C000144498

1,000,000 Books

are available to read at

---◆---

www.ForgottenBooks.com

---◆---

Read online
Download PDF
Purchase in print

ISBN 978-1-5277-7080-5
PIBN 10889822

This book is a reproduction of an important historical work. Forgotten Books uses
state-of-the-art technology to digitally reconstruct the work, preserving the original format
whilst repairing imperfections present in the aged copy. In rare cases, an imperfection in
the original, such as a blemish or missing page, may be replicated in our edition. We do,
however, repair the vast majority of imperfections successfully; any imperfections that
remain are intentionally left to preserve the state of such historical works.

Forgotten Books is a registered trademark of FB &c Ltd.
Copyright © 2018 FB &c Ltd.
FB &c Ltd, Dalton House, 60 Windsor Avenue, London, SW19 2RR.
Company number 08720141. Registered in England and Wales.

For support please visit www.forgottenbooks.com

1 MONTH OF
FREE
READING

at

www.ForgottenBooks.com

By purchasing this book you are eligible for one month membership to ForgottenBooks.com, giving you unlimited access to our entire collection of over 1,000,000 titles via our web site and mobile apps.

To claim your free month visit:

www.forgottenbooks.com/free889822

* Offer is valid for 45 days from date of purchase. Terms and conditions apply.

English
Français
Deutsche
Italiano
Español
Português

www.forgottenbooks.com

Mythology Photography **Fiction**
Fishing Christianity **Art** Cooking
Essays Buddhism Freemasonry
Medicine **Biology** Music **Ancient
Egypt** Evolution Carpentry Physics
Dance Geology **Mathematics** Fitness
Shakespeare **Folklore** Yoga Marketing
Confidence Immortality Biographies
Poetry **Psychology** Witchcraft
Electronics Chemistry History **Law**
Accounting **Philosophy** Anthropology
Alchemy Drama Quantum Mechanics
Atheism Sexual Health **Ancient History**
Entrepreneurship Languages Sport
Paleontology Needlework Islam
Metaphysics Investment Archaeology
Parenting Statistics Criminology
Motivational

CANADIAN WOODWORKER

and

Furniture Manufacturer

C.M.C.

OUR "BIG DRIVE"

Our big offensive this year is on the woodworking trade. We hope to capture more machinery orders than ever before. The **C.M.C.** line of woodworking machinery is the most complete on the market, including every machine required by the up-to-date woodworking plant.

C.M.C.
MACHINES

Matchers
Surfacers
Moulders
Sanders
Tenoners
Boring Machines
Clamps
Mortisers

C.M.C.
MACHINES

Grinders
Rip Saws
Cross Cut Saws
Band Saws
Shapers
Stickers
Dovetailers
Re-Saws

The machine illustrated is our No. 504 Tenoner, single or double head.

Write us for information on any or all of our machines.

CANADA MACHINERY CORPORATION
LIMITED

GALT - ONTARIO

Alphabetical Index to Advertisers, Page 8

Put Good Belting

on your

Good Machines

Your machines are expensive, and naturally you expect them to turn out the quality and quantity of work commensurate with their cost.

Then why harness them with belting which cannot begin to do them justice?

Use

"AMPHIBIA"
Planer Belting

and get the most work from your machines, in the quickest time at the lowest cost per day of service.

Try a sample run of AMPHIBIA Planer and prove its merits.

"Leather like gold has no substitute."

Sadler & Haworth

Established 1876

Tanners and Manufacturers

For 40 years Tanners and Manufacturers of the Best Leather Belts.

Toronto	St. John, N.B.	MONTREAL	Winnipeg	Vancouver
38 Wellington St. E.	159 Prince William St.	511 William St.	Galt Building	217 Columbia Ave.

Shell Box Equipment

and

General Wood Working Tools

We have in stock at Toronto for immediate shipment, for use on Fixed Ammunition Boxes :—

Cowan Special Dado Machines, for grooving sides
J. & C. 12 Spindle Dovetailers
J. & C. Case Clamp for squeezing

also a Used Advance 15 Spindle Dovetailer—in very good condition

We have installed several plants for shell box work.
Let us co-operate with you in selecting and laying out your tool equipment.
Our Services are at your disposal.

We handle all kinds of wood-working machinery, including the complete line of the

AMERICAN WOODWORKING MACHINERY CO.
ROCHESTER, N.Y.

and the line of special boring machines of

M. L. ANDREWS & CO., CINCINNATI, OHIO

Garlock - Machinery

197 Wellesley Street   -   Toronto

Telephone North 6849

Wood Steaming Retort

Wood Bending Manufacturers:

This is one of our

Perfection Retorts

which we guarantee will save you

50% Less Breakage

in your bending department than your present process; that your stock will dry in your forms or presses in one-third less time; that you will have no stained stock; that your stock will retain its shape much better after being bent; that it will dry in your dry-kiln in one-half less time and that your steam consumption will be reduced at least 90 per cent.

The door can be opened and closed in ten seconds, and it is steam and water tight and for this reason can be placed anywhere in your factory.

Compare this IMPROVED RETORT with your present steam boxes, then write us for our Booklet on Progressive Wood Steaming.

SHIPPED ON 30 DAYS TRIAL

Perfection Wood Steaming Retort Co.

PARKERSBURG - WEST VIRGINIA

The Atlantic Lumber Co.

Manufacturers and Wholesale Dealers

American and Canadian Hardwood Lumber and Mahogany, Quartered and Plain, White and Red Oak a Specialty.

110 Manning Chambers - **TORONTO**
Phone Main 6386

HOME OFFICE, BOSTON MASS.

Mills: KNOXVILLE, TENN.,   WALLAND, TENN.
FRANKLIN, VA.

Immediate Shipment

West Virginia Stock

Thoroughly Dry and Well Manufactured

1 car 4/4 1s and 2s, Plain White Oak.
4 cars 4/4 No. 1 Common Plain White Oak.
2 cars 4/4 No. 2 Common Plain White Oak.
1 car 4/4 No. 1 Common and Better Plain White
     Oak. Worm holes no defect.
1 car 6/4 No. 1 Common Plain White Oak.
6 cars 4/4 S. W. Chestnut.
1 car 6/5 S. W. Chestnut.
4 cars 6/4 S. W. Chestnut.
1 car 8/4 S. W. Chestnut
3 cars 4/4 No. 3 Common Chestnut.
1 car 4/4 No. 1 Common and Better Chestnut.
½ car 8/4 No. 1 Common and Better Chestnut.
1 car 4/4 Saps and Select Poplar.
2 cars 4/4 No. 2 Common and Better Poplar.
1 car 8/4 No. 2 Common Poplar.
1 car 6/4 No. 2 Common Poplar.
1 car 4/4 No. 2 Common and Better Birch.

We solicit your inquiries on special cut bill Oak.

Mill at Holly Junction, W. Va.

Write at once for Quotations   The Price is Right

The Imperial Lumber Co.
COLUMBUS - OHIO

We'll Demonstrate this Machine—FREE—
On Your Own Work!

Built-up pieces, large or small, are sanded on the 427.

Curved surfaces and irregular shapes are sanded equally well.
Special forms and devices can be furnished
for holding the work.

IF you will send us samples of such work as you are still sanding by hand, those most expensive operations, we will show you how much you should be saving with this latest Berlin hand-block belt sander.

On millwork, furniture or special jobs, the saving will return the investment many times. Work you've here-to-fore considered impossible to sand by machine is now easily handled.

You may have belt sanders now—good machines for some purposes—but not as labor-saving on your work as the Berlin "427."

No matter how efficient you consider your present methods, here is a real profit-saver with better quality finish thrown in.

Free details on construction mailed
on request.   Ask for circular 427.

Berlin Machine Works, Ltd.
Hamilton, Ont.

Either
DIRECT MOTOR
Or BELT
Drive

One Piece Sides and Ends for

SHELL BOXES

From Narrow (Cheaper) Lumber at
a Low Rate per 1000 feet on the

LINDERMAN AUTOMATIC DOVETAIL MATCHER

*Samples
on
Request*

THESE CONCERNS USE LINDERMAN MACHINES:

W. C. Edwards & Co.	Ottawa, Ont.	Dominion Box & Package Co.	Montreal, Que.
(2 Machines)		(2 Machines)	
Czerwinski Box Co.	Winnipeg, Man.	C. P. R. Angus Shops	Montreal, Que.
Cummer-Dowswell Co.	Hamilton, Ont.	J. W. Kilgour & Bro.	Beauharnois, Que.
Canada Furniture Mfrs.	Woodstock, Ont.	Durham Furniture Co.	Durham, Ont.
Martin Freres & Cie	Montreal, Que.	Brunswick-Balke-Collender Co.	Toronto, Ont.
Alberta Box Co.	Calgary, Alta.	Singer Mfg. Co.	St. Johns, Que.

Canadian Linderman Company
Woodstock, Ontario

Muskegon, Mich.   Knoxville, Tenn.   New York, N. Y.

Shell Box Machinery

Special Two Spindle Boring Machine

Designed for boring the large holes in the internal fittings of the 18-lb. 4-round Ammunition Box.

Machine is heavily constructed, having large steel spindles and a heavy steel gear.

With the use of this machine you can turn out a large number of blocks a day. We have over 50 of these Borers in operation on Ammunition Boxes.

Note the special clamp for holding the blocks, also the positive rack feed and the stops for gauging depth of cuts.

Increase Your Output of Shell Boxes

Particulars and prices gladly furnished for this and other shell box machines.

We also manufacture a complete line of

Boring and Dovetailing Machines
for Shell Box Work

Jackson, Cochrane & Company
BERLIN - ONTARIO

Alphabetical List of Advertisers

Atlantic Lumber Company   4
Ault & Wiborg Company 31

Baird Machinery Works 42
Berlin Macinne Works   5

Canada Macinnery Corporation 1
Canadian Boomer & Boschert Mfg.
   Company 43
Canadian Morehead Mfg. Company
Carborundum Company   47
Central Veneer Company   37
Chapman Double Ball Bearing Co.   43
Clark Veneer Co., Walter. 35
Clipper Belt Lacer Company 10
Coignet Chemical Products Co. 31

Delany & Pettit 12

Evansville Veneer Company 32

Foster, Merriam & Company   32

Garlock Machinery Limited   3
Grand Rapids Dry Kiln Company ..   38

Hartzell, Geo. W. 35
Hay & Company   31
Hay Knife Company, Peter 41
Hoffman Bros. Company   35

Imperial Lumber Company   4

Jackson, Cochrane & Company   7

Kersley, Geo. 41
Klotz Manufacturing Company 10
Knechtel Bros. 46
Kraetzer-Cured Lumber Company..

Linderman Machine Company   6
Lightfoot, Stanley 42
Luehrmann Hardwood Lumber Co.,
   C. L. 12

Mattison Machine Works, C. 48

Nash, J. M. 43
National Securities Corporation ... 42
Neilson & Company, J. L.   8

Ober Manufacturing Company   43
Ohio Veneer Company   35

Perfection Wood Steaming Retort
   Company   4
Penrod Walnut & Veneer Company   37
Perkins Glue Company   9
Perrin, William R. 11

Radcliff Saw Company 41
Reynolds Pattern & Macine Works   9

Saranac Machine Company   47
Sadler & Haworth   2
Shawver Company 11
Sheldons Limited ... ...
Simonds Canada Saw Company ...
Spencer, C. A. 41

Toomey, Frank, Inc.   42

Vokes Hardware Company 39

Waetjen & Company, G. L. 35
Walter & Company, B.   47
Weeks Lumber Company, F. H. ...   11
Whitney & Son, Baxter D. ... .

| Lock Corner Outfits
Box Makers Sawing Sets
Resaws and Filing Outfits
Box Board Matchers, etc. | We Specialize

in furnishing | Dovetail Box Outfits
Nailing Machinery
Box Board Printers
Surfacers, Borers, etc. |

BOX MACHINERY
(for LOCK-CORNER, NAILED and DOVETAILED BOXES)
Also all other kinds of
WOOD-WORKING MACHINERY

TWENTY years actual, first hand, working experience in box factories, sash and door factories and planing mills enables us to
give you the intelligent service all buyers desire.

Buy from people who will understand
your requirements

J. L. NEILSON & CO. - Winnipeg, Man.

The Only Way to Drive Screws

There is a stage-coach method and a modern method of doing everything. When it comes to driving screws the only modern way is with an Automatic Screw Driving Machine, which means greater and better production and greater profit.

The Reynolds
Automatic Screw-Driving Machine

is guaranteed to produce results. It is no experiment. The No. 2 Standard Type, shown herewith, is particularly suited for placing the screws in the straps of shrapnel boxes. It may be fitted to drive, without change or adjustment, any three consecutive sizes flat or round head, or two sizes filister head, No. 6 to 14 inclusive, wood or machine screws, or three sizes flat or round head, or two sizes filister head machine screws up to No. 20. All the above up to 1½ inches long if smaller than No. 10, or up to 1¾ inches long, No. 10 or larger. Fitted with boring attachment if desired.

Many Machines in Use in Canada.

Ask for our Catalogue.

REYNOLDS PATTERN & MACHINE CO.
101 and 103 Third Ave., Moline, Ill., U.S.A.

No. 2 Standard Type Automatic Screw-driving Machine.

Perkins Vegetable Glue
The Best for Built-up Stock and Veneer Laying

Dinner wagon manufactured by the D. Hibner Furniture Co., Limited, Berlin, Ont., in which Perkins Glue was used.

Perkins Vegetable Glue sticks and holds. It is unaffected by dampness, extremely dry atmosphere or severe sanding. We guarantee to effect a 20 per cent. saving in your glue room by replacing hide glue with Perkins Vegetable Glue. Along with this saving you can have a glue room comfortable as to temperature and sweet as to smell. The glue is absolutely uniform in quality. Every shipment exactly the same. Let us demonstrate in your own factory and on your own stock just what can be accomplished with Perkins Vegetable Glue.

Perkins Glue Company, South Bend, Ind.

Here It Is!

The Shaping Saw Machine

That Saves Time

THE Klotz Shaping Saw Machine is specially made and designed for cutting out Shapes or Blanks for Handles, Spokes and similar work, from planks or slabs after they are cut from the log.

The frame is made from hard wood and firmly bolted together. The table is made of well-seasoned ash strips, glued together, and runs on large turned rolls; with wrought iron axles and metal bearing. A chuck is attached to the table for holding the timber; the same is adjustable for different lengths of work, and is very quickly operated by a suitable handle. The mandrel is of steel, and the balance-wheel is turned so as to run perfectly true. The drop-table is arranged so that it may be easily lifted out of position, in order to get at nut on mandrel to take off the saw. The saw furnished with the machine is 26 inches in diameter, and pulley on mandrel is 8 inch diameter, 6½ inch face, and will make about 1400 revolutions per minute. Power required, 2 H.P. Shipping weight 900 lbs. Floor space 4 x 7 feet.

Write us for prices and details.

Klotz Machine Co.

Sandusky, Ohio

"Clipper"

30 Days' Trial

The Rapid Belt Lacer
For Shell Box Machinery

In these rush days when every mill making shell boxes is working at full capacity, even minor delays are costly. The "Clipper" Belt Lacer is the quickest remedy for a stretched or broken belt. The "Clipper" machine can be taken to the belt and the lacing completed in three minutes without removing the belt from shaft.

Clipper Belt Lacer Co.

1010 Front Ave.
Grand Rapids, Mich.

30 Days' Trial

We will send you a "Clipper" Belt Lacer on 30 days' trial and if you are not satisfied with results your money will be refunded.

For Sale by
All the Leading Supply Houses in Canada

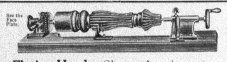

See the Face Plate.

Fluting Heads (Shaper Attachment)

For fluting table legs, chair legs, bed posts, columns, balusters, etc., etc.
Swings any diameter up to 8 in. and may be extended to any length.
Spaces automatically numbers 1 to 48.

Ask for circulars of Fluting Heads, Twist Machines and Twist Sanders.

SHAWVER COMPANY - - - - - **Springfield, Ohio**

Solid Cutters

Made of high-grade Crucible Steel. Tempered hard to grind only. Sharpened from under throat side. These Cutters will hold their original form until worn out.

4 to 8 cutting edges means smooth work, long life.

ONE OPERATOR
SIXTY SETS
SIXTY MINUTES

BRICK HOUSE JAMBS

FRAME HOUSE JAMBS

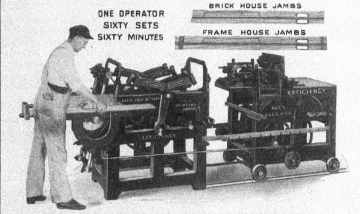

THE EFFICIENCY WINDOW FRAME MACHINE

A new style machine which **REDUCES LABOR COST FULLY 60%.** Works automatically—one man can turn out sixty sets of pulley stiles an hour. Cross cuts the ends of pockets, routs the pulley mortises and dadoes the head and sill gain, automatically, while the operator slits the pockets. You take no chance—write us.

THE F. H. WEEKS LUMBER COMPANY - **AKRON, OHIO**

The Canadian Woodworker and Furniture Manufacturer covers a specialized field. It represents maximum value to its subscribers.

Subscription price $1.00 per year.

PRESSES

For Veneer and Veneer Drying

Made in Canada

William R. Perrin

Limited

Toronto

Chas. F. Luehrmann Hardwood Lumber Co.

148 Carroll Street

ST. LOUIS, MO.

Manufacturers of the original
St. Francis Basin Gum

We have on hand a complete stock of Red Gum and Unselected Gum, in thicknesses from 1″ to 3″. Nice dry stock of 1″ to 2″ Quartered Red Gum.

We also carry a complete stock of Oak, Quartered and Plain, Red and White, 1″ to 5″ thick, and large stocks of Ash, soft Maple, soft Elm, Cypress, Yellow Pine and Cottonwood.

We make a specialty of Yellow Pine Structural Timber, and can furnish either Long Leaf or Short Leaf.

We carry a full and complete stock of all kinds of Veneers, in Red Gum, Mahogany, Oak, Poplar, Yellow Pine and Birch.

Joint
and
Veneer
Glue

Always
Uniform

Garnet
Paper
and
Cloth

Delany & Pettit, Limited

Office:—133 Jefferson Ave.   **Toronto, Ont.**   *Works:*–105-131 Jefferson Ave.
106-118 Atlantic Ave.

Canadian Woodworker

A Monthly Publication in the Interest of the Woodworking Industry.
Reaches the factories producing interior finish, doors, sash, flooring,
woodenware, furniture, pianos, boxes, and general mill products.

Subscription, $1.00 a year; foreign $1.50.

Woodworker Publishing Company, Limited

Branches: Montreal, Vancouver,
Chicago, New York, London, Eng.

345 Adelaide St. West, Toronto

Phone Ade. 2700

Authorized by the Postmaster General for Canada, for transmission as second class matter.
Entered as second class matter July 18th, 1914, at the Post Office at Buffalo, N.Y., under the Act of Congress of March 3, 1879.

Vol. 16	January, 1916	No. 1

Shell Box Patronage in a Cabinet Minister's Constituency

The publicity we have recently given to the manner in which many shell box contracts were let by the Shell Committee has met the approval of our readers in all parts of Canada. From letters received, we are convinced that nearly everyone interested in the production of boxes of any kind has come across instances similar to those we have described, in which contracts have been let to parties who are not in any way entitled to them, while other parties well equipped, with their machinery idle and their men out of employment, have been unable to get contracts.

However, although we have published the facts in connection with several of these cases and have disclosed a state of affairs that is intolerable, namely, the giving of contracts for the production of shell boxes to political friends who are in many cases without the necessary experience or equipment, while firms which are justly entitled to secure contracts because of their experience and equipment, have been neglected in spite of their repeated applications for the business, nothing has been done by the Government in the way of investigating the situation. In times like the present surely the first consideration in the giving of contracts by the Government should be economy and efficiency. Political consideration should never be permitted to enter into the letting of contracts for shell boxes or for anything else. Yet we find example after example in which only the most searching investigation would convince us that political considerations have not been the chief factors. Another instance which comes to our attention is similar to others already reported; we have enquired carefully into the circumstances and believe the following is a plain account of the facts:

A contract for 5,000 shell boxes to hold four 18-lb. shells and fuses, at $2.50 per box, was secured by C. W. Burgoyne, of Fenelon Falls, Ont., which is located in the Dominion riding of Victoria, which included shipment to the point designated for delivery. C. W. Burgoyne, who secured this contract, is in partnership with his father in the firm of W. Burgoyne & Company, which carries on a general store business at Fenelon Falls. Not being in a position to manufacture the boxes himself, we are informed that he was willing to sub-contract with anyone who would undertake to handle the contract or parts of it. The contract was finally allotted to Mr. Alfred Tiers, of Fenelon Falls, who conducts a small planing mill at that place. This contract has now been completed and the boxes were delivered some time ago.

While this contract was being worked out, an additional contract for 25,000 boxes was secured, according to our information, by C. W. Burgoyne, and this also was taken over by Alfred Tiers. We understand that this contract was known as an optional one, which we suppose means that Burgoyne was at liberty to take it or leave it, as he desired. It included the stipulation that no more than 3,000 boxes were to be paid for in any one month. The price for this box was a trifle less than the price for the box covered by the first contract. By the end of the year 3,000 boxes had been produced under this contract. We are informed further that Tiers was not able to fill the first contract entirely by himself, but purchased the tops and bottoms for a number of the boxes, securing them at 14c per pair, and that in connection with the second contract he is purchasing the ends ready to be dovetailed, at 5c. each, securing them from a manufacturer at Lindsay.

We cannot see how such methods of handling the shell box business can be justified. Even in ordinary times the letting of contracts in this way could not be defended, and at a time like the present it is particularly indefensible. The suspicion that this contract was handed over by Burgoyne to Tiers for a definite monetary consideration is strengthened by the fact that these two parties formed a partnership under the name of C. W. Burgoyne & Company. Evidently they wished to give the transaction a more respectable appearance; but we cannot see how this can be considered as lending the affair any respectability. It amounts to the same thing in the end, namely, that a party quite unable to produce the boxes, secured the contracts and turned them over to another party. The fact that the first party entered into a partnership with Tiers indicates rather clearly that he did so in order to make it appear that he was entitled to a share of the profits which he was securing and towards which he had contributed nothing. When one considers that throughout Canada there are hundreds of planing mill plants which have suffered heavily through business depression and would be greatly benefited by securing shell box contracts, one is thoroughly justified in criticizing a government which has permitted the letting of contracts in such a manner.

What the public wants to know is whether the letting of shell box contracts has been used to promote the interests of a political party. We believe that many contracts have been so let and that it is incumbent upon the Government to throw light upon the whole situation. Such information as we have gathered upon the subject is at the service of the Government any time it desires to open an investigation which shall be thorough and impartial. Our object is to help in clearing up a most unsatisfactory intolerable situation. The cases we have already referred to cover a large part of Ontario and part of Quebec, and they strongly suggest party patronage in the giving out of contracts.—Canada Lumberman.

What Do We Mean by 'Rods'?

The writer has been a constant reader of trade papers for quite a number of years and having gathered considerable knowledge from that source wishes to offer something which has not yet appeared in any trade paper, but which is taught by some of the largest schools of furniture designing in the States.

We all know what an important part modern machinery and modern methods play in present day woodworking factories, also that the science of draughting has given us the most intelligible method of presenting our ideas for other men to work from.

In most of the progressive factories of today the usual method of procedure is to have a full size drawing, showing all details of construction. The stock bill is made up from the drawing and all necessary pieces of material having been cut up in the roughing out department of the machine room, the detail drawing accompanies the material to the finishing machine room, and from there to the cabinet room.

You who have had experience in draughting know what condition your drawing is in by the time it makes one trip through the machine room. To overcome this objectionable feature of having drawings on which the details cannot be distinguished after the first job has been completed, and drawing to assist the machine man who has no place to lay out a drawing unless on a truck load of material, a method has been adopted by some of the largest and most up-to-date furniture factories manufacturing medium and high grade work, of which I propose to give an outline.

Take a piece of board, preferably nice clear soft maple about six feet long, ten inches wide and five-sixteenths inch thick and having sanded two sides and jointed both edges, proceed to lay out, with the assistance of a marking gauge and sharp knife, as per diagrams. After making knife cut run the point of a sharp pencil along the cut.

This board is known in the trade as a rod and after receiving a coat of white shellac and having a hole bored in the end for hanging it up, is about the handiest thing you can have in your machine room, as it can be placed on a truck or any convenient place to take measurements from.

It will be seen that this rod is nothing more nor less than broken plan and sections, arranged so that any measurement may be had from them. In the factories which use this system of rods you will also find the following system in connection with their drawings:—

Two copies of each drawing are made, one being filed in the cabinet room in a cupboard built for that purpose, the other one goes to a man who makes all necessary patterns, shaper forms, etc., for the job and is then filed away in a cupboard in the machine room.

This system when properly carried out with all drawings and rods numbered is pretty hard to beat, and while the rod serves for almost all machine work, the drawing is always at hand for a reference.

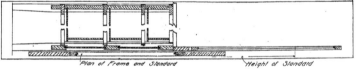

Plan of Frame and Standard   Height of Standard

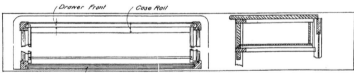

Drawer Front   Case Rail

Panel

Upper section shows height of case, etc. Lower drawings show plan and cross section. These drawings are placed on opposite sides of the 'rod.'

"Punch"

The British Empire, and Canada its most important unit, faces the new year with unimpaired
vigor, uncountable resources, unshaken confidence and redoubled determination.

The French Period Styles—Louis XIV

The first of a series of illustrated articles dealing with developments in furniture design during this important era.

By James Thomson

Of French furniture styles of historic periods those representative of the 17th and 18th centuries undoubtedly are of the greatest interest to the furnishers of today and incidentally to designers and manufacturers. The Louis XIV, XV and XVI periods were responsible for styles that, though in a measure distinct and separate, are usually classed together. Precisely as we co-ordinate the names Chippendale, Heppelwhite and Sheraton so do we bring the styles of the three Louis into partnership.

Louis the fourteenth the so called "Grand Monarque" ruled France from 1643 to 1715 by his exaction, extravagance and waste, sowing the seeds that a century later brought forth the bitter fruit of revolution and terror. With the national financial system bankrupt here was a king that maintained exalted place as the first sovereign of Europe, only in the end to have his burial equipage rounded by curses deep,

Fig. 1

vehement and prolonged. Unlimited expenditure of money wrung from the pockets of a hard working peasantry was the means by which the mare was kept magnificently going. The many were persistently exploited in order that the few might pass their lives

in luxuriant idleness. Never was king more lavish with means wrung from helpless people. It mattered not that money was scarce, means had to be found to pay for the erection of magnificent Versailles and other royal residences. The furniture and interior

Fig. 2   Fig. 3

decorations of the palaces and chateaus of the France of the time of the Grand Monarch was as sumptuous and extravagantly elegant as could possibly be imagined. With the means at hand and grand edifices in process of erection the most talented workers of other lands flocked to France. Architectural and decorative genius had here opportunity such as seldom was offered, and the output of the period certainly carries evidence of the fact. No botcher had a hand in the development of this really splendid style. On the erection of Versailles alone there was expended a thousand million livres. Fit dwelling was it for strutting majesty in cinnamon colored coat, diamond embroidered wig adding half a dozen inches to height, and red shoes that lifted the wearer four inches from the ground. All this splendor bought with the sweat, blood and tears of a crushed and cowed people, it is little wonder the king's carriage was more than once stoned when returning from some function in the twilight hour. What the peasantry dared not do by day they did by night.

Outside of the product of Andre Charles Boule—

LOUIS XIV.

Fig. 4

Louis XIV

Fig. 5

Fig. 6

Fig. 7

Fig. 8

Fig. 9

Fig. 10

Fig. 11

whose output was in manner separate and distinct from all others—the dominating feature in the style is theatrical ostentation. Spectacular effect was deemed of main importance, and as rooms were large and lofty the spectator was enabled to take in a large section at a glance. The work was so excellently executed, details could well bear close examination. Never was carving done better.

The style doubtless had rise in exterior decorative features of building, being of Italian origin. Palladio, an Italian architect of the 16th century reduced to a workable basis a style of building in which the five orders of antiquity were utilized. His style had great vogue and carried far and no doubt in the beginning the Louis XIV style was based upon it. But it soon partook of a character essentially its own, and in direction novel in the extreme. As it developed there was disclosed something the world had never looked upon before. It was a style having dignity, proportions stately, symmetry of arrangement in ornamental features, decorative features though often extremely rich and elaborate in detail yet handled with rare restraint. It was only near the end of the reign the tendency to flamboyance and departure from rule and reason was observable and not then to any detrimental extent.

Paris enthusiastically took up with the new style. It was such as appealed to a society where the pursuit of pleasure was paramount over every other consideration; a time when female frailty was ennobled and downright rascality exalted. Paris made the style its own but eventually gave it another identity and at once a wonderful brilliancy. Though in general the scheme was classical, details making no pretence to being classical made appearance, shells, flowers, fruit, birds, heads human and otherwise, scrolls in plain guise or in that of the accanthus, heads and faces of maids of honor, ribbons, shields, straps, etc., all judiciously and symmetrically mingled; the arrangements always having regard for the possibilities of light and shade. Nothing could be more splendid and enchantingly attractive than a large and lofty room in this brilliant style. For this reason in the embellishment of the theatre, the ballroom and similar places of public assembly it is the style par excellence. No prentice hand had part in its development. It was designed by master draughtsmen, as the arrangement of parts and the surety of line discloses. The men too, who did the manual part of the work were masterful craftsmen each in his special line. A jewelry-like quality pervades the ormolu mounts especially such as help adorn the product of Boule of which I take pleasure in submitting a few examples.

In Figure 1 is shown a splendid example in the way of a bedroom of the period. Imagine room after room of palace or chateau furnished in this ornate yet dignified manner. Here is a style demanding a lofty room and adequate space in order that justice may be done it. Here was a time when building operations of kingly character were conceived on a grand scale. The Gallerie des Glacis at Versailles in which the King was wont to receive, was two hundred and forty-two feet long by forty high. Notwithstanding the immensity of this room it did no more than accommodate the gaudily garbed crowd of courtiers daily presenting themselves before royalty.

In figure 4 we have a handsome example in the shape of a sideboard. This is characteristic of the style in its earlier and formal aspect, dignified, broad

in treatment, excellently proportioned and with admirable restraint in decoration.

In the three score years of this reign the style showed many changes. As may be noted in panel ornament S and C forms appear with amazing frequency. Take an S, the upper and lower limbs separate features, maybe they are joined together by leafage and make an attractive figure. And as time went on the S and C forms were utilized as structural features in the manner exemplified in figures 2 and 5. Peculiarities of the style became exaggerated as time went on until when the end came, balance was many times entirely eschewed, the two sides of a mirror say, being unlike.

Figure 6 shows a chair designed in pure style before the curved leg became a feature. As time went on, sprawling legs called cancan, after the vile dance of the name became the fashion. In figure 7 we have a beautiful example from the transitional period.

Boule work was one of the peculiar cabinet-making manifestations of the period. Of the unique product

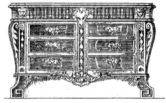

Louis the fourteenth style
as interpreted by Chippendale.

Fig. 12

of this maker, a great deal of it went to furnish the new royal chateau at Versailles the inventor afterwards becoming the head of the royal furniture making department. As to the boulework, its peculiar quality is in the marquetry, the surface embellishment. It is composed of tortoise shell and brass and on occasion ivory and enamelled metals. As at first made, the product was very expensive but as time went on, by sawing two or more thicknesses at a time the cost was much lessened though at best it was most expensive and only for kings. The technical name of the new method was "boule and counter". By counter changing, two or more designs were obtained by one sawing. In one piece of marquetry the shell would form the pattern, in the other, form the border. In the later work the shell is laid on a vermillion ground or perhaps gold, the brass as usual being sharply and elaborately chased with the graver. By and by brass mounts of the usual order were employed, edgings, brackets, claw feet, egg and dart mouldings and the like. Very brilliant and beautiful is this latest work as a reference to figures 3, 8 and 9 will disclose.

One of the finest Louis XIV chairs I have reserved for the last, and because of the beauty of the carved back I have in figure 11 added a larger scaled and better rendered drawing of it. On the same sheet is shown two characteristic panels. In these last examples the C and S tendencies may be noted.

In Chippendale's book are many more or less bald copies from the French. In some instances he made no changes whatsoever, but frankly lifted the patterns and claimed them for his own. In figure 12 are shown two sideboards adapted from the French. A reference to figure 4 makes plain the source of the upper one, while the lower one is frankly Louis fourteenth at its best; the possibly original creation of the English craftsman. As an example of the style I cannot see how it could be bettered.

In conclusion I would remark that here was a style in interior decoration excellently suited where great display and splendor was a desideratum—the style of state occasion and parade. Walls were panelled and in these panels were costly paintings and tapestries. The furniture coverings also were of a most brilliant character. Examination in detail of the chair coverings here shown will give an idea of the beauty of design in textile fabrics.

Some Problems of Shell Box Making in a Furniture Factory
By Foreman

The shell box business has undoubtedly proved the most important feature of the year 1915 in the woodworking industry.

Furniture manufacturers have been severely hit by the war and in order to keep their factories running have turned their attention to shell box making, and though the prices are low, yet at this particular time, when the furniture business is a little dull they are glad to keep their factories moving by making shell boxes.

The writer, in conversation with a workman was told that a furniture factory was no place to make shell boxes; this man was certainly wrong. There is, of course, a great difference between making shell boxes and high grade furniture, and while the above workman was a good mechanic he was not willing to adapt himself to new conditions. If he had only taken time to consider the question thoroughly no doubt he would have come to a different conclusion.

the right kind of lumber and carelessness in handling. Not only is the right kind of lumber necessary in view of the government inspection and test, but it helps considerably in the manufacture. This goes to show that it is good business to buy lumber of a fairly good grade as it requires less labor to work it and there is a smaller percentage of waste. Other difficulties that will arise come from the handling during the different operations such as: bad gaining or dadoing of the sides of the boxes; the holes in the bridges or cradles not being in perfect alignment; screwing on the bottoms and handles and bad gluing. These can be overcome by proper attention and careful supervision.

The bridge work is where the most difficulty usually occurs. A boring machine specially built for this work is made by Jackson, Cochrane & Co., Berlin. A deviation of one sixteenth of an inch may mean the rejection of some bridges and refilling the boxes.

Some specifications call for a dovetailed corner

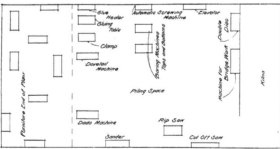

Fig. 1.—Machine Room, first floor.

There is no reason whatever why shell boxes cannot be manufactured in a furniture factory. Shell box making is just as easy a process as furniture making if you have the right tools and men. Adaptation is the keynote, not for the manufacturers alone but for the workmen too.

There are difficulties to be encountered in making shell boxes, due to the strict specifications, scarcity of while others call for nailing. Whether the box be dovetailed or nailed the work should be systematized. The dovetailed box calls for more labor in the machine room, while the nailed box calls for more labor in the cabinet room. It is wise in making the dovetailed box to do as many operations as possible on the one floor.

There are manufacturers who have rearranged

their machine rooms, installing new machines to meet the requirements of this particular box. In arranging their machine rooms some have placed the machines down one side of the room, doing all the operations here down to the assembling and sanding of the

machines are installed or changes made, if the men are careless or slipshod in their particular operation, shell box making will be anything but satisfactory.

The nailed box calls for more work in the cabinet room. In this box forms should be made for the dif-

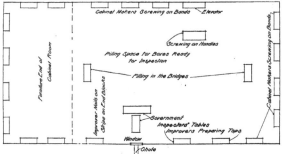

Fig. 2.—Cabinet room, second floor..

box. The boxes are then trucked into the shipping room for the screwing on of the iron bands and filling in of the bridge work, the box being then ready for inspection, and placing in the car. Other manufacturers have placed the shell box machines practically in a square at one end of their machine room as shown in Fig. 1. From the kiln the lumber is taken to the cut off saw; and so on around the square until the box parts are ready for the cabinet room when they are placed on the elevator and sent up to that department, for the screwing on of the iron bands, and the filling in of the bridge work, after which the box is inspected, placed in a chute and slides down into the car, Fig. 2.

The output of the whole factory largely depends upon the results of the machine room, also the cost of production, hence the importance of the different operations of the machine room. Whatever new

ferent operations such as preparing the tops, preparing the bottoms, and the assembling of the box. These forms will enable a man to work away regardless of measurements, the forms of course being inside dimensions.

Sectional work is the most preferable way of handling boxes in the cabinet room. It is wise to give the work out and watch results for a few days and then arrange your workmen accordingly, giving each man that particular operation with which he can show best results.

The arrangement of the whole factory so that the lumber comes in at one end in the unfinished state and goes out at the other in the form of a shell box without creating confusion and with the least possible handling of the material is the essential problem of shell box manufacture.

How I Make My Paper Last When Sanding Gummy Woods
By Edward Lee*

Sanding had always been given a pretty hard name around the plant where I worked. The boss claimed he bought more sandpaper and got less use out of it than any sash and door manufacturer in the business. The result was that somebody was always coming around and asking questions or giving me advice about running my machine. One time it would be to put the paper tighter on the drums, and the next time they'd say to try a different size or make of paper. One would think I ought to take a lighter cut, or maybe say the drums were running too fast, while another would blame the oscillators. Anyway, they only fussed around and made things worse.

Well, I was like the rest. I wouldn't admit it was my fault, but laid the blame to the stock I was run-

*Berlin Machine Works.

ning, and I really thought so too. You see, we mostly made veneered panel doors, and used nothing but hemlock and Douglas fir, just like any other do. Now both these woods are gummy like and wear out paper in no time. The fir clogs and wears out paper the fastest because the wood itself is harder, so that it takes just that much more pressure to make paper cut. Still I wanted to make my paper last better. But there isn't anything that will clean sandpaper. I tried to figure it because there's a fortune in it. I tried running chalk over the paper, because a fellow down south once told me it would help. I tried kerosene oil, and running in a hard piece of dry wood at an angle, and holding a stiff brush against the running drums, but there's nothing works.

I said there's nothing works; there is one thing, but

its different. After I'd tried everything and messed up more stock than I'd saved, I finally got onto the right idea. It's this. Don't try to find out how to clean the paper after its clogged, but prevent its clogging in the first place.

Now both these woods I work are mostly curly grained with very close pores. When fine or dull grit is used it settles right against these gummy pores, especially so as the woods are hard and heavy pressure is used.

The result is that the paper doesn't get a chance to clean itself and it soon runs in a layer of gum. This soon kills the cut of the grit and the only thing to do then is start with fresh paper. But the better way, and the way that got me out of this hole, is to use only coarse paper that is real sharp. The proper size is 2, 1, ½ garnet. This is large enough so that just the cutting part of the grit strikes the surface, and the gum can't get a good hold.

It's just as important after you've got the right paper to always make sure your adjustments are right enough not to give a heavy pressure, and that your bed and drums are lined up so that the sanding is even.

Ever since doing this I've been able to make paper last 6 or 8 hours, where it used to wear out in 2 or 3 hours. But about this hour business a fellow can't tell. After he's done his part, it all depends on how his doors are made up. So that when gummy woods are used, it also pays in sanding time and paper expense to get a pretty regular surface before the sanding is done.

Safety in the Machine Room
By W. J. Beattie

The prevention of accidents has been, and is getting, the attention that should have been given years ago, but better late than never. A fellow feeling, for your associates should always be considered in preference to the mere dollars and cents costs of damage suits. Had many manufacturers been more considerate and not try to "get past" with doing as little as possible in the way of safety devices, the present compensation law might have been more to the liking of

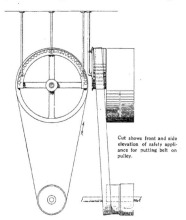

Cut shows front and side elevation of safety appliance for putting belt on pulley.

some of them. Even from a selfish point of view, where or how can a manufacturer see a profit in neglecting these things, when the time lost by even one good man for six or eight weeks would go a considerable way towards safety devices that would have rendered his accident impossible.

We cannot excuse the workmen altogether, either, as many are notoriously careless both of themselves and their fellows, but as the employer is the one to take the initiative on account of his position, he is slightly more to blame.

However, as we are working out an equitable compensation law it behooves us all to feel that our interests are one, and to do all we can to prevent the loss and suffering that have been all too common in the past.

Many accidents, some fatal, have been caused by employees running belts on to moving pulleys, using the hand and arm, not even taking the trouble to have the engine slowed down. Old employees, or rather men of long experience, are the greatest offenders in this respect, as familiarity breeds contempt of belts and line shafts as well as of other things much less dangerous. The accompanying sketch will give a fair idea of a method of running a belt off and on a line shaft pulley when in motion, without the slightest chance of any accident occurring, and much less trouble than usual.

The frame that carries the rollers is somewhat less in radius than the pulley and is fastened in place so that the "on" side is in line with the face of the pulley and the "off" side is slightly lower, so that when the belt is thrown over the tension is lessened.

It is very easy to push the belt on to the face of the revolving pulley with a stick having the end notched. The rollers used are of wood 1¼ in. in diameter, they are let into the two pieces that form the frame, so that they project slightly and engage the belt the moment it starts to leave the pulley.

The rollers are turned with a slight flange on the outside end and are held in place by the piece shown in the sketch.

The two pieces that form the frame are 1¼ in. thick and are held in place the proper distance apart by bolts having jam nuts attached; the whole thing is suspended to the ceiling by 1¼ in. pipes and flanges, making a neat, substantial job.

The sketch shows the belt run off on to the rollers, and the centres of the shafts are proportionately closer together than in actual use, to save space, but to convey the idea fully as well.

The writer knows that this is a practical method and where a machine may be only used one half of the time, the saving in power by being able to run the belt off and on so easily is worth consideration also. Had occasion a short time ago to put up one of these

Continued on page 38.

Possible Trade Development for the Woodworking Industry

To the owners of woodworking plants in Canada the prospects of securing sufficient business to keep their plants running and staffs employed looked rather dull only a few short months ago. They found themselves face to face with a situation which called for prompt action on their part and those who took the initiative and adapted their plants to the manufacture of shell boxes found that they were not only able to keep their plants running and staffs employed, but were able to do so with considerable profit to themselves.

A lot of discussion has arisen as to how long this business will continue and what the prospects for woodworking factories will be after sufficient shell boxes have been manufactured. An interesting contribution to this discussion comes in the form of a possible development of trade in portable houses for France and Belgium, which promises to open up a new avenue of business for those engaged in woodworking. Judged by the manner in which they are handling the shell box orders there should be no doubt but that they will be equal to the occasion. Progress is already shown in that the Canadian Timber Products Association has recently been formed with the object of promoting the export trade of Canada in manufactured wood products and at a meeting held in Ottawa on December 14th, with Mr. F. E. Barker of W. C. Edwards & Company, Ltd., Ottawa, in the chair, a resolution was passed to send a representative to Paris, France, whose duties it will

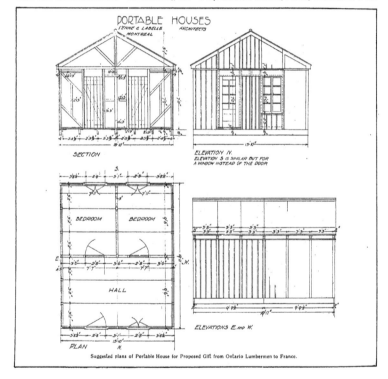

Suggested plans of Portable House for Proposed Gift from Ontario Lumbermen to France.

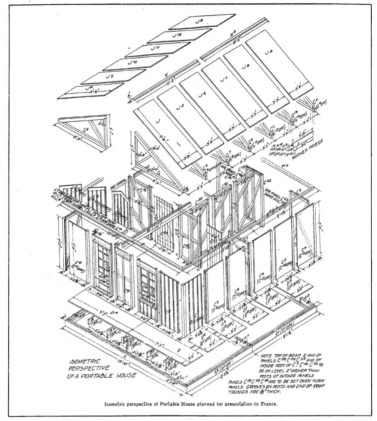

Isometric perspective of Portable House planned for presentation to France.

be to get into touch with the market for portable houses and other Canadian wood products and seek to work up trade for the Association. Another important result of the meeting was the passing of a resolution to the effect that each member of the Association would do the milling for one or two portable houses, free of charge, if the lumbermen of Ontario will contribute the lumber and the other trades will contribute the balance of the material, such as beaver board, roofing, etc.

Plans of the portable houses which Ontario firms proposed to present are published herewith. The specifications which have been worked out by the firm of Venne and Labelle, architects, Montreal, P. Q.,

call for about 3,500 feet of lumber worked to special sizes. The house shown is 15 ft. 10 in. by 19 ft. 11 in. and contains three rooms, one 9 ft. 6 in. by 15 ft. 5 in. and two 7 ft. 7 in. by 9 ft. 6 in. with a uniform height of storey 8 ft. The roof represents 395 sq. ft. which will require about 450 feet of roofing paper or such other material as may be used. For ceiling and wall lining the following quantities will be required: Ceiling 292 sq. ft.; walls 481 sq. ft.; inside partitions 155 sq. ft. on each side.

It will be seen from the plans that there will be a considerable amount of mill work on these houses. They can be accurately machined, packed and shipped, and will occupy practically the same space that the roug1 lumber would occupy. T1us our manufacturers could. by mac1ining t1em in large quantities, deliver at lower cost t1an if t1ey were mac1ined in Europe.

It can also be appreciated t1at t1e proposed gift, apart from a patriotic standpoint, would tend to create a favorable impression wit1 regard to Canadian pro-

ducts and s1ould be the means of future large orders being placed in t1is country. The importance of t1is opportunity s1ould not be overlooked as United States manufacturers will not be slow to recognize it and are almost certain to put fort1 a strong effort to secure the business for t1emselves t1at may develop in t1is particular field.

The Box Industry of British Columbia

Output rapidly expanding to cope with the steadily increasing home and foreign demand

The statement may appear paradoxical, yet it is nevert1eless true, t1at t1e depression w1ic1 overtook t1e B. C. lumber industry in 1913 seems to 1ave contributed materially to t1e upbuilding of t1e box industry of t1e province. The business has made great strides in t1e last two years, and may now be said to be firmly established on profitable lines. The output has more t1an doubled in t1e period mentioned, w1ile plant betterments now under way or contemplated would seem to be fully justified by t1e constantly increasing volume of business now offering.

The explanation of t1e apparent paradox is t1at t1e fruit and vegetable growers of t1e province saved the box industry at a critical juncture. Many new orc1ards came into bearing in 1914-15, and t1e insistent prairie demand for fruits provided a market for everyt1ing not consumed locally. New box factories were established in t1e Okanagan and ot1er interior

least one B. C. factory is manufacturing largely for t1e United States market, and American capital is being invested in a concern now getting ready to

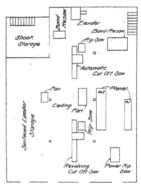

Ground floor plan of factory.

operate in Prince Rupert district, t1e intention being to market t1e product across t1e border.

One of Vancouver's Busy Box Plants

The factory of t1e Vancouver Box Co., Ltd., one of t1e best in t1e province, is located on Dufferin street west, Vancouver, a s1ort distance from t1e sout1ern approac1 to Connaug1t Bridge. The site occupied is 250 x 250 ft., 1aving t1ree street frontages and railway switc1. The factory building is 100x120 ft., t1ree storeys, so t1at t1ere is ample space available for new buildings. The company was organized in 1912 by A. Y. Jo1nston and A. M. S1arpe, as successor to t1e B. C. Box Co., Ltd., w1ic1 had been in business since 1903. Mr. Jo1nston becoming president and Mr. S1arpe retaining t1e position in w1ic1 1e had contributed so muc1 to t1e success of t1e old company—t1at of manager. The erection of t1e new factory and t1e introduction of modern mac1inery meant an increase in capacity of 40 per cent., thus en-abling t1e new management to reac1 out after new business and 1andle orders of almost any size. Gradually t1ey began to specialize in certain lines of work, and by installing t1e best equipment possible t1e fac-

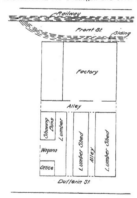

General layout Vancouver Box Co.'s plant.

fruit raising districts, yet orders continued to reac1 t1e Coast plants for large quantities of box material. Following t1e war a considerable export trade in box s1ooks was developed wit1 t1e United Kingdom Australia and New Zealand, and now t1at t1e manufactories of t1e province are again becoming busy t1ey are demanding more containers t1an formerly. At

tory speedily acquired a reputation for good work, prompt delivery and reasonable prices.

On the saw floor are two McGregor & Gourlay planers, (daily capacity 25,000 feet) ; two cut-off saws, two trimmer saws, one 60-in. band resaw, one lumber

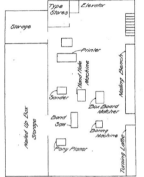

Second floor plan of factory.

ripsaw, (Canadian Machinery Co.) six small cut-off saws, and one bundling machine. As will be noted, the various machines are placed to the best advantage to handle the lumber, which is delivered on trucks from the alleys east of the sawing floor.

The second floor, served by a freight elevator, is given over to the printing, nailing, sanding and gluing

Third floor plan of factory.

departments, dado and boxing machines, work benches, etc. The printing outfit comprises one upright color press and one color rotary.

The third floor accommodates the filing room, belt storage room, combination saw and jointer, roomy

work benches and ample space for the assembling and setting up of special order work not marked for speedy delivery.

While the company have made a specialty of catering to the demands of the Coast wholesale trade, they also do a large volume of business every spring with the salmon canneries, and later with the fruit and vegetable growers of the Fraser Valley, carload lots of boxes and crates for apples, plums, peaches, strawberries, raspberries, cherries, tomatoes, rhubarb, etc., being shipped at intervals to Mission, Hatzic, Hammond, Chilliwack and other Valley points during the past summer. In addition the company found it possible to make several large shipments of butter boxes to Ireland and Liverpool, as well as a few carloads of various types of containers to Winnipeg and other prairie wholesalers.

The company's plans for 1916 cover the erection of a sawmill to cut their own box material. They own a considerable area of hemlock and spruce timber up the Fraser and elsewhere, and by selling the clear

Corner of sawing floor, Vancouver Box Co.

lumber they will be able to make an additional profit that will look well on the yearly balance sheet.

During 1915 an average of 35 men appeared on the company's payroll, and for part of the season the working time had to be increased to thirteen hours per day to cope with the rush orders.

The Difference Between Success and Failure

The manager whose time is continually occupied with details cannot originate, or plan or carry out, methods of increasing the efficiency or of cutting out leaks in his factory. The most successful managers shift details to the shoulders of less expensive employees, and devote their time to the big things. They aim to keep their time so free from the small things that they always have leisure to investigate a good thing; to follow up a good idea; and to think of plans to increase business, to increase efficiency and to decrease cost of manufacture. In a small organization the manager will always find it difficult to get rid of details, because the business is not large enough to stand the organization required to handle all detail matter thoroughly; but there is no excuse for the manager of a medium sized business or a large business not having time at his disposal.

Making Foremen out of Journeymen

By G. D. Keller*

Sometimes a sudden rush of business emphasizes the need not only of additional equipment. but of additional men; and not so much additional machine hands as additional foremen and others capable of instructing, supervising and managing men and machinery. The plant which has in every journeyman a possible foreman is well provided for against emergencies of the kind described. Napoleon declared that every soldier of France carried a marshal's baton in his knapsack; and the modern need is that every employe in every wood-working shop shall carry under his cap the gray matter belonging to a foreman or a superintendent.

Not long ago a concern in a special branch of the wood-working business got a tremendous order on account of the war. Instead of operating one plant, it found it desirable. almost over night, to set up additional ones at this point and at that, and to create facilities for manufacturing six or eight times the amount of material which had formerly come out of its plant. Getting the men and machinery for these additional factories was easy enough, but getting superintendents and foremen capable of running them was a horse of another color. One of the greatest difficulties which the concern experienced was in "organizing" these new departments of its business; and "organization," in this as in other cases, was principally a question of securing the right kind of supervision.

This matter of producing foremen, at least potential foremen, who, in turn, can easily be developed into factory superintendents, has been tackled from a good many angles during the past few years. There is none more promising than that of the night manual training school. The idea is to give the factory hand, who may be doing something of a routine and monotonous nature, a broader outlook, a more thorough understanding of the theoretical as well as the practical side of the work, and the mental equipment by means of which he can go forward.

In cities where the experiment has been tried—and the idea is new enough to be called still an experiment—excellent results have been obtained. The very fact that so many men are anxious to obtain the training that will fit them for something better is in itself a splendid indication of the morale among employees of wood-working plants. It also shows a commendable ambition, based on willingness to go through the necessary preparation to "get there."

Know as Much as the Boss

A big industrial plant in a western city has given a lot of publicity to the inside workings of its business. Among the things which it has said about its force is: "Every man is supposed to know as much as his boss." This is a rather clever and suggestive way of saying that every man is expected to equip himself to step into the shoes of the boss. whenever that may be necessary. And certainly the man at the machine. who knows not only the details of his own job. but those of the men at his side. is better fitted to handle the supervision of that part of the factory than anybody else on the same footing with him.

The night manual training school gives factory hands an excellent opportunity to take up work which is right in line with what they are already doing. Mechanical drawing, for instance, is something that every wood-worker ought to know about and be able to do. It will open the way to designing, if he is in a furniture factory. The employee who can appreciate draughtsmanship and who can turn out a creditable piece of work himself, has all sorts of opportunities. He can see what is being produced and can emulate it, with the prospect of developing workable ideas.

The man who understands this subject and who can read blueprints—an art which everybody in industry should master, if he has not already acquired it —is in line for all sorts of jobs, if he is employed in a planing mill. The desk of the estimator is one of the most important in such a plant, and the employe who is thoroughly familiar with and can interpret the drawings of the architects in connection with their specifications, has the foundation on which to build the proper equipment for this task, though of course experience is required before anybody can be regarded as able to hold down this position.

The machine hand who is able to assist the foreman or superintendent who may be working over blueprints, and figuring out the requirements for certain kinds of mill work, is a more valuable employe than the chap who is content to plod along in his own little groove, without ever attempting to learn anything outside of the small compass of his particular job.

The Manual Training School

This is only one feature of the possibilities of the night manual training school. The student at such an institution, who has the run of the shop and can use all of the equipment provided, can obtain a better working knowledge of the function of each machine than otherwise. He can study each by itself, and each in relation to the others. Thus he will not only be able to do his own work better, but will appreciate its importance, and will understand the part it plays in producing the finished goods.

It usually happens, in a big wood-working plant, that it is most convenient for the superintendent to have his men specialize. In other words, efficiency in handling one particular operation is developed by having the same man work at that machine all the time. This is a good thing for the business, but it tends to narrow the operative of the machine. He does not learn anything new, and he ceases to grow unless he takes pains on the outside to keep up his development, and to continue to learn things about the business other than those which come within his vision in the factory.

When he operates the various machines in the shop at the manual training school, when he understands to build from start to finish a piece of cabinet work, he necessarily handles every process, from the machine work to the joining and finishing. He even selects the material which is to be used, and is in charge of its production absolutely. On a small scale he becomes a "complete wood-worker," and inasmuch as his work is reinforced by the practical knowledge of machinery which he gets at his daily task, he is bound to become an abler artisan.

*The Woodworker.

QUESTION BOX

A Practical Department of Reference for the Practical Man

Ventilation of Dry Kilns
(Continued from December number.)

Q.—The answers to your questions concerning our present kiln are as follows:—

Length of tunnel—100 feet.
Width—4 feet.
Height—3 feet.
Number of pipes for exhaust steam in tunnel— One 4 inch pipe.
Length of pipes—Full length of tunnel.
Size of pipes—4 in. 2½ in. 1 in.
Size of openings for hot air—18 in. x 9 in., 2 to each compartment.
Size of fan outlet—5 feet.
Speed of fan—550 r.p.m.
Temperature of kiln; charging end—150 deg.; discharge end, 150 deg.
Kind of lumber dried—Both hard and soft.
Thickness—3, 2, 1½, 1 in.
Are you satisfied with the quality of your dried lumber—Not always.
If not, what are its defects?—Case hardening, if not sufficiently seasoned.
Is there room at the back of the kiln at the loading end for the building of a steaming chamber large enough to hold at least one car of lumber—Yes.

While we have answered your questions above briefly, to give you intelligent information, we will require to take up some of the questions more fully.

Our kiln is divided into six compartments, each being as you intimate, a self contained kiln.

We require to dry both hard and soft lumber and of various thicknesses from 3 in. to 1 in.; stock used for manufacturing pianos.

The timber that gives us the greatest trouble is Gumwood which, unless it is thoroughly air dried before entering the kiln, will case harden. Oak acts somewhat similarly, but any timber that is green when placed in the kiln gives us more or less trouble.

The exhaust steam is conveyed to the kiln through 4 inch pipes running the entire length of the kilns through the tunnel before referred to. This pipe is tapped at each compartment by a 2½ in. pipe extending up above the floor of the kiln, which conveys the steam into a coil of eight 1 inch pipes running under the cars the entire length of each compartment, which is 30 feet, or the full width of the kiln proper.

I have given you 150 deg. as the temperature of the kiln at both charging and discharging end, which I understand to mean the charging and discharging of the lumber cars, but the temperature varies in the different compartments, the ones nearest the fan get a more intense heat than those farther away. The compartment farthest away is generally 60 deg. lower in temperature than the one nearest."

A—Having before us the very complete information regarding your dry kiln with which you have favored us, we are no longer at a loss to understand why it is that your results have been unsatisfactory. A hot blast kiln like this, while generally satisfactory for soft wood if it is hot enough, is seldom of much use with hard wood, on account of the fact that it drys the surface hard before the interior moisture has a chance to escape. The only hope of success with a kiln like this is to maintain a high degree of humidity at the start and gradually reduce it as the drying process continues. This it has been impossible for you to do, for all the hot air that really contributed to the drying was forced into the kiln from the outside and the humid air was forced out through the crevices, etc., making it impossible to either achieve a high humidity or a high enough temperature, owing to the fact that only as much air as could escape from the kiln could be forced in. no matter how large the fan or how hot the coils. Further than this, all the openings from the tunnel into the various compartments are the same size, and as the air pressure drops rapidly as the distance from the fan increases, not as much would be forced into the end compartment as into the first one, resulting in the difference in temperature you mention. What we must achieve first of all if we are to accomplish anything is to secure an even distribution of heat from the fan to each one of the compartments, and this may be done by making the opening nearest the fan small and increasing the size in each successive compartment until the end one is reached. Leaving the width of the openings the same as they are now, this would call for the lengths

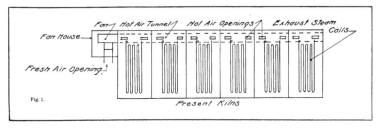

Fan House　Fresh Air Opening　Fan—Hot Air Tunnel　Hot Air Openings　Exhaust Steam Coils

Fig. 1.　　　Present Kilns

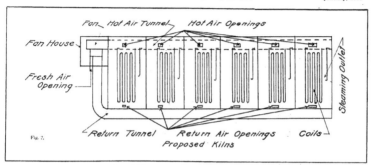

Fig. 2.

of the openings to be from the first to the last consecutively 18 in., 25 in., 32 in., 39 in., 46 in., and 54 in. The total area of these openings equals the area of the tunnel, and the pressure at all openings should be the same; if not the size of the openings may be changed until the temperatures are the same in all compartments.

Having secured a uniform temperature, (which will probably be too low, but which we will change later) we must provide for a proper circulation of the air to prevent too much of its heat and moisture from being lost before we are through with it, and this will probably be best accomplished by returning a certain percentage of it through the fan and heating coils. To this end, construct another tunnel at the back of the kiln the same as the one at the front end, but only 2 ft. x 3 ft. This may be inside the kiln at the back end or outside altogether, so long as its top is under the tracks. This tunnel must be tight and well protected to prevent the loss of the heat through useless radiation, and should have openings into each compartment the same length as the ones at the other end, but only half as wide. These openings as well as the ones at the other end must be provided with dampers. One end of this new tunnel must be connected into the intake of the fan, so as to create a suction, leaving the rest of the fan intake communicating with the outside air as before. The result of this construction will be to permit a larger total volume of air to enter the kiln at the hot end, and half of this air, after passing through the lumber and becoming moisture laden enters the ducts at the cool end and is returned after reheating to the hot end, mingled with a certain percentage of fresh air. It is plain to be seen that as twice as much air is going in at the hot end as can be withdrawn through the openings at the cool end, the balance must escape somewhere and this escape will take place through the natural openings around the doors. etc., as before, thus providing the ventilating desired, but under proper control.

Eight one inch pipes for exhaust steam in each of these compartments is not sufficient. and we would suggest that you install eight more above these so as to provide twice the heating surface you now have. When you make the connection for this additional coil, cut into the supply pipe with a branch running half way down the kiln, terminating in an open elbow turned inward as shown in Fig. 2. This will be controlled by a separate valve and will serve to admit

wet exhaust steam for saturating your lumber before the drying process begins.

With your kiln fitted out in this manner, your process will be as follows:—

Treat each compartment as a box kiln, which is to be fully loaded, closed, dried, and opened only after all the lumber in it is dry. Load only one kind and thickness of lumber in one compartment as far as possible.

After loading the compartment full, close the doors and admit exhaust steam through the open pipe. The openings at each end leading to and from the fan should be closed by sliding dampers. Leave steam on until lumber is saturated, (experience will teach the proper time) then shut off steam from the open pipe and admit steam to the coils in the compartment, leaving the kiln in this condition until the temperature is about 130 deg. and the air and lumber soaking wet. This may require 24 hours or more.

It is now safe to open the air ducts at both ends, one half for the first day and full after that, and the lumber should be left in this way until sufficiently dry, which will probably be about four days for 1 in. lumber and longer for thicker. When the process is completed, the lumber should be removed at once to a storage shed which contains no heat and the lumber left here in open piling for at least forty-eight hours before working up.

You will probably find that with this arrangement you will get too much heat, and this you must reduce by reducing the speed of the fan, bearing in mind that the less rapid the circulation, beyond a certain limit, that you have through the lumber, the better will be the quality of the drying. The speed of the fan constitutes part of the very delicate control of which this kiln will be capable, and should be regulated to just maintain the desired temperature, without in any way approaching a hot blast, which causes surface drying and case hardening.

If it is found that the temperature in the various compartments varies it will be because the sizes of the openings are not exactly right, and they should be changed, reducing the ones in the hot compartments. and increasing the ones in the cool compartments until the temperature is uniform throughout, not forgetting to change the openings in the other end in the same proportion.

This is about the best arrangement of your kilns we can suggest without spending more than the value of new equipment. What we have advocated is based on sound reasoning and facts, but the intelligent interpretation and careful execution of these suggestions is absolutely necessary for success. Having never seen the plant and with so litttle actual knowledge of conditions, we have been obliged to assume much, but fully believe that the kilns we have outlined will be far better than the ones you have, if carefully installed and given good supervision. It is sometimes customary to leave this detail in the hands of a common laborer and Dame Chance, but it is well to remember that the value of the material is high, and intelligent and even expert supervision will earn for your company a lot of perfectly good money.

Why Does a Belt Cling to a Crowned Pulley?

The Editor,   Haileybury, Ont., Dec. 8, '15·
Canadian Woodworker:

Dear Sir:—In your November issue, on the Question Box page, I notice an explanation as to a belt travelling on a pulley. The writer appears to me to be right in only one part of his answer, where he states that the belt will run off on the lower side. I think that the answer would have been much clearer if it had been stated in the following way. If the writer had said that the belt will run off the side where it first comes in contact with the pulley he would have covered the whole subject.

When the writer states that there is only one way to make the belt travel straight on a pulley, that is to move the ends of the shaft towards which the belt is running off, he is correct, but he makes a mistake in explaining the travelling of a belt on a crown faced pulley. No doubt a belt will run towards the side which first comes in contact with the pulley in the case of a crown faced pulley also. Let me give the following explanations:

1. Take two straight shafts parallel to each other with two straight faced pulleys in line; here the belt will travel straight on the pulley. Now move one end of the shaft nearer to the other, and you will find that the belt will run off to the side. Then see which edge of the belt first comes in contact with the pulleys.

2. In order to prove that the writer was in error in his explanation concerning the crown faced pulleys, and also when he stated that a crown faced pulley was two separate pulleys out of line with the shaft, you have only to take a crown faced pulley in line on each shaft. Put your belt about two inches from the pulley and run your shaft. It will then be apparent that the belt will climb over the crown without the other half of the pulley. In that case you will see that the two edges of the belt do not pull in opposite directions, and also that when the belt reaches the crown of the pulley it will make an even contact in the centre. This is the explanation of how a belt keeps on the pulley. An experiment will show that a belt will climb over a crown; the belt will not run either on the low side or on the high side, but will run off the side which comes in contact with the pulley.

3. Take a band saw on a pulley, stand in front of it with the teeth of the saw towards you, tilt the top pulley to you, and it will be seen that the saw will come in the same direction. Then notice which edge of the saw will first come in contact with the pulley, and you will find it will be the front edge.

4. Take a board feeding through a planer roller, line your roller at an angle greater than 45 degrees with the fence, and observe which edge of the board will first come in contact with the roller. It will be the edge against the fence and the board will feed through tight against the fence.

5. Line your rollers at an angle less than 45 degrees, run the board through and it will feed through the machine in working away from the fence. The edge of the board which will first come in contact with the rollers will be the edge away from the fence.

6. Now reverse the speed of the rollers, and the board will try to work towards the fence. In this instance the edge of the board which will first come in contact with the rollers on the other side will be the one against the fence.

7. In the case of a belt with a tightener on it, line your tightener pulley a little crossways with the belt, so that one edge of the pulley will meet the running of the belt before the other. Your belt will run off that side. If you reverse the speed of your belt the other edge of the pulley on the other side will meet the running of the belt first and it will run off that side.

In the same issue a question appears as to why a band saw stays on the wheels. The answer was that the saw was held on the wheels by a kind of wedge action, resulting from the front edge being shorter than the back. Now if this answer be right what would be the result with a double cutting band which is the same on both edges? According to the answer a double cutting band would not stay on a pulley. I am afraid the writer has ignored the large number of successful single cutting band saw fitters who never used a back gauge.

The writer must understand that the saw is held on the pulley by the tension in the saw, which makes a greater contact with the pulley on both edges, providing the two pulleys are well in line. If a single cutting band can stand a little more feed than a double cutting band it is due to the fact that there is in the front of a single cutting band what is known as a tire; that is to say, a strap about an inch and a half without tension will make a greater contact on that edge with the wheel and permit the saw to run straighter and harder to push back off the pulley.

Yours truly,

J. A. CHEVRETTE.

The Nicholson Lumber Company

The Nicholson Lumber Company, Limited, Burlington, Ont., carry on an extensive lumber trade, in connection with which they operate a well-equipped planing mill, which, in normal times, employs about fifty men and carries a stock of from 1,500,000 feet to 2,000,000 feet. The equipment of the planing mill, which is thoroughly up-to-date, includes 14 machines, among them being a Cowan & Company self-feed band resaw, a Berlin No. 108 fast-feed planer and matcher, a Ballantine sticker and moulder, a Preston Woodworking Machinery fast-feed power rip-saw, etc. The product of the plant includes flooring, ceiling, siding, mouldings, etc. The second floor of the plant is equipped, especially in the main department, for the manufacture of frames, stairs, newels, columns and all detail work. On the second floor there is also a box factory, which is thoroughly equipped.

On the first floor especial attention is given to the work of milling-in-transit. Special facilities are included for unloading direct from cars to the machines, the cars running on a siding through the mill.

THE FINISHING ROOM

Overcoming Shortage in Aniline Dyes

Mr. Walter K. Schmidt, an analytical chemist, explains what is being done to provide efficient substitute.[*]

THE present market conditions of aniline stains is understood pretty thoroughly by everyone consuming aniline colors. At the present time there is an absolute cessation of activities in many industries, and if this should come to be the case in the furniture industry, which at its best has not been in a very flourishing condition, it would work a hardship, unless a relief of the situation were found. There are three finishes that practically represent the selling shades—golden oak, fumed oak, and mahogany. The latter is the one finish that practically depends upon the imported colors.

When brown mahogany was introduced there was no war abroad, and it is a peculiar coincidence that this color should come on the market and meet with such universal favor just at this crucial time. Golden oak is made mostly of asphaltum gums, and the color brightened by the addition of an oil soluble yellow. In some cases an oil soluble black is used. These two give more snap to the color. They penetrate the flake, but, should we suddenly be deprived of these colors, the writer would suggest that the stain be made of a high grade turpentine and 50 per cent. naphtha. If yellowish color is desired, it can be produced by the use of fustic, which is a vegetable dye. Make a strong extract by boiling the ground fustic, and give the work a water coat, sand, and then apply the asphaltum coat. To overcome the expense of the additional coat, the natural filler can be substituted. The asphaltum stain should be put on heavy enough to give it a decided brown appearance, as though it were varnished with a thin coat of asphaltum varnish. After this has set for about twenty minutes apply the natural filler in the usual manner. This will lift the excess asphaltum, and in so doing color itself. The pores will be colored brown, as they should be, in golden oak, whereas the flake will have the brightness that is desired.

American Products Satisfactory for Fuming

Fumed oak when made in a fuming box, with dependence only upon the fuming process, if augmented by the use of tannic and pyrogallic acid, is not dependent upon foreign markets, while until about last November practically all the pyrogallic acid was imported. There has been, however, a large factory which has undertaken the manufacture of this acid, and its product is on the market in competition with the imported article. While for real fine chemical work these manufacturers may not have attained much efficiency as yet, the product costs less than the imported article, and is plenty good enough for the fuming process. What is more, it is made in our own country. The manufacturers who rely on stains to produce fumed oak can easily change their method of pro-

[*] Furniture Manufacturer and Artisan.

duction and use the chemical fuming process, which gives magnificent results.

The results are so good that an expert has difficulty in distinguishing between the actually fumed wood and that which may be produced with the following formula. The first coat, which is also used as the sponging coat, is made up of one-half ounce of tannic acid and one ounce of pyrogallic acid; water, one gallon. Apply very freely. Then sand and apply the following coat: carbonate of soda, 8 ounces; bichromate of potash, 2 ounces. Dissolve in one gallon of water. To this gallon add the following mixture: 1 ounce of sulphate of copper, and 4 ounces of hot water. Allow this to cool, and add stronger water of ammonia, which as soon as it is poured into the copper solution forms a precipitate. Keep on adding ammonia until the solution is clear and of a dark blue color; then add this solution to the soda and bichromate mixture, which will give an olive green ammoniacal smelling mixture. This, when completed, constitutes the second coat.

After the first coat has stood for twenty-four hours and has been thoroughly dried and sanded, apply the second coat, when the wood will begin to turn a regular fumed color, which process will continue until it has ripened, usually requiring twenty-four hours.

An Even Shade

The great advantage of this method is in the evening of the shade. The application of the two acids produces an excess on the entire piece of furniture. By this I mean that there is more than enough tannin present to produce the desired brownish effect. In a measure this helps to overcome the deficiencies of some of the timber, so that the second growth and first growth, and white and red oak, have a more uniform appearance than would be the case were we to rely upon the tannin naturally present in the work.

If the effect depended entirely upon tannin and the other constituents of the wood, and their texture had nothing to do with the result in color, the tannin alone would give an absolutely uniform shade; but, as far as this is concerned, the color is uniform because, as the chemicals in the second coat, is sufficient to completely neutralize the action of the fuming coat in cases where chemicals are employed, or the fumes of ammonia are employed, when the same piece of furniture is subjected to fuming. Both the actually fumed furniture and that which is chemically fumed can be greatly enriched if the life of the wood is replaced by oiling. It must be understood that when wood is subjected either to the strong fumes of ammonia or to the chemicals a certain amount of life is taken from the wood, which is replaced with the oil coat. It gives a

THREE-PLY MAPLE VENEER
SHEETS

carried in stock in the following sizes:

36 x 36 x 3/16　　　　　42 x 42 x 3/16

48 x 48 x 3/16

WRITE FOR PRICES

Hay & Company, Limited

Woodstock, Ontario

WE HAVE BEEN MAKING THE

Best French Glue
in the World

SINCE 1818; ALMOST A CENTURY

In competition at the numerous Expositions abroad, the quality of our glues has been found so high that we are usually declared "Beyond Competition." Our trade in this country makes the same declaration. Consistency in quality and service has brought this result and we are proud of it.

Perfection in manufacture from the best raw material makes the quality of COIGNET FRENCH GLUES consistently the highest.

Large stocks at Montreal assure the best of service.

Why not take advantage of these most important factors by using our glues?

At your service for prices and samples.

Coignet Chemical Products Co.

17 State Street,　N. Y. CITY, N.Y.

Factories at {St. Denis / Lyons} France

ANGLO
RUBBING and POLISHING

Works free and easy and can be rubbed in two days.

WRITE FOR SAMPLE

The
Ault & Wiborg Co. of Canada, Ltd.
Varnish Works

Montreal　　　Toronto　　　Winnipeg

THE GLUE BOOK

WHAT IT CONTAINS

Chapter 1—*Historical Notes*
Chapter 2—*Manufacture of Glue*
Chapter 3—*Testing and Grading*
Chapter 4—*Methods in the Glue Room*
Chapter 5—*Glue Room Equipment*
Chapter 6—*Selection of Glue*

WOODWORKER PUBLISHING CO., LIMITED

345 Adelaide St. West, Toronto

certain depth to the finish that makes fumed oak so beautiful—that wholesome appearance which is due to this peculiar style of finish, which makes the wood appear as though the color was through and through. It removes any possibility of its having a color and finish of a superficial appearance.

Thus we find that we can take care nicely of two of our popular finishes with products made at home. Early English, while it is no longer a good seller, can be handled also. Although no actual tests have been made, the following suggestions will be sufficient for any foreman finisher, should he find himself lacking in the aniline stains. Take a weak solution of iron, and a solution made of two ounces of logwood extract, and four ounces of fustic. Boil these in a gallon of water, and add thereto sufficient iron solution to produce the shade desired. On oaks this will make a beautiful Early English. The shade will be in absolute control by the changing of the amounts of material employed.

The mahoganies such as are on the market to-day, especially those of the stronger reds, will give us more difficulty. In the making of the brown mahogany there should be little or no difficulty. The production of a brown could be made by the use of catechue or japonica, logwood extract, and fustic, and to produce a weak tint of red, madder or alkanet root can be employed. The entire proposition depends upon the extraction of the color values, and the standardizing of each. The beginning of this would necessitate a bit of careful manipulation. A given quantity would have to be boiled, in a given quantity of water, until all the color value was extracted. Then this water must be brought back to a standard quantity, and naturally it would have to be strong. When the strength has been produced, it then depends upon the quantities of each required to produce the same color of stain that had been employed when made of aniline dyes. Each batch would have to be standardized and compared with the original standard prepared for the first experiment. After that the foreman has his formula.

In the case of the red mahoganies, the excess would be with the red extracts of alkanet root or madder, these to be toned back with the brown solution, omitting the use of fustic. The filler material could be augmented by the use of reds in the filler, so as to aid the stain in producing the desired colors. A cochineal solution could be applied and set with a mordant of alum solution. It would be quite necessary for the foreman to do some experimenting, but any of these stains, thus produced, when given a coat containing some alum, or a weak solution of sugar of lead, will be quite permanent. It must be remembered that any water stain, when applied hot, will be less apt to fade than when applied cold.

It is the opinion of the writer that the supply of anilines will be short for at least three or four months, but that there is a sufficient amount of stocks held by some manufacturers, if spread around, which would see every factory through the crisis. I know of one stock of over 4,000 pounds, carried as red and brown mahogany, and while this factory is absolutely secure from any contingency that may arise, there are other factories that are not so fortunate. A few pounds of this large stock would see these through the season. Dealers are pretty well cleaned up, and importers are in no position to fill any orders. New York warehouses are absolutely cleaned out.

This consignment of African Mahogany is the selection from a million feet, which accounts for its superiority.

Mahogany and Cabinet Woods—Sawed and Sliced

Quartered INDIANA White Oak, Red Oak,
Figured Red Gum, American Walnut, Etc.

Rotary Cut Stock in Poplar and Gum for Cross
Banding, Back Panels, Drawer Bottoms and Panels

THE EVANSVILLE VENEER CO.
Evansville, Indiana

The Importance of Veneers and Built-up Stock

By Albert Hudson (Continued from December)

The storing of veneers is very important. The stock room for veneers should have an earthen floor if possible. The finer grades of face veneer are nearly always very thin and if all moisture is dried out of them they are certain to crack; even thicker veneers will become brittle if dried, so that they are easily broken in handling. The object of having a stock room with an earthen floor is to ensure that the air will be kept moist, since dry air is one of the worst enemies of veneers while stored away before use. An earthen floor always keeps the humidity high in an enclosed space even in the summer months.

Now a word as regards the selection of veneers and the proper handling of the same preparatory to gluing up into the finished product.

All veneers should be selected according to color and texture as well as for figure. In high grade furniture manufacture there is a call for modest display of figure. At one time, especially in the quartered oak, the call was for a large splashy figure irrespective of color or texture. There are many well figured veneers that are unfit for use in the manufacture of high grade built up stock because of two reasons. (1) A soft spongy texture has a tendency to take up much color in the staining and looks flat under the finish. (2) Because of the natural alternation of smooth, hard and soft streaks, as the result of which the veneer will show light and dark streaks after staining. This is quite as undesirable as the soft spongy wood. The most artistic results are obtained from veneer of a good even figure and texture rather than from the splashy figured stock.

Stock Carefully Redried

Knowing the tendency of veneers to absorb moisture while in the bulk in the storeroom, both in the fancy goods and the cheap rotary stock which is used mostly for crossbanding, etc., great care should be taken to see that such stock is carefully redried before it is glued up into the finished product. Redrying veneers does away with the possibility of damp or green stock being used. It is the use of undried stock that causes such a lot of trouble in the way of checks, loose veneers, open joints, etc. One would not think of laying veneer without matching; neither should one think of laying veneer without redrying. Do not be afraid of having veneer good and dry; it may be easier to handle when soft and pliable, but to lay in that condition means trouble. Probably 50 per cent. of the poor results obtained from gluing up comes from improperly dried stock—hence the importance of redrying.

Some manufacturers have trouble with their veneer joints, especially fancy woods, such as crotch mahogany and circassian. It should not be beyond the ability of a good workman to make good joints either with straight veneers or with crotch and cir-

cassian veneers. Ordinary veneers are, for the most part, of even texture, and their response to moisture in the way of expansion and contraction is practically uniform throughout the whole veneer. For this reason the workman is mostly successful with his joints on this class of veneers. Crotch mahogany and circassian are practically the opposite; hence the veneer man should study the texture of the two veneers in question, ordinary plain veneers and fancy veneers. He would then be successful in getting good joints on the fancy veneers as well as on the ordinary plain veneers.

Laying Veneers

The fancy veneers cannot be laid in the same manner as the ordinary veneers. This is where some men fail, as they do not consider the effect of moisture on the different textures and therefore proceed to lay the fancy veneers just as they would the plain veneer, the result being uneven expansion and poor joints.

In laying crotch mahogany, etc., prepare the core first by toothing, then the glue is spread on and after allowing the surplus moisture to enter the core, the crossband is laid on and after remaining in the press long enough to dry is piled away and the crossband allowed to season. After seasoning, the crossbanded material is toothed and glue spread on as before; the face veneer is then applied and the whole again put in the press and the hot cauls put in place. By this method the glue will become fairly congealed, but this will not hurt as the hot cauls will remedy this.

By this way of handling, the face veneer does not have a chance to absorb moisture to sufficient extent to damage the joint before the pressure is on, the pressure being once applied, there is no further danger.

Some may say that the process is a long one. "I admit that it is," but in the end it pays and you have the finished product O.K. if handled properly. In carrying out a system so many persons fail just because they do not carry it out in its entirety, sometimes one little thing is left undone that spoils the whole thing. Personally I believe it is better to be safe than sorry.

Again, do not attempt to joint too many pieces at once, or glue up too many pieces at once. Six pieces is quite sufficient. When taken out of the presses, fill up any small holes with lithage colored with a little red ochre and give a coat of glue size.

The secret of laying end wood veneers is to treat and handle them properly. If the veneer is properly treated and correctly laid it will stand up indefinitely without checking unless the veneered piece is allowed to become filled with moisture from long storage in a damp warehouse or storeroom. Take time to let the work dry out thoroughly before starting to finish, also allow plenty of time between the different processes and the results should be satisfactory.

Don't Lay the Blame on the Veneer Man When the Glue is at Fault

By M. A. Burt

Blaming the veneer man for poor veneering is a fault of many manufacturers of furniture and panels, yet these same men will try to beat the glue salesman down a half or whole cent a pound on the price of his glue, and chuckle with satisfaction to think they are getting eleven cent glue for ten cents. Here is probably the cause of considerable poor veneering that the veneer man gets blamed for. The manager thinks he is getting eleven cent glue and uses it as such, while in reality he is getting and using just what he is paying for—ten cent glue. The result is a batch of panels, or other stock, goes through the shop with blisters and loose veneer, and the veneer man gets blamed for not having his cauls hot enough, or else too hot; or not putting on enough pressure, or else putting on too much pressure; while in reality the cause of it all is the manager's bargain glue.

This sort of thing happens so frequently that suspicions are aroused when a new brand of glue comes in, for with it always comes the same old story, that is, a trial order that will stand a certain percentage of water, will spread so many feet per square inch, etc., usually appearing more economical than the glue previously used. Experiences with this sort of thing, however, usually convince one that ten cent glue is ten cent glue, no matter where bought, and its value is determined by its strength and spreading qualities. Glue manufacturers have established values whereby one is not likely to get fifteen cent glue for ten cents.

One veneer department used nine cent glue with good results by mixing 45 pounds of dry glue per 100 pounds solution. This was not what the glue salesman claimed the glue would do, nor what the manager wanted it to do. The glue salesman claimed that 36 pounds of dry glue per 100 pounds of solution would give a glue with sufficient body for any kind of wood, used at 175 degrees F., and the spread would be 35 square feet per pound of glue. Tried in these proportions it was found that while it would hold on hardwood veneers fairly well, under most favorable conditions, it was of no use whatever on the softer woods. Hence the first mentioned solution was decided on after considerable experimenting, and this grade of glue is never used with less body than forty per cent. dry glue, and at this proportion it is used on hardwood core bodies with hardwood cross banding, usually soft maple and mahogany face veneers. On this class of goods it is the custom to spread about 28 square feet per pound of glue. This may seem to be pretty thick, but this plant figures that glue is cheaper than repairs. When the glue salesman's advice was followed one man was always busy on repairs.

Going Back for Repairs

This sort of thing was more expensive than it appeared to be, for in addition to the repair man's wages, there was the extra glue used in the case of reveneering, also glue and labor. But the annoyance did not cease with the loss occasioned by the repairs or reveneering, for in the cabinet department jobs would be held up that were partly benched, waiting for parts that had to go back for repairs. While it may seem a very simple matter to return tops, drawer fronts, etc., to be reveneered, the trouble and extra work for the bench in such cases is a bigger item than is usually supposed. Take, for instance, a man starting to bench a batch of fifty sideboards. He will have his stock in the bench room and the different parts separated and in such order as will be most convenient for work. He will then commence to clean up such parts as require to be cleaned before setting up, but if he finds some of the gables require veneering, it simply means that as many cases as these gables represent cannot be set up. The delay caused by reveneering amounts to a week or more, while in the meantime the balance of the job is likely to be benched, and the remaining parts have to be set to one side to await the return of the gables. When these arrive the stock has to be handled again. This extra work is too often looked upon as a matter of course, and its effect on production and profits is usually underestimated. Sometimes these things are unavoidable, but usually a little common sense investigation will reveal the cause. and it is generally in the veneer department, but not always the fault of the veneer man.

Taking the Salesman's Word

To illustrate, let us consider the case where the veneer man handed the manager a requisition for glue, stating that on hand would last until a certain time. The manager desired to know the why and wherefore, and went to the veneer department and put the hydrometer in the glue dissolver containing the glue from which the spreader was being supplied. He found it contained 43 pounds of dry glue per 100 pounds of solution. This glue was out of a shipment that the glue salesman had said would produce a good body at 36 pounds. The manager had the solution reduced to the salesman's standard, but it happened that it was an extra solution, made a little heavier for a special job. The manager was so informed, but it didn't make any difference the solution was reduced, and 500 pieces of three ply basswood were glued up. A week later, when the stock was taken off the strips, it was found that only 230 pieces were fit for use.

The majority of manufacturers of veneered stock seem to be more willing to pay $2.00 for labor than $1.00 for glue. While it is, doubtless, true that many veneer men waste a certain amount of glue, by making solutions stronger than is necessary, it is safer and more economical to err on the side of safety than otherwise. A glue solution a little stronger than is necessary will not do any harm to the stock, or the man who produces it, but a solution a little weaker than necessary is apt to injure the reputation of the firm. Defects from poor or light glue do not always present themselves before the goods leave the factory. Small blisters may develop, and an examination show that the veneer may be peeled off very easily, which is evidence that the glue used was too light.

Proprietors and managers of furniture factories will find it more profitable to adopt proportions for their glue solutions to suit their particular class of work, rather than take for granted all the glue salesman may state, for it usually happens that the veneer foreman knows from experience what is best for the work at hand.

If outstanding contracts are filled, and the war continues throughout 1916, it seems clear that during 1915 and 1916 there will have been spent in Canada for war supplies considerably more than $500,000,000.—Sir Edmund Walker, at annual meeting Bank of Commerce.

OAK MAHOGANY CIRCASSIAN MAPLE

VENEERS

BIRCH POPLAR CHERRY WALNUT GUM

THE OHIO VENEER COMPANY

Importers and Manufacturers

Foreign and Domestic Veneers and Hardwood Lumber

We always carry a large and assorted stock of Mahogany, Circassian
Walnut, Sawed and Sliced Quartered Oak.

Send us your enquiries and orders. We guarantee good service.

2624 to 2644 Colerain Avenue, CINCINNATI, OHIO.

George W. Hartzell

Piqua, Ohio

Wholesale Manufacturer

AMERICAN WALNUT

LUMBER and VENEERS

Large Stocks Dry Lumber
Ready for Prompt Shipment

We can also supply

Clear Dimension Stock

Write Us For Prices
Send Us Your Orders

*Special Attention Given to the
Requirements of Furniture Manufacturers*

Hoffman Bros. Co.

Estab. 1867, Incor. 1904

800 West Main Street,
FORT WAYNE, INDIANA

Manufacturers of

VENEERS and LUMBER

In the Domestic Hardwoods

ANY THICKNESS.

1/24 and 1/30 Slice Cut
(Dried flat with Smith Roller Dryer)

1/20 and thicker Sawed Veneers,
Band Sawn Lumber.

—SPECIALTY—

Indiana Quartered Oak

VENEERS & PANELS

CARRIED IN STOCK

VENEERS

Red and White Oak, Birch, Poplar, Gum,
Yellow Pine, Cypress, Etc.

PANELS

Three and Five Ply—All Woods.

GEO. L. WAETJEN & CO.
Milwaukee, Wisconsin

Veneers and Panels

LARGE STOCKS FOR
QUICK SHIPMENTS

Send for Lists

Walter Clark Veneer Co.

GRAND RAPIDS, MICH.

Veneer for Door-making
By O. B. Henley*

Veneer for door-making is becoming a more important item in the veneer world every year, and it has reached the point now where this class of stock is deserving of special study, with a view to getting the best results both for veneer manufacturers and for the doors themselves, structurally and in appearance.

The more common method of making panels for the regulation doors in birch, gum and other native woods is to make them of three-ply ⅛-in. stock. This produces a ⅜-in. panel, and it means that in cutting the veneer stock there must be two faces to one centre. In this respect the door panel differs somewhat from many furniture panels and from other mill work panels, where, quite commonly, there is just one face, a back and a centre. In panel work calling for only one face it is not essential that more than one-third of the stock be good enough for faces. In door work, however, the kind referred to here, there must be two good faces for every centre or filler. In other words, two-thirds of the stock should be practically clear.

Given comparatively good timber in either birch or gum, it may be easy enough to get this percentage of

The Imperial Munitions Board have given orders in Canada for 22,800,000 shells, having a value of $282,000,000. If we add to this the orders for cartridge cases, primers, forgings, friction tubes, etc., a total of $303,000,000 is reached.
—Sir Edmund Walker, at annual meeting of Bank of Commerce.

faces, but it will call for good timber, and, also, to put up panels in this way, calls for a thickness out of keeping with thicknesses used for face stock in the cabinet world.

Generally, face stock in the cabinet world runs from 1/20 to 1/28-in. in thickness, and even the cross-banding under the thin face stock is now commonly cut to 1/20-in. One strong reason for doing this is that good face stock is valuable and it is a waste of material to cut it ⅛-in. thick. It is using the thin material for purposes that may just as well be served with less valuable wood. Moreover, it is contended that the thin face stock can be cut with less rupturing of the wood than thick stock and makes a better job of veneering.

When we get into some of the finer jobs of door-making, where doors are made to order or to architect's specifications, we find thin face stock being used. It may be figured gum, oak, mahogany or walnut, but it is thin face stock, and the work partakes of the same nature as furniture panels, and involves the use of cross-banding under the face stock to secure satisfactory results.

We are, then, confronted with the question in connection with veneer for door-making of whether it will be wise to practically reorganize the whole system of work, using thin face veneer and the built-up core body underneath. It seems a lot of extra work and expense to make panels for the ordinary four or six-panel door, five-ply—that is, to build them up with a three-ply body and face on each side with thin veneer. It is possible that there is room to compromise here and to make a four-ply job of it, make the centre body one

* In "Veneers."

of two-ply common veneer, faced on each side with thin face veneer.

Still, it only requires a little more work to produce a five-ply job, which is much better, and, when it comes to the doors with larger panels, with only one or two panels in a door, it becomes essential to make a five-ply job of the panel. It matters not what the thickness or what the core body, planing mill men and door manufacturers have long since found out that it is not practicable to make three-ply door panel of large size and have it stand up as it should. There will be wrinkling and buckling, and the only way to secure a satisfactory job is to cross-band the core or centre on each side and then use a thin face veneer. The indications are that in the course of time smaller panels for the better grade doors will be made in the same way. For cheaper doors, those to be stained or painted, it may be practicable to make them up of either two or three-ply plain veneer, but in the highest grades of work, where the natural beauty of the wood is to be brought out, the tendency at present is strongly toward five-ply panels with thin face veneer.

With thin face veneer used in the panels there will naturally follow a disposition to use thin face veneer on the stiles and rails. This, to insure a good job, will perhaps call for cross-banding under the face veneer on stiles and rails.

It will not be necessary for the cross-banding to literally cross the other in grain. It will be more likely a case of laying a cushion of 1/20 or 1/16-in. plain veneer on the solid or built-up stock for stiles and rails and applying the face veneer to this. It will cost a little more, but it will make a better door, and that is where the tendency is to-day—toward the making of doors higher in the quality of finish and in which is displayed the natural beauty of the wood.

The plain five-cross-panel doors may continue for some little time in the old way, the panels being made up of three-ply ⅛-in. stock and the stiles and rails faced with ⅛-in. stock, especially when they are made in birch and gum; but the big two-panel and single-panel doors will not prove satisfactory when made up in this manner. They call for five-ply work and will naturally take on the thinner and finer face veneer. So, in the course of time, the veneer for door-making, instead of being almost universally ⅛-in. stock, will be of the same stock as to thickness for face body as is used in cabinet work, and the door construction will have to be altered in whatever manner is necessary to conform to the needs of this kind of veneering.

The Tillson Company, Limited

The Tillson Company, Limited, Tillsonburg, Ont., carry on quite an extensive woodworking business. Their plant covers a floor area of about 18,000 square feet. On the ground floor all the heavy machinery is located, including the following:—one S. A. Woods 12-in. inside moulder, one Cowan 26-in. revolving bed surfacer, one American Woodworking Machinery Company power-feed rip saw, one Cowan band resaw, one Goldie McCulloch swing cut-off saw. On the second storey the equipment includes one four-sided sticker, one universal woodworker, one Cowan chain mortiser, one 2-spindle shaper, one elbow sander, one Cowan tenoner for doors, sash, etc., one railway cut-off saw, one four drum sander, one Cowan 26-in. surfacer, one sash and door sticker, one rip saw, one band saw, one hollow chisel mortiser and relisher, one stair router, one Cowan universal saw table, one turning lathe, door clamps, sash clamps, etc.

OAK — MAHOGANY — CIRCASSIAN — MAPLE

VENEERS

BIRCH — POPLAR — CHERRY — WALNUT — GUM

PENROD WALNUT & VENEER CO.

Manufacturers

——

Exclusively Walnut Lumber and Veneers

——

KANSAS CITY, MO.

Penrod, Jurden & McCowen

Manufacturers of

Hardwood Lumber and Veneers

announce the opening of

GENERAL OFFICES

in

MEMPHIS, TENNESSEE

| Veneer Mills | Band Mills |
| HELENA, ARK. | BRASFIELD, ARK. |

We specialize on

Poplar

cross-banding and backing

Veneer

Highest Grade — Right Prices

The largest Poplar Veneer Mill in the world.

The
= Central Veneer Co.

Huntington, West Virginia

The Glue Book

What It Contains:

Chapter 1—Historical Notes
Chapter 2—Manufacture of Glue
Chapter 3—Testing and Grading
Chapter 4—Methods in the Glue Room
Chapter 5—Glue Room Equipment
Chapter 6—Selection of Glue

Price 50 Cents

Woodworker Publishing Co.
Limited
345 Adelaide St. West, Toronto

Breaking-out or Cutting Stock to the Clear·
By A. G. S.

This work has been and is still done in a great many of the piano, furniture, chair and casket factories, on the old type of hand feed rip saw, and at best is a very slow and, consequently, very expensive method of doing this work.

The excessive cost of labor, however, in doing this work is not the most important item. The amount of lumber that is yearly wasted by this method is enormous.

Recently the writer had an interview with a casket manufacturer who installed one of the new chain feed rip saws. This gentleman stated that by the new method of doing this work in their plant they were saving two feet of lumber on each casket manufactured, which amounts to about 9 per cent. of the total lumber used in the building of a casket.

Considerably better results were obtained when the chain feed band rip saw was placed on the market. There were, however, and are to-day, some serious objections to the installation of band rip saws, especially so in the smaller factories where only one of these machines would be used and where no band resawing is done. In a factory of this kind, it becomes necessary when installing a band rip saw, to also purchase a filing room equipment, in order to dress the band rip saws properly.

Next, is the question of room to place this filing room equipment properly.

And last, but by no means least, the necessity of

· American Furniture Manufacturer.

securing a filer who is capable of taking care of wide band saw blades. These men are scarce, as is a well-known fact, also the small factory is more or less at the mercy of a filer of this kind, because if the filer leaves them, it means that the band rip saw will have to stand idle until another man is secured, who is capable of taking care of the blades properly, and the small manufacturer found at the end of the year

The Customs receipts of the city of Toronto for the month of December amounted to $2,128,050—the largest in the history of the city.

after installing a band rip saw that the additional floor space required, extra cost for band saw blades and extra wages paid the band saw filer, more than ate up the saving which was effected by the machine proper.

Since the overcut chain feed rip saw has been placed on the market, all these objections have been removed, as no extra floor space for filing room equipment is necessary, neither is an expert filer necessary.

The saw used in this machine costs no more than those used on any other rip saw, and the capacity of the machine is limited only to what the operator can get through the machine.

Some of the manufacturers who have installed a machine of this type claim that one ripper and off-bearer do as much as two hand rippers formerly did.

Another manufacturer has told the writer that his ripper produced as much on a chain feed rip saw now as formerly three rippers produced, and the operators are able to do this absolutely without any danger of getting cut.

The great saving, however, obtained from this machine is not in labor, it is in lumber.

First: The machine will rip absolutely straight, and this means that a very much lighter cut can be taken on the glue jointer. This refers to hard wood face joints only.

All core stock joints can be made directly from the saw and the saving effected in this manner means from a quarter of an inch to three-eighths of an inch on joints made, saving the jointing operation entirely, which is eliminated.

It seems that in these days of efficiency, and also the days of small profits, that a great many furniture manufacturers could inject a little efficiency in their method of breaking out their stock to very good advantage:

Safety in the Machine Room
(Continued from page 21.)

attachments in connection with a shaper, the belt runs almost perpendicular and is crossed, and the consumption of power and oil when the belt was running idle was a dead loss, and owing to the danger and delay of throwing off the belt and putting it on again when needed, necessitated the better method.

If this gives any of the readers of the Woodworker some help, the writer will consider the time well spent.

Don't build up a foundation for a dull rubbed or polished surface with a "finishing varnish."

Casehardening ?

You are casehardening your lumber if you don't use a **Grand Rapids Vapor Kiln.** The vapor is a dense hot fog surrounding the lumber and keeps the surface soft and moist while the center dries. Lumber free from casehardening will show this test.

Ask for "dope" on how to prevent casehardening.

1200 kilns in successful use.

Canada's Premier Hardware House
(Established 1885)

Shell Box Hardware
Bands, Handles (wire and rope)
Screws, Nails, Staples, Houseline,
Corrugated Joint Fasteners

Large Stock—Approved and Inspected Goods
Write for Prices and Samples

Sole Selling Agents for the British American Hardware Manufacturing Co.

Tools – Tools – Tools

Write for Catalogue

FACTORY and MILL SUPPLIES, CABINET HARDWARE

BARTON'S GARNET PAPER

Canadian Agents: UNIVERSAL CASTERS

TRY US FOR SPECIAL MADE-TO-ORDER HARDWARE

 The
Vokes Hardware Co., Ltd.
40 Queen Street East
Toronto – Canada

News of the Trade

The H. Chatillon Furniture Company, Ottawa, have made an assignment.

Lindsay Woodworker's Limited, Lindsay, Ont., have obtained a charter.

The Gold Medal Furniture Company will erect 15 residences at Uxbridge, Ont.

The London, Ont., School Board are contemplating the erection of a Technical School.

Semmens & Evel Casket Company, Hamilton, damaged by fire, loss estimated about $1,000.

The Henry Wise Company, of Palmerston, Ont., are running overtime making shell boxes.

Mr. David Earl is considering the erection of a woodworking plant either in Berlin or Waterloo.

William Rudd, of the firm of William Rudd & Company, Dresden, carriage manufacturers, died recently of pneumonia.

William Gray & Sons, Limited, of Chatham, Ont., carriage builders, had their dry kiln damaged by fire recently.

Mr. J. Kerr Osborne, for many years vice-president of the Massey-Harris Company, died recently in Bournemouth, Eng.

Mark Dadson, an artificial limb-maker, for over thirty years with Authors & Cox, died recently at the age of 73.

The picture frame factory of Thomas Crowley, Toronto, was damaged by fire. Loss about $8,000; covered by insurance.

The factory of the Jamestown Table Company, Jamestown, Ont., was damaged by fire, loss about $2,000, covered by insurance.

Frank Nyberg, cabinet and antique furniture manufacturer, St. John, N. B., suffered loss by fire recently but was fully insured.

Mr. Frank Storey, of Little Current, Ont., has increased the capacity of his shop in order to cope with the demand for fish boxes.

Lambert & Frere, 324 Prefontaine Street, Montreal, are erecting a small addition to their sawmill to manufacture sash and doors.

The toy department of the Toronto Relief Office, employing crippled men, is now in full swing making toys formerly imported.

The Perkins Glue Company, Lansdale, Pa., have leased a factory in Hamilton and will commence operations about the middle of February.

The Sussex Manufacturing Company have completed a contract for 20,000 shell boxes and have under consideration another contract for 100,000.

The millmen of British Columbia are planning to urge on the shell commission the use of hemlock in place of spruce for manufacturing shell boxes.

A floor in the factory of the Malcolm Furniture Company, which was heavily weighted down with shell boxes, collapsed; fortunately no one was injured.

Mr. T. Z. Pariseau, of Pariseau Freres, Limited, box manufacturers, Outremont, Que., has been elected a member of the Montreal Board of Trade.

Mr. Alexander Ramsay, president of the A. Ramsay & Son, Company, paint and varnish makers, died suddenly on January 14 at his residence, Argyle Avenue, Westmount, aged 76. Mr. Ramsay was a native of Glasgow, but was edu-

cated in Montreal, and in 1867 acquired the paint and varnish business established by his father. Under his direction the trade very largely increased.

J. Lewis & Sons, Limited, Truro, N. S., have been incorporated to carry on the business of lumbering and manufacturing of wood and other materials.

Mr. J. J. Thomas, for 29 years superintendent of the Bell Piano factory, passed away recently after an illness that had extended over several months.

The Windsor Casket Company, Limited, Windsor, Ont., has been incorporated for the purpose of manufacturing coffins, caskets and undertaking apparatus.

G. A. Roy, Limited, St. Pie, P.Q., has been incorporated to carry on the manufacture of coffins, caskets, hearses, furniture, sash and doors, toys and turned articles.

The National Furniture Manufacturers' Association meeting in Grand Rapids recently, expressed the view that furniture would advance from 20 to 50 per cent. at the beginning of the year.

The village of Stirling will erect a factory to cost $1,500 which will be used by Messrs. Wallace, Chapman and Marshall for the purpose of manufacturing cheese boxes, berry boxes, etc.

The annual "Made in Stratford" furniture exhibit was held in that city from January 10th to the 15th. Almost all the local concerns were represented and many new lines were shown.

The A. R. Williams Machinery Company, of Vancouver, have entered a claim for $9,000 of the insurance collected by Mr. John Graham as assignee of the Westminster Woodworking Company.

Mr. Thomas W. Leslie has been appointed by the Manitoba Government as an expert to choose the furniture for the new legislative buildings in Winnipeg. Canadian furniture will be used exclusively.

Mr. Charles Powell, for fourteen years a foreman in the George McLagan Furniture Company's factory, Stratford, Ont., was instantly killed a few weeks ago, being struck in the head by a broken band-saw.

The Frontenac Moulding Company, Limited, has been incorporated at Toronto, with capital stock of $250,000, to manufacture and import mouldings, frames, pictures, picture frames, mantels, wood fittings, furniture, cabinet-woodwork.

A new company, to be known as the Nanoose Shingle Company, Limited, has been incorporated at Victoria, B.C., with a capital of five thousand dollars. Among other interests, the company will acquire sawmills and factories for the manufacture of any article of which wood forms a component part.

Messrs. A. Lindstrom and Nelson, of Kenora, for several years foremen with the Scott & Hudson Building Company, have formed a partnership under the firm name of Lindstrom and Nelson. They have rented the Sharp Building and will install modern machinery to manufacture sash, doors and other articles of woodwork.

Mr. John Wesley died recently at his home in Toronto after four days' illness. Mr. Wesley was the organizer of the Dominion Piano & Organ Company, and for some years held the position of secretary-treasurer with that company. In 1890 he started the Berlin Organ and Piano Company, and eighteen years ago removed to Toronto and has been business manager of the Mendelsohn Piano Company ever since.

The Arionola Manufacturing Company of Canada, Limited, have been incorporated in the province of Ontario "to carry on trade throughout Canada and other countries as wholesale or retail dealers in and manufacturers of musical instruments of every description and all parts thereof, and

To Manufacturers of
SHELL BOXES

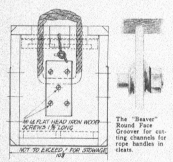

№ 14 FLAT HEAD IRON WOOD SCREWS 1⅝ LONG

NOT TO EXCEED ⅞ FOR STOWAGE 10⅞

The "Beaver" Round Face Groover for cutting channels for rope handles in cleats.

You Should Use These
RADCLIFF SPECIALS

On the left is the Dado head for cutting grooves in the sides of shell boxes and slots in the end pieces of shrapnel boxes. These dados cut absolutely smooth and clean.

In the centre is our special saw for cutting holes in the top and bottom trays of the boxes. It is made of the best Swedish bandsaw steel with special coil spring in centre for ejecting the cores.

On the right is our combination tooth planer saw, which cuts with or across the grain so smoothly that it eliminates any extra finishing.

It will pay you to get our prices on these saws.

THE RADCLIFF
SAW MFG. CO., LIMITED
550 Dundas Street
TORONTO - ONTARIO

Dry Spruce
and Birch

Good Stocks, Prompt Shipments, Satisfaction

C. A. SPENCER, Limited
Wholesale Dealers in Rough and Dressed Lumber

Eastern Townships Bank Building
MONTREAL - - Quebec

 TEAK

Write for prices on all

D O M E S T I C

Plain and Fancy
Lumbers and Veneers

Mahogany, Oak, Birch, Walnut, Ash, Poplar, etc., rotary, sliced or sawed in all thicknesses

SHELL BOX PARTS
in birch and spruce
Bridges and Ends

Write for Quotations

L U M B E R

George Kersley
224 St. James Street, Montreal

 MAHOGANY

Peter Hay Machine Knives

Peter Hay knives are made of special quality steel tempered and toughened to take and hold a keen cutting edge. Hay knives are made for all purposes.

Have you our Catalogue?

Made in Canada

The Peter Hay Knife Co., Limited
GALT, ONTARIO

all furniture. fixtures and appurtenances of every kind used in connection with musical instruments; to manufacture, buy, sell, print and deal in music and musical text books of every description, places and musical records of every description and all articles of every kind used in connection with musical instruments of every description." Capital stock $50,000; head office Toronto.

Second-Hand Woodworking Machines

Band Saws—26-in., 36-in., 40-in.
Jointers—12-in., 16-in.
Shapers—Single and double spindle.
Re-Saw—42-in. American.
Lathes—20-in. with rests and chisels.
Surfacers—16 x 6 in. and 18 x 8 in.
Mortisers—No. 5 and No. 7 New Britain.
Rip and Cut-off Saws; Molders; Sash Relishers, etc.
Over 100 machines, New and Used, in stock.

BAIRD MACHINERY COMPANY,
123 Water Street, Pittsburgh, Pa.

Second-Hand Woodworking Machinery

Baluster Machine, Mattison No. 8.
Band Saws, 36 in. and 38 in.
Boring Machine, 3 Spindle, horizontal, H. B. Smith.
Boring Machine, 4 Spindle, horizontal, Garden City.
Dovetailing Machine, 10 Spindle, Advance.
Dowell Machine, double, Crescent, power feed.
Dowell Machine, six heads, Fay & Egan, power feed.
Glue Heater, instantaneous, Zimmerman.
Glue Heater, 45 gal., Power Agitator, Howard.
Glue Spreader, 2 Francis 31 in. x 37 in., double rolls.
Jointers, 16 in. Oliver, 16 in. Smith, 16 in. Colladay.
Planer, Single, 24 in. x 6 in., Daisy.
Planer, Single, 24 in. x 8 in., Valley City.
Planer, Single, 20 in. x 8 in., Rowley & Hermance.
Planer, Single, 32 in. x 7 in., L. Power Co.
Planer, Single, 24 in. x 8 in., No. 4 Williamsport.
Planer, double, 30 in. x 4 in. Whitney.
Planer, double, 20 in. x 14 in., endless bed, L. Power Co.
Planer and Matcher, 14 in. x 4 in., Hoyt Bros.
Sander, triple drum 42 in., Invincible.
Sander, triple drum 42 in., Young Bros.
Sander, single drum 30 in. Surface, American.
Sander, Bell, No. 173 with No. 174 attachment, Wysong & Niles.
Sander, Universal disc and spindle, Fay & Egan.
Shapers, double spindle, 1 Clement, 2 American, 1 Fay & Co.
Swing Saw, parallel, No. 8 Pryible.
Tenoner, No. 2½ American, single heads, copes and cut-off.
Tenoner, double end, single heads, 2 copes.
Sticker, single head, heavy type, American.
Universal Woodworker, No. 30, Famous.

FRANK TOOMEY, Inc.
131 N. Third Street,   -   PHILADELPHIA, PA.

PATENTS

STANLEY LIGHTFOOT
Registered Patent Attorney
Lumsden Building (Cor. Adelaide & Yonge Sts.) TORONTO, CAN.
The business of Manufacturers and Inventors who wish their Patents systematically taken care of is particularly solicited. Ask for new free booklet of complete information.

7% INVESTMENT 7%   PROFIT SHARING
Sums of $100, $500, $1000 and up. Safe as any mortgage.
Business at back of this Security established 28 years.

Send for special folder to

National Securities Corporation, Ltd., Confed. Life Bldg. TORONTO

"ACME" Casters
Outlast The Furniture

The choice of trimmings and casters may be a deciding factor in the classification of your furniture. Use "Acme" Ball-Bearing or Non Ball-Bearing Casters and you will be known as a maker of the best furniture.

"Acme" Casters usually outwear the furniture they carry. They never bend, loose their wheels or develop flat sided wheels.

We solicit your enquiries and will be pleased to have our Canadian representative call and discuss any special Casters or Period Trimming you may require.

Foster, Merriam & Co.
Meriden, Conn.

Canadian Representative:—F. A. Schmidt, Berlin, Ont.

"ACME"
For All Furniture

TAKE ITS COST OUT OF THE PAY ROLL

It saves $3.00 to $4.00 a day

Let me show you how

J. M. NASH, MILWAUKEE, Wis.

Work your waste into small handles and other small turnings.

The Ober Lathes will help you.

We also make Lathes for turning Axe, Pick, Sledge and other irregular handles, Whiffletrees, Neckyokes, Etc., Sanders, Rip Saws, Cut Off Saws, Chucking Machines, Etc.

Write for free catalogue No. 31

THE OBER MFG. COMPANY

CHAGRIN FALLS, O., U.S.A.

VENEER PRESSES

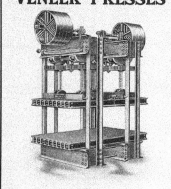

WRITE FOR BULLETIN C 1.

Canadian Boomer & Boschert Press

Company, Limited

18 Tansley Street, MONTREAL

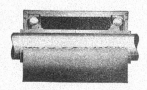

Your Loose Pulley Trouble Disappears

when installed with

CHAPMAN BALL BEARING LOOSE PULLEYS

A little vaseline once a year is all the attention and lubrication required.

Our loose Pulleys cannot run hot and do not cut the shaft.

Write us for particulars and price list.

The

Chapman Double Ball Bearing Co.

of Canada, Limited

339-351 Sorauren Ave., Toronto, Canada

The "Canadian Woodworker" Buyers' Directory

For Alphabetical Index see page 20

ABRASIVES
The Carborundum Co., Niagara Falls, N.Y.

BALL BEARINGS
Chapman Double Ball Bearing Co., Toronto.

BALUSTER LATHES
Ober Mfg. Company, Chagrin Falls, Ohio.
Whitney & Son, Baxter D., Winchendon, Mass.

BAND SAW FILING MACHINERY
Berlin Machine Works, Ltd., Hamilton, Ont.
Fay & Egan Co., J. A., Cincinnati, Ohio.
Prybil Machine Company, P., New York.
Shawver Company, Springfield, O.

BAND SAWS
Berlin Machine Works, Ltd., Hamilton, Ont.
Jackson, Cochrane & Company, Berlin, Ont.
Petrie, H. W., Toronto.
Prybil Machine Co., P., New York City.
Shawver Company, Springfield, O.

BAND SAW MACHINES
Berlin Machine Works, Hamilton, Ont.
Fay & Egan Co., J. A., Cincinnati, Ohio.
Petrie, H. W., Toronto.
Williams Machinery Co., A. R., Toronto, Ont.

BAND SAW STRETCHERS
Berlin Machine Works, Ltd., Hamilton, Ont.
Fay & Egan Co., J. A., Cincinnati, Ohio.

BASKET STITCHING MACHINE
Saranac Machine Co., Benton Harbor, Mich.

BELT LACER
Clipper Belt Lacer Co., Grand Rapids, Mich.

BENDING MACHINES
Fay & Egan Co., J. A., Cincinnati, Ohio.
McKnight & Son Co., L. G., Gardner, Mass.
Perfection Wood Steaming Retort Company,
Parkersburg, W. Va.

BELTING
Fay & Egan Co., J. A., Cincinnati, Ohio.

BLOWERS
Invincible Blow Pipe Company, Toronto.
Sheldons, Limited, Galt, Ont.

BLOW PIPING
Invincible Blow Pipe Company, Toronto.
Sheldons, Limited, Galt, Ont.

BORING MACHINES
Berlin Machine Works, Ltd., Hamilton, Ont.
Canada Machinery Corporation, Galt, Ont.
Fay & Egan Co., J. A., Cincinnati, Ohio.
Garlock Machinery Company, Toronto, Ont.
Jackson, Cochrane & Company, Berlin, Ont.
Nash, J. M., Milwaukee, Wis.
Prybil Machine Co., P., New York.
Reynolds Pattern & Machine Co., Moline, Ill.
Shawver Company, Springfield, O.

BOX MAKERS' MACHINERY
Berlin Machine Works, Ltd., Hamilton, Ont.
Canada Machinery Corporation, Galt, Ont.
Fay & Egan Co., J. A., Cincinnati, Ohio.
Garlock Machinery Company, Toronto, Ont.
Jackson, Cochrane & Co., Berlin, Ont.
Neilson & Company, J. L. Winnipeg, Man.
Reynolds Pattern & Machine Co., Moline, Ill.
Whitney & Son, Baxter D., Winchendon, Mass.
Williams Machinery Co., A. R., Toronto, Ont.

BRUSHES
Meakins & Sons, Hamilton.

CABINET PLANERS
Berlin Machine Works, Ltd., Hamilton, Ont.
Fay & Egan Co., J. A., Cincinnati, Ohio.
Jackson, Cochrane & Co., Berlin, Ont.
Prybil Machine Co., P., New York City.
Shawver Company, Springfield, O.

CARS (Transfer)
Sheldons, Limited, Galt, Ont.
Standard Dry Kiln Co., Indianapolis, Ind.

CARBORUNDUM PAPER
Carborundum Co., Niagara Falls, N.Y.

CARVING MACHINES
Curtis Machinery Corporation, Jamestown, N.Y.
Fay & Egan Co., J. A., Cincinnati, Ohio.
Jackson, Cochrane & Company, Berlin, Ont.

CASTERS
Foster, Merriam & Co., Meriden, Conn.

CLAMPS
Fay & Egan Co., J. A., Cincinnati, Ohio.
Grand Rapids Hand Screw Co., Grand Rapids,
Mich.
Jackson, Cochrane & Company, Berlin, Ont.
Palmer & Sons, A. E., Owosso, Mich.
Simonds Canada Saw Company, Montreal, P.Q.

COLUMN MACHINERY
Fay & Egan Co., J. A., Cincinnati, Ohio.

COMBINATION WOODWORKER
Hutchison Woodworker Co., Toronto, Ont.

CUT-OFF SAWS
Berlin Machine Works, Ltd., Hamilton, Ont.
Fay & Egan Co., J. A., Cincinnati, Ohio.
Jackson, Cochrane & Co., Berlin, Ont.
Knechtel Bros., Limited, Southampton, Ont.
Ober Mfg. Co., Chagrin Falls, Ohio.
Prybil Machine Co., New York.
Shawver Company, Springfield, O.
Simonds Canada Saw Co., Montreal, Que.

CUTTER HEADS
Berlin Machine Works, Hamilton, Ont.
Fay & Egan Co., J. A., Cincinnati, Ohio.
Jackson, Cochrane & Company, Berlin, Ont.
Mattison Machine Works, Beloit, Wis.
Shimer & Sons, Samuel J., Galt, Ont.

DADO HEADS
Elliott, W. A., Toronto.
Huther Bros. Saw Mfg. Co., Rochester, N.Y.

DOVETAILING MACHINES
Canadian Linderman Machine Company, Wood-
stock, Ont.
Fay & Egan Co., J. A., Cincinnati, Ohio.
Garlock Machinery Company, Toronto, Ont.
Jackson, Cochrane & Company, Berlin, Ont.

DOWEL MACHINES
Fay & Egan Co., J. A., Cincinnati, Ohio.
Ober Mfg. Co., Chagrin Falls, Ohio.

DRY KILNS
Grand Rapids Veneer Works, Grand Rapids,
Mich.
Morton Dry Kiln Co., Chicago, Ill.
Sheldons, Limited, Galt, Ont.
Standard Dry Kiln Co., Indianapolis, Ind.

DOWELS
Knechtel Bros., Limited, Southampton, Ont.

DUST COLLECTORS
Invincible Blow Pipe Company, Toronto.
Sheldons, Limited, Galt, Ont.

DUST SEPARATORS
Invincible Blow Pipe Company, Toronto.
Sheldons, Limited, Galt, Ont.

EDGERS (Single Saw)
Fay & Egan Co., J. A., Cincinnati, Ohio.
Simonds Canada Saw Co., Montreal, Que.

EDGERS (Gang)
Berlin Machine Works, Ltd., Hamilton, Ont.
Fay & Egan Co., J. A., Cincinnati, Ohio.
Petrie, H. W., Toronto.
Simonds Canada Saw Co., Montreal, P.Q.
Williams Machinery Co., A. R., Toronto, Ont.

END MATCHING MACHINE
Berlin Machine Works, Ltd., Hamilton, Ont.
Fay & Egan Co., J. A., Cincinnati, Ohio.
Jackson, Cochrane & Company, Berlin, Ont.

EXHAUST FANS
Invincible Blow Pipe Company, Toronto.
Sheldons, Limited, Galt, Ont.

FILES
Simonds Canada Saw Co., Montreal, Que.

FLOORING MACHINES
Berlin Machine Works, Ltd., Hamilton, Ont.
Fay & Egan Co., J. A., Cincinnati, Ohio.
Petrie, H. W., Toronto.
Whitney & Son, Baxter D., Winchendon, Mass.

FLUTING HEADS
Fay & Egan Co., J. A., Cincinnati, Ohio.
Prybil Machine Company, P., New York.
Shawver Company, Springfield, O.

FLUTING and TWIST MACHINE
Prybil Machine Company, P., New York.
Shawver Company, Springfield, O.

FURNITURE TRIMMINGS
Foster, Merriam & Co., Meriden, Conn.

GARNET PAPER AND CLOTH
The Carborundum Co., Niagara Falls, N.Y.
Delany & Pettit, Limited, Toronto, Ont.

GRAINING MACHINES
Berlin Machine Works, Ltd., Hamilton, Ont.
Fay & Egan Co., J. A., Cincinnati, Ohio.
Williams Machinery Co., A. R., Toronto, Ont.

GLUE
Coignet Chemical Product Co., New York,
N.Y.
Delany & Pettit, Limited, Toronto, Ont.
Perkins Glue Company, South Bend, Ind.

GLUE CLAMPS
Jackson, Cochrane & Company, Berlin, Ont.

GLUE HEATERS
Fay & Egan Co., J. A., Cincinnati, Ohio.
Jackson, Cochrane & Company, Berlin, Ont.

GLUE JOINTERS
Berlin Machine Works, Ltd., Hamilton, Ont.
Chas. E. Francis Co., Rushville, Ind.
Jackson, Cochrane & Company, Berlin, Ont.

GLUE SPREADERS
Fay & Egan Co., J. A., Cincinnati, Ohio.
Jackson, Cochrane & Company, Berlin, Ont.

GLUE ROOM EQUIPMENT
Boomer & Boschert Company, Montreal, Que.
Perrin & Company, W. R., Toronto, Ont.

GLUING CAMP CARRIERS
Billstrom, Nels. J., Rockford, Ill.

GRINDERS (Cutter)
Fay & Egan Co., J. A., Cincinnati, Ohio

GRINDERS, HAND AND FOOT POWER
The Carborundum Co., Niagara Falls, N.Y.

GRINDERS (Knife)
Berlin Machine Works, Ltd., Hamilton, Ont.
Fay & Egan Co., J. A., Cincinnati, Ohio.
Petrie, H. W., Toronto.

GRINDERS (Tool)
Fay & Egan Co., J. A., Cincinnati, Ohio.
Jackson, Cochrane & Company, Berlin, Ont.
Pryibil Machine Company, P., New York.
Shawver Company, Springfield, O.

GRINDING WHEELS
The Carborundum Co., Niagara Falls, N.Y.

GROOVING HEADS
Fay & Egan Co., J. A., Cincinnati, Ohio.
Shimer & Sons, Samuel J., Galt, Ont.

GUARDS (Saw)
Pryibil Machine Company, P., New York.
Shawver Company, Springfield, O.

GUMMERS, ETC.
Fay & Egan Co., J. A., Cincinnati, Ohio.

HAND PROTECTORS
Fay & Egan Co., J. A., Cincinnati, Ohio.

HAND SCREWS
Black Bros. Machinery Co., Mendota, Ill.
Fay & Egan Co., J. A., Cincinnati, Ohio.
Grand Rapids Hand Screw Co., Grand Rapids, Mich.

HANDLE THROATING MACHINE
Klotz Machine Company, Sandusky, Ohio.

HANDLE TRIMMING MACHINE
Klotz Machine Company, Sandusky, Ohio.

HANDLE & SPOKE MACHINERY
Fay & Egan Co., J. A., Cincinnati, Ohio.
Klotz Machine Co., Sandusky, Ohio.
Nash, J. M., Milwaukee, Wis.
Ober Mfg. Company, Chagrin Falls, Ohio.
Whitby & Son, Baxter D., Winchendon, Mass.

HARDWARE SPECIALTIES
Reflector & Hardware Mfg. Co., Chicago, Ill.

HARDWOOD LUMBER
Atlantic Lumber Company, Toronto.
Blair & Rolland, Limited, Montreal, Que.
Robert Bury & Company, Toronto, Ont.
Duhlmeier Bros., Cincinnati, Ohio.
Imperial Lumber Company, Columbus, Ohio.
Geo. Kersley, Montreal, Que.
Kraetzer-Cured Lumber Co., Moorehead, Miss.
Luehrmann Hardwood Lumber Co., St. Louis, Mo.
Mowbray & Robinson, Cincinnati, Ohio.
Penrod Walnut & Veneer Co., Kansas City, Mo.
Powell Myers Lumber Co., South Bend, Ind.
Prendergast Company, Cincinnati, Ohio.
Spencer, C. A., Montreal, Que.

HUB MACHINERY
Fay & Egan Co., J. A., Cincinnati, Ohio.

HYDRAULIC VENEER PRESSES
Perrin & Company, Wm. R., Toronto, Ont.

JOINTERS
Canadian Linderman Machine Company, Woodstock, Ont.
Berlin Machine Works, Ltd., Hamilton, Ont.
Fay & Egan Co., J. A., Cincinnati, Ohio.
Jackson, Cochrane & Company, Berlin, Ont.
Knechtel Bros., Limited, Southampton, Ont.
Petrie, H. W., Toronto, Ont.
Pryibil Machine Company, P., New York.
Shawver Company, Springfield, O.

KNIVES (Planer and others)
Berlin Machine Works, Ltd., Hamilton, Ont.
Fay & Egan Co., J. A., Cincinnati, Ohio.
Huther Bros. Saw Mfg. Co., Rochester, N.Y.
Shimer & Sons, Samuel J., Galt, Ont.
Simonds Canada Saw Co., Montreal, P.Q.

LUMBER
Austin & Nicholson, Chapleau, Ont.
Blair & Rolland, Limited, Montreal, Que.
Gillies Bros., Limited, Braeside, Ont.
Imperial Lumber Company, Columbus, Ohio.
Wistar, Underhill & Nixon, Philadelphia, Pa.

LATHES
Fay & Egan Co., J. A., Cincinnati, Ohio.
Jackson, Cochrane & Company, Berlin, Ont.
Klotz Machine Company, Sandusky, Ohio.
Ober Mfg. Company, Chagrin Falls, Ohio.
Petrie, H. W., Toronto, Ont.
Pryibil Machine Company, P., New York.
Shawver Company, Springfield, O.
Whitney & Son, Baxter D., Winchendon, Mass.

MACHINE KNIVES
Berlin Machine Works, Ltd., Hamilton, Ont.
Galt Knife Company, Galt, Ont.
Peter Hay Knife Company, Galt, Ont.
Simonds Canada Saw Co., Montreal, P.Q.

MITRE MACHINES
Fay & Egan Co., J. A., Cincinnati, Ohio.

MITRE SAWS
Fay & Egan Co., J. A., Cincinnati, Ohio.
Simonds Canada Saw Co., Montreal, P.Q.

MORTISING MACHINES
Berlin Machine Works, Ltd., Hamilton, Ont.
Fay & Egan Co., J. A., Cincinnati, Ohio.

MOULDINGS
Knechtel Bros., Limited, Southampton, Ont.

MULTIPLE BOXING MACHINES
Fay & Egan Co., J. A., Cincinnati, Ohio.
Nash, J. M., Milwaukee, Wis.
Reynolds Pattern & Machine Co., Moline, Ill.

PAINTS & VARNISHES
Marietta Paint & Color Co., Marietta, Ohio.

PATENT SOLICITORS
Dennison, H. J. S., Toronto, Ont.
Lightfoot, Stanley, Toronto, Ont.

PATTERN SHOP MACHINES
Berlin Machine Works, Ltd., Hamilton, Ont.
Fay & Egan Co., J. A., Cincinnati, Ohio.
Jackson, Cochrane & Company, Berlin, Ont.
Whitney & Son, Baxter D., Winchendon, Mass.

PICTURE FRAME MACHINERY
Black Bros. Machinery Co., Mendota, Ill.
Chicago Machinery Exchange, Chicago, Ill.

PLANERS
Berlin Machine Works, Ltd., Hamilton, Ont.
Fay & Egan Co., J. A., Cincinnati, Ohio.
Jackson, Cochrane & Company, Berlin, Ont.
Petrie, H. W., Toronto, Ont.
Pryibil Machine Company, P., New York.
Whitney & Son, Baxter D., Winchendon, Mass.
Williams Machinery Co., A. R., Toronto, Ont.

PLANING MILL MACHINERY
Berlin Machine Works, Ltd., Hamilton, Ont.
Fay & Egan Co., J. A., Cincinnati, Ohio.
Jackson, Cochrane & Company, Berlin, Ont.
Petrie, H. W., Toronto, Ont.
Pryibil Machine Company, P., New York.
Shawver Company, Springfield, O.
Shimer & Sons, Samuel J., Galt, Ont.
Whitney & Son, Baxter D., Winchendon, Mass.
Williams Machinery Co., A. R., Toronto, Ont.

PRESSES (Veneer)
Black Bros. Machinery Co., Mendota, Ill.
Jackson, Cochrane & Company, Berlin, Ont.
Perrin & Co., Wm. R., Toronto, Ont.

PULLEYS
Fay & Egan Co., J. A., Cincinnati, Ohio.
Pryibil Machine Company, P., New York.

RESAWS
Berlin Machine Works, Ltd., Hamilton, Ont.
Fay & Egan Co., J. A., Cincinnati, Ohio.
Jackson, Cochrane & Company, Berlin, Ont.
Petrie, H. W., Toronto, Ont.
Simonds Canada Saw Co., Montreal, P.Q.
Williams Machinery Co., A. R., Toronto, Ont.

RIM AND FELLOE MACHINERY
Fay & Egan Co., J. A., Cincinnati, Ohio.

RIP SAWING MACHINES
Berlin Machine Works, Ltd., Hamilton, Ont.
Fay & Egan Co., J. A., Cincinnati, Ohio.
Klotz Machine Company, Sandusky, Ohio.
Jackson, Cochrane & Company, Berlin, Ont.
Ober Mfg. Company, Chagrin Falls, Ohio.
Pryibil Machine Company, P., New York.

RUBBING MACHINES
Bagg's Export Combine, Chicago, Ill.

SANDERS
Berlin Machine Works, Ltd., Hamilton, Ont.
Duplex Moulding Sander Co., Hornell, N.Y.
Elliot Woodworker Company, Toronto, Ont.
Fay & Egan Co., J. A., Cincinnati, Ohio.
Klotz Machine Company, Sandusky, Ohio.
Jackson, Cochrane & Company, Berlin, Ont.
Nash, J. M., Milwaukee, Wis.
Ober Mfg. Company, Chagrin Falls, Ohio.
Pryibil Machine Company, P., New York.
Shawver Company, Springfield, O.

SANDPAPER
Black Bros. Machinery Co., Mendota, Ill.

SASH, DOOR & BLIND MACHINERY
Berlin Machine Works, Ltd., Hamilton, Ont.
Fay & Egan Co., J. A., Cincinnati, Ohio.
Jackson, Cochrane & Company, Berlin, Ont.
Shimer & Sons, Samuel J., Galt, Ont.

SAWS
Disston & Sons, Henry, Philadelphia, Pa.
Fay & Egan Co., J. A., Cincinnati, Ohio.
Radcliff Saw Mfg. Company, Toronto, Ont.
Simonds Canada Saw Co., Montreal, P.Q.
Huther Bros. Saw Mfg. Co., Rochester, N.Y.

SAW GUARDS
Huther Bros. Saw Mfg. Co., Rochester, N.Y.
Pryibil Machine Company, P., New York.
Shawver Company, Springfield, O.

SAW GUMMERS, ALOXITE
The Carborundum Co., Niagara Falls, N.Y.

SAW SWAGES
Fay & Egan Co., J. A., Cincinnati, Ohio.
Simonds Canada Saw Co., Montreal, Que.

SAW TABLES
Berlin Machine Works, Ltd., Hamilton, Ont.
Fay & Egan Co., J. A., Cincinnati, Ohio.
Jackson, Cochrane & Company, Berlin, Ont.
Knechtel Bros., Limited, Southampton, Ont.
Pryibil Machine Company, P., New York.

SAW JOINTERS
Huther Bros. Saw Mfg. Co., Rochester, N.Y.

SCRAPING MACHINES
Whitney & Son, Baxter D., Winchendon, Mass

SCROLL CHUCK MACHINES
Klotz Machine Company, Sandusky, Ohio.

SCREW DRIVING MACHINES
Reynolds Pattern & Machine Co., Moline, Ill.

SCREW DRIVING MACHINERY
Reynolds Pattern & Machine Works, Moline, Ill.

SCROLL SAWS
Berlin Machine Works, Ltd., Hamilton, Ont.
Fay & Egan Co., J. A., Cincinnati, Ohio.
Jackson, Cochrane & Company, Berlin, Ont.
Pryibil Machine Company, P., New York.
Simonds Canada Saw Co., Montreal, P.Q.

SECOND-HAND MACHINERY
Pryibil Machine Company, P., New York.
Williams Machinery Co., A. R., Toronto, Ont.

SHAPERS
Berlin Machine Works, Ltd., Hamilton, Ont.
Fay & Egan Co., J. A., Cincinnati, Ohio.
Jackson, Cochrane & Company, Berlin, Ont.
Ober Mfg. Company, Chagrin Falls, Ohio.
Petrie, H. W., Toronto, Ont.
Pryibil Machine Co., P., New York.
Shawver Company, Springfield, O.
Shimer & Sons, Samuel J., Galt, Ont.
Simonds Canada Saw Co., Montreal, P.Q.
Whitney & Sons, Baxter D., Winchendon, Mass.

SHARPENING TOOLS
The Carborundum Co., Niagara Falls, N.Y.

SHAVING COLLECTORS
Invincible Blow Pipe Company, Toronto, Ont.
Sheldons Limited, Galt, Ont.

SHELL BOX HANDLES
Yokes Hardware Co., Toronto, Ont.

SHELL BOX EQUIPMENT
Yokes Hardware Co., Toronto, Ont.

SINGLE SPINDLE BOXING MACHINES
Fay & Egan Co., J. A., Cincinnati, Ohio.
Ober Mfg. Company, Chagrin Falls, Ohio.

SPRAYING APPARATUS
De Vilbiss Mfg. Co., Toledo, Ohio.

STAINS
Marietta Paint & Color Co., Marietta, Ohio.

STAVE SAWING MACHINE
Whitney & Sons, Baxter D., Winchendon, Mass.

STEAM TRAPS
Canadian Morehead Mfg. Co., Woodstock, Ont.
Standard Dry Kiln Co., Indianapolis, Ind.

STEEL SHELL BOX BANDS
Metallic Roofing Co., Toronto, Ont.
Yokes Hardware Co., Toronto, Ont.

SURFACERS
Berlin Machine Works, Ltd., Hamilton, Ont.
Fay & Egan Co., J. A., Cincinnati, Ohio.
Jackson, Cochrane & Company, Berlin, Ont.
Petrie, H. W., Toronto, Ont.
Whitney & Sons, Baxter D., Winchendon, Mass.
Williams Machinery Co., A. R., Toronto, Ont.

SWING SAWS
Berlin Machine Works, Ltd., Hamilton, Ont.
Fay & Egan Co., J. A., Cincinnati, Ohio.
Jackson, Cochrane & Company, Berlin, Ont.
Ober Mfg. Co., Chagrin Falls, Ohio.
Simonds Canada Saw Co., Montreal, P.Q.

TABLE LEG LATHES
Berlin Machine Works, Ltd., Hamilton, Ont.
Fay & Egan Co., J. A., Cincinnati, Ohio.
Ober Mfg. Co., Chagrin Falls, Ohio.
Whitney & Sons, Baxter D., Winchendon, Mass.

TABLE SLIDES
Walter & Co., B., Wabash, Ind.

TENONING MACHINES
Berlin Machine Works, Ltd., Hamilton, Ont.
Fay & Egan Co., J. A., Cincinnati, Ohio.
Jackson, Cochrane & Company, Berlin, Ont.
Pryibil Machine Company, P., New York.

TRIMMERS
Berlin Machine Works, Ltd., Hamilton, Ont.
Fay & Egan Co., J. A., Cincinnati, Ohio.

TRUCKS
Standard Dry Kiln Co., Indianapolis, Ind.
Sheldons Limited, Galt, Ont.

TURNINGS
Knechtel Bros., Limited, Southampton, Ont.

TURNING MACHINES
Berlin Machine Works, Ltd., Hamilton, Ont.
Fay & Egan Co., J. A., Cincinnati, Ohio.
Klotz Machine Company, Sandusky, Ohio.
Ober Mfg. Company, Chagrin Falls, Ohio.
Pryibil Machine Company, P., New York.
Shawver Company, Springfield, O.
Whitney & Sons, Baxter D., Winchendon, Mass.

UNDER-CUT SELF-FEEDING FACE
PLANER
Fay & Egan Co., J. A., Cincinnati, Ohio.
Jackson, Cochrane & Company, Berlin, Ont.

VARNISHES
Ault & Wiborg Company, Toronto, Ont.
Marietta Paint & Color Co., Marietta, Ohio.

VENEERS
Robert Bury & Company, Toronto, Ont.
Central Veneer Co., Huntington, W. Virginia.
Walter Clark Veneer Co., Grand Rapids, Mich.
Evansville Veneer Co., Evansville, Ind.
Hay & Company, Woodstock, Ont.
Hoffman Bros. Company, Fort Wayne, Ind.
Huddleston-Marsh Lumber Co., Chicago, Ill.
Geo. Kersley, Montreal, Que.
Ohio Veneer Company, Cincinnati, Ohio.
Penrod Walnut & Veneer Co., Kansas City, Mo.
Robbins Veneer & Lumber Co., Buffalo, N. Y.
Waetjen & Co., Geo. L., Milwaukee, Wis.
Whisler, C. L., Hillsboro, Ohio.

VENTILATING APPARATUS
Sheldons Limited, Galt, Ont.

VENEERED PANELS
Hay & Company, Woodstock, Ont.

VENEER PRESSES (Hand and Power)
Canadian Boomer-Boschart Co., Montreal, Que.
Jackson, Cochrane & Company, Berlin, Ont.
Perrin & Company, Wm. R., Toronto, Ont.

VISES
Fay & Egan Co., J. A., Cincinnati, Ohio.
Huther Bros. Saw Mfg. Co., Rochester, N.Y.
Simonds Canada Saw Company, Montreal, Que.

WAGON AND CARRIAGE MACHINERY
Berlin Machine Works, Ltd., Hamilton, Ont.
Fay & Egan Co., J. A., Cincinnati, Ohio.
Ober Mfg. Company, Chagrin Falls, Ohio.
Whitney & Sons, Baxter D., Winchendon, Mass.

WOOD FINISHES
Ault & Wiborg, Toronto, Ont.
Marietta Paint & Color Co., Marietta, Ohio.

WOOD TURNING MACHINERY
Berlin Machine Works, Ltd., Beloit, Wis.
Shawver Co., Springfield, Ohio.

WORK BENCHES
Fay & Egan Co., J. A., Cincinnati, Ohio.

WOODWORKING MACHINES
Canada Machinery Corporation, Galt, Ont.
Cowan & Company, Galt, Ont.
Garlock Machinery Company, Toronto, Ont.
Jackson, Cochrane & Co., Berlin, Ont.
Reynolds Pattern & Machine Works, Moline,
Ill.
Williams Machinery Co., A. R., Toronto, Ont.

WHEELS, Grinding, Carborundum and Aloxite
The Carborundum Co., Niagara Falls, N.Y.

WOOD SCREWS
Robertson & Company, P. L., Milton, Ont.

WOOD SPECIALTIES
Knechtel Bros., Limited, Southampton, Ont.

WOODWORKING CUTTERS
Galt Knife Company, Galt, Ont.

YELLOW PINE
Prendergast Company, Cincinnati, Ohio.

DOWELS

Plain and Spiral Grooved

Our fifteen years' experience in this line at your disposal.
Let us send you prices.

WE ALSO MANUFACTURE A FULL LINE OF

Embossed and Carved Dentil Mouldings, Turnings and Balusters

*We will be pleased to furnish information
and prices on your requirements.*

KNECHTEL BROS., LIMITED
SOUTHAMPTON　　-　　-　　-　　-　　ONTARIO

Making TABLE-SLIDES is a Specialty Business

For more than TWENTY-FIVE YEARS we have made TABLE SLIDES exclusively. Our Factory is equipped with Special Machinery which enables us to make SLIDES.— BETTER and CHEAPER than the furniture manufacturer.

Canadian Table makers are rapidly adopting WABASH SLIDES

Because { They ELIMINATE SLIDE TROUBLES Are CHEAPER and BETTER

Reduced Costs   BY USING
Increased Out-put   **WABASH SLIDES**

MADE BY

B. Walter & Company
Wabash, Ind.

The Largest EXCLUSIVE TABLE-SLIDE Manufacturers in America
ESTABLISHED-1887

Just What You Need

There's no necessity now for you to wish for a corrugated joint fastener machine which will rise fully to your expectations; you can get one.

The Saranac Corrugated Joint Fastener Machine

is the absolute realization of your wish. This illustration gives you an idea of what it looks like, but you can have no conception of its **efficiency, economy, speed** and **reliability** until you get it working for you.

It is substantially constructed, and the workmanship and materials are of the very best. Particularly suitable for manufacturers of sash, doors, blinds, screens, porch columns, boxes, etc., etc.

Made in various types for all conditions.

Write for further particulars and prices

Saranac Machine Co., SOLE MAKERS
Benton Harbor, Mich.

INCREASED production and greater economy in your sanding department can be obtained by using

CARBORUNDUM BRAND GARNET PAPER AND CLOTH

They show a saving in labor, quicker production and better results. Carborundum Brand Garnet Paper and Cloth is made from the best garnet grains. Free from all impurities. Uniformly coated upon flexible backings. The best materials united with superior methods of manufacture make Carborundum Brand Garnet Paper and Cloth the ideal abrasive for the woodworking trade.

Working samples upon request

THE CARBORUNDUM COMPANY
NIAGARA FALLS, N. Y.

New York   Chicago   Grand Rapids
          Boston   Cincinnati

MATTISON
NEW MODEL
Turning and Shaping Lathe

Period Style Furniture at Rock Bottom Cost—
This is the Machine that does it

It is the *turned work* wherein the manufacture of Period Styles differs from others. If you solve the problem of cost in your Period Turnings, you also solve it for your Period Furniture.

The Mattison New Model Turning and Shaping Lathe not only produces them on a strict manufacturing basis of cost, but in as small lots as required *without sacrificing economy.*

It is a machine on which you can turn the complicated and delicate patterns of *today:* of infinite variety and in a wide range of sizes, easily, quickly, and economically, in large or small lots.

It is a machine on which you can produce your *future* turnings with equal facility, by reason of its adaptability. It's the great furniture turning machine of both today and tomorrow. And it's the only standardized method.

*There is no other Lathe like the Mattison
and it is built only by*

C. MATTISON MACHINE WORKS, BELOIT, WIS., U.S.A.
883 FIFTH STREET

*We have an interesting folder entitled "Why are Period Styles
in Demand, and How Long Will They Last. Ask for a copy.*

Volume Sixteen　　　　　TORONTO, FEBRUARY, 1916　　　　　Number Two

CANADIAN WOODWORKER
and Furniture Manufacturer

Where You Profit Most

Whether you make duplicate turnings as a specialty or to utilize waste—the proper gauge lathe insures success.

A Whitney Back-Knife Gauge Lathe

has a big working capacity—its speed means dollars earned—its accuracy means perfect reproductions of the back-knife pattern—the variety of turnings possible depends only upon the number of knives with which your Whitney Gauge Lathe is equipped.

The test of a gauge lathe is not on a few mechanical devices alone—but the machine's complete adaptability to your work.

We make the Whitney in four different sizes—20, 30, 40 and 50 inch. They turn stock up to 3 inches square.

The guiding of tools is done mechanically. The head and tail spindles are made of cast steel. Bearings are oil-flooded but oil cannot work out of boxes. The head center, instead of the tail center, is moved to clamp the work which saves stock.

A quick way to find out how much you need this profitable Whitney Gauge Lathe is to send today to our nearest office for booklet.

Baxter D. Whitney & Son, Winchendon, Mass.

California Office:—Berkeley, California.　　　　Selling Representatives:—Henry Kelly & Co.
H. W. Petrie, Ltd., Toronto, Ontario, Agents for Ontario and Quebec.　　　26 Pall Mall, Manchester, England.

Put Good Belting

on your

Good Machines

Your machines are expensive, and naturally you expect them to turn out the quality and quantity of
work commensurate with their cost.

Then why harness them with belting which cannot
begin to do them justice?

Use

"AMPHIBIA"
Planer Belting

and get the most work from your machines, in the
quickest time at the lowest cost per day of service.

**Try a sample run of 'AMPHIBIA' Planer
and prove its merits.**

"Leather like gold has no substitute."

Sadler & Haworth

Established 1876

Tanners and Manufacturers

For 40 years Tanners and Manufacturers of the Best Leather Belts.

| Toronto | St. John, N.B. | MONTREAL | Winnipeg | Vancouver |
| 38 Wellington St. E. | 139 Prince William St. | 511 William St. | Galt Building | 217 Columbia Ave. |

Shell Box Equipment

and

General Wood Working Tools

We have in stock at Toronto for immediate shipment, for use on Fixed Ammunition Boxes:—

Cowan Special Dado Machines, for grooving sides
J. & C. 12 Spindle Dovetailers
J. & C. Case Clamp for squeezing

also a Used Advance 15 Spindle Dovetailer—in very good condition

We have installed several plants for shell box work.
Let us co-operate with you in selecting and laying out your tool equipment.
Our Services are at your disposal.

We handle all kinds of wood-working machinery, including the complete line of the
AMERICAN WOODWORKING MACHINERY CO.
ROCHESTER, N.Y.

and the line of special boring machines of
M. L. ANDREWS & CO., CINCINNATI, OHIO

Garlock - Machinery
197 Wellesley Street   -   Toronto
Telephone North 6849

For your Files

We are not any interested in any sanders now, but send us the book for our files.

Remark...

Name...

Address...

For Mr...

FILL OUT AND MAIL THIS COUPON—NOW

Clip the Coupon Get the Book.

The Latest Views on Sanding Operations—

FREE !

48 **pages of the most interesting and instructive information ever compiled on this important finishing method.**

For the convenience of you who give close attention to every sanding operation and the machines best adapted for your particular work and conditions, this book will offer invaluable suggestions.

No one type of sander is suitable for all classes of work. The peculiar size and shape of your particular stock may best be handled with greatest economy and perfect workmanship on the latest Roll-feed Sander, either with one, two, three, four or six drums.

Then again, an Endless-bed feed may best serve your purpose. Or perhaps a Flexible-belt Sander may save considerable by finishing such work as you can now only do by hand.
Each type of Berlin Sander is adapted for a definite purpose, yet all "overlapping" one another in usefulness to insure highest efficiency of operation on the work in hand.

To judge such usefulness requires first-hand information. The book above supplies a definite knowledge.

THE BERLIN MACHINE WORKS, LTD.

Hamilton, Ont.

U. S. Plant,   -   Beloit, Wis.

Our Service Department is in position to supply you, without expense or obligation, with data on the latest sanding methods used in every kind of factory where sanders are used. Why not avail yourself of such help?

MADE RIGHT—RIGHT IN CANADA

Dry Spruce
and Birch

Good Stocks, Prompt Shipments, Satisfaction

C. A. SPENCER, Limited
Wholesale Dealers in Rough and Dressed Lumber

Eastern Townships Bank Building
MONTREAL   -   -   Quebec

 TEAK

Write for prices on all

D

Plain and Fancy
Lumbers and Veneers

L

O
M

Mahogany, Oak, Birch,
Walnut, Ash, Poplar, etc.,
rotary, sliced or sawed
in all thicknesses

U

E
S

M

SHELL BOX PARTS
in birch and spruce
Bridges and Ends

B

T
I

E

Write for Quotations

C

R

George Kersley
224 St. James Street, Montreal

 MAHOGANY

Peter Hay Machine Knives

Peter Hay knives
are made of special
quality steel temper-
ed and toughened
to take and hold a
keen cutting edge.
Hay knives are made
for all purposes.
Have you our Catalogue?
Made in Canada

The Peter Hay Knife Co., Limited
GALT, ONTARIO

The Atlantic
Lumber Co.

Manufacturers and
Wholesale Dealers

American and Can-
adian Hardwood Lum-
ber and Mahogany,
Quartered and Plain,
White and Red Oak
a Specialty.

110 Manning Chambers - **TORONTO**
Phone Main 6386

HOME OFFICE, BOSTON MASS.

Mills: KNOXVILLE, TENN.,   WALLAND, TENN.
FRANKLIN, VA.

Immediate Shipment

West Virginia Stock

Thoroughly Dry and Well Manufactured

1 car 4/4 1s and 2s. Plain White Oak.
6 cars 4/4 No. 1 Common Plain White Oak.
2 cars 4/4 No. 2 Common Plain White Oak.
1 car 4/4 No. 1 Common and Better Plain White
       Oak. Worm holes, no defect.
1 car 6/4 No. 1 Common Plain White Oak.
6 cars 4/4 S. W. Chestnut.
1 car 5/4 S. W. Chestnut.
4 cars 6/4 S. W. Chestnut.
1 car 8/4 S. W. Chestnut
3 cars 4/4 No. 3 Common Chestnut.
1 car 4/4 No. 1 Common and Better Chestnut.
½ car 8/4 No. 1 Common and Better Chestnut.
1 car 4/4 Saps and Select Poplar.
2 cars 4/4 No. 2 Common and Better Poplar.
1 car 5/4 No. 2 Common Poplar.
1 car 6/4 No. 2 Common Poplar.
1 car 4/4 No. 2 Common and Better Birch.

We solicit your inquiries on special car bill Oak.

Mill at Holly Junction, W. Va.

Write at once for Quotations   The Price is Right

The Imperial Lumber Co.
COLUMBUS   -   OHIO

One Piece Sides and Ends for

SHELL BOXES

From Narrow (Cheaper) Lumber at
a Low Rate per 1000 feet on the

LINDERMAN AUTOMATIC
DOVETAIL MATCHER

Samples
on
Request

THESE CONCERNS USE LINDERMAN MACHINES:

W. C. Edwards & Co.	Ottawa, Ont.	Dominion Box & Package Co.	Montreal, Que.
(2 Machines)		(2 Machines)	
Czerwinski Box Co.	Winnipeg, Man.	C. P. R. Angus Shops	Montreal, Que.
Cummer-Dowswell Co.	Hamilton, Ont.	J. W. Kilgour & Bro.	Beauharnois, Que.
Canada Furniture Mfrs.	Woodstock, Ont.	Durham Furniture Co.	Durham, Ont.
Martin Freres & Cie	Montreal, Que.	Brunswick-Balke-Collender Co.	Toronto, Ont.
Alberta Box Co.	Calgary, Alta.	Singer Mfg. Co.	St. Johns, Que.

Canadian Linderman Company
Woodstock, Ontario

Muskegon, Mich.   Knoxville, Tenn.   New York, N. Y.

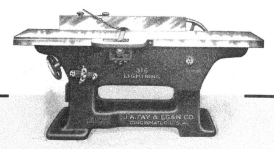

FOOL PROOF

No matter how careless the operator, he can not injure himself or the machine if he is operating the new Fay-Egan No. 316 Hand Planer.

The head is the Fay-Egan Patented Safety Type, with thin air-hardened steel knives. The knives require only a very slight projection and the steel-lipped tables set up snug—no possibility of getting fingers cut off. **Note too, that cylinder is adjustable for alignment with tables.** Tables are mounted on long continuous ways, and could not be gotten out of alignment even by abuse. Infeed table adjusts by large hand wheel at rear, out of way, but easily accessible and very easy to turn—**think of the convenience and time saved by being able to adjust table with one hand while you are feeding your work up to the cylinder.**

Notice especially the design, the clearance for operator's feet, the rabbetting arm, the lever adjustment of outfeed table for making hollow or spring joints, and the shaving chute cast integral with the base, and arranged for convenient blowpipe connection.

Made in 12, 16, 20, 24 and 30-in. sizes.

No. 316 IS THE "LAST WORD" IN HAND PLANERS.
Write for bulletin H-10, or fill in and mail attached coupon.

J. A. FAY & EGAN CO.
West Front Street
CINCINNATI, OHIO

Date................................

J. A. FAY & EGAN CO.,
    Cincinnati, Ohio.

Please tell me more about your "Fool Proof" No. 316 Hand Planer and Jointer.

Mr. ....

Company Name ....

Street Address ....

City .... State ....

Alphabetical List of Advertisers

Atlantic Lumber Company   5
Ault & Wiborg Company 33

Baird Machinery Works 45
Berlin Machine Works   4

Canada Machinery Corporation ....
Canadian Boomer & Boschert Mfg.
  Company 47
Canadian Linderman Co. 6
Canadian Morehead Mfg. Company   43
Carborundum Company   51
Central Veneer Company 39
Chapman Double Ball Bearing Co.   47
Clark Veneer Co., Walter 37
Clipper Belt Lacer Company ... ..
Coignet Chemical Products Co. 33

Delaney & Petit 12

Evansville Veneer Company 34

Fay & Egan Company, J. A.   7
Foster, Merriam & Company 46

Gage Mfg. Company, E. E. 11
Garlock-Machinery   3

Globe Soap Company 45
Grand Rapids Dry Kiln Company .. 44

Hartzell, Geo. W. 37
Hay & Company 33
Hay Knife Company, Peter   5
Hoffman Bros. Company 37

Imperial Lumber Company   5

Jackson, Cochrane & Company 12

Kersley, Geo.   5
Klotz Manufacturing Company 10
Knechtel Bros. 50
Kraetzer-Cured. Lumber Company..

Lightfoot, Stanley 45

Mattison Machine Works 52

Nash, J. M. 47
National Dry Kiln Company 44
National Securities Corporation ... 45
Neilson & Company, J. L.   8

Ober Manufacturing Company 47
Ohio Veneer Company 37

Perfection Wood Steaming Retort
  Company ... ...
Penrod Walnut & Veneer Company 39
Perkins Glue Company   9
Perrin, William R. 11
Petrie, H. W. 45

Radcliff Saw Company 43
Reynolds Pattern & Machine Works   9

Saranac Machine Company 51
Sadler & Haworth   2
Shawver Company 46
Sheldons Limited 46
Spencer, C. A.   5
Stow Mfg. Company 10

Toomey, Frank, Inc. 45

Vokes Hardware Company 41

Waetjen & Company, G. L. 37
Walter & Company, B. 51
Weeks Lumber Company, F. H. 11
Whitney & Son, Baxter D.   1

| Lock Corner Outfits
Box Makers Sawing Sets
Resaws and Filing Outfits
Box Board Matchers, etc. | **We Specialize**

in furnishing | Dovetail Box Outfits
Nailing Machinery
Box Board Printers
Surfacers, Borers, etc. |

BOX MACHINERY

(for LOCK-CORNER, NAILED and DOVETAILED BOXES)

Also all other kinds of

WOOD-WORKING MACHINERY

TWENTY years actual, first hand, working experience in box factories, sash and door factories and planing mills enables us to
give you the intelligent service all buyers desire.

Buy from people who will understand
your requirements

J. L. NEILSON & CO. - Winnipeg, Man.

The Only Way to Drive Screws

There is a stage-coach method and a modern method of doing everything. When it comes to driving screws the only modern way is with an Automatic Screw Driving Machine, which means greater and better production and greater profit.

The Reynolds
Automatic Screw-Driving Machine

is guaranteed to produce results. It is no experiment. The No. 2 Standard Type, shown herewith, is particularly suited for placing the screws in the straps of shrapnel boxes. It may be fitted to drive, without change or adjustment, any three consecutive sizes flat or round head, or two sizes filister head, No. 6 to 14 inclusive, wood or machine screws, or three sizes flat or round head, or two sizes filister head machine screws up to No. 20. All the above up to 1½ inches long if smaller than No. 10, or up to 1¾ inches long, No. 10 or larger. Fitted with boring attachment if desired.

Many Machines in Use in Canada.

Ask for our Catalogue.

REYNOLDS PATTERN & MACHINE CO.
101 and 103 Third Ave., Moline, Ill., U. S. A.

No. 2 Standard Type Automatic
Screw-driving Machine.

Are Your Products
Perkins Glued?

Ladies' Desk manufactured
by The Berlin Furniture
Co., Berlin, Ont., in which
Perkins Glue was used.

The live dealer realizes the value of selling his customers Perkins Glued products because they never come apart. Many of Canada's largest furniture factories use Perkins Vegetable Glue. It costs 20 per cent. less than hide glue, is applied cold, antiquating the obnoxious and wasteful cooking process, is absolutely uniform in quality and will not blister in sanding.

We will be glad to demonstrate in your own factory and on your own stock what can be accomplished with Perkins Vegetable Glue.

Write us today for more detailed information.

Perkins Glue Co.
South Bend, Ind.

Here It Is!

The Shaping Saw Machine

That Saves Time

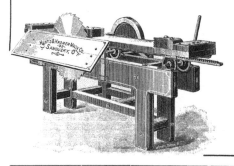

THE Klotz Shaping Saw Machine is specially made and designed for cutting out Shapes or Blanks for Handles, Spokes and similar work, from planks or slabs after they are cut from the log.

The frame is made from hard wood and firmly bolted together. The table is made of well-seasoned ash strips, glued together, and runs on large turned rolls, with wrought iron axles and metal bearing. A chuck is attached to the table for holding the timber; the same is adjustable for different lengths of work, and is very quickly operated by a suitable handle. The mandrel is of steel, and the balance-wheel is turned so as to run perfectly true. The drop-table is arranged so that it may be easily lifted out of position, in order to get at nut on mandrel to take off the saw. The saw furnished with the machine is 26 inches in diameter, and pulley on mandrel is 8 inch diameter, 6½ inch face, and will make about 1400 revolutions per minute. Power required, 2 H.P. Shipping weight 900 lbs. Floor space 4 x 7 feet.

Write us for prices and details.

Klotz Machine Co.
Sandusky, Ohio

Efficiency—
Increased Production
"STOW"
Motor Driven Screw Drivers

Built in Several Sizes   Suspended or Pedestal Design

Immediate Shipment — *Write us your requirements to-day*

STOW Tools are an INVESTMENT and not an EXPENSE

STOW MANUFACTURING CO.
Binghamton, N. Y., U. S. A.
Oldest Portable Tool Manufacturers in America
Electric and Belt Driven Drills, All Sizes
INVENTORS AND LARGEST MANUFACTURERS OF STOW FLEXIBLE SHAFTING

CUTTER HEADS for Making SHELL BOXES

Our Cutter Heads are especially suitable for the production of shell boxes. Will cut through Pine or Hardwood lumber up to 1¼ inches in thickness.
We also manufacture Machine Knives of all kinds, made from the best English Steel and Swedish Iron.

Send us your order, or write for prices

E. E. GAGE MFG. CO.
41 Main St., GARDNER, MASS.
Largest Makers of Back Knives in New England

ONE OPERATOR
SIXTY SETS
SIXTY MINUTES

BRICK HOUSE JAMBS

FRAME HOUSE JAMBS

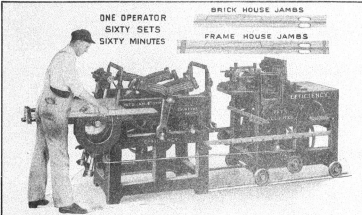

THE EFFICIENCY WINDOW FRAME MACHINE

A new style machine which **REDUCES LABOR COST FULLY 60%**. Works automatically—one man can turn out sixty sets of pulley stiles an hour. Cross cuts the ends of pockets, routs the pulley mortises and dadoes the head and sill gain, automatically, while the operator slits the pockets. You take no chance—write us.

THE F. H. WEEKS LUMBER COMPANY   —   AKRON, OHIO

The Canadian Woodworker and Furniture Manufacturer covers a specialized field. It represents maximum value to its subscribers.

Subscription price $1.00 per year.

PRESSES
For Veneer and Veneer Drying
Made in Canada

William R. Perrin
Limited
Toronto

Gang Dovetailer

for Ammunition Boxes

The Best Machine Yet Produced

Can be made to do either the closed end drawer joint or the open end box joint, and can be changed from one joint to the other at a very small cost.

The frame is well proportioned and very strong. All belts are now done away with except the belt drive from the line shaft. This is important, as it occupies only one-third the floor space usually given to a machine of this class and reduces belt cost to the minimum.

A large number of these machines, installed during the past three years, are giving excellent satisfaction. Our guarantee goes with each machine.

Have you our Bulletins?

Jackson, Cochrane & Co.
BERLIN, CANADA

Our No. 90 Gang Dovetailing Machine

Joint
and
Veneer
Glue

*Always
Uniform*

Garnet
Paper
and
Cloth

Delany & Pettit, Limited

Office:—133 Jefferson Ave.   **Toronto, Ont.**   *Works:*–105-131 Jefferson Ave.
106-118 Atlantic Ave.

Canadian Woodworker

A Monthly Publication in the Interest of the Woodworking Industry.
Reaches the factories producing interior finish, doors, sash, flooring,
woodenware, furniture, pianos, boxes, and general mill products.

Subscription, $1.00 a year; foreign $1.50.

Woodworker Publishing Company, Limited

Branches: Montreal, Vancouver,
Chicago, New York, London, Eng.

345 Adelaide St. West, Toronto

Phone Ade. 2700

Authorized by the Postmaster General for Canada, for transmission as second class matter.
Entered as second class matter July 18th, 1914, at the Post Office at Buffalo, N.Y., under the Act of Congress of March 3, 1870.

Vol. 16   February, 1916   No. 2

The Furniture Situation

Letters received from leading furniture manufacturers state that the demand for furniture during the latter part of 1915 was very much better than for the same months in the preceding year. The general opinion is that the demand during 1916 will continue much the same as in the past few months, but some manufacturers expect that the first six months will be better than the last six. A large number of men are enlisting for overseas service. A few of them are breaking up their homes, but on the other hand thousands of men are engaged in making munitions and many of them are earning better wages than they ever earned before. These men will be buying furniture for which they and their families have long expressed a desire. The demand, although it may not be anything like that of normal times, will at least keep the wheels turning as long as the war lasts.

Canadian furniture manufacturers will certainly have to follow the example of their neighbors across the line and increase their prices, as the cost of practically everything which enters into the manufacture of furniture has increased and in some cases is still increasing. The most noticeable increases have been in lumber, hardware, glue and particularly in upholstering materials. The cost of labor has also been increased, as the factories making munitions are offering fancy wages, and furniture manufacturers are obliged to offer an increase or lose some of their best men. There is also a scarcity of labor owing to the large number of men enlisting. Another feature which adds to the cost of manufacture is that owing to the limited demand it is necessary to manufacture in much smaller quantities. The importance of this feature can hardly be appreciated unless one is engaged in the business.

The furniture which is finding most ready sale is the cheap and medium grades, with very little demand for the higher priced lines. As a means of developing trade it has been suggested that to provide the most attractive designs at the lowest prices consistent with local conditions is about the only means in sight at present. We believe a large volume of export business could be secured by the manufacturers of cheap tables and chairs which could be shipped knocked down. A large volume of export trade might be secured in manufacturing other articles which are not exactly in the furniture line but which could be very nicely taken care of in furniture factories. In connection with this business it would be advisable for a practical man to visit the importing countries to report on the market and secure samples from which prices could be quoted.

Practically every plant in Great Britain of any size has been fitted up to make munitions and when the war is over it will take some time to dismantle and reorganize them for the purposes for which they were formerly used. Canadian factories have not been disorganized to any great extent and manufacturers here will be in a position to meet the demand which will undoubtedly arise for manufactured goods of every description.

The Importance of Glue

Glue continues to receive a great deal of attention from the manufacturers of wood products, and not without reason. The improper handling of glue is the cause of more bad work than almost any other article which enters into woodworking. After lumber and machine work, glue is next in importance, and should be treated accordingly. Intelligent superintendents have recognized this importance, but some of them confine their efforts to the purchasing end and think that when they have purchased a good grade of glue their responsibility has ended. There are other superintendents who, after purchasing a good grade, give the handling of it the same careful supervision which is bestowed on the kiln drying and machining of the lumber.

There is no place in the woodworking factory

where an intelligent man can be employed to better advantage than in charge of the glue department. Men have realized that glue should not be cooked, and the majority of them are faithfully attending to this end of it; but the man in the gluing department cannot accomplish very much without the co-operation of the foreman and superintendent.

There are other things to be taken care of in the handling of glue besides the cooking part. The superintendent should see that a proper supply of hot water is available in the glue room. If he doesn't, he can't blame the men if they dip the hot water out of the glue heater. They know as well as he does that it contains boiler compound, but the responsibility is his. An ordinary kitchen boiler installed in the boiler room and piped to the various glue heaters in the factory and taps provided to draw the water from will eliminate the trouble. If too much piping is required, probably different arrangements could be made to suit local conditions. The difficulty with having long stretches of piping is that quite a lot of cold water has to be drawn off to get the hot. Another important thing is not to make up any more glue than is required for the day, and the pots should be cleaned out every morning before making up the glue.

We know of a factory which has glue heaters located in several different departments, and each heater has half a dozen different glue pots, one for each man. This factory has a double supply of glue pots, and it is the duty of one man to change all the pots every morning. A large pot of glue is made up in the veneer room, and this man puts enough glue in each of the small pots to last a man for the day, judging by his work and the amount he uses. The pots are then placed on a truck and distributed to the different departments. The other pots are collected, brought back to the veneer room, and emptied into the large pot. This glue is only a day old and there is not much of it, so it is quite safe to mix it with the glue in the large pot to be used that day. The small pots are then all put in a tank of hot water, installed for the purpose, and thoroughly cleaned. and are then ready for the next morning's distribution. The labor cost of this is a trifle; there is no waste, and it insures nothing but fresh glue and clean pots being used. System! And it proved worth while.

Demand in Britain for Broom Handles

Broom handles are in demand in Great Britain, according to information the trade and commerce department has sent out. Harrison Watson, trade commissioner, London, writes the department that the situation has recently been eased by the receipt of supplies from the United States and that prices have dropped considerably from the extreme figures current last spring.

Large quantities of broom handles are exported to Great Britain from Finland, Sweden and the United States. Although Swedish whitewood and redwood squares 1 1-8 inch by 1 1-8 inch by 51 inch, free from knots, are being imported at £14 7s. 6d. per standard c. i. f., since the outbreak of the war, Douglas fir has also been used and found satisfactory. The greater part of the broom handle supply, however, is imported ready manufactured. The varieties imported are spruce, basswood, maple or beech. Spruce and basswood are preferred because of their light weight. Broom handles must be smooth. The only complaint against Canadian broom handles imported up to the present time has been that they were not finished sufficiently smooth, and that consequently they had to be graded as 2 and 3.

Canadian Firms Likely to Get Portable House Trade

The Canadian Timber Products Association recently received through the Department of Trade and Commerce, a communication from Mr. Philippe Roy, General Commissioner for Canada, Paris, France, stating that the French Government has authorized him to make enquiries into the possibility of obtaining a large number of shelter houses in Canada, to be used in the devastated portions of France which have been recovered from the enemy. The communication states that if the houses can be turned out at a suitable price the French Government will place an initial order for a very large quantity and that further orders are likely to follow.

Working upon this enquiry the Canadian Timber Products Association has appointed a committee to consult with architects, and already have, in conjunction with Mr. John M. Lyle, who represents the French National Relief Society at Toronto, made very definite progress towards the preparation of plans for a couple of houses of different design. Each of these houses is to contain three rooms, two being bed rooms and one a kitchen and dining room. Mr. Lyle is now completing the plans and as soon as they are in shape, one of the members of the Canadian Timber Products Association will have one sample of each house built. The Canadian Government has promised, through Sir Geo. E. Foster, Minister of Trade and Commerce, that in connection with this business it will guarantee that transportation facilities will be available.

The other propositions previously taken up by the Canadian Timber Products Association, relating to the donation of a number of houses by Canadian lumbermen, woodworkers, and the Government, is for the time being in abeyance. The Ontario woodworking firms represented in the Canadian Timber Products Association have given definite undertakings to do their share in this work and when the other provinces have been heard from the matter is likely to be brought to completion.

Gothic Church Work

By Walter J. Allen, R.S.A.

Among the many and varied branches of wood work practised in this country none have suffered from an ignorance of its technique as has Gothic work.

One would have thought that a craft which has been eulogized for ages by all sorts and conditions of men—antiquaries, historians, architects, philosophers, scientists, travellers, etc.—would have reached, in some sense, the exalted plane it rightly occupies in all civilized countries of the world.

Our neighbors across the line have long gained the pseudonym of "globe-trotters"; but the result of this "trotting" around the ancient cathedral churches of the Continent and Britain is reflected in some of the fairly magnificent churches of the United States—notably, perhaps, in the cathedral church of St. John, New York.

There are not wanting signs that Canada is slowly awakening to the peculiar richness and beauty of the true Gothic style. This is shown by the accompanying illustration, which, though small, sufficiently conveys the idea referred to. It is known as a "reredos," and was constructed by L. Rawlinson, Limited, for St. Simon's Church, Toronto. It is in the "Decorated Style" (or 13th-14th Century work). The three subject panels are surmounted by pierced canopy work, which finishes with a cap mould, the chief member of which is pierced enrichment, and finally the cresting, also pierced, and carries the dedicatory "Sanctus." This, and a few other examples, were wholly executed in Toronto.

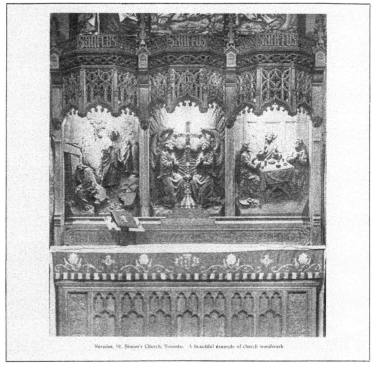

Reredos, St. Simon's Church, Toronto. A beautiful example of church woodwork

A Handsome Calendar

We are in receipt of a very handsome calendar entitled "Lilies of the Valley" from the Central Veneer Company of Huntington, W. Va. "Lilies of the Valley" is a hand-painted water-color after the original painting by William Haskell Coffin.

The Circulation of Air in the Dry Kiln

By Harry Donald Tiemann, Madison, Wis,[*]

The kiln drying of lumber is recognized by all who are engaged in Woodworking as a very important subject. Mr. Tiemann after considerable experimenting for the U. S. Government recommends the use of the Humidity Dry Kiln. It may be that other men through experience have found a different kind of kiln better suited to their work. We are always pleased to receive communications from practical men engaged in the trade and some valuable ideas on kiln drying might be brought out in this way. May we hear from you?—[Editor.]

During the last year concentrated study has been made of the manner in which the air moves within the

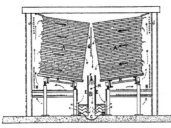

Fig. 1.—Diagrammatic section of improved humidity regulated kiln truck with spray flue in centre.

dry kiln, and particularly within the pile of lumber. Temperatures of the lumber at various points have also been measured in this connection. The observations have developed some valuable conclusions. The importance of this circulation within the pile has been greatly emphasized. In fact, it is evident upon consideration that the humidities and temperatures inside the piles of lumber are largely dependent thereon. That is to say, the temperature and humidity within the kiln taken alone are no criterion of the conditions of drying within the pile of lumber if the circulation in any portion be deficient. It is possible to have an extremely rapid circulation of the air within the dry kiln itself and yet have stagnation within the pile, the air passing chiefly through open spaces and channels. Wherever stagnation exists or the movement of air is too sluggish there the temperature will drop and humidity increase, perhaps to the point of saturation. As evaporation is a cooling process it is evidently necessary to have a continual supply of hot air to keep up the temperature. For example, suppose the air around a pile of lumber is at a temperature of 140 degrees F. and a humidity condition of 51 per cent. Then if the temperature at any point inside the pile be 115 degrees (which may easily happen) at that point saturated conditions will exist. No evaporation will take place there, and the timber may even gain in weight, while the outer parts of the boards are drying rapidly. In this connection it should be observed that so long as evaporation is taking place the temperature of the

[*] In charge of section of Timber Physics, Forest Products Laboratory.

lumber itself (which may be measured by a thermometer inserted in a hole bored in the wood) is always less than that of the air. When the lumber is very wet its temperature is almost identical with the wet-bulb of a hygrometer, but as it becomes drier its temperature will lie somewhere between that of the wet-bulb and that of the actual air temperature.

The extraction of heat from the air which comes in contact with the lumber, through evaporation of moisture, causes it to cool and to tend to descend. This tendency is considerable and may be sufficient to counteract the tendency of the air, heated by steam pipes beneath to rise through the piles. It has been found sufficient to produce a reversed circulation causing it to pass downward instead of upward from steam pipes placed beneath the piles. This condition is particularly manifest when wet lumber is placed in the kiln, and especially so when it is cold or frozen. In the former humidity regulated kiln, in which inclined piling was used, the arrangement was such as to cause the circulation to be forced upward through the pile of lumber and downward in the spray chambers on either side.

When flat piling is used it has been found that even with a good circulation in the kiln itself stagnation sometimes may exist within the center portion of the piles to so great an extent that mold may form between the layers of lumber. With the inclined piling as originally designed for use with this kiln the effect of the cooling is not nearly so great as with the flat piling, but it may be sufficient to cause unnecessarily slow drying in portions of the pile. In a large sized opera-

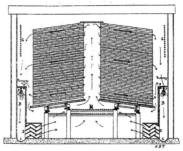

Fig. 2.—Diagrammatic section of improved dry kiln with spray chambers on sides double truck form.

tion when the forced circulation is in the opposite direction from that induced by the cooling of the air by the lumber there is always more or less uncertainty as to the movement of the air through the piles. Even with the boards placed edgewise, the stickers running vertically, with the heating pipes beneath the lumber, it was found that although the air passed upward through most of the spaces it was actually descending through others, and very unequal drying resulted.

While edge piling would at first thought seem ideal for the freest circulation in an ordinary kiln with steam pipes below it, in fact it produces an indeterminate condition; air columns may pass downward through some channel as well as upward through others, and probably stagnate in others. Nevertheless, edge piling is greatly superior to flat piling where the heating system is below the lumber.

The idea has gradually been developed from these experiments and from a study of conditions in commercial kilns so to arrange the parts of the kiln and the pile of lumber that advantage may be taken of this cooling of the air to assist the circulation instead of being a counteracting influence. That this can readily be accomplished without doing away with the present features of regulation of humidity by means of a spray of water is clear from the accompanying diagram, Fig. 1, which represents a cross section of the improved humidity regulated dry kiln.

In the simplest typical form, shown in the sketch,

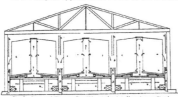

Fig. 3.—Improved Tiemann Dry Kiln, six truck battery form without partitions.

Note.—In the progressive type the condensers G are placed at the dry end only, and sprays F at moist end only.

the spray chambers formerly placed on the sides of the kiln are done away with and are replaced by a single chamber, or flue B, which runs through the center near the bottom. This flue is only about six or seven feet in height and together with the water spray F and the baffle plates DD constitutes the humidity control feature of the kiln. This control of humidity is effected in the same manner as in the former kiln by the temperature of the water used in the spray. This spray completely saturates the air in the flue B at whatever predetermined temperature is required. The baffle plates DD are to separate all entrained particles of water from the air, so that it is delivered to the heaters in a saturated condition at the required temperature. This temperature is, therefore, the dew point of the air when heated above, and the method of humidity control may therefore be called the dew-point method. It is a very simple matter by means of the Humidity Diagram designed for the purpose*, or by a hydrodeik, to determine what dew point temperature is needed for any desired humidity above the heaters. This spray F also accomplishes another thing besides the humidity regulation; namely, it acts as an ejector and forces a circulation of air through the flue B. The heating system H is concentrated near the outer walls so as to heat the rising column of air. The temperature within the drying chamber is controlled by means of any suitable thermostat, actuating a valve on the main steam line. The lumber is piled in such a way that the stickers slope downward toward the center.

*Forest Service Bulletin 104, "Principles of Drying Lumber and Humidity Diagram." Superintendent of Documents, Government Printing Office, Washington, D.C., 5 cents.

Referring to the remaining letters on the diagram, M is an auxiliary steam spray pointing downward for use at very high temperatures. C is a gutter to catch the precipitation and conduct it back to the pump, the water being recirculated through the sprays. JJ are auxiliary heating coils, for maintaining the temperature of the rising column of hot air, but may be omitted. G is a pipe condenser for use toward the end of the drying operation. K is a baffle plate for diverting the heated air and at the same time shielding the under layer of boards from direct radiation of the steam pipes.

The operation is extremely simple. The heated air rises in the spaces HJ at the sides of the piles of lumber. As it comes in contact with the piles portions of it are cooled and pass downward and inward through the layers of boards into the space G. Here the column of cooled air descends into the spray-flue B, where its velocity is increased by the force of the water spray. It then passes out from the baffle plates to the heaters and repeats the cycle.

Various modifications of this arrangement may be made. For example, a single track kiln may be used. This form would be represented by simply dividing the diagram vertically into two parts by extending the line E (on the left side) upward to represent the outer wall, and erasing the part to the left of this line. Or, again, the spray chambers may be kept on the sides, just as in the former kiln, as shown in figure 2. The lumber would then slope in the opposite direction with respect to the center of the kiln, the air would rise in the center and descend on the sides. With this arrangement flue B may be extended upward to the top of the pile as in the former kiln. In fact, the former kiln may be used to accomplish this reverse circulation, by merely sloping the piles of lumber in the direction shown in figure 2, without any alteration of the parts, except that the steam pipes should be concentrated toward the center or so baffled as to throw the heat toward the center of the pile. Experiment indicates that the method will work satisfactorily with this arrangement, the air rising in the center, descending outwardly through the lumber, rising to the top of the partitions and again descending in the spray chambers. It is advantageous, however, to lower the partitions and also the spray system so that the tops of the spray chambers are but slightly above the bottom of the pile of lumber.

One of the greatest advantages of this reversed circulation method is that the colder the lumber when placed in the kiln the greater is the movement produced, under the very conditions which call for the greatest circulation—just the opposite of the direct circulation method. This is a feature of the greatest importance in winter, when the lumber is put into the kiln in the frozen condition. One truck load of lumber at 60 percent moisture may easily contain over seven thousand pounds of ice. Think of the circulation of air needed to melt three and a half tons of ice, before any heating or drying can begin to take place, and this for every truck load placed in the kiln.

The result is, in fact, self-regulatory. The colder the lumber the greater the circulation produced; and, moreover, the effect is increased toward the cooler and wetter portions of the pile.

In figure 3 is illustrated diagrammatically how a battery of six trucks, equivalent to three ordinary kilns, may be placed under a single roof without any partitions between.

The French Period Styles—Louis XV

The second of a series of illustrated articles dealing with developments in furniture design during this important era

By James Thomson

Nothing in the way of furnishing can be more imposing, and to the cultivated eye more artistically satisfying than a capacious lofty-ceilinged room done in best Louis XIV. fashion. Seldom is there to be found here aught likely to offend artistic sensibility.

Fig. 1.

Fig. 2.

Quite otherwise is it with the French style we owe to the reign that followed. Beginning fairly well, before it ended the Louis XV. style had defied all laws of regularity and orderly arrangement, and was only saved from being contemptible by the high quality of the craftsmanship—by the genius of designers, artists, cabinet-makers, decorators, workers in metals, weavers of textiles, responsible for its development. Contemplating the high quality of the workmanship one is apt to overlook the lawlessness of conception, the tendency to senseless arrangement and taudriness in motive.

Louis XV. was but a child when called to the throne. The Duke of Orleans, a depraved debauchee, became Regent, and his assumption of power was the signal for an irregularity of morals that went on increasing up to the end of the reign of the monarch he was representing.

It is singular, but a fact, that furniture of a given period very adequately reflects the spirit of the age. In the style developed during the reign of Louis XV.

Fig. 3.

the inward feeling of moral looseness and debased feeling find outward expression. Louis XIV. to the day of his death was absolute master, nor brooked opposition. The style of furnishing developed in his reign is a reflection of his taste and desire. When Louis XV., however, came to the throne, he may be said to have "reigned but did not rule." He as a rule allowed the reigning favorite to have all the say. Pompadour in her day ruled France, and after her death Dubany had her turn. In matters of taste they were dictators. Competent designers, artists and decorators had to swallow their pride and cater to the desires of these depraved mistresses of the King. If the reigning favorite was a woman of taste, good, as affected furnishings at Court, might result. If a woman of vicious taste and vulgar ideas, the results from an artistic point of view might be deplorable. When, therefore, we find in the style incoherence, abundance of meaningless detail, want of definite plan, and vulgar display in the product, we can easily understand how it happened to come about.

The style at the beginning had many good features, but in process of time came that phase of it to which has been attached the appellation "Rococo." Besides rocks and shells as decorative features, are grottos, roses, cornucopias, vases with spilling contents, scrolls, weeds, sprawling accanthus, all seemingly thrown together haphazard. Everything evidently was grist that came to the decorative mill. In some of the decoration one is reminded of plant growth of a rank kind; vegitation that springs into plenitude from the green scum of stagnant pools. The one idea of the decorator seems to have been to get effect through deep shadow and high light. The depth in shell and rock supplied

Fig. 4.

shade and shadow, the prominences, brilliant points of light. Here was multitudinous detail, ridged and broken outlines in constant succession, the whole forming an aspect richness of effect. When we come to the style that prevailed in the latter part of the reign (now known as the rococo) we find all formal treatment eschewed and symmetrical arrangement of not the smallest account; in fact a thing to be avoided. To have the two sides of an object to entirely differ was the aim. Chatearoux was mistress of the King in 1744. Pompadour in 1764, Dubany until 1774, and each left impression of taste—from an individual point of view—on the furniture and decoration of the reign.

A process belonging to this time, called after the inventor Veinis-Martin, had great vogue. He was a carriage painter. His varnish—a transparent lac—was possibly derived from Japan. The work associated with his name is to be found in the shape of fans, snuff boxes, needle cases, as well as in furniture of domestic

Fig. 5.

character. The decoration is usually upon a ground of gold dust.

While he declared his secret should die with him, as a matter of fact he was succeeded by many imitators.

Later in the century came the names of Riesener, David and Gauthiere, who gained great repute, the first two as makers of marquetry, the last named as a founder and chaser of metal mounts such as edgings and lock-plates. Specimens of this work now bring fabulous prices.

I am fortunately in position to submit some very fine examples as illustrations of this article, and should be pleased to submit more did the exigencies of space permit.

In Figs. 1 and 2 are beautiful and characteristic

Fig. 6.

Chippendale treatment of the Louis fifteenth style.

Fig. 7.

chairs of an early period. In Fig. 3 the tendency to greater elaboration is observable, though symmetry is still an end to be sought. A reference to Fig. 4 discloses the "hit or miss" unsymmetrical tendency that later ran riot. In the frame surrounding the faun's head we have the rococo in its purity. Take note of the difference in arrangement of festoons, and the nondescript nature of the materials supplying the decorative motives. See also how unsymmetrical are the frame corners. With designs such as these the sole appeal is in the high quality of the craftsmanship.

Fig. 8.

When such carving is done in a clumsy, inartistic manner the result is deplorable.

In Fig. 5 are shown excellent specimens of the rococo now in the Fontinbleau Palace. A recessed bed from the Chateau Lausun, Paris, is depicted in Fig. 6. These recessed beds were a favorite feature in this

Chippendale, who flourished during this period, was a bald pilferer of Parisian designs. Many designs published under his name are in fact absolutely French in origin. In Fig. 7 are shown British examples of the style, though in all particulars they might be classed as Parisian.

In Figs. 8 and 9 are shown examples from Chippendale's published works. They are so thoroughly Parisian it has been surmised that this master on occasion was not above incorporating the work of others in his book, thus leading the public to believe it to be his own. At any rate, many designs attributed to this master are so thoroughly French as to lead one to the conclusion that he borrowed them from others.

In conclusion I would remark that taking it in the mass, the Louis XV. style is one of great splendor. In white and gold or all gold, nothing can be more effective. It is a style for the luxurious, as to give it full play requires that the walls be panelled and that in these panels are paintings or tapestries of bright har-

Fig. 9.

monius coloring. Fragonard and Baucie were noted court painters of the period and the panellings of the gay interiors in palaces of French royalty and nobility were made gay with their work.

Will Use Canadian Hardwoods

In order to encourage the use of Canadian hardwoods for interior decoration, Baron Shaughnessy has issued instructions to use nothing but Canadian forest products in the sleeping, parlor, dining and observation cars and in the offices and hotel buildings of the Canadian Pacific Railway Company. This decision was made only after careful consideration and experiment. Baron Shaughnessy had samples of all Canadian hardwoods treated at the Angus shops in Montreal, where selected specimens were tested with polishes, stains, etc., and the results showed that the Canadian woods compared very favorably with imported varieties. In view of the fact that the interiors of the Canadian Pacific sleeping, dining, parlor and observation cars and hotels are recognized as setting the very highest standard in their class, the encouragement given to Canadian lumber and manufacturing industries will undoubtedly be very great.

Safety in the Machine Room

By W. J. Beattie

The one danger in the Safety First crusade is, that the thing will be overdone. This is rather a peculiar statement to make, as most people would argue that you could not possibly overdo it.

The way it can be overdone is by too much talk, more than by actual safeguarding of machines. Factory men know well that once a man loses confidence in his ability to operate a machine, his days of usefulness are over, at least so far as that particular machine is concerned.

Let a man get a few blows in the stomach from a shaper and his confidence will be sorely tried, he will get as nervous as a child the first day at school.

There are good safeguards made for shapers, but I once saw a man get knocked down on account of the guard, but it was his own fault. He was shaping a heavy cove moulding around the edges of panels and the knives were necessarily very large and murderous looking, he screwed the rocker guards down so tight that he could not shove the work into the knives easily, the result was that he gave a quick lunge at the panel which sent it into the knives with a rush and it was thrown back with great force, with the result that he was carried home on a stretcher, and was lucky to be only temporarily injured. The fact of him having the spring guard too tight was discovered after the accident. This fact is not advanced against the use of shaper guards, but to show that when not used properly they can be made a source of danger. The man mentioned, so far as I know, would never run a shaper again, his confidence was gone.

It goes without contradiction that if you frighten a man sufficiently by too many warnings and he takes them seriously enough, he becomes a useless incumbrance.

A careful watching of men who are running woodworking machines and checking anything that is

‡ Piping used as an effective guard around planer.

careless on their part, and using all necessary safeguards, is all that any one can be expected to do.

An habitually reckless or careless man should never be employed to run woodworking machinery. In this issue the writer presents a couple of sketches of machine guards that help to protect the machine as well as the worker, even the "poor" machine needs protection quite frequently. The one sketch illustrates

piping as used to guard a planer. This machine stands near a pillar which was used to give the guard rigidity, note that one piece of pipe runs between the pulley and the "on" side of the feed belt, thus making it impossible for anything or any person to be drawn under the belt. The top is bent around and is close to the cylinder belt thus protecting that point. This makes a very effectual guard.

The other sketch illustrates the protection of the

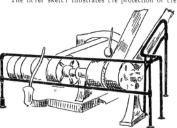

Protecting counter shaft of sticker with piping.

counter shaft of a sticker by using iron piping; it explains itself.

Any guard made of iron piping can be made to remove easily by employing telescope joints. Practically every factory has quantities of worn out piping that can be utilized for making machine guards, and by painting it a vermillion color the effect is noticeable and pleasing.

It is unnecessary to state that the solidity and appearance are much ahead of any wooden railings that could be used for such a purpose.

These are sketches of actual installations and were made from old piping, put together as chance occurred, and at a minimum cost, replacing unsightly and dirt harboring wooden constructions.

Cotton Wood for Butter Boxes

The following press dispatch was recently sent from Madison, Wis.:

"The peculiar and penetrating odor in factories where cottonwood is being worked naturally led to the belief that this species of wood could not be used in the manufacture of butter boxes because it would contaminate the butter. One of the large box companies recently requested the Forest Products laboratory here to determine the truth of this theory and furnished a number of boxes for trial.

"In co-operation with the dairy department of the University of Wisconsin tests were conducted as follows: Butter wrapped in waxed paper and enclosed in a paper carton, some wrapped in waxed paper only and some unwrapped were placed in the cottonwood boxes and these were then placed in storage. At the end of one week they were removed and the butter tested by five graders. The butter wrapped in waxed paper and that enclosed in a paper carton indicated no contamination. That which had not been wrapped absorbed only a slight taste from the wood and that on the surface only. It was the opinion of the five judges that the butter which had been packed in cottonwood boxes was fully as good as that packed in boxes of other species.

The Air Compressor in Furniture Factories

A New Factor Coming into Use that Requires Sensible Methods in Operation

By L. O. Hanson

Compressed air is gradually coming into more common use in furniture factories, and for certain work it is being found quite indispensable—notably in the finishing room for spray and rubbing machines. Some plants, once equipped with a compressed air installation find many practical uses that meet individual requirements. The writer has in mind one factory making upholstered goods where compressed air is used in thirty-eight different operations.

Numerous accidents occur in connection with the use of high-pressure air, and explosions of various parts of the apparatus are likely to be most serious in their results, both with regard to property damage and personal injury. Air-compressor explosions, considered in proportion to the number of compressors in use, are perhaps even more frequent than explosions of steam boilers; and all possible precautions should therefore be taken to safeguard the compressing apparatus by designing and constructing the various parts with suitable factors of safety, by providing safety-valves, and by operating the installations in an intelligent and careful manner.

Air-compressor explosions, like boiler explosions, are caused by over pressure. This excess pressure may be due to the failure of the safety-valve to operate properly, or it may be caused by the spontaneous ignition of explosive gases of some kind within the compressor cylinders, or in the pipe lines or receivers. Lubricating oil is the only combustible substance that can be present in the cylinders of an air compressor under normal conditions, and it is therefore responsible for the explosions that take place in them. Oil is combustible in its liquid state, or when vaporized, or when partially carbonized and deposited in solid form upon the walls of the compressor cylinders, or upon the inner surfaces of the piping.

There is reason to believe that there is little danger to be feared from explosions of the oil when it is in the form of either liquid or vapor, because during the normal operation of the compressors it is hardly possible for oil to accumulate, in either of these forms, in sufficient amount to cause such explosion. We must therefore look, for the cause of the trouble, to the soot-like deposits that are formed when a poor grade of oil is used, or when the oil is fed in excessive quantities.

Chemical analysis shows that a combination of one part of finely-divided carbon and 11½ parts of air, both by weight, forms a mixture with explosive possibilities. One pound of carbon gives off 14,500 British thermal units when burned, and by the explosion of a mixture such as we have just assumed, there would be 979 units of heat liberated, per cubic yard of air. Naturally all this heat could not be expended in the production of mechanical energy, but we may perhaps assume that ten per cent. of it could be so expended. That is, 98 British thermal units, or over 76,000 foot pounds of mechanical energy, would be available for every cubic yard of air. This shows the enormous destructive energy that may be stored up in the air cylinders of a large compressor, and the corresponding importance of guarding against the possibility of explosions.

Finely-divided carbon ordinarily ignites at a temperature of about 6000 Fahr., but if it is subjected to continuous heat and pressure the igniting point is much lower. It is therefore highly important to keep the cylinder temperature low, and this may be done by providing adequate water jacketing, and by keeping the water circulating freely and continuously.

As already stated, carbon is usually the result of an excessive oil supply, or of using oil of poor quality for lubricating the air cylinders. An excess of oil causes the parts and valves to become obstructive by carbon deposits, and this is likely to bring about an increased temperature in the cylinder. Experience shows that from one to three drops of oil every five minutes is sufficient to lubricate cylinders of ordinary size. Mineral oil of medium viscosity is to be preferred, and it should be free from bituminous and other foreign matter.

The air cylinders should be washed out frequently with soft soap and water, or with a solution of caustic soda or caustic potash, to remove slight accumulations of carbon deposit. Under no circumstances, however, should kerosene, gasoline, or other similar inflammable liquids or substances be used for cleaning purposes. If care is taken to prevent carbon deposits, and an adequate and constant circulation of cooling water is maintained, and safety-valves are connected to the cylinders, explosions of this part of the equipment will be of very infrequent occurrence.

Explosions of the air receivers of compressor installations are more frequent than those of the air cylinders. In many cases receiver tanks are installed that were never designed for such service—even ordinary kitchen boilers being used in some instances. Sometimes, also no safety-valves are provided, and the safety element seems to be entirely neglected throughout the entire installation. Receiver tanks should be designed especially for the purpose for which they are to be used, and they should have an adequate factor of safety. The material and workmanship should be of the best quality, and the tanks should be thoroughly tested before being placed in service.

In order to remove oil that may accumulate in a receiver, a flange may be riveted to the receiver at its lowest point. A 4-inch nipple, about one foot long, may be screwed into the flange, and this may be bushed to receive a stop-cock. A pocket or trap is thus formed, in which the oil will collect and from which it may be blown by opening the cock. A manhole or handhole should be located in each receiver, to afford adequate facilities for cleaning.

Explosions occur frequently in the air lines, or pipes because of defective air valves which permit the compressed air to return to the cylinder. As the air is re-compressed in consequence of such leakage, the temperature constantly increases until it reaches a point high enough to vaporize the oil that has accumulated on the interior of the pipes, or even to ignite the oil vapor, and thus cause an explosion. If care is taken to keep the valves in good working order trouble of this kind may be averted in large measure.

The following additional recommendations are suggested for the safety of the entire installation.

Accidents are likely to occur if open lights are placed in proximity to a cylinder head that is being removed, and under no circumstances should lighted matches, candles, kerosene lamps or other open-flame lights be used. Serious explosions have been caused by escaping gases when removing cylinder heads, these gases resulting from the use of kerosene or gasoline for cleaning the cylinders, or from vaporized lubricating oil.

If it becomes necessary to replace piston rings, or to do other work in the cylinders, the compressor should be shut down some time in advance, to allow it to cool off thoroughly before the work is begun.

It is highly desirable to use an aftercooler and a separator. The purpose of the aftercooler is to condense the moisture and oil so that these will be retained by the separator, from which they may be removed. It cannot take place in the pipes or receiver. Both of these devices should be used, as they work independently of each other.

The valves should be regularly inspected and tested, and the receiver and pipes into which the compressor discharges should be blown out periodically to remove any oil and water that may have accumulated.

It is extremely important that the discharge valves be regularly inspected and cleaned, because if they stick, hot, compressed air will return into the air cylinders, thereby increasing the temperature of the air to such a point that ignition of the lubricating oil may take place on the next stroke.

Recording thermometers, placed on the air lines close to the compressors, are highly to be recommended, for showing the temperature of the outgoing air.

Automatic electrical emergency shutdown devices, operated by the thermometers, are also in use in some plants, and these are also of great value, but they must be given careful attention and be tested frequently to make sure that they are working properly. Another effective warning device consists in a gong which sounds whenever the temperature of the outgoing air rises sufficiently to close an electrical contact on the thermometer or thermostat.

Fusible signal plugs have been devised for use with air compressors and these should be generally adopted where no other warning device is provided.

These plugs are similar to those used in boilers, and may be placed in the discharge spaces of single-stage compressors, on the high-pressure side of the two-stage machines, or in the receivers, or other containers or tanks. In case a dangerous temperature develops, the fusible metal melts and the opening so formed allows the air to escape with a shrill, whistling sound, thereby giving warning to the engineer. By varying the composition of the fusible metal in these plugs they are made suitable for use at different temperatures.

The chief objection to the use of air as a motive power is its dependence upon some other form of power to furnish and maintain the required pressure; and as the first cost of an air-compressing installation is a considerable item, and its subsequent maintenance must be considered in addition to the expense of the equipment and of the power necessary to operate the compressors, some other form of motive power is usually preferred to compressed air.

The Moulder Man a Specialized Expert

By E. H. Barnes

With the possible exception of the shaper, the moulder, or "sticker," as it is called most of the time, is least understood of any machine used in the furniture factory. There is so much of real skill necessary in its efficient operation that the average man would rather take up any other machine than to study out the best methods of accomplishing the things that he is called upon to do in almost any factory. This is not giving the machine operators very much credit for being progressive, as many of them are, but when it is taken into consideration that many of the things to be done by the moulder operator require downright hard study, it is not strange that when a man can he will select another machine to take care of rather than the moulder. Much of the knowledge required of the moulder man is also needed by the shaper operator, but the shaper man stops before he gets to the majority of things that the moulder operator feels are just starting points in moulder operation.

It is a thing very much to be regretted that there is so little opportunity for men to learn what they should know of machine operation before they take on real jobs in some furniture factory, for as it is they simply have to learn the necessary points by butting in against them or by asking questions of the older men who have had their turns at the moulder. Not many of the industrial schools in the country are equipped with even small stickers, so that they could give instruction in their operation, and it seems to the writer that they miss one of the greatest opportunities to be helpful just for this reason. A small sticker would cost so little that any school in a good-sized town could afford one, and to help make up the cost of the instruction in its operation a great deal of work about town could be done for carpenters, contractors and builders generally. To be sure it would hardly be up to all towns with industrial or manual training schools to install a sticker for instruction purposes, but where there are any woodworking industries at all it would certainly pay the manufacturers well to see that this sort of instruction is given. It would surely be much more to the point than the teaching of how to operate hand jointers, as is done in almost all well equipped schools.

To see some moulder operators go after a machine in setting it up, with hammer or piece of babbitt metal, you might think that they had got their training in somebody's kitchen. Their wrenches are always mislaid. The right size wrench is never at hand, it generally being covered up with shavings under the machine. And as to a screw driver for turning in screws on the hold downs, it is so much easier to keep the universal hammer hanging up on the post near at hand, or laying it down near where it was used last. Tool box? A darn nuisance, always filled with shavings, and you have to dig for a tool every time you want anything. I have seen this sort of thing so much in different places where I have been that I have reached the point where I expect it. In fact, if I were to get into a place where everything required at the moulder was right where it should be, I should think that they had just started up things and hadn't yet had time to strew them around.

As a matter of fact, a hammer is not the most useful tool at the moulder, but is perhaps seen more than any other. And at that, nine times out of ten, it will be a claw hammer instead of a mechanic's tool. When you have to use a hammer or a piece of babbitt to shift a chipbreaker or a guide or a hold down, I think that a better policy would be to keep the machine clean enough to enable you to shift parts more easily, or to use the proper wrench enough to loosen them so that they could be moved without a hammer.

Whenever I have had charge of any machine, whether it has been a moulder or some other machine, the first thing I have done was to arrange a tool box where it was handy to the operating side of the machine. A box fifteen or sixteen inches long and about a foot wide and perhaps eight or nine inches deep serves the purpose very well. Such a box might be made up of ⅝-inch stuff and be able to stand all the battering up it will get from the tools it holds. It is better to have the bottom of this tool box of heavy screening with one inch mesh. Then the shavings will all sift through, except the big ones, and they won't give any trouble because the tools are used so much at the moulder that they are naturally thrown out when taking out the tools.

It may not seem necessary that a man have any particular skill in order to use a few common tools, and yet the slip-shod way that many men have of using the commonest of tools shows that a bit of instruction would help a great deal. And a bit of discipline would also teach them to take care of their tools as they should be.

After a moulder has been properly set up and everything ready to run, it is always a mighty good scheme to see that everything is really ready. For example, are the belts tight enough, providing that gravity tighteners are not used on the main drive belts, to stand whatever pull will be required of them throughout the run? Start up your heads first and be sure that they are running perfectly before feeding a thing through the machine. Some men are not careful enough about this. A box may have been tightened too far or some dirt may have sifted down into the bearing and caused hard running, and the first thing the operator knows his belt has burned enough to have done considerable damage. It doesn't take long, you know, to burn a high-speed belt. Be sure always that the bearings are in first-class condition. A hot journal does a lot of harm without necessarily burning out the box. If a box seems worn enough to allow any wobble or play, get to it and pour a new bearing and don't say to yourself that you will run until noon or night to get out an order. It is more essential that an expensive machine should be kept in good order than that some department head be pleased with quick delivery of something that he wants in order to make a nice showing for himself.

Scraping in a bearing is no fool's job. It takes even a good intelligent man some time to get the knack of scraping a bearing that will run well under average conditions. Some men still pour their boxes on the regular journal. They may get away with this practice for a time without springing the journal, but it is a dangerous practice, one that will cost a nice little sum of cash when a journal does spring a bit.

The best methods of setting up moulders have been given many times in this and other practical papers, and it will not be necessary for me to enter into a discussion of them here. It is enough to say that the setting up of a moulder requires a great deal of that specialized knowledge of which I spoke at the beginning of this article. Even with the milled-to-pattern cutters with which many, many moulders are being operated nowadays, a good bit of skill is needed in the man that gets out the very best quality of work, in order to make quick, good set-ups. And with the thick cutters a man must be a real mechanic in order to do anything at all. Too much will never be said about the necessity of carefully balancing knives. Use a good knife balance by all means.

I have always kept my moulder knives on special shelves, and I put them away in sets, numbered, so that when I come to use them again I know what pairs were used opposite each other, and although I always balance all my knives before using them, when they are numbered I can do it quicker than when they are not, because I know where to start and about which knives are rightly balanced. In putting away knives I always put a piece of the pattern they cut in front of them on the shelves so that I could find what I want even in the dark. Place different kinds of patterns on different shelves so that their filing has some sort of classification.

The moulder man is a dozen machine men all done up in one hide. He must know all that the planer man knows, all that the shaper man and the jointer operator and all the rest of them together know about machine work before he is a real first-class moulder operator. And when he graduates from the moulder he is mighty well fitted to be a machine room foreman, if not superintendent of the whole factory, for his moulder work has made him think, and thinking men are needed in the superintendent's place in every furniture factory.—American Furniture Manufacturer.

An Unusual Chip

This picture represents a piece of steel which was cut by a 27½ in. Chipper Knife in use at the Brown Corporation, La Tuque, Quebec. There were three pieces about alike cut by the same knife and it is still in use. The knife was a Simonds make.

Piece of Steel Cut with a Chipper Knife.

The Value of Industrial Democracy

A plea for closer relationship between employer and employee—Better co-operation means bigger profits

By A. W. Wishart*

"We read history," said Wendell Phillips, "not through our eyes, but through our prejudices." That is the way most of us hear discussions of the labor problem or read books on the relations between employer and employee.

We approach the consideration of the labor problem with preconceived ideas and prejudices, more or less fixed, and which are largely the result of the atmosphere in which we live, think and work. The result is that the average workingman looks with suspicion upon every interpretation of the labor problem which comes from the manufacturing or employing class. The employer reads the opinion of labor leaders through the colored glasses of his own interests and is antecedently suspicious of such professional students of the labor problem as college professors and preachers. It is this unconscious influence of the personal equation and group-interest which makes it hard for all of us to put ourselves in the other man's place and to recognize whatever truth there may be in his point of view. It may, therefore, facilitate a more impartial consideration of what I have to offer on this question if I disclaim at the outset any superior knowledge of this most perplexing topic or any more earnest desire to see justice prevail than the average employer possesses. I have no final solution of the labor problem to offer. I believe there is no such finality in life touching any of the great industrial, political and religious issues of this or any other age. Advancing knowledge and broadening experience are inevitably accompanied by higher ideals and new interpretations of old principles. Life is dynamic, not static. We never finally arrive, but are always on the way to new adjustments, necessitated by changing conditions. We indulge in vain delusions when we hope for the time when mal-adjustments. inequality of capacity or of opportunity and the class of interests will disappear from the world. The best that any right-minded employer may hope to do is to understand as far as he may be able the tendency in the industrial world toward better conditions, and to grasp those industrial and ethical principles derived from experience, which, when applied to the problems which face every employer, will gradually remove unnecessary friction, produce increased efficiency and promote a higher degree of co-partnership than now exists between capital and labor.

Competent Leaders

"I have a profound conviction," says N. C. Gilman, that a true and natural aristocracy—the leadership of the competent—is to endure in the industrial world, and elsewhere, for an indefinite period."

It is the privilege and the duty of the competent, whether in the ranks of the employees or the employers, to exert their powers of leadership to obtain the best possible relations between labor and management. The elimination of the incompetent, who, in many cases, are also morally inefficient, is one of the great needs of our industrial age. When the incompetent without high moral principles gain power, either in

the ranks of the employers or in the ranks of the workers, there is bound to be an increase in the obstacles to an amicable adjustment of industrial relations.

The labor problem is generally defined in terms of human relationship or in terms of profit. On the one hand the labor problem is the question of so adjusting the relations between those who work with their hands and those who manage the industry that industrial peace will prevail and some measure of contentment will be obtained because the spirit of co-partnership has been aroused through executive skill and just dealings.

On the other hand the problem has been defined in terms of profit. For example, Seth Low has said. "There has grown up very widely among employees a feeling that the men who put labor into a railroad system or into any other vast industrial plant, help to create that system just as truly as the men who put their money into it, and out of this belief has grown and is growing a constant strengthening conviction that those who work for such an enterprise acquire a property right in it just as real as the property right of those who embark capital in it."

A Complex Problem

The evolution of industry has been marked, in some cases, by a very great increase in the number of owners or stockholders, in a more complex organization which has multiplied the number of managers of of all sorts who stand between the employees and the owners. This fact creates new difficulties in the way of a just and equitable recognition of the two claims to property right in industry—the one arising from capital and the other arising from labor. The adjustment between these relations of man to man and of these conflicting claims over property right can be promoted only by those who keep in mind the nature of the problem.

I have no panacea to offer for the solution of this problem. All that I hope to do is to make clear a few fundamental principles which must guide in the consideration of the issue and to discuss briefly some of the applications of those principles which give rise to concrete efforts to establish better relations and to deal justly in the distribution of the profits of industry.

It has been said that the first thing necessary to the workman is that his employer shall be a man of intellectual ability and general force of character. It is far more important to a workman that his employers shall be financially successful than that he shall be kind or generous in his dealings. If the truth of this statement is not pushed too far, it is of value. An incompetent employer is always an enemy of the workingman, for his incompetency is liable to lead to low wages, bad conditions and unemployment. The economically successful employer is a man of power and it is his duty to use that power for the good of the worker—not only because the interests of industry require this, but also because the employer is more than an economic producer. He is a citizen among citizens and his ideals with their resulting conduct must have

*Extracted from address before National Furniture Manufacturers' Association.

a profound influence upon the lives of his workers and their families, and, therefore, upon the whole community in which he lives. In short, the intelligence and efficiency of the employer need to be supplemented by conscience. Power over material forces needs to be restrained and guided by those moral and humane considerations, which, if ignored, produce misery and social strife.

The treatment of men as machines or as economic instruments to produce profit for the investor is a distinct violation of moral obligation and in the end will be found to be economically unprofitable. The competent, industrial leader should take a higher attitude toward the workingman than that of contempt for organized labor, or negative criticism of the weaknesses of those who toil. The industrial world needs practical and positive suggestions and experiments in the direction of more humane relations between men, more justice in the distribution of profits and more kindly co-operation in production.

Modification of Wage System

Mr. Gilman, whose work on profit-sharing is a standard authority, believes that the wage-system in one form or another will probably endure for a long while even though it may be greatly modified in the process of time. He also believes that co-operative production in the strict sense of the term, or the socialistic state, is a long way off, but he says that employers should be in the front ranks of the advocates for legislation that will bring those who are unwilling to conform to high standards up to the mark.

"Labor laws are absolutely essential to the protection of the unprotected workpeople and absolutely essential to maintain standards, which those best informed and most willing among the employers have themselves maintained or are trying to maintain. Opposition to the injustices of trade unions, or what are regarded as injustices by employers, should not take the form exclusively of opposition to the principle of organization itself, but it should be accompanied by a just, frank and generous recognition of the good that labor organizations have achieved, of the right of the workers to organize, of the necessity of such organization in view of the character of modern industrial life."

Frequent objection is expressed to what is called the appeal to the class spirit, but no thoughtful and fair-minded man can deny that in a sense there exists in this country a capitalistic and employing class on the one hand and industrial classes on the other. In each of these groups will be found certain class standards and class opinions which are the inevitable result of a community of interests and organization for the promotion of those interests. The class struggle is not all bad. The chief evil of the class consciousness is when loyalty to the class dominates loyalty to broader interests whether they be social, political, industrial or religious. Nationalism, in which you all believe, is, in a sense, class consciousness. Nationalism has its place and its value. It only becomes a curse when the broader interests of humanity are subordinated to a narrow nationalism. It is perfectly natural that the workers at the bench, the profits of whose industry come to them in the form of wages, should possess a certain community of feeling arising from their community of interests. It is only natural that there should be some attempt to organize this class just as those who are in the employing class find it to their advantage to organize for the promotion of their

industrial welfare. But when this principle of class co-operation is pushed to an extreme then ensue those unhappy classes which imperil the existence of democracy. When industry is autocratically controlled with a minimum of democracy in a country where democracy is recognized as a legitimate and valuable principle of political life, there is bound to be serious friction. Therefore, one of the serious phases of the labor problem is the excessive self-assertion of organizations and groups; the struggle for the dominance of one group over the other, which, to the extent it is achieved, means the triumph of autocracy. That autocracy may mean the despotism of labor as well as the despotism of capital. Industrial democracy is the harmonious co-operation of these groups in what has been called co-partnership production. The plain truth is that the operation of social forces in the field of industry has produced all sorts of inequality in station, power and property. Some of these are useful and some are harmful, as Professor Charles H. Cooley of the University of Michigan has pointed out. He says: "A true freedom, a reasonable equality, aims to conserve the useful differences and to limit the harmful differences. The idea of freedom prevalent in the managing class, is, however, somewhat narrow and hardly hospitable to the group self-assertion of the less privileged classes. The labor movement has made its way by energy and reasonableness in the face of a rather general distrust and opposition—sometimes justified by its aberrations—on the part of the masters of industry."

As Professor Cooley adds elsewhere: "The great mass of wealth is accumulated by solid qualities, energy, tenacity, shrewdness and the like, which may co-exist with great moral refinement or with its opposite."

This success has been achieved by a considerable degree of autocratic control, necessitated by the condition of modern life and, therefore, to that extent not at all blameworthy. But these capable and efficient leaders in industry should remember that the time may have arrived when there should be a curtailment of autocracy and an increase in the application of the democratic principle.

The Appeal to Self Interest

Henry D. Lloyd has said: "The system which comes nearest to calling out all the self-interest and using all the faculties and sharing all the benefits will out-compete any system that strikes a lower level of motive, faculty and profit."

It cannot be denied that an increasing number of employers are actively and conscientiously engaged in endeavoring to discover, through experiment and study, industrial devices in management and production that will promote real democracy and industry. On the other hand the evils of labor organization must not blind the eyes of the employer to whatever good labor organizations have achieved or to the part which they have played in pushing these new devices of industry into the foreground of popular consideration. We must all recognize the fact, as Professor Cooley has well said, that we want a promising industrial plant but have not yet learned to make it turn out the desired product of righteousness and happiness. Wherever there is power that has outstripped the growth of moral and legal standards there is sure to be some kind of anarchy. And so it is with our commerce and finance.

But since employers are "neither heroes, nor saints,

nor philanthropists," the ideals which are presented for their acceptance must be within their reach. They must be such as can be fitted into the actual conditions of modern industry. The next desirable step is never the destruction of existing systems and the building of new ones from the ground up. This was attempted in the French Revolution. It ended disastrously, and men had to begin again the slow reconstruction of government, industry and social conditions. Therefore, every employer should distinguish between those ideals which look forward to a new age, more or less remote from the present, and those ideals which have to do with the immediate improvement of industrial relations. In this field many interesting experiments are being tried which are characterized as welfare work, insurance against accident, illness or old age and death, profit-sharing, bonus and premium systems, employees' stockholding, and democratic management, meaning thereby various experiments in the direction of giving the workmen some sort of representation in

the management of the industry. And to these experiments must be added the general effort to promote efficiency and welfare through scientific management and industrial education.

As Mr. A. W. Burrett, of Bridgeport, Conn., has recently said: "If the employer desires to get enthusiastic service, a high degree of loyalty from his entire force, from the management to the least important employee, he must create good personal relations which can be felt all along the line and good working conditions; he must pay the full market wages; he must know the labor cost of each item of his product and the producing value of every employee; he must be quick to recognize improvement and ability and acknowledge it by increased pay or advancement. If in addition to these things he wishes to practise some wisely devised plan of profit-sharing or industrial partnership, it should stimulate that enthusiastic loyalty which results in the highest degree of efficiency so necessary to meet the demands of the time.

System in Designing Mouldings and Taking Care of Knives
By J. H. Walker

How often does the designer or superintendent of a furniture factory design a moulding to harmonize with some particular style of case, and after two or three jobs of this case have gone through the factory he finds the moulding is entirely different from the one he originally designed.

This is a very common occurrence and is due to lack of care in grinding the knives as some men in setting up a shaper or sticker and finding their knives to be a trifle dull will proceed to touch them up on the points only, as this is the part of the knives which do the most cutting and therefore get dull before the other parts. It will be seen that if this action is repeated a few times, instead of having a graceful, nicely proportioned moulding, as the designer intended, you will have one which looks stumpy and may spoil the appearance of the whole case, also if the moulding was designed to show a certain margin when finished, you will find you have a wider margin than should be and this may be sufficient to make the whole case look out of proportion.

As designer and draughtsman in a large furniture factory the writer has adopted a system which proved worth the trouble. With a piece of tissue paper trace the exact shape of the moulding from the drawing, then by means of glue or shellac fasten the paper to a thin piece of wood. A zinc pattern is then made by placing a piece of zinc between two pieces of wood, one piece having the paper pattern attached and bandsawing to shape. After band-sawing, the pattern should be tried on the drawing and touched up with a file to get the correct shape, or if it is a small moulding the bandsawing may be dispensed with and just the file used.

A large board is hung up beside the draughting table, having hooks arranged in rows and a number above each hook. The zinc pattern is then numbered and hung on the hook bearing corresponding number. If a new piece of furniture is being designed and a moulding required, a look at the patterns will show if one of the stock mouldings will be suitable. If so the pattern can be used for drawing it and the number

specified on the stock bill and then on the tag which accompanies the job.

In connection with this system another pattern is made for the machine room (the making of which I will deal with later) for the purpose of grinding the knives and bears a corresponding number to the pattern in the draughting room.

The knives are then ground and trued up accurately to this pattern and after being used are placed in a cupboard made for the purpose, having shelves, conveniently arranged for keeping the knives in place and a number corresponding to the patterns is placed beside the knives. When another job comes through which requires this particular moulding, all that is necessary is to take the knives and if they require sharpening they should be ground accurately to the pattern as when first made.

It may require a little more time but it may also save considerable trouble later on and the finished piece of furniture will have the desired appearance. It is system in caring for details such as this which leads to that final efficiency for which all manufacturers are striving, and the fact should not be overlooked that although you may scimp these details in order to save manufacturing costs, when your furniture is placed on the dealer's floor it will have to compare favorably in appearance and structural details with that of your competitor who may be giving special attention to just such details as these.

In making a knife pattern the most important point is to determine the correct amount of knife projection necessary to cut the different members of the mould. The remaining parts may be drawn in freehand. The writer has found that about the most convenient method of throwing out these projections is by means of a scale, which has been described in the columns of this and other trade journals, but which it may be advisable to take up again for the benefit of those who may not be familiar with the method.

Make an accurate drawing of the head of the machine for which the knives are intended to be used, as at Fig. I of the accompanying sketch, then lay out the projection of a straight knife past the lip of the cutter head as at (B), mark the exact centre of the cutter head, which will be at (C), then draw a line from centre (C) to a few inches past the point of knife (B) and at right angles to line (CD) draw line (E). This line represents the cutting line for straight knives and also the first projection of moulding knives and is the basis for calculating projections. At this point on line (C.D.) proceed to mark off the measurements necessary for determining the corresponding projections. In the accompanying sketch two inches are laid off in eighths which are shown at (F). Extend the line of knife to distance shown at (G) and with a compass, one leg of which is set at the centre of cutter head (C) the measurements are transferred from line (D) to line (G) as at (H). The measurements on line (D) represent the depth of cut and the corresponding measurements on line (G) represent the amount of knife projection required and will be referred to hereafter as the projection scale. The accompanying sketch shows a sticker head but a scale for shaper knives may be made in the same manner.

For shaper knives where the knives may be used on two or three different sized collars, take an average of the different sizes and make pattern accordingly. To make a knife pattern, using the projection scale, mark off an outline of the mould as at (1) Fig. 2. The outline of the mould is the heavy line. At the thickest part of the moulding, draw a line parallel to its base as at (2), then at points where projections occur draw perpendicular lines as at (3), (4), (5), (6), and (7). The measurements from the horizontal line (2) to the outline of the mould on the perpendicular lines may be changed into projections by the use of the projection scale and the parts of the moulding between the projections can be drawn in freehand. The outline of the pattern is shown by the dotted lines.

The pattern may be made for the full size of knife showing bolt slots, etc., according to size of cutter head.

South Africa Open to Buy B. C. Doors

South Africa offers a good market for Canadian made doors, according to information received from the trade and commerce department at Ottawa, which department is in receipt of a communication from H. R. MacMillan, special trade commissioner, stating that the largest firm of timber importers in South Africa import several thousand doors each year from the United States. These doors are nearly all made in New York state, largely from lumber supplied from Canada.

Several years ago, this company imported Canadian doors which were found to be satisfactory but, during the past few years, their New York agents have been supplying them with American doors. It is possible that there are manufacturers in Canada who are in a position to compete with American manufacturers both in quality and price.

This firm is willing to take trial orders of a hundred 6-foot 8-inch by 2-foot 8-inch by 1½ inch panel double moulded No. 3 pine doors, provided that these doors can be shipped before the end of March, for a price of 10s each, c. i. f. Durban. They desire that these doors should be shipped by a reputable manufacturing firm prepared to enter into the export trade. The firm shipping the doors may draw upon the company in question at sight for payment in full.

When the shipment is made the South African firm will be glad to receive a statement from the shipper dealing with the possibility of the development of an export trade from Canada in doors, giving information regarding grades, prices and conditions of shipment.

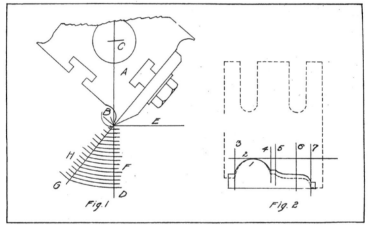

Making a Projection Scale for Sticker Knives.

Letters to the Editor

Why Does a Belt Cling to a Crowned Pulley?

Bridgetown, N. S., Jan. 27, '16

To the Editor, Canadian Woodworker:

Dear Sir—I have been reading with no small degree of interest the letters recently published in the January issue also in issue of November, on the subject of "Why does a belt cling to a crowned pulley." Now there is such a thing as an editor getting tired of too much controversy on one subject, but I hope that you will allow me to give the readers what I think is the real reason why a belt will run to the highest part of a pulley (when the shafting is parallel.)

The reason is summed up in one word—lead. A slight pressure with a belt shifter against the edge of a belt where it is approaching a pulley, will suffice to push it over the other pulley, fast or loose. But all mill men know that it is useless to place a belt shifter on the part of belt that is leaving the pulley. Fig. 1 shows a belt running free on a single round face pulley.

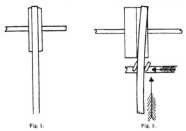

Fig. 1.　　　　　Fig. 2.

Fig. 2 shows a belt running on a fast and loose pulley with the shifter commencing to push against one edge of belt. You will notice that the shifter is pushing towards the left, which puts a slight bend in the belt edgeways. You will then notice that the belt no longer leads straight around the pulley but is approaching at an angle, same as if one were winding a rope on a revolving cylinder and holding the rope on an angle with the cylinder, it would wind in a spiral and quickly travel to the other end of cylinder. In the same way the belt being led on to the pulley at an angle, quickly runs over until it is in line with the shifter, where it will remain.

Now this may seem "off" the subject, but it leads up to the subject, for exactly the same thing happens when a belt is first put on one edge of a crowning pulley as at Fig. 3, which shows belt on right side of a crowning pulley. Then the belt being approximately at right angles with shaft and the pulley being higher in centre, the tension will be much greater on the left side of the belt, which tends to stretch it on that side and throw it over on a curve, same as when pushed over by a shifter, causing it to "lead" towards the left or towards the centre of pulley. As soon as the highest point has been reached and the tension is same on both edges, the belt will become straight and not run either way.

As to the cause of a belt running to the slack side of pulley when shafts are not parallel, the same rule holds good. Fig. 4 shows a shaft with one end dropped down so as to be nearer the lower shaft than

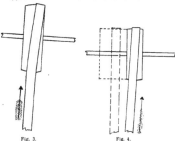

Fig. 3.　　　　　Fig. 4.

at the other end. You again see the belt leading not square on the pulley but on an angle or spiral and if the pulley were widened out as shown in dotted lines the belt would travel over until it became square with the shaft as shown in dotted lines to the left. That is provided the belt would stay in original position on the lower shaft, but as the top end shifting over would also lead the lower end of belt over, the result is that the belt runs half off both pulleys, or off far enough to leave all the tension on the side that remains on the pulleys which curves the belt as in Fig. 5, until each end leads square on the pulley. Fig. 5 will show as well as Fig. 4 why a band saw runs to the front if you tilt the upper wheel towards you.

The quarter twist belt shows what the "lead" does to a belt. On a tenoner for instance, owing to the coper belt being short and pulleys quite large the

Fig. 5.

angle of leaving is quite large but as the belt "leads" square on each pulley it runs as well as if not twisted.

Yours truly,

R. W. W. Purdy.

If you read anything in the columns of this paper that you do not agree with or if there is any subject on which you would like to comment, you may feel at liberty to do so. There is nothing we like better than a discussion of some practical subject.—[Editor.]

THE FINISHING ROOM

The Aeron System of Finishing*

Getting the Most in Quality, Speed, Economy and Efficiency out of the Finishing Room

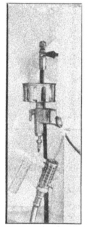

Showing the Auto-Electric Air-Heater Connected to Separate Container, Type Aeron.

The man today at the head of the wood-working plant has confronting him such questions as, how can I save time here and labor there, how can I offer my workmen better protection, how can I increase my output and hold my overhead down—how can I do all of these things and at the same time hold my own in the market by producing a better looking, longer lasting and quicker selling article?

A variety of labor and time saving and safety first devices have modernized his factory to an appreciable extent. To produce, however, a better looking, longer lasting and quicker selling article, there is but one new idea, process or system yet to be applied to his business, and that is the installation of the Aeron System of finishing.

The Aeron System of applying varnishes, enamels, shellac, lacquers, and practically every kind of finishing material on wood products with compressed air, is the result of extensive painting and mechanical experience, and a comprehensive study of finishing problems. This system with its complete accessory equipment successfully meets every angle and problem of the application of finishing materials with a thoroughness and positiveness, and a certainty of greater quality, speed, economy and efficiency. It not only accomplishes this, but it also offers protection to the workmen, in that it makes the work clean and healthful.

In by-gone days the slow-going, hand brush method of applying finishing materials held sway, in fact it was the only method available. In those days it was perhaps a case of overlooking the unevenness of surface, sags, runs, streaks, thin spots, fatty edges and brush marks, and turning out the job as a first class one, finding justification in the fact that the finisher did the best he could with what he had to work with. That day is past. The Aeron System has completely revolutionized the finishing room and its processes. There is now no legitimate reason why any manufacturer should turn out his product bearing the marks of the antiquated brush method of finishing.

In describing the Aeron System from an equipment standpoint, it is important to mention that the Aeron—in every type—and all accessory equipment are mechanically correct in basic principles, and are complete in every detail. This must of necessity be the case to insure the procuring of maximum results.

There are two styles of Aerons, one with the attached fluid cup, and the other without the cup, receiving the fluid from a container placed overhead. Of these two styles there are several sizes and different types to meet every requirement. Then there is the air compressor and the air receiver, and also the air transformer set, together with air duster, for regulating the air pressure and purifying the air, and for cleaning the parts to be finished.

There is also the electric air heater which is attached to the Aeron at the last possible point of contact, and which supplies the only practical way of heating the air and keeping it heated till it reaches the work, and of raising the temperature of the material.

To complete the equipment there is the fire-proof indestructible steel fumexer in which the aeroning or spraying is done. The back of the fumexer is funnel-shaped clear to the floor. This scientifically correct style of back together with the large fan opening and arrangement and the short exhaust pipe combine to insure the height of exhausting efficiency. The fumexer is made in a variety of sizes, ranging from 3 to 16 feet in width, with the proper number of fans installed in each size. A turntable, which is also supplied, greatly facilitates the handling of the work. The autocool electric exhaust fan installed in the fumexer has a protected and automatically cooled motor; it can be swung inward for cleaning; each fan is a self-contained unit and can be adapted to any kind of work; it requires no belts, nor mill wright work, and takes up no valuable floor space; it has a 1/12 h. p. motor and can be attached to any electric light socket, using 1/10th to 1/20th the current to do the same work as other style fans. The autocool fan is made in one size only, the number of fan units being increased to two or more for fumexers above five feet in width. The big advantage of the fan arrangement is that a better distribution of exhaust is secured, and the vapor is quickly moved at low pressure.

In using the Aeron System, for which the air pressure required varies from 30 to 80 pounds, it is not necessary to finish separately different parts of any

* By R. D. Waltz, of The DeVilbiss Mfg. Co., Toledo, Ohio.

particular job, allowing time for one part to set up before coating another, in order to obtain a full bodied application. All surfaces that are to be finished can be coated in one operation, and it is also quite possible to put on a heavier coat than with the brush without the danger of sags or runs. This is of particular advantage in the application of coach varnish, where a full body and high gloss must be obtained with as few coats as possible.

One supreme test of the efficiency of the Aeron System is to apply a full coat of material on a vertical surface. The Aeron produces a uniform surface, one that is of even thickness throughout, and because of this uniformity will set evenly and dry more quickly than a brushed surface. Such uniformity of finish eliminates all sags and runs—there being no thin spots over which heavier portions of the coat may run or sag, as is usually the case in brushing.

In brushing carved or ornamental relief work the finisher encounters the difficulty of the low spots collecting more varnish than the high places and the finish is far from perfect. Using the Aeron on identically the same work results in a surface that is covered uniformly and evenly, it all being accomplished with one sweep of the machine, at an even speed.

In order to have the finishing material of the same consistency when applied upon work with the Aeron as it is when brushed it is sometimes better and in other cases necessary to reduce the material slightly with turpentine or other solvents. This is also logical because of the slight evaporation of the solvent resulting from the action of the air on the material.

This matter of reducing materials is naturally governed more or less by the conditions existing in different plants; generally speaking, however, it is safe to say that in using a free flowing coach varnish no reducer is necessary; rubbing varnishes are usually

Showing Steel Fumexer with Two Exhaust Fan Installation. Here the Work is Shown placed on the Turn-table.

reduced slightly as they do not flow as readily as coach varnishes. The aeroned coat of rubbing varnish is more easily and quickly rubbed than when brushed, and the undercoat more easily sanded.

In aeroning or spraying pigment primers or first coaters it is better to have the pigment content less than for brushing—it should be finely ground.

Because of greater uniformity of surface, a flat finish more closely resembles a rubbed finish than when brushed. This to the finisher is an obvious advantage.

The most experienced and dexterous brush hands

find it extremely difficult to brush shellac uniformly and without laps. It is undoubtedly in the application of shellac that the Aeron System shows to greatest advantage. The aeroned coat of shellac is perfectly smooth and uniform and, comparatively speaking, requires no sanding; the gum is usually cut 2½ pounds to the gallon of alcohol.

It is with surprising ease that white undercoats and enamels are handled with the Aeron. In brushing white finishes several coats are necessary to properly cover the job, while in aeroning—owing to uni-

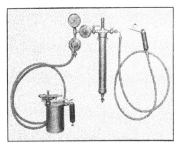

Air Transformer Set, with Attached Cup Aeron and Air Duster Connected.

formity of surface and absence of brush marks—one or more coats can usually be saved.

As further evidence of the superiority of the Aeron System it might be pointed out that one coat of white primer when aeroned will completely cover a piece of work as well or better than two coats brushed. Highly presentable product is turned out in two and three coats satin finish with this System; highest grade rubbed work is done in four coats.

There are factories which are successful in the aeroning of stains and fillers. However this process is not so applicable to these materials as to the others mentioned.

As might be expected the saving of time is more or less regulated by the materials applied, the class of goods finished, and the factory process of handling the goods. It is safe to say, however, that one man operating the Aeron can do a given amount of work in from one-half to one-tenth of the time required to brush the same job.

Economy of time is much greater with the shellacs and enamels—one man being able to do the work of from four to eight men on ordinary case goods or flat surfaces. On carved work there is a still greater saving. On any work with many different angles and surfaces the saving is most notable.

As one might reasonably understand, varnish and enamel coats cannot be handled as rapidly as the undercoats, notwithstanding the differences in time required with the Aeron System and with the brush are nearly as marked as in the case of the undercoats.

Although the Aeron System is of the highest type of mechanical construction, embodying every fundamental that makes for the success of the system, it is extremely simple and easy to operate. The average man will find that a week's handling of the system is sufficient time in which to become a proficient and expert operator.

The complete removal of all vapors and fumes from the finishing room by the exhaust fans installed in the steel booth makes the work clean, pleasant and, what is more, healthful and sanitary.

Under normal conditions there is usually some loss of material when the Aeron System is used. This is scarcely noticeable except on small work where it may amount to fifteen or twenty per cent. In many cases there is no waste whatever, in others an actual saving this where coats are eliminated by the use of the system. The slight loss that is recorded is largely of the solvent used in reducing the material. The loss of material is offset to such large extent by the many savings the Aeron System effects that it is of negligible character.

The installation of the Aeron System in any finishing room where wood products are to be finished makes possible such compelling economies as the 50 to 90 per cent. saving of time and labor; the great conservation of floor space, which during the busy seasons, especially, is at a premium; the remarkable ease and speed in handling the work; the higher grade and quality of work done; and the absolute protection of the workmen, which has a tendency to arouse a greater working spirit.

The all-around and general advantage and convenience of the Aeron System of applying finishing materials surely is obvious even to the casual observer, and without question should receive careful attention from every manufacturer who has a desire to better his own individual factory conditions, as well as to give recognition to something which is perhaps contributing more than all else to the uplift and betterment of the finishing industry.

The Supply of Competent Finishers
By H. K. Stanford

In the early days of furniture making it was necessary for a cabinet maker to understand his trade thoroughly and be able, to place a piece of furniture on the floor complete even to the finshing. In these days of keen competition it is necessary for a firm manufacturing cabinet work to handle their work in an entirely different manner and in order to secure greater efficiency towards reducing costs the factory is divided into departments, each department handling the work in a different stage of its manufacture. Each department is again divided, each man handling a certain portion of the work. This is undoubtedly an efficient way of doing the work, as it only requires a few skilled men to do the most important parts of the work and men with a very limited amount of knowledge and experience to do the balance, but it is a handicap for the young man who is ambitious to qualify as a foreman or superintendent. However, a man who is engaged in a cabinet room or a machine room and has the ambition to push himself to the front has the advantage of night classes in technical schools and correspondence schools by which he can learn mechanical drawing, designing, etc., which will give him a good knowledge of constructional details and enable him to understand the need for the different operations in the machine and cabinet rooms and will thus make him a much better mechanic and open up avenues for progress, perhaps leading to a much better position and higher salary.

With the boy or young man working in the finishing room, however, it is different, as he has very few chances to learn anything outside of the factory about the art of mixing colors, making stains, etc., and the scientific side of the finishing part of the business, and he has very little chance of getting the knowledge inside of the factory as the foreman does all the experimenting, mixing, etc. The boy or young man is generally put to work when he starts in the finishing room, wiping off filler, sandpapering, and various jobs requiring about the same degree of skill.

The skilled tradesmen who are paid the best wages are expected to turn out just as much work as possible for the time they are at work so the beginner will get very little information or instruction from that source, and the foreman usually has his hands full filling the demands of the shipping department, and attending to his various duties so that he has very little time to spare for the beginner.

Foremen finishers are usually the subject of a lot of criticism with regard to carrying their formulas and secrets around in their pockets and being very backward in parting with them, but I must say that during my short experience I have never met any of this kind as I always found them quite willing to help me out when I appealed to them for information.

The practice in the finishing room as in other departments used to be that a certain number of apprentices were kept according to the number of journeymen. The apprentice was then advanced stage by stage through all the different branches of the work until he could take up the duties of a journeyman. This method seems to have been abandoned, and since in the very small shops and as a consequence the supply of competent finishers is diminishing year by year and the time does not seem to be very far distant when first-class finishers will be able to demand almost any wage.

It is the opinion of the writer that manufacturers having finishing departments in their factories should get together and appeal to the heads of the technical education movement to inaugurate a department in the various technical schools throughout the country whereby instruction might be given which would assist those engaged in this important branch of the woodworking business. Manufacturers in other lines of business have appealed to this body and have had their applications attended to and the woodworking industry is sufficiently important in this country to warrant such a department being introduced in these schools.

Greater demands are being made on the finishing department every year for new styles of finish, some of them do not remain in fashion long, but the finishing department has to be able to produce them while they are in demand, and if classes were started in this branch of the work it might be the means of stimulating a great deal of interest in finishing processes and bring out a great many new ideas. It is also the opinion of the writer that classes established to give instruction in this line of work would prove very beneficial to a great many foremen finishers as well as finishing room employees. Pretty near every

THREE-PLY MAPLE VENEER
SHEETS

carried in stock in the following sizes:

36 x 36 x 3/16   42 x 42 x 3/16

48 x 48 x 3/16

WRITE FOR PRICES

Hay & Company, Limited

Woodstock, Ontario

WE HAVE BEEN MAKING THE

Best French Glue
in the World

SINCE 1818; ALMOST A CENTURY

In competition at the numerous Expositions abroad, the quality of our glues has been found so high that we are usually declared "Beyond Competition." Our trade in this country makes the same declaration. Consistency in quality and service has brought this result and we are proud of it.

Perfection in manufacture from the best raw material makes the quality of COIGNET FRENCH GLUES consistently the highest.

Large stocks at Montreal assure the best of service.

Why not take advantage of these most important factors by using our glues?

At your service for prices and samples.

Coignet Chemical Products Co.

17 State Street, N. Y. CITY, N.Y.

Factories at { St. Denis } France { Lyons }

ANGLO
RUBBING and POLISHING

Works free and easy and can be rubbed in two days.

WRITE FOR SAMPLE

The
Ault & Wiborg Co. of Canada, Ltd.
Varnish Works

Montreal   Toronto   Winnipeg

THE GLUE BOOK

WHAT IT CONTAINS

Chapter 1—*Historical Notes*
Chapter 2—*Manufacture of Glue*
Chapter 3—*Testing and Grading*
Chapter 4—*Methods in the Glue Room*
Chapter 5—*Glue Room Equipment*
Chapter 6—*Selection of Glue*

WOODWORKER PUBLISHING CO., LIMITED
345 Adelaide St. West, Toronto

other branch of the woodworking industry is receiving attention in some of these schools, carpentry, cabinet making, machine work, carving, etc., and while the pupils may be given sufficient instruction to finish the piece of work which they make at school, it does not deal with the practical methods used in the large factories.

A great deal of time and money is wasted through experiments being made in factories, owing to the fact that quite a number of men have not the slightest idea where to start from to make their experiments and just simply blunder along until they get what they want or something like it, more by good luck than through any intelligent idea of the materials they are experimenting with.

Now if classes were started to teach practical finishing to men and boys engaged in the trade, a good instructor could be secured and these men and boys given instruction in chemistry pertaining to finishes, and basic principles established from which to work, so they would not be groping in the dark when called upon to produce a certain finish.

The majority of firms making finishing materials maintain laboratories and are constantly making experiments with the result that when their salesmen call on the foreman finisher and proceed to explain to him some wonderful discovery they have made and which is far ahead of anything that has ever been used for the purpose before and finally he leaves a sample for the foreman to try. Now if the foreman had a knowledge of chemistry pertaining to finishing he could tell by a simple analysis whether the sample was any use or not, while without this knowledge all he can do is try it on a piece of board and if it stands

alright it is then generally decided from the standpoint of price, and then after it has been in use for some time it is found to have gone back in some manner and either lacks the correct color or something else which makes it undesirable.

Now these instances which the writer has cited would not occur with some foremen finishers as they are thoroughly grounded on finishing materials and their uses but on the other hand a great many are called upon to do work with which they have had absolutely no previous experience.

The foreman finisher can usually get the proper materials for the work in hand by buying from a reliable firm, but even so a thorough knowledge of the business will make him a more intelligent buyer and may perhaps lead to a better understanding and closer co-operation between him and the house he buys from.

Now, dear reader, these are a few random thoughts of one who has had a little experience in the trade, and who would be very pleased to hear the opinion of some practical men expressed in the columns of this paper.

In recent years the subject of wood-waste has received considerable attention from those engaged in the woodworking industry, and various ways and means of using this waste have been suggested. In a book recently published by Scott, Greenwood and Son, 8 Broadway, Ludgate, E. C., London, England, entitled "The Utilisation of Wood-Waste," the author Mr. Ernst Hubbard, deals very extensively with the subject, explaining various uses that the waste might be put to, also appliances for affecting the different uses.

This consignment of African Mahogany is the selection from a million feet, which accounts for its superiority.

Mahogany and Cabinet Woods—Sawed and Sliced

Quartered INDIANA White Oak, Red Oak,
Figured Red Gum, American Walnut, Etc.

Rotary Cut Stock in Poplar and Gum for Cross
Banding, Back Panels, Drawer Bottoms and Panels

THE EVANSVILLE VENEER CO.
Evansville, Indiana

Handling Butt Joint Veneers

By T. H. Wilson

The writer has had considerable experience in the handling of what are known in the trade as butt joint veneers, and during that experience has picked up a few points which may be of assistance to fellow workers in the trade.   Before going any further a short explanation as to what butt joint veneers are may be in order for the benefit of those not familiar with the term.

When lumber was a great deal more plentiful than it is now a man cutting down a tree never thought of cutting it off close to the ground but cut it off at the height which was the most convenient for himself. In this manner stumps ranging from twenty-five to thirty-six inches high were left standing. This applied to what are today some of the most expensive cabinet woods.   When the demand for these woods increased with the growth of the furniture industry the great forests which it was thought at one time

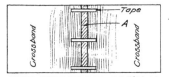

Crossband with Piece in Centre to come under Butt Joint. A is
Piece of Mahogany or Walnut 1½ in. wide Cut on Bias.

would last forever were fast disappearing. As the trees disappeared the price increased and finally some shrewd persons realized that the stumps which were left standing were worth a fortune and they were then cut off close to the ground.   After being cut the stumps are made into veneers and are called butt joint veneers, from the fact that it generally takes two pieces jointed end to end (called a butt joint) to veneer a piece of core stock.   When these stumps are made into veneers they are worth from $100 to $200 apiece.

Butt joint veneers are generally used on pianos and high-class furniture and when the grain or figure is carefully matched, present a very beautiful appearance.

The writer, in this article, will deal with the handling of the veneers between the time they are taken from the storage cellar and laid on the core stock, and will assume that they are going to be used for piano cases. although the same method of handling will be found suitable when the veneers are used for furniture.   In the first place, ascertain the number of feet contained in the butt and estimate how many cases it will make, allowing one hundred feet to a case.   In order to obtain the best results as well as reduce the waste to a minimum it is advisable to cut up the whole

butt at the same time and for the same style of case. For instance, if twelve cases are required and the butt contains enough veneer for eighteen, cut the eighteen, and six may be placed away carefully till some more cases of the same style are required. The next move is to lay out the veneer to the very best advantage before cutting it, using the best figured sheets for the most prominent parts of the case. This will be greatly facilitated if the sheets are all numbered in rotation before commencing to lay out.

No definite rules can be followed for laying out the veneer, as a man has to use his own judgment, but the writer always commences with the widest pieces required.   It is sometimes necessary to sacrifice a little in the figure to get the desired width.   After laying it all out it may be found impossible to get the number of cases figured on owing to holes in the veneer and this is where the numbering comes in handy, as you can put the veneer back in its place and start over again with a couple of cases less.   Sometimes if a very wide piece is required it may be found necessary to make a joint lengthwise as well as the butt joint, but this adds to the labor cost and should be avoided if possible.   In order to make the veneer match perfectly care should be taken that the two pieces to be jointed are the pieces which were next to each other in the log.

Before jointing, the veneer should be perfectly flat, and this will be best accomplished by damping, then drying and straightening it between hot cauls.   This is the point where a lot of men fall down on the job as they soak it with water, put it between hot cauls in a press and screw it down as tight as their strength will permit and then wonder why their veneer comes out all split in pieces.

The writer damps his veneer with a sponge and puts it between hot cauls, about two or three between each caul, but instead of screwing it down tight just lays a couple of extra cauls on top and leaves it over night.   The next day it is put between hot cauls again and the weight increased slightly.   This is repeated for three or four days and while it may seem a slow process it pays as there is very rarely any trouble with veneer splitting and when there is it is nothing serious and is easily repaired.

The reason for damping the veneer is to make it pliable, but with the heat of the cauls it stands to reason that it is going to shrink, and if the press is screwed down tight it can't shrink, so there is nothing else for it to do but split.   It will not be necessary to go into details about the actual jointing as this is done in the usual manner on a shooting board with a sharp plane.

A point in the preparation of the crossband merits the consideration of any person handling this class of work.   The crossband is always cut through the centre on the bandsaw and the piece marked A in the accompanying diagram is placed in the centre, where it will come under the butt joint and is fastened to the crossband with a few short pieces of tape.

Walnut or mahogany cuttings, bandsawed on the bias can be used for A, as it is not necessary to make a perfect joint with the crossband. It will be seen that this method of preparing the crossband cannot

be used when two veneers are laid at once but the writer would not attempt to lay two veneers at once with this kind of veneer.

This method of procedure may seem slow and commonplace to some men in the trade but the writer can say in its favor that in the factory where it is employed open joints in veneer are almost entirely unknown.

Some Effects of Rushing the Work
By Edgar Arnold

Recently the writer was consulted regarding the peculiar condition of some table tops that had been veneered about three months and had been finished about two months. They had been cross-banded and faced with very thin veneer.

Taking one top as a sample of the others, it was as follows: About 2 ins. from the front edge was a slight depression, about ½ in. wide. Then about 3 ins. behind that was another depression about 2½ ins. wide. These depressions extended from one end of the top to the other. All the way across the top at irregular intervals were narrow and wide depressions.

The finisher was up against it. These defects had not been discovered until the goods had been varnished and the rubbers were working on them. Some of the first ones had been shipped out, and those yet in stock seemed to be getting worse.

Everybody disclaimed all responsibility for the condition of the tops—all except the finisher, and he could not say anything. The machine foreman said the cores were properly dressed, and the veneer foreman said the stock was alright when it left him, and the cabinet foreman said the stock was O.K. when he sent it to the finishing rooms; if not, why did the finisher foreman not object? The finisher foreman could put up no defense and make no explanation further than that he was satisfied the cause of the trouble was not connected with his department.

"Perhaps the varnish was not put on evenly," suggested one. "Maybe the rubbers rubbed deeper some places than others," said another.

To these the finisher made no reply, but only gave them a saddened look, as though he were sorry for them. But the saddened look did not alter the fact that here were some expensive tables gradually developing defects, and if those that had been shipped out were doing the same thing there would be some disappointed customers. These were the things about which the manager was concerned. He was also afraid of a repetition of the trouble, and, if such should be the case, where would it end? These were the thoughts that disturbed the manager's mind, and, while the various foremen were anxious to shelve the blame on some one else, he was anxious to get at the cause, no matter where it was.

The different foremen having viewed the defects and cleared themselves of all blame in the matter, had left the finishing room and returned to their respective departments, and the manager and foreman finisher were left alone.

"It looks as though it is up to you to get at the bottom of this trouble," said the manager. "The goods seem to have been O.K. when they reached this department, and as the defects developed here, I see no other course open but to start here to find the cause."

The following day the foreman finisher got busy. He did not agree with the manager that the only place

to start an investigation was in the vicinity where the trouble was discovered. He knew from his experience as a finisher that many troubles which do not manifest themselves until some considerable time after the goods are finished may have their inception in the early stages of the work. He was also satisfied that the trouble in hand did not originate in the finishing room, but was in the wood or the manner of preparing it for the finish. For the next day or two it was noticed that he spent more than his usual amount of time out of his department, and he might have been seen talking to the men in the veneer and glueing departments.

At the end of two days he went to the cost accountant and asked to see the time cards used in the veneer and machine departments on the job that had given the trouble, and this is what he discovered: The table tops had been glued up in the morning and planed off in the afternoon, then veneered the following day, both cross-banding and face being laid at the one time. Four days later they were put through the scraper, and a week after being glued up they were in the finishing room. Here was the solution of the problem. Here was the cause of the trouble. Away back in the early history of these tables, in fact at the very beginning, the seeds of death had been sown, and had but been waiting for favorable conditions to show their power.

Someone may ask what was wrong with the way the work was done. In the first place, the stock was planed too soon after being glued up. The glue swells the wood along the edge of the joint, and if this is not allowed to dry out and shrink back to normal before planing, it will shrink below the level after planing, hence these narrow depressions extending along the top from end to end.

The wider depression was caused by the use of two kinds of wood in the core, and when the glue was put on, one kind had swelled up above the other; and the stock being put through the scraper before it was dry, had been scraped off level, cutting down these raised parts, which shrank below the level when dry.

As a result of the investigation no more stock will be rushed through in like manner, no matter how big the hurry. The manager says he doesn't want to take another chance like that.—Veneers.

The sawed veneer and drying departments of the Evansville Veneer Company, of Evansville, Ind., were destroyed by fire early in February. The company advise us that they were fortunate in having ample stocks of sawed and sliced veneer in their warehouse, which was not damaged in the slightest, and from which they will be able to take care of orders for this material. They have a considerable stock of rotary cut veneers dry that they will cut to size to fill orders. It is their intention to re-build in such a way that a fire will be practically impossible.

OAK   MAHOGANY   CIRCASSIAN   MAPLE
VENEERS
BIRCH   POPLAR   CHERRY   WALNUT   GUM

THE OHIO VENEER COMPANY

Importers and Manufacturers

Foreign and Domestic Veneers and Hardwood Lumber

We always carry a large and assorted stock of Mahogany, Circassian
Walnut, Sawed and Sliced Quartered Oak.

Send us your enquiries and orders.   We guarantee good service.

2624 to 2644 Colerain Avenue,   -   -   CINCINNATI, OHIO.

We operate exclusively in

AMERICAN WALNUT

Our entire supply of raw material is obtained
from Ohio and Indiana, which produces the best
walnut in the United States.

We have the most modern and best equipped
plant in the United States for producing

WALNUT LUMBER and VENEERS

Special attention given to the requirements of
Furniture Manufacturers

George W. Hartzell,   PIQUA, Ohio

(HB) **Hoffman Bros. Co.** (HB)
Estab. 1867, Incor. 1904

800 West Main Street,
FORT WAYNE, INDIANA

Manufacturers of

VENEERS and LUMBER

In the Domestic Hardwoods
ANY THICKNESS,
1/24 and 1/30 Slice Cut
(Dried flat with Smith Roller Dryer)

1/20 and thicker Sawed Veneers,
Band Sawn Lumber.

—SPECIALTY—

(HB) **Indiana Quartered Oak** (HB)

VENEERS & PANELS

CARRIED IN STOCK

VENEERS

Red and White Oak, Birch, Poplar, Gum,
Yellow Pine, Cypress, Etc.

PANELS

Three and Five Ply—All Woods.

GEO. L. WAETJEN & CO.
Milwaukee,   Wisconsin

Veneers and Panels

LARGE STOCKS FOR
QUICK   SHIPMENTS

Send for Lists

Walter Clark Veneer Co.
GRAND RAPIDS, MICH.

Rising Cost of Raw Materials

At a meeting held in Grand Rapids on January 25th by the National Furniture Manufacturers' Association, to discuss the rising cost of raw materials, it was estimated that the price of mahogany and walnut veneer had increased 25 to 30 per cent. Veneer manufacturers are reporting an increased demand and a shortage of logs, and although these figures may not hold in all cases it is certain that the price of veneer is firm, with a strong tendency to advance. There is a pronounced scarcity of logs in gum, walnut, and mahogany. The floods in the south have seriously hindered the getting out of gum logs, and stocks are low. The war contractors are buying up all available walnut for manufacturing into rifle stocks. A large quantity of the mahogany used in Canada and the United States is imported from Africa and the West Indies and manufactured into veneer. The scarcity of ships and extremely high ocean freight rates have made importing difficult.

At a meeting of the Indiana oak manufacturers, held in Indianapolis in January of 1915, an inventory of the stocks of seventeen manufacturers of oak veneers showed that there were some 19,970,000 ft. of 1/20 and 1/16 in. oak veneers in the various warehouses, in excess of orders, and this amount was probably increased considerably before the veneers began to move freely. At a meeting held in January of this year this excess stock had been reduced to 4,250,000 ft.

This shows that while the veneer factories have been producing to capacity for the past few months, they have lacked some 15,720,000 ft. of keeping pace with the demand, and quite a number have had to buy oak veneers to fill their orders.

Now the effect of the curbing of logging operations is being felt, great difficulty is being experienced in obtaining oak logs and flitches. Weather conditions in the south have also made it impossible to transport logs of any kind to the mills, which increases the shortage.

The inevitable result will be a sharp raise in the prices of oak veneers, and those consumers who were wise enough to lay in a good supply during the depression, when prices were low, will be in an enviable position and in line to realize a nice profit on their forethought.

The Power Driven Screw Driver in the Woodworking Factory

One of the greatest steps leading towards efficiency and reduced costs in the woodworking factory has been made possible through the application of power-driven machines to the old tedious and slow operation of driving screws.

The belt-driven power machine illustrated on this page is manufactured by the Stow Manufacturing Company of Binghamton, N. Y., and is not only a screw-driver, but is applicable to a variety of operations, such as drilling, grinding, buffing and polishing.

It will be seen from the illustration, that this machine is attached to the ceiling, so that it is out of the way when not in use, and when required for use all that is necessary is to reach up and pull it down—just as simple as lifting a tool off the bench.

Referring again to the illustration, it will be observed that a very cleverly arranged device up close to

the ceiling, together with the flexible shaft, enables a man to perform any of the above-named operations almost any place within a radius of twenty feet and at practically any angle or in any corner that a hand drill or hand screw-driver may be used.

Almost every woodworking factory has a certain amount of boring and sometimes some grinding which cannot be done at the stationary machines used for this work, and for such work this machine is particularly well adapted.

If used for this purpose alone the machine would soon pay for itself by labor saved. But when these operations can be performed as well as using the machine it can easily be appreciated the enormous saving which would be effected. On the other hand if a screw-driver were required to be

Belt driven screw driving machine attached to ceiling.

in constant use, two machines could be used to good advantage.

Another admirable feature in connection with this machine is an appliance by which the motion of the tool which is being used can be stopped instantly without the necessity of throwing up the head of the frame. Another feature is that the machine stops automatically, so that there is no wear when not in use.

The machine is furnished in different sizes to suit the requirements of the customer.

The Stow Manufacturing Company are makers of a full line of this class of machinery, including a motor-driven machine covering practically the same range of operations as the belt-driven machine. The motor-driven machine, however, lends itself admirably as a portable tool, and hooks may be arranged at various desirable locations and it is only necessary to attach it to the hook by means of a ring placed on the machine for that purpose.

This firm is also the inventors and manufacturers of the Stow Flexible Shaft.

Alder logs are being taken out in Langley, B. C., in connection with some interesting experiments with alder wood carried out two years ago, when the suitability of that wood for furniture-making was tested. It was found that after a lengthy period of air-drying alder could be beautifully polished and stained without any sign of warping in the boards. It is also claimed that the wood makes a good veneer.

If you are in the market for either lumber or veneers in walnut, "the aristocrat of American hardwoods," it will pay you to get in touch with us, for we are

SPECIALISTS IN WALNUT

We have been in this business for thirty years; and in addition to knowing exactly how to make walnut lumber and thin stock, and in addition to having a stock which for variety and size is unequaled anywhere else in the United States, we have on hand a big bunch of

Unique Figured Veneers

from the Southwest

These veneers are highly figured, and of unusual color, and for special work are absolutely unrivaled. Ask us for samples.

Penrod Walnut & Veneer Co.

KANSAS CITY, Mo.

We specialize on

Poplar

cross-banding and backing

Veneer

Highest Grade — Right Prices

The largest Poplar Veneer Mill in the world.

The Central Veneer Co.

Huntington, West Virginia

The Glue Book

WHAT IT CONTAINS

Chapter 1—Historical Notes

Chapter 2—Manufacture of Glue

Chapter 3—Testing and Grading

Chapter 4—Methods in the Glue Room

Chapter 5—Glue Room Equipment

Chapter 6—Selection of Glue

PRICE 50 CENTS

Woodworker Publishing Co. Limited

345 Adelaide St. West, Toronto

High Pressure Steaming

It has taken a long time to get modern efficiency ideas introduced into the work of steaming and bending wood, but we are now getting improved appliances for steaming as well as for bending, which are well worth investigating by those who have occasion to bend wood for their work.

In days gone by the preparatory process for bending wood consisted of boiling or steaming it in some wooden vat or open tank. There is some of this kind of work to-day, in fact too much of it. There are other appliances to do this work. There are modern steaming retorts made of metal, with opening and closing doors that can be put in place and tightened up easily so that not merely live steam may be used, but actual steam pressure may be applied in the preparation of wood for bending.

With the high pressure process the manufacturer is able to steam his stock without using any more steam than is required by the old low pressure processes. The high pressure method of steaming has been tested out very thoroughly for the past several years and has proved superior to the low pressure method for several reasons.

In the first place, it softens the stock uniformly throughout and does not over steam it on the outside.

When the stock is bent it does not break on the outside or buckle on the inside, as is the case when steamed by the low pressure. Second, the high pressure drives the sap and acids out of the stock, leaving the pores open so that the air can get into them quickly. By the low pressure process the sap and acids are not taken out in steaming and it is necessary to water soak the stock and put more water in it than by the high pressure steaming.

The reason that high pressure steaming was not a success until recent years was that no manufacturer was able to build a steam box that would withstand a high enough pressure to give sufficient heat to soften the stock quickly enough to be practical for use. However, these boxes have been perfected by the manufacturer and are being used by many of the large bending concerns in preference to the old low pressure steaming.

It has been found, however, that to steam the stock properly under high pressure, it is necessary to use a pressure of 35 to 40 or over, as the pressure below this amount does not contain enough heat to put the stock in proper bending condition. It has also been found that it is very necessary to use cast iron for the holding of this pressure, from the fact that the cast iron is not affected by the acids in the timber and that there is no possible chance of the cast iron giving away after a few years of service; also the manufacturers of these cast iron retorts make the walls sufficiently heavy to retain enough heat to keep the stock in proper bending condition for a considerable length of time without closing the doors, which makes them very convenient for taking the stock out where same is being taken out a few pieces at a time.

The Snyder Brothers Upholstering Company, of Waterloo, have purchased the business of John B. Snider, for many years a manufacturer of office furniture, etc., in that town.

To Test Wooden Boxes

The Northwestern Association of Box Manufacturers, with headquarters at Portland, Ore., has forwarded samples of standard cannery cases for testing at the Forest Products Laboratory, Madison, Wis. These cases are made according to association specifications and will be loaded with the goods designed to be shipped therein. The experiments will be made with a view of endeavoring to fix a standard of manufacture for wooden boxes, and will be done in co-operation with the committee D-10 of the Society for Testing Materials, a national organization, composed of manufacturers of every class of containers.

Turnings for William and Mary Designs

The Gage Manufacturing Company, of Gardner, Massachusetts, makers of machine knives and cutter heads, are making cutter heads suitable for cutting large holes in wood. Heads of this description are used for making shell boxes and will cut through pine or hardwood lumber up to 1¼ inches thick. They have been used successfully in cutting out seats for children's chairs. The company some time ago built a machine for cutting round holes in chair seats for what are known as cobbler seat chairs. This machine

Back Knife for William and Mary Turnings.

could be used to advantage in making shell box partitions. On this page is an illustration of a back knife made by the same company. These knives are being extensively used on turnings for the new William and Mary designs in furniture.

Forest Service Bureau

Mr. R. H. Campbell, Director of Forestry, Ottawa, speaking before the B. C. Forest Club, Victoria, recently, said:—

"We have now established laboratories in connection with McGill University on somewhat similar lines to those followed by the United States Forest Service. The object will be to study every phase of all possible uses of Canadian woods and the properties which make them suitable for different manufactures. The testing of the mechanical properties of timber was most pressing, and has been taken up first. McGill University has a testing plant which is as good as any on the continent, which is placed at the disposal of the Department, and we are now testing Douglas fir both from the coast and the interior of British Columbia and from Alberta, so as to try it from all conditions. The work has proceeded well, and results, which are now about ready for publication, show this timber in a very favorable light. We are now engaged on the testing of mining timber from Nova Scotia, and we hope to reach all Canadian species in time."

Canada's Premier Hardware House
(Established 1885)

A Wide Range at Right Prices

TOOLS

Let Us Send You Our Catalog

V O K E S

Shell Box Hardware

Everything in Hardware for the Shell Box Manufacturer.

Bands, Handles (Wire and Rope,) Screws, Nails, Staples, Houseline, Corrugated Joint Fasteners

A large assortment in stock—Approved and inspected goods.

Write for Samples and Prices

Sole Selling Agents for the British American Hardware Manufacturing Co.

**FACTORY and MILL SUPPLIES, CABINET HARDWARE
BARTON'S GARNET PAPER**
Canadian Agents   **UNIVERSAL CASTERS**

Try Us for Special Made-to-order Hardware

The **Vokes Hardware Co.**
Limited
40 Queen St. East, Toronto, Can.

The Lumber Market

Domestic Woods

The greatest demand for domestic lumber comes from the manufacturers of shell boxes. Large quantities of pine and birch are being used for this purpose. The Government specifications call for boards nine and ten inches wide, but owing to the difficulty of securing lumber of this width the specifications have been changed and parts of the boxes may now be glued up providing a suitable joint is used. This will give manufacturers an opportunity of getting rid of the narrow stock. Lumbermen are strongly advocating the use of spruce for shell boxes. Large quantities of spruce have been exported to Great Britain for the erection of soldiers' huts, etc.

The Dominion Government's trade commissioner has been sending in satisfactory reports with regard to the market for Canadian lumber and wood products abroad, but it will be some time before this business can be developed sufficiently to affect prices to any extent. The recent action of Baron Shaughnessy of the C. P. R. in giving instructions that only Canadian woods should be used by that company will no doubt be followed by many others, and will stimulate the use of domestic woods.

Imported Woods

Owing to the limited demand for hardwood lumber for factory consumption since the beginning of the war, lumbermen have let their stocks dwindle down about as low as they dare. A scarcity is now said to exist in the staple woods such as oak, ash, gum, poplar, maple and birch. Reports show that the unsold stock at northern mills is very light, especially in upper grades. About one-half the supply of maple has been disposed of, and the same is true of birch, basswood, and brown ash.

Recent reports indicate that practically one-third of the total stock of Wisconsin and Northern Michigan hardwoods was unsold around the middle of January.

Prices on birch are quoted at about $42 for firsts and seconds inch; on elm $38 for firsts and seconds inch, and on basswood $37 for firsts and seconds inch. Hard maple is selling around $21 to $38, according to grade.

The following, taken from the recent address of President Jurden before the Commercial Rotary Gum Association, deals with the gum situation:—

"One very important feature to which I wish to call your attention is the gum log situation in what I would term the Memphis territory or the Mississippi valley. Sixty or ninety days ago general business conditions did not seem to warrant the production of large supplies of gum logs, and the manufacturers who logged their own timber and the loggers who produced logs for the market did not perhaps produce more than 40 to 50 per cent. of a normal fall crop of logs. The result today is, with this revival of business, the wildest scramble for logs ever known in this section. Weather conditions are here which retard operation, and only a small amount of logs are coming out or can be produced at this time. The result is that gum logs in this section have advanced at loading points $2 to $3 per 1,000 ft. in the last two to four weeks, and unquestionably will reach a still higher level in the next thirty days."

Southern woods other than gum are firm as to price; and plain oak is quoted at from $50 to $55. The southern hardwood situation is complicated by the floods, which are apt to leave much of the lumber in a condition which will depreciate its value; besides logging will be practically impossible until the floods recede, which may not be for a couple of months. Operations at the mills in this district have been at a standstill. It is estimated that a reduction of fully 50 per cent. will be made in the output of the hardwood mills in the Mississippi Valley.

Reports show that orders placed at the recent furniture expositions in Grand Rapids showed an increase of from 25 to 50 per cent. over those of January, 1915. The woods most in evidence are mahogany, oak, and American walnut, with gum, maple, and birch showing some activity.

Keeping a Level Head

The ability to maintain mental equilibrium is an excellent thing to acquire. By this means the mind is kept clear to consider means to solve the problems that may suddenly be presented, enabling one to act to the best advantage in emergencies. The bookkeeper has it who hears the fire alarm and quickly puts his books in the safe and closes it before seeking to save his own life by flight. The man who sees financial disaster overtaking him and faces the trial, keeping his mind clear to help others go through the wreck and save the remains, realizing upon the assets, is another who is deserving of credit for his calmness in an emergeney. Few there are who can stand by and see the labor of a lifetime go for naught without being seriously distressed and losing their poise. Many persons have been able to conquer adversity and get a fresh start until eventually led to restoring their fortunes by preserving their calmness in adverse circumstances. It is a valuable trait and well worth cultivating. The man who loses his head in emergencies or when business reverses overtake him labors under disadvantages that materially affect him and militate against his interests. It is said that one of the advantages of playing poker is that players learn to maintain an impassive countenance regardless of the fortunes of the game. Win or lose, they are calm and their adversary cannot tell from their demeanor how the game is going. While this is a valuable faculty to cultivate it is not altogether essential to gamble in order to do so.

Wood Laboratory Work

The Forest Products laboratory, operated at Madison, Wis., by the United States Government, has enabled the manufacturers of the nation to save thousands of dollars during the past year. Over 112,000 tests were made upon 95 different kinds of American timber. They have been the basis of the revision of many building codes, as they provide dependable strength formulas for each kind of wood. The laboratory has made improved designs for many articles that will mean a tremendous annual saving in material. Chemical discoveries at the laboratory have improved the processes of treating hardwood, and increased the production of wood alcohol. The latest announcement is that a yellow dye may be extracted from wood. This will help to tide over the shortage caused by the suspension of the importation from Germany.

RADCLIFF
Special Shell Box Saws and they are very useful in your regular business too

In the centre is our special saw for cutting holes in the top and bottom trays of the boxes. It is made of the best Swedish steel with special coil spring in centre for ejecting the cores.

Recent improvements on this saw have overcome its former disadvantages and it does the work thoroughly and with absolutely no dissatisfaction.

On the left is the Dado head for cutting grooves in the sides of shell boxes and slots in the end pieces of shrapnel boxes. These dados cut absolutely smooth and clean.

On the right is our combination tooth planer saw, which cuts with or across the grain so smoothly that it eliminates any extra finishing.

It will pay you to get our prices on these saws.

WE CAN SUPPLY YOU WITH
Aloxite Saw Gumming Wheels, Hoe Bits and Shanks, Inserted Tooth Saws, Shingle Saws, Machine Knives, etc.

THE RADCLIFF SAW MFG. CO., LIMITED
TORONTO   550 Dundas Street   ONTARIO

Practical Pointers on Boiler Room Efficiency

IN THE NEW "Morehead" book you will find real ready-to-use information on the handling of your steam equipment which may save you thousands of dollars a year.

Besides a general discussion on the drainage of kilns and steam heated machinery generally, there are interesting

Pictures and Data

which will show you how representative plants everywhere are turning their condensation into golden profit.

Write for the book today

Morehead
> Back to Boiler >
SYSTEM

Feed Your Boilers Pure Water HOT !

Canadian Morehead Manufacturing Co.
Woodstock Dep. "H" Ont.

NATIONAL TRUCKS

Easy Running
Strong   **Durable**
Always in stock

National Cross Piling Truck—Fig. 5

All National Trucks are Equipped with Malleable Iron Wheels

IN all National Trucks the sides are made of channel steel; the axles and rollers of cold rolled shafting; and the spreaders of malleable iron. The axles have milled ends, which does away with riveting. All our trucks are held together with bolts and can be dismantled without injury. Sides are cut to exact length and punched on a template, which insures uniformity.

Write today for catalog and prices.

THE NATIONAL DRY KILN CO.

1118 East Maryland St.   Indianapolis, Ind.

Manufacturers of brick, lumber and cooperage stock
dry kilns, steel roller bearing trucks, wheels, etc.

GRAND RAPIDS

VAPOR DRY KILN

GRAND RAPIDS
MICHIGAN

129

Grand Rapids Vapor Kilns sold in the last

122

days of the year 1915.

54

of these were Repeat Orders.

Repeat orders represent satisfaction. We can guarantee experienced engineering ability and efficient service.

Over 1300 Grand Rapids Vapor Kilns in use.

Write us regarding better drying

News of the Trade

The Northern Veneer Company, of Winona, Ont., has made an assignment.

J. Kreiner, Sr., of J. Kreiner & Company, furniture manufacturers, Berlin, Ont., is dead.

W. Sanders, Rosebank, has succeeded to the woodworking business of Joseph Maughan.

The Canada Casket Company, Montreal Que., have dissolved. John St. Jean continuing.

The death occurred recently of Mr. George Rowley, a well-known boat builder of St. John, N. B.

The death occurred recently of Mr. Robert Parker, who was for many years a carriage builder at Guelph.

Cushing Bros., Limited, of Saskatoon, Canada, are building a new box plant at that place. It will be equipped on the individual motor system.

Mr. George D. Stothart, who for many years operated a sash and door factory at Newcastle, N. B., passed away recently after an illness of about three weeks.

The box mill owned by Messrs. Haley & Son, St. Stephen, N. B., and located at Calais, Maine, was recently destroyed by fire. The loss is estimated at $5,000, and is covered by insurance.

At the annual meeting held in Montreal, the financial statement of Carriage Factories Limited, showed earnings at the rate of 22.1 per cent. on the preferred stock and 15.3 per cent. on the common stock.

The box factory belonging to the Roebuck Cheese and Butter Factory, Roebuck, Ont., was completely destroyed by fire last month. The loss was partly covered by insurance. Mr. Lindsay is the proprietor.

A. L. Hampton, a well-known Vancouver real estate dealer, is reported to have organized a company for the manufacture of ammunition boxes and to have arranged for the erection of two factories at Hall's Lake, near Seattle, Washington, the output of which will be 30,000 boxes per month.

The Beamsville Basket and Veneer Company, Limited, Beamsville, Ont., have been granted a charter. The company is capitalized at $40,000, and will manufacture fruit baskets, veneer, woodenware and lumber. Mr. Aquilla Walsh Reid and Mr. Samuel Gibbs Near, basket manufacturers, are provisional directors.

The Valley City Seating Company, Limited, Dundas, Ont., are making an attractive line of Mission store equipment. All units are shipped knocked down, thus effecting a considerable saving in freight charges. A recent contract completed by the Valley City Company was the installation of a quantity of carved work in St. Paul's Anglican Church, Toronto, which is one of the city's finest edifices.

Mr. McMillan, a contractor of North Sydney, Cape Breton is erecting a new woodworking factory which is expected to be ready for occupation in the early spring. The building, which will occupy a floor space of 80 x 40 feet, will be two storeys high, the lower part being built of concrete, with large concrete and fireproof basement. Nothing but the most modern and improved machinery will be installed. All waste and refuse lumber, shavings, etc., will be carried by a large fan into the fire box. Adjoining the factory are two cottages nearly completed, which were built by Mr. McMillan and will likely be occupied by employees.

Box Machinery Bargain

One Dovetail Machine. One Setting-up Machine.
One Four-Saw Trimmer—self feeding; made by Dovetail Machine Co., St. Paul, Minn.
One No. 22 Box Trimmer, made by Fischer Machine Works, Chicago, Illinois.
All cost $3,300. Will guarantee machines and sell at a bargain.

Inquire,　The GLOBE SOAP COMPANY,
2-3-4　　　　　　　　　　　　Cincinnati, Ohio.

Second - Hand Woodworking Machinery

Baluster Machine, Mattison No. 8.
Band Saws, 36 in. and 38 in.
Boring Machine, 3 Spindle, horizontal, H. B. Smith.
Boring Machine, 4 Spindle, horizontal, Garden City.
Dovetailing Machine, 10 Spindle, Advance.
Dowell Machine, double, Crescent, power feed.
Dowell Machine, six heads, Fay & Egan, power feed.
Glue Heater, instantaneous, Zimmerman.
Glue Healer, 45 gal., Power Agitator, Howard.
Glue Spreader, 2 Francis 31 in. x 37 in., double rolls.
Jointers, 16 in. Oliver, 16 in. Smith, 16 in. Colladay.
Planer, Single, 24 in. x 6 in., Daisy.
Planer, Single, 24 in. x 8 in., Valley City.
Planer, Single, 26 in. x 8 in., Rowley & Hermance.
Planer, Single, 32 in. x 7 in., L. Power Co.
Planer, double, 24 in. x 8 in., No. 4 Williamsport.
Planer, double, 30 in. x 4 in. Whitney.
Planer, double, 26 in. x 14 in., endless bed, L. Power Co.
Planer and Matcher, 14 in. x 4 in., Hoyt Bros.
Sander, triple drum 42 in., Invincible.
Sander, triple drum 42 in., Young Bros.
Sander, single drum 30 in. Surface, American.
Sander, Belt, No. 173 with No. 174 attachment. Wysong & Niles.
Shapers, double spindle, 1 Clement, 2 American, 1 Fay & Co.
Swing Saw, parallel, No. 5 Pryible.
Tenoner, No. 2½ American, single heads, copes and cut-off.
Tenoner, double end, single heads, 2 copes.
Sticker, single head, heavy type, American.
Universal Woodworker, No. 90, Famous.

FRANK TOOMEY, Inc.
131 N. Third Street,
PHILADELPHIA, PA.

7% Investment 7% Profit Sharing

Sums of $100, $500, $1000 and up. Safe as any mortgage.
Business at back of this Security established 28 years.

Send for special folder to

National Securities Corporation,
Limited
Confederation Life Bldg., Toronto

PATENTS
STANLEY LIGHTFOOT
Registered Patent Attorney
Lumsden Building - - TORONTO, CAN.
(Cor. Adelaide & Yonge Sts.)
The business of Manufacturers and Inventors who wish their Patents systematically taken care of is particularly solicited. Ask for new free booklet of complete information.

For Sale

Two 54 in. x 12 ft. Stationary Boilers, in use only 8 months, complete with full arch fronts, smoke box and stack, and all regular fixtures and fittings, working pressure 100 lbs.

WILLIAMS & WILSON, LTD.,
2　　　　　　　　　　　Montreal, Que.

Machinery For Sale

One McGregor Gourlay second hand Resaw, 54-in. wheels, carries 5-in. saws, machine in first-class condition, only used on light work. Located in Central British Columbia.

Two second hand McGregor Gourlay Q. Y. Planers and Matchers, one 9-in., one 15-in. wide, in good condition, ran every day till October, when fast feed Matcher was installed. Located in Saskatchewan. Box 20, Canadian Woodworker, Toronto, Ont.　　　　　　　　　　　　　2

Second-Hand Woodworking Machines

Band Saws—26-in., 36-in., 40-in.
Jointers—12-in., 16-in.
Shapers—Single and double spindle.
Re-Saw—42-in. American.
Lathes—20-in. with rests and chisels.
Surfacers—16 x 6 in. and 18 x 8 in.
Mortisers—No. 5 and No. 7 New Britain.
Rip and Cut-off Saws; Molders; Sash Relishers, etc.
Over 100 machines, New and Used, in stock.

Baird Machinery Company,
123 Water Street,
PITTSBURGH, PA.

POSITIONS VACANT

WANTED

Veneer Man for Furniture Factory; must be experienced in cutting, jointing and laying quartered oak and mahogany veneer and to take full charge of veneer room with six men. Apply, stating experience and wages wanted, to Box 683, Globe Office, Toronto.　　　　　　　　　　　　2

It doesn't pay to be stingy with glue, nor does it pay in any way to use too much glue. The best way is to get a man who knows just how much to use for every purpose and then keep him eternally satisfied. Money doesn't keep men satisfied, by the way. Proper working conditions and pleasant people to work with and for, do more than money ever can.

There seems to be a ready opportunity for the man who can invent a machine that will sand veneers before they are glued on the centre stock. Much loss results from sanding veneered stock. The veneer is easily sanded through if the centre stock is warped or planed unevenly with varying thicknesses on different planers. It is practically impossible to patch stock so damaged and when it cannot be cut down to a smaller size it must be thrown away. In either case the result is a loss.

PETRIE'S
MONTHLY LIST
of
NEW and USED
WOOD TOOLS
in stock for immediate delivery

Wood Planers
30″ American double surfacer.
30″ Whitney pattern single surfacers.
26″ Revolving bed double surfacers.
26″ Goldie & McCulloch single surfacer.
24″ MacGregor Gourlay planer & matcher.
24″ Major Harper planer and matcher.
24″ Single surfacers, various makes.
20″ Dundas pony planer.
18″ Little Giant planer and matcher.
16″ Galt jointer.
12″ Crescent jointer.

Band Saws
42″ Fay & Egan power feed rip.
38″ Atlantic tilting frame.
36″ Crescent pedestal.
30″ Ideal pedestal.
28″ Rice 3-wheel pedestal.
28″ Jackson Cochrane bracket.

Saw Tables
Preston variable power feed.
Cowan heavy power feed.
No. 5 Crescent sliding top.
No. 3 Crescent universal.
No. 2 Crescent combination.
No. 17 Clark-Demill dimension.
12″ Defiance automatic equalizer.
MacGregor Gourlay railway cut-off.
7½′ Martin iron frame swing.
6½′ Crescent iron frame swing.

Mortisers
No. 5 New Britain, chain.
Cowan hollow chisel.
Fay upright, graduated stroke.
Galt upright.
Smart foot power.

Moulders
10″ Clark-Demill four side.
10″ Houston four side.
6″ Dundas sash sticker.

Sanders
48″ Berlin 3-drum.
36″ Egan double drum.
12″ C.M.C. disk and drum.
18″ Crescent disk.
Ober double belt automatic.

Miscellaneous
Fay & Egan 12-spindle dovetailer.
No. 10 Ober automatic handle lathe.
16″ and 18″ Ideal turning lathes.
No. 235 Wysong & Miles post boring machine.
MacGregor Gourlay 2-spindle shaper.
Elliott single spindle shaper.
No. 51 Crescent universal woodworker.
40″ MacGregor Gourlay band resaw.
Rogers vertical resaw.
Cowan sash clamp.
Egan sash and door tenoner.
6-nail box nailer, Meyers patent.
20″ American wood scraper.
4-head rounding machines.
Cowan spindle carver.

Prices, Descriptions and Full Particulars on Request

H. W. PETRIE, LTD.
Front St. W., Toronto, Ont.

Twist Machine

New Cutter Head for Twist Machine

This machine will produce twist and spiral Columns, Balusters, Table Legs and Pedestals, Chair Posts and Spindles, etc., right and left turn.

The dividing plates will space 2, 3, 4, 6 and 8-in. Special plates furnished on application.

The feed gear set gives one turn in 1½, 3, 4½, 6, 7½ and 9-in.

The centres will swing any diameter up to 12-in., 3 ft. between centers.

For prices and further description write

Eight Knives, 30 Degrees Angle.

The Shawver Company
Springfield, Ohio

Woodworkers
and
Lumbermen

The Sheldon Exhaust Fans have characteristics that adapt themselves to the service of the Planing Mill or Woodworking Plant. They are specially designed for this kind of work, having a saving in power and speed of 25% to 40%.

Write for Industrial Booklet.

SHELDONS LIMITED - Galt, Ont.
Toronto Office, 609 Kent Building
— Agents —

Messrs. Ross & Greig, 412 St. James St., Montreal, Que.
Messrs. Walker's Limited, 259-261 Stanley St., Winnipeg, Man.
Messrs. Gorman, Clancey & Grindley Ltd., Calgary and Edmonton, Alta.
Messrs. Robt. Hamilton & Co., Ltd., Bk. of Ottawa Bldg, Vancouver, B.C.

"ACME" CASTERS

"Made since 1835—the oldest in the business"

The choice of trimmings and casters help to identify your furniture. Use "ACME" Ball-Bearing or Non Ball-Bearing Casters and you will be known as a maker of the best furniture. "ACME" Casters usually outwear the furniture they carry.

"ACME" Casters have been giving satisfaction for eighty years.

We solicit your inquiries and will be pleased to have our Canadian representative call and discuss any Special Casters or PERIOD TRIMMINGS you may require.

Foster, Merriam & Co.
Meriden, Conn.
Canadian Representative:—F. A. Schmidt, Berlin, Ont.

"ACME"
of
Reliability for all Furniture.

A BOY

can run the

NASH HANDLE SANDER

and do the work of *five men* with hand belts

SAVING $3.00 to $4.00 a day

J. M. NASH, MILWAUKEE, Wis.

Reduce the Cost of Your Sanding and Improve the Finish

of your Handles by using an

"OBER" DOUBLE BELT SANDER

for sanding broom, mop, fork, hoe, and rake handles, pike poles, etc.

We also make LATHES, for turning this work, also other Lathes for turning irregular work, RIP SAWS, BOLSTERS, Etc.

Write for free catalogue No. 31.

THE OBER MFG. COMPANY

CHAGRIN FALLS, Ohio., U.S.A.

VENEER PRESSES

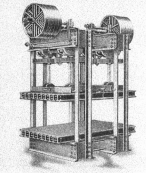

WRITE FOR BULLETIN C 1.

Canadian Boomer & Boschert Press
Company, Limited
18 Tansley Street, MONTREAL

Your Loose Pulley Trouble Disappears

when installed with

HAPMAN BALL BEARING LOOSE PULLEYS

A little vaseline once a year is all the attention and lubrication required.

Our loose Pulleys cannot run hot and do not cut the shaft.

Write us for particulars and price list.

The
Chapman Double Ball Bearing Co.
of Canada, Limited
339-351 Sorauren Ave., Toronto, Canada

The "Canadian Woodworker" Buyers' Directory

For Alphabetical Index see page 8

ABRASIVES
The Carborundum Co., Niagara Falls, N.Y.

BALL BEARINGS
Chapman Double Ball Bearing Co., Toronto.

BALUSTER LATHES
Garlock-Machinery, Toronto, Ont.
Ober Mfg. Company, Chagrin Falls, Ohio.
Whitney & Son, Baxter D., Winchendon, Mass.

BAND SAW FILING MACHINERY
Berlin Machine Works, Ltd., Hamilton, Ont.
Fay & Egan Co., J. A., Cincinnati, Ohio.
Garlock-Machinery, Toronto, Ont.
Shawver Company, Springfield, O.

BAND SAWS
Berlin Machine Works, Ltd., Hamilton, Ont.
Garlock-Machinery, Toronto, Ont.
Jackson, Cochrane & Company, Berlin, Ont.
Petrie, H. W., Toronto.
Shawver Company, Springfield, O.

BAND SAW MACHINES
Berlin Machine Works, Hamilton, Ont.
Fay & Egan Co., J. A., Cincinnati, Ohio.
Garlock-Machinery, Toronto, Ont.
Petrie, H. W., Toronto.
Williams Machinery Co., A. R., Toronto, Ont.

BAND SAW STRETCHERS
Berlin Machine Works, Ltd., Hamilton, Ont.
Fay & Egan Co., J. A., Cincinnati, Ohio.

BASKET STITCHING MACHINE
Saranac Machine Co., Benton Harbor, Mich.

BELT LACER
Clipper Belt Lacer Co., Grand Rapids, Mich.

BENDING MACHINES
Fay & Egan Co., J. A., Cincinnati, Ohio.
Garlock-Machinery, Toronto, Ont.
Perfection Wood Steaming Retort Company,
Parkersburg, W. Va.

BELTING
Fay & Egan Co., J. A., Cincinnati, Ohio.

BLOWERS
Sheldons, Limited, Galt, Ont.

BLOW PIPING
Sheldons, Limited, Galt, Ont.

BORING MACHINES
Berlin Machine Works, Ltd., Hamilton, Ont.
Canada Machinery Corporation, Galt, Ont.
Fay & Egan Co., J. A., Cincinnati, Ohio.
Garlock-Machinery, Toronto, Ont.
Jackson, Cochrane & Company, Berlin, Ont.
Nash, J. M., Milwaukee, Wis.
Reynolds Pattern & Machine Co., Moline, Ill.
Shawver Company, Springfield, O.

BOX MAKERS' MACHINERY
Berlin Machine Works, Ltd., Hamilton, Ont.
Canada Machinery Corporation, Galt, Ont.
Fay & Egan Co., J. A., Cincinnati, Ohio.
Garlock-Machinery, Toronto, Ont.
Globe Soap Company, Cincinnati, Ohio.
Jackson, Cochrane & Co., Berlin, Ont.
Neilson & Company, J. L. Winnipeg, Man.
Reynolds Pattern & Machine Co., Moline, Ill.
Whitney & Son, Baxter D., Winchendon, Mass.
Williams Machinery Co., A. R., Toronto, Ont.

CABINET PLANERS
Berlin Machine Works, Ltd., Hamilton, Ont.
Fay & Egan Co., J. A., Cincinnati, Ohio.
Garlock-Machinery, Toronto, Ont.
Jackson, Cochrane & Co., Berlin, Ont.
Shawver Company, Springfield, O.

CARS (Transfer)
Sheldons, Limited, Galt, Ont.

CARBORUNDUM PAPER
Carborundum Co., Niagara Falls, N.Y.

CARVING MACHINES
Fay & Egan Co., J. A., Cincinnati, Ohio.
Garlock-Machinery, Toronto, Ont.
Jackson, Cochrane & Company, Berlin, Ont.

CASTERS
Foster, Merriam & Co., Meriden, Conn.

CLAMPS
Fay & Egan Co., J. A., Cincinnati, Ohio.
Garlock-Machinery, Toronto, Ont.
Jackson, Cochrane & Company, Berlin, Ont.
Simonds Canada Saw Company, Montreal, P.Q.

COLUMN MACHINERY
Fay & Egan Co., J. A., Cincinnati, Ohio.

CUT-OFF SAWS
Berlin Machine Works, Ltd., Hamilton, Ont.
Fay & Egan Co., J. A., Cincinnati, Ohio.
Garlock-Machinery, Toronto, Ont.
Jackson, Cochrane & Co., Berlin, Ont.
Knechtel Bros., Limited, Southampton, Ont.
Ober Mfg. Co., Chagrin Falls, Ohio.
Shawver Company, Springfield, O.
Simonds Canada Saw Company, Montreal, Que.

CUTTER HEADS
Berlin Machine Works, Hamilton, Ont.
Fay & Egan Co., J. A., Cincinnati, Ohio.
Gage Mfg. Company, E. E., Gardner, Mass.
Jackson, Cochrane & Company, Berlin, Ont.
Mattison Machine Works, Beloit, Wis.

DOVETAILING MACHINES
Canadian Linderman Machine Company, Woodstock, Ont.
Fay & Egan Co., J. A., Cincinnati, Ohio.
Garlock-Machinery, Toronto, Ont.
Jackson, Cochrane & Company, Berlin, Ont.

DOWEL MACHINES
Fay & Egan Co., J. A., Cincinnati, Ohio.
Ober Mfg. Co., Chagrin Falls, Ohio.

DRY KILNS
Grand Rapids Veneer Works, Grand Rapids,
Mich.
National Dry Kiln Co., Indianapolis, Ind.
Sheldons, Limited, Galt, Ont.

DOWELS
Knechtel Bros., Limited, Southampton, Ont.

DUST COLLECTORS
Sheldons, Limited, Galt, Ont.

DUST SEPARATORS
Sheldons, Limited, Galt, Ont.

EDGERS (Single Saw)
Fay & Egan Co., J. A., Cincinnati, Ohio.
Garlock-Machinery, Toronto, Ont.
Simonds Canada Saw Co., Montreal, Que.

EDGERS (Gang)
Berlin Machine Works, Ltd., Hamilton, Ont.
Fay & Egan Co., J. A., Cincinnati, Ohio.
Garlock-Machinery, Toronto, Ont.
Petrie, H. W., Toronto.
Simonds Canada Saw Co., Montreal, P.Q.
Williams Machinery Co., A. R., Toronto, Ont.

END MATCHING MACHINE
Berlin Machine Works, Ltd., Hamilton, Ont.
Fay & Egan Co., J. A., Cincinnati, Ohio.
Garlock-Machinery, Toronto, Ont.
Jackson, Cochrane & Company, Berlin, Ont.

EXHAUST FANS
Garlock-Machinery, Toronto, Ont.
Sheldons Limited, Galt, Ont.

FILES
Simonds Canada Saw Co., Montreal, Que.

FLOORING MACHINES
Berlin Machine Works, Ltd., Hamilton, Ohio.
Fay & Egan Co., J. A., Cincinnati, Ohio.
Garlock-Machinery, Toronto, Ont.
Petrie, H. W., Toronto.
Whitney & Son, Baxter D., Winchendon, Mass.

FLUTING HEADS
Fay & Egan Co., J. A., Cincinnati, Ohio.
Shawver Company, Springfield, O.

FLUTING AND TWIST MACHINE
Shawver Company, Springfield, O.

FURNITURE TRIMMINGS
Foster, Merriam & Co., Meriden, Conn.

GARNET PAPER AND CLOTH
The Carborundum Co., Niagara Falls, N.Y.
Delany & Pettit, Limited, Toronto, Ont.

GRAINING MACHINES
Berlin Machine Works, Ltd., Hamilton, Ont.
Fay & Egan Co., J. A., Cincinnati, Ohio.
Williams Machinery Co., A. R., Toronto, Ont.

GLUE
Coignet Chemical Product Co., New York,
N.Y.
Delany & Pettit, Limited, Toronto, Ont.
Perkins Glue Company, South Bend, Ind.

GLUE CLAMPS
Jackson, Cochrane & Company, Berlin, Ont.

GLUE HEATERS
Fay & Egan Co., J. A., Cincinnati, Ohio.
Jackson, Cochrane & Company, Berlin, Ont.

GLUE JOINTERS
Berlin Machine Works, Ltd., Hamilton, Ont.
Jackson, Cochrane & Company, Berlin, Ont.

GLUE SPREADERS
Fay & Egan Co., J. A., Cincinnati, Ohio.
Jackson, Cochrane & Company, Berlin, Ont.

GLUE ROOM EQUIPMENT
Boomer & Boschert Company, Montreal, Que.
Perrin & Company, W. R., Toronto, Ont.

GRINDERS (Cutter)
Fay & Egan Co., J. A., Cincinnati. Ohio.

GRINDERS, HAND AND FOOT POWER
The Carborundum Co., Niagara Falls, N.Y.

GRINDERS (Knife)
Berlin Machine Works, Ltd., Hamilton, Ont.
Fay & Egan Co., J. A., Cincinnati, Ohio.
Garlock-Machinery, Toronto, Ont.
Petrie, H. W., Toronto.

GRINDERS (Tool)
Fay & Egan Co., J. A., Cincinnati, Ohio.
Jackson, Cochrane & Company, Berlin, Ont.
Shawver Company, Springfield, O.

GRINDING WHEELS
The Carborundum Co., Niagara Falls, N.Y.

GROOVING HEADS
Fay & Egan Co., J. A., Cincinnati, Ohio.

GUARDS, (Saw)
Shawver Company, Springfield, O.

GUMMERS, ETC.
Fay & Egan Co., J. A., Cincinnati, Ohio.

HAND PROTECTORS
Fay & Egan Co., J. A., Cincinnati, Ohio.

HAND SCREWS
Fay & Egan Co., J. A., Cincinnati, Ohio.

HANDLE THROATING MACHINE
Klotz Machine Company, Sandusky, Ohio.

HANDLE TRIMMING MACHINE
Klotz Machine Company, Sandusky, Ohio.

HANDLE & SPOKE MACHINERY
Fay & Egan Co., J. A., Cincinnati, Ohio.
Klotz Machine Co., Sandusky, Ohio.
Nash, J. M., Milwaukee, Wis.
Ober Mfg. Company, Chagrin Falls, Ohio.
Whitney & Son, Baxter D., Winchendon, Mass.

HARDWOOD LUMBER
Atlantic Lumber Company, Toronto.
Imperial Lumber Company, Columbus, Ohio.
Kersley, Geo., Montreal, Que.
Kraetzer-Cured Lumber Co., Moorhead, Miss.
Penrod Walnut & Veneer Co., Kansas City, Mo.
Spencer, C. A., Montreal, Que.

HUB MACHINERY
Fay & Egan Co., J. A., Cincinnati, Ohio.

HYDRAULIC PRESSES
Can. Boomer & Boschert Press, Montreal, Que.

HYDRAULIC PUMPS & ACCUMULATORS
Can. Boomer & Boschert Press, Montreal, Que.

HYDRAULIC VENEER PRESSES
Perrin & Company, Wm. R., Toronto, Ont.

JOINTERS
Canadian Linderman Machine Company, Woodstock, Ont.
Berlin Machine Works, Ltd., Hamilton, Ont.
Fay & Egan Co., J. A., Cincinnati, Ohio.
Garlock-Machinery, Toronto, Ont.
Jackson, Cochrane & Company, Berlin, Ont.
Knechtel Bros., Limited, Southampton, Ont.
Petrie, H. W., Toronto, Ont.
Pryibil Machine Company, P., New York.
Shawver Company, Springfield, O.

KNIVES (Planer and others)
Berlin Machine Works, Ltd., Hamilton, Ont.
Fay & Egan Co., J. A., Cincinnati, Ohio.
Simonds Canada Saw Co., Montreal, P.Q.

LUMBER
Hartzell, Geo. M., Piqua, Ohio.
Imperial Lumber Company, Columbus, Ohio.

LATHES
Fay & Egan Co., J. A., Cincinnati, Ohio.
Garlock-Machinery, Toronto, Ont.
Jackson, Cochrane & Company, Berlin, Ont.
Klotz Machine Company, Sandusky, Ohio.
Ober Mfg. Company, Chagrin Falls, Ohio.
Petrie, H. W., Toronto, Ont.
Shawver Company, Springfield. O.
Whitney & Son, Baxter D., Winchendon, Mass.

MACHINE KNIVES
Berlin Machine Works, Ltd., Hamilton, Ont.
Peter Hay Knife Company, Galt, Ont.
Simonds Canada Saw Co., Montreal, P.Q.

MITRE MACHINES
Fay & Egan Co., J. A., Cincinnati, Ohio.

MITRE SAWS
Fay & Egan Co., J. A., Cincinnati, Ohio.
Garlock-Machinery, Toronto, Ont.
Simonds Canada Saw Co., Montreal, P.Q.

MORTISING MACHINES
Berlin Machine Works, Ltd., Hamilton, Ont.
Fay & Egan Co., J. A., Cincinnati, Ohio.
Garlock-Machinery, Toronto, Ont.

MOULDINGS
Knechtel Bros., Limited, Southampton, Ont.

MULTIPLE BOXING MACHINES
Fay & Egan Co., J. A., Cincinnati, Ohio.
Nash, J. M., Milwaukee, Wis.
Reynolds Pattern & Machine Co., Moline, Ill.

PATENT SOLICITORS
Lightfoot, Stanley, Toronto, Ont.

PATTERN SHOP MACHINES
Berlin Machine Works, Ltd., Hamilton, Ont.
Fay & Egan Co., J. A., Cincinnati, Ohio.
Garlock-Machinery, Toronto, Ont.
Jackson, Cochrane & Company, Berlin, Ont.
Whitney & Son, Baxter D., Winchendon, Mass.

PLANERS
Berlin Machine Works, Ltd., Hamilton, Ont.
Fay & Egan Co., J. A., Cincinnati, Ohio.
Garlock-Machinery, Toronto, Ont.
Jackson, Cochrane & Company, Berlin, Ont.
Petrie, H. W., Toronto, Ont.
Whitney & Son, Baxter D., Winchendon, Mass.
Williams Machinery Co., A. R., Toronto, Ont.

PLANING MILL MACHINERY
Berlin Machine Works, Ltd., Hamilton, Ont.
Fay & Egan Co., J. A., Cincinnati, Ohio.
Garlock-Machinery, Toronto, Ont.
Jackson, Cochrane & Company, Berlin, Ont.
Petrie, H. W., Toronto, Ont.
Shawver Company, Springfield, O.
Whitney & Son, Baxter D., Winchendon, Mass.
Williams Machinery Co., A. R., Toronto, Ont.

PRESSES (Veneer)
Jackson, Cochrane & Company, Berlin, Ont.
Perrin & Co., Wm. R., Toronto, Ont.

PULLEYS
Fay & Egan Co., J. A., Cincinnati, Ohio.

RESAWS
Berlin Machine Works, Ltd., Hamilton, Ont.
Fay & Egan Co., J. A., Cincinnati, Ohio.
Garlock-Machinery, Toronto, Ont.
Jackson, Cochrane & Company, Berlin, Ont.
Petrie, H. W., Toronto, Ont.
Simonds Canada Saw Co., Montreal, P.Q.
Williams Machinery Co., A. R., Toronto, Ont.

RIM AND FELLOE MACHINERY
Fay & Egan Co., J. A., Cincinnati, Ohio.

RIP SAWING MACHINES
Berlin Machine Works, Ltd., Hamilton, Ont.
Fay & Egan Co., J. A., Cincinnati, Ohio.
Garlock-Machinery, Toronto, Ont.
Klotz Machine Company, Sandusky, Ohio.
Jackson, Cochrane & Company, Berlin, Ont.
Ober Mfg. Company, Chagrin Falls, Ohio.

SANDERS
Berlin Machine Works, Ltd., Hamilton, Ont.
Fay & Egan Co., J. A., Cincinnati, Ohio.
Garlock-Machinery, Toronto, Ont.
Klotz Machine Company, Sandusky, Ohio.
Jackson, Cochrane & Company, Berlin, Ont.
Nash, J. M., Milwaukee, Wis.
Ober Mfg. Company, Chagrin Falls, Ohio.
Shawver Company, Springfield, O.

SASH, DOOR & BLIND MACHINERY
Berlin Machine Works, Ltd., Hamilton, Ont.
Fay & Egan Co., J. A., Cincinnati, Ohio.
Garlock-Machinery, Toronto, Ont.
Jackson, Cochrane & Company, Berlin, Ont.

SAWS
Fay & Egan Co., J. A., Cincinnati, Ohio.
Radcliff Saw Mfg. Company, Toronto, Ont.
Simonds Canada Saw Co., Montreal, P.Q.

SAW GUARDS
Shawver Company, Springfield, O.

SAW GUMMERS, ALOXITE
The Carborundum Co., Niagara Falls, N.Y.

SAW SWAGES
Fay & Egan Co., J. A., Cincinnati, Ohio.
Simonds Canada Saw Co., Montreal, Que.

SAW TABLES
Berlin Machine Works, Ltd., Hamilton, Ont.
Fay & Egan Co., J. A., Cincinnati, Ohio.
Garlock-Machinery, Toronto, Ont.
Jackson, Cochrane & Company, Berlin, Ont.
Knechtel Bros., Limited, Southampton, Ont.

SCRAPING MACHINES
Garlock-Machinery, Toronto, Ont.
Whitney & Sons, Baxter D., Winchendon, Mass.

SCROLL CHUCK MACHINES
Klotz Machine Company, Sandusky, Ohio.

SCREW DRIVING MACHINES
Reynolds Pattern & Machine Co., Moline, Ill.
Stow Manufacturing Co., Binghamton, N.Y.

SCREW DRIVING MACHINERY
Reynolds Pattern & Machine Works, Moline, Ill.
Stow Manufacturing Co., Binghamton, N.Y.

SCROLL SAWS
Berlin Machine Works, Ltd., Hamilton, Ont.
Fay & Egan Co., J. A., Cincinnati, Ohio.
Garlock-Machinery, Toronto, Ont.
Jackson, Cochrane & Company, Berlin, Ont.
Simonds Canada Saw Co., Montreal, P.Q.

SECOND-HAND MACHINERY
Petrie, H. W.
Williams Machinery Co., A. R., Toronto, Ont.

SHAPERS
Berlin Machine Works, Ltd., Hamilton, Ont.
Fay & Egan Co., J. A., Cincinnati, Ohio.
Garlock-Machinery, Toronto, Ont.
Jackson, Cochrane & Company, Berlin, Ont.
Ober Mfg. Company, Chagrin Falls, Ohio.
Petrie, H. W., Toronto, Ont.
Shawver Company, Springfield, O.
Simonds Canada Saw Co., Montreal, P.Q.
Whitney & Sons, Baxter D., Winchendon, Mass.

SHARPENING TOOLS
The Carborundum Co., Niagara Falls, N.Y.

SHAVING COLLECTORS
Sheldons Limited, Galt, Ont.

SHELL BOX HANDLES
Vokes Hardware Co., Toronto, Ont.

SHELL BOX EQUIPMENT
Vokes Hardware Co., Toronto, Ont.

SINGLE SPINDLE BOXING MACHINES
Fay & Egan Co., J. A., Cincinnati, Ohio.
Ober Mfg. Company, Chagrin Falls, Ohio.

STAVE SAWING MACHINE
Whitney & Sons, Baxter D., Winchendon, Mass.

STEAM TRAPS
Canadian Morehead Mfg. Co., Woodstock, Ont.
Standard Dry Kiln Co., Indianapolis, Ind.

STEEL SHELL BOX BANDS
Yokes Hardware Co., Toronto, Ont.

SURFACERS
Berlin Machine Works, Ltd., Hamilton, Ont.
Fay & Egan Co., J. A., Cincinnati, Ohio.
Garlock-Machinery, Toronto, Ont.
Jackson, Cochrane & Company, Berlin, Ont.
Petrie, H. W., Toronto, Ont.
Whitney & Sons, Baxter D., Winchendon, Mass.
Williams Machinery Co., A. R., Toronto, Ont.

SWING SAWS
Berlin Machine Works, Ltd., Hamilton, Ont.
Fay & Egan Co., J. A., Cincinnati, Ohio.
Garlock-Machinery, Toronto, Ont.
Jackson, Cochrane & Company, Berlin, Ont.
Ober Mfg. Co., Chagrin Falls, Ohio.
Simonds Canada Saw Co., Montreal, P.Q.

TABLE LEG LATHES
Berlin Machine Works, Ltd., Hamilton, Ont.
Fay & Egan Co., J. A., Cincinnati, Ohio.
Ober Mfg. Co., Chagrin Falls, Ohio.
Whitney & Sons, Baxter D., Winchendon, Mass.

TABLE SLIDES
Walter & Co., H., Wabash, Ind.

TENONING MACHINES
Berlin Machine Works, Ltd., Hamilton, Ont.
Fay & Egan Co., J. A., Cincinnati, Ohio.
Garlock-Machinery, Toronto, Ont.
Jackson, Cochrane & Company, Berlin, Ont.

TRIMMERS
Berlin Machine Works, Ltd., Hamilton, Ont.
Fay & Egan Co., J. A., Cincinnati, Ohio.
Garlock-Machinery, Toronto, Ont.

TRUCKS
Sheldons Limited, Galt, Ont.
National Dry Kiln Co., Indianapolis, Ind.

TURNINGS
Knechtel Bros., Limited, Southampton, Ont.

TURNING MACHINES
Berlin Machine Works, Ltd., Hamilton, Ont.
Fay & Egan Co., J. A., Cincinnati, Ohio.
Garlock-Machinery, Toronto, Ont.
Klotz Machine Company, Sandusky, Ohio.
Ober Mfg. Company, Chagrin Falls, Ohio.
Shawver Company, Springfield, O.
Whitney & Sons, Baxter D., Winchendon, Mass.

**UNDER-CUT SELF-FEEDING FACE
PLANER**
Fay & Egan Co., J. A., Cincinnati, Ohio.
Garlock-Machinery, Toronto, Ont.
Jackson, Cochrane & Company, Berlin, Ont.

VARNISHES
Ault & Wiborg Company, Toronto, Ont.

VENEERS
Central Veneer Co., Huntington, W. Virginia.
Walter Clark Veneer Co., Grand Rapids, Mich.
Evansville Veneer Co., Evansville, Ind.
Hartzell, Geo. W., Piqua, Ohio.

Hay & Company, Woodstock, Ont.
Hoffman Bros. Company, Fort Wayne, Ind.
Geo. Kersley, Montreal, Que.
Ohio Veneer Company, Cincinnati, Ohio.
Penrod Walnut & Veneer Co., Kansas City, Mo.
Waetjen & Co., Geo. L., Milwaukee, Wis.

VENTILATING APPARATUS
Sheldons Limited, Galt, Ont.

VENEERED PANELS
Hay & Company, Woodstock, Ont.

VENEER PRESSES (Hand and Power)
Can. Boomer & Boschert Press, Montreal, Que.
Garlock-Machinery, Toronto, Ont.
Jackson, Cochrane & Company, Berlin, Ont.
Perrin & Company, Wm. R., Toronto, Ont.

VISES
Fay & Egan Co., J. A., Cincinnati, Ohio.
Simonds Canada Saw Company, Montreal, Que.

WAGON AND CARRIAGE MACHINERY
Berlin Machine Works, Ltd., Hamilton, Ont.
Fay & Egan Co., J. A., Cincinnati, Ohio.
Ober Mfg. Company, Chagrin Falls, Ohio.
Stow Manufacturing Co., Binghamton, N.Y.
Whitney & Sons, Baxter D., Winchendon, Mass.

WOOD FINISHES
Ault & Wiborg, Toronto, Ont.

WOOD TURNING MACHINERY
Berlin Machine Works, Ltd., Beloit, Wis.
Garlock-Machinery, Toronto, Ont.
Shawver Co., Springfield, Ohio.

WORK BENCHES
Fay & Egan Co., J. A., Cincinnati, Ohio.

WOODWORKING MACHINES
Canada Machinery Corporation, Galt, Ont.
Garlock-Machinery, Toronto, Ont.
Jackson, Cochrane & Co., Berlin, Ont.
Reynolds Pattern & Machine Works, Moline, Ill.
Williams Machinery Co., A. R., Toronto, Ont.

WHEELS, Grinding, Carborundum and Aloxite
The Carborundum Co., Niagara Falls, N.Y.

WOOD SPECIALTIES
Knechtel Bros., Limited, Southampton, Ont.

DOWELS
Plain and Spiral Grooved

Our fifteen years' experience in this line at your disposal.
Let us send you prices.

WE ALSO MANUFACTURE A FULL LINE OF

Embossed and Carved Dentil Mouldings, Turnings and Balusters

*We will be pleased to furnish information
and prices on your requirements.*

KNECHTEL BROS., LIMITED
SOUTHAMPTON   —   —   —   —   ONTARIO

Making TABLE-SLIDES is a Specialty Business

For more than TWENTY-FIVE YEARS we have made TABLE SLIDES exclusively. Our Factory is equipped with Special Machinery which enables us to make SLIDES,—BETTER and CHEAPER than the furniture manufacturer.

Canadian Table makers are rapidly adopting WABASH SLIDES

Because { They ELIMINATE SLIDE TROUBLES
Are CHEAPER and BETTER

Reduced Costs | BY USING
Increased Out-put | **WABASH SLIDES**

MADE BY

B. Walter & Company
Wabash, Ind.

The Largest EXCLUSIVE TABLE-SLIDE Manufacturers in America
ESTABLISHED-1887

Just What You Need

There's no necessity now for you to wish for a corrugated joint fastener machine which will rise fully to your expectations; you can get one.

The Saranac Corrugated Joint Fastener Machine

is the absolute realization of your wish. This illustration gives you an idea of what it looks like, but you can have no conception of its efficiency, economy, speed and reliability until you get it working for you.

It is substantially constructed, and the workmanship and materials are of the very best. Particularly suitable for manufacturers of sash, doors, blinds, screens, porch columns, boxes, etc., etc.

Made in various types for all conditions.

Write for further particulars and prices

Saranac Machine Co., SOLE MAKERS
Benton Harbor, Mich.

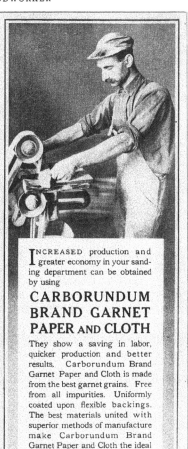

INCREASED production and greater economy in your sanding department can be obtained by using

CARBORUNDUM BRAND GARNET PAPER AND CLOTH

They show a saving in labor, quicker production and better results. Carborundum Brand Garnet Paper and Cloth is made from the best garnet grains. Free from all impurities. Uniformly coated upon flexible backings. The best materials united with superior methods of manufacture make Carborundum Brand Garnet Paper and Cloth the ideal abrasive for the woodworking trade.

Working samples upon request

THE CARBORUNDUM COMPANY
NIAGARA FALLS, N. Y.

New York   Chicago   Grand Rapids
          Boston   Cincinnati

These
Miniature
Turnings
tell a story
of machine
work that
will surprise
you.

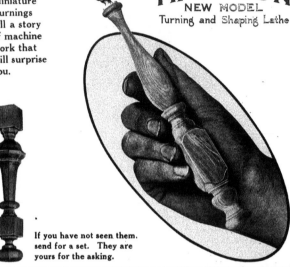

MATTISON
NEW MODEL
Turning and Shaping Lathe

If you have not seen them,
send for a set. They are
yours for the asking.

Period Styles Demand Efficient Turning Methods

The old-fashioned hand lathe or band-saw methods are too slow and wasteful to be considered in the manufacture of Turnings for Period Style Furniture.

But the popularity of Period Patterns demands that ornamental turned work be used to a large extent on all classes and grades of furniture. It is no longer a question of making Period Furniture, but of making it "at a price."

The Mattison Lathe solves the problem for you —it puts your turned work on a systematic manufacturing basis. It produces the artistic round, oval, square or octagon shaped turnings at from

one-fifth to one-twentieth the cost of hand-method costs.

The Mattison Lathe is always on the job—insures uninterrupted production. Any ordinary machine man can operate it. The cutters are no more difficult to grind than sticker or shaper knives.

This machine method enables you to compete on an equal basis with the leading manufacturers of Period Style Furniture.

If you need a Mattison Lathe for this season's work, now is the time to get posted. It's a pleasure to explain the improvements on the Model "B."

There is no other Lathe like the Mattison
and it is built only by

C. MATTISON MACHINE WORKS, BELOIT, WIS., U.S.A.
883 FIFTH STREET

Volume Sixteen   TORONTO, MARCH, 1916   Number Three

CANADIAN WOODWORKER
and Furniture Manufacturer

The illustration shows our

No. 561 Hollow Chisel Mortiser

which is especially suited for light mortising in sash, doors, blinds and general cabinet work. It has an extra wide base and runs without vibration.

New Feature

One of the essential points of a machine of this type is the bit and chisel holder. The bit and chisel holder on this machine is our new improved design, being the result of long practical experience, and is used on our machines exclusively. It is quickly and easily adjusted. Saves wear on the bits and saves time and money in operating and upkeep ; an improvement worthy of investigation.

Full Information and illustrations on request.

CANADA MACHINERY CORPORATION
LIMITED

Head Office - GALT, ONT.

Toronto Office and Showrooms - - Brock Avenue Subway

Put Good Belting

on your

Good Machines

Your machines are expensive, and naturally you expect them to turn out the quality and quantity of work commensurate with their cost.

Then why harness them with belting which cannot begin to do them justice?

Use

"AMPHIBIA"
Planer Belting

and get the most work from your machines, in the quickest time at the lowest cost per day of service.

Try a sample run of AMPHIBIA Planer and prove its merits.

"Leather like gold has no substitute."

Sadler & Haworth

Established 1876

Tanners and Manufacturers

For 40 years Tanners and Manufacturers of the Best Leather Belts.

Toronto	St. John, N.B.	MONTREAL	Winnipeg	Vancouver
38 Wellington St. E.	139 Prince William St.	511 William St.	Galt Building	217 Columbia Ave.

What are Your Requirements

for

Wood Working Machinery

We handle all kinds—for every purpose

including full line of

American Wood Working Machinery Co.

ROCHESTER, N.Y.

M. L. Andrews & Co.,

CINCINNATI, OHIO

We have installed several plants for SHELL BOX work.

Let us co-operate with you in selecting and laying out your tool equipment.

Our Services are at your disposal.

Garlock - Machinery

197 Wellesley Street   -   Toronto

Telephone North 6849

A page in the book on handle factory efficiency

Here is an all-parts perfect Splitting or Bolting Saw Machine which offers every feature necessary for fast, strictly accurate splitting and ripping of lumber. On its first job it will change your present idea of handle factory efficiency. You should see it in operation to appreciate its value.

Klotz Splitting or Bolting Saw

Saw is hardwood frame, securely bolted; steel mandrel, true-running flywheel, heavily plated; a light, thoroughly seasoned ash table operated on turned rolls with steel axles and metal bearings; an instantly adjustable chuck attached to table for holding timber. Made in four standard sizes. Ask us which size is suitable for your particular needs.

Turn a Page With Us Next Issue

FOR YOU

who have a thought of manufacturing handles, or those desiring maximum production at correct cost, get the Klotz Machine Company's idea of a money-making, perfectly

MODERN HANDLE FACTORY

One with such dependable machines in it as their Blocking Saw, Shaping Saw, Lathe, Throater, Polisher, Knobber or Scroll Machine, Disc Wheel and End Shaper, Splitting or Bolting Saw. All have a place in an efficient handle factory. Write

Klotz Machine Co., Sandusky, O.

Efficiency—
Increased Production
"STOW"
Motor Driven Screw Drivers

Built in Several Sizes   Suspended or Pedestal Design

Immediate Shipment   —   Write us your requirements to-day

STOW Tools are an INVESTMENT and not an EXPENSE

STOW   MANUFACTURING   CO.
Binghamton, N. Y., U. S. A.
Oldest Portable Tool Manufacturers in America
Electric and Belt Driven Drills, All Sizes
INVENTORS AND LARGEST MANUFACTURERS OF STOW FLEXIBLE SHAFTING

MADE RIGHT—RIGHT   IN CANADA
Berlin
NEW TURNING AND SHAPING MACHINE

As installed in prominent furniture plants—leaders in their respective lines.

A booklet illustrating many popular Period Patterns, and describing this latest and most economical method of producing them, may be had free on request.

Prepared to Produce
the Patterns That Dominated the Furniture Shows

MORE than ever do furniture makers realize the need of this Berlin Turner. Period Patterns will continue "all-the-go."

To make them at rock-bottom cost this Berlin Turner is absolutely necessary—the simplest and most efficient machine method in use.

Any mechanic can operate it. Most any size and shape can be made at but a fraction of the cost of other methods.

Be prepared. Make your patterns economically. Those at the shows selling at attractive prices got the business. The time to get in line is NOW. There is a machine of just the size to fit your requirements. Equip it to it your work in particular.

Begin to look it up by asking for folder of description and details.

P. B. YATES MACHINE COMPANY, LIMITED
Successors to

U.S. Plant—BELOIT, Wis.   The Berlin Machine Works, Limited   HAMILTON, ONT.

One Piece Sides and Ends for
SHELL BOXES

From Narrow (Cheaper) Lumber at
a Low Rate per 1000 feet on the

LINDERMAN AUTOMATIC
DOVETAIL MATCHER

*Samples
on
Request*

THESE CONCERNS USE LINDERMAN MACHINES:

W. C. Edwards & Co.	Ottawa, Ont.	Dominion Box & Package Co.	Montreal, Que.
(2 Machines)		(2 Machines)	
Czerwinski Box Co.	Winnipeg, Man.	C. P. R. Angus Shops	Montreal, Que.
Cummer-Dowswell Co.	Hamilton, Ont.	J. W. Kilgour & Bro.	Beauharnois, Que.
Canada Furniture Mfrs.	Woodstock, Ont.	Durham Furniture Co.	Durham, Ont.
Martin Freres & Cie	Montreal, Que.	Brunswick-Balke-Collender Co.	Toronto, Ont.
Alberta Box Co.	Calgary, Alta.	Singer Mfg. Co.	St. Johns, Que.

Canadian Linderman Company
Woodstock, Ontario

Muskegon, Mich.   Knoxville, Tenn.   New York, N. Y.

The Only Way to Drive Screws

There is a stage-coach method and a modern method of doing everything. When it comes to driving screws the only modern way is with an Automatic Screw Driving Machine, which means greater and better production and greater profit.

The Reynolds
Automatic Screw-Driving Machine

is guaranteed to produce results. It is no experiment. The No. 2 Standard Type, shown herewith, is particularly suited for placing the screws in the straps of shrapnel boxes. It may be fitted to drive, without change or adjustment, any three consecutive sizes flat or round head, or two sizes filister head, No. 6 to 14 inclusive, wood or machine screws, or three sizes flat or round head, or two sizes filister head machine screws up to No. 20. All the above up to 1½ inches long if smaller than No. 10, or up to 1¾ inches long, No. 10 or larger. Fitted with boring attachment if desired.

Many Machines in Use in Canada.

Ask for our Catalogue.

REYNOLDS PATTERN & MACHINE CO.
101 and 103 Third Ave., Moline, Ill., U.S.A.

No. 2 Standard Type Automatic Screw-driving Machine.

Perkins Vegetable Veneer Glue

IS A

Uniform, Guaranteed Patented Glue

¶ We have, during the past ten or more years, built up a large volume of business and maintained it, because we sell only

HIGHEST QUALITY

¶ Our list of customers proves the statement that a very large per cent. of the highest quality furniture, including office and household furniture, also doors, interior trim, etc., are built up with Perkins Glue.

¶ Be sure you have *Perkins Quality*—absolutely uniform.

¶ We will demonstrate to your satisfaction in your own factory

Ladies' Desk manufactured by The D. Hibner Furniture Co., Limited, Berlin, Ont., in which Perkins Glue was used.

Originator and Patentee of Perkins Vegetable Glue.

PERKINS GLUE COMPANY, LIMITED

Address all inquiries to Sales Office,
Perkins Glue Company, SOUTH BEND, Indiana, U.S.A.

Hamilton, Ont.

Alphabetical List of Advertisers

Atlantic Lumber Company 10
Ault & Wiborg Company 33

Baird Machinery Works 45

Canada Machinery Corporation 1
Canadian Boomer & Boschert Mfg.
  Company 47
Canadian Linderman Co. 6
Canadian Morehead Mfg. Company
Carborundum Company 51
Central Veneer Company 39
Chapman Double Ball Bearing Co. 47
Clark Veneer Co., Walter 37
Clipper Belt Lacer Company ... ...
Coignet Chemical Products Co. 33

Darling Bros... 42
Delaney & Pettit 12

Evansville Veneer Company 34

Fay & Egan Company, J. A.... ... 9
Foster, Merriam & Company 8

Garlock-Machinery 3
Globe Soap Company 45
Grand Rapids Dry Kiln Company .. 47

Hartzell, Geo. W. 37
Hay & Company 33
Hay Knife Company, Peter 10
Hoffman Bros. Company 37

Imperial Lumber Company 10

Jackson, Cochrane & Company 11

Kersley, Geo.... 10
Klotz Manufacturing Company .. 4
Knechtel Bros. 50
Kraetzer-Cured Lumber Company..

Lightfoot, Stanley 45

Mattison Machine Works 52

Nash, J. M. 44
National Dry Kiln Company 44
National Securities Corporation ... 45
Neilson & Company, J. L. 43

Ober Manufacturing Company 47
Ohio Veneer Company 37

Perfection Wood Steaming Retort
  Company 43
Penrod Walnut & Veneer Company 39
Perkins Glue Company 7
Perrin, William R 46
Petrie, H. W. 45

Radcliff Saw Company 12
Reynolds Pattern & Machine Works 7

Saranac Machine Company 51
Sadler & Haworth 2
Shawver Company 46
Sheldons Limited 46
Spencer, C. A. 10
Stow Mfg. Company 4

Vokes Hardware Company 41

Waetjen & Company, G. L. 37
Walter & Company, B. 51
Weeks Lumber Company. F. H. ... 46
Whitney & Son, Baxter D. ... ...

Yates Machine Co., P. B. 5

THE COMPLETE LINE

For eighty years ACME CASTERS have been reliable casters, the use of which will add prestige to your furniture.

Discriminating manufacturers appreciate the perfection of ACME BALL BEARING or non-ball bearing CASTERS.

The adoption of ACME CASTERS insures the best and quickest service.

We solicit your inquiries and will be pleased to have our Canadian representative call and discuss any special casters or period trimmings you may require.

Foster, Merriam & Co.
Meriden, Conn.
Canadian Representative: F. A. Schmidt, Berlin, Ont.

" ACME "
The
RELIABLE CASTER

—a new small four-side matcher
—that is *really* new

FAY-EGAN "LIGHTNING" No. 369, Planer, Matcher and Moulder

—Surfaces two sides, 24" x 8", matches 15" wide, cuts 1½" moulds.

—Enclosed cut gear and chain drive—powerful and silent.

—Feeds 75' a minute, with two additional rates.

—New "gap" type frame, makes all parts accessible.

—Four 5" driven feed rolls, spring pressures. (No weights.)

—Infeed table removable and separate counter-shaft make matcher easily portable.

—Side wing angle clamp bearings.

—Gravity belt tightener

—Just what many a mill owner has been looking for—a small general purpose matcher with the economy of the big fast feeders—well, here it is.

Bulletin C-10, Free for Asking, or fill in and Mail the Coupon

J. A. FAY & EGAN CO.
153-173 WEST FRONT STREET   **CINCINNATI, OHIO**

J. A. Fay & Egan Co., Cincinnati, Ohio, U. S. A.

Please send Bulletin C-10 and quote price.

Mr. ....

Street ....

City .... Province ....

Dry Spruce
and Birch

Good Stocks, Prompt Shipments, Satisfaction

C. A. SPENCER, Limited

Wholesale Dealers in Rough and Dressed Lumber

Eastern Townships Bank Building
MONTREAL   -   -   Quebec

 TEAK

D
O
M
E
S
T
I
C

Write for prices on all

Plain and Fancy
Lumbers and Veneers

Mahogany, Oak, Birch,
Walnut, Ash, Poplar, etc.,
rotary, sliced or sawed
in all thicknesses

SHELL BOX PARTS
in birch and spruce
Bridges and Ends

Write for Quotations

L
U
M
B
E
R

George Kersley

224 St. James Street, Montreal

 MAHOGANY

Peter Hay Machine Knives

Peter Hay knives
are made of special
quality steel temper-
ed and toughened
to take and hold a
keen cutting edge.
Hay knives are made
for all purposes.
Have you our Catalogue?
Made in Canada

The Peter Hay Knife Co., Limited
GALT, ONTARIO

The Atlantic Lumber Co.

**Manufacturers and
Wholesale Dealers**

American and Canadian Hardwood Lumber and Mahogany, Quartered and Plain, White and Red Oak a Specialty.

110 Manning Chambers - **TORONTO**
Phone Main 6386
HOME OFFICE, BOSTON MASS.
Mills: KNOXVILLE, TENN.,   WALLAND, TENN.
FRANKLIN, VA.

Immediate Shipment

West Virginia Stock

Thoroughly Dry and Well Manufactured

1 car 4/4 Is and 2s, Plain White Oak.
4 cars 4/4 No. 1 Common Plain White Oak.
2 cars 4/4 No. 2 Common Plain White Oak.
1 car 4/4 No. 1 Common and Better Plain White
   Oak. Worm holes no defect.
1 car 6/4 No. 1 Common Plain White Oak.
6 cars 4/4 S. W. Chestnut.
1 car 5/5 S. W. Chestnut.
4 cars 6/4 S. W. Chestnut.
1 car 8/4 S. W. Chestnut
3 cars 4/4 No. 3 Common Chestnut.
1 car 4/4 No. 1 Common and Better Chestnut.
½ car 8/4 No. 1 Common and Better Chestnut.
1 car 4/4 Saps and Select Poplar.
2 cars 4/4 No. 2 Common and Better Poplar.
1 car 5/4 No. 2 Common Poplar.
1 car 6/4 No. 2 Common Poplar.
1 car 4/4 No. 2 Common and Better Birch.

We solicit your inquiries on special cut bill Oak.

Mill at Holly Junction, W. Va.

Write at once for Quotations   The Price is Right

The Imperial Lumber Co.
COLUMBUS   -   OHIO

Rips and Cross Cuts Lumber— and Reduces Costs

Our No. 164 Double Rip and Cross Cut Saw deserves a place in every woodworking plant. It is designed for cutting off to accurate lengths all kinds of materials used in woodworking factories and will be found **particularly profitable in the shell box industry.** The frame is a single cast iron structure. The tables are made of iron planed perfectly true and grooved to receive the gauges.

Mandrels are arranged to carry two saws, one on each end, and are adjustable up and down by hand wheel shown in cut. Each mandrel is fitted with an expansion collar to accommodate varying size holes in saws.

Belt Tighteners. This is a new feature on this machine and greatly improves the work of the saw.

Equipment regularly furnished is 2 ripping guides, 2 mitre guides, 3 extension guides and 1 guide for short stock, also two 14 inch cross cut saws and necessary wrenches.

Ask for Bulletin

Jackson, Cochrane & Co.
Berlin, Ontario

"Beaver" Saws do better work and give you better service.

When you get a saw made of a very fine quality of steel and tempered to the exact degree—not too tough or not too brittle—to give it the longest life and best service, a saw that seems to do better work every day it is used, look at the name. Nine times out of ten it's a "BEAVER."

Write us for prices on our Labor Saving Saws

Special groover for cutting grooves in shell boxes for wire handles. It cuts with a smoothness that makes further finishing unnecessary.

Radcliff Saw Mfg. Company
Limited
TORONTO   **550 Dundas Street**   **ONTARIO**

Agents for—L. & J. White Company Machine Knives

Joint
and
Veneer
Glue

Always
Uniform

Garnet
Paper
and
Cloth

Delany & Pettit, Limited

Office:—133 Jefferson Ave.   **Toronto, Ont.**   *Works:*—105-131 Jefferson Ave.
106-118 Atlantic Ave.

Canadian Woodworker

A Monthly Publication in the Interest of the Woodworking Industry.
Reaches the factories producing interior finish, doors, sash, flooring,
woodenware, furniture, pianos, boxes, and general mill products.
Subscription, $1.00 a year; foreign $1.50.

Woodworker Publishing Company, Limited

Branches: Montreal, Vancouver,
Chicago, New York, London, Eng.

345 Adelaide St. West, Toronto
Phone Ade. 2700

Authorized by the Postmaster General for Canada, for transmission as second class matter.
Entered as second class matter July 18th, 1914, at the Post Office at Buffalo, N.Y., under the Act of Congress of March 3, 1879.

Vol. 16   March, 1916   No. 3

Competition in Prices of Shell Boxes

Now that the Imperial Munitions Board are adopting the English practice of sending out specifications to many firms likely to be able to make boxes, they are no doubt getting the lowest possible quotations. This is in many ways a good thing. Enquiries into exorbitant prices and fat contracts do not look well in print, nor are they pleasant for the parties concerned. It seems opportune to inquire, however, how far the Munitions Board are justified in accepting some of the quotations they get. How far is a purchaser responsible to see that the price he accepts goods at is sufficient for the seller to trade honestly?

There are certain conditions about estimating that are common to all. (a) The cost of raw material is very nearly alike to all; especially is this true of hardware (for our hardware friends know their business). (b) Labor. This differs with location somewhat, though in the average it does not affect the price a great deal. (c) Plant. At first sight it looks as though this will affect prices considerably. Really, it is output more than cost that is affected by this condition or extent of your plant. If the facilities are meagre the cost of help will be up and overhead low, and if this plant is extensive and efficient, labor will be down but upkeep and administration relatively higher. (d) Office depreciation, supervision and interest charges. (e) Profits.

Recently an order was placed for shell boxes, not a thousand miles from Montreal, at a price which looks low. As nearly as can be judged the price worked out this way: Lumber cost 44 per cent., hardware 23 per cent., labor 20 per cent., leaving 13 per cent. for depreciation, supervision and profit. What probably happened was that the box maker figured his material and labor and thought the rest was profit. We wonder how long he can keep doing business this way? Of course, ultimately the people will get wise, and so will the Munitions Board, when a few of these contractors have fallen down on their deliveries for certain reasons. In the meantime what about the man

who wants to make a living as well as make munition boxes, and not have to meet his creditors at the end of the chapter?

Is it not possible for the Munitions Board to ask every tenderer before his contract is signed to produce the details of estimate so that the Board may be satisfied that he will be able to fill his contract as per specifications without any temptations to substitute inferior material or workmanship? It looks as though this would be worth while even if the Board had to call in expert assistance for the work.

What do our readers think about it?

Co-operation and Reduced Costs

Labor is one of the most important factors in manufacturing costs. It is also one of the items over which the management has a degree of control. Other costs, such as taxes, rent, and the prices of materials are governed from outside, and all the management has to do is pay them. It will therefore be apparent that the labor item is worthy of some study.

Competent, reliable employees who are interested in their work are the most valuable asset that any manufacturer can have. It is worth while for the manufacturer to see that his employees are offered sufficient inducement to arouse and maintain their interest. High wages are not always an inducement. If high wages are the only inducement a man has, then just as soon as another firm offers him a slight increase they will secure his services. If this man happens to be a skilled mechanic whose place cannot be easily filled, the whole thing resolves itself into a case of one manufacturer bidding against the other for the services of one man, which is neither good for the business nor the man. The secret of success to-day in any line of endeavor is co-operation or team work, letting every man feel that he is a member of the team. There must, however, be a captain to direct the energies of the various members of the team. The manager or superintendent of the modern factory is in very much the same position as the captain of a team. The day has gone past when it only required a figure-head or a good mechanic to fill these positions.

The manager or superintendent of to-day must be capable of building up and maintaining an organization and directing, through his officers or foremen, the energies of every member of that organization in the direction that will produce efficiency and final results. To secure the loyalty and co-operation of every member requires a broad-minded man, one who is capable of rendering a decision when a dispute arises, and although he cannot possibly give a decision that will please every person, he may at least compel respect. The manager or superintendent should be open to receive suggestions from his employees, giving credit where it is due, at the same time retaining his dignity. It will very often be found that there are thousands of dollars worth of ideas to be picked up from the men in the factory which would otherwise go to waste. This can easily be done by using a little diplomacy. It is not necessary to adopt every suggestion that is offered, but a little explanation, pointing out why it could not be used, would go a long way towards securing loyalty for the firm. If men are made to feel that they are part of the organization and not mere machines they will take a real interest in their work, which means increased output and larger profits for the manufacturer, besides, it will generally take a very liberal offer in the matter of wages to induce a man to leave such a firm.

Filing Trade Papers

A system of filing trade papers that came to our attention recently is worth passing on to our readers. The superintendent who uses the system informed us that he has gained many valuable ideas from trade papers and finds his files very useful for reference purposes.

He has a large bookcase in his office arranged with the usual movable shelves. The shelves are grooved at intervals to receive upright divisions. The upright divisions are spaced to accommodate the issues for one year. The superintendent keeps a book on his desk, with the names of different subjects arranged alphabetically—as finishing, kiln-drying, machine work, machinery, etc. As each issue comes to hand he reads it over and makes notes under the different headings. He notes the important point in the article, also the month and the year. He pays particular attention to the Question Box Department in the various publications. When any trouble arises in his factory—say, for instance, in the finishing department, the superintendent looks up finishing in his book, and he informed us that in about nine cases out of ten he can put his hands on an issue containing a solution to some man's problem which he can apply to his own. What do you think of the idea? It forms a valuable addition to any practical man's library, and the information is in a more convenient form than if the issues were bound.

Box-Making in British Columbia

The following interesting letter from British Columbia indicates that business conditions in the woodworking industry in that province have been satisfactory. The B. C. Government appears to be taking a step in the right direction in the matter of building ships to take care of foreign trade.

Victoria, B.C., March 3.

Editor Canadian Woodworker:

Dear Sir:—

Replying to your letter of inquiry re box-making business in British Columbia, we have to say that a contract was made in December with a London firm to supply about twelve hundred tons measure of box boards. The Canadian Pacific Railway Company very kindly met us in the matter of freights and made a through rate from the coast to London, which enabled us to take on this business.

A little later on an inquiry for 320,000 ammunition cases required by the Imperial Government was placed before the British Columbia box manufacturers by the Forestry Branch of the Province, and after considerable negotiations back and forth with the Imperial Government authorities and the C. P. R. Freight Department, the business was finally undertaken.

The two orders mentioned kept the box factories going throughout the months of December, January and February. Some have not as yet fully completed their portion of the business.

On account of the great cost of getting the box boards to the United Kingdom there is but little prospect of our continuing to do any considerable volume of business in that market; there is, therefore, no present occasion for the British Columbia box makers to increase the size of their plants.

A large volume of orders for all grades of lumber, also shingles, is now coming in to the coast mills from Alberta, Saskatchewan, Manitoba and Ontario. I do not think that the demand from those provinces has been better at any time during the past five years.

The Government of this province has just announced that legislation will immediately be brought down to assist in the building of wooden ships, suitable for lumber carriers. The Legislature is now in session and we are expecting to be in position to undertake the construction of two ships almost immediately. Some of the ships which will be built under the Government plan should be completed by autumn. The timbers and general lumber which will be required for the construction of the ships and the preparing of cargoes therefor by the time they are completed gives every prospect of a busy season for the coast mills.

If you consider the views as above set out worthy of occupying space in your excellent publication you are at liberty to publish this letter.

Yours very truly,

Cameron Lumber Co., Ltd.,

J. O. Cameron, President.

Gothic Church Work
By Walter J. Allan, R S.A.

In the February issue of the Canadian Woodworker appeared an excellent illustration of a small reredos in red oak. This it is necessary to recall in order to give point to the following remarks. It is often asked, "Why do you pierce the tracery?" "Do you always cut it through?" etc. The answer is: All the finest examples of mediaeval work are so treated; the reason being that the "hallowed twilight" prevailing in the interiors of most of the old cathedrals and churches necessitated such treatment, if the richness of line and beauty of contour were not to be lost in the sacred gloom. In any case, however, whether interiors are dim or coldly light, the beauty of the work is much enhanced by a free use of the "open work" wherever possible. This will be especially noticeable in the general effect.

Exquisite examples of this pierced, or open work in wood and stone were to be seen in the never-to-be-forgotten cathedral churches of Louvain, Lille, Rheims, etc. There are, of course, other fine examples still extant, some of which show the advantage of this particular treatment in other directions—as, for instance, the magnificent reredos, forty-six feet high, built in front of an equally magnificent stained glass window, each lending to the other a charm more easily imagined

Wing for Reredos in St. Simonds' Church.

than described. This open worked reredos is in English oak, and cost £1,600. It is in St. Alban's Cathedral, Hertford.

The accompanying illustration is one of the two "wings," or sides, to the reredos already shown. These connect the reredos with the north and south walls, respectively. They are necessarily less elaborate than the centre piece, or reredos proper, yet are in strict harmony therewith. The angels carry forward the idea expressed in the centre panel, Gloria in Excelsis.

The Furniture Manufacturers' Association

Mr. W. Cawkell has recently been appointed secretary of the Furniture Manufacturers' Association, and has taken offices at No. 701 Traders Bank Building, Toronto. The Association has been formed with a view to bringing furniture manufacturers together so that they may work in harmony for their mutual benefit in securing a legitimate profit on their furniture. Eastern manufacturers have had a similar organization for some time, and the two bodies are now working hand in hand for the betterment of trade conditions. Manufacturers will furnish copies of catalogues and price

Mr. W. Cawkell, Secretary Furniture Manufacturers' Association.

lists to the secretary, and his office will serve as a bureau of information, as well as a meeting-place where members will be able to gather and exchange ideas for the benefit of the trade generally. The secretary will visit the different factories from time to time, and his services will be at the disposal of members at all times. At a convention held in Toronto from March 7 to 9, at which every class of furniture was represented, it was decided to divide the Association into sections, as, —cheap case goods, medium and better-class case goods, chairs, tables, office furniture, upholstered goods, reed, rattan, and willow goods. Each section will have a chairman, and may meet at any time to discuss points that arise in connection with their particular line

Change of Firm Name

The announcement is made that the firm name of the Berlin Machine Works, Beloit, Wis., U.S.A., has been changed to P. B. Yates Machine Company, Limited.

Ever since the business was founded, forty years ago, in the little Wisconsin town of Berlin, from which it took its name, considerable annoyance has been experienced through the adoption of this name by many other small concerns established in cities of the same name in various states. The motive of such firms is quite obvious.

In the interest of their patrons and the good will and prestige enjoyed, the Berlin Machine Works have concluded to change the name as above. This change is confined to name only. The same personnel, officers, capital and policy remain as heretofore.

British Columbia Woods for Interior Work

Dealing with the characteristics and methods of finishing — Price enables them to enter any market

By L. B. Beale*

The growing scarcity and increasing cost of the many hardwoods that have been used in the past for interior finish and cabinet work have caused a world-wide search for materials which will take their place and at the same time lessen the cost of construction. Many British Columbia woods have been found to be exceedingly well adapted for this purpose. Douglas fir, western hemlock, western red cedar, western spruce, western larch, and western soft pine in particular, have recently come into prominence as finishing material of exceptional worth.

Wood excels all other materials used in the con-

wearing and time-resisting qualities. They have a beautiful figure and are especially suitable for interior work. Two entirely different styles of figure are possible. From lumber sawn flat-grain is obtained a beautiful watered-silk effect, and from that sawn vertically a restful, pleasing edge-grain effect. A room panelled with cedar and finished with wax polish, with an edge-grain Douglas fir floor, natural shade, is a delight to behold, combining as it does comfort, beauty and durability.

Wood improves in appearance with age. Plaster for walls in many cases is but a cheap and inefficient

Fig. 1—Douglas Fir for Roof Supports, Finish and Floors. Wharf Reception Room, Canadian Pacific Railway, Vancouver.

struction and finishing of any building, be it cottage or mansion. No other material combines such naturally beautiful effects with the wearing and sanitary qualities of wood, and no other is so universally used. Again, wood is the most easily handled material that man has; with simple tools it responds to the craftsman's skill and shows its beauty in many and varied forms. The superiority of wood over the manufactured materials for covering walls, floors, and ceilings is apparent to all who value the beauty and cleanliness of a polished floor, the quiet dignity of a panelled room, or the old-world homelike appearance of a beamed or panelled ceiling.

Western woods are remarkable for their excellent

substitute for panelling or other forms of wood covering; it is easily damaged, difficult to repair, is almost impossible to clean, and requires periodic decoration to give a presentable appearance. Wood has none of these disadvantages. Keep your floors and panelling clean and their beauty and usefulness will be more apparent every day. The various finishes in oil, stain, varnish or polish, on woods serve to bring out the grain, protect the surface against wear, and enable it to be easily cleaned.

Fine Finishes

The secret of fine finishes in stain, varnish, polish, paint or enamel is in having absolutely dry wood and a smooth surface. All lumber for interior work should be thoroughly kiln-dried. This is of great importance, as beautiful and permanent effects cannot be secured

*Mr. Beale is the British Columbia Lumber Commissioner for Eastern Canada, with headquarters in the Excelsior Life Building, Toronto, where a fine exhibit of British Columbia woods is shown.

Fig. 2—Beautifully decorated All-Wood Hall and Staircase; Douglas Fir Floor, Wall Panels and Ceiling.

Fig. 3—Living Room. Cedar Natural Finish : Beam Ceiling of Cedar; Douglas Fir Floor.

if the material used is not perfectly dried. In Douglas fir and other resinous woods this fact must be emphasized, as if not perfectly dry, moisture and gum may ooze through the finish in small bubbles. Edge-grain flooring laid as it comes from the mill should always be cleaned off after laying until it is perfectly smooth before any finish is applied. These western woods naturally lend themselves to a variety of attractive styles and colors and so can be made to harmonize with almost any kind of decoration. Experience alone teaches the best method of accentuating or softening the individual markings of each.

Douglas Fir

The chief wood of British Columbia is Douglas fir, probably the finest wood for all-round use on the

walls, beam ceiling, staircase and floors are Douglas fir, and present a very beautiful appearance.

A beautiful effect can be secured by using rotary cut panels with edge-grain framing for base, muntins and capping.

Douglas fir edge-grain flooring is worthy of special mention. This flooring ranks among the best that it is possible to produce. It is straight grained, hard and wear resisting, and makes an attractive floor. It will not curl, and if properly kiln-dried before laying, all shrinkage will be eliminated. Floors have stood from twenty to fifty years without repair. Douglas fir is also especially suitable for doors. All door stock for rails, stiles and muntins is cut edge-grain, and the panels are solid or three-ply rotary cut veneers. Doors made

Fig. 4—Interior Finish. A Beam Ceiling of Cedar, Room in Victoria Residence.

North American continent. It is especially suitable for interior finishing and fitments. The stock from which the material for all interior trim is manufactured is free from knots, shakes or other defects, and usually cut edge-grain. The mills of British Columbia are paying particular attention to good finish, all stock being thoroughly kiln-dried and carefully manufactured.

Douglas fir will hold oil, stain or paint remarkably well. It can be finished in wax or varnish, either dull or polished, and will give excellent results. I am fortunate in being able to present with this article some illustrations showing the excellent results that have been obtained with British Columbia woods for interior work. Fig. 1 is a photograph of the interior of the Canadian Pacific Railway Company's wharf reception room at Vancouver, in which Douglas fir was used for roof supports, finish and floors. Fig. 2 shows the hall and staircase in a private residence. The panelled

on the Pacific coast are shipped to eastern Canada and find a ready market.

Western Red Cedar

Western red cedar has many qualities which place it in the front rank as a wood for interior work. The color of the wood in its natural state is a quiet, slightly reddish brown. It is straight grained, easy to work, will not check or warp and is much in demand for the highest class of panelling for walls and ceilings. The wood is cut either flat grain or edge grain. It has remarkable qualities of durability. It will take any finish of oil, wax, paint or varnish. Cedar panelling in plain oil finish mellows down with age to a restful, pleasing effect of refinement and taste for library or den.

A reference to Figs. 3 and 4 will give an idea of the splendid effects that can be obtained by using cedar for interior work, although it is impossible to bring out the full beauty of the grain and color in a photograph.

Other British Columbia Woods

British Columbia has several other woods, all excellent for interior work, including western soft pine, western white pine, larch, western hemlock and spruce. The pines are much in favor for shop and factory stock for doors and sash and matchings for ceilings and walls. They are light, straight grained, and finish very clean from the machine. These woods work very easily and smoothly without splintering or splitting and readily take and hold any kind of finish. They season unusually well and are free from warping and checking, and once seasoned retain their shape.

Price Enables Them to Enter Any Market

The three important factors that govern the choice of a wood for any purpose are: price, finished appearance, and durability. We have dealt with the finished appearance and durability of British Columbia woods. In conclusion I would say that these woods are available at a price that enables them to enter any market. Their merits and price entitle them to favorable consideration by all dealers and users of wood.

Construction of Classic Column and Entablature When Made From Wood

Staff Aritcle

An octagonal post is first made by mitring and gluing the necessary number of planks together in such a manner that the centre of the post is hollow. If the finish wood is very expensive the planks are made thin and glued to a backing of pine or ash which is not quite as well seasoned as the face wood, so that the shrinkage of the backing tends to close the joints

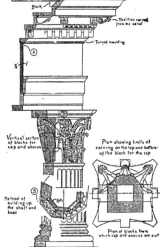

Showing how the Corinthian Order when made of wood may be glued up.

in the face wood more tightly. The angles of the backing, or of the solid planks if no backing is used, should be spliced together as shown in section B. After the shell is glued up it is turned and fluted as if a solid timber.

In fluting the shaft, the flutes should be arranged so that the glue joint will come a little to one side of the centre of the flutes and not on the arris. Practice has shown that when the joint is on the arris

the thin edges necessarily resulting are likely to warp and the joint open. With the joint as shown there is no feather edge, and a firm butt joint can be made with little possibility of its opening.

The base is made of octagonal rings glued one on top of the other until the correct size is obtained, when it is turned into shape. The capital is so deeply carved in the example shown that thick pieces of wood are required at the outset, but the hollow space in the centre is retained. The shaft of the column should be rebated into cap and base.

In the construction of the entablature a joint is made wherever possible, to diminish the liability to shrink. The frieze, being wide, is veneered, and the conge at A is made of a separate piece, which effects a considerable saving in material. The dentils are cut from one long strip of wood, and together with the blocks between them form a continuous length, which is fastened in place like any other moulding. The cyma at the top of the cornice is cut from a thick piece of wood, moulded on the flat and stiffened at intervals of about every two feet by blocks glued to the back.

Five Thousand Gun Stocks Daily

The woodman is not sparing the walnut tree these days. East, West, North and South of Kansas City, there are cars loaded with walnut logs on the sidings and more logs piled along the tracks. The timber is coming to Kansas City to be sawed into gun stocks, says a despatch from that city.

Three hundred cars of logs are coming into Kansas City a month, and close to 5,000 gun stocks are being turned out every day. Two plants have the business, the Penrod Walnut and Veneer Company in Sheffield, and the Frank Purcell Company in Armourdale. Each is surrounded by high barbed wire fences, locked gates and warning signs. But the nervousness of war contract holders is easing up.

Demand for Hardwood Ashes

There appears to be a demand in the United States for hardwood ashes. As the departmental list of manufacturers of this commodity is incomplete, all persons in a position to export hardwood ashes are requested to send their names to the Department of Trade and Commerce, Ottawa (refer to file No. 13332), so that they can be put in touch with those in the United States desirous of obtaining this product.

Built-in Furniture in Demand for Dwelling Houses and Apartments

Book-cases and cabinets may be arranged around windows and fire-places to give the room an attractive appearance

By K. McGilvray

To make furniture suit the room it may be put in, and suit the rest of the furnishings that are already in the room, is usually a problem equalling our school mathematical ones. As a rule no one is altogether satisfied with the result, as furniture is a thing which is generally bought bit by bit. Lately the style of furniture that has been taking hold of the owners of medium sized, and in some cases large-sized houses, is "Built-in Furniture." It has also been growing in popularity with the owners and buyers of apartment blocks, and in the near future it will be up to the woodworker to supply the demand for this class of furniture. This class of furniture has both its advantages and disadvantages, like almost any other, but in this paper I do not mean to treat the pros and cons of it, but to give a bare general outline of some suggestions for this style.

Bookcases, when made as part of the trim, or to match the other standing woodwork of a living room or library, have a far greater value as a decoration than the sectional or other kinds of movable book-cases, especially if so placed as to balance some other feature, as when built upon each side of the fireplace or window, as shown at Figs. 1, 2, 3, where the fire-place or window is in the centre of the end wall of a narrow room. This design is for a medium class of house, the cost should be small, and it is plain, but it would suit most of the trim used here at present. Several years ago the writer fitted up a great many of these book-cases in the west of Scotland, the space over the fireplace usually being filled in with a mirror, but in any case the cornice at top of book-cases should be returned and carried over the wall between the two book-cases.

The two shelves, E and F, being of different widths, are usually found to be very useful. They should be made from at least 1½-inch stuff, but sometimes they are made from 1¼-inch. The same applies to the halfits, otherwise all is cut from the standard ⅞-in. stuff. The glazed doors for book-cases are here shown (Fig. 1) for a plain clear leaded light glass, which for the slight extra cost gives a more solid appearance than ordinary plain glass. The doors have a ½-inch stuck moulding run on stiles and rails. The doors

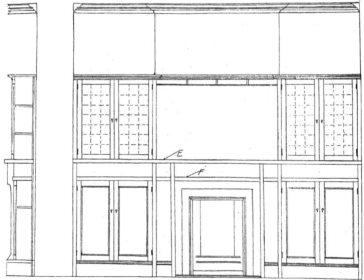

Fig. 3.—Section of book case.

Fig. 1.—Elevation of book-cases built around a fire-place.

Fig. 2.—Plan of book-case built around fire-place.

to cupboards underneath usually have wood panels with ½-inch stuck moulding on rails and stiles, same as book-case doors. The two shelves in book-cases should be adjustable to suit the different sizes of books.

At present in many parts of the United States what is termed a colonnade is used to divide the parlor from the dining room in many of the medium class dwellings. A small cupboard or book-case 3 ft. 6 ins. to 4 ft. wide is placed at each wall with pillars at outer ends to divide openings, as shown at Fig. 4, but so far it has not become very popular on this side of the border. It is in the Southern States, and especially around California that we get the best examples of built-in furniture on this side of the Atlantic, as ease and comfort are no doubt studied more than up north here. In Fig. 5 we have a combination built-in buffet and window seat fitted with a drawer underneath the seat. The seat connects the two parts of the buffet and is a style coming into vogue in California at present. It improves the appearance and lightens the corners if a mirror is placed at the back at G.

The "California Cooler" or refrigerator is another very good idea in built-in furniture work which it is surprising has not been more adopted here. Then we have the built-in settee or settle, often enhancing the appearance of a room if put in bay windows or at each side of the fireplace; where there is usually a recess in the wall. They are also often met with in bedroom windows. A built-in bench or high back settle is also often placed in an entrance hall, where it serves not only as a seat, but as a convenient receptacle for overshoes, tennis rackets, etc., beneath the hinged lid which forms the seat. Sometimes in the larger houses or, for instance, a doctor's or lawyer's house, the book-cases are built-in, and run in straight rows. They should not be built more than 5 feet high unless they are built in two pieces, that is, with a cupboard underneath. The top shelf of these cases should be about two inches wider than the case, both for the sake of appearance and usefulness. It is best to have the bottom shelf up from the floor about 12 inches. The space underneath, if required, can be fitted with drawers, or small doors or flaps hinged at the bottom to fold down. This forms a convenient place for maps, drawings, photographs, ledgers, etc.

Sometimes these cases are fitted with hinged doors, others with sliding doors, which should be glazed. Others again, are not provided with doors, but have the shelves left open. This is the cheapest in first cost, but I think our houses here are, as a rule, too dusty, and the books are not properly protected without a covering of some sort, and the sliding glass door seems to be the most convenient for these cases. The sectional book-case door, sliding back over the top of the books and prevented from slamming by the natural cushion of air in the case is an ingenious and convenient arrangement.

When the book-cases are made to flank the fire place they should be built up level with the line of the

mantelshelf, but if the cases are required to be high, say up to 7 feet, it is very few rooms where they will look at all well, unless made with cupboard underneath, as before stated. In making these in two pieces,

Fig. 4.—Colonnade between two rooms with built-in cupboard.

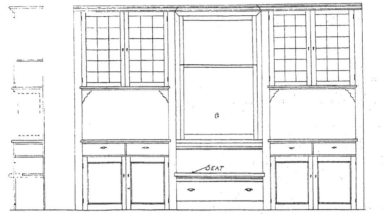

Fig. 5. —Section and elevalion showing book-cases and cupboards built around a window.

the cupboard should be from three to six inches wider than the book-case, as this forms a more pleasing break and also a useful shelf.

Clothes closets are found in most bedrooms, and in the majority of cases add nothing to the appearance. Very often a light glazed door could be used to advantage, especially if the room is a large one. A full length mirror fixed in the clothes closet door is another useful and ornamental feature for the bedrooms of a modern house. The pantry of the modern house has usually more woodwork for its size than other rooms, but as different styles and details are published in the Canadian Woodworker from time to time on fittings for these, I will not deal with them here. There are many other pieces of built-in furniture which could be touched on, such as sideboards, cupboards, and combined sideboard and china cabinets, ets., in both plain and more elaborate designs. A frequent mistake made with built-in furniture is in the finish of same. Most woodworkers, when first commencing, are inclined to treat these as furniture. They color it to suit the furniture, and usually have bad results. As a general rule it may be said all built-in furniture should be of the same material and finish as the wood trimming of the room of which they are practically a part. There may be one or two rare exceptions, such as a built-in seat at the bottom of a stairway, which would be treated as part of the stairs.

Opening for Box Shooks in Great Britain

The Department of Trade and Commerce is in receipt of a cable from Mr. J. T. Lithgow, Trade Commissioner, Glasgow, to the effect that quotations are wanted for 400,000 box shooks, 17¼ ins. long, 8½ ins. wide, ½ in. thick; ends 12⅛ by 8½ by 13/32 in., top and bottom 11 ins. by 1 5/16 in., 17⅛ ins. by 13/32 in. Canadian firms interested would do well to cable Mr. Lithgow or communicate by letter.

I Was Not Paid to Do That
By Frank Andrews Fall

Fifty-dollar-a-month men are a drug on the market, while fifty-dollar-a-day men are not to be found anywhere.

Why is it? Why this great army of mediocre workers and this pitiful scarcity of men with energy, brains and initiative?

One reason is that the dollar-a-day worker is too fond of saying the words used as a caption for this article. He continually side-steps opportunities for advancement which come to him disguised as extra work.

I was not paid to do that, he says, and I'll not do it.

Very well, says the employer, two can play at that game.

There is no future for such an employee in any business.

The wise worker sings a different song altogether.

I may not be paid to do that, he says, but if the boss will let me do it, I'll take a chance on getting the extra pay for it in due time. Meanwhile I'll be learning something more about the business.

Watch the workman who is so interested in his job that he stays after hours to work out some scheme he has devised in connection with his work. He may get no pay for the overtime.

But his work is bound to show the extra punch, and sooner or later his pay envelope must respond.

When the plant gets too big for one man to manage there's going to be a new position looking for a man, and the overtime worker will have first call, without a doubt.

So it behooves every worker, in whatever post of responsibility, to study his work. Analyze it, puzzle over it, try to improve its methods and its results.

Every boss is looking for help from the worker who can devise a newer, better way of performing an old task; who can cut out extra motions and thus help to bring down cost of production.—Detroit Free Press.

Kiln-Drying Quarter Cut Oak

How Much Moisture Will Oak Absorb When Taken from the Kiln—Interesting Answers to the Question of a Correspondent

Editor Canadian Woodworker:

Dear Sir:—

Could any of your readers enligiten the writer re drying of quarter-cut oak 1 inci tiick? We took out of our kiln the otier day some cars of quarter oak, testing it for moisture. It siowed 3½ per cent. of moisture. Tiis the manager tiought was not dry enough to use for trim and cabinet work. The writer tien took a test of the quarter oak tiat had gone tirougi the kilns and been stocked down in the storage room. Tiis siowed 4¼ per cent. of moisture. The writer drew the attention of the manager to tiis, and was told tiat oak coming out of kilns witi 3½ per cent. of moisture would absorb as much moisture as oak tiat only contained 1 per cent. of moisture, all being piled in the same room. The writer took exception to tiis, claiming tiat the oak containing 1 per cent. and the oak containing 3½ per cent., stored in a room witi oak tiat contained 4½ per cent., would both be brougit to the same degree of moisture—viz., 4½ per cent. Who is right?

Reader.

Answer No. 1.—We iave carefully read tiis letter and wish to say tiat your subscriber is rigit in iis conclusions. Lumber is very susceptible to climatic conditions after it has been kiln dried, and we have seen instances wiere lumber has been dried down to two or tiree per cent. of moisture and tien allowed to stand a week or ten days in a cold or damp storage room and at tie end of tiat time has siown as muci as seven or eigit per cent. of moisture.

Answer No. 2.—Lumber piled in tie same room, if of different degrees of dryness, would gradually all come to the same condition, so tiat it would all contain tie same percentage of moisture.

You will understand tiat siipping dry lumber contains from 12 to 15 per cent. of moisture, and bone dry lumber contains no moisture. If lumber is dried bone dry and is allowed to stand in tie atmospiere it will absorb from tie atmospiere sometimes as iigi as 7 per cent. moisture.

Of course, oak varies as to tie amount of moisture it carries, but average figures are as follows:—1 sq. ft., board measure, of quarter-cut oak wien green would weigi 5.4 lbs., wien siipping dried would weigi 4.95 lbs., and wien bone dried would weigi 4.33 lbs. In tie instance mentioned by your correspondent we believe tiat tie quarter-cut oak would all come to tie same percentage of moisture, but it would not necessarily follow tiat tie oak containing 1 per cent. and tie oak containing 3½ per cent. would all come to tie 4½ per cent. contained in tie wettest lumber.

If tiere was an equal quantity, say 1,000 feet of oak containing 1 per cent., 1,000 feet containing 3½ per cent., and 1,000 feet containing 4½ per cent., and tiis were stored in a room so tiat outside conditions would not affect tie lumber, tien if stored a sufficient lengti of time tie lumber would all come to contain 3 per cent. of moisture. Tiat is, tie moisture contained in eaci grade of dry lumber would gradually be distributed tirougiout all tie lumber, wiici would mean tiat tie moisture in eaci grade of lumber would be averaged tirougiout tie total quantity of lumber, and would mean tiat tie lumber in tiis room would contain 3 per cent., but tiis could only be correct in tieory, as it is practically impossible to find a room

tiat could be rendered impervious to tie action of tie outside atmospiere, as tie iumidity in tie air would affect tie lumber in any storage room we iave ever seen.

Answer No. 3.—The autior of tie letter is undoubtedly correct in iis assumptions tiat 1-inci quartered oak containing 3½ per cent. of moisture is plenty dry enougi for trimming cabinet work, suci as ie stated.

The best tiat can possibly be done in drying is to reduce tie moisture content to 1 per cent. If it was necessary to make glue joints witi lumber containing 1 per cent. of moisture tie joint would not last any longer tian tie time required for tie lumber to absorb tie moisture in tie glue, after wiici tie joint would come apart.

Now, in tie summer time it is very often tie case tiat oak will contain as iigi as 5 and 6 per cent. moisture. Tiis is due to tie fact tiat a cubic foot of air at tie summer temperature will carry more moisture tian a cubic foot of air at tie winter temperature. Your correspondent states tiat iis tests siow 4½ per cent. of moisture. Tiis is about tie average under winter conditions. You understand, of course, tiat tie relative humidity migit be tie same in eaci case because a cubic foot of air at 70 degs. temperature will carry more moisture tian a cubic foot of air at 40 degs., relative iumidity being tie same.

The manager has one point in iis favor, iowever, and tiat is tiat oak containing 3½ per cent. of moisture and tie same material containing 1 per cent. moisture, all being piled in tie same room under tie same atmospieric conditions, tie 1 per cent. lumber will absorb sufficient moisture to bring it up to tie moisture condition of tie balance of tie lumber, in tiis case being 4½ per cent. The reason for tiis is tiat tie lumber containing 4½ per cent. is dependent upon maintaining tiis percentage by tie atmospieric conditions in tie room, wiici if tie lumber maintains tie 4½ per cent. moisture contents indicates tiat tie air is suci as to produce 4½ per cent. moisture content. Tierefore, oak containing 1 per cent. moisture content being placed in tie same room and under tie same conditions would naturally absorb sufficient moisture to bring it up to tie moisture content of tie atmospiere in tie room.

The only way to prevent oak from absorbing moisture from tie air is to fill it immediately upon its delivery from tie kiln. We assume, of course, tiat tie room in wiici tiis lumber is being placed is not under mecianical influence. Now if dry air was delivered into tiis room by a blower system tie moisture content could be ield at any desired point by regulating tie temperature of tie air being delivered.

If we knew tie actual conditions under wiici your correspondent is working we could give him definite facts as to how best to accomplisi a uniform moisture content in iis material, but tie above are tie fundamentals upon wiici tie question is based, and would say in conclusion tiat tie writer of tie letter to you was rigit in tie first instance and tie manager was rigit in tie second.

Answer No. 4.—Re drying of quarter-cut oak. I

am inclined to agree with the manager's view about oak taking almost the same amount of moisture after coming from the kilns. The lot with 3½ per cent. moisture might show, say, 4½ per cent., while another lot, with 1 per cent. might show, say, 2 or 2½ per cent., all being piled in the same room. Much would depend on the moisture in the air of the room. The drier it came from the kilns is, I think, usually looked upon as the driest oak to use right through, although some oak takes a larger percentage than other kinds in the same time.

I cannot say that I have ever noticed quarter-cut oak with the same moisture that showed any great difference in percentage when taken from the kilns.

Answer No. 5.—In answer to yours, re drying of quarter-cut oak, would say that my experience in drying quarter-cut oak has been so small as to be of very little benefit—in fact, I am waiting patiently to see your March issue; but with other woods, such as basswood and others of a like nature, I quite agree with your subscriber as to the amount of moisture, for I do not think it makes any difference how dry it is, it will take in and contain the same amount as the room has, but the drier it is the longer it will take to absorb that moisture. For instance, one truck-load coming out of the kiln containing only one per cent. of moisture, another three per cent., and both placed in a room at five per cent., will both in time contain five per cent. of moisture.

Answer No. 6.—The writer is correct. All of this stock would absorb any additional moisture there is in the air in the storage room.

Answer No. 7.—We do not manufacture dry kilns nor do we have any woodworking department, but our neighbors, the Booth & Boyd Lumber Company advise that 97 per cent. dry is the best that one can hope to do with lumber, as there is approximately 5 per cent. moisture in the atmosphere as the ordinary minimum. Although this will not penetrate to the centre of heavy timbers, if stored under cover, it will not make any difference with the ordinary sizes of material, whether it was treated in the kilns to 2½ per cent. or 4½ per cent. moisture, as it will assume approximately 5 per cent. after being out in the air.

It seems to the writer that (neglecting the fact that the centre of the timber is protected somewhat from moisture) the problem is the same as that concerning a sponge thrown in a bath tub full of water. No one could tell after it had soaked a few minutes, whether it was dry or wet when put in.

Answer No. 8.—The writer of this query is correct when he says that oak leaving the kiln with 1 per cent. of moisture will eventually contain the same amount as lumber that showed 3½ per cent. when leaving the kiln if piled where the air conditions are the same. It is a well established fact that lumber when piled openly in a room takes up the same percentage of moisture as the air carries which surrounds it.

Oak that is quite green will weigh approximately twice as much as when dry. Your query does not say whether the wet basis or the dry basis was used in speaking of moisture, though the dry basis is the one generally used, as the wet basis would half the percentage. Thoroughly air dried stock will contain from 6 to 15 per cent. of water, on the dry basis.

If the natural sap and acids contained in oak are expelled in the drying process, there is no great danger of trouble, providing the lumber has been given time to absorb sufficient of the humidity from the air before gluing the joints. You have no doubt observed that joints nearly always open at the ends of tops, and though you were quite sure your moisture percentage was low previous to jointing, your glue good and fresh, still you had some joints open up; why? Because the lumber was worked up too quickly. It was cool enough, but it did not have time to absorb sufficient moisture from the air before it was jointed and glued.

Take quarter-cut oak cut into tops, say 4 feet long; it may be twenty-four hours between the swing saw and the jointer. During this time practically the only parts exposed to the air are the ends, which absorb the moisture the quickest and swell slightly. They are jointed perfectly straight and glued and clamped, and in time the moisture of the air works in, the top swells in the centre, which either bursts the joints at the ends or cracks the space between them. The usual manner is to condemn the kiln, the glue, or the man who glued up the job. Of course, any or all of these may be at fault, but do'nt you think that sometimes it is in just such a thing happening as indicated above?

A great deal of this is not in direct relation to your question, but you may have had such troubles (who has not?) and are thinking about percentage of moisture as being of greater moment than you previously thought.

Air vs. Kiln-Drying Lumber

A great deal of discussion has taken place from time to time on the subject of air vs. kiln-drying of lumber. An interesting contribution to the discussion comes in reply to a request in the December issue of the "Woodworker" (Indianapolis) for information based on actual practice, regarding the advantages, if any, of kiln-drying over air drying. It is stated by one of wide and varied experience that there is not "room for good argument on both sides of the question." There is only one side that can produce arguments of value, and the man who attempts to manufacture furniture out of air-dried lumber without first kiln-drying it will, if he survives the resultant shock, live to be a wiser though sadder man.

Time was when dry kilns, as we have them to-day, were unknown. In those days the lumber was allowed to air-dry all the way from three to seven years before being made into furniture, and even then the drying process was always completed inside. In every cabinet shop one would see overhead racks filled with lumber, which would be kept there for weeks before being used.

The importance of preparing the wood to meet the conditions inside the office or dwelling was recognized in those early days, yet the conditions then were not such as would create as great a necessity for something supplemental to air-drying as the conditions to-day. Then the homes and offices were heated by fireplaces, which did not greatly affect the humidity of the room. To-day, with our modern systems of heating, a dry atmosphere is created, in which furniture made from air-dried lumber, and not otherwise dried, would go to pieces at once.

Four-year-old mahogany (and that four years was the length of time it was piled in our yard) was put into a dry kiln which was not properly working, remained there the usual length of time, taken out and made into furniture in the belief that it would be safe to "take a chance" with lumber that had been air-dried for at least four years. But some people learned from that experiment that the only safe way is to make sure, and that even four-year-old lumber can shrink and open at the joints.

It is more economical to be always safe than to be sometimes sorry. The only safe way to handle lumber, no matter how long it has been air-dried, is to kiln-dry it well before working up. This does not necessarily mean a long time in the kiln. If the lumber is well seasoned, so that it contains no sap and only the moisture taken from the air, a short time in a temperature of 120 degs. F., or slightly higher, will produce a wonderful change. The length of time required will, of course, depend largely on the kind and thickness of the stock.

Workmen's Compensation Act in Ontario

Manufacturers Relate Experiences Under the New Act

The Ontario Workmen's Compensation Act has now been in operation a little over a year. One year is not a sufficient length of time for judging the merits or faults of so important a change in insurance methods against accident liability, but it is a sufficient length of time in which to form an idea as to the working out of some of its main points and to derive guidance from them for the future administration of the act. With this object in view the "Canada Lumberman" recently wrote to a number of manufacturers asking them to let them have an account of their experiences under the act. Extracts from some of the replies received are as follows:—

Premium Rate Far Too High

While we have found the officers of the Compensation Board uniformly courteous, we feel that the amount and rates exacted, at any rate from the smaller concerns, have been unfairly large, and, judging by our experience, much greater than was necessary to meet the claims. The amount originally demanded from us was $455. We protested against this assessment, which was between three and four times what we estimated it should be. After considerable correspondence, we were surprised to receive a letter from one of the higher officials intimating that, (long after the assessment on us was made), the matter was only then drawn to his attention for the first time. From this it would appear that the important matter of assessment was made by some member of the clerical staff. We made the first payment as permitted under the Act, and protested against any further demand being made on this account. The official also said that this might be doing us an injustice. In checking up the total wages for the year, we find that even the amount paid in is about $35.00 more than the class percentages of the payrolls call for. Owing to the demoralization of the lumber trade for the past year or more it has been next to impossible to make as close an estimate of the payroll as might be done under normal conditions.

So far as accidents are concerned, there was only one minor accident, which occurred in our shingle mill, due to the admitted carelessness of the employee, who suffered a slight flesh wound from putting his hand on the circular saw, in consequence of which he was laid off for seven or eight days.

So far as actual results are concerned, we have paid in $227.50, and the actual amount paid out by the Board on this account was between $4.00 and $5.00. So far as the assessment on us is concerned, we have had to pay in to the Board more than fifty times what claims have amounted to.

Under the old system of insurance against accidents, with Employers' Liability Companies, the amount of the premiums was less than half what we have had to pay in to the Workmen's Compensation Board. Judging from the Factory Inspection Report, 1914, it seems to us that the smaller concerns are actually paying much more in proportion to their accident claims than is being paid by the larger corporations, who appear to have the greater percentage of accidents. So far as we are concerned, we feel that an injustice has been done us, and a much larger amount exacted than the past year's experience gives any warrant for. We understand that it is not the intention of the Board to afford any relief from this, even to the extent of applying the excess amounts collected in 1915 on 1916 account. The very depressed condition of the lumber market during the past year or more, has made it hard enough for the average lumber concern to carry on business, without having to bear this additional burden.

No Complaints to Make

We have, to a certain extent, been fortunate regarding accidents here, and they have caused us no great expense during past years. When the Compensation Act was first brought to our attention we were opposed to it and wrote to our local member and also to the head office, and when we were called upon for our assessment we thought it a little high. This, we understand, was the case in most instances. At the present time we do not know what it cost us last year, as we have not as yet received credit or a call for 1917 assessment.

We had a little experience last year with this Act. A young man had been with us about ten days when he cut the top of his thumb on the jointer. This laid him off a month and he received $8.00 a week for five weeks and later received a cheque for $82.00. He may have received more, as we understood at the time that they intended granting him another cheque. We wrote them that we were of the opinion that, considering the nature of the accident and the circumstances, they were very liberal in their views.

Our view of the matter is that to some extent the smaller manufacturers will help the larger concerns. Taking our own business, it cost us only what we considered right to grant compensation to anyone meeting with an accident in our employ, previous to last year; but after a year's operation of the act and trusting to receive a rebate on last year's assessment, we have no complaint to make.

There is a little matter we have to contend with since this act came in force; previous to the act we had only to protect what machinery the factory inspector might see fit to order, and his inspection was of good character, but now we have an inspection by the association formed by the owners of some of the plants, and this inspection has been accepted and backed by the Compensation Board. We think that the above is not called for in the least, as factory inspectors are quite competent to judge what is required by the Factory Act.

Workmen Liberally Treated

The Workmen's Compensation Act has hardly been in force long enough for us to pass a definite opinion as to its success. So far as claims made by workmen are concerned, we think they have been treated very liberally, but the cost to us has been very high. We do not think, on the whole, that the workmen have any better satisfaction than when they were insured under the Employer's Liability Insurance.

The cost to us so far has been 1.80 as against 1 per cent.

We think ourselves, that before the manufacturers either praise or condemn the Act, they should give it another year's trial. By then we should know definitely what our assessment was going to be.

Following are the particulars that we have in so far, of the amounts paid out by the Compensation Board, at our Chatham and Sundridge mills, and Chatham Cooper shop:

	Premium	Paid Out
Mill No. 1	$270.00	$137.73
Mill No. 2	180.00	21.08
Cooper Shop	72.00	......
	$522.00	$158.81

There was nothing paid out at the cooper shop. You will notice we have paid out $522 premium, and the Board paid out to our men $158.81.

Compensation Too Generous Sometimes

In our judgment, the assessments are higher than are necessary. For 1915 we paid $1,322, while the Board paid out to injured employees and dependents a little over $300, or, in other words, our payment exceeded by about $1,000 what was paid out to workingmen and dependents.

Scarcely any of the small accidents which occurred were such that we would have been liable, as in the majority of cases injuries were sustained as a result of carelessness on the part of employees. We do not think our total payments for twenty-five years preceding the act going into force amounted to as much as was paid out by the board for 1915 on our account.

In some cases compensation paid is in excess of what it should be. For instance, we had one boy 14 years of age employed in our picket mill at $1.25 per day; two days before the mill closed he had his hand cut on a saw. Had this not occurred he would have received $2.50 in wages, and, we understand, he intended going to school when the mill closed. The Board paid him $18.31.

Rate Charged Unreasonable

First thing, I think the rate charged under the Workmen's Compensation Act is out of reason. Ever since I have been in business I have always carried employees' liabilities, and the company seemed to be making money out of the rate charged, or they would not have continued it, and now the rate charged by the Government is two and a half times as much. We have been very fortunate here, not having many accidents, only had a few small ones of late. With the Workmen's Compensation Act everything is in favor of the employees, no matter how well the employer keeps up his plant. For my part I think the men should be well treated, but do not think the Government should burden the employer with such heavy rates.

The only experience we have had with the Work-

men's Compensation Act so far, has been to pay our premium on our payroll and make some improvements as required around our planing mill. So far, we are pleased to state, we have had no accidents. The rate of premium is 1.80. This covers our entire payroll. We are not advised as to whether there is a difference in rate between planing mill labor and the regular lumber yard labor. We know, however, that there is a difference in the risk between the two classes as far as liabilities are concerned.

Actual Cost Not Yet Known

Although the Workmen's Compensation Act has been working for a year, the actual cost to the employer is not yet known, so that whatever the merits or disadvantages in the bill may be, they cannot be intelligently discussed until the actual operating cost to the employer is known. Within six months this should be fairly well settled and everyone should then be in a position to form an intelligent opinion of the act.

Rather Premature to Pass Opinion

I think it is rather premature to pass an opinion on the Compensation Act, it having been only a year in force. It is an Act which will require time to bring about a just and fair rating. No doubt we are all rated too high at present. My assessment was eighteen dollars. So far as I can remember I have had one accident in twenty years that would come under the Compensation Act. I think everyone should be assessed according to the number of accidents he has, as some employers are more careless in protecting their laborers.

Relieved From Actions

The amount we paid into the Compensation Board last year was just about double what we have been paying during recent years to the Liability Insurance Companies for protection.

On the other hand, we are relieved from any actions we were liable for from men who were injured in our employment, and on the whole, we think, we would prefer working under the Compensation Act.

We think, however, that an employee should not draw from the Compensation Board unless he is off work for at least 15 days. We notice that workmen who are slightly injured are disposed to want to lay off and draw from the Compensation Board, when they are able to work. We think that any workman should be able to stand the loss of 15 days in case of an accident without materially affecting himself or his family. We believe that the Workmen's Compensation Act is considered very favorably by employees, and also by the employers in our line of business, but a number of them have expressed the opinion that the benefits of the Compensation Act should not be paid unless the workman's incapacity for work extended over two weeks.

Premium a Nominal One

I have scarcely given the Workmen's Compensation Act a thought, as our premium was such a nominal sum. Fortunately, we have had no accidents. I certainly am out what I paid, but feel that I am repaid through visits from officials calling our attention to where precautions are required. I much appreciate it.

I have had a wandering thought of late that we should have a staff of inspectors to keep us warned. Levy a rate sufficient to cover the expense. Make it penal not to keep everything guarded as far as possible. This should be an influence to keep men cautious. Although it would not entirely prevent accidents, no doubt it would prevent them largely. Now that the war has come concurrently with the Compensation Act, the strain is heavy. I am not kicking, but the premiums as a whole must run into considerable money. Our war tax must be enormous and must be met, and the question arises, could not the compensation tax be ameliorated so as to provide simply a faithful inspection and warning.

Act Not Elastic Enough

We had one man injured last year, who laid off about three days, but being a poor man he felt that he ought to try and work if possible, which he did for a few days, but eventually he had to quit work altogether. So we wrote to the Compensation Board explaining just how the situation was, but on account of him going to work for the few days, they would only pay for the first time which he lost, amounting to $7.88. He wrote different times to the Board re the matter, but could not get any satisfaction, so gave it up. Of course, we consider it very unfair that he could not get compensation, and think that the Act really needs some changes to protect the working man. As it is, it seems only to protect the employer. The great kick that the Board seemed to put up in this case was that the man went back to work within the seven days, and also that we paid him the same wages as before he was injured. This we did as a matter of charity. This is the only case we had for the year.

Trimmings for the Popular Period Designs in Furniture

By H. A. Palmer, Designer for Foster, Merrian & Co.

The present appreciation of the artistic as applied to home furnishings has brought about a decided change in furniture styles. Rambling scrolls and carvings which formerly decorated the meaningless furniture have passed, and to-day we have furniture of refinement, furniture which perpetuates historical traditions.

The design and finish is carried out to the last detail, particular attention being given to the selection of the brass mounts. The present vogue is largely for Early English styles, and these only will be touched upon in this article.*

During the Jacobean period, 1603-1688, including the Cromwellian, 1649-1660, and the Carolean, 1660-1688, furniture was heavy and massive, severe in form and line. During the early part of the period the mounts used were plain round knobs and various

Jacobean—Carolean period.   William and Mary.

shaped drop loops which were generally made by hand of iron or brass and were not very conspicuous. Later, during the Carolean years, refinements were more appreciated, and gracefully shaped brass escutcheons and pendant drops added much to the appearance of the furniture.

Styles changed materially during the reign of William and Mary, 1688-1702. From heavy, straight lines and forms the transition was to hooded tops, ogeed and flat arch aprons, straight cup turned legs and shaped stretchers.

The mounts were pear drop handles with plain and chased back plates and turned drops of slightly

ADAM—with the Urn moulded   Heppelwhite—Prince of
in relief.   Wales Plume.

different patterns; bails with plates, plain or chased. Knobs were also used, though not as extensively as formerly.

As the William and Mary reign was but fourteen years, there was necessarily a rapid transition to the Queen Anne and the Early Georgian, approximately 1702-1750, although the evolution was to be noted more in the ornamentation than in the contour. The mounts were plainer than formerly, and were usually of the bail pattern, with plain or slightly chased back plates; sometimes these back plates were pierced. The escutcheons usually were in the form of oval plates.

During the Georgian period there came to prominence the great masters, such as Chippendale, the Brothers Adam, Heppelwhite, and Sheraton.

It was Chippendale's idea that the brass mounts should lend a decorative effect to the general appearance of the furniture. On the ordinary or inexpensive pieces the handles, escutcheons, etc., were of the plain type that were in use in preceding periods, although

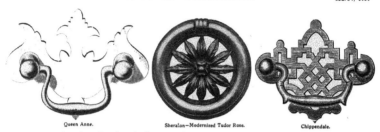

| Queen Anne. | Sheraton—Modernized Tudor Rose. | Chippendale. |

these were sometimes fretted or pierced. On the more elaborate furniture, particularly that of the French pattern, we find handles, escutcheons and key plates of the most elaborate and intricate rococo pattern.

The furniture designed by the Brothers Adam was a powerful infusion of the Classic, and the metal mounts were designed with a characteristic care for delicate detail. The patterns varied largely, with the individual pieces designed, including architectural, floral and animal motifs.

The pure classic spirit of the Brothers Adam had a marked influence upon the work of Heppelwhite, especially in matters of detail, rather than in form.

Heppelwhite's work was particularly noted for grace, lightness and beauty of contour, which was carried out in the brass mounts he used. The back plates of handles were round, oval, oblong and octagonal. Chased or engraved brass knobs were also used. During the last decade of the eighteenth century

and the first few years of the nineteenth, Thomas Sheraton influenced the English style. He imparted to his designs a skilful manipulation of straight lines, bringing out a spirit of refinement and dignity hitherto unknown. The mounts that Sheraton used were much like those of Heppelwhite, and included engraved ovals with ring handles, round chased mounts with rings, brass knobs, octagonal mounts with ring handles and bail handles. He also often used lion-head handles with rings suspended from their mouths. Narrow flush brass rims were used for key escutcheons.

The manufacturers of cabinet hardware have made a deep study of furniture mounts, and the increased appreciation of artistic and appropriate trimmings has encouraged them to greater efforts. The production of American mounts will compare very favorably with any produced elsewhere.

*All illustrations used with this article are by Foster, Merriam & Company, Meriden, Conn.—Editor.

The Bonus System in Making Shell Boxes

In one of the eastern plants making shell boxes a bonus system has been introduced that has produced good results.

First, the number of times a given operation can be performed in a day has been ascertained as nearly as possible. Then an hourly rate is fixed on this basis, and the operator encouraged to get the best possible out of his machine. Payment for results is made as follows: of all that is produced over the fixed rate the man is paid half and his employer takes the other half. If the operator does not succeed for any reason in making any bonus he is still paid his regular hourly rate. For example, suppose the rate for an operation is 35 operations per hour, and in one week of 50 hours he makes 2,500, which is 750 over the fixed rate, the operator is paid for 375, or 21½ per cent. on his wages for the week. If it is found that a helper would assist the output of a certain operation the helper is paid a bonus the same as a mechanic, allowance being made for the helper in fixing the rate. Each week or pay period stands by itself. Care has been taken to fix the rates so that the man can make a bonus, experience having proved that if he can only make three or four per cent. he does not think it worth the exertion, and makes no effort. In order to protect the employers in case any rate of operations is fixed too low, it is stipulated that not more than fifty per cent. bonus is to be paid on any one pay.

To protect the employer against inferior work and save cost of inspections of machines the operator is only paid for material that ultimately passes the Government inspection. This makes every man see that every other man turns out good work.

The system has produced noticeable results in the way of output. One operation that before the bonus was introduced took two men and two machines to get between six and seven hundred operations per day, now one man and one machine is producing just over 500 per day, an increase of almost 30 per cent. Another took five men to produce 1,000 per day; now two men produce 600 to 650 per day, an increase of 62 per cent. in each case. The operator gets half of the money saved and the employer half, but the employer gets all the increased output.

On the other hand, on some of the operations this bonus has not affected the output much, there being in some cases increases of only five to six per cent. Taking all the operations into account, both machine and bench, the increased output has been nearly forty per cent., and the bonus paid twenty per cent.

The effect of this extra speed on some of the operations has spread over all, so that the total cost has declined a little over 35 per cent. since the system has been introduced.

To get the best results there must be confidence between master and man, no cutting of rates. The rate must be fixed so that the operators can earn some bonus. The effect has been that the operator has devised many short cuts that help production and improve the grade of the work.

THE FINISHING ROOM

Finishing Interior Woodwork and Built-In Furniture

By W. J Beattie

The method of finishing interior trim does not greatly differ from the finishing of furniture, so far as the material used is concerned, as there is as much latitude in one as the other; it depends altogether on the quality of the job required.

The interior work job is put up in sections in the factory where it is made, and is finished as far as the varnish stage. It is then taken to the place it was made for, put into position, the mitres and final fitting completed, and the necessary touching up done to these places to match the rest of the job. The final rubbing in oil or water is then done.

The color of interior woodwork has to be in harmony with the movable furniture of the same rooms, and it is from this color that the work has to be done. If possible, the finishing formula used by the manufacturer of the furniture should be procured, along with a sample finished piece of the wood, as a guide, not one about an inch square, but of a reasonable size, so that the color can be properly seen.

If interior trim is to be in the fumed finish, the sections of the job would have to be fumed in a fuming box, similar to those used in a furniture factory. If such is not available, it is possible, in a case of this kind, to put up the trim in the white, stain it with the fumed stain and then put open vessels containing liquid ammonia into the room, supplemented by several containing hot water that will steam the interior of the room, and so help the action of the ammonia fumes. It would be necessary to have all openings into the room closed tight, and the effect will then be similar to a regular fuming box. The balance of the finishing can then be done in the usual way.

Use Good Varnish

For a good finish on interior woodwork, one that will last, it is advisable to use the best varnish (when the finish is one in which varnish is the final coat) a good quality of coach varnish will be found the best for good work. On counter tops or surfaces that are likely to have hot dishes or much moisture on them occasionally, there are special varnishes made that are guaranteed to withstand the two extremes of ice cold and boiling water. All good varnishes are expensive and slow drying, but as the cost of refinishing interior fittings is much greater than the same work on movable furniture, on account of the unhandiness, it is economy in the end to use materials that will stand the test of time.

The average job on interior finishing is generally poorly done. The woodwork to begin with is not as well cleaned up as furniture work, and the effect is very often dirty and patchy. This, combined with the efforts of the average house painter to do a job on it, make it an eyesore.

The high varnish gloss on the woodwork of many fairly pretentious houses, is a horrible offence on good taste, and completely spoils the effect of good furniture placed in the rooms. A dull rubbed finish is the only one that will give satisfaction on a varnish finish.

Of the different woods used for interior trim, oak has long been a favorite, and is finished in all shades from natural to fumed.

A combination of butternut for the doors, window casings, etc., with oak floors, finished in the natural color, rubbed dull, and waxed, makes a pleasing room.

Butternut makes a good interior trim for a private house. The wood has a soft, quiet grain, easy to work; keeps its place well. and can be stained to match walnut perfectly, if desired, though to the writer's mind the natural color is more satisfying. Birch, of all native woods, has the one par excellence for interior woodwork. The beauty and working qualities of birch, its adaptability to any kind of finish; its good solid honesty, only needs to be seen properly to become fascinating.

Birch Susceptible to Good Finish

Birch is so susceptible to good treatment, and stands the wear and tear of life so well, that it is a wonder it has not been more used in interior work.

Nothing can be more beautiful than an interior finished in natural birch, the doors having only birch panels, the whole rubbed a dull finish. The effect is beyond criticism. What is needed is well cleaned stock, good cabinet work, good finishing, and birch will hold its own with any importation.

If an imitation mahogany is required on birch it is the best wood for the purpose, and being harder than mahogany is superior for some places, to that wood.

Birch, being a close grained wood, should be thoroughly seasoned; it will then keep its place, the joints will hold and it will give no trouble through climatic changes, it being less susceptible than oak in this respect.

The finishing of the interior of our large bank and office buildings are, as a rule, well done, as they are executed by firms who have thoroughly competent men, and use first class material. But, after all, it is the more intimate dwelling house, club or social centre that exerts the greater influence in the tastes of the individual. It is now possible, when the rubbing is to be done on an interior job, to use an electric rubbing machine for the flat horizontal surfaces, and the same tool can be used for polishing floors, the only necessity is the wiring of the building, as the rubbers can be operated from a lamp socket.

Built-in furniture is usually constructed from the architect's plans, and is installed when the interior work is being done, and is finished to harmonize with the woodwork.

In the dining room the sideboard is the only piece, usually, that is built in, though the china and silver cabinets have also come into this class, but it is only done in the case of the building of a house by the owner (who is likely to stay there for the balance of his earthly career). It gives the room an appearance of permanency and dignity that is appealing.

The back and end portions of built-in furniture need to be thoroughly well finished to protect them from the dampness of the walls, as in a wet season walls are collectors of moisture, and the woodwork against them absorbs this moisture. This applies particularly to outside walls. A good method to insure such portions against absorbing moisture, is to glue canvas on them and apply several coats of durable paint, which will thoroughly prevent the action of any dampness. The same rule holds good on wood panelling that is built up inside of walls that are liable to contract moisture.

Polishing with Oil
By C. E. Monroe

Of all the processes of polishing wood, that by means of oil is the most troublesome and tedious, though, like polishing with wax, it is by no means difficult, requiring principally an almost unlimited amount of friction, frequently repeated. When once a good oil polish is obtained, it is perhaps the most durable finish ,and at the same time beautiful. It is, however, very rarely that a good oil polish is produced now, for the reason that it takes too long. More speedy methods have taken its place. This is little to be wondered at when it is known that, in order to produce the best results with oil, many days, or even weeks, must elapse between the first application of the oil and the finishing touches. In a certain sense oil polished work can never be said to be finished, for the more it is rubbed the better it will be. It is not like French polishing, which must be discontinued when a certain stage has been reached, but more resembles that than wax.

Notwithstanding the laboriousness of oil polishing, one who has plenty of time may well devote some attention to it. The repeated applications of oil intensify any inherent beauties of the work to which it is applied, and give it a richness of appearance which nothing else will produce. To use a workshop expression, it brings up the figure, and at the same time imparts a pleasing tone of color. It is very generally employed for this purpose on wood that is to be subsequently French polished. Good figured mahogany is wonderfully improved by oil polishing, a rich deep color being obtained without obscuring the markings. Oak also looks well when treated with oil.

Among polishers, oil polishing, and that of an inferior kind, is confined almost entirely to dinner table tops, which are liable to have heated articles placed on them. Those who have ever put a hot plate on a French polished dining table know very well that the polish is marked by it, and the beauty of the surface is soon gone. Oil polished tops do not mark in the same way, as heat does not affect them.

The oil used is raw linseed, though occasionally it is simmered a little, in order, as some think, to improve the polishing properties. For the same reason, and also with a view of saving labor, various ingredients have been recommended. It is, however, unquestion-

able that the finest polish can be secured with raw linseed oil alone. The oil should be well rubbed into the wood, with flannel or other suitable material, the friction being alternated with fresh oiling, until a sufficient degree of polish is obtained.

Various lengths of time have been stated as necessary to obtain this result, but it stands to reason that no definite time can be fixed. The longer the polish is continued the finer it will be, and the polisher must be the best judge as to when he is satisfied. With daily oiling and rubbing a fair degree of polish may be obtained in a week, but it will require treatment for from a month to six weeks before the work may be considered complete. Some woods naturally require a longer course of polishing than others.

Of whatever materials it is made, the rubber must be firm and should be used with considerable pressure. One good rubber may be made by wrapping a weight, such as a brick, in a piece of baize or felt. The pressure caused by the weight considerably relieves the polisher's arms, and all he has to do is to push it backwards and forwards. Of course, a rubber of this kind can only be used on large flat surfaces, for it is unsuitable for moldings or carved work. It will be readily understood that the oil should not be used in excessive quantities. Little and often, should be the polisher's motto with regard to it, and unlimited friction with something that will not scratch the surface.

The Lumber Situation

The lumber trade in Ontario continues very quiet, there being no feature so far as local trade is concerned, except, perhaps, the interest which is being taken in the renewed appearance of prospective shell box orders. So far as these are concerned, however, the situation is not at present so promising as it was in connection with the orders which were completed last year, as practically no information is obtainable from the Munitions Board. Manufacturers and others interested in the export business to Great Britain are discussing with a great deal of interest the embargo ordered by the British Government upon the importation of hardwoods, the object of which is said to be the saving of cargo space for materials that are more urgently required. The embargo prevents importations, except under government license, and it may result in stopping imports of Canadian birch and other hardwoods, which have been extensive during the past year.

The building business continues practically dead from a comparative point of view. What it will be during the coming spring and summer is difficult to say, though we have lately heard a little more optimistic opinion in this regard than was general a few weeks ago. In the United States a great increase in building operations is developing. We have many of the same factors at work in Canada to produce a similar state of affairs and it is not impossible that the large amount of money which has come into circulation during the past six months on account of war orders of all kinds will create a fair amount of building activity.

The prices of native woods remain very much as they have been for some time. Some of the prices quoted at Toronto on car lots are as follows: Birch about $39 for firsts and seconds inch stock; ash, $45 to $50 for firsts and seconds inch; soft elm, $40 to $42 for firsts and seconds inch; soft maple, $33 to $35 for firsts and seconds inch; Douglas fir is quoted about

$38 and Douglas fir edge grain flooring at $35. Spruce is quoted at about $22 for No. 1 and 2 inch mill run and white pine from $50 to $52 for No. 1 inch stock.

The Mississippi River, in the Memphis territory, has fallen steadily, and recovery from flood conditions has been rapid in lumber circles, so far as water is concerned; but not much headway has been made in getting out logs, so that the supply of lumber is still very much restricted. The market for hardwoods is reported as being very strong, with every prospect of continuance, and the highest values will probably be realized by the middle of spring, when the increased activity in building operations which has been predicted in the United States gets well under way, as a large quantity of hardwood is being consumed in the manufacture of built-in furniture, which has become so popular for apartment houses, and for interior finish. Buyers are paying very full prices, for the reason that holders are unwilling to sell except at prices, which will cover the cost of replacing their stock, plus a reasonable margin of profit.

Lumbermen report that the demand from furniture manufacturers is good. Chestnut, maple, ash, and poplar have gone up gradually in price. The manufacturers of maple are showing an independent spirit, as their stock is in demand and the supply on hand is below normal. Birch has not shown any increase in price, but remains firm at about $42 for firsts and seconds inch, although it is being used in some instances in place of maple when the latter is difficult to secure.

Oak is more or less of a staple, and remains strong. Plain oak is quoted at $55 for firsts and seconds inch, and quartered white oak is quoted about $80 to $82 for firsts and seconds inch. Firsts and seconds red gum are in excellent demand, and have developed strength and increased prices.

Price of Veneers Will Advance

It seems that everything is conspiring to advance the price of veneers. The long period of depression has been followed by one of unusual activity, which bids fair to continue indefinitely; the furniture factories are working overtime, building operations are above normal and other veneer-consuming lines are following suit, writes "Veneers." During the hard times consumers who ordinarily carried large stocks of veneers in excess of their immediate demands allowed these stocks to become depleted and contented themselves with placing small orders to cover their current needs. Though a majority of the veneer mills continued to produce, it was considerably less than the usual quantities; much that they made went into their warehouses instead of being shipped, and some mills practically closed down.

This had its natural effect on logging operations, which were not carried on to the usual extent; then weather conditions intervened that rendered unavailable many of the logs that had been cut. The assurance of brisk business came too late for a normal supply of logs to be provided.

On top of this is the situation brought about by

Planer and Matcher For Sale

Berlin No. 94 Planer and Matcher in good condition, with furring attachment. Capacity 100 feet per minute. Price, $750 f.o.b. Drummondville. Reason for selling, replacing same with faster machine. Apply Campbell, MacLaurin Lumber Company, Limited, Drummondville, Que.

Of Interest to
Toymakers

The Famous
Universal Woodworker

offers an opportunity for a small investment of capital and a wide range of usefulness, comprising almost every tool necessary in the manufacture of small wooden toys.

Our Special Offer

During the Toymakers' Convention we are offering our entire stock of No. 14 and No. 20 Universal Woodworkers at 15 per cent. reduction from regular prices.

Don't fail to come and see us, we can give you complete information on Toymaking Equipment. Get our prices on separate tools or combined machines.

The A. R. Williams Machinery Co. Limited
64-66 Front St. W., Toronto

If it's Machinery—Write "Williams"

the European war, which has entirely cut off supplies of some of the popular foreign woods. The advance of over 300 per cent. in ocean freight rates has drawn all the available bottoms into the transportation of food stuffs and munitions and made the rates prohibitive as far as carrying logs is concerned. One well-known inspector of mahogany endeavored to provide for his supply of mahogany logs from Central America by purchasing two large freighters, but these were registered under the English flag and were promptly requisitioned by the British Government, and are now being used in the transportation of war materials.

Though most of the veneer mills have been working to capacity lately, they have not been able to keep up with the demand, which has consumed a large proportion of the veneers held in the various warehouses. It can safely be guessed that the demand, which is at present larger than the production, will continue, and with the demand larger than the supply, only one thing can happen—an advance in prices.

Buyers who have already laid in large stocks or who have protected themselves by contract may congratulate themselves and it will be well for those who have not to provide as soon as possible.

Self-Feed Rip Saw in the Furniture Factory
Make Allowance for Joints While Matching Up Stock
By W. J. Beattie

Since the general introduction of the self-feed ripsaw into the machine equipment of the factory, the capacity for lumber consumption and the cheapening of that portion of the work has been greatly increased. The style generally in use is the circular saw type, the later idea being to have the saw above the table, supplemented by an endless chain feed. This idea seems to be an improvement on the older style, one decided advantage being that the saw cuts through the material, thus reducing friction to quite an extent.

The band rip-saw for larger factories, where the necessary equipment and help to keep the saw in order are available, is doubtless the better tool, as the lesser waste of material, combined with greater speed, are strong points in their favor.

There are still some plants that cling to the old style railroad rip-saw for use in matching up tops or large surfaces, especially on quartered oak goods, in many cases carefully matching up the material, which may consist of six or more pieces, laying the whole lot on the table and shoving them through the saw, cutting off the surplus from the edge.

This is a slow and laborious method, and can just as well be done on a self-feed machine with a great saving of labor.

Assuming that the light is good, and it certainly should be, the matching up can be done on a table behind the rip saw, and the one piece only that requires ripping is passed back and sawn to the necessary width. Of course, it is understood that the stock has already had the necessary edging done to it after leaving the swing saw.

The expert matcher is behind the saw in this case, and not in front of it, and sorts the material for figure as it comes to him, on a table of the required size, which can be made several inches lower than the saw table, and so placed that it is in the handiest position. A rule, such as is illustrated, will be found very handy for this work, made of maple of the necessary length; the inches are divided into quarters and thirds, which will correspond with the various lengths for each joint. For instance, if the tops are 46 inches long and 20 inches wide, finished width, and there are four joints, the width when leaving the matching table will be 27¼ inches, which allows 1 inch for the four joints plus ¼ inch for buzzing the one edge and the final ripping to exact width.

For longer stock the larger divisions of the inch would likely be necessary; it all depends on the accuracy of the rip saw and the minimum allowance for the jointer. At any rate it is possible to have only

the same amount of final ripping waste from a top having as many as six or seven joints as one having only three, if the proper allowance is made for each joint. The maple rule referred to is just as feasible for work that is not matched for figure. The material goes through the saw and is laid on the table until it is within a few inches of the desired width. The "tailer" can then tell just what he wants to complete the requirement in inches and it is immediately delivered to him and the material placed on the truck. The same thing goes on until that size and quantity is completed.

Another help in this connection is to have the rule bored at the lime intersections, into which are inserted

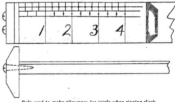

Rule used to make allowance for joints when ripping stock.

small steel pins which correspond with the number of joints that may be necessary. If the top is to finish 20 inches wide the first pin would be inserted at the 20¼ in. measurement, the next ¼ in. more, etc., the idea being that it is easier to count the pins corresponding to the joints than if they were not used.

It will be found handier to have, say, three rules made of various lengths, to use according to the material requirements of the factory. In making a rule secure a piece of white maple 1½ in. wide, 7/16 in. thick, bevelled on the under corners, having a stop on the left end facing; lay off the inches and divisions with a knife, bore the small holes for the pins, give a coat or two of white shellac, sand, ink on the figures; then give a coat of good tough varnish, and you will have a good bright serviceable article, and one that will be found of great help in the rip saw work. The accompanying illustration will give an idea of the rule referred to. The figures necessarily begin at one inch, though this is not needed, particularly, at least so far as the divisions of the inches are concerned.

Save Money by Purchasing

"Made in Canada"

Veneered Panels

Maple - Elm - Basswood - Birch - Plain Oak

Quartered Oak - Mahogany - Black Walnut

Single 3, 5 or 7 Plies.

Hay & Company, Limited   Woodstock Ontario

WE HAVE BEEN MAKING THE

Best French Glue in the World

SINCE 1818; ALMOST A CENTURY

In competition at the numerous Expositions abroad, the quality of our glues has been found so high that we are usually declared "Beyond Competition." Our trade in this country makes the same declaration. Consistency in quality and service has brought this result and we are proud of it.

Perfection in manufacture from the best raw material makes the quality of COIGNET FRENCH GLUES consistently the highest.

Large stocks at Montreal assure the best of service.

Why not take advantage of these most important factors by using our glues?

At your service for prices and samples.

Coignet Chemical Products Co.

17 State Street, N. Y. CITY, N.Y.

Factories at { St. Denis } France { Lyons }

ANGLO
RUBBING and POLISHING

Works free and easy and can be rubbed in two days.

WRITE FOR SAMPLE

The
Ault & Wiborg Co. of Canada, Ltd.
Varnish Works

Montreal   Toronto   Winnipeg

THE GLUE BOOK

WHAT IT CONTAINS

Chapter 1—*Historical Notes*
Chapter 2—*Manufacture of Glue*
Chapter 3—*Testing and Grading*
Chapter 4—*Methods in the Glue Room*
Chapter 5—*Glue Room Equipment*
Chapter 6—*Selection of Glue*

WOODWORKER PUBLISHING CO., LIMITED
345 Adelaide St. West, Toronto

Lightning Hollow Chisel Mortiser

Mortising sash stiles, rails and muntins and relishing out the tenon and cutting away that part of the tenon which is under the mold, while closely related operations, have heretofore always been done on separate machines.

The cost of three separate machines together with space required have been too much for the jobbing

No. 351 Lightning Hollow Chisel Mortiser and Relisher.

shops and smaller manufacturers of window sash, and they have, therefore, always had to do most of this work by hand.

Such people will eagerly welcome a new machine recently placed on the market as illustrated herewith.

It is in the first place, a high grade general purpose hollow chisel mortiser, working material up to 6 in. by 6 in., 3 in. deep or by reversing through, 6 in. It will carry chisels up to ¾ in. square and any larger

or irregular shaped mortise can be made quickly and easily by overlapping cuts.

It has a travelling relishing table, relishing head being carried on the same spindle with the bit for the hollow auger.

At right angles to the relishing head is a boring bit that cuts away that part of the tenon which is under the mold.

This machine is entirely self-contained, takes up small floor space, uses very little power and does clean cut and accurate work. It is so simple to adjust and easy to operate that a boy can run it. No pounding or jarring; mortises are perfect corners, square and sharp; no chips left to be picked out by hand.

The advantages and economy of this small combination machine, which is being made to sell at a very attractive price, can be learned and further information regarding it can be had upon application to the manufacturers, J. A. Fay & Egan Co., 152-173 W. Front St., Cincinnati. Ohio.

The latest statement of Dominion finances, covering the period from April 1, 1915, to February 1, 1916, shows an increase in revenue from all sources of approximately $30,000,000 as compared with the same period a year ago. Also, expenditures have been reduced by $18,500,000. Our revenue position for 10 months is therefore improved to the extent of $48,500,000 as compared with the same period of 1915, or at the rate of $58,200,000 a year.

We don't merely buy logs; we select them. The above "Simon Pure" Indiana Forked Leaf White Oak Logs are merely average logs with us. This is one reason why our Oak Veneer is accepted standard.

Importers and Manufacturers

Mahogany and Cabinet Woods—Sawed and Sliced

Quartered INDIANA White Oak, Red Oak, Figured Red Gum, American Walnut, Etc.—Rotary Cut Stock in Poplar and Gum for Cross Banding, Back Panels, Drawer Bottoms and Panels.

The Evansville Veneer Company, Evansville, Ind.

Panelled Elliptical Arch of Three-ply Veneer

Simple Method of Handling What Looks Like a Difficult Piece of Work
Staff Article

There are a great many problems that confront the manufacturer of high-class interior woodwork from time to time.

One that came to the attention of the writer and was dealt with satisfactorily may be of interest to some of our readers. The problem was the making of a panelled elliptical arch as illustrated in Fig. 1. Now, there may be different methods of constructing such an arch, but the method described and illustrated in this article will be found as simple and practical as any. The first step is to take the width of the opening and the height of the arch. Then construct the form as shown in Fig. 2 with common dry lumber. For the sake of strength, at least four ribs should be made and nailed together and placed at each side, as shown in the section drawing, Fig. 4, with cross pieces top and

Fig. 1.—Showing panelled arch in place.

bottom. The bevelled strips should then be nailed to the ribs and the nails set well in. The whole thing should then be trued up to the desired ellipse. In laying out the width be very careful to allow for the thickness of the bevelled strips. The pieces which form the false stiles are first steamed and bent around the form. The false stiles shown in Fig. 4 are shown thicker than they would be in actual practice. The false rails are then fitted in between the stiles. Before proceeding to bend the veneer the centre piece, which is cross-grained, should be put in the heater and allowed to get fairly hot. The face veneer which forms the panels should be glued on one side and along the edges on the other side, where it glues on the back of the stiles. The other outside veneer is glued on one side only. The three veneers are then placed together, the hot one in the centre, and a small nail put through them at each end to prevent slipping when the glue

starts to loosen up. Taking for granted that the bevelled pieces for the backing have been prepared before gluing the veneer, you should then proceed to bend the veneer around the form, starting at one end and fastening it securely to the form. After being fastened at one end the veneer should be worked around the form and fastened down temporarily in two or three places. You should then start at one end and put

Fig. 2.—Making a panelled elliptical arch.

on the backing strips. These are glued down to the veneer, and also glued edge to edge. They are then screwed to the form, the screws passing the outside edge of the stile, as shown in the section drawing, Fig. 4. After all the backing strips are in place the glue is allowed to dry thoroughly. The screws may then be taken out and the whole thing lifted off the

Fig. 3.—Backing strips left off to show veneer glued on false stiles.

Fig. 4.—Section through A A showing method of screwing on backing.

form and taken to the band saw, where the protruding ends of the backing strips are cut off. You will find that by this method you will have a first-class job and one that will last for all time.

Preparation of Core Stock and Cross-Banding

By "Veneer Room Foreman"

Veneering is one of the big items of expense in the manufacture of high-grade furniture, but it is well worth the cost if the work is well done. The only way to keep the cost down to its worth is to do the work right the first time. Some managers have attempted to reduce the cost by giving less attention to details in the laying of the veneer, but the attempt has, usually, if not always, resulted in an increase in cost and much inferior work. Others have tried the experiment of using inferior materials and giving less attention to the selection and preparation of the core stock, but the increased cost in the cabinet room, resulting from the repairs rendered necessary by the inferior work, proved the experiment to be an absolute failure. It is merely shifting the cost from one department to another.

The up-to-date manufacturer is one who not only experiments for himself, but is willing to learn from the experience of others. The more ready he is to adopt ideas that others have tried and found satisfactory, the more successful he will be. The only way to keep the cost of the veneering down is to keep the quality of the work up. Never let the idea creep in that the excessive cost results from the work being done too well. On the contrary, excessive cost usually, if not always, results from poor work. In the first place, a little more care and more attention to details will save many dollars in repairs, to say nothing of the enhanced value of the goodwill of the concern, because of its reputation for strictly high-class work. If the goods are worth the time and material required to veneer them they are certainly worth the time required to do them right.

Some men contend that it does not pay to have work too well done. A little blister here, a rough spot there or a glue joint open, would not do any harm, but all manufacturers do not think the same as that. If a few cents' worth of time and care will increase the value of the article almost as many dollars, it seems to me to be the part of wisdom to make the investment, as with the enhanced reputation for quality will come the business and the increased price.

Work started smooth and kept smooth will keep smooth to the end. This is evident and requires no argument, but I sometimes think that the matter of preparing core stock for veneer is altogether too simple for some people to fully understand. You often see the rolls on expensive beds as if they had been intended for wash-boards. These parts are run through the sticker, and without having the knife-marks removed are taken to the veneer room, with the idea that, because they are being hidden for the time, they will never be discovered. However, the marks are there and will show themselves at a time when it is impossible to remove them. Hundreds of dollars are spent every year in some finishing rooms applying extra coats of varnish and extra rubbing in an effort to make this kind of work passable, thereby spending a dollar's worth of time and material to save a dime in some other department. This is a piece of nonsense, but it is very often the case. If work is to be finished right, it must be smoothed off to remove knife marks, either from the sticker or planer. If the core stock is of hardwood sanding will remove the marks, but if of softwood it should be sponged before sanding, as the knife-marks or bruises will be much deeper in soft-wood than in hardwood, and the sponging will swell out the bruised portions. It is better to do the sponging immediately after the stock comes through the machine and before the bruised portions become set.

Having prepared the core stock, the question is whether the cross-banding and face veneer should both be laid at the same time. There are various reasons why cross-banding and face veneer should not be laid at the same time, but we will mention only the most important ones.

Much of the stock used for cross-banding and face veneer contains creases where the wood broke away during the process of slicing, and it is usually too thin to dress before laying. If the cross-band and face veneer are laid at the same time these creases will fill up with glue. When the stock is allowed to dry thoroughly the glue in the creases will shrink and draw the veneer down. This accounts for the ridges which so often mar the appearance of sideboard, dresser and table tops. To insure a perfect job, cross-banding should always be laid some time before the face veneer and stripped on trucks to dry thoroughly. This gives it a chance to do all its shrinking. It should then be run through the sander, using coarse paper, before the face veneer is applied.

Our high-grade veneers are cut so thin that they will not stand the amount of scraping and sanding necessary to remedy the defects of cross-banding and core stocks and you should not expect it. High grade lumber is getting scarce, and high grade veneer is expensive, and there is no reason why core stock or cross-banding should be left in such a condition that the veneer should cut through before a smooth surface can be obtained. It is up to the user of veneer to adjust his factory and pick his men so that these modern conditions can be met.

Overlapping of cross-band is a frequent source of trouble, especially where it and face veneer are laid at the same time. To prevent this, the cross-band must be thoroughly dried. Lay it first and get it into the press as soon as possible after it touches the glue, to prevent swelling.

Veneer cannot expand once pressure is applied. If there is no expansion, there will be no overlapping, no contraction and no checking of the face veneers from these causes.

D. Herscovitch and A. Leibovitch have registered at Montreal, and will carry on business as furniture manufacturers under the style of Herscovitch & Leibovitch.

DARLINGS STEAM APPLIANCES

DARLING BROTHERS
LIMITED
Engineers and Manufacturers
MONTREAL, CANADA

Branches: Agents:
Toronto and Winnipeg   Halifax St. John, Calgary, Vancouver

OAK   MAHOGANY   CIRCASSIAN   MAPLE

VENEERS

BIRCH   POPLAR   CHERRY   WALNUT   GUM

THE OHIO VENEER COMPANY

Importers and Manufacturers

Foreign and Domestic Veneers and Hardwood Lumber

We always carry a large and assorted stock of Mahogany, Circassian
Walnut, Sawed and Sliced Quartered Oak.

Send us your enquiries and orders. We guarantee good service.

2624 to 2644 Colerain Avenue,   ⋈   ⋈   **CINCINNATI, OHIO.**

We operate exclusively in

AMERICAN WALNUT

Our supply of raw material is obtained from
Ohio and Indiana, which produces the best
walnut in the United States.

We have the most modern and best equipped
plant in the United States for producing

WALNUT LUMBER and VENEERS

Special attention given to the requirements of
Furniture Manufacturers

George W. Hartzell,   **PIQUA, Ohio**

(HB) **Hoffman Bros. Co.** (HB)

Reg. U.S. Pat. Off.   Estab. 1867, Incor. 1904   Reg. U.S. Pat. Off.

800 West Main Street,
FORT WAYNE, INDIANA

Manufacturers of

VENEERS and LUMBER

In the Domestic Hardwoods
ANY THICKNESS,

1/24 and 1/30 Slice Cut
(Dried flat with Smith Roller Dryer)

1/20 and thicker Sawed Veneers,
Band Sawn Lumber.

—SPECIALTY—

(HB) **Indiana Quartered Oak** (HB)

Reg. U.S. Pat. Off.   Reg. U.S. Pat. Off.

VENEERS & PANELS

CARRIED IN STOCK

VENEERS

Red and White Oak, Birch, Poplar, Gum,
Yellow Pine, Cypress, Etc.

PANELS

Three and Five Ply—All Woods.

GEO. L. WAETJEN & CO.
Milwaukee,   Wisconsin

Veneers and Panels

LARGE STOCKS FOR
QUICK SHIPMENTS

Send for Lists

Walter Clark Veneer Co.
GRAND RAPIDS, MICH.

A Portable Woodworking Machine

One of the most important features in the cost of a building is the labor cost in putting the interior trim in place. A great deal of that cost would be eliminated if the work could be done by machinery, but the trouble has been to secure a machine that would cover the various operations required and at the same time be of a size that it could be taken in and out of the doors of the ordinary building. A machine which has all these features is illustrated on this page, and is known as the Elliot Woodworker No. 3. Observe the jointer placed to one side of the machine, on

Elliot Woodworker.

which the edges of all material may be straightened in a fraction of the time required to do it by hand. On the other side of the machine is a table on which a great variety of operations may be performed, such as crosscutting, dadoing, boring, housing stair strings, etc., simply by changing the tool on the mandrel. Another valuable feature is that there are two mandrels —one for the jointer and the other for saws, dado heads, or drill chuck—and two men can work on the machine at one time. Either mandrel may be stopped for adjusting purposes without interfering with the other. The saw table is stationary, but the machine may be swung around at any angle. This machine is manufactured by the Elliot Woodworker, Limited, corner College and Bathurst Street, Toronto.

Wooden Boxes Coming into Their Own

J. B. Knapp, manager of the Northwestern Association of Box Manufacturers, Portland, Ore., has been carrying out some investigations along the line of the value of wooden boxes compared with fiber, as shipping containers. It is shown that, damage to merchandise in transit has increased enormously since the fiber container came into use. A summary of Mr. Knapp's findings follows:

The people of the United States are annually confronted with a minimum waste amounting to fifty cents per capita by reason of damage claims annually paid by transportation lines. These claims, approximating thirty-seven million dollars annually, are paid by the railroad companies and are a net loss to these important agencies of communication. In the year 1900 the total damage claims paid by American railways amounted to seven million, as compared with thirty million during the year 1914, an increase of 471 per cent.

In the ten years from 1904 to 1914, the increase in damage claims was 314 per cent., while the increase in railroad revenues was only 100 per cent. During these ten years a great change in the form of shipping containers has taken place in this country, and the

very large portion of the increase in damage claims may be directly attributed to the introduction of containers made from so-called substitutes for wood.

In years previous to 1914 practically all of the important shipping containers were in the form of wooden boxes, barrels and crates. In the past ten years it is estimated that thirty per cent. of the trade formerly using wooden boxes has turned to the use of substitute fiber containers; therefore this substitute is undoubtedly responsible for the large percentage of increase in damage claims.

The principal appeal of the fiber container is its low first cost and the saving to the shipper in the difference of its weight as compared with the weight of a similar container made of wood. In these two points of merit the fiber package appeals particularly strong to that element of our shipping class which considers its packing cases as an overhead expense without proper consideration of the protection and security which it affords to the commodities packed.

Strong Agitation for Better Containers

The wood box manufacturers has allowed his package to suffer in an attempt to meet the competition of the substitute containers and is himself to blame for some portion of the recent increase in damage claims. The wood box is the modern efficient container when properly constructed and its real merit is only becoming understood by comparison with those substitutes which are advanced to displace it.

There is a strong agitation for better containers under way and an active investigation of the merit of all kinds of shipping containers is being conducted by American railroad interests, entirely with a view to their own protection against losses and claims for which they are largely not responsible. In this endeavor the railroads are supported by the Interstate Commerce Commission and by government agencies seeking to advance the market for American goods in foreign countries.

The manufacturers of wooden boxes are strongly in sympathy with this movement. They appreciate the reflection on their containers by reason of the practice of some manufacturers in skimping material to the detriment of the efficiency of the package in order to meet competition from inferior substitute materials.

This entire matter is one of great importance and has attracted the attention of all parties having to do with the manufacture, use, transportation, handling and consumption of all forms of packed commodities and the packages for them. It has resulted in particular activities on the part of the joint committee of the American Society for Testing Materials, known as committee D-10, made up of representatives of all agencies, industries and trades concerned in promoting standard efficient shipping containers.

This committee in co-operation with the National Association of Box Manufacturers and the United States Forest Service has recently developed a new form of testing machine for determining the standard of construction for developing the highest efficiency as compared with boxes made of substitute materials.

These tests are now being inaugurated and samples of boxes representing the standard followed in the several districts of the United States will be tested and a national standard for each class of box is intended as the result.

If you are in the market for either lumber or veneers in walnut, "the aristocrat of American hardwoods," it will pay you to get in touch with us, for we are

SPECIALISTS IN WALNUT

We have been in this business for thirty years; and in addition to knowing exactly how to make walnut lumber and thin stock, and in addition to having a stock which for variety and size is unequaled anywhere else in the United States, we have on hand a big bunch of

Unique Figured Veneers

from the Southwest

These veneers are highly figured, and of unusual color, and for special work are absolutely unrivaled. Ask us for samples.

Penrod Walnut & Veneer Co.
KANSAS CITY, Mo.

We specialize on

Poplar

cross-banding and backing

Veneer

Highest Grade — Right Prices

The largest Poplar Veneer Mill in the world.

The
═ Central Veneer Co.
Huntington, West Virginia

The Glue Book

WHAT IT CONTAINS

Chapter 1—Historical Notes   Chapter 4—Methods in the Glue Room

Chapter 2—Manufacture of Glue   Chapter 5—Glue Room Equipment

Chapter 3—Testing and Grading   Chapter 6—Selection of Glue

PRICE 50 CENTS

Woodworker Publishing Co. Limited
345 Adelaide St. West, Toronto

Market for Furniture in South Africa

A market exists in South Africa for furniture in a knocked-down state, and inquiries have been addressed to Mr. W. J. Egan, Trade Commissioner, Cape Town, re the opening up of furniture agencies. If Canadian manufacturers are in a position to export the class of goods illustrated below, complete details should be forwarded Mr. Egan, who will bring the information to the notice of importing houses:—

Fig. 1.   Fig. 5.

Fig. 4.

Fig. 1 is an American-made chair, cane seat, packed one dozen in case, weight 118 pounds. This chair is quoted at $6.50 f.o.b. New York, less 2½ per cent. discount.

Fig. 2 is made in bentwood, black or brown, with a cane seat, packed one dozen in case, weighing 118 pounds. Fig. 3 is a bentwood chair, with a brown embossed wood seat. These are packed one dozen in case, weighing 120 pounds to case. The pre-war prices were 40s. a dozen net c.i.f. Durban. Figs. 2 and 3 are of Austrian manufacture.

The illustration shown in Fig. 4 indicates the cheaper class of bureau imported in the knocked-down state. This is an American make and is a four drawer bureau,

with or without mirror, packed four to a crate, two light and two dark. The crate without mirror weighs 340 pounds and with mirror 540 pounds. The bureau is sold at $5.95 f.o.b. New York, less 2 per cent. discount.

The washstand shown in Fig. 5 is American make and indicates the single washstand of the cheaper grade. The single stands are packed one dozen in case and double washstands one-half dozen to a case.

Tables are also in demand. The best selling sizes are 4 feet by 2 feet 6 inches, 5 feet by 3 feet, and 6 feet by 3 feet. The weights per case of tables without drawers are 240 pounds, 310 pounds and 360 pounds, respectively. The same size tables with one drawer weigh 250 pounds, 330 pounds, and 400 pounds.—Department of Trade and Commerce Weekly Bulletin.

Fig. 2.   Fig. 3.

In a recent bulletin on hemlock, written by E. H. Frothingham, of the U. S. Forest Service, he says that a very small amount of hemlock—less than one per cent.—is used annually for veneer manufacture. Hemlock in 1909 ranked twenty-fourth among veneer-producing species, with an annual consumption of 207,000 board feet of logs. Most of the hemlock veneer is made in New York, while Maine, Pennsylvania, Ohio, North Carolina, and the Lake States contribute small amounts. It is employed chiefly in the manufacture of shipping packages of various kinds, laminated or built-up lumber, etc. Because of heat defect (knots, shake and decay) hemlock cores left after veneer production are of little value for anything but fuel.

Canada's Premier Hardware House
(Established 1885)

A Wide Range at Right Prices

TOOLS

Let Us Send You Our Catalog

Shell Box Hardware

Everything in Hardware for the Shell Box Manufacturer.

Bands, Handles (Wire and Rope,) Screws, Nails, Staples, Houseline, Corrugated Joint Fasteners

A large assortment in stock—Approved and inspected goods.

Write for Samples and Prices

Sole Selling Agents for the British American Hardware Manufacturing Co.

**FACTORY and MILL SUPPLIES, CABINET HARDWARE
BARTON'S GARNET PAPER**
Canadian Agents   **UNIVERSAL CASTERS**

Try Us for Special Made-to-order Hardware

The
Vokes Hardware Co.
Limited
40 Queen St. East, Toronto, Can.

VOKES

Small Belt Matters

Barrel and box factories are big users of belts, and since all kinds of belts are expensive as to first cost—leather, rubber, canvas, or camel hair—it is plain that belts warrant considerable attention in their selection, writes N. G. Near in "Barrel and Box."

Belts in barrel and box factories are also subject to pretty severe usage, and since this is so the quality and cost must again be considered, for if a belt must be renewed often the cost multiplies rapidly. . . . . . .

The trouble with most superintendents and managers, though, is that they do not look farther than the belt. When laying in a new supply the cost of that supply may be $1,000 or more, and the superintendent naturally "pricks up his ears." He makes up his mind that the matter is important enough for him to give it his own personal attention, which he does. He buys or supervises the buying of a belt that he believes is the "best" for his purpose. But right there he stops. He does not think of the lacings that must be used to tie these expensive belts together, nor the dressing that must be applied to keep the belt in good pulling shape. These latter details cost only a couple of dollars or so, and he is willing to leave so small an amount to the discretion of the purchasing agent, or to any mechanic he feels like sending to the store after it.

This is where the superintendent makes a mistake, however. He does not stop to consider that the kind of lacing used often determines the life of the belt. And a poor dressing can also quickly destroy a belt. So, all his pains quickly come to naught. His efforts to save money are wrecked because the right kind of lacing and dressing were not purchased and applied.

Let us be a little more definite, now. What is the best lacing? Leather or wire? I am not going to answer that directly because if I should you would probably say that I am prejudiced. You know, though, the endless belt is by far the best. The endless belt has no "joints" at all; it is pliable in every direction; it does not make a noise while running aside from an efficient-sounding "swishing." I always prefer the endless belt. It costs more than the others, of course, because a short length of the belt is spoiled in making the joint, and because more time is required in making the splice. But after it is made, you have something of which you can well be proud. You cannot make anything better.

The best lacing, then, is the one that most closely approaches the cemented joint—the one that is most pliable in every way—the one that is almost as strong. The lacing must not be bulky for if it is it will make a noise as it runs around the pulleys, it lifts the belt up so as to destroy contact, and it may cause slipping and running off. The holes punched in the leather or other belt must be very small as not to cut away the strength of the belt, and the lacing must be strong.

About the first thing that comes to mind after reading these specifications is "Wire lacing." It does make a little lump at the joints, etc., but when properly put together the joint can be made a pretty good approach to the endless belt joints. I am not opposed to either wire or rawhide; I am in favor of the best. If you can make a more flexible, more noiseless, etc., joint out of rawhide, go ahead and use it.

If you have had any belting experience at all you know well that laced joints always give more trouble than cemented joints. The lacing breaks, pulls out, catches, tears the belt, and takes up valuable time for re-fixing. The belt runs off the pulleys, catches, tears, and you have to cut out the torn part, replace it, and in go two more laced joints. Your troubles then start all over again. Your cheap lacing is therefore the direct cause of the destruction of your expensive belting.

In order to lace with heavy rawhide lacing you have to punch large holes into the belt. The loss of strength varies directly as the area cut away by these large holes, and as we all know, that loss in strength often amounts to considerable. The fact that belt failures usually occur at the joint (in the case of laced joints) is proof that there is where the belt is weakest. Many an expensive belt has been ruined in this way, but the belt men do not seem to know it. In mending the belt they do the same thing over again.

Belt Dressings

Next we come to belt dressings. There are many dope dressings on the market that should not be used at all because they ruin belts. I am not going to be explicit here because I do not want to knock anything. I just want to warn belt men against the stuff that might ruin several thousand dollars' worth of belts in a year's time. By spending a few dollars more per year for a good dressing the belts can be saved and made to do their work much better for many years.

For instance, I have seen boiled linseed oil recommended for belts as a good dressing because it makes the surface hard and smooth. Now, if there is anything that will cause belts to slip it is a hard smooth surface. It stands to reason that the belt pulling surface is a soft, grippy surface. Note the great difference in tractive power between rubber on steel and steel on steel. Again, we find soap, resin, grease, or any old kind of oil recommended for the cure of temporary belt slip. I myself have used shellac with what some people would call success, but I would not use it again because I have too much regard for my pocketbook.

Neatsfoot oil and castor oil are perhaps recommended more than any other oils for this purpose, and they are pretty good. They are better than some of the dopes I have mentioned, and I would use them myself if I could not get anything better, but there are better dressings on the market, and those are the ones I would use. I am not boosting any particular dressing any more than I am boosting any particular lacing. I am just pointing out the importance of the small details in connection with belt drives.

I trust I have emphasized my point so strong that hereafter you will look into these small matters yourself. I can name instances where belts have run for twenty years in service that is severe enough to put the same size belt out of commission in a year's time if it is not properly cared for—if it is doped—if it is cut full of big holes for lacing. These belts run nice and slack. Their surfaces are not hard and smooth. They are nice, soft, pliable, and grippy, just as good belts should be.

The best advice I can offer along this line is: Investigate some of these belts that has endured so many years. Learn how it was done. In what way was the belt, laced? What is the stress per inch of width? What dressings are used? Is the belt tight or slack? What sizes of pulleys are used? Get as much information on it as you can, and coupling it together with what I have told you now you should have no trouble in saving much money per year hereafter. The principal point is—make the whole belt pliable. Make the joints pliable and keep them so.

One dollar per year, in this way, will save hundreds per year.

OUR SPECIALTY

All Machinery Supply Firms like Machinery Manufacturers have lines which they specialize. This is because they know more about these particular lines than any other. We know more about woodworking machinery than anything else and can give you a greater selection and more intelligent service in this line than any of our competitors. Therefore

Our Specialty is

WOOD-WORKING

MACHINERY

Send us Your Inquiries and See if We Have not Something Better in Both Machinery and Service.

J. L. NEILSON & CO.

Winnipeg, Man.

"Furniture in England"

This book is an encyclopaedia of artistic suggestion—an education in itself.

Upwards of 400 exquisite half-tone and color plates on pages 14 ins. x 10 ins.

Among the subscribers to this book are Her Majesty Queen Mary, His Majesty the King of Greece, and many of the foremost manufacturers the world over.

Price $12.00

Delivered to any address in Canada

Copies may be obtained from

The Woodworker Publishing Co.

Limited

Toronto　　　Ontario

Wood Steaming Retort

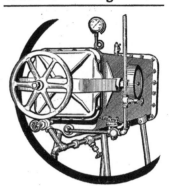

Wood Bending Manufacturers:

This is one of our

Perfection Retorts

which we guarantee will save you

50% Less Breakage

in your bending department than your present process; that your stock will dry in your forms or presses in one-third less time; that you will have no stained stock; that your stock will retain its shape much better after being bent; that it will dry in your dry-kiln in one-half less time and that your steam consumption will be reduced at least 90 per cent.

The door can be opened and closed in ten seconds, and it is steam and water tight and for this reason can be placed anywhere in your factory.

Compare this IMPROVED RETORT with your present steam boxes, then write us for our Booklet on Progressive Wood Steaming.

SHIPPED ON 30 DAYS TRIAL

Perfection Wood Steaming Retort Co.

PARKERSBURG · WEST VIRGINIA

QUALITY AND QUANTITY OF PRODUCT AND LOW COST OF OPERATION DISTINGUISH THE NASH SANDER

For long straight handles, curtain poles, chair stock, dowel rods, shade rollers and similar work, this Sander can save its cost in three or four months.

It will pay you to investigate.

J. M. NASH, MILWAUKEE, Wis.

NATIONAL TRUCKS

Easy Running
Strong   Durable
Always in stock

National Cross Piling Truck—Fig. 5

All National Trucks are Equipped with Malleable Iron Wheels

IN all National Trucks the sides are made of channel steel; the axles and rollers of cold rolled shafting; and the spreaders of malleable iron. The axles have milled ends, which does away with riveting. All our trucks are held together with bolts and can be dismantled without injury. Sides are cut to exact length and punched on a template, which insures uniformity.

Write today for catalog and prices.

THE NATIONAL DRY KILN CO.

1118 East Maryland St.   Indianapolis, Ind.

Manufacturers of brick, lumber and cooperage stock dry kilns, steel roller bearing trucks, wheels, etc.

News of the Trade

William Travis contemplates the erection of a planing mill at Wyoming, Ont.

Joseph Landreville, of the firm of J. Landreville & Son, coopers, Toronto, is dead.

The Toronto Wood Specialty Works suffered loss by fire recently; loss partly insured.

Alfred Gallagher, furniture manufacturer, Montreal, was burnt out recently; loss partly insured.

A storehouse belonging to the Goderich Planing Mill Company, at Goderich, Ont., was recently destroyed by fire.

Nathan Calder is planning to install equipment in his planing mill at Listowel, Ont., to be operated by Hydro power.

Mr. F. W. Colpitts and Mr. George McEachern have started up a box and woodworking factory in Moncton, N.B.

The planing mill belonging to S. L. Lambert at Welland, Ont., was recently destroyed by fire. Loss $23,000, partly insured.

William Gardner, who was for many years partner in a wagon and carriage-making business in Woodstock, died recently.

The planing mill at 141 Main Street, East Ottawa, owned by Alderic Charpentier was recently damaged by fire to the extent of about $3,000.

The planing mill belonging to the Buchanan Planing Mills Company, of Goderich, Ont., was recently damaged by fire. Damage estimated at $6,000.

The Barrie Planing Mill resumed operations on Wednesday, after having their plant overhauled. They report having a large number of orders on hand.

Mr. William Burke died recently at his home in Welland, Ont., as the result of a stroke. He was for many years manager of Waugh's planing mill at Welland.

The Jeffries Furniture Company, of Welland, Ont., had their factory destroyed by fire. Mr. Jeffries estimates his loss at about $4,800, partly covered by insurance. He will rebuild as soon as possible.

The death occurred recently of Mr. Samuel Running, Sr., at his home in Port Elmsley. He was engaged for many years in the sawmill business and the manufacture of cheese boxes in different parts of Ontario.

The Diamond Lumber and Shingle Company, Limited, has recently been incorporated at Vancouver, B. C., with capital of $15,000, to carry on the business of sawmilling and manufacturing shingles, sash, and doors.

F. W. & S. Mason, cabinet makers and upholsterers, St. Andrews, N. B., have dissolved partnership, F. W. Mason retiring. A co-partnership has been formed between Sam Mason and Walter F. McMullon, the firm name remaining F. W. & S. Mason.

Mr. J. A. Orton, of Orillia, Ont., who recently had his planing mill destroyed by fire, will carry on the business temporarily in the factory which he recently purchased from Mr. W. H. Crawford, also of Orillia, who was in the same line of business. Mr. Orton will erect an up-to-date plant during the coming summer.

The M. S. Glassco Company, Limited, have been incorporated with a capital of $50,000, to carry on the business of manufacturing and dealing in furniture. The place of business will be located in Hamilton, Ont. Malcolm S. Glassco, merchant, William C. Hammond, salesman, and Reginald D. Glassco, banker, all of Hamilton, are interested.

Box Machinery Bargain

One Dovetail Machine. One Setting-up Machine.
One Four-Saw Trimmer—self feeding; made by Dovetail Machine Co., St. Paul, Minn.
One No. 22 Box Trimmer, made by Fischer Machine Works, Chicago, Illinois.
All cost $3,300. Will guarantee machines and sell at a bargain.
Inquire, The GLOBE SOAP COMPANY,
2-3-4 Cincinnati. Ohio.

Woodworking Machinery For Sale

1—36" Crescent band saw, A1 condition	$75.00
1—24" Cowan double surfacer, A 1 condition	375.00
1—8" Crescent jointer	75.00
1—Variety saw table, iron	85.00
1—Wood frame, 8 ft. saw table	60.00
1—Combination drill and mortiser, McGregor-Gourlay, 4" stroke, table ¾ x 5½, raise and lower 12", side movement 10", knife 1" wide	50.00

Pollard Manufacturing Company, Ltd.,
3 Niagara Falls, Ont.

Second-Hand Woodworking Machines

Band Saws—20-in., 86-in., 40-in.
Jointers—12-in., 16-in.
Shapers—Single and double spindle.
Re-Saw—42-in. - American.
Lathes—20-in. with rests and chisels.
Surfacers—16 x 6 in. and 18 x 8 in.
Mortisers—No. 5 and No. 7 New Britain.
Rip and Cut-off Saws; Molders; Sash Relishers, etc.
Over 100 machines, New and Used, in stock.

Baird Machinery Company,
123 Water Street,
PITTSBURGH, PA.

7% Investment 7% Profit Sharing

Sums of $100, $500, $1000 and up. Safe as any mortgage.
Business at back of this Security established 28 years.

Send for special folder to

National Securities Corporation,
Limited
Confederation Life Bldg., Toronto

PATENTS
STANLEY LIGHTFOOT
Registered Patent Attorney
Lumsden Building - - TORONTO, CAN.
(Cor. Adelaide & Yonge Sts.)

The business of Manufacturers and Inventors who wish their Patents systematically taken care of is particularly solicited. Ask for new free booklet of complete information.

Wire-Stitch Jointing

One of the problems of three-ply built-up work, where narrow pieces are used for the center ply or filler, is to keep this filler properly jointed together, so that there may neither be overlapping of the pieces nor open spaces in the veneer. This is enough of a problem that at times it causes people to hesitate about using a series of narrow pieces to make up the filler in three-ply work and prefer to let this go to waste and use full-size stock.

The wire stitching machine offers some help to the solution of this problem in many classes of work—a machine somewhat of the same type, though differing a little in detail from those used in making light packages of various kinds and stitching them together. The idea is that two or three or more wire stitches can be made right across the joint of the pieces of veneer, making up the center or filler, and in this way the center can be made into practically a solid sheet from a series of narrow strips, making it practical to utilize waste and at the same time hold the filler stock firmly and closely jointing together without overlapping.

Comparatively light wire should be used for stitching of this kind and the stitcher carefully adjusted so that it will imbed the wire in the work, both on the face and in the clinches. The number of stitches will be determined by the size of the stock to be stitched, and of course one should avoid stitching near the outer edge, where, in trimming, the saws are likely to come in contact with the stitches. Stitching machines are made that will do this work rapidly, and the idea suggests itself as a good one for experiment in the utilization of narrow stock, that might otherwise go to waste, for fillers in three-ply work.—Veneers.

The Machinery Man

The machinery man in the woodworking industry has often been complained of as a sort of pest because he is always after the manufacturer's money with something new in the machinery line, writes "Veneers." Sometimes machinery men individually may prove objectionable, and some of their methods may be open to question, but, after all, it is the machinery man who puts the wheels on progress. In veneers improved machinery has done much to help us make progress. Jointing and taping machines have reduced to a fraction the cost of matching up veneers; modern presses have done much to reduce the cost of making built-up work, and everywhere we can find where the machinery men have devised and put on the market machines and appliances for reducing cost, increasing capacity and improving the quality of the work. The machinery man is really a harbinger of progress. Wherever he goes with his new offerings and his song of salesmanship there are signs of progress left in his wake, and we find the industry moving forward and seldom ever going backward again to older methods after having introduced machines.

While a manufacturer here and there may feel irritable now and then because machinery men are insistent and their offerings come too rapidly, we must, after all, take off our hats to the machinery man as being the instigator of ideas that lead to real progress.

PETRIE'S
MONTHLY LIST
of
NEW and USED
WOOD TOOLS
in stock for immediate delivery

Wood Planers
36" American double surfacer.
30" Whitney pattern single surfacers.
26" Revolving bed double surfacers.
26" Goldie & McCulloch single surfacer.
24" MacGregor Gourlay planer & matcher.
24" Major, Harper planer and matcher.
24" Single surfacers, various makes.
30" Dundas pony planer.
18" Little Giant planer and matcher.
16" Galt jointer.
12" Crescent jointer with safety head.

Band Saws
42" Fay & Egan power feed rip.
38" Atlantic tilting frame.
36" Crescent pedestal.
30" Ideal pedestal.
28" Rice 3-wheel pedestal.
28" Jackson Cochrane bracket.

Saw Tables
Preston variable power feed.
Cowan heavy power feed. M138.
No. 8 Crescent sliding top.
No. 3 Crescent universal.
No. 2 Crescent combination.
12" Defiance automatic equalizer.
MacGregor Gourlay railway cut-off.
7½" Martin iron frame swing.
6½" Crescent iron frame swing.

Mortisers
No. 5 New Britain, chain.
Cowan hollow chisel. M.190.
Fay upright, graduated stroke.
Galt upright, compound table.
Smart foot power.

Moulders
10" Clark-Demill four side.
10" Houston four side.
6" Dundas sash sticker.

Sanders
48" Berlin 3-drum.
36" Egan double drum.
12" C.M.C. disk and drum.
18" Crescent disk.

Miscellaneous
Fay & Egan 12-spindle dovetailer.
16" and 18" Ideal turning lathes.
No. 285 Wysong & Miles post boring machine.
MacGregor Gourlay 2-spindle shaper.
Elliott single spindle shaper.
No. 51 Crescent universal woodworker.
40" MacGregor Gourlay band resaw.
Rogers vertical resaw.
Cowan sash clamp.
Pedestal tenoner double heads and copes.
Egan sash and door tenoner.
No. 6 Lion universal woodtrimmer.
Rivall box nailer, Meyers patent.
20" American wood scraper.
4-head rounding machines.
Cowan spindle carver. M63.
Cowan veneer press screws.

Prices, Descriptions and Full Particulars on Request

H. W. PETRIE, LTD.
Front St. W., Toronto, Ont.

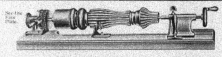

Fluting Heads (Shaper Attachment)

For fluting table legs, chair legs, bed posts, columns, balusters, etc., etc.
Swings any diameter up to 8 in. and may be extended to any length.
Spaces automatically numbers 1 to 48.

Ask for circulars of Fluting Heads, Twist Machines and Twist Sanders.

Solid Cutters

Made of high-grade Crucible Steel. Tempered hard to grind only. Sharpened from under throat side. These Cutters will hold their original form until worn out. 4 to 8 cutting edges means smooth work long life.

SHAWVER COMPANY   -   -   -   -   Springfield, Ohio

ONE OPERATOR
SIXTY SETS
SIXTY MINUTES

BRICK HOUSE JAMBS

FRAME HOUSE JAMBS

THE EFFICIENCY WINDOW FRAME MACHINE

A new style machine which REDUCES LABOR COST FULLY 60%. Works automatically—one man can turn out sixty sets of pulley stiles an hour. Cross cuts the ends of pockets, routs the pulley mortises and dadoes the head and sill gain, automatically, while the operator slits the pockets. You take no chance—write us.

THE F. H. WEEKS LUMBER COMPANY   -   AKRON, OHIO

The Canadian Woodworker and Furniture Manufacturer covers a specialized field. It represents maximum value to its subscribers.

Subscription price $1.00 per year.

PRESSES

For Veneer and Veneer Drying

Made in Canada

William R. Perrin
Limited
Toronto

GRAND RAPIDS
VAPOR DRY KILN
GRAND RAPIDS
MICHIGAN

129

Grand Rapids Vapor Kilns sold in the last

122

days of the year 1915.

54

of these were Repeat Orders.

Repeat orders represent satisfaction. We can guarantee experienced engineering ability and efficient service.

Over 1300 Grand Rapids Vapor Kilns in use.

Write us regarding better drying

Reduce the Cost of Your Sanding and Improve the Finish

of your Handles by using an

"OBER" DOUBLE BELT SANDER

for sanding broom, mop, fork, hoe, and rake handles, pike poles, etc.

We also make LATHES, for turning this work, also other Lathes for turning irregular work, RIP SAWS, BOLSTERS, Etc.

Write for free catalogue No. 31.

THE OBER MFG. COMPANY
CHAGRIN FALLS, Ohio., U.S.A.

VENEER PRESSES

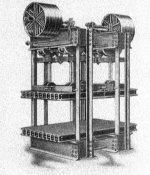

WRITE FOR BULLETIN C 1.

Canadian Boomer & Boschert Press
Company, Limited
18 Tansley Street, MONTREAL

Your Loose Pulley Trouble Disappears

when installed with

CHAPMAN BALL BEARING LOOSE PULLEYS

A little vaseline once a year is all the attention and lubrication required.

Our loose Pulleys cannot run hot and do not cut the shaft.

Write us for particulars and price list.

The

Chapman Double Ball Bearing Co.
of Canada, Limited

339-351 Sorauren Ave., Toronto, Canada

The "Canadian Woodworker" Buyers' Directory

For Alphabetical Index see page 8

ABRASIVES
The Carborundum Co., Niagara Falls, N.Y.

BALL BEARINGS
Chapman Double Ball Bearing Co., Toronto.

BALUSTER LATHES
Garlock-Machinery, Toronto. Ont.
Ober Mfg. Company, Chagrin Falls, Ohio.
Whitney & Son, Baxter D., Winchendon, Mass.

BAND SAW FILING MACHINERY
Berlin Machine Works, Ltd., Hamilton, Ont.
Fay & Egan Co., J. A., Cincinnati, Ohio.
Garlock-Machinery, Toronto. Ont.
Shawver Company, Springfield, O.

BAND SAWS
Berlin Machine Works, Ltd., Hamilton, Ont.
Garlock-Machinery, Toronto. Ont.
Jackson, Cochrane & Company, Berlin, Ont.
Petrie, H. W., Toronto.
Shawver Company, Springfield, O.

BAND SAW MACHINES
Berlin Machine Works, Hamilton, Ont.
Fay & Egan Co., J. A., Cincinnati, Ohio.
Garlock-Machinery, Toronto. Ont.
Petrie, H. W., Toronto.
Williams Machinery Co., A. R., Toronto, Ont.

BAND SAW STRETCHERS
Berlin Machine Works, Ltd., Hamilton, Ont.
Fay & Egan Co., J. A., Cincinnati, Ohio.

BASKET STITCHING MACHINE
Saranac Machine Co., Benton Harbor, Mich.

BELT LACER
Clipper Belt Lacer Co., Grand Rapids, Mich.

BENDING MACHINES
Fay & Egan Co., J. A., Cincinnati, Ohio.
Garlock-Machinery, Toronto, Ont.
Perfection Wood Steaming Retort Company, Parkersburg, W. Va.

BELTING
Fay & Egan Co., J. A., Cincinnati, Ohio.

BLOWERS
Sheldons, Limited, Galt, Ont.

BLOW PIPING
Sheldons, Limited, Galt, Ont.

BORING MACHINES
Berlin Machine Works, Ltd., Hamilton, Ont.
Canada Machinery Corporation, Galt, Ont.
Fay & Egan Co., J. A., Cincinnati, Ohio.
Garlock-Machinery, Toronto. Ont.
Jackson, Cochrane & Company, Berlin, Ont.
Nash, J. M., Milwaukee, Wis.
Reynolds Pattern & Machine Co., Moline, Ill.
Shawver Company, Springfield, O.

BOX MAKERS' MACHINERY
Berlin Machine Works, Ltd., Hamilton, Ont.
Canada Machinery Corporation, Galt, Ont.
Fay & Egan Co., J. A., Cincinnati, Ohio.
Garlock-Machinery, Toronto, Ont.
Globe Soap Company, Cincinnati, Ohio.
Jackson, Cochrane & Co., Berlin, Ont.
Neilson & Company, J. L. Winnipeg, Man.
Reynolds Pattern & Machine Co., Moline, Ill.
Whitney & Son, Baxter D., Winchendon, Mass.
Williams Machinery Co., A. R., Toronto, Ont.

CABINET PLANERS
Berlin Machine Works, Ltd., Hamilton, Ont.
Fay & Egan Co., J. A., Cincinnati, Ohio.
Garlock-Machinery, Toronto, Ont.
Jackson, Cochrane & Co., Berlin, Ont.
Shawver Company, Springfield, O.

CARS (Transfer)
Sheldons, Limited, Galt, Ont.

CARBORUNDUM PAPER
Carborundum Co., Niagara Falls, N.Y.

CARVING MACHINES
Fay & Egan Co., J. A., Cincinnati, Ohio.
Garlock-Machinery, Toronto, Ont.
Jackson, Cochrane & Company, Berlin, Ont.

CASTERS
Foster, Merriam & Co., Meriden, Conn.

CLAMPS
Fay & Egan Co., J. A., Cincinnati, Ohio.
Garlock-Machinery, Toronto, Ont.
Jackson, Cochrane & Company, Berlin, Ont.
Simonds Canada Saw Company, Montreal, P.Q.

COLUMN MACHINERY
Fay & Egan Co., J. A., Cincinnati, Ohio.

CUT-OFF SAWS
Berlin Machine Works, Ltd., Hamilton, Ont.
Fay & Egan Co., J. A., Cincinnati, Ohio.
Garlock-Machinery, Toronto, Ont.
Jackson, Cochrane & Co., Berlin, Ont.
Knechtel Bros., Limited, Southampton, Ont.
Ober Mfg. Co., Chagrin Falls, Ohio.
Shawver Company, Springfield, O.
Simonds Canada Saw Co., Montreal, Que.

CUTTER HEADS
Berlin Machine Works, Hamilton, Ont.
Fay & Egan Co., J. A., Cincinnati, Ohio.
Gage Mfg. Company, E. E., Gardner, Mass.
Jackson, Cochrane & Company, Berlin, Ont.
Mattison Machine Works, Beloit, Wis.

DOVETAILING MACHINES
Canadian Linderman Machine Company, Woodstock, Ont.
Fay & Egan Co., J. A., Cincinnati, Ohio.
Garlock-Machinery, Toronto, Ont.
Jackson, Cochrane & Company, Berlin, Ont.

DOWEL MACHINES
Fay & Egan Co., J. A., Cincinnati, Ohio.
Ober Mfg. Co., Chagrin Falls, Ohio.

DRY KILNS
Grand Rapids Veneer Works, Grand Rapids, Mich.
National Dry Kiln Co., Indianapolis, Ind.
Sheldons, Limited, Galt, Ont.

DOWELS
Knechtel Bros., Limited, Southampton, Ont.

DUST COLLECTORS
Sheldons, Limited, Galt, Ont.

DUST SEPARATORS
Sheldons, Limited, Galt, Ont.

EDGERS (Single Saw)
Fay & Egan Co., J. A., Cincinnati, Ohio.
Garlock-Machinery, Toronto, Ont.
Simonds Canada Saw Co., Montreal, Que.

EDGERS (Gang)
Berlin Machine Works, Ltd., Hamilton, Ont.
Fay & Egan Co., J. A., Cincinnati, Ohio.
Garlock-Machinery, Toronto, Ont.
Petrie, H. W., Toronto.
Simonds Canada Saw Co., Montreal, P.Q.
Williams Machinery Co., A. R., Toronto, Ont.

END MATCHING MACHINE
Berlin Machine Works, Ltd., Hamilton, Ont.
Fay & Egan Co., J. A., Cincinnati, Ohio.
Garlock-Machinery, Toronto, Ont.
Jackson, Cochrane & Company, Berlin, Ont.

EXHAUST FANS
Garlock-Machinery, Toronto, Ont.
Sheldons Limited, Galt, Ont.

FILES
Simonds Canada Saw Co., Montreal, Que.

FLOORING MACHINES
Berlin Machine Works, Ltd., Hamilton, Ont.
Fay & Egan Co., J. A., Cincinnati, Ohio.
Garlock-Machinery, Toronto, Ont.
Petrie, H. W., Toronto.
Whitney & Son, Baxter D., Winchendon, Mass.

FLUTING HEADS
Fay & Egan Co., J. A., Cincinnati, Ohio.
Shawver Company, Springfield, O.

FLUTING AND TWIST MACHINE
Shawver Company, Springfield, O.

FURNITURE TRIMMINGS
Foster, Merriam & Co., Meriden, Conn.

GARNET PAPER AND CLOTH
The Carborundum Co., Niagara Falls, N.Y.
Delany & Pettit, Limited, Toronto, Ont.

GRAINING MACHINES
Berlin Machine Works, Ltd., Hamilton, Ont.
Fay & Egan Co., J. A., Cincinnati, Ohio.
Williams Machinery Co., A. R., Toronto, Ont.

GLUE
Coignet Chemical Product Co., New York, N.Y.
Delany & Pettit, Limited, Toronto, Ont.
Perkins Glue Company, South Bend, Ind.

GLUE CLAMPS
Jackson, Cochrane & Company, Berlin, Ont.

GLUE HEATERS
Fay & Egan Co., J. A., Cincinnati, Ohio.
Jackson, Cochrane & Company, Berlin, Ont.

GLUE JOINTERS
Berlin Machine Works, Ltd., Hamilton, Ont.
Jackson, Cochrane & Company, Berlin, Ont.

GLUE SPREADERS
Fay & Egan Co., J. A., Cincinnati, Ohio.
Jackson, Cochrane & Company, Berlin, Ont.

GLUE ROOM EQUIPMENT
Boomer & Boschert Company, Montreal, Que.
Perrin & Company, W. R., Toronto, Ont.

GRINDERS (Cutter)
Fay & Egan Co., J. A., Cincinnati, Ohio

GRINDERS, HAND AND FOOT POWER
The Carborundum Co., Niagara Falls, N.Y.

GRINDERS (Knife)
Berlin Machine Works, Ltd., Hamilton, Ont.
Fay & Egan Co., J. A., Cincinnati, Ohio.
Garlock-Machinery, Toronto, Ont.
Petrie, H. W., Toronto.

GRINDERS (Tool)
Fay & Egan Co., J. A., Cincinnati, Ohio.
Jackson, Cochrane & Company, Berlin, Ont.
Shawver Company, Springfield, O.

GRINDING WHEELS
The Carborundum Co., Niagara Falls, N.Y.

GROOVING HEADS
Fay & Egan Co., J. A., Cincinnati, Ohio.

GUARDS (Saw)
Shawver Company, Springfield, O.

GUMMERS, ETC.
Fay & Egan Co., J. A., Cincinnati, Ohio.

HAND PROTECTORS
Fay & Egan Co., J. A., Cincinnati, Ohio.

HAND SCREWS
Fay & Egan Co., J. A., Cincinnati, Ohio.

HANDLE THROATING MACHINE
Klotz Machine Company, Sandusky, Ohio.

HANDLE TRIMMING MACHINE
Klotz Machine Company, Sandusky, Ohio.

HANDLE & SPOKE MACHINERY
Fay & Egan Co., J. A., Cincinnati, Ohio.
Klotz Machine Co., Sandusky, Ohio.
Nash, J. M., Milwaukee, Wis.
Ober Mfg. Company, Chagrin Falls, Ohio.
Whitney & Son, Baxter D., Winchendon, Mass.

HARDWOOD LUMBER
Atlantic Lumber Company, Toronto.
Imperial Lumber Company, Columbus, Ohio.
Kersley, Geo., Montreal, Que.
Kraetzer-Cured Lumber Co., Moorhead, Miss.
Penrod Walnut & Veneer Co., Kansas City, Mo.
Spencer, C. A., Montreal, Que.

HUB MACHINERY
Fay & Egan Co., J. A., Cincinnati, Ohio.

HYDRAULIC PRESSES
Can. Boomer & Boschert Press, Montreal, Que.

HYDRAULIC PUMPS & ACCUMULATORS
Can. Boomer & Boschert Press, Montreal, Que.

HYDRAULIC VENEER PRESSES
Perrin & Company, Wm. R., Toronto, Ont.

JOINTERS
Canadian Linderman Machine Company, Woodstock, Ont.
Berlin Machine Works, Ltd., Hamilton, Ont.
Fay & Egan Co., J. A., Cincinnati, Ohio.
Garlock-Machinery, Toronto, Ont.
Jackson, Cochrane & Company, Berlin, Ont.
Knechtel Bros., Limited, Southampton, Ont.
Petrie, H. W., Toronto, Ont.
Pryibil Machine Company, P., New York.
Shawver Company, Springfield, O.

KNIVES (Planer and others)
Berlin Machine Works, Ltd., Hamilton, Ont.
Fay & Egan Co., J. A., Cincinnati, Ohio.
Simonds Canada Saw Co., Montreal, P.Q.

LUMBER
Hartzell, Geo. W., Piqua, Ohio.
Imperial Lumber Company, Columbus, Ohio.

LATHES
Fay & Egan Co., J. A., Cincinnati, Ohio.
Garlock-Machinery, Toronto, Ont.
Jackson, Cochrane & Company, Berlin, Ont.
Klotz Machine Company, Sandusky, Ohio.
Ober Mfg. Company, Chagrin Falls, Ohio.
Petrie, H. W., Toronto, Ont.
Shawver Company, Springfield, O.
Whitney & Son, Baxter D., Winchendon, Mass.

MACHINE KNIVES
Berlin Machine Works, Ltd., Hamilton, Ont.
Peter Hay Knife Company, Galt, Ont.
Simonds Canada Saw Co., Montreal, P.Q.

MITRE MACHINES
Fay & Egan Co., J. A., Cincinnati, Ohio.

MITRE SAWS
Fay & Egan Co., J. A., Cincinnati, Ohio.
Garlock-Machinery, Toronto, Ont.
Simonds Canada Saw Co., Montreal, P.Q.

MORTISING MACHINES
Berlin Machine Works, Ltd., Hamilton, Ont.
Fay & Egan Co., J. A., Cincinnati, Ohio.
Garlock-Machinery, Toronto, Ont.

MOULDINGS
Knechtel Bros., Limited, Southampton, Ont.

MULTIPLE BOXING MACHINES
Fay & Egan Co., J. A., Cincinnati, Ohio.
Nash, J. M., Milwaukee, Wis.
Reynolds Pattern & Machine Co., Moline, Ill.

PATENT SOLICITORS
Lightfoot, Stanley, Toronto, Ont.

PATTERN SHOP MACHINES
Berlin Machine Works, Ltd., Hamilton, Ont.
Fay & Egan Co., J. A., Cincinnati, Ohio.
Garlock-Machinery, Toronto, Ont.
Jackson, Cochrane & Company, Berlin, Ont.
Whitney & Son, Baxter D., Winchendon, Mass.

PLANERS
Berlin Machine Works, Ltd., Hamilton, Ont.
Fay & Egan Co., J. A., Cincinnati, Ohio.
Garlock-Machinery, Toronto, Ont.
Jackson, Cochrane & Company, Berlin, Ont.
Petrie, H. W., Toronto, Ont.
Whitney & Son, Baxter D., Winchendon, Mass.
Williams Machinery Co., A. R., Toronto, Ont.

PLANING MILL MACHINERY
Berlin Machine Works, Ltd., Hamilton, Ont.
Fay & Egan Co., J. A., Cincinnati, Ohio.
Garlock-Machinery, Toronto, Ont.
Jackson, Cochrane & Company, Berlin, Ont.
Petrie, H. W., Toronto, Ont.
Shawver Company, Springfield, O.
Whitney & Son, Baxter D., Winchendon, Mass.
Williams Machinery Co., A. R., Toronto, Ont.

PRESSES (Veneer)
Jackson, Cochrane & Company, Berlin, Ont.
Perrin & Co., Wm. R., Toronto, Ont.

PULLEYS
Fay & Egan Co., J. A., Cincinnati, Ohio.

RESAWS
Berlin Machine Works, Ltd., Hamilton, Ont.
Fay & Egan Co., J. A., Cincinnati, Ohio.
Garlock-Machinery, Toronto, Ont.
Jackson, Cochrane & Company, Berlin, Ont.
Petrie, H. W., Toronto, Ont.
Simonds Canada Saw Co., Montreal, P.Q.
Williams Machinery Co., A. R., Toronto, Ont.

RIM AND FELLOE MACHINERY
Fay & Egan Co., J. A., Cincinnati, Ohio.

RIP SAWING MACHINES
Berlin Machine Works, Ltd., Hamilton, Ont.
Fay & Egan Co., J. A., Cincinnati, Ohio.
Garlock-Machinery, Toronto, Ont.
Klotz Machine Company, Sandusky, Ohio.
Jackson, Cochrane & Company, Berlin, Ont.
Ober Mfg. Company, Chagrin Falls, Ohio.

SANDERS
Berlin Machine Works, Ltd., Hamilton, Ont.
Fay & Egan Co., J. A., Cincinnati, Ohio.
Garlock-Machinery, Toronto, Ont.
Klotz Machine Company, Sandusky, Ohio.
Jackson, Cochrane & Company, Berlin, Ont.
Nash, J. M., Milwaukee, Wis.
Ober Mfg. Company, Chagrin Falls, Ohio.
Shawver Company, Springfield, O.

SASH, DOOR & BLIND MACHINERY
Berlin Machine Works, Ltd., Hamilton, Ont.
Fay & Egan Co., J. A., Cincinnati, Ohio.
Garlock-Machinery, Toronto, Ont.
Jackson, Cochrane & Company, Berlin, Ont.

SAWS
Fay & Egan Co., J. A., Cincinnati, Ohio.
Radcliff Saw Mfg. Company, Toronto, Ont.
Simonds Canada Saw Co., Montreal, P.Q.

SAW GUARDS
Shawver Company, Springfield, O.

SAW GUMMERS, ALOXITE
The Carborundum Co., Niagara Falls, N.Y.

SAW SWAGES
Fay & Egan Co., J. A., Cincinnati, Ohio.
Simonds Canada Saw Co., Montreal, Que.

SAW TABLES
Berlin Machine Works, Ltd., Hamilton, Ont.
Fay & Egan Co., J. A., Cincinnati, Ohio.
Garlock-Machinery, Toronto, Ont.
Jackson, Cochrane & Company, Berlin, Ont.
Knechtel Bros., Limited, Southampton, Ont.

SCRAPING MACHINES
Garlock-Machinery, Toronto, Ont.
Whitney & Sons, Baxter D., Winchendon, Mass.

SCROLL CHUCK MACHINES
Klotz Machine Company, Sandusky, Ohio.

SCREW DRIVING MACHINES
Reynolds Pattern & Machine Co., Moline, Ill.
Stow Manufacturing Co., Binghamton, N.Y.

SCREW DRIVING MACHINERY
Reynolds Pattern & Machine Works, Moline, Ill.
Stow Manufacturing Co., Binghamton, N.Y.

SCROLL SAWS
Berlin Machine Works, Ltd., Hamilton, Ont.
Fay & Egan Co., J. A., Cincinnati, Ohio.
Garlock-Machinery, Toronto, Ont.
Jackson, Cochrane & Company, Berlin, Ont.
Simonds Canada Saw Co., Montreal, P.Q.

SECOND-HAND MACHINERY
Petrie, H. W.
Williams Machinery Co., A. R., Toronto, Ont.

SHAPERS
Berlin Machine Works, Ltd., Hamilton, Ont.
Fay & Egan Co., J. A., Cincinnati, Ohio.
Garlock-Machinery, Toronto, Ont.
Jackson, Cochrane & Company, Berlin, Ont.
Ober Mfg. Company, Chagrin Falls, Ohio.
Petrie, H. W., Toronto, Ont.
Shawver Company, Springfield, O.
Simonds Canada Saw Co., Montreal, P.Q.
Whitney & Sons, Baxter D., Winchendon, Mass.

SHARPENING TOOLS
The Carborundum Co., Niagara Falls, N.Y.

SHAVING COLLECTORS
Sheldons Limited, Galt, Ont.

SHELL BOX HANDLES
Vokes Hardware Co., Toronto, Ont.

SHELL BOX EQUIPMENT
Vokes Hardware Co., Toronto, Ont.

SINGLE SPINDLE BOXING MACHINES
Fay & Egan Co., J. A., Cincinnati, Ohio.
Ober Mfg. Company, Chagrin Falls, Ohio.

STAVE SAWING MACHINE
Whitney & Sons, Baxter D., Winchendon, Mass.

STEAM TRAPS
Canadian Morehead Mfg. Co., Woodstock, Ont.
Standard Dry Kiln Co., Indianapolis, Ind.

STEEL SHELL BOX BANDS
Vokes Hardware Co., Toronto, Ont.

SURFACERS
Berlin Machine Works, Ltd., Hamilton, Ont.
Fay & Egan Co., J. A., Cincinnati, Ohio.
Garlock-Machinery, Toronto, Ont.
Jackson, Cochrane & Company, Berlin, Ont.
Petrie, H. W., Toronto, Ont.
Whitney & Sons, Baxter D., Winchendon, Mass.
Williams Machinery Co., A. R., Toronto, Ont.

SWING SAWS
Berlin Machine Works, Ltd., Hamilton, Ont.
Fay & Egan Co., J. A., Cincinnati, Ohio.
Garlock-Machinery, Toronto, Ont.
Jackson, Cochrane & Company, Berlin, Ont.
Ober Mfg. Co., Chagrin Falls, Ohio.
Simonds Canada Saw Co., Montreal, P.Q.

TABLE LEG LATHES
Berlin Machine Works, Ltd., Hamilton, Ont.
Fay & Egan Co., J. A., Cincinnati, Ohio.
Ober Mfg. Co., Chagrin Falls, Ohio.
Whitney & Sons, Baxter D., Winchendon, Mass.

TABLE SLIDES
Walter & Co., E., Wabash, Ind.

TENONING MACHINES
Berlin Machine Works, Ltd., Hamilton, Ont.
Fay & Egan Co., J. A., Cincinnati, Ohio.
Garlock-Machinery, Toronto, Ont.
Jackson, Cochrane & Company, Berlin, Ont.

TRIMMERS
Berlin Machine Works, Ltd., Hamilton, Ont.
Fay & Egan Co., J. A., Cincinnati, Ohio.
Garlock-Machinery, Toronto, Ont.

TRUCKS
Sheldons Limited, Galt, Ont.
National Dry Kiln Co., Indianapolis, Ind.

TURNINGS
Knechtel Bros., Limited, Southampton, Ont.

TURNING MACHINES
Berlin Machine Works, Ltd., Hamilton, Ont.
Fay & Egan Co., J. A., Cincinnati, Ohio.
Garlock-Machinery, Toronto, Ont.
Klotz Machine Company, Sandusky, Ohio.
Ober Mfg. Company, Chagrin Falls, Ohio.
Shawver Company, Springfield, O.
Whitney & Sons, Baxter D., Winchendon, Mass.

UNDER-CUT SELF-FEEDING FACE PLANER
Fay & Egan Co., J. A., Cincinnati, Ohio.
Garlock-Machinery, Toronto, Ont.
Jackson, Cochrane & Company, Berlin, Ont.

VARNISHES
Ault & Wiborg Company, Toronto, Ont.

VENEERS
Central Veneer Co., Huntington, W. Virginia.
Walter Clark Veneer Co., Grand Rapids, Mich.
Evansville Veneer Co., Evansville, Ind.
Hartzell, Geo. W., Piqua, Ohio.
Hay & Company, Woodstock, Ont.
Hoffman Bros. Company, Fort Wayne, Ind.
Geo. Kersley, Montreal, Que.
Ohio Veneer Company, Cincinnati, Ohio.
Penrod Walnut & Veneer Co., Kansas City, Mo.
Waetjen & Co., Geo. L., Milwaukee, Wis.

VENTILATING APPARATUS
Sheldons Limited, Galt, Ont.

VENEERED PANELS
Hay & Company, Woodstock, Ont.

VENEER PRESSES (Hand and Power)
Can. Boomer & Boschert Press, Montreal, Que.
Garlock-Machinery, Toronto, Ont.
Jackson, Cochrane & Company, Berlin, Ont.
Perrin & Company, Wm. R., Toronto, Ont.

VISES
Fay & Egan Co., J. A., Cincinnati, Ohio.
Simonds Canada Saw Company, Montreal, Que.

WAGON AND CARRIAGE MACHINERY
Berlin Machine Works, Ltd., Hamilton, Ont.
Fay & Egan Co., J. A., Cincinnati, Ohio.
Ober Mfg. Company, Chagrin Falls, Ohio.
Stow Manufacturing Co., Binghamton, N.Y.
Whitney & Sons, Baxter D., Winchendon, Mass.

WOOD FINISHES
Ault & Wiborg, Toronto, Ont.

WOOD TURNING MACHINERY
Berlin Machine Works, Ltd., Beloit, Wis.
Garlock-Machinery, Toronto, Ont.
Shawver Co., Springfield, Ohio.

WORK BENCHES
Fay & Egan Co., J. A., Cincinnati, Ohio.

WOODWORKING MACHINES
Canada Machinery Corporation, Galt, Ont.
Garlock-Machinery, Toronto, Ont.
Jackson, Cochrane & Co., Berlin, Ont.
Reynolds Pattern & Machine Works, Moline, Ill.
Williams Machinery Co., A. R., Toronto, Ont.

WHEELS, Grinding, Carborundum and Aloxite
The Carborundum Co., Niagara Falls, N.Y.

WOOD SPECIALTIES
Knechtel Bros., Limited, Southampton, Ont.

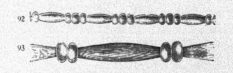

Bead and Turned Mouldings

We manufacture a very complete line of Rope and Turned Mouldings. If the design you want is not found in our line we can make it to your order.

Write us for information and prices

KNECHTEL BROS., LIMITED
SOUTHAMPTON, ONT.

Making TABLE-SLIDES is a Specialty Business

For more than TWENTY-FIVE YEARS we have made TABLE-SLIDES exclusively. Our Factory is equipped with Special Machinery which enables us to make SLIDES.— BETTER and CHEAPER than the furniture manufacturer.

Canadian Table makers are rapidly adopting WABASH SLIDES

Because { They ELIMINATE SLIDE TROUBLES / Are CHEAPER and BETTER

BY USING

Reduced Costs | **WABASH**

Increased Out-put | **SLIDES**

MADE BY

B. Walter & Company
Wabash, Ind.

The Largest EXCLUSIVE TABLE-SLIDE Manufacturers in America
ESTABLISHED-1887

The Machine Supreme

for wire stitching Climax Baskets, is undoubtedly the

SARANAC
Climax Basket
Stitching Machine

It is the newest and most perfect machine of its kind. Its use insures a higher degree of efficiency, greater capacity, absolute accuracy, easier operation and bigger savings in material and labor.

If more profit means anything to you—instal this machine.

MAY WE SEND YOU FURTHER PARTICULARS?

Saranac Machine Co.
Makers of high-grade fruit stapling machines for all purposes
Benton Harbor, Mich.

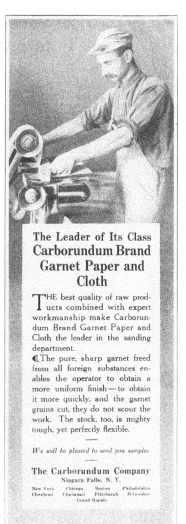

The Leader of Its Class
Carborundum Brand Garnet Paper and Cloth

THE best quality of raw products combined with expert workmanship make Carborundum Brand Garnet Paper and Cloth the leader in the sanding department.

The pure, sharp garnet freed from all foreign substances enables the operator to obtain a more uniform finish—to obtain it more quickly, and the garnet grains cut, they do not scour the work. The stock, too, is mighty tough, yet perfectly flexible.

We will be pleased to send you samples.

The Carborundum Company
Niagara Falls, N. Y.

New York   Chicago   Boston   Philadelphia
Cleveland   Cincinnati   Pittsburgh   Milwaukee
Grand Rapids

Sand Your Square and Octagon Turnings
Faster and Better
on the

ATTISON
EDGE SANDER
With Oscillating Belt

The stationary steel roll at the left may be used to advantage for reaching nearly all the surfaces of patterns made up largely of ornamental members — and the sanding is done lengthways of the grain.

By means of the narrow-nosed form, the curved members, as well as the plain portions of square turnings and similar shaped work, may be sanded with one handling— accurately and economically.

The Secret is in the Combination

And the key to this combination is "Freedom from Vibration."

It is practically out of the question to attempt to sand work like the above against small, revolving rolls — the vibration due to high speed is too great. And to use a stationary roll or narrow form is just as useless if the machine itself chatters and vibrates; for every vibration leaves its mark in the work.

That is why the Mattison Edge Sander is built like a machine tool—rigid and accurate—free from vibration.

Combining this kind of construction with the use of nonvibrating sanding forms; together with our special tensioning device that insures free and smooth travel to the sandbelt, has put this machine in a class by itself.

Compare it with other edge sanders and you'll quickly see the difference.

Our Circular Describes its Construction in Detail.
Send for a Copy.

Volume Sixteen   TORONTO, APRIL, 1916   Number Four

CANADIAN WOODWORKER
and
Furniture Manufacturer

Where You Profit Most

Whether you make duplicate turnings as a specialty or to utilize waste—the proper gauge lathe insures success.

A Whitney Back-Knife Gauge Lathe

has a big working capacity—its speed means dollars earned—its accuracy means perfect reproductions of the back-knife pattern—the variety of turnings possible depends only upon the number of knives with which your Whitney Gauge Lathe is equipped.

The test of a gauge lathe is not on a few mechanical devices alone—but the machine's complete adaptability to your work.

We make the Whitney in four different sizes—20, 30, 40 and 50 inch. They turn stock up to 3 inches square.

The guiding of tools is done mechanically. The head and tail spindles are made of cast steel. Bearings are oil-flooded but oil cannot work out of boxes. The head center, instead of the tail center, is moved to clamp the work, which saves stock.

A quick way to find out how much you need this profitable Whitney Gauge Lathe is to send today to our nearest office for booklet.

Baxter D. Whitney & Son, Winchendon, Mass.

California Office: Berkeley, California.   Selling Representatives:—Henry Kelly & Co.,
H. W. Petrie, Ltd., Toronto, Ontario, Agents for Ontario and Quebec.   26 Pall Mall, Manchester, England.

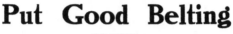

Put Good Belting

on your

Good Machines

Your machines are expensive, and naturally you expect them to turn out the quality and quantity of work commensurate with their cost.

Then why harness them with belting which cannot begin to do them justice?

Use

"AMPHIBIA"
Planer Belting

and get the most work from your machines, in the quickest time at the lowest cost per day of service.

Try a sample run of AMPHIBIA Planer
and prove its merits.

"Leather like gold has no substitute."

Sadler & Haworth

Established 1876

Tanners and Manufacturers

For 40 years Tanners and Manufacturers of the Best Leather Belts.

| Toronto | St. John, N.B. | MONTREAL | Winnipeg | Vancouver |
| 38 Wellington St. E. | 139 Prince William St. | 511 William St. | Galt Building | 217 Columbia Ave. |

—a new small four-side matcher
—that is *really* new

FAY-EGAN "LIGHTNING" No. 369, Planer, Matcher and Moulder

—Surfaces two sides, 24" x 8", matches 15" wide, cuts 1½" moulds.

—Enclosed cut gear and chain drive—powerful and silent.

—Feeds 75' a minute, with two additional rates.

—New "gap" type frame, makes all parts accessible.

—Four 5" driven feed rolls, spring pressures. (No weights.)

—Infeed table removable and separate counter-shaft make matcher easily portable.

—Side wing angle clamp bearings.

—Gravity belt tightener

—Just what many a mill owner has been looking for—a small general purpose matcher with the economy of the big fast feeders—well, here it is.

Bulletin C-10, Free for Asking, or fill in and Mail the Coupon

J. A. FAY & EGAN CO.
153-173 WEST FRONT STREET   **CINCINNATI, OHIO**

J. A. Fay & Egan Co., Cincinnati, Ohio, U. S. A.

Please send Bulletin C-10 and quote price.

Mr. ....

Street ....

City .... Province ....

Second page in the book on handle factory efficiency

Because of the nature of the work and possibilities for speed in cutting blocks, a Blocking Saw should be so simple in construction and easily operated that accurate work is readily accomplished. The Blocking Saw made by the Klotz Machine Company is another necessity for a money-making, one hundred percent efficient handle factory.

KLOTZ BLOCKING SAW MACHINE

has a sloping table so waste falls quickly out of the way; raised platform for operator to stand on; steel mandrel and turned balance wheel. Saw is 20-in. in diameter. Easy access to bearings and nut on mandrel by lifting table which is hung on hinges. This machine, like others of the Klotz line, soon pays for itself.

Turn a Page With Us Next Issue

FOR YOU

who have a thought of manufacturing handles, or those desiring maximum production at the correct cost, get the Klotz Machine Company's idea of a perfectly modern handle factory. One with such dependable machines in it as their

BLOCKING SAW

SHAPING SAW

LATHE

THROATER

BOLTING SAW

POLISHER

DISC WHEEL OR END SHAPER

KNOBBER OR SCROLL MACHINE

All have a place in an efficient handle factory. Write

Klotz Machine Co., Sandusky, O.

Efficiency—
Increased Production
"STOW"
Motor Driven Screw Drivers

Built in Several Sizes   Suspended or Pedestal Design

Immediate Shipment   —   *Write us your requirements to-day*

STOW Tools are an INVESTMENT and not an EXPENSE

STOW MANUFACTURING CO.
Binghamton, N. Y., U. S. A.

Oldest Portable Tool Manufacturers in America

Electric and Belt Driven Drills, All Sizes

INVENTORS AND LARGEST MANUFACTURERS OF STOW FLEXIBLE SHAFTING

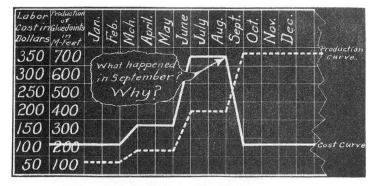

Chart showing relation of cost and production during increased output

COMPARE THESE GLUE JOINTING CURVES

Suppose you're paying a man approximately $85 a month to make 75,000 feet of glue-joints, and the business boom has landed you some big orders for quick delivery. How'll you meet the demand? Increasing your men and machines won't increase your output in proportion to your labor costs. A study of the chart during March and June shows that it's impossible for you to safely increase your output that way.

September 1st you install a New Yates Continuous Feed Glue Jointer. Since the machine can be operated by any mechanic and a boy, the cost curve falls lower than in the beginning, and because the Yates Glue Jointer will do as much as eight hand jointers the production curve goes "sky high." Every joint is perfect. The machine is simple, convenient, and adaptable to many uses other than glue-jointing.

Photographs, Specifications and Full Details of this New Machine sent upon request

The first to adopt an unbreakable roller bearing steel chain feed provided with take-up and single hand-wheel adjustment for pressure bars

Feed tables adjustable at either end of machine; original and distinctive Berlin construction.

P. B. Yates Machine Co. Ltd.

HAMILTON, ONT. CANADA

" Treat your machine as a living friend."

No. 162

High-Speed Ball-Bearing Shaper

FIRST BALL-BEARING SHAPER MADE IN CANADA

Speed 7,000 Revolutions per minute

OUR CLAIM:—Double the amount of work can be done on this HIGH SPEED BALL BEARING SHAPER than can be done on an ordinary shaper, with much more ease to the operator.

DIMENSIONS, WEIGHTS and SPEEDS

Floor Space, 120 in. x 58 in. Net Weight, 2650 lbs. Shipping Weight, 2750 lbs. Tight and Loose Pulleys on Countershaft, 10 in. x 6 in. Driving Pulleys on Countershaft, 18 in. x 5 in. Idler-Pulleys, 14 in. x 5 in. R.P.M. Countershaft, 1200. Pulley on Spindle, 3½ in. x 6 in. R.P.M. Spindle, 7000. Diameter Spindle, 2 3/16 in. Spindle Centres, 24 in. Table, 58 in. x 41 in. Diameter Spindle, Top, 1¼ in. Horse Power Required, 3. Front Projection, 17 in. Height, 35 in.

REPEAT ORDERS BEST PROOF OF MERIT

The Chevrolet Motor Car Company, Limited, of Oshawa, bought three of these machines November 26, 1915. On March 22, 1916, they wrote us: "The three Shapers which we purchased from you several months ago are giving every satisfaction. How soon could you supply us with two more and what allowance could you make for two old shapers?" On March 28 they ordered these two additional shapers, making five in all. They further expressed their opinion in a letter dated April 5, 1916, in which they say: "We cannot speak too highly of the design and workmanship of the machines, or the results which we are able to get from them. **They are, in our opinion, one hundred per cent. perfect.**"

¶ If you are interested in these machines, write us.

The Preston Woodworking Machinery Co. Limited

¶ PRESTON, ONT.

"Treat your machine as a living friend."

No. 184
New "Preston" Tenoning Machine

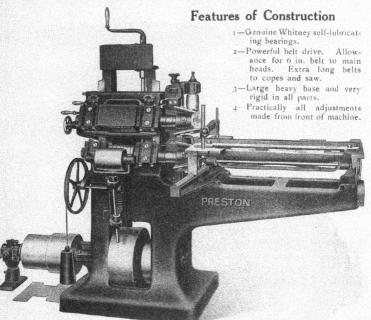

Features of Construction

1—Genuine Whitney self-lubricating bearings.

2—Powerful belt drive. Allowance for 6 in. belt to main heads. Extra long belts to copes and saw.

3—Large heavy base and very rigid in all parts.

4—Practically all adjustments made from front of machine.

This machine is designed for Door, Sash, Blind, Furniture and Piano work. It is the most up-to-date, modern and improved Tenon Machine made to-day. Its design is the very latest, its adjustments are the most convenient, and it embodies all known improvements.

Write us for detailed description. You can reduce costs and increase profits.

Tell us your requirements and let us submit you a proposition.

The Preston Woodworking Machinery Co. Limited
PRESTON, ONT.

Rips and Cross Cuts Lumber— and Reduces Costs

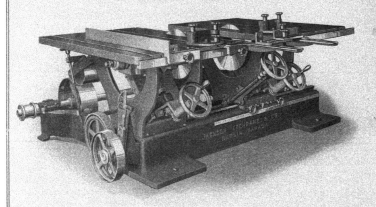

Our No. 164 Double Rip and Cross Cut Saw deserves a place in every woodworking plant. It is designed for cutting off to accurate lengths all kinds of materials used in woodworking factories and will be found particularly profitable in the shell box industry. The frame is a single cast iron structure. The tables are made of iron planed perfectly true and grooved to receive the gauges.

Mandrels are arranged to carry two saws, one on each end, and are adjustable up and down by hand wheel shown in cut. Each mandrel is fitted with an expansion collar to accommodate varying size holes in saws.

Belt Tighteners. This is a new feature on this machine and greatly improves the work of the saw.

Equipment regularly furnished is 2 ripping guides, 2 mitre guides, 3 extension guides and 1 guide for short stock, also two 14 inch cross cut saws and necessary wrenches.

Ask for Bulletin

Jackson, Cochrane & Co.
Berlin, Ontario

The Only Way to Drive Screws

There is a stage-coach method and a modern method of doing everything. When it comes to driving screws the only modern way is with an Automatic Screw Driving Machine, which means greater and better production and greater profit.

The Reynolds
Automatic Screw-Driving Machine

is guaranteed to produce results. It is no experiment. The No. 2 Standard Type, shown herewith, is particularly suited for placing the screws in the straps of shrapnel boxes. It may be fitted to drive, without change or adjustment, any three consecutive sizes flat or round head, or two sizes filister head, No. 6 to 14 inclusive, wood or machine screws, or three sizes flat or round head, or two sizes filister head machine screws up to No. 20. All the above up to 1½ inches long if smaller than No. 10, or up to 1¾ inches long, No. 10 or larger. Fitted with boring attachment if desired.

Many Machines in Use in Canada.

Ask for our Catalogue.

REYNOLDS PATTERN & MACHINE CO.
101 and 103 Third Ave., Moline, Ill., U. S. A.

No. 2 Standard Type Automatic Screw-driving Machine.

Perkins Vegetable Glue
The Best for Built-up Stock and Veneer Laying

Perkins Vegetable Glue sticks and holds. It is unaffected by dampness, extremely dry atmosphere or severe sanding. We guarantee to effect a 20 per cent. saving in your glue room by replacing hide glue with Perkins Vegetable Glue. Along with this saving you can have a glue room comfortable as to temperature and sweet as to smell. The glue is absolutely uniform in quality. Every shipment exactly the same. Let us demonstrate in your own factory and on your own stock just what can be accomplished with Perkins Vegetable Glue.

Dinner wagon manufactured by the D. Hibner Furniture Co., Limited, Berlin, Ont., in which Perkins Glue was used.

PERKINS GLUE COMPANY LIMITED,   HAMILTON, ONT.

"Address all inquiries to Sales Office, Perkins Glue Company, South Bend, Indiana, U. S. A."

Alphabetical List of Advertisers

American Hardwood Lumber Co., 14
American Woodworking Mach. Co., 12
Atlantic Lumber Company 14
Ault & Wiborg Company 37

Baird Machinery Works 51

Canada Machinery Corporation ...
Canadian Boomer & Boschert Mfg.
  Company 55
Canadian Linderman Co. 11
Canadian Moreland Mfg. Company 15
Carborundum Company 59
Central Veneer Company 43
Chapman Double Ball Bearing Co. 55
Churchill, Milton Lumber Co. 14
Clark Veneer Co. Walter 41
Clipper Belt Lacer Company ... ..
Coignet Chemical Products Co. ... 37

Darling Bros. 50

Evansville Veneer Company 38
Elgie & Jarvis Lumber Co. 51

Fay & Egan Company. J. A. 3
Foster, Merriam & Company 10

Garlock-Machinery 13
Globe Soap Company 51
Grand Rapids Dry Kiln Company .. 55

Hartzell, Geo. W. 41
Hay & Company 37
Hay Knife Company, Peter 54
Hoffman Bros. Company 41

Jackson, Cochrane & Company ... 8

Kersley, Geo. 42
Klotz Manufacturing Company ... 4
Knechtel Bros. 58
Kraetzer-Cured Lumber Company.. 49

Lightfoot, Stanley 51

Metallic Roofing Co. 15
Mattison Machine Works 60
Mowbray & Robinson 14

Nash, J. M. 52
National Dry Kiln Company 54
National Securities Corporation ... 51
Neilson & Company, J. L. 52

Ober Manufacturing Company 55
Ohio Veneer Company 41

Perfection Wood Steaming Retort
  Company ... ...
Perkins Glue Company 9
Penrod Walnut & Veneer Company 43
Perrin, William R. 54
Petrie, H. W. 51
Preston- Woodworking Mach. Co.,. 6-7

Radcliff Saw Company 16
Reynolds Pattern & Machine Works 9
Roberts, John N, ... 42
Saranac Machine Company 59
Sadler & Haworth 2
Shawver Company 52
Sheldons Limited 52
Simonds Canada Saw Co. 53
Spencer, C. A. 14
Stow Mfg. Company 4
Toronto Blower Company 16
Vokes Hardware Company 47
Walter & Company, B. 59
Weeks Lumber Company, F. H. ... 54
Whitney & Son, Baxter D. 1
Wood Mosaic Company 42

Yates Machine Co., P. B. 5

THE COMPLETE LINE

For eighty years ACME CASTERS have been reliable casters, the use of which will add prestige to your furniture.

Discriminating manufacturers appreciate the perfection of ACME BALL BEARING or non-ball bearing CASTERS.

The adoption of ACME CASTERS insures the best and quickest service.

We solicit your inquiries and will be pleased to have our Canadian representative call and discuss any special casters or period trimmings you may require.

Foster, Merriam & Co.
Meriden, Conn.

Canadian Representative:—F. A. Schmidt, Berlin, Ont.

"ACME"
The
RELIABLE CASTER

One Piece Sides and Ends for
SHELL BOXES

From Narrow (Cheaper) Lumber at a Low Rate per 1000 feet on the

LINDERMAN AUTOMATIC DOVETAIL MATCHER

Samples on Request

THESE CONCERNS USE LINDERMAN MACHINES:

W. C. Edwards & Co. (2 Machines)	Ottawa, Ont.	Dominion Box & Package Co. (2 Machines)	Montreal, Que.
Czerwinski Box Co.	Winnipeg, Man.	C. P. R. Angus Shops	Montreal, Que.
Cummer-Dowswell Co.	Hamilton, Ont.	J. W. Kilgour & Bro.	Beauharnois, Que.
Canada Furniture Mfrs.	Woodstock, Ont.	Durham Furniture Co.	Durham, Ont.
Martin Freres & Cie	Montreal, Que.	Brunswick-Balke-Collender Co.	Toronto, Ont.
Alberta Box Co.	Calgary, Alta.	Singer Mfg. Co.	St. Johns, Que.

Canadian Linderman Company
Woodstock, Ontario

Muskegon, Mich.   Knoxville, Tenn.   New York, N. Y.

Direct Attached Motor

Mr. Woodworker,
Mr. Furniture Man,

here are three new ones— American No. 444 Cabinet Surfacer; American No. 26 Fast Feed Outside Moulder, and American Combination Roll Feed and Endless Feed Sander.

All Slip-On Heads

Nothing on the market can compare with them. Write for the big pictures, or consult our Canadian Representatives:

Garlock-Machinery

506 Foy Bldg.

32 Front St. West

Toronto   -   Ont.

Telephone Main 5346

Equally effective on heavy or light stock, on short or long pieces—Two machines in one

American Woodworking Machinery Co.

Executive and General Sales Office

Rochester, N. Y.

Woodworking Machinery

We handle all kinds, including full line of

American Wood Working Machinery Co.
ROCHESTER, N.Y.

M. L. Andrews & Co.
CINCINNATI, OHIO

Let us show you the machines covered by cuts in the advertisement of the American Woodworking Machinery Co. on opposite page in operation here in Toronto.

We have a number of American machines for sale—slightly used only — at bargain prices. Write us regarding such tools as you may need.

Let us co-operate with you in selecting and laying out your tool equipment.

Our services are at your disposal.

Garlock - Machinery
32 Front Street West   -   -   · Toronto
Telephone Main 5346

Dry Spruce
and Birch

Good Stocks, Prompt Shipments, Satisfaction

C. A. SPENCER, Limited
Wholesale Dealers in Rough and Dressed Lumber

Eastern Townships Bank Building
MONTREAL   -   -   Quebec

Churchill Milton
Lumber Co.

Louisville, Ky.

We carry at
NEW ALBANY, IND.

Genuine Indiana Plain and
Quartered Oak, Poplar

At GLENDORA, MISS.

Cypress, Gum,
Cottonwood, Oak

SEND US YOUR INQUIRIES

*We have dry stock on hand and
can make quick shipment*

American Hardwood
Lumber Co.
St. Louis, Mo.
Large stock of—

Dry Ash, Quartered Oak
Plain Oak and Gum

Shipments from — NASHVILLE, Tenn.,
NEW ORLEANS, La., and BENTON, Ark.

The Atlantic
Lumber Co.

Manufacturers and
Wholesale Dealers

American and Can-
adian Hardwood Lum-
ber and Mahogany,
Quartered and Plain,
White and Red Oak
a Specialty.

110 Manning Chambers - TORONTO
Phone Main 6386

HOME OFFICE, BOSTON MASS.

Mills: KNOXVILLE, TENN.,   WALLAND, TENN.
FRANKLIN, VA.

Plain and Quartered
Red and White Oak
and Other Hardwoods

Even Color   Soft Texture

We have **35,000,000 feet dry stock**
all of our own manufacture, from our own
timber grown in Eastern Kentucky

MADE (MR) RIGHT

OAK FLOORING

PROMPT SHIPMENTS

The Mowbray & Robinson Co., Inc.
GENERAL OFFICES
CINCINNATI, OHIO
Mills: Quicksand, Ky., West Irvine, Ky., Viper, Ky.

Canadian Representative:
N. H. FARNHAM
601 Elmwood Ave., BUFFALO, N. Y.

STEEL BANDS FOR SHELL BOXES

We are making excellent bands, officially inspected and approved. We are making deliveries as agreed.

You can save real money, also a lot of worry and trouble by linking up with us for your band supply. Ask some of our customers. (Names on request.)

The Metallic Roofing Co., Limited, Toronto, Ont.
ALSO MANUFACTURERS OF "QUALITY FIRST"
Sheet Metal Roofing, Siding, Ceiling, Skylights, Ventilators, Corrugated Iron

Morehead
⟩ Back to Boiler ⟩
SYSTEM

Feed
Your Boilers
Pure Water
HOT !

Practical Pointers on
Boiler Room Efficiency

IN THE NEW "Morehead" book you will find real ready-to-use information on the handling of your steam equipment which may save you thousands of dollars a year.

Besides a general discussion on the drainage of kilns and steam heated machinery generally, there are interesting

Pictures and Data

which will show you how representative plants everywhere are turning their condensation into golden profit.

Write for the book today

Canadian Morehead
Manufacturing Co.

Woodstock Dep. "H" Ont.

BOOKS FOR SALE

The following books are offered at special prices subject to previous sale

Carpentry and Joinery, by Gilbert Townsend. Published in 1913 by the American School of Correspondence. 258 pages, illustrated. Price $1.00.

Saw Fitting Manual, a treatise on the care of saws and knives. Deals with everything in the saw and knife alphabet, from adjustments to widths. 144 pages. Price $2.00.

Common-sense Handrailing, by Fred T. Hodgson. Published by Frederick J. Drake & Company, Chicago. 114 pages, illustrated. Price 50c.

Roof-Framing Made Easy, by Owen B. Maginnis. Published by The Industrial Publication Company, New York. 104 pages, illustrated. Price 50c.

Handrailing Simplified, by An Experienced Architect. Published by William T. Comstock, New York. 52 pages, illustrated. Price 50c.

How to Join Mouldings; or, The Arts of Mitering and Coping, by Owen B. Maginnis. Published by William T. Comstock, New York. 72 pages, illustrated. Price 50c.

Popular Mechanics Shop Notes. Published by Popular Mechanics, Chicago. Easy Ways to do Hard Things, etc. Years 1905-1906. Price 40c. each.

Cabinet Making, by J. H. Reid. Published by Grand Rapids Furniture Record Company. 200 pages, illustrated. Price $1.50.

Utilization of Wood-Waste (Second Revised Edition), by Ernst Hubbard. Published in 1916 by Scott, Greenwood & Sons. 192 pages, illustrated. Price $1.50.

Woodworker Publishing Company, Limited
345 Adelaide Street West, Toronto, Ontario

Special "BEAVER" Cylinder Saw

for cutting holes in shell boxes and in furniture making

Cuts Clean and Fast

The Beaver Cylinder Saw is made of 11 gauge steel and fitted with a heavy crucible steel spring to eject the core. It has a standard shank and will fit any ordinary boring machine. We fully guarantee this saw as it is "fool-proof" and exceptionally strong. Made in practically any size desired.

Illustration shows three views of the "Beaver" Cylinder Saw, also a sample of its work. Note the heavy coil spring for ejecting cores.

Write for prices

Radcliff Saw Mfg. Co., Limited

TORONTO   **550 Dundas Street**   **ONTARIO**

Agents for—L. & I. J. White Company Machine Knives

Read this letter—it proves why

you should write us about our

Improved Automatic Furnace Feeders

Special

Hoods For Woodworking Machines

Slow Speed and Low Power Exhaust Systems

for removing shavings, sawdust, emery dust, lint, smoke and odors

Full Particulars Upon Request

PHONE JUNCTION 1224   DETAIL WORK AND VENEERED DOORS A SPECIALTY

T. H. HANCOCK

SASH, DOORS, FRAMES, BAND SAWING, SHAPING, PLANING, MOULDING, ROUGH AND DRESSED LUMBER

TORONTO   March 14, 1916.

The Toronto Blower Co.,
City.

Dear Sirs:—

I have pleasure in writing this letter in regard to the Blower System you installed in my factory last year. Both the shaving and the separate sawdust systems are working very satisfactorily and I know we have as good a job of piping as it is possible to put up. This work was done under the direction of your Mr. Daniels, whom I consider a very good mechanic.

Yours truly,
J. H. Hancock

THE TORONTO BLOWER COMPANY

TORONTO   156 DUKE ST.   **ONTARIO**

Canadian Woodworker

A Monthly Publication in the Interest of the Woodworking Industry.
Reaches the factories producing interior finish, doors, sash, flooring,
woodenware, furniture, pianos, boxes, and general mill products.

Subscription, $1.00 a year; foreign $1.50.

Woodworker Publishing Company, Limited

Branches: Montreal, Vancouver,
Chicago, New York, London, Eng.

345 Adelaide St. West, Toronto
Phone Ade. 2700

Authorized by the Postmaster General for Canada, for transmission as second class matter.
Entered as second class matter July 18th, 1914, at the Post Office at Buffalo, N.Y., under the Act of Congress of March 3, 1879.

Vol. 16   April, 1916   No. 4

Technical Training

In another section of this issue appears a descriptive article dealing with the woodworking department of the Toronto Technical School. The woodworking department, although complete in itself, occupies only one corner of this immense building, which has been erected for the purpose of giving technical training to the young men and women of Toronto, in order that they may be better qualified to deal with the problems encountered in their daily work. The cry to-day in almost every branch of the manufacturing industry is for skilled workers. Present-day competition demands efficiency in manufacturing, and in order to secure efficiency skilled workers are necessary. That the manufacturers of Toronto have recognized the value of technical training is evidenced by some of the shops in the school, which have been equipped by trade associations who wanted a course of instruction provided in their particular line of work.

Although it would not be possible for the smaller manufacturing towns to erect as expensive a school as the one in Toronto, yet a great deal could be done in the way of providing technical training, particularly pertaining to the principal industries of the town—as, for instance, a town in which furniture or woodworking was the chief industry could provide courses of instruction in cabinet-making, the use of woodworking machinery, and drafting. There is not much doubt but what manufacturers would give such a proposition substantial support. The mechanic who seeks a position today in the furniture or woodworking factories and does not possess a knowledge of detail drawings and blue-prints will never rise above the plane of an ordinary mechanic, who simply does what he is told, and has to be told every move to make. The skilled mechanic who has a knowledge of making and reading drawings is in line for promotion, and there are very few manufacturers who are not looking for a good man to fill some responsible position.

While it is not possible for every man to be a foreman or superintendent, there are some pretty good jobs to be picked up around the woodworking plant. Take for instance, that machine called a "sticker"—it is a sticker (literally speaking) to a great many men, and when a man comes along who can handle it he will not have to seek very long for a position. There are very few sticker men out of work, and the good ones draw much better wages than the ordinary machine hand or cabinet-maker. The demand for good sticker men is always increasing—for the reason that styles in furniture and woodwork in general are becoming more standardized all the time, and the man who can figure out how he can do in one operation on the sticker what formerly required perhaps two or three on other machines is the man manufacturers are looking for.

Even special work, such as that executed from architects' drawings, requires a good man on the sticker; so it must be apparent that the man who possesses mechanical ability and has a fair working knowledge of detail and scale drawings would do well to give the sticker real, serious consideration. More superintendents have graduated from the machine and cabinet rooms than from any other departments. The ones from the cabinet room will usually be found to possess a good working knowledge of machines, and practically all of them will be found to possess a knowledge of drawings—which is sufficient evidence that technical training is an important factor in the woodworking industry.

Know Your Costs

On another page of this issue appears an article on the cost of making shell boxes, which should be particularly interesting to those manufacturers who are submitting tenders with a view to receiving orders. The author of the article has taken for an example the well known 18-pounder 6-round box, and in a thorough and practical manner reduced the operations necessary for its manufacture, into money.

There are manufacturers tendering for shell box orders who lack a practical knowledge of box making.

These men are very liable to overlook important features that will seriously affect the cost. An instance that came to the writer's attention some time ago will illustrate the point. A large furniture factory secured an order for shell boxes. They were making furniture and shell boxes at the same time. It occurred to them that by purchasing their shell box lumber kiln-dried, they would not interfere with the drying of their furniture lumber. While placing an order for a large quantity of lumber for shell boxes the manager asked the superintendent if it would be advisable to have the lumber ripped to the finished width. The superintendent replied "certainly have it ripped to the finished width." Well, to make the story short, the lumber was not thoroughly kiln-dried, and after it was received at the factory shrunk sufficiently to make it too narrow. Now the superintendent was a capable man in the furniture business, but he overlooked a point that a practical box manufacturer would have observed in an instant and made provision for.

There are numerous other points that may be overlooked, with the result that inexperienced men will submit tenders that are too low. It will therefore be to the advantage of any manufacturer who contemplates tendering for shell box orders to read the article "The Cost of Making Shell Boxes" which has been written specially for the Canadian Woodworker, by a man thoroughly conversant with the subject. When you have read the article tell us whether you agree with the statements.

Should a Man Be Discharged for Making a Mistake?

One of the most noticeable improvements in the attitude of employer toward employee, and one that is coming to be a strong bond between them, is the abolishment of the plan of discharging for every slight offence. The time has gone past when this was supposed to be an incentive to better work. With the advancement of the times it has come to be recognized that interest in one's work is the greatest incentive to securing the best result and that much greater interest is taken by employees when they are made to feel that they are a permanent part of the business and that advancement is sure to come as a result of good work.

No man can do his best work with the thought on his mind that he may receive a discharge with his next pay envelope. It is the thought of being a part, however small and inferior it may be, in building up a great organization which is the spur to greater endeavor. It will do much towards banishing discouragement and giving one an optimistic outlook to feel that he has the opportunity to make a better place for himself by doing his best to build up the part of the business with which he is concerned. Where slow promotions are the rule this takes away much of the interest, but firms which are

sufficiently advanced to encourage the workers to greater endeavor through the thought of being a part of the concern should also be alert to the advantage of making promotions as rapidly as one is ready and a place can be made.

Managers differ in regard to this subject, but one successful manager of a large woodworking plant said that it was his opinion that the firm lost time in training new men to fill the place of one discharged. He found it much better to point out mistakes in a way that would prevent their being repeated instead of discharging the one that had made them, for a new employee would have to be trained to succeed him and after several weeks spent in doing this his general qualities might not be equal to the one that was discharged, when the other, if given the opportunity of overcoming that one fault might have been far better.

Some managers and superintendents contend that it is far better to shift employees to other positions if it is found that they do not do acceptable work in a department, rather than to discharge, that is unless it is something more than the regular incompetency, for one might do excellent work in one position and be a failure in another. This method of trying to find the right place for the people whom they engage, lead their employees to remark to others that are in search of a position, "If you succeed in becoming located with that firm you will be sure of a position as long as you want it," if you do good work, for they never discharge an employee except for an excellent reason. The result of such a reputation is that a firm which has it can always secure the best class of employees. Its members have learned that they have a duty to employees, other than paying them their salary and one that will react to the benefit of the concern.

Efficiency

The average man is apt to get a little tired of having the word "efficiency" constantly brought before him, but we cannot overlook the fact that "efficiency" plays an important part in the success of any modern business enterprise. Even the laborer who sweeps the floor can be systematic in his work. System simply means planning your work and then accomplishing it with the least possible number of motions. In this way a man can do a great deal more work in the time at his disposal, with the same amount of labor. The word "co-operation" explains itself and co-operation plus system equals efficiency. Efficiency increases profits. Increased profits mean or should mean increased wages. Thus we have a problem in arithmetic and although it presents some difficulties, by giving it the amount of study it deserves, a solution can be worked out that will give satisfaction to both employer and employee.

A Well String for a Circular Winding Stair

Simple and Practical Method for Use in Woodworking Shop

By W. Graham

The following is a very simple and practical method of developing the well string for a circular winding stair. The first thing to do is. to draw a full-sized part plan of the well and winders, as in Fig. 1, keeping treads 10 and 15 back about two inches from the springing line, and dividing the remaining space equally. By keeping treads 10 and 15 back, as stated, you

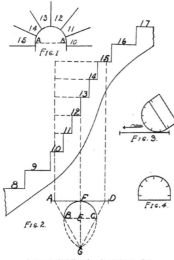

Fig. 1

Fig. 2

Fig. 3.

Fig. 4.

Laying out a Well String for a Circular Winding Stair.

obtain more tread for the winders, which is a great advantage in winding stairs. We need now to secure a development of the half circle with twelve-inch diameter. The following is a simple method. Draw a line B. C. (Fig. 2) twelve inches long—that is, equal to the diameter of the semi-circle that forms the plan of the well. With the point E. as centre midway between B. and C. draw the semicircle B. F. C. Through E. draw F. G. at right angles with B. C. Through F. at right angles with F. G. draw A. D., which will, of course, be parallel with B. C. With C. as centre and radius C. B., draw the arc B. G. With B. as centre and radius B. C. draw the arc C. G. From G. the intersection of the two arcs, draw G. C. and continue the line to D. Then draw G. B. A. in a similar manner. The length of the straight line A. D. is equal to the length of the semicircle B. F. C. Having obtained this length, we can proceed to mark out positions of each riser. To do this take a piece of thin cardboard, Fig. 4, and draw a semicircle twelve inches in diameter and cut

out cardboard to semicircular shape. Then place over Fig. 1, marking out positions of risers. Having obtained this, proceed as in Fig. 3 (which is self-explanatory) to mark the positions on a straight line. This straight line will be equal in length to A. D. It will then be necessary to transfer these positions to A. D., Fig. 2. Draw perpendiculars from 10 and 15 (Fig. 2), and draw in the risers 10, 11, 12, 13, 14, and 15.

To obtain the curved line of the string, draw in treads 8, 9, 16, and 17. From straight parts of string at top and bottom, draw an easy curved line, leaving on sufficient material to give strength to string. All this should be drawn full size on stout paper and cut out to use as a template for transferring to lumber. The curved portion of string will be made of a thin board from one-quarter to three-eighths thick. Steam the thin board well, and proceed to bend around the drum (the making of which we will deal with later), having previously marked positions of springing lines on the drum. Now clamp the thin board on the drum, and be sure to have the springing lines coincide. Prepare the bevelled pieces as shown in Fig. 5, and glue them to back side of veneer. This being done, a piece of stout canvas or other suitable material should be glued over the strips to bend all of them together, making a sound job. The drum should be at least three or four feet wide, consisting of three or four semicircular pieces of seven-eighths pine, good and dry, and the bevelled strips well secured to them. The whole should then be worked to a true semicircular sweep of twelve inches diameter. When the glue is thoroughly dry, the well string may be taken from the cylinder and the paper template applied to the inside of the curve, keeping the springing lines in agreement with each other. Mark on the string the steps and under edge of easement, using the paper template for the purpose. The notches and easement

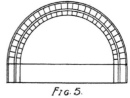

Fig. 5.

Method of building up curved portion of stair string.

lines should now be cut out level with landing—that is to say, be sure to keep the line of saw teeth radial from the axis of the cylinder whilst sawing the notches and easement lines.

It will be understood that treads one to nine form the straight portion of the stair.

Do not be ashamed to shift belts with a stick instead of the hand, and for overhead work always use a stick reaching nearly to the floor.

Revival in Use of Turned Wooden Boxes

There is what looks like a splendid chance to revive and considerably enlarge the trade in turned wooden boxes or small round packages that are familiar in the drug trade and several special lines.

Many of these that were familiar in days gone by have been replaced with glass bottles or pasteboard or fiber boxes and with tins. Perhaps in some instances the substitutes were cheaper, but the indications are that a large part of this substitution has come because the substitute people have pushed their product, exploited it extensively, while the wood workers stood by and let the game go on.

Always something can be accomplished by extensive exploitation and right now the opportunities look better than usual because the parcel post is a factor in the business. There is much complaint being heard of articles shipped by parcels post being damaged in transit because of flimsy packing. The corrugated pasteboard which has shock absorbing qualities will not stand the strain of traffic and is quite often damaged. The ideal plan is to use wooden packages and line them with the corrugated stuff to help lessen the danger of breakage from shocks and have the strength of the wooden package as a safeguard against damage in handling.

It is difficult to figure out specifically many lines of turned wooden boxes that might be advanced by proper effort. It is easily evident, however, that there are many lines, especially boxes in which to pack bottles and other frail goods and also boxes in which to pack medicine in the form of tablets, pills and many other chemical products shipped in small quantities on mail orders.

Those who have machinery for making this line of goods or have material out of which they can be made, by adding some machines, may well make a special study of the matter with a view to seeing something of the field of possibilities. There is evidently room for some good work and the chance to materially enlarge the consumption of turned wooden products of this class by proper effort in the right direction.— Wood Turning.

The Cost of Making Shell Boxes

Is There Any Profit in Present Prices?—Under Favorable Conditions the 18-Pounder 6-Round Shell Box Costs 83 Cents to Produce—Items of Expense Which Cannot be Overlooked

Of all the various details of supplies required in the present war, the boxes for transporting shells and munitions from the factories to the front have been popularly supposed to represent the greatest possible simplicity of manufacture, and therefore a most lucrative form of business requiring no experience, in which various lumber mills, box factories, furniture factories and other concerns who have no factories at all could engage.

It has been said upon good authority, however, that the box business normally requires more brains and experience than most other lines of business, on account of the very small margins between costs and selling prices, and the very large quantities of material that must be handled to insure a good volume of business, and a fair profit. This is all very well for those who understand the game, but it will readily be seen that a slight reversal of conditions, such as a drop in the selling price or an advance in raw materials, may mean as much or more of a loss than there was profit before, and the larger the output the greater the loss.

Then again, there are many expenses entering into the cost of a box besides the bare cost of material and labor, and it is a mistake to overlook or under-estimate them. Especially dangerous is it to assume that larger savings may be effected in the labor, for on a box on which the total labor charge is but 10 cents, no saving can be made which will justify, say, even a five-cent reduction in price. Lumber prices do not vary materially, neither does labor or the component parts of boxes, such as nails, screws, etc., and if any variations in quoted prices of component box men occur, these variations are bound to be small, and depend upon change in conditions which would be too small to figure in most other lines of manufacture.

In the case of the box under consideration, the original contracts were let to box makers at box makers' prices, and were a safe venture under those conditions, but no bonanza, the profit per box probably amounting to about 10 cents. As more experience was gained and the orders were issued in larger lots, prices could be and were reduced, but on the other hand, the costs of all materials entering into their construction began to advance, so that the maintenance of the safe margin called for still further refinements of manufacture, which were secured by the building or purchase of improved machinery and the purchase of material in quantities sufficient to secure lower prices. In order to do this it became necessary to purchase far ahead of orders in hand, and the uncertainty as to the continuation of the business on the same basis added another factor tending to make the straight and narrow path anything but safe. The margin has steadily narrowed, until for the past year it has been impossible to do more than carry the overhead expense of the plant, and at the present time the manufacture of shell boxes is a most precarious and uncertain line of business, requiring, in order to succeed, resources in capital, equipment and supplies which far exceed the popular conception of this line of work. There have been, and still are many concerns allied with the lumber business, who when asked for quotations and estimates on these various boxes have submitted prices which are so low as to give rise to a suspicion concerning their familiarity with the real problems in this line of manufacture, and this article is written with the idea of making as many as possible familiar with the actual facts of the case, particularly as concerns the cost of producing these boxes, leaving the problems of manufacture for the consideration of the individuals who are concerned, as the saving possible by further refinements in manufacturing do not account for the discrepancies in the prices.

It would be impossible to do justice to more than one box in one short discussion, therefore we have selected as being most suitable for this purpose at the present time, the old familiar 18 Pr. 6 Rd. Shrapnel and H. E. Box, which has already been manufactured in very large quantities, and with which all are more or less familiar. The figures and details of manufacture concerning this box which follows are, however, applicable to all other styles of shell boxes at present being manufactured, with the possible exception of the 4 Rd. Fixed Ammunition Box, which is another proposition altogether.

A drawing of the 6 Rd. 18 Pr. Box is given herewith. The lumber, according to the specification, is spruce, finished to ⅞ in. thickness for the sides and cover. Birch, finished to ⅞ in. thickness for the ends

box only, and who have no outlet for their small cuttings and ends, this figure for waste is altogether too conservative, and, depending upon the quality of the lumber, the skill of the operatives, and the wisdom with which the lumber has been bought, so far as securing the proper widths and lengths is concerned, the figure for waste might well run as high as twenty per cent. It seems, therefore, entirely fair in a case like this where we are seeking the rock bottom figures to leave the waste allowance at 15 per cent., so that the actual amount of spruce required to cut one of these boxes would be 6.321 board feet, and the amount of hardwood 1.82 board feet.

Of course, there is considerable labor to be expended upon this lumber before it is in condition for the various special operations which make this box different from an ordinary packing case. These preliminary operations are common to all shooks of any kind, being the resawing, planing, cross-cutting, rip-

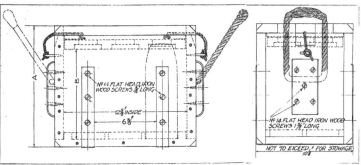

18-Pounder 6-round Shell Box— Carrier Case.

and handle cleats, and pine or spruce 1⅜ in. in thickness for the bottom. The amount of lumber required for this box is as follows:—

2 sides	14⅜ x 11 11/16 x ⅞	2.36	
1 cover	14⅜ x .9⅞ x ⅞	.985	
1 bottom	14⅜ x 8⅛ x 1⅜	1.21	Spruce
2 cover cleats	8 x 1½ x ½	.10	
2 tray cleats	7⅛ x ¾ x 5/16	.035	
1 tray	12⅜ x 7⅞ x ⅞	.683	
		5.373	
2 ends	11 11/16 x 8⅛ x ⅞	1.33	Birch
2 handle cleats	5½ x 2¾ x ¾	.21	
		1.54	

making the total lumber in this box 5.373 board feet of spruce; and 1.54 board feet of hardwood.

In the manufacture of boxes with regular companies engaged in the business and fitted up in every way both with the proper tools for manufacturing, and with a market for the various cuttings and waste, it is an accepted fact that 15 per cent. waste over and above the actual contents in board feet of the box is only a fair and safe working margin. With furniture factories and companies who are working on one size

ping, grooving and trimming operations which give us the shook for the box cut and dressed to the proper size, such as might be furnished by a shook mill in the case of a box for ordinary shipping purposes. Assuming that spruce 1 inch thick and of a quality suitable for the manufacture of these boxes can be bought at $22.50 per thousand feet board measure, and allowing $7.50 per thousand feet board measure to cover the handling of the lumber in the yard, kiln drying what portion is necessary, and manufacturing the shook up to the point we have just indicated, we arrive at a price for the finished shook of $30 per thousand feet, or .03 cents per foot for the spruce, and $40 per thousand feet, or .04 cents per foot for the hardwood, making the cost of our lumber for this box .1896 for the spruce, and .0728 for the hardwood, or a total of .2624. These figures on the cost of lumber and manufacturing of the shook are low, and can only be duplicated at the present time by those who are using lumber bought long ago. In placing the cost of manufacturing the shook at $7.50 per thousand feet, we have named a figure which is as low as this shook could be turned out for by a concern possessing exactly the right equipment and experience for this work, allowing no profit; in fact, in order to purchase the lumber and produce the shook at the figures above named, same would have to be handled in very large quantities, and by people who know every move of the game.

It is very probable that the prices named are too low to apply to the average manufacturer, but as it is our desire to show a minimum price at which this box can be manufactured, we will allow these figures to stand, as they can be adjusted by individuals to suit individual requirements.

In addition to the lumber in these boxes, there is required a certain amount of hardware, a list of which follows, with the price per box which is in force at the present time:—

2 Galvanized iron staples	.004
10 screws, 1½ x No. 14	.021
16 screws, ¾ x No. 14	.025
4 screws, 1¾ x No. 14	.0095
2 rope handles	.10
8 nails, 1¼ in. long; 32 nails, 2¼ in. long; 6 nails, 1 in. long	.0043
2 pieces houseline	.006
2 bands	.06
Corrugated fasteners	.0005
	.2303

Allowing 2 per cent. additional for waste, shortage in count, defective material, etc., total cost of hardware would be .235.

The special operations which are required to convert this shook and hardware into the finished munition box, and deliver same within a reasonable radius as called for by the Imperial Munitions Board, are as follows:—

Ends:		
Cut notches for cover.		
Cut groove for string.		
Drill two holes for staples.		
Drive staples.		
Print or stamp one end.		
Place handles and block	.01	
Cover:		
Cut notches for ends.		
Cut and tie string.		
Cut groove for string.		
Cleat and insert string.	.004	
Bottom:		
Bore six holes 3⅜ ins.		
Cut out corners.	.006	
Tray:		
Bore six 2½-in. holes.		
Countersink twelve 2½-inch holes (on both sides)		
Cleat tray.	.015	
Handle Blocks:		
Cut off.		
Groove		
Drill.	.01	
Nail (frame and bottom)	.01	
Band.	.01	
Insert tray and apply cover	.005	
Inspector's helper.	.005	
Ship.	.005	
Deliver.	.02	
Figuring the cost of our lumber at		.2624
Hardware.		.235
Labor.		.10
Brings the actual cost of this box for material and labor to		.5974

It is necessary, however, in figuring the cost of any article to allow a certain percentage to cover overhead expenses, which overhead includes such fixed charges as rent, taxes, insurance, salaries, repairs to machinery and plant, and depreciation of machinery, or an allowance for the purchase of new tools such as may be required for this work, and for these expenses, an allowance of 20 per cent. is no more than sufficient, bringing the total cost of our box to .746. It will be noted that thus far we have added nothing for profit, and we think that there will be no profit or overhead either at the present price. However, in order to arrive at a fair and just price for this box, we would assume that we are willing to operate at a profit of 10 per cent., which brings the total cost of our box to .83. It has been our experience in fitting up for and manufacturing a large number of these boxes that it was necessary to spend a very considerable amount of money in the purchase of new machines such as were required in the manufacture of these boxes in large quantities, and also in the rearrangement and repair of the various units from time to time in order to maintain and safeguard the very low manufacturing price (without profit) which we have named above, and we are absolutely sure that any individuals or concerns who undertake to manufacture any quantity of these boxes without giving this matter of overhead expense very serious consideration are simply laying themselves open to a disaster sooner or later.

Even considering the bare cost of the lumber, labor and material with no allowance for either overhead and profit, the difference between .5974 which we consider to be the rock bottom price of these three items, and .65c. at which at least one contract was allotted, is too small to justify the large investment required, or for that matter any investment at all.

There are always in any line of business certain individuals who, lacking the necessary knowledge and experience, are attracted by the glitter of a large output and come in with prices which not only mean loss to themselves, but throw discredit upon those already engaged in the work, in the eyes of the purchasers of the product. In some instances, however, the previous record of the old manufacturers and the knowledge of conditions which every good purchasing agent should possess combine to detect an impossible condition like this, and the incautious bidder is turned down even though his price may be less. In the present instance, however, these boxes are being bought at the lowest quoted prices, and manufacturers have no protection against loss except their own knowledge of the situation, and a determination to stand up for their just and legitimate returns, which are absolutely necessary for the best interests of all concerned.

It should be borne in mind that the costs shown above are about as low as can be secured, and that it will be absolutely impossible to lower this figure by trimming either material or labor.

In the manufacture of shells, there is still a considerable margin represented by the abnormal wages being paid to machine operatives all over the country. When a $2 a day mechanic or helper can go into a shell shop and take away $4 to $6 per day in piece work prices, there is certainly a liberal margin for reduction, but there is no such margin in the manufacture of boxes at the present time. If the labor charge could be eliminated altogether, the reduction in price of this box could not possibly be more than 10 cents, and certainly we cannot make the box without labor, any more than we can ignore our fixed charges or overhead.

A Speedy and Accurate Method of Bandsawing Round Table Tops

By W. R. Gould

·The writer was employed for some time in a furniture factory where they manufactured a large number of round tables, such as card tables and extension dining tables. The method of cutting these was to have a boy nail two together and then mark them with a pattern, the man on the bandsaw cutting them by following the line in the usual manner.

One of the men working in the factory suggested making an attachment for the bandsaw which would eliminate the marking and make a nicer job. The suggestion was adopted and the attachment made, with the result that a considerable reduction was made in the cost of the tables and there was absolutely no chance of running off the mark and gouging the edge of the top.

The tops which we had to saw for our extension tables were four feet in diameter, and, of course, were made in two halves. The attachment can be made to saw tops of almost any diameter, and in the hope that it may offer a suggestion to some person in the trade the writer is submitting a description and two drawings, which will convey the idea quite clearly.

The first step is to secure a piece of maple, shown at A, which on our attachment was three feet six inches long by twelve inches wide and one and one-half inches thick. The bandsaw table has two holes bored in it, and the piece A has two holes bored to correspond, and is fastened to the bandsaw table by means of two bolts, H. A reference to the section drawing will show that the bolts H have their heads sunk flush with the surface of the piece A. A piece of board was then glued up four feet long, two and one-half feet wide, and seven-eighths of an inch thick; this piece is shown at B, and it will be noticed that it was made wider than half of a top. This was done

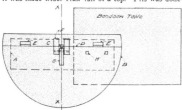

Plan of attachment for bandsawing round table tops.

in order that the stops E and the blocks F might be attached to it for the purpose of holding the tops in place while they were being sawn. By referring to the drawing it will be seen that piece B is attached to piece A by means of bolt C, which has the head set in flush with the surface of piece B. The stops E may be made with two pieces of wood or two pieces of angle iron about three-eighths of an inch thick. Various methods might be adopted for the purpose of holding the tops in place, but the one shown in the drawing will be found as simple and convenient as any. Now all that is necessary to cut the tops is to lay them on the piece B, shove them up against the

stops E, then fasten them down by means of the lever G, and revolve the attachment on the bolt C. It may be found possible to saw more than two at once, but the writer found they could be handled quicker by sawing two . If it is desired to saw round tops which are not made in halves, it will be necessary to make the piece B the full size of the top which it is desired to cut, and in order to hold the tops secure while sawing, nails could be driven into the piece B and then filed off to a point, leaving them to project about one-eighth of an inch. The top would then be pressed down on the nail points, and would be held in place

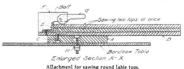

Attachment for sawing round table tops.

while being sawn. The holes caused by the nail points would not matter on the bottom side of the table tops.

This attachment can be very quickly put in place for sawing or very quickly removed when not required.

Motor-Driven Rip Saws

Because of the wide variation from time to time in the power requirements the rip saw is one woodworking machine that presents quite a complex problem when it comes to equipping it with the individual motor drive. It is not only that there are light rip saws and heavy ones, but a rip saw of given size or pattern may be called upon during the day for a wide range of work. It may rip up inch poplar boards one hour and the next hour two inch or three inch oak. Certainly there is .quite a difference in the amount of power required. This puts a man between two fires in figuring on the matter, for we are told on one hand that to get the best order of efficiency out of motor drives the motor should be of such size that it is fairly well loaded all the time. On the other hand, we know that to equip a machine with a motor too light for the work means trouble and probable damage to the motor. The opinions expressed by those who have had experience in equipping motor-driven rip saws indicate that quite commonly people err by making these motor drives too light for the first cost. To the writer it seems that the best order of efficiency may be secured by using direct motor drives only on the heavier rip saws which have a comparatively steady duty at heavy work and making the motors of ample size. Then for the lighter saws used intermittently and on stock varying in thickness from one-half to three inches it is better where practical to make a group motor drive of several light machines on the theory that all of them will not be in commission at the same time and less motor power will be required for the group than if each machine was driven with an individual motor. This should save both in first cost and in current.

The French Period Styles–Louis XVI

The Third of a Series of Illustrated Articles Dealing with Developments in Furniture Design During this Important Era

By James Thomson

On the death of Louis XV.—his grandson the dauphin becoming king—a great change for the better in the moral atmosphere of the court circle was soon observed—a change that soon found reflection in fashions. Furniture began to take form in lines of

Fig. 1.

greater refinement and correctness to classic ideal. Discoveries at Herculanium and Pompeii no doubt had great influence in forming the new style. Decorations of a mural kind, furniture, metal articles and the like had been brought to light in excavating. All

Fig. 2.

such gave impetus to classic idea in designing for modern needs. As a result of this there was soon observable both in France and Britain a reversion to classic forms and a purer taste in design. As a consequence we find in the Louis XVI. style a great deal

that is borrowed from the Pompean. There is, in fact, a strong likeness between the Pompean style and the Louis XVI.

Room decoration in this reign reached the highest point of elegance that French art of a sumptuous kind had touched since the sixteenth century. Rooms panelled from floor to ceiling in straight lines had mouldings richly cut up, enclosing carvings done in the highest styles of the art. Panels were divided by elaborately decorated pilasters, the quills in the flutes en-

Louis XVI.

Fig. 3.

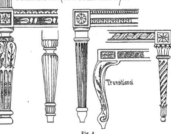

Characteristic examples in the Louis XVI. style.

Fig. 4.

riched by beads or similar subdivision. Panels were exquisitely and delicately carved in arabesque designs, the enclosing mouldings being also enriched with carving. Houses built by the members of the court of Marie Antoinette were filled with exquisite work in this manner, chaste in design and of excellent execution. Where the mural decoration was not gilt, the wood was painted in white or in delicate tones of pale green, rose, or green. The names of a few eminent cabinet-makers of the period have come along the centuries to us, the chief of which are Riesener and Roentgen. They were what is known as ebonisters, which is to say, producers or fabricators of fine cabinet work.

In addition to carving as an embellishment great stress was placed on marquetry and ormolu. Classic and costly woods were employed in the inlay—the world being drawn upon for the rarest. The working of brass had been brought to a high state of perfection, the product of Gouthier being of masterly quality. Splendidly designed, excellently moulded, every piece was finished to a jewelry-like sharpness by tooling with the graver. Such castings were made liberal use of as mouldings, pendants, panel ornaments, key plates and the like on the furniture of Riesener and Roentgen. In the Boston Museum of Fine Arts are a set of beautiful oaken panels from the palace of Montmorency. Exquisitely carved in the best Louis

Fig. 5.

XVI. manner they are a revelation of what can be done with so hard a material as oak when a master workman shows his skill on it. There are eight of these panels meant to decorate from floor to ceiling. The quality of carving is of the same class as is here shown in Fig. 5. Carving all cut from the solid, nothing being glued on.

Figs. 1 and 2 show characteristic historic examples in the way of arm chairs. In Fig. 3 we have a chair representative of the transitional period. Here we discern an effort to get away from the waywardness and flamboyance that characterizes so much of late Louis XV. design. In the covering of seat and back we find a most refined feeling, which leads me to direct readers to the differences to be found in designs of textiles of the various periods, of this series. Compare, if you will, the chair coverings of the examples submitted as representative of the Louis XIV. and XV. with those herewith submitted. It constitutes in itself a liberal education in style, which is the very reason for my being so particular as to include them

Not usually in chair delineation are the designers of coverings so definitely detailed.

Fig. 4 shows chair legs of the period as usually to be found. The curved leg is representative of the transitional period. Once the style was established the cabriole or curved leg went into the discard. It is never to be met with in pure examples of the style. Fig. 7 exhibits the high quality of much of the arabesque carving and running ornament of the times.

Fig. 6.

Salembin, a noted designer of the day, is responsible for these exquisite examples of designing fancy. In Fig. 6 is shown a table typical of the style. While a great deal of the furniture and interior woodwork of the time was coated with plaster of paris and then gilt, or if not gilt, painted in white, pale green, delicate rose and the like, with gold trimmings. Many examples were in natural wood finished with the greatest brilliancy. When finished in green after the Vernis-Martin manner the embellishment was by paintings of the shepherd and shepherdess order. Brass mounts of the utmost delicacy of execution supplied needed emphasis.

Most of the wall painting of the time was in accordance with the "simple life" idea. Marie Antoinette

Fig. 7.

and her court were wont to retire to the little Trianon and play at being villagers. At the dairy the queen herself perhaps made butter, while others played at herding cattle and sheep. The artists of the time taking their cue from this circumstance devoted their talents to depicting scenes of rural felicity, though in a sumptuous way. The paintings in the panelling of walls have for motive, rustic scenes. Textile fabrics and tapestry hangings follow the same fashion. Where furniture of the Vernis-Martin order had painted decoration the subjects were of the shepherd and shepherdess order.

In Fig. 7 is shown a recessed bed to which attaches peculiar interest, historical and otherwise. While a splendid example, designed in best Louis XVI spirit, it in reality indicates his reign by a considerable number of years. This beautiful bed is in the apartment of Madame de Maintenon at the Fontainbleau Palace, France. Madame de Maintenon was the wife of Louis XIV. She was very religious, made an excellent wife and her refined and quiet tastes are shown by the quality of the furnishing by which she chose to be surrounded. Louis XIV. died in 1715, while Louis

XVI. did not begin to reign until 1774, yet here is a room furnished in true Marie Antoinette manner devised at a time when the prevailing style was entirely different. It all goes to show that Madame de Maintenon was a woman of most refined tastes, and one to whom the pretentious Louis XIV. style had no appeal. The Louis XVI. is a style that is chaste and refined. Based upon the classic order of the ancient Greek, there has been engrafted upon it a sumptuous splendor well calculated to appeal to cultivated modern femininity.

In concluding the present series I would remark that my endeavor has been to, in succinct manner, place before readers examples in the various styles that should render it easy to differentiate. A simple presentation of the three styles such as has here been done should be of educational value.

Qualifications a Foreman Should Have

He Should be a Competent Workman—Able to Make His Department Pay—and Popular with His Men

By Lewis Taylor

The superintendent is undoubtedly the brains of any industrial establishment, but it is the foreman or (literally the first man) who gives the verbal instructions to the men and, therefore, comes into closer contact with the employes. His is a dual duty; his primal duty is to discharge his obligations to the firm by whom he is employed—that is, to see that the work is carried out profitably, expeditiously and above suspicion regarding durability and appearance. The cost of producing any article in the wood-working line, like any other, is of chief importance, and justly so, for only by paying attention to that point can any business be conducted successfully. If he is a mill foreman and has the necessary tact and qualifications of his office, he will watch and study the men under his control to see if they are specially fitted for such and such a machine, for there is no doubt that the efficiency of a man is greatly increased if he is put on work that he is adapted for. I have known men who were lost on one machine, put on another and they became experts at it. Then again there are foremen who show a lamentable lack of tact in dealing with their men. There are careful workers and there are careless workers. When the conscientious and careful workman errs, his mistake should be pointed out to him certainly, but in the words of the old saw: "a word to the wise is as good as a sermon." But there are a certain class of workers who have to be disciplined like balky horses or they will be continually doing bad work, and it should be up to the foreman to judge accordingly. There are some men, however, who are incorrigible and of course there is only one way to deal with them.

Then again there is another phase of the question. An incompetent foreman who has a bad and imperfect system in operation in his department is continually finding fault with the results of a faulty system. I recollect working in a small factory years ago with a firm that shall be nameless, as they are still in existence. They had just started in business and the foreman had a batch of stock veneered. As far as any one knew the work was done all right, but—they piled them in a room below the surface of the street with only the floor between them and the damp earth,

with the result the joints parted and the veneer peeled off—results, three men who did the gluing were fired and the foreman kept on. That was certainly a clear case of a foreman's incompetency. A foreman, to be true to his employers, should have an open mind and be ever on the alert for new ideas. Some foremen are loath to move beyond a fixed circle prescribed to them by their old-fashioned way of working, and look with dread, even terror, on any new departure in their department.

I have known foremen who had new ideas forced on them by an astute and enterprising superintendent go about to execute them with a sullen unwillingness, almost akin to despair, with a sort of "Oh I know this thing will never work out right" and then after finally getting into the new way would not go back to the old way at any price.

I stated at the beginning of this article that a foreman had two duties; one to his employers, which no one will deny as being of primary importance, but his duty to his own men is by no means of slight importance, because he has, like any other man in a position of power, the opportunity of making their working day a hell or the reverse. There is always a happy medium between the two obligations, which any foreman who desires the respect of both his employers and his own men can aspire to if he only gives the subject the attention it is entitled to.

It has been noticed that invariably a department takes its tone from the man in charge. If he is not punctual, or is careless, the men will be likewise. The foreman is looked upon as a leader and guide and if he fails in this, the department naturally suffers. There is no sadder sight in a factory than where the foreman is held in silent contempt through inefficiency. It also re-acts on the men, making them lose that spirit that makes for efficiency. What a difference in going into a department where the foreman is in every respect capable, everyone moves smartly and everything hums. In conclusion I would say that a qualified foreman means a paying department both for the employers and for the men, while an unqualified one suits neither. This fact all men who have served as privates in the great industrial army will agree to.

A Well-Equipped Woodworking Plant in the Toronto Technical School

Expert Instruction Provided for Men Who Seek a Livelihood in Woodworking Trades

Staff Article

The impression one receives on entering the woodworking shops of the Toronto Technical School is that he is in a splendidly equipped and up-to-date commercial woodworking plant.

The mill room is 57 feet by 28 feet and equipped with a roller table cut-off saw, power feed rip saw, 16-

Fig. 1—View in Mill Room—Toronto Technical School.

inch jointer, bandsaw, a two-spindle shaper and large belt-sander, all made by the Preston Woodworking Machinery Company, of Preston, Ont. There is also a 30-inch cabinet surfacer of the American Woodworking Machinery Company's make, an automatic

grinder for sharpening saws and an appliance for grinding the jointer and surfacer knives without taking them out of the machines, which makes them absolutely true. Each machine is connected to a blower system. A small but well equipped dry-kiln occupies one end of the mill room. Lumber is purchased in the rough state and when material is required for use in any of the various woodworking shops, a stock bill must be made out for use in the mill room. Fig. 1 is a view of the mill room, and shows the layout of the machinery. The dry kiln, located at one end of the mill room, is also shown.

The cabinet room is 57 feet by 24 feet and the equipment includes a vertical mortiser of variable speeds from 760 to 2,000 r.p.m., made by the American Woodworking Machinery Company, a Cowan single end tenoner, a 12-in. jointer and a variety saw by the Preston Woodworking Machine Company. There are benches and small tools to accommodate twenty students. This class is restricted to those actually engaged in the trade, and the work taken up includes the laying out and construction of furniture, mantels, etc. The student is required to have a fair working knowledge of scale drawings and period styles. Fig. 2 shows the lathes, boring machine, and grinder, arranged at one end of the cabinet room; and the variety saw is shown in Fig. 6. Note the arrangement of the motor, and the fuse box placed at the front of the machine.

There are two pattern-making shops, each 57 x 30 feet. One of them is equipped with a Cowan vertical mortiser, a Fay and Egan 8-in. jointer, an American

Fig. 2.—Machine End of Cabinet Room—Toronto Technical School.

Woodworking Machinery Company's grinder, a variety saw and 11 lathes by the Preston Woodworking Machinery Company, as well as 20 benches and sets of small tools. The equipment of the other room consists of a 20-inch motor head block pattern making lathe with power feed, six 12-inch motor head block lathes, one Preston band saw, one universal saw table, a 6-inch jointer, a Fay and Egan disk and spindle sander, a trimmer and 20 benches and sets of small tools. The instruction given in pattern making includes allowances for shrinkage of castings, the necessary taper for withdrawing a pattern from the mould, allowance for subsequent machining of castings, patterns made in halves, use of cores, core boxes, and core prints, work built up in segments to avoid shrinkage and give strength, parts dovetailed or pinned on to allow of

Fig. 3—Machine End of Pattern Making Room—Toronto Technical School.

Fig 4 -Elementary Wood-working Shops—Toronto Technical School.

withdrawal from the mould and lathe work in wood. Fig. 3 is a view of the machine end of one of the pattern-making rooms. The disk and spindle sander is located in the corner by the band-saw, but is hardly discernible.

One of the real features of the school, and one of which the directors are justly proud, is the construction room, which is 80 feet long, 37 feet wide and 30 feet high, affording ample space for the erection of two moderate sized dwelling houses at one time. A carpenter shop, 50 feet by 28 feet, equipped with a variety saw with mortising attachment, a jointer, and 20 benches and sets of small tools, is located next to the construction room. The arrangement of the carpenter shop is shown in Fig. 5. The students receive instruction in the use of the steel square, laying out and construction of complete hip and valley roof frames, with use of machines for making doors,

sash, and window frames. The more advanced work includes laying floors, laying out and construction of newels, hand rails, strings, treads and risers for stair cases, hardwood trimming, construction of complete stair cases, roof trusses or other problems and the use of woodworking machinery. An Elliot portable woodworking machine is provided which can be used in either the carpenter shop or the construction room, as plugs are located at different places in the walls for the purpose of connecting the machine.

These plugs are also used to connect up the lights after the students learning electrical work have wired the houses. The houses will be finished complete, even to the brick laying, plumbing and heating, as all the different trades have been provided for in the matter of shops and instruction. At the time the Canadian Woodworker's representative visited the school there was a house framed complete, with studding, rafters and winding stair cases in place. The elementary or manual training shops, of which there are two, are equipped with a Preston band saw, an American Woodworking Machinery Company's tool grinder, 6

Fig. 5—Carpenter Shop—Toronto Technical School.

lathes, a Fay and Egan 6-inch jointer and 24 benches and sets of small tools. There is also a pneumatic sanding block by the Cleveland Pneumatic Tool Company and a mortising and tenoning machine supplied by the H. W. Petrie Company, Limited, Toronto.

The work taken up in the elementary classes includes typical problems in construction and the principal joints used in carpentry and joinery. In connection with the bench work, attention is paid to design, and students may design, construct, assemble, and finish a piece of furniture. They are taught the use and care of tools, kinds of woods and their properties, cutting and seasoning of lumber, prevention of checking and warping, varnishing and waxing. They are also taught the use of wood-turning lathes and lathe tools. The two elementary, or manual training shops are located next to each other, with a communicating door, and the arrangement of the lathes is shown in Fig. 4.

All machines are motor-driven by individual motors, with the exception of the lathes in one or two of the shops, which are arranged in rows. The motor

Fig. 6 –Variety Saw in Cabinet Room—Toronto Technical School.

is placed on the floor at the end of the row and the line shaft runs along underneath the lathes quite close to the floor. All danger from belts is eliminated. With junior pupils small classes are necessary in order that the instructor may constantly be in touch with each one to give instruction and avoid accidents.

One point worth emphasizing is that the instructors in charge of the different shops are all experts in their particular line and have been picked from some of the leading industrial plants of the country. It is also gratifying to know that the men in the trade have not been slow in taking advantage of the instruction offered, although the classes have been very much depleted of late, owing to the number of men that have enlisted.

The Shoe Peg Business

Shoe pegs are bought and sold by the pint, peck, and bushel, while the narrow ribbons of wood of which the pegs are made are bartered in rolls. It is therefore apparent that the pegs with which soles are fastened on shoes appear on the market in two forms. In one, the pegs are in separate pieces, each shaped and pointed ready for driving. The other form is a ribbon of wood as wide as the peg is long, and thick enough

to form the body of the peg. The first form is intended for the use of cobblers and shoe makers who work with hand tools. They punch holes in the shoe with a pegging awl, and drive in the pegs, one at a time, by light taps with a hammer. That was the old style method of making shoes, and it is still followed by the worker in the small shop.

The large factory follows another method. It makes shoes by machinery. Instead of buying pegs each individually complete, as the country cobbler does, it purchases them only partly manufactured in the form of the peg ribbons above described. These ribbons are in rolls and are fed to the machine as thread comes from a spool. The machine points the pegs, slices them off one by one very rapidly, and drives them in the sole, all without human assistance except to keep the machine in working order and well supplied with peg rolls.

The pegs are nearly all of paper birch. That wood possesses properties which cause it to be given preference over all other woods. This is the wood of which most thread spools are made. It is a white color, is hard, strong, and tools are not dulled rapidly in cutting it. The peg material is first cut from the log as rotary veneer in proper thickness for the peg that is to be made. The veneer is then further manufactured, either into ribbons or finished pegs.

Pegs are several sizes, ranging in length from those about one-third of an inch to nearly an inch. The short pegs are for light, thin soles, the long ones for heels and the heaviest soles. Pegs are not used in the manufacture of all kinds of shoes. Fine ones may have none, the soles being sewed on. The largest use is in heavy shoes and boots. The pegged sole makes the shoe stiff and rigid and it is apt to be less comfortable than the sewed sole, on the wearer's foot.

It is not possible to determine exactly how much wood is required annually to furnish the country with shoe pegs. Available statistics include pegs, shanks, and wooden heels all under one head, and these articles call for 7,483,000 feet of paper birch; but it cannot be stated how much of this goes into pegs. In New Hampshire the yearly consumption of wood for pegs totals 2,512,000 feet, of which a little more than one-third is yellow birch, the remainder paper birch.

In some factories shoe pegs are split from blocks instead of being cut from veneer. The disks are cut the length of the peg, grooves are cut across the end in both directions forming squares the size of the peg, and at the same process each peg is pointed.

Trade Inquiries

The following trade inquiries have been received by the Department of Trade and Commerce. Those interested can obtain the names and addresses of the firms by writing to the Department and referring to the number of the inquiry.

292. **Dummy rifles.**—Of hardwood, from 3-inch to ¾-inch, turned 2 feet down from handle, remainder rough turned; 35,000 required. (For drawing refer to File A-1665.

302. **Wood coat hangers.**—A Glasgow firm wishes to obtain quotations for wood coat hangers to substitute for continental supplies. Must be good quality and finish.

335. **Broom and hoe handles.**—A firm of Hull importers wishes to be placed in touch with shippers prepared to supply broom and hoe handles.

Points in the Manufacture of Leather Belting

Qualities Essential in Good Belts—Study the Methods of Manufacture to Know How to Choose a Belt

In a paper presented at a meeting of the Providence Association of Mechanical Engineers, affiliated with the American Society of Mechanical Engineers. F. H. Small, chemist of the Graton & Knight Mfg. Co., Worcester, Mass., discussed the manufacture of leather belting, from the selection of the hide to the final operations in the belt shop. After calling attention to the qualities that are inherent in the hide, the speaker described the tanning process at length and in closing enumerated the essentials of good belting, as follows:

First.—We need good driving surface, sufficient friction between belt and pulley to eliminate slippage as completely as may be and to enable the belt to carry its load under minimum tension, thus avoiding useless waste of power at the bearings.

Second.—Lateral stiffness coupled with pliability, the stiffness to keep the belt from twisting and waving and to prevent its curling at the edges when shifted; the pliability to enable it to hug the pulley, wrapping itself round and so securing large arc of contact, and to enable it to alter its shape with the minimum of internal resistance as it travels round the pulley.

Third.—Good tensile strength that it may carry its load without breaking.

Fourth.—Little stretch but considerable elasticity; the former so that it will need to be shortened as seldom as may be, i. e., will do its work uninterruptedly;

up into belting and less than 40 per cent. will go into first quality belting. First, we reject the bellies, cropping or cutting them off at the flank; this loses us 25 per cent. of the hide. Next we cut off the shoulder at a point 4 feet 4 inches from the tail or, if the hide be exceptionally small, at a less distance; this loses us 25 per cent. more, leaving us for cutting into belting a "bend" which constitutes less than 50 per cent. of the original hide.

This bend has now to be curried, i. e., given a supplementary grease-tannage; set out—to give a smooth flat piece of leather—and then stretched. The stretching is done on frames in which the wet leather may be clamped and subjected to as much tension as desired, the leather being allowed to dry under tension on the frames so that an additional stretch resultant upon the natural shrinkage of the leather in drying is imparted to the leather. Before being stretched the bend is usually cut into a center and two side pieces, inasmuch as the center portion being more close fibred will not stretch as much as the side pieces and by being divided as above the leather can be stretched more in accordance with its capacity. After stretching, the leather is rolled and glassed to improve its looks and is then ready to go to the stock room.

When the leather is received in the stock room, it is sorted according to weight or thickness—which in this connection are practically synonymous terms—into extra heavy, heavy, medium and light, and then

---

Can an 18-pounder shell box be made for 83 cents?
Read the article on page 18 and write us about it.

---

the latter so that it may easily take up and let go its load as it travels round the pulley.

Fifth.—Firmness or stability, much the same as lateral stiffness. so that the leather springs little when cut. holds its shape, remains straight and runs true on the pulleys.

Sixth.—Resistance to external conditions, such as heat. moisture, chemicals, etc.; that it may do its work in any place, at any time, and enduringly.

Seventh.—Low initial cost.

It is apparent that no leather can have all these qualities in the highest degree, for some of them are incompatible. Sole leather would do admirably as far as lateral stiffness, little stretch, and firmness are concerned, but would fail lamentably to satisfy the requirements of pliability. tensile strength, elasticity, etc. The best result we can achieve is bound to be somewhat of a compromise. We must aim to get the largest measure possible of the most desirable qualities in our leather, with the least necessary sacrifice of others. Quality belting demands then—first, suitable hides—second, suitable tannage.

We now have the leather and may proceed to prepare it for making into belting. Of the hides which we have tanned only 50 per cent. may legitimately be cut packed down for future use. The sides are packed in square piles with alternate layers at right angles, to

keep the stock flat and straight and allow of a circulation of air through the pile, thus aiding and hastening the seasoning process. This seasoning is an often neglected but most desirable operation, for the use of well seasoned leather is as important in the manufacture of belting as the use of well-seasoned lumber is in building. Belts made from well-seasoned stock stretch less and more uniformly, retain their elasticity and wear longer than belts from green stock.

From the stock room the leather goes to the belt shop where its manufacture into belt takes place. The first step in this process is to straighten one edge of the leather. Next it is cut into strips of various widths by passing between a rapidly revolving circular knife and a guide, the strips being graded for width and roughly cut for quality as they come from the knife, and then stored in racks. From these racks the leather goes to the sorters, by whom it is most carefully graded, both for thickness and quality, and on their expertness depends the maintenance of the standard set for each brand of belt. Accurate judgment of quality depends on wide experience in handling leather and a good all-round knowledge of the specific characteristics of leather in different parts of the hide, for it is a fact that no two square inches of hide are precisely alike.

After the strips have been sorted, they then go to

fitters to be matched and have the laps marked. Pieces cut from the right side of the hide are matched with pieces cut from the left side, because all strips which are not backbone center pieces will stretch in a curve if subjected to a sufficient strain. Narrow strips from a properly stretched side, merely as a result of the stripping, contract to a slight curve. Belts made by joining alternately rights and lefts will roll out in a curve on the floor but will run true upon the pulleys, while belts in the construction of which no attention is paid to the matching of rights and lefts will roll out straight on the floor but will invariably stretch crooked if subjected to sufficient tension on the pulleys. Not all belts show these characteristics noticeably because not all belts are required to transmit sufficient load to develop them early enough in the life of the belt for them to attract attention.

The laps are marked according to the thickness of the stock and usually range between 4 inches and 10 inches in length. Laps must be longer on the shoulder end of the piece because hides become thinner, taper off faster, near the shoulder than near the rump, and longer laps are needed therefore to maintain a uniform thickness of belt. Shoulder ends are joined to shoulder ends and butt to butt because the length of the laps match better, the thickness is more uniform, the stock is similar in quality and the component parts will therefore wear and stretch more uniformly. The laps are scarfed.

The usual cement employed to stick the laps has for its basis animal glue. Each manufacturer is likely to have his own pet formula calling for certain additions to the glue solution and particular methods of compounding, but they all look much alike. By this it is not meant to disparage the cement, for a cement that will hold as long as the leather will wear, is a necessary component of a quality belt. About twelve years ago saw the beginnings of the now indispensable waterproof cement. Laps stuck with this are absolutely unaffected by water, either cold or hot. Whatever cement is used the process of stitching is the same in its essentials. The surfaces of the leather to be joined are coated with the cement, put together in their final position, placed between the plates of a hydraulic press and subjected to heavy pressure. From the presses the belt goes to the inspector and then to stock.

The above description applies more particularly to single belting, but the processes are much the same if double or three-ply belting is to be made. A liberal quantity of stock which has been scarfed and the flesh side of which has been cleaned up with a scraper to remove grease, loose flesh, etc. (this being the side which is cemented) is placed upon the fitter's bench. He matches these together on a smooth surface against wooden blocks which are his standard of thickness, the pieces being matched to secure as uniform thickness as possible and so that the laps of one ply come about half way between the laps of the opposite ply.

Musical Instrument Trade in the West Indies

The import of musical instruments into the West Indies last year was valued at about £25,000. Although not a large trade, it is a growing one, and important enough to warrant the attention of manufacturers of pianos in Canada, as about three-quarters of the total trade in musical instruments is carried on in pianos. The number imported is not given in available statistics, but it is probably between seven and eight hundred. Cabinet organs are not so much in favor, only a few being sold, as the moist climate tends to swell the various parts of the action, making it difficult to keep the instrument in good condition. This cause does not appear to affect adversely the action of pianos to the same extent. Quite a number of phonographs are imported. Imports also occur occasionally of brass instruments for bands or musical societies. The music trade as a whole, however, is not well organized; the stores do not carry, as in other countries, any attractive stock, nor is there an atmosphere in these islands that encourages musical sales. That this could probably be overcome and a much better condition created is evidenced by the success obtained by an American firm in selling its sewing machines through making an attractive display of its products in the various Islands, and by thoroughly canvassing from house to house and selling on the easy payment plan.

Canadian Piano Manufacturers Interested

It is gratifying to note that an interest is beginning to be taken by Canadian piano manufacturers in this trade, due partly, no doubt, to the preference given to Canadian pianos under the Canada-West Indies Agreement, and partly to the growth and necessity for expansion of the Canadian piano trade itself. One Canadian piano manufacturer has intimated that he intends sending down a representative to work up the trade, as he is now making a piano that appears in most particulars to meet the demand in these Islands for a small piano at a moderate price. Several Canadian firms, indeed, have made inquiry for fullest particulars in regard to the requirements of the market, and for a description of the styles and general make-up of the pianos sold, and one Canadian firm has already been able to place a few pianos in Barbados. In view of the fact that some attention has been centred on the piano trade of these Islands, it appears desirable to set out as fully as possible the kind of piano obtaining readiest sale.

Style and Finish of Piano

The style of case preferred, speaking generally, is one that is small in size and simple in ornament. The majority of houses in the West Indies being of the cottage type, the piano suitable for such houses would necessarily require to be small to harmonize with the surroundings. Mahogany wood appears to be most in demand. Occasionally pianos in ebony are imported. One English firm is placing small pianos in the market in solid mahogany cases, the mahogany being probably of the cheapest grade and approximating to cedar. Mahogany alone of all the woods from which piano cases are constructed resists the destructive effects of the wood ant, which eventually eats away and destroys all cabinet work made from any kind of wood except mahogany or cedar. Piano cases that are started to be "made for the Tropics" are secured against warping by screws inserted at certain points along the sides and front where warping is likely to occur. Whether this gives any additional strength to the case or not, it has at least that appearance, and as it is done at small expense it would be advisable for Canadian firms making a small piano for these markets to add this feature to the case, demonstrating at least that the piano was intended for tropical countries. Nearly all pianos were formerly imported with brass candle-sticks on the front, but this does not appear to be necessary, except from the point of view of ornamentation, as most of the better class of houses are very well lighted at night. It would be well, however, to have the front designed with two square end panels, so that if found

Fig. 1

necessary candlesticks could be afterwards attached. The illustration, Fig. 1, will show the style and wood of the case, with measurements, of an English piano now sold in Barbados.

Toy Exhibition

From March 28 to April 11 there was on exhibition in Toronto an extensive display of toys, collected by buyers who were sent to New York by the Department of trade and Commerce at the instance of the Minister, Sir George Foster.

The purpose of the exhibition was two-fold—first, to bring buyers together for the purpose of broadening the demand for Canadian-made toys, and, second, to afford Canadian manufacturers generally a demonstration of the kind of toys heretofore supplied by countries with which we are now at war.

There was also on exhibition a large number of wooden toys and novelties which are manufactured in Canada at the present time, and there is no doubt that, owing to the changed conditions in the relation of the countries, European toys will in future be excluded, thus affording Canadian manufacturers an opportunity of adding to their present lines and securing a market for their products in their own country.

One of the most noticeable differences between the Canadian-made toys and those formerly imported from foreign countries is that the Canadian toys are made with the idea of usefulness as well as amusement, while the foreign toy is apt to aim solely at amusement, and is soon discarded.

The Noah's ark, with its grotesque animals, is a typical foreign-made toy—it is amusing, but not useful; the child's sled, nicely painted, and strongly constructed of good hardwood, is a typical Canadian toy; it is a toy from which the child will get much pleasure as well as good healthy exercise. The toy church, with its spire and colored windows, represents another European idea of a toy; the child's wagon, that is strong enough to haul as big a load as the child is strong enough to pull, represents the Canadian idea.

One point which the manufacturers of toys should always bear in mind, and one which, judging by their exhibits, Canadian manufacturers have realized, is that children are always deeply interested in whatever they see grown-up people do; they like to imitate their elders; they want toy ships and boats to match real water-craft; toy wagons, sleds, trains, houses, and furniture in imitation of these things in real life, about them.

Give the boy a good set of building blocks and a few designs to work from, and watch how long it takes him to construct all the designs; then watch him start to work out new designs for himself, and you will be surprised at the ingenuity and resourcefulness of the little mind. The same is true of the girl: give her a small house with real rooms and furniture, and she will spend many hours arranging and rearranging the furniture.

Toy manufacturing in Canada is only in its infancy, and if manufacturers continue to make toys of a useful and educational nature they deserve the whole-hearted support of the Canadian people—and a conference of women representing the many organizations of Toronto which met for the purpose of uniting in a movement to give preference to Canadian toys leads us to believe that they will get that whole-hearted support. Another point worthy of notice is the effort that has been made to provide interesting amusement in the matter of games for the boys and girls who have grown beyond the toy stage. In this connection the exhibit of Schultz Brothers, of Brantford, Ont., was particularly interesting, with a display of miniature billiard and bagatelle tables, crokinole boards, and others of like interest.

A conference of the manufacturers and buyers at which over three hundred were present was presided over by Sir George Foster, Minister of Trade and Commerce. A committee composed of six manufacturers and six buyers was formed for the purpose of arranging to hold an annual toy exhibition. Altogether the wooden toy and novelty industry seems to be limited only by the amount of thought and energy which is put into it.

Designing and Draughting

"Designing and Draughting" is the title of a book by Alvan Crocker Nye. The book is for the use of designers, draughtsmen, and others who at times find it desirable to make drawings for furniture. It deals with the theory of design, methods of proportioning and rendering, and the dimensions and construction of different articles of furniture. There are twenty-five plates, carefully drawn by the author, who was for some years designer for several of the important furniture makers of New York City. The book is about 10 ins. by 7 ins., nicely bound in dark red cloth. The publishers are the William T. Comstock Company, 23 Warren Street, New York, and the book will be sent post paid on receipt of price, $2.

Increasing the Diameter of an Iron Pulley

The following question was received from a correspondent. An answer is given below, but we should like to hear further from our readers on this subject:—

Carnduff, Sask., March 6, 1916.

Editor Canadian Woodworker:—

Dear Sirs,

I have a 12-in. diameter by 5-in. face iron clutch pulley on an engine, belonging to one of my customers, and he wants to make it larger in diameter. Can you give me an idea how to do this and to fix it on to the pulley, to make a job that will stand up?

Yours truly,

W. H. Cottis.

A method which has been used successfully for increasing the diameter of an iron pulley is shown

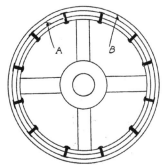

Increasing the diameter of an iron pulley.

by the accompanying illustration. A piece of thick leather belting is attached to the pulley by means of glue and rivets as shown, making a bevel joint at A. Layers of leather may then be glued one on top of the other, breaking joints as shown at B, until the required diameter is secured. In gluing on the first thickness be sure and heat the pulley well and hammer the rivets flat. The number of rivets required would depend on the size of the pulley. A pulley twelve inches in diameter would only require about one-third of the number shown in the illustration.—The Editor.

The Machinery Market

A man who has apparently made a study for years of conditions in the iron and steel industries says that the world is now short some 70,000,000 tons in the production of iron, that the warring nations are wasting iron at the rate of 10,000,000 tons a year, and that since the opening of hostilities there has scarcely been a locomotive, car or rail made by any of the belligerents, except for purposes directly connected with the war.

He estimates that these nations produced fully 55 per cent. of the world's supply, that with their forces of skilled laborers decimated, with new rolling stock to be built, new rails to be laid, new bridges to be built, etc., that at least two or three years must elapse before they will be able to supply the iron and steel

necessary for their domestic consumption, and that with production doubled, the United States cannot meet the demand.

Already there have been sharp advances in the prices of iron and steel and there is no doubt says "Veneers" but they will go much higher within the near future. Those who need new machinery now will be wise to order at once and it will also be the part of wisdom to anticipate future needs in the machinery line as far as possible, because the indications are that it will soon be difficult to secure wood-working machinery at any price. A recent letter, received from a large manufacturer of veneer machinery, says "We have been compelled to increase the price of our machines at the present time running the factory day and night and we find that it is now a matter of obtaining material at any price, and whereas our customers, or some of them, put up a kick, at the same time they, like ourselves, have to take their medicine."

Keeping the Shafting Clean
By Richard Newbecker

Some woodworking plants pay little, if any, attention to keeping their shafting clean, and very often enough oil-soaked sawdust will accumulate and work its way into the boxes to cause trouble. Other plants have their oiler clean off the shafting at regular intervals. This is done by placing a ladder against the revolving shaft and rubbing it with oil-soaked waste to remove any accumulations of sawdust, and at the same time keep the shaft from rusting. This procedure is rather dangerous, and requires a very careful man. If some portion of his clothing happens to catch on a projecting set-screw, the chances are he will be fatally injured before the motor or engine can be shut down.

A far safer and more effective way to clean shafting is by the aid of a simple device constructed as

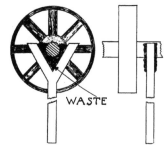

"Safety First" method of cleaning shafting.

shown in the accompanying illustration. It takes only a short time to construct, and with a handful of oily waste fastened as shown, sawdust can be removed and the shafting oiled in a very short time. It saves the oiler from the danger of getting wound up with the shafting. The device should be made to suit individual requirements; no definite sizes can be given, as the height of shafting from the floor varies in different factories.

Woodworking Machinery

The Preston Woodworking Machinery Company of Preston, Ont., are, as the name implies, manufacturers of a complete line of woodworking machinery, in fact they specialize on machines for all kinds of woodwork. Their power feed rip saws No. 129 and 130 will be particularly interesting to those manufacturers who are trying to increase their production and decrease the labor cost on their product. The No. 130 is a short centre machine and especially suitable for furniture factories where it is necessary to work up a great deal of short stock which otherwise would be wasted.

When one considers that in a woodworking factory the waste on lumber averages anywhere from 30 to 40 per cent., it is apparent that a machine that will reduce this waste to a considerable extent is a worth while investment. On the other hand the proprietor of a woodworking factory has to be very careful not to get too much money tied up in machinery or he immediately runs up his over-head expense. It behooves him them to choose wisely and he cannot do better than consult a firm who has made a study of his particular requirements and have embodied in their machines all the latest features that make for efficiency in manufacturing. They will give him expert advice as well as the very best of service.

The Preston Woodworking Machinery Company recently installed a number of machines in the woodworking department of the Toronto Technical School (a descriptive article of which appears on another page of this issue.) This fact in itself speaks volumes for their machines, as nothing but the very best was purchased for this school. One machine worthy of special mention is their No. 162 high speed ball bearing shaper, one of which was purchased for the mill department of the Technical School. This machine has a heavy base over which the weight is evenly distributed, thereby eliminating any vibration in the table. The speed of the machine is 7,000 R. P. M. These two features together with the ball bearing feature commend themselves to those manufacturers who require a machine that will cut without leaving knife marks, as work of this kind requires very little sand papering.

One of their latest designs is their No. 184 new Preston Tenoning Machine, which is particularly suited for door, sash, blind, furniture and piano work. This machine possesses a number of excellent features. A back view of the machine is shown herewith, and a front view is shown in their advertisement in this issue.

The number of testimonial letters the company are

Back view of No. 184 New Preston Tenoning Machine.

receiving with reference to their No. 126 high speed ball bearing shaper indicates that its good qualities are appreciated.

Preservation of Woodwork.

Generally speaking, any wood will last a long time when protected from moisture. It would, however, lose its agreeable color and surface unless some means are used to protect it. The use of chemical preservations is confined almost entirely to underground timbers and so has little use in millwork. One of the places millwork would be benefitted by chemical treatment is at the base of porch columns, where treated wood might well be used where water is apt to penetrate. Creosote is the best of practically all preservatives, and there are a number of methods of using it.

Should the Foreman Finisher Always be Blamed for Defects in the Finish?

By A. Williams

If a piece of furniture shows defects in the finish after being placed on the dealer's floor, and is returned to the factory, the foreman of the finishing department is invariably the man that is called upon to walk the carpet and explain the situation. Very often it is impossible for him to convince the manager that the fault may lie with some other department. The manager's answer is generally to the effect that his (the foreman finisher's) was the last department to handle the work and that if the defects had been caused by inferior work in the machine or cabinet departments they would have been noticed before he received the goods.

In very few cases does the fault rest with the finishing room; it will invariably be found to originate in either the machine or cabinet department. A short review of the progress of the furniture, from the dry kiln to the finishing room, will show where most of the defects have their beginning. Each foreman receives instructions to the effect that he must conduct his department on a percentage of the cost of the finished product. Now, it is to be expected that no foreman will allow an operation to be performed in his department that he can shift on to "the other fellow." The machine room foreman is the first man on the job, and he is generally the man who looks after putting the stock in the kilns. His first job, then, is to hand over the stock bills to the stock-cutter.

The necessary cars having been brought out of the kiln, and the lumber allowed to cool, the stock-cutter then proceeds to work up his stock to the best advantage. An operation very closely related to sawing the stock that once came to the writer's attention might be worth taking up at this point. The dry kilns were located about fifty feet in a direct line from the breaking-out end of the factory, and the intervening space covered with a good roof. A planer which had been replaced by a new one in the factory, instead of being scrapped was moved out and placed about half way between the kilns and the factory. When the lumber was taken from the kilns it was run through this machine, cleaned on one or two sides (depending on what it was going to be used for), and put back on the car. The car was then run into the stock-cutter, who was able to cut the stock to much better advantage, as he could plainly see any defects, and it would not be necessary to get out extra pieces to replace faulty ones.

One point where trouble very often starts, but does not show up till it reaches the finishing room, is when sufficient stock is not taken out of the kiln to complete the cutting list and another lot is rushed right into the factory and cut up without allowing time for it to cool. The machine room foreman will very often give instructions for this to be done, to avoid keeping his men idle, thereby keeping his percentage down. The trouble that is liable to develop by rushing lumber through the factory without allowing it to cool cannot be estimated; it shows up later on, and the finishing department generally gets the blame. Another point that should be chalked up against the machine room foreman is hiring cheap help for some of the most important machines. Some foremen figure out that they can reduce their percentage by having one or two good men to set up the machines and letting any Tom, Dick or Harry put the stock through. This is certainly a mistake, as these men put the stock through without looking to see which way the grain runs, or anything else, and consequently the machine has not a chance to make a clean cut—with the result that the cabinet room foreman and finishing room foreman have to patch up the work between them.

There is only one way to cope with a machine room foreman of this kind—and that is to send all the stock back to him and compel him to do the work right. On the other hand, the machine room foreman sometimes has such a large department to look after that he cannot possibly supervise all the details of the work. In that case he should have reliable assistants to look after certain machines and see that they are turning out satisfactory work. This would prevent any overlapping, and give him more time to devote to something else.

We will now assume that all the machine work has been completed and the parts sent to the cabinet room for assembling. The cabinet room is also generally conducted on a percentage basis, and the slogan is, "Put the stuff together and get rid of it." When it goes to the finishing room the foreman is afraid to examine it too closely, because he knows he will find glue in the corners of the panels which should have been cleaned off, small splits in the ends of some of the solid oak tops which should have had a little glue run in them and then drawn up with a clamp, little blisters in the veneer, spots that should have been sandpapered, mouldings that should have been sponged before cleaning, and a dozen other things that should have been attended to in the cabinet room.

These are only a few of the things the foreman finisher is up against. He is between two fires, because, in a great many cases, he is getting bad work from the cabinet room, and at the same time the office will be chasing him up for goods to fill their orders. The only relief he can hope for is a good superintendent who understands the business and can place the blame where it belongs, and one who will make a careful examination of all the assembled work before it is passed on to the finishing room. It has often been said that the superintendent should pay a good deal of attention to the shipping room to see that the stuff is going out. It is all right to watch the quantity going out, but he should also spend some time around the machine room in order to be sure that he is getting quality as well as quantity. If the lumber is started right, and given careful supervision all the way through the factory, the finishing department will have a fair chance, and there is not likely to be much furniture returned owing to defects in the finish.

Good Results can be Secured with the Spray

By Herbert Brettman

Four years ago, when the spray first came into this shop, the manager said to me: "Herbert, here is a paint sprayer; see if you can find it useful in our shop. It is said to be a great labor-saver."

I was rather pleased that the boss picked me out to test the machine, because I was one of the youngest brush hands in the shop. Among my associates were men who had years of experience. Naturally, they were inquisitive when they saw the sprayer in operation, and to one of them I thoughtlessly repeated the remark of the boss, viz., "It is said to be a great labor-saver." That remark was really a declaration of war between the rest of the shop hands and myself, for their combined cunning, ingenuity, diplomacy, criticism, etc., were used against me and the spray. If there was any trick, device, or act that they did not resort to in order to down the spray proposition, it was because none of their number thought of it.

The foreman was with me until I began to make successful use of the spray, when he evidently saw the handwriting on the wall and figured that the possibilities were that he would lose a considerable percentage of the men he had working for him, and, therefore, he would not be as "big" a foreman in the eyes of the outside world as he was. On the other hand, I felt that the foreman's value was rated, not by the number of men he supervised, but by his cost of producing a certain article.

When I could spray a coat of shellac on a small cabinet in one-fourth the time that it took me to brush it on, and do it in such a way as to require hardly any sanding afterward; when I could blow dust out of the drawers by compressed air, instead of trying to coax it out with a bristle duster; when I could get into the carving without having to poke and stab at it with a brush, I knew I had a friend in the spray and did not care if the foreman liked it or not; in fact, I already had designs on his position.

However, what resulted in the future was another story, but suffice to say we are using eight of these spray devices in this shop today. We had four brush hands when the first machine came in. Now we use one brush hand for miscellaneous work that cannot be sprayed. This fact is remarkable, especially when one considers that although our shop is three times as large now as it was when the first spray came in, it is a fact that we use only one brush hand today for miscellaneous work.

At first we exhausted the fumes caused by the spray, by means of a blower type exhaust fan, but this fan got clogged quickly and its operation soon became very unsatisfactory. The next improvement was to connect up the fan without having to push the exhaust out into the open. This was done by making a "Y" pipe connection. While this kept the blower blades clean, it required a larger blower and took twice the power to operate it. Our first installation was operated by a 2-h.p. motor, and the wide pipe arrangement compelled us to use a 5-h.p. motor to do the same work formerly done by the 2-h.p.

Today we are using an especially designed electric exhaust fan, having consigned the blower type fan to the junk heap. This we were able to do because we are using a three-unit electric exhaust fan set, in the back of an exhaust cabinet, side by side, the three having a combined power consumption of ½-h.p. If this were the only advantage, it would have paid us to make the change to electric, but it happened that we secured another decidedly distinct advantage that the manufacturers of the spray equipment had not themselves realized—a gentle exhaust. What an operator really wants is not a powerful exhaust that will almost pull off his hat, but one that will exhaust the fumes gently and efficiently over a wide area.

My experience proved conclusively to me that it was wrong to try to pull or push the bottom of a 12-in. diameter cone, which is really the shape the spray becomes as it leaves the gun, the cone's apex being at the nozzle, through an 8-in. pipe. Therefore, I am pleased with the electric exhaust, which has an 18-in. opening for the 12-in. cone of varnish or shellac, whichever the case may be. This relative area of opening would, of course, be proportionately increased in a fumexer for handling, say, 12-ft. tables. The proportionate exhaust areas in our installation would be an 8-in. circle under the old condition and three times an 18-in. circle under the electric system.

Pitting

In the early days we had, of course, more or less trouble with varnish due to "pitting." We tried to get away from this by heating the air inlet pipe, but could not get any uniform results, as the air heated in this way would vary a great deal in temperature and would nearly always be cold, due to irregular radiation. At that time the quality of our finish, while a little bit better than that produced by the brush, was not altogether satisfactory. We were not able to get what we would call highly satisfactory results until we installed an electric heater. With the aid of this heater we are now able to produce the most uniform results, perfectly free from pits.

Pinholes

It might be well to mention something about pinholes, a trouble which usually appears in mahogany. I have heard a great many finishers lay the cause of this trouble to the spray, but my experience has been that there is no way that you could manipulate the spray to make it produce pinholes, should you so desire. We used to have the same trouble before the spray ever came into the finishing room, and it is my firm conviction that the whole question of pinholes is due to improper filling of mahogany in the first place.

In using the spray I take three factors into consideration: 1—The air pressure; 2—the consistency of the material; 3—the distance the spray is held away from the work to be finished.—Veneers.

BUILT-UP VENEER PANELS

Ready for Immediate Shipment April 15, 1916 **Subject to Prior Sale**

Good Two Sides, 5-ply.

1 Qrt. Oak, 2s,	72 x 24 x ½	11 Pl. Oak, 2s,	60 x 30 x ⅜
17 Pl. Oak, 2s,	60 x 20 x ⅜	22 Qt. Oak, 2s,	60 x 30 x ⅜
9 Qrt. Oak, 2s,	60 x 20 x ⅜	12 Birch, 2s	72 x 20 x ⅜
32 Pl. Oak, 2s,	72 x 24 x ⅜	8 Birch, 2s,	60 x 30 x ⅜
41 Qt. Oak, 1s, Pl. Oak, 1s,	60 x 30 x ⅜	2 Qt. Oak, 1s, Birch, 1s,	60 x 30 x ⅜
		57 Pl. Oak, 1s, Birch, 1s,	60 x 30 x ⅜

Good Two Sides, 3-ply.

22 Birch, 2s,	60 x 30 x ¼	10 Pl. Oak, 2s,	60 x 30 x 3/16
20 Birch, 2s,	60 x 30 x 3/16	19 Pl. Oak, 1s, Birch, 1s.	60 x 30 x 3/16
5 Qt. Oak, 1s, Birch, 1s,	60 x 24 x ¼	18 Qt. Oak, 1s, Birch, 1s,	60 x 18 x ¼

Good One Side, 5-ply.

23 Qrt. Oak, 1s,	60 x 24 x ½	30 Qrt. Oak, 1s,	60 x 18 x ⅜
10 Qrt. Oak, 1s,	48 x 18 x ⅜	140 Qrt. Oak, 1s,	60 x 30 x ⅜
1 Mahogany, 1s,	72 x 24 x ⅜	47 Qrt. Oak, 1s,	60 x 24 x ⅜
4 Mahogany, 1s,	72 x 20 x ⅜	61 Qrt. Oak, 1s,	60 x 20 x ⅜
3 Mahogany, 1s,	60 x 20 x ⅜	20 Pl. Oak, 1s,	60 x 30 x ⅜

Good One Side, 3-ply.

6 Mahogany, 1s,	72 x 24 x ¼	6 Pl. Oak, 1s,	60 x 30 x ¼
10 Mahogany, 1s,	60 x 30 x ¼	4 Birch, 1s,	60 x 30 x ¼
9 Mahogany, 1s,	60 x 20 x ¼	6 Birch, 1s,	60 x 14 x ¼
		7 Qrt. Oak, 1s,	60 x 30 x 3/16

Single Ply Veneer

1/28 in. MAPLE VENEER, 60 in. to 72 in. long and 6 in. to 36 in. wide, averaging 24 in. to 30 in.
1/20 in. MAPLE VENEER, 72 in. long and 6 in. to 38 in. wide.
1/16 in. MAPLE VENEER, 60 in. to 72 in. long and 6 in. to 36 in. wide.
ALSO CULL VENEER 1/8 IN. AND THICKER, SUITABLE FOR CRATING.
Write for Prices

HAY & COMPANY, Limited - - WOODSTOCK, ONT.

WE HAVE BEEN MAKING THE

Best French Glue in the World

SINCE 1818; ALMOST A CENTURY

In competition at the numerous Expositions abroad, the quality of our glues has been found so high that we are usually declared "Beyond Competition." Our trade in this country makes the same declaration. Consistency in quality and service has brought this result and we are proud of it.

Perfection in manufacture from the best raw material makes the quality of COIGNET FRENCH GLUES consistently the highest.

Large stocks at Montreal assure the best of service.

Why not take advantage of these most important factors by using our glues?

At your service for prices and samples.

Coignet Chemical Products Co.

17 State Street, N. Y. CITY, N.Y.

Factories at {St. Denis / Lyons} France

ANGLO
RUBBING and POLISHING

Works free and easy and can be rubbed in two days.

WRITE FOR SAMPLE

The Ault & Wiborg Co. of Canada, Ltd.

Varnish Works

Montreal Toronto Winnipeg

THE GLUE BOOK

WHAT IT CONTAINS

Chapter 1—*Historical Notes*
Chapter 2—*Manufacture of Glue*
Chapter 3—*Testing and Grading*
Chapter 4—*Methods in the Glue Room*
Chapter 5—*Glue Room Equipment*
Chapter 6—*Selection of Glue*

WOODWORKER PUBLISHING CO., LIMITED

345 Adelaide St. West, Toronto

The Manufacture of Jewelry Cases

An interesting branch of the woodworking business which has come to our attention recently is the manufacture of fine jewellery cases and cutlery cabinets, etc., as carried on by the J. Coulter Company, of Toronto. These cases are made in walnut, mahogany, and oak, and present some very fine cabinet work. The company do not use any veneer in the making of their products. Everything is made from the solid wood; even serpentine fronts are bandsawed out of the solid piece.

The tops for some of the cases are framed up with pieces about 3½ or 4 inches wide, mitred at the corners, then moulded on the shaper. It requires a very heavy shaper to do this work, as the knives project from three to four inches.

The frames for the tops are shaped by placing them on a heavy hardwood form and fastening them securely. The form is then run against the collar, and requires careful manipulating when the mitred corner is reached. An interesting feature in this company's plant is a dust-proof sanding room, where the sanding machines are located. All dust is drawn away by means of a fan, leaving a good wholesome atmosphere for the workmen. At the same time it prevents the dust from penetrating to other parts of the factory and making conditions unpleasant for the other employees.

The other machines in use are saws, planers, jointers, etc., common to woodworking factories, and a variety of small machines adapted to small case work.

The cases are all lined with velvet or silk in a separate department. This work is done by skilled female help, the most of whom received their experience in

Birmingham, England. The lining gives the case a nice finished appearance. We are pleased to be able

Cutlery Cabinet.

to reproduce a cut of the company's product, which will no doubt prove of interest to wood-workers.

The wearing of long coats by workmen should be discouraged. Aprons not ragged and fastened with strings that will not break easily under strain are better.

If you will shift a belt by hand, use the palm with thumb and fingers extended, and look out for metal splicings.

We don't merely buy logs; we select them. The above "Simon Pure" Indiana Forked Leaf White Oak Logs are merely average logs with us. This is one reason why our Oak Veneer is accepted standard.

Importers and Manufacturers

Mahogany and Cabinet Woods—Sawed and Sliced

Quartered INDIANA White Oak, Red Oak, Figured Red Gum,
American Walnut, Etc.—Rotary Cut Stock in Poplar and Gum
for Cross Banding, Back Panels, Drawer Bottoms and Panels.

The Evansville Veneer Company, Evansville, Ind.

How One Man Solved Veneer Troubles

By L. O. Hanson

That the average run of veneer troubles can be overcome is an established fact, and all that is required to do it is proper facilities, a little common sense and good judgment. An incident to prove this happened recently in a plant manufacturing high-grade furniture.

Like many others in the business, these people were having troubles in the veneer department, or rather most of them were discovered in the finishing room. They were of the ordinary variety, such as blisters, checks and bad joints. The finisher was blamed for some of it and the veneer man for the balance. The plant was owned and operated by a man who had spent nearly forty years at the business, but the competition of recent years was fast forcing him to the wall.

The outfit in the veneer room consisted of a variety of hand presses, some of which were constructed of wood, and these, through constant use, had become so strained and weak that it was impossible to get sufficient pressure for some kinds of work. There was also a home-made glue-spreader that would sometimes put on enough glue, but more times it wouldn't, and as a rule about half of the stock had to be re-spread with a brush. The cauls were a collection of several years, and as a result there was quite a variety—some of zinc of various thicknesses, while others were of tin, and a few, recently added, were from a tinsmith's stock of stovepipe material. A large caul-heater occupied one corner of the room, and it was quite evident that there was steam in the pipes, as it was conspicuous at nearly every joint.

The quality of work that came from the veneer room was just about on a par with the facilities they had to do it with. A few months ago the man who had been foreman of the factory for years, resigned, partly through disgust and partly on account of old age, and the proprietor was searching for a capable man to take his place. There were several applications, but only one appealed to the proprietor as being fit for the position. He was well recommended by a firm in a neighboring town as a first-class man.

The proprietor was rather close on the salary question, but finally engaged the man at a moderate salary and a percentage of the net profits of the concern. He didn't reckon the new man would get rich quick by his share of the profits, but as he pressed for a five-year contract, with absolute authority to make any changes he saw fit, the proprietor agreed. The new man fully realized that the veneering was the one great drawback to the successful operation of the plant, and at once got busy in that department. He had the place enlarged and properly ventilated and added some new facilities, the first of which was a fine big hydraulic press; next was an up-to-date glue-spreader and glue-dissolver. The matter of cauls

also received consideration, and a supply of retaining forms were made to use in connection with the new press, and in the course of a month everything in the veneer room had been arranged to the new man's taste. While the proprietor had very little to say, it was quite evident he thought he had given the new man too free a hand; however, he kept his thoughts to himself and awaited results.

Unlike many other plants, most of the trouble here was on flat work, and it was a common occurrence to find, on starting the machine work, that a good percentage of the job had to be returned to the veneer room to be re-veneered; sometimes for nothing more than a lap joint in the face veneer, but more frequently it was a case of blisters and split veneers. As the class of goods would not permit of too conspicuous a patch, the only alternative was to plane off the veneer and do the job over again.

The new man had taken note of these things and had given them careful consideration, and figured that the re-organization of the veneer department would remedy the evil; and it had the desired effect, for it was only a short time until the work began to go through the veneer department in a more mechanical way. As fast as the stock was veneered it was placed in a drying room, arranged by the new man, and under no consideration was it allowed to go to the machine department until it was in proper shape. The result was that when work was started on a job everything was ready for it and no parts had to go back for repairs. This was quite different from the former regime, for the veneer room could just nicely supply the machine room when the work was satisfactory; but with the amount of veneer repairs and the delays caused by it in the machine room, the output of that department was 25 per cent. less than it was capable of doing under favorable conditions. With the new facilities in the veneer department, however, a supply of veneered stock was fast accumulating, and instead of waiting for repaired parts and for another job to come from the veneer room, the machine department found itself well supplied with work that would pass inspection at every stage of the game.

The beneficial effect of the new order of things was just as noticeable in the cabinet room as it was in the machine room, and here, as elsewhere, it doesn't take much to waste a few hours trying to doctor up a poor job of veneering. In the finishing department a similar effect was noted, and right here is probably where the old order of things was the most annoying, for it was practically an everyday occurrence to start to rub a case and find it in such a state, through poor veneering, that it would be necessary to body it up again with varnish before it could be shipped out. Even after it was re-varnished there was no assurance that it was going to be satisfactory, and in not a few cases the atmospheric changes in the furniture warerooms revealed the defects that had been covered up by an extra coat or two of varnish before the goods had left the factory. After the new process of veneering had been inaugurated, however, all that was necessary when the goods reached the finishing room was to put them through the ordinary process of finishing.

Many furniture factory proprietors do not fully realize to what extent poor veneering affects the cost of production. In this particular case the proprietor thought he couldn't afford the expense of a new veneer room equipment, or the salary to a competent man to superintend the factory, while in reality he couldn't afford to do otherwise. The new man in this case didn't consider the salary offered as an inducement, but the percentage offered in case the business paid fairly well, when added to his salary, would net him a good income. It has proved as he surmised, and at the present writing his percentage amounts to a fraction over 50 per cent. of his total salary, and the proprietor's dividends and business reputation are correspondingly better.

Care in the Use of Glue
By "Veneer Room Foreman"

Almost every woodworker has something to do with glue. The majority are more or less interested in the subject, and the interest is to some extent intensified by the fact that to many it is most uncertain in its effects. Some men, when they use a batch of glue, worry about whether it is going to hold or not, while others have become accustomed to a certain percentage of their work going bad and have ceased to worry, living in hopes that the present lot will not be worse than previous ones. This should not be necessary, as when a man starts a batch of glue he should be certain of the result. Assuming that the glue is O.K. to start with, there should be no chance work about it. The manufacture of glue is a science, and the same may be said of its use. To follow a definite course in the use of glue insures definite results at all times. Frequently this is not done, and the results are not satisfactory.

A man prepares a batch of glue, using part of it on one lot of goods and the balance on another lot. One lot turns out O.K., while the other is disappointing. What is the cause? Why should the same batch of glue make a good job on one lot of goods and the very opposite on the other lot? The reason is that both lots of goods were not made of the same kind of wood, and there is more difference in woods than glue. No glue manufacturer will send out glue that is unfit for use, so we may take for granted that the glue has been properly made.

Let us consider the different grades of glue, and the uses for which they are adapted. It is not possible for all animal glue to be of the same quality, and it would not be wise to make only one grade, as by doing so, only a medium grade would be produced, which would be unfit for high-grade work. The different grades of glue are made at different temperatures. This is an important point to remember when preparing glue. The gelatinous matter which makes the highest grade of glue is readily released from the raw material at a comparatively low temperature, and when once released is skimmed off and becomes the highest grade. Other grades are made by adding more water and slightly raising the temperature. The

various grades of glue have their own uses. It does not always take the highest grade for the highest grade of work. The fibre and texture of the wood must be very carefully considered. High grade glue should always be used for thin and hard woods. Cross-grained woods require a thin glue in order that it may penetrate the surface and secure a firm grip. For wood that is soft and porous a medium grade glue will do just as well, and it should be made fairly thick to prevent absorption. In gluing hard and soft woods together, make the glue to suit the soft wood. The fibre of the hard wood must be raised in order for the glue to get a permanent grip.

Glue should not be heated over 160 degs. Fair., and should be used as fresh as possible, not leaving it to soak over three hours.

Is Veneer Dry Because It Is Wrinkled?
By W Alexander

Recently the writer read an article which said that wrinkles in veneer "are a pretty good sign of thorough dryness." This is a fallacy, and the man who lays veneer without redrying, under the impression that because it was wrinkled it was dry, will certainly find out his mistake when too late.

Veneer becomes wrinkled during the process of drying and it is an indication of unevenness in the formation of the wood which made it dry unevenly. Crotch and other fancy figured woods are more liable to wrinkle than are the plain woods. This is because much of the surface is made up, in varying degrees, of end wood. This end wood dries out more rapidly than the parts that have not the open pores exposed so completely to the air. When these parts begin to dry and shrink they create a strain on the slower drying parts which causes them to twist. But this twisting is done long before the veneer is thoroughly dry.

But even suppose the twist did not come into the veneer until it was thoroughly dry, that would not constitute a valid reason for believing that the veneer was dry at the time it was being laid. Because veneer will wrinkle when drying does not mean that it will straighten out again when it becomes moist. Veneer may be made thoroughly dry and become badly wrinkled, and if stored in a damp basement or other place where it might obtain moisture, it will lose its dryness without losing its wrinkles.

If one wishes to be sure that his veneer is dry before being laid he had better have some surer sign than wrinkles. If veneer has been kept for some considerable time in a warm, dry room, it will, unless it has been packed closely, be dry enough to use.

But if it is dry and badly wrinkled it would not be well to lay it in that condition, because to do so would result in badly broken veneer. If one has a redrier it would be well to moisten and use the machine; but if there is no redrier, it is better to moisten and place between hot boards to hold it straight and extract the moisture in this way.—"Timber."

It is dangerous to wear finger rings while at work, especially those handling boxes, or who work on drill presses and milling machines.

Do not leave material, tools or work lying on the floor where anyone might stumble over them, nor in overhead trestles or platforms, where they might fall or be jarred off.

THE OHIO VENEER COMPANY
Importers and Manufacturers

Foreign and Domestic Veneers and Hardwood Lumber

We always carry a large and assorted stock of Mahogany, Circassian
Walnut, Sawed and Sliced Quartered Oak.

Send us your enquiries and orders. We guarantee good service.

2624 to 2644 Colerain Avenue,   ￭   ￭   CINCINNATI, OHIO.

We operate exclusively in

AMERICAN WALNUT

Our supply of raw material is obtained from
Ohio and Indiana, which produces the best
walnut in the United States.

We have the most modern and best equipped
plant in the United States for producing

WALNUT LUMBER and VENEERS

Special attention given to the requirements of
Furniture Manufacturers

George W. Hartzell,   PIQUA, Ohio

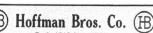

(HB) Hoffman Bros. Co. (HB)

Estab. 1867, Incor. 1904

800 West Main Street,
FORT WAYNE, INDIANA

Manufacturers of

VENEERS and LUMBER

In the Domestic Hardwoods
ANY THICKNESS,

1/24 and 1/30 Slice Cut
(Dried flat with Smith Roller Dryer)

1/20 and thicker Sawed Veneers,
Band Sawn Lumber.

– SPECIALTY—

(HB) Indiana Quartered Oak (HB)

The Canadian Woodworker and Furniture Manufacturer covers a specialized
field. It represents maximum value to
its subscribers.

Subscription price $1.00 per year.

Veneers $_a^{nd}$ Panels

LARGE STOCKS FOR
QUICK SHIPMENTS
Send for Lists

Walter Clark Veneer Co.
GRAND RAPIDS, MICH.

Wood Mosaic Company

Main Office, New Albany, Ind.

Are you interested in

Quartered Oak Veneers?

We have the best timber and the best plant for the manufacture of this stock to be found anywhere. We make our own flitches, using only Kentucky and Indiana stock, which is famous for mildness, color and texture. We can make immediate deliveries of Quartered Oak Veneers on orders received now.

Get acquainted with our Famous Indiana and Kentucky QUARTERED WHITE OAK

TEAK

DOMESTIC

Write for prices on all
**Plain and Fancy
Lumbers and Veneers**
Mahogany, Oak, Birch,
Walnut, Ash, Poplar, etc.,
rotary, sliced or sawed
in all thicknesses

SHELL BOX PARTS
in birch and spruce
Bridges and Ends

Write for Quotations

George Kersley
224 St. James Street, Montreal

LUMBER

MAHOGANY

American Walnut Veneers
Figured Wood
Sliced - Half Round - Stump Wood

5 Essential Qualifications:

1.—Unlimited supply of Walnut Logs.
2.—Logs carefully selected for figure.
3.—All processes of manufacture at our own plant under personal supervision.
4.—Dried flat in Philadelphia Textile Dryer.
5.—3,000,000 feet ready for shipment.

JOHN N. ROBERTS,   New Albany, Indiana
The Pioneer Walnut Veneer Producer of America

If you are in the market for either lumber or veneers in walnut, "the aristocrat of American hardwoods," it will pay you to get in touch with us, for we are

SPECIALISTS IN WALNUT

We have been in this business for thirty years; and in addition to knowing exactly how to make walnut lumber and thin stock, and in addition to having a stock which for variety and size is unequaled anywhere else in the United States, we have on hand a big bunch of

Unique Figured Veneers

from the Southwest

These veneers are highly figured, and of unusual color, and for special work are absolutely unrivaled. Ask us for samples.

Penrod Walnut & Veneer Co.
KANSAS CITY, Mo.

We specialize on

Poplar

cross-banding and backing

Veneer

Highest Grade — Right Prices

The largest Poplar Veneer Mill in the world.

The
═ **Central Veneer Co.**
Huntington, West Virginia

The Glue Book

WHAT IT CONTAINS

Chapter 1—Historical Notes   Chapter 4—Methods in the Glue Room
Chapter 2—Manufacture of Glue   Chapter 5—Glue Room Equipment
Chapter 3—Testing and Grading   Chapter 6—Selection of Glue

PRICE 50 CENTS

Woodworker Publishing Co. Limited
345 Adelaide St. West, Toronto

The Automatic Glue Jointer in Box Work

The adaptability of the automatic glue jointer spreads itself into more fields than its name would indicate, writes H. Van Earl, in "Woodworker" (Indianapolis). Furniture men are not the only ones to benefit through the greater accuracy and capacity this tool gives them over the old hand glue jointer methods. Other special branches of woodworking have come to know this tool also, as a machine far more practical for their work than various "special-purpose" machines which it has displaced.

Take the box manufacturer, for instance. The installation of the continuous-feed glue jointer has not only improved on the work of both buzz planer and boxboard matcher, but has lowered the cost of work formerly produced on these machines as much as 85 per cent. This sounds pretty big, but let us back it up by an actual instance. In the box factory department of an eastern manufacturer, jointing a certain amount of 1-in. square box cover stock was costing in the neighborhood of $2 on a buzz planer; later, with the continuous-feed glue jointer on the job, this same work was done at a cost of less than 25 cents, or less than 15 per cent. of the former expense.

The method was to run the stuff through the glue jointer on two sides and finish the other two sides on the single surfacer. The saving was due to a combination of advantages which all traced back to the method of feed. Owing to power-feed, stock was worked at where each separate piece had to be followed through a continuous feed, with far less effort and time than by hand, as on the buzz planer. Small, narrow stock, that formerly was regarded as waste, was put to very profitable use, owing to the safety of the automatic feed. Absolutely straight joints were secured, owing to the stock being rigidly held to a chain which passes through planed ways. As a regular thing, such accuracy is impossible on a buzz planer, while if the boards are a trifle warped it often takes three or four cuts to get them straight, as a man cannot hold a board steady that does not ride on at least two places.

On the buzz planer the actual feeding or handling of stock, moreover, takes practically twice as long as is required by the automatic glue jointer method. Instead of having to stack stock that has been jointed on one edge, on the truck, and then feed and stack it all over again when the other edge is straightened, a boy merely stands at the out-feeding end of the automatic machine and turns the out-coming stock over on placing it in the return feedway. The continuous-feed glue jointer has also the additional advantage over the buzz planer of being able to make any kind of mold or tongue-and-groove joint on the edge of stock, by simply changing heads.

In regard to the continuous-feed glue jointer as a boxboard matcher, a prominent casket manufacturer states that he matches all of his box stock on the continuous-feed machine, two men thereby accomplishing more in one hour than one man could in a half day in the old way, doing much better work. While such increased capacity cannot be claimed where the best boxboard matchers having sixteen rolls are working, far better work can always be claimed.

A boxboard matcher, when stock is run on edge, cannot be a successful machine unless it has a series of four pairs of rolls. Any of these machines, with only two pairs of end-feed rolls and two pairs of outside feed rolls, will not hold the stock down properly, so that it will ride over the cutters, riding high on the knots and dubbing at the ends. Even these machines

with sixteen rolls, can only take a light cut, as the smooth rolls do not take the place of a hold-down, as do the corrugated blocks on the continuous feedway of the continuous-feed glue jointer. Owing to this, the continuous-feed machine's joint is always absolutely tight on the first run, and is often made, therefore, in a day's run. Furthermore, adjustments hold far more accurately and upkeep is much lower on the continuous-feed machine, as it has about six times the strength and rigidity of the boxboard matcher.

Ball Bearings in Woodworking Machines

During the last few years mechanical and electrical improvements have brought about something of a revolution in the woodworking industry. Competition, that incentive to improvements, constantly urges the production of specialties, and the machine-building industry as a whole is rapidly developing, so that manufacturers of woodworking machines are daily receiving fresh suggestions for the improvement of the details of their machines.

One of the noteworthy improvements introduced in the last decade is the use of ball bearings. This improvement is noteworthy because the woodworking industry, as an industry, did not always recognize the importance of avoiding unnecessary friction. The cost of fuel alone was considered. The use of sawdust and scrap for fuel under the boilers discounted the attention which the power consumption problem should have received. The far-sighted woodworker, however, discovered three serious mistakes in this method of operation.

In the first place, by using ball-bearing-equipped machines, the power requirements are less, and the scrap thus saved can be sold, or the excess steam, if this sawdust is used, can be employed for factory heating, dry kilns, etc. The better market for well-dried lumber, as compared with sappy or half-dried stock, shows how uneconomical it is to waste steam in unnecessary power consumption.

Second, an excessive wastage of power is inevitable where plain bearings are used. Friction is there, and with it wear, for friction always means wear—wear on bearings, wear on shaft. The bearings wearing down, permit the shaft to drop from its proper setting. Not only do the bearings themselves wear down, but the shafts also become scored and worn to eccentric form. This is why chatter takes place, why cutters require frequent setting, and the quality of the work is not kept up unless constant care is used.

Third, where plain bearings are used, hot bearings are a serious menace. The time lost in shutting down the machine to repair or replace a bearing should be charged as additional cost of plain bearings.

To give any specific statement on power saving which would cover all cases would be misleading. Users have reported savings ranging from 20 to 60 per cent. of the friction wasted by plain bearings—the amount varying with the character of the work, the size of shaft, speed, conditions of temperature, weight of rotating parts, etc. It should be noted especially that the starting effort by the use of ball bearings is very much reduced. This is because the hardened steel balls in a ball bearing roll on hardened steel races; and, in starting the bearing, only the effort to make the balls start rolling is required. Compare this with a plain bearing. When the machine is stopped for any considerable length of time, the oil separating the shaft from the bearing lining is squeezed out, and the two metals come into contact. In starting a plain-bearing machine, therefore, the friction of metal-to-metal contact must be overcome. This is why "cold" plain-bearing machines are so hard to start, and even in the case where the machines are frequently started, the coefficient of friction is considerably higher in starting than when running.—Timber, London.

Efficiency in the Packing Department
By Albert Hudson

Great attention is being given to-day to the product of the furniture factory as it progresses from one department to another in its manufacture, with a view to securing the very best in quality and efficiency. There is one department, however, where a slackening off is very often noticed, especially in small factories. That is the department where the trimming and packing is done. It is wrong to assume that because every other department is being conducted on a systematic basis that "anything" will do in the trimming and packing room.

How easy it would be to extend the system a little further and give some time and thought to the equipment of this department. With up-to-date equipment and reliable men, the management will have the satisfaction of knowing that their product will reach the customer in a condition that will go a long way in bringing repeat orders. Satisfied customers are one of the best assets that any manufacturer can have.

Goods are often packed so carelessly that they are damaged in transit, and sometimes so carelessly that when delivered they are found short of casters or keys. Sometimes a mirror is loose or drawers running tight. They are often crated in such light crates that is is a matter of impossibility for them to arrive at their destination without being damaged. The writer has seen goods returned owing to poor crating and packing. In transit they have worked loose, and sometimes the tops or gables have been damaged, or feet have been broken off. Crates should be made strong, with a good body, instead of trying to see how light they can be made. There may be a little saved in making them light, but in the end "prevention is better than cure," and good, strong crates are a preventative. In making crates, first make the ends, nail well together; then nail on the back, leaving off the front slats until they are required.

When packing furniture, first see that everything is complete, drawers and doors open and shut properly, then be sure to lock and fasten them. After this is done, cover up with a sheet of paper. Then place in the crate and fasten with screws at the back to hold the furniture firm. Turn the crate over and nail on the front slats.

All furniture should be wadded on the ends and front with wads made from excelsior or wood wool, to prevent rubbing against the crate and help keep the furniture in place. All tops should be protected, as sometimes these get damaged by careless handling on the railway, or by something heavy being placed upon the crate. With china cabinets, buffets, etc., where glass is used, this should be protected by strips of wood. Some pieces that are very fragile and liable to have feet or legs broken, should be placed on a slat or swing in the crate, the feet or legs about six inches from the bottom of the crate. Too much care cannot be taken in crating furniture. The writer once saw a man put a screw right through the back of a crate into a mirror. The result was a broken mirror. Mirrors should be covered with paper and the frames screwed to the crate from the back, then wadded to protect them from rubbing.

Chairs should be wrapped with paper before crating, and when packed two in a crate, should be properly fastened together. It will be found very convenient for future reference if the sizes of the materials used for crating the different articles are placed on a stock bill. When packing new goods it is advisable to make a note of the cubic feet of space required, and the weight of the article, for shipping information, especially when it is required to ship in car lots. The writer once went into a packing room where the packers were using a bandsaw to cut up the crating. This took up a lot of time and was very inconvenient. A cut-off saw or a small circular saw in the packing room will pay for itself in a very short time. Failing this the crating lumber could be cut in the machine room to the sizes specified on the stock bills and stacked in racks in the packing room, built for the purpose.

Often a piece of furniture gets damaged by careless handling, either by the trimmer or packer, and for this reason should be examined by the foreman before it is put in the crate. Sometimes the trimmer may find difficulty in putting a mirror in a frame or in a wardrobe door, or the glass in a china cabinet. Frames for mirrors usually come through in large batches, and if one is first tried in the machine room the rest could be made accordingly. This would stop a lot of patching and manœuvering in the trimming department. Sometimes a trimmer will damage a piece of furniture by his screwdriver slipping, or by placing something on the top of the case, or by chipping or scratching the silver off the mirror. Accidents will happen sometimes, but by exercising care and good judgment they can be reduced to a minimum. A good idea for the trimmer when putting in mirrors is to place a piece of thin cardboard between his hammer and the glass. This will prevent the silver from being scratched or damaged.

Speaking about a screwdriver slipping; a word may be said in favor of the Robertson socket head screws. These screws were introduced in the factory where the writer is foreman and were found very suitable. There is no danger of the screwdriver slipping and they are preferable to the slit screws for the reason that no ragged heads appear. One objection to this screw may be advanced by the workmen, and that is the necessity of purchasing special screwdrivers. However, the manufacturer furnished us with screwdrivers, as well as bits for the Yankee spiral screwdrivers.

A word about waste in the packing room may be in order. How often one may see nails scattered around the floor, casters and sockets, locks and keys lying about. These get swept up and very often find their way into the fire. All packing rooms are not like this, but a good plan is to have all trimmings and packing material under the direct supervision of the shipper.

The Lumber Market

Domestic Woods

There are several features in connection with the lumber situation in Canada which makes it almost impossible at this time to estimate with any degree of accuracy the prices that will be most likely to prevail in the near future. The prohibition of the import of furniture wood, hardwoods and veneers into Great Britain will affect Canadian lumbermen to some extent. Some of the woods prohibited that were formerly shipped from Canada are maple, beech, birch and elm. The embargo recently announced by the Intercolonial Railroad against lumber shipments and the scarcity of cars on American railroads will prevent Canadian lumbermen from disposing of any appreciable portion of their stock in the United States.

Canadian mills will commence cutting very shortly (if some have not already commenced) and as the quantity of logs ready for cutting is estimated to be about the same as 1915, lumbermen will have, fairly large stocks on hand. It looks at this time as though they will be compelled to seek a market in their own country. The quantity of lumber being consumed for building purposes is practically negligible, so the only market in sight seems to be the wood-working factories.

In this connection Mr. George Kersley, a wholesale lumber and veneer merchant of Montreal, recently paid a month's visit to the principal furniture and piano centres in Ontario and reported that business, particularly in the latter industry, was good and some furniture manufacturers were in a much better position as regards orders than they had been for some time.

Another branch of the woodworking industry that is showing signs of rapid development is the manufacture of toys and novelties. These were formerly imported in large quantities from foregn countries, but will now be manufactured in Canada and orders are being placed for stocks for the Christmas trade. The amount of lumber that will be consumed by this branch of the industry is as yet unknown. It has been suggested on different occasions that cuttings could be used to good advantage for this work, but it is very doubtful as the labor cost of sorting and handling cuttings very often makes their use prohibitive, even when used for manufacturing small articles.

In the majority of cases it is much cheaper to run a long board over the saw a few times and cut it into strips and then put the strips through the sticker than to work up short pieces as each piece has to be handled separately. Therefore, in order to place their products on the market at prices that will be attractive to the average buyer, it is almost certain that manufacturers of toys and novelties will buy their lumber wholesale the same as other woodworking factories.

It has been reported that on March 17th the Imperial Munitions Board allotted contracts for a little over a million shell boxes. The contracts were for the production of 6 round 18 pounder boxes, 50 round 4.5 C. C. boxes, 2 round 60 pounder boxes and 2 round 6 inch boxes. These contracts will require roughly about 12 or 13 million feet of lumber, the varieties being spruce and pine in 1 inch stock, 4 inches and up in width for the tops, bottoms and sides and birch and

maple in 1 inch stock varying from 7 inches to 9 inches in width for the ends. Spruce boards 1 inch thick are reported scarce. This fact taken with the quantity required will have a tendency towards keeping prices firm, but weighing everything in the balance it looks as though manufacturers will be able to secure lumber at the present prices for some time to come. Some of the prices quoted at Toronto for firsts and seconds inch, car lots are as follows:—Ash $50, birch $42, basswood $40, soft elm $40, soft maple $33, hard maple $40, hickory $70, white pine $50 to $52, spruce $22, Douglas fir $46.

Imported Woods

From the reports of actual sales of hardwood lumber compiled by the Hardwood Manufacturers' Association of the United States, the hardwood trade is showing a continuance of the increased sales noted in recent months. There has been a gradual increase in prices, and orders are said to be plentiful, with stocks at the mills very much lower than normal. There has been some improvement in conditions with regard to the manufacturing of lumber, and some of the mills in the district recently visited by floods have resumed operations, but it will be some time before production becomes normal.

From information received at the offices of the Association it is seen that oak, gum, poplar, and chestnut are in strong demand, and prices are reported to have an upward tendency. There is also a demand for maple, birch, basswood, and ash. Automobile manufacturers are now large users of elm and maple, and have been doing some pretty heavy buying during the last few weeks.

Furniture factories in the United States are busy, and are taking all of the desirable birch offered, and are ready buyers of fancy hardwoods. Owing to decided activity in the building business in the United States, sash and door factories and interior finish factories are taking greater interest in the market than they have for some time, and are taking all the desirable birch, maple, and oak offered them. Great quantities of basswood are being used, and prices are up a notch because of the comparatively light stocks of dry materials.

Large quantities of gum of No. 2 stock in suitable lengths are being consumed by the cooperage interests. Some sections of the southern producing mills are unable to make shipments on account of car shortage. Cars are increasingly scarce, and the shipping problem is becoming serious—not only because of the car shortage itself, but because of the embargoes in effect on many lines to eastern destinations. Thus it would appear that furniture manufacturers and others using imported woods would be well advised in looking over their stocks on hand and placing their orders accordingly.

As much as $65 has recently been paid for firsts and seconds inch, plain oak, and nothing is offered at less than $62. Quotations at Boston on some of the active woods, firsts and seconds inch, are as follows: basswood, $44 to $46; maple, $41 to $44; plain oak, $62 to $65; quartered oak, $90 to $91; red birch, $56 to $59; sap birch, $47 to $50.

When operating a machine, give the work your undivided attention. It may save spoiling a job, if not more serious damage.

Never reach across the revolving oiler to regulate the oil feed.

V O K E S

A Wide Range at Right Prices

TOOLS

CABINET HARDWARE
FACTORY AND MILL SUPPLIES

BARTON'S GARNET
and
SAND PAPER

UNIVERSAL CASTERS
(Canadian Agents)

*Write for prices on Garden Tools, Lawn Mowers, Lawn
Rollers and Hose*

PRICES RIGHT — PROMPT SHIPMENT

Let Us Send You Our Catalogs

The **Vokes Hardware Co.**
Limited
40 Queen St. East, Toronto, Can.

HARDWARE

A New Belt Sander

The accompanying illustration shows a new belt sander recently placed on the market. This machine is capable of a wide range of work, and the price places it within the reach of those who require a belt sander and yet have not sufficient work to warrant purchasing one of the more expensive types. It is of sturdy and simple construction, and will give highly satisfactory service. No overhead rigging or tighteners are required. The machine is made in two styles —No. 412, on which the travelling table has no vertical adjustment, and No. 412-B, on which vertical adjustment, fence, and end table rest are provided.

In many woodworking factories the thickness of the stock to be sanded does not vary enough to require any change in the height of the table. In some factories, however, large pieces of cabinet work are sanded after assembling, and for such factories No. 412-B type is made. The frame is made of selected hardwood and is mortised, tenoned, and bolted in a very substantial manner. The table is 4 ft. 6 ins. long, made of strips of hardwood, glued up and made open pattern to permit the sanding dust to drop through. This keeps the table clean at all times.

The table is mounted on frictionless rollers, and has a horizontal movement backward and forward, of 30

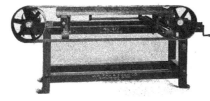

A new Fay & Egan Belt Sander

ins. (this travel can be extended on special order). A handle is provided underneath the table so that the operator's hands can never come into contact with the abrasive material. The table, furnished on "B" type, has a vertical adjustment of 3 ins. by lever movement on inclines, raising concentric with diameter of pulley at in-feed table end. A fence, adjustable to any point on the table, is provided, and also an end table, so that the sand belt may be reversed from its position in the 412-A type, and large pieces sanded from above. Exhaust connection can be made underneath the end table. A locking device is provided for the traveling table.

The sand belt is carried on pulleys, which are made for any width of belt up to 8 inches. These pulleys are carefully turned and balanced, and mounted on shafts in long, self-oiling bearings. One pulley is provided with a screw adjustment in dovetail slides to accommodate the varying lengths of belts and to take up the natural belt stretch. The crank operating this adjusting screw is extended sufficiently to prevent the operator's hand getting in contact with the belt. On either type of machine, the belt is easily put on or taken off, without dismantling the machine in any way, and as the sanded or cutting surface faces the pulleys, there is not the danger of accidents

that is always present where the sanded surface is turned out. The length between the pulleys makes the belt very flexible, and prevents uneven wear when sanding irregular shapes. Users can make hand hold-down blocks out of wood to suit the different kinds of work they have to do.

This machine is manufactured by the J. A. Fay & Egan Company, 153 and 173 West Front Street, Cincinnati, Ohio, and they will gladly furnish full information to anyone interested, upon receipt of their request.

Birch Dowels Required in Great Britain

The Department of Trade and Commerce has received a communication with reference to the opening in Great Britain for birch dowels. The firm making the inquiry are a very important organization, manufacturing veneers in many forms and doing a large business in tea chests. Originally manufacturing in Russia, they opened a large factory outside of London, and the practical closing of transportation from Russia and interruption with Scandinavia presumably necessitate their obtaining supplies of dowels from new sources. This company submitted to Mr. Harrison Watson, Trade Commissioner, London, the particulars of an average specification, together with prices. These are as follows:—12 inch : $\frac{1}{4}$-inch, 4s.; 5-16-inch, 4s. 10d.; $\frac{3}{8}$-inch, 5s. 10d.; 7-16 inch, 6s. 11d.; $\frac{1}{2}$-inch, 8s. 36 inch : 15s. 2d. per 1,000; 5/16-inch, 17s. 3d. per 1,000; $\frac{3}{8}$-inch 20s. 3d. per 1,000 ; 7/16-inch 23s. 11d. per 1,000; $\frac{1}{2}$-inch 21s. 2d. per 1,000.

These prices are c.i.f. London, including war risk, and are based on a freight to London of $1 per 100 pounds. It is stated that Americans were exporting these dowels into London previous to the war at much lower prices, e.g., the size $\frac{3}{8}$-inch by 36-inch was being offered at 14s. 2d. per 1,000 c.i.f. London. Canadian firms who are in a position to cater for the large quantities of birch dowels required may obtain the name and address of the firm making the inquiry on application to the Department of Trade and Commerce, Ottawa. (Refer Trade Inquiry No. 285. File A-2008).

Be sure stops are set and that everything will run clear during the entire cut before throwing in feeds.

Thinking

(By Walter D. Wintle, in Credit Men's Bulletin)

If you think you are beaten, you are,
　If you think you dare not, you don't.
If you'd like to win, but you think you can't,
　It's almost a cinch you won't.
If you think you'll lose, you're lost,
　For out of the world we find
Success begins with a fellow's will.
　It's all in the state of mind.

If you think you're outclassed, you are;
　You've got to think high to rise,
You've got to be sure of yourself before
　You can ever win a prize.
Life's battles don't always go
　To the stronger or faster man;
But soon or late the man who wins,
　Is the one who thinks he can.

Cut Straight
Stuck Straight
Cured Straight
And Dried Straight...

THATS WHY-WE CAN GUARANTEE EVERY·BOARD

The Most Modern Hardwood Operation In The World

IT NEVER LEAVES THE TRUCK UNTIL IT IS --LOADED IN THE -CAR·EVERY BOARD -IS BRANDED THE--THE KRAETZER -CURED LUMBER CO·&MARKED WITH THE -GRADE·WE STAND-BEHIND THE GUAR-ANTEE

WE are specializing in Gum Lumber. We believe our Gum Lumber is different from, and superior to anything in this line that has ever been put on the market.

It is cut from as fine timber as ever grew, it is all properly Kraetzer-cured and treated and handled by our own peculiarly effective methods.

Over 70% of our business during 1915 was with people who had previously used Gum, and abandoned it as unsatisfactory and impractical. **Without exception they are now regular and enthusiastic users of our product.** Need more be said?

Our prices on the rough lumber may be slightly higher than others are asking, but the ultimate cost, measured by the results obtained, is much less than lumber selling at a lower price. "The best is always the cheapest." A trial will convince you.

Our Quartered and Plain Oak and Ash, while it comprises only a small part of our cut is of the same high quality as our Gum.

LOOK FOR THE BRAND

That's your guarantee THE KRAETZER-CURED LUMBER CO. *Our mutual protection*

ON EVERY BOARD

THE KRAETZER-CURED LUMBER CO--

Mills—MOORHEAD, Miss.   Sales Office—CINCINNATI, Ohio

WRITE TODAY FOR A SAMPLE CAR LOAD — —

News of the Trade

The Canada Builders Ltd., will rebuild their planing mills at Orillia, Ont.

The Skedden Brush Company are erecting a new factory at Hamilton, Ont.

The Beaver Toy Manufacturing Company (Thomas Evans and Walter Royston), Montreal, Que., have registered.

A factory to cost $20,000 will be erected at Montreal, P. Q., by L. Wisintainer & Son, 58 St. Lawrence Blvd., for the purpose of manufacturing picture frames.

The M. F. P. Aeroplanes Ltd., will manufacture aeroplanes, hydroplanes, flying boats and air craft of all kinds at Toronto, Ont. The company will have a capital of $500,000.

Mr. E. W. Forsberg of the Stratford Davenport Company has recently returned from a business trip to Chicago and reports that the furniture trade is booming in the American cities.

The Flint Varnish & Color Works of Canada, Ltd., will have their head office at Toronto. The company is capitalized at $250,000 and will manufacture oils, paints, varnishes, enamels, dyes and colors.

The North Bay Toy Company is a new concern, but according to the "Times" of that town it has a very promising future. The company manufacture some ten different lines and are continually adding new ones.

The furniture factory belonging to D. H. Langlois and Co. at St. John, Que., was recently destroyed by fire. The factory will be replaced by a one storey building costing $8,000, on which work has been started.

A toy making project in aid of returned soldiers has been put into operation at Winnipeg. The committee in charge have announced that premises have been secured for a factory in the Conway Building and are now being put into shape.

The Peerless Weaving and Belting Company, Limited, has been incorporated with capital of $150,000 to manufacture leather and reinforced belting. The chief place of business will be at Hamilton, Ont. John Selwyn Rhodes, broker, of Hamilton, Ont., is interested.

The Plaola Piano Company Ltd., will manufacture pianos, player pianos, piano actions and keys and all kinds of automatic musical instruments at Oshawa, Ont. The capital of the company is $40,000 and the provisional directors are L. A. Litlico, J. G. Leckie, J. A. Kent, all of Toronto, Ont.

The offices of Robert Bury & Co., Lumber Merchants, and the Dominion Mahogany & Veneer Co., Limited, have been removed to 456 King St. West, Toronto. The veneer warehouse of the Dominion Mahogany & Veneer Co. Limited, is also now located at the above address, but the lumber yard of Robert Bury & Co. is still at foot of Spadina Avenue.

The Acme Steel Goods Company of Canada, Limited, has been incorporated to take over the Canadian business of the Acme Steel Goods Co., Chicago, manufacturers of box strapping and fasteners. Mr. J. E. Beauchamp, who previously represented the Chicago company, is the manager of the Canadian company, with offices at 20 Nicholas Street, Montreal.

A company to be known as the Commercial Motor Bodies and Carriages, Limited, has been formed with capital of $40,000 to take over the carriage woodworking business of C. Kloepfer, Limited, of Toronto. The chief place of business of the company will be at Guelph, Ont., and they will carry on the business of making motor bodies and carriages.

Bauman & Letson, Elmira, Ont., recently lost their sash and door factory by fire. The amount of the loss is estimated at $10,000 which was partly covered by insurance. The owners will rebuild at once. We understand they are in the market for a full line of woodworking machinery. Communications should be addressed to Noah Beringer, Arthur Street, Elmira, Ont.

The "Hillcrest Lumber Company, Limited" has recently been incorporated with capital of $400,000 to carry on a planing mill and general lumbering business, Renfrew, Ont. Some of those interested are James McNiece Austin and George Bremner Ferguson, lumbermen; Daniel Watson Stewart, accountant, John Geale and Stanley Thorn Chown, solicitors—all of Renfrew, Ont..

The Menard Motor Truck Company, Limited, has been incorporated with capital stock of $150,000 to take over as a going concern the business of Moses L. Menard in the city of Windsor, and to carry on the business of manufacturing motor trucks, wagons, carriages, etc. The place of business will be at Windsor, Ont., and those interested include William Nordwell Gatfield, of Sandwich, Ont., manager, Frederick Hurst Neal, William Hughes Neal, and Moses Louis Menard, manufacturers, all of Windsor, Ont.

The Dominion Casket Company has been incorporated with capital stock of $100,000 to carry on business as manufacturers of caskets, coffins, boxes, hearses, ambulances, carriages, novelties, etc. Their place of business will be at Guelph, Ont., and those interested are William Pears, manufacturer, of Toronto, William George Whitehead, manager, James Mason Arnold, superintendent, and Walter Ellis Buckingham, barrister, all of Guelph, Ont.

Mr. George Kersley, wholesale lumber and veneer merchant, Montreal, recently returned from a month's successful visit to the principal piano and furniture centres in Ontario. He reports that business, particularly in the former branch of industry has revived in a marked degree; some furniture manufacturers are also in a much better position so far as orders are concerned as compared with a few months ago.

A company has been formed at Toronto, Ont., to be known as the Crowley Manufacturing Company, Limited, with capital of $100,000, to manufacture all kinds of wood products, as—mouldings, picture frames, furniture, wood novelties, and specialties. Those interested are J. F. Coughlin, barrister; T. W. Pinnell, and A. W. Gilmour, tinsmiths; F. P. O'Hearn, accountant, and J. L. Waggoner, bookkeeper, all of Toronto, Ont.

DARLINGS STEAM APPLIANCES

DARLING BROTHERS
LIMITED
Engineers and Manufacturers
MONTREAL, CANADA

Branches:
Toronto and Winnipeg

Agents:
Halifax, St. John, Calgary, Vancouver

Box Machinery Bargain

One Dovetail Machine. One Setting-up Machine.
One Four-Saw Trimmer—self feeding; made by Dovetail Machine Co., St. Paul, Minn.
One No. 22 Box Trimmer, made by Fischer Machine Works, Chicago, Illinois.
All cost $1,300. Will guarantee machines and sell at a bargain.
Inquire, The GLOBE SOAP COMPANY,
2-3-4 Cincinnati, Ohio.

Second-Hand Woodworking Machines

Band Saws—26-in., 36-in., 40-in.
Jointers—12-in., 16-in.
Shapers—Single and double spindle.
Re-Saw—42-in. American.
Lathes—20-in. with rests and chisels.
Surfacers—16 x 6 in. and 18 x 8 in.
Mortisers—No. 5 and No. 7 New Britain.
Rip and Cut-off Saws; Moulders; Sash Relishers, etc.
Over 100 machines, New and Used, in stock.

Baird Machinery Company,

123 Water Street,
PITTSBURGH, PA.

1", 1¼" and 2" White Pine and Spruce
⅞", 1", 2" x 4 and up Pine and Spruce Crating Lumber

Write for our stock sheet

The Elgie & Jarvis Lumber Co., Ltd.

18 Toronto St., Toronto, Canada

Imitation Ebony

The method of polishing wood with charcoal, now much used by French cabinetmakers, gives furniture a beautiful dead black color and a smooth surface, the wood seeming to have the density of ebony. Compared with furniture rendered black by stain and varnish, the difference is marked. In charcoal polishing every detail in carving is respected, while paint and varnish will clog up the holes and widen the ridges.

Only carefully selected woods of a close and compact grain are used, and they are first covered with a coat of camphor dissolved in water, and almost immediately afterward another coat, composed chiefly of sulphate of iron and nutgall. The two compositions in blending penetrate the wood and give it an indelible tinge. When these two coats are dry, the wood is first rubbed with a very hard brush, and then with charcoat of substance as light and friable as possible, because if a single hard grain remained in the charcoal this alone would scratch the surface. The flat parts are rubbed with natural stick charcoal, the indented portions and crevices with charcoal powder. Alternately with the charcoal the piece of furniture is rubbed with flannel soaked in linseed oil and essence of turpentine.

Wood Decay is Contagious

A circular prepared at the Forest Products Laboratory, Madison, Wis., has been distributed by the Department of Agriculture. It deals with the sanitation of lumber yards and places where wood products are stored. The circular states that decaying lumber scatters germs which cause decay in other wood where they fall. For that reason no decaying wood should be permitted to lie about a lumber yard or place where wood is stored. The usual sources of danger are rotting foundations on which piles are placed; decaying tramway tracks, and unsound pieces scattered about the premises. A growth of weeds is also conducive to decay because it keeps the premises moist and in the proper condition to receive the germs of decay. Yards ought to be kept clear of decaying wood and also of weeds and grass.

Men Scarce in Canada

The Sutherland-Innes Co., Ltd., slack cooperage stock manufacturers, Chatham, Ont., report that business has been exceptionally good for this season of the year, and everything is opening up well for a good brisk season. We have not been able to stock up all of our mills on account of the poor winter and the lack of labor. All our mills that are running have been stocked up full, so that our output in Canada will only be reduced about twenty-five per cent. this year. The problem that is facing us this spring, is to get sufficient men of the right kind, to run our mills, as most of the able-bodied men in the country have enlisted for the war, and we will have to depend on men over the age of forty-five and the men who are not accepted by the military authorities to do our work. It is going to incapacitate us to a considerable extent, and increase the price of production this year.

7% Investment 7% Profit Sharing

Sums of $100, $500, $1000 and up. Safe as any mortgage.

Business at back of this Security established 28 years.

Send for special folder to

National Securities Corporation,

Limited
Confederation Life Bldg., Toronto

PATENTS

STANLEY LIGHTFOOT

Registered Patent Attorney
Lumsden Building - - TORONTO, CAN.
(Cor. Adelaide & Yonge Sts.)

The business of Manufacturers and Inventors who wish their Patents systematically taken care of is particularly solicited. Ask for new free booklet of complete information.

PETRIE'S

MONTHLY LIST

of

NEW and USED

WOOD TOOLS

in stock for immediate delivery

Wood Planers

36" American double surfacer.
30" Whitney pattern single surfacer.
26" Revolving bed double surfacer.
26" Goldie & McCulloch single surfacer.
24" MacGregor Gourlay planer & matcher.
24" Major Harper planer and matcher.
24" Single surfacers, various makes.
20" Dundas pony planer.
18" Little, Giant planer and matcher.
16" Galt Jointer.
12" Crescent Jointer with safety head.

Band Saws

42" Fay & Egan power feed rip
38" Atlantic tilting frame.
36" Crescent pedestal.
30" Ideal pedestal.
28" Rice 3-wheel pedestal.
28" Jackson Cochrane bracket.

Saw Tables

Weston variable power feed.
Cowan heavy power feed. M128.
No. 5 Crescent sliding top.
No. 3 Crescent universal.
No. 2 Crescent combination.
12" Defiance automatic equalizer.
MacGregor Gourlay railway cut-off.
1½" Martin iron frame swing.
6½" Crescent iron frame swing.

Mortisers

No. 5 New Britain, chain
Cowan hollow chisel, M.190.
Fay upright, graduated stroke.
Galt upright, compound table.
Smart foot power.

Moulders

10" Clark-Demill four side.
10" Houston four side.
6" Dundas sash sticker.

Sanders

40" Berlin 3-drum.
30" Egan double drum.
12" C.M.C. disk and drum.
18" Crescent disk.

Miscellaneous

Fay & Egan 12-spindle dovetailer.
16" and 18" Ideal turning lathes.
No. 235 Wysong & Miles post boring machine.
MacGregor Gourlay 2-spindle shaper.
Elliott single spindle shaper.
No. 51 Crescent universal woodworker.
40" MacGregor Gourlay band resaw.
Rogers vertical resaw.
Cowan sash clamp.
Pedestal tenoner double heads and copes.
Egan sash and door tenoner.
No. 6 Lion universal woodtrimmer.
6-nail box nailer, Meyers patent.
20" American wood scraper.
4-head rounding machines.
Cowan spindle carver M62.
Cowan veneer press screws.

Prices, Descriptions and Full Particulars on Request

H. W. PETRIE, LTD.

Front St. W., Toronto, Ont.

Twist Machine

New Cutter Head for Twist Machine

This machine will produce twist and spiral Columns, Balusters, Table Legs and Pedestals, Chair Posts and Spindles, etc., right and left turn.

The dividing plates will space 2, 3, 4, 6 and 8-in. Special plates furnished on application.

The feed gear set gives one turn in 1½, 3, 4½, 6, 7½ and 9-in.

The centres will swing any diameter up to 12-in. 3 ft. between centers.

For prices and further description write

Eight Knives. 30 Degrees Angle.

The Shawver Company
Springfield, Ohio

YOU CAN
cut down your cost of sanding round work 85%

How? By installing a Nash Sander

Let me send you details

J. M. NASH, MILWAUKEE, Wis.

OUR SPECIALTY

All Machinery Supply Firms like Machinery Manufacturers have lines which they specialize. This is because they know more about these particular lines than any other. We know more about woodworking machinery than anything else and can give you a greater selection and more intelligent service in this line than any of our competitors. Therefore

Our Specialty is

WOOD-WORKING
MACHINERY

Send us Your Inquiries and See if We Have not Something Better in Both Machinery and Service.

J. L. NEILSON & CO.
Winnipeg, Man.

Woodworkers
and
Lumbermen

The Sheldon Exhaust Fans have characteristics that adapt themselves to the service of the Planing Mill or Wood-working Plant. They are specially designed for this kind of work, having a saving in power and speed of 25% to 40%.

Write for Industrial Booklet.

SHELDONS LIMITED - Galt, Ont.
Toronto Office, 609 Kent Building
— Agents —

Messrs. Ross & Greig, 412 St. James St., Montreal, Que.
Messrs. Walker's Limited, 259-261 Stanley St., Winnipeg, Man.
Messrs. Gorman, Clancey & Grindley Ltd., Calgary and Edmonton, Alta.
Messrs. Robt. Hamilton & Co., Ltd., Bk. of Ottawa Bldg, Vancouver, B.C.

It is always to the advantage of mill owners and operators to use Simonds Solid Tooth and Inserted Tooth Saws, as well as Simonds Band Saws and Planer Knives. The owner profits more by larger production, better lumber, and fewer delays. The Employee gains by working with safe and sure saws.

Write for our 1916 Catalog and prices

Simonds Canada Saw Co., Limited

ST. JOHN, N. B.

Factory:
St. Remi Street & Acorn Ave., MONTREAL

VANCOUVER, B. C.

GREEN LUMBER PROTECTED

Receiving End

NOTE that when lumber first enters a National Progressive Drier equipped with **Vertical** piping, (receiving end) it does not stand directly over the hot coils, which would cause too rapid surface drying. In a National **Vertical** Piping System the coils are all confined to the **discharging** end of the kiln.

Our catalog explains in detail this and many **other** advantages offered by the National System of **Vertical** Piping. The catalog is free. Write for it today.

Discharging End

THE NATIONAL DRY KILN CO.   1117 E. Maryland St.   INDIANAPOLIS, IND.

ONE OPERATOR
SIXTY SETS
SIXTY MINUTES

BRICK HOUSE JAMBS

FRAME HOUSE JAMBS

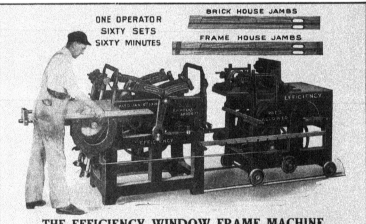

THE EFFICIENCY WINDOW FRAME MACHINE

A new style machine which **REDUCES LABOR COST FULLY 60%**. Works automatically—one man can turn out sixty sets of pulley stiles an hour. Cross cuts the ends of pockets, routs the pulley mortises and dadoes the head and sill gain, automatically, while the operator slits the pockets. You take no chance—write us.

THE F. H. WEEKS LUMBER COMPANY   -   AKRON, OHIO

Peter Hay Machine Knives

Peter Hay knives are made of special quality steel tempered and toughened to take and hold a keen cutting edge. Hay knives are made for all purposes.

Have you our Catalogue?

Made in Canada

The Peter Hay Knife Co., Limited
GALT, ONTARIO

PRESSES

For Veneer and Veneer Drying

Made in Canada

William R. Perrin
Limited
Toronto

GRAND RAPIDS

VAPOR DRY KILN

GRAND RAPIDS
MICHIGAN

129
Grand Rapids Vapor Kilns sold in the last

122
days of the year 1915.

54
of these were Repeat Orders.

Repeat orders represent satisfaction. We can guarantee experienced engineering ability and efficient service.

Over 1300 Grand Rapids Vapor Kilns in use.

Write us regarding better drying

Work your waste into small handles and other small turnings

The OBER LATHES will help you

We also make lathes for turning Axe, Pick Sledge and other irregular Handles, Whiffletrees Neckyokes, etc. Sanders, Rip Saws, Cut Off Saws, Chucking Machines, etc.

Write for free Catalogue No. 31

THE OBER MFG. CO.
CHAGRIN FALLS, O., U. S. A.

VENEER PRESSES

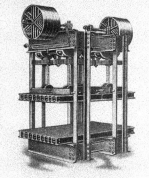

WRITE FOR BULLETIN C 1.

Canadian Boomer & Boschert Press
Company, Limited
18 Tansley Street, MONTREAL

Your Loose Pulley Trouble Disappears

when installed with

CHAPMAN BALL BEARING LOOSE PULLEYS

A little vaseline **once a year** is all the attention and lubrication required.

Our loose Pulleys cannot run hot and do not cut the shaft.

Write us for particulars and price list

The
Chapman Double Ball Bearing Co.
of Canada, Limited
339-351 Sorauren Ave., Toronto, Canada

The "Canadian Woodworker" Buyers' Directory

ABRASIVES
The Carborundum Co., Niagara Falls, N.Y.

BALL BEARINGS
Chapman Double Ball Bearing Co., Toronto.

BALUSTER LATHES
Garlock-Machinery, Toronto, Ont.
Ober Mfg. Company, Chagrin Falls, Ohio.
Whitney & Son, Baxter D., Winchendon, Mass

BAND SAW FILING MACHINERY
Berlin Machine Works, Ltd., Hamilton, Ont.
Canada Machinery Corporation, Galt, Ont.
Fay & Egan Co., J. A., Cincinnati, Ohio.
Garlock-Machinery, Toronto, Ont.
Shawver Company, Springfield, O.

BAND SAWS
Berlin Machine Works, Ltd., Hamilton, Ont.
Canada Machinery Corporation, Galt, Ont.
Garlock-Machinery, Toronto, Ont.
Jackson, Cochrane & Company, Berlin, Ont.
Pettie, H. W., Toronto.
Shawver Company, Springfield, O.

BAND SAW MACHINES
Berlin Machine Works, Hamilton, Ont.
Canada Machinery Corporation, Galt, Ont.
Fay & Egan Co., J. A., Cincinnati, Ohio.
Garlock-Machinery, Toronto, Ont.
Pettie, H. W., Toronto.
Williams Machinery Co., A. R., Toronto, Ont.

BAND SAW STRETCHERS
Berlin Machine Works, Ltd., Hamilton, Ont.
Fay & Egan Co., J. A., Cincinnati, Ohio.

BASKET STITCHING MACHINE
Saranac Machine Co., Benton Harbor, Mich.

BENDING MACHINES
Fay & Egan Co., J. A., Cincinnati, Ohio.
Garlock-Machinery, Toronto, Ont.
Perfection Wood Steaming Retort Company,
Parkersburg, W. Va.

BELTING
Fay & Egan Co., J. A., Cincinnati, Ohio.

BLOWERS
Sheldons, Limited, Galt, Ont.
Toronto Blower Company, Toronto, Ont.

BLOW PIPING
Sheldons, Limited, Galt, Ont.

BORING MACHINES
Berlin Machine Works, Ltd., Hamilton, Ont.
Canada Machinery Corporation, Galt, Ont.
Fay & Egan Co., J. A., Cincinnati, Ohio.
Garlock-Machinery, Toronto, Ont.
Jackson, Cochrane & Company, Berlin, Ont.
Nash, J. M., Milwaukee, Wis.
Reynolds Pattern & Machine Co., Moline, Ill.
Shawver Company, Springfield, O.

BOX MAKERS' MACHINERY
Berlin Machine Works, Ltd., Hamilton, Ont.
Canada Machinery Corporation, Galt, Ont.
Fay & Egan Co., J. A., Cincinnati, Ohio.
Garlock-Machinery, Toronto, Ont.
Globe Soap Company, Cincinnati, Ohio.
Jackson, Cochrane & Co., Berlin, Ont.
Neilson & Company, J. L. Winnipeg, Man.
Preston Woodworking Machinery Company,
Preston, Ont.
Reynolds Pattern & Machine Co., Moline, Ill.
Whitney & Son, Baxter D., Winchendon, Mass.
Williams Machinery Co., A. R., Toronto, Ont.

CABINET PLANERS
Berlin Machine Works, Ltd., Hamilton, Ont.
Canada Machinery Corporation, Galt, Ont.
Fay & Egan Co., J. A., Cincinnati, Ohio.
Garlock-Machinery, Toronto, Ont.
Jackson, Cochrane & Co., Berlin, Ont.
Preston Woodworking Machinery Company,
Preston, Ont.
Shawver Company, Springfield, O.

CARS (Transfer)
Sheldons, Limited, Galt, Ont.

CARBORUNDUM PAPER
Carborundum Co., Niagara Falls, N.Y.

CARVING MACHINES
Canada Machinery Corporation, Galt, Ont.
Fay & Egan Co., J. A., Cincinnati, Ohio.
Garlock-Machinery, Toronto, Ont.
Jackson, Cochrane & Company, Berlin, Ont.

CASTERS
Foster, Merriam & Co., Meriden, Conn.

CLAMPS
Fay & Egan Co., J. A., Cincinnati, Ohio.
Garlock-Machinery, Toronto, Ont.
Jackson, Cochrane & Company, Berlin, Ont.
Simonds Canada Saw Company, Montreal, P.Q.

COLUMN MACHINERY
Fay & Egan Co., J. A., Cincinnati, Ohio.

CUT-OFF SAWS
Berlin Machine Works, Ltd., Hamilton, Ont.
Canada Machinery Corporation, Galt, Ont.
Fay & Egan Co., J. A., Cincinnati, Ohio.
Garlock-Machinery, Toronto, Ont.
Jackson, Cochrane & Company, Berlin, Ont.
Knechtel Bros., Limited, Southampton, Ont.
Ober Mfg. Co., Chagrin Falls, Ohio.
Shawver Company, Springfield, O.
Simonds Canada Saw Co., Montreal, Que.

CUTTER HEADS
Berlin Machine Works, Hamilton, Ont.
Canada Machinery Corporation, Galt, Ont.
Fay & Egan Co., J. A., Cincinnati, Ohio.
Gage Mfg. Company, E. E., Gardner, Mass.
Jackson, Cochrane & Company, Berlin, Ont.
Mattison Machine Works, Beloit, Wis.

DOVETAILING MACHINES
Canada Machinery Corporation, Galt, Ont.
Canadian Linderman Machine Company; Wood-
stock, Ont.
Fay & Egan Co., J. A., Cincinnati, Ohio.
Garlock-Machinery, Toronto, Ont.
Jackson, Cochrane & Company, Berlin, Ont.

DOWEL MACHINES
Canada Machinery Corporation, Galt, Ont.
Fay & Egan Co., J. A., Cincinnati, Ohio.
Ober Mfg. Co., Chagrin Falls, Ohio.

DRY KILNS
Grand Rapids Veneer Works, Grand Rapids,
Mich.
National Dry Kiln Co., Indianapolis, Ind.
Sheldons, Limited, Galt, Ont.

DOWELS
Knechtel Bros., Limited, Southampton, Ont.

DUST COLLECTORS
Sheldons, Limited, Galt, Ont.

DUST SEPARATORS
Sheldons, Limited, Galt, Ont.

EDGERS (Single Saw)
Canada Machinery Corporation, Galt, Ont.
Fay & Egan Co., J. A., Cincinnati, Ohio.
Garlock-Machinery, Toronto, Ont.
Simonds Canada Saw Co., Montreal, Que.

EDGERS (Gang)
Berlin Machine Works, Ltd., Hamilton, Ont.
Canada Machinery Corporation, Galt, Ont.
Fay & Egan Co., J. A., Cincinnati, Ohio.
Garlock-Machinery, Toronto, Ont.
Simonds Canada Saw Co., Montreal, P.Q.
Williams Machinery Co., A. R., Toronto, Ont.

END MATCHING MACHINE
Berlin Machine Works, Ltd., Hamilton, Ont.
Canada Machinery Corporation, Galt, Ont.
Fay & Egan Co., J. A., Cincinnati, Ohio.
Garlock-Machinery, Toronto, Ont.
Jackson, Cochrane & Company, Berlin, Ont.

EXHAUST FANS
Garlock-Machinery, Toronto, Ont.
Sheldons Limited, Galt, Ont.

FILES
Simonds Canada Saw Co., Montreal, Que.

FLOORING MACHINES
Berlin Machine Works, Ltd., Hamilton, Ont.
Canada Machinery Corporation, Galt, Ont.
Fay & Egan Co., J. A., Cincinnati, Ohio.
Garlock-Machinery, Toronto, Ont.
Pettie, H. W., Toronto.
Whitney & Son, Baxter D., Winchendon, Mass.

FLUTING HEADS
Fay & Egan Co., J. A., Cincinnati, Ohio.
Shawver Company, Springfield, O.

FLUTING AND TWIST MACHINE
Shawver Company, Springfield, O.

FURNITURE TRIMMINGS
Foster, Merriam & Co., Meriden, Conn.

GARNET PAPER AND CLOTH
The Carborundum Co., Niagara Falls, N.Y.

GRAINING MACHINES
Berlin Machine Works, Ltd., Hamilton, Ont.
Canada Machinery Corporation, Galt, Ont.
Fay & Egan Co., J. A., Cincinnati, Ohio.
Williams Machinery Co., A. R., Toronto, Ont.

GLUE
Coignet Chemical Product Co., New York,
N.Y.
Perkins Glue Company, South Bend, Ind.

GLUE CLAMPS
Jackson, Cochrane & Company, Berlin, Ont.

GLUE HEATERS
Fay & Egan Co., J. A., Cincinnati, Ohio.
Jackson, Cochrane & Company, Berlin, Ont.

GLUE JOINTERS
Berlin Machine Works, Ltd., Hamilton, Ont.
Canada Machinery Corporation, Galt, Ont.
Jackson, Cochrane & Company, Berlin, Ont.

GLUE SPREADERS
Fay & Egan Co., J. A., Cincinnati, Ohio.
Jackson, Cochrane & Company, Berlin, Ont.

GLUE ROOM EQUIPMENT
Boomer & Boschert Company, Montreal, Que.
Perrin & Company, W. R., Toronto, Ont.

GRINDERS (Cutter)
Fay & Egan Co., J. A., Cincinnati, Ohio.

GRINDERS, HAND AND FOOT POWER
The Carborundum Co., Niagara Falls, N.Y.

GRINDERS (Knife)
Berlin Machine Works, Ltd., Hamilton, Ont.
Canada Machinery Corporation, Galt, Ont.
Fay & Egan Co., J. A., Cincinnati, Ohio.
Garlock-Machinery, Toronto, Ont.
Petrie, H. W., Toronto.

GRINDERS (Tool)
Fay & Egan Co., J. A., Cincinnati, Ohio.
Jackson, Cochrane & Company, Berlin, Ont.
Shawver Company, Springfield, O.

GRINDING WHEELS
The Carborundum Co., Niagara Falls, N.Y.

GROOVING HEADS
Fay & Egan Co., J. A., Cincinnati, Ohio.

GUARDS (Saw)
Shawver Company, Springfield, O.

GUMMERS, ETC.
Fay & Egan Co., J. A., Cincinnati, Ohio.

HAND PROTECTORS
Fay & Egan Co., J. A., Cincinnati, Ohio.

HAND SCREWS
Fay & Egan Co., J. A., Cincinnati, Ohio.

HANDLE THROATING MACHINE
Klotz Machine Company, Sandusky, Ohio.

HANDLE TRIMMING MACHINE
Klotz Machine Company, Sandusky, Ohio.

HANDLE & SPOKE MACHINERY
Fay & Egan Co., J. A., Cincinnati, Ohio.
Klotz Machine Co., Sandusky, Ohio.
Nash, J. M., Milwaukee, Wis.
Ober Mfg. Company, Chagrin Falls, Ohio.
Whitney & Son, Baxter D., Winchendon, Mass.

HARDWOOD LUMBER
American Hardwood Lumber Co., St. Louis, Mo.
Atlantic Lumber Company, Toronto.
Kersley, Geo., Montreal, Que.
Kraetzer-Cured Lumber Co., Moorhead, Miss.
Mowbray & Robinson, Cincinnati, Ohio.
Penrod Walnut & Veneer Co., Kansas City, Mo.
Spencer, C. A., Montreal, Que.

HUB MACHINERY
Fay & Egan Co., J. A., Cincinnati, Ohio.

HYDRAULIC PRESSES
Can. Boomer & Boschert Press, Montreal, Que.

HYDRAULIC PUMPS & ACCUMULATORS
Can. Boomer & Boschert Press, Montreal, Que.

HYDRAULIC VENEER PRESSES
Perrin & Company, Wm. R., Toronto, Ont.

JOINTERS
Berlin Machine Works, Ltd., Hamilton, Ont.
Canada Machinery Corporation, Galt, Ont.
Fay & Egan Co., J. A., Cincinnati, Ohio.
Garlock-Machinery, Toronto, Ont.
Jackson, Cochrane & Company, Berlin, Ont.
Knechtel Bros., Limited, Southampton, Ont.
Petrie, H. W., Toronto, Ont.
Preston Woodworking Machinery Company, Preston, Ont.
Prybil Machine Company, P., New York.
Shawver Company, Springfield, O.

KNIVES (Planer and others)
Berlin Machine Works, Ltd., Hamilton, Ont.
Canada Machinery Corporation, Galt, Ont.
Fay & Egan Co., J. A., Cincinnati, Ohio.
Simonds Canada Saw Co., Montreal, P.Q.

LUMBER
American Hardwood Lumber Co., St. Louis, Mo.
Churchill, Milton Lumber Co., Louisville, Ky.
Elgie & Jarvis Lumber Co., Toronto, Ont.
Hartzell, Geo. W., Piqua, Ohio.

LATHES
Canada Machinery Corporation, Galt, Ont.
Fay & Egan Co., J. A., Cincinnati, Ohio.
Garlock-Machinery, Toronto, Ont.
Jackson, Cochrane & Company, Berlin, Ont.
Klotz Machine Company, Sandusky, Ohio.
Ober Mfg. Company, Chagrin Falls, Ohio.
Petrie, H. W., Toronto, Ont.
Preston Woodworking Machinery Company, Shawver Company, Springfield, O.
Whitney & Son, Baxter D., Winchendon, Mass.

MACHINE KNIVES
Berlin Machine Works, Ltd., Hamilton, Ont.
Canada Machinery Corporation, Galt, Ont.
Peter Hay Knife Company, Galt, Ont.
Simonds Canada Saw Co., Montreal, P.Q.

MITRE MACHINES
Canada Machinery Corporation, Galt, Ont.
Fay & Egan Co., J. A., Cincinnati, Ohio.

MITRE SAWS
Canada Machinery Corporation, Galt, Ont.
Fay & Egan Co., J. A., Cincinnati, Ohio.
Garlock-Machinery, Toronto, Ont.
Simonds Canada Saw Co., Montreal, P.Q.

MORTISING MACHINES
Berlin Machine Works, Ltd., Hamilton, Ont.
Canada Machinery Corporation, Galt, Ont.
Fay & Egan Co., J. A., Cincinnati, Ohio.
Garlock-Machinery, Toronto, Ont.

MOULDINGS
Knechtel Bros., Limited, Southampton, Ont

MULTIPLE BOXING MACHINES
Fay & Egan Co., J. A., Cincinnati, Ohio.
Nash, J. M., Milwaukee, Wis.
Reynolds Pattern & Machine Co., Moline, Ill.

PATENT SOLICITORS
Lightfoot, Stanley, Toronto, Ont.

PATTERN SHOP MACHINES
Berlin Machine Works, Ltd., Hamilton, Ont.
Canada Machinery Corporation, Galt, Ont.
Fay & Egan Co., J. A., Cincinnati, Ohio.
Garlock-Machinery, Toronto, Ont.
Jackson, Cochrane & Company, Berlin, Ont.
Whitney & Son, Baxter D., Winchendon, Mass.

PLANERS
Berlin Machine Works, Ltd., Hamilton, Ont.
Canada Machinery Corporation, Galt, Ont.
Fay & Egan Co., J. A., Cincinnati, Ohio.
Garlock-Machinery, Toronto, Ont.
Jackson, Cochrane & Company, Berlin, Ont.
Petrie, H. W., Toronto, Ont.
Preston Woodworking Machinery Company, Whitney & Son, Baxter D., Winchendon, Mass.
Williams Machinery Co., A. R., Toronto, Ont.

PLANING MILL MACHINERY
Berlin Machine Works, Ltd., Hamilton, Ont.
Canada Machinery Corporation, Galt, Ont.
Fay & Egan Co., J. A., Cincinnati, Ohio.
Garlock-Machinery, Toronto, Ont.
Jackson, Cochrane & Company, Berlin, Ont.
Petrie, H. W., Toronto, Ont.
Preston Woodworking Machinery Company, Shawver Company, Springfield, O.
Whitney & Son, Baxter D., Winchendon, Mass.
Williams Machinery Co., A. R., Toronto, Ont.

PRESSES (Veneer)
Canada Machinery Corporation, Galt, Ont.
Jackson, Cochrane & Company, Berlin, Ont.
Perrin & Co., Wm. R., Toronto, Ont.

PULLEYS
Fay & Egan Co., J. A., Cincinnati, Ohio.

RESAWS
Berlin Machine Works, Ltd., Hamilton, Ont.
Canada Machinery Corporation, Galt, Ont.
Fay & Egan Co., J. A., Cincinnati, Ohio.
Garlock-Machinery, Toronto, Ont.
Jackson, Cochrane & Company, Berlin, Ont.
Petrie, H. W., Toronto, Ont.
Preston Woodworking Machinery Company, Preston, Ont.
Simonds Canada Saw Co., Montreal, P.Q.
Williams Machinery Co., A. R., Toronto, Ont.

RIM AND FELLOE MACHINERY
Fay & Egan Co., J. A., Cincinnati, Ohio.

RIP SAWING MACHINES
Berlin Machine Works, Ltd., Hamilton, Ont.
Canada Machinery Corporation, Galt, Ont.
Fay & Egan Co., J. A., Cincinnati, Ohio.
Garlock-Machinery, Toronto, Ont.
Klotz Machine Company, Sandusky, Ohio.
Jackson, Cochrane & Company, Berlin, Ont.
Ober Mfg. Company, Chagrin Falls, Ohio.

SANDERS
Berlin Machine Works, Ltd., Hamilton, Ont.
Canada Machinery Corporation, Galt, Ont.
Fay & Egan Co., J. A., Cincinnati, Ohio.
Garlock-Machinery, Toronto, Ont.
Klotz Machine Company, Sandusky, Ohio.
Jackson, Cochrane & Company, Berlin, Ont.
Nash, J. M., Milwaukee, Wis.
Ober Mfg. Company, Chagrin Falls, Ohio.
Shawver Company, Springfield, O.

SASH, DOOR & BLIND MACHINERY
Canada Machinery Corporation, Galt, Ont.
Berlin Machine Works, Ltd., Hamilton, Ont.
Fay & Egan Co., J. A., Cincinnati, Ohio.
Jackson, Cochrane & Company, Berlin, Ont

SAWS
Fay & Egan Co., J. A., Cincinnati, Ohio.
Radcliff Saw Mfg. Company, Toronto, Ont.
Simonds Canada Saw Co., Montreal, P.Q.

SAW GUARDS
Shawver Company, Springfield, O.

SAW GUMMERS, ALOXITE
The Carborundum Co., Niagara Falls, N.Y.

SAW SWAGES
Fay & Egan Co., J. A., Cincinnati, Ohio.
Simonds Canada Saw Co., Montreal, Que.

SAW TABLES
Canada Machinery Corporation, Galt, Ont.
Berlin Machine Works, Ltd., Hamilton, Ont.
Fay & Egan Co., J. A., Cincinnati, Ohio.
Garlock-Machinery, Toronto, Ont.
Jackson, Cochrane & Company, Berlin, Ont.
Knechtel Bros., Limited, Southampton, Ont.

SCRAPING MACHINES
Canada Machinery Corporation, Galt, Ont.
Garlock-Machinery, Toronto, Ont.
Whitney & Sons, Baxter D., Winchendon, Mass

SCROLL CHUCK MACHINES
Klotz Machine Company, Sandusky, Ohio.

SCREW DRIVING MACHINES
Canada Machinery Corporation, Galt, Ont.
Reynolds Pattern & Machine Co., Moline, Ill.
Stow Manufacturing Co., Binghamton, N.Y.

SCREW DRIVING MACHINERY
Canada Machinery Corporation, Galt, Ont.
Reynolds Pattern & Machine Works, Moline, Ill.
Stow Manufacturing Co., Binghamton, N.Y.

SCROLL SAWS
Canada Machinery Corporation, Galt, Ont.
Berlin Machine Works, Ltd., Hamilton, Ont.
Fay & Egan Co., J. A., Cincinnati, Ohio.
Garlock-Machinery, Toronto, Ont.
Jackson, Cochrane & Company, Berlin, Ont.
Simonds Canada Saw Co., Montreal, P.Q.

SECOND-HAND MACHINERY
Petrie, H. W.
Williams Machinery Co., A. R., Toronto, Ont.

SHAPERS
Canada Machinery Corporation, Galt, Ont.
Berlin Machine Works, Ltd., Hamilton, Ont.
Fay & Egan Co., J. A., Cincinnati, Ohio.
Garlock-Machinery, Toronto, Ont.
Jackson, Cochrane & Company, Berlin, Ont.
Ober Mfg. Company, Chagrin Falls, Ohio.
Petrie, H. W., Toronto, Ont.
Preston Woodworking Machinery Company, Preston, Ont.
Shawver Company, Springfield, O.
Simonds Canada Saw Co., Montreal, P.Q.
Whitney & Sons, Baxter D., Winchendon, Mass.

SHARPENING TOOLS
The Carborundum Co., Niagara Falls, N.Y.

SHAVING COLLECTORS
Sheldons Limited, Galt, Ont.

SHELL BOX BANDS
Metallic Roofing Company, Toronto, Ont.

SHELL BOX HANDLES
Vokes Hardware Co., Toronto, Ont.

STEAM APPLIANCES
Darling Bros., Montreal.

SHELL BOX EQUIPMENT
Vokes Hardware Co., Toronto, Ont.

SINGLE SPINDLE BOXING MACHINES
Canada Machinery Corporation, Galt, Ont.
Fay & Egan Co., J. A., Cincinnati, Ohio.
Ober Mfg. Company, Chagrin Falls, Ohio.

STAVE SAWING MACHINE
Whitney & Sons, Baxter D., Winchendon, Mass.

STEAM TRAPS
Canadian Morehead Mfg. Co., Woodstock, Ont.
Standard Dry Kiln Co., Indianapolis, Ind.

STEEL SHELL BOX BANDS
Vokes Hardware Co., Toronto, Ont.

SURFACERS
Berlin Machine Works, Ltd., Hamilton, Ont.
Canada Machinery Corporation, Galt, Ont.
Fay & Egan Co., J. A., Cincinnati, Ohio.
Garlock-Machinery, Toronto, Ont.
Jackson, Cochrane & Company, Berlin, Ont.
Pettie, H. W., Toronto, Ont.
Preston Woodworking Machinery Company,
Preston, Ont.
Whitney & Sons, Baxter D., Winchendon, Mass.
Williams Machinery Co., A. R., Toronto, Ont.

SWING SAWS
Berlin Machine Works, Ltd., Hamilton, Ont.
Canada Machinery Corporation, Galt, Ont.
Fay & Egan Co., J. A., Cincinnati, Ohio.
Garlock-Machinery, Toronto, Ont.
Jackson, Cochrane & Company, Berlin, Ont.
Ober Mfg. Co., Chagrin Falls, Ohio.
Simonds Canada Saw Co., Montreal, P.Q.

TABLE LEG LATHES
Berlin Machine Works, Ltd., Hamilton, Ont.
Fay & Egan Co., J. A., Cincinnati, Ohio.
Ober Mfg. Co., Chagrin Falls, Ohio.
Whitney & Sons, Baxter D., Winchendon, Mass.

TABLE SLIDES
Walter & Co., B., Wabash, Ind.

TENONING MACHINES
Berlin Machine Works, Ltd., Hamilton, Ont.
Canada Machinery Corporation, Galt, Ont.
Fay & Egan Co., J. A., Cincinnati, Ohio.
Garlock-Machinery, Toronto, Ont.
Jackson, Cochrane & Company, Berlin, Ont.
Preston Woodworking Machinery Company,
Preston, Ont.

TRIMMERS
Berlin Machine Works, Ltd., Hamilton, Ont.
Canada Machinery Corporation, Galt, Ont.
Fay & Egan Co., J. A., Cincinnati, Ohio.
Garlock-Machinery, Toronto, Ont.
Preston Woodworking Machinery Company,
Preston, Ont.

TRUCKS
Sheldons Limited, Galt, Ont.
National Dry Kiln Co., Indianapolis, Ind.

TURNINGS
Knechtel Bros., Limited, Southampton, Ont.

TURNING MACHINES
Berlin Machine Works, Ltd., Hamilton, Ont.
Fay & Egan Co., J. A., Cincinnati, Ohio.
Garlock-Machinery, Toronto, Ont.
Klotz Machine Company, Sandusky, Ohio.
Ober Mfg. Company, Chagrin Falls, Ohio.
Shawver Company, Springfield, O.
Whitney & Sons, Baxter D., Winchendon, Mass.

UNDER-CUT SELF-FEEDING FACE
PLANER
Fay & Egan Co., J. A., Cincinnati, Ohio.
Garlock-Machinery, Toronto, Ont.
Jackson, Cochrane & Company, Berlin, Ont.

VARNISHES
Ault & Wiborg Company, Toronto, Ont.

VENEERS
Central Veneer Co., Huntington, W. Virginia.
Walter Clark Veneer Co., Grand Rapids, Mich.
Evansville Veneer Co., Evansville, Ind.
Hartzell, Geo. W., Piqua, Ohio.
Hay & Company, Woodstock, Ont.
Hoffman Bros. Company, Fort Wayne, Ind.
Geo. Kersley, Montreal, Que.
Ohio Veneer Company, Cincinnati, Ohio.
Penrod Walnut & Veneer Co., Kansas City, Mo.
Roberts, John N., New Albany, Ind.
Wood Mosaic Co., New Albany, Ind.

VENTILATING APPARATUS
Sheldons Limited, Galt, Ont.

VENEERED PANELS
Hay & Company, Woodstock, Ont.

VENEER PRESSES (Hand and Power)
Canada Machinery Corporation, Galt, Ont.
Can. Boomer & Boschert Press, Montreal, Que.
Garlock-Machinery, Toronto, Ont.
Jackson, Cochrane & Company, Berlin, Ont.
Perrin & Company, Wm. R., Toronto, Ont.

VISES
Fay & Egan Co., J. A., Cincinnati, Ohio.
Simonds Canada Saw Company, Montreal, Que.

WAGON AND CARRIAGE MACHINERY
Berlin Machine Works, Ltd., Hamilton, Ont.
Canada Machinery Corporation, Galt, Ont.
Fay & Egan Co., J. A., Cincinnati, Ohio.
Ober Mfg. Company, Chagrin Falls, Ohio.
Stow Manufacturing Co., Binghamton, N.Y.
Whitney & Sons, Baxter D., Winchendon. Mass.

WOOD FINISHES
Ault & Wiborg, Toronto, Ont.

WOOD TURNING MACHINERY
Berlin Machine Works, Ltd., Beloit, Wis.
Canada Machinery Corporation, Galt, Ont.
Garlock-Machinery, Toronto, Ont.
Shawver Co., Springfield, Ohio.

WORK BENCHES
Fay & Egan Co., J. A., Cincinnati, Ohio.

WOODWORKING MACHINES
Canada Machinery Corporation, Galt, Ont.
Garlock-Machinery, Toronto, Ont.
Jackson, Cochrane & Co., Berlin, Ont.
Preston Woodworking Machinery Company,
Preston, Ont.
Reynolds Pattern & Machine Works, Moline,
Ill.
Williams Machinery Co., A. R., Toronto, Ont.

WHEELS, Grinding, Carborundum and Aloxite
The Carborundum Co., Niagara Falls, N.Y.

WOOD SPECIALTIES
Knechtel Bros., Limited, Southampton, Ont.

92

93

Bead and Turned Mouldings

We manufacture a very complete line of Rope
and Turned Mouldings. If the design you want
is not found in our line we can make it to your
order.

Write us for information and prices

KNECHTEL BROS., LIMITED
SOUTHAMPTON, ONT.

55   58

Making TABLE-SLIDES is a Specialty Business

For more than TWENTY-FIVE YEARS we have made TABLE SLIDES exclusively. Our Factory is equipped with Special Machinery which enables us to make SLIDES.— BETTER and CHEAPER than the furniture manufacturer.

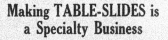

Canadian Table makers are rapidly adopting WABASH SLIDES

Because { They ELIMINATE SLIDE TROUBLES — Are CHEAPER and BETTER

BY USING

Reduced Costs | **WABASH**

Increased Out-put | **SLIDES**

MADE BY

B. Walter & Company
Wabash, Ind.

The Largest EXCLUSIVE TABLE-SLIDE Manufacturers in America

ESTABLISHED-1887

The Machine Supreme

for wire stitching Climax Baskets, is undoubtedly the

SARANAC Climax Basket Stitching Machine

It is the newest and most perfect machine of its kind. Its use insures a higher degree of efficiency, greater capacity, absolute accuracy, easier operation and bigger savings in material and labor.

If more profit means anything to you—install this machine.

MAY WE SEND YOU FURTHER PARTICULARS?

Saranac Machine Co.
Makers of high-grade fruit stapling machines for all purposes

Benton Harbor, Mich.

WITH spring many new homes will be built, new furniture will be required. This means an increase in your sanding department. Are you prepared to meet the demand?

CARBORUNDUM BRAND GARNET PAPER AND CLOTH

Produce a better, quicker and finer finish. They are made from selected garnet—free from all impurities. Uniformly coated on the best paper and cloth stock.

Working Samples Upon Request

The Carborundum Company, Niagara Falls, N. Y.

MATTISON
BELT SANDERS
DIFFERENT KINDS for DIFFERENT WORK

Turn Those Costly Hand-Sanding Jobs Over to This Mattison "131"

It will save enough in time and labor, on odd jobs alone, to pay for itself in a short time. There is undoubtedly a lot of your work being sanded by hand, no matter how many "Special" sanding machines you may have. And that's what "raises Ned" with your finishing costs. Most of the time so spent is wasted, because this Mattison "131" will do the work in less than half the time, with cheap help.

The Mattison Hand-Block Sander is the most practical mechanical method for finishing flat surfaces or molded shapes. Its wonderful flexibility, obtained through the long reach of the sand belt and the way it is tensioned, permits of handling unlimited sizes and shapes of work.

To accomplish convenience of operation and quick dependable adjustments, a substantial foundation, especially for the table, is most necessary.

While this style of sander is not subjected to severe strains or excessive vibrations, the different parts must keep in proper alignment under continuous operation. If the adjustments stick or the table ceases to roll easily, the efficiency of the machine is lost.

The operator goes over the whole surface being sanded, by applying hand-block over sand belt, at the same time moving the table back and forth. So the importance of having every part work freely and with greatest convenience is quite apparent.

Talk over these points with your practical man and verify their value as applied to your finishing problems.

Our Bulletin Shows 29 Views of this Sander on Labor-Saving Operations. Send for a Copy.

C. MATTISON MACHINE WORKS
BUILDERS OF BELT SANDERS, ELECTRIC RUBBING MACHINES SHAPING LATHES AND SPECIAL WOODWORKING MACHINERY

883 Fifth Street, BELOIT, WISCONSIN, U.S.A.

Volume Sixteen   TORONTO, MAY, 1916   Number Five

CANADIAN WOODWORKER
and
Furniture Manufacturer

C.M.C.

CHAIN   MORTISERS

The only Builders in
Canada of the original
Chain Saw Mortisers
by special arrangement
with the New Britain
Machine Co.

BUILT IN
THREE |
SIZES.

Strong, rapid machine
for mortising Sash,
Doors, Blinds, Furni-
ture and Pianos, etc.

*Bulletins
and
Full Details
sent on
request.*

CANADA MACHINERY CORPORATION
LIMITED
GALT   -   ONTARIO

Toronto Office and Showrooms   ·   ·   ·   ·   Brock Avenue Subway

Put Good Belting
on your
Good Machines

Your machines are expensive, and naturally you expect them to turn out the quality and quantity of work commensurate with their cost.

Then why harness them with belting which cannot begin to do them justice?

Use
"AMPHIBIA"
Planer Belting

and get the most work from your machines, in the quickest time at the lowest cost per day of service.

Try a sample run of AMPHIBIA Planer
and prove its merits.

"Leather like gold has no substitute."

Sadler & Haworth

Established 1876

Tanners and Manufacturers

For 40 years Tanners and Manufacturers of the Best Leather Belts.

| Toronto | St. John, N.B. | MONTREAL | Winnipeg | Vancouver |
| 33 Wellington St. E. | 139 Prince William St. | 511 William St. | Galt Building | 217 Columbia Ave. |

Woodworking Machinery

We handle all kinds, including full line of

American Wood Working Machinery Co.
ROCHESTER, N.Y.

M. L. Andrews & Co.
CINCINNATI, OHIO

Let us show you the machines covered by cuts in the advertisement of the American Woodworking Machinery Co. on opposite page in operation here in Toronto.

We have a number of American machines for sale— slightly used only — at bargain prices. Write us regarding such tools as you may need.

Let us co-operate with you in selecting and laying out your tool equipment.

Our services are at your disposal.

Garlock - Machinery

32 Front Street West   -   Toronto

Telephone Main 5346

Page 3 in the book on handle factory efficiency

The Klotz Machine Company has spared neither expense nor effort in designing and manufacturing handle factory equipment of undoubtable excellence, and naturally, they would strain a point to perfect a lathe which would maintain the example of efficiency set by its other makes. So, because the lathe is the mainspring of a money-making handle factory,

KLOTZ IMPROVED AXE HANDLE LATHE

is all a lathe could possibly be. Has a surprising capacity for turning any kind of irregular shapes. Good mechanical construction has practically eliminated any upkeep expense. Automatic fast or slow feed; revolving rest; sprocket wheel and chain operate pattern and handle preventing any slipping; safety shield over the 12-in. live gauge saw furnished with each lathe. Pattern regulates shape to be turned—size regulated by raising and lowering revolving rest. Turns lengths from 8-in. to 42-in. Larger sizes to order. Useful in wood-working plants to turn waste stock into cash.

Turn a Page With Us Next Issue

FOR YOU

who have a thought of manufacturing handles, or those desiring maximum production at the correct cost, get the Klotz Machine Company's idea of a perfectly modern handle factory. One with such dependable machines in it as their

BLOCKING SAW

SHAPING SAW

LATHE

THROATER

BOLTING SAW

POLISHER

DISC WHEEL OR END SHAPER

KNOBBER OR SCROLL MACHINE

All have a place in an efficient handle factory. Write

Klotz Machine Co., Sandusky, O.

Gang Dovetailer

for Ammunition Boxes

The Best Machine Yet Produced

Can be made to do either the closed end drawer joint or the open end box joint, and can be changed from one joint to the other at a very small cost.

The frame is well proportioned and very strong. All belts are now done away with except the belt drive from the line shaft. This is important, as it occupies only one-third the floor space usually given to a machine of this class and reduces belt cost to the minimum.

A large number of these machines, installed during the past three years, are giving excellent satisfaction. Our guarantee goes with each machine.

Have you our Bulletins?

Jackson, Cochrane & Co.

BERLIN, CANADA

Our No. 90 Gang Dovetailing Machine

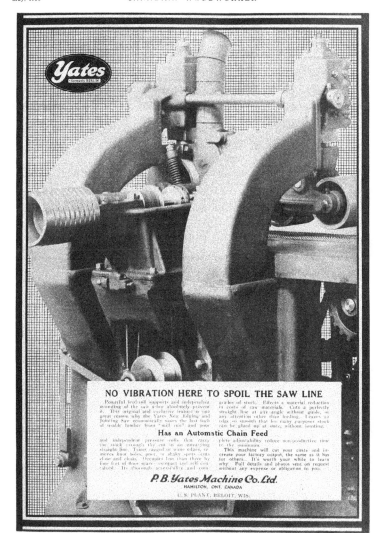

NO VIBRATION HERE TO SPOIL THE SAW LINE

Powerful feed-roll supports and independent mounting of the saw a-bar absolutely prevent it. This original and exclusive feature is one great reason why the Yates New Edging and Jointing Saw economically saves the last inch of usable lumber from "mill run" and poor grades of stock. Effects a material reduction in costs of raw materials. Cuts a perfectly straight line at any angle without guide, or any attention other than feeding. Leaves an edge so smooth that for many purposes stock can be glued up at once, without jointing.

Has an Automatic Chain Feed

and independent pressure rolls that carry the stock through the cut in an unvarying straight line. Trims ragged or wane edges, removes knot holes, poor, or shaky spots cuts close and clean. Occupies less than three by four feet of floor space compact and self contained. Its thorough accessibility and complete adjustability reduce non-productive time to the minimum.

This machine will cut your costs and increase your factory output, the same as it has for others. It's worth your while to learn why. Full details and photos sent on request without any expense or obligation to you.

P. B. Yates Machine Co. Ltd.

HAMILTON, ONT. CANADA

U. S. PLANT, BELOIT, WIS.

Are you one of those who could make a big profit on this machine?

¶ Not every business can make a profit on a "FAY-EGAN" "LIGHTNING" No. 291 Automatic. ¶ But if your product requires large quantities of equalized lumber, i.e., stuff cut-off or trimmed to exact length, then this Hopper Fed Machine can and will make a big profit for you. ¶ Its capacity is practically unlimited. ¶ It will cut off both ends as fast as your operator can throw stock into the hopper by the armful and off-bearer clear it away. ¶ Does as much work as six old style trim saws. ¶ And every piece is **exactly** the length of all the others. ¶ If you equalize in quantities, use the enclosed post card to get Bulletin Q-28 and prices.

J. A. FAY & EGAN CO.

153-173 West Front St.   -   -   -   -   Cincinnati, Ohio

—this new 230 page book
on Woodworking Machinery free!

Our new handy reference catalog is just off the press. Shows all of the newest models in woodworking machinery.

Every owner and user of woodworking machinery, every superintendent and foreman should have a copy of this new book, which is really a text on modern woodworking machinery.

A copy will be sent to you, free of charge upon receipt of your request.

Fill in and mail the coupon to-day.

J. A. Fay & Egan Co.,
153-173 W. Front St.,
Cincinnati, Ohio, U.S.A.

Please send your new Book on Woodworking Machinery free of charge to:

Mr. ....

Position ... (Give name of company you are with) ...

Street ....

City ....

Province ....

" Treat your machine as a living friend."

What is the problem in Canada today which is the cause of most anxiety to all manufacturers?

"THE SCARCITY OF LABOR"

If you have trouble of this nature, install one of our

No. 162 High-Speed Ball-Bearing Shapers

and keep your production up to top notch.

Speed 7,000 Revolutions per minute

HERE IS WHAT WE CLAIM:— Double the amount of work can be done on this HIGH SPEED BALL BEARING SHAPER that can be done on an ordinary shaper, with much more ease to the operator. LET US PROVE IT.

DIMENSIONS, WEIGHTS and SPEEDS

Floor Space, 120 in. x 58 in. Net Weight, 2650 lbs. Shipping Weight, 2750 lbs. Tight and Loose Pulleys on Countershaft, 10 in. x 6 in. Driving Pulleys on Countershaft, 18 in. x 5 in. Idler Pulleys, 14 in. x 5 in. R.P.M. Countershaft, 1200. Pulley on Spindle, 3⅛ in. x 6 in. R.P.M. Spindle, 7000. Diameter Spindle, 2 3/16 in. Spindle Centres, 28 in. Table, 58 in. x 41 in. Diameter Spindle Top, 1¼ in. Horse Power Required, 3. Front Projection, 17 in. Height, 35 in.

REPEAT ORDERS BEST PROOF OF MERIT

The Chevrolet Motor Car Company, Limited, of Oshawa, bought three of these machines November 26, 1915. On March 22, 1916, they wrote us: "The three Shapers which we purchased from you several months ago are giving every satisfaction. How soon could you supply us with two more and what allowance could you make for two old shapers?" On March 28 they ordered these two additional shapers, making five in all. They further expressed their opinion in a letter dated April 5, 1916, in which they say: "We cannot speak too highly of the design and workmanship of the machines, or the results which we are able to get from them. They are, in our opinion, one hundred per cent. perfect." We specialize in Labor Saving Machines. If you are interested in these machines, write us.

The Preston Woodworking Machinery Co. Limited, Preston, Ont.

Peter Hay Machine Knives

Peter Hay knives are made of special quality steel temper-ed and toughened to take and hold a keen cutting edge. Hay knives are made for all purposes.

Have you our Catalogue?

Made in Canada

The Peter Hay Knife Co., Limited
GALT, ONTARIO

PRESSES

For Veneer and Veneer Drying

Made in Canada

William R. Perrin
Limited
Toronto

A New Drying Principle
That Has "Made Good"

Only a little more than a year ago the National Dry Kiln Company announced a **Vertical** Piping System for the drying of hardwoods. Today there are National **vertically-piped** hardwood kilns in actual use in practically every section of the United States. Yes—and following the success of this new System in the States, it has been adopted as the most efficient and economical dry kiln equipment by leading manufacturers in Canada—among them the

Electric Planing Mills
Welland, Ontario

Isn't a system that has, because of its unusual merit, been so generally adopted in so short a time, at least worth investigating by every alert, progressive manu-facturer? It surely is! Then why not write for our catalog, giving complete information about **Vertical** piping today? No obligation.

The National Dry Kiln Co.
1117 Maryland Street, INDIANAPOLIS

Do You
Want This?

1/3 RD ORIGINAL COST

(1) An outfit of E. B. Hayes Machine Company's Dowel Door Machin-ery.
Combined Stile and Rail Borer.
Dowel Sticker, Cut-off and Pointer.
Dowel Gluer and Driver. Capa-city 125 doors.

(2) An 8-ft. Linderman Dovetail Joint-er. Used one year.

(3) A Mattison Block and Spindle Machine.

(4) A McGregor-Gourlay 12-in. Four-Side Moulder.

Write NOW.

J. L. NEILSON & CO.
Winnipeg, Man.

BOOKS FOR SALE
The following books are offered at special prices subject to previous sale:

Carpentry and Joinery, by Gilbert Townsend. Published in 1913 by the American School of Correspondence. 258 pages, illus-trated. Price $1.00.

Saw Fitting Manual, a treatise on the care of saws and knives. Deals with everything in the saw and knife alphabet, from adjustments to widths. 144 pages. Price $2.00.

Common-sense Handrailing, by Fred T. Hodgson. Published by Frederick J. Drake & Company, Chicago. 114 pages, illus-trated. Price 50c.

Roof Framing Made Easy, by Owen B. Maginnis. Published by The Industrial Publication Company, New York. 164 pages, illustrated. Price 50c.

Handrailing Simplified, by An Experienced Architect. published by William T. Comstock, New York. 52 pages, illustrated. Price 50c.

How to Join Mouldings; or, The Arts of Mitering and Coping, by Owen B. Maginnis. Published by William T. Comstock, New York. 72 pages, illustrated. Price 50c.

Popular Mechanics Shop Notes. Published by Popular Mechanics, Chicago. Easy Ways to do Hard Things, etc. Years 1905-1906. Price 40c. each.

Cabinet Making, by J. H. Rudd. Published by Grand Rapids Fur-niture Record Company. 210 pages, illustrated. Price $1.50.

Utilization of Wood-Waste (Second Revised Edition), by Ernst Hubbard. Published in 1915 by Scott, Greenwood & Sons. 192 pages, illustrated. Price $1.50.

Woodworker Publishing Company, Limited
345 Adelaide Street West, Toronto, Ontario

The Only Way to Drive Screws

There is a stage-coach method and a modern method of doing everything. When it comes to driving screws the only modern way is with an Automatic Screw Driving Machine, which means greater and better production and greater profit.

The Reynolds
Automatic Screw-Driving Machine

is guaranteed to produce results. It is no experiment. The No. 2 Standard Type, shown herewith, is particularly suited for placing the screws in the straps of shrapnel boxes. It may be fitted to drive, without change or adjustment, any three consecutive sizes flat or round head, or two sizes filister head, No. 6 to 14 inclusive, wood or machine screws, or three sizes flat or round head, or two sizes filister head machine screws up to No. 20. All the above up to 1½ inches long if smaller than No. 10, or up to 1¾ inches long, No. 10 or larger. Fitted with boring attachment if desired.

Many Machines in Use in Canada.

Ask for our Catalogue.

REYNOLDS PATTERN & MACHINE CO.
101 and 103 Third Ave., Moline, Ill., U.S.A.

No. 2 Standard Type Automatic
Screw-driving Machine.

Are Your Products
Perkins Glued?

Ladies' Desk manufactured by The Berlin Furniture Co., Berlin, Ont., in which Perkins Glue was used.

The live dealer realizes the value of selling his customers Perkins Glued products because they never come apart. Many of Canada's largest furniture factories use Perkins Vegetable Glue. It costs 20 per cent. less than hide glue, is applied cold, antiquating the obnoxious and wasteful cooking process, is absolutely uniform in quality and will not blister in sanding.

We will be glad to demonstrate in your own factory and on your own stock what can be accomplished with Perkins Vegetable Glue.

Write us today for more detailed information.

Perkins Glue Co.
Hamilton, Ont.   Limited

Address all inquiries to Sales Office, Perkins Glue Company,
South Bend, Indiana, U. S. A.

Alphabetical List of Advertisers

American Hardwood Lumber Co... 11
Atlantic Lumber Company 11
Ault & Wiborg Company 33

Baird Machinery Works 43
Brown & Co., Geo. C. 46

Canada Machinery Corporation ... 1
Canadian Boomer & Boschert Mfg.
　Company 47
Canadian Morehead Mfg. Company 45
Carborundum Company 51
Central Veneer Company 39
Chapman Double Ball Bearing Co. 47
Churchill, Milton Lumber Co. 11
Clark Veneer Co., Walter 37
Clark & Son, Edward 10
Coignet Chemical Products Co. ... 33

Darling Bros. 46

Evansville Veneer Company 34
Elgie & Jarvis Lumber Co. 43

Fay & Egan Company, J. A. 6
Foster, Merriam & Company 50

Garlock-Machinery 3
Globe Soap Company 51
Grand Rapids Dry Kiln Company .. 47

Hartzell, Geo. W. 37
Haughton Veneer Company 39
Hay & Company 33
Hay Knife Company, Peter 5
Hoffman Bros. Company 37

Jackson, Cochrane & Company ... 4

Kersley, Geo. 38
Klotz Manufacturing Company ... 4

Lightfoot, Stanley 43

Mattison Machine Works 52
Mowbray & Robinson 11

Nartzik, J. J. 37
Nash, J. M. 45
National Dry Kiln Company 8
Neilson & Company, J. L. 8

Ober Manufacturing Company 47
Ohio Veneer Company 37

Perfection Wood Steaming Retort
　Company 45
Penrod Walnut & Veneer Company 39
Perkins Glue Company 9
Perrin, William R. 8
Petrie, H. W. 43
Preston Woodworking Mach. Co.. 7

Radcliff Saw Company 12
Reynolds Pattern & Machine Works 9
Roberts, John N. 38

Saranac Machine Company 51
Sadler & Haworth 2
Sheldons Limited ... ...
Simonds Canada Saw Co. ... .
Spencer, C. A. 11

Toronto Blower Company 12

Vokes Hardware Company 41

Walter & Company, B. 51
Weeks Lumber Company. F. H. ... 46
Whitney & Son, Baxter D. ... .
Wood Mosaic Company 38

Yates Machine Co., P. B. 9

Edward Clark & Sons

Toronto

CANADIAN HARDWOODS—winter cut Birch, Bass, Maple

10,000,000 ft. on hand, shipment June 1st
10,000,000 ft. sawing, shipment August 1st

ONTARIO and **QUEBEC** stocks, East or West, rail or water shipment, direct from
　mill to factory.

Put Up for Shell Boxes
　3,000,000 ft. 4/4 Birch, No. 1 and 2 Common.
　2,000,000 ft. 6/4 Birch, No. 1 and 2 Common.
　1,000,000 ft. 12/4 Birch, No. 1 and 2 Common.

For Furniture Manufacturers
　1,000,000 ft. each 10/4, 12/4, 16/4 Birch, 75
　per cent. No. 1 and 2.

For Trim Trade
　500,000 ft. 4/4 Birch, No. 1 and 2, 30 per cent. 14/16 ft., Red all in.
　300,000 ft. 5/4 Birch, No. 1 and 2, ⎫ 9 in. average. Red all in.
　700,000 ft. 6/4 Birch, No. 1 and 2, ⎬ 40 to 60 per cent. 14/16 ft. long.
　500,000 ft. 8/4 Birch, No. 1 and 2, ⎭
　200,000 ft. 4/4 to 8/4 Brown Ash, No. 1 and 2, 40 per cent. 14/16 ft.

We Commend to Your Notice:
　Our lumber is **reliable**, our grades **uniform**, our shipments **prompt**, and receive personal supervision;
our prices **consistent**. We have had **long experience**. We desire to serve.

Dry Spruce
and Birch

Good Stocks, Prompt Shipments, Satisfaction

C. A. SPENCER, Limited
Wholesale Dealers in Rough and Dressed Lumber

Offices—500 McGill Building

MONTREAL   -   -   Quebec

Churchill Milton Lumber Co.
Louisville, Ky.

We carry at
NEW ALBANY, IND.

Genuine Indiana Plain and Quartered Oak, Poplar

At GLENDORA, MISS.

Cypress, Gum, Cottonwood, Oak

SEND US YOUR INQUIRIES

We have dry stock on hand and can make quick shipment

American Hardwood Lumber Co.
St. Louis, Mo.

Large stock of—

Dry Ash, Quartered Oak Plain Oak and Gum

Shipments from — NASHVILLE, Tenn., NEW ORLEANS, La., and BENTON, Ark.

The Atlantic Lumber Co.

Manufacturers and Wholesale Dealers

American and Canadian Hardwood Lumber and Mahogany, Quartered and Plain, White and Red Oak a Specialty.

110 Manning Chambers   - TORONTO
Phone Main 6386

HOME OFFICE, BOSTON MASS.

Mills: KNOXVILLE, TENN.,   WALLAND, TENN.
FRANKLIN, VA.

Plain and Quartered Red and White Oak

and Other Hardwoods

Even Color   Soft Texture

We have **35,000,000 feet dry stock** all of our own manufacture, from our own timber grown in Eastern Kentucky

MADE (MR) RIGHT

OAK FLOORING

PROMPT SHIPMENTS

The Mowbray & Robinson Co., Inc.
GENERAL OFFICES
CINCINNATI, OHIO
Mills: Quicksand, Ky., West Irvine, Ky., Viper, Ky.

Canadian Representative:

N. H. FARNHAM
601 Elmwood Ave., BUFFALO, N. Y.

Beaver   Brand

Reduce the Cost of
Making Shell Boxes

by using a

Special
"Beaver" Dado

This special "Beaver" Dado was designed to meet the requirements of shell box manufacturers. This particular construction eliminates the trouble previously experienced by many manufacturers in cutting across the grain. It leaves a clean, smooth surface when cutting with or across the grain.

Illustration shows one complete order recently shipped by us to a shell box manufacturer. It consists of five "Beaver" Dados from 1½ in. to 2 in. wide; also two special Planer Saws for cutting of box stock; the small groover shown at centre-top is to cut slot for wire handle or for string for the cover.

Get our prices on special shell box saws—we can save you money.

Radcliff Saw Manufacturing Company, Limited
TORONTO   550 Dundas Street   ONTARIO

Agents for L. & I. J. White Company Machine Knives

Read this letter—it proves why

you should write us about our

Improved Automatic
Furnace Feeders

PHONE JUNCTION 1234

DETAIL WORK AND
VENEERED DOORS A SPECIALTY

T. H. HANCOCK

SASH, DOORS, FRAMES, BAND SAWING, SHAPING, PLANING, MOULDING,
ROUGH AND DRESSED LUMBER

TORONTO   March 14, 1916.

The Toronto Blower Co.,

City.

Dear Sirs:-

I have pleasure in writing this letter in regard to the Blower Systems you installed in my factory last year. Both the shaving and the separate sawdust systems are working very satisfactorily and I know as good a job of piping as it is possible to put up. This work was done under the direction of your Mr. Daniels, whom I consider a very good mechanic.

Yours truly,

J. H. Hancock

Special

Hoods For Wood-
working Machines

Slow Speed
and Low Power
Exhaust Systems

for removing shavings, sawdust, emery dust, lint,
smoke and odors

Full Particulars Upon Request

THE TORONTO BLOWER COMPANY
TORONTO   156 DUKE ST.   ONTARIO

Canadian Woodworker

A Monthly Publication in the Interest of the Woodworking Industry.
Reaches the factories producing interior finish, doors, sash, flooring,
woodenware, furniture, pianos, boxes, and general mill products.
Subscription, $1.00 a year; foreign $1.50.

Woodworker Publishing Company, Limited

Branches: Montreal, Vancouver,
Chicago, New York, London, Eng.

345 Adelaide St. West, Toronto
Phone Ade. 2700

Authorized by the Postmaster General for Canada, for transmission as second class matter.
Entered as second class matter July 18th, 1914, at the Post Office at Buffalo, N.Y., under the Act of Congress of March 3, 1879.

Vol. 16   May, 1916   No. 5

Stock Taking

Undoubtedly the most important feature in connection with any business is to know whether it is showing a profit or a loss. This information is obtained by means of cost systems, etc., worked out by competent cost accountants, but it is generally necessary for a business house carrying any appreciable amount of stock to check it up at least once a year. With a firm carrying a large and varied stock, however, such as a woodworking plant, it is very often necessary to check up or "take stock" (as it is more commonly called) twice a year, that the manufacturer may know just exactly what stock he has on hand and its value. Stock taking simply means making an inventory of everything on the premises, but this should be done under two distinct headings, namely, stock and equipment.

Stock taking is regarded as rather a disagreeable job in a large number of factories. In fact, it has been the custom to look on it as a sort of an annual house-cleaning, but our friend, "Mr. System," that powerful factor in modern business, has been making some wonderful strides in the woodworking plant as elsewhere, and the result is that there are very few factories that require an annual house-cleaning. Consequently stock taking is a much more simple matter than it used to be. A great many plants shut down for one and two weeks for the purpose of taking stock, and although part of this time is very often taken up in making repairs to the machinery and boilers, etc., we believe some particulars of an up-to-date method of taking stock will be welcomed by a large number of manufacturers. We have accordingly devoted some space to it in this issue. Our original intention was only to deal with stock taking, but we have received some interesting information on keeping track of stock. Our information has therefore been put under two headings—Stock Taking in the Woodworking Plant, and The Importance of Keeping Track of Stock.

The purpose of this journal is to present practical and timely information to men engaged in the woodworking business, and the information which we offer under these headings represents no idle theory, but has been gained through experience by men prominent in their particular branch of woodworking. Their experience in many cases has cost real money and yet we find them ready and willing to give out this information for the benefit of others. Surely this is sufficient evidence that manufacturers are willing to cooperate for the benefit of the industry. It also proves the value of the "Canadian Woodworker and Furniture Manufacturer" as an effective medium through which they may exchange ideas. The men who concentrate their minds on their own particular plants without keeping in touch with developments in their line of work may find that they have missed some important idea which would have helped them to reduce their manufacturing costs. On the other hand, their business may demand so much attention that they have very little time to spare to keep in touch with what other manufacturers are doing. These are the men who will find it to their advantage to subscribe to the trade paper which specializes on their particular field.

Labor Situation Serious

During a recent trip to Berlin, Galt, and other woodworking centres, the Editor of the "Canadian Woodworker" found business conditions in the woodworking and furniture industries in a healthy condition. Furniture factories without an exception report that they are fairly well fixed for orders, but they cannot get men to make the furniture. Upholsterers are particularly scarce; in fact, it is next to impossible to get them without taking them away from other manufacturers. The labor situation is a serious one and there does not appear to be any relief in sight at present. Recruiting officers are still active and the manufacturers of munitions are offering inducements in the shape of big wages for men with some mechanical skill to operate machines for turning out shells, etc.

The average woodworking mechanic can be "broken in" to do this work in a very short time and consequently a great many who are not enlisting are drifting into the munition work in order to take advantage of the high wages that are offered. While they cannot be blamed for making the shift, it is rough on the manufacturers who kept their staffs on when they had hardly any work for them to do.

It has recently been brought to our attention that some of the furniture manufacturers located at different points in Ontario have employed women in some of the departments. There may be some jobs in furniture factories that women can handle, such as sandpapering stock in the finishing room and applying the finish on some of the cheaper goods, but there are many reasons why they would not be able to take the place of machine hands or cabinet makers. They might be able to do upholstering, but by the time they had learned how to do the work, the war would be over. So it looks as though the only place a woman could be used in a furniture factory would be in some of the departments where there is no machinery and on boys' work.

The planing mills and general woodworking plants are not so handicapped for labor as the furniture manufacturers, as there has not been any decided activity in the building trades and they find that they are able to come out just about even, with enough men to take care of the work that comes their way.

Finishing Interior Trim

Should sash, doors, and interior trim for dwelling houses be finished before it leaves the factory? The writer discussed this question with the proprietor of a large planing mill at Berlin, Ontario. This party intends adding a finishing department to his plant in the near future, as he is convinced from long experience in supplying finish for all classes of buildings that all woodwork for interior use should be finished before it leaves the factory to be erected in the building.

This looks like a step in the right direction and we would like to hear the opinions of some of our readers on the subject. When we consider that the lumber from which interior finish is manufactured is kiln dried and after remaining for some time in the dry atmosphere of the factory is taken to the job and very often piled outside with only a few rough boards for a covering and then put up in a building which is minus windows and doors, it is evident that some treatment of the wood is necessary.

The lumber will absorb sufficient moisture to make it practically as wet as before it entered the kiln. Then, after the painters have finished their work the building is closed up and the furnace started, with the result that the wood will shrink and show unsightly cracks and streaks of unfinished wood where

door panels, base, stair treads, etc., have shrunk away from other portions of the work. Complaints then begin to reach the planing mill man that the lumber was not properly dried, when, as a matter of fact, he had performed his part of the work honestly and faithfully.

To us it looks like a splendid opportunity for planing mill men to create business by preparing plans of dwelling houses and samples of finished wood so that they will be in a position to give "service" in the fullest sense of the word, to prospective home-builders. They would have no difficulty in securing the co-operation of the manufacturers of finishing materials and some good colored pictures showing interiors of rooms would also be in order. The prospective home builder could then compare the picture with the finished sample of wood. The satisfaction of knowing just what his home was going to look like when completed and that the woodwork was not going to shrink like it had in other houses he had seen would take a load off his mind.

These features would also make a strong appeal to the man who builds houses to sell. It would be a relief to him to know that he was going to have lumber in his houses that would "stay put" and that he could choose the style and color of finish he wanted at the same time that he ordered his lumber. There does not appear to be any room for objections on the planing mill man's side. The stock is all machined to patterns, so that all the work required on the job is to cut off the necessary lengths.

Box Manufacturers' Convention

The Board of Governors of the National Association of Box Manufacturers of the United States, after careful consideration of the proposition, have decided to hold the mid-summer meeting of the organization at the Belmont Hotel, New York, Aug. 23, 24 and 25. It was originally intended to have the convention in June, but because of the fact that practically all box manufacturers make up their inventories during the latter part of June, it was decided to hold the convention in August so as to have as good an attendance as possible. One of the subjects to be taken up at the convention is the uniform cost, price and discount sheets which were considered at the last annual session. In all likelihood definite action will be taken in the matter.

Veneer and Panel Manufacturers Meet

The semi-annual meeting of the National Veneer and Panel Manufacturers' Association will be held at Indianapolis, Ind., on Tuesday and Wednesday, June 13th and 14th. The Association is endeavoring to improve conditions in the veneer and panel industry and a cordial invitation is extended to members and non-members alike to be present at this meeting and help along the Association work by expressing their views on the subjects to be discussed.

Figure Your Profits on Sales Instead of on Costs

By W. A. McKinnon, President and Manager Western Planing Mills Company, Calgary, Alta.

A few years ago a newspaper man asked one of the most prominent business men in the United States how he thought the "trusts" in that country could be best controlled and was surprised to receive the following brief reply: "Make it a legal offence to sell goods below cost plus a profit." Undoubtedly he had given the subject considerable study before arriving at that conclusion, for it is a known fact that trusts and kindred organizations are always started with a view to protecting legitimate business against the individuals or firms who do not know the true value of their goods and sell at a loss. It seems to be a coincidence that the only idea of salesmanship which

Mr. W. A. McKinnon.

this class of people possess is "cut prices below the other fellow."

I was always of the opinion that the proper methods of determining costs and profits were universally understood in the woodworking and lumber business until a few years ago when I was appointed one of a committee of several to compile a "price list," allowing a fair profit and also sufficient margin to allow 10 per cent. discount for cash payment. Imagine my surprise on finding men with years of business experience totally ignorant of the correct methods of figuring costs and profits. This led me to believe that the subject was not given the attention it is entitled to by the trade journals and other sources of education. The object of this article is to bring the matter to the attention of readers of the "Canadian Woodworker" in the hope that other manufacturers will speak out and give us the benefit of their knowledge and experience, as different minds run in different channels of thought, and we should not regard knowledge of this kind as trade secrets.

If all firms in our line of business were educated to properly understand this important subject, the usual trouble experienced in getting a fair price for

our goods would be practically eliminated. I am quite sure that fully three-quarters of the failures in the woodworking business are due either directly or indirectly to ignorance of the correct methods of figuring costs and adding profits. The dealer who sells his goods below cost not only harms himself, but his competitors, and both their creditors are placed in jeopardy by his action. Just as soon as the public learns of his prices they naturally flock to him and use his prices as a weapon to reduce the prices of the legitimate dealer with whom they have been doing business in the past.

The legitimate dealer, seeing his business slipping away from him, will in desperation cut his prices lower and lower, until, sooner or later, he finds himself not master of his business, but in the same state as his competitor, who does not know the value of his goods.

You may say, "Why doesn't he stand still until the firm doing business in such a foolish manner has gone broke?" but that is where the difficulty arises. Just as soon as the legitimate dealer allows his customers to leave him and purchase from the competitor the volume of the competitor's business will increase by leaps and bounds and he is immediately recognized by his bank as a "live and aggressive business man" and has no trouble in getting credit. In fact, he is encouraged to extend his operations, which greatly assists in prolonging the evil day, as the increased volume of business will to a certain extent decrease his overhead expense. The bank never dreams that they are fostering a commercial leper and, glorying in their good fortune in having such a "live account," never think of investigating his business methods or asking themselves the question, "Does he know the value of his goods?" They are satisfied with his honest and aggressive reputation.

The legitimate dealer who has lost, or is losing his trade through the actions of this "live wire," is faced with one of two methods of business suicide. He can stand still and be devoured by his overhead expense which he now finds has been increased by decreased volume of business and considerable unrequired equipment which was necessary to handle his former business, or meet Mr. Live Wire at his own game and slide merrily down on the toboggan to bankruptcy with him.

I was in conversation a short time ago with one of the largest wholesalers (now an old man) on the continent. He owns a line of stores all over the country which he had been obliged to take over to save himself. He was commenting on the large percentage of such businesses that failed, while he had no difficulty in making them pay and stated that he found in 75 per cent. of the cases ignorance of how to estimate their costs and profits the cause of their failure. It has often seemed strange to me that this subject is not more fully explained in our schools and business colleges. You will quite frequently find in their arithmetic books such examples as this: A man buys a horse for $50.00 and sells it for $75.00. What per-

SHEET No. 0

EXPLANATION & ILLUSTRATIONS
TO SHOW WHY
PROFITS SHOULD BE FIGURED ON SALES
INSTEAD OF ON COSTS

PERCENTAGES OF EXPENSES ARE USUALLY, AND MOST SAFELY, CALCULATED ON THE YEARS SALES.

THE SALARY OR OTHER REMUNERATION OF SALESMEN, IS ALWAYS RECKONED ON THE SALE, AND THE AMOUNT IS ALWAYS BASED MORE OR LESS ON A PERCENTAGE OF THE SALES TOTAL.

MERCANTILE OR OTHER TAXES OF A SIMILAR NATURE ARE ASSESSED ON A CERTAIN PERCENTAGE OF THE ANNUAL SALES: ALSO IF ANY SPECIAL TAXES ARE LEVIED BY GOVERNMENTS, THE AMOUNT IS ALWAYS A CERTAIN PERCENTAGE OF THE SALES, AND NOT A PER:CENTAGE OF THE COST.

THIS ILLUSTRATION SHOWS THE GREATER SAFETY IN FIGURING ON SALES, ESPECIALLY WITH UNTRAINED MINDS WHO DO NOT PROPERLY DISCRIMINATE.

A man sold an article which cost 80 cents, for $1.00 and basing his percentage of profit on the cost, figured he was making 25 per cent. At the end of a given period, the Sales totalled $20.000.00 The manager told a stockholder the amount of Sales and also the percentage of profit. The presumption was that a profit of $5,000.00 had been realised, while the books showed a profit of $4,000, or 20% on the sales.

Cases have been known where arrests for defaulting have been made in such instances.

ILLUSTRATION No 2.

A firm purchased an article which cost them $20.00 F.O.B. their warehouse. To this amount they added 20%. This firm's overhead expense is 15% of their total sales. What profit did they make on this sale

PROCESS:

20% OF $20.00 IS 16⅔% ON SELLING PRICE = $4.00
15% OF 24.00 THE SELLING PRICE 3.60
.40

ANSWER:- 40¢ OR 1⅔% PROFIT

Note:- Always figure your profits on the Sale, as shown on Sheets Nos. 1 & 2 – which is the correct method; then you will be on the safe side

W.A.McK.

Fig. 1.

centage of profit does he make? Answer: 50 per cent.

No more misleading answer was ever penned. They lead us to think of the percentage of profit from an unbusinesslike standpoint and cause many business men to think they are making much larger profits than they really are. This makes them careless of expenses and often causes failure, where, if they had had a proper knowledge of percentages their failure could have been avoided. Supposing a sale was transacted on the above basis, the man who had purchased the horse for $50.00 decided he would sell it for $75.00 and, not having the time to attend to the matter, called in a dealer who informed him he would assume all responsibility in effecting a sale for a commission of 33 1/3 per cent. The owner, having 50 per cent. profit in sight agrees to this and when a sale is completed he receives a statement from the dealer as follows:—

Sold one horse for	$75.00
Commission for selling same, 33 1/3 per cent..	25.00
Balance due	$50.00

The owner's books would show a profit of 50 per cent. entirely eaten up by a commission of 33 1/3 per cent.

In Card No. 0, Fig. 1, appears an explanation and illustration to show why profits should be figured on sales instead of on costs. If you carefully read the explanation and check over the figures in the illustration you will recognize the accuracy of this method. The table on Card No. 1, Fig. 2, enables a man of average intelligence to very quickly determine the correct selling price of any article. The value of this feature will be apparent to any person who has charge of a business establishment, as it is impossible for the manager to give personal attention to every detail of the business. The only course open to him is to train men to think as he thinks and the best method of accomplishing this is to have "his thoughts" in such a

SHEET No 1.

: TABLE FOR FINDING THE SELLING PRICE OF ANY ARTICLE :

COST TO DO BUSINESS	: NET PER CENT PROFIT DESIRED :																				
	1	2	3	4	5	6	7	8	9	10	11	12	13	14	15	20	25	30	35	40	50
15%	84	83	82	81	80	79	78	77	76	75	74	73	72	71	70	65	60	55	50	45	35
16	83	82	81	80	79	78	77	76	75	74	73	72	71	70	69	64	59	54	49	44	34
17	82	81	80	79	78	77	76	75	74	73	72	71	70	69	68	63	58	53	48	43	33
18	81	80	79	78	77	76	75	74	73	72	71	70	69	68	67	62	57	52	47	42	32
19	80	79	78	77	76	75	74	73	72	71	70	69	68	67	66	61	56	51	46	41	31
20	79	78	77	76	75	74	73	72	71	70	69	68	67	66	65	60	55	50	45	40	30
21	78	77	76	75	74	73	72	71	72	69	68	67	66	65	64	59	54	49	44	39	29
22	77	76	75	74	73	72	71	70	69	68	67	66	65	64	63	58	53	48	43	38	28
23	76	75	74	73	72	71	70	69	68	67	66	65	64	63	62	57	52	47	42	37	27
24	75	74	73	72	71	70	69	68	67	66	65	64	63	62	61	56	51	46	41	36	26
25	74	73	72	71	70	69	68	67	66	65	64	63	62	61	60	55	50	45	40	35	25

RULE
FOR USING SELLING PRICE TABLE

Multiply the cost price (Invoice price with freight added or Material and labour added together), by 100 & divide by the figure in the column of Net rate per cent profit desired on the line with per cent Cost to do business.

Example:

A gasoline engine costs F.O.B. Winnipeg $50.00
Freight W'peg to Cal'y...... 5.00
$55.00

You desire to make a net profit of.............10%
It costs to do business.......20%

Take the figure in Col. 10 on line with 20, – which is 70

70 | 5500.00 | 78.57⅐ THE SELLING PRICE
490
600
560
400
350
500
490
10/70 = 1/7

: W.A.McKINNON
: CALGARY : ALTA :

Fig. 2.

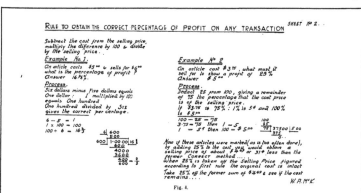

Fig. 4.

form that they may be consulted without worrying him.

Card No. 2, Fig. 3, shows two examples of the correct method of obtaining the percentage of profit on a transaction.

Card No. 3, Fig. 4, shows how to arrive at the per cent. cost of doing business. The items included under the different headings in Fig. 4 will be found on Card No. 4, Fig. 5. A few examples of the percentages which will have to be added to the cost price to give a desired percentage on the selling price are shown on Card No. 5, Fig. 6.

I read in an excellent article written in one of our leading magazines on "Why Business Firms Fail," a summary of avoidable causes, as follows: (1) Lack of Skill, which includes unpreparedness for the business, inefficiency, inaptitude, and the general subject of buying and selling; (2) Lack of Capital; (3) Over Extension; (4) Unwise Credits; (5) Speculation; (6) Dishonesty.

I would like to have seen the writer define the first cause more closely. If he had taken the trouble to do so he would have come to the conclusion that at least 90 per cent. of the first cause was "not knowing what price goods should sell for to make a profit."

Our Governments, both Dominion and Provincial, have formed departments and employed experts to assist citizens in agriculture, stock-raising of all kinds, mining, forestry, railroading, fisheries, etc., and the advancement of these lines is proof of their wisdom in doing so, but the citizens "who sell things" or "manufacture things" are overlooked entirely (except when there is an opportunity to pass legislation to add to their burdens). Whereas, if there was a department of experts in commercial affairs who could assist the manufacturer or merchant in analyzing problems in cost systems, salesmanship, credits, factory organization, financing, etc., they would soon find a field where as much could be accomplished for the public benefit as in any other department they have organized.

Lack of cost keeping knowledge is a serious problem, especially in a new country, where a great number of the business heads are made up of men not trained in practical business methods, as they are in

: COST OF DOING BUSINESS :		
: YEARLY STATEMENT :		
1 TAXES.		
2 INSURANCE.		
3 FUEL, LIGHT, WATER ETC.		
4 RENT.		
5 SALARIES.		
6 EXTRA CLERK HIRE.		
7 ADVERTISING.		
8 EXPRESS, TELEGRAPH & PHONES.		
9 OFFICE SUPPLIES, POSTAGE ETC.		
10 SHIPPING SUPPLIES.		
11 LIVERY DRAYAGE ETC.		
12 HORSES AND WAGONS.		
13 REPAIRS.		
14 DEPRECIATION.		
15 DEDUCTIONS.		
16 DONATIONS AND SUBSCRIPTIONS.		
17 LOSSES.		
18 MISCELLANEOUS EXPENSE.		
19 INTEREST ON TOTAL INVESTMENT.		
TOTAL EXPENSE.		
TOTAL SALES.		
PER CENT COST TO DO BUSINESS.		

RULE : MULTIPLY TOTAL EXPENSE BY 100 AND DIVIDE BY TOTAL SALES — RESULT IS PER CENT COST TO DO BUSINESS.
TO FIX SELLING PRICE SEE SELLING PRICE TABLES.
NOTE : THIS FORM IS AN EXAMPLE ONLY AND CAN BE CHANGED TO SUIT ANY BUSINESS.

: SEE EXPLANATION SHEET :
W. A. McK.

Fig. 4.

SHEET Nº 4.

:COST OF DOING BUSINESS

:EXPLANATION OF YEARLY STATEMENT:

1 TAXES: Include all Taxes & Licenses
2 INSURANCE: Fire, Employers Liability & all insurance except Life Insurance.
3 FUEL: Fuel, light & water.
4 RENT: Include rent of all property used in the business.
5 SALARIES: Include amounts to cover Salaries of Proprietor, Partners, Accountants and all services not included in Prime Cost.
6 EXTRA CLERK HIRE: Include all Canvassers, Auditors, Etc.
7 ADVERTISING: Include all money expended on advertising or expenses in connection with promoting business.
8 EXPRESS, TELEPHONE & TELEGRAPH: Include all amounts expended for these items where not added to invoice price of goods or charged to customer.
9 OFFICE SUPPLIES, POSTAGE etc. Include all bills for Stationery, Ink, Pencils, Postage stamps, etc.
10 SHIPPING SUPPLIES: Include all bills for wrapping Paper, Twine, Boxes, Crating, etc.
11 LIVERY, DRAYAGE etc.: All expenses when hired from others.
12 HORSES & WAGONS: If owned, figure all expenses of upkeep.
13 REPAIRS: This item should include all amounts paid to keep buildings & equipment in repair.
14 DEPRECIATION: Include a proper deduction from all Buildings & Equipment subject to decline from wear and tear also all goods that cannot be sold at regular prices.
15 DEDUCTIONS: Include amounts allowed customers for damaged goods or any cause whatever.
16 DONATIONS: Include money or goods donated to Charity.
17 LOSSES: Include Notes, Accounts which are uncollectable also amounts paid lawyers & others for collections and goods lost or stolen etc.
18 MISCELLANEOUS EXPENSE: Include all expenses not provided for above.
19 INTEREST ON TOTAL INVESTMENT: Figure interest on total assets at the beginning of Business Year. (Cash, Notes, Accounts, Merchandise etc) If this is done it ensures profits at least equal to interest had capital been loaned instead of invested.

W. A. McK.

Fig. 5.

SHEET Nº 5

TABLE
SHOWING NET PROFIT ON SELLING PRICE OF CERTAIN PERCENTAGES ADDED TO COST PRICE

5 %	ADDED TO	COST IS	4½ %	PROFIT ON	SELLING	PRICE
7½	"	"	7	"	"	"
10	"	"	9	"	"	"
12½	"	"	11 1/9	"	"	"
15	"	"	13	"	"	"
16⅔	"	"	14 2/7	"	"	"
17½	"	"	15	"	"	"
20	"	"	16⅔	"	"	"
25	"	"	20	"	"	"
30	"	"	23	"	"	"
33⅓	"	"	25	"	"	"
35	"	"	26	"	"	"
37½	"	"	27½	"	"	"
40	"	"	28½	"	"	"
45	"	"	31	"	"	"
50	"	"	33⅓	"	"	"
55	"	"	35½	"	"	"
60	"	"	37½	"	"	"
65	"	"	39½	"	"	"
66⅔	"	"	40	"	"	"
70	"	"	41	"	"	"
75	"	"	42 6/7	"	"	"
80	"	"	44½	"	"	"
85	"	"	46	"	"	"
90	"	"	47½	"	"	"
100	"	"	50	"	"	"

EXAMPLE:—
An article cost \$100.00 and you add 20% for profit—which equals \$120.00 (the selling price). What percentage of the selling price is the profit of \$20.00?

RULE:—
Subtract the COST from the SELLING price, multiply the difference by 100, and divide by the selling price.

120.00 (Selling)
100.00 (Cost)
20.00 x 100 equals
120.00) 2000.00 (16⅔ %
12000
80000
72000
8000 / 12000 = 2/3

W. A. McK.

Fig. 6.

the older countries. They think because they have been good mechanics, taught school or sold real estate, that they are qualified to shine in the commercial world and their first brilliant act in their ignorance of costs is to start cutting prices in order to get business, thinking by their way of figuring that there are enormous profits in the business and they are not going to be hoggish. They will be benefactors to the "poor dear public." Instead, they are the beginning of all evil in the business world. If you can prevent or reduce business failures, you prevent or reduce "waste which is wicked." This in turn will reduce interest rates, insurance rates, credit losses, legal and court expenses and other wastes too numerous to mention, which will eventually mean that prices of all commodities will be reduced and the purchasing power of a dollar increased.

Progress Made in Bending Wood

Wood bending is almost entirely confined to hardwoods, for the reason that nearly every article for which bent wood is used requires hardwood to give the necessary strength—as, for instance, rims for vehicles, shafts, and chair braces. The chief difficulty to be overcome in bending wood is to prevent the crushing or breaking of the fibres, in order that the wood will retain its strength.

In the progress of the bending industry many machines and appliances have been invented to facilitate the work and reduce the percentage of breakage. Until recently most of the attention has been centered on appliances for clamping and holding and bending the wood, but it seems now as though the attention has been centered "on the wrong side of the fence."

In other words, more satisfactory progress seems to have been made by paying attention to the preparation of the wood to be bent.

For many years the practice has been to boil wood in open vats, or steam it in home-made wooden boxes to prepare it for bending. The steaming idea was all right, but it was impossible to concentrate enough steam in one place to make the wood pliable, even with a good head of steam on, as it all leaked out, and another thing was that the steaming box had to be located outside the factory building, which made it very inconvenient for the man doing the bending.

These difficulties have been overcome through the introduction of closed metal retorts for steaming the stock in. These retorts can be placed in any part of the factory where it is desired to do the bending and they soon pay for themselves by the amount of steam saved, besides ensuring a smooth, even bend and almost eliminating breakage. The retorts are made by the Perfection Wood Steaming Retort Company, Parkersburg, W. Va., and they merit the serious consideration of the manufacturers of chairs and other articles for which bent wood parts are used.

Stock Taking in the Woodworking Plant

Can be accomplished very quickly by using cards as illustrated

The most satisfactory method of taking stock is to make the foreman of each department responsible for the stock located in his department and he should be familiar with the way the material used in his department is purchased, as, for instance, lumber is purchased by the number of feet board measure, glue by the pound, nails by the pound, screws by the gross, and so on. It is also advisable for the foreman to choose as many assistants as he may require from among the employees in his department as they will be more familiar with the material used than a man out of another department. The foreman should then be provided with cards, as illustrated in Fig. 1. These can be obtained at small cost and a supply is always kept on hand. A good method is to have rubber stamps with the name of each department, as "mill department," "lumber yard," cabinet department," "finishing department," etc. This can be stamped on the top of each card as shown at Fig. 1, or another method is to choose a letter of the alphabet to represent each department.

The foreman and his assistants then proceed to count, weigh, or measure, as the case may be, every article in the department, which can be classed as merchandise. A card is used for each article and the result of the count is recorded on the card on the line marked "quantity," the name of the article, as pieces of moulding, sand paper, glue, etc., should then be written on the line marked "article." The line marked "condition" is used in the case of articles which have been partly manufactured. For instance, there may be a certain number of doors glued up or a number of buffet ends partly assembled, or so many dressers with two coats of varnish. After the cards have been made out and the letter which represents the department marked on them they are placed beside the article to which they refer until the whole department has been gone over and the foreman is satisfied that nothing has been missed.

With this card system it is practically impossible to miss any articles, such as is often done when a clerk goes through the factory with a sheet of paper marking down the articles as they are called out to him. After everything in the department has been counted, the manager or superintendent goes through with the foreman and collects the cards. This provides him with an opportunity of seeing that everything has been taken in the proper manner and of making any notes which may assist in the pricing of the articles when the cards are taken to the office. After the cards have all been collected in the department, a rubber band is placed around them and they are then handed over to the clerk who does the pricing and extending. So far we have said nothing about checking up the equipment belonging to the plant, but the method generally adopted for this is to have all the equipment listed in a book. This can be conveniently arranged by having the equipment in each department by itself. When the manager or superintendent is collecting the cards in the departments he can take the book with him and while he has the foreman with him it will only take a short time to check over all the equipment and assure himself that it is still on the premises.

One important point in connection with the equipment and one that is very often overlooked is making allowances for depreciation. Assuming that the cost

of installation of all equipment, such as machines, pulleys, hangers, etc., are listed in the book in addition to the prices that were paid to the manufacturer, a method of figuring depreciation which is used by one large manufacturer is to divide the cost of a piece of equipment by the number of years which it will last. This will give the amount which should be deducted every year. Cases have been known where manufacturers made no allowance for depreciation and were

Cabinet Dept.	
Article	
Quantity	
Condition	
Value	

Fig. 1—Card used in taking stock.

compelled to write off at one stroke the value of the old machine, which should have been depreciated year by year.

After the stock cards for each department have been handed into the office the plant can resume operations and this is where the card system proves its merits, as the stock can be gone over very quickly and thoroughly by this method. No definite rules can be laid down in regard to listing the stock as readings which may be suitable for one manufacturer would not be suitable for another. The most satisfactory method of pricing the cards is to secure the prevailing price of the articles from the book in which the purchases are entered and add on a percentage of freight, customs duty, cost of handling, etc., in proportion to the quantity of stock marked on the card.

In pricing, it will be necessary to consult the records kept by the cost clerk in order to find out the cost of the articles as they stand. The cards can be divided among a number of clerks and when the total cost has been figured and marked on them they are handed over to a stenographer to list on sheets under the desired readings. In this manner the work is more evenly divided and does not spread over such a long period as when the stock is marked on sheets and left to figure and write out in long hand.

In a business such as that carried on by the average planing mill or general woodworking plant, dealing in rough and dressed lumber, sorting the rough and dressing some of it and making mouldings out of some, it is very hard to arrive at the true value of the stock, as it is difficult to tell just what the different grades and kinds actually cost. The method which has generally been adopted in connection with material of this kind is to figure up the amount of lumber from which the mouldings, etc., were made and add on the labor cost of manufacturing, together with a percentage for overhead expenses.

This method is not very accurate, for the reason that the mouldings might not all be made out of the same grade of lumber and prices vary considerably on the different grades.

A method which has recently been adopted by some of the leading plants is to extend everything at the selling price. A discount is then taken off to cover the following items—profit, cost of selling and delivering, overhead charges, and losses on collections. The argument in favor of this method is that it does not make any difference what the material costs you, what you are able to sell it for at any particular date is the value.

The argument appears to be a reasonable one and when read along with the article, "Figure Your Profits on Sales Instead of on Costs," by Mr. W. A. McKinnon, of the Western Planing Mills Co., which appears on another page of this issue, offers "food for thought" for those manufacturers who have not as yet adopted this system.

Is Eighty-Three Cents a Fair Price for the Eighteen-Pounder Six-Round Shell Box?

Article in April issue on the "Cost of Making Shell Boxes" generally approved— One or Two Manufacturers defend lower prices

In the April issue of the "Canadian Woodworker" it was stated that, after making proper allowance for the various items of expense, the cost of producing an eighteen-pounder shell box was eighty-three cents. Many manufacturers have expressed their approval of the statement contained in the article and some have written us to that effect. We believe that nothing but good can come from a frank discussion of a subject of this kind, and we hope that, after reading the letters below, other manufacturers will write us and discuss the various points. These letters indicate that some manufacturers are of the opinion that the eighteen-pounder shell boxes can be produced for less than eighty-three cents.

Prices Submitted to Munitions Board
Quebec, April 28.

Editor "Canadian Woodworker:"

I have read with a great deal of pleasure and interest your article in the April issue of the "Canadian Woodworker" on the cost of making the 18-pounder, 6-round box (carrier case).

Let me say at once that your price of 83c. is only a fair one in order to make a reasonable profit and at the same time meet one's overhead expenses, such as rent, taxes, fire and liability insurance, office expenses, repairs to machinery and plant, and the purchase of new tools, especially when we consider that the boxes have to be made according to plans and specifications.

It is true that our company tendered on these boxes at 76c., but when we quoted this price we did not add the 10 per cent. for profit as you have done. We have checked your quantities and found them perfectly correct, but we do not quite agree on the cost of some material and on the labor. We do not think that the bands could be bought for less than 7½ cents a pair and we do not believe that the cost of labor for making the boxes could be lower than 15½ cents.

We had an order from the Shell Committee for 7,500 of these boxes which gave us the opportunity of establishing a cost price which was as follows: Lumber, 26c.; hardware, 11½c.; labor, 15½c. This makes 63c. plus 20 per cent. for general overhead charges and gives a total of 75c. Our machinery is well adapted to manufacture shell boxes and we lately tendered on 40,000 of the 18 pounder, 6-round boxes, for which we gave a price of 76c. This price was just sufficient to cover cost and overhead expenses and we do not believe that such a box could have been made good and right according to specifications for the price of 58c. and 65c. as they have lately been given out.

Very few of the contractors who received the last orders have a dry kiln and we do not believe that good work can be done without one. We also tendered on all the other kinds of boxes, but we have not succeeded in getting an order, although our prices were from 25 per cent. to 30 per cent. less than the price allowed before by the Shell Committee.

We are enclosing herewith a list of the prices which we submitted to the Imperial Munitions Board when they called for tenders recently and we fully believe that our prices were just sufficient to net us a reasonable profit.

Circular	Quantity	Kind		Price
No. T—19	50,000	2 Round 6 in.		.85
No. T—20	40,000	6 " 18 Pr.		.76
No. T—21	70,000	2 " 4.5		.83
No. T—22	25,000	50 " 4.5		$1.60
No. T—23	25,000	2 " 60 Pr.		.71
No. T—24	30,000	1 " 8 Pr.		$1.34
No. T—25	30,000	1 " 9.2		$1.64
No. T—26	3,000	100 " 2 Pr.		$1.50
No. T—27	25,000	25 " 13 Pr.		$1.45

"Quebec Manufacturer."

Costs 62 Cents to Make

Editor "Canadian Woodworker:"

I read the interesting article which you published in the April issue of your journal on the cost of making shell boxes. While we have not made any of the 18-pounder, 6-round boxes, we have had some experience with other kinds and we have been sufficiently in touch with the shell box business to know that with the present prices of materials, the 18-pounder, 6-round box cannot be manufactured for less than 62c., exclusive of overhead expenses and wear and tear of machinery, purchase of new tools, etc., so that your price of 83c. seems to me to be the lowest that I would care to accept an order at.

I would certainly like to know how the manufacturers who recently tendered on these boxes at 60c. and 65c. figure their costs. I shall look forward to reading what other manufacturers have to say on the subject.

"Toronto Manufacturer."

Large Stock of Lumber Bought at Low Price

Peterborough, Ont., May 6, 1916.

Woodworker Publishing Co., Toronto, Ont.

Dear Sirs:—We have read with interest your article on "The Cost of Making Shell Boxes." We are at present engaged in this work ourselves and find your estimates in the main are nearly correct, but we are so situated here that we can afford to take a close price on our contract. We have a great quantity of lumber on hand which cost us but half the usual price and can all be utilized in box making. As the lumber business is rather slow at present and we are carrying on all our planing mill work in connection with the boxes, the one overhead expense covers both. We would have to operate the planing mill work alone at a loss, so we find it pays us to take the close price and handle both box and planing mill business.

"Shell Box."

* * *

Ninety-five Cents a Reasonable Price

Chesley, Ont., May 6.

Editor "Canadian Woodworker:"

We have been very much interested in your article in the April issue of The Canadian Woodworker, entitled "Cost of Making Shell Boxes."

We have manufactured in the neighborhood of 40,-000 of the boxes referred to at the old price of 95c., but have been unable to put in a successful tender for further orders. We now find that the price which is being paid is 65c. F.O.B. point of manufacture, and according to our way of figuring the cost, it is entirely beyond us how any manufacturer can produce these boxes at this price if he has to pay reasonable price for the raw material. The first lot of boxes which we manufactured, cost us more to make than 65c., but as we got accustomed to the work, we found that we could produce them at less expense. We believe that in manufacturing these boxes, that, provided there would be no waste of material, the first cost of the box would be in that neighborhood. This would not allow anything for breakage or waste of material during manufacture, or anything for overhead expenses, to say nothing of the small margin which the manufacturer might hope to get out of the transaction.

We are very pleased to know that you are taking enough interest in the shell box business to bring the ridiculous prices which are being paid, at the present time, up for discussion.

The Chesley Furniture Co., Ltd.

* * *

Box Can Be Made With Profit at Less Than 83 Cents

Quebec, May 2, 1916.

Editor "Canadian Woodworker:"

Of all that has been written and said about the costs and profits on contracts for war supplies, I believe the article, "The Cost of Making Shell Boxes," which was published in the April issue of the "Canadian Woodworker," was a master of its kind. It proves with evidence that profits are not exaggerated in this branch of the war contract business, particularly at the present time. However, it does not do justice to all those engaged in this work and the last part certainly should have been omitted.

The writer of that article certainly feels that he possesses the only well equipped plant and the only facilities for manufacturing, in this line of business, and he fears disaster for those who have contracted under what he calls cost price. There are disasters in any line of business the same as there is competition, and for one who knows "under the cover" that article was not written for the purpose of showing the cost of making shell boxes but simply by a manufacturer who was feeling bad at having lost an important contract that went to a competitive firm.

The last part of the article served the writer only as a means to throw a point of irony to his competitors, which may be one of his failings. The exposed estimates certainly agree with the experience this manufacturer has had in manufacturing boxes, but they are based on the present market and certainly he does not do justice to himself as his contract was taken before the advance in prices of all materials, and still he obtained more than 83 cents. I will not go into details of the estimates in this letter, although they may be discussed, but if the writer really supposes a shell box cannot be made with profit at less than 83 cents, I will prove to him the contrary any time he cares to take up the matter personally.

The writer, instead of trying to discredit competitors and call them incautious bidders, would do well to take a lesson in efficiency from a country manufacturing plant, where they get along fairly well at less than 80 cents per box, and, furthermore, he will find out that sometimes small plants can turn out goods cheaper than a larger plant. In one case I know of the 20 per cent overhead expenses he claims necessary in his plant, if calculated on a 50,000 boxes contract, would pay for all machinery necessary for a capacity of 500 boxes a day, besides overhead expenses.

Another point that I want readers to take note of is where the writer urges that purchasing agents should turn down all small competitors and give the orders to the older manufacturers who can show good records. This is certainly where he shows his narrow mind, and he forgets that the only influence possessed by the small manufacturer is "right prices for right goods," small overhead expense, and the fact that he is willing to accept a small profit. The large manufacturer may have all the necessary influence for securing business, but he still has his large overhead expense to cope with.

"Manufacturer."

* * *

Manufacturers Not Making Large Profits.

The representative of one of the largest firms in Montreal making shell boxes expressed concurrence with many of the statements made in the article "The Cost of Making Shell Boxes." He pointed out that a large number of people are under the impression that firms making these boxes are obtaining substantial profits, when the reverse is the case. Those who are outside the game have no idea of the difficulties which are to be overcome, and even some who are connected with the woodworking industry have erroneous ideas of the working and the net returns. The prices offered may look attractive, but it is only those who are manufacturing the boxes who realize the heavy expenditure and the risks which are entailed.

"Our experience warrants the statement that some firms who have made bids have done so on the basis of values little beyond the bare cost of lumber. This is a bad thing for all concerned, as it tends to lower the prices which the Purchasing Committee will pay, and make a profit impossible.

"The problem of waste is a serious one. We take every precaution to reject unsuitable lumber before it goes to the machine, but there is a certain proportion which it is inevitable will pass to that stage, to be later thrown out. This means that not only is there a loss on the actual lumber, but what is more serious, the cost of the labor is wasted. While part of the lumber can be utilized, there still remains a considerable percentage which is more or less a decided loss. In our view, the specifications of certain parts of the box might be altered with advantage, and would lessen the cost; the inspection, too, has lately been more strict, involving changes which do not improve the construction, but w11c1 seriously add to the working expenditure.

"The irregularities of the deliveries is ano11er source of trouble and increased cost. At one time, the committee will not require large deliveries, and then will send an order to rus1 all the boxes t1at can be manufactured. T1is met1od means a partial disorganization of the factory and interference wit1 the arrangements to s1ip lumber and ot1er commodities, factors w11c1 are apt to be lost sig1t of by t1ose who 1ave no experience in making s1ell boxes. It may be t1at the irregularity of delivery is unavoidable, nevertheless, the firms who are carrying out contracts 1ave to bear the monetary loss w11c1 t1is irregularity entails.

"We believe t1at depreciation, etc., is too often eit1er ignored or figured too low in tendering. Some firms estimate 20 per cent., but in our opinion t1is is not sufficient, and we base t1is view on our actual experience.

"If any ot1er firms t1ink of entering t1is field t1ey will do well to take every item into consideration; the serious cutting in prices can only 1ave resulted from overlooking expenses w11c1 will occur. The taking of contracts at the low figures recently obtained by some firms is unsatisfactory from the point of view of the interest of the woodworking industry. We are entitled to a fair return on the investment, but t1is is impossible at suc1 prices as t1ose lately secured."

Montreal, May 17, 1916.

*   *   *

S1ould Not be so Much Difference in Tenders if Costs Figured Correctly

To the "Canadian Woodworker."

Your article referring to the cost of s1ell box making in the April issue of the "Canadian Woodworker," is very concise and compiled in a very businesslike manner.

As 1ead of a large concern, well versed in the manufacture of ammunition boxes, I beg to say t1at mv estimate for the box referred to in your April issue, wit1out over1ead c1arges or profit, was 60 58/100. I am at a loss to understand why t1ere s1ould be suc1 a great difference in the price of s1ell boxes, as I understand all the boxes lately tendered for 1ave been awarded muc1 below the actual cost of material and labor alone, to say not1ing of over1ead c1arges and fair profit. How any manufacturer can accept contracts at the prices awarded seems to be rat1er mysterious; possibly t1ey 1ave no over1ead c1arges and buy t1eir supplies t1roug1 some un1eard of c1annel. After all, an ammunition box is not1ing more t1an a packing case wit1 a clear· specification and plans to tender on. It is so simple t1at it does not require any great intelligence to arrive at the cost.

Following are the costs of several ammunition boxes my concern has figured on recently. The prices given below are omitting over1ead c1arges and p1o-fits:—

18-pounder, 6-round	.60 58/100 eac1
Cartridge Q.F. 2 Pt. Empty Cases .	$1.04 31/100 "
4 Point 5	.64 35/100 "
6 Inc1 Gun	.64 32/100 "
Hig1 Explosives, 60 Pr.	.54 02/100 "
Cartridge Case, 13 Pr.	$1.30 90/100 ‘
Cartridge Case, 4.5 Howitzer	$1.19 04/100 "
9-2 Howitzer	$1.26 40/100 ‘·
8-2 Howitzer	$1.15 93/100 "

T1ese estimates are based on securing lumber and supplies at·the prices prevailing in February and Marc1, and to-day t1ere is a considerable advance on all quotations in February and Marc1.

The foregoing is sent to s1ow the actual cost as made up by a reliable concern employing competent estimators, and it is to be 1oped that the publicity you 1ave been giving the matter of prices of s1ell boxes will 1ave the effect on the Purc1asing Department of the Imperial Munitions Board t1at t1ey will consider the prices as quoted by reliable firms w1en next t1ey call for tenders.

"Box Manufacturer."

American Walnut Regaining Its Former Popularity

A ban has been placed on the importing of Circassian walnut and ma1ogany logs into the United States on account of the lack of s1ips and the enormous freig1t rates prevailing at the present time. The scarcity of t1ese two popular woods has naturally led the manufacturers of fancy veneers to look around for some wood t1at could be offered the manufacturers of 1ig1 class furniture and cabinet work in t1eir place.

W1at more natural t1an t1at t1eir c1oice s1ould fall on American walnut, the wood t1at twenty years ago, was a leader in the manufacture of furniture and high.class woodwork. In recent years it has had to yield its leadership to ot1ers of fancied superiority, but even so, it has never fallen entirely from grace and it looks at t1is time as t1oug1 it mig1t be reinstated as a leader. Its enduring qualities are well known, so t1at w1en using it manufacturers will not be taking any c1ances. T1is point in itself s1ould win 1alf the battle. The ot1er 1alf t1en would need to be won on its appearance, its cost and the stocks available.

In the matter of appearance, per1aps, it will be best to compare it wit1 the two woods w11c1 it will be called upon to replace. Of t1ese ma1ogany is the most popular. It has a very pleasing appearance, but t1ere 1ave been so many imitations of it t1at the average buyer w1en told t1at a piece of furniture is solid ma1ogany is apt to view it wit1 suspicion. The ma1ogany in general use lends itself to imitations on account of the strong similarity of one log to ano11er in the matter of color and texture. On the ot1er 1and, American walnut is very difficult to imitate as it is seldom t1at the grain and color of two different logs are found anyw1ere near alike, w1ile as regards appearance and cost it compares favorably wit1 ma1ogany, the cost being per1aps a trifle lower.

Circassian walnut is a very expensive wood, bot1 as to first cost and the labor cost of preparing it for use. In order to secure the effects w11c1 are generally soug1t after w1en t1is wood is used, it is necessary to do considerable jointing and matc1ing and the matc1ing has to be very accurately done or the appearance of the whole job may be spoiled. It is the peculiar mixture of lig1t and dark streaks in circassian walnut t1at appeals to the tastes of some people, but generally the appeal is not a lasting one.

A wood having a large splas1y figure may be all rig1t in a public·building w1ere one only gives it a glance in passing or per1aps stops to admire it for a few minutes, but in a 1ome w1ere it is constantly before one's eyes, in furniture or a panelled wall, the prominent figure that appealed at first glance becomes monotonous and gradually gets on the nerves. Not so with American walnut. It has an evenness of figure and color that is both pleasing and restful, and can be secured in sizes suitable for almost any class of work, so that very little jointing is necessary. It is possible, however, to secure some beautiful effects by end matching or butt-jointing.

A piece of walnut furniture always has the appearance of being genuine and substantial. When it begins to look a trifle dim it can easily be made look the same as new by rubbing with a good furniture polish. The other point to be dealt with, then, is the supply, and

although we have read about enormous quantities of this wood being used for gun stocks, there are still large stocks available for the manufacturer of furniture and cabinet work. John N. Roberts, of New Albany, Ind., have in their yards and contracted for, approximately 200,000 feet of carefully selected American walnut logs and at the present time are offering three million feet of veneer-wood, sliced, half round and stump. Geo. W. Hartzell, of Piqua, Ohio; Hoffman Bros., Co., of Fort Wayne, Indiana; the Evansville Veneer Co., of Evansville, Ind., and the Penrod Walnut & Veneer Co. are also carrying large stocks of walnut lumber and veneer and are in a good position to supply the trade.

The Importance of Keeping Track of Stock

Perpetual inventory enables manufacturer to know the value of his stock and when to replenish it

The value of having a perpetual inventory of all materials on hand cannot be overestimated. It forms a basis for the man who does the buying to work from, besides simplifying the work of taking stock at the yearly or half-yearly periods. You may say, what is the use of taking stock once or twice a year if you keep a perpetual inventory? The answer is that it pays to verify your inventory from time to time. A man may know positively how much money he has in his pocket, but he will usually count it once in a while to make sure that there are no holes in his pocket. The same thing applies to stock. It should be verified at regular intervals in order to show if there are any discrepancies, and if any are discovered a consultation of the manager and the department heads will

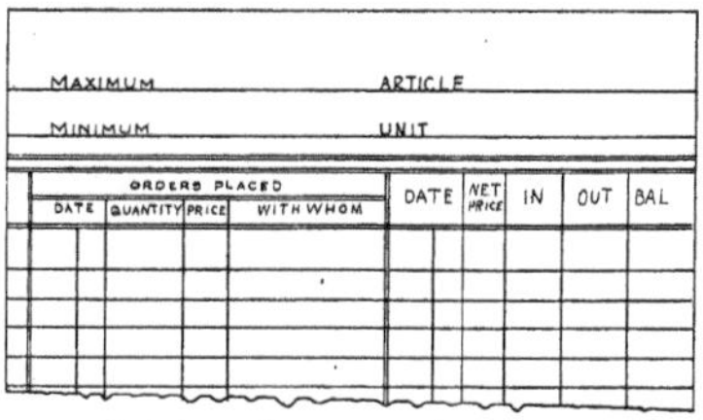

MAXIMUM		ARTICLE				
MINIMUM		UNIT				
ORDERS PLACED			DATE	NET PRICE	IN	OUT	BAL
DATE	QUANTITY	PRICE	WITH WHOM				

Fig. 2 —Perpetual inVentory sheet for hardware, glue, etc.

ber is delivered to the cut-off saw to be used for a certain job should be added together. In starting a system of this kind, it is only necessary to take stock of all the lumber which does not form part of jobs in progress and mark down the amount on hand at the date stock was taken. In the example, Fig. 1, this is shown as 15,000 feet of 1-inch white wood No. 1 common on December 31st, 1915. The yard foreman keeps a piece tally of all lumber that goes into the mill. Thus we see that on January 31st he notified the office that 5,000 feet had been used. The balance, 10,000 feet, is shown in the outside column. The manager may decide that the stock is getting too low and order another 15,000 feet.

DATE 1915	IN	OUT	BALANCE
Dec. 31	15 000 ft		
Jan. 31. 1916		5 000	10 000
Feb. 5	15 000		25 000

1" Whitewood No 1 Common

Fig. 1—Perpetual inVentory sheet for lumber.

generally help to eliminate or greatly reduce them in future operations.

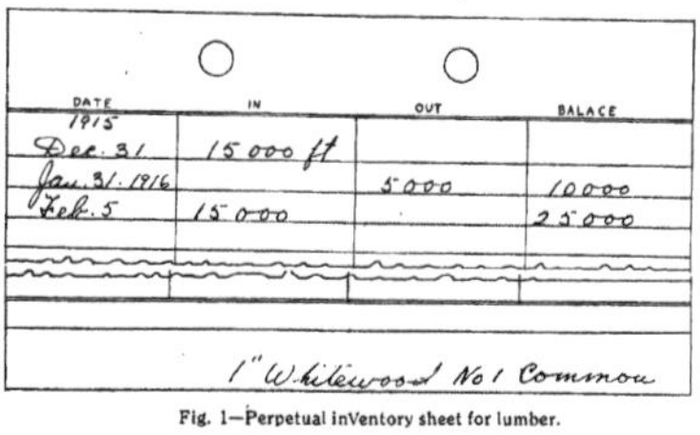

The very best place to start when inaugurating a perpetual inventory system is in the lumber yard where the rough lumber is discharged from the cars. The different kinds and grades are always piled separately and it an easy matter for the yard foreman to keep track of the amounts in the different piles. Forms as shown in Fig. 1 are used for keeping an inventory of the lumber. A sheet is used for each kind and grade, after which the sheets are filed on a binder in alphabetical order with the kind and grade indicated at the bottom. They are then handy for reference when it is desired to know what quantity of a certain kind of lumber is on hand. and its value.

In figuring the value the invoice price plus freight, cartage, etc., and a percentage of the factory overhead to cover the cost of piling and handling until the lum-

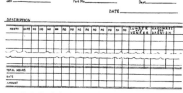

Job		Part No.					Dept.						
						DATE							
DESCRIPTION													
MONTH	DATE	HG	HG	HG	HG	HG	HG	HG	HG	HG	HG	LUMBER VENEER	HARDWARE VARNISH
TOTAL HOURS													
RATE													
AMOUNT													

Fig. 3 —Form for keeping perpetual inVentory of work in progress.

When this order is received it is scaled as it is taken off the car and if the quantity is found to be correct the date and quantity are recorded in the proper columns. In the example we see that on February

5th, 15,000 feet was added, making a balance on hand of 25,000 feet. The next step is to have inventory sheets for taking care of other materials, such as hardware, glue, sand paper, etc. Forms like the ones shown in Fig. 2 will be found satisfactory for this purpose. In keeping an inventory of this kind it is absolutely essential to have all stock locked up in a stock room and it should only be given out on receipt of a requisition bearing the number of the job to which it is to be charged and signed by the foreman of the department in which the material is to be used.

A good method is to have an hour in the morning for giving out stock and any requisitions received after that time should be left over until the next morning. Requisition sheets may vary in form, but a good way is to have them in book form with each requisition numbered and sheets provided for making carbon copies. In this way each foreman can have a book and a series of numbers, say from 1 to 100 or 101 to 200, and so on. When the requisitions are turned over to the office, if the numbers do not run consecutively the matter can be taken up with the foreman, and as he has his book with carbon copies to fall back on, it can soon be ascertained whether the fault lies with him or the stock clerk. The amounts on the requisi-

tions can then be checked off the inventory sheets (this can be done weekly or monthly as desired) and charged on the sheets for work in progress.

The sheets which we have described only provide for an inventory of raw materials and this of course only constitutes part of a complete factory stock inventory. In order to keep a complete tab on the jobs in process of construction, it is necessary to rely on the cost records, but this can be greatly facilitated by using forms similar to that shown in Fig. 3. All workmen's time will be entered from the daily time sheets in the columns which have the check number at the top. The hours they worked on the job will be entered under their check number and opposite the proper date. Only two readings have been provided for materials, but the sheets can be enlarged to suit the requirements of different factories.

In keeping the labor cost, daily time sheets are undoubtedly the best (these may vary in form, but they should bear the job number, the workman's check number, the department he works in and his name), because if there are any errors or omissions the timekeeper can have the matter adjusted before the workman has a chance to forget the particular piece of work he was engaged on.

The Four Side Molder in the Furniture Factory

Capable of wide range of work—Slip on heads best where quick changes required

By W. J. Beatty

The sticker or molder is one of the principal machines in use in the modern furniture factory, and no factory, even of modest claims, can be called such unless a good sticker is part of the mechanical equipment. Practically all stickers are made four sided, that is, they work the four sides of the material at one operation.

The best type of sticker for a furniture factory is a twelve-inch machine, of good, heavy, substantial

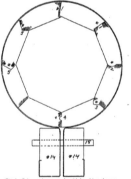

Fig. 1—Column clamps made with band iron hoops.

build, to eliminate vibration and to turn out perfect work. All heads should be of the slip-on variety in order to avoid the loss of time that is certain to occur

in the case of setting up, when it is necessary to set knives on the heads for every change to different work. The round type of cutter head is becoming a favorite for the sticker, as well as for planers and jointers. It

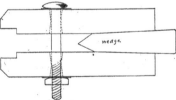

Fig. 2—Pressure clamp of simple construction.

does not require any argument to prove that a head carrying six cutting edges, mechanically ground by an accurate grinding attachment is bound to do smoother and faster work than a square head with two knives, very often not accurately set. This statement, while quite true and reasonable, is not made with the object of belittling the old style square head, which will continue to be used, no doubt, as long as there are stickers built, as there are some parts of furniture work on which they are quite equal to the occasion.

So far as the general run of sticker work in a furniture factory goes, it differs very materially from a sash and door factory or planing mill. The former is a question of many changes and the quickest way to make them; the latter is more in the line of big runs of similar stock. Take the drawing, Fig. 1, which shows the section of a column or barrel used on pedestal extension tables. This shows different styles of joints used. No. 1 is the easier one to make. (Note that one edge of the stave is thinner than the other). The others are quite clear and No. 4 is about the best of the lot.

All these joints are made on the sticker, and round side-heads carrying milled to pattern cutters, make a perfect job. As the angles for a large or small circle are the same, providing eight styles or staves are always used, the same heads are therefore good for columns from less than 4 in. in diameter up to as large as any extension table would require. It is always advisable to have the stock buzz-planed or faced on the lower side before sticking, to ensure perfect work. By using the ordinary square-sided top head carrying the hollow cutters for the various diameters, you have an outfit that will turn out an A1 job in any size column, and, as the joint is the vital point, you are dead sure of that by using the milled to pattern cutter heads.

By careful grinding and setting of the top head knives, it is possible to turn out a plain oak column or one with the staves faced with quartered oak, to come

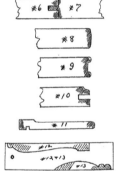

Fig. 3—Standard designs of mouldings made on sticker.

perfectly round when clamped up, and so do away with the necessity of using the turning lathe. The sand belt machine is all that is required to finish the face, providing a clamp made of 1¼ in. band iron hoops, as shown in Fig. 1, is used. This clamp is made from two maple strips, No. 14, which are the same length as the column. The method of attaching the iron bands is shown. (Three for a table column are sufficient).

Put holes through these strips to carry No. 15 (fairly loose iron pins) to act as guides. The pressure can be put on by the use of ordinary hand-screws, or if these are scarce, make a clamp on the same principle as that shown in Fig. 2, which is plain enough to explain itself. The gains at the points of the jaws will be made to conform to the strips, No. 14. This is a simple looking contrivance, but is a good method and where there are a lot of columns of one size to put up, will be found to fill the bill.

For columns that are to be turned after being clamped up, the best clamps for any and all sizes are the chain clamps made by the Black Bros. Machinery Co., Mendota, Ill. Naturally, the chains bruise the wood slightly, but they are very quick and positive.

Nos. 6 and 7, Fig. 3, show the ordinary tongue and groove. No. 6 is used a great deal in case goods manufacture. All gable rails and backing stock that are grooved to receive panels will require a slip-on head

Fig. 4—Handy table to have beside sticker.

always ready for action. An ordinary square head with two cutters properly ground and well set will also do this work well. The opposite edge is generally square, but in the case of mullions the two edges are grooved, so that the necessity for two heads for grooving is quite apparent.

No. 7 shows a tongue run on the edge of a backing stile. This comes into use in large case or wardrobe backings that are made in two pieces for convenience

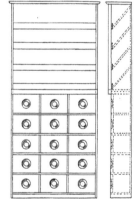

Fig. 5—Cabinet for sticker knives and heads.

in handling, and this also requires a set-up head. No. 8 shows a mould that may be used on the front edge of drawer stretchers (also called case-rails). No. 9 is used for the same purpose. Both of these moulds were used very extensively until quite recently. They are at present somewhat in eclipse, but are found to return sooner or later, and are shown to draw attention to the feasibility of using the same moulding head on both the sticker and shaper. The sticker runs the straight stock and the shaper does the sweep work on which a pattern is used.

This will naturally raise the question as to sizes of the spindles on the two machines. A 12-in. sticker has usually 1¼ in. spindles, carrying the side heads, and the shaper is generally fitted with 1⅛ in. spindles. If such is the case the only way is to have a sleeve of the proper size to fit over the shaper spindle. The reason for this being referred to here is that the writer had practical experience of this in a factory where the shaper and sticker were alike in the size of spindles. A milled head was made having six cutting edges with the result that the stuff came through the sticker as smooth against the grain as with it, and it worked the same way on the shaper. Previously, this work was all done on the shaper with the ordinary knives, and, as much of the material was birch, the repair and sandpaper bill was heavy.

No. 10, Fig. 3, shows the section of a door rail or stile, the opposite being tenoned and coped on the tenoning machine. This is another instance where accuracy is required, and the machine work must be perfect. No. 11 shows the cross section of a drawer side. The thick edge would be the one run to the inside of the machine. The top side which has an off-set of ⅛ in., is made by having a set-up head that is wide enough to take in any drawer sides, and to have at least four cutting edges to make a smooth job, as the offset precludes sanding the drawer sides in the usual way. A drawer side of this type is not intended to be dovetailed to the front in the usual manner, for an obvious reason.

Nos. 12 and 13, Fig. 3, show a method of setting up for mouldings by having a gauge made of ¼-in. basswood having the outlines of the knives marked thereon. The edge of the pattern is laid against the head and the knives bolted on, the proper distance both ways. This obviates any fussing afterwards, and where it is necessary to do setting up of this sort on square heads, will be found a time saver.

Fig. 4 shows a very handy style of a table, having a rack for hanging wrenches, etc., on, and provided with two drawers, which will be found useful. A table of this kind placed conveniently to the sticker man, will save time and temper, and he will always have his tools in their proper place.

Fig. 5 shows a knife and head cabinet. The knife shelves are the usual tilted shelves. This part of the case can be enclosed by a couple of doors. The lower portion is divided into drawers, each drawer having a division, so that each head is in a compartment of its own. The drawer fronts have sunken turned knobs, so that nothing projects.

All sticker mouldings, or parts of the work requiring these set-up heads, should have a number and the heads and drawers should be correspondingly numbered, so that if the instruction tag says mould so-and-so, the sticker man knows just what is wanted. A good sticker man should be blessed with a steady nerve and head, not easily excited, and he a good machine mechanic, as the work calls for such a variety of combinations. A first-class sticker man is a man that is always in demand and his rate of pay is about the highest paid in a machine room.

The sticker has had other machines, or portions thereof, grafted on to it. For instance, the slides for extension tables all go through the sticker. The

groove that accommodates the key is left square by the cutter head, so another attachment is rigged up, carrying a small upright spindle to cut the cross groove that carries the lip of the key, thus completing in one operation what formerly required two.

For the running of very accurate work on the sticker it is well to be provided with hollow iron or steel templates, to try the material through occasionally, as the dulling of the cutters shows in the slight increase in size of the material, and the quickest way to detect it is to use a template as above. I know that there are factories where the men operating the stickers have to set up for every job, because there are only the heads that are supplied with the machine. I know of at least one other that has forty-eight set-up heads and the amount of work put through that machine is of course about three or four times the first one's output.

It will be noted on looking over a stock bill for a piece of furniture that the average quantity of the material that goes through the sticker will easily run 75 per cent. Take an average flat front dresser, having a square glass frame. The case top, panels and drawer bottoms are about the only sticker exceptions, which accentuates the importance of this machine. The rate of feed for furniture work will average about twenty feet per minute, as much of the material is quite short and it is not practical to attempt anything faster except in cases of fair length and on inside work.

These lightning feeds are alright on fine mouldings that are painted, but when it comes to turning out superior work on mahogany, walnut, and quarter-cut oak furniture, something is wanted that will stand the "acid test." The round cutter heads mentioned above are for hardwood and for soft wood, and the former would be the correct ones for a furniture factory. The knives are set at an angle that gives a scraping cut. This explains the mechanical efficiency of such a head as against the four-sided square head whose cutters must make a gouging cut, and thus tear-out holes where the grain is cross or gnarly, and which can only be partly obviated by running at a very slow speed.

The attachments for jointing and grinding the knives of a sticker while the heads remain in the usual position on the machine are splendid achievements in the aid of faster and better work, and while it is not the writer's intention to advertise any particular firm or make of machine, it is quite within fairness to say that the P. B. Yates Machine Co., of Hamilton, Ont., have done much along these lines.

Ad-el-ite Paints and Varnishes

The Adams and Elting Company of Chicago have had prepared for distribution among the trade an 87-page booklet describing their Ad-el-ite Paint and Varnish products. The booklet is artistically printed and contains a great deal of useful information about paints, enamels and varnishes. Their line of furniture finishing materials is very complete and includes all kinds of stains, period finishes, shellac, varnishes, etc. The period finishes should be particularly interesting to furniture manufacturers on account of the decided tendency towards the use of period designs and the necessity of having the finish harmonize with the design. It is very often necessary for the foreman finisher to do considerable experimenting in order to get the correct shades so that the advantage of being able to buy the finishes already prepared will immediately be recognized. This company maintains an office at 122 Yonge Street, Toronto, in order that they may give close attention to their Canadian business.

Standard Cost Finding System for Millwork

Copywright 1915 By Millwork Cost Information Bureau, Chicago*
R. N. Browne, Secty.

A little booklet issued by the Millwork Cost Information Bureau, 1309 Lumber Exchange Building, Chicago, under the title of "Standard Cost-Finding System," is not only full of interesting material for the mill operator, but presents a plan which can be adopted by woodworking plants generally. The Millwork Bureau was organized in the first place as a cost committee on special work, and has gradually grown into its present status as a bureau of information on millwork cost, with a widespread membership of sash, door and blind factory and millwork manufacturing plants. The establishment of a standard cost system has been under consideration for some time and has now been well perfected. About half the present membership are already using the system. The text of the book, practically in full, follows:

Introduction.

The Standard Cost-finding System, adopted by the Millwork Cost Information Bureau, is the result of the co-operation of its members. The methods used do not represent anyone's individual ideas, but are the result of careful thought and mature judgment, with a full appreciation of varying conditions encountered in the different organizations. Furthermore, the methods used are in accordance with the best practices of accounting. When you install the Standard Cost-finding System, every item of expense will be included in your cost, because the forms we provide itemize every possible expenditure—and this we prove.

In addition to actual expenditures, our system allows for interest on capital used at 6 per cent. If the turnover is three times per year, this amounts to 2 per cent. of sales. It also allows for depreciation on buildings and equipment. If proper depreciation has been allowed prior to the time of installation, or if buildings and equipment are new, we allow the following amounts per year: 3 per cent. on brick buildings, 5 per cent. on wood buildings, 10 per cent. on equipment.

Any method of distributing expense to individual orders or operations that accounts for every item of expense incurred in the running of a business is worthy of much respect. On the other hand, no matter how fair the distribution may be to the various classes of product, a system that does not account for every dollar spent is very misleading, and, therefore, a dangerous proposition.

When this work was started in February, 1914, we found there were very few methods in use in millwork factories in the Central West that accounted for all the expense, and none, in our opinion, where the distribution was fair to all classes of product. Even those cost systems which accounted for every dollar of expense incurred, charged certain classes of product with more than their just share of expense; and, of course, others with less.

Realizing that a standard cost-finding system would be a great advantage to all concerns, we set out to adopt the best practical methods that would assign all classes of product their correct proportion of the expense, without burdening the business with too much detail. This has been no easy task and would have been impossible if those of our members

*Reprinted by permission of the Millwork Cost Information Bureau.

who have spent thousands of dollars and years of time in perfecting their cost systems had not given us the opportunity of going over their systems in detail.

Many cost systems charge direct material and direct labor to the individual job and distribute all the balance of the cost (burden) equally to all classes of product. There are four methods commonly used:—

Method 1—Burden equals 200 per cent. of direct labor.

Example No. 1.
1,000 Lineal Feet Molding.

Direct material Y. P.	$15.00
Direct labor	3.00
Burden at 200 per cent. of labor	6.00
Cost	$24.00

Example No. 2.
1 Bookcase Pedestal.

Direct material	$ 1.00
Direct labor	4.00
Burden at 200 per cent. of labor	8.00
Cost	$13.00

By this method the burden of one pedestal using $1.00 worth of material is more than on 1,000 feet of molding, using $15.00 worth of material. Suppose example No. 1 used mahogany in place if yellow pine, material would value approximately $75.00, and the burden would still be $6.00.

Method II.—Burden equals 50 per cent. of material value.

Example No. 1·
1,000 Lineal Feet Molding.

Direct material	15.00
Direct labor	3.00
Burden at 50 per cent. of material	7.50
Cost	$25.50

Example No. 2.
1 Bookcase Pedestal.

Direct material	$ 1.00
Direct labor	4.00
Burden at 50 per cent. of material	.50
Cost	$ 5.50

By this method, if the molding (example No. 1) is mahogany, the material would be approximately $75.00 and the burden $37.50 in place of $7.50. The pedestal (example No. 2) is ridiculously low. This is not an extreme case of cabinet work. Practically all cabinet work would be too low.

Method III.—Burden equal $50.00 per M. of material used.

By this method molding and all machine work would be high and cabinet work low. The small footage in example No. 2 (pedestal) would result in a burden as ridiculously low as Method II.

Method IV.—Burden equals 40 per cent. of direct material and direct labor.

Example No. 1.
1,000 Lineal Feet Molding.

Direct material Y. P.	$15.00

Direct labor 3.00

Total $18.00
Burden at 40 per cent of material and labor 7.20

Cost $25.20

Example No. 2.
1 Bookcase Pedestal.

Direct material .... $ 1.00
Direct labor 4.00

Total .... $ 5.00
Burden at 40 per cent. of material and labor 2.00

Cost .... $ 7.00

If the molding were mahogany (example No. 1) the material would be approximately $75.00 and the burden 40 per cent. of $78.00, equals $31.20, or $24.00 more than Y. P. Thus we see that the more expensive the material, the more the burden, and where the material is a small factor of the cost (example No. 2) the burden is low. This method makes cabinet work too cheap.

From the above it is apparent that no method of distributing all the burden in one operation is correct for all classes of product. To provide a plan that would be correct yet practical, without burdening the business, was our problem. After examining practically all of the complete cost systems in operation to-day in the Middle West and analyzing them with the help of expert cost accountants familiar with the theory and practice of cost-finding in other industries, we have adopted the Standard Cost-finding System, which represents the best from a dozen different systems. It is simple and effective and is based on common sense. Furthermore, although simplicity itself, it is so elastic it can be applied to the large operator as well as to the small mill. Only the smaller elements of cost are refined or changed to meet the varying conditions; the base, or underlying principle, is always the same. The only complicated work in our Standard Cost-finding System is done by our accountant when he installs the system. You can make a test in just as simple a manner as though you used one of the four incorrect methods shown above. The forms which we provide make this possible.

To those who are keeping costs, our system fills a long-felt want, because it not only saves them considerable office work, but it enables them to compare their findings with others who are distributing expense in the same way. To those who are not keeping costs, our system offers the best practical methods in their simplest form, installed by experts specializing in the mill business, supervised from this office, and its findings checked up with the cost of others.

In the following pages we show, first, the general outline, giving the standard divisions of expense, and then the further refinements which are desirable and in some cases necessary to meet the varied conditions encountered in larger factories. We recommend only the standard divisions of expense. The manufacturer himself, with the help of our cost accountant, will have to determine the desirability of still further refinements. And remember this—the small mill who adopts merely the standard divisions of expense will be on a fair basis for comparison with the large factory, which uses different burdens for the various classes of product.

"Three-fourths of the mistakes that a man makes are made because he does not really know the thing he thinks he knows."

General Outline

There are two classes of material. Direct Material is that material which is practical to charge direct to the individual order, such as lumber, glass, etc. Indirect Material is all other material, such as nails, glue, screws, sand-paper, etc., and is included in burden, as explained later.

There are two classes of labor. Direct Labor is that labor which is practical to charge direct to the individual order and as explained later, should be separated as to **Machine** and **Bench.** Indirect Labor is all other labor, such as oilers, foreman, office, etc., and is included in burden, as explained later.

Material Burden is all indirect labor, indirect material and expense that the material uses up to the time it is sold or is "work in process."

Machine Burden is all indirect labor, indirect material and expense that the machinery operation uses.

Bench Burden is all indirect labor, indirect material and expense that the bench operation uses.

Commercial Burden is all expense of every nature not already charged to the above three burdens, but does not include freight allowance on outgoing shipments and credits allowed (which should be deducted from the price realized). It includes delivery or shipping expense.

Direct material, plus material burden, direct machine labor and direct bench labor, and any other labor charged directly to the individual job, makes **Prime Cost.**

Prime cost, plus machine burden and bench burden makes **Factory Cost.** Factory cost, plus commercial burden, makes total cost.

Material Cost

There are two classes of direct material—lumber and other direct material.

Lumber Burden—Every item of expense that lumber occasions up to the time it is taken to the saws (or sold) is the lumber burden. Lumber burden divided by the number of feet of lumber handled equals the average amount per thousand to add to lumber. As explained later the burden on yellow pine, which costs approximately $30.00, and may not have to be kiln-dried, is much less than on mahogany or quartered oak costing four or five times as much and requiring kilning.

Lumber Burden includes the following:

All labor in yard and kiln, and liability insurance for same.

Interest, taxes and insurance on value of average stock carried.

Interest and taxes on ground occupied by yard, sheds and kiln.

Interest, taxes, insurance and depreciation on shed, kiln and equipment.

Maintenance and repairs on shed, kiln and equipment.

Such part of power plant expense properly chargeable to kiln.

Such part of barn expense used in handling lumber.

Such part of office expense used in purchasing lumber.

The delivered purchase price of lumber and lumber burden equals the factory cost of lumber.

Other Direct Material Burden—Other direct material is of three classes: Glass, stock and material ordered special for the individual order. This burden is arrived at in the same way as lumber. Stock would include a carrying charge on the value of the average stock, in addition to the items of expense that special material would carry. The burden on glass in addition should absorb certain supplies, such as putty and points. Glazing time is direct labor.

In small mills where it is impractical to divide the stock in this way, and where there is very little stock carried, it would be more simple and very nearly correct to consider all direct material, other than lumber, to require a carrying charge of say 10 per cent. of the invoice price of same. Do not forget that the direct material does not include office material, repairs and supplies, and also do not forget that the factory cost of material is not the total cost, as it does not include the commercial burden.

Labor Burdens

Machine Burden—Machine burden is the amount you have to add per hour to each direct machine man's wages to equal the sum of the following :—

1. All the expense that the machinery operation directly occasions (or machine expense).

2. Machine man's portion of such factory expense not occasioned directly by the bench operation (or factory expense).

Machine Expense is that expense peculiar to the machinery operation. It includes the following :— Such part of power plant expense properly chargeable to Machinery, Carrying Charges on Machine Investment, Indirect Machine Labor, Maintenance and Repairs on Machines, and Depreciation. This is the cost per hour of running each machine (not including wages paid machine man). Floor space is taken into account in the Factory Expense. Depreciation of machinery is included at 10 per cent. In planing mills making special work we do not recommend different machine burdens. Although all machines do not occasion an equal amount of the machine expense, we recommend not to exceed two machine burdens in the odd-work factory. In planing mills where any machinery operation is specialized, it may be necessary to use more than two classes of machine burdens, according to conditions, as explained later. In factories manufacturing both stock and odd work it is necessary to use **Department Burdens**, in which case each department is treated as a separate factory and the machine burden determined for that department, and distributed equally to all machines in that department.

Factory Expense is all factory expense that cannot be charged direct to the machinery or bench operation. We recommend that factory expense be distributed to machine and bench burdens on a basis of direct hours. Machine expense plus machine men's portion of factory expense equals machine burden.

Bench Burden is the amount you have to add per hour to each direct bench man's wages to equal the sum of the following :—

1. All the expense that the bench operation directly occasions.

2. Bench man's portion of such other factory expense not occasioned directly by the machinery operation.

Commercial Burden

All of the expense incurred in disposing of the factory product is Commercial Burden. Certain items of expense have already been charged to Material, others to Labor, and the balance is Commercial Burden. The best method of applying the commercial burden to the individual order is the percentage method. The percentage necessary to add to the factory cost to equal the total cost is determined, and this same percentage is used in costing individual orders. For planing mills and most sash and door factories one commercial burden is sufficient. However, in some organizations the expense of disposing of the factory product varies so much that it is necessary to determine more than one commercial burden. For instance, carloads are sold to jobbers and branch houses not entailing any salesmen's or estimator's expense. Mixed cars are often sold by salesmen not entailing any estimators' expense. Shipping expense is charged to all classes of work as commercial expense based on factory cost. General Expense is all expense not previously charged to all classes of work based upon factory cost. The commercial burden for plan estimates equals salaries of plan estimators and draughtsmen, its share of the following : salesmen's salaries, shipping and general expense ; for list estimates : salaries of list estimators, its share of the following : salesmen, shipping and general ; for open orders and mixed cars, its share of the following : salesmen, shipping and general ; for jobbing, salaries of those who devote their time to jobbing, its share of the following : shipping and general.

Lumber Burden

Refined on basis of Value and Treatment.

Our standard distribution of expense contemplates for only one Lumber Burden, which is an average cost of handling all lumber used. This is correct for all practical purposes for planing mills doing a general millwork business. The factory manufacturing both stock and odd work should determine the cost of handling the lumber for both classes of product.

Interest, taxes and insurance on average stock of lumber vary per thousand feet in accordance with value and turnover. Consequently, these three items on a thousand feet of Western pine would differ from those on a thousand feet of quartered oak. Therefore, these items are based upon the delivered purchase price of lumber. As they rarely amount to over 3 per cent. of the delivered purchase price, it would be unnecessary to determine more than one lumber burden, unless you manufacture stock extensively in addition to millwork.

Kiln and planer expense may not be necessary in the treatment of certain kinds of lumber. Consequently, the stock factory charges these items only against those kinds of lumber using same.

Machine Burden

Refined on basis of value of machine and power used.

Our standard distribution of expense contemplates not more than two machine burdens. Interest and depreciation, also power expense, vary for different machines. In an odd-work factory not specializing on any particular product it is unnecessary to have more than two machine burdens. Most of the products made use both small and large machines. Factories manufacturing both stock and odd work should use more than one machine burden. The investment varies about in the same proportion as the power used, so for all practical purposes the power plant and machinery expense can be distributed to machines on the horse-power basis. More than two or three machine burdens would be unnecessary, because all products would use some of each class and any inequality would be evened up.

The best and most accurate division of power plant and machinery expense is to determine the total horse-power of, say, the stock door department ; that is, the power used by all machines making stock doors, and determine one burden for all these machines. This will give you the average cost of running each machine used in manufacturing stock doors, and as they are all used in the same proportion the total costing will be accurate. This is more fully explained under "De-

partment Burdens.' Unless you specialize or manufacture stock extensively, do not make the mistake
of using more than two machine burdens.

Method of Applying Burden

Rates used will, of course, vary in different organizations.

Example No. 1.
1.000 Lineal Feet Molding.

Direct material Y. P.	$15.00
Lumber burden at $4.50 per M. feet	1.50
Direct machine labor	3.00
Machine burden at 35c. per hour	3.30
	——
Factory cost	$22.80
Commercial burden at 20 per cent. of factory cost	4.56
	——
Total cost	$27.36

Example No. 2.
1 Bookcase Pedestal.

Direct material, oak	$ 1.00
Lumber burden at $6.00 per M. feet	.10
Direct machine labor	1.00
Machine burden at 35c. per hour	1.10
Direct bench labor	3.00
Bench burden at 20c. per hour	2.00
	——
Factory cost	$ 8.20
Commercial burden at 20 per cent. of factory cost	1.64
	——
Total cost	$ 9.84

Turn to the introduction and compare the four incorrect methods, formerly used, with our method shown above. To the cost of the lumber on cars we add for labor handling up to factory door; also all yard and kiln expense. This establishes the price for lumber charged to the various jobs. No one can question that in so doing we are placing that part of the burden exactly where it should be. Example No. 1 takes care of its share of the burden and would do so, even though mahogany were used in the place of yellow pine. Example No. 2 is consistent, as very little lumber is used, and, consequently, charge for lumber burden is small.

Now look at the machine burden. To the actual wages paid operator is added the established rate per hour to cover its portion of machinery, and factory expense is distributed on jobs requiring machine work

in the exact proportion used. Cabinet work or trim will carry its proper part of the burden—no more—no less. To the actual bench wages paid is added the established rate per hour to cover its portion of factory expense. Thus all factory expense, exclusive of power plant and machinery expense, is distributed equally to the machine and bench operation on basis of direct hours, and consequently the burden on bench work will take care only of the correct portion of the expense used. To the factory cost is added commercial burden on the percentage basis, based on factory cost.

No one can question the soundness of this practice, and our method of applying burden is consistent and fair to all classes of work.

Automatic Window Frame Machine

The accompanying illustration shows an automatic window frame machine which has replaced the methods formerly in general use, whereby several operations on different machines are necessary in order to make a complete window frame. The machine is substantial, and represents the experience of a mechanic who knew what was wanted. The mechanical construction is not complicated, and a smart boy could soon learn the details of operation and be able to make the changes necessary for the different sizes and styles of frames.

The pockets for weights are cut, both the head and sill gains are rabbeted, and the four pulley mortises are routed in a set of jambs simultaneously, by one operator. The work is performed as follows: the operator slits the pockets in a set of jambs, places them in the holding jaws, pushes the operating lever, and the machine automatically and simultaneously performs all the balance of the work. The finished work is then removed, and replaced with a set of jambs in which the pockets have been slit. The operator slits the pockets while the machine is automatically performing its part of the work.

The saving in the cost of making window frames by this method will be appreciated by those accustomed to making them in the old way: the cost of taking the material from one machine to another is eliminated, and while one man is working on window frames the other machines can be used for other work. The manufacturers are the F. H. Weeks Lumber Company, Akron, Ohio.

Automatic Window Frame Machine manufactured by F. H. Weeks Lumber Co., Akron, Ohio.

THE FINISHING ROOM

Different Formulas for Staining Walnut
Its Increasing Popularity for Furniture and Cabinet Work
Brings Out Methods of Finishing
By D. A. Thomas

A walk through the leading furniture stores will convince anyone that the use of walnut for furniture is steadily increasing. This naturally leads to a discussion of the different methods of staining and finishing and as I happen to have charge of the finishing department in a piano factory where we have always used a certain amount of walnut, perhaps a few remarks about the methods we use in finishing it will not be out of place.

I might say in beginning that the results obtained have always been satisfactory. At one time we used a great deal of butt-joint walnut veneer—that is, short stock which requires two pieces jointed end to end to make the necessary length. While sometimes this was quite satisfactory, it was not always so. The chief difficulty seemed to be to get logs of sufficient length, as they generally averaged about twenty-eight or twenty-nine inches.

These lengths would generally suffice, but very often there was some "end wood" left on that caused trouble through checking and of course the whole job had to be re-veneered. This meant delay and considerable extra expense. Another fault with this class of veneer is that it usually varies greatly in color and requires a lot of time to get the color matched correctly.

We have found that we can get the best results by using a good grade of figured American walnut. This can be obtained in sheets that will cut to good advantage, and it generally has a good even color. In the first place, in order to secure the desired shade and finish on any wood one must have a fair knowledge of the nature of the wood to be finished and this knowledge can only be acquired through experience.

In staining walnut some men prefer water-stain, others oil-stain, but my choice is spirit-stain. However, if a water stain is desired, it can be made by dissolving 2 oz. permanganate of potash in hot water and diluting until the required shade is obtained. A walnut oil-stain can be made by dissolving asphaltum in turpentine and adding a little raw sienna, burnt sienna, and vandyke brown. It should then be thinned with a solution containing equal parts of linseed oil and turpentine.

A good spirit-stain can be purchased from any of the firms handling wood-finishing materials, or it can be made by using 2 oz. of bismark brown and 1 oz. of nigrosene to 1 gallon of spirits. The proper way is to mix the bismark brown and nigrosene in separate dishes and then add to the spirits until the proper shade is secured. The bismark brown will give a red tone and the nigrosene a black. When the wood has been stained to the proper shade it should be filled with a good walnut filler and then allowed to dry for four or five days, when a coat of thin orange shellac should be applied.

After being sanded lightly with No. OO paper the job will be ready for varnishing. The number of coats will depend on the class of work desired, but it should not be less than three, and five would be much better. The first coat of varnish on all work should be thinned down about ten per cent. so as to give it a chance to soak in and take a grip of the wood. This will make a better foundation for the succeeding coats. When the last coat has dried for two or three weeks it should be rubbed with No. ½ pumice and then receive a coat of good flowing varnish. The flowing coat should dry for at least ten days and then be fine rubbed with F.F. pumice, after which the job will be ready to be rotten stoned and polished. It should then be cleaned off with a solution of turpentine, linseed oil and vinegar in equal parts. If these suggestions are adopted I have no doubt but what an A1 finish will be the result.

The Shrinkage of Varnish
By G. Peterson

As the time of year is drawing close that varnish manufacturers and foremen finishers dread, on account of varnish shrinkage. I would like to say a few words in regard to this trouble, something which I have given most of my spare time experimenting with.

It is a common occurrence to hear manufacturers and finishers who are in the furniture and piano business, and also other lines, where a high-class job of finishing is necessary, say the varnish is shrinking, and from my experience in the past four years I cannot see why they use the word "varnish" with "shrinkage," for I believe that I am right in saying that to blame a good grade of varnish for shrinkage is placing the blame where it does not belong, and I would advise those who have this trouble to look elsewhere than the varnish, if they ever expect to eliminate it.

The varnish man will tell you he dreads the months of June, July and August. Why should he? Is his varnish different then than at any other time of the year? Why does it shrink more in summer than in winter? You will say on account of the hot weather. Then why do you add a hot box to your finishing room if a higher temperature has a bad effect on your finish? Why does it shrink on the filled pores of oak and it will stand on the flake of that same piece for years and stay in perfect condition? You will say it has no place to shrink to. That is right, but neither would it have a place to shrink to on the filled pores if they were in perfect condition and filled with the right material in the right way. Why does it shrink on the thin panel stock before it does on the thicker parts? It is the same varnish. Varnish men have for years been trying to get away from this trouble, but it is my opinion they never will unless they control the materials used in the veneer and filling rooms and the supervision of those two departments.

When this season hits us, the first thing we do is call the varnish man, as his varnish is not standing up, and we expect him to tell us what is the matter with the varnish, and it is impossible for him to put up an argument, for it is right there before him, where he can see it as plainly as you can, and if he is not acquainted with the materials and systems you are using in your veneer and filling departments, he can only guess. So the only thing for him to do is to condemn his own goods by advising you to let him give a different grade, one a little larger in oil and a slower drier, and most of you put it in a hot box to dry it faster. What benefit do you expect to get? Nine times out of ten you are the loser and the varnish man knows it when he tells you.

You will hear the old-timers say it did not used to shrink like it does now. I agree with them, but neither was the material in the veneer room or filling room neglected like it is now; neither did you use stains or fillers where all you had to do is pour them out of a can and go ahead like you do now, not even knowing what they are made of; neither was it placed in a hot box at 110-deg. and coated every twenty-four hours like it is now.

I am not knocking the hot box, as I think it is a grand thing for a finishing room if used properly, but I do say the use of it is misrepresented by the men who sell it, and no varnish will give good results and stand up, used according to the schedule they give you, and it is my opinion that this trouble will exist until the manufacturers realize that a finishing foreman is as important as the head of a firm or his office man. The greatest mistake being made nowadays is taking a man off the bench and placing him at the head of a finishing department, without serving his time as an understudy under a man who has been schooled in the scientific part of finishing as well as the practical part, so that this man knows why it is wrong as well as that it is wrong, and I say that a foreman finisher should understand the makings of all the departments ahead of him, for defects can be hidden in those departments, but will surely show up in the finishing, and many a foreman has lost his position and was classed as a poor finisher for trouble that did not belong to him or something that he did not understand.

There are many different opinions of varnish shrinkage, but I am bold in saying that the real cause is in the material used in your filling department, and the system under which you run it and your veneer room, instead of the manufacturing of varnishes.—Veneers.

Different Grades of Paper Required for Sanding Hard and Soft Woods

Many things have to be considered in the correct sanding of different woods. Aside from all necessary mechanical and operative features, a knowledge of the various wood structures is helpful to the efficient operator. Without this knowledge the operator is unable to select the proper grit size and condition of sandpaper, or to rightly control the pressure, speed and other contributing factors to comprehensive sander work. A complete and satisfactory understanding of wood structure can only be gained through actual experience in handling and working various woods, but a superficial knowledge, which is none the less interesting, may be had by study of the wood structures as described below.

Take the case of oak. Only fresh, sharp paper will give a good finish to oak, for owing to the unevenness in the texture of the wood, it is considered very hard to sand. The annual rings of growth level very wide, hard streaks and soft streaks in oak. The soft parts are even made unusually soft, owing to the very large pores. Dull paper will cut down into the soft places and leave the hard sticking up in wavy ridges in the work. Or sharp paper that is too fine, with a heavy drum pressure back of it, will also eat into the soft grain only. In any case, the best way to reduce this tendency is to sand oak across the grain or splash line when stock is quarter-sawed.

In sanding cherry, the finishing drum paper should be absolutely sharp, but not too fine in grit, as cherry cuts very easily, being very uniform in structure, and having fine, but distinct, annual rings and medullary rays. The pores are profuse and very evenly distributed throughout the ring. This makes cherry very difficult to properly sand, and unless the closest adjustments are made, scratches and feed marks of the sanding will show up. About the same grain and

figure and easy cutting qualities are noticeable in apple, pear, plum, baywood, cedar, and mahogany.

Elm is a very soft or "tender" wood, owing to its numerous large pores, which arrange themselves in separate rings. The rays are "fine" and invisible to the unassisted eye. When dried, elm is apt to become "dusty" in sanding, so that it is well to not use too fine a paper, and see that the brush is doing its work well, otherwise dust will accumulate and mash into the soft wood, spoiling the sanding. Elm is very durable in contact with water, so it is a great favorite of ship-builders, and most of Venice is built on piles of this wood. It is also in some request for furniture and interior decoration purposes, and much resembles white pine, whitewood, butternut and basswood.

Poplar is very similar in construction to elm, being equally soft and mild, and, when seasoned, sands well. It is in great request for cabinet work and general fit tings, most of it being consumed by the plywood factories.

The grain of hardwood is easily distinguished from softer woods, as being very close-grained, with few pores. Take birch, for example. This wood, as well as maple, is very hard on sandpaper, as the grit is soon worn away, but owing to its close grain a fine polish with fairly coarse paper can be given.

Rosewood and red gum are also close-grained hardwoods, but, owing to the fancy figures of the grain, fine sandpaper is required to give the best polish. The tough, curling grain of rosewood makes this wood so popular in fine cabinet work. The peculiar nature of the red gum pores makes it easy to understand why sandpaper clogs and is made sticky.

These woods, as described, represent most of the types a sander operator has to do with, and a study of their peculiarities should help in the intelligent selec-

BUILT-UP VENEER PANELS

Ready for Immediate Shipment April 15, 1916 **Subject to Prior Sale**

Good Two Sides, 5-ply.

1 Qrt. Oak, 2s,	72 x 24 x ½	11 Pl. Oak, 2s,	60 x 30 x ⅜
17 Pl. Oak, 2s,	60 x 20 x ⅜	22 Qt. Oak, 2s,	60 x 30 x ⅜
9 Qrt. Oak, 2s,	60 x 20 x ⅜	12 Birch, 2s	72 x 20 x ⅜
32 Pl. Oak, 2s,	72 x 24 x ⅜	8 Birch, 2s,	60 x 30 x ⅜
41 Qt. Oak, 1s, Pl. Oak, 1s,	60 x 30 x ⅜	2 Qt. Oak, 1s, Birch, 1s,	60 x 30 x ⅜
		57 Pl. Oak, 1s, Birch, 1s,	60 x 30 x ⅜

Good Two Sides, 3-ply.

22 Birch, 2s,	60 x 30 x ¼	10 Pl. Oak, 2s,	60 x 30 x 3/16
20 Birch, 2s,	60 x 30 x 3/16	10 Pl. Oak, 1s, Birch, 1s,	60 x 30 x 3/16
5 Qt. Oak, 1s, Birch, 1s,	60 x 24 x ¼	18 Qt. Oak, 1s, Birch, 1s,	60 x 18 x ¼

Good One Side, 5-ply.

23 Qrt. Oak, 1s,	60 x 24 x ½	30 Qrt. Oak, 1s,	60 x 18 x ⅜
10 Qrt. Oak, 1s,	48 x 18 x ⅜	140 Qrt. Oak, 1s,	60 x 30 x ⅜
1 Mahogany, 1s,	72 x 24 x ⅜	47 Qrt. Oak, 1s,	60 x 24 x ⅜
4 Mahogany, 1s,	72 x 20 x ⅜	61 Qrt. Oak, 1s,	60 x 20 x ⅜
3 Mahogany, 1s,	60 x 20 x ⅜	20 Pl. Oak, 1s,	60 x 30 x ⅜

Good One Side, 3-ply.

6 Mahogany, 1s,	72 x 24 x ¼	6 Pl. Oak, 1s,	60 x 30 x ¼
10 Mahogany, 1s,	60 x 30 x ¼	4 Birch, 1s,	60 x 30 x ¼
9 Mahogany, 1s,	60 x 20 x ¼	6 Birch, 1s,	60 x 14 x ¼
		7 Qrt. Oak, 1s,	60 x 30 x 3/16

Single Ply Veneer

1/28 in. MAPLE VENEER, 60 in. to 72 in. long and 6 in. to 36 in. wide, averaging 24 in. to 30 in.
1/20 in. MAPLE VENEER, 72 in. long and 6 in. to 36 in. wide.
1/16 in. MAPLE VENEER, 60 in. to 72 in. long and 6 in. to 36 in. wide.
ALSO CULL VENEER 1/8 IN. AND THICKER, SUITABLE FOR CRATING.

Write for Prices

HAY & COMPANY, Limited - - WOODSTOCK, ONT.

WE HAVE BEEN MAKING THE

Best French Glue in the World

SINCE 1818; ALMOST A CENTURY

In competition at the numerous Expositions abroad, the quality of our glues has been found so high that we are usually declared "Beyond Competition." Our trade in this country makes the same declaration. Consistency in quality and service has brought this result and we are proud of it.

Perfection in manufacture from the best raw material makes the quality of COIGNET FRENCH GLUES consistently the highest.

Large stocks at Montreal assure the best of service.

Why not take advantage of these most important factors by using our glues?

At your service for prices and samples.

Coignet Chemical Products Co.

17 State Street, N. Y. CITY, N.Y.

Factories at {St. Denis / Lyons} France

ANGLO
RUBBING and POLISHING

Works free and easy and can be rubbed in two days.

WRITE FOR SAMPLE

The Ault & Wiborg Co. of Canada, Ltd.
Varnish Works

Montreal Toronto Winnipeg

THE GLUE BOOK

WHAT IT CONTAINS

Chapter 1—*Historical Notes*
Chapter 2—*Manufacture of Glue*
Chapter 3—*Testing and Grading*
Chapter 4—*Methods in the Glue Room*
Chapter 5—*Glue Room Equipment*
Chapter 6—*Selection of Glue*

WOODWORKER PUBLISHING CO., LIMITED

345 Adelaide St. West, Toronto

tion of sandpaper and machine adjustments best suited to the particular wood sanded.—"Timber."

George M. Mason, Limited

Located in the heart of the city of Ottawa and in the centre of the Ottawa Valley lumber district is the large and modern plant of George M. Mason, Limited, of which Mr. George M. Mason is the president and general manager. For sixteen years the plant has been running under the above firm name, and year by year its operations have broadened.

The plant covers over four acres and is located at the corner of Wellington Street and Bayswater Avenue, consisting of planing mill, sash and door factory and lumber yard. To meet the demands of their growing trade a branch yard and planing mill is now running at Woodroffe, a suburb of the city of Ottawa.

The main plant has a capacity for one hundred men, and turns out all kinds of house finish, doors, columns, sash, hardwood flooring and interior trim. The planing mill is equipped with the most modern machinery and has a capacity for handling 150,000 feet of lumber per day. The sash and door factory has also a large capacity, and is equipped with all the latest machinery.

A feature of the mill, of interest to our readers, is its splendid facilities for the dressing of lumber-in-transit, in which end of the business the firm is constantly expanding. The transportation facilities are splendid, the C. P. R. and G. T. R. passing by the door.

In the yards lumber of all kinds is stocked, including dimension timber, scantling, rough and dressed lumber, lath and shingles, and all kinds of hardwood.

The George M. Mason Limited are already fully equipped to take care of any order, large or small, in the portable house line. For some time past they have been specializing in the manufacture of ready-cut houses, and have made and shipped quantities of these all over Canada.

Speed of Woodworking Machinery

There seems to be small effort made to instruct the mechanics in railroad planing mills anything about the science of their business, yet there are many interesting things to be learned concerning woodworking machinery, things that would astonish many master mechanics.

Woodworking machines require great power to drive them owing to the enormous capacity for doing work. Wood is more easily worked than metal, but, the material is cut so rapidly that it represents immense concentration of power.

A properly driven circular saw has a peripheral speed of 7,000 feet per minute—nearly a mile and a half. A band saw is run at about half that speed. Planing machine cutters have a speed at the edge of 6,000 feet per minute, as the cutters of molding machinery trims out material at about 4,000 feet per minute. Wood carving drills run 5,000 revolutions per minute. Augers one and one half inch diameter are run 900 revolutions per minute and those half that size are run 1,200 revolutions per minute. Mortising machine cutters make about 300 strokes per minute. Two men operating a cross cut saw, regulate their speed by the eyes of the foreman.—Railway and Locomotive Engineering.

This shows the possibility of FIGURED QUARTERED RED GUM as used in our private office. We specialize this wood

Importers and Manufacturers

Mahogany

and

Cabinet Woods

Sawed and Sliced

Quartered INDIANA White Oak, Red Oak, Figured Red Gum, American Walnut, Etc.

Rotary Cut Stock in Poplar & Gum for Cross Banding, Back Panels, Drawer Bottoms & Panels.

The Evansville Veneer Company, Evansville, Ind.

Metal Versus Wood Cauls for Veneering
Superintendents and Veneer Room Foremen Reply to Question of a Correspondent

Winnipeg, April 10.

Editor "Canadian Woodworker":

Dear Sir:—We have quite a lot of veneering to do in our factory and about two years ago we were advised to adopt the use of aluminum cauls for veneering, but they have not given us as good satisfaction as the wood cauls which we previously used. Since then we have heard that a number of manufacturers have discarded metal cauls and gone back to wooden ones. We would like very much to have the opinion of some of your readers on this subject.

"Manufacturer."

Answer No. 1.—Replying to yours re cauls for veneering, would say, I have had no experience with aluminum cauls, but have had experience with zinc. I discarded zinc for 3-ply wood cauls and I find them much better, as they give better results. They do not have to be so hot as zinc and they retain the heat much longer. I also find that the glue will set or harden better and much more quickly than it did with zinc cauls. I have taken five years' use out of the batch of 3-ply cauls which I made and they are still in good condition. They are made of rotary birch, one-quarter inch core, and three-sixteenths inch face, sanded down to a smooth surface and coated with an oil filler. They are kept greased with a heavy cup grease and I have no trouble with mahogany or walnut veneer sticking to them, so I stake my luck on good 3-ply birch cauls.

Subscriber.

Answer No. 2.—In reply to your request for information on the subject of aluminum versus wood cauls, I gladly give the benefit of my experience. It depends to a certain extent on what and how the process of veneering is being done to make a comparison between the two kinds, as both have their good points. If the idea is to have a caul that will hold a great deal of heat and hold it for a long time, then the wood caul would be preferable, but that is only required where the glue has got "tacky" before laying the veneer. To my mind that and the fact that they are reasonably cheap is the only advantage the wood cauls have over the metal cauls. The method or process of veneering which seems to be gaining favor is to get the stock under pressure as quickly as possible after gluing to prevent swelling or absorbing of moisture, in which case aluminum cauls would be the best. With this method of veneering the caul would not require to be hot, just warm, and it is not necessary that it should hold the heat for any length of time. With a spreader and hydraulic press a few minutes is all that is required to put in a press. I have used one lot of aluminum cauls for five years and every one of them is still good, in fact, nearly as good as new. They stand moderate usage but will not stand rough usage. They should be kept clean, but never scraped, and they will not get thin in places as zinc cauls do. If a light application of some good wax is given a couple of times a week they will leave the stock free and in splendid condition. The first cost of aluminum cauls is greater than wood cauls

and I know of no reason why with moderate care they should stand for years. They are much more convenient to handle, take less space in the press and in the caul box, and in my estimation they are an ideal caul.

Piano Veneer Man.

Answer No. 3.—In the matter of wood cauls vs. aluminum cauls, my choice certainly lies in favor of wood cauls for various reasons. Wood cauls when left in the caul box a reasonable length of time will retain enough heat at the proper temperature to ensure the safe gluing of a large batch of stock without any danger of it being chilled. Their surfaces are easily kept in good condition and free from blemishes or cavities which would be likely to spoil the stock they came in contact with.

On the other hand, metal cauls—as aluminum—are not so good. Their surfaces cannot be kept so free from blemishes as wood and any dent or hollow they may sustain is not easily remedied, so they have to be discarded. Again, they receive while in the caul box preparatory to gluing a quick fierce heat, that is emitted as fast as it is acquired, thereby doing double damage to the stock. In laying a batch of veneer, when the metal caul is first taken out of the heater and laid on the stock it is quite liable to burn the glue and yet not have sufficient heat to keep the glue warm until the final process of clamping down is commenced. Therefore, in my opinion it has two faults; a too primal heat and not a sustaining one.

It is quite true that more stock can be veneered at one time with the aluminum cauls owing to their thinness, but after all it is the last analysis that counts in this as in any operation. I worked in a factory in the Old Country where they used zinc cauls of about the same thickness as the aluminum, and as far as I could learn they had good success with them, but then of course zinc is a totally different metal to the aluminum. It is not so ductile and its density is much greater, therefore it is able to hold heat. In conclusion, I may state that a large firm that I am acquainted with have discarded the aluminum cauls entirely and reverted to the wood cauls again as being more satisfactory and enduring.

Lewis Taylor.

Answer No. 4.—In my opinion, wood cauls are undoubtedly the best, but their great drawback is that they are easily damaged, and only a small amount of stock can be put in the press. The metal cauls are favored on account of their thinness, which makes it possible to put more stock in the press than with wood cauls and they are more easily kept clean. Of the different kinds of metal cauls the aluminum cauls are the best. They last longer than zinc, do not wear thin in places, remain straight and hold the heat well. They are also light to handle and they leave the work free and clean.

Metal cauls, either zinc or aluminum, should never

be scraped. They should be cleaned with some soft material such as waste. I have known both wood and aluminum cauls to be used so hot that the very life or nature was burnt out of the glue, which of course caused loose veneers. There is reason in all things and it should be applied in heating cauls. I have come to the conclusion, however, that the cold glue system is the best for veneering, after having

seen it worked. Cauls are not used by this method.

I have pointed out the so-called advantages of aluminum cauls and the reasons why they are used; but if it is desired to stick to the old method of using hot glue, then for good, honest, enduring work, give me wood cauls, as they will hold an even heat for a much longer period than metal cauls.

Furniture Veneer Man.

Properly Laid Veneer Will Enable the Sanding Machines to Do Better Work

By Lewis Taylor

With the advent of new machines in the woodworking industry, for almost every conceivable use, some of them almost human in their wonderful mechanism, we are apt to overlook the fact that a machine can only turn out good work when the material operated on by it is in proper shape and condition for the operation.

To illustrate what I mean I may say that sanding machines (either belt or stroke sanders) will be the source of a great deal of trouble and expense when operating on stock veneered with the usual thin rotary cut or sliced veneer if, firstly, the stock has not been thoroughly dried out after the veneer was laid, and, secondly, if the operator presses too hard on one place too long, for the heat generated by this friction is very great and is apt to cause blisters in the veneer. This is not an idle theory, because we have had this same trouble and could put it down to nothing else. Hand cleaning of the veneer is not one bit better, if as good, as the work the machines turn out. The least we can do is to give our machines and their operators good stock to work on, as loose veneer is one of the most annoying things that can be found in any factory.

The trouble with some men seems to be that they do not realize the extreme thinness of the veneer and think that it will stand almost anything. On the other hand, we have made the advent from hand work to machine work so rapidly that we have failed to adapt the stock to meet the different conditions. Of course, this is only true regarding certain machines, but no one doubts that gluing and sanding machines touch the heart of things and are vital to the finished product.

I have great faith in hand toothing, and I think it would pay to hand tooth the cores of the most important parts; for if there is any hole or depression on the surface of the core where any glue could find lodgment the toother would soon find it out and then it could be rectified instead of being an eye-sore in the shipping room.

Machinery is increasing in effectiveness all the time and any one that "knocks" a machine is like one trying to stop the steady march of progress or put back the clock of time. But at the same time we might as well realize that a machine has no judgment. It will only do its part faithfully if those who provide the stock for it and operate it are faithful also.

I was watching a gluing machine in operation the other day, but their method of working was opposite to the principles of veneering that I was taught when we did the gluing by hand. After the crossbanding was put through the gluing machine, they laid the piece with the hot glue on both sides of it on the core stock, then immediately laid the outside veneer on the still steaming glue. Now, this is opposed to the true principles of veneering, as the glue should be chilled before the veneer is placed on the core stock, on account of the action the hot glue has on the veneer. The difficulty experienced when using the gluing machine could be easily overcome by having a series of horizontal racks to place the veneers on after they have passed through the gluing machine. Then after they had cooled a little they would be ready to lay and put in the press.

There is no doubt that with the increasing number of mechanical and labor-saving machines in the woodworking trade that "inside cabinet making" will be a lost art in the near future and then again there is great danger that little details that the old cabinet maker was so careful to observe will be forgotten amidst the roar of the machinery.

I read an article in a recent edition of The Canadian Woodworker about some table tops that had developed depressions in the finishing room. In my opinion that trouble would never have occurred if the core stock had been crossbanded and then after standing a while, had been carefully hand toothed with a fine toother. Surely it is worth that much trouble for first-class work, especially when the first cost is trifling compared with the surety of durability.

We must all bow to the might of machinery, but in so doing we must not lose sight of the fact that there are certain parts of the work where the old hand method will work out the cheapest in the long run and also sustain the reputation of the firm for enduring work. I might say in conclusion that I worked for a long time in a small shop where the old methods were used and during that time loose veneers and blisters were unknown, and I think, therefore, that the closer we keep to the old principles, even with the up-to-date methods, the better will be the results obtained.

506 Foy Building, 32 Front Street West, Toronto, and manufacturers intending to place orders for machinery can rely on getting the very best in quality and service at this address.

Garlock - Walker Machinery Co., Limited

Mr. Wm. Garlock, Jr., who has for some years been in charge of the Canadian business of the American Woodworking Machinery Company, and Mr. A. B. Walker, who has been associated with him, have recently formed the Garlock-Walker Machinery Company, Limited. The company will carry a full line of woodworking and metal working machines. Mr. Garlock will look after the woodworking machinery end of the business and Mr. Walker the metal working machinery. Both men are particularly well qualified to handle their respective lines, as their experience in the machinery business extends over a number of years. The company will have their offices at Room

THE OHIO VENEER COMPANY

Importers and Manufacturers

Foreign and Domestic Veneers and Hardwood Lumber

We always carry a large and assorted stock of Mahogany, Circassian
Walnut, Sawed and Sliced Quartered Oak.

Send us your enquiries and orders.   We guarantee good service.

2624 to 2644 Colerain Avenue,   ⋈   ⋈   CINCINNATI, OHIO.

American Walnut

in

Lumber - Veneers
Dimension Stock

Our sole product is well manufactured
Soft Ohio and Indiana Walnut. As specialists'in this high class lumber we are able to
offer you the best band sawn and equalized
stock in America.

Special attention given to furniture manufacturers.

Ask us for samples and prices.

Geo. W. Hartzell
Piqua, Ohio

 ## Hoffman Bros. Co.
Reg. U.S. Pat. Off.   Estab. 1867, Incor. 1904   Reg. U.S. Pat. Off.

800 West Main Street,
FORT WAYNE, INDIANA

Manufacturers of

VENEERS and LUMBER
In the Domestic Hardwoods

ANY THICKNESS,

1/24 and 1/30 Slice Cut
(Dried flat with Smith Roller Dryer)

1/20 and thicker Sawed Veneers,
Band Sawn Lumber.

– SPECIALTY—

Indiana Quartered Oak

Reg. U.S. Pat. Off. Reg. U.S. Pat. Off.

Veneers and Panels
PANELS

Stock Sizes for Immediate Shipment
All Woods   All Thicknesses   3 and 5 Ply
or made to your specifications

VENEERS

5 Million feet for Immediate Shipment
Any Kind of Wood   Any Thickness

J. J. NARTZIK
1986–78 Maud Ave.,   ⋈   ⋈   CHICAGO, ILL.

Veneers and Panels

LARGE STOCKS FOR
QUICK SHIPMENTS

Send for Lists

Walter Clark Veneer Co.
GRAND RAPIDS, MICH.

OAK · MAHOGANY · CIRCASSIAN · MAPLE · VENEERS · BIRCH · POPLAR · CHERRY · WALNUT · GUM

Wood Mosaic Company

Main Office, New Albany, Ind.

Are you interested in

Quartered Oak Veneers?

We have the best timber and the best plant for the manufacture of this stock to be found anywhere. We make our own flitches, using only Kentucky and Indiana stock, which is famous for mildness, color and texture. We can make immediate deliveries of Quartered Oak Veneers on orders received now.

Get acquainted with our Famous Indiana and Kentucky QUARTERED WHITE OAK

TEAK

Write for prices on all

**Plain and Fancy
Lumbers and Veneers**

Mahogany, Oak, Birch, Walnut, Ash, Poplar, etc., rotary, sliced or sawed in all thicknesses

SHELL BOX PARTS
in birch and spruce
Bridges and Ends
Write for Quotations

George Kersley
224 St. James Street, Montreal

MAHOGANY

D O M E S T I C

L U M B E R

American Walnut Veneers
Figured Wood
Sliced ‑ Half Round ‑ Stump Wood

5 Essential Qualifications:

1.—Unlimited supply of Walnut Logs.
2.—Logs carefully selected for figure.
3 —All processes of manufacture at our own plant under personal supervision.
4.—Dried flat in Philadelphia Textile Dryer.
5.—3,000,000 feet ready for shipment.

JOHN N. ROBERTS,   New Albany, Indiana
The Pioneer Walnut Veneer Producer of America

Trying It "On the Dog"

A "new beginner," as Potash would say, must experiment before he learns how, just as a young doctor who has just passed his state board examinations must have somebody to practice on before he becomes expert enough to command much confidence. Most of us prefer to let him get his experience out of somebody else.

It's the same way with lumber and veneers: to be specific, with walnut lumber and veneers. We have been in this business for thirty years; our Mr. J. N. Penrod is generally regarded as the most expert manufacturer of figured walnut stock in the United States, and the leading consumers rely upon him for service of an exceptional character.

You may be able to get what you want from the "new beginner", but why take chances, when you can come to headquarters and get the real thing at the same price?

Penrod Walnut & Veneer Co.
KANSAS CITY, Mo.

We specialize on

Poplar
cross-banding and backing

Veneer
Highest Grade — Right Prices

The largest Poplar Veneer Mill in the world.

The
= Central Veneer Co.
Huntington, West Virginia

MAHOGANY VENEERS
Notwithstanding the scarcity, we are in a position to take care of our customers.

CIRCASSIAN
Some stocks on hand suitable for interior finish and furniture manufacturers.

FIGURED AMERICAN WALNUT
Over two million feet, nice striped and curley wood.

QUARTER SAWED WHITE AND RED OAK
1-20 to 1-8, 6 to 15-in. wide

GRAND RIM POPLAR
Cut to size. We can take care of your 1916 requirements from choice Yellow Poplar logs.

FIGURED WALNUT BUTTS
Two hundred twenty-three thousand feet newly cut Butts, some highly colored stock in this lot.

MAHOGANY LUMBER
This is something we make a specialty of, representing in Chicago and the Northwest, one of the largest importers.

HAUGHTON VENEER COMPANY
1152 W. LAKE STREET   CHICAGO, ILLINOIS

Vertically Piped Dry Kilns

The Electric Planing Mills Company of Welland, Ont., have recently installed a dry kiln equipped with the vertical system of piping, which has been perfected by the National Dry Kiln Company of Indianapolis, U.S.A., after considerable experimenting. The merits of the system are shown by the fact that it has been adopted by a great many leading manufacturers. The advantages of vertical piping are given as follows: —First, costs less to operate; second, joints and pipes can be easily reached; third, geen lumber just entering the kiln is fully protected; fourth, the risk of fire is minimized.

The chief difference between the vertical piping system and the old style horizontal system lies in the arrangement of the coils of pipes. In a vertical piping system the coils are placed in vertical tiers with one pipe above the other, instead of in horizontal tiers with each pipe beside the other as in the horizontal system. More pipe can be installed in vertical tiers above one another than in horizontal tiers and the convenience of the vertical arrangement is much greater. In a National kiln, with its system of vertical piping, enough pipe can be installed to operate the kiln on exhaust steam by day and low pressure steam at night, or, exhaust steam by day, and high pressure steam at night, with part of the coils shut off. The saving in this is obvious.

If it is desired to operate a horizontally piped kiln on exhaust steam the following condition as immediately presented. The piping required will more than cover the floor of the kiln. There is only one way to "beat" this situation and that is to place the pipes in two or more rows, one above the other. When this is done the horizontally piped kiln becomes in reality a vertically piped kiln, for that is all the vertical piping system is. The placing of the pipes above one another in vertical tiers instead of side by side in horizontal tiers.

The difference between the two systems, however, is very great; for instance, suppose it is desired to reach a joint or pipe in the bottom row when the pipes are laid horizontally. Think of the inconvenience of reaching it with two or more rows of pipes forming a net work above it. There is hardly a time when such a contingency arises that it is not necessary to run one or more loaded cars out of the kiln. This not only results in loss of time, but is hard on the partially dried lumber. In the National system the vertical coils are placed twenty-five or more inches apart, and the coils are below the track line so that if it is necessary for a workman to reach any joint or pipe he can do so easily and conveniently, as there is sufficient room to walk or work between the coils and under the cars, thus making it unnecessary to move out loaded cars or to shut down the kiln to make repairs.

There is another important advantage offered by the National system of vertical piping and that is the protection afforded to green lumber. There are three types of kilns in general use for drying hardwoods; the progressive, the compartment, and the box types. In the progressive type the lumber is pushed in at one end of the kiln and advances each day through stages that are progressively hotter and drier until it is finally taken out thoroughly dried at the opposite end of the kiln. This type is ideal when the output of the factory is such that it is desired to dry a certain amount of one kind of stock every day—or different kinds of stock requiring the same drying period. It will be realized as of the utmost importance that the heating pipes in a progressive kiln be confined very largely to the discharging end, as otherwise the green lumber upon entering the kiln would be exposed to the direct heat of the pipes and case-hardening, checking, and warping would result.

With the progressive drier equipped with vertical piping enough pipe can be installed to operate the kiln with exhaust steam if this is desired and still keep most of the coils confined in the discharging end of the kiln. It is this fact together with the system of natural draught circulation concentrating the moisture at the receiving end of the kiln where it envelopes and protects the green lumber that assures perfect results with the National progressive drier. The same protection is afforded green lumber in a compartment or box kiln equipped with vertical piping as in the progressive type of kiln. In the compartment or box kiln the lumber cannot be moved along through varying stages of moisture and temperature as in a progressive drier, so it is necessary that some means of regulating these elements be provided. The common practice is to start the kiln with a low temperature and a high humidity, then as the drying progresses increase the heat and reduce the humidity.

Each row of pipe is supplied with steam independently of the others, thus all that is necessary to vary the temperature of the kiln is to turn on or shut off one or more of the coils. The kiln can be started with one or two coils the first day, a coil can be added the second day, two more the third day and another the fourth day—gradually increasing the heat each day and meeting easily and simply every requirement of perfect drying. The moisture in a compartment or box kiln is regulated by a steam spray and the National system of natural draught circulation in this type of kiln makes only the minimum of regulation of both heat and moisture necessary.

All heating systems spring leaks occasionally, but with the vertical system of piping, if the leak is discovered in one coil, that one coil can be shut off and the kiln can continue to run until Saturday or Sunday when it is generally convenient to make repairs. Another advantage with the vertical piping is that bark and rubbish cannot fall on the hot pipes as with the horizontal system. This greatly reduces the danger of fire and the kiln can be cleaned out regularly without any trouble.

The National Dry Kiln Company, manufacturers of the National dry kiln, also manufacture a full line of dry kiln accessories, such as steel roller bearing trucks, malleable iron wheels and door carrier equipment.

Lumbermen's Convention

The 19th annual convention of the National Hardwood Lumber Association will be held in Chicago on June 15th and 16th. It is expected that the attendance will exceed anything in the history of the organization. The business and entertainment programmes have been arranged and members can look forward to hearing some interesting addresses and spending an enjoyable time.

When working around a dangerous machine, be sure good footing is provided and never remove guards or safety devices unless absolutely necessary in order to do the work.

Before throwing in an automatic feed, be sure the stops are set so that no damage will be done either to the work or machine by its use.

V O K E S

A Wide Range at Right Prices

TOOLS

CABINET HARDWARE
FACTORY AND MILL SUPPLIES

BARTON'S GARNET
and
SAND PAPER

UNIVERSAL CASTERS
(Canadian Agents)

Write for prices on Garden Tools, Lawn Mowers, Lawn Rollers and Hose

PRICES RIGHT — PROMPT SHIPMENT

Let Us Send You Our Catalogs

The **Vokes Hardware Co.**
Limited
40 Queen St. East, Toronto, Can.

HARDWARE

The Lumber Market

Domestic Woods

The prices of birch, basswood, maple, elm and ash are $7 or $8 a thousand higher than they were a year ago. This is due in part to the fact that hardwood stocks are decidedly low. American buyers have been taking all the hardwoods that were offered and furniture manufacturers, implement manufacturers and other users of hardwood have been making steady purchases. There does not appear to be much relief in sight at present, as mill men have been very conservative in getting out logs for their spring cut, which is estimated to be about 25 per cent. less than formerly. Manufacturers would be well advised in buying up any hardwoods that are offered them, as it will be some time before the present cut is in condition for shipping.

There is no feature of particular interest in connection with soft woods unless it be spruce for shell boxes, and the millmen seem to have made provision for this business in their estimates, as there have been one or two unusually heavy cuts of spruce reported.

The building business remains practically at a standstill except for several large buildings that are under way or contemplated, and with the exception of heavy timbers for construction purposes the lumber required for these will be mostly hardwood for floors, trim, etc.

Some of the current prices quoted at Toronto on car lots are as follows: White ash, firsts and seconds, inch, $60; white ash, No. 1 common, $45; brown ash, firsts and seconds, inch, $50; birch, firsts and seconds, inch, $42; birch, No. 1 common, $35; basswood, firsts and seconds, inch, $40; basswood, No. 1 common, $34; soft elm, firsts and seconds, inch, $40; soft elm, No. 1 common, $32; soft maple, firsts and seconds, inch, $33; soft maple, No. 1 common, $25; hard maple, firsts and seconds, inch, $40; hard maple, No. 1 common, $30.

Imported Woods

The situation in regard to imported woods does not change materially from month to month. Prices are holding very firm, particularly on oak, maple, and ash, due to the marked shortage of hardwood stocks. The weather in the Southern States has been unfavorable for getting out new stocks and in the Northern States the embargoes in effect against lumber shipments on the railroads has prevented the distribution of such stocks as were cut.

The report issued by the Hardwood Manufacturers' Association of the United States of stocks of hardwood lumber on hand April 1st, indicates more clearly than anything else the seriousness of the shortage. The report includes figures from twenty-four mills which are as follows: plain white oak, 1,597,000 feet; plain red oak, 1,287,000 feet; quartered white oak, 908,000 feet; quartered red oak, 176,000 feet; poplar 857,000; quartered poplar, 38,000; chestnut, 822,000. In plain white oak the greatest decreases are in firsts and seconds, Nos. 1, 2, and 3, common.

The same grades decreased most in plain red oak. The reports for quartered white oak show decreases in firsts and seconds, Nos. 1 and 2, common, and for quartered red oak in Nos. 1 and 2, common. For poplar the greatest decreases are in Nos. 2 and 3, common. Chestnut decreases are heaviest in No. 2 wormy, and No. 2 common. Stocks in practically all grades of gum

are short and the prices are accordingly firm. Owing to the impossibility of importing African mahogany, the price has jumped to $200 to $250 per thousand. American walnut is rather active at the present and is quoted around $140 to $145 for 1 inch, and $190 for 1½ inch. No. 1 common, which is the grade mostly used by furniture factories, sells at $65 to $70. Oak still holds its own as regards sales, and the prices quoted are: 1 inch plain oak, $65; quartered oak, $90.

Woods Used for Furniture

According to figures compiled by the United States Forest Service the furniture industry of the United States uses in one year 944,677,807 feet of lumber, board measure, costing $28,717,873.46. Following is the table showing the principal woods used:

Woods Used for Furniture in United States

Kind of wood	Quantity Used annually (Ft. b. m.)	Per cent	Average cost per 1,000 ft.	Total Cost f. o. b. factory
Oak	431,053,289	45.63	$34.11	$14,704,675.58
Red gum	101,866,767	10.78	20.15	2,052,423.40
Maple	87,571,456	9.27	26.17	2,291,852.80
Birch	54,677,450	5.79	25.02	1,367,919.24
Yellow poplar	53,374,580	5.65	24.29	1,296,715.65
Chestnut	44,734,180	4.73	20.70	926,042.00
Basswood	33,146,276	3.51	26.13	866,132.43
Beech	21,103,204	2.24	19.74	417,712.11
Yellow pine	18,926,400	2.00	15.69	296,916.90
Ash	15,868,588	1.66	27.21	426,270.59
Mahogany	15,697,125	1.66	133.91	2,093,990.08
Elm	12,154,102	1.29	24.50	297,706.79
Douglas fir	11,387,790	1.21	24.24	276,042.16
White pine	9,832,808	.99	25.33	236,579.30
Hemlock	7,053,446	.75	13.38	94,347.50
Cottonwood	5,158,309	.55	22.48	115,964.00
Cypress	3,477,800	.37	18.01	62,619.00
Tupelo	2,900,100	.31	22.32	64,735.00
Spruce	2,270,500	.24	26.06	59,170.36
Cedar	1,856,100	.20	57.60	106,919.20
Western yellow pine	1,806,985	.19	25.00	45,173.56
Black walnut	1,689,967	.18	109.90	185,732.02
Sycamore	1,474,882	.16	23.99	35,381.66

The position which oak occupies in the manufacture of furniture stands out prominently. Its total is more than four times that of its nearest competitor.

Cooperage Men Meet

The nineteenth annual meeting of the National Slack Cooperage Manufacturers' Association, which was held in Chicago, May 9, 10, 11, resulted in this old and important organization affiliating with the National Coopers' Association and Tight Barrel Stave Manufacturers' Association, thus completing the amalgamation of cooperage industries.

The majority of the leading men in the business were present at the meeting and their vote was unanimous. Some time in the near future the present Executive Committee of the National Slack Cooperage Manufacturers' Association will meet with the Executive Committee chosen at the formation of Cooperage Industries, in St. Louis, when plans for work under the new amalgamation will be laid out.

The old officers were re-elected and, of course, the organization will retain its identity as will the National Coopers' Association and the Tight Barrel Stave Manufacturers' Association until such a time as it may be deemed advisable to discontinue the separate organizations. Some interesting papers were read at the convention and Mr. W. A. Fraser, of the Trenton Cooperage Mill, of Trenton, Ont., gave a talk on the baby barrel in Canada, which was listened to with much interest. Mr. Fraser said these packages were used largely among apple growers and shippers, and grocerymen are finding a big demand among the housewives. Some of these baby barrels are painted in different colors, others have just the natural stave and head colors with painted hoops.

Machinery for Sale

1—18 x 24 in. Slide Valve Engine.
1—10 ft. Fly-wheel and 9 ft. drive pulley, 24 in. face.
1—Wood Pulley 69 in. x 24 in. face, 4½ in. bore.
1—Wood Pulley 84 in. x 18 in. face, 4 in. bore.
1—Wood Pulley 90 in. x 18 in. face, 3½ in. bore.
1—Jack-ladder chain and sprockets, 300 ft. long.
1—Boiler, 72 in. diam., 14 ft.
1—Boiler 52 in. diam., 14 ft.

Can be inspected at our mill at Callander.

CALLANDER SAW MILLS
Callander, Ont.

For Sale

Woodworking Machinery

The following second-hand machines have been placed with us for sale:

1 Woods No. 32 Matcher; 1 Goldie 10-in. Moulder; 1 Goldie 24-in. Matcher; 1 Cowan 42-in. Sander; 1 Jackson-Cochrane 36-in. Sander; 1 Egan 36-in. Single Surfacer; 1 McGregor-Gourlay 2-drum 30-in. Boss Sander; 1 Egan 30-in. 2-drum Sander; 1 CMC 8-in. 4-side Sticker.

P. B. Yates Machine Co., Limited
Hamilton, Ontario

Second-Hand Woodworking Machines

Band Saws—26-in., 36-in., 40-in.
Jointers—12-in., 16-in.
Shapers—Single and double spindle.
Re-Saw—42-in. American.
Lathes—20-in. with rests and chisels.
Surfacers—16 x 6 in. and 18 x 8 in.
Mortisers—No. 5 and No. 7 New Britain.
Rip and Cut-off Saws; Molders; Sash Relishers, etc.
Over 100 machines, New and Used, in stock.

Baird Machinery Company,
123 Water Street,
PITTSBURGH, PA.

File Efficiency

Experts who test things from all sides and measure up analytically the final results, have put a black mark against the idea of economizing in files by saving them up and having them re-cut. It is claimed that careful tests have demonstrated that re-cut files will only do about half as much work as new files, and that for this reason the loss in time and efficiency amounts to more than the saving in re-cutting. If they do nothing else, these experiments serve to call attention to the fact that quality in files is worth something and worth paying more for. To buy files that are a little cheaper and yet are deficient in cutting capacity, which are rough and will not cut smoothly, or are defective in some way, is false economy. In selecting files, as in the buying of saws, knives and other tools, it is easily evident that it is economy in the end to pay the higher price and get the better quality. The time of the man using the file, and the results accomplished, are of much more importance than a little difference in the first cost. As to what had best be done with worn files, at the present price for good steel and other scrap metal, the answer seems to be, preserve and sell them to those who want the metal.—Woodworker, Indianapolis.

Making Bungs and Faucets

The bung's business is to hold liquid in a barrel or keg, the faucet's is to draw it out. The making of these two small articles is treated as a single industry in statistical compilations, though they are not generally the product of the same factory. They are so intimately connected with cooperage that they are naturally classed with that industry. Though the articles are small, they call for a large amount of wood for their production. Their yearly total needs in the United States, according to latest figures, exceeds 21,000,000 feet, and includes 17 kinds of wood, 12 of which are hardwoods, five soft. In quantity, yellow poplar amounts to more than all others combined.

Special Packing Methods

An American consular officer has called attention to a special style of packing that has been adopted by certain European exporting houses with a view to the prevention of pilfering.

The report states that the merchandise is first placed in an inside container, which is then bound with wire. Burlap is wrapped around the entire package, and this wrapping is again bound with wire transversely, with the result that the entire wrapping must be removed to gain access to the inner package. The package prepared in this way is then crated or boxed, as may be necessary. In the outer crating or box special nails of serrated form are used. These nails can not be readily removed without evidence of the removal appearing plainly on the case. This method of packing is not adapted to heavy and bulky merchandise, but it is this class of exports that suffers least from pilfering.

1", 1½" and 2" White Pine and Spruce
⅝", 1", 2" x 4 and up Pine and Spruce
Crating Lumber

Write for our stock sheet

The Elgie & Jarvis Lumber Co., Ltd.
18 Toronto St., Toronto, Canada

PATENTS
STANLEY LIGHTFOOT
Registered Patent Attorney
Lumsden Building - - TORONTO, CAN.
(Cor. Adelaide & Yonge Sts.)

The business of Manufacturers and Inventors who with their Patents systematically taken care of is particularly solicited. Ask for new free booklet of complete information.

PETRIE'S
MONTHLY LIST
of
NEW and USED
WOOD TOOLS
in stock for immediate delivery

Wood Planers

36" American double surfacer.
30" Whitney pattern single surfacers.
26" Revolving bed double surfacers.
26" Goldie & McCulloch single surfacer.
24" MacGregor Gourlay planer & matcher.
24" Major Harper planer and matcher.
24" Single surfacers, various makes.
20" Dundas pony planer.
18" Little, Giant planer and matcher.
16" Galt jointer.
12" Crescent jointer with safety head.

Band Saws

42" Fay & Egan power feed rip.
38" Atlantic tilting frame.
36" Crescent pedestal.
30" Ideal pedestal.
28" Rice 3-wheel pedestal.
28" Jackson Cochrane bracket.

Saw Tables

Preston variable power feed.
Cowan heavy power feed. M138.
No. 5 Crescent sliding top.
No. 3 Crescent universal.
No. 2 Crescent combination.
12" Defiance automatic equalizer.
MacGregor Gourlay railway cut-off.
7½' Martin iron frame swing.
6½' Crescent iron frame swing.

Mortisers

No. 5 New Britain, chain.
Cowan hollow chisel. M.100.
Fay upright, graduated stroke.
Galt upright, compound table.
Smart foot power.

Moulders

10" Clark-Demill four side.
10" Houston four side.
6" Dundas sash sticker.

Sanders

40" Berlin 3-drum.
30" Egan double drum.
12" C.M.C. disk and drum.
18" Crescent disk.

Miscellaneous

Fay & Egan 12-spindle dovetailer.
40" and 18" Ideal turning lathes.
No. 285 Wysong & Miles post boring machine.
MacGregor Gourlay 2-spindle shaper.
Elliott single spindle shaper.
No. 51 Crescent universal woodworker.
40" MacGregor Gourlay band resaw.
Rogers vertical resaw.
Cowan sash clamp.
Pedestal tenoner double heads and copes.
Egan sash and door tenoner.
No. 6 Lion universal woodtrimmer.
6-nail box nailer, Meyers patent.
20" American wood scraper.
4-head rounding machines.
Cowan spindle carver, M63.
Cowan veneer press screws.

Prices, Descriptions and Full Particulars on Request

H. W. PETRIE, LTD.
Front St. W., Toronto, Ont.

News of the Trade

The Eagle Lumber Company, Quebec, have increased their capital from $90,000 to $500,000.

The planing mill at Simcoe, Ont., owned by Lewis Fick & Son, was destroyed by fire early in May.

Canadian Hardwoods, Limited, Deseronto, Ont., have changed their style to Dominion Hardwoods, Limited.

The Preston Car Company was awarded the contract for the construction of 13 bodies for street cars for the Toronto civic car lines.

The Tudhope Carriage Company, a subsidiary company of Carriage Factories. Ltd., is reported to be operating double time to keep up with their orders.

The Amherst Piano Company, Amherst, N.S., recently held their annual meeting. They paid their half yearly dividends and placed $30,000 to reserve.

Chas. J. Anderson, for many years secretary-treasurer of the Dominion Match Company of Deseronto, recently passed away at the Wellesley Hospital, Toronto.

The Elkhorn Trading Co., Elkhorn, Man., has been incorporated with capital stock of $45,000, to carry on business as manufacturers of sash and doors and to sell shingles.

William Travis, of Wyoming, Ont., is erecting a new planing mill at that place. The building will be forty by fifty and fourteen feet high. The owner has machinery.

Furniture factories throughout Ontario are beginning to employ female help. The Malcolm Furniture Company, of Kincardine, have a number of women on their staff now.

Mr. E. Monaghan, who recently died at his home in Glencoe, Ont., was for several years foreman of the woodworking department of the E. Leonard & Sons Foundry, of London, Ont.

The Onward Mfg. Company, manufacturers of the Onward Sliding Furniture Shoe, will build a new factory on East King Street, Berlin, Ont. Mr. T. A. Witzel is manager of the company.

A toy exhibition was held recently in the Morrin College Hall, Quebec, in order to demonstrate that Canadian-made toys are equal, if not superior to those formerly imported from foreign countries.

Reeve Alfred Tier's planing mill at Fenelon Falls, Ont., was destroyed by fire on April 18th. There was no insurance. Since the outbreak of the war this factory has been continually working on extensive contracts for ammunition boxes.

J. H. Moffatt, superintendent of the Gibbard Furniture Co., Napanee, Ont., had his hand badly lacerated in an exhaust fan in the factory recently. Mr. Moffat was for some years draughtsman and designer for the Geo. McLagan Furniture Co., Stratford, Ont.

The Nor'west Farmers' Co-operative Lumber Company is the name of a new company which has been formed with capital of $100,000, to manufacture lumber and lumber products. The name indicates the field of their operations and the head office will be at Vancouver.

A new company has been incorporated with place of business at Toronto, to manufacture all kinds of finish, fixtures, furniture and fittings, for public and private buildings. The name will be the McLean-Simpson Company, Ltd., and they will operate on a capital of $50,000. Mr. J. W. McLean will be the manager.

The White Sewing Machine Company has acquired the plant and business of the Raymond Mfg. Company, of Guelph, Ont. The new company will double the capacity of the present plant and is already erecting a new factory to cost $250,000. About 760 men will be employed when the plants are in operation.

The Studebaker Corporation of Canada, Ltd., has been formed to take over the assets and liabilities of the former company bearing the same name, which was incorporated under an Ontario charter. The company will manufacture automobiles and motor boats at Walkerville, Ont., and they will operate on a capital of $400,000.

The Canadian branch of Samuel J. Shimer & Sons, manufacturers of the well-known Shimer Cutter Heads, has been taken over by a new company, the Shimer Cutter Head Company of Canada, Limited. Mr. Robert Pollock, who for many years represented Samuel J. Shimer & Sons on the road, is manager of the new company.

An interesting indication of the high price of wood goods in Great Britain is afforded by the fact that a Toronto importer of cotton goods from Manchester has received notice that in future he will be charged 11s. 6d. for each box. Formerly these boxes, which are about 3 ft. 6 in. square and made of a good grade of lumber, were furnished free by the exporters in Manchester.

A downward revision of freight rates on sash, doors, and other lumber products from the Pacific Coast to Eastern Canada and the Eastern States, is being sought from the Interstate Commerce Commission of the United States by manufacturers of British Columbia, California, Oregon and Washington. The petitioners complained that rates were advanced last October from 1 to 22 per cent. and seek to recover the excess already paid.

Owing to the inactivity in building operations, the Galbraith Company, of Owen Sound, who formerly conducted a planing mill business, have gone into the manufacture of wood novelties. They have taken over the small turnings business of Knechtel Bros., of Southampton, Ont., and are prepared to supply the trade with special order turnings. The company have designed their products along special lines such as toys to sell at 10c. and 15c. retail, wooden mechanical toys to sell at 20c. and 25c., and parlor games and kitchen ware of various kinds. Mr. G. C. Seibert is the manager of the company.

Bauman & Letson, of Elmira, who recently lost their planing mill by fire, are erecting a new concrete building, 70 feet by 90 feet, one storey high. The building will be fireproof as far as can be conveniently arranged. The boiler is in a separate building and will be used only for heating purposes. They have installed a 20 h.p. motor and have placed their orders for the necessary machinery. A joint stock company will be formed and the business carried on under the name of Elmira Planing Mills Company, Ltd. They will manufacture the same line of goods as formerly, with the exception that they will manufacture a complete line of silo material.

The Machinery Company of America

The Machinery Company of America has been organized with general offices at 505 Sixth Street Grand Rapids, Mich., to act as distributors of the saw and knife fitting machinery, tools and supplies of Baldwin, Tuttill & Bolton, of Grand Rapids, Mich.; the Covel Manufacturing Company, Benton Harbor, Mich.; and the Hanchett Swage Works, Big Rapids, Mich. The business will be under the direction and control of men who are fully conversant with trade conditions and requirements. The lines represented have been on the market for a number of years and have been sold extensively in the United States and other countries.

Wood Steaming Retort

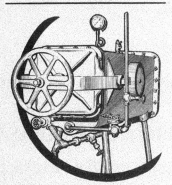

Wood Bending Manufacturers:

This is one of our

Perfection Retorts

which we guarantee will save you

50% Less Breakage

in your bending department than your present process; that your stock will dry in your forms or presses in one-third less time; that you will have no stained stock; that your stock will retain its shape much better after being bent; that it will dry in your dry-kiln in one-half less time and that your steam consumption will be reduced at least 90 per cent.

The door can be opened and closed in ten seconds, and it is steam and water tight and for this reason can be placed anywhere in your factory.

Compare this IMPROVED RETORT with your present steam boxes, then write us for our Booklet on Progressive Wood Steaming.

Made in Preston, Ontario

Perfection Wood Steaming Retort Co.

PARKERSBURG - WEST VIRGINIA

For the Economical Sanding of Round Work

My Sanders are very efficient

If you have any kind of round work to sand write me

J. M. NASH, MILWAUKEE, Wis.

Morehead

Back to Boiler

SYSTEM

They Saved 25% in Fuel and 50% in Repairs—

So writes Mr. William H. Turner, Secretary and Treasurer of the Easton Furniture Company, of Easton, Md., after giving the "Morehead" System a thorough trial.

□ **Rejuvenate Your Steam Plant**

by making the inexpensive "Morehead" System a part of it. Save fuel by bringing the pure condensation directly back to the boiler HOT!

Save repairs by eliminating forever the necessity for the waste-ful and constantly out-of-order steam pump.

The simple, easily installed "Morehead" System will much more than pay for itself the first year of use.

The Morehead Book shows actual photographs of interesting installations. Write for your copy now.

Canadian Morehead Manufacturing Company

Dept. H
WOODSTOCK, ONTARIO

GEO. C. BROWN & COMPANY

Manufacturers St. Francis Basin Hardwoods

Band Mills—PROCTOR, ARK.   General Offices—Bank of Commerce Bldg.—MEMPHIS, TENN.

STOCK FOR SALE

SAP GUM (Kraetzer Cured)
June   July
4-4 1s and 2s   51,965 105,740
4-4 No. 1 Common   120,000

SELECTED RED GUM
Plain Sawed—Kraetzer-Cured
4-4 No. 1 Common   41,405
5-4 No. 1 Common   40,800   44,020
6-4 No. 1 Common   59,100

SOFT ELM
10-4 No. 2 Common   39,570

SELECTED RED GUM
Quartered—Kraetzer-Cured
June   July
10-4 No. 1 Com. and Bet. 110,505
12-4 No. 1 Com. and Bet.   9,750

PLAIN RED OAK
4-4 No. 1 Common   78,000 140,556
4-4 No. 2 Common   19,500   39,030

PLAIN WHITE OAK
4-4 No. 1 Common   63,708   18,744
4-4 No. 2 Common   15,027   4,436

QUARTERED RED OAK
June   July
4-4 No. 1 Com., 6 in. and
   wider   93,592
4-4 No. 2 Com., 6 in. and
   wider   22,148
5-4 Nos. 1 and 2 Com. 6
   in. and wider   17,040

QUARTERED WHITE OAK
4-4 1s and 2s   63,785   24,500
4-4 No. 1 Common   33,767   21,280

TENNESEE AROMATIC RED CEDAR SPECIALISTS   OUR TRAFFIC DEPARTMENT TRACES EVERY SHIPMENT

30 Sets of Jambs in 28 Minutes

Here's the Machine that will do it!

ONE OPERATOR
SIXTY SETS
SIXTY MINUTES

BRICK HOUSE JAMBS

FRAME HOUSE JAMBS

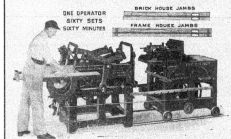

Every manufacturer wants more profit, more economy in production, more output per man. Unequalled results are to be obtained from the

"Efficiency"

Window Frame Machine

One operator cuts the pockets, routs the pulley mortises, rabbets head and sill gain simultaneously. Guaranteed to reduce the time to work pulley stiles 50 to 60 per cent. Try the "Efficiency" Machine and be convinced.

The F. H. Weeks Lumber Co. - Akron, Ohio

The Canadian Woodworker and Furniture Manufacturer covers a specialized field. It represents maximum value to its subscribers.

Subscription price $1.00 per year.

DARLINGS STEAM APPLIANCES

DARLING BROTHERS
LIMITED
Engineers and Manufacturers
MONTREAL, CANADA

Branches: Agents:
Toronto and Winnipeg   Halifax St. John Calgary Vancouver

GRAND RAPIDS
VAPOR DRY KILN
GRAND RAPIDS
MICHIGAN

129

Grand Rapids Vapor Kilns sold in the last

122

days of the year 1915.

54

of these were Repeat Orders.

Repeat orders represent satisfaction. We can guarantee experienced engineering ability and efficient service.

Over 1300 Grand Rapids Vapor Kilns in use.

Write us regarding better drying

Work your waste into small handles and other small turnings

The OBER LATHES will help you

We also make lathes for turning Axe, Pick Sledge and other irregular Handles, Whiffletrees Neckyokes, etc. Sanders, Rip Saws, Cut Off Saws, Chucking Machines, etc.

Write for free Catalogue No. 31

THE OBER MFG. CO.
CHAGRIN FALLS, O., U. S. A.

VENEER PRESSES

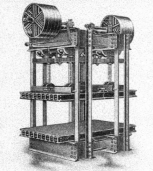

WRITE FOR BULLETIN C 1.

Canadian Boomer & Boschert Press
Company, Limited
18 Tansley Street, MONTREAL

Your Loose Pulley Trouble Disappears

when installed with

CHAPMAN BALL BEARING LOOSE PULLEYS

A little vaseline **once a year is all the attention** and lubrication required.

Our loose Pulleys cannot run hot and do not cut the shaft.

Write us for particulars and price list.

The

Chapman Double Ball Bearing Co.
of Canada, Limited

339-351 Sorauren Ave., Toronto, Canada

The "Canadian Woodworker" Buyers' Directory

ABRASIVES
The Carborundum Co., Niagara Falls, N.Y.

BALL BEARINGS
Chapman Double Ball Bearing Co., Toronto.

BALUSTER LATHES
Garlock-Machinery, Toronto, Ont.
Ober Mfg. Company, Chagrin Falls, Ohio.
Whitney & Son, Baxter D., Winchendon, Mass.

BAND SAW FILING MACHINERY
Berlin Machine Works, Ltd., Hamilton, Ont.
Canada Machinery Corporation, Galt, Ont.
Fay & Egan Co., J. A., Cincinnati, Ohio.
Garlock-Machinery, Toronto, Ont.
Preston Woodworking Machinery Company,
Preston, Ont.

BAND SAWS
Berlin Machine Works, Ltd., Hamilton, Ont.
Canada Machinery Corporation, Galt, Ont.
Garlock-Machinery, Toronto, Ont.
Jackson, Cochrane & Company, Berlin, Ont.
Petrie, H. W., Toronto.
Preston Woodworking Machinery Company,
Preston, Ont.

BAND SAW MACHINES
Berlin Machine Works, Hamilton, Ont.
Canada Machinery Corporation, Galt, Ont.
Fay & Egan Co., J. A., Cincinnati, Ohio.
Garlock-Machinery, Toronto, Ont.
Petrie, H. W., Toronto.
Preston Woodworking Machinery Company,
Preston, Ont.
Williams Machinery Co., A. R., Toronto, Ont.

BAND SAW STRETCHERS
Berlin Machine Works, Ltd., Hamilton, Ont.
Fay & Egan Co., J. A., Cincinnati, Ohio.

BASKET STITCHING MACHINE
Saranac Machine Co., Benton Harbor, Mich.

BENDING MACHINES
Fay & Egan Co., J. A., Cincinnati, Ohio.
Garlock-Machinery, Toronto, Ont.
Perfection Wood Steaming Retort Company,
Parkersburg, W. Va.

BELTING
Fay & Egan Co., J. A., Cincinnati, Ohio.

BLOWERS
Sheldons, Limited, Galt, Ont.
Toronto Blower Company, Toronto, Ont.

BLOW PIPING
Sheldons, Limited, Galt, Ont.

BORING MACHINES
Berlin Machine Works, Ltd., Hamilton, Ont.
Canada Machinery Corporation, Galt, Ont.
Fay & Egan Co., J. A., Cincinnati, Ohio.
Garlock-Machinery, Toronto, Ont.
Jackson, Cochrane & Company, Berlin, Ont.
Nash, J. M., Milwaukee, Wis.
Preston Woodworking Machinery Company,
Preston, Ont.
Reynolds Pattern & Machine Co., Moline, Ill.

BOX MAKERS' MACHINERY
Berlin Machine Works, Ltd., Hamilton, Ont.
Canada Machinery Corporation, Galt, Ont.
Fay & Egan Co., J. A., Cincinnati, Ohio.
Garlock-Machinery, Toronto, Ont.
Jackson, Cochrane & Co., Berlin, Ont.
Neilson & Company, J. L. Winnipeg, Man.
Preston Woodworking Machinery Company,
Preston, Ont.
Reynolds Pattern & Machine Co., Moline, Ill.
Whitney & Son, Baxter D., Winchendon, Mass.
Williams Machinery Co., A. R., Toronto, Ont.

CABINET PLANERS
Berlin Machine Works, Ltd., Hamilton, Ont.
Canada Machinery Corporation, Galt, Ont.
Garlock-Machinery, Toronto, Ont.
Jackson, Cochrane & Co., Berlin, Ont.
Preston Woodworking Machinery Company,
Preston, Ont.

CARS (Transfer)
Sheldons, Limited, Galt, Ont.

CARBORUNDUM PAPER
Carborundum Co., Niagara Falls, N.Y.

CARVING MACHINES
Canada Machinery Corporation, Galt, Ont.
Fay & Egan Co., J. A., Cincinnati, Ohio.
Garlock-Machinery, Toronto, Ont.
Jackson, Cochrane & Company, Berlin, Ont.

CASTERS
Foster, Merriam & Co., Meriden, Conn.

CLAMPS
Fay & Egan Co., J. A., Cincinnati, Ohio.
Garlock-Machinery, Toronto, Ont.
Jackson, Cochrane & Company, Berlin, Ont.
Preston Woodworking Machinery Company,
Preston, Ont.
Simonds Canada Saw Company, Montreal, P.Q.

COLUMN MACHINERY
Fay & Egan Co., J. A., Cincinnati, Ohio.

CUT-OFF SAWS
Berlin Machine Works, Ltd., Hamilton, Ont.
Canada Machinery Corporation, Galt, Ont.
Fay & Egan Co., J. A., Cincinnati, Ohio.
Garlock-Machinery, Toronto, Ont.
Jackson, Cochrane & Co., Berlin, Ont.
Ober Mfg. Co., Chagrin Falls, Ohio.
Preston Woodworking Machinery Company,
Preston, Ont.
Simonds Canada Saw Co., Montreal, Que.

CUTTER HEADS
Berlin Machine Works, Hamilton, Ont.
Canada Machinery Corporation, Galt, Ont.
Fay & Egan Co., J. A., Cincinnati, Ohio.
Jackson, Cochrane & Company, Berlin, Ont.
Mattison Machine Works, Beloit, Wis.

DOVETAILING MACHINES
Canada Machinery Corporation, Galt, Ont.
Fay & Egan Co., J. A., Cincinnati, Ohio.
Garlock-Machinery, Toronto, Ont.
Jackson, Cochrane & Company, Berlin, Ont.

DOWEL MACHINES
Canada Machinery Corporation, Galt, Ont.
Fay & Egan Co., J. A., Cincinnati, Ohio.
Ober Mfg. Co., Chagrin Falls, Ohio.

DRY KILNS
National Dry Kiln Co., Indianapolis, Ind.
Sheldons, Limited, Galt, Ont.

DUST COLLECTORS
Sheldons, Limited, Galt, Ont.

DUST SEPARATORS
Sheldons, Limited, Galt, Ont.

EDGERS (Single Saw)
Canada Machinery Corporation, Galt, Ont.
Fay & Egan Co., J. A., Cincinnati, Ohio.
Garlock-Machinery, Toronto, Ont.
Simonds Canada Saw Co., Montreal, Que.

EDGERS (Gang)
Berlin Machine Works, Ltd., Hamilton, Ont.
Canada Machinery Corporation, Galt, Ont.
Fay & Egan Co., J. A., Cincinnati, Ohio.
Garlock-Machinery, Toronto, Ont.
Petrie, H. W., Toronto.
Simonds Canada Saw Co., Montreal, P.Q.
Williams Machinery Co., A. R., Toronto, Ont.

END MATCHING MACHINE
Berlin Machine Works, Ltd., Hamilton, Ont.
Canada Machinery Corporation, Galt, Ont.
Fay & Egan Co., J. A., Cincinnati, Ohio.
Garlock-Machinery, Toronto, Ont.
Jackson, Cochrane & Company, Berlin, Ont.

EXHAUST FANS
Garlock-Machinery, Toronto, Ont.
Sheldons Limited, Galt, Ont.

FILES
Simonds Canada Saw Co., Montreal, Que.

FLOORING MACHINES
Berlin Machine Works, Ltd., Hamilton, Ont.
Canada Machinery Corporation, Galt, Ont.
Fay & Egan Co., J. A., Cincinnati, Ohio.
Garlock-Machinery, Toronto, Ont.
Petrie, H. W., Toronto.
Preston Woodworking Machinery Company,
Preston, Ont.
Whitney & Son, Baxter D., Winchendon, Mass.

FLUTING HEADS
Fay & Egan Co., J. A., Cincinnati, Ohio.

FLUTING AND TWIST MACHINE
Preston Woodworking Machinery Company,
Preston, Ont.

FURNITURE TRIMMINGS
Foster, Merriam & Co., Meriden, Conn.

GARNET PAPER AND CLOTH
The Carborundum Co., Niagara Falls, N.Y.

GRAINING MACHINES
Berlin Machine Works, Ltd., Hamilton, Ont.
Canada Machinery Corporation, Galt, Ont.
Fay & Egan Co., J. A., Cincinnati, Ohio.
Williams Machinery Co., A. R., Toronto, Ont.

GLUE
Coignet Chemical Product Co., New York,
N.Y.
Perkins Glue Company, South Bend, Ind.

GLUE CLAMPS
Jackson, Cochrane & Company, Berlin, Ont.

GLUE HEATERS
Fay & Egan Co., J. A., Cincinnati, Ohio.
Jackson, Cochrane & Company, Berlin, Ont.

GLUE JOINTERS
Berlin Machine Works, Ltd., Hamilton, Ont.
Canada Machinery Corporation, Galt, Ont.
Jackson, Cochrane & Company, Berlin, Ont.

GLUE SPREADERS
Fay & Egan Co., J. A., Cincinnati. Ohio.
Jackson, Cochrane & Company, Berlin, Ont.

GLUE ROOM EQUIPMENT
Boomer & Boschert Company, Montreal, Que.
Perrin & Company, W. R., Toronto, Ont.

GRINDERS (Cutter)
Fay & Egan Co., J. A., Cincinnati. Ohio.

GRINDERS, HAND AND FOOT POWER
The Carborundum Co., Niagara Falls, N.Y.

GRINDERS (Knife)
Berlin Machine Works, Ltd., Hamilton, Ont.
Canada Machinery Corporation, Galt, Ont.
Fay & Egan Co., J. A., Cincinnati, Ohio.
Garlock-Machinery, Toronto, Ont.
Petrie, H. W., Toronto.
Preston Woodworking Machinery Company,
Preston, Ont.

GRINDERS (Tool)
Fay & Egan Co., J. A., Cincinnati, Ohio.
Jackson, Cochrane & Company, Berlin, Ont.

GRINDING WHEELS
The Carborundum Co., Niagara Falls, N.Y.

GROOVING HEADS
Fay & Egan Co., J. A., Cincinnati, Ohio.

GUMMERS, ETC.
Fay & Egan Co., J. A., Cincinnati, Ohio.

GUARDS (Saw)
HAND PROTECTORS
Fay & Egan Co., J. A., Cincinnati, Ohio.

HAND SCREWS
Fay & Egan Co., J. A., Cincinnati, Ohio.

HANDLE THROATING MACHINE
Klotz Machine Company, Sandusky, Ohio.

HANDLE TRIMMING MACHINE
Klotz Machine Company, Sandusky, Ohio.

HANDLE & SPOKE MACHINERY
Fay & Egan Co., J. A., Cincinnati, Ohio.
Klotz Machine Co., Sandusky, Ohio.
Nash, J. M., Milwaukee, Wis.
Ober Mfg. Company, Chagrin Falls, Ohio.
Whitney & Son, Baxter D., Winchendon, Mass.

HARDWOOD LUMBER
American Hardwood Lumber Co., St. Louis, Mo.
Atlantic Lumber Company, Toronto.
Brown & Company, Geo. C., Memphis, Tenn.
Clark & Son, Edward, Toronto, Ont.
Kersley, Geo., Montreal, Que.
Kraetzer-Cured Lumber Co., Moorhead, Miss.
Mowbray & Robinson, Cincinnati, Ohio.
Penrod Walnut & Veneer Co., Kansas City, Mo.
Spencer, C. A., Montreal, Que.

HUB MACHINERY
Fay & Egan Co., J. A., Cincinnati, Ohio.

HYDRAULIC PRESSES
Can. Boomer & Boschert Press, Montreal, Que.

HYDRAULIC PUMPS & ACCUMULATORS
Can. Boomer & Boschert Press, Montreal, Que.

HYDRAULIC VENEER PRESSES
Perrin & Company, Wm. R., Toronto, Ont.

JOINTERS
Berlin Machine Works, Ltd., Hamilton, Ont.
Canada Machinery Corporation, Galt, Ont.
Fay & Egan Co., J. A., Cincinnati, Ohio.
Garlock-Machinery, Toronto, Ont.
Jackson, Cochrane & Company, Berlin, Ont.
Knechtel Bros., Limited, Southampton, Ont.
Petrie, H. W., Toronto, Ont.
Preston Woodworking Machinery Company,
Preston, Ont.

KNIVES (Planer and others)
Berlin Machine Works, Ltd., Hamilton, Ont.
Canada Machinery Corporation, Galt, Ont.
Fay & Egan Co., J. A., Cincinnati, Ohio.
Simonds Canada Saw Co., Montreal, P.Q.

LUMBER
American Hardwood Lumber Co., St. Louis, Mo.
Brown & Company, Geo. C., Memphis, Tenn.
Churchill, Milton Lumber Co., Louisville, Ky.
Clark & Son, Edward, Toronto, Ont.
Elgie & Jarvis Lumber Co., Toronto, Ont.
Hartzell, Geo. W., Piqua, Ohio.

LATHES
Canada Machinery Corporation, Galt, Ont.
Fay & Egan Co., J. A., Cincinnati, Ohio.
Garlock-Machinery, Toronto, Ont.
Jackson, Cochrane & Company, Berlin, Ont.
Klotz Machine Company, Sandusky, Ohio.
Ober Mfg. Company, Chagrin Falls, Ohio.
Petrie, H. W., Toronto, Ont.
Preston Woodworking Machinery Company,
Shawver Company, Springfield, O.
Whitney & Son, Baxter D., Winchendon, Mass.

MACHINE KNIVES
Berlin Machine Works, Ltd., Hamilton, Ont.
Canada Machinery Corporation, Galt, Ont.
Peter Hay Knife Company, Galt, Ont.
Simonds Canada Saw Co., Montreal, P.Q.

MITRE MACHINES
Canada Machinery Corporation, Galt, Ont.
Fay & Egan Co., J. A., Cincinnati, Ohio.

MITRE SAWS
Canada Machinery Corporation, Galt, Ont.
Fay & Egan Co., J. A., Cincinnati, Ohio.
Garlock-Machinery, Toronto, Ont.
Simonds Canada Saw Co., Montreal, P.Q.

MORTISING MACHINES
Berlin Machine Works, Ltd., Hamilton, Ont.
Canada Machinery Corporation, Galt, Ont.
Fay & Egan Co., J. A., Cincinnati, Ohio.
Garlock-Machinery, Toronto, Ont.

MOULDINGS
Knechtel Bros., Limited, Southampton, Ont.

MULTIPLE BOXING MACHINES
Fay & Egan Co., J. A., Cincinnati, Ohio.
Nash, J. M., Milwaukee, Wis.
Reynolds Pattern & Machine Co., Moline, Ill.

PATENT SOLICITORS
Lightfoot, Stanley, Toronto, Ont.

PATTERN SHOP MACHINES
Berlin Machine Works, Ltd., Hamilton, Ont.
Canada Machinery Corporation, Galt, Ont.
Fay & Egan Co., J. A., Cincinnati, Ohio.
Garlock-Machinery, Toronto, Ont.
Jackson, Cochrane & Company, Berlin, Ont.
Preston Woodworking Machinery Company,
Preston, Ont.
Whitney & Son, Baxter D., Winchendon, Mass.

PLANERS
Berlin Machine Works, Ltd., Hamilton, Ont.
Canada Machinery Corporation, Galt, Ont.
Fay & Egan Co., J. A., Cincinnati, Ohio.
Garlock-Machinery, Toronto, Ont.
Jackson, Cochrane & Company, Berlin, Ont.
Petrie, H. W., Toronto, Ont.
Preston Woodworking Machinery Company,
Whitney & Son, Baxter D., Winchendon, Mass.
Williams Machinery Co., A. R., Toronto, Ont.

PLANING MILL MACHINERY
Berlin Machine Works, Ltd., Hamilton, Ont.
Canada Machinery Corporation, Galt, Ont.
Fay & Egan Co., J. A., Cincinnati, Ohio.
Garlock-Machinery, Toronto, Ont.
Jackson, Cochrane & Company, Berlin, Ont.
Petrie, H. W., Toronto, Ont.
Preston Woodworking Machinery Company,
Whitney & Son, Baxter D., Winchendon, Mass.
Williams Machinery Co., A. R., Toronto, Ont.

PRESSES (Veneer)
Canada Machinery Corporation, Galt, Ont.
Jackson, Cochrane & Company, Berlin, Ont.
Perrin & Co., Wm. R., Toronto, Ont.
Preston Woodworking Machinery Company,
Preston, Ont.

PULLEYS
Fay & Egan Co., J. A., Cincinnati, Ohio.

RESAWS
Berlin Machine Works, Ltd., Hamilton, Ont.
Canada Machinery Corporation, Galt, Ont.
Fay & Egan Co., J. A., Cincinnati, Ohio.
Garlock-Machinery, Toronto, Ont.
Jackson, Cochrane & Company, Berlin, Ont.
Petrie, H. W., Toronto, Ont.
Preston Woodworking Machinery Company,
Preston, Ont.
Simonds Canada Saw Co., Montreal, P.Q.
Williams Machinery Co., A. R., Toronto, Ont.

RIM AND FELLOE MACHINERY
Fay & Egan Co., J. A., Cincinnati, Ohio.

RIP SAWING MACHINES
Berlin Machine Works, Ltd., Hamilton, Ont.
Canada Machinery Corporation, Galt, Ont.
Fay & Egan Co., J. A., Cincinnati, Ohio.
Garlock-Machinery, Toronto, Ont.
Klotz Machine Company, Sandusky, Ohio.
Jackson, Cochrane & Company, Berlin, Ont.
Ober Mfg. Company, Chagrin Falls, Ohio.
Preston Woodworking Machinery Company,
Preston, Ont.

SANDERS
Berlin Machine Works, Ltd., Hamilton, Ont.
Canada Machinery Corporation, Galt, Ont.
Fay & Egan Co. J. A., Cincinnati, Ohio.
Garlock-Machinery, Toronto, Ont.
Klotz Machine Company, Sandusky, Ohio.
Jackson, Cochrane & Company, Berlin, Ont.
Nash, J. M. Milwaukee, Wis.
Ober Mfg. Company, Chagrin Falls, Ohio.
Preston Woodworking Machinery Company,
Preston, Ont.

SASH, DOOR & BLIND MACHINERY
Canada Machinery Corporation, Galt, Ont.
Berlin Machine Works, Ltd., Hamilton, Ont.
Fay & Egan Co., J. A., Cincinnati, Ohio.
Garlock-Machinery, Toronto, Ont.
Jackson, Cochrane & Company, Berlin, Ont
Preston Woodworking Machinery Company,
Preston, Ont

SAWS
Fay & Egan Co. J. A., Cincinnati, Ohio
Radcliff Saw Mfg. Company, Toronto, Ont.
Simonds Canada Saw Co., Montreal, P.Q.

SAW GUMMERS, ALOXITE
The Carborundum Co., Niagara Falls, N.Y.

SAW SWAGES
Fay & Egan Co. J. A., Cincinnati, Ohio.
Simonds Canada Saw Co., Montreal, Que

SAW TABLES
Canada Machinery Corporation, Galt, Ont.
Berlin Machine Works, Ltd., Hamilton, Ont.
Fay & Egan Co. J. A., Cincinnati, Ohio.
Garlock-Machinery, Toronto, Ont.
Jackson, Cochrane & Company, Berlin, Ont
Preston Woodworking Machinery Company,
Preston, Ont.

SCRAPING MACHINES
Canada Machinery Corporation, Galt, Ont.
Garlock-Machinery, Toronto, Ont.
Whitney & Sons, Baxter D., Winchendon, Mass

SCROLL CHUCK MACHINES
Klotz Machine Company, Sandusky, Ohio

SCREW DRIVING MACHINES
Canada Machinery Corporation, Galt, Ont.
Reynolds Pattern & Machine Co., Moline, Ill

SCREW DRIVING MACHINERY
Canada Machinery Corporation, Galt, Ont.
Reynolds Pattern & Machine Works, Moline, Ill

SCROLL SAWS
Canada Machinery Corporation, Galt, Ont
Berlin Machine Works, Ltd., Hamilton, Ont.
Fay & Egan Co. J. A., Cincinnati, Ohio.
Garlock-Machinery, Toronto, Ont.
Jackson, Cochrane & Company, Berlin, Ont.
Simonds Canada Saw Co., Montreal, P.Q.

SECOND-HAND MACHINERY
Petrie, H. W.
Williams Machinery Co., A. R., Toronto, Ont

SHAPERS
Canada Machinery Corporation, Galt, Ont.
Berlin Machine Works, Ltd., Hamilton, Ont
Fay & Egan Co., J. A., Cincinnati, Ohio.
Garlock-Machinery, Toronto, Ont.
Jackson, Cochrane & Company, Berlin, Ont.
Ober Mfg. Company, Chagrin Falls, Ohio
Petrie, H. W., Toronto, Ont.
Preston Woodworking Machinery Company,
Preston, Ont.
Simonds Canada Saw Co., Montreal, P.Q.
Whitney & Sons, Baxter D., Winchendon, Mass

SHARPENING TOOLS
The Carborundum Co., Niagara Falls, N.Y

SHAVING COLLECTORS
Sheldons Limited, Galt, Ont.

SHELL BOX HANDLES
Vokes Hardware Co., Toronto, Ont.

STEAM APPLIANCES
Darling Bros., Montreal.

SHELL BOX EQUIPMENT
Yokes Hardware Co., Toronto, Ont.

SINGLE SPINDLE BOXING MACHINES
Canada Machinery Corporation, Galt, Ont.
Fay & Egan Co., J. A., Cincinnati, Ohio.
Ober Mfg. Company, Chagrin Falls, Ohio.

STAVE SAWING MACHINE
Whitney & Sons, Baxter D., Winchendon, Mass

STEAM TRAPS
Canadian Morehead Mfg. Co., Woodstock, Ont.
Standard Dry Kiln Co., Indianapolis, Ind.

STEEL SHELL BOX BANDS
Yokes Hardware Co., Toronto, Ont.

SURFACERS
Berlin Machine Works, Ltd., Hamilton, Ont.
Canada Machinery Corporation, Galt, Ont.
Fay & Egan Co., J. A., Cincinnati, Ohio.
Jackson, Cochrane & Company, Berlin, Ont.
Pettie. H. W., Toronto, Ont.
Preston Woodworking Machinery Company,
Preston, Ont.
Whitney & Sons. Baxter D., Winchendon, Mass.
Williams Machinery Co., A. R., Toronto, Ont.

SWING SAWS
Berlin Machine Works, Ltd., Hamilton, Ont.
Canada Machinery Corporation, Galt, Ont.
Fay & Egan Co., J. A., Cincinnati, Ohio.
Garlock-Machinery, Toronto, Ont.
Jackson, Cochrane & Company, Berlin, Ont.
Ober Mfg. Co., Chagrin Falls, Ohio.
Preston Woodworking Machinery Company,
Preston, Ont.
Simonds Canada Saw Co., Montreal, P.Q.

TABLE LEG LATHES
Berlin Machine Works, Ltd. Hamilton, Ont.
Fay & Egan Co., J. A., Cincinnati, Ohio.
Ober Mfg. Co., Chagrin Falls, Ohio.
Whitney & Sons, Baxter D., Winchendon, Mass.

TABLE SLIDES
Walter & Co., B., Wabash, Ind.

TENONING MACHINES
Berlin Machine Works, Ltd., Hamilton, Ont.
Canada Machinery Corporation, Galt, Ont.
Fay & Egan Co., J. A., Cincinnati, Ohio.
Garlock-Machinery, Toronto, Ont.
Jackson, Cochrane & Company, Berlin, Ont.
Preston Woodworking Machinery Company,
Preston, Ont.

TRIMMERS
Berlin Machine Works, Ltd., Hamilton, Ont.
Canada Machinery Corporation, Galt, Ont.
Fay & Egan Co., J. A., Cincinnati, Ohio.
Garlock-Machinery, Toronto, Ont.
Preston Woodworking Machinery Company,
Preston, Ont.

TRUCKS
Sheldons Limited, Galt, Ont.
National Dry Kiln Co., Indianapolis, Ind.

TURNING MACHINES
Berlin Machine Works, Ltd., Hamilton, Ont.
Fay & Egan Co., J. A., Cincinnati, Ohio.
Garlock-Machinery, Toronto, Ont.
Klotz Machine Company, Sandusky, Ohio.
Ober Mfg. Company, Chagrin Falls, Ohio.
Whitney & Sons, Baxter D., Winchendon, Mass.

UNDER-CUT SELF-FEEDING FACE PLANER
Fay & Egan Co., J. A., Cincinnati, Ohio.
Garlock-Machinery, Toronto, Ont.
Jackson, Cochrane & Company, Berlin, Ont.

VARNISHES
Ault & Wiborg Company, Toronto, Ont.

VENEERS
Central Veneer Co., Huntington, W. Virginia.
Walter Clark Veneer Co., Grand Rapids, Mich.
Evansville Veneer Co., Evansville, Ind.
Hartzell, Geo. W., Piqua, Ohio.
Haughton Veneer Company, Chicago, Ill.
Hay & Company, Woodstock, Ont.
Hoffman Bros. Company, Fort Wayne, Ind.
Geo. Kersley, Montreal, Que.
Nartsik, J. J., Chicago, Ill.
Ohio Veneer Company, Cincinnati, Ohio.
Penrod Walnut & Veneer Co., Kansas City, Mo.
Roberts, John N., New Albany, Ind.
Wood Mosaic Co., New Albany, Ind.

VENTILATING APPARATUS
Sheldons Limited, Galt, Ont.

VENEERED PANELS
Hay & Company, Woodstock, Ont.

VENEER PRESSES (Hand and Power)
Canada Machinery Corporation, Galt, Ont.
Can. Boomer & Boschert Press, Montreal, Que.
Garlock-Machinery, Toronto, Ont.
Jackson, Cochrane & Company, Berlin, Ont.
Perrin & Company, Wm. R., Toronto, Ont.

VISES
Fay & Egan Co., J. A., Cincinnati, Ohio.
Simonds Canada Saw Company, Montreal, Que.

WAGON AND CARRIAGE MACHINERY
Berlin Machine Works, Ltd., Hamilton, Ont.
Canada Machinery Corporation, Galt, Ont.
Fay & Egan Co., J. A., Cincinnati, Ohio.
Ober Mfg. Company, Chagrin Falls, Ohio.
Whitney & Sons, Baxter D., Winchendon, Mass.

WOOD FINISHES
Ault & Wiborg, Toronto, Ont.

WOOD TURNING MACHINERY
Berlin Machine Works, Ltd., Beloit, Wis.
Canada Machinery Corporation, Galt, Ont.
Garlock-Machinery, Toronto, Ont.

WORK BENCHES
Fay & Egan Co., J. A., Cincinnati, Ohio.

WOODWORKING MACHINES
Canada Machinery Corporation, Galt, Ont.
Garlock-Machinery, Toronto, Ont.
Jackson, Cochrane & Co., Berlin, Ont.
Preston Woodworking Machinery Company.
Preston, Ont.
Reynolds Pattern & Machine Works, Moline,
Ill.
Williams Machinery Co., A. R., Toronto, Ont.

WHEELS, Grinding, Carborundum and Aloxite
The Carborundum Co., Niagara Falls, N.Y.

THE COMPLETE LINE

Long experience and absolute perfection in the many details of design and workmanship is the result of eighty years of CASTER manufacturing.

The adoption of ACME CASTERS insures the stability of your furniture.

We solicit your inquiries and will be pleased to have our Canadian representative call and discuss your CASTER proposition.

We also make a complete line of PERIOD TRIMMINGS.

Foster, Merriam & Co.
Meriden, Conn.

Canadian Representative:—F. A. Schmidt, Berlin, Ont.

" ACME "
The
RELIABLE CASTER

Making TABLE-SLIDES is a Specialty Business

For more than TWENTY-FIVE YEARS we have made TABLE SLIDES exclusively. Our Factory is equipped with Special Machinery which enables us to make SLIDES,—BETTER and CHEAPER than the furniture manufacturer.

Canadian Table makers are rapidly adopting WABASH SLIDES

Because { They ELIMINATE SLIDE TROUBLES
Are CHEAPER and BETTER }

BY USING

Reduced Costs

**WABASH
SLIDES**

Increased Out-put

MADE BY

B. Walter & Company
Wabash, Ind.

The Largest EXCLUSIVE TABLE-SLIDE Manufacturers in America

ESTABLISHED-1887

Just What You Need

There's no necessity now for you to **wish** for a corrugated joint fastener machine which will rise fully to your expectations; you can **get** one.

The Saranac Corrugated Joint Fastener Machine

is the absolute realization of your wish. This illustration gives you an idea of what it looks like, but you can have no conception of its **efficiency, economy, speed** and **reliability** until you get it working for you.

It is substantially constructed, and the workmanship and materials are of the very best. Particularly suitable for manufacturers of sash, doors, blinds, screens, porch columns, boxes, etc., etc.

Made in various types for all conditions.

Write for further particulars and prices

Saranac Machine Co., SOLE MAKERS
Benton Harbor, Mich.

WITH spring many new homes will be built, new furniture will be required. This means an increase in your sanding department. Are you prepared to meet the demand?

CARBORUNDUM BRAND GARNET PAPER AND CLOTH

Produce a better, quicker and finer finish. They are made from selected garnet—free from all impurities. Uniformly coated on the best paper and cloth stock.

Working Samples Upon Request

The Carborundum Company, Niagara Falls, N. Y.

MATTISON
NEW MODEL
Turning and Shaping Lathe

Cost of Period Turnings Reduced to Bed Rock
This is the Machine That Does the Trick

Period styles can't be made on a true manufacturing basis unless a standard machine method is used for producing the turned work.

If your outlay for labor on the turnings—round, square or octagon—is excessive, the manufacturing cost for your period furniture as a whole will be unusually high.

Slow and wasteful hand methods are inadequate. They must be replaced by equipment that provides the needed capacity at lowest cost.

There's the reason for the Mattison New Model Turning and Shaping Lathe—*it cuts deep into the cost of turned work.*

For the delicate and complicated patterns of to-day, varying greatly in design, and in a wide range of sizes, this Mattison Lathe is indispensable. And its adaptability makes it equally essential for future turnings.

Let us help you investigate the equipment best suited for low-cost production of your particular turned work.

*There is Only One Mattison Lathe
and it is built only by*

C. MATTISON MACHINE WORKS
883 Fifth Street BELOIT, Wisconsin, U.S.A.

Volume Sixteen   TORONTO, JUNE, 1916   Number Six

CANADIAN WOODWORKER
and Furniture Manufacturer

Cutting — Sawing — Planing

The Planing Counts Most

A planer that does its duty often swings the business up to success—buying a planer merits your deep study.

Fully and frankly, we tell you about

Whitney Planers
Single and Double Surfacers

We detail the reasons why the installation of a Whitney will increase the volume of your planing output and better the quality—why the upkeep and operation costs are carried down to the lowest possible notch—how Whitney Planers stand up sturdily to the hardest kind of day-by-day service—we also suggest the most efficient equipment for your work of square cylinders and two or four ordinary knives or round cylinders and four thin high-speed steel knives and jointing, grinding and setting devices.

Just write us a line

Notice how Whitney Planers are found in wood-working shops that are considered representative.

No. 4 Whitney Single Surface Planer

BAXTER D. WHITNEY & SON, Winchendon, Mass.

California Office:
Berkeley,
California

Canadian Representatives:
H. W. Petrie, Limited
Toronto, Ontario

Put Good Belting

on your

Good Machines

Your machines are expensive, and naturally you expect them to turn out the quality and quantity of work commensurate with their cost.

Then why harness them with belting which cannot begin to do them justice?

Use

"AMPHIBIA"
Planer Belting

and get the most work from your machines, in the quickest time at the lowest cost per day of service.

Try a sample run of AMPHIBIA Planer
and prove its merits.

"Leather like gold has no substitute."

Sadler & Haworth

Established 1876

Tanners and Manufacturers

For 40 years Tanners and Manufacturers of the Best Leather Belts.

Toronto	St. John, N.B.	MONTREAL	Winnipeg	Vancouver
38 Wellington St. E.	139 Prince William St.	511 William St.	Galt Building	217 Columbia Ave.

Mr. Box Manufacturer:

Are you preparing to build the so-called "Bethlehem" 18 pr. Box?

If so, let us tell you about our proposition of boring all Eighteen holes in the cross bridges of this box at one operation. Our machine will pay for itself and save you money during the life of a single contract.

We can also furnish you such other machinery as you require for this particular box or any of the other types being called for by the Imperial Munitions Board.

Tell us your needs in the way of Machinery or Equipment. We can offer you something for your every requirement.

Garlock-Walker Machinery, Limited
32 Front Street West - - Toronto
Telephone Main 5346

Canadian Representatives

American Wood Working Machinery Co.
ROCHESTER, N.Y.

M. L. Andrews & Co.
CINCINNATI, OHIO

Shell Box Machinery

Special Two Spindle Boring Machine

Designed for boring the large holes in the internal fittings of the 18-lb. 4-round Ammunition Box.

Machine is heavily constructed, having large steel spindles and a heavy steel gear.

With the use of this machine you can turn out a large number of blocks a day. We have over 50 of these Borers in operation on Ammunition Boxes.

Note the special clamp for holding the blocks, also the positive rack feed and the stops for gauging depth of cuts.

Increase Your Output of Shell Boxes

Particulars and prices gladly furnished for this and other shell box machines.

We also manufacture a complete line of

Boring and Dovetailing Machies
for Shell Box Work

Jackson, Cochrane & Company
BERLIN   -   ONTARIO

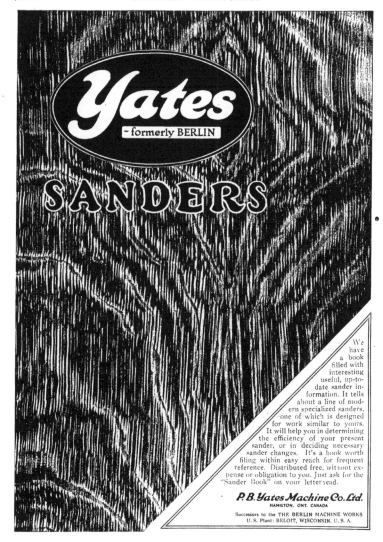

Yates
— formerly BERLIN
SANDERS

We have a book filled with interesting useful, up-to-date sander information. It tells about a line of modern specialized sanders, one of which is designed for work similar to yours. It will help you in determining the efficiency of your present sander, or in deciding necessary sander changes. It's a book worth filing within easy reach for frequent reference. Distributed free, without expense or obligation to you. Just ask for the "Sander Book" on your letterhead.

P. B. Yates Machine Co. Ltd.
HAMILTON, ONT. CANADA

Successors to the THE BERLIN MACHINE WORKS
U. S. Plant: BELOIT, WISCONSIN, U. S. A.

The "Beaver" Circular Planer Saw

This style of saw is specially adaptable to manufacturers of Shell Boxes and Toys. It is hollow ground, so as to run without set and will cut either with or across the grain. The work is done so smoothly that there is no need of sanding.

Size	Gauge at Collar Line	Gauge at Teeth and Hole
4 inches.	21	18
5 "	20	17
6 "	20	17
7 "	19	16
8 "	19	16
9 "	18	15
10 "	18	15
11 "	17	14
12 "	17	14
14 "	16	13
16 "	16	13
18 "	15	12
20 "	15	12
22 "	14	11
24 "	14	11

Choose your sizes and write for prices.

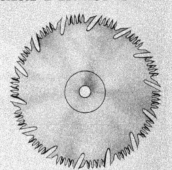

Radcliff Saw Manufacturing Co., Limited

TORONTO   550 Dundas Street   ONTARIO

Agents for L. & I. J. White Company Machine Knives

We can save money for you on your
BLOWER SYSTEM

The "FOSTER" BLOWER

The

Foster Blower

is specially designed for handling shavings, sawdust, excelsior, or any conceivable material. It is equipped with three bearings, shaft extending through, with bearing on inlet side. This eliminates the over-hung wheel and avoids heating, clogging and shaking. It is the only non-centre suction fan in existence. The Foster Fan is so constructed that the material entering the fan is immediately discharged without passing through or around the wheel.

Our Automatic Furnace Feeders will feed your material under boilers or into any desirable receptacle.

Consult us on your exhaust system troubles.

THE TORONTO BLOWER COMPANY
TORONTO   156 DUKE ST.   ONTARIO

" Treat your machine as a living friend."

Do You need Men in Your Veneer Room?

If so

Buy a "Black Bros." Patented Veneer Taping Machine

and

We guarantee that it will save 50% of your present labor cost

That's strong talk, but we stand back of every word of it, and can refer you to Furniture Manufacturers in Ontario today for proof of what we say.

If you have eight men now you could do with four if you had this **Machine,** and that should be a particularly appealing argument in view of the very great scarcity of labor existing at the present time.

The Machine will interest you if you do any Veneered work. It will tape together all kinds of Veneer, no matter whether the Veneer be thick or thin, handling both kinds equally well.

It will use the ordinary gummed tape or it will use plain paper tape and put on its own gum or glue just before it is laid.

The Machine is built in two sizes—24 in. and 36 in.

Write us for prices and particulars.

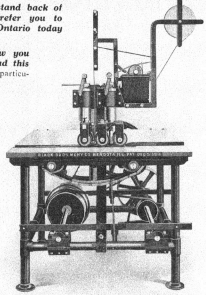

We control the Canadian Patent on this Machine

The Preston Woodworking Machinery Co.

Preston, Ont.   Limited

Peter Hay Machine Knives

Peter Hay knives are made of special quality steel tempered and toughened to take and hold a keen cutting edge. Hay knives are made for all purposes.

Have you our Catalogue?

Made in Canada

The Peter Hay Knife Co., Limited
GALT, ONTARIO

PRESSES

For Veneer and Veneer Drying

Made in Canada

William R. Perrin
Limited
Toronto

A New Drying Principle That Has "Made Good"

Only a little more than a year ago the National Dry Kiln Company announced a **Vertical** Piping System for the drying of hardwoods. **Today** there are National vertically-piped hardwood kilns in actual use in practically every section of the United States. Yes—and following the success of this new System in the States, it has been adopted as the most efficient and economical dry kiln equipment by leading manufacturers in Canada—among them the

Electric Planing Mills
Welland, Ontario

Isn't a system that has, because of its unusual merit, been so generally adopted in so short a time, at least worth investigating by every alert, progressive manufacturer? It surely is! Then why not write for our catalog, giving complete information about **Vertical** piping today? No obligation.

The National Dry Kiln Co.
1117 Maryland Street, INDIANAPOLIS

Do You Want This?

1/3 RD ORIGINAL COST

(1) An outfit of E. B. Hayes Machine Company's Dowel Door Machinery.
Combined Stile and Rail Borer.
Dowel Sticker, Cut-off and Pointer.
Dowel Gluer and Driver. Capacity 125 doors.

(2) An 8-ft. Linderman Dovetail Jointer. Used one year.

(3) A Mattison Block and Spindle Machine.

(4) A McGregor-Gourlay 12-in. Four-Side Moulder.

Write NOW.

J. L. NEILSON & CO.
Winnipeg, Man.

BOOKS FOR SALE

The following books are offered at special prices subject to previous sale:

Carpentry and Joinery, by Gilbert Townsend. Published in 1913 by the American School of Correspondence. 268 pages, illustrated. Price $1.90.

Saw Fitting Manual, a treatise on the care of saws and knives. Deals with everything in the saw and knife alphabet, from adjustments to widths. 144 pages. Price $2.00.

Common-sense Handrailing, by Fred T. Hodgson. Published by Frederick J. Drake & Company, Chicago. 114 pages, illustrated. Price 50c.

Roof Framing Made Easy, by Owen B. Maginnis. Published by The Industrial Publication Company, New York. 104 pages, illustrated. Price 50c.

Handrailing Simplified, by An Experienced Architect. Published by William T. Comstock, New York. 82 pages, illustrated. Price 50c.

How to Join Mouldings; or, The Arts of Mitering and Coping, by Owen B. Maginnis. Published by William T. Comstock, New York. 72 pages, illustrated. Price 50c.

Popular Mechanics Shop Notes. Published by Popular Mechanics, Chicago. Easy Ways to do Hard Things, etc. Years 1905-1906. Price 50c. each.

Cabinet Making, by J. H. Rudd. Published by Grand Rapids Furniture Record Company. 210 pages, illustrated. Price $1.50.

Utilization of Wood-Waste (Second Revised Edition), by Ernst Hubbard. Published in 1915 by Scott, Greenwood & Sons. 192 pages, illustrated. Price $1.50.

Woodworker Publishing Company, Limited
345 Adelaide Street West, Toronto, Ontario

The Only Way to Drive Screws

There is a stage-coach method and a modern method of doing everything. When it comes to driving screws the only modern way is with an Automatic Screw Driving Machine, which means greater and better production and greater profit.

The Reynolds
Automatic Screw-Driving Machine

is guaranteed to produce results. It is no experiment. The No. 2 Standard Type, shown herewith, is particularly suited for placing the screws in the straps of shrapnel boxes. It may be fitted to drive, without change or adjustment, any three consecutive sizes flat or round head, or two sizes filister head, No. 6 to 14 inclusive, wood or machine screws, or three sizes flat or round head, or two sizes filister head machine screws up to No. 20. All the above up to 1½ inches long if smaller than No. 10, or up to 1¾ inches long, No. 10 or larger. Fitted with boring attachment if desired.

Many Machines in Use in Canada.

Ask for our Catalogue.

REYNOLDS PATTERN & MACHINE CO.

101 and 103 Third Ave., Moline, Ill., U.S.A.

No. 2 Standard Type Automatic Screw-driving Machine.

Perkins Vegetable Veneer Glue

IS A

Uniform, Guaranteed Patented Glue

¶ We have, during the past ten or more years, built up a large volume of business and maintained it, because we sell only

HIGHEST QUALITY

¶ Our list of customers proves the statement that a very large per cent. of the highest quality furniture, including office and household furniture, also doors, interior trim, etc., are built up with Perkins Glue.

¶ Be sure you have *Perkins Quality*—absolutely uniform.

¶ We will demonstrate to your satisfaction in your own factory

Ladies' Desk manufactured by The D. Hibner Furniture Co., Limited, Berlin, Ont., in which Perkins Glue was used.

Originator and Patentee of Perkins Vegetable Glue.

PERKINS GLUE COMPANY, LIMITED

Address all inquiries to Sales Office, Perkins Glue Company, SOUTH BEND, Indiana, U.S.A.

Hamilton, Ont.

Alphabetical List of Advertisers

Acme Steel Goods Company 49
American Hardwood Lumber Co... 12
Atlantic Lumber Company 13
Ault & Wiborg Company 37

Baird Machinery Works 45
Brown & Co., Geo. C. 12

Callander Saw Mills ... ..., 45
Canada Machinery Corporation ...
Canadian Boomer & Boschert Mfg.
 Company 51
Canadian Morehead Mfg. Company 50
Carborundum Company 55
Central Veneer Company 43
Chapman Double Ball Bearing Co. 51
Churchill, Milton Lumber Co. 12
Clark Veneer Co., Walter 41
Clark & Son, Edward 10
Coignet Chemical Products Co. ... 37

Darling Bros. 50
Dominion Mahogany & Veneer Co.. 14

Evansville Veneer Company 38

Fay & Egan Company, J. A. 11
Foster, Merriam & Company 54

Garlock-Walker-Machinery 3
Gillies Bros. ... ...'... 44

Hart & McDonagh 13
Hartzell, Geo. W. 13
Haughton Veneer Company 43
Hay & Company 37
Hay Knife Company, Peter 8
Hoffman Bros. Company 41

Jackson, Cochrane & Company ...· 4

Kersley, Geo. 42
Klotz Manufacturing Company ... 50

Mattison Machine Works 56
Mowbray & Robinson 12

Nartzik, J. J. 41
Nash, J. M. 51
National Dry Kiln Company 8
Neilson & Company, J. L. 8

Ober Manufacturing Company 51
Ohio Veneer Company 41

Paepcke Leicht Lumber Co. ... ·. 14
Perfection Wood Steaming Retort

Company ... ...
Penrod Walnut & Veneer Company 43
Perkins Glue Company :.. 9
Perrin, William R. 8
Petrie, H. W. 45
Preston Woodworking Mach. Co... 7

Radcliff Saw Company 6
Reynolds Pattern & Machine Works 9
Roberts, John N. 42

Saranac Machine Company 55
Sadler & Haworth 2
Sheldons Limited 13
Simonds Canada Saw Co. 47
Spencer, C. A. 12

Toronto Blower Company 6
Toronto Veneer Co. 44

Walter & Company, B. 55
Wardwell Mfg. Company 49
Weeks Lumber Company, F. H. ... 49
Whitney & Son, Baxter D. 1
Wood Mosaic Company 43

Yates Machine Co., P. B. 5-45

Edward Clark & Sons
Toronto

CANADIAN HARDWOODS—winter cut **Birch, Bass, Maple**

10,000,000 ft. on hand, shipment June 1st
10,000,000 ft. sawing, shipment August 1st

ONTARIO and **QUEBEC** stocks, East or West, rail or water shipment, direct from mill to factory.

Put Up for Shell Boxes
　3,000,000 ft. 4/4 Birch, No. 1 and 2 Common.
　2,000,000 ft. 6/4 Birch, No. 1 and 2 Common.
　1,000,000 ft. 12/4 Birch, No. 1 and 2 Common.

For Furniture Manufacturers
　1,000,000 ft. each 10/4, 12/4, 16/4 Birch, 75 per cent. No. 1 and 2.

For Trim Trade
　500,000 ft. 4/4 Birch, No. 1 and 2,
　300,000 ft. 5/4 Birch, No. 1 and 2,
　700,000 ft. 6/4 Birch, No. 1 and 2,
　500,000 ft. 8/4 Birch, No. 1 and 2,
　200,000 ft. 4/4 to 8/4 Brown Ash, No. 1 and 2, 40 per cent. 14/16 ft.

30 per cent. 14/16 ft., Red all in.
9 in. average. Red all in.
40 to 60 per cent. 14/16 ft. long.

We Commend to Your Notice:
　Our lumber is reliable, our grades uniform, our shipments prompt, and receive personal supervision; our prices consistent. We have had long experience. We desire to serve.

Are you one of those who could make a big profit on this machine?

¶ Not every business can make a profit on a "FAY-EGAN" "LIGHTNING" No. 291 Automatic. ¶ But if your product requires large quantities of equalized lumber, i.e., stuff cut-off or trimmed to exact length, then this Hopper Fed Machine can and will make a big profit for you. ¶ Its capacity is practically unlimited. ¶ It will cut off both ends as fast as your operator can throw stock into the hopper by the armful and off-bearer clear it away. ¶ Does as much work as six old style trim saws. ¶ And every piece is **exactly** the length of all the others. ¶ If you equalize in quantities, use the enclosed post card to get Bulletin Q-28 and prices.

J. A. FAY & EGAN CO.

153-173 West Front St.   -   -   -   -   Cincinnati, Ohio

—this new 230 page book
on Woodworking Machinery free!

Our new handy reference catalog is just off the press. Shows all of the newest models in woodworking machinery.

Every owner and user of woodworking machinery, every superintendent and foreman should have a copy of this new book, which is really a text on modern woodworking machinery.

A copy will be sent to you, free of charge upon receipt of your request.

Fill in and mail the coupon to-day.

J. A. Fay & Egan Co.,
153-173 W. Front St.,
Cincinnati, Ohio, U.S.A.

Please send your new Book on Woodworking Machinery free of charge to:

Mr. ....

Position .... (give name of company you are with) ....

Street ....

City ....

Province ....

GEO. C. BROWN & COMPANY

Manufacturers St. Francis Basin Hardwoods

Ban Mills—PROCTOR, ARK. General Offices—Bank of Commerce Bldg.—MEMPHIS, TENN.

STOCK FOR SALE

SAP GUM (Kraetzer Cured)

	June	July
4-4 1s and 2s	51,005	105,740
4-4 No. 1 Common	150,000	

SELECTED RED GUM

Plain Sawed—Kraetzer-Cured

4-4 No. 1 Common	41,905	
5-4 No. 1 Common	40,860	44,030
6-4 No. 1 Common	39,190	

SOFT ELM

10-4 No. 2 Common	30,570	

SELECTED RED GUM

Quartered—Kraetzer-Cured

	June	July
10-4 No. 1 Com. and Bet.	110,595	
12-4 No. 1 Com. and Bet.	9,750	

PLAIN RED OAK

4-4 No. 1 Common	78,000	146,556
4-4 No. 2 Common	19,500	36,639

PLAIN WHITE OAK

4-4 No. 1 Common	63,708	18,744
4-4 No. 2 Common	15,927	4,430

QUARTERED RED OAK

	June	July
4-4 No. 1 Com. 6 in. and wider	93,592	
4-4 No. 2 Com., 6 in. and wider	22,148	
5-4 Nos. 1 and 2 Com. 6 in. and wider	17,640	

QUARTERED WHITE OAK

4-4 1s and 2s	63,785	24,500
4-4 No. 1 Common	33,767	21,280

OUR TRAFFIC DEPARTMENT
TRACES EVERY SHIPMENT

TENNESEE AROMATIC RED CEDAR SPECIALISTS

Churchill Milton Lumber Co.

Louisville, Ky.

We carry at
NEW ALBANY, IND.

Genuine Indiana Plain and Quartered Oak, Poplar

At GLENDORA, MISS.

Cypress, Gum, Cottonwood, Oak

SEND US YOUR INQUIRIES

We have dry stock on hand and can make quick shipment

Plain and Quartered Red and White Oak

and Other Hardwoods

Even Color Soft Texture

We have **35,000,000 feet dry stock**
all of our own manufacture, from our own
timber grown in Eastern Kentucky

MADE (MR) RIGHT

OAK FLOORING

PROMPT SHIPMENTS

The Mowbray & Robinson Co., Inc.

GENERAL OFFICES
CINCINNATI, OHIO

Mills- Quicksand, Ky., West Irvine, Ky., Viper, Ky.

Canadian Representative:
N. H. FARNHAM
601 Elmwood Ave., BUFFALO, N. Y.

American Hardwood Lumber Co.

St. Louis, Mo.

Large stock of—

Dry Ash, Quartered Oak Plain Oak and Gum

Shipments from — NASHVILLE, Tenn.,
NEW ORLEANS, La., and BENTON, Ark.

Dry Spruce and Birch

Good Stocks, Prompt Shipments, Satisfaction

C. A. SPENCER, Limited

Wholesale Dealers in Rough and Dressed Lumber

Offices—500 McGill Building

MONTREAL - - Quebec

The Atlantic Lumber Co.

Manufacturers and Wholesale Dealers

American and Canadian Hardwood Lumber and Mahogany, Quartered and Plain, White and Red Oak a Specialty.

110 Manning Chambers - **TORONTO**
Phone Main 6386

HOME OFFICE, BOSTON MASS.

Mills: KNOXVILLE, TENN.,   WALLAND, TENN.
FRANKLIN, VA.

Shell Box Stock

1 x 6 birch, in No. 1, 2, and 3 common.

1 x 7 and 8 birch, in No. 1, 2, and 3 common.

1 x 8½ birch, in No. 1, 2, and 3 common

1 x 9 and wider, birch, in No. 1, 2, and 3 common.

1 x 4 and wider, spruce. 2 x 4 and wider, spruce.

For the Furniture Trade

4/4, 5/4, 6/4, 8/4, 10/4, 12/4, 16/4, birch, maple, elm, basswood and black ash.

Elm, Basswood and Black Ash

in No. 1 common and better grades—dry stock in assorted sizes.

Hart & McDonagh

Continental Life Bldg.

TORONTO   -   -   **ONTARIO**

FOR SALE

Black Walnut Lumber

15,000 feet 4/4 No. 1 Common		
20,000 " 4/4 No. 2	"	
15,000 " 5/4 No. 1	"	
15,000 " 5/4 No. 2	"	
15,000 " 6/4 No. 2	"	
40,000 " 8/4 No. 1	"	
60,000 " 8/4 No. 2	"	
30,000 " 9/4 No. 1	"	

Will make special low prices on all of the above stock to move it quick. All Band Sawn, Equalized, and all thoroughly dry, ready for immediate shipment. Write for prices.

Geo. W. Hartzell

Piqua, Ohio

Woodworkers
and
Lumbermen

The Sheldon Exhaust Fans have characteristics that adapt themselves to the service of the Planing Mill or Woodworking Plant. They are specially designed for this kind of work, having a saving in power and speed of 25% to 40%.

Write for Industrial Booklet.

SHELDONS LIMITED - Galt, Ont.

Toronto Office, 609 Kent Building

— Agents —

Messrs. Ross & Greig, 412 St. James St., Montreal, Que.
Messrs. Walker's Limited, 259-261 Stanley St., Winnipeg, Man.
Messrs. Gorman, Clancey & Grindley Ltd., Calgary and Edmonton, Alta.
Messrs. Robt. Hamilton & Co., Ltd., Bk. of Ottawa Bldg., Vancouver, B.C.

14   CANADIAN WOODWORKER   June, 1916

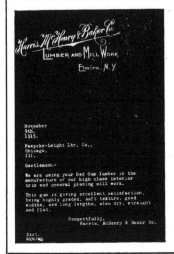

Of course it is true that

RED GUM

is America's finest cabinet wood—but
Just as a poor cook will spoil the choicest
viands while the experienced chef will turn
them into prized delicacies, so it is true that

The inherently superior qualities
of Red Gum can be brought
out only by proper handling.

When you buy this wood, as when you buy a new
machine, you want to feel that you have reason for
believing it will be just as represented.

We claim genuine superiority for our Gum. The
proof that you can have confidence in this claim is
shown by the letter reproduced herewith.

Your interests demand that you remember this
proof of our ability to preserve the wonderful qualities of the wood when you again want RED GUM.

Paepcke Leicht Lumber Co.

Conway Building   111 W. Washington St.
CHICAGO, ILL.

Band Mills: Helena and Blytheville, Ark.; Greenville, Miss.

Buy Made in Canada Veneers

Dominion Mahogany
& Veneer Co., Limited

Head Office and Factory   Toronto Office and Warehouse
Montreal West, P.Q.   455 King St. West

Mahogany, Walnut,
and all Fancy Woods—Lumber

SAWN, SLICED and ROTARY CUT VENEERS

Mahogany, Walnut, Quartered and Plain Oak, Ash, Maple,
Birch, Elm, Poplar, Gum.

Gum and Poplar Cross Banding.   Hard Maple Pin Block Stock.

SEND US YOUR ENQUIRIES.

Canadian Woodworker

A Monthly Publication in the Interest of the Woodworking Industry.
Reaches the factories producing interior finish, doors, sash, flooring,
woodenware, furniture, pianos, boxes, and general mill products.

Subscription, $1.00 a year; foreign $1.50.

Woodworker Publishing Company, Limited

Branches: Montreal, Vancouver,
Chicago, New York, London, Eng.

345 Adelaide St. West, Toronto
Phone Ade. 2700

Authorized by the Postmaster General for Canada, for transmission as second class matter.
Entered as second class matter July 18th, 1914, at the Post Office at Buffalo, N.Y., under the Act of Congress of March 3, 1879.

Vol. 16	June, 1916	No. 6

Furniture Manufacturers Must Advance Prices

The enormous advance in price of every item of raw material which enters into the manufacture of furniture, together with the scarcity of labor and the higher wages demanded, makes it imperative that manufacturers take some immediate and definite action. Furniture prices must advance, and the fact that some of the large buyers have recently been endeavoring to secure contracts for rather large quantities is sufficient evidence that dealers can see the trend of events.

The Editor was in conversation a couple of weeks ago with a manufacturer who had to turn down an order for $2,000 worth of buffets. He could not fill the order from stock, and as prices on raw materials are constantly advancing and he cannot get men, he was not going to take any chances. The fact that the particular buyer who attempted to place the order is always met at the depot when on a buying trip by the manager, with a "big car," makes the act more significant. It seems rather strange that dealers are preparing for increased prices and yet furniture manufacturers have not made any united effort to advance prices.

The Furniture Manufacturers' Association, recently formed, opens up an avenue by which its members can compare notes and adjust prices, so as to protect themselves, and it is to the advantage of every furniture manufacturer to be a member of the Association. As an illustration, we cannot do better than quote a paragraph from the address of Mr. Robert W. Irwin, past president of the Federation of Furniture Manufacturers of the United States, at their meeting in Chicago: "Ours is a manufactured product, and I believe there are few men in the business who have the same courage to advance their selling price, that the dealer in raw material has. Competition in manufactured goods, and in the furniture trade particularly, is very keen, and I think that we are all prone to wonder at times like these just what our competitors are going to do. It is my judgment that the greatest value that comes from association work is the opportunity that it affords for manufacturers to get together at just such times as this. By having an organization with regular dates for meetings which brings us together from time to time we are sure of an opportunity for interchange of experiences at the very time it is most needed. I have gathered more courage to advance prices since I have been in the furniture business as a result of these meetings than in any other way. We have had these conditions before, and we are generally fearful of the approaching market. We realize more money is needed, but we hesitate, not knowing whether our competitors fully realize it or not. When we get together in an open meeting, the importance of the changed conditions comes to us more clearly and with it more strength and courage to ask for our product that which we know we need to have if we are to keep up with advancing costs."

"Break Out" One Job at a Time

The question has often been raised as to which is the most economical and satisfactory method of breaking out stock in a furniture factory. Some superintendents contend that the best results can be secured by loading a number of cars with one kind of lumber, say plain oak or quartered oak, as the case may be, drying them altogether and then having the men work on that one kind until it is all cut up.

By this method it is necessary to pick out all the jobs requiring the particular kind of wood which is being cut, and this is where trouble begins. There will be some parts of the job for which a different kind of wood will be required and as it will not be possible to keep the bulk of the job in the breaking out department until these parts are gotten out, it will have to be machined at different times. The result will be that the parts will not all reach the cabinet room together and this means loss of time in assembling the work.

The method which seems to give the most satisfaction is to determine the amount of each kind of lumber required for a job and then place it on cars and put the cars together in the kiln. This is continued until the kiln is full, and then when the drying has been completed, the jobs can be taken to the breaking out department in rotation. The whole job can then be cut out complete and placed on trucks so that the different parts will not get mixed in with other jobs and after being machined the parts can all be sent to the cabinet room together for assembling.

The time that can be saved by having all the different parts of the job reach the cabinet room together will more than offset any little saving which might be effected in the machine room by having a "big cut" on one kind of lumber. The quantities of lumber required for the different jobs can be determined pretty accurately after a little experimenting, and it is better to have a few boards over than a few short. A dry shed in which the left-over boards can be stored will be found to be a very useful addition to the plant, as in case one or two pieces develop defects on their journey through the machine room, they can be replaced from this source.

By keeping a piece tally of the lumber which is put on the different cars, it is possible to keep very accurate account of the amount of lumber to be charged to each job.

Lumber Consumed for Manufacturing Purposes

The total lumber consumption for manufacturing purposes in the province of Ontario is given in a special bulletin issued by the Department of the Interior, Ottawa, for the year 1912, at 807,456,000 board feet. Of this, 82 per cent., or 662,113,920 board feet, was supplied by Ontario; 21 per cent., or 30,424,590 feet was purchased from eastern Canada; 2,897,580 feet from British Columbia, and 111,556,830 feet from for-

The total annual lumber consumption of the maritime provinces is approximately 200,000,000 feet, of which only about 12.3 per cent. is purchased outside of those provinces. Of this 12.3 per cent., the other Canadian provinces supply 2,017,000 feet, and 23,145,000 feet is imported from other countries. The total annual consumption of the prairie provinces is about 68,439,000 feet, of which British Columbia supplies 32,766,000, or about one-half of the total consumption. Eastern Canada supplies 12,291,000 feet, or about 20 per cent. of the total, and 242,000 feet is imported by foreign countries, while the balance is supplied by the prairie provinces.

Enormous Increase in Cost of Raw Materials

The following figures compiled by the secretary of the Furniture Manufacturers Association of Canada speak more plainly than words as to why furniture manufacturers should increase the selling price of their products. The figures cover the period from April, 1915, to May, 1916:—

Lumber	10 to 25 per cent.
Veneer	25 to 50 " "
Mirrors	100 " "
Stains	300 to 400 " "
Enamels	25 " "
Oils	25 " "
Shellac	100 " "
Varnish	25 " "
Hardware	50 to 100 " "
Brass Trimmings	25 " "
Packing Paper	100 " "
Burlap	100 " "
Catalogues, Photos, etc.	25 " "
Twine	50

Scarcity of Tool Steel

There is in Canada an alarming scarcity of tool steel, due to the fact that the British Government are using every ounce of steel that can be produced in

---

Can the new "Bethlehem" shell box be made for $1.23? Read the article on page 18. What is your opinion?

---

eign countries. The greatest consumption of lumber is in the manufacture of sash, doors and interior trim. 250,000,000 feet is used in this industry, of which 85.5 per cent., or 216,250,000 feet is supplied by Ontario. Hard pine, oak and cypress form the bulk of the imports for this industry, but the imports amount to less than 10 per cent. of the whole, or about 24,500,000 feet.

The figures given for the furniture industry covers those factories engaged in the manufacture of household furniture, office furniture, lodge and school furniture, pianos, organs, etc., and ranks next in importance, with an annual consumption of about 80,000,000 feet, of which about 27 per cent., or 21,600,000 feet is imported. The imported woods are chiefly gum, sycamore, red cedar, butternut, walnut and mahogany.

the British Isles for the manufacture of munitions of war, and the American manufacturers of munitions are using all that can be produced in the States.

Sheffield, England, is undoubtedly the home of tool steel, and the Canadian representative of one of the largest Sheffield companies told the "Canadian Woodworker" that unless he got some shipments immediately he would be forced out of business. One or two of the steel companies were fortunate in having large stocks on hand at their local warehouses, but these stocks are being rapidly depleted. A great many of the sizes cannot now be had and steel that formerly sold at 85c a lb., is now $3.50 a lb. It looks as though Canadian manufacturers may have to appeal to the British Government to permit a few tons to be shipped to this country.

Reduction in Workmen's Compensation Rates
Favorable Results in Woodworking Industries Lead to
Lower Premiums for 1916

The Workmen's Compensation Board has furnished the "Canadian Woodworker" with special information about the results of the first year's operation of the act and exclusive information regarding changes to be made in the rates charged in connection with a number of lines of industry during 1916. The most important feature of this information is that in relation to the new rates. In advising the "Canadian Woodworker" of the changes, the Workmen's Compensation Board say:—

"The aim of the Board is to make the rates for each kind of industry conform to the actual cost for such industry.

"The rates fixed for 1915 before any experience had been had in the actual operation of the new Act were necessarily based upon other tables of rating with allowances for the difference between the provisions of the Ontario Act and the covering of the other ratings. We have now had one year's experience, but the actual figures for the year are still incomplete, inasmuch as at the close of the year many claims for accidents happening in 1915 had not yet been filed, or, by reason of continuing disability or incompleteness of reports, had not been finally dealt with, and inasmuch as the assessments, which were

amounts distributed to employees and their families amounted to $1,186,221. There was thus a surplus of assessments over distributions amounting to $395,026.

Compensation was given in 9,829 cases. Among these the parties benefiting included 65 widows, 116 children, 35 mothers, and 10 fathers. The cost of administration of the Act for 1915 was $100,846. Cheques to the number of 150, on the average, were sent out from the Board's offices daily, amounting to about $3,600.

In its report for the year the Board says:— "Though, in view of the unavoidable incompleteness of the actual figures, the first year's experience cannot be regarded as conclusive, it indicates that the financial burden of the new law will be less than was anticipated. The assessments levied for 1915 are in most cases more than sufficient to meet the requirements. Though in a very few instances the bad accident record has called for an increase in rate, the Board has felt warranted in making substantial reductions in most classes of industry. In a number of classes, or parts of classes, where the experience was good and the surplus large, reductions are being applied retroactively for 1916."

Examine the drawings and specifications for the new shell box on page 18, then write and tell us what you think it can be made for.

---

based upon estimated payrolls, had not been adjusted to the actual payrolls.

"With the help of estimates, however, the situation has been gauged as nearly as possible. This has been done not only with respect to each class of industry but also, as far as could be, with respect to each particular line of industry in the class, and the rates for 1916 have been fixed accordingly.

"While in a few cases it has been deemed necessary to increase the rate, there is for 1916 in most cases a reduction from the 1915 rate, in many cases a very substantial one. In cases where the surplus was large it has been reduced by lowering the rates for 1916 retroactively.

"Separate account is kept of all the moneys received and paid out in respect of each class of industry, and the standing of each class is considered separately. It is very essential that within the class each kind of industry shall bear its proper proportionate rate, but it is immaterial to employers in one class what the rate may be in another class. If there is a surplus in the class it means a subsequent lowering of rates, while if there is a deficit it will mean an increase."

The results of the operation of the act during 1915 show many interesting points. The total number of accidents reported was 17,033. Assessments collected from employers amounted to $1,539,492. The

Reductions in the Rates

The rates for 1915 and the new rates for 1916 in connection with woodworking, etc., are as follows:—

Industry	1915 rate	1916 rate
Manufacture of		
Veneer or excelsior	$1.60	$1.50
Staves, spokes or headings	1.80	1.50
Planing mills, sash and door factories, and lumber yards (including delivery) in connection with planing mills or sash and door factories..	1.50	1.30
Manufacture of		
Wooden boxes, cheese boxes and mouldings	1.50	1.30
Ammunition shell boxes	1.50	1.00
Cigar boxes	1.50	.70
Window and door screens	1.20	1.00
Wooden articles or wares or baskets, or wooden toys	1.20	.1.00
Vehicle woodwork (retroactive) ..	1.20	1.00
Furniture, interior woodwork, fixtures, etc.	.80	1.00

Benefits of the Act

"The Act has worked smoothly and satisfactorily. The benefits of the new system of law to both workmen and employers are recognized and appreciated. Claims are expeditiously and inexpensively disposed of. Employers are immune from the expense and annoyance of litigation. The intricacies and hardships upon workmen and their families of the old doctrines of negligence and assumed risk are eliminated. The facts to be determined by the Board are usually few and simple. There is no longer the need for payment of legal fees either by workmen or employers."

Evidence that employers as a whole have been satisfied with the working of the Act in its first twelve months is shown by the fact that many employers have asked that industries or operations carried on by them which are not now under the main schedule of the Act should be included. Although the Act introduced a rather radical alteration in compensation methods, the co-operation shown by all parties in the first year's experience has been excellent and steadily improves.

The New "Bethlehem" Shell Box

Brief Outline of the Construction of the Box and the Materials Necessary to Make It.

The Imperial Munitions Board have issued plans and specifications of a box to be made for the new 18-pounder "Bethlehem" shell. The outside over all dimensions of the box are 27¼ ins. x 10 ⅜ ins. x 10⅜ ins. It will hold four complete rounds, including fuzes and fuze covers. The kinds of wood to be used are spruce, pine or basswood for the top, bottom, sides and ends, and birch, beech and maple for the top and bottom battens and diaphragms. All metal fittings are to be of mild steel galvanized.

The construction of the box is somewhat different from the other boxes that manufacturers are acquainted with. It has a removable end which is held in place by a central spindle that passes through the centre of the box from end to end, fastened with a wing nut. Two small galvanized steel plates are screwed to the bottom and top of the box to keep the head of the bolt and wing nut from cutting into the wood.

Fig. 3 shows a side elevation and end elevation of the box and on the side elevation can be seen the central spindle, wing nut and small steel plates in dotted lines. It will be noticed that the wing nut and head

the box are made of 3-ply hardwood glued up. The centre ply is ¼ of an inch thick and the two outside plys are 5/16 inches, making a total thickness of ⅞ of an inch. The specifications state that these thicknesses are nominal, which we take to mean that the total thickness of the diaphragms may vary slightly from the measurements given. The diaphragms are for

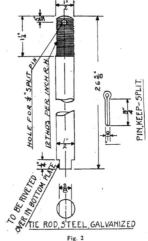

Fig. 2

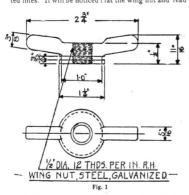

½ DIA. 12 THDS. PER IN. R.H.
— WING NUT, STEEL, GALVANIZED. —
Fig. 1

of the bolt do not project past the ends of the box. This is to allow of the boxes being piled on top of each other. The battens that go on the ends are mitred and project ⅜ of an inch over the end pieces, thus forming a rebate for the end to fit into the box.

The three diaphragms or divisions that go inside

the purpose of holding the shells in place. Each one has five holes bored in it, the centre one being a ⅝ inch hole for the central spindle to pass through. The other four holes are for the shells and are a different size in each diaphragm. Those in the diaphragm nearest the removable end are 3 13/16 inches in diameter, those in the centre diaphragm are 3 13/32 inches in diameter, while those in the diaphragm nearest the closed end are 3⅛ inches tapered down to 2 15/16 inches.

A piece of soft gray felt ¼ inch thick is fastened to the inner side of the removable end as shown in Fig. 4 to keep the shells in place and to keep the ends from

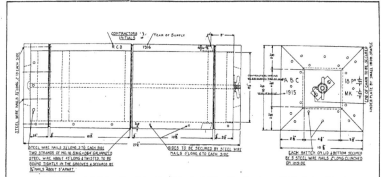

Fig. 3—Side and End Elevation of New "Bethlehem" Shell Box, Showing Central Spindle [in dotted lines]
that Holds Removable End in Place.

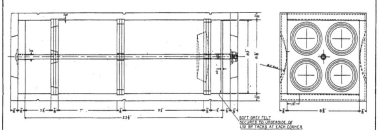

Fig. 4—Plan and End View of "Bethlehem" Shell Box. Note the Hand Holes in Dotted Lines.

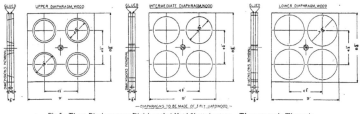

Fig 5—Three Diaphragms or Divisions that Hold Shells in place. They are made Three-ply.

coming in contact with the wood. The felt is fastened to the lid by means of tacks at each corner. The box has two hand holes for lifting purposes, the shape and depth of which are shown in Fig. 4. The width and angle at which they are cut is shown in Fig. 3.

The wing nut that holds the removable end in

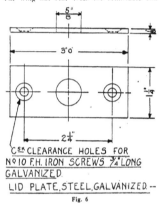

C'S'K CLEARANCE HOLES FOR
N° 10 F.H. IRON SCREWS ¾"LONG
GALVANIZED.
LID PLATE, STEEL, GALVANIZED. --

Fig. 6

place has a ring and chain attached to it to keep it from getting mislaid. The ring has an opening 1 inch in diameter and is made of ring steel No. 13 S.W.G., .092 of an inch wide. One end of the chain is attached to the ring and the other end secured under the lid

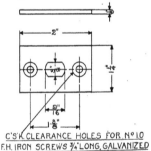

C'S'K. CLEARANCE HOLES FOR. N° 10
F.H. IRON SCREWS ¾"LONG, GALVANIZED

BOTTOM PLATE, STEEL GALVANIZED

Fig. 7

plate by a small staple. The dimensions and shape of the wing nut are shown in Fig. 1 and the ring and chain in Fig. 8.

The size and location of the holes in the lid and bottom plates are shown in Figs. 6 and 7 and the dimensions of the central spindle and the split pin that

goes in the end of the spindle are given in Fig. 2. The box has three banding wires of two strand galvanized steel wire, No. 16 S. W. G., .064 of an inch tightly bound around the four sides in grooves provided for the purpose and secured by ¾ inch wire nails about 5 inches apart. After the box is assembled and wired, it is marked in two places, (as indicated in Fig. 3) with the contractor's initials and the year of supply.

We understand that some contracts have recently been let for this box to be made at $1.23, and we have been told that it cannot be made for less than $1.50.

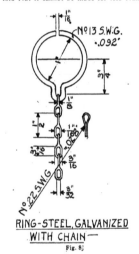

RING-STEEL, GALVANIZED
WITH CHAIN ―
Fig. 8;

We would like our readers to tell us what they think the box can be made for, as we believe that a discussion of the probable cost would be advisable at the present time when tenders are being called for and contracts given.

Should any action be taken to bring the manufacturers together to discuss the prices and the problems of shell box making?

Kiln for Drying B. C. Spruce

Namu, B. C., June 5, 1916.
Editor "Canadian Woodworker":
Dear Sir:—

I want to kiln dry B. C. spruce for boxes and do not know whether it has ever been tried out or not. Will you kindly ask the question in the "Canadian Woodworker," also how large a kiln would have to be built for a capacity of fifteen thousand boxes a day, and what is the best style of kiln for the purpose?

Yours truly,
W. S. Ball.

We shall be very glad to hear from our readers on this subject.—Editor.

The Cost of Making Shell Boxes

Article in April Issue "Written by Man Who Knows" was the Verdict of Many Men with Whom the Subject was Discussed

The "Canadian Woodworker" recently had the pleasure of discussing with a number of owners of woodworking plants in Hamilton the article that was published in our April issue on the cost of making the 18-pounder 6-round shell box. Without an exception they expressed the opinion that the article was written by a man who "really knew" what it cost to make a shell box. We regard this as a compliment to our journal, because it has always been our aim, and always will be, to have our articles written by practical men.

There seems to have been a misunderstanding on the part of some people as to how much money was being made on shell box orders. This misunderstanding no doubt led a number of firms to think when tenders were called for, "well those people cleaned up a lot of money on their order at such and such a price, I will take a chance and put in a bid 10 or 15 cents less than the price they received." This is undoubtedly how some of the recent tenderers arrived at the figures which they submitted, because if each firm had figured their costs without taking into consideration the prices the "other fellow" received, there would not have been such a wide difference in the figures submitted, even allowing for the different methods of figuring used by different men.

Before the first orders were given out by the Shell ing and experimenting to eat up a large share of the profit.

The firms that really made money on shell boxes were those that secured orders ranging from 50,000 to 125,000 of one kind of box. They had everything in their favor. They knew just exactly how much lumber, hardware, iron bands, handles, etc., to contract for and were thus in a position to take advantage of prices prevailing at that time, with full knowledge that they could run their plant to capacity for six or seven months. The only thing they could not contract for was labor and although a few of them may have been hampered on this score the majority of them came through with a large balance in the bank. The orders given out by the Shell Committee generally ranged from 10,000 to 50,000, but the writer knows of one instance (and doubtless there were others) where an order was placed (or an understanding given that sufficient orders would be forthcoming to make up the quantity), for 125,000 boxes. From the fact that the firm in question made preparation for six months' work, it can be reasonably deduced that they either had the order or the understanding.

In regard to the boxes that have been manufactured for the 4.5 "High Explosive" shells, the statement has been made that the price formerly paid for

---

> ## We have been told that the new "Bethlehem" shell box cannot be made for less than $1.50. Do you agree? How do your costs work out?

Committee, a representative of that committee visited the plant of each party that was endeavoring to secure an order. The visit was made for the purpose of determining whether the size of the plant was adequate for a reasonably large output. And from a business point of view, no person can deny that the plan was a wise one, although it may not have appeared just to the smaller manufacturers.

The amount of money that would be required to pay extra freight rates and inspectors' salaries and travelling expenses if the orders had been distributed indiscriminately to all comers, regardless of the size of the output of which their plants were capable, or the location of the plant, would be a considerable item on the cost of the boxes. There is not the slightest doubt that a number of firms were turned down by the Shell Committee who were in a better position to manufacture boxes and turn them out in suitable quantities than some of the firms who received the orders. But on the other hand a number of the firms who received orders from the Shell Committee did not make much money on them, for the simple reason that they did not receive large enough orders for one particular kind of box. They would probably secure an order for 10,000 of one kind and after establishing a cost price by getting their machines adjusted and getting their men into the swing of the work, they would get another order for 10,000 or 15,000 of an entirely different kind of box. This meant just sufficient readjust-

this box was $1 and that the present price is 80c. This statement is rather misleading, as there were two different kinds of boxes made for the 4.5 "High Explosive" shells. One was made to accommodate two shells, two cartridge cases, two plugs, and handles. The other was only made to hold two shells. The first mentioned one was the box for which the Shell Committee paid $1 and for the latter one they paid 80c. It is quite possible that both these boxes have been tendered on at a figure below these prices, but the fact was brought to our attention that some manufacturers who had not made either of the 4.5 boxes were not aware of the two different kinds, and consequently they thought that the $1 box was being made for 80c.

The Munitions Board have recently been calling for tenders for a quantity of the 4-round fixed ammunition boxes, and as the majority of the manufacturers that cut prices below the point where there was any possibility of making a profit will likely spend a little more time figuring their costs before they put in further tenders, we may soon look for a period when more favorable manufacturing conditions will prevail for those legitimate woodworking firms who wish to participate in the shell box business. All due respect should be given to the man who figures his costs to the best of his ability and bases his tenders on intelligent figuring, but guessing at their costs is liable to put the manufacturers that "guess" out of business, besides spoiling the business for others.

How to Make and Read Working Drawings

An Illustrated Article Dealing With the Method of Teaching Draughting in Technical Schools.

By W. Alexander

I read with a great deal of interest an editorial which appeared in a recent issue of the "Canadian Woodworker," pointing out the advantages of technical training for those men and boys who hope some day to occupy responsible positions in the woodworking business.

Technical training certainly includes a knowledge of draughting and as the editor has pointed out, it is one of the qualifications absolutely essential to the present-day mechanic. It matters not whether he works in a planing mill, a furniture factory, a box factory, or in fact any kind of plant that is manufacturing wood products, a knowledge of how to make and read working drawings will be a valuable asset.

The question then, naturally, arises where is he going to secure that knowledge? There are correspondence schools and courses of instruction provided

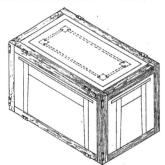

Fig. 1.—Box with Glass Sides Used for Teaching Draughting.

in night schools for the purpose of imparting the necessary knowledge, but there may be many reasons why a man cannot spare the time or the money to take such a course, although he would be amply repaid by doing so.

It has often occurred to me that one or two articles dealing with the basic principles of making working drawings would be appreciated by men engaged in the trade and while I fully realize that there are men better qualified to deal with the subject than myself, yet I have been tempted to offer a little information and can only hope that my efforts will not be criticized too severely by those men who are already draughtsmen.

There are two kinds of drawings in general use, namely, perspective drawings and working drawings. The reason I mention perspective drawings is because I have found that to the average man in the shop a "drawing" is a drawing no matter what kind it is. Perspective drawings are only used to give a representation of how an article will look when completed. The best illustration that one can give of a per-

spective drawing is a photograph. Everything is seen from one point. In other words, a man standing looking at a house or a piece of furniture would probably see the front and one side (depending on just where he stood) and he would draw it just exactly as it appeared to him. Hence the lines of the object would diminish the further away they were from the eye. This can be clearly seen in a photograph, where the lens represents the eye.

It will be apparent, then, that a man could not build a house or a piece of cabinet work (unless it were very simple in design and construction) from a perspective drawing or a photograph unless he had all the necessary measurements given to him and even then he would have to do a certain amount of guessing.

A working drawing of an article may consist of any number of different views (depending on the size of the article and whether the construction is complicated), called plans, elevations, and sections. This is where the average workman gets, to use a shop expression, "all balled up." A short time ago I had occasion to take a drawing out in the factory to a shaper man to get him to make the knives for a new moulding. He could see the moulding all right when I pointed it out to him, but he wanted to know what the drawing represented.

He said he could understand a drawing of a chair when it was hung up on the wall because it was just like seeing the chair standing up in front of him (he was referring to the side elevation of a chair), but when it came to a buffet or a china cabinet he could not make head or tail of it. Perhaps the reason he could understand the drawing of a chair was that it only takes six or eight lines to make the side elevation of a chair and as the seat projects out from the back it would be reasonably clear to the workman of average intelligence. The difficulty he had with the drawing of a buffet, or china cabinet, was that the lines appeared to be run in one on top of the other.

I spent about fifteen minutes one day giving that man a little demonstration of how they teach draughting in the schools and he proved a very apt pupil. For some time after that he used to spend his noon hours examining all the drawings that he could get hold of and it was not long before he could understand them as well as any person around the place, so I do not think I can do better than try and repeat the demonstration and I sincerely hope that I will be able to make the subject clear.

A box is used for the demonstration, but instead of having solid pieces for the ends, sides, top and bottom like an ordinary box, it is made up of a number of frames with glass in them like windows. The frames are hinged so that they can be folded up to represent a box or they can be opened up so as to lie flat on the table. A reference to Figs. 1 and 2 will convey the idea quite clearly. For the purpose of demonstrating we will assume that it is desired to make a working drawing of an ordinary kitchen table.

When making a working drawing of an article you must be able to picture in your mind the article you wish to draw. Then, if you understand the principles of draughting and have sufficient mechanical knowledge to know how the article should be constructed, it is an easy matter to draw it out on paper. We now come to the point where draughting differs clearly and distinctly from perspective drawing and if you succeed in grasping this point the rest is only a matter of practice.

Instead of seeing two or more sides of an object from one point and drawing it just as it appears to us as in perspective drawing, we have to draw it as it really is, or as we would construct it. For instance, in the case of the kitchen table, a man would make the two sides and then the ends and then the top, or he might reverse the order of procedure and make the top first and then the sides and ends. In making a drawing he would go about it in just exactly the same way. He would draw the top view or plan, then the ends or end elevations, then the sides or side elevations.

In making the plan we have to imagine that we are directly above the object looking down on it. This is where the box with the glass sides is used in the demonstration. The object to be drawn is placed inside the box. In this case we will assume that the kitchen table is inside.

A bottle of white fluid and a small brush for marking on the glass are then secured. Imagining ourselves looking down at the table through the glass top of the box, we would see the four points which represent the corners of the top. We mark these four points and draw lines to join them together. Now we know that underneath that top there are the tops of four legs. The next thing to do, then, is to locate how far those legs are from the edges and ends of the top. This being done, we draw the little squares to represent the tops of the legs.

Right here might be a good place to say that anything which "we know is there, but which cannot be seen" should be drawn with dotted lines. These are called invisible lines, meaning for parts which are invisible. We also know that underneath that top there are four rails which join the legs together. The thickness of these should be determined and then they should also be drawn in dotted lines the proper distance from the edges and ends of the top. If the rails are going to be tenoned into the legs the tenons should also be shown. This completes the plan so we will next draw the side elevation.

Now we have to imagine ourselves looking squarely at the side of the table. This will be drawn on the side of the glass box. The first thing to do is to get the height and the length of the top and then draw two lines to represent the edge of the top. We have already determined, on the plan, how far the legs are in from the end, also that they are going to be straight on the outside and tapered part way down on the inside, so we can proceed to draw them accordingly. The only thing necessary now is a line to show the width of the rail which connects the legs and the side elevation is complete. Both sides of the table are exactly alike so that only one side elevation is necessary.

For the end elevation we have to imagine that we are looking squarely at the end of the table. This will

Editor's Note.—Mr. Alexander has expressed his willingness to take up more advanced work in draughting and taking off stock bills from drawings, etc., if sufficient of our readers are interested. We shall be guided by the number of communications we receive. If you like this article let us hear from you, either by letter or post card.

be drawn on the end of the glass box. The end elevation is drawn in the same manner as the side elevation. The only difference will be that the legs will be closer together and the width of the top will be shown instead of the length. Both ends are alike so that only one end elevation is necessary. The glass box can now be opened up and laid flat so that we have our drawing in front of us, as shown in Fig. 1.

All that remains to be done now is to put a sheet of paper on a drawing board and draw the plan and elevations on paper, arranging them in the most convenient manner in order to save space.

In making working drawings, then (which is draughting), the thing to bear in mind is that each side of the object is drawn on a different plane. (A plane is simply a level surface such as a piece of glass which can either be laid flat or stood on edge). The glass box represents the three planes:—First, the horizontal plane on which the plan is drawn; second, the perpendicular plane on which the side elevations are drawn; third, the perpendicular plane on which the end elevation is drawn.

In practical work the different views are drawn on one plane or sheet of drawing paper. The glass box opened out flat gives a representation of how this can

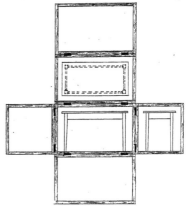

Fig. 2.—Box opened out to lie flat on table.

be done. When making a drawing a man can picture the different views in his mind. He can then proceed to put them on paper and arrange them to the best advantage the same as he would move the men on a checker board.

If you can get these points securely fixed in your mind and then examine all the drawings you can get your hands on and study them out you will soon have no difficulty in taking off a stock bill or constructing a job from a drawing.

I have tried to make the subject as clear as possible, and if this article is of assistance to any person in the trade I shall be pleased to take up more advanced work in future articles.

Hamilton as a Woodworking Centre

Many and Varied are the Products Manufactured in the Mountain City

The city of Hamilton, although it has not as many woodworking plants, according to population, as some of the other towns in Canada, is, nevertheless, worthy of special mention as a woodworking centre. Literally, everything that can be made from wood is manufactured in Hamilton, so it will be evident that the operations extend over a pretty broad field. Business conditions, while not all that can be desired, are much better than last year and the general opinion among the manufacturers of Hamilton is that they will continue to improve. There are one or two exceptions to the general rule, as in the case of firms who make a specialty of built-to-order fixtures and high-grade interior woodwork. These have found that the amount of business offering in their line is very limited.

The woodworking factories in Hamilton have manufactured large quantities of shell boxes and two or three firms are at the present time working on large orders. The main feature in connection with this business (outside of reduced prices and the increased cost of raw materials) is the seriousness of the labor situation. Apart from the fact that Hamilton has contributed liberally to the forces of the Empire there are a great many steel working plants in the city, and these are practically all turning out shells for the Government. Some are working night and day and the number of men required is abnormal. This demand for men by the shell manufacturers, with the wages that they can afford to pay, means that some of the other industries must go short. The writer was informed by the proprietor of a box-making factory that youths who a short time ago would have jumped at an offer of $8 per week now laugh if they are offered $2 per day.

In describing the woodworking industries of Hamilton, our description must necessarily be brief, as it would be impossible to do justice to the numerous plants in one article. We will, therefore, give the names of some of the largest firms, together with the products manufactured, and one or two of the principal features noticed at their plant. We hope to be able to give a detailed description of one or two of the leading plants in a future issue of The Canadian Woodworker.

The Cummer-Dowswell Company undoubtedly have the best equipped and most up-to-date woodworking plant in Hamilton. Their equipment includes a Linderman automatic dove-tail jointer and the very latest in machinery of every description. The company manufacture clothes wringers, washing machines, barrel churns, lawn mowers, clothes driers, clothes mangles, etc.

The woodworking plant of D. Aitchison and Company is very favorably situated in the central portion of the city. The company manufacture a general line of planing mill products and supply everything in the nature of wood which may be required for building purposes. Part of their plant is devoted to the making of boxes and packing cases, which are consumed locally. Apart from the usual machinery common to woodworking plants, a feature of this company's plant is a Dutch oven for burning shavings. The machines are piped to a blower system in the usual manner, but instead of the main pipe leading horizon-

tally (or on an angle) into the fire box and discharging the shavings, as in the ordinary method, the fire box is placed below the floor level and projects out about two feet past the front of the boiler (which is level with the floor), and the pipe discharges vertically into the fire box. The fire box has an opening somewhat larger than the diameter of the pipe. The pipe just reaches to the top of the opening and the shavings catch fire as soon as they are discharged from the pipe. The draft created through the opening being larger than the diameter of the pipe, carries the "streak of fire" back under the boiler. This method of burning shavings is said to give much greater heat than the ordinary method and it practically eliminates smoke.

The Kilgour Mfg. Company manufacture boxes and packing cases in a factory well equipped for the purpose. The company have "done their bit" in the manufacture of shell boxes, having completed an order for 15,000 25-round boxes, and they are at present engaged on an order for 25,000 4.5 boxes which they expect to have completed in the course of a few weeks.

All kinds of interior finish, sash, doors, water and acid tanks, and silos, are manufactured at the plant of the Patterson-Tilley Company. The plant is admirably situated and equipped to take care of large orders for anything in the line of manufactured woodwork. A large portion of the woodwork for interior purposes turned out by this company is executed to architect's details. This class of work calls for a high degree of mechanical skill in order to secure the maximum of efficiency necessary to make the business profitable. The company have also taken care of a portion of the shell box business. They completed an order for 10,000 4.5 boxes, but their plant is large enough to handle a much larger order. Mr. Patterson has charge of the mechanical end of the business and Mr. Tilley the office end.

The manufacture of caskets as carried on by the Semmens and Evel Company and the Evel Casket Company, is a highly interesting branch of the woodworking industry. One of the features of this class of work is the heavy mouldings which have to be run on the shaper. Some of them require a knife projection of nearly 4 inches, and this makes it rather a dangerous operation. The knives have to be very accurately adjusted in order to turn out fine work. Casket making involves some very fine cabinet work and the finish is equal to that used in high class furniture and pianos.

The furniture manufacturing industry is represented in Hamilton by the Malcolm and Souter Company. They manufacture a complete line of medium grade furniture, but for some time they have been devoting their energies to the manufacture of shell boxes as well as furniture. They are at present engaged on an order for 50,000 4-round fixed ammunition boxes and expect that it will be six weeks or two months before this order is completed. Previous to this they manufactured 10,000 of the 60-pounder boxes.

The Burton and Baldwin Company manufacture all kinds of interior woodwork and built-to-order fixtures. They have a factory which is splendidly laid out and equipped. On entering the machine room of this factory the first thing that attracted the attention of the Canadian Woodworker was the splendid lighting conditions that prevailed. Working on the principle that a machine room cannot have too much light, this company have carried out an idea, which, to say the least, gives splendid results. The line shaft and all countershafts, pulleys, hangers, etc., are located in the basement and the belts pass up through the floor, This makes it very easy for the man who attends to the oiling and examining of the belts to perform his duties as the line shaft and countershafts are attached to the basement floor. The Burton and Baldwin Company have also taken care of some of the shell box business, having manufactured 25,000 of the 4.5 boxes.

The principal products of the Newbigging Cabinet Company are phonograph cases and record cabinets. When we consider that these have to take their place in the home beside a piano or the living room furniture, we can understand the class of cabinet work required on them. The cases vary slightly in construction, according to the particular make of machine for which they are to be used, but the workmanship is of the same high quality on the different cases.

The Present Necessity for Good Wood Turners
Popular "William and Mary" Designs Call for Skilled Workmen Where Quantity Required Does Not Warrant a Turning Machine
By W. J. Beattie

The wood turning craft is one of the oldest in the fashioning of wood into things of beauty and utility, and the importance of good woodturning in the furniture industry is greater at present than for some years back, in fact, it is questionable if it ever was so important in the history of Canadian factories, as at present.

There have been times when it was quite common to see a number of turners in some factories, in the days of the old spindle beds, but the slogan was quantity rather than quality, and old timers are fond of telling how many bed-posts So-and-So could do in a ten-hour day, and the quantity does not grow less by repetition.

Until quite recently it was hard enough to keep even one turner going in a good sized plant, and as this state of affairs lasted some years, the supply of wood turners became pretty nearly exhausted, and now we have on us a demand, which is growing, for the William and Mary style in house furniture, which calls for turning and for quality work.

It is true that the machinery men have perfected machine-turning lathes that turn out really good work, but their use is largely confined to factories that manufacture patterns in large quantities, and the large majority of plants that make reasonably good grade goods in small and medium batches are not within reach of the benefit. For instance, if a plant was making only twelve of a pattern, it would not pay to go to the trouble of setting up a machine to turn out twenty-four or forty-eight pieces, so that the hand turner is still the necessary man.

The William and Mary style, now so popular, originated in England, the name gives the date in history, which was from 1688 to 1702. The style was largely influenced by Dutch furniture which was brought into England at that time. Some of the old patterns were very fine examples, and others have quite the opposite, if one is to judge by illustrations that claim to be copies of the originals, at least so far as beauty and proportion are concerned, though there is an apparent honesty and goodness evident in all of them. The William and Mary style was the immediate predecessor of the Queen Anne, and forms the furniture style link between the Jacobean and Queen Anne. It is only necessary to study the various styles mentioned to see the evolution.

The William and Mary style is made, generally, in oak and walnut, the latter being a favorite. The lighter and daintier patterns look really beautiful in walnut, and there is an increasing pleasure in looking

at them. And all this leads up to the importance of the turning being executed well, as a clumsily turned leg will ruin the appearance of a table, or the poor turning of a toilet stand will spoil a dressing case, etc. It takes mighty little to make the difference between a pleasing effect and the opposite, when it comes to the turning on a William and Mary design.

As some examples of the two kinds of work referred to, see Fig. 1 and Fig. 2. At first glance there does not appear to be any great difference, but it can soon be seen that Fig. 1 is much superior to Fig. 2.

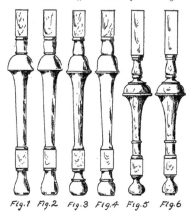

Fig.1 Fig.2 Fig.3 Fig.4 Fig.5 Fig.6

Good and bad examples of William and Mary turnings.

Give a turner a perfect drawing of Fig. 1 and he may present you with Fig. 2.

Figs. 3 and 4 show the same differences in execution of detail. These show very plain patterns, small detail being absent, and yet if they are well and crisply turned, they will be pleasing for cheaper patterns.

Fig. 5 shows a more elaborate pattern, having more shape and style to it. Fig. 6 shows the same pattern not so neatly or crisply turned out, and it lacks the style of Fig. 5. It is not so easy to show these differences in small sketches, where half the thickness of a line makes a decided change, but they are sufficiently clear to show the importance of taste and skill on the part of the turner.

Of course, all the foregoing comments assume that the turner has been provided with a properly drawn detail to work from. The writer's experience is that a good turner should have the same taste and skill as a good carver, a good eye for pleasing lines and proportions, in fact, an artist in his way. Behind all this is the need for technical training so as to be master of his trade, and so turn out work that will be a pleasure and credit to himself and to the furniture business.

Saving Money in the Boiler Room

Skill and Efficiency can be Applied in this Department as Elsewhere

By M. A. Burt

It is customary in the conduct of most kinds of business to give special attention to those departments or branches of a business in which the largest amount of material is used. This is done to discover as soon as possible any opportunity for loss to occur through preventable causes. All careful business men are constantly on the lookout for such sources of loss and they endeavor to maintain conditions favorable to the economical use of materials. The same care is often exercised in the use of oil, nails, brooms, etc., in the factory, and in the office the clerks are reminded that pens, ink, lead pencils, and pins cost considerable in the course of a year.

When it comes to the power plant all this care and painstaking methods of preventing waste are thrown to the four winds. This, of course, is not invariably the case, but such conditions are found much more frequently than the average person is willing to admit, and oftentimes in establishments where it would be least expected. It is both surprising and amusing to note the methods some persons employ in an endeavor to economize, and doubtless they wonder why their methods do not yield greater profit.

Ordinarily, fuel represents the largest single item of expense, excepting labor, in establishments run by steam power, and there are many persons in charge of such establishments who invariably begin to economize at the wrong end. Probably in the majority of cases the engine represents the spigot and the boiler the bung. Untiring effort is put forth to save at the spigot, even to throwing out an old-style, but good, engine, and replacing it with one strictly up-to-date. The term efficiency has been worked so hard in connection with steam-using machines that many persons are disposed to believe it applies only to steam-using and not to steam-generating machines.

Perfect indicator diagrams, high vacuum, close regulation of speed and copious lubrication are much to be desired. They have their places in the list of items that influence economy, but they are the branches of the tree, so to speak, while the boiler and furnace represent the roots, the source of energy and motion, and a factor of tremendous importance in determining the cost of power. It is difficult to imagine why raw material should be wasted in one department and not in another—why the cost of minor supplies should be scanned and the cost of coal allowed to go unheeded.

Economy consists simply in avoiding waste. A man who economizes to the utmost at one point and wastes at another, is not, on the whole, economical. It simply amounts to an endeavor to offset uneconomical methods. The buying of expensive engines for the purpose of reducing the steam consumption without putting the boiler plant in equally good trim, is in effect saving steam at the engine to cover up needless waste at the boiler, which, on the whole, is not true economy.

As every engineer knows, there are many ways of economizing fuel and saving money without buying new engines, condensers, and other engine room equipment. The source of the energy is the place to begin to save. When steam is generated as cheaply as modern appliances will permit, then it is time to overhaul the steam-using machines and install new ones wherever it will pay to do so. Two of the most important ways of saving coal are, keeping the boiler clean inside and outside, and supplying it with hot water.

It requires no small amount of brains and hard work to keep a boiler plant in strictly first-class condition. It represents a continual study of conditions to be sure that one is not saving at the spigot and wasting at the bung. It is easier to save a little by installing new grates or a new heater than to save a like amount by keeping a boiler clean inside and out. But it requires both to save all that can be saved—to avoid waste. It is for the engineer in charge to determine which is the most important. Much can be accomplished in the saving of coal by attacking the problem systematically. Begin with the coal, then the grates, furnace and draft, and then the products of combustion, and lastly the setting.

The next step is to obtain a rapid absorption of heat, which requires heating surfaces free from soot on the outside and free from scale on the inside. There is the quality and temperature of the water fed to the boiler, and means of getting it to the boiler economically. If these things are carefully attended to and the best results which the conditions will warrant are obtained, the cost of steam in any particular plant will have been reduced to the minimum, and any further saving of expense will have to be made in the labor charges and in the steam-using machinery.

Much more attention is being given to the burning of fuel, or combustion, and the wastefulness of admitting an unnecessary quantity of air to the furnaces is more generally understood. The installation of enormous boiler plants in which a fraction of one per cent of the fuel would run a fairly good sized establishment has resulted in careful investigations of the losses incident to poor combustion. Engineers generally have profited by these investigations and much more care in many plants is being taken of the boiler furnaces and in the adoption of boiler room methods. The value of the damper regulator is not understood, even now, as generally as it should be and as it will be when the price of steam coal climbs a few notches higher.

Economizing in heat is comparatively easy when the proper appliances are employed. In very many instances much more heat could be gotten out of the exhaust steam and returned to the boiler if it were not for a penny-wise policy, which is responsible for the installation of heaters that are too small. First cost often blinds men to those methods which yield the greatest returns in the long run. The benefits to be derived by maintaining a uniform water level, thus avoiding heavy strain on the boiler and unnecessary cleaning of fires, are better appreciated now. The near future will undoubtedly bring a much greater demand for simple, reliable feed water regulators.

Those who view the feed water regulator from the viewpoint of saving labor, must necessarily fail to understand that is one of the least of the advantages it offers.

It is important to save labor, especially in large plants. The human element in power plant operation should be, and is being, eliminated as far as practicable. It lessens expense and relieves those in charge of the machinery of much unnecessary care and fatigue. It also places the engineers in a better position to view the plant as a business undertaking, as well as a place demanding the exercise of mechanical skill. Since to err is human, a plant equipped with modern labor-saving appliances should be safer, as well as more economical, than one wholly dependent upon eyes, ears and memory.

Some More About Figuring Profits on Sales Instead of on Costs

, West Toronto, June 2, 1916.

Editor "Canadian Woodworker":

Dear Sir,—I read a very interesting article in the May issue of the "Canadian Woodworker," by W. A. MacKinnon, entitled "Figure Your Profits on Sales Instead of on Costs." On page 16, Mr. MacKinnon gives a "table for finding the selling price of any article." I have worked out several prices according to his table, and find it to be very unreliable.

I enclose an example taken from the table and compare it with the correct method of finding the selling price of an article. The difference is so very great that it cannot be let go unchallenged. Again, on page 17, sheet No. 2, example No. 1, Mr. MacKinnon says if an article cost $5.00 and sells for $6.00, the percentage of profit is 16 2/3%. Surely this is altogether wrong. The profit is $1.00, and $1.00 is 1/5 of $5.00, and 1/5 of anything is 20%. Mr. MacKinnon's method in this case (example No. 1, sheet No. 2) is wrong, and for the benefit of your readers, I give the correct method of finding percentage of profit.

Example According to Mr. MacKinnon's Table

Cost of a car of lumber $670
Net per cent. profit desired 35%

Cost to do business 23 per cent. of sale price.

According to the table, we multiply by 100 and divide by 42.

$$670 \times 100 = 67000$$

42)67000(1595
  42
  ───
  250
  210
  ───
  378
  ───
  220
  210
  ───
   10
  400

The selling price is $15.95, showing a profit of 83% on cost price, after deducting 23% of sale price for working expenses (but the profit desired is only 35%.)

Correct Method of Working Same Example

Cost of car of lumber $670
Profit of 35% on $670 234
                                            ───
Cost of material and profit $904

Now, the selling price must be that sum, which, when 23 per cent. of it is deducted, we have a remainder of $904 (our cost and profit). Therefore, $904 must equal 77 per cent. of the selling price. Thus we have the problem reduced to a simple rule of 3:

If 77% = $904, what does 100% equal? Dividing by 77 gives you 1% and multiplying by 100 gives you 100%, which is the correct selling price.

$$904 \times 100$$

                  77
77)90400(1174
= 77
   ───
   134
    77
   ───
   570
   539
   ───
   310
   308
   ───
     2

Selling price, $1,174.
Proof:—
23% of 1174=$270.
35% of 1174= 234
Cost price = 670

Selling price $11 74

Correct Method of Obtaining Percentage of Profit

An article costs $35 and sells for $42, what is the percentage of profit? A profit of $7 is made.

$42
 35
 ──
 $7

Now the question is, what percentage of 35 is 7? or, in more simple words, if $35 has gained $7, what would $100 have gained in the same transaction? Again we have the simple rule of 3:

Dividing by 35 and multiplying by 100. Answer, 20%.

$$7 \times 100$$

                  35
35)700(20
   70
   ──

Take a simple example: If an article cost $100 and sold for $150, it is very evident you make 50% profit on your outlay. Yet by Mr. Mackinnon's rule you only make 33 1 3%.

Yours truly,
Wm. Graham.

Reply by Mr. Mackinnon

Calgary, Alta., June 13th, 1916.
Editor "Canadian Woodworker":

Dear Sir—Replying to Mr. Wm. Graham's criticism of my article entitled "Figure Your Profits on Sales Instead of Costs," I am sorry to see that Mr. Graham has not given it the study necessary to thoroughly understand the principle involved.

He starts out by using the table with a profit of 35 per cent. and an overhead expense of 23 per cent., which he states brings a profit of 83 per cent. on "cost price," and in his second example, which he terms correct, he uses the same percentage of profit on cost as he did in his first example, although he calculates one on sales and the other on cost.

Now he knows that sales are a greater amount than costs: consequently his net profit will bear a lower ratio to sales than it does to cost and it is unreasonable to think that you should use the same figures for profit when figuring on a basis of sales as when figuring on a basis of cost.

For example, we will suppose that we pursue an article, the prime cost of which is $100. Our overhead expense (as per card No. 3, Fig. 4 on page 17 of May issue) is 15 per cent. of sales and we desire a profit on sales of 5 per cent. Now, by using the table on page 16 we will have to divide the prime cost, after multiplying by 100, by 80 which makes a selling price of $125.

Example:—

80) 10000.00 ($125.00 — Selling Price
    80
   ——
   200
   160
   ——
   400
   400

Now if you wish to use the prime cost as a factor on which to base your overhead expense and profit, you would have to make the former 18¾ per cent. and the latter 6¼ per cent., but this method is very cumbersome and impractical in business.

Mr. Graham's last example states "an article costs $100 and is sold for $150," and he correctly states that this is a profit of 50 per cent. on the outlay. But my article is not based on outlay, but on profits based on the sale, which is one-third of the sale, or 33 1/3 per cent.

I am very thankful to Mr. Graham for his attention to this article, as there cannot be too much interchange of ideas on business methods. His illustrations of the problems are superb from an arithmetical standpoint, but I consider his system of application entirely wrong, and if he will apply his intelligence and education to a close study of the different cards in my article, I feel sure we will soon have a booster for the system of "Figure Your Profits on Sales Instead of Costs."

Yours truly,
W. A. Mackinnon.

The "Canadian Woodworker" is a practical man's paper.

Carelessness

I am more powerful than the combined armies of the world.

I have destroyed more men than all the wars of the nations.

I am more deadly than bullets, and I have wrecked more homes than the mightiest siege guns.

I steal, in one country alone, over $300,000,000 each year.

I spare no one, and I find my victims among the rich and poor alike, the young and old, the strong and weak. Widows and orphans know me.

I loom up to such proportions that I cast my shadow over every field of labor, from the turning of the grindstone to the moving of every railroad train.

I massacre thousands upon thousands of wage earners a year.

I lurk in unseen places, and do most of my work silently. You are warned against me, but you heed not.

I am relentless.

I am everywhere—in the house, on the streets, in the factory, at railroad crossings, and on the sea.

I bring sickness, degradation and death, and few seek to avoid me.

I destroy, crush, or maim; I give nothing but take all.

I am your worst enemy.

I AM CARELESSNESS.

Canadian Box Shooks Required in S. Africa

The Weekly Bulletin of the Department of Trade and Commerce in a recent issue, states that a South African merchant is open to import box shooks from Canada. He will purchase not less than $10,000 worth annually in small sizes. His trade is chiefly with fruit growers.

The following details are submitted for the information of intending exporters:—

Wood must be white, free from knots, and fine sawn. Sizes mainly required are: 18 x 12 x 2½ in.; 18 x 12 x 3 in.; 18 x 12 x 3½ in.

Thicknesses of wood for the above are:—

Two ends	½ in.
Two tops................	3/16 in.
Two bottoms	3/16 in.
Two sides	3/16 in.

An air space of one-half inch is to be allowed clear on each side. and on tops, bottoms, and sides. Thus, to cover the lid there are required two pieces 5½ inches wide and bottoms 5½ inches wide. Sides will also be one-half inch short of the actual depth of ends.

Shooks should be packed in bundles in such a manner that 14 bundles make 100 boxes.

Large quantities of boxes size 18 by 12 by 5½-inch are also required.

The following are the thicknesses of wood for the above:—

Two end pieces	⅝ in.
Two tops	3/16 in.
Two bottoms	3/16 in.
Two sides	3/16 in.

The same number of pieces and the same air space referred to above are required.

Orange Boxes

The export of oranges is calculated in a few years to use millions of boxes. The size of the cases required is 26 by 12½ by 12½-inch.

These boxes are made up of eleven pieces, consisting of two tops, two bottoms, four sides, all of equal size, with ends and one centre piece of equal size. The ends and the centre piece should be made solid by patent jointing process. Steel clips are of no use.

The ends and centre piece should be 12 by 12 by ⅝-inch, and tops, bottoms, and sides 26 by 5½ by

Goods should be shipped through bankers against documents Cape Town, and prices quoted c.i.f. Cape Town are preferred.

Wood Used for Artificial Limb Making
In No Other Way Can it Play More Helpful Part to Humanity

With the rapid developments of recent years in the manufacture of furniture and general wood products, the average woodworker is very apt to have overlooked the modest industry of artificial limb making. And with all due respect to artificial limb makers we would have preferred to see the industry remain in obscurity. But it is very gratifying to find at this time when so many of our soldiers are returning from the battlefields of France and Flanders, where they have lost perhaps a leg or an arm, that the manufacturers of artificial limbs have been achieving some wonderful results along mechanical lines in

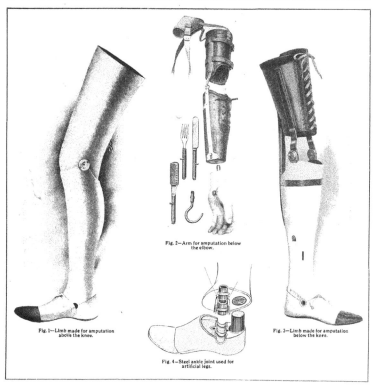

Fig. 1—Limb made for amputation above the knee.

Fig. 2—Arm for amputation below the elbow.

Fig. 3—Limb made for amputation below the knee.

Fig. 4—Steel ankle joint used for artificial legs.

the production of limbs to replace those supplied by nature. It is also gratifying to know that limbs equal to (if not superior to) those produced in any other part of the world, are manufactured right here in Canada, and it is to be hoped that our Government will be as generous in providing artificial limbs for returned soldiers as those soldiers were in offering their lives in the service of their country. It is also to be hoped that they will take advantage of the highest degree of mechanical skill that is obtainable.

The very least that can be done is to give the parties requiring limbs an order on some good, reliable manufacturer where they can secure individual attention, which carries with it the privilege of returning to have any slight adjustments made which may be necessary for the comfort of the wearer. This is one instance where the Government should forego the system of calling for tenders and having the work done wholesale.

Artificial limbs are made from white English willow, which grows in abundance in some parts of Canada and the United States. Great care must be exercised in the selection of the wood, however, as it must meet certain rather rigid requirements. It will not do to manufacture an artificial limb block from a small tree containing the heart in the centre of the piece, because of the greater liability to checking. A tree of 12 inches diameter, or over, is therefore required, although occasionally smaller trees are utilized.

Only the straight, clear portion of the trunk is usable, which often means less than 10 feet of merchantable material from the ordinary tree. The logs are split into quarters as soon as cut, in order to avoid possibility of checks in drying. Sometimes the ends of the sticks are painted to prevent too rapid seasoning, or a hole is bored in the end of the piece, which tends to prevent end checking by distributing the seasoning process farther back into the piece. Most of the manufacturers of artificial limbs believe the wood should be cut in winter and seasoned for about eighteen months or two years.

The seasoning or preparation of the wood for use is a very important operation and a keen appreciation of its fibrous qualities is necessary in order that the best results may be secured. The wood, though tough, responds readily to cutters. Its fibres will spread to admit rivets for the rivetting of metals and it has sufficient spring in its texture to prevent crushing or disorganization of the grain. Fig. 1 shows what is known as a thigh leg and is made for an amputation above the knee. It is made entirely of wood, covered with rawhide, has metal knee and ankle joints and the foot has a joint across the toes.

One is apt to think from looking at the illustration that the party wearing the limb would not have sufficient control over the knee joint to prevent it from swinging right around when coming down stairs.

This is not so, however, as the mechanism that controls the knee is almost human. A strap controlling the mechanism passes over the shoulders of the wearer and if he feels the knee bending over too far he can stop it by slightly raising his shoulder. To a reader this may seem complicated, but in reality the whole thing is so simple that it only takes the wearer a short time to secure almost human control over it. The limb is finished in flesh-colored enamel. Fig. 4 shows the ankle joint used by Authors & Cox, of Toronto in their artificial legs. Fig. 3 is a leg made for an amputation below the knee. The lower part is made of wood with ankle joint, and the top portion is made of leather with steel for the knee joint. Fig. 5 shows an arm and hand of wood and having an

Fig. 5—Arm for amputation above the elbow.

elbow joint. The hand is made so that it will grip firmly a pencil or any other article necessary to perform light work. For heavy work, of course, a hook would have to be substituted for the hand. The change from the hand to the hook can be very quickly made by means of a socket in the end of the arm. In Fig. 2 is shown an arm and hand, part wood, and part leather. The spring in the hand which permits the thumb to close tightly against the fingers to hold the pencil or other light object can be plainly seen in the illustration.

Handy Draughting Table for a Woodworking Factory
Staff Article

The writer had the pleasure of visiting a draughting room of a woodworking factory recently which seemed to be about the last word in arrangement and equipment. The draughting table used is worthy of special mention.

The accompanying drawings and a short explanation will be sufficient to point out the most important features. The end of the table at which the drawers are placed is panelled, and the other end is open. Six drawers are provided on each side of the table for drawing materials or papers. The large drawer at the bottom contains divisions which may be spaced to suit individual requirements. The two large open shelves are used for drawings, T squares, etc. The two brackets on the panelled end have a piece of $\frac{3}{4}$-inch piping placed across them to hold a roll of drawing paper. The piping acts as a roller on which the roll of paper can be revolved. A small brass plate

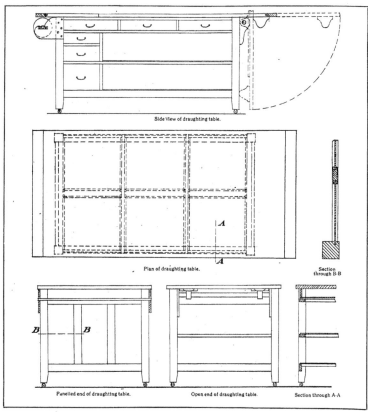

Side View of draughting table.

Plan of draughting table.

Section through B-B

Panelled end of draughting table.   Open end of draughting table.   Section through A-A

screwed on the outside of each bracket keeps the roller in place. Slides are screwed on the under side of the top by which it may be extended to the position indicated by the dotted lines, and then swung so that it is straight up and down. This swinging top is a very useful feature, especially for the man who is making large drawings, as it enables him to get an idea of the proportion of the article he is about to construct. It is also very useful for the man who is designing furniture. The top itself is eight feet long, forty-five inches wide, and one and one-quarter inches thick. It is made of basswood, and has cleats four inches wide fastened to the ends by means of a slip-feather to prevent it from warping. The method of arranging the slides will be seen on the end elevation. The top

is swung by means of a wooden roller which passes through two brackets screwed to the posts and also through two blocks which are attached to the slide. The top is not fastened to the table by any other means than the roller, as its own weight holds it perfectly flat. The height from the floor to the top of the table is thirty-nine inches, and the width was arranged to accommodate a roll of paper forty-two inches wide. The body of the table is made of elm. It is finished with one coat of varnish and trimmed with brass handles, and presents a very handsome appearance.

The man whose middle name is work, gets there.

A Solution of the Waste Problem

Closer Co-operation Between Lumber Manufacturers and the Owners of Woodworking Plants

By James E. Dewey*

I have been requested to address you to-day on the problem of waste in the manufacture of your products, but do not intend to confine my remarks strictly to this subject, as it seems to me this offers an opportunity to cover a much wider field of operations concerning your various problems, and I trust you will regard my comments and suggestions to you from an impartial standpoint.

The problem of waste, without doubt, is the most vital question that confronts every manufacturing institution in existence to-day, not only in the consumption of lumber alone, but it applies to any raw material that enters into the process of manufacturing your products. When you consider that statistics show all of the principal materials have increased in value from 25 to 150 per cent., which, I presume, is a conservative maximum, then you appreciate, to some extent at least, the great necessity for conserving the waste in these materials. Modern machinery and improved methods absorb a reasonable percentage of this increase in cost of materials by reducing the handling and manufacturing costs and otherwise creating a saving in production. We must then give our attention to other means of waste reduction, and I now refer particularly to your lumber requirements.

A great deal of abuse and criticism has been heaped upon the heads of the lumber industry over the question of establishing a set of inspection rules which will conform more closely to the requirements of lumber consumers. While it is probable that the present rules can be improved upon in this respect, until this past year no united effort has been made by the consumers to encourage an interchange of ideas on the subject. The Federation of Furniture Associations has finally appointed a committee to confer with the rules committee of the National Hardwood Lumber Association, and I am confident that a set of rules will be drafted acceptable to all concerned.

The elimination of waste is a factor which appeals to me more each day as I study manufacturing problems, and it offers unlimited possibilities toward the solution of our difficulties. This, again, suggests the importance of closer co-operation between the consumers and manufacturers of lumber. Those among you who are not familiar with the operation of a sawmill cannot appreciate the remarkable improvement in manufacturing methods of an up-to-date plant. To some extent this has been due to a natural evolution and the necessity of reducing costs, but the idea of conservation has been paramount in the development of these methods. Why, then, should not the consumer derive a benefit from these developments through a closer communion with the manufacturer? It is a surprising fact that a large number of lumber consuming industries of the present day make no attempt to compute their waste, or estimate their cost of production. The important feature with them is the total production of their finished product within a specified time. Quantities of stock are burned which could be utilized in the production of useful articles

*Abstracted from an address delivered before the Manufacturers' Cost Association, Chicago, by James E. Dewey of the Stearns Salt & Lumber Co.

at a profit. I consider this actually a crime, and some means of punishment should be meted out by our government for such gross and needless waste of this country's resources.

We have come in contact with one case through our service department which will serve to illustrate this condition. This company uses a large amount of lumber each year, and one of our salesmen who calls on them regularly explained the idea of our service plan. The buyer was mildly interested and agreed to consult our special representative, so that an interview was arranged. Our representative has now held three meetings with that buyer, but so far has been unable to interest him. On the first visit the buyer resented his suggestions and seemed to consider them a reflection on his ability. The succeeding visits were a little more productive, but so far the privilege of investigating their manufacturing operations has been refused. We know positively that we can offer suggestions which will save money, not only in the purchase of material, but also the waste in their production. This instance is only one of many and emphasizes my former remarks regarding the closer co-operation between the purchasing and operating departments.

The principal factor in the waste elimination is in the selection of your raw material, and this again reverts to the purchase of your requirements. In this connection we must first consider the kind of material best adapted. It is necessary to know what physical properties in the wood are most essential for the use intended. This is in regard to strength, texture, grain, weight, etc. Next, the choice of grade or grades which will produce the maximum of product with the minimum of waste at the least cost, and then the selection of widths, lengths and thickness most economical for the sizes of your cuttings. In this connection, where narrow widths or short lengths are desirable, it immediately reduces the initial cost of your material.

Now we come to the products of timber waste which may be utilized to advantage. This at once offers opportunities for conservation that are unlimited. Considering these facts, who is better qualified to offer constructive ideas and furnish specific information in the proper understanding of these factors than the lumber manufacturer whose knowledge from years of study and experience enables him to proffer most valuable knowledge? It is a deplorable fact that we have never been given the proper consideration in this respect, while in practically every other line of materials you immediately call in the producer to confer on your problems.

Our mutual interests have millions of dollars invested that are dependent on the great timber resources of this country, so I implore your earnest co-operation and assistance to curtail this wholesale slaughter of nature's bounty for the sake of our own existence and the benefit of posterity.

Do not fail to read about the New Shell Box on page 18.

The Lumber Market

Domestic Woods

The situation in regard to domestic woods has not shown any developments since the time of our last writing. There is still a decided scarcity of hardwood stocks, but the demand coming from the furniture, automobile and implement industries is not as pronounced as it was a couple of months ago. This may be on account of the regular slackness that occurs at this season of the year when so many people are preparing to spend the summer away from home. Or on the other hand, it may be due to the great difficulty that manufacturers are experiencing in getting skilled mechanics to do the work.

However, once the hot weather is past there will undoubtedly be a revival in the furniture and other industries, and as hardwood stocks are liable to be a great deal more scarce then than they are now, it looks as though it would be a good plan for manufacturers to purchase any hardwood lumber that may be offered to them in the meantime. If they hold back, American lumbermen will not be slow in picking up any cars that are offered, as practically every industry in the United States in which lumber is used is booming and their stocks are none too plentiful.

The prices quoted at Toronto on domestic woods for car lots are as follows: white ash, firsts and seconds, inch, $60; white ash, No. 1 common, $45; brown ash, firsts and seconds, inch, $50; No. 1 common, $40; birch, firsts and seconds, inch, $42; No. 1 common, $35; basswood, firsts and seconds, inch, $40; No. 1 common, $34; Nos. 2 and 3 common, $18.50; soft elm, firsts and seconds, inch, $40; No. 1 common, $43; No. 2 and 3 common, $17.50; rock elm, firsts and seconds, 1½ inch and 2 inch, $55; No. 2 common, 1½ inch and 2 inch, $37; soft maple, firsts and seconds, inch, $32; No. 1 common, $25; hard maple, firsts and seconds, inch, $40; No. 1 common, $30; plain red oak, firsts and seconds, inch, $66; No. 1 common and better, $40; No. 2 common, $32; plain white oak, firsts and seconds, inch, $68; No. 1 common, $40; No. 2 common, $32; quarter cut white oak, firsts and seconds, inch, $89; No. 1 common, $57; hickory, firsts and seconds, inch, $70.

Imported Woods

Prices of imported woods remain practically the same as they have been for some time. The demand has not been as great during the past month as it was earlier in the spring. This is no doubt on account of manufacturers stocking up their yards when prices were slightly lower. The announcement has been made that the embargoes that have been in effect on the railroads will likely be lifted in a few days. This will enable some of the manufacturers who were unable to secure shipments on account of the embargoes to replenish their stocks, but it is not expected that it will materially affect the price of lumber.

Indications show that the different factories continue busy. The factories in the United States manufacturing sash, doors, and interior trim, are particularly busy, due to the increased activity in building operations. In some parts of the United States where statistics are available, these show that the building permits issued this year exceed anything for the same period during the last ten years; this means that large

quantities of hardwood lumber will be required for the better class of buildings.

The report issued by the Hardwood Manufacturers' Association of the United States covering the period from April 1st to May 1st, show that the stocks on hand in the different yards are somewhat better than might be expected. The greatest decrease is in plain white oak and amounts to 2,628,000 feet for the month. Plain red oak is next with a decrease of 520,000 feet for the month. Taking the others in order, they are, quartered white oak, 1,845,000 feet; poplar, 588,000 feet; chestnut, 728,000 feet. The stocks of quartered red oak have increased by 241,000 feet, and quartered poplar by 11,000 feet.

It was pointed out at a recent meeting of the Louisville Hardwood Club that the advance on the price of hardwoods going to the furniture trade have been proportionately less than that of any other raw material used in the manufacture of furniture. The following percentages of advances for lumber are shown in tables prepared by the furniture men. Figured mahogany veneer, 45 per cent.; mahogany lumber, 21 per cent.; birch, 7.5 per cent.; gum, 21 per cent.; log run elm, 8 per cent.; crating material, 20 per cent. The advance in plain oak has been only 6 per cent., and quartered oak, 2½ to 4 per cent.

There is at the present time a strong demand for poplar, which is quoted at $60 to $63 for firsts and seconds, inch, and maple in all thicknesses up to 3 inch, is in demand. The demand of the automobile manufacturers for elm has placed that wood in a pretty strong position as regards price. Fancy woods like mahogany and walnut have been somewhat in demand of late. The prices on Honduras and Mexican mahogany run from $135 to $175 per thousand, 1-inch stock, f.o.b. New York, according to grade. Cuban mahogany is slightly lower in price, being quoted at $130 per thousand for a pretty good grade. The price of Philippine mahogany ranges from $100 to $140 per thousand for 1-inch stock, depending on color, figure and freedom from pin holes.

The demand for oak has not been quite so great during the past month, but the price remains about the same. Quartered white oak is being sold at $88 to $93 per thousand in New York. Plain oak still remains around the $60 mark for firsts and seconds, inch. The price of walnut is about $100 to $105 per thousand for firsts and seconds, inch. Chestnut is still used in large quantities for furniture and piano factories for core stock, and the price averages from $48 to $52 for firsts and seconds, inch.

Alder for Gunpowder

Years ago the British Government caused to be preserved extensive areas of alder as a reserve supply of wood from which to make gunpowder, and this foresight is now bringing results. Much willow is also used, but the willow is chiefly utilized for smokeless powder. The wood is reduced to charcoal in specially constructed retorts, and the best charcoal is made from small twigs, though all parts of trunks are used. American grown willow is likewise supplying much of the material for powder. Large areas in Delaware and other eastern States were privately planted in willow some years ago, and the osiers grown on the land now bring good prices at the powder works. American alder in the Eastern States is too small for any other use than powder making, but is just right for that. Large quantities of it grow in swamps and along the banks of streams. Let it be hoped that no other chance like the present to sell gunpowder material will ever come.

System in Oiling

This thing of the careless and happy-go-lucky use of the oil can around machines is not in keeping with modern methods and it endangers the safety of machinery bearings, says Wood Turning.

For a generation or more the usual practice has been to provide every machine operator with one or more oil cans. With these conveniently placed by his machine he squirts a little oil into a bearing here and there whenever the spirit moves him, most of the oil working out immediately and dripping on to the floor somewhere, and sometimes when he is busy he forgets about his journals and they may suffer for want of oil until they become hot and send out a danger signal in the shape of an odor calling attention to their needs. Then it requires a lot of oil and perhaps a loss of time to get them in shape, and quite often serious damage is done either to the journal or to the box.

What we need is to systematize this business of oiling and to reduce it to a more safe, sane and sensible basis. Economy in oil is a small matter, but small things count, and besides there is involved the safety of the journals and the continuous running of machines as well as the matter of economy and the proper utilization of oil.

In so far as practical, the journals on machines and shafting should be standardized in the matter of oiling. What is meant by this is that in so far as practical they should be so arranged that each will require oil with substantially the same frequency.

A splendid thing if it can be worked out is that of having each journal and box so arranged and fitted that an application of oil once a day will serve its needs. This is what is meant by systematizing the work. When it is put on a basis of that kind any operator can make his round of oiling, say at starting time in the morning, and this will become such a fixed habit that it will not be neglected and overlooked and it gives a measure of insurance of proper attention.

There are various devices and ideas being promulgated in connection with lubrication. Some with grease body attachments may require attention only once a month or more. This is good, the only weak spot in it being that one is more likely to forget when the attention is needed than if it is a matter of daily occurrence and part of the daily task. In fact this is the one weak spot in the grease system of lubrication, the chance that it may be overlooked because there is such a long time between the calls for attention.

There are many devices and make-ups of journal boxes both for the use of oil and grease, and each institution may thrash out the problem for itself and decide on which it prefers, but some idea should be standardized in a way so that the attention given to oiling will be systematic and sure.

For example, it is disturbing to system to have a machine in use with a few bearings that a little oil once a day will suffice, other bearings that need oil once an hour and others that need it two, three, or four times a day. Efforts should be made so that as nearly as practical all the journals in the factory will require oil attention with the same frequency.

In some instances it may be advisable to have a division, say to have certain journals requiring daily attention, those on the machines for example, while on the line shafting when grease cups or other system is used it may be weekly or monthly. The division may have to be extended to cover the needs in some factories and the ideas suggested here may have to be amended in various ways. Still, the arguments hold good that system in oiling makes for safety. Also that to get any kind of a systematic basis to work from one must in so far as practical standardize the time needs of attention to the various journals about a machine. This may call for different types of boxes or oiling devices for journals of varying sizes and speeds but it should be practical to get away from the old haphazard method of oiling and to so systematize the work as to insure proper lubrication every hour of the day with a minimum of waste time and waste oil.

Wooden Goods for South Africa

Mr. W. J. Egan, Trade Commissioner, Cape Town, has forwarded the accompanying illustrations to the

Faucets   Flour Pails

Butter Tubs

Department of Trade and Commerce, as indicative of kinds of wooden flour pails, butter tubs and faucets in demand in South Africa.

Easy Way to Remove Splinters

A correspondent of Woodworker (Indianapolis), describes a simple and safe surgical appliance which is worth trying, if occasion calls for something along that line. The writer says that one of the most annoying minor accidents to which every wood-worker is subjected at times, is to have a sliver of wood stuck in his hand; and its removal is painful if not properly performed. If the splinter is softwood, it cannot be removed very easily with a needle or other sharp instrument.

A very easy and effective way of removing a splinter without pain or inconvenience is to take a wide-mouthed bottle (such as a milk bottle), filled nearly full of water as hot as the glass will stand. Place the injured part over the mouth of the bottle, pressing down slightly, thus preventing any steam from escaping. This will cause the flesh to be drawn down, and in a minute or two the steam will extract the splinter, at the same time preventing the inflammation which usually follows an injury of this kind.

THE FINISHING ROOM

Hints on the Proper Handling of Varnish

By "Foreman Finisher"

Cracking of Varnish

When varnish, after being on for a few days, begins to show cracks, the blame is generally laid on the varnish, but it is not always the varnish that is at fault. If a cheap grade of varnish is applied over a quick drying foundation it may stand for quite a long time without showing any cracks, but in time it will turn white, that is to say, all the life will die out of it. If this same varnish is put on a foundation having a certain amount of elasticity, it will perhaps crack in a day or two, but in any event will crack before turning white. The reason is that a low grade of varnish is short and brittle and will work to better advantage when used on a surface that is likewise hard and brittle. If a high grade of oil varnish is used on a hard, brittle surface, the chances are that it will soon show cracks. The deduction to be made from these facts is that a varnish will do its very best when used on a surface that is as nearly like the composition of the varnish as possible, but in all events the surface should be perfectly dry.

Sweating of Varnish

One of the first questions usually put to the varnish salesman is "does your varnish sweat?" More than likely he will respond that it does not sweat, when, as a matter of fact, any varnish will sweat out under certain conditions. When a varnish sweats it is a sign that it still has life. A varnish that will not sweat does not possess any life whatever. If you will take a piece of varnished work, preferably one that has been varnished for some time and try it for sweating, you will, in many cases, find, on rubbing it, that it will sweat, which is a sign that sufficient moisture or oil is still in the varnish to give it proper elasticity. This means that it will be good for a long period of wear and tear.

The writer has been asked, "does the sweating of a rubbing coat injure the flowing coat over it?" It certainly does, because the fact that it sweats shows that it was not dry enough. Yet a person will say, "I thought all varnishes would sweat." They will, but there are different degrees of sweating, and the sweating of an old or well dried coat is not injurious to the finish. It is the soft and fresh coat that causes trouble.

Wrinkling of Varnish

Wrinkling of varnish is caused by a too heavy flow. Of course, in some cases it may be due to the fact that the varnish is thin and new. It is always advisable to buy varnish that is sufficiently aged, and if you are dealing with a reliable house it is not likely that they will try to sell you anything but properly aged varnish, as their reputation will be at stake. Flowing varnish in particular requires to be well aged, for on it to a very large degree the finished appearance and lasting qualities of the work depend. Some of the faults in varnished work may occasionally be attributed to the way in which it is applied, but if it is put on by an experienced brush hand there will not likely be any trouble on that score. The brush itself, however, should be suited to the work. It should be pliable enough to allow the varnish to be properly worked over the surface of the job being coated and the varnish should be applied as quickly as possible to avoid the setting of the same before it is properly spread out.

Slow drying varnish contains a larger proportion of oil and is therefore more liable to run and streak than one that is heavier in body and takes longer to dry. It will therefore require more dexterity and careful handling by the brush hand in order to prevent running or lapping of one "stroke" over the other. When left standing for a long time all varnish will deposit settlings. For this reason it is advisable when the larger portion of a can has been used to put the balance away for some less particular work. Varnish will improve with age up to a certain point, but after that it is liable to become fatty and very often so much so that it will not be fit for use. If kept in an airtight can, however, it will keep perfectly, but very often the can may appear to be air-tight and at the same time the air will be getting in and oxidizing the varnish.

Proper Staining of Birch

By W. S. Thompson

I would like to offer a little information on the proper staining of birch, as I have seen so many jobs where the finished appearance was not what it should have been and this on account of the proper stains not being used. As is well known, there are three kinds of stains, spirit, oil, and water. Stains of nearly any shade desired can be purchased ready to apply. Samples and color schemes are freely supplied by manufacturers. The only thing to be careful of is to see that the stains are of tested quality. Spirit stains are not recommended for birch since the alcohol evaporates so rapidly that it is difficult to apply it uniformly. Oil stains give splendid effects, but not as clear and transparent colors as are produced with water stains.

Water or acid stains are evenly and quickly applied and permit any method of finishing over them. The objection that water stain sometimes raises the grain of the wood can be disposed of by sponging the wood with clear water and sanding it before applying the stain. Varnish shows up more clearly the natural appearance of the wood, so that defects are magnified as well as natural beauty. This makes it important to have the surface smooth and free from defects, dirt, and dust before the varnish is put on. Varnish works best at a temperature of about 70 degrees and in a dry atmosphere.

Rubbed or rubbed and polished finishes give the

best results with birch. An excellent silver gray can be produced on birch by treating it with one coat of acid stain, followed by sand papering and then finishing with one coat each of white shellac and wax and then rubbing it. The user of birch should always remember that he is handling a high class hardwood and that its peculiar richness of tone and figure are worthy of his most painstaking efforts. Birch has proved its merits for such uses as store counters, show cases and fixtures and the interior trim of some very handsome apartments, offices and hotels. Its reasonable cost places it within the reach of people who desire handsome woodwork and yet cannot afford mahogany or walnut.

A few of the shades which show up particularly well on birch are silver gray, walnut, natural, and brown. These colors will harmonize well with almost any scheme of interior decoration, but if the user desires some other shade of finish, he can still use birch, as it will take and hold any color of stain if it is properly applied. One of the best features of birch is that it harmonizes so well with other woods. Maple or oak floors with birch trim make an excellent combination, and birch veneered doors can be used with trim of other woods or white enamelled trim.

Chip Marks Often Affect Finish
By J. Crow Taylor

When there are chip marks on dressed lumber, that is marks that indicate chips and shavings have remained on the board and passed through under the pressure bar or some of the out feeding rolls, it is a pretty sure sign that from one cause or another the blower system is not performing its duty.

It is a comparatively easy thing to handle the chips from a planer cutting a narrow stock like flooring, but when it comes to surfacing wide boards it is a bigger problem and it is there we are most likely to find chip marks on lumber.

Sometimes these may be due to a sort of over-burden, to a thick place in a board which calls for an unusual amount of cutting on the part of the head so that the chips and shavings produced are in abnormal quantity and not all of them will be carried away.

When this is not the cause, the trouble is either in the hooding or lack of speed or power capacity on the part of the fan, or due to a low dragging speed all around, which results in the cutterhead itself not giving the shavings a good enough start, and so low is the air pressure that they are not carried away properly.

When the fan is of a good type and kept up to the proper speed so that there is enough air pressure for ordinary requirements, it may be taken that chip marks mean lack of proper fitting of the hood over the cutterhead.

It is impossible to get efficient service out of any blower system unless the hoods are carefully and properly fitted to the machines. This is a work that should never be done by an inexperienced man. Either those who specialize in erecting blower pipe systems and fitting hoods should be called in to do the work, or else the manufacturers of planers themselves should be asked to have hoods properly fitted to the machines before they are sent out.

There are some manufacturers who with certain machines make it a point to send the hood along with the machine, properly fitted, which helps some, especially where it is not convenient to secure a high order of expert service for making and fitting hoods. However, where this expert service is available one should

be able to secure just as good results by having the hoods fitted after the machine arrives.

One should make sure, however, that the service is competent, because on this depends a lot of the success of the blower system. Even a concern that has been long in the business may send out a comparatively new workman on the job and to safeguard against errors from this practice, one should ask for a guarantee of service or performance from the concern putting in the blower system, and then if the trouble develops, put it up to them to come and correct it.

Chip marks on surfaced lumber are not only aggravating things but they certainly depreciate the value of the lumber, and means should be taken to prevent them happening in any regular manner. There may be an accidental happening now and then but no continuation of chip marks should be tolerated. The suggestions given above are enough to point out where and how to look for the trouble, and when chip marks appear persistently the trouble should be chased down and gotten rid of.—Yates Quality.

Turpentine Must Not Be Adulterated

Furniture factories are large users of turpentine and will no doubt be interested in the following standards which have been established by the Laboratory of the Inland Revenue Department at Ottawa.

A very large number of complaints have been received in recent years by the laboratory from users of turpentine and the results of recent investigations carried out by the department prove that ample grounds existed for the complaints. The chief adulterant was found to be a petroleum product, but resin oil has also been used for the purpose.

The standards were adopted after a careful and exhaustive inspection of 212 samples which were purchased throughout the Dominion in July, August and September of last year, and were specially obtained from dealers of paint materials. Fifty of the samples fail to meet the requirements and must be described as adulterated under the Act.

Standards for Turpentine

I. When turpentine is sold as a drug, for medicinal purposes or to the order of a physician, it shall meet pharmacopoeia requirements as provided by section 7 of the said Act.

II. When turpentine is sold under any other conditions than for purposes of medicine, as above, it shall meet the following requirements:—

(1) It shall be entirely free from mineral oil.

(2) Unless sold as wood turpentine, it shall absorb not less than 340 times its weight of iodine (Hubl Solution and Method). If sold as wood turpentine it shall absorb not less than 240 times its weight of iodine by same method.

(3) The undissolved (unpolymerized) residue on treatment of 10cc with 40cc of a sulphuric acid containing 20 per cent. of the fuming acid, shall not exceed 10 per cent. by volume of the sample.

(4) The refractive index of this residue shall be not less than 1.4950 at 20°C.

(5) The refractive index of the sample at 20°C. shall lie between 1.4680 and 1.4730.

(6) The specific gravity of the sample at 20°C. shall not be less than 0.860.

(7) The initial boiling point shall not be lower than 150°C. under ordinary atmospheric pressure.

(8) At least 75 per cent. by volume shall distil below 160°C.

(9) The residue on evaporation over a steam bath shall not exceed two (2) per cent.

Single Ply Veneer

suitable for

Crating Purposes

Basswood　　　　Maple　　　　Elm

Various Widths and Lengths.

Write for prices

Hay & Company, Limited, Woodstock, Ont.

WE HAVE BEEN MAKING THE

Best French Glue in the World

SINCE 1818; ALMOST A CENTURY

In competition at the numerous Expositions abroad, the quality of our glues has been found so high that we are usually declared "Beyond Competition." Our trade in this country makes the same declaration. Consistency in quality and service has brought this result and we are proud of it.

Perfection in manufacture from the best raw material makes the quality of COIGNET FRENCH GLUES consistently the highest.

Large stocks at Montreal assure the best of service.

Why not take advantage of these most important factors by using our glues?

At your service for prices and samples.

Coignet Chemical Products Co.

17 State Street, N. Y. CITY, N.Y.

Factories at { St. Denis } France
{ Lyons }

ANGLO
RUBBING and POLISHING

Works free and easy and can be rubbed in two days.

WRITE FOR SAMPLE

The
Ault & Wiborg Co. of Canada, Ltd.
Varnish Works

Montreal　　　　Toronto　　　　Winnipeg

THE GLUE BOOK

WHAT IT CONTAINS

Chapter 1—*Historical Notes*
Chapter 2—*Manufacture of Glue*
Chapter 3—*Testing and Grading*
Chapter 4—*Methods in the Glue Room*
Chapter 5—*Glue Room Equipment*
Chapter 6—*Selection of Glue*

WOODWORKER PUBLISHING CO., LIMITED
345 Adelaide St. West, Toronto

Hydraulic Press of Large Dimensions

The hydraulic press shown in the accompanying illustration was built by the Hydraulic Press Mfg. Company, of Mount Gilead, Ohio, for the Ahnapee Veneer and Seating Company, to be used for veneering purposes. It is of the double platen type with four cylinders, and has a double pressure capacity of

This Press Will Hold Nearly Fifty Men.

920 tons. Each of the four cylinders has a 13-inch ram working in it. The four rams may be operated in unison when it is desired to raise the two platens

or whole pressing service, or they may be operated in two pairs, each pair of rams raising one platen. When the two platens are in use the press will accommodate stock 192 inches long and 96 inches wide, or when the platens are operated separately, they will each accommodate stock 96 inches square. The daylight space between the platen and pressure head is 64 inches, which is sufficient to accommodate a truck and clamping plate, together with a package of stock 36 inches high.

On each side of the platen four crickets are provided to support the wheels of the truck until the platen begins to rise, when the wheels will be received into the pockets in the platen track, thus permitting the bolsters of the truck to come firmly against the platen. Each pair of rams is controlled by two hydraulic valves of the globe type. One valve admits the pressure to a pair of cylinders and the other controls the outlet to the return line. The use of this type of valve limits the pressure to whatever amount is required and holds the pressure on the stock until the retaining rods are secure in place.

Steel is used throughout in the construction of the press. Double steel I-beams are used for the head and base, which prevents any tendency to spring in the centre of the press and insures the veneered stock being perfectly flat. Some idea of the size of the press can be ascertained from the picture, which shows nearly 50 men comfortably seated inside.

Enthusiasm is what you need. Unless you have it you are only marking time in the business world. Lack of progress is equivalent to loss of ground. No one stands still. He goes either forward or back.

This shows the possibility of FIGURED QUARTERED RED GUM as used in our private office. We specialize this wood

Importers and Manufacturers

Mahogany

and

Cabinet Woods

Sawed and Sliced

Quartered INDIANA White Oak, Red Oak, Figured Red Gum, American Walnut, Etc.

Rotary Cut Stock in Poplar & Gum for Cross Banding, Back Panels, Drawer Bottoms & Panels.

The Evansville Veneer Company, Evansville, Ind.

The Cause of Uneven Surfaces in Veneered Work

Staff Article

The superintendent of a woodworking plant in which a veneer department had recently been installed put the following problem up to the writer a few days ago: "The work being turned out by our veneer department is not satisfactory. Instead of having a smooth, flat surface, it is all tiny hills and hollows and we cannot understand why. We use a chestnut core and run it through the sander, using coarse paper before laying the cross banding, which is 1/20 in. rotary cut poplar. Two days after the cross banding is laid we put the stock through the sander and apply the face veneer. The unevenness is not noticed until after we get a coat of varnish on in the finishing room and then it becomes worse as the time passes."

The whole trouble is undoubtedly due to the fact that they only allow two days before taking the cross banded stock out and sanding it. It should stand for at least two weeks to give it time to dry out thoroughly before it is sanded and the face veneer laid. A review of the methods of handling core stock and veneer, as carried out in one or two large factories may prove interesting to readers who have experienced trouble on this or similar points. The core stock after being accurately thicknessed in the machine room is run through a planer, which, instead of being fitted with ordinary planer knives, has tooted knives in their place. The idea of the tooted knives is to scratch up the hard, smooth surface of the core stock so that the glue will have a better chance to penetrate and take a grip on the wood. After being tooted the core stock is sent to the veneering department to be cross-banded.

The gluing of the core stock can be done either by hand or with a gluing machine, as long as a satisfactory body of glue can be put on. After the glue is put on it is allowed to set until, when touched with the end of the finger it feels almost as if it were too hard. Then the cross banding is put on and the stock put away in the press. It is generally allowed to remain in the press overnight and on being taken out in the morning is piled away between strips (of equal thickness) and allowed to stand at least two weeks.

When the two weeks are up, if the stock is needed it is taken to the sander and sanded with coarse paper in order that every particle of grease, etc., that may have adhered to it will be removed. It is then ready for the application of the face veneer, which has in the meantime been prepared in another department. The method of preparing the face veneer consists of cutting it into pieces suitable for the different parts of the work and marking it so that it can be used for the right parts, then gluing tape across the ends to prevent it from splitting when the pressure is put on the press. After the ends are taped the veneer is placed in a rack (built for the purpose) and the rack

placed in a caul-box or other suitable heater. The rack is constructed so that the heat will pass freely through and dry all parts of the veneer. The length of time the veneer should stay in the heater depends on the amount of heat present. However, it should stay in long enough so that when taken out it will be perfectly dry. After it has been dried the veneer is sent to the department where it is to be laid on the cross-banded core stock.

In laying the face veneer the process of gluing and putting in the press, previously mentioned, is repeated. After the stock is removed from the press it is piled away between strips as before and allowed to remain for at least 18 or 20 days, after which it is taken to the cabinet room and made up into the job for which it is intended.

Veneer Press Constructed Under Difficulties

A few years ago the writer was engaged to take charge of the veneering of the interior woodwork and fixtures for a large Government building. The building was being erected in an out of the way part of the country and the Government built and equipped a small shop and dry kiln and did the work themselves. They, however, overlooked a very necessary part of the veneering equipment, namely, a press. Some of

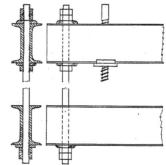

Fig. 1—Veneer Press Made with Channel Irons and Shafting.

the stock to be veneered was 12 feet long, so we required a substantial one, and although money was no object, time was, and we had to rig up something. We looked around the shop and could not find anything from which we could build a suitable press, so we took a walk up around the building where they were erecting the steel work. An idea occurred to us and the accompanying illustration shows one section of the press that was constructed from the idea. The bed and top parts were made of channel irons. For the uprights we managed to secure some pieces of 1¼ inch shafting. The channel irons were held together by bolts, as indicated at A, Fig. 1. The washers, B and C, and the nuts, D, Fig. 1, held the top channel irons in place. The screws and the sockets in which they worked were made at a machine shop in a nearby town.

Fig. 2 shows a plan and section of the casting that was made to work on the lips of the channel irons

and hold the threaded socket in which the screw worked. The press gave good service and turned out

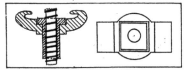

Fig. 2—Casting Made for Veneer Press.

work that would compare favorably with that turned out by any hydraulic press.

Chair Drawing Made on Piece of Three-ply Birch

By J. Walker

In a previous issue of the "Canadian Woodworker," the writer gave a brief outline of the method usually employed in furniture factories in regard to their drawings, and described in detail what is known in the trade as a "rod" and its uses so far as the manufacturing of case goods is concerned.

This article will deal with the method adopted by some of the leading chair factories. First, an accurate

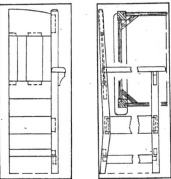

Chair Drawing Made on Piece of Three-ply Birch

drawing is made, showing every detail of construction. All patterns are then traced from the drawing by the draughtsman, and the drawing and paper patterns are handed over to a man who makes the finished patterns and shaper forms required. The finished patterns are usually made of wood for straight parts and zinc for bent parts. The drawing is then returned to the draughting room and remains there so that in case of a mistake or a dispute it is always at hand for a reference. A "rod" is also made for use in chair factories, which I will describe later.

Owing to the fact that there is very little straight work in the manufacture of chairs, there being a considerable number of angle cuts, bevels, etc., it is necessary to make almost constant reference to the drawing. It will therefore be seen that the drawing would soon get worn out, or at least would get so dirty that it would be almost impossible to distinguish anything on it. The "rod" will be found much more convenient in the machine and cabinet rooms than the ordinary drawing on paper.

About the best material that can be used for these rods is three-ply birch, which can be secured in sheets 42 inches by 36 inches. This is about the most convenient size, as it can be ripped in half and makes two pieces. The advertising section of The Canadian Woodworker will put you in touch with manufacturers of this material.

After the pieces are ripped, they are jointed on two edges and sanded on two sides, finishing about 42 in. by 17½ in., and are then piled away between strips to keep them straight. When it is required to make a "rod" one of these pieces is taken from the pile and placed on the draughting table or a bench and held down with a couple of thumbscrews. It can then be laid out with the assistance of T-square, set-squares, etc., in the same manner as the ordinary drawing.

It will be seen from the accompanying diagrams that on one side of the "rod" is placed half of the front elevation and on the other side is placed a broken side elevation and half of plan. I have chosen a chair which is very simple in design and construction, as in a small-scale drawing it is difficult to show all the details. As, for instance, in the seat plan I have only shown one section through the back post, while on a full-size drawing, sections could be shown through the back post at the stretcher, seat-rail, and arm, thus giving the accurate length of the stretchers and rails and showing where the arm is checked to fit on the back post. The rod will be found very useful in the machine room if the stretchers, rails, etc., are sized on the sticker.

It will be seen from the drawing that the only parts which are bandsawed are the back posts, top back rail, and arm. The patterns for these are numbered and kept at the bandsaw, where they will be convenient for use when the parts are brought to this machine for sawing. The different parts of the chairs then go to the cabinet room, where the machines for performing the various operations necessary for the assembling of the chairs are located. In this department the rod will be found indispensable, as the lengths and bevel cuts of the rails, stretchers, front legs, etc., may be had from it, also the location and size of all tenons and mortises.

OAK — MAHOGANY — CIRCASSIAN — MAPLE

VENEERS

BIRCH — POPLAR — CHERRY — WALNUT — GUM

THE OHIO VENEER COMPANY

Importers and Manufacturers

Foreign and Domestic Veneers and Hardwood Lumber

We always carry a large and assorted stock of Mahogany, Circassian
Walnut, Sawed and Sliced Quartered Oak.

Send us your enquiries and orders. We guarantee good service.

2624 to 2644 Colerain Avenue,   -   -   **CINCINNATI, OHIO.**

Buyers of Veneers and Panels

will find it to their advantage to purchase from the manufacturers and dealers whose advertisements appear in the "Canadian Woodworker." They are progressive firms —the leaders in the business, which is a guarantee of good service and prompt attention to orders.

Give your business to the man who will spend his time and money to get in touch with you. He deserves it—if his stock and prices are right.

Hoffman Bros. Co.

Estab. 1867, Incor. 1904

**800 West Main Street,
FORT WAYNE, INDIANA**

Manufacturers of

VENEERS and LUMBER

In the Domestic Hardwoods
ANY THICKNESS.

1/24 and 1/30 Slice Cut
(Dried flat with Smith Roller Dryer)

1/20 and thicker Sawed Veneers,
Band Sawn Lumber.

— SPECIALTY —

Indiana Quartered Oak

Veneers and Panels

PANELS

Stock Sizes for Immediate Shipment
All Woods   All Thicknesses   3 and 5 Ply
or made to your specifications

VENEERS

5 Million feet for Immediate Shipment
Any Kind of Wood   Any Thickness

J. J. NARTZIK

1966-76 Maud Ave.,   -   -   CHICAGO, ILL.

Veneers and Panels

**LARGE STOCKS FOR
QUICK   SHIPMENTS**

Send for Lists

Walter Clark Veneer Co.

GRAND RAPIDS, MICH.

OAK MAHOGANY CIRCASSIAN MAPLE

VENEERS

BIRCH POPLAR CHERRY WALNUT GUM

Wood Mosaic Company

Main Office, New Albany, Ind.

Are you interested in

Quartered Oak Veneers?

We have the best timber and the best plant for the manufacture of this stock to be found anywhere. We make our own flitches, using only Kentucky and Indiana stock, which is famous for mildness, color and texture. We can make immediate deliveries of Quartered Oak Veneers on orders received now.

Get acquainted with our Famous Indiana and Kentucky QUARTERED WHITE OAK

TEAK

Write for prices on all

**Plain and Fancy
Lumbers and Veneers**

Mahogany, Oak, Birch, Walnut, Ash, Poplar, etc., rotary, sliced or sawed in all thicknesses

SHELL BOX PARTS
in birch and spruce
Bridges and Ends

Write for Quotations

George Kersley
224 St. James Street, Montreal

MAHOGANY

D O M E S T I C

L U M B E R

QUALITY

American Walnut Veneers
Figured Wood
Sliced - Half Round - Stump Wood

5 Essential Qualifications:

1.—Unlimited supply of Walnut Logs.
2.—Logs carefully selected for figure.
3 —All processes of manufacture at our own plant under personal supervision.
4.—Dried flat in Philadelphia Textile Dryer.
5.—3,000,000 feet ready for shipment.

JOHN N. ROBERTS, New Albany, Indiana
The Pioneer Walnut Veneer Producer of America

OAK — MAHOGANY — CIRCASSIAN — MAPLE
VENEERS
BIRCH — POPLAR — CHERRY — WALNUT — GUM

Trying It "On the Dog"

A "new beginner," as Potash would say, must experiment before he learns how, just as a young doctor who has just passed his state board examinations must have somebody to practice on before he becomes expert enough to command much confidence. Most of us prefer to let him get his experience out of somebody else.

It's the same way with lumber and veneers: to be specific, with walnut lumber and veneers. We have been in this business for thirty years; our Mr. J. N. Penrod is generally regarded as the most expert manufacturer of figured walnut stock in the United States, and the leading consumers rely upon him for service of an exceptional character.

You may be able to get what you want from the "new beginner", but why take chances, when you can come to headquarters and get the real thing at the same price?

Penrod Walnut & Veneer Co.
KANSAS CITY, Mo.

We specialize on

Poplar

cross-banding and backing

Veneer

Highest Grade — Right Prices

The largest Poplar Veneer Mill in the world.

The
= Central Veneer Co.
Huntington, West Virginia

MAHOGANY VENEERS
Notwithstanding the scarcity, we are in a position to take care of our customers.

CIRCASSIAN
Some stocks on hand suitable for interior finish and furniture manufacturers.

FIGURED AMERICAN WALNUT
Over two million feet, nice striped and curley wood.

QUARTER SAWED WHITE AND RED OAK
1-20 to 1-8, 6 to 15-in. wide

GRAND RIM POPLAR
Cut to size. We can take care of your 1915 requirements from choice Yellow Poplar logs.

FIGURED WALNUT BUTTS
Two hundred twenty-three thousand feet newly cut Butts, some highly colored stock in this lot.

MAHOGANY LUMBER
This is something we make a specialty of, representing in Chicago and the Northwest, one of the largest importers.

HAUGHTON VENEER COMPANY
1152 W. LAKE STREET   CHICAGO, ILLINOIS

Laying Crotch Mahogany Veneers

The writer had occasion a short time ago to prepare and lay some crotch mahogany veneer that was to be used for piano cases. Any person who has ever handled this stuff knows that it will almost split if you breathe on it. This is owing to the fact that there is so much short grain or end wood in it. If you attempt to put any weight on it, it will crack and split in every conceivable direction.

After one or two unsuccessful attempts, we decided that the best way to lay it was to glue some cotton on the back of it. We accordingly secured a few yards of very cheap factory cotton, and after gluing the entire surface of the veneer laid the cotton on and worked it into contact with the veneer. The experiment was a success, so we treated all the veneer in the same manner and after the stock was taken from the press and the cotton removed, we found we had a beautiful surface. If any of the readers of the "Canadian Woodworker" are ever called on to use crotch mahogany they may find the idea worth adopting.

Cutting Magnolia Veneers

The evergreen magnolia (Magnolia foetida) of the South is a tree so attractive as an ornament that it is to be regretted that it must be cut for lumber, writes Hardwood Record; but it contains good material, and it goes to mills along with its associates. The lumber sometimes passes as yellow poplar, but it cannot do so if closely inspected. It is largely used in the manufacture of rotary-cut veneers, such as are made into berry and small fruit boxes. Some veneer mills report difficulties in cutting magnolias. Concerning this matter a correspondent writes that there are three factors involved here, either of which may contribute to the trouble. They are: the knife, the pressure-bar and the boiling. Smooth cutting calls for a keen knife, carefully whetted, also for a pressurebar, carefully adjusted, with enough pressure on it to keep the stock smooth. With the knife sharp and the pressurebar properly adjusted, the only problem remaining is that of boiling. There are many woods which, if steamed or boiled for any great length of time, become woolly and difficult to cut smooth. Try some experimenting with the boiling and you will probably find that you are overdoing it. The chances are that when your magnolia is good and hot through and through it is ready to cut, and if left longer develops woolly tendencies.

David Gillies President	J. S. Gillies Vice-Pres. and Man. Dir.	D. A. Gillies Sec.-Treas.
Established 1873	**GILLIES BROS.** LIMITED	Mills and Head Office BRAESIDE, ONT.

— Manufacturers of —

White Pine - Red Pine - Spruce

We offer Lumber for Box and Crating:

WHITE PINE MILL RUN		RED PINE LOG RUN	
4/4 x 4/10 x 6/9 ft.	500 M	4/4 to 8/4 x 4/5/6/7/8	
4/4 to 8/4 x 6-7-8 x		in. x 10/17	1500 M
10/16	900 M	RED PINE BOX & CULLS	
WHITE PINE BOX		4/4 to 8/4 x 4 and up	
4/4 to 8/4 x 4 and up		x 6/16	1000 M
x 6/16	400 M	SPRUCE MILL RUN	
WHITE PINE MILL CULLS		2 x 4 and up x 6/22	25 M
4/4 to 8/4 x 4 and up		SPRUCE BOX & CULLS	
6/16	200 M	4/4, 5/4, 8/4 x 4 and	
4/4 x 10/11/12 x 6/16	300 M	up x 6/16	500 M

The above stock is dry and much of it is assorted widths.

Toronto Veneer Company

V E N E E R S

Exclusive Importers of High Grade Veneers and
Fancy Woods

MAHOGANY
AMERICAN WALNUT
QUARTERED OAK
FIGURED GUM
CURLY BIRCH

We have a large stock of carefully selected veneers and fancy woods. It is our desire to have our prospective customers, when convenient, call and make their own selections from this varied assortment. This will enable you to choose stocks most desirable.

We have just received from our mill a new cutting of American Walnut; this lot comprises some very good values very moderately priced. If you are looking for some nice large, clean stocks, we have them.

A card will bring our representative with a full line of samples.

93-99 Spadina Avenue

TORONTO   ONTARIO

F A N C Y   W O O D S

FOR SALE

In good condition, a

Shavings Collector
(Cyclone)

suitable to connect with a 35-in. Planing Mill Exhauster. Blue Prints and further information furnished on request.

CUMMER-DOWSWELL, Limited,
6-7-8　　　　　　　　　Hamilton, Ont.

Machinery for Sale

1—18 x 24 in. Slide Valve Engine.
1—10 ft. Fly-wheel and 9 ft. drive pulley, 24 in. face.
1—Wood Pulley 60 in. x 24 in. face, 4½ in. bore.
1—Wood Pulley 84 in. x 18 in. face, 4 in. bore.
1—Wood Pulley 90 in. x 18 in. face, 3½ in. bore.
1—Jack-ladder chain and sprockets, 300 ft. long.
1—Boiler, 72 in. diam., 14 ft.
1—Boiler 52 in. diam., 14 ft.

Can be inspected at our mill at Callander.

CALLANDER SAW MILLS
Callander, Ont.

For Sale

Woodworking Machinery

The following second-hand machines have been placed with us for sale:
1 Woods No. 32 Matcher; 1 Goldie 10-in. Moulder; 1 Goldie 24-in. Matcher; 1 Cowan 42-in. Sander; 1 Jackson-Cochrane 36-in. Sander; 1 Egan 36-in. Single Surfacer; 1 McGregor-Gourlay 2-drum 30-in. Boss Sander; 1 Egan 30-in. 2-drum Sander; 1 CMC 8-in. 4-side Sticker.

P. B. Yates Machine Co., Limited
Hamilton, Ontario

Second-Hand Woodworking Machines

Band Saws—26-in., 36-in., 40-in.
Jointers—12-in., 16-in.
Shapers—Single and double spindle.
Re-Saw—42-in. American.
Lathes—20-in. with rests and chisels.
Surfacers—16 x 6 in. and 18 x 8 in.
Mortisers—No. 5 and No. 7 New Britain.
Rip and Cut-off Saws; Molders; Sash Relishers, etc.
Over 100 machines, New and Used, in stock.

Baird Machinery Company,
123 Water Street,
PITTSBURGH, PA.

Carbuilders Return to Wood

The shortage of steel and the consequent high prices have resulted in a change of specifications for 2,250 freight cars which were recently decided upon by a United States railway company. The specifications are changed from steel to wooden underframes, because of the price problem, and also because of the inability to get deliveries in steel in time to get prompt completion of the cars. Other similar car contracts, calling for wooden underframes, include 1,000 freight cars for the Northwestern system, and 300 stock cars for the Illinois Central. This change in specifications is unusual, for the railroads have, even when ordering most of their cars with steel underframes, even when the bodies were of wood. The change in specifications will mean millions of feet of additional contracts for dimension timbers.—Barrel and Box.

Mahogany

The normal receipts of all sorts of mahogany in the United States, according to the Lumberman's Review, are about 50,000,000 feet a year, though 70,-000,000 feet were received in the banner year of 1907. But on the authority of this same journal it is said that there are great developments pending which will put this country into the lead as a mahogany centre, with a consequent growth in receipts.

Newark Bay is now the site of an 11-acre yard which is to be devoted solely to the distribution of mahogany, and there is said to be more than 2,000,000 feet now on hand. Boston was once the real mahogany city of this country, but New York has taken its place and there are mahogany sawmills on Long Island.

Woods Used for Specialties

The making of toy furniture is a line that has been growing fast during the past few years. Some special machinery has been devised to turn this work out and it means much work to make the things which mean play for the children. Nail-driving machines are used and the articles are dipped in a vat and not hand painted. They are decorated by hand and thoroughly inspected. Most of these are made in the New England States.

Most of the bamboo broom handles used in the United States come from Amsterdam, N. Y., and are imported from Japan. When they were first brought here they proved unsatisfactory, as they cracked and split. The newer methods prevent this. George H. Maus, of Amsterdam, imports most of them and he has started a hardwood broom handle factory in Japan.

The writer has been through several of the pencil slat factories, used to make pencils and it is a big industry. Trade has been cut off from Germany and Austria and there is a good deal of secrecy about the work after it reaches the pencil slat condition. Most of the wood is red cedar from the South, some incense cedar from California and for cheap pencils white pine and basswood. The United States exports more pencils than it imports. There are eleven pencil factories in the United States and a new one in Toledo to be built, to cost $200,-000. About 200,000,000 lead pencils are made in the United States each year.

PETRIE'S
MONTHLY LIST
of
NEW and USED
WOOD TOOLS
in stock for immediate delivery

Wood Planers
30″ American double surfacer.
30″ Whitney pattern single surfacers.
26″ Revolving bed double surfacers.
26″ Goldie & McCulloch single surfacer.
24″ MacGregor Gourlay planer & matcher.
24″ Major Harper planer and matcher.
24″ Single surfacers, various makes.
20″ Dundas pony planer.
18″ Little Giant planer and matcher.
16″ Galt jointer.

Band Saws
42″ Fay & Egan power feed rip.
38″ Atlantic tilting frame.
36″ Major Harper.
30″ Ideal pedestal.
28″ Jackson Cochrane bracket.

Saw Tables
Preston variable power feed.
Cowan heavy power feed. M138.
No. 5 Crescent sliding top.
No. 3 Crescent universal.
No. 2 Crescent combination.
Ballantine dimension.
12′ Defiance automatic equalizer.
MacGregor Gourlay railway cut-off.
6½′ Crescent iron frame swing.

Mortisers
No. 5 New Britain, chain.
Cowan hollow chisel. M190.
Fay upright, graduated stroke.
Galt upright, compound table.
Smart foot- power.

Moulders
12″ Cowan four side.
10″ Clark-Demill four side.
10″ Houston four side.
6″ Cowan four side.
6″ Dundas sash sticker.

Sanders
36″ Egan double drum.
24″ Fay double drum.
12″ C.M.C. disk and drum.
18″ Crescent disk.

Miscellaneous
Fay & Egan 12-spindle dovetailer.
No. 285 Wysong & Miles post boring machine.
MacGregor Gourlay 2-spindle shaper.
Elliott single spindle shaper.
No. 51 Crescent universal woodworker.
40″ MacGregor Gourlay band resaw.
Rogers vertical resaw.
Cowan sash clamp.
Pedestal tenoner double heads and blade.
Egan sash and door tenoner.
No. 6 Lion universal woodtrimmer.
6-nail box nailer, Meyers patent.
20″ American wood scraper.
4-head rounding machines.
Cowan spindle carver. M63.
Cowan veneer press screws.

Prices, Descriptions and Full Particulars on Request

H. W. PETRIE, LTD.
Front St. W.,　　Toronto, Ont.

Motor Driven Saw-Filing Machine

The Wardwell Manufacturing Company, of Cleveland, Ohio, manufacturers of saw and knife sharpening machinery and tools, have recently adapted their seven different types of filers, combination filers and setters, ranging in price from $40 to $100, so that they may be direct connected to motors. The steadily increasing demand for direct connected motor-driven machinery makes this a desirable feature. The machine illustrated herewith will file and set all kinds of band-

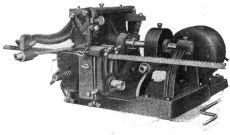

Motor Driven Saw-Filing Machine.

saws at one operation. It is direct connected to a motor, which eliminates the long stretches of belts usually required and permits of the machine being placed in what would otherwise be waste space. The machine has several other features which are worthy of mention. All wearing parts are of steel. The vice through which the saw passes is steel lined. The setter points are a high grade hardened tool steel. The filing arm works between heavy adjustable steel slides which makes it very accurate in the work which it performs.

The company last year designed a circular saw filer (model J) for a large motor manufacturing firm to cut brass and copper tubing. The saws, which it is called upon to sharpen, have 32 teeth to the inch, and these can be sharpened at a speed of 90 per minute. The filing arm that was put on the machine exactly resembles that on the filing and setting machines, so one can appreciate the accurate work that can be accomplished. The filing arms have an elliptical movement and are guided by vertical rods set at an angle, which permits of the teeth being filed with a hook. All the teeth can be set as little or as much as desired, and every tooth set in such a way as to leave proper clearance behind the cutting edge. The smallest teeth can be sharpened without a particle of burr. The motor is connected by a flexible universal coupling and can be removed, should occasion require. From this coupling a shaft extends to a cut worm and gear, the gear being on the drive shaft of the machine. The machine and motor are supported on a cast iron, ribbed base and the motor used is of the "all enclosed" type, which is built to meet the requirements demanded by the United States Government.

A company to be known as the Sorel Shipbuilding and Coal Company has been formed to engage in shipbuilding and ship repairing. The capital will be $100,000, and the place of business will be at Montreal, P.Q.

Automatic "One-Man" Band Saw Mill

The accompanying illustration shows an automatic "one man" bandsaw mill, which is used by manufacturers of "woodenware" for cutting small short logs into box shooks, chair stock, spokes, bobbins, handle blocks, etc. As the name indicates, the machine only requires one man to operate it. A lever is provided by which the operator dogs the log. Another lever then starts the carriage, which automatically off-sets, returns and moves forward into the cut, continuing until the entire log is cut up. In the meantime the operator can have the next log ready and no time need be wasted.

The mill can be set to automatically cut logs into any dimensions and there is no waste because the special dogging device on the carriage holds the log until the last half inch. The machine is portable and should interest manufacturers that use large quantities of small stock, as the automatic feature does away with a lot of hand labor. The manufacturers of the machine, the J. A. Fay & Egan Company have recently issued a new book on "Band Sawing Machinery." The book is known as catalogue 96 and will be sent to any manufacturer upon request.

Automatic "One-Man" Band Saw Mill.

SIMONDS

Knives
For Planing and Moulding Mills, Box Factories, etc.

Saws
For All Purposes

Beader Bits   Groover Bit   Straight Bit

Beader Bit   Milled Matcher Bits   Tongue and Groove Bits

It pays to use the better grade of cutting tools such as Saws or Planer Knives.

Planer Knife

Write to us about any size or style. Remember we make Grooving Saws, Novelty Saws, Narrow and Wide Band Saws, Web and Scroll Saws, Swages, Saw Sets, Files, Hack Saws, etc. May we send you a copy of our catalog describing the complete line we manufacture?

Straight Knife

Untempered Moulding Blank

Moulding Knives

Moulding Knife

Simonds Canada Saw Co., Limited

Factory: St. Remi St. and Acorn Ave.,
Montreal

St. John, N. B.   Vancouver, B. C.

News of the Trade

· The National Wood Mfg. Co., Ltd., South River, Ont., has obtained a charter.

The death occurred recently of Mr. E. C. Wood, who for many years carried on a carriage making business at Mount Forest, Ont.

The damage caused by fire recently at the plant of L. W. Fick & Son, Simcoe, Ont., amounted to $8,000. Upon this there was insurance of $4,595.

The Preston Woodworking Machinery Company recently presented wrist watches to three of their employees who enlisted for overseas service.

The sash and door factory, and stock therein, belonging to J. G. Morton, at Victoria, B.C., was recently damaged by fire. The loss is fully covered by insurance.

The large woodworking factory and grist mill belonging to Mr. John Gaudet, at Memramcook, N.B., has been destroyed by fire. The total loss is estimated at nearly $20,000.

Fire destroyed Chrysler's barrel heading and custom sawmill at Windham Centre on May 27th. The loss included about $1,500 worth of heading. Stock to the value of $2,000 in the yard and some custom lumber was saved.

The Kilgour Davenport Company, Limited, have obtained a charter and will manufacture furniture, principally davenports, at Toronto. The provisional directors are H. P. MacGregor, A. MacGregor, A. E. Wilson and W. Keenan, all of Toronto.

The by-law to grant exemption of taxes for ten years to the Elmira Planing Mills Company, of Elmira, Ont., was carried unanimously. The business was formerly carried on by Messrs. Bauman and Letson, both of whom are interested in the new company.

Mr. J. F. Wildman, for many years general manager of the Office Specialty Manufacturing Company, was presented with a handsome diamond pin and illuminated address by the head office and factory employees recently, on resigning his position with that company.

The sash and door factory belonging to the Peter McLaren Lumber Company at Brockville, Ont., was seriously damaged by fire. The machinery and equipment suffered the most damage. Two of the employees were seriously injured in making their exit from the burning building.

The Ontario Lumber Company, Limited, has been incorporated with head office at Vancouver, B.C., and capital stock of $10,000. to carry on business as manufacturers of and dealers in lumber of all kinds, including wood pulp, paper mill refuse, and to own sawmills, planing mills, etc.

The Maple Ridge Lumber Company have purchased the mills owned by Abernethy & Lougheed at Port Moody and Port Haney, B.C., and will operate both plants at full capacity. The Maple Ridge Lumber Company is a subsidiary of the B. C. Box Company of New Westminster. Their box factory is located at Queensborough, B.C. The head office is at New Westminster, B.C., and the capital stock is $25,000.

The Wright Furniture Company has been incorporated at Port Arthur, Ont., to carry on the business of manufacturing, buying, and selling furniture. The company is capitalized at $40,000, and the provisional directors are W. A. Wright, W. F. Langworthy, and A. J. McComber, of Port Arthur.

The Baetz Bros.' Furniture Company, Limited, will take over and carry on the furniture manufacturing business of Baetz Bros. and Company, of Berlin, Ont. The company will operate on a capital of $75,000. C. J. Baetz, J. H. Baetz and J. A. Scellen are the provisional directors of the new company.

A group of Meaford manufacturers are negotiating with the Town Council of Stayner, Ont., for the erection of a planing mill and woodworking plant at that place. The manufacturers propose erecting a building to cost $25,000 and are asking for a loan of $15,000 from the town. If the deal goes through 35 men will be employed the year round.

"McArthur Beltings, Limited" is the name of the company that has been formed to take over the business of J. D. McArthur & Company, belt manufacturers, of Brockville, Ont. The capital stock of the company is $40,000, and those interested are J. D. McArthur and J. S. McArthur, manufacturers, and J. M. Palmer and W. M. Moran, belt makers, all of whom reside at Brockville, Ont.

The Toronto Veneer Company have opened an office and stockrooms at 93 to 99 Spadina Avenue, Toronto, Ont., with Mr. J. A. Houde and Mr. S. Hamilton in charge. The company are specializing in fancy woods, such as mahogany, American walnut, quartered oak, figured gum, curly birch, etc. They carry a large stock at their warehouse from which customers can make their own selections.

At the annual meeting of the Montreal branch of the Canadian Manufacturers' Association, held on June 6, Mr. William Rutherford, of William Rutherford and Sons Company, Limited, was elected vice-chairman, while Mr. J. H. A. Acer, Laurentide Co., Ltd.; Mr. S. H. B. Roland, Rolland Paper Co., Ltd., and Lieut.-Col. W. J. Sadler, Sadler and Haworth, Ltd., were elected members of the executive committee. A resolution of regret at the death of Mr. George Esplin, was passed.

Mr. George Esplin, who died at Lachine, P.Q., on May 28, aged 70, was senior member of the firm of G. and J. Esplin, box manufacturers, Duke Street, Montreal. Mr. Esplin had been ill for a considerable period, and recently had only attended business at intervals. He was never a prominent citizen, but took a deep interest in educational and hospital work, being chairman of the Lachine School Board and governor of three or four hospitals. The firm is one of the largest manufacturers of boxes in Montreal.

Reduced Manufacture of Carriages and Wagons

The inroads of the automobile into the carriage and wagon industry during the period from 1909 to 1914 were greater in respect to pleasure vehicles than those used for business purposes. In 1909, according to statistics gathered by the United States Bureau of the Census, carriages represented 53.2 per cent. and wagons 39.7 per cent. of all vehicles made, but in 1914 the proportion represented by carriages had declined to 47 per cent., while that represented by wagons had increased to 48.2 per cent.

In the preparation of the 1914 census of manufactures for the manufacture of carriages and wagons and of bodies, tops, cushions, hubs, felloes, spokes, wheels, and other materials used in the production of the complete vehicles, reports were received from 5,320 establishments, which manufactured 1,817,002 vehicles of all classes, valued at $72,283,898. At the 1909 census there were reported 5,613 establishments, with an output of 1,584,571 vehicles, valued at $94,-037,000. The number of establishments thus decreased during the five-year period by 293, the number of vehicles by 397,569, and the value by $21,754,-002.—Hardwood Record.

MOST IMPROVED FILING AND SETTING MACHINES MADE, THEY FILE, SET AND JOINT ALL BAND SAWS IN ONE OPERATION

The only machines made that do it.

WRITE FOR DETAILS

THE WARDWELL MFG. CO.

Specialty Manufacturers
Combination Filing and Setting Machines,
Circular Saw Filers and Grinders,
Band Saw Grinders, Etc.

116 Hamilton Ave., Cleveland, Ohio

---

30 Sets of Jambs in 28 Minutes

Here's the Machine that will do it!

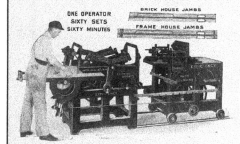

ONE OPERATOR
SIXTY SETS
SIXTY MINUTES

BRICK HOUSE JAMBS

FRAME HOUSE JAMBS

Every manufacturer wants more profit, more economy in production, more output per man. Unequalled results are to be obtained from the

"Efficiency"
Window Frame Machine

One operator cuts the pockets, routs the pulley mortises, rabbets head and sill gain simultaneously. Guaranteed to reduce the time to work pulley stiles 50 to 60 per cent. Try the "Efficiency" Machine and be convinced.

The F. H. Weeks Lumber Co. - Akron, Ohio

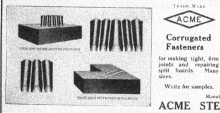

Acme Divergent Saw Edge Fasteners

TRADE MARK

ACME

Corrugated Fasteners

for making tight, firm joints and repairing split boards. Many sizes.

Write for samples.

Box Strapping and Cold Rolled Steel Bands
FOR SHELL BOXES

Manufactured by   Box Strapping

ACME STEEL GOODS CO., LIMITED
Coristine Building, MONTREAL, Quebec

DARLINGS STEAM APPLIANCES

DARLING BROTHERS
LIMITED
Engineers and Manufacturers
MONTREAL, CANADA

Branches: Agents:
Toronto and Winnipeg   Halifax, St. John, Calgary, Vancouver.

The Canadian Woodworker and Furniture Manufacturer covers a specialized field. It represents maximum value to its subscribers.

Subscription price $1.00 per year.

Morehead Back to Boiler
SYSTEM

Read What The Wabash Cabinet Co. Say!

"With reference to the Morehead Trap, it is everything that Mr. MacLachlan said it was. Before we made this change, our watchman was obliged to put in coal every fifteen minutes. Now he puts in a fire on an average of once every hour and WE ARE SAVING OVER 50 PER CENT. ON THE COST OF OUR FUEL."

You can do the same. The first step is to

Write for the "Back-to-Boiler" Book

In it you will find the condensation question discussed from several angles—and interesting pictures of installations under every possible condition.

The economic principle of the "Morehead" System will save you dollars. Write for the book TO-DAY.

Canadian Morehead Mfg. Co.
WOODSTOCK   Dept. "H"   ONTARIO

Fourth page in the book on handle factory efficiency

Real success in making handles won't come with an incomplete equipment. A Handle Throater is an indispensable machine which removes irregularities left by the saw. It's another of the Klotz all-parts-perfect machines and possesses the same high degree of mechanical superiority.

THE KLOTZ HANDLE THROATING MACHINE

has a 12-in., fine-tooth cut-off saw on one end of mandrel and throater on other end. Throating and trimming handles to proper length without leaving your tracks saves considerable time in a day. Sliding table and adjustable gauge permit cutting off at any desired angle.

This machine is the result of thirty years concentration perfecting handle machinery that has no equal.

Turn a Page With Us Next Issue

FOR YOU

who have a thought of manufacturing handles, or those desiring maximum production at the correct cost, get the Klotz Machine Company's idea of a perfectly modern handle factory. One with such dependable machines in it as their

     BLOCKING SAW

     SHAPING SAW

     LATHE

     THROATER

     BOLTING SAW

     POLISHER

     DISC WHEEL OR END SHAPER

     KNOBBER OR SCROLL MACHINE

All have a place in an efficient handle factory. Write

Klotz Machine Co., Sandusky, O.

Saves from $3 to $4 a day

That is what the

NASH SANDER HAS DONE

on such
work as

Chair Stock,
Dowel Rods,
Curtain Poles,
Hoe, Fork,
Rake, Broom
and Shovel
Handles,
Whip Stocks,
Flag Sticks,
Veneered
Columns,
etc.

Wouldn't you like to have it do
that for you?

J. M. NASH, MILWAUKEE, Wis.

There is MONEY in the HANDLE BUSINESS.
Write for free catalogue No. 31 of

OBER LATHES

For turning Axe, Pick, Sledge, Hammer and
Hatchet Handles, Fork, Hoe, Rake, Broom
Handles, Whiffletrees, Spokes, etc.

SANDERS, RIPSAWS
and other Machinery

The

Ober Mfg. Co.

Chagrin Falls, O., U.S.A.
31 Bell Street

VENEER PRESSES

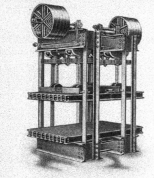

WRITE FOR BULLETIN C 1.

Canadian Boomer & Boschert Press

Company, Limited
18 Tansley Street, MONTREAL

Your Loose Pulley Trouble Disappears

when installed with

CHAPMAN BALL BEARING LOOSE PULLEYS

A little vaseline once a year is all the
attention and lubrication required.
Our loose Pulleys cannot run hot and do
not cut the shaft.

Write us for particulars and price list.

The

Chapman Double Ball Bearing Co.

of Canada, Limited

339-351 Sorauren Ave., Toronto, Canada

The "Canadian Woodworker" Buyers' Directory

ABRASIVES
The Carborundum Co., Niagara Falls, N.Y.

BALL BEARINGS
Chapman Double Ball Bearing Co., Toronto.

BALUSTER LATHES
Garlock-Machinery, Toronto, Ont.
Ober Mfg. Company, Chagrin Falls, Ohio.
Whitney & Son, Baxter D., Winchendon, Mass.

BAND SAW FILING MACHINERY
Canada Machinery Corporation, Galt, Ont.
Fay & Egan Co., J. A., Cincinnati, Ohio.
Garlock-Machinery, Toronto, Ont.
Preston Woodworking Machinery Company, Preston, Ont.
Yates Machine Co., P. B., Hamilton, Ont.

BAND SAWS
Canada Machinery Corporation, Galt, Ont.
Garlock-Machinery, Toronto, Ont.
Jackson, Cochrane & Company, Berlin, Ont.
Petrie, H. W., Toronto.
Preston Woodworking Machinery Company, Preston, Ont.
Yates Machine Co., P. B., Hamilton, Ont.

BAND SAW MACHINES
Canada Machinery Corporation, Galt, Ont.
Fay & Egan Co., J. A., Cincinnati, Ohio.
Garlock-Machinery, Toronto, Ont.
Petrie, H. W., Toronto.
Preston Woodworking Machinery Company, Preston, Ont.
Williams Machinery Co., A. R., Toronto, Ont.
Yates Machine Co., P. B., Hamilton, Ont.

BAND SAW STRETCHERS
Fay & Egan Co., J. A., Cincinnati, Ohio.
Yates Machine Co., P. B., Hamilton, Ont.

BASKET STITCHING MACHINE
Saranac Machine Co., Benton Harbor, Mich.

BENDING MACHINES
Fay & Egan Co., J. A., Cincinnati, Ohio.
Garlock-Machinery, Toronto, Ont.
Perfection Wood Steaming Retort Company, Parkersburg, W. Va.

BELTING
Fay & Egan Co., J. A., Cincinnati, Ohio.

BLOWERS
Sheldons, Limited, Galt, Ont.
Toronto Blower Company, Toronto, Ont.

BLOW PIPING
Sheldons, Limited, Galt, Ont.
Toronto Blower Company, Toronto, Ont.

BORING MACHINES
Canada Machinery Corporation, Galt, Ont.
Fay & Egan Co., J. A., Cincinnati, Ohio.
Garlock-Machinery, Toronto, Ont.
Jackson, Cochrane & Company, Berlin, Ont.
Nash, J. M., Milwaukee, Wis.
Preston Woodworking Machinery Company, Preston, Ont.
Reynolds Pattern & Machine Co., Moline, Ill.
Yates Machine Co., P. B., Hamilton, Ont.

BOX MAKERS' MACHINERY
Canada Machinery Corporation, Galt, Ont.
Fay & Egan Co., J. A., Cincinnati, Ohio.
Garlock-Machinery, Toronto, Ont.
Jackson, Cochrane & Co., Berlin, Ont.
Neilson & Company, J. L. Winnipeg, Man.
Preston Woodworking Machinery Company, Preston, Ont.
Reynolds Pattern & Machine Co., Moline, Ill.
Whitney & Son, Baxter D., Winchendon, Mass.
Williams Machinery Co., A. R., Toronto, Ont.
Yates Machine Co., P. B., Hamilton, Ont.

CABINET PLANERS
Canada Machinery Corporation, Galt, Ont.
Fay & Egan Co., J. A., Cincinnati, Ohio.
Garlock-Machinery, Toronto, Ont.
Jackson, Cochrane & Co., Berlin, Ont.
Preston Woodworking Machinery Company, Preston, Ont.
Yates Machine Co., P. B., Hamilton, Ont.

CARS (Transfer)
Sheldons, Limited, Galt, Ont.

CARBORUNDUM PAPER
Carborundum Co., Niagara Falls, N.Y.

CARVING MACHINES
Canada Machinery Corporation, Galt, Ont.
Fay & Egan Co., J. A., Cincinnati, Ohio.
Garlock-Machinery, Toronto, Ont.
Jackson, Cochrane & Company, Berlin, Ont.

CASTERS
Foster, Merriam & Co., Meriden, Conn.

CLAMPS
Fay & Egan Co., J. A., Cincinnati, Ohio.
Garlock-Machinery, Toronto, Ont.
Jackson, Cochrane & Company, Berlin, Ont.
Preston Woodworking Machinery Company, Preston, Ont.
Simonds Canada Saw Company, Montreal, P.Q.

COLUMN MACHINERY
Fay & Egan Co., J. A., Cincinnati, Ohio.

CORRUGATED FASTENERS
Acme Steel Goods Company, Chicago, Ill.

CUT-OFF SAWS
Canada Machinery Corporation, Galt, Ont.
Fay & Egan Co., J. A., Cincinnati, Ohio.
Garlock-Machinery, Toronto, Ont.
Jackson, Cochrane & Company, Berlin, Ont.
Ober Mfg. Co., Chagrin Falls, Ohio.
Preston Woodworking Machinery Company, Preston, Ont.
Simonds Canada Saw Co., Montreal, Que.
Yates Machine Co., P. B., Hamilton, Ont.

CUTTER HEADS
Canada Machinery Corporation, Galt, Ont.
Fay & Egan Co., J. A., Cincinnati, Ohio.
Jackson, Cochrane & Company, Berlin, Ont.
Mattison Machine Works, Beloit, Wis.
Yates Machine Co., P. B., Hamilton, Ont.

DOVETAILING MACHINES
Canada Machinery Corporation, Galt, Ont.
Fay & Egan Co., J. A., Cincinnati, Ohio.
Garlock-Machinery, Toronto, Ont.
Jackson, Cochrane & Company, Berlin, Ont.

DOWEL MACHINES
Canada Machinery Corporation, Galt, Ont.
Fay & Egan Co., J. A., Cincinnati, Ohio.
Ober Mfg. Co., Chagrin Falls, Ohio.

DRY KILNS
National Dry Kiln Co., Indianapolis, Ind.
Sheldons, Limited, Galt, Ont.

DUST COLLECTORS
Sheldons, Limited, Galt, Ont.
Toronto Blower Company, Toronto, Ont.

DUST SEPARATORS
Sheldons, Limited, Galt, Ont.
Toronto Blower Company, Toronto, Ont.

EDGERS (Single Saw)
Canada Machinery Corporation, Galt, Ont.
Fay & Egan Co., J. A., Cincinnati, Ohio.
Garlock-Machinery, Toronto, Ont.
Simonds Canada Saw Co., Montreal, Que.

EDGERS (Gang)
Canada Machinery Corporation, Galt, Ont.
Fay & Egan Co., J. A., Cincinnati, Ohio.
Garlock-Machinery, Toronto, Ont.
Petrie, H. W., Toronto.
Simonds Canada Saw Co., Montreal, P.Q.
Williams Machinery Co., A. R., Toronto, Ont.
Yates Machine Co., P. B., Hamilton, Ont.

END MATCHING MACHINE
Canada Machinery Corporation, Galt, Ont.
Fay & Egan Co., J. A., Cincinnati, Ohio.
Garlock-Machinery, Toronto, Ont.
Jackson, Cochrane & Company, Berlin, Ont.
Yates Machine Co., P. B., Hamilton, Ont.

EXHAUST FANS
Garlock-Machinery, Toronto, Ont.
Sheldons Limited, Galt, Ont.
Toronto Blower Company, Toronto, Ont.

EXHAUST SYSTEMS
Toronto Blower Company, Toronto, Ont.

FILES
Simonds Canada Saw Co., Montreal, Que.

FILING MACHINERY
Wardwell Mfg. Company, Cleveland, Ohio.

FLOORING MACHINES
Canada Machinery Corporation, Galt, Ont.
Fay & Egan Co., J. A., Cincinnati, Ohio.
Garlock-Machinery, Toronto, Ont.
Petrie, H. W., Toronto.
Preston Woodworking Machinery Company, Preston, Ont.
Whitney & Son, Baxter D., Winchendon, Mass.
Yates Machine Co., P. B., Hamilton, Ont.

FLUTING HEADS
Fay & Egan Co., J. A., Cincinnati, Ohio.

FLUTING AND TWIST MACHINE
Preston Woodworking Machinery Company, Preston, Ont.

FURNITURE TRIMMINGS
Foster, Merriam & Co., Meriden, Conn.

GARNET PAPER AND CLOTH
The Carborundum Co., Niagara Falls, N.Y.

GRAINING MACHINES
Canada Machinery Corporation, Galt, Ont.
Fay & Egan Co., J. A., Cincinnati, Ohio.
Williams Machinery Co., A. R., Toronto, Ont.
Yates Machine Co., P. B., Hamilton, Ont.

GLUE
Coignet Chemical Product Co., New York.
Russia Glue Company, South Bend, Ind.

GLUE CLAMPS
Jackson, Cochrane & Company, Berlin, Ont.

GLUE HEATERS
Fay & Egan Co., J. A., Cincinnati, Ohio.
Jackson, Cochrane & Company, Berlin, Ont.

GLUE JOINTERS
Canada Machinery Corporation, Galt, Ont.
Jackson, Cochrane & Company, Berlin, Ont.
Yates Machine Co., P. B., Hamilton, Ont.

GLUE SPREADERS
Fay & Egan Co., J. A., Cincinnati, Ohio.
Jackson, Cochrane & Company, Berlin, Ont.

GLUE ROOM EQUIPMENT
Boomer & Boschert Company, Montreal, Que.
Perrin & Company, W. R., Toronto, Ont.

GRINDERS (Cutter)
Fay & Egan Co., J. A., Cincinnati, Ohio.

GRINDERS, HAND AND FOOT POWER
The Carborundum Co., Niagara Falls, N.Y.

GRINDERS (Knife)
Canada Machinery Corporation, Galt, Ont.
Fay & Egan Co., J. A., Cincinnati, Ohio.
Garlock-Machinery, Toronto, Ont.
Petrie, H. W., Toronto.
Preston Woodworking Machinery Company, Preston, Ont.
Yates Machine Co., P. B., Hamilton, Ont.

GRINDERS (Tool)
Fay & Egan Co., J. A., Cincinnati, Ohio.
Jackson, Cochrane & Company, Berlin, Ont.

GRINDING WHEELS
The Carborundum Co., Niagara Falls, N.Y.

GROOVING HEADS
Fay & Egan Co., J. A., Cincinnati, Ohio.

GUMMERS, ETC.
Fay & Egan Co., J. A., Cincinnati, Ohio.

GUARDS (Saw)
HAND PROTECTORS
Fay & Egan Co., J. A., Cincinnati, Ohio.

HAND SCREWS
Fay & Egan Co., J. A., Cincinnati, Ohio.

HANDLE THROATING MACHINE
Klotz Machine Company, Sandusky, Ohio.

HANDLE TRIMMING MACHINE
Klotz Machine Company, Sandusky, Ohio.

HANDLE & SPOKE MACHINERY
Fay & Egan Co., J. A., Cincinnati, Ohio.
Klotz Machine Co., Sandusky, Ohio.
Nash, J. M., Milwaukee, Wis.
Ober Mfg. Company, Chagrin Falls, Ohio.
Whitney & Son, Baxter D., Winchendon, Mass.

HARDWOOD LUMBER
American Hardwood Lumber Co., St. Louis, Mo.
Atlantic Lumber Company, Toronto.
Brown & Company, Geo. C., Memphis, Tenn.
Clark & Son, Edward, Toronto, Ont.
Kersley, Geo., Montreal, Que.
Kraetzer-Cured Lumber Co., Moorhead, Miss.
Mowbray & Robinson, Cincinnati, Ohio.
Penrod Walnut & Veneer Co., Kansas City, Mo.
Spencer, C. A., Montreal, Que.

HUB MACHINERY
Fay & Egan Co., J. A., Cincinnati, Ohio.

HYDRAULIC PRESSES
Can. Boomer & Boschert Press, Montreal, Que.

HYDRAULIC PUMPS & ACCUMULATORS
Can. Boomer & Boschert Press, Montreal, Que.

HYDRAULIC VENEER PRESSES
Perrin & Company, Wm. R., Toronto, Ont.

JOINTERS
Canada Machinery Corporation, Galt, Ont.
Fay & Egan Co., J. A., Cincinnati, Ohio.
Garlock-Machinery, Toronto, Ont.
Jackson, Cochrane & Company, Berlin, Ont.
Knechtel Bros., Limited, Southampton, Ont.
Petrie, H. W., Toronto, Ont.
Preston Woodworking Machinery Company, Preston, Ont.
Yates Machine Co., P. B., Hamilton, Ont.

KNIVES (Planer and others)
Canada Machinery Corporation, Galt, Ont.
Fay & Egan Co., J. A., Cincinnati, Ohio.
Simonds Canada Saw Co., Montreal, P.Q.
Yates Machine Co., P. B., Hamilton, Ont.

LUMBER
American Hardwood Lumber Co., St. Louis, Mo.
Brown & Company, Geo. C., Memphis, Tenn.
Churchill, Milton Lumber Co., Louisville, Ky.
Clark & Son, Edward, Toronto, Ont.
Dominion Mahogany & Veneer Co., Montreal, Que.
Hart & McDonagh, Toronto, Ont.
Hartzell, Geo. W., Piqua, Ohio.
Paepcke Leicht Lumber Company, Chicago, Ill.

LATHES
Canada Machinery Corporation, Galt, Ont.
Fay & Egan Co., J. A., Cincinnati, Ohio.
Garlock-Machinery, Toronto, Ont.
Jackson, Cochrane & Company, Berlin, Ont.
Klotz Machine Company, Sandusky, Ohio.
Ober Mfg. Company, Chagrin Falls, Ohio.
Petrie, H. W., Toronto, Ont.
Preston Woodworking Machinery Company, Shawver Company, Springfield, O.
Whitney & Son, Baxter D., Winchendon, Mass.

MACHINE KNIVES
Canada Machinery Corporation, Galt, Ont.
Peter Hay Knife Company, Galt, Ont.
Simonds Canada Saw Co., Montreal, P.Q.
Yates Machine Co., P. B., Hamilton, Ont.

MITRE MACHINES
Canada Machinery Corporation, Galt, Ont.
Fay & Egan Co., J. A., Cincinnati, Ohio.

MITRE SAWS
Canada Machinery Corporation, Galt, Ont.
Fay & Egan Co., J. A., Cincinnati, Ohio.
Garlock-Machinery, Toronto, Ont.
Simonds Canada Saw Co., Montreal, P.Q.

MORTISING MACHINES
Canada Machinery Corporation, Galt, Ont.
Fay & Egan Co., J. A., Cincinnati, Ohio.
Garlock-Machinery, Toronto, Ont.
Yates Machine Co., P. B., Hamilton, Ont.

MOULDINGS
Knechtel Bros., Limited, Southampton, Ont.

MULTIPLE BOXING MACHINES
Fay & Egan Co., J. A., Cincinnati, Ohio.
Nash, J. M., Milwaukee, Wis.
Reynolds Pattern & Machine Co., Moline, Ill.

PATTERN SHOP MACHINES
Canada Machinery Corporation, Galt, Ont.
Fay & Egan Co., J. A., Cincinnati, Ohio.
Garlock-Machinery, Toronto, Ont.
Jackson, Cochrane & Company, Berlin, Ont.
Preston Woodworking Machinery Company, Preston, Ont.
Whitney & Son, Baxter D., Winchendon, Mass.
Yates Machine Co., P. B., Hamilton, Ont.

PLANERS
Canada Machinery Corporation, Galt, Ont.
Fay & Egan Co., J. A., Cincinnati, Ohio.
Garlock-Machinery, Toronto, Ont.
Jackson, Cochrane & Company, Berlin, Ont.
Petrie, H. W., Toronto, Ont.
Preston Woodworking Machinery Company,
Whitney & Son, Baxter D., Winchendon, Mass.
Williams Machinery Co., A. R., Toronto, Ont.
Yates Machine Co., P. B., Hamilton, Ont.

PLANING MILL MACHINERY
Canada Machinery Corporation, Galt, Ont.
Fay & Egan Co., J. A., Cincinnati, Ohio.
Garlock-Machinery, Toronto, Ont.
Jackson, Cochrane & Company, Berlin, Ont.
Petrie, H. W., Toronto, Ont.
Preston Woodworking Machinery Company,
Whitney & Son, Baxter D., Winchendon, Mass.
Williams Machinery Co., A. R., Toronto, Ont.
Yates Machine Co., P. B., Hamilton, Ont.

PRESSES (Veneer)
Canada Machinery Corporation, Galt, Ont.
Jackson, Cochrane & Company, Berlin, Ont.
Perrin & Co., Wm. R., Toronto, Ont.
Preston Woodworking Machinery Company,
Preston, Ont.

PULLEYS
Fay & Egan Co., J. A., Cincinnati, Ohio.

RESAWS
Canada Machinery Corporation, Galt, Ont.
Fay & Egan Co., J. A., Cincinnati, Ohio.
Garlock-Machinery, Toronto, Ont.
Jackson, Cochrane & Company, Berlin, Ont.
Petrie, H. W., Toronto, Ont.
Preston Woodworking Machinery Company,
Preston, Ont.
Simonds Canada Saw Co., Montreal, P.Q.
Williams Machinery Co., A. R., Toronto, Ont.
Yates Machine Co., P. B., Hamilton, Ont.

RIM AND FELLOE MACHINERY
Fay & Egan Co., J. A., Cincinnati, Ohio.

RIP SAWING MACHINES
Canada Machinery Corporation, Galt, Ont.
Fay & Egan Co., J. A., Cincinnati, Ohio.
Garlock-Machinery, Toronto, Ont.
Klotz Machine Company, Sandusky, Ohio.
Jackson, Cochrane & Company, Berlin, Ont.
Ober Mfg. Company, Chagrin Falls, Ohio.
Preston Woodworking Machinery Company,
Preston, Ont.
Yates Machine Co., P. B., Hamilton, Ont.

SANDERS
Canada Machinery Corporation, Galt, Ont.
Fay & Egan Co., J. A., Cincinnati, Ohio.
Garlock-Machinery, Toronto, Ont.
Klotz Machine Company, Sandusky, Ohio.
Jackson, Cochrane & Company, Berlin, Ont.
Nash, J. M., Milwaukee, Wis.
Ober Mfg. Company, Chagrin Falls, Ohio.
Preston Woodworking Machinery Company,
Preston, Ont.

SASH, DOOR & BLIND MACHINERY
Canada Machinery Corporation, Galt, Ont.
Fay & Egan Co., J. A., Cincinnati, Ohio.
Garlock-Machinery, Toronto, Ont.
Jackson, Cochrane & Company, Berlin, Ont.
Preston Woodworking Machinery Company,
Yates Machine Co., P. B., Hamilton, Ont.

SAWS
Fay & Egan Co., J. A., Cincinnati, Ohio.
Radcliff Saw Mfg. Company, Toronto, Ont.
Simonds Canada Saw Co., Montreal, P.Q.

SAW GUMMERS, ALOXITE
The Carborundum Co., Niagara Falls, N.Y.

SAW SWAGES
Fay & Egan Co., J. A., Cincinnati, Ohio.
Simonds Canada Saw Co., Montreal, Que.

SAW TABLES
Canada Machinery Corporation, Galt, Ont.
Fay & Egan Co., J. A., Cincinnati, Ohio.
Garlock-Machinery, Toronto, Ont.
Jackson, Cochrane & Company, Berlin, Ont.
Preston Woodworking Machinery Company,
Preston, Ont.
Yates Machine Co., P. B., Hamilton, Ont.

SCRAPING MACHINES
Canada Machinery Corporation, Galt, Ont.
Garlock-Machinery, Toronto, Ont.
Whitney & Sons, Baxter D., Winchendon, Mass.

SCROLL CHUCK MACHINES
Klotz Machine Company, Sandusky, Ohio.

SCREW DRIVING MACHINES
Canada Machinery Corporation, Galt, Ont.
Reynolds Pattern & Machine Co., Moline, Ill.

SCREW DRIVING MACHINERY
Canada Machinery Corporation, Galt, Ont.
Reynolds Pattern & Machine Works, Moline, Ill.

SCROLL SAWS
Canada Machinery Corporation, Galt, Ont.
Fay & Egan Co., J. A., Cincinnati, Ohio.
Garlock-Machinery, Toronto, Ont.
Jackson, Cochrane & Company, Berlin, Ont.
Simonds Canada Saw Co., Montreal, P.Q.
Yates Machine Co., P. B., Hamilton, Ont.

SECOND-HAND MACHINERY
Petrie, H. W.
Williams Machinery Co., A. R., Toronto, Ont.

SHAPERS
Canada Machinery Corporation, Galt, Ont.
Fay & Egan Co., J. A., Cincinnati, Ohio.
Garlock-Machinery, Toronto, Ont.
Jackson, Cochrane & Company, Berlin, Ont.
Ober Mfg. Company, Chagrin Falls, Ohio.
Petrie, H. W., Toronto, Ont.
Preston Woodworking Machinery Company,
Preston, Ont.
Simonds Canada Saw Co., Montreal, P.Q.
Whitney & Sons, Baxter D., Winchendon, Mass.
Yates Machine Co., P. B., Hamilton, Ont.

SHARPENING TOOLS
The Carborundum Co., Niagara Falls, N.Y.

SHAVING COLLECTORS
Sheldons Limited, Galt, Ont.

SHELL BOX HANDLES
Vokes Hardware Co., Toronto, Ont.

STEAM APPLIANCES
Darling Bros., Montreal.

SINGLE SPINDLE BOXING MACHINES
Canada Machinery Corporation, Galt, Ont.
Fay & Egan Co., J. A., Cincinnati, Ohio.
Ober Mfg. Company, Chagrin Falls, Ohio.

STAVE SAWING MACHINE
Whitney & Sons, Baxter D., Winchendon, Mass.

STEAM TRAPS
Canadian Morehead Mfg. Co., Woodstock, Ont.
Standard Dry Kiln Co., Indianapolis, Ind.

STEEL BOX STRAPPING
Acme Steel Goods Co., Chicago, Ill.

STEEL SHELL BOX BANDS
Acme Steel Goods Co., Chicago. Ill.

SURFACERS
Canada Machinery Corporation, Galt, Ont.
Fay & Egan Co., J. A., Cincinnati, Ohio.
Garlock-Machinery, Toronto, Ont.
Jackson, Cochrane & Company, Berlin, Ont.
Petrie, H. W., Toronto, Ont.
Preston Woodworking Machinery Company,
    Preston, Ont.
Whitney & Sons, Baxter D., Winchendon, Mass.
Williams Machinery Co., A. R., Toronto, Ont.
Yates Machine Co., P. B., Hamilton, Ont.

SWING SAWS
Canada Machinery Corporation, Galt, Ont.
Fay & Egan Co., J. A., Cincinnati, Ohio.
Garlock-Machinery, Toronto, Ont.
Jackson, Cochrane & Company, Berlin, Ont.
Ober Mfg. Co., Chagrin Falls, Ohio.
Preston Woodworking Machinery Company,
    Preston, Ont.
Simonds Canada Saw Co., Montreal, P.Q.
Yates Machine Co., P. B., Hamilton, Ont.

TABLE LEG LATHES
Fay & Egan Co., J. A., Cincinnati, Ohio.
Ober Mfg. Co., Chagrin Falls, Ohio.
Whitney & Sons, Baxter D., Winchendon, Mass.
Yates Machine Co., P. B., Hamilton, Ont.

TABLE SLIDES
Walter & Co., B., Wabash, Ind.

TENONING MACHINES
Canada Machinery Corporation, Galt, Ont.
Fay & Egan Co., J. A., Cincinnati, Ohio.
Garlock-Machinery, Toronto, Ont.
Jackson, Cochrane & Company, Berlin, Ont.
Preston Woodworking Machinery Company,
    Preston, Ont.
Yates Machine Co. P. B., Hamilton, Ont.

TRIMMERS
Canada Machinery Corporation, Galt, Ont.
Fay & Egan Co., J. A., Cincinnati, Ohio.
Garlock-Machinery, Toronto, Ont.
Preston Woodworking Machinery Company,
    Preston, Ont.
Yates Machine Co., P. B., Hamilton, Ont.

TRUCKS
Sheldons Limited, Galt, Ont.
National Dry Kiln Co., Indianapolis, Ind.

TURNING MACHINES
Fay & Egan Co., J. A., Cincinnati, Ohio.
Garlock-Machinery, Toronto, Ont.
Klotz Machine Company, Sandusky, Ohio.
Ober Mfg. Company, Chagrin Falls, Ohio.
Whitney & Sons, Baxter D., Winchendon, Mass.
Yates Machine Co., P. B., Hamilton, Ont.

**UNDER-CUT SELF-FEEDING FACE
    PLANER**
Fay & Egan Co., J. A., Cincinnati, Ohio.
Garlock-Machinery, Toronto, Ont.
Jackson, Cochrane & Company, Berlin, Ont.

VARNISHES
Ault & Wiborg Company, Toronto, Ont.

VENEERS
Central Veneer Co.,' Huntington, W. Virginia.
Dominion Mahogany & Beneer Company, Mont-
    real, Que.
Walter Clark Veneer Co., Grand Rapids, Mich.
Evansville Veneer Co., Evansville, Ind.
Hartzell, Geo. W., Piqua, Ohio.
Haughton Veneer Company, Chicago, Ill.
Hay & Company, Woodstock, Ont.
Hoffman Bros. Company, Fort Wayne, Ind.
Geo. Kersley, Montreal, Que.
Nartzik, J. J., Chicago, Ill.
Ohio Veneer Company, Cincinnati, Ohio.
Penrod Walnut & Veneer Co., Kansas City, Mo.
Roberts, John N., New Albany, Ind.
Toronto Veneer Company, Toronto, Ont.
Wood Mosaic Co., New Albany, Ind.

VENTILATING APPARATUS .
Sheldons Limited, Galt, Ont.

VENEERED PANELS
Hay & Company, Woodstock, Ont.

VENEER PRESSES (Hand and Power)
Canada Machinery Corporation, Galt, Ont.
Can. Boomer & Boschert Press, Montreal, Que.
Garlock-Machinery, Toronto, Ont.
Jackson, Cochrane & Company, Berlin, Ont.
Perrin & Company, Wm. R., Toronto, Ont.

VISES
Fay & Egan Co., J. A., Cincinnati, Ohio.
Simonds Canada Saw Company, Montreal, Que.

WAGON AND CARRIAGE MACHINERY
Canada Machinery Corporation, Galt, Ont.
Fay & Egan Co., J. A., Cincinnati, Ohio.
Ober Mfg. Company, Chagrin Falls, Ohio.
Whitney & Sons, Baxter D., Winchendon, Mass.
Yates Machine Co., P. B., Hamilton, Ont.

WOOD FINISHES
Ault & Wiborg, Toronto, Ont.

WOOD TURNING MACHINERY
Canada Machinery Corporation, Galt, Ont.
Garlock-Machinery, Toronto, Ont.
Yates Machine Co. P. B., Hamilton, Ont.

WORK BENCHES
Fay & Egan Co., J. A., Cincinnati, Ohio.

WOODWORKING MACHINES
Canada Machinery Corporation, Galt, Ont.
Garlock-Machinery, Toronto, Ont.
Jackson, Cochrane & Co., Berlin, Ont.
Preston Woodworking Machinery Company,
    Preston, Ont.
Reynolds Pattern & Machine Works, Moline,
    Ill.
Williams Machinery Co., A. R., Toronto, Ont.

WHEELS, Grinding, Carborundum and Aloxite
The Carborundum Co., Niagara Falls, N.Y.

THE COMPLETE LINE

ACME CASTERS made by FOSTER MERRIAM & CO. have
the backing of eighty years of experience which means that ACME
CASTERS are better casters.

They are the choice of thousands of discriminating users both in
the BALL BEARING and NON BALL BEARING styles.

We solicit your inquiries and will be pleased to have our
CANADIAN representative call and assist you in your Caster
and period trimmings problems.

Foster, Merriam & Co.
Meriden, Conn.

Canadian Representative:—F. A. Schmidt, Berlin, Ont.

" ACME "
The
RELIABLE CASTER

Making TABLE-SLIDES is a Specialty Business

For more than TWENTY-FIVE YEARS we have made TABLE SLIDES exclusively. Our Factory is equipped with Special Machinery which enables us to make SLIDES—BETTER and CHEAPER than the furniture manufacturer.

Canadian Table makers are rapidly adopting WABASH SLIDES

Because { They ELIMINATE SLIDE TROUBLES
Are CHEAPER and BETTER

BY USING

Reduced Costs

Increased Out-put

WABASH SLIDES

MADE BY

B. Walter & Company
Wabash, Ind.

The Largest EXCLUSIVE TABLE-SLIDE Manufacturers in America

ESTABLISHED-1887

Just What You Need

There's no necessity now for you to **wish** for a corrugated joint fastener machine which will rise fully to your expectations; you can get one.

The Saranac Corrugated Joint Fastener Machine

is the absolute realization of your wish. This illustration gives you an idea of what it looks like, but you can have no conception of its **efficiency, economy, speed** and **reliability** until you get it working for you.

It is substantially constructed, and the workmanship and materials are of the very best. Particularly suitable for manufacturers of sash, doors, blinds, screens, porch columns, boxes, etc., etc.

Made in various types for all conditions.

Write for further particulars and prices

Saranac Machine Co., SOLE MAKERS
Benton Harbor, Mich.

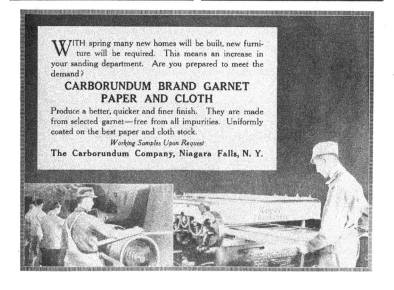

WITH spring many new homes will be built, new furniture will be required. This means an increase in your sanding department. Are you prepared to meet the demand?

CARBORUNDUM BRAND GARNET PAPER AND CLOTH

Produce a better, quicker and finer finish. They are made from selected garnet—free from all impurities. Uniformly coated on the best paper and cloth stock.

Working Samples Upon Request

The Carborundum Company, Niagara Falls, N. Y.

Sand Your Square and Octagon Turnings
Faster and Better
on the

MATTISON
EDGE SANDER
With Oscillating Belt

The stationary steel roll at the left may be used to advantage fur reaching nearly all the surfaces of patterns made up largely of ornamental members—and the sanding is done lengthways of the grain.

By means of the narrow-nosed form, the curved members, as well as the plain portions of square turnings and similar shaped work, may be sanded with one handling—accurately and economically.

The Secret is in the Combination

And the key to this combination is "Freedom from Vibration."

It is practically out of the question to attempt to sand work like the above against small, revolving rolls — the vibration due to high speed is too great. And to use a stationary roll or narrow form is just as useless if the machine itself chatters and vibrates; for every vibration leaves its mark in the work.

That is why the Mattison Edge Sander is built like a machine tool—rigid and accurate—free from vibration.

Combining this kind of construction with the use of nonvibrating sanding forms; together with our special tensioning device that insures free and smooth travel to the sandbelt, has put this machine in a class by itself.

Compare it with other edge sanders and you'll quickly see the difference.

Our Circular Describes its Construction in Detail.

Send for a Copy.

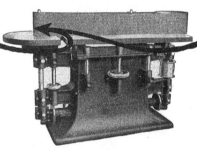

C. Mattison Machine Works
223 FIFTH STREET   BELOIT, WISCONSIN

Sash Relisher and Mortiser

This machine is particularly designed for relishing and mortising sash and blinds. Its cost is moderate, its capacity high, and a variety of work can be rapidly obtained from it by convenient adjustments.

Ask us for bulletin, which fully describes the use of this machine.

CANADA MACHINERY CORPORATION
LIMITED

GALT - ONTARIO

Toronto Office and Showrooms   -   -   Brock Avenue Subway

Alphabetical Index to Advertisers, Page 10

"Leather like gold has no substitute."

A Severe Test

Here is the real test of any belt—a high speed heavy
duty run over small pulleys. It is the performance of
Amphibia Planer Belting in just such runs that
has placed it foremost among leather belting.

By the use of Amphibia Planer Belting you not only
get belting service but you command higher production and better work from your machines.

Perfect leather belting on good machines makes for
the profitable operation of woodworking plants.

For over 40 years we have been tanners and manufacturers of the best leather belts.

Try a sample run of Amphibia Planer

Sadler & Haworth

MONTREAL, 511 William Street

TORONTO	**ST. JOHN**	**WINNIPEG**	**VANCOUVER**
38 Wellington St. E	149 Prince William St.	Galt Building	107-111 Water St.

Mr. Manufacturer

Does your equipment "balance" for properly manufacturing Shell Boxes?

Wouldn't the addition of a machine or two help in allowing the boxes to move more smoothly through your plant, thus cutting down your manufacturing costs?

Let us offer you our services.

Tell us what your troubles are.

We can furnish you with such machines as you require.

Anything in the way of WOOD WORKING MACHINERY for every purpose.

Garlock-Walker Machinery Co., Limited
32 Front Street West - - Toronto
Telephone Main 5346

Canadian Representatives

American Wood Working Machinery Co.
ROCHESTER, N.Y.

M. L. Andrews & Co.
CINCINNATI, OHIO

Is it true that you have finally "got wise" to the fact that people of the finest taste have really been educated up to the unequalled beauty of good **RED GUM** furniture?

And that those dealers who sell it are doing a mighty satisfactory business, selling all the **RED GUM FURNITURE** they can get from the manufacturers who are smart enough to make it?

It is ALSO being used in a lot of the most recently designed BANKS for Interior Woodwork *in place of Mahogany.*

A wood that is good enough for a fine bank interior is worthy of the most careful application of the finest art of the best *Furniture Maker.* **RED GUM** gets it.

Why not *get in line* by getting in a stock of perfectly seasoned **RED GUM** from a thoroughly responsible manufacturer—one who takes personal pride in his product? (Beware of GUM from "hurry-up" manufacturers—who lack both capital and facilities for *properly* and *safely* producing this Finest of America's Hardwoods.)

Write us for Samples, Particulars and General Information.
Our reply will be Prompt, Personal and Dependable.

Gum Lumber Manufacturers' Association
1314 Bank of Commerce Building, Memphis, Tennessee

Gang Dovetailer
for Ammunition Boxes
The Best Machine Yet Produced

Can be made to do either the closed end drawer joint or the open end box joint, and can be changed from one joint to the other at a very small cost.

The frame is well proportioned and very strong. All belts are now done away with except the belt drive from the line shaft. This is important, as it occupies only one-third the floor space usually given to a machine of this class and reduces belt cost to the minimum.

A large number of these machines, installed during the past three years, are giving excellent satisfaction. Our guarantee goes with each machine.

Have you our Bulletins?

Jackson, Cochrane & Co.
BERLIN, CANADA

Our No. 90 Gang Dovetailing Machine

Get the Output of Six Hand Sanders At the Cost of One

That's what you can do with the Yates Flexible Belt Sander, because it finishes at machine speed so much of the work that, formerly, was sanded by hand. Its versatility opens brand new fields for economic machine sanding. Can be applied to any flat surface, and almost any round or concave surface by hand blocks, cut to fit. Any length of belt giving any degree of flexibility desired can be easily obtained. Quality of finish can be made equal to hand finish in every way.

WITH A YATES FLEXIBLE BELT SANDER

you can banish hand sanding except for corners and other out-of-the-way places. The investment required puts it within reach of even the smallest woodworking plants. It's certainly worth your while, at least, to learn more about it and its possible influence on your margin of profit. Complete information yours for the asking.

The Yates Sander line—the standard by which sanders are measured—includes a sander for every purpose. Our Sander Book describes them. Why not get it now—free?

P. B. Yates Machine Co. Ltd.
HAMILTON, ONT. CANADA

Successors to THE BERLIN MACHINE WORKS, LTD.
U. S. Plant: BELOIT, WISCONSIN, U. S. A.

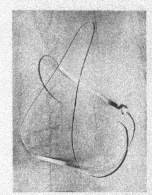

Here's a long story told in one look
about
RADCLIFF BAND SAWS

The accompanying photo speaks volumes for the toughness of our band saws. This saw was furnished to a Toronto manufacturer by us. The operator pushed blade off the wheel through forcing work too much, with result that rim was torn off the wheel, two spokes torn out and operator's hand severely cut. The saw was returned to us "to be straightened." On examination not a single crack was found either in blade or braze

This Manufacturer's name sent on request

When you need a Band Saw get a good one—a Radcliff

Write for prices

Radcliff Saw Manufacturing Co., Limited
TORONTO   550 Dundas Street   ONTARIO

Agents for L. & I. J. White Company Machine Knives

SLOW SPEED LOW POWER
BLOWER SYSTEM

The
Foster Blower

The "FOSTER" BLOWER

is specially designed for handling shavings, sawdust, excelsior, or any conceivable material. It is equipped with three bearings, shaft extending through, with bearing on inlet side. This eliminates the over-hung wheel and avoids heating, clogging and shaking. It is the only non-centre suction fan in existence. The Foster Fan is so constructed that the material entering the fan is immediately discharged without passing through or around the wheel.

Our Automatic Furnace Feeders will feed your material under boilers or into any desirable receptacle.

Consult us on your exhaust system troubles.

THE TORONTO BLOWER COMPANY
TORONTO   156 DUKE ST.   ONTARIO

Are you one of those who could make a big profit on this machine?

¶ Not every business can make a profit on a "FAY-EGAN LIGHTNING" No. 291 Automatic. ¶ But if your product requires large quantities of equalized lumber, i.e., stuff cut-off or trimmed to exact length, then this Hopper Fed Machine can and will make a big profit for you. ¶ Its capacity is practically unlimited. ¶ It will cut off both ends as fast as your operator can throw stock into the hopper by the armful and off-bearer clear it away. ¶ Does as much work as six old style trim saws. ¶ And every piece is **exactly** the length of all the others. ¶ If you equalize in quantities, use the enclosed post card to get Bulletin Q-28 and prices.

J. A. FAY & EGAN CO.

153-173 West Front St.　　　　·　　　·　　　·　　　·　　　·　　　Cincinnati, Ohio

—this new 230 page book
on Woodworking Machinery free!

Our new handy reference catalog is just off the press. Shows all of the newest models in woodworking machinery.

Every owner and user of woodworking machinery, every superintendent and foreman should have a copy of this new book, which is really a text on modern woodworking machinery.

A copy will be sent to you, free of charge upon receipt of your request.

Fill in and mail the coupon to-day.

J. A. Fay & Egan Co.,
153-173 W. Front St.,
Cincinnati, Ohio, U.S.A.

Please send your new Book on Woodworking Machinery free of charge to.

Mr. ....

Position .... (Give name of company you are with)

Street ....

City ....

Province ....

Peter Hay Machine Knives

Peter Hay knives are made of special quality steel tempered and toughened to take and hold a keen cutting edge. Hay knives are made for all purposes.

Have you our Catalogue?

Made in Canada

The Peter Hay Knife Co., Limited
GALT, ONTARIO

PRESSES

For Veneer and Veneer Drying

Made in Canada

William R. Perrin
Limited
Toronto

The Electric Planing Mills
Welland, Ontario,

is only one of many widely-known and progressive Canadian manufacturers who use

VERTICALLY-PIPED
DRY KILNS

Manufactured by

THE NATIONAL DRY KILN CO.
1117 E. Maryland St., Indianapolis, Ind., U.S.A.

You
CONTRACTORS
and
SMALL MILL MEN
Just the machine you need
—ANY KIND OF POWER—

A Whole Mill in Itself
A GREAT TOOL
Write for Catalogue and Prices NOW

J. L. NEILSON & CO.
Winnipeg, Man.

BOOKS FOR SALE
The following books are offered at special prices subject to previous sale:

Carpentry and Joinery, by Gilbert Townsend. Published in 1913 by the American School of Correspondence. 268 pages, illustrated. Price $1.00.

Saw Fitting Manual, a treatise on the care of saws and knives. Deals with everything in the saw and knife alphabet, from adjustments to widths. 144 pages. Price $2.00.

Common-sense Handrailing, by Fred T. Hodgson. Published by Frederick J. Drake & Company, Chicago. 114 pages, illustrated. Price 50c.

Roof Framing Made Easy, by Owen B. Maginnis. Published by The Industrial Publication Company, New York. 164 pages, illustrated. Price 50c.

Handrailing Simplified, by An Experienced Architect. Published by William T. Comstock, New York. 62 pages, illustrated. Price 50c.

How to Join Mouldings; or, The Arts of Mitering and Coping, by Owen B. Maginnis. Published by William T. Comstock, New York. 72 pages, illustrated. Price 50c.

Popular Mechanics Shop Notes. Published by Popular Mechanics, Chicago. Easy Ways to do Hard Things, etc. Years 1905-1906. Price 40c. each.

Cabinet Making, by J. H. Rudd. Published by Grand Rapids Furniture Record Company. 210 pages, illustrated. Price $1.50.

Utilization of Wood-Waste (Second Revised Edition), by Ernst Hubbard. Published in 1915 by Scott, Greenwood & Sons. 192 pages, illustrated. Price $1.50.

Woodworker Publishing Company, Limited
345 Adelaide Street West, Toronto, Ontario

The Only Way to Drive Screws

There is a stage-coach method and a modern method of doing everything. When it comes to driving screws the only modern way is with an Automatic Screw Driving Machine, which means greater and better production and greater profit.

The Reynolds
Automatic Screw-Driving Machine

is guaranteed to produce results. It is no experiment. The No. 2 Standard Type, shown herewith, is particularly suited for placing the screws in the straps of shrapnel boxes. It may be fitted to drive, without change or adjustment, any three consecutive sizes flat or round head, or two sizes filister head, No. 6 to 14 inclusive, wood or machine screws, or three sizes flat or round head, or two sizes filister head machine screws up to No. 20. All the above up to 1½ inches long if smaller than No. 10, or up to 1¾ inches long, No. 10 or larger. Fitted with boring attachment if desired.

Many Machines in Use in Canada.

Ask for our Catalogue.

REYNOLDS PATTERN & MACHINE CO.
101 and 103 Third Ave., Moline, Ill., U. S. A.

No. 2 Standard Type Automatic Screw-driving Machine.

Perkins Vegetable Glue
The Best for Built-up Stock and Veneer Laying

Perkins Vegetable Glue sticks and holds. It is unaffected by dampness, extremely dry atmosphere or severe sanding. We guarantee to effect a 20 per cent. saving in your glue room by replacing hide glue with Perkins Vegetable Glue. Along with this saving you can have a glue room comfortable as to temperature and sweet as to smell. The glue is absolutely uniform in quality. Every shipment exactly the same. Let us demonstrate in your own factory and on your own stock just what can be accomplished with Perkins Vegetable Glue.

Dinner-wagon manufactured by the D. Hibner Furniture Co., Limited, Berlin, Ont., in which Perkins Glue was used.

PERKINS GLUE COMPANY, LIMITED,   HAMILTON, ONT.

"Address all inquiries to Sales Office, Perkins Glue Company, South Bend, Indiana, U. S. A."

Alphabetical List of Advertisers

Acme Steel Goods Company 47
American Hardwood Lumber Co... 12
Atlantic Lumber Company 13
Ault & Wiborg Company 37

Baird Machinery Works 45
Brown & Co., Geo. C. 12

Callander Saw Mills 45
Canada Machinery Corporation ... 1
Canadian Boomer & Boschert Mfg.
Company 49
Canadian Morehead Mfg. Company 49
Carborundum Company 55
Central Veneer Company 43
Chapman Double Ball Bearing Co. 51
Churchill, Milton Lumber Co. 12
Clark Veneer Co., Walter 41
Clark & Son, Edward 10
Coignet Chemical Products Co. ... 37

Darling Bros. 13
Dominion Mahogany & Veneer Co... 14

Evansville Veneer Company 38

Fay & Egan Company, J. A. 7
Foster, Merriam & Company 54

Garlock-Walker-Machinery Co. ... 3
Gillies Bros. 13
Gum Lumber Manufacturers' Assn. 4

Hart & McDonagh 13
Hartzell, Geo. W. 13
Haughton Veneer Company 43
Hay & Company 37
Hay Knife Company, Peter 8
Hoffman Bros. Company 41

Jackson, Cochrane & Company ... 4

Kent-McClain Ltd. 47
Kersley, Geo. 42
Klotz Manufacturing Company ... 50
Kraetzer-Cured Lumber Co. 34

Mattison Machine Works 56
Mowbray & Robinson 12

Nartzik, J. J. 41
Nash, J. M. 51
National Dry Kiln Company 8
Neilson & Company, J. L. 8

Ober Manufacturing Company 51
Ohio Veneer Company 41

Paepcke Leicht Lumber Co. 14

Perfection Wood Steaming Retort
Company 49
Penrod Walnut & Veneer Company 43
Perkins Glue Company 9
Perrin, William R. 8
Petrie, H. W. 45
Preston Woodworking Mach. Co... 11

Radcliff Saw Company 6
Reynolds Pattern & Machine Works 9
Roberts, John N. 42

Saranac Machine Company 55
Sadler & Haworth 2
Sheldons Limited ... ...
Simonds Canada Saw Co. ... .
Spencer, C. A. 12

Toronto Blower Company 6
Toronto Veneer Co. 44

Walter & Company, B. 55
Wardwell Mfg. Company 47
Whitney & Son, Baxter D. ... ...
Wood Mosaic Company 42

Yates Machine Co., P. B. 5-45

Edward Clark & Sons

Toronto

CANADIAN HARDWOODS—winter cut Birch, Bass, Maple

10,000,000 ft. on hand, shipment June 1st
10,000,000 ft. sawing, shipment August 1st

ONTARIO and **QUEBEC** stocks, East or West, rail or water shipment, direct from mill to factory.

Put Up for Shell Boxes

3,000,000 ft. 4/4 Birch, No. 1 and 2 Common.
2,000,000 ft. 6/4 Birch, No. 1 and 2 Common.
1,000,000 ft. 12/4 Birch, No. 1 and 2 Common.

For Furniture Manufacturers

1,000,000 ft. each 10/4, 12/4, 16/4 Birch, 75 per cent. No. 1 and 2.

For Trim Trade

500,000 ft. 4/4 Birch, No. 1 and 2,   30 per cent. 14/16 ft., Red all in.
300,000 ft. 5/4 Birch, No. 1 and 2,   9 in. average. Red all in.
700,000 ft. 6/4 Birch, No. 1 and 2,   40 to 60 per cent. 14/16 ft. long.
500,000 ft. 8/4 Birch, No. 1 and 2,
200,000 ft. 4/4 to 8/4 Brown Ash, No. 1 and 2, 40 per cent. 14/16 ft.

We Commend to Your Notice:

Our lumber is reliable, our grades uniform, our shipments prompt, and receive personal supervision; our prices consistent. We have had long experience. We desire to serve.

" Treat your machine as a living friend."

Do You need Men in Your Veneer Room?

If so

Buy a "Black Bros." Patented Veneer Taping Machine

and

We guarantee that it will save 50% of your present labor cost

That's strong talk, but we stand back of every word of it, and can refer you to Furniture Manufacturers in Ontario today for proof of what we say.

If you have eight men now you could do with four if you had this Machine, and that should be a particularly appealing argument in view of the very great scarcity of labor existing at the present time.

The Machine will interest you if you do any Veneered work. It will tape together all kinds of Veneer, no matter whether the Veneer be thick or thin, handling both kinds equally well.

It will use the ordinary gummed tape or it will use plain paper tape and put on its own gum or glue just before it is laid.

The Machine is built in two sizes—24 in. and 36 in.

Write us for prices and particulars.

We control the Canadian Patent on this Machine

The Preston Woodworking Machinery Co.

Preston, Ont.　　　　　Limited

GEO. C. BROWN & COMPANY

Manufacturers St. Francis Basin Hardwoods

Band Mills—PROCTOR ARK.   General Offices—Bank of Commerce Bldg.—MEMPHIS, TENN.

STOCK FOR SALE—JULY AND AUGUST SHIPMENT

150 M' 4/4 No. 1 and No. 2 Com. Plain Red Oak.	200 M' 4/4 1st and 2nds Sap Gum (straight and flat).
150 M' 4/4 No. 1 and No. 2 Com. Plain White Oak.	50 M' 8/4 1st and 2nds Sap Gum (straight and flat).
70 M' 4/4 1st and 2nds. Qtd. White Oak (15 to 20 per cent. 10 in. and up).	150 M' 4/4 1st and 2nds Sel. Red Gum (straight and flat) 50 M' 4/4 No. 1 Com. and Bet. Sel. Red Gum (quarter-sawed).
70 M' 4/4 No. 1 and No. 2 Com. Qtd. White Oak.	50 M' 8/4 No. 1 Com. and Bet. Sel. Red Gum (quarter-sawed).
20 M' 5/4 No. 1 and No. 2 Com. Qtd. Red Oak (6 in. and up).	60 M' 10/4 No. 1 Com. and Bet. Sel. Red Gum (quarter-sawed).

"Our Gum is all Kraetzer-Cured"—Splendid Condition.

TENNESSEE AROMATIC RED CEDAR SPECIALISTS   OUR TRAFFIC DEPARTMENT TRACES EVERY SHIPMENT

Churchill Milton Lumber Co.

Louisville, Ky.

We carry at
NEW ALBANY, IND.

Genuine Indiana Plain and Quartered Oak, Poplar

At GLENDORA, MISS.

Cypress, Gum, Cottonwood, Oak

SEND US YOUR INQUIRIES

We have dry stock on hand and can make quick shipment

Plain and Quartered Red and White Oak

and Other Hardwoods

Even Color   Soft Texture

We have **35,000,000 feet dry stock**
all of our own manufacture, from our own
timber grown in Eastern Kentucky

MADE (MR) RIGHT

OAK FLOORING

PROMPT SHIPMENTS

The Mowbray & Robinson Co., Inc.

GENERAL OFFICES
CINCINNATI, OHIO

Mills: Quicksand, Ky., West Irvine, Ky., Viper, Ky.

Canadian Representative:
N. H. FARNHAM
601 Elmwood Ave., BUFFALO, N. Y.

American Hardwood Lumber Co.

St. Louis, Mo.

Large stock of—

Dry Ash, Quartered Oak Plain Oak and Gum

Shipments from — NASHVILLE, Tenn.,
NEW ORLEANS, La., and BENTON, Ark.

Dry Spruce and Birch

Good Stocks, Prompt Shipments, Satisfaction

C. A. SPENCER, Limited

Wholesale Dealers in Rough and Dressed Lumber

Offices—500 McGill Building
MONTREAL - - Quebec

David Gillies
President
J. S. Gillies
Vice-Pres. and Man. Dir.
D. A. Gillies
Sec.-Treas.

Established 1873
GILLIES BROS.
LIMITED
Mills and
Head Office
BRAESIDE, ONT.

— Manufacturers of —

White Pine - Red Pine - Spruce

We offer Lumber for Box and Crating:

WHITE PINE MILL RUN
4/4 x 4/10 x 6-9 ft.   500 M
4/4 to 8/4 x 6-7-8 x
10/16   900 M
WHITE PINE BOX
4/4 to 8/4 x 4 and up
x 6/16   400 M
WHITE PINE MILL CULLS
4/4 16 8/4 x 4 and up x
6/16   x 2/0 M
4/4 x 10/11/12 x 6/16   300 M

RED PINE LOG RUN
4/4 to 8/4 x 4/5/6/7/8
in. x 10/17   1550 M
RED PINE BOX & CULLS
4/4 to 8/4 x 4 and up
x 6/16   1000 M
SPRUCE MILL RUN
2 x 4 and up x 6/22   25 M
SPRUCE BOX & CULLS
4/4, 5/4, 8/4 x 4 and
up x 6/16   500 M

The above stock is dry and much of it is assorted widths.

The Atlantic Lumber Co.

Manufacturers and Wholesale Dealers

American and Canadian Hardwood Lumber and Mahogany, Quartered and Plain, White and Red Oak a Specialty.

110 Manning Chambers - TORONTO
Phone Main 6386

HOME OFFICE, BOSTON MASS.

Mills: KNOXVILLE, TENN.,   WALLAND, TENN.
FRANKLIN, VA.

DARLINGS STEAM APPLIANCES

DARLING BROTHERS
LIMITED
Engineers and Manufacturers
MONTREAL, CANADA

Branches:
Toronto and Winnipeg

Agencies:
Halifax, St. John, Calgary, Vancouver

Shell Box Stock

1 x 6 birch, in No. 1, 2, and 3 common.

1 x 7 and 8 birch, in No. 1, 2, and 3 common.

1 x 8½ birch, in No. 1, 2, and 3 common

1 x 9 and wider, birch, in No. 1, 2, and 3 common.

1 x 4 and wider, spruce. 2 x 4 and wider, spruce.

For the Furniture Trade

4/4, 5/4, 6/4, 8/4, 10/4, 12/4, 16/4, birch, maple, elm, basswood and black ash.

Elm, Basswood and Black Ash

in No. 1 common and better grades—dry stock in assorted sizes.

Hart & McDonagh

Continental Life Bldg.

TORONTO   -   -   ONTARIO

FOR SALE

Black Walnut Lumber

15,000 feet 4/4 No. 1 Common
20,000   "   4/4 No. 2   "
15,000   "   5/4 No. 1   "
15,000   "   5/4 No. 2   "
15,000   "   6/4 No. 2   "
40,000   "   8/4 No. 1   "
60,000   "   8/4 No. 2   "
30,000   "   9/4 No. 1   "

Will make special low prices on all of the above stock to move it quick. All Band Sawn, Equalized, and all thoroughly dry, ready for immediate shipment. Write for prices.

Geo. W. Hartzell

Piqua, Ohio

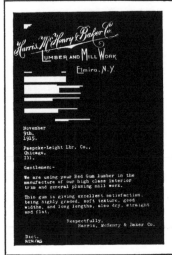

Of course it is true that

RED GUM

is America's finest cabinet wood—but
Just as a poor cook will spoil the choicest
viands while the experienced chef will turn
them into prized delicacies, so it is true that

The inherently superior qualities
of Red Gum can be brought
out only by proper handling.

When you buy this wood, as when you buy a new
machine, you want to feel that you have reason for
believing it will be just as represented.

We claim genuine superiority for our Gum. The
proof that you can have confidence in this claim is
shown by the letter reproduced herewith.

Your interests demand that you remember this
proof of our ability to preserve the wonderful qualities of the wood when you again want RED GUM.

Paepcke Leicht Lumber Co.

Conway Building   111 W. Washington St.
CHICAGO, ILL.

Band Mills: Helena and Blytheville, Ark.; Greenvil Miss.

Buy Made in Canada Veneers

Dominion Mahogany
& Veneer Co., Limited

Head Office and Factory   Toronto Office and Warehouse
Montreal West, P.Q.   455 King St. West

Mahogany, Walnut,
and all Fancy Woods—Lumber

SAWN, SLICED and ROTARY CUT VENEERS

Mahogany, Walnut, Quartered and Plain Oak, Ash, Maple,
Birch, Elm, Poplar, Gum.

Gum and Poplar Cross Banding.   Hard Maple Pin Block Stock.
SEND US YOUR ENQUIRIES.

Canadian Woodworker

A Monthly Publication in the Interest of the Woodworking Industry.
Reaches the factories producing interior finish, doors, sash, flooring,
woodenware, furniture, pianos, boxes, and general mill products.

Subscription, $1.00 a year; foreign $1.50.

Woodworker Publishing Company, Limited

Branches: Montreal, Vancouver,
Chicago, New York, London, Eng.

345 Adelaide St. West, Toronto
Phone Ade. 2700

Authorized by the Postmaster General for Canada, for transmission as second class matter.
Entered as second class matter July 18th, 1914, at the Post Office at Buffalo, N.Y., under the Act of Congress of March 3, 1879.

Vol. 16   July, 1916   No. 7

Advantages of Piece Work

The question of day work or piece work is of vital interest to both the employer and employee, and although a great many workmen have strong objections to any form of piece work, yet it is undoubtedly the fairest system for both parties if it is properly worked out. Piece work gives the honest workman a chance to prove his metal and shows up the "slacker" who under a day work system might be drawing the same wages. But even the honest and conscientious workman in a great many cases gets "cold feet' when he hears that a piece work system is being introduced into the shop. This is largely due to the fact that some employers take advantage of their men by reducing prices once they get the piece work system established.

The prices are generally set by having a man do the job day work and then making a slight reduction from the day work price, it being assumed that a man will work harder when working piece work. The method of setting the prices is alright, but when men put forth extra efforts and devise short cuts for doing the work, only to find that the prices are cut as soon as they make a little more money, they lose interest in their work and either leave at the first opportunity or develop into chronic grumblers. On the other hand, fair treatment aided by a little diplomacy will make them into loyal, contented workers, a valuable asset to the firm.

In introducing a piece-work system the first aim of the employer should be to establish confidence between himself and his employees. Let them feel that they can rely on getting a "square deal," and in nine cases out of ten the system can be operated to the benefit of both parties.

The bonus system of piece work that has recently been adopted by a number of manufacturers has many advantages over the old system of straight piece work. In the first place, a price is fixed for the job and then the workman is told that he is sure of his week's wages and that everything he makes over and above the set price will be divided between him and the firm. The satisfaction of knowing that he is sure of his week's wages is a great stimulant for a man to think and figure out ways and means of turning out a greater quantity of work, whereas when he is not sure of his wages, he figures out that he has to do so much work an hour, regardless of quality. He may have it sent back to him, but the delay in the other departments costs the firm more than it would cost them to pay the workman a better price for the job.

Another disadvantage with straight piece work is that when a man finishes a job and has to wait for another (maybe through no fault of his own) he loses the time that he has to wait. Of course, a good foreman will see that the next job is beside the man's bench ready for him to start on when the first one is finished, but it sometimes occurs that some parts of the job are not ready and the workman has to take another one. This means delay, and as the workman himself is not responsible he should not be the loser. A little consideration by both the employer and employee, of each other's problems should go a long way in working out an equitable piece-work system.

Mechanical Skill Versus Executive Ability

What qualifications should a man possess in order to be advanced to the position of superintendent? The editor recently discussed this question with a man who had had a number of years' experience in the woodworking business. He had learned his trade as a cabinet-maker and then after spending his spare time for a number of years in perfecting his knowledge of draughting he occupied a position in the draughting room of a furniture factory, where he had charge of all draughting, stockbilling, etc. Then when there was a vacancy in the factory for a superintendent and he applied for it he was informed by the manager that the position was only open to a man who was an expert machine hand, capable of operating (and repairing if necessary) every machine in the factory. The result was that the man in question severed his connection with the woodworking industry and drifted into an entirely different line of business, where he is at the present time holding a very responsible position.

The incident opens up an interesting subject. The fact that the man had a good general knowledge of woodworking and that he stepped right into another position requiring considerable executive ability as well as the ability to handle men, would seem to indicate that he had the necessary qualifications to make a successful superintendent and the question naturally suggests itself: can the woodworking industry afford to lose men of this calibre? It is undoubtedly essential that a superintendent should have a first class working knowledge of manufacturing, but it seems rather a large order to ask him to be an expert at setting up and operating every machine, and where a competent machine room foreman is kept it should hardly be necessary. If a factory is large enough to warrant engaging a superintendent then his services might be employed to better advantage planning how he could reduce the cost of production and organizing his forces accordingly, than tinkering around a machine.

We are of the opinion that the first quality necessary in a superintendent is an excellent knowledge of human nature and he should be a "man" in the true sense of the word—above all petty jealousies and favoritism—and capable of maintaining discipline, together with the respect of his men. He should be thoroughly familiar with every operation (but not necessarily an expert at performing it) required in the manufacture of the firm's products. He should know the uses of every machine and the output of which it is capable. Besides this he should possess sufficient executive ability to keep the factory running to the best advantage as regards quality, quantity and cost of output.

Another Toy Exhibition

The following letter has been sent out by the Canadian Toy Association and the opportunity afforded of exhibiting samples of their products at the Canadian National Exhibition, free of charge, should be taken advantage of by those woodworking firms that have gone into the manufacture of toys and novelties and those who contemplate doing so. The value of such exhibits can be realized when it is considered that the Canadian National Exhibition attracts visitors from every part of the country and when within about four short months after it is over, with that holiday season approaching when toys reign supreme, it is only natural to suppose that people's thoughts will turn to the wonderful examples seen at the exhibition and resolve to buy "made in Canada" toys.

Canadian Toy Association,
53 Yonge Street, Toronto.

Gentlemen.—The Executive Committee of the Canadian National Exhibition were so favorably impressed with the Toy Fair which was held in the Royal Bank Building, Toronto, during the past spring, that they asked for a repetition of it for the Canadian Exhibition. Their proposal had the approval of the Canadian Toy Association, which was formed at the time of the Toy Fair, and we now have the assurance of Hon. Geo. E. Foster, Minister of Trade and Commerce, that the Department of Trade and Commerce will co-operate with the Canadian Toy Association in making a success of the Fair, if the toy makers will exhibit and go into the matter with a will. We have been able to get the following conditions for the manufacturers, which look very reasonable:—

(a) All space at the Exhibition will be given free by the Canadian Toy Exhibition.

(b) The Department of Trade and Commerce will pay the cost of constructing the stands necessary for the exhibition of the toys, and also for the general decorations.

(c) The exhibitors will be expected to pay the cost of the decorating of their own stands, and the transportation of their goods to and from the Exhibition.

(d) The Department of Trade and Commerce will furnish a general supervisor who will give the exhibitors assistance as far as possible; however, each exhibitor will be responsible for the care of his own toys during the Exhibition.

(e) All toys must be in place before the date of the opening of the Fair.

As secretary of the Canadian Toy Association I shall be obliged if you will let me know as soon as possible if you will exhibit samples of your toys, and if so, what space you will require, as the management of the Canadian National Exhibition wish to know at once what space is to be set aside for our toy fair, and the Department of Trade and Commerce must make an immediate arrangement for the construction of stands and general decorations.

While the Toronto Exhibition does not open until near the end of August, space must be immediately allotted to all the different classes of exhibits at once, and unless the carpentry work in connection with the construction of the stands is arranged immediately it will possibly be difficult to get carpenters to do the work later on. You will appreciate the importance of letting me know your decision as quickly as possible.

I have no doubt you will be pleased to exhibit your toys there, as there are usually about a million visitors at the Toronto Exhibition.

If you have not mailed in your dollar for membership to the Canadian Toy Association, kindly do so at once.

Yours faithfully,
Canadian Toy Association
Per L. G. Beebe.

Speaking of the need for sharp tools to work cypress and gum, what about the splash line in quartered oak? If there is anything in native wood harder to cut than that, it has not been heard from.

How to Properly Adjust the Fast Feed Matcher

Application of Oil on Bearings Will Often Prevent Waves—Knives Must be Ground Straight—Should Show Contact With Jointing Stone

By H. H Shumaker

To those who have to deal with the fast-feed machines in the modern planing mill, the care of the matcher, which catches the widest range of work, is of the first importance. The man who has been used to the slow-feed machines finds that, in taking hold of the late machines, they require more thoughtful attention. I say thoughtful, because a man had better be pretty sure when he sarts after a "trouble" that he is on the right track.

In days gone by, when a wave or uneven knife mark showed on the dressed surface, the machinist would tighten down on the bearing caps, with perfect assurance that he was doing one of the few things necessary to get a perfect cut. If the bearings happened to be drawn a little too tight, and heating, without especial damage to the bearings, resulted, still the knife marks usually showed regular. On the late type of machine possibly more trouble results from the effort to correct an uneven cut in this way, than any other. Then, if the trouble is not in the loose bearing, where is it?

To illustrate: We have a machine in our mill, built on splendid. lines, which was equipped with square cylinder, with clamps to hold thin-steel knives. It took constant work and care to get anything resembling a perfect cut with the thin knives at 100-ft. per minute. But remove the clamps and use the slotted knives, set up with rule or gauge, and feed the stock at 50-ft. per minute, and we would get what would have been called in the old days a perfect cut. This shows that the machine must be heavier, and adjustments more perfect, to get first-class work with round cylinders and jointed knives.

Improper lubrication is a common cause of trouble. In the perfect-running bearing the metal of shaft and babbitt is not in contact, but separated by a thin coat of oil. This coat must extend the full length of bearing, obtained by free oil-roads cut in babbitt, and these, when possible, filled with wicking or felt to draw the oil to extreme ends of bearing. I recently had trouble with a bearing that, to my mind, was nearly perfect in construction and size for the work required. There was no sign of heating, so a light jointing of the knives was resorted to, with no desired result. I unscrewed one bearing cap, and noticed a streak about 1 in. wide on bearing, close to cylinder. that was dry enough to show a dark gumming. I cleaned this, then opened up oil-roads, poured on some oil for starting, replaced cap, and the trouble disappeared. I have frequently noticed a wave developing on the face of stock coming from the machine, when another application of oil to the bearings entirely effaced the trouble. Had I resorted to the first idea usually occurring to the novice, I would have tightened down the bearing caps and brought on more trouble.

Knives not ground perfectly straight are another source of trouble. When set in the cylinder the knife must show contact with the jointing stone. To get this, the straight knife will naturally have to be jointed heavily, in order that the low place in the crooked knife will show a joint. Then the knives will have an uneven heel. and first-class work under these conditions is hardly possible, for experience soon teaches

that all knives should show an equal joint. Neither can the equal joint be obtained when the bevels ground are not the same. Don't hurry the grinding; you not only bring trouble to the machine operator, but naturally shorten the life of the knife. It is a well-proven fact that the best-finished stock and the lightest-running cylinder come with the very lightest-jointed knives.

Another feature which goes toward injuring the fast-feed machines is the jointing device. While all the parts controlling the stone should work freely, be sure there is no lost motion, or the stone will have too much vibration when coming in contact with the revolving knives. Point the jointer stone so that the point coming in contact with knives is not over ⅛ by ¾ in. wide. A blunt stone may joint perfectly a lightly-jointed knife, but will be unsteady when the heel is heavier. Move the stone slowly, then it will put a straight cutting edge to knife and avoid the diagonal knife marks on finished stock.

In concluding this article I will repeat old advice: Don't neglect repair work; when you see a loose key or setscrew, tighten it at once. If a cylinder or cutting head shows signs of cutting up, go right after it; it can be fixed much easier when first detected, while the journals are still true, and so with any weakened part of the matcher. "Don't wait" is the very best motto, unless you are absolutely certain that damage will not result from a little delay in repairs. I have seen many good men lose their positions simply by neglecting small troubles. As a result, these accumulate until the low grade of work produced calls a protest from the man higher up. The sequel to this we all know.

Proper Attention Increases Usefulness of Belts

By N. G. Near

If manufacturers and dealers in belting could be assured that the belting would receive proper attention and care while in the hands of the buyer they could often sell at a much lower figure.

I know of a manufacturer, for instance, who generally uses a "factor of safety of two" in selling his belting, because he fears the belting will be expected to "take care of itself."

The rule this manufacturer of rubber belting uses is: "Each inch in width, running 1,000 ft. per min., will transmit one horse power."

With proper care this man states that the figure 500 could be used just as well, just as safely, just as durably, but he will not risk it.

In buying belts, then, it is well to tell the dealer or manufacturer just what kind of care you expect to give the belts.

The outside of a board is apt to be more or less gritty; or, if not, is a trifle harder than the inside. A knife will dull less quickly if the cut is deep enough so that it gets under the outside shell at the start. It dulls a knife less to lift up a bit of grit than to cut down through it.

Fig. 1—Front Elevation of House.

Modern Dwellings Call for Large Quantity of Manufactured Woodwork

The Small Town Planing Mill Gets Large Share of this Business. House Illustrated has Considerable Exterior as Well as Interior Work

One of the most noticeable features in connection with the woodworking industry in Canada is the constantly increasing demand for high grade planing mill work in the small towns and rural districts. A great many people have been prone to regard the farmer as a "tightwad," but a short journey into the country in almost any direction will reveal the fact that he is

Fig. 2—Side Elevation.

spending his money generously in providing himself and his family with a home built on modern lines. The same is true of some of the small towns and cities that have gradually evolved from sleepy villages into manufacturing districts of considerable importance.

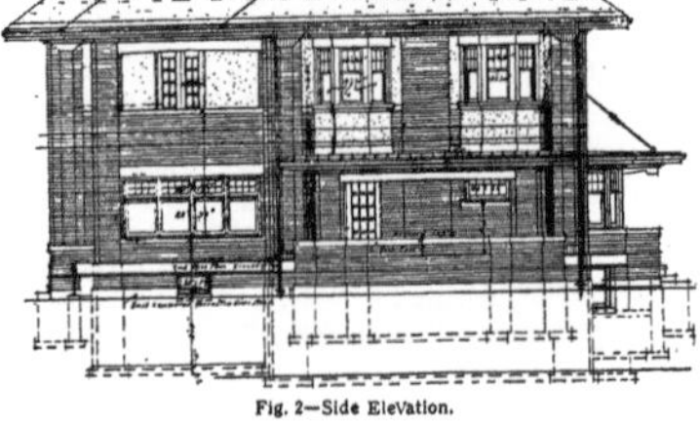

When a man finds himself becoming prosperous, about the first thing his thoughts turn to is a new home suitable to his new station in life. His artistic tastes may not always develop in proportion to his bank account, but nevertheless, he generally wants something large for his money and something that no person else has—no stock designs for him. He outlines his dream of a home, together with a few practical details, such as the number of rooms, etc., it must contain, to the architect, who proceeds to work out something that will look well and at the same time embody the necessary features, as outlined by his client. This means that the planing mill man, if he wants the business, must be able to furnish all the necessary woodwork, manufactured according to the detail drawings and at the same time furnish it at a price that will enable him to compete with others in the business.

The house which we have selected as an illustration was designed and erected under the supervision of S. B. Coon & Son, Architects, Toronto, for an out-of-town client. The details of the doors, trim, etc., Figs. 3 and 5, should be interesting to planing mill men as being a good example of the class of work that they are from time to time called upon to figure on, and one of the first things to be considered when figuring on work of this kind is how many operations will be required on the various machines. It should, therefore, be worth while to examine closely the details of the doors, newel post, beam in dining room and trim and determine how many operations would be required to make them. One handling of a batch of stock may mean the difference between profit and loss.

Another thing that has to be made at the planing mill is the staircases. These are generally made in sections, according to the number of landings, winders, etc., to each flight. The number of steps and widths of landings has to be taken from the floor plan, see Fig. 4, and sufficient left on the ends of the strings to allow for fitting on the job. Then besides the standing finish and stairs there is usually in dwellings considerable work in the way of fixtures and fittings for pantries, closets and built-in furniture, all of which means work for the planing mill.

Now it will be evident that in order to turn out this class of work the planing mill must be up-to-date both as regards machinery and the mental equipment of the men in charge.

A point that was brought up recently by a writer in one of the American journals to the effect that the

owners of planing mills, instead of delivering the materials to the job and having outside carpenters put it in place should have charge of the work until the building is turned over to the owner, is worthy of con-

Fig. 3—Detail of Beams in Dining Room.

sideration by those engaged in the trade. In order to suggest that the woodworker who is making building materials owes it to himself to see that his product is properly handled after it leaves the factory, a few cases in point may be examined. Of course, it is true that the millwork man who sells his product to a con-

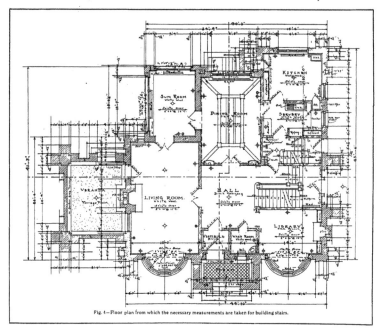

Fig. 4—Floor plan from which the necessary measurements are taken for building stairs.

tractor, whether handling the whole work or a detail, has no other customer, and loses control over the material after it is delivered; but it would nevertheless be a good thing to offer service in insuring the proper installation of the work. Wood is the one material that goes into a building that needs positive protection from the weather. That is to say, the iron, stone,

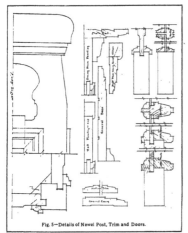

Fig. 5—Details of Newel Post, Trim and Doors.

cement, brick, etc., are not greatly affected and can be handled with little regard to the weather. Therefore, as long as the roof is on and his men can work inside, the contractor does not often stop to consider whether or not his woodwork has been exposed long enough to take up moisture nor whether the building has dried out sufficiently to prevent the millwork from being damaged. His aim is to finish up the building as quickly as possible, turn it over to the architect or owner and collect his money. Once he gets his money, if any defects appear in the woodwork the man that supplied the millwork gets the blame, even though he may have attended to his end of the work faithfully and honestly.

Any person knows that if a door is exposed to the weather for any length of time, or the trim is put in place without any windows in the building, the wood will absorb sufficient moisture to make it practically as wet as before it entered the kiln. Then after the painters have finished their work and the building is occupied the wood starts to shrink and show unsightly cracks and streaks of unfinished wood where door panels, base, stair treads, etc., have shrunk away from other portions of the work. The effects of moisture will be even more noticeable if the woodwork has been glued up.

The writer was recently told that some of the leading flooring manufacturers in the United States who have a reputation to maintain, and who are determined that every job in which their material is used shall be handled right, use the plan of supervising all of this

work, even where their interest consists only in the sale of the flooring. In the first place, they keep their stock in heated warehouses, both at the factory and at the warehouse from which it is distributed to the contractors. Rooms in which the flooring is laid down are heated, so that there will be no excess moisture to be driven off later and to cause gaps in the floor or cupping of the flooring. They watch all of these fine points and the result is floors of beauty.

Machines for Sandpapering Small Jobs

The sandpapering machine illustrated on this page is manufactured by the Downer Grinder Company, 83 Jarvis Street, Toronto, and is about one of the handiest machines for sandpapering small stock, either flat or curved, that has come to our attention. It is of the disc and spindle type, having two discs (one at each side) with the spindle or drum at the back and between the discs, but differs greatly from others of this kind, in that, instead of having wooden discs with the sandpaper fastened at the outer edge, permitting it to "bag" in the centre, it has steel discs, planed absolutely true, with the sandpaper securely attached to the surface of both sides of the discs with shellac. It only requires a couple of minutes to remove the discs and reverse them to use the opposite side.

The spindle or drum has a vertical and revolving motion, which prevents scratching of the stock and can be removed and attached to the main arbor in place of one of the discs if desired. The tables on both the discs and the spindle can be raised and low-

The Downer and Sanding Machine.

ered so as to use every portion of the paper. They can also be tilted and accurately adjusted at any angle up to forty-five degrees, either up or down. The discs are attached to the arbor by means of four bolts with screw heads. The upright shafts on which the tables are raised and lowered, instead of being round, with slots for the set screws, are square, preventing the screws from working loose.

When the sandpaper is worn out it can be very quickly removed from the discs with a blunt chisel, and as two extra discs are provided with the machine no time need be lost while working on a job. An assortment of spindles or drums is also provided. The machine takes up very little floor space and three men can work on it at the same time without interfering with each other.

Hints on the Hardening and Tempering of Steel

Table of Temperatures and Corresponding Colors as Indicated by a Reliable Pyrometer

By W. Brown

Milled to pattern knives and cutters are being used more extensively in woodworking factories as the advantages of using them become known. The old method was to have the shaper and sticker men make their own knives, but in the steady march toward the goal "efficiency," this method has been passed up and the more modern one of having the knives made by firms who specialize on that particular work has taken its place. It sometimes happens, however, that a pair of knives are required in a hurry, and they have to be made in the factory. In that case it will be necessary that the man whose duty it is to make them shall know how to harden and temper them properly. The majority of the experienced shaper and sticker hands possess this knowledge, so that the object of this article is to point out for the benefit of the younger generation, the principles of and reasons for hardening and tempering steel.

It might not be out of place to state at the beginning that steel is nothing more or less than iron mixed with different elements (one of which is carbon) to give it the necessary toughness and hardness. There are three different kinds of tool steel in general use; carbon steel, high speed steel and self-hardening steel. Each kind has varying quantities of different elements such as carbon, tungsten, valadium, etc. Carbon steel is the kind that is most widely used for woodworking machine cutters, so we will deal with it first.

The percentage of carbon in carbon steel is generally very low and depends on the kind of work the steel is being used for. That used for woodworking machine cutters, hand chisels, plane irons, etc., contains from .85 per cent. to 1.00 per cent. A slight difference in the percentage of carbon makes a vast difference in the percentage of which is generally designated by the manufacturers as being of such and such a temper. The different tempers are numbered and if the manufacturer is informed what use the steel is required for he will furnish steel of the correct temper. The word "temper" as used in this connection, however, simply means the percentage of carbon and should not be confused with the act of tempering, (which we will explain later).

Incorrect treatment in the heating process and the selection of an unsuitable quenching liquid are more often the cause of faulty tools than anything else. Oil is generally used for quenching because it has a milder effect on the steel than water and does not make it as brittle. But almost any desired result can be secured by quenching in water, if it is properly treated. The water may be perfectly cold when the tool is simple in shape and required to be glass hard. For more intricate tools it should be heated to 60 or 70 degrees Fahrenheit. If it be previously saturated with calcium chloride—which raises the boiling point very considerably—it may be heated to 176 degrees Fahr. and every advantage of oil quenching and others not obtainable with oil may be secured. When oil is used, opinions differ as to the best kind, but the writer has always found cottonseed oil and fish oil to be satisfactory, both as regards price and results. After the quenching medium, the next thing that deserves consideration is the fire.

It has often been pointed out that a clean charcoal fire is best for heating steel, but good forge coal will give equally satisfactory results. Green coal should be avoided, because, if any impurities, such as sulphur or phosphorous are present, the steel will be ruined. The heating of the tool is important and the best method is to start with a bed of good live coals, placing the knife so the thin cutting edges will not be heated first and burned before the remainder of the knife reaches the proper temperature. When heating a knife it is advisable to place it on the coals face up, letting the thin cutting edges and the narrow points project a little past the hot blast. Start a slow, even draft and keep the fiercest heat concentrated along the knife just back of the bevel where it is of full thickness, until the low, dull red color appears, then force the draft until the color brightens somewhat and

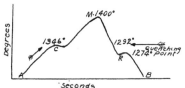

Diagram showing how heat rises and falls when heating steel.

quickly withdraw and plunge the knife point downward into the bucket of oil or water.

By uniform and careful heating each part is expanded without strain and at relatively the same rate and without oxidizing and decarbonizing the outer surface of the steel. Decarbonization means soft places on the surface of the knife. Irregular and hasty heating will cause one part of the knife to become hotter than another and then nothing can prevent it expanding more. The result will be that parts adjacent will become permanently strained, or even bend or break, and on the tool being quenched, both warping and cracking will take place. When hardening the knife great care should be taken to soak the steel thoroughly, that is, to see that the necessary quenching heat is diffused to the core, but on no account must the steel be roasted. Increase the heat slowly and preheating previous to placing the tools in the hardening fire is very desirable.

The correct temperature to which the steel has to be heated for hardening is fixed by the manufacturer and published by him in the the degrees (centigrade or Fahrenheit), at which a given temper or grade of steel should be quenched. If the degrees are given centigrade they can be changed to Fahrenheit by multiplying by nine-fifths and adding thirty-two. Quenching should never be done from a rising heat, but from a falling one, and an explanation together with the accompanying diagram will show why.

On heating a piece of steel (taking a piece of plain carbon tool steel as an example) in a uniform way at the rate, say, of one degree per second, there will oc cur in the region of 1,346 degrees Fair., a period of seconds during which, although the furnace is getting hotter, the temperature of the steel lags or does not rise at all. After this period of seconds has elapsed, the steel begins again to rise regularly and the procedure is represented by the line A. C. M. with the lag indicated by the kink C. Now, if instead of quenching the piece of steel at the heat represented at the point M, we allow it to cool at the same rate at which it was heated (i.e., one degree per second) we shall observe at a temperature of about 1,274 degrees Fair. again a period of seconds during which the steel does not cool at all. It may even during this period get hotter, although, all the time the furnace is cooling down.

This cooling process is a reversal of the changes we observed in heating and is indicated on the diagram by the line M. R. B. with the kink at R. . The points R and C are of equal importance, because they indicate to the operator that only when a piece of steel is heated above the change point C, can it be hardened on quenching and that from any temperature above C the piece of steel will harden when quenched. It will also harden at any temperature during cooling before the temperature indicated by the change point R is reached. The aforementioned change points occur at different temperatures in different steels and being pre-determined it remains only to heat the tools in the uniform manner to some 10, 20 or 30 degrees (according to the size, etc.) above the change point C. Allow them to cool evenly to a temperature 5 to 10 degrees above the change point R and then quench. This quenching on the falling heat prevents cracking and represents the best practice. Quenching, no matter how simple or how small the tool may be, will always set up strains. The best mode of quenching will attain the needful hardness with minimum strain.

The heating for hardening should be as uniform as possible and should not be more than the degree fixed by the manufacturer for each separate temper or number. Overheating does not augment the hardness, but it increases brittleness. It is better to risk a failure by giving too low a heat for hardening, as in one case you can re-harden and in the other you may have spoilt the steel. The tool should be kept in motion while in the quenching fluid. It might be well to remark that to determine by experiment the lowest temperature at which different grades of steel will harden thoroughly is always worth a great deal more than the trouble thereby involved and the workman who combines in himself patience, observation, and a love of experiment, will notice that hardness, minimum brittleness and the finest possible fracture are always found together.

Modern progress in shop work has developed the necessity of reliable instruments for measuring high temperatures. Pyrometers are now found in many steel hardening shops and their use obviates the uncertainty as to the proper degree of heat required to harden steel. It should, however, be always remembered that when using a pyrometer the instrument must be frequently tested to make sure that it is recording correctly and it is also important that the heat of the furnace as recorded by the pyrometer be uniformly communicated to the steel all the way through. This heat in the steel must be given by a thorough soaking of the piece in the furnace until it is without doubt the same in the centre as the outside.

One important point in using a pyrometer is to place it in the furnace as close to the steel being heated as possible. Otherwise it might be in a hotter portion of the fire and the steel would not be at the heat registered by the pyrometer. A pyrometer costs around the $100 mark, so it would hardly pay to have one in a shop that was only making a pair of knives once in a while. The only other course is to rely on the color test and after a few experiments with it the average mechanic will be able to get fairly accurate results. Below is a table of heats and corresponding colors for steel as indicated by a reliable pyrometer. 1,292 degress Fair., dark red; 1,472 degrees Fair., dull cherry red. Average heat for hardening carbon steel, 1,652 degrees Fair., cherry red; maximum heat for carbon steel, 1,832 degrees Fair., bright cherry red.

2,012 deg. Fair., orange red $\left.\begin{array}{l}\end{array}\right\}$ Minimum and maximum heats for high speed steels.
2,192 deg. Fair., orange yellow
2,372 deg. Fahr., yellow white

After being hardened the knife will have to be tempered, which means that it has to be reheated and the temperature gradually raised until sufficient of the hardness has been drawn or taken out to make the knife suitable for the work that it is to do. Correct tempering depends on the ability of the workman to raise the temperature to the proper degree before cooling. The method of determining the temperature is by the colors that will appear as the tool is being heated. The first one, a faint yellow, will appear at about 430 degrees Fair., and will quickly change to an orange color and then purple and blue. Faint yellow indicates a very high temperature and would only be used for steel engraving tools and planing tools for steel. For woodworking machine cutters, a reddish purple will give good results. There are two methods of heating the knife; one is to heat a piece of iron to a dull red and lay the knife on it, or the knife may be held over a flame. When the proper color appears the knife should be cooled. Before starting to draw the temperature, the surface of the knife along the cutting edge should be brightened up so that the colors can be seen. A beginner may have some difficulty in seeing the colors at first, but they are very plain in good daylight.

Some mechanics use a method of hardening and tempering steel at one heat. When the tool is hot enough for quenching it is plunged edge downward in oil and left for a few seconds until the edge is black and then quickly withdrawn and finished in water. With some brands of steel this will give the right degree of hardness, but when too hard it can be tempered in the usual way.

High speed steel is being used quite extensively in woodworking factories where fast feeds and large quantities are the dominating factors. The treatment of high speed steel differs considerably from that of the ordinary carbon steel. When hardening the knife it should be heated slowly to bright red, then heated up quickly until the point appears to be melting and then quench in oil, air blast, or in the open air. With high speed steel it is better to melt the tip of the tool nose than to get it insufficiently hot. After the tool is hardened the surface should be ground sufficient to remove the outer layers that have been exposed to the direct action of the fire.

Self hardening steel is very little used now that high speed steel has made its appearance. When used, however, it should be worked at full red heat gradually but thoroughly applied. It should be laid down to cool off itself as it does not require hardening and water must not touch it.

Boring for Extension Table Tops and Leaves

A Steel Gauge as Illustrated Insures Accuracy When New Batches of Tables are Being Made

By W. J. Beattie

If a vote was taken of manufacturers, as to their favorite article of manufacture, the extension table would not be at the top of the list, and the same verdict would result from a vote of the men in the factory. The extension table is a thing to be tolerated with the best grace possible. This short article will deal with one only of the troubles that beset the making of extension tables, namely the boring of the tops and leaves, or fillers, as some call them.

To keep the tops and leaves of each width of extension table absolutely the same, batch after batch, and year after year, is the goal to be aimed at. The only sure way to do this is to have iron gauges made for each width of table, similar to the sketch shown. Three pins are all that are necessary for tables up to and including 48 inches in width. Make your gauges all the same, so far as the distance from the ends to the outside pins is concerned; two inches to the centre of the hole will be right.

It is quite evident that an iron gauge made of ⅝-inch square material, cut exactly the length required and having iron tapered pins the exact duplicate of the wooden ones used in the table inserted, and having pieces screwed across the ends as shown, will make a perfect and unchangeable standard to work from.

The idea of having the end pieces project back on the opposite side to the pins is to form a gauge for cutting off the tops and leaves by. The gauge therefore serves a dual purpose. If the table is finished with a shaper mould you still require the gauge to keep an absolutely correct and unchanging length to work from. A gauge that the work has to drop into to fit, is away ahead of one that is laid on the work and then the hand passed over it to see if it fits.

The common practice is to keep one leaf of each length to work from when new batches are being machined. This is not accurate enough, and is sure to bring trouble sooner or later. The woodworking trades are not, as a rule, particular enough in such matters, it being quite rare to see the necessary pains taken to provide gauges and patterns that are perfect. These iron patterns would, of course, be made accurately in a machine shop.

It may be a debatable point with some as to which is the best for extension tables, the old style tapered wooden ⅜-inch dowel or the steel pin and socket made for the same purpose. The writer was at one time a convert to the steel pin and socket, but after considerable experience with it decided to return to the wood pins.

The steel pins and sockets necessitate two handlings of the leaves for boring, as one edge is bored for the socket and the other for the pins, so this is one count against them. As the pins must necessarily be almost the size of the hole in the socket, they have to meet very accurately or the socket will be displaced. This happens if there is a slight wind in the leaves (which is very frequent) and has often been noted by the writer.

The cost is also greater than the wood pins. In fact it remains to be shown that the steel pins and sockets are superior in any one way to the wooden pins. The advantage of wooden pins is that in the boring

there are only one-half of the holes to be "plugged," and if the boring is of just sufficient depth to stop the pin at the beginning of the taper, the parts slip together much easier than any metal to metal pins and sockets do. At first glance a person is likely to be favorably impressed by the metal as against wood, but time and trouble teach a whole lot of things!

The writer can personally assure any reader of the "Canadian Woodworker" that metal gauges similar to

Iron cut-off and boring gauge for extension table tops and leaves.

the one shown will prove a source of profit, and though the sketch is not drawn to any scale, it is sufficiently clear to explain itself to a practical man.

Cost of Machinery

In connection with factory costs, there is a difficult problem for every manufacturer to work out, and that is the cost of the machinery. What is meant by this is not the original cost of any given machine, but the proportional cost of the item of machinery for doing any given work. This must, necessarily, include the cost of machinery as compared to the work it does, the life of a machine or the amount of depreciation in value each year, and the chance, meantime, of some other more up-to-date machine being invented, necessitating the replacing of the machine before it has really served its time.

It is considered that machinery depreciates about ten per cent. each year. This is, in substance, that the life of a machine should be reckoned at about ten years. This may be a fair average, but it is not conclusive. There are machines in use to-day that are seventy-five years old. Other machines are sometimes ruined through accident or something of the kind inside of five years. Some machines wear out rapidly. Others last as long as a man. Estimates are complicated by the claims made for some new machinery that it will pay for itself in a very limited time.

The matter of machinery paying for itself raises a complicated question. It goes without saying that every machine must pay for itself in a certain length of time. Otherwise, it is a loss. But to make it pay for itself, a certain amount each year must be allowed to the machine for its share of the work, just as so much is allowed for the machinist. This is one way of looking at it. There is another way. There are those who say that improved machinery will do more work than others previously in use at the same labor cost. This is frequently true, but it, like lots of other truisms, depends, to a certain extent, on other things.

In the first place, if it effects a saving over the old machine, you will have an advantage in first cost until your competitor puts in the new machine also. Then the situation changes. You haven't any more advantage over the other man in the business than you formerly had, and the benefits that accrue from the saving go to the world in general by cheapening the articles produced. This is one of the inevitable results of competition, and is a point which must be considered in deciding whether or not a machine will pay for itself on the basis of its work compared to the work of the older machines.

Another factor is the possibility of some other machine being invented soon after this new one is installed. This other machine may offer such advantages over the new one that it will become necessary to make a change before the machine has had time to pay for itself. In this case, if you don't buy the new machine, you are left behind by your competitor, who will get it and then have the advantage over you. It is easy to see, therefore, that this matter of the cost of machinery, and of machines paying for themselves, is a subject in which one may get tangled.—Hardwood Record.

Activity in the Manufacture of Shell Boxes

Contracts for Bethlehem Boxes Let for $1.24 per Box—Munitions Board Issues Plans and Specifications of Three New Boxes

Large orders have been placed for the Bethlehem boxes, drawings and specifications of which were published in our last issue. We understand the price for this box is $1.24, which appears to be a very small figure when the amount of material and work in the box is considered. At least one Toronto firm tendered on these boxes at $1.65, and although this may be a trifle high the firm tendering at this price based their figures on first-class material and workmanship.

They told the editor of the Canadian Woodworker that the price quoted to them on the tie rod, complete with wing nut, ring and chain and the two small plates for the ends of the box was 35 cents. The price quoted on the felt for the inside of the lid worked out at 10c. per box. The quantities of lumber were 9 ft. of pine and 4 ft. of birch. Allowing for kiln drying of the birch for the three-ply diaphragms and gluing, the cost of the diaphragms worked out at 14 cents. In figuring on the birch for the diaphragms the firm figured on resawing their birch, as rotary birch of the required thickness would be hard to get. The three strands of wire that bind the box worked out at 3 cents per box. Then allowing 4 cents for nails, 2 cents for galvanized screws and 55 cents to cover total productive labor and overhead cost the cost of the box would work out at $1.40.

Some slight reduction in the cost might be possible, however, in the case of a firm securing a very large order. We understand that one Montreal firm is installing hydraulic presses to make the three-ply glued-up diaphragms, and that the same firm secured a quotation on the tie rod, complete, for 22 cents. This, of course, would make the cost of the box work out very much less.

There is a rumor current that the Munitions Board have waived the three-ply diaphragm in favor of solid wood. If such is the case, this would also reduce the cost of the box. The reduction would be particularly noticeable in the productive labor cost, but it does not seem hardly fair to waive the matter of three-ply diaphragms after the tenders have been figured on that basis. We would like to hear the opinion of our readers on the cost of this box and what quotations they were able to receive.

We reproduce herewith the drawings and specifications of three new boxes which have been issued by the Munitions Board. The quantities in which contracts are being let would seem to indicate that the "big drive" has commenced on the Western battlefront, and that Canada is going to continue to do her share in turning out large quantities of ammunition. The boxes do not differ materially from some of those that our readers are familiar with, but it will

no doubt be necessary to make some slight readjustments of machinery and working conditions to secure the maximum of efficiency in manufacturing.

The B. 310 box is for the 4.5 shells and will hold three shells instead of two, as formerly. This will mean that wider material will be required for the tops and bottoms, which will necessitate considerable jointing. The specifications call for spruce, pine, or basswood for the tops, bottoms, sides and cleats, and birch, beech, elm or maple for the ends, blocks, and handle cleats. Each box is to be bound with two bands of No. 16 gauge iron bands, one and one-quarter inches wide. The bands will project into the sides of the box ¼ inch, five inches up from the bottom, and are to be secured by No. 12 wood screws, ¾ inch long, seven to each band. The lid will be secured to the body of the box with six No. 12 wood screws 1¾ inches long. The blocks that support the small end of the shell will be secured with 1¾ inch, No. 12 screws, while the blocks that support the large end will be secured with 1¾ inch nails.

The handle blocks will be held in place by five screws. The handles will be of spliced gummets of ½ in. diameter tarred manilla rope of the best quality obtainable. The cleats on the under side of the lid are to be secured by No. 12 wood screws, five to each cleat, and the bevelled blocks that go on top of the cleats will be fastened by two No. 12 wood screws, 1¼ inches long.

The contractor's initials, together with the year of supply, and the name of the ammunition the box is to hold, are to be stamped on the end of each box with letters ½ inch high. All sharp corners on the outside of the box are to be nicely rounded off.

The B. 304 box is for the 6-inch howitzer shells, and will accommodate two. The specifications for this box are practically the same as for the B. 310. It is also similar in construction to the B. 310, and calls for stock fifteen inches wide for the tops and bottoms.

The other box is known as the No. 1071, and is to be used for transporting cordite. It will hold 100 lbs. and will have stamped on the lid in black letters ¾ inch high, "Smokeless Powder for Cannon." The No. 1071 is made entirely of "any suitable soft wood." Two tongue and groove glued joints are allowed in sides, ends and bottom, but no piece may be less than three inches in width. One tongue and groove joint is allowed in the top, but no piece in the top may be less than six inches in width. The box is dovetailed at the corners and the lid is secured by 1½ inch No. 12 screws. The bottom is secured with 3½ inch wire nails and the number of the box must be stamped on the top and one end in black letters.

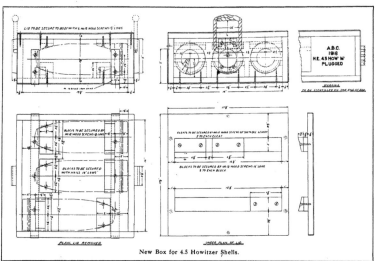

New Box for 4.5 Howitzer Shells.

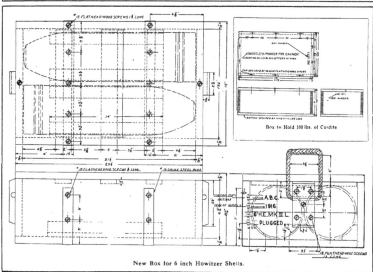

Box to Hold 100 lbs. of Cordite

New Box for 6 inch Howitzer Shells.

Electric Power in the Woodworking Industry

Equipment of the Brunette Saw Mills at New Westminster, B.C.—Modern Drive has Greatly Increased Mill Capacity

By D. Penzer Dunn, Assoc. A.I.E E.

The Brunette saw mills do a general lumber business, both domestic and export, having commenced operations in the year 1878, as the outcome of a demand for manufactured lumber around the larger towns in British Columbia.

At that time the amount called for was very small as compared to that of the present day, the business having increased considerably up to the present time. The plant now consists of a saw mill with a daily capacity of 100 M., planing mill 100 M., a box factory turning out five thousand boxes per day, and a shingle mill with a daily capacity of 200,000 shingles.

The site chosen was particularly favorable from the point of view of a manufacturing centre, being situated on the Fraser River about 15 miles from the Gulf of Georgia, and within the corporative limits of New Westminster, thus affording a convenient handling of export and import trade. The company owns extensive timber limits, the most important one being on Vancouver Island, where the logs are cut and towed direct to the mills by steam tugs belonging to the company.

The original motive power for the various wood working and planing machines was steam until the year 1910, when the output had increased to the extent that it was found necessary to greatly enlarge the factory in order to

View of Electrically Driven Planing Mill

cope with the increased business; it was at this time that the company seriously considered the economy of maintenance and operation in connection with steam drive, and after due consideration, decided to adopt electric power.

From the first the company decided to generate their own power owing to the waste lumber, saw dust, etc., that could be used for fuel, as had been the case hitherto with the steam power. Three phase induction motors were considered the best applicable to the various mill machines, owing to their constant speed characteristics, etc., and to this end a turbo-generator set was installed of 500 kw. capacity. This machine took care of all the power load of the factory, which consists of 43 induction motors with horsepowers varying from 3 to 75 h.p., making a total of 900 h.p. connected.

A series of tests made on the plant prior to its initial operation, gave the following results: The average power factor was approximately 73 per cent, at full load 76 per

cent., at maximum overload 67 per cent.; the drop at the farthest motor was approximately 8 volts. Several motors have been added since these initial tests were carried out and a more recent test of overload showed that it would require a synchronous motor condenser set of 250 kw.a. to correct the power factor to 90 per cent. The question as to the advisability of installing a synchronous condenser as against an additional power unit being brought up at this time with a final decision in favor of the latter.

After four years of service, the installation of an additional power unit of 300 kw. capacity, owing to the fact that the turbo-generator was subject almost continuously during the latter time to an overload of 40 per cent, was discussed. In making the decision as to increase of capacity, however, it was decided to buy from the Western Canada Power Co., the chief reason for this being standby service in the case of a breakdown of the mill power plant and to enable repairs to be made on the 500 kw. unit when necessary without totally crippling the service.

The Power House

The power house for containing the plant is a reinforced concrete building and contains controlling and generating equipment as well as condensers, etc., in the basement, the capacity in all, including transformers, being 800 kw. There are 7 boilers of 513 h.p. capacity total for generating the steam, being of the return tubular type. These are equipped with natural draught and use as a fuel the waste products from the lumber mills; this is fed by means of a conveyor into the boiler room and fired automatically. To feed the various boiler furnaces, the wood shavings, etc., are led through a 30 inch galvanized iron pipe, No. 14 gauge, through which it is drawn by suction fans; the feed water for the boilers and condensers is taken from the Brunnette River.

Turbine

This machine is of Allis Chalmers Parsons type, steam pressure 110 lbs. to the square inch, vacuum 25 inches, having three stages of expansion; B.h.p.; 600; speed, 6,300 r.p.m.; governor oil pressure, 30 lbs. to the sq. inch.

Generator

This machine is Allis Chalmers type, being a three-phase, 60 cycle, 480 volt, amperes 602, 3,600 r.p.m., two pole machine, delta connected.

Exciters

These are two in number, No. 1 being direct connected through the switch board to the generator, while the other is kept as a reserve and also to supply direct current power and lighting to the factory; exciter No. 1 is driven by a Curtis steam turbine with a speed of 4,500 r.p.m.. It is C. G. E. type, constant current, 125 volt machine, with a full load rating of 120 amperes (15 kw.). Exciter No. 2 is a 125 volt machine of 35 kw. capacity, and is driven by a single cylinder reciprocating engine, 300 r.p.m., size 9 x 10, Goldie & McCulloch, Ltd., direct coupled. This machine now takes care of the factory lighting, and a small d. c. power load consisting of a crane of 15 h.p. and a 10 h.p. shunt motor connected to a band saw. The lighting load consists of about 15 kw.

Switchboards

The main panel contains the following: 1 Tirrill voltage regulator, generator and exciter instruments, 1 ammeter and volt meter for each exciter, 1 generator field ammeter, 1 machine a. c. volt meter and ammeter. The main oil switch and field switches are stationed at the base of this panel. The field rheostats regulator handles for generator and exciter are located next to this panel and operate the resistances which are located in the basement. The direct current board is marble, with angle iron supports and was supplied by the Hinton Electric Co., Vancouver, B.C.. It contains one main and 5 feeder switches, and d. c. volt and ammeters. Reference to the connection diagram herewith will make this layout clear. This diagram includes the complete wiring layout of the plant from generation to point of supply.

Transformers

The 300 kw. bank of 3 phase transformers is contained on a built-up wooden platform made level with the engine room and located outside the building. These consist of six

switches, enabling the power supply to be drawn from the lumber company's own plant or the power company's mains. (See diagram).

Feeders

These are 4 in number as just mentioned. Each is controlled by a type H Westinghouse oil switch, 3300 amps., 600 volts. Feeders are 19 strand No. 10 B. & S. rubber covered wire; no voltage and overload release equipments are provided with these oil switches. The feeders are run in conduit, 2 in. diameter, ⅛ in. thick, to the outside of the power house wall, where they emerge from condulet outlets onto glass insulators supported on wooden cross arms; these circuits feed the following: box factory, planing mill, and saw mill proper.

The circuits are run to the centre of distribution in the various buildings where they are enclosed in conduit and

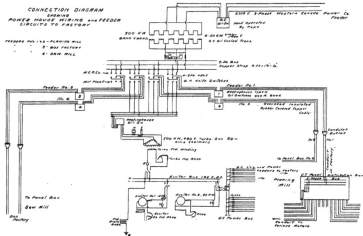

Wiring Diagram of Electrical Equipment, Brunette Mills, New Westminster, B.C.

Can. Gen. Elec. 50 kw. units of the oil-cooled, out-door type, delta connected on both high and low tension sides, the voltage being 2,200/440 v. They are fed by a No. 4 copper line about one and a half miles long, from the Western Canada Power Co., sub-station. They are run in conduit under ground on the mill company's premises and thence through a pot-head unto a short pole line of about 800 ft. to the factory power house. The transformers may be disconnected from the line by an oil switch outside the building, operated by a rope from the ground below. Reference to the diagram will make this clear. The kw. hour meters in connection with these transformers are mounted on the inside wall of the factory power house together with series and shunt transformers, the power being metered on the secondary side at a reasonably low cost per unit with a guaranteed minimum from the mill company. The secondary leads from the transformers consist of 19 No. 10 B. & S. stranded copper wire, rubber covered, and enter the building in conduit, where they connect with a 3-phase copper bus bar 4 ft. long, 1½ in. deep and ¼ in., mounted at the back of slate switchboard: the front of this board contains four 3-phase change over knife

enter a distribution panel box. From there the various motor circuits are tapped off a 3-phase bus and run to their individual motors, each circuit being protected by fuses of the required rating. There are 4 of these distribution panels altogether, the greatest number of circuits from one panel being 18 in the planing factory. The motors themselves are Westinghouse semi-enclosed type, with squirrel cage rotors. The control equipment consists of a compensator and disconnecting oil switch in all cases. The compensators are equipped with overload trip coils and in some of the larger machines a no-volt release device has also been added. This apparatus is handled easily by the employes of the factory and has so far given satisfactory service, the only troubles having been due to a few burnouts on the compensator coils. The position of some of the motors is such that they need a protective covering owing to the large accumulation of saw dust within the building.

The distance from the power house to the farthest motor being about 500 ft., this gives a measured voltage drop at full load of 5 per cent. at the motor terminals. One of the biggest motors is in the box factory, a 75 h.p. unit. The compensator has overload and no-voltage control equipment mounted on it, the speed being 850 r.p.m. This motor is direct connected to a box planer.

The following tables include a list of the various motors with the machines attached thereto, giving the speed, output, and location; they may be sub-divided as belonging to the following: (1) box factory; (2) saw mill; (3) planing factory.

Box Factory

H.P.	Speed, r.p.m.	Drive	Use
75	850	Direct	Box planer.
10	850	Direct	Tumbling cut-off saw.
50	850	Belt	Exhaust fan.
50	580	Belt	Roller band re-saw.
30	1120	Belt	Counter shaft running several small machines.
15	1120	Belt	Dovetailing plant.
5	1700	Belt	3 rip saws.

Saw Mill

15	1120	Belt	Conveyor lumber.
50	1120	Direct	Pony edger.
20	850	Belt	Sorting table.
75	850	Belt	Roller band re-saw.
10	1120	Belt	Transfer chain to roller band re-saw.

Planing Factory

50	850	Direct	4-sided timber planer.
30	850	"	Matcher.
40	850	"	Matcher.
30	850	"	Moulding machine.
7½	1700	"	Rip saw.
7½	1120	Belt	Scroll hand saw.
50	1120	"	Fast feed matcher.
50	1120	"	Double 50-in. fan.
15	1120	"	Bevel siding splitter.
30	1120	"	Single 50-in. fan.

The motor that works the crane at the mill end is a 15 h.p., 125 v., d.c. machine, with reversible drum controller, Westinghouse type, used for piling and loading timber at the mill end. This crane has a capacity of 10 tons.

A few particulars with regard to the operation of two of the larger machines might be of interest in the box fac-

50 H.P. Motor Installation.

tory. The 75 h.p. box planer is capable of dressing lumber four sided at the rate of 200 lineal ft. per minute, while the 50 h.p. timber planer in the planing factory, is capable of dressing lumber four sided at a speed of 250 lineal ft. per minute.

The management of the factory, is organized as follows: one general supervisor, each department being under a foreman; the labor employed is mixed, there being a number of Japs and Hindoos along with the white men, the latter, of course, holding the most responsible jobs. Upwards of 200 men are employed.

The wood chiefly used is first spruce, and cedar is handled also; the shipping is done by rail and water; domestic and export in New Westminster by rail (C.P.R.). The engineering staff consists of the following: chief engineer, electrician, blacksmith and helper, machinist, and mill-wrights. The company own their own machine shop and repairs are done right on the job. The equipment includes 2 lathes and a shaper, drill press, and blacksmith shop. The average mill cut for 275½ day year works out at 25,723,853, while the maximum for the last five years was 29,349,358.

In conclusion, the writer wishes to thank Mr. Duby, general superintendent of the plant, and Mr. Bennett, chief engineer, in supplying the necessary information required in the preparation of this article; the illustrations shown, as well as the information regarding the electrical tests of the plant were supplied by the kindness of Messrs. Mather & Yuill, consulting and electrical engineers, Vancouver, B. C.

Lacing Belts on the Pulleys

By N. G. Near

The modern way to adjust a belt is to do it "right on the pulley." It is the quickest way, the safest way, the most economical way. That is why it is the modern way.

About twenty years ago it was thought that belts and pulleys had reached their maximum of efficiency. We thought that there was no more to know—leather was "King" of belt materials; rawhide lacing, was IT. Almost "any old pulley" would do the trick; better bearings were undreamed of, unheard of, and so forth.

To-day there are many belt men who will therefore look upon the modern method of adjusting belts with contempt. They will say, "It's a useless extravagance," without looking into the matter thoroughly. For, just as we did twenty years ago, belt men to-day are prone to think that we have reached the maximum of efficiency and that there is "nothing new under the sun."

Well, lacing a belt while on the pulley is a new stunt, and it is a good one. It has every argument in its favor.

To do this lacing properly requires a machine—a couple of clamps—one on each belt end—and a rack arrangement for cranking the two ends together right where you want them. The machine holds the two ends in correct position for lacing and leaves your hands free to do the work most efficiently.

Even where the belt is to be run fairly slack or easy, it is plain that a device like this will prove beneficial. It takes the belt's measure, helps do the work, and leaves the belt right where you want it—all at once.

When a cast pulley doing hard service at a high speed develops cracks in either the rim or spokes, the best way to repair it, and the cheapest way in the end, is by replacing it with a new pulley.

When there is a rattling and shaking of machines or the shafting, look to the balance of running parts. Lots of trouble with machinery comes from lack of balance. It is one of the first conditions that should be looked after when machinery does not run smoothly.

Handling Lumber in Drying

P. D. Thomas, in The Woodworker, Indianapolis

In transferring lumber from railroad car to factory, it is of course advisable to cut out all unnecessary movements, as every time a board is handled it adds to the cost of manufacture. Therefore, in laying out a system of lumber storage for both air-drying and kiln-drying stock, it is essential to carefully diagnose the factory requirements and arrange unloading and yard storage tracks and dry kilns and dry storage so as to meet those requirements in the most efficient manner. In other words, to achieve with the fewest possible interruptions a cycle from the railroad car of raw material to the railroad car of finished product.

A typical installation for the economical handling trucks are removed to dry kilns as required, and from there, when kiln-dried, either to dry lumber storage or direct to factory. By this system a very considerable amount of rehandling of stock is eliminated, thus insuring a corresponding decrease in cost of production.

Convention of the Box Trade

Preparations for the summer meeting of the National Association of Box Manufacturers, which is to be held at the Hotel Belmont, New York City, August 23 and 25, are now complete. The programme will include subjects of vital importance to every member of the organization and will call for the discussion and decision by the box manufacturers of questions

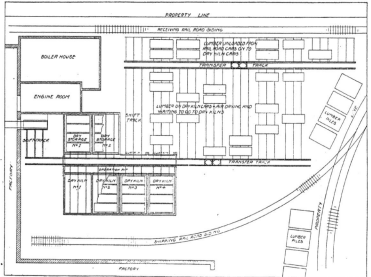

Lumber is unloaded from railroad cars on receiving siding and cross-piled on kiln trucks ready for kilns.

of lumber is shown in the sketch herewith, designed and installed especially for the P. A. Starck Piano Company, of Chicago, by the Grand Rapids Veneer Works, Grand Rapids, Mich. The sketch is self-explanatory, but it may not be out of place to call attention to the fact that by this arrangement lumber is unloaded from railroad cars on receiving siding and cross-piled on kiln trucks ready for kiln. By means of a transfer car running on tracks crosswise between loading tracks and air-drying lumber storage, the loaded trucks are sorted and shifted to the tracks whereon similar kinds and thicknesses of lumber are stored.

At the reverse end of the storage tracks is another transfer track and car, by means of which the loaded which go to the very heart of their business. Of immediate interest and consequence will be the reports that will be submitted showing the progress made with the case of the box industry that is to be presented to the Interstate Commerce Commission in connection with the investigation by that body into the question of a readjustment of railroad rates and classifications as applying to lumber and lumber products.

———

Ash is one of our easiest native woods to dry, and when dry does very little swelling and shrinking. Consequently it is a good wood to use where it is desired to guard against coming and going with the weather.

The Importance of Knowing True Costs

Address delivered before the Veneer and Panel Manufacturers' Association by Robert E. Belt, United States Federal Trade Commissioner

In my opinion the day will come when the success of a manufacturer will depend upon the information a complete and accurate cost system gives him. It will be as essential as good workmen and a good organization.

The Federal Trade Commission of the United States is urging manufacturers to give this subject the attention it deserves. In its work it has found that a majority of our business failures can be attributed to poor accounting and inadequate business information. It has found that bad office methods, inadequate and unreliable costs of production and distribution cause a great deal of unfair competition and a heavy business death rate.

Undoubtedly ruinous prices are due more to the fact that many manufacturers do not know what their actual costs are than to a desire on their part to sell at prices which do not yield a fair return on the money invested. At times most manufacturers, as a matter of expediency, are obliged to take orders which return but little, if any, profit; but in the great majority of cases we find that unprofitable prices are made by manufacturers as a result of their not knowing their total and complete costs of production and distribution.

There is another phase of this subject which manufacturers should not lose sight of, and that is the attitude of banks in the matter of granting credit. Banks are paying more and more attention to the accounting methods used by the manufacturer to whom they extend credit. They are willing to give larger loans and very often more liberal terms to the business man who keeps his books in a way that enables him to show the bank at any time the true condition of his business. A business man who can give a bank this information will receive more consideration with the same amount of assets than the one who cannot. Even if he is successful, but cannot conclusively show it because of poor accounting methods, the bank will not consider him a desirable risk.

The number of unsuccessful business corporations in the country is staggering. There are in the United States 230,000 corporations exclusive of banking, railroad and public utility corporations. Of this number, over 100,000, or 40 per cent., do not earn a penny for their stockholders. This shows clearly that there is an unduly large proportion of unsuccessful business corporations in the country.

What is the remedy for this condition? The remedy lies in a large measure in co-operative effort to increase efficiency and to eliminate unintelligent competition. It is the belief of the Federal Trade Commission that if merchants and manufacturers would agree on some standard practical system of accounting, it would go a long way towards eliminating the unintelligent and destructive competition that is so manifest in many of our industries to-day.

It is of great importance to the manufacturers of an industry that they agree on the elements or fundamentals of a cost practice, especially to make sure that all the elements of cost are included in their calculations. This is of vital importance. Make sure that you have included in your cost an adequate provision for depreciation and every expense of every description incurred during the year. The test of the accuracy of a cost system is that it dovetails in and co-ordinates with the financial books of account. The costs then, in the aggregate at least, must be correct.

Excellent results are being attained by trade associations appointing a man to devote his entire time to visiting the concerns that are engaged in their industry, getting acquainted, preaching the gospel of good accounting practice, reconciling differences, and in standardizing methods. When the person engaged for this work has the subject to heart, is congenial and persuasive and well grounded in the principles of cost accounting, the results in improving the general welfare of the industry are surprising. There should be no selfish motives back of such service. There should be the same willingness and free spirit to help and to aid and bring in line the non-member of an organization as there is manifested in the member. The non-member is really the trouble-maker.

Business men are looking upon activities for the general welfare of their industry in a new light. They are being governed by the single purpose to promote the best interests of the community and of the industry in which they are engaged. They are putting aside sectional differences, and there is the broadest and most general co-operation for the general good. As an example, in the paint industry one of the largest and most successful manufacturers has a standing invitation to its competitors to inspect its cost records. to adopt its efficiency methods and to profit by its long period of business experience.

The Federal Trade Commission is aiding and co-operating with trade associations and individual merchants and manufacturers in improving accounting methods and business practice. If the appropriations of the commission will permit, it is contemplated to conduct an educational campaign upon this subject among merchants and manufacturers. It outlines what a cost system is, describes the method of handling material and labor, and defines overhead expense and shows a logical method of distributing it. It will also show the books and entries necessary to properly control the cost accounts and fit them for the general books of account.

Nearly every business man is perfectly willing to include in his costs all items for which he pays out actual money, but he is inclined to overlook those which do not require a visible outlay, and depreciation is one of these, although it unquestionably exists. A number of manufacturers do not charge any depreciation, and give as a reason that they keep their plant in first-class condition and it does not depreciate. This is one of the most fallacious ideas in business to-day. Every machine, building and apparatus, like every man, has a period of life, no matter how much care you take of the machine, or how much medicine you give the man, death is bound to come.

It is evident that where profits are divided without adequate provision being made for depreciation, they are really taken from capital, and when the plant is finally worn out the manufacturer is forced to raise more capital, not for the purpose of enlarging his plant, but actually to put it back in the same condition as when he began operations. Many manufacturers hesitate to charge off depreciation because of the mistaken idea that by so doing they are spending money. Nothing could be farther from the truth. You have not lost one penny by making an adequate charge for depreciation; on the contrary, you have strengthened your business and you are playing fair with yourself and with your competitors.

The Federal Trade Commission believes that with the co-operation of business men it can be of constructive help in assisting to bring about better trade conditions; conditions which will lessen our business death rate and enable our merchants and manufacturers to compete more intelligently with each other, as well as make them more formidable competitors in the markets of the world.

Manufacturing Pipes from Maple and Cherry

Could the Industry be Established on a Large Scale ?—Plentiful Supply of Raw Material in Canadian Forests

By P. B. Walmsley

A tobacconist in Huntsville, Ont., is making pipes out of black cherry root, maple knots and bird's-eye maple. His factory acreage is about a yard of counterspace next his window, and he is now on his fifty-fourth dozen of his hand-made pipes. He has a diminutive lathe, worked by a treadle. His tools are simple; some penknives, brace and bits, taper drills and reamers, with a small vice at the end of the counter. He saws out a block of the required size, and then saws out the rough square of the pipe, works the wood on the lathe, and bores the stem. The bowl he hollows out with brace and bit, giving it a touch with the taper-drill before the reamer, so as not to split the wood. Then he puts it in the lathe and sandpapers the inside and edge of the bowl. He uses three grades of sandpaper; the finest one being for the cherry root.

The polishing of the outside is done by a little pressure with a piece of wood while on the lathe and then coated with shellac. Afterwards, vulcanite and amber mouthpieces are attached. The price of the pipes depends mostly on the mouthpiece. The tobacconist in question ventured the assertion, that in regard to the "French briar," "the men who make them do not smoke them, otherwise they would never bring the bore of the stem out in the bottom of the bowl." He makes his come out a little above the bottom, and claims the following advantages from so doing: If the pipe is clogged, and a pipe-cleaner is inserted, it will stir up the tobacco, and fit the pipe for smoking, and if cleaning an empty pipe, the cleaner will not stick in the bowl, but turn the corner and come up. Moreover, it leaves a sort of well or trap which stops nicotine from being drawn up the stem.

The other advantages of the pipes arise from the nature of the wood. Both cherry and maple are lighter in weight than the briar, but the cherry is lighter than the maple. Both are absorbent, but the maple is more so than the cherry. More of the maple pipes are sold than the cherry, but the cherry is good for the man who is looking for an especially light pipe. The maple pipe will color like the best meerschaum, and, being a sweet, absorbent wood, absorbs all the juices as well as a clay pipe would do, but it is without the unpleasant taste of a clay pipe. As it is free from oils, it is never bitter, and is not as liable to burn or crack as the briar. It may, however, burn if smoked in a gale. Another thing in favor of maple is that no putty has to be placed in it to stop holes, because there are no holes to stop. The wood, of course, must be dry before it is worked, or it will warp.

On my inquiring what others were doing in the way of pipe-making, and whether it could not be carried out on a larger scale, the tobacconist said he knew of one firm in the United States making pipes, and there was a manufacturer in a town in Ontario who imported the briar and then made it into pipes here, but he did not know of any other man making pipes from the maple. There was certainly plenty of raw material in the neighborhood. His stove wood happened to be bird's-eye maple, and he found no difficulty in procuring the black cherry root. The Hydro-Electric Commission is installing its system in Huntsville, and the tobacconist hopes shortly to take up some of this power and use it to run a proper pipe lathe, but he will probably still be only single-handing it.

Now that we are considering how we can organize our resources, and import less and manufacture and export more goods, and the consumer is being exhorted to purchase "Made-in-Canada" goods, the men who make things should also be urged to further efforts by another slogan—"Make-it-in-Canada." Apply this to the pipes. Why should we go to Europe for the materials for our pipes or for the pipes themselves? Of course, custom is strong, and prejudices would have to be overcome. The briar-root has had a great vogue. It is, we are told, the name given to the roots and knots of the tree-heath Erica arborea, a plant belonging to the Ericeceae, and abounding in countries bordering on the Mediterranean, and the word itself is a corruption of the French bruyere, "heath."

The war has sent up the price of the bruyere pipe. Why not seize the opportunity to make pipes from our own woods for home and perhaps export trade, too? Here surely is a healthy "infant industry" which only needs a proper up-bringing to become of great service to the community. I think it is worthy the attention of some manufacturer. One can imagine a pipe factory located somewhere in the midst of the maple and black cherry. Is this vision of pipes merely a "pipe-dream"? I leave it with the woodworkers of Canada.

Trade Inquiries

607. **Box shooks.**—A London manufacturing firm asks to be placed in communication with Canadian manufacturers who can supply box shooks.

609. **Handles.**—A London company wishes to get into touch with Canadian firms manufacturing on a large scale, wooden handles (hickory and ash) for shovels, spades, pickaxes, axes, hammers, etc., which are required for shipment to China.

635. **Box shooks.**—A Cape Province firm, shippers of green and manufacturers of preserved fruit, ask for quotations on two sizes of boxes. Require 100,000 boxes a year. Sizes and other particulars may be obtained on application to the Commercial Intelligence Branch, Department of Trade and Commerce, Ottawa.

Eliminating Prejudice on Veneer

Manufacturers of the better class of veneered goods, particularly furniture, have in the past been, and still continue to be, handicapped by a consumer prejudiced against this class of goods. Many people are unable to get away from the literal definition of the word veneer—"A thin strip of superior wood for overlaying"—outside show, pretence.

It is too true that years ago the definition correctly described most of the veneered goods that were put upon the market, and the effect is still felt by those engaged in the business. The superficial time has passed, however, and, with few exceptions, the present makers of veneered goods are turning out stock that will stand the test of time. Yet the prejudice still remains in many places, and for the purpose of further advocating a previous suggestion this article is written.

The makers of veneered goods have been on the defensive. Their articles have been in the "just as good" class, and this is never very convincing. The time has come for the manufacturers to drop the defensive attitude and take a more aggressive or progressive stand. The attention that has been given this work of late years has so perfected the methods of construction that when properly made it will stand the test of time, and nothing but fire or a severe water accident can destroy it.

Keeping these facts in mind, the word "veneered" when applied to furniture is a misnomer and entirely misrepresents its true character. This suggests the feasibility for a new name, one that will combine the chief characteristic of this class of goods. One word might well be "reinforced." It is true that the word veneer might change in time, if its use were continued, and the change would apply mostly to furniture, but the word veneer will always be associated with "outside show," etc., which would still keep the makers on the defensive.

With the adoption of the word "reinforced" there is a description of the construction. The word means to supply with new strength, and when a well made core has been crossbanded and faced with veneer, this five-ply construction, with the grain of each ply running in the opposite direction from the other, and each acting as a support to the other, we have an article that is as indestructible as anything in wood can be.

Reinforced goods should be made with two distinct objects in view—durability and appearance. The core stock and the crossbanding should both be selected for strength, and be of a material to which glue will readily adhere. Soundness is a proper requirement, but the poorest figure may be used. When these two are properly combined and given a face veneer selected for its beauty, one will have an article hard to excell in appearance, and of such strength and durability that one may reasonably expect it to find an honored place among the antiques of generations yet unborn.

It may be said that in all probability the time is coming when all the best grades of furniture will be made on the reinforced principle. When wood becomes so scarce that the more common class of furniture will be made of substitutes for wood it will be necessary to cut the better figured woods very thin to conserve the supply. The trend is more and more towards such conservation all the time. Unless manufacturers of reinforced goods begin to educate the public to the merits of this class of goods, metal furniture makers are liable to hold the inside track. A national campaign would soon eliminate the prejudice against these goods, provided the makers were careful to have the goods equal to the claims made

The educational feature should be extended to include the salesman who reaches the consumer. He should be given an opportunity to visit some factory and see the process by which these goods are made. The care with which the core is selected, and then reinforced with another layer placed at right angles and firmly cemented to the core under tremendous pressure; the various operations in detail, of preparing the wood and glue, the piling away to dry until the danger of warping and twisting has passed, and the overlaying of this reinforced body with a face of some wood selected for its beauty present a story so interesting that if witnessed by a salesman will enable him to talk most entertainingly of the goods he has to sell, and generally increase his selling power.—Hardwood Record.

Sanding and Scraping Flooring

The hardwood flooring people introduced the idea of fancy floors, and flooring to be proud of, instead of floors covered up with carpets. Ever since that time sanding or scraping has been an item of more or less importance in connection with the manufacture of flooring, says the Hardwood Record.

Some manufacturers have featured the finish as a talking point; some have praised their sanding, and some their scraping. Lately there has been some argument about the relative merits of each process of finishing. One man, evidently in favor of sanding, says that flooring manufacturers have experienced trouble with scraping because of a case-hardened finish it puts upon their stock, and that a large number of mills have abandoned scraping flooring after it is matched because of habitaual complaints. One might well add to this that it is a good sign to see hardwood floor manufacturers abandon both scraping and sanding to finish their stock.

The real finish is put on a floor after it is laid down. This applies to a straight, plain floor as well as to fancy laid floors and parquetry. It is impossible to get a floor to lie so evenly that it will not require some kind of finishing after it is laid. Therefore it seems a waste of time to polish a piece of flooring before it is laid.

Appearance cuts some figure and as a selling point it is well to have a flooring that is finely finished and presents a pleasing appearance. This has been fairly well taken care of, however, in the modern flooring machines, and really the death knell to the scraping of hardwood floors might very well have been sounded when the first flooring machine was turned out with a device for jointing the knives right on the cutterhead so as to produce a smooth finish on stock fed through at a much higher rate of speed than ever was dreamed of in the early days.

Practically all modern flooring machines are now so equipped that knives can be jointed or ground and jointed right on the cutterhead, and with proper attention any modern machine will turn out flooring as smoothly finished on its face as need be. It seems impractical to get a sub-floor so even that flooring can be laid without some smoothing of the top surface afterward.

As to the relative merits of sanding and scraping, when it comes to the final smoothing off of laid floors, it is largely a matter of machines and methods available for different kinds of work.

The Lumber Market

Domestic Woods

A wholesale lumber dealer informed the editor the other day that he had all the business he could handle and he was quite elated because one of his customers had asked him to defer shipping a car of lumber until August, as he said it was going to give him some trouble to fill all the orders he had on his books. This would seem to indicate that some branches at least of the woodworking industry are busy.

Prices on most of the popular hardwoods have stiffened somewhat during the past week or two and this is particularly true of the thicker than inch stocks. Lumber dealers are no doubt anticipating a steady demand and as their stocks are below normal they are going to hold out for their price.

The large number of shell box contracts that have been let during the past few weeks has caused quite a little stir in lumber circles and may be the cause of advanced prices unless manufacturers can secure an option on the lumber before they tender for the boxes. The boxes are of a new type so far as Canada is concerned and one kind calls for a considerable quantity of three-ply hardwood for the diaphragms that hold the shells in place. A rumor has been circulated to the effect that the three-ply diaphragms have been waived in favor of solid softwood, but this has not been verified. If the three-ply hardwood is adhered to it will undoubtedly mean that there will be a big demand for hardwoods to be manufactured into rotary cut stock.

The Imperial Munitions Board has also called for tenders on a large quantity of rifle boxes, the specifications for which are said to be very stiff. A contract for shooks for 400,000 boxes for a British firm, provisionally placed in Montreal, was subsequently cancelled on the ground that the price was too high. Many inquiries have been received in the Montreal district from Ontario, asking for quotations on suitable box lumber, but it is said that there is some difficulty in quoting in view of the continued scarcity of certain descriptions. The building trade is much about the same and there is no prospect of any substantial improvement. Most of the permits being issued are for small houses.

Prices at Toronto, on car lots, for some of the most active woods, are as follows: white ash, firsts and seconds, inch, $60; white ash, No. 1 common, $45; brown ash, firsts and seconds, inch, $50; No. 1 common, $40; birch, firsts and seconds, inch, $44; birch, firsts and seconds, 1½ and 2 inches, $46; birch, firsts and seconds, 2 and 3 inches, $60; basswood, firsts and seconds, inch, $40; No. 1 common, $35; No. 2 common and better, $28; soft elm, firsts and seconds, inch, $40; No. 1 common, $43; No. 2 and 3 common, $17.50; rock elm, firsts and seconds, 1½ and 2 inch, $55; No. 2 common, 1½ and 2 inch, $37; soft maple, firsts and seconds, $32; No. 1 common, $25; hard maple, firsts and seconds, inch $40; No. 1 common, $33; hard maple, firsts and seconds, 3 and 4 inch, $50; plain red oak, firsts and seconds, inch, $66; No. 1 common and better, $40; No. 2 common, $32; plain white oak, firsts and seconds, inch, $68; No. 1 common, $40; No. 2 common, $32; quarter cut white oak, firsts and seconds, inch, $89; No. 1 common, $57; hickory, firsts and seconds, inch, $70.

* * *

Imported Woods

Lumber prices in the United States have shown a slight tendency to go down hill during the last month. This may be due to a number of causes; the possibility of war with Mexico being perhaps the principal one. But since the smoke of battle seems to have cleared away from the horizon and both parties are conversing in more friendly tones, a general resumption of all lines of business may be looked for.

The "American Lumberman," discussing the situation, says, "Road salesmen report that everywhere there are indications that industrial activity is going to continue into the fall at least, upon a high plane. With southern hardwoods oak is reviving notably and there is hope that this wood will be back in a stronger position this fall. Northern hardwoods, especially birch, are in fair demand, and prices are also fair. In Southern hardwoods, aside from the revived conditions of oak, a fair demand exists for maple and elm. Cottonwood continues a good item.

Poplar continues one of the hardwoods that still find themselves in a fairly good position. For many weeks movement was very good and then for two or three weeks there was a lapse. However, the wood has now recovered again and the demand is good. Reports from the mills indicate that not enough stock remains in the manufacturers' hands to make them eager to market the dry lumber they have left.

Reports from the United States show that the decrease in oak consumption in furniture manufacture has been very marked of late, and in addition, the substitution of cheaper woods for speculative building interiors and the development of better finishes for other woods, resulting in their broadened use in high class interiors is decreasing the amount of oak formerly used for this work.

The monthly report of stocks of oak, poplar, chestnut and cottonwood lumber on hand at mills reporting to the Hardwood Manufacturers' Association is as follows:

In plain white oak there were stocks on hand May 1 at the eastern mills of 22,606,000 feet, and at the southern mills of 15,425,000 feet;-a total of 38,031,000 feet. On June 1 there were stocks at the eastern mills aggregating 7,798,000 feet, and at the southern mills aggregating 15,350,000 feet; a total of 39,468,000 feet. This shows an increase of 2,226,000 feet, or about 7 per cent.

In plain red oak there were 3,406,000 feet on hand May 1 at the eastern mills and 13,508,000 feet on hand at the southern mills, making a total of 16,914,000 feet. On June 1 there were on hand 3,623,000 feet at the eastern mills and 12,866,000 feet at the southern mills, making a total of 16,489,000 feet. This shows an increase on June 1 of 6.3 per cent. in the eastern mills and a decrease of 4.4 per cent. at the southern mills.

Quartered white oak shows an increase of one-tenth of 1 per cent. at the eastern mills and 4.4 per cent. at the southern mills. Quartered red oak shows a decrease of 1.7 per cent. at the southern mills and 4.4 per cent. at the eastern mills.

Poplar, manufactured at the eastern mills, shows a decrease in stocks of 806,000 feet, or 5.7 per cent. In chestnut, manufactured at eastern mills, the report shows a decrease of 5½ per cent. in stocks.

Our Method of Sticking Insures
Straight, Flat, Properly Dried Lumber

Some months ago the demand for our **Kraetzer-Cured** Gum Lumber became so great that much to our regret, we were unable to supply all the wants of our regular patrons, and were obliged to turn down all new business.

We are now pleased to announce that owing to a material increase in our capacity we are again able to supply the requirements of all our old patrons and also to take care of the increased demand that has been created by the quality and character of our product and the service we render.

To those who have tried our lumber and are familiar with our methods, it is unnecessary to speak of either and the above announcement will suffice.

To those who have not used any of our **Kraetzer-Cured** Gumwood we express the hope that a trial will be granted us. Our stock will be found different from and superior to any other Gum Lumber that the market has ever afforded, and a trial and a comparison with other material will convince the most skeptical of the merit of our product.

Our method of manufacture, sticking and handling, together with our special **Kraetzer-Curing** treatment, all of which is at variance from the usual saw-mill practice, enables us to put out Gum lumber that is **Flat, Straight and Bright.**

We are manufacturers exclusively, and not dealers. Every board that we ship is taken from our own timber, sawn on our own mill and handled in our extremely careful manner. In this way we are able to offer a supply that is dependable, reliable, uniform and of absolutely maximum quality.

ᴛʜᴇ KRAETZER-CURED LUMBER Cọ--
MILLS—MOORHEAD, MISS.
SALES OFFICE—CINCINNATI, OHIO.

WRITE TODAY FOR A SAMPLE CAR LOAD -- --

A LUMBER YARD ON WHEELS

THE FINISHING ROOM

Filling and Coloring Different Kinds of Woods
By James Culverwell

The subject of staining and finishing woods is a very broad and varied one, and one worthy of special study. We claim to have in this country forty different kinds of oak, most of which may be used for some special purpose. But the domestic oaks, as they might rightly be termed, are but very few. The principal ones being:

White oak (Quercus rulia).
Rock oak (Quercus primas murticola).
Black oak (Quercus tindoris).

All are capable of being handsomely finished, and they stand in merit respectively, white, black, red; but the red oak is not nearly as good a wood for staining as the white.

Most all woods need a filler before a satisfactory job can be obtained, and oak, being one of our coarse or open grained woods, necessitates a good filler. There are a number of fillers of different materials used. There is no part of wood finishing more important than the filling, and great care should be exercised, both in selecting and applying the filler, as a wood may lose its lustre and beauty if filled with a muddy colored filler, and it is wise, as far as possible, to choose the transparent colors for staining the filler.

It is therefore important to get conversant with the colors that are used for staining. I will refer to the more important transparent colors and how they are applied, later on in this article. Most of our home-made fillers are based on a silicas earth, such as bolted whiting, silver white, China clay, terra alba, calcined plaster, and others. All more or less have a whitish color and therefore necessitate staining to match the wood it is intended to fill. Corn starch makes a good filler, but has one fault, that of never getting hard. China clay—the English is the best—makes an exceedingly good filler. It is light, almost white in color and drys very hard.

Calcined magnesia, calcined plaster and bolted whiting, mixed together dry, with sufficient boiled oil to make it into a stiff putty, and with patent or japan dryer added, also makes a good filler. White japan should be used for a light wood and brown or ordinary japan for a stained wood. It must then be thinned down to a creamy consistency with turpentine. Should this not dry quick enough, napitha may be added, and should it dry too quickly increase the quantity of oil. It is always advisable to keep the oil away as much as possible as oil has a tendency to darken the wood and also to swell the grain or fibres, and there is then a constant shrinking of the wood until the oil has either dried or evaporated.

Before proceeding with the filling, all dust must be thoroughly cleaned out of the grain of the wood. The dust is composed chiefly of wood dust, but it is all inimical to grain lustre if it gets mixed up with filler. One of the best fillers on the market to-day is a preparation manufactured from a basic form of mineral silica in an atomic shape, which when pulverized to dust form assumes a pointed shape. In its use as a filler these points fasten themselves into the fibres or pores of the wood, giving the work a good flat finish.

This filler is known as Wheeler's wood-filler. It is more costly than the home-made fillers, but gives more satisfaction. There are many cheap ways and methods of filling, but they are not to be recommended, unless for the very cheap work. I would not refer to these only for the fact that there is a great demand for cheaper articles, and I feel of necessity bound to mention one or two. Many wood finishers content themselves by giving the work two or three coats of a glue size, which has been stained to match the wood being finished. For mahogany, venetian red is added, and for walnut, add burnt umber and raw umber, or Vandyke brown, and for the lighter wood, raw sienna is used.

This size stain must be applied to the wood while hot and lightly rubbed over with a cloth, first across the grain and then finish the way of the grain. This method of filling and rubbing applies with every kind of filler. Work that has been sized in this way does not require further filling. All woods, as ash, oak, mahogany, walnut, cherry, etc., should have the grain filled. Ash and oak are known in the trade as being "hungry woods," meaning that they take a considerable amount of filler to fill them and it is on that account that judgment should be used in using the filler that shrinks least.

While giving these formulas for home-made fillers or silicas earth fillers, I do not recommend them against the many patent fillers that can be purchased to-day, which dry perfectly hard in a few hours. Better results are obtained, however, by letting it stand from eight to twenty-four hours, and it should always be borne in mind that the substance of wood consists of a multitude of small tubes or pores lying side by side. These cells or pores are not continuous from the top of the tree to the bottom, but are comparatively short and taper out so that the widest portion is in the middle. Owing to this the wood absorbs a great quantity of oil or spirits, and it is owing to this fact that the filler must be well rubbed into the cells. There is not the time and labor spent by using the patent fillers as by using a filler made by silver white or china clay, etc.

A few of the home-made formulas are as follows:
Filler for light wood—

    5 lbs. of bolted English whiting.
    3 lbs. calcined plaster.
    1 lb. corn starch.
    3 ozs. Calcined magnesia.
    ½ gallon raw linseed oil.
    1 qt. spirits of turpentine.
    1 qt. brown Japan.

Sufficient French ochre should be added to stain the

white. Mix well and apply with a brush; rub well in with tow or waste across the grain and finish with a clean rag with the grain.

Filler for oak—

5 lbs. bolted English whiting.
2 lbs. calcined plaster.
1 oz. burnt sienna.
1 qt. raw linseed oil.
½ oz. French yellow.
1 pint benzine.

Rub in well with excelsior or tow and clean off with rags.

Walnut filler—

5 lbs. bolted whiting.
1 lb. calcined plaster.
6 ozs. calcined magnesia.
1 oz. burnt umber.
1 oz. French yellow.

Mix well; rub well in and clean off with rags.

Some prefer to give the walnut a very thin coating of white shellac and let it stand for a few hours, then rub down, clean off and apply the filler. There are many ways of making up the fillers from these silicas earths and all are stained according to the color of the wood.

It will now be necessary to consider the staining of woods in oil, water, spirit and fumigating, which will have special attention in another issue of the "Canadian Woodworker".

Painting the Interior Woodwork White

By A. Ashmun Kelly

The priming must be adapted to the wood, white pine requiring simply white lead thinned down with raw oil, a little turps, and japan drier. But much of the white pine contains sap and hard places which require special treatment. Touch up the sap with an extra coat, and use heavier priming for the hard parts, one with less oil, and brush it out well. It may also be necessary to apply a coat of white shellac, in order to get a foundation that will not make the white finish look spotty. Sandpaper each coat and keep the surface smooth. Lay a cloth down to catch the lead dust, and when done shake the cloth outdoors. Dust off woodwork after sandpapering it. By making the sandpaper slightly moist with turps or benzine no dust will escape or fly. Many prefer steel wool to sandpaper, and in many cases it is best, but sandpaper does all right. If you have any small places to get into, split the sandpaper in two, making a thin sheet. Or use steel wool.

Putty Work

Puttying well done is half the finish, they say. Putty up all imperfections, make it smooth. Never use store putty for a white job, for the putty will show through several coats of white. Make a white putty with lead and enough whiting to stiffen it, this putty drying hard and sandpapering nicely. If you were doing a tinted job then white lead added to store putty would answer. White lead putty will not shrink. Apply a very thin coat of white shellac on the bare wood; some prefer to apply it to the priming coat, but there does not seem to be any advantage in doing this, while it has long been the practice to coat the bare wood.

What About the Brushes?

New brushes will not do a smooth job; a brush that has been broken in on other work will do, and if about three-fourths worn will do better still. Use as good a brush as you can buy. Flat or turpentine paint causes the bristles of the brush to show marks on the surface, and it is difficult to avoid this unless your brush is very smooth, and even then you have to lay the paint on carefully. With oil paint it is different, the oil in the paint causes it to flow out and hide any brush marks.

Use an 8-0 or 10-0 round brush, and remember that a brush should be adapted to the work, a larger brush for large surface, and smaller one for smaller surface. But always use the larger tool when it will do the work right, for with the large brush you can get over ground faster. I knew of an old boss painter who took a sash tool from a journeyman's hand and flung it away as far as he could, because the workman was using it where he ought to have been using a large brush. This is a point worth remembering, for the main cost in doing painting is in the time for labor.

The Right Paint

To insure a good job of painting the right kind of paint must be used, and with all due respect to ready prepared paint of the best quality, I prefer to mix my own, for any particular purpose, such as flat inside work. There is a difference even in pure carbonate of lead, some is much finer and whiter than others, all due to the care in the making. As white lead has more body than zinc white, it is used for the finish because of its superior whiteness and quality of maintaining its whiteness in the presence of sulphur gas, etc. There is also a difference of quality in zinc, there being several grades.

The best paint is that which allows of much rubbing out, and both zinc and lead admit of this, when of the right grade.

Dead Flat Work

Meaning without lustre, both terms, dead and flat. When no oil is used in the paint there can be no lustre. But the dead effect will be more marked if the white lead, having been ground in oil, has that little oil taken out. It might be asked, why not use dry or powdered lead? It would not have the requisite bond, that is to say, the little oil remaining in dead flat paint is enough to bind the paint but not sufficient to affect the deadness. To get the oil out of white lead it is mixed with turpentine and left to stand over night; benzine also will do. By morning the oil will be found floating on top, and it may then be skimmed off. Repeat this operation until no more oil will arise. As it is the oil that causes inside white paint to yellow, the less oil we allow to remain the whiter the job will be.

A Good Rubbed White Job

A very little driers will dry flat paint, yet that little must be light enough in color not to darken the paint the least shade. White japan drier will do, though mostly patent paste drier is used, this being simply a compound of lead acetate, or sugar of lead, with whiting to form a paste.

After the priming coat very little oil is used, say one-third in the second coat, then none at all in subsequent coats. Let each coat dry well. Use lead up to and including the third coat, after which use zinc white. The finish should be the best French zinc ground in damar varnish, thinned down with white copal rubbing varnish of the best grade, adding enough turps to make the paint flow easily under the brush, and level down right. Rub down this coat.

Old woodwork, well painted and smooth, will not require as many coats as new wood, nor will new wood need as many coats as it otherwise would if it is kept smooth.

In doing white interior painting do not make the

Single Ply Veneer

suitable for

Crating Purposes

Basswood   Maple   Elm

Various Widths and Lengths.

Write for prices

Hay & Company, Limited, Woodstock, Ont.

WE HAVE BEEN MAKING THE

Best French Glue
in the World

SINCE 1818; ALMOST A CENTURY

In competition at the numerous Expositions abroad, the quality of our glues has been found so high that we are usually declared "Beyond Competition." Our trade in this country makes the same declaration. Consistency in quality and service has brought this result and we are proud of it.

Perfection in manufacture from the best raw material makes the quality of COIGNET FRENCH GLUES consistently the highest.

Large stocks at Montreal assure the best of service.

Why not take advantage of these most important factors by using our glues?

At your service for prices and samples.

Coignet Chemical Products Co.

17 State Street, N. Y. CITY, N.Y.

Factories at { St. Denis } France
{ Lyons }

ANGLO
RUBBING and POLISHING

Works free and easy and can be rubbed in two days.

WRITE FOR SAMPLE

The
Ault & Wiborg Co. of Canada, Ltd.
Varnish Works

Montreal   Toronto   Winnipeg

THE GLUE BOOK

WHAT IT CONTAINS

Chapter 1—*Historical Notes*
Chapter 2—*Manufacture of Glue*
Chapter 3—*Testing and Grading*
Chapter 4—*Methods in the Glue Room*
Chapter 5—*Glue Room Equipment*
Chapter 6—*Selection of Glue*

WOODWORKER PUBLISHING CO., LIMITED
345 Adelaide St. West, Toronto

mistake of using benzine or turpentine substitutes, for they will not do the work as pure turpentine does it. Benzine will yellow the paint. Turpentine bleaches paint white. Just how benzine affects white in the manner stated does not appear to be known, but its action was first discovered by a maker of oil cloth, who suspected it was the benzine, and tried turpentine, and the yellowing ceased.

While speaking about flat paint, it may be in order to say something about another form of flat paint, known as flat wall paint, and it is very different from the flat paint we have been considering. Flat wall paint is indeed quite different from any other paint ever used. Without going into the details of its composition, let us note its use. The wall to which it is to be applied should be made perfectly non-porous, with ordinary oil paint. The last coat before putting on the flat paint ought to dry with a gloss. This paint is not applied like regular oil paint, but is flowed on, with a full brush, and not rubbed out; the work must be done quickly. Don't try to spread the paint nor lay it off, but use a stippler always. Use a 4-inch flat brush for applying the wall paint. Use the stippler gently, don't pound the wall with it. Don't use benzine in it, but turps. Benzine will cause the paint to dry streaky. Don't stipple in one place twice and don't miss stippling any part, for it will surely show.

These flat paints are on the market, under various brands. Painters criticize this form of paint, saying it does not wear well, and that it is difficult to apply; etc. They make a very nice looking job, however.

Some painters make their own flat wall paint. The old-time way was to simply prepare a flat paint with lead and turpentine. This makes the very best form of flat wall paint, one that will stand and give satisfaction, and not cost any more than the modern flat wall paint, perhaps not as much. There is a little trick painters use in preventing the turpentine in the flat wall paint from drying out too quickly, namely, by adding a cup of water to a can of paint; the best way is to mix the water and turpentine together, as these two fluids mix readily; then the mixture can be mixed into the flat paste paint. This causes the paint also to work cool, or, easy, and of course the water afterwards dries out. You need no driers in this paint. Another way is to mix up the lead and oil color with turps, then stir in about one pint of raw oil to the gallon of flat paint, then add two or three handfuls of whiting, which will absorb the oil, stir all up together, and your paint will dry flat, provided enough whiting has been added. Use no driers. Such are a few of the formulas which have originated with inventive painters. I might tell of many more, of various kinds, some of which are protected by letters patent, but won't. What I have told in this article is for the carpenter and builder contractor, who ought to know as much as any painter, seeing that he often has to be responsible for the painting, and ought to know when it is done right.—National Builder.

The search for new uses of wood is widespread and far-reaching; but the Turks have scored last. The army of 13,000 men which they sent to Egypt to fight 80,000 British and Colonials at the Suez Canal, was supplied, in part, with wooden bullets. They had been painted a metallic lustre, and the unfortunate soldiers doubtless supposed the cartridges were standard. The British captured large quantities of these wooden pegs which passed for bullets when they came from the contractor's factory. The Turk gets it coming and going.

This shows the possibility of FIGURED QUARTERED RED GUM as used in our private office. We specialize this wood

Importers and Manufacturers

Mahogany

and

Cabinet Woods

Sawed and Sliced

Quartered INDIANA White Oak, Red Oak, Figured Red Gum, American Walnut, Etc.

Rotary Cut Stock in Poplar & Gum for Cross Banding, Back Panels, Drawer Bottoms & Panels.

The Evansville Veneer Company, Evansville, Ind.

The Use of Vulcanized Fibre Cauls for Veneering

By Wm. Eves

Vulcanized fibre cauls, or Vul-cot cauls to give them their proper name, have been used successfully by manufacturers for a number of years, but their use has come to the attention of veneer layers more and more in the past few months on account of the prohibitive advances in the prices of aluminum and zinc cauls. The advantages of Vul-Cot Cauls over those of metal and wood are many, but it frequently requires the advantage of price to introduce a different product to a man who does not care to make a change.

The manufacturer of veneered stock who considers operating costs, desires a caul material which is thin, light, tough, and of sufficient strength to withstand hard usage. By using a thin caul a greater percentage of panels can be built up in the press at one time, and in the case of thin cores this percentage becomes large. In addition, less storage space is required for the unused cauls and also for the clamped stacks of panels while the glue is setting. This tends to reduce overhead costs.

Vul-Cot cauls appeal to the man who does the veneering on account of their lightness, which makes them very easy to handle. Another thing in their favor is the fact that they are very tough and pliable and cannot be bent out of shape or broken. They have a specific gravity varying from 1.2 to 1.3 as compared with 2.7 for aluminum, 7.0 for zinc and 0.7 for woods such as are used for cauls. Therefore, a Vul-Cot caul would have but half the weight of an aluminum caul and one-sixth the weight of a zinc caul if they were of the same thickness.

Although the specific gravity of the Vul-Cot caul is nearly twice that of a wood caul, wood cauls cannot be used in thicknesses less than ⅛ inch to ¼ inch, so that the Vul-Cot caul compares favorably here also. They are made in any thicknesses down to and including 1/32 inch and the thickness should be as low as is consistent with the type of veneer work that is being done. For very large cores, cauls from 1/16 inch to ⅛ inch may be used, but by far the greater proportion of veneer work requires the use of a 1/32 inch caul.

Many manufacturers have made the mistake, when purchasing Vul-Cot cauls, of ordering cauls of the same thickness as the metal or wood cauls that they had been using in the past, and have not obtained satisfactory results because the cauls did not keep their shape. A thin caul is kept flat and in good shape by the weight of the work, and it possesses the other advantages due to thinness and light weight which have been previously mentioned.

To the man who uses animal glue, the question of a caul holding its heat is of great importance. The caul material should have a high specific heat and a low thermal conductivity. Such a caul would have a high heat capacity and would lose its heat more slowly by conduction. Vul-Cot cauls, being made from vulcanized fibre, possess these desired qualities. A metal caul loses its heat so rapidly that it is often cold by the time the pile has been built up in the press, which, of course, causes trouble. On the other hand, Vul-Cot cauls will retain their heat practically all night, and moreover, because of their good heat insulating qualities, they do not chill such glues as are not heated, and in such cases do not require heating to prevent chilling.

Many times, blisters, loose edges, rough spots and other defects of the veneer can be traced directly or indirectly to the caul. If the caul is not uniform in thickness, blisters are likely to result. If the edges of the caul wear, the veneer will not hold and loose edges will be the result. If the surface of the caul becomes rough, due to splintering or to corrosion caused by acids in the glue or wood, the veneer will be marred. Vul-Cot cauls possess great resistance to wear and last for years. For this reason, loose edges will not be found. Furthermore, when being manufactured, these cauls are all rolled in large calendar rolls, designed to give not only uniform thickness, but also a hard, smooth ivory-like surface which is essential in high class veneering. They are not pitted by the glue or wood acids, and this adds to the quality of the veneer. One large item in veneer work is the labor required in sanding, scraping, cleaning and oiling certain kinds of cauls. Because of the smooth, uniform surface of Vul-Cot cauls, this is not necessary as the glue does not stick to them.

Their pliability makes them very useful for veneering curved surfaces. They can be bent to fit the curvature of the surface to be veneered and after the process is completed, they can be flattened out again, ready for another surface. A caul compound has been prepared, which is especially adapted for use with Vul-Cot cauls. It should be used more frequently at the start than is necessary later on, in order that the cauls may become well coated. They will "season" and keep in very good condition if this care is taken at the outset.

A careful investigation and trial of this type of caul may pay the user many times. The quality of the work depends to such a degree on the characteristics of the caul that much time and money can be saved by using the right one. This information in regard to Vul-Cot cauls has been given with the hope that it may prove interesting to those who build veneers, and valuable to many who are considering new cauls or changes in their present equipment.

Experiences in Handling Veneered Work

Study of Which May Help Some Workman to Avoid Trouble When Called Upon To Do the Same Class of Work

By Lewis Taylor

There is nothing more humiliating to a workman than to have his work go wrong and sent back to him to do over again (especially if he is working piecework). This happened to me one time while veneering the end wood of some stock in a certain factory. I had done this operation on the same kind of stock dozens of times before and never had any trouble with it, but this was the first time I had ever used sliced oak veneer. Now, where I made the mistake was in using the same method of operation with sliced oak veneer, that I had previously used with sawn oak veneer. The result was, all the stock that I veneered with sliced oak was back on my hands to do over again at my own expense, while the mahogany and walnut that I had done was perfect.

After the work had been repaired I commenced to reason the matter out with a view to eliminating future trouble. I examined a piece of the veneer closely and found that it was very "shelly" and that it appeared to have absorbed nearly all the glue. I realized then that I had made two mistakes; in the first place I should have used a thicker glue size, and in the second I should have used a thick glue; or, again, I might have given the veneer itself a coating of glue size. If I had taken these precautions the first time, as I did the second, I would have saved a lot of trouble for the department and humiliation for myself.

Mistakes like these generally occur when a man is accustomed to a certain routine in doing his work and does not take proper care to distinguish between different woods and alter his system accordingly. The best way to avoid trouble is to locate the cause.

Veneering Mouldings

Some factories have a lot of trouble in veneering mouldings and there are various causes contributory to this. If the mouldings are stuck correctly and the template corresponds with the caul, there should apparently be no trouble. But even with these points perfect there is trouble at times and the chief source of trouble that I found was in the centre of pressure—the pressure not bearing directly on the centre of the moulding. The consequence is that instead of the glue being squeezed out it remains there, and when it dries out the veneer becomes loose.

When zincs are used, allowance should be made of course for the double thickness of zinc, one thickness on each side of the centre of the moulding with only one thickness in the centre. Neglect of this leads naturally to blisters, and when blisters are continually occurring in the centre of the round or hollow of the mouldings, look to the cauls; see that allowance is made for the double ply of zinc. With this precaution the chances are that the difficulty will be overcome.

In bending veneers (cross grained) for veneering mouldings, after dampening and bending them with pressure in the cauls there should be taken to relax the pressure to give them a chance to shrink, as otherwise they will shrink and split. At the factory where I am employed, we have had good success in veneering ogee mouldings with cross-grained veneer by having the cauls well soaked in beeswax and not using zincs.

While looking over some walnut veneer a short time ago just before it went into the finishing room, I noticed at the centre or butt joint where the strips of cotton had been glued on the veneer, that this strip was entirely free from cracks, while immediately to the side of where the cotton had been, the veneer was full of cracks. The inference I drew from this was that if the cotton had covered the entire surface of the veneer, these cracks would not have been there.

There is no doubt that the strips of cotton protected the veneer during the process of laying and drying out. The explanation of cracks in short grained wood is that veneer of this kind is easily parted by extreme pressure when it has no protection, and thus forms into cracks or splits. Then when the material goes to the finishing room it receives in most cases a stain with potash in it, which eats the glue from the minute cracks and leaves it worse than ever.

I certainly think it would pay to protect high-class veneer (it is usually the best figured veneer, such as good walnut and crotch mahogany, that cracks), and this could easily be done by gluing cotton over the entire surface of the veneer. It would no doubt save future annoyance and complaints from the dealers, for no one likes to pay high prices for veneered stock that is faulty.

Preparation of Core Stock

It is surprising how careless some factories are regarding the surface of the core stock to be veneered. I have seen core stock plastered with charcoal and glue, or sawdust and glue—with more glue than anything else—and then before the fillings had thoroughly shrunk, they were toothed or rasped down and the veneer applied. The consequence was that the shrinking continued after the veneer was laid, and in good light the depressions could be plainly seen. The extreme thinness of the veneer should always be considered and we should never attempt to lay veneer on a surface until we are absolutely sure it is free from blemishes. In the case of fillings in the core stock, care should be taken to see that these are thoroughly dried in.

The way we do when veneering is to mark the edge of the core-stock with a V mark, the point of the V to the best side. The marking is very quickly done with a small saw, and when the face veneer is being applied, it is not necessary to turn the piece of core-stock over two or three times trying to decide which is the best side.

You can't get too good a surface for the face veneer when it has to stand the test of the public gaze, generally in a good light. There is one thing that can always be depended upon and that is that a finished piece of veneered stock will be as smooth and as free from depressions as the core stock upon which the veneer is laid (providing the operation of laying has been properly handled). On the other hand, if no care is taken with the core-stock and it is full of small holes, the finished stock will be likewise, for there is a continuous shrinking going on, like water seeking its own level, and where there is a hollow it is going to be shown up.

THE OHIO VENEER COMPANY

Importers and Manufacturers

Foreign and Domestic Veneers and Hardwood Lumber

We always carry a large and assorted stock of Mahogany, Circassian
Walnut, Sawed and Sliced Quartered Oak.

Send us your enquiries and orders. We guarantee good service.

2624 to 2644 Colerain Avenue,   ⋈   ⋈   CINCINNATI, OHIO.

Buyers of Veneers and Panels

·will find it to their advantage to purchase
from the manufacturers and dealers whose
advertisements appear in the "Canadian
Woodworker." They are progressive firms
—the leaders in the business, which is a
guarantee of good service and prompt attention to orders.

Give your business to the man who will
spend his time and money to get in touch
with you. He deserves it—if his stock and
prices are right.

 Hoffman Bros. Co.

Estab. 1867, Incor. 1904

**800 West Main Street,
FORT WAYNE, INDIANA**

Manufacturers of

VENEERS and LUMBER

In the Domestic Hardwoods
ANY THICKNESS,
1/24 and 1/30 Slice Cut
(Dried flat with Smith Roller Dryer)

1/20 and thicker Sawed Veneers,
Band Sawn Lumber.

—SPECIALTY—

 Indiana Quartered Oak

Veneers and Panels

PANELS
Stock Sizes for Immediate Shipment
All Woods   All Thicknesses 3 and 5 Ply
or made to your specifications

VENEERS
5 Million feet for Immediate Shipment
Any Kind of Wood   Any Thickness

J. J. NARTZIK

1966-76 Maud Ave.,   ⋈   ⋈   CHICAGO, ILL.

Veneers aⁿᵈ Panels

LARGE STOCKS FOR
QUICK SHIPMENTS
Send for Lists

Walter Clark Veneer Co.

GRAND RAPIDS, MICH.

Wood - Mosaic Co.

NEW ALBANY,   ·   ·   IND.

WALNUT

Lumber – Dimensions – Veneers

If you are in the market for anything in Walnut,
let us figure with you.   We have cut three million feet of Walnut in the last year, and have a
good stock of all thicknesses dry.   We can also
furnish Walnut dimension—**Walnut Veneers.**
We have a large stock of veneers, plain, stripe
figured and stump wood.

Wood-Mosaic Co., Inc.

NEW ALBANY,   ·   ·   IND.

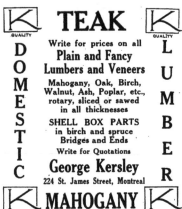

TEAK

D O M E S T I C

Write for prices on all

Plain and Fancy
Lumbers and Veneers

Mahogany, Oak, Birch,
Walnut, Ash, Poplar, etc.,
rotary, sliced or sawed
in all thicknesses

SHELL BOX PARTS
in birch and spruce
Bridges and Ends

Write for Quotations

George Kersley
224 St. James Street, Montreal

L U M B E R

MAHOGANY

American Walnut Veneers

Figured Wood

Sliced – Half Round – Stump Wood

5 Essential Qualifications:

1.—Unlimited supply of Walnut Logs.
2.—Logs carefully selected for figure.
3.—All processes of manufacture at our own plant
   under personal supervision.
4.—Dried flat in Philadelphia Textile Dryer.
5.—3,000,000 feet ready for shipment.

JOHN N. ROBERTS,   New Albany, Indiana

The Pioneer Walnut Veneer Producer of America

OAK MAHOGANY CIRCASSIAN MAPLE

VENEERS

BIRCH POPLAR CHERRY WALNUT GUM

Dear Mr. Consumer:

If you're using walnut, you can't afford to take chances with anything but the best.

Properly manufactured, properly matched and properly laid, American Black Walnut is the most beautiful of woods.

But you must have the right foundation in the material.

When it comes from Penrod, it's right. Just remember that, when you need Walnut Veneers or Lumber, Plain Stock or Figured.

Penrod Walnut & Veneer Co.

"WALNUT SPECIALISTS FOR THIRTY YEARS"

KANSAS CITY, Mo.

We specialize on

Poplar

cross-banding and backing

Veneer

Highest Grade — Right Prices

The largest Poplar Veneer Mill in the world.

The Central Veneer Co.

Huntington, West Virginia

MAHOGANY VENEERS
Notwithstanding the scarcity, we are in a position to take care of our customers.

CIRCASSIAN
Some stocks on hand suitable for interior finish and furniture manufacturers.

FIGURED AMERICAN WALNUT
Over two million feet, nice striped and curley wood.

QUARTER SAWED WHITE AND RED OAK
1-20 to 1-8, 6 to 15-in. wide.

GRAND RIM POPLAR
Cut to size. We can take care of your 1916 requirements from choice Yellow Poplar logs.

FIGURED WALNUT BUTTS
Two hundred twenty-three thousand feet newly cut Butts, some highly colored stock in this lot.

MAHOGANY LUMBER
This is something we make a specialty of, representing in Chicago and the Northwest, one of the largest importers.

HAUGHTON VENEER COMPANY

1152 W. LAKE STREET   CHICAGO, ILLINOIS

Perkins Vegetable Glue

During the past winter the Perkins Glue Company of Pennsylvania, have incorporated the Perkins Glue Company, Limited, of Canada, with a capital stock of $40,000, and have equipped a factory in Hamilton, Ont., to supply their Canadian patrons who have formerly been supplied from the Pennsylvania factory.

The Canadian factory is running to full capacity and if business in Canada continues as during the past three months they will feel warranted in having made the investment. Perkins Glue is patented in the United States, Canada and many European countries. The patents have been tested out in the United States Courts and pronounced valid, and on the strength of this decision the Perkins Company feel justified in making the investment in Canada, believing that the Canadian Courts will uphold their patent rights the same as the United States Courts have done.

The Perkins Company are this year celebrating their tenth anniversary, and while some years of experimenting had occurred previous to this period, they consider that their first established business in high class furniture factories was in the spring and summer of 1906. During these ten years they have sold a great many millions of pounds of glue to factories all over the United States and Canada, and during the past three or four years they have almost doubled the output of the previous year. The Perkins Company sell service as well as their customers and their engineers are fully capable of giving advice on any problem which comes up in the building up of built-up stock, whether the problem is in the glue-room or at the lathe or in any other department. The manufacturers of Perkins glue have had to combat two of the greatest enemies that any new product ever met with. These are, custom, and prejudice, and the fact that their product is being used in factories making high grade furniture signifies that it has won on its merits.

The Perkins Company are always striving to adapt their product to new uses to satisfy the ever-increas-

Perkins glue being applied in a North Carolina factory.

ing requirements of the many branches of the woodworking industry represented by their customers. They also pay particular attention to shipments so that customers can rely on having their orders promptly delivered. The accompanying illustration shows Perkins glue being applied in a North Carolina factory.

You Don't Have to Wait for Veneers
ORDERED HERE

Notwithstanding the scarcity of MAHOGANY Veneer we have a good stock on hand and another large consignment which will be ready for shipment in about three weeks.

We carry the largest and best stock of AMERICAN BLACK WALNUT Veneers in Canada. We have some very highly figured stump wood, good straight sliced stripe and curly figured stocks, also semi rotary and full rotary cut stock of all kinds.

QUARTERED OAK we carry in all thicknesses in both sawed and sliced. In this we have received from our mill a new cutting a few days ago, including some very choice extra wide 1/16″ and 1/20″ sawed Oak.

Let us show you what we have to offer in FIGURED GUM, in this we have the goods and prices.

Get our prices on WALNUT LUMBER

TORONTO VENEER COMPANY
TORONTO   93-99 Spadina Avenue   ONTARIO

Machinery
For Sale—Woodwork

New 12-Spindle Dovetail Machine;
Band Saw, 30 and 32 inch; Cross Cut
Saw; Swing Saw; Panel Planer, 20
inch; Gig Saw; and Berlin 3-Drum
Sander, 36 inch.

Apply D. Picard

202 Papineau Ave., Montreal, P.Q.

Machinery for Sale

1—18 x 24 in. Slide Valve Engine.
1—10 ft. Fly-wheel and 9 ft. drive pulley, 24 in.
    face.
1—Wood Pulley 60 in. x 24 in. face, 4½ in. bore.
1—Wood Pulley 84 in. x 18 in. face, 4 in. bore.
1—Wood Pulley 90 in. x 18 in. face, 3½ in. bore.
1—Jack-ladder chain and sprockets, 300 ft. long.
1—Boiler, 72 in. diam., 14 ft.
1—Boiler 52 in. diam., 14 ft.
Can be inspected at our mill at Callander.

CALLANDER SAW MILLS
Callander, Ont.

For Sale
Woodworking
Machinery

The following second-hand machines
have been placed with us for sale:
1 Woods No. 32 Matcher; 1 Goldie 10-
in. Moulder; 1 Goldie 24-in. Matcher; 1
Cowan 42-in. Sander; 1 Jackson-Cochrane
36-in. Sander; 1 Egan 36-in. Single Surfacer; 1 McGregor-Gourlay 2-drum 30-in.
Boss Sander; 1 Egan 30-in. 2-drum Sander; 1 CMC 8-in. 4-side Sticker.

P. B. Yates Machine Co., Limited
Hamilton, Ontario

Second-Hand
Woodworking Machines

Hand Saws—26-in., 36-in., 40-in.
Jointers—12-in., 16-in.
Shapers—Single and double spindle.
Re-Saw—42-in. American.
Lathes—20-in. with rests and chisels.
Surfacers—16 x 6 in. and 18 x 8 in.
Mortisers—No. 5 and No. 7 New Britain.
Rip and Cut-off Saws; Molders; Sash Relishers,
    etc.
Over 100 machines, New and Used, in stock.

Baird Machinery Company,
123 Water Street,
PITTSBURGH, PA.

Wooden Sporting Articles

The making of sporting articles of
wood is carried on to a large extent.
Base ball bats are made of ash, maple
and hickory. Elm goes into gymnasium
goods and red oak into horizontal bars.
White cedar and white pine are used
for fish net floats, polo sticks and hockey
sticks of white oak, ash and hickory.
English beech was once used for the
heads of golf sticks, but now they are
made mostly of dogwood and persim-
mon. Yellow poplar, and chestnut are
used much for billiard and pool table
tops. Hickory is used for ladder rungs
and spring boards; sugar maple for ten-
pins, lignum vitae for bowling balls,
and dogwood is sometimes used for the
latter. The legs of billiard tables are
made of red oak. The billiard and pool
table racks are of white and red oak,
cushion rails of oak and ash. Bowling
alleys consume shortleaf pine, also a lot
of sugar pine, hemlock and spruce.
The inside frame work of billiard and
pool tables is elm, white pine, spruce,
chestnut, yellow poplar and basswood.
The billiard cues are made of surgar
maple, rosewood, Circassian walnut,
black walnut and ebony. In some sections beech, ironwood and willow are
used for cheaper baseball bats. The triangle used to set pool balls is made of
cherry. White ash is used for polo and
hockey sticks and tennis racket frames
and hammock frames of birch, and black
walnut for gun stocks.

For woodbending the wood was once
immersed in hot water vats, but now it
is steamed under high pressure in retorts. It is used largely for rims, shafts,
wheels, etc., and must reduce breakage
to a minimum.

For bicycle rims maple is generally
used, and it has advanced considerably
in price. It is resilient and adds to the
life of the tire.

Large quantities of canes, umbrella
and parasol handles are made in Pennsylvania and New York. For the shanks
sugar and soft maple are generally used,
and beech for the handles. The crooked sticks are straightened by placing in
hot sand. Many high priced canes are
made of malacca wood, some selling as
high as $100. Beech, mahogany and
ebony are also used for the umbrella
handles and many imported woods are
used for these lines.

Woodenware such as pails, buckets,
snow shovels, traps, etc., are getting to
be quite a line of work. The mouse
traps are made of beech, maple, red
gum, white elm and yellow poplar.
Bowls, dishes. butter cups, tubs, plates,
rolling pins and spoons are made of
red spruce, white pine, balsam, fir and
hemlock, also some mahogany, black
walnut, cherry, sycamore, white ash and
white oak. Sugar maple, beech and
paper birch are used much for plates
and dishes. Breadboards are made of
basswood, cottonwood, white cedar,
maple and birch. Elm is used in some
of these lines.

PETRIE'S
MONTHLY LIST
of
NEW and USED
WOOD TOOLS
in stock for immediate delivery

Wood Planers

36" American double surfacer.
26" Revolving bed double surfacers.
20" Goldie & McCulloch single surfacer.
24" Major Harper planer and matcher.
24" Single surfacers, various makes.
20" Dundas pony planer.
18" Little Giant planer and matcher.
16" Galt jointer.

Band Saws

42" Fay & Egan power feed rip.
38" Atlantic tilting frame.
34" Major Harper pedestal.
32" Ideal pedestal.
28" Jackson Cochrane bracket.

Saw Tables

Ballantine variable power feed.
Preston variable power feed. M13S.
Cowan heavy power feed. M13S.
No. 3 Crescent universal.
No. 2 Crescent combination.
Ballantine dimension.
Ideal variety.
12' Defiance automatic equalizer.
MacGregor Gourlay railway cut-off.
6½' Crescent iron frame swing.

Mortisers

No. 5 New Britain, chain.
Cowan hollow chisel. M.190.
Fay upright, graduated stroke.
Galt upright, compound table.
Smart foot power.

Moulders

13" Clark-Demill four-side.
12" Cowan four-side.
10" Clark-Demill four side.
10" Houston four side.
6" Cowan four side.
6" Dundas sash sticker.

Sanders

24" Fay double drum.
12" C.M.C. disk and drum.
18" Crescent disk.

Clothespin Machinery

Humphrey No. 8 giant slab resaw.
Humphrey gang slitter.
Humphrey cylinder cutting-off machine.
Humphrey automatic lathes (6).
Humphrey double lathes (4).

Miscellaneous

Fay & Egan 12-spindle dovetailer.
No. 285 Wysong & Miles post boring
    machine.
MacGregor Gourlay 2-spindle shaper.
Elliott single spindle shaper.
No 51 Crescent universal woodworker.
40" MacGregor Gourlay band resaw.
Rogers vertical resaw.
Cowan sash clamp.
Egan sash and door tenoner.
No. 6 Lion universal woodtrimmer.
6-nail box nailer, Meyers patent.
20" American wood scraper.
16" Ideal wood lathes.
4-head rounding machines.
Cowan spindle carver. M63.
Cowan veneer press screws.

Prices, Descriptions and Full Particulars on Request

H. W. PETRIE, LTD.
Front St. W.,   Toronto, Ont.

News of the Trade

The St. Lawrence Wagon Company, 34 King Street, Montreal, are erecting a factory to cost $7,000.

The Victor Saw Works are erecting a $6,000 addition to their factory at Hamilton, Ont.

The planing mill at Harrow, Ont., belonging to C. F. Smith, destroyed by fire, will be rebuilt at a cost of $10,000.

A short course in woodworking will shortly be introduced into a number of the public schools in Manitoba.

Cheese factories in the districts of St. Thomas, Norwich and Simcoe are experiencing some difficulty in getting boxes, due to lack of experienced help in the box factories.

It is reported that the furniture factory of J. W. Kilgour and Bros. at Beauharnois, Que., is at present working overtime on an order for shell boxes.

The Dupont Fabrikoid Company, Toronto, manufacturers of artificial leather, are contemplating the erection of a new factory at New Toronto.

The Ontario Manufacturers' Corporation, of Toronto, will take over a vacant factory building at New Hamburg, Ont., and manufacture matches and other articles.

The Canada Broom and Brush Company has been incorporated, with capital of $15,000. The place of business will be at Ridgetown, Ont.

The Windsor Lumber Company, Windsor, Ont., are erecting a small addition to their planing mill and are contemplating the erection of a warehouse.

Fire recently damaged the factory of the Stratford Davenport Company, Stratford, Ont., to the extent of about $4,000. The damage will be repaired immediately.

An employee of the Wilson Box Manufacturing Company at Westfield, N.B., was killed by an electric storm while handling lumber in the company's yard.

The box factories of British Columbia are busy turning out berry crates. The increase in acreage of small fruits is estimated to be about 50 per cent., which makes a large demand for crates.

The factory at Deseronto, Ont., owned by the Dominion Hardwoods, Limited, was last week destroyed by fire. The loss is estimated at $100,000, partly covered by insurance. In all probability the company will rebuild immediately.

The Military Hospitals Commission, Traders Bank Building, Toronto (Mr. Riddell, secretary), are equipping a large house in Toronto to manufacture artificial limbs for returned soldiers. Seven or eight limb makers will be employed under a skilled superintendent.

The toy exhibition held recently at the Soldiers' Club, St. John, N.B., attracted many visitors. A table of wooden toys which had been designed to show how cheaply and attractive wooden toys could be made was particularly interesting.

The plant of the National Furniture Company, at Berlin, Ont., has been purchased by Mr. E. O. Weber, owner of the Waterloo Furniture Company, Waterloo, Ont. It is expected that the new factory will employ seventy-five or eighty men. Mr. A. Edwards will be the manager.

A Toy Exhibition held at the Whitehall Art Gallery, London, England, demonstrated the fact that just as good toys could be made by British workmen as by the Huns. It is expected that in future German toys will be replaced by toys made in Britain and America.

The firm of V. E. Traversy, Montreal, has been turned into a limited liability company, under the name of Traversy, Limited, with a capital of $100,000. The powers cover the business of manufacturers and dealers in mill work and lumber, box makers and general contractors.

The Dominion Mahogany and Veneer Company, Limited, Montreal West, report that business is very brisk. The company has purchased three additional rotary lathes, which will be installed in a factory in Ontario. This will greatly facilitate the operations of the company, which of late has secured some good orders.

Lieut. J. G. Shearer, 1st Regiment, Grenadier Guards, president and managing director of the James Shearer Company, Limited, dealers in lumber and sashes and doors, Montreal, has been presented by Col. J. A. Fages with a parchment testimonial from the Canadian Royal Humane Society in recognition of his gallant services in February last in stopping a runaway horse in Montreal. The presentation was made at the close of an inspection of the Guards.

Montreal firms were very successful in securing orders for the Bethlehem shell boxes. One contract for 300,000 boxes was awarded to Henry Morgan and Company, and another, which will probably exceed this number, went to the Dominion Box and Package Company; contracts of smaller amounts were allotted to other Montreal box makers. Henry Morgan and Company are installing a large amount of new machinery for the Bethlehem boxes, including nailing machines and plant for making the three-ply glued-up diaphragms.

Replacing Guards

Following are six rules for replacing guards, which ought to be studied by the operator:—

1. A good guard out of place is worse than no guard at all. The value of a guard depends on the enforcement of its use.

2. Avoid, if possible, arbitrarily imposing guards on men. Consult the man on the job—get his suggestions—especially on guards which cover tools. such as guards for punch presses, circular saws, wood jointers and shapers. In a certain woodworking plant the operator of a shaper refused to use any guard. The foreman encouraged him to work out a guard of his own, with the result that he designed a model guard, and afterwards took great pride in using it.

3. If a guard is not used, there is a reason. Either the guard is wrong or the man is wrong. Before condemning the man, go over the situation with him and find out his objections. It may help both of you. There are more bad guards than bad men.

4. The use of guards, like all other forms of safety work, will never be carried out successfully until the workmen have been brought to an intelligent appreciation of their value. If the workmen are right-minded, you can secure their co-operation.

5. Men who wilfully refuse to use guards should be disciplined. To allow efficient guards to be treated with indifference breeds contempt for all safety work.

6. Repair men should be carefully instructed, and should be made responsible for replacing guards which they have removed.—Bulletin of National Safety Council.

Actual cost, plus a percentage of profit, is a mighty safe plan for doing odd jobs of work; but, for all that, it does not appeal strongly to the average man. He prefers an uncertainty, with a chance for large profits, to a certainty limited down to narrow margins.

MOST IMPROVED FILING AND SETTING MACHINES MADE, THEY FILE, SET AND JOINT ALL BAND SAWS IN ONE OPERATION

The only machines made that do it.

WRITE FOR DETAILS

THE WARDWELL MFG. CO.

Specialty Manufacturers
Combination Filing and Setting Machines,
Circular Saw Filers and Grinders,
Band Saw Grinders, Etc.

116 Hamilton Ave., Cleveland, Ohio

FOR SALE
Shell Box Machinery

2 Screw Drivers, electric power

2 Dovetailers, 12 Spindle
2 doz. extra cutters

3 Spindle Borers with Counter Shafts
3 sets heads and knives
4 bits for boring end blocks

1 Boring Machine, 18 Spindle, with mechanical feed

1 Boring Machine, 6 Spindle

1 Boring Machine, 5 Spindle, for handle cleats

1 Printing Machine

1 Sander, Double Drum, for bevelling edges

1 Special Dado for Sides, power feed, cut 5 grooves at once

1 Automatic Numbering and Initialing Machine

1 Gauge for testing blocks

These machines have been used ONE MONTH and are in FIRST-CLASS condition

KENT-McCLAIN, LIMITED

Carlaw Avenue   -   -   -   TORONTO

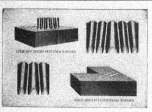

TRADE MARK

ACME

Corrugated Fasteners

for making tight, firm joints and repairing split boards. Many sizes.

Write for samples.

Box Strapping and Cold Rolled Steel Bands
FOR SHELL BOXES

Manufactured by   Box Strapping

ACME STEEL GOODS CO., LIMITED
Coristine Building, MONTREAL, Quebec

Acme Divergent Saw Edge Fasteners

Cross Cut Saws

Circular Saws

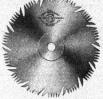
Circular Mitre Saws

The better the equipment, The bigger the *PROFIT*

DISSTON

Saws, Tools, Files

mean minimized repair, replacement and out-of-service expense and maximum output of a high grade product.

Get the Catalog

Narrow Band Saws

Keystone Groover

Henry Disston & Sons
Limited
2-20 Fraser Ave.   TORONTO, ONT.

Circular Grooving Saws

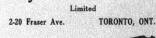

Planer Knives   Hand, Rip and Panel Saws

"We are Saving over 40% of our Coal!"

"I need only clean the boiler once in four weeks!"

MR. O. H. HALE, of the King Furniture Company, Warren, Ohio, is so enthusiastic about the

Morehead
Back to Boiler
SYSTEM

that he has written an account of his experiences which will profit his brother engineers everywhere.

For the Morehead "Back-to-Boiler" Book— with its many pages of data and interesting illustrations—and a complete copy of Mr. Hale's letter, just write

Canadian Morehead Mfg. Co.
WOODSTOCK   Dept. "H"   ONTARIO

"You never have to think about low water in the Boiler!"

"I have worked other Traps but I never saw one as simple as the Morehead!"

VENEER PRESSES

WRITE FOR BULLETIN C 1.

Canadian Boomer & Boschert Press
Company, Limited
18 Tansley Street, MONTREAL

Wood Steaming Retort

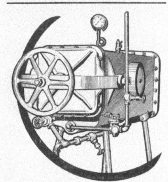

Wood Bending Manufacturers:

This is one of our

Perfection Retorts

which we guarantee will save you

50% Less Breakage

in your bending department than your present process; that your stock will dry in your forms or presses in one-third less time; that you will have no stained stock; that your stock will retain its shape much better after being bent; that it will dry in your dry-kiln in one-half less time and that your steam consumption will be reduced at least 90 per cent.

The door can be opened and closed in ten seconds, and it is steam and water tight and for this reason can be placed anywhere in your factory.

Compare this IMPROVED RETORT with your present steam boxes, then write us for our Booklet on Progressive Wood Steaming

Made in Preston, Ontario

Perfection Wood Steaming Retort Co.
PARKERSBURG - WEST VIRGINIA

Page 5 in the book on handle factory efficiency

Producing handles in the rough at minimum cost is ably taken care of with the Klotz Splitting Saw, Blocking Saw, Lathe and Throater. Then is the operation of finishing which has a great bearing on the prices you will get if your handles have that unusual smooth, glossy polish achieved on a

KLOTZ HANDLE POLISHING MACHINE

To tell you it is a durable machine with "stand-up" quality built in every angle it is only necessary to assure you that every part in it gets the careful scrutiny of men who "know handle machinery" by the only grade The Klotz Machine Company will let go out to its trade—the best.
It has steel spindles; self-oiling, and run in adjustable hollow bearings; perfectly balanced pulleys. Tension of sand belts is regulated by screws and hand wheels. Three sizes, 8-in. to 12-in. sand belt, double or single.

Turn a Page With Us Next Issue

FOR YOU

who have a thought of manufacturing handles, or those desiring maximum production at the correct cost, get the Klotz Machine Company's idea of a perfectly modern handle factory. One with such dependable machines in it as their

BLOCKING SAW

SHAPING SAW

LATHE

THROATER

BOLTING SAW

POLISHER

DISC WHEEL OR END SHAPER

KNOBBER OR SCROLL MACHINE

All have a place in an efficient handle factory. Write

Klotz Machine Co., Sandusky, O.

The Glue Book

WHAT IT CONTAINS

Chapter 1—Historical Notes

Chapter 2—Manufacture of Glue

Chapter 3—Testing and Grading

Chapter 4—Methods in the Glue Room

Chapter 5—Glue Room Equipment

Chapter 6—Selection of Glue

PRICE 50 CENTS

Woodworker Publishing Co. Limited
345 Adelaide St. West, Toronto

IT SOLVES THE
LABOR PROBLEM

One boy can operate the NASH SANDER and do as much work as 5 men with sand belts. Better work too.   See the saving?

J. M. NASH, MILWAUKEE, Wis.

"OBER" signifies

DURABILITY in Construction
SIMPLICITY in Operation
EFFICIENCY in Production
of
HANDLES and other
TURNED WORK

We make LATHES for turning handles, spokes, whiffletrees, neck yokes and variety work.

SANDERS, RIPSAWS, CHUCKS, Etc.

Write for free Catalogue No. 31

THE OBER MFG. CO.
Chagrin Falls, O., U.S.A.

L OSS of power is one of the greatest drawbacks to modern manufacturing. It is constantly harassing the profits of the manufacturer. Oftentimes it is the real factor that stands between profits and losses.

Eliminate that loss of power caused by friction by installing Chapman Double Ball Bearings on your line and counter shafts.

Installation of the above bearings will reduce your power costs from 15% to 40%. The cash you paid for waste power during the past year and a half would have purchased a complete installation of Chapman Ball Bearings. Why not install them now, and cut out this unnecessary expense in the future?

No trouble to change over our bearings suit any standard hanger

The Chapman Double Ball Bearing Co. of Canada, Limited
TORONTO   347 Sorauran Ave.   ONTARIO

The "Canadian Woodworker" Buyers' Directory

ABRASIVES
The Carborundum Co., Niagara Falls, N.Y.

BALL BEARINGS
Chapman Double Ball Bearing Co., Toronto.

BALUSTER LATHES
Garlock-Walker Machinery Co., Toronto. Ont.
Ober Mfg. Company, Chagrin Falls, Ohio.
Whitney & Son, Baxter D., Winchendon, Mass.

BAND SAW FILING MACHINERY
Canada Machinery Corporation, Galt, Ont.
Fay & Egan Co., J. A., Cincinnati, Ohio.
Garlock-Walker Machinery Co., Toronto, Ont.
Preston Woodworking Machinery Company,
    Preston, Ont.
Yates Machine Co., P. B., Hamilton, Ont.

BAND SAWS
Canada Machinery Corporation, Galt, Ont.
Garlock-Walker Machinery Co., Toronto, Ont.
Jackson, Cochrane & Company, Berlin, Ont.
Petrie, H. W., Toronto.
Preston Woodworking Machinery Company,
    Preston, Ont.
Yates Machine Co., P. B., Hamilton, Ont.

BAND SAW MACHINES
Canada Machinery Corporation, Galt, Ont.
Fay & Egan Co., J. A., Cincinnati, Ohio.
Garlock-Walker Machinery Co., Toronto, Ont.
Petrie, H. W., Toronto.
Preston Woodworking Machinery Company,
    Preston, Ont.
Williams Machinery Co., A. R., Toronto, Ont.
Yates Machine Co., P. B., Hamilton, Ont.

BAND SAW STRETCHERS
Fay & Egan Co., J. A., Cincinnati, Ohio.
Yates Machine Co., P. B., Hamilton, Ont.

BASKET STITCHING MACHINE
Saranac Machine Co., Benton Harbor, Mich.

BENDING MACHINES
Fay & Egan Co., J. A., Cincinnati, Ohio.
Garlock-Walker Machinery Co., Toronto, Ont.
Perfection Wood Steaming Retort Company,
    Parkersburg, W. Va.

BELTING
Fay & Egan Co., J. A., Cincinnati, Ohio.

BLOWERS
Sheldons, Limited, Galt, Ont.
Toronto Blower Company, Toronto, Ont.

BLOW PIPING
Sheldons, Limited, Galt, Ont.
Toronto Blower Company, Toronto, Ont.

BORING MACHINES
Canada Machinery Corporation, Galt, Ont.
Fay & Egan Co., J. A., Cincinnati, Ohio.
Garlock-Walker Machinery Co., Toronto, Ont.
Jackson, Cochrane & Company, Berlin, Ont.
Nash, J. M., Milwaukee, Wis.
Preston Woodworking Machinery Company,
    Preston, Ont.
Reynolds Pattern & Machine Co., Moline, Ill.
Yates Machine Co., P. B., Hamilton, Ont.

BOX MAKERS' MACHINERY
Canada Machinery Corporation, Galt, Ont.
Fay & Egan Co., J. A., Cincinnati, Ohio.
Garlock-Walker Machinery Co., Toronto, Ont.
Jackson, Cochrane & Co., Berlin, Ont.
Neilson & Company, J. L. Winnipeg, Man.
Preston Woodworking Machinery Company,
    Preston, Ont.
Reynolds Pattern & Machine Co., Moline, Ill.
Whitney & Son, Baxter D., Winchendon, Mass.
Williams Machinery Co., A. R., Toronto, Ont.
Yates Machine Co., P. B., Hamilton, Ont.

CABINET PLANERS
Canada Machinery Corporation, Galt, Ont.
Fay & Egan Co., J. A., Cincinnati, Ohio.
Garlock-Walker Machinery Co., Toronto, Ont.
Jackson, Cochrane & Co., Berlin, Ont.
Preston Woodworking Machinery Company,
    Preston, Ont.
Yates Machine Co., P. B., Hamilton, Ont.

CARS (Transfer)
Sheldons, Limited, Galt, Ont.

CARBORUNDUM PAPER
Carborundum Co., Niagara Falls, N.Y.

CARVING MACHINES
Canada Machinery Corporation, Galt, Ont.
Fay & Egan Co., J. A., Cincinnati, Ohio.
Garlock-Walker Machinery Co., Toronto, Ont.
Jackson, Cochrane & Company, Berlin, Ont

CASTERS
Foster, Merriam & Co., Meriden, Conn.

CLAMPS
Fay & Egan Co., J. A., Cincinnati, Ohio.
Garlock-Walker Machinery Co., Toronto, Ont.
Jackson, Cochrane & Company, Berlin, Ont.
Preston Woodworking Machinery Company,
    Preston, Ont.
Simonds Canada Saw Company, Montreal, P.Q.

COLUMN MACHINERY
Fay & Egan Co., J. A., Cincinnati, Ohio.

CORRUGATED FASTENERS
Acme Steel Goods Company, Chicago, Ill.

CUT-OFF SAWS
Canada Machinery Corporation, Galt, Ont.
Fay & Egan Co., J. A., Cincinnati, Ohio.
Garlock-Walker Machinery Co., Toronto, Ont.
Jackson, Cochrane & Co., Berlin, Ont.
Ober Mfg. Co., Chagrin Falls, Ohio.
Preston Woodworking Machinery Company,
    Preston, Ont.
Simonds Canada Saw Co., Montreal, Que.
Yates Machine Co., P. B., Hamilton, Ont.

CUTTER HEADS
Canada Machinery Corporation, Galt, Ont.
Fay & Egan Co., J. A., Cincinnati, Ohio.
Jackson, Cochrane & Company, Berlin, Ont.
Mattison Machine Works, Beloit, Wis.
Yates Machine Co., P. B., Hamilton, Ont.

DOVETAILING MACHINES
Canada Machinery Corporation, Galt, Ont.
Fay & Egan Co., J. A., Cincinnati, Ohio.
Garlock-Walker Machinery Co., Toronto, Ont.
Jackson, Cochrane & Co., Berlin, Ont.

DOWEL MACHINES
Canada Machinery Corporation, Galt, Ont.
Fay & Egan Co., J. A., Cincinnati, Ohio.
Ober Mfg. Co., Chagrin Falls, Ohio.

DRY KILNS
National Dry Kiln Co., Indianapolis, Ind.
Sheldons, Limited, Galt, Ont.

DUST COLLECTORS
Sheldons, Limited, Galt, Ont.
Toronto Blower Company, Toronto, Ont.

DUST SEPARATORS
Sheldons, Limited, Galt, Ont.
Toronto Blower Company, Toronto, Ont.

EDGERS (Single Saw)
Canada Machinery Corporation, Galt, Ont.
Fay & Egan Co., J. A., Cincinnati, Ohio.
Garlock-Walker Machinery Co., Toronto, Ont.
Simonds Canada Saw Co., Montreal, Que.

EDGERS (Gang)
Canada Machinery Corporation, Galt, Ont.
Fay & Egan Co., J. A., Cincinnati, Ohio.
Garlock-Walker Machinery Co., Toronto, Ont.
Petrie, H. W., Toronto.
Simonds Canada Saw Co., Montreal, P.Q.
Williams Machinery Co., A. R., Toronto, Ont.
Yates Machine Co., P. B., Hamilton, Ont.

END MATCHING MACHINE
Canada Machinery Corporation, Galt, Ont.
Fay & Egan Co., J. A., Cincinnati, Ohio.
Garlock-Walker Machinery Co., Toronto, Ont.
Jackson, Cochrane & Company, Berlin, Ont.
Yates Machine Co., P. B., Hamilton, Ont.

EXHAUST FANS
Garlock-Walker Machinery Co., Toronto, Ont.
Sheldons Limited, Galt, Ont.
Toronto Blower Company, Toronto, Ont.

EXHAUST SYSTEMS
Toronto Blower Company, Toronto, Ont.

FILES
Simonds Canada Saw Co., Montreal, Que.

FILING MACHINERY
Waidwell Mfg. Company, Cleveland, Ohio.

FLOORING MACHINES
Canada Machinery Corporation, Galt, Ont.
Fay & Egan Co., J. A., Cincinnati, Ohio.
Garlock-Walker Machinery Co., Toronto, Ont.
Petrie, H. W., Toronto.
Preston Woodworking Machinery Company,
    Preston, Ont.
Whitney & Son, Baxter D., Winchendon, Mass.
Yates Machine Co., P. B., Hamilton, Ont.

FLUTING HEADS
Fay & Egan Co., J. A., Cincinnati, Ohio.

FLUTING AND TWIST MACHINE
Preston Woodworking Machinery Company,
    Preston, Ont.

FURNITURE TRIMMINGS
Foster, Merriam & Co., Meriden, Conn.

GARNET PAPER AND CLOTH
The Carborundum Co., Niagara Falls, N.Y.

GRAINING MACHINES
Canada Machinery Corporation, Galt, Ont.
Fay & Egan Co., J. A., Cincinnati, Ohio.
Williams Machinery Co., A. R., Toronto, Ont.
Yates Machine Co., P. B., Hamilton, Ont.

GLUE
Coignet Chemical Product Co., New York,
    N.Y.
Perkins Glue Company, South Bend, Ind.

GLUE CLAMPS
Jackson, Cochrane & Company, Berlin, Ont.

GLUE HEATERS
Fay & Egan Co., J. A., Cincinnati, Ohio.
Jackson, Cochrane & Company, Berlin, Ont.

GLUE JOINTERS
Canada Machinery Corporation, Galt, Ont.
Jackson, Cochrane & Company, Berlin, Ont.
Yates Machine Co., P. B., Hamilton, Ont.

GLUE SPREADERS
Fay & Egan Co., J. A., Cincinnati, Ohio.
Jackson, Cochrane & Company, Berlin, Ont.

GLUE ROOM EQUIPMENT
Boomer & Boschert Company, Montreal, Que.
Perrin & Company, W. R., Toronto, Ont.

GRINDERS (Cutter)
Fay & Egan Co., J. A., Cincinnati, Ohio.

GRINDERS, HAND AND FOOT POWER
The Carborundum Co., Niagara Falls, N.Y.

GRINDERS (Knife)
Canada Machinery Corporation, Galt, Ont.
Fay & Egan Co., J. A., Cincinnati, Ohio.
Garlock-Walker Machinery Co., Toronto, Ont.
Petrie, H. W., Toronto.
Preston Woodworking Machinery Company,
    Preston, Ont.
Yates Machine Co., P. B., Hamilton, Ont.

GRINDERS (Tool)
Fay & Egan Co., J. A., Cincinnati, Ohio.
Jackson, Cochrane & Company, Berlin, Ont.

GRINDING WHEELS
The Carborundum Co., Niagara Falls, N.Y.

GROOVING HEADS
Fay & Egan Co., J. A., Cincinnati, Ohio.

GUMMERS, ETC.
Fay & Egan Co., J. A., Cincinnati, Ohio.

HAND PROTECTORS
Fay & Egan Co., J. A., Cincinnati, Ohio.

HAND SCREWS
Fay & Egan Co., J. A., Cincinnati, Ohio.

HANDLE THROATING MACHINE
Klotz Machine Company, Sandusky, Ohio.

HANDLE TRIMMING MACHINE
Klotz Machine Company, Sandusky, Ohio.

HANDLE & SPOKE MACHINERY
Fay & Egan Co., J. A., Cincinnati, Ohio.
Klotz Machine Co., Sandusky, Ohio.
Nash, J. M., Milwaukee, Wis.
Ober Mfg. Company, Chagrin Falls, Ohio.
Whitney & Son, Baxter D., Winchendon, Mass.

HARDWOOD LUMBER
American Hardwood Lumber Co., St. Louis, Mo.
Atlantic Lumber Company, Toronto.
Brown & Company, Geo. C., Memphis, Tenn.
Clark & Son, Edward, Toronto, Ont.
Kersley, Geo., Montreal, Que.
Kraetzer-Cured Lumber Co., Moorhead, Miss.
Mowbray & Robinson, Cincinnati, Ohio.
Paepcke Leicht Lumber Co., Chicago, Ill.
Penrod Walnut & Veneer Co., Kansas City, Mo.
Spencer, C. A., Montreal, Que.

HUB MACHINERY
Fay & Egan Co., J. A., Cincinnati, Ohio.

HYDRAULIC PRESSES
Can. Boomer & Boschert Press, Montreal, Que.

HYDRAULIC PUMPS & ACCUMULATORS
Can. Boomer & Boschert Press, Montreal, Que.

HYDRAULIC VENEER PRESSES
Perrin & Company, Wm. R., Toronto, Ont.

JOINTERS
Canada Machinery Corporation, Galt, Ont.
Fay & Egan Co., J. A., Cincinnati, Ohio.
Garlock-Walker Machinery Co., Toronto, Ont.
Jackson, Cochrane & Company, Berlin, Ont.
Knechtel Bros., Limited, Southampton, Ont.
Petrie, H. W., Toronto, Ont.
Preston Woodworking Machinery Company, Preston, Ont.
Yates Machine Co., P. B., Hamilton, Ont.

KNIVES (Planer and others)
Canada Machinery Corporation, Galt, Ont.
Fay & Egan Co., J. A., Cincinnati, Ohio.
Simonds Canada Saw Co., Montreal, P.Q.
Yates Machine Co., P. B., Hamilton, Ont.

LUMBER
American Hardwood Lumber Co., St. Louis, Mo.
Brown & Company, Geo. C., Memphis, Tenn.
Churchill, Milton Lumber Co., Louisville, Ky.
Clark & Son, Edward, Toronto, Ont.
Dominion Mahogany & Veneer, Co., Montreal, Que.
Gillies Bros., Braeside, Ont.
Hart & McDonagh, Toronto, Ont.
Hartzell, Geo. W., Piqua, Ohio.
Paepcke Leicht Lumber Company, Chicago, Ill.

LATHES
Canada Machinery Corporation, Galt, Ont.
Fay & Egan Co., J. A., Cincinnati, Ohio.
Garlock-Walker Machinery Co., Toronto, Ont.
Jackson, Cochrane & Company, Berlin, Ont.
Klotz Machine Company, Sandusky, Ohio.
Ober Mfg. Company, Chagrin Falls, Ohio.
Petrie, H. W., Toronto, Ont.
Preston Woodworking Machinery Company,
Shawver Company, Springfield, O.
Whitney & Son, Baxter D., Winchendon, Mass.

MACHINE KNIVES
Canada Machinery Corporation, Galt, Ont.
Peter Hay Knife Company, Galt, Ont.
Simonds Canada Saw Co., Montreal, P.Q.
Yates Machine Co., P. B., Hamilton, Ont.

MITRE MACHINES
Canada Machinery Corporation, Galt, Ont.
Fay & Egan Co., J. A., Cincinnati, Ohio.

MITRE SAWS
Canada Machinery Corporation, Galt, Ont.
Fay & Egan Co., J. A., Cincinnati, Ohio.
Garlock-Walker Machinery Co., Toronto, Ont.
Simonds Canada Saw Co., Montreal, P.Q.

MORTISING MACHINES
Canada Machinery Corporation, Galt, Ont.
Fay & Egan Co., J. A., Cincinnati, Ohio.
Garlock-Walker Machinery Co., Toronto, Ont.
Yates Machine Co., P. B., Hamilton, Ont.

MOULDINGS
Knechtel Bros., Limited, Southampton, Ont.

MULTIPLE BOXING MACHINES
Fay & Egan Co., J. A., Cincinnati, Ohio.
Nash, J. M., Milwaukee, Wis.
Reynolds Pattern & Machine Co., Moline, Ill.

PATTERN SHOP MACHINES
Canada Machinery Corporation, Galt, Ont.
Fay & Egan Co., J. A., Cincinnati, Ohio.
Garlock-Walker Machinery Co., Toronto, Ont.
Jackson, Cochrane & Company, Berlin, Ont.
Preston Woodworking Machinery Company, Preston, Ont.
Whitney & Son, Baxter D., Winchendon, Mass.
Yates Machine Co., P. B., Hamilton, Ont.

PLANERS
Canada Machinery Corporation, Galt, Ont.
Fay & Egan Co., J. A., Cincinnati, Ohio.
Garlock-Walker Machinery Co., Toronto, Ont.
Jackson, Cochrane & Company, Berlin, Ont.
Petrie, H. W., Toronto, Ont.
Preston. Woodworking Machinery Company,
Whitney & Son, Baxter D., Winchendon, Mass.
Williams Machinery Co., A. R., Toronto, Ont.
Yates Machine Co., P. B., Hamilton, Ont.

PLANING MILL MACHINERY
Canada Machinery Corporation, Galt, Ont.
Fay & Egan Co., J. A., Cincinnati, Ohio.
Jackson, Cochrane & Company, Berlin, Ont.
Petrie, H. W., Toronto, Ont.
Preston Woodworking Machinery Company,
Whitney & Son, Baxter D., Winchendon, Mass.
Williams Machinery Co., A. R., Toronto, Ont.
Yates Machine Co., P. B., Hamilton, Ont.

PRESSES (Veneer)
Canada Machinery Corporation, Galt, Ont.
Jackson, Cochrane & Company, Berlin, Ont.
Perrin & Company, Wm. R., Toronto, Ont.
Preston Woodworking Machinery Company, Preston, Ont.

PULLEYS
Fay & Egan Co., J. A., Cincinnati, Ohio.

RED GUM LUMBER
Gum Lumber Manufacturers' Association.

RESAWS
Canada Machinery Corporation, Galt, Ont.
Fay & Egan Co., J. A., Cincinnati, Ohio.
Garlock-Walker Machinery Co., Toronto, Ont.
Jackson, Cochrane & Company, Berlin, Ont.
Petrie, H. W., Toronto, Ont.
Preston Woodworking Machinery Company, Preston, Ont.
Simonds Canada Saw Co., Montreal, P.Q.
Williams Machinery Co., A. R., Toronto, Ont.
Yates Machine Co., P. B., Hamilton, Ont.

RIM AND FELLOE MACHINERY
Fay & Egan Co., J. A., Cincinnati, Ohio.

RIP SAWING MACHINES
Canada Machinery Corporation, Galt, Ont.
Fay & Egan Co., J. A., Cincinnati, Ohio.
Garlock-Walker Machinery Co., Toronto, Ont.
Klotz Machine Company, Sandusky, Ohio.
Jackson, Cochrane & Company, Berlin, Ont.
Ober Mfg. Company, Chagrin Falls, Ohio.
Preston Woodworking Machinery Company, Preston, Ont.
Yates Machine Co., P. B., Hamilton, Ont.

SANDERS
Canada Machinery Corporation, Galt, Ont.
Fay & Egan Co., J. A., Cincinnati, Ohio.
Garlock-Walker Machinery Co., Toronto, Ont.
Klotz Machine Company, Sandusky, Ohio.
Jackson, Cochrane & Company, Berlin, Ont.
Nash, J. M., Milwaukee, Wis.
Ober Mfg. Company, Chagrin Falls, Ohio.
Preston Woodworking Machinery Company, Preston, Ont.

SASH, DOOR & BLIND MACHINERY
Canada Machinery Corporation, Galt, Ont.
Fay & Egan Co., J. A., Cincinnati, Ohio.
Garlock-Walker Machinery Co., Toronto, Ont.
Jackson, Cochrane & Company, Berlin, Ont.
Preston Woodworking Machinery Company, Preston, Ont.
Yates Machine Co., P. B., Hamilton, Ont.

SAWS
Disston & Sons, Henry, Toronto & Philadelphia
Fay & Egan Co., J. A., Cincinnati, Ohio.
Radcliff Saw Mfg. Company, Toronto, Ont.
Simonds Canada Saw Co., Montreal, P.Q.

SAW GUMMERS, ALOXITE
The Carborundum Co., Niagara Falls, N.Y.

SAW SWAGES
Fay & Egan Co., J. A., Cincinnati, Ohio.
Simonds Canada Saw Co., Montreal, Que.

SAW TABLES
Canada Machinery Corporation, Galt, Ont.
Fay & Egan Co., J. A., Cincinnati, Ohio.
Garlock-Walker Machinery Co., Toronto, Ont.
Jackson, Cochrane & Company, Berlin, Ont.
Preston Woodworking Machinery Company, Preston, Ont.
Yates Machine Co., P. B., Hamilton, Ont.

SCRAPING MACHINES
Canada Machinery Corporation, Galt, Ont.
Garlock-Walker Machinery Co., Toronto, Ont.
Whitney & Sons, Baxter D., Winchendon, Mass

SCROLL CHUCK MACHINES
Klotz Machine Company, Sandusky, Ohio.

SCREW DRIVING MACHINERY
Canada Machinery Corporation, Galt, Ont.
Reynolds Pattern & Machine Works, Moline, Ill

SCROLL SAWS
Canada Machinery Corporation, Galt, Ont.
Fay & Egan Co., J. A., Cincinnati, Ohio.
Garlock-Walker Machinery Co., Toronto, Ont.
Jackson, Cochrane & Company, Berlin, Ont.
Simonds Canada Saw Co., Montreal, P.Q.
Yates Machine Co., P. B., Hamilton, Ont.

SECOND-HAND MACHINERY
Petrie, H. W.
Williams Machinery Co., A. R., Toronto, Ont.

SHAPERS
Canada Machinery Corporation, Galt, Ont.
Fay & Egan Co., J. A., Cincinnati, Ohio.
Garlock-Walker Machinery Co., Toronto, Ont.
Jackson, Cochrane & Company, Berlin, Ont.
Ober Mfg. Company, Chagrin Falls, Ohio.
Petrie, H. W., Toronto, Ont.
Preston Woodworking Machinery Company, Preston, Ont.
Simonds Canada Saw Co., Montreal, P.Q.
Whitney & Sons, Baxter D., Winchendon, Mass
Yates Machine Co., P. B., Hamilton, Ont.

SHARPENING TOOLS
The Carborundum Co., Niagara Falls, N.Y.

SHAVING COLLECTORS
Sheldons Limited, Galt, Ont.

SHELL BOX HANDLES
Vokes Hardware Co., Toronto, Ont.

STEAM APPLIANCES
Darling Bros., Montreal.

SINGLE SPINDLE BOXING MACHINES
Canada Machinery Corporation, Galt, Ont.
Fay & Egan Co., J. A., Cincinnati, Ohio.
Ober Mfg. Company, Chagrin Falls, Ohio.

STAVE SAWING MACHINE
Whitney & Sons, Baxter D., Winchendon, Mass

STEAM TRAPS
Canadian Morehead Mfg. Co., Woodstock, Ont.
Standard Dry Kiln Co., Indianapolis, Ind.

STEEL BOX STRAPPING
Acme Steel Goods Co., Chicago, Ill.

STEEL SHELL BOX BANDS
Acme Steel Goods Co., Chicago, Ill.

SURFACERS
Canada Machinery Corporation, Galt, Ont.
Fay & Egan Co., J. A., Cincinnati, Ohio.
Garlock-Walker Machinery Co., Toronto, Ont.
Jackson, Cochrane & Company, Berlin, Ont.
Petrie, H. W., Toronto, Ont.
Preston Woodworking Machinery Company,
Preston, Ont.
Whitney & Sons, Baxter D., Winchendon, Mass.
Williams Machinery Co., A. R., Toronto, Ont.
Yates Machine Co., P. B., Hamilton, Ont.

SWING SAWS
Canada Machinery Corporation, Galt, Ont.
Fay & Egan Co., J. A., Cincinnati, Ohio.
Garlock-Walker Machinery Co., Toronto, Ont.
Jackson, Cochrane & Company, Berlin, Ont.
Ober Mfg. Co., Chagrin Falls, Ohio.
Preston Woodworking Machinery Company,
Preston, Ont.
Simonds Canada Saw Co., Montreal, P.Q.
Yates Machine Co., P. B., Hamilton, Ont.

TABLE LEG LATHES
Fay & Egan Co., J. A., Cincinnati, Ohio.
Ober Mfg. Co., Chagrin Falls, Ohio.
Whitney & Sons, Baxter D., Winchendon, Mass.
Yates Machine Co., P. B., Hamilton, Ont.

TABLE SLIDES
Walter & Co., B., Wabash, Ind.

TENONING MACHINES
Canada Machinery Corporation, Galt, Ont.
Fay & Egan Co., J. A., Cincinnati, Ohio.
Garlock-Walker Machinery Co., Toronto, Ont.
Jackson, Cochrane & Company, Berlin, Ont.
Preston Woodworking Machinery Company,
Preston, Ont.
Yates Machine Co., P. B., Hamilton, Ont.

TRIMMERS
Canada Machinery Corporation, Galt, Ont.
Fay & Egan Co., J. A., Cincinnati, Ohio.
Preston Woodworking Machinery Company,
Preston, Ont.
Yates Machine Co., P. B., Hamilton, Ont.

TRUCKS
Sheldons Limited, Galt, Ont.
National Dry Kiln Co., Indianapolis, Ind.

TURNING MACHINES
Fay & Egan Co., J. A., Cincinnati, Ohio.
Garlock-Walker Machinery Co., Toronto, Ont.
Klotz Machine Company, Sandusky, Ohio.
Ober Mfg. Company, Chagrin Falls, Ohio.
Whitney & Sons, Baxter D., Winchendon, Mass.
Yates Machine Co., P. B., Hamilton, Ont.

**UNDER-CUT SELF-FEEDING FACE
PLANER**
Fay & Egan Co., J. A., Cincinnati, Ohio.
Garlock-Walker Machinery Co., Toronto, Ont.
Jackson, Cochrane & Company, Berlin, Ont.

VARNISHES
Ault & Wiborg Company, Toronto, Ont.

VENEERS
Central Veneer Co., Huntington, W. Virginia.
Dominion Mahogany & Veneer Company, Montreal, Que.
Walter Clark Veneer Co., Grand Rapids, Mich.
Evansville Veneer Co., Evansville, Ind.
Hartzell, Geo. W., Piqua, Ohio.
Haughton Veneer Company, Chicago, Ill.
Hay & Company, Woodstock, Ont.
Hoffman Bros. Company, Fort Wayne, Ind.
Geo. Kersley, Montreal, Que.
Nartzik, J. J., Chicago, Ill.
Ohio Veneer Company, Cincinnati, Ohio.
Penrod Walnut & Veneer Co., Kansas City, Mo.
Roberts, John N., New Albany, Ind.
Toronto Veneer Company, Toronto, Ont.
Wood Mosaic Co., New Albany, Ind.

VENTILATING APPARATUS
Sheldons Limited, Galt, Ont.

VENEERED PANELS
Hay & Company, Woodstock, Ont.

VENEER PRESSES (Hand and Power)
Canada Machinery Corporation, Galt, Ont.
Can. Boomer & Boschert Press, Montreal, Que.
Garlock-Walker Machinery Co., Toronto, Ont.
Jackson, Cochrane & Company, Berlin, Ont.
Perrin & Company, Wm. R., Toronto, Ont.

VISES
Fay & Egan Co., J. A., Cincinnati, Ohio.
Simonds Canada Saw Company, Montreal, Que.

WAGON AND CARRIAGE MACHINERY
Canada Machinery Corporation, Galt, Ont.
Fay & Egan Co., J. A., Cincinnati, Ohio.
Ober Mfg. Company, Chagrin Falls, Ohio.
Whitney & Sons, Baxter D., Winchendon, Mass.
Yates Machine Co., P. B., Hamilton, Ont.

WOOD FINISHES
Ault & Wiborg, Toronto, Ont.

WOOD TURNING MACHINERY
Canada Machinery Corporation, Galt, Ont.
Garlock-Walker Machinery Co., Toronto, Ont.
Yates Machine Co., P. B., Hamilton, Ont.

WORK BENCHES
Fay & Egan Co., J. A., Cincinnati, Ohio.

WOODWORKING MACHINES
Canada Machinery Corporation, Galt, Ont.
Garlock-Walker Machinery Co., Toronto, Ont.
Jackson, Cochrane & Co., Berlin, Ont.
Preston Woodworking Machinery Company,
Preston, Ont.
Reynolds Pattern & Machine Works, Moline,
Ill.
Williams Machinery Co., A. R., Toronto, Ont.

WHEELS, Grinding, Carborundum and Aloxite
The Carborundum Co., Niagara Falls, N.Y.

THE COMPLETE LINE

ACME CASTERS are the development of long experience, they are always dependable and reliable.

We solicit your inquiries and will be pleased to have our Canadian Representative call and assist you in your Caster and Period Trimmings problems.

Foster, Merriam & Co.

Meriden, Conn.

Canadian Representative:—F. A. Schmidt, Berlin, Ont.

"ACME"
The
RELIABLE CASTER

Making TABLE-SLIDES is a Specialty Business

For more than TWENTY-FIVE YEARS we have made TABLE SLIDES exclusively. Our Factory is equipped with Special Machinery which enables us to make SLIDES.—BETTER and CHEAPER than the furniture manufacturer.

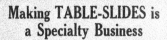

Canadian Table makers are rapidly adopting WABASH SLIDES

Because { They ELIMINATE SLIDE TROUBLES
Are CHEAPER and BETTER }

Reduced Costs   } BY USING

Increased Out-put } **WABASH SLIDES**

MADE BY

B. Walter & Company
Wabash, Ind.

The Largest EXCLUSIVE TABLE-SLIDE Manufacturers in America
ESTABLISHED-1887

Just What You Need

There's no necessity now for you to wish for a corrugated joint fastener machine which will rise fully to your expectations; you can get one.

The Saranac Corrugated Joint Fastener Machine

is the absolute realization of your wish. This illustration gives you an idea of what it looks like, but you can have no conception of its **efficiency, economy, speed and reliability** until you get it working for you.

It is substantially constructed, and the workmanship and materials are of the very best. Particularly suitable for manufacturers of sash, doors, blinds, screens, porch columns, boxes, etc., etc.

Made in various types for all conditions.

Write for further particulars and prices

Saranac Machine Co., SOLE MAKERS
Benton Harbor, Mich.

Real Sanding Service is Production with Economy

TO get real sanding service you are dependent upon your garnet paper and cloth. If you are getting paper or cloth that is uniformly coated with clean, hard, sharp garnet—if the paper, cloth and glue used are of the best possible quality—if you are getting Carborundum Brand Garnet Products—the paper and cloth that cut clean and fast, that hold the grain and show long life—*then you are getting real sanding service.*

Cut in all Standard Sizes in all Grits
Working Samples on Request

THE CARBORUNDUM COMPANY
NIAGARA FALLS, N. Y.

Grand Rapids   New York   Chicago   Boston   Cleveland
Philadelphia   Cincinnati   Pittsburgh   Milwaukee

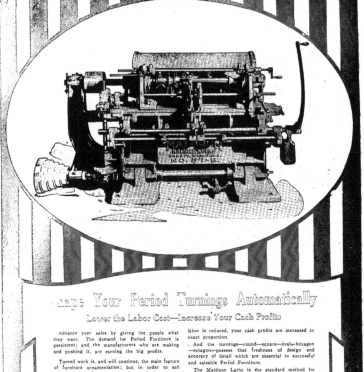

Shape Your Period Turnings Automatically

Lower the Labor Cost—Increase Your Cash Profits

Advance your sales by giving the people what they want. The demand for Period Furniture is persistent; and the manufacturers who are making and pushing it, are earning the big profits.

Turned work is, and will continue, the main feature of furniture ornamentation; but in order to sell this Period stuff at a substantial profit, your turnings must be made on a low-cost basis of manufacture. You can not afford to whittle out these patterns by hand, or purchase them from outside at high prices.

What you need is the Model "B" Mattison Automatic Shaping Lathe—a practical, economical machine turner. It provides unlimited capacity and minimizes the labor cost. Every time an item of labor is reduced, your cash profits are increased in exact proportion.

And the turnings—round—square—oval—hexagon—octagon—possess that freshness of design and accuracy of detail which are essential to successful and saleable Period Furniture.

The Mattison Lathe is the standard method for automatically shaping Period turned work. It is efficient and dependable—requires but an ordinary mechanic to operate—gives your designer unlimited scope in the creation of elaborate or complicated patterns—places your production on an equal basis with that of your competitors.

Investigate this up-to-date machine method. Write us today and we shall gladly give you further interesting facts. This doesn't obligate you in the least.

No Other Shaping Lathe like the Mattison
It is built only by the
Mattison Machine Works
883 Fifth St., Beloit, Wis.

Volume Sixteen   TORONTO, AUGUST, 1916   Number Eight

CANADIAN WOODWORKER
and Furniture Manufacturer

Why AND *How*

We can help you, is told in our new catalog on

The Whitney Wood Scraping Machine

For Final Finishing of Flat Hardwood Surfaces

The booklet should be in your hands—it shows you, clearly and concisely, the "reason why" scraping with a Whitney Wood Scraping Machine is better, quicker, and more economical than any other method you can employ, sanding or hand scraping, in finishing-up flat solid hardwood or veneering.

Not only will we mail you this catalog when you write, but we will send you samples of hardwood, scraped on a Whitney Wood Scraping Machine, to demonstrate the high quality work a Whitney will do.

Baxter D. Whitney & Son, Winchendon, Mass.

Pacific Coast Office: Representatives for Ontario:
Berkeley, Cal.   H. W. Petrie, Limited, Toronto, Ontario

THE A.R. WILLIAMS MACHINERY CO., LTD.
ST. JOHN, N.B. TORONTO WINNIPEG VANCOUVER
Canada's Leading Machinery House

Our Exhibit in
Government Building

Watch **For** *It*

Of Interest to Toy
Manufacturers

IN THIS GREAT NEW CANADIAN INDUSTRY — Toy-Making—
there are three principles, all-important, and linked up by a mutual beneficial object
—You—the Government—A. R. Williams Machinery Co.

AT THE CANADIAN NATIONAL EXHIBITION, in the Government
Building, there will be a Toy display under Government supervision. In connection with this display our Service department has risen to the occasion and will
show a few sample lines of machines which are largely in use in the manufacture
of cheap wooden toys.

YOU MAY CONSIDER our representatives at this exhibit entirely at your
disposal. They will be pleased to go into the details of toy making. They will
advise you in regard to equipment and give you the benefit of our experience in
meeting problems presented by this new industry.

Service and Co-operation

are the two great objects which we have set before us. If you cannot
solve your problem let us get together and discuss them fully.
Two heads are better than one—you know what you wish to achieve
—our Service department will help you to attain that object.

64-66 Front Street West - - TORONTO

GARLOCK-WALKER MACHINERY CO.
LIMITED
32 FRONT ST. WEST.   **TORONTO**   **TELEPHONE MAIN 5346**

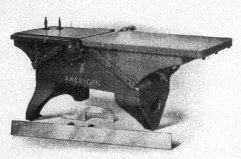

Made in 8″, 12″,
16″, 24″, 30″,
36″ sizes.

Also with
Power Feed.

NOTE
Adjustable Boxes,

Long Gibbed In-
cline ways for
raising and low-
ering table.

Steel Lips,

7′ Table,

Angular adjustable
Fence.

American Buzz Planer and Hand Jointer

A FULL LINE OF

MACHINERY

FOR ALL PURPOSES

Canadian Representatives

American Wood Working Machinery Co.
ROCHESTER, N.Y.

M. L. Andrews & Co.
CINCINNATI, OHIO

METAL and WOODWORKING MACHINERY of all Kinds

"AMERICA'S FINEST CABINET WOOD"

RED GUM

furniture "has come into its own at last." Whereas a few short years ago the alert furniture manufacturer (and dealer), with a finger on the public pulse, showed RED GUM only when it was insisted upon, and then with a half-apologetic manner, HE NOW SAYS: "COME RIGHT UP HERE IN FRONT —I WANT TO SHOW YOU MY RED GUM LINES." (Proud of it— and well he may be.)

WHY DON'T YOU, MR. FURNITURE MAKER, GET YOUR RED GUM LINE INTO THE SPOTLIGHT OF POPULAR FAVOR?

There's a lot of honorable profit in meeting public taste. Do you realize what our advertising to the general public (the real market) is doing for manufacturers who feature GUM? "Come on in—the selling's fine" with "America's Finest Cabinet Wood."

WRITE US FOR SAMPLES, PARTICULARS AND GENERAL INFORMATION. OUR REPLY WILL BE PROMPT, PERSONAL AND DEPENDABLE.

GUM LUMBER MANUFACTURERS' ASS'N
1314 Bank of Commerce Building, Memphis, Tennessee

Sixth page in the book on handle factory efficiency

A manufacturer is mostly known by the quality of product he makes. Put a smooth finish on the beveled end of your handles and you will find a ready market, and get a price for your handles which will repay the small expense of installing a

KLOTZ DISC WHEEL OR END SHAPER

It is an extremely serviceable machine for any handle factory and has never failed to cement the good will of users of Klotz line of handle machinery—in other words, it holds up the Klotz reputation for good machines and the users' reputation for good handles. The disc wheel is turned and polished on inside face; provided with knives; adjustable gauge holds handles in proper position while ends are being finished.

Let us diagram for you the exact position for this machine and the other machinery for a thoroughly efficient handle factory.

Turn a Page With Us Next Issue

FOR YOU

who have a thought of manufacturing handles, or those desiring maximum production at the correct cost, get the Klotz Machine Company's idea of a perfectly modern handle factory. One with such dependable machines in it as their

BLOCKING SAW
SHAPING SAW
LATHE
THROATER
BOLTING SAW
POLISHER
DISC WHEEL OR END SHAPER
KNOBBER OR SCROLL MACHINE

All have a place in an efficient handle factory. Write

Klotz Machine Co., Sandusky, O.

Absolute
Accuracy!

The nature of the work the equalizing saw must perform puts accuracy above all other considerations. The work must have both ends trimmed at absolute right angles to the length or it is worthless. Perfect parallel saw cuts are invariably made

With the Yates Automatic Double Cut-Off Saw

Two endless chains of the automobile type run over sprocket wheels and carry the stock past the saws. Slipping of chains is impossible. Stock is maintained perpendicular to the line of travel at all times. Absolute accuracy of cut is thereby assured.

In addition, the machine is equipped with hopper feed and is capable of the quantity of output that hopper feed induces. The operator piles stock in the hopper by armfulls. The machine automatically makes the cut and delivers the equalized stock to the off-bearer. Thus this machine has both the accuracy and output that makes a perfect double cut-off saw.

Circular 229, describing this machine in detail will be sent upon request without expense or obligation in any way.

"Furniture Fundamentals"

Is the name of a folder we have recently published, dealing with the principal primary manufacturing units as found in the average furniture factory. A copy is yours for the asking.

P. B. Yates Machine Co. Ltd.
HAMILTON, ONT. CANADA
U.S. PLANT—BELOIT, WIS.

Beaver Brand

Reduce the Cost of Making Shell Boxes

by using a

Special
"Beaver" Dado

This special "Beaver" Dado was designed to meet the requirements of shell box manufacturers. This particular construction eliminates the trouble previously experienced by many manufacturers in cutting across the grain. It leaves a clean, smooth surface when cutting with or across the grain.

Illustration shows one complete order recently shipped by us to a shell box manufacturer. It consists of—five "Beaver" Dados from 1¾ in. to 2 in. wide; also two special Planer Saws for cutting of box stock; the small groover shown at centre top is to cut slot for wire handle or for sizing for the cover.

Get our prices on special shell box saws—we can save you money.

Radcliff Saw Manufacturing Company, Limited

TORONTO   550 Dundas Street   ONTARIO

Agents for L. & I. J. White Company Machine Knives

SLOW SPEED LOW POWER
BLOWER SYSTEM

The
Foster Blower

The "FOSTER" BLOWER

is specially designed for handling shavings, sawdust, excelsior, or any conceivable material. It is equipped with three bearings, shaft extending through, with bearing on inlet side. This eliminates the over-hung wheel and avoids heating, clogging and shaking. It is the only non-centre suction fan in existence. The Foster Fan is so constructed that the material entering the fan is immediately discharged without passing through or around the wheel.

Our Automatic Furnace Feeders will feed your material under boilers or into any desirable receptacle.

Consult us on your exhaust system troubles.

THE TORONTO BLOWER COMPANY

TORONTO   156 DUKE ST.   ONTARIO

Are you one of those who could make a big profit on this machine?

❡ Not every business can make a profit on a "FAY-EGAN LIGHTNING" No. 291 Automatic. ❡ But if your product requires large quantities of equalized lumber, i.e., stuff cut-off or trimmed to exact length, then this Hopper Fed Machine can and will make a big profit for you. ❡ Its capacity is practically unlimited. ❡ It will cut off both ends as fast as your operator can throw stock into the hopper by the armful and off-bearer clear it away. ❡ Does as much work as six old style trim saws. ❡ And every piece is **exactly** the length of all the others. ❡ If you equalize in quantities, use the enclosed post card to get Bulletin Q-28 and prices.

J. A. FAY & EGAN CO.

153-173 West Front St.   ·   ·   ·   ·   ·   Cincinnati, Ohio

—this new 230 page book
on Woodworking Machinery free!

Our new handy reference catalog is just off the press. Shows all of the newest models in woodworking machinery.

Every owner and user of woodworking machinery, every superintendent and foreman should have a copy of this new book, which is really a text on modern woodworking machinery.

A copy will be sent to you, free of charge upon receipt of your request.

Fill in and mail the coupon to-day.

J. A. Fay & Egan Co.,
153-173 W. Front St.,
Cincinnati, Ohio, U.S.A.

Please send your new Book on Woodworking Machinery free of charge to:

Mr. ....

Position .... (Give name of company you are with) ....

Street ....

City ....

Province ....

Are you acquainted with
the superior qualities of these four
British Columbia Woods

for Shop and Factory work?

If not, let us send you a sample and
descriptive booklet.

| Douglas Fir | Western Larch | Western Soft Pine | Western Red Cedar |

The best softwoods in the world for Shop and Factory are found in British Columbia

Write to
British Columbia Lumber Commissioner,
Excelsior Life Building, Toronto and Adelaide Streets,
Toronto, Ontario

for samples and booklets.

British Columbia Has a Wood for Every Purpose

"Treat your machine as a living friend."

What is the problem in Canada today which is the cause of most anxiety to all manufacturers?

"THE SCARCITY OF LABOR"

It will be worse before it is better and if you are up against this proposition at all
why not install one of our

No. 162 High-Speed Ball-Bearing Shapers

and do away with some of your worry?

Speed 7,000 Revolutions per minute

HERE IS WHAT WE CLAIM:—Double the amount of work can be done on this HIGH-SPEED BALL-BEARING SHAPER that can be done on an ordinary shaper, with much more ease to the operator. LET US PROVE IT TO YOU AS WE HAVE PROVED IT TO OTHERS.

DIMENSIONS, WEIGHTS and SPEEDS

Floor Space, 120 in. x 58 in. Net Weight, 2650 lbs. Shipping Weight, 2750 lbs. Tight and Loose Pulleys on Countershaft, 10 in. x 6 in. Driving Pulleys on Countershaft, 18 in. x 5 in. Idler Pulleys, 14 in. x 5 in. R.P.M. Countershaft, 1200. Pulley on Spindle, 3½ in. x 6 in. R.P.M. Spindle, 7000. Diameter Spindle, 2 3/16 in. Spindle Centres, 28 in. Table, 58 in. x 41 in. Diameter Spindle Top, 1¼ in. Horse Power Required, 3. Front Projection, 17 in. Height, 35 in.

REPEAT ORDERS BEST PROOF OF MERIT

The Chevrolet Motor Car Company, Limited, of Oshawa, bought three of these machines November 26, 1915. On March 22, 1916, they wrote us: "The three Shapers which we purchased from you several months ago are giving every satisfaction. How soon could you supply us with two more and what allowance could you make for two old shapers?" On March 28 they ordered these two additional shapers, making five in all. They further expressed their opinion in a letter dated April 5, 1916, in which they say: "We cannot speak too highly of the design and workmanship of the machines, or the results which we are able to get from them. They are, in our opinion, one hundred per cent. perfect." Since then we have sold them still another.

We specialize in Labor Saving Machines. If you are interested in these machines, write us.

The Preston Woodworking Machinery Co. Limited, Preston, Ont.

Alphabetical List of Advertisers

Acme Steel Goods Company   50
American Hardwood Lumber Co...   12
Atlantic Lumber Company   46
Ault & Wiborg Company   35

Baird Machinery Works   47
Barnaby, Chas. A.   41
B. C. Lumber Commissioner   8
Brown & Co., Geo. C.   12

Canada Machinery Corporation ...
Canadian Morehead Mfg. Company   49
Carborundum Company   55
Central Veneer Company   43
Chapman Double Ball Bearing Co.   51
Churchill, Milton Lumber Co.   12
Clark Veneer Co., Walter   41
Clark & Son, Edward   10

Darling Bros.   46
Dominion Mahogany & Veneer Co..   14

Fay & Egan Company, J. A.   7
Foster, Merriam & Company ...` ...   54

Garlock-Walker-Machinery Co. ...   3
Gillies Bros.   46

Gorham Bros. Co.   45
Gum Lumber Manufacturers' Assn.   4

Hart & McDonagh   46
Hartzell, Geo. W.   46
Haughton Veneer Company   43
Hay & Company   35
Hay Knife Company, Peter   50
Hoffman Bros. Company   41

Jackson Cochrane & Company ...   11

Kersley, Geo.   42
Klotz Manufacturing Company ...`   4
Kraetzer-Cured Lumber Co. ... ...

Long-Knight Lumber Co.   45

Mattison Machine Works   56
Mowbray & Robinson   12

Nartzik, J. J.   41
Nash, J. M.   51
National Dry Kiln Company   50
Neilson & Company, J. L.   50

Ober Manufacturing Company   51
Ohio Veneer Company   50

Paepcke Leicht Lumber Co. ...`. ..   14

Perfection Wood Steaming Retort
  Company ... ...
Penrod Walnut & Veneer Company   43
Perkins Glue Company   13
Perrin, William R.   50
Petrie, H. W....   47
Preston Woodworking Mach. Co...   9

Radcliff Saw Company   6
Reynolds Pattern & Machine Works   13
Roberts, John N.   42

Saranac Machine Company   55
Sadier & Haworth   45
Sheldons Limited   49
Simonds Canada Saw Co.   48
Spencer, C. A.   12

Toronto Blower Company   6
Toronto Veneer Co.   36

Walter & Company, B.   55
Wardwell Mfg. Company   49
Whitney & Son, Baxter D.   1
Williams Machinery Co., A. R.   2
Wood, Mosaic Company   42

Yates Machine Co.,.P. B. 5-47

Edward Clark & Sons

Toronto

CANADIAN HARDWOODS—winter cut Birch, Bass, Maple

10,000,000 ft. on hand, shipment June 1st
10,000,000 ft. sawing, shipment August 1st

ONTARIO and QUEBEC stocks, East or West, rail or water shipment, direct from mill to factory.

Put Up for Shell Boxes
3,000,000 ft. 4/4 Birch, No. 1 and 2 Common.
2,000,000 ft. 6/4 Birch, No. 1 and 2 Common.
1,000,000 ft. 12/4 Birch, No. 1 and 2 Common.

For Furniture Manufacturers
1,000,000 ft. each 10/4, 12/4, 16/4 Birch, 75 per cent. No. 1 and 2.

For Trim Trade
500,000 ft. 4/4 Birch, No. 1 and 2,   30 per cent. 14/16 ft., Red all in.
300,000 ft. 5/4 Birch, No. 1 and 2, { 9 in. average. Red all in.
700,000 ft. 6/4 Birch, No. 1 and 2, } 40 to 60 per cent. 14/16 ft. long.
500,000 ft. 8/4 Birch, No. 1 and 2;
200,000 ft. 4/4 to 8/4 Brown Ash, No. 1 and 2, 40 per cent. 14/16 ft.

We Commend to Your Notice:
Our lumber is reliable, our grades uniform, our shipments prompt, and receive personal supervision; our prices consistent. We have had long experience. We desire to serve.

The
Herzog Self-Feed Jointer
means

Increased Production	Simplicity of Operation
Small Floor Space	Safety to Employees

Does Four Times the Work of the Hand Jointer.

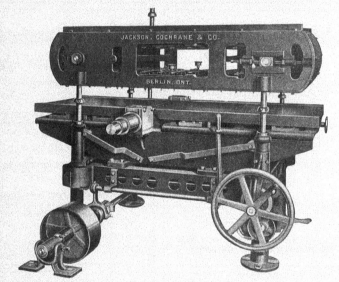

Our No. 34 Herzog Jointer, illustrated above, is one of the most efficient machines on the market to-day. It is appreciated by the manufacturer and employees alike, because, while it will produce from three to five times as much work as the hand jointer, it does not require skilled operators, but eliminates the danger so common to other makes. It can be operated by two boys. It will handle stock varying in width from 1 inch to the full width of the jointer, will feed fast or slow, takes only one-fourth the floor space of hand jointers, and requires only one-fifth of the sharpening of the knives. It is fitted with power feed raising and lowering attachment, with cylinder double belted and driven from both ends.

If interested in reducing your costs, write us.

Jackson, Cochrane & Company
BERLIN   -   CANADA

GEO. C. BROWN & COMPANY

Manufacturers St. Francis Basin Hardwoods

Band Mills—PROCTOR, ARK.   General Offices—Bank of Commerce Bldg.—MEMPHIS TENN.

STOCK FOR SALE—JULY AND AUGUST SHIPMENT

150 M' 4/4 No. 1 and No. 2 Com. Plain Red Oak.	200 M' 4/4 1st and 2nds Sap Gum (straight and flat).
150 M' 4/4 No. 1 and No. 2 Com. Plain White Oak.	50 M' 8/4 1st and 2nds Sap Gum (straight and flat).
70 M' 4/4 1st and 2nds. Qtd. White Oak (15 to 20 per cent. 10 in. and up).	150 M' 4/4 1st and 2nds Sel. Red Gum (straight and flat)
70 M' 4/4 No. 1 and No. 2 Com. Qtd. White Oak.	50 M' 8/4 No. 1 Com. and Bet. Sel. Red Gum (quarter-sawed).
20 M' 5/4 No. 1 and No. 2 Com. Qtd. Red Oak (6 in. and up).	60 M' 10/4 No. 1 Com. and Bet. Sel. Red Gum (quarter-sawed).

"Our Gum is all Kraetzer-Cured"—Splendid Condition.

TENNESSEE AROMATIC RED CEDAR SPECIALISTS   OUR TRAFFIC DEPARTMENT TRACES EVERY SHIPMENT

Churchill Milton Lumber Co.

Louisville, Ky.

We carry at
NEW ALBANY, IND.

Genuine Indiana Plain and Quartered Oak, Poplar

At GLENDORA, MISS.

Cypress, Gum, Cottonwood, Oak

SEND US YOUR INQUIRIES

We have dry stock on hand and can make quick shipment

Plain and Quartered Red and White Oak

and Other Hardwoods

Even Color   Soft Texture

We have **35,000,000 feet dry stock** all of our own manufacture, from our own timber grown in Eastern Kentucky

MADE (MR) RIGHT

OAK FLOORING

PROMPT SHIPMENTS

The Mowbray & Robinson Co., Inc.

GENERAL OFFICES
CINCINNATI, OHIO

Mills: Quicksand, Ky., West Irvine, Ky., Viper, Ky.

Canadian Representative:
N. H. FARNHAM

601 Elmwood Ave., BUFFALO, N. Y.

American Hardwood Lumber Co.

St. Louis, Mo.

Large stock of—

Dry Ash, Quartered Oak Plain Oak and Gum

Shipments from — NASHVILLE, Tenn., NEW ORLEANS, La., and BENTON, Ark.

Dry Spruce and Birch

Good Stocks, Prompt Shipments, Satisfaction

C. A. SPENCER, Limited

Wholesale Dealers in Rough and Dressed Lumber

Offices—500 McGill Building
MONTREAL   - -   Quebec

The Only Way to Drive Screws

There is a stage-coach method and a modern method of doing everything. When it comes to driving screws the only modern way is with an Automatic Screw Driving Machine, which means greater and better production and greater profit.

The Reynolds
Automatic Screw-Driving Machine

is guaranteed to produce results. It is no experiment. The No. 2 Standard Type, shown herewith, is particularly suited for placing the screws in the straps of shrapnel boxes. It may be fitted to drive, without change or adjustment, any three consecutive sizes flat or round head, or two sizes filister head, No. 6 to 14 inclusive, wood or machine screws, or three sizes flat or round head, or two sizes filister head machine screws up to No. 20. All the above up to 1½ inches long if smaller than No. 10, or up to 1¾ inches long, No. 10 or larger. Fitted with boring attachment if desired.

Many Machines in Use in Canada.

Ask for our Catalogue.

REYNOLDS PATTERN & MACHINE CO.
101 and 103 Third Ave., Moline, Ill., U. S. A.

No. 2 Standard Type Automatic
Screw-driving Machine.

Are Your Products
Perkins Glued?

Ladies' Desk manufactured
by The Berlin Furniture
Co., Berlin, Ont., in which
Perkins Glue was used.

The live dealer realizes the value of selling his customers Perkins Glued products because they never come apart. Many of Canada's largest furniture factories use Perkins Vegetable Glue. It costs 20 per cent. less than hide glue, is applied cold, antiquating the obnoxious and wasteful cooking process, is absolutely uniform in quality and will not blister in sanding.

We will be glad to demonstrate in your own factory and on your own stock what can be accomplished with Perkins Vegetable Glue.

Write us today for more detailed information.

Perkins Glue Co.
Hamilton, Ont.   Limited

Address all inquiries to Sales Office, Perkins Glue Company,
South Bend, Indiana, U. S. A.

Of course it is true that

RED GUM

is America's finest cabinet wood—but

Just as a poor cook will spoil the choicest viands while the experienced chef will turn them into prized delicacies, so it is true that

The inherently superior qualities of Red Gum can be brought out only by proper handling.

When you buy this wood, as when you buy a new machine, you want to feel that you have reason for believing it will be just as represented.

We claim genuine superiority for our Gum. The proof that you can have confidence in this claim is shown by the letter reproduced herewith.

Your interests demand that you remember this proof of our ability to preserve the wonderful qualities of the wood when you again want RED GUM.

Paepcke Leicht Lumber Co.

Conway Building   111 W. Washington St.

CHICAGO, ILL.

Band Mills: Helena and Blytheville Ark.; Greenville, Miss.

Buy Made in Canada Veneers

Dominion Mahogany
& Veneer Co., Limited

Head Office and Factory
Montreal West, P.Q.

Toronto Office and Warehouse
455 King St. West

Mahogany, Walnut,
and all Fancy Woods—Lumber

SAWN, SLICED and ROTARY CUT VENEERS

Mahogany, Walnut, Quartered and Plain Oak, Ash, Maple, Birch, Elm, Poplar, Gum.

Gum and Poplar Cross Banding.   Hard Maple Pin Block Stock.

SEND US YOUR ENQUIRIES.

Canadian Woodworker

A Monthly Publication in the Interest of the Woodworking Industry.
Reaches the factories producing interior finish, doors, sash, flooring,
woodenware, furniture, pianos, boxes, and general mill products.

Subscription, $1.00 a year; foreign $1.50.

Woodworker Publishing Company, Limited

Branches: Montreal, Vancouver,
Chicago, New York, London, Eng.

345 Adelaide St. West, Toronto
Phone Ade. 2700

Authorized by the Postmaster General for Canada, for transmission as second class matter.
Entered as second class matter July 18th, 1914, at the Post Office at Buffalo, N.Y., under the Act of Congress of March 3, 1879.

Vol. 16 — August, 1916 — No. 8

Making British Columbia Woods Better Known

The time appears to be opportune to say a few words regarding the publicity campaign which has been organized and is being successfully carried out by the British Columbia Government for the purpose of making known to the people of Eastern Canada the many desirable qualities of the woods grown in that province.

The campaign was opened some months ago, when there was very little activity in the woodworking industries. but the lack of activity seems to have been more of a help than a detriment. This can be understood when it is remembered that the mission of the British Columbia Government is not to sell British Columbia lumber, but simply to point out its good qualities and the many uses to which the different kinds can be put.

A man's mind is in a much more receptive mood when he has time to think than when he is rushed with business, and consequently the literature (advertisements and booklets) circulated by the British Columbia Government has been read and absorbed.

There is also a great deal to be said about the businesslike manner in which the publicity work is being carried out. For instance, in Toronto a large office is maintained, where there is a splendid exhibit of the wood products of British Columbia, where all kinds of samples can be seen and information secured. The office and exhibit is in charge of Mr. L. B. Beale, a very capable and enthusiastic exponent of the merits of British Columbia woods. Mr. Beale does not confine his efforts entirely to the office, but goes out and gets into personal touch with the users of lumber, taking with him a number of samples and a great deal of information. The field which he has to cover is a large one, however, and it may be interesting to the proprietors of out-of-town woodworking factories to know that they can meet Mr. Beale and see a beautiful exhibit of British Columbia woods (detailed mention of which appears elsewhere in this issue) at the Canadian National Exhibition, which opens at Toronto on August 26.

There is a limit, though, to the work that can be done by the British Columbia Government. They can introduce the product to the user and point out its merits, but after that it is up to the lumber manufacturers of British Columbia to demonstrate that they are entitled to supply the lumber. This they will have to do by means of salesmanship and giving "service." The giving of "service" may mean having stocks of lumber located at different points in order that factories may be supplied promptly. Of course, problems such as these would have to be worked out, and this could be best accomplished by the lumbermen getting together and holding a general discussion. One thing is certain: they will have to follow up the pioneer work being done by their Government or that work will have been in vain, and there is no time like the present, when a great many planing mill men and proprietors of woodworking factories have been favorably impressed with samples of British Columbia woods that have been shown to them.

Exhibits of Interest to Woodworkers

By the time this issue of the "Canadian Woodworker" is off the press the opening day of the Canadian National Exhibition will be close at hand, and on account of the large number of articles that are being manufactured in Canada that were formerly imported from foreign countries (and it is really surprising how numerous the imported articles were when one commences to investigate) there will be many exhibits of particular interest to woodworkers.

The Dominion Government have arranged for the manufacturers of toys and novelties to exhibit their products in a portion of the Government Building, and as no charge is being made for the space, it is hardly possible that any manufacturer will fail to avail himself of the opportunity of showing his goods. The manufacturers of wooden toys and novelties have all been asked to exhibit their goods, and we hope to see such a well-chosen assortment that Canadian people will in future demand them in preference to all others. We are of the opinion that quite a number of articles made of wood that were formerly imported in large quantities can be manufactured in Canada by Canadian manufacturers.

Occasionally we hear about some enterprising woodworker that has gone after orders (which in a great many cases wholesale houses, department stores, etc., were only too willing to place), and applied himself to the making of some particular line, with the result that he is now in a fair way to become independent. Just as an instance, the editor was in conversation a few days ago with a man who, previous to the war, conducted a builders' supply business, in connection with which he had a woodworking shop. This man, when he found his trade in building supplies had dwindled down past the point where there was any profit, immediately set about to find a new line. The line he chose was cheap rulers—the ones that school children buy for one cent and lose or break the next day—and yardsticks for advertising purposes, etc. He said there was approximately $100,000 worth of business per year in Canada in this one line alone that could be secured if the prices were right. By concentrating his mechanical ability on the one line he has been able to get his prices right. This party also said that since then he has had other lines offered to him, but he will not touch them because he already has all he can handle.

One of the lines that he was asked to make—by a buyer of a large departmental store — was wooden forms for drying stockings on. These forms are simply flat pieces of board about one-quarter of an inch thick, band sawn to the shape of a stocking. The edges are rounded on the shaper and small holes bored around the outside, about three-eighths of an inch from the edge. They are, of course, made in different sizes, and we have been told that thousands of them are used by the knitting factories. They are also sold retail in some of the dry goods stores for use in the home. Yet our informant assures us that they are not manufactured in Canada, and the firm that wanted them had to get them from the States. There should be no difficulty in making them. All the finishing they require is sanding smooth so that they will not tear the stockings.

We believe that if some of our manufacturers who find business dull would "nose around" a little outside of the "beaten track" in woodwork they would discover some very attractive line for which their plants could be utilized.

Any of our readers contemplating taking up a new line are invited to send us enquiries. We may be able to tell you, without loss of time, whether the amount of business to be had makes the line worth considering. We are also glad to offer our services in putting you in touch with firms that are in a position to furnish the most up-to-date machinery for manufacturing any class of woodwork.

B. C. Woods at Toronto Exhibition

In view of the growing hold of British Columbia woods on the important market afforded by Eastern Canada, the exhibit to be placed on view this year will be on a more ambitious scale. Space has been secured in the Government Building, and in front of this space will be erected a handsome facade, composed of pillars of Douglas fir in the rough, behind which the smaller exhibits will be arranged. The walls will bear panels of the different commercial woods of British Columbia, together with photographic enlargements showing forest stands, operations of the logging and sawmill industries, etc. The walls will be divided into six sections by means of pilasters of various woods, and to each of these will be attached, in such a manner that they can be readily swung, doors in various styles and finishes. Exhibits of articles made from Douglas fir, Western soft pine, Western hemlock, Western larch, Western red cedar, Western white pine, and Western spruce will be shown, with samples of flooring and veneer work.

Interior trimmings in the form of specimens of mouldings, casing, bases, etc., will be included, and among other features, silo stock, oars, and paddles, a fully equipped boat, boxes and fruit packages, and samples of spruce showing the qualities of that wood for aeroplane construction. The whole will be in charge of the British Columbia Lumber Commissioner in Eastern Canada, Mr. L. B. Beale, whose headquarters are at the Excelsior Life Building, Toronto.

Visit Us When at the Exhibition

To the many woodworkers and furniture manufacturers who intend visiting the Toronto Exhibition from out-of-town points we extend a cordial invitation to visit the office of the Canadian Woodworker.

It may be that you will have occasion to dictate some letters or use a telephone, or you might wish to do some writing away from the noise and bustle of your hotel. All of these facilities are at your disposal at our office, and we shall be glad to have you make use of them. Again, you may wish to discuss some phase of the woodworking business with us, or there may be some information you want that we can secure for you. In any case, we will be glad to see you, so do not forget the address—345 Adelaide Street West.

Why a Saw Cuts

Of course we all know that the cutting action of the saw is due to the shape of the teeth. But it has remained for Henry Disston Company, Philadelphia, Pa., to put all of the information, showing exactly how the teeth enter the wood and tear it apart, in a most interesting booklet, which they are prepared to send to our readers on request. The various kinds of saws are shown, with their relatively shaped teeth, many of these drawings greatly enlarged to show to better advantage.

The Practical Manufacture of Wooden Toys

Factory Should be Located Where Cuttings can be Obtained—Beginners Advised to go Slow When Purchasing Equipment.

By "Experience"

The fact that German toys can no longer be brought into this country has made the market for wooden toys and novelties a large one, and the manufacture of them an attractive proposition, provided it is gone into in the right way. The mere fact of it being attractive, however, is liable to induce men into it that will give the industry a "black eye" right at the beginning. There are a number of ways in which this might be done. For instance, a number of men might start in without having had any previous experience in manufacturing and fail through not being able to achieve the utmost in economy.

The result of this would be that other men who were contemplating going into the business would be "scared off." This would be "black eye" number one. Then, again, there is the mechanic who is anxious to "earn money for himself instead of a boss." There are, of course, a great many of these who have gone into business for themselves and made it pay. But, on the other hand, there are many more who do not understand even the rudiments of conducting a business. They do not know how to buy their material to advantage, and they forget to include all the necessary items in their "overhead expense." There is also the man who, to make money, will turn out an inferior article, and this is probably the worst "black eye" that the industry could possibly get. For, after all, the buying public are only human, and they are not going to purchase an inferior article, even if it is "made in Canada," if they can get a better foreign-made article for the same price.

A man contemplating going into the manufacture of wooden toys and novelties should, in the first place, select the class of toy he intends to manufacture. He should endeavor to keep his line as small as possible. In fact, the very best way to make money on toys is to pick out one line, and stick to it. You will always be discovering "short cuts" and thereby saving money, whereas the man who equips a factory with the ordinary run of woodworking machines and then proceeds to make some "nifty" samples and solicit orders is going to meet with some difficulties when he gets the orders and starts to fill them. Also the man who starts to make building blocks and wheel goods, such as small wagons, doll carriages, etc., in one factory will very soon find out that he cannot manufacture them economically. The same thing will apply in making small wheel goods and large wheel goods in one factory, and so on down the list.

Men when playing a game of checkers will generally do some figuring before they "move," but the same men will, in many cases, start and manufacture articles without any consideration as to whether they will work in together or not. If they will only remember that business is the biggest and most real game in life, and that there are more competitors in it than in any other game, they will get down to "brass tacks," and figure out every move and every item of cost before they start.

To make a success of the toy business the factory must be located where cuttings can be bought from other manufacturers, because to buy whole lumber at the price it is selling at to-day is out of the question.

There has been a great number of toy samples exhibited that are out of the range of manufacture. That is, they require so much "fiddling" work that in order to make a profit on them they would have to be sold at a price that would make them prohibitive to the buyer. The class of toy that seems to have the greatest sale and are the best educators are building blocks or construction toys. There are quite a number already on the market—some secured by patents—but the field is large, and new designs are constantly coming out. There is still plenty of room for new models, and toys along this line will always be good sellers. They are best for manufacturers because they quickly find their way to the wood-box, and if we would take a leaf out of the German manufacturers' book we would study this point most particularly.

Another important point, and one that cannot be emphasized too strongly, is if the man who contemplates manufacturing toys is not a practical mechanic he should secure the services of a practical mechanic. One with a financial interest in the business is best, because where the pocket is affected the eye is apt to be very keen in finding out "leaks" and the brain very alert in figuring out how the leaks can be avoided. It is one of the most pitiful things in business to see a man trying to run a business when he does not know anything about it. In a number of cases he can do nothing but sit still and lose money hand over fist, while, by getting a practical man to take charge, the business could be put on a paying basis.

The margin of profit in the toy and novelty business is so small that every detail of manufacture must be worked out. The best methods of cutting up the stock must be decided on. The machines must be arranged so as to save as much time and labor as possible in handling the stock from one machine to another. All adjustments of knives, etc., must be accurately made, because in the manufacture of this class of goods bench work must be eliminated. The labor to be used is also worthy of some consideration by the manufacturer. It is a mistake to suppose that any class of picked-up labor will do. Boys and girls should be used wherever they can do the work, but where machine hands are required it is very often better to hire a "green" man than an experienced machine hand. You may have a reason for wanting a certain operation on a machine performed in a certain way. But when you attempt to tell the experienced machine hand how to do it he is apt to regard it as a reflection on his intelligence and get offended. On the other hand, a "green" man will generally adapt himself to do the work the way you want him to do it. When your men have been "broken in" to do certain parts of the work it is advisable to keep them, if possible, because it takes time to "break in" others.

The writer would strongly advise anyone contemplating the manufacture of toys and novelties to go very slow in the matter of purchasing equipment. There is no use loading up a plant with expensive machinery until you find out if you really need it. The majority of the successful toy and novelty makers have been able, by a study of their needs and application of their mechanical ability, to work out machines to do their own particular work and cut their costs very low.

The French Period Styles—Empire

The Continuation of a Series of Illustrated Articles Dealing with Developments in Furniture Design During this Important Era

By James Thomson

The "Empire," a development of the reign of Napoleon the Great, was the last of the French historic styles that have stood the test of time. Styles come, styles go, only those that are really worth while persisting. When (the Louis Nineteenth meanwhile having been temporarily thrown on the scrap heap by the Revolution), after the passing of the Reign of Terror, Napoleon began to rule, there was developed a style in furnishing that has been handed along to us under the name of "Empire," or, to give it its full title, "the style of the first empire."

In December, 1799, Napoleon was made consul, or,

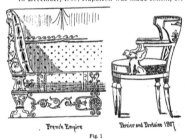

French Empire   Percier and Fontaine 1807

Fig. 1

to speak more definitely, first consul; in August, 1802, he was made consul for life, and by May, 1804, had persuaded a deluded people to make him emperor, though so shortly before they had waded in blood in order to do away with kings, emperors, and all such rulers by divine right. Well; here were they of their own volition forging the chains once more and giving indications of immensely enjoying the unflattering distinction.

Having thrown off the republican mask, Napoleon must needs have imperial trappings to accord with assumption of divine right to rule. In obedience to imperial desire, the artist and artizan talent of the time was drawn upon to supply the need. And as France had been for long a leader in matters of taste, there was small difficulty in getting talent of highest order to meet requirements. The talented designers, decorators, and painters of first rank, who had under the direction of Marie Antoinette, developed the beautiful, refined and satisfying Louis Sixteenth style, were at the service of the new master.

Among the chief developers of the style were the architects, Percier and Fontaine, who have left evidences of their skill as furniture designers in a book published in 1807. The designs are very beautiful, but unless greatly modified, almost useless for practical purposes. A cast-iron inflexibility, as a rule, pervades the product. Heavily conceived arm chairs appear as if cast from a mould, the upholstery bulging out in likeness to rubber filled with air. One is driven to the conclusion that only as accessories to imperial pretentions would the articles deliniated (when made up) be

of real use. As a matter of fact, it is quite unlikely that many of the designs in the book were ever put to practical purpose.

There were other designers, however, whose product (under modification) has accorded with modern needs. The style as exemplified by best examples is dependent for effect on beautifully figured branch mahogany, the ornaments being principally of brass, the quality of which has always been accorded high praise. The design of the ormolu decoration is chaste and refined, the casting of unexampled workmanship.

The Napoleonic conquests had an influence in turning the minds of Frenchmen to the conquests of antiquity. Closest study was made of ancient architecture, monuments, metal and stone work, vases and amphora, particularly as regards the decoration thereof. Robert Adam had half a century before given to the world the results of four years' study of the ruins of Diocletian's villa at Spalatro. The classic taste that had dominated England was soon to be equally dominant in France. But while both based on classic formula, in reality the Adam style and that of the Empire radically differed. In the one was neatness, refinement, lightness of effect, in the other, heaviness, in some instances, absolute clumsiness, an aspect of stone and refinement only as to decoration. As a matter of fact, Empire furniture's sole appeal was through the beautifully figured mahogany of which made, sweetness of line in carved ornamentation, and the jewelry-like ormolu mounts. Omit the brasses and in many an instance the charm goes.

Though the English were so bitter against Napoleon, they were thereby not deterred from adopting a style in furnishing to which he had given impetus. England's designers, however, with perhaps one or two exceptions, made a sad mess of it, one of the worst being the renowned Sheraton. The possible exceptions were Thomas Hope and George Smith, the former publishing an "Empire" book in 1807, the latter a similar work

Ormolu Mounts

French Empire

Fig. 2

in 1808. The chairs shown in Fig. 4 are characteristic examples of English handling of the style, and in my opinion may be attributed to Smith. Hope, having no practical knowledge of the trade, is not quite—from a practical point of view—so happy, though his designs in a pictorial way, are most pleasing.

Soon did the Empire fashion take possession of

French Empire.
Fig. 3

Chairs in the English Empire Style
Originals in London Museum
Fig. 4

American Empire
Fig. 5

American Empire
Fig. 6

American Empire
Fig. 7

American Empire.
Fig. 8

New England, where the mother country was patterned, rather than the French. In the American emanations, carving on casework was rare, the grain of the crotch veneer being considered embellishment sufficient. Brasses also were tabooed, in some cases the handles being of glass. Heaviness, extending even to absolute clumsiness, characterized much of the American product. French work, though heavy, was relieved by the beautiful brass mounts. British Empire furniture was often elaborately carved in Pompeian manner (see Fig. 4); the American had no carving or other decoration to lighten the effect. Bald and clumsy, therefore, many American examples were, as

American Empire.

Fig. 9

is exemplified in the chest of drawers shown in Fig. 7. As a rule, the scroll carving on the console supports was absent.

It was a sad day artistically for England and America when the mad Empire venture was made. It marked the beginning of a decadence from which there was no emergence until three-quarters of the century had gone. With the advent of the sham classic came the practice of using mahogany branch veneer to the exclusion of all kinds else. Veneering across the grain upon a pine base became the common practice in using smallest bits, lots of the work being done with the hammer. Many supposedly solid mahogany pieces, valued beyond price because of the possession of that quality, are to-day in the possession of New Englanders of old stock. Enlightenment comes when they try to sell to a person conversant with the facts.

The designs submitted speak for themselves, but a few explanatory remarks may be in order. Extreme French Empire examples have been avoided as tending to give a wrong impression. The sofa and chair, as representative of the work of Percier and Fontaine, and depicted in Fig. 1, are more adapted for present day purposes. The French Empire sofa was made high to the top of upholstery because the custom of using a footstool was commonly followed by the ladies.

In Fig. 4 we have the English Empire. It differs from the French in the quality of the carving, wherein Pompeian motives, very heavily treated, are employed. So late as the sixties of the last century the writer, as an apprentice in the old country, worked on furniture in which carving in this identical manner was to be found.

In Figs. 5 and 6 we have the American Empire at its best. Patterns such as this are well worthy of perpetuation. As a matter of fact, I to-day find replicas of the bureau shown, in the American market.

In Fig. 7 is shown a type of American Empire, which, though common enough, is not to be commended. Chipping of the veneer from these ogee surfaces was a common occurrence. No special fault can be found with the design as here submitted. The trouble came when people of no taste and less knowledge, in the interest of cheapness, altered the design. Some desperately bad examples of such botching are to-day to be found in New England.

Chairs shown in Fig. 8, and the sofa in Fig. 9, are characteristic American examples. People with a craze for antiques are scouring the country for sofas of this character, though, as a matter of fact, they represent the early Victorian period. The carving on some of them looks as if done with a meat axe, yet the people who buy think it immense.

The little drop-leaf table is a good sort. This is a design being reproduced to-day. One in my possession, made to order about the year 1830, has a top of solid San Domingo mahogany that by its weight astonishes people who essay to lift the table.

Modified French Empire furniture can be of great charm, as is here exemplified in the few samples the space at command admits of showing.

The Empire is the last of the great historic styles that came from the French. Much as they have tried, they have not in the last hundred years been able to invent another of a persisting kind. Modern attempts at originality always ends in the "freak," which nobody will have.

Clever Method of Framing Dormer Window

The accompanying sketch illustrates a clever method of framing a dormer window. The piece, A, which forms the lookouts is spiked to the plates B, and the short uprights, C, are spiked in place directly above the studding G, so as to form a straight wall line for the plaster. Then the piece, A, is sawn off as indicated by the lines D, to allow for head room.

Framing a dormer window.

The pieces, E, C, F, form a cantilever and cannot be beaten for strength. There is no waste of material as the piece that is sawn out is used up above for a collar tie and forms the ceiling.

Canada's Foreign Trade in Wood Goods

Canada's imports of wood and manufactures of wood for the twelve months ending April, 1916, were valued at $8,696,065, as against $13,465,189 for the preceding twelve months. The 1916 imports were divided as follows: From the United States, $8,294,666; from the United Kingdom, $151,531.

Cost Accounting in the Jobbing Planing Mill

Ascertaining Cost of Operating Individual Machines for Each Job—Distribution of Expense—Records of Finished Stock Kept for Future Reference

By J. D. Clark

The operator of a planing mill, dimension mill, or other woodworking plant where "jobs" rather than regular work are handled, naturally finds the cost-keeping problem more difficult than is confronted in the furniture factory which is turning out the same goods right along. Of course, the problems of the furniture man are not simple, because the number of items produced is great enough to suggest the necessity of discriminating in figuring costs, and it is needful to know just how much expense is attached to the production of each item.

On the other hand, the planing mill which gets work by bidding on it in competition with others must find out to the last cent how much the production of the job involves, for the reason that it is only logical to learn the results of each operation, if for no other reason than to determine whether the original estimate was correct, and as a guide against possible repetition of errors. So also with the dimension factory which is cutting lumber to size for use in some other woodworking plant. If it has a large cutting order there is a good chance of overlooking certain items and of assuming that a price which is really insufficient to pay all the charges will cover the requirements.

The dimension proposition is newer than the planing mill; and, in view of the rapid growth of this kind of business, it may be of interest to other woodworkers to explain how a model system of this kind is operated. The plan is not put forward as ideal, but is certainly a big improvement over the rough-and-tumble, hit-or-miss methods which often prevail, especially when the dimension business is operated in connection with a sawmill, as is usually the case.

In the first place, the correct basis of the proposition is obtained by figuring the cost of the lumber at the price which it had on the market for it. In a lot of cases where low-grade lumber (which is not easily sold) is used in dimension manufacturing, the tendency is to disregard the value of the material. Obviously this is not a correct method of procedure, since the object should be to sell the product at a profit, and a price which does not take account of the original intrinsic value of the raw material is not properly ascertained.

If the material is kiln dried, as is usually the case when dimension stock is manufactured to order, the cost of drying is included, this averaging about $3 per 1,000 feet. That is the usual custom charge; and, while an individual manufacturer operating his own kilns should be able to do the work for something less, it will not miss this figure much by the time the cost of handling, which is a big item in kiln-drying operations, is taken into account.

The system referred to is handled by means of a loose-leaf ledger, each job being figured on its own page. The idea is to enable the manufacturer to go over the whole proposition and see just how much he has made or lost on each job. This is a big advantage, compared with the ordinary system of grouping expenses without reference to their application to various jobs.

The first item on the page is rough lumber. The date and quantity of each kind and grade of lumber delivered for dimension manufacturing are shown, together with prices, enabling the value of the material to be calculated accurately. As suggested, the cost of kiln-drying is included, this being added to the market price of the stock, without the K. D. feature being taken into account.

The labor costs involved in the work are then figured in detail, columns being provided for "Cross-cutting labor," "Ripping labor," "Resaw labor," "Inspecting labor," and "Bundling labor." This, it should be noted, applies to labor expense only, a separate calculation being made to cover machine expense. The labor costs are determined by the use of time-cards for each employe, whose records show just how much time was put in on each job, and therefore what proportion of his wages is chargeable to that job. The jobs, it should be added, are handled in the dimension plant by number only, so that none of the hands is advised regarding the destination of the material.

The labor expense is handled more easily in this particular plant by reason of the fact that the machine hands and inspectors are paid by the piece, so there is no difficulty about figuring the actual cost which is to be included on each job. The inspection feature is, of course, important in this kind of work, as one of the troubles which many plants have in making good with customers is due to the fact that a lot of material is shipped which is not up to specifications. This means that, in order to get results, the inspection work must be properly taken care of. Inspection in this case does not mean technical inspection for the purpose of determining grade, but is used to learn whether the material is clear and whether the general character and appearance of the stock is up to the specifications. The plan of paying so much per thousand pieces for inspection suggests a possible tendency to speed at the expense of care, but, of course, any complaints regarding the stock are immediately put up to the inspector, who for this reason is bound to be careful in going through the stock before it is bundled and shipped.

Possibly the most interesting feature of this cost sheet is the figuring of machinery expense. In many plants no attempt is made to analyze this, power and machinery expense being grouped and applied against all of the work done by means of some general or overhead charge. But there is as much individuality in machines as there is in men, and the plan used in the plant now under consideration, whereby the cost of operating each machine for each job is carefully calculated, seems to be by far the better system.

In the first place, the rate is determined by figuring the total capacity of the machine per year—that is, the normal number of hours it may be expected to operate. This is based on actual tests, and takes into account loss of time for repairs, adjustments, etc. Then the value of the machine is considered, and the expense represented by interest, depreciation, and other charges is figured. Tests are also made to determine the power consumption; and, if electric motors are used, the current consumed can be actually measured and definite figure secured for this. If steam power is used, the power being transmitted from the engine by mechanical means, the distribution of the expense would, of course, be harder to figure, and might have to be applied on a pro rata basis.

The point to be kept in mind, however, is that the object of cost accounting should be to get away from the use of general charges as much as possible, and to substitute individual items, because it is in the calculation of those general figures that errors are introduced, due to the actual variations in the capacity, efficiency, etc., of various machines. Putting each machine, each man, and each job on a basis determined by the facts in each case is the only logical plan.

In the plant referred to the rate per hour for each machine is calculated on the basis suggested, so that by having a record of the number of hours each machine works on each job the cost of the machine time applicable to it may be entered with the assurance that it will be very close to the correct data.

As pointed out above, the rate for machine time is based on the assumption that the time used will be normal. If the plant happens to run overtime, this, of course, reduces the cost to the manufacturer, but the normal rate is used, as it would be unfair to allow a customer whose work was done at night a lower rate than that applied to the manufacture of goods in regular hours. Such an unearned increment as this, it is also worth noting, might serve to offset unexpected losses of time due to breakdowns of more than the usual importance.

The record for the finished stock is analyzed with reference to the number of pieces produced, the amount of lumber used, the waste, and hence the net results in thousands of feet. The sales price is indicated on this calculation, with the sales value in terms of feet.

In this particular mill an effort is made to dispose of the waste and offal of the plant, and a careful record is also kept in this respect. It shows the character and quantity of all the material, with the inventory value, this being credited to the job against which it is figured. These figures are unusually valuable, because they serve not only as an inventory of the waste (which otherwise would be hard to secure), but also provide a basis for determining the average waste to be allowed for in the manufacture of the finished product.

The whole job is recapitulated on the same sheet. The gross sales value of the finished stock is shown, together with the allowances made for freight and other items. Space is provided to enable the freight to be figured, as many of the sales made by this concern are on a basis of delivered prices, and it is necessary that an allowance be shown in order to enable the net results of the job to be determined. The net sales value, after making the deductions indicated, is then entered.

The next item, determined from the figures already entered on the sheet, is cost. This includes lumber, which is analyzed as follows: Total cost of lumber sent to sawyers, less rough lumber laid out or left over; gross lumber actually used, less shorts and reusable offal produced. The result of this calculation is the net cost of the lumber. The manufacturing expense is then figured, the items here being as follows: Cross-cutting labor, cross-cutting machine cost and burden; ripping labor, ripping machine cost and burden; resawing labor, resawing machine cost and burden; inspecting labor; bundling labor; miscellaneous labor; loading labor; hauling labor and expense. The total thus secured gives the total manufacturing and shipping cost, including the cost of the lumber shown above.

The total cost of the product shipped, which means with the freight allowance included, is then secured by adding the lumber and manufacturing cost, and the result is compared with the net sales value realized (already calculated) to determine the gross profit or loss on the order. This is gross for the reason that the general selling expenses, which are figured in percentage on the total sales, must be deducted. After this is figured the result is the net profit or loss on the job.

One other calculation which is made—and one that would be interesting if available to every woodworker —is to show the correct price based on these cost figures. The total cost shipped is entered, plus selling expense, plus cash discounts, plus freight allowance, the result being the correct delivered price. That enables the sheet to be referred to from time to time when repeat orders are secured, to insure the right quotations being made on the job.—The Woodworker (Indianapolis).

Manufacturing Shoe Lasts in a Montreal Plant

The shoe last as we have it to-day has been a gradual development. It is now possible to turn out with machinery lasts from models which heretofore required much work by hand. Thousands can now be made in the time it took to make hundreds. A recent visit to the last factory of Robin Brothers, Carriere Street, Montreal, verifies the latter statement. The rough blocks of maple 13 inches long and smaller (the length varies according to size of last required), about like a large shoe carton, come in from the Canadian Last Block Company, Iberville, P.Q., of which Mr. Robin is president. At Iberville the maple logs are cut into lengths as above indicated, and are then cut, pie like, into a number of pieces, the number depending upon the circumference of the log. The machine used in cutting these blocks was invented by the Canadian Last Block Company. The blocks are neat "roughed," making them slightly cylindrical, but smaller at one end. They are then stored and allowed to season, or air dry, for at least six months before shipping to the last factory.

On reaching the factory the rough blocks are either stored or placed in dry kilns. They remain in these kilns at least six weeks, losing about one-third of their weight. The roughed turned kilns in the factory are of Mr. Robin's own design, the result of his many years' experience and observation. The proper heat is automatically maintained by mechanical devices, thus preventing checking. From the dry kilns the blocks, still in the rough shape, are sorted according to size and placed in large bins ready to be drawn upon to fill orders as required. They have then to go to the lathe, to be finally shaped to the model desired.

The model and pattern room is the centre of this business. New style lasts are continually being obtained from the shoe style centers, and new models are also designed in this factory. When a manufacturer requiring new lasts visits the model room he may find what he wants or he may ask for alterations. He may want a lower or a higher toe, or something changed in some other place. A model is cut down, built or patched up, all by hand of course, until the exact shape he wants is obtained. This then becomes the model from which his lasts for that particular style are made. From this model a sole pattern is made, and then from this original one a set is graded for every size of the lasts to be made, and all the lasts are turned according to these patterns.

The model is then placed in one side of the turning lathe, with a rough block, which has been kiln dried, placed on the other end of the lathe. As the lathe turns, a guide wheel follows accurately every curve of the model, and the machine is so adjusted that the cutter heads, or knives, are cutting out of the rough block an exact duplicate of the model. The machine needs no attention other than taking off the last and putting in another rough block. The machine can be adjusted like a pantograph, so that several sizes can be made from the same model. From the turning lathe the lasts are passed on to other operators. In all there are some 90 operations before some of the lasts are ready for the manufacturer.

Last making may be divided into five departments —model making, ironing (placing a metal bottom or a heel), shaving (i.e., shaping heels and toes by hand or with the help of special machinery), hingeing, and scouring or finishing. The hinge is to allow the last to be withdrawn without breaking the shoe, and the Robin hinge is Mr. Robin's own design, for which he claims strength, durability, simplicity, and economy. The big demand for these special hinged lasts is certainly a testimonial.

All the machinery in the Robin factory is electrically driven. A boiler in the basement, fed with the wood refuse, supplies heat for the building and the kilns. All by-products are used by the firm or sold.

Wooden Spokes for Wagons and Automobiles

There are to-day about four divisions of the spoke business—two divisions as to wagon spokes proper, one division of hickory spokes for buggies, carriages, and similar vehicles, and one division of automobile spokes.

In the wagon spoke business, which until the coming of the automobile was the big thing in the spoke trade, there are two general divisions. One is in the manufacture of finished spokes for the blacksmith shops and smaller wagon makers throughout the country and for the iron stores which supply the blacksmith shops. The other division is that of making club-turned spokes for the larger wagon factories which prefer to finish up their own spokes—that is, to give them their final shape as to throat and contour and to do the tenoning.

Every hardwood man who becomes interested in the subject of spokes should make a study of these divisions before entering into the matter, and decide upon what branch or division he is in best shape to serve or how many of them, because they call for differences in equipment and also in the manner of handling and distribution.

The equipping of a complete spoke factory for turning finished spokes such as are used in blacksmith shops and by the small wagon maker, and are purchased and distributed by the iron stores, calls for quite a string of machines and also for skilled operators to run them. It calls for enough equipment to make one hesitate about going into it unless there is a somewhat permanent location at some assembling point, and that the length of time the business may be operated will justify the installation.

The equipment for making finished wagon spokes may also be used for making buggy and carriage spokes, with some changes and different sets of patterns. Moreover, they may be made alternately, and thus a man with one equipment can serve these two branches of the trade.

It may be remarked in passing that a man with an equipment of this kind can make club-turned spokes and auto spokes, but if either club-turned oak spokes or auto spokes are to be the feature of manufacture there is no occasion for all the equipment required for the finishing plant.

A club-turned spoke is simply a spoke blank that has been put into a lathe and turned to rough size of whatever dimension spoke it is intended to make. This is the kind of spoke that big wagon manufacturers usually use, because they like to do their own finishing. The reason for buying the club-turned spokes instead of the rough blocks is that by this turning process the freight is materially reduced and the spokes are uniform and will dry out more quickly.

A factory for making club-turned spokes is a comparatively simple institution, which almost any hardwood sawmill can operate as a side line. It consists of nothing more than an equalizing saw and a lathe for roughly turning the spokes. Some factories of this type are operated as a side line to sawmilling, and others are operated as spoke factories, pure and simple, buying their bolts from various sources, roughing turning them, and shipping them. This probably represents the best system of handling spokes for the bigger wagon factories. A wagon factory can take these spokes and either pile them under a shed and let them air dry or kiln dry them and then finish them to suit its own peculiar requirements.

The automobile spoke is a distinct development of the past ten years. It is short and heavy and generally made from hickory, but the requirements in the way of toughness and quality do not seem so severe as those for the old buggy and carriage spokes, which were made very slender and of considerable length. The auto spoke can be made of almost any kind of solid, clear hickory, and it is made in enormous quantities, so that labor-saving machinery not heretofore used in spoke work has been introduced, such as double equalizing saws, chain conveyors, hopper feeds, and other automatic handling equipment.

The auto spoke differs from the old-time spoke in that it has no tenon. The inner end is shaped so that the spokes themselves form the hub, or solid centre. So they do serve pretty much the same purpose as the old tenon and the mortise in the hub, but instead of being filled into mortises they are fitted to match together, one upon the other.

This idea really did not originate with the automobile, but it is the development of a patented wheel that came into use before the automobile. This wheel was a type that seemed to fit the needs of the automobile industry, and it has become practically the exclusive method of constructing auto wheels.

Perhaps some day, as progress is made in standardization, we will find small portable plants and single lathes as additions to hardwood mills scattered here and there turning out small quantities of auto spokes.
　　　　　　　　　　　　　　　　—Hardwood Record.

Proper Method of Laying Oak Flooring

Stock should be examined to ascertain if it contains moisture—Filler should be applied as soon as possible after laying is completed

By W. L. Claffey

TO-DAY, by improved machinery, equipment, and quantity manufacture, the cost of making flooring has been so reduced that beautiful oak floors are now within reach of everyone.

Oak flooring is generally laid by a profession commonly known as floor-layers, who specialize in the laying of hardwood floors. These floor-layers may be divided into two classes — good workmen and a class that are careless. The expert floor-layer obtains his reputation by the high class and perfect work that he turns out. It is practically his only asset in the game. Many large and prosperous floor-laying concerns have reached their prosperous condition chiefly through conscientious workmanship in their earlier days. The floor-layer who is careless in his work will never succeed.

It is not necessary to be an expert to produce a good floor-laying job, but it is very essential that considerable care should be exercised, and all the details, from the very start to the finish, should be carefully studied before the floor-laying work is taken in hand.

Before starting to lay oak flooring the stock should be examined to ascertain if it has absorbed any moisture while at the lumber yard, on the wagon, or at the job, as usually during rainy weather oak flooring will absorb considerable moisture, mostly at the ends — thereby causing it to swell as much as one-sixteenth of an inch. If this condition is not discovered before the floor is laid, unsightly crevices will appear in the floor. The sub-floor, as well as the plaster work, should be thoroughly dry before starting to lay oak floors. If in winter, the rooms should have a temperature of about 70 degrees to insure the best results, and the oak flooring bundles should be in the rooms at least ten days to thoroughly dry out, in case the stock has been subjected to any moisture, before the main work is started.

Oak flooring leaves the mill in perfect physical condition, but is very often abused by improper handling before it reaches the job. There are many lumber yards and contractors that almost treat oak flooring like rough lumber. This is a mistake.

The sub-floor should be thoroughly swept, and it is well to use a damp proof paper, and where sound-proof results are desired a heavy, deadening felt is recommended.

The sub-floor should be of serviceable wood, but not less than ⅞ in. thick, dressed one side to an even thickness. Sub-floors should be nailed securely to the joists, but not driven too tight together so as to

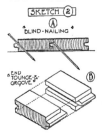

permit it to swell, then bulging; 4-inch to 6-inch strips are preferred widths for sub-floors.

When starting with the first oak flooring strip it is well to leave at least ⅜ in. for expansion space between the first strip and the baseboard, and likewise at the other end of the room, as there is more or less expansion and contraction in all kiln-dried oak flooring.

Oak flooring should always be laid at an angle to the sub-floor, and, after laying and nailing three or four pieces, use a short piece of hardwood, 2 in. x 4 in., placed against the tongue and drive it up with a heavy hammer.

The nailing of oak flooring is very important. All tongued and grooved oak flooring should be blind nailed. The best flooring made can be spoiled by the use of improper nails. The steel cut variety is recommended for 13/16 in. stock—use 8-penny nails every 16 inches; for ⅜-inch flooring use 3-penny wire finishing nails every 10 inches. If even better results are desired the nails can be driven closer.

A floor-layer should use discretion in regard to certain strips that do not blend in color with the majority of strips. A few badly discolored pieces in a room will mar the appearance greatly. Bad discolored

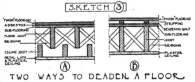

TWO WAYS TO DEADEN A FLOOR

pieces should always be set aside and used in closets and other out-of-the-way places. Where there is a wide variation in color it is good policy to separate the pieces before they are nailed down. This insures a more regular run of color, and blends better together than if scattered throughout all the rooms. Every floor-layer should watch this feature of his work closely, as it is the appearance of the floor after laid that counts.

Oak floors, with some care, should last a lifetime, and it is for this very reason that all floor-layers should be very particular when they lay oak flooring. The wood itself is practically never permitted to wear —that is, in the better grades that are used in homes. It is the wax or varnish finish that wears, which is always replenished. Honest and careful workmanship on the part of the floor-layer spells success. A good job of floor-laying is the best advertising, while a poor job gets nothing but kicks and no reward.

Scraping oak floors is always done in the better grades, or in all homes where people dwell. In order to get the best results for a nicely finished surface it is best to scrape it. This scraping process can be done by the ordinary scrapers, such as used by cabinetmakers, or by one of the many types of power or hand machines that are generally used by contractors and carpenters. Always scrape lengthwise of the wood, and not across the grain. A floor properly scraped looks very smooth, but it should be thoroughly gone over with No. 1½ sandpaper to obtain the best results in finishing. After this the floor should be swept clean and the dust removed with a soft cloth.

The floor is then ready for the filler, which should be put on as soon as possible after the laying work is finished, as the filler fills up the pores of the wood and keeps it from shrinking.

Utilizing the Waste in Box Factories[*]

A Machine for Making Sawdust from Shavings Gives Good Results, but is Costly to Operate

A box maker located in the West told how he had solved the waste problem in his factory, as follows:

We are not overly burdened with edgings such as many of you may have to deal with. Conditions have greatly changed during the past fifteen or twenty years, and I think it may be worth mentioning that fifteen or twenty years ago the average mill cut their lumber thicker than they do to-day, and also there were not nearly so many of them that had planers, and you could buy the lumber cheaper rough than surfaced. In those days we used to pay about $12 or $15 a thousand. We worked the problem out in our mill by figuring how much we would save if we used our waste for fuel, and if that proved to be the case, we would sell our waste and buy fuel.

After I had arrived at the point as to how much we could get for our sawdust, selling it at a good profit. I found out that it was often advisable to sell it in preference to using it for fuel, and I will say that to-day we are selling all that we can spare, because we can get more out of it that way than by using it for fuel. We have raised the price of sawdust about 50 per cent., and I find that we are getting increased sales for this product right along. You can make money on it, and good money, by selling it and buying your fuel.

I might say that in the last few years our fuel bill has been about $1,800, in addition to what we sold, and I started to work last year to see if I could not reduce the fuel bill, and in my opinion I thought it would be worth while to take our old boilers out and put in a different type, steel encased, asbestos lined. I figured that the saving would pay for the boiler in about two years, and I have found that I was right—that we are saving a good percentage. I took the old boilers out and threw them away.

A New England member: It seems to me that in the East the problem of the distribution of waste is largely a local matter. I think that the waste will be much heavier with those who are working around dead stock. That is obvious; they have more sawdust. There is a machine that will make sawdust from shavings, and this is beneficial, as there is a greater market for sawdust than for shavings. I think if you bale the shavings you do not get enough out of them. I would like to ask whether any of the gentlemen here are using the machine which makes sawdust out of shavings whether it is practical. I believe that that would solve the question of shavings if it is practical to make them into sawdust.

In reply to a question as to baling the shavings, one member said he had tried it, but found there was too much expense attached to this method, as it costs more than it is worth.

Regarding a machine for making shavings into sawdust, the discussion developed that there is such a machine, which takes, perhaps, 40 horsepower to operate, and will make sawdust to any size required, from the powdered right up to the larger sizes.

A Western member: We have one of these machines working at our place, and it is costing from $1.50 to $2 a ton to make a satisfactory sawdust for the machine, which is pretty high, but we get a good price for our sawdust. Three or four years ago I thought we might make a little money by putting in a scale, as we were selling everything by measure. After putting in the scale and selling the sawdust by weight, we found that it was one of the best investments we ever made, as the scale paid for itself in about two months. Our customers are very glad to get this sawdust and pay us good prices for it.

We dispose of all the sawdust we have at $6 a ton. We sell it at 30 cents a bag. We used to peddle sawdust and sell it at so much a barrel, but one man would have a large barrel and the next one a small one and then probably the next a hogshead. We, therefore, started to standardize a package, and we now use a 25-inch bag holding from 47 to 50 pounds. Our name is on the bag, and we allow 120 bags to the load. We have a regular delivery day three times a week, and get 35 cents a bag and allow the customer a rebate if he returns the bag in perfect condition. The bag weighs close to 50 pounds, and our customers order the sawdust in quantities of 25, 50, or 100 bags at a time.

Waterproof Belting Cement
By N. Richard

While a waterproof belting cement is not exactly a necessity in the average run of woodworking plants, nevertheless there are times when one can be used to good advantage. Therefore, a formula for making a good waterproof belting cement may not come amiss.

Dissolve a quantity of gutta percha in bisulphide of carbon until it becomes the consistency of molasses. Slice down and thin the ends of the belt to be jointed, warm the laps, and apply the cement; then submit the joint to as heavy pressure as possible, and you will obtain as perfect a waterproof joint in your belt as it is possible to make.

Remedy for Greasy Emery Wheels
By R. Newbecker

It will often occur that parts of machinery which have broken down will be dressed up on the emery wheels used for grinding knives and making moulding cutters. This procedure will in time result in a greasy or glass-like finish of the wheel face, making it almost impossible to do any grinding.

All that is necessary to remove this oily condition is to wash the emery wheel with bisulphide of carbon, and after a few applications it will cut good as new.

[*]At a meeting of the National Association of Box Manufacturers the question of utilizing the waste in box factories was discussed.

Securing Maximum Service from Sanding Machines

Feeding Mechanism Must be in Perfect Adjustment—Drums Must Not be Too High—Vibration of Machine Will Cause Oscillators to Make Waves

By A. T. Deinzer

Sanding is one of the things that ordinarily cannot receive too much attention. In most factories a very high finish is demanded. As the quality of the finish depends on the way the wood is cleaned up, the necessity for good sanding is at once apparent. The manufacturer whose sander work is not giving full satisfaction will do well to visit woodworking plants in his vicinity or to correspond or talk with manufacturers of sanding machines, for I am certain he will gain information that will help him out considerably and prove profitable in every way. One hears operators say: "We have the same make of machine as Jones & Co. They are having no trouble sanding stock perfectly, but we have no end of trouble with this machine. The manufacturer certainly unloaded a lemon on the old man when he ordered and paid for that piece of junk." My argument in a case of that kind is, if your machine is of the same make and style as that used by Jones & Co. it ought to do the work as perfectly as does the machine your friends are using. There is no doubt that you are handling your machine entirely different than they, and, as a consequence, your results are also different.

Too little attention is given the feeding mechanism of many sanders. It is not uncommon to see sanders feeding the stock to the second roll and there become stuck. The operator will push on the feed end, do more or less cussing, and blame the boss for buying such a "d—— machine."

The feeding mechanism must be in perfect adjustment if it is expected that the stock will travel through the machine. The operator should see that his drums are not too high nor his idle rolls too low. Indeed, neglect to take this precaution has often resulted in a ruined drum covering, in addition to spoiled paper. It is by far the safest plan to raise the drums to the desired cut and lower the idle rolls to the desired pressure while the first piece is passing through the machine. Pressure rolls that are down too heavily will cause hollow places.

One of the most common causes of scratchy stock is that of not having the front drum low enough, so that it cuts too heavily, while the back drum is not raised enough. However, all the drums may be too low and the pressure rolls too heavy, so that as stock slides along over the bars grains of sand get in between stock and bars and cause scratches.

If waves appear, look to lost motion in the oscillators for the trouble. Vibration while oscillators are in the centre of their stroke causes a slight depression, which disappears at the end of every stroke, resulting in close-set, regular waves. When stock or paper is burned and the feed is jerky, the rolls may be too high or the drums too low, so that the pressure soon wears out the paper and burns the stock. The same thing happens when rolls are slipping because of oil having been spattered over them, as a result of oiling up while in motion.

An irregular line left standing above the stock is usually the result of a defect or depression in the sandpaper, which is not cutting at that point. The oscillating motion of the drum causes the line to "snake" lengthwise over the stock. A nail or foreign substance in the wood will cause the same mark, by tearing out a streak of grit in the paper. The only way to correct this is to re-cover the drum.

If drums are run too slowly, sandpaper will wear more quickly, and will not cut as fast and smooth as when running at their proper speed. Here is where the manufacturer of sandpaper frequently comes in for unjust criticism of his brand of paper, whereas, if the truth were known the fault would often be found in the slow speed of the machine. Remember this: To perform its work most satisfactorily the mechanism of the sander must be driven at a specific rate of speed. We already explained what will happen if driven too slow. If, on the other hand, the machine is driven too fast, the paper will glaze over and burn the stock.

Another thing worthy of some consideration is the proper care of sandpaper before being used. It is all too common in many plants to see rolls of sandpaper hung in some damp corner of the shop until used. We know, or should know, what dampness will do to sandpaper. The paper must be thoroughly dry so that the sand will not come loose from the warmth and strain of service. One will find some who will favor putting paper into a hot box before using. An objection to this is that the paper will become too dry if left in the hot box for an unreasonable length of time. Sandpaper too dry will cause trouble, for the reason that it has a tendency to break and fly, for the sand will not hold as well as it would if there were enough moisture to relieve it of its harsh dryness.

The care of felt is something that receives too little attention, for proper attention will give surprisingly large returns, not alone in the saving of garnet paper, but in the quality of finish turned out. The felt should be brushed with a good quality bristle brush each time the papers are changed.

In sanding birch, oak, elm, or any other hardwood with a drum sander, much depends upon the condition of the lumber as it comes to the machine. If the lumber is thoroughly dried and smoothly and evenly planed so that it does not require much stock to be removed in the polishing of the same on the sander, a much finer grade of paper can be used throughout the machine and a better finish obtained.

Most manufacturers test glues, varnishes, oils, coal, etc., when purchasing these for their plants. Why not test the sandpaper? Indeed, this is possible. Experiments will not only show wearing quality, but one can determine what paper will be the cheapest to use, i.e., quality and price considered. Records should be kept of these tests, and I am certain the little time required will be well paid for both in saving of money and improved quality of the work.

The sander needs a certain amount of lubrication. So much can be said about proper lubrication that we have neither the time nor the space to properly deal with the subject in this article. Graphite and graphite greases are excellent lubricants. In fact, any lubricant that has body and can be spread out comparatively thin will do the trick. Be careful when oiling the sander not to smear up a lot of stock, possibly on trucks near the machine about to be or already sanded.

Last, but by no means least, inspect the sander frequently, and see that it is leveled up properly, and see that it is firmly anchored to the floor. We know only too well that vibration is the natural enemy of all machinery, especially machinery possessing fast revolving parts. So do not place the machine on a shaky floor. If it is absolutely necessary to do this, then at least support the floor with timbers directly beneath, which may help some.—Yates Quality.

The Grading of Lumber

The Federation of Furniture Manufacturers of the United States certainly started something when, under the influence of a few radicals, it refused to accept certain rules for the grading of lumber adopted by the National Hardwood Lumber Association at its meeting held in 1914, and recommended to the furniture manufacturers that they refuse to buy lumber except upon the rules of 1912, writes "The Furniture Manufacturer and Artisan." This action resulted in the largest meeting of the hardwood lumbermen of the country ever held, which occurred in Chicago during June, and a long and educational debate. But it resulted in little else.

It should be understood for the fullest understanding of the situation that there are two associations among the hardwood lumbermen. One of these—the association which held the meeting herein referred to —the National Hardwood Lumber Association—made up of manufacturers, jobbers, and buyers of hardwood lumber—and the other the Hardwood Manufacturers' Association, the membership of which is confined to the manufacturers of hardwood lumber. Both associations have inspection rules, differing but slightly, but enough to create confusion and cast doubt on every transaction in hardwood lumber. It has been the aim of every individual looking to the best interests of the industry to secure a system of inspection which would be uniform and universal.

Two reports were made by the committee on inspection rules at the Chicago convention—a majority and a minority report. The majority report outlined conferences which had been held with a committee from the Federation of Furniture Manufacturers and a committee from the Hardwood Manufacturers' Association, at which a substantial agreement had been reached on changes to be made in the inspection rules. Concessions had been made on both sides, and the changes recommended by the majority of the committee were to be accepted by both associations of lumbermen, and the amended result was furthermore to be accepted by the furniture manufacturers as represented by the Federation. The minority of the committee reported against any changes, which was the original position of the Federation. The majority report commanded only 140 votes, with 213 against, and as it takes a two-thirds vote to make changes in the rules, the adoption of the recommendations of the majority fell far short of carrying. But since the things contended for, it seems to us, were right, and in the interests of the entire industry—and particularly the furniture manufacturers and the manufacturers of lumber—we are sure will eventually prevail.

The proposal contained in the report of the majority of the committee, and which excited the long and illuminating debate, contemplated closer grading. A grade of selects was provided for, and No. 1 and No. 2 common was to be subdivided into A and B grades.

The A grade of No. 1 was to be inspected from the poor face to give the manufacturer who required clear cutting lumber what would suit his needs, and the B grade of No. 1 common was to be inspected from the best face, which would give a grade that could be used for one side cuttings. The same subdivisions were to be made in the No. 2 common.

These proposed changes are in line with the proposal frequently advocated in these pages by Mr. Crain, that the furniture manufacturers be placed in position so as to buy the stock best suited to their purposes, and avoid the waste which is inseparable from the present system. It is unfortunate that the majority report did not prevail, and the day of one system of grading rules ushered in. These rules, rendered thus universal, might then become the basis for further but slight amendment from time to time, for it is a mistaken idea that rules once made should not be changed as the changes in the character of the raw material dictate. It was this feature of the action of the Federation to which we took exception. The later committee appointed by President Irwin, less radical than the originators of the objection to the action of the National Hardwood Lumber Association, reached this point of view, judging from their acquiescence in the majority report. But the radicals are the ones who bring about reforms after all; and, had the majority report prevailed, such would have been the result in this instance.

Nearly twenty years have been employed in the attempt to bring about universal inspection. The meeting held in Chicago last month was the largest and most representative yet held at which the attempt was made to bring this about. But all is not lost. The understanding of the question is more general than it ever has been, and we are certain that if the right spirit is permitted to prevail the result will be accomplished in the near future.

Planer Shavings Should Curl

One day, while standing beside a planer, my attention was attracted by the noise the knives were mak-

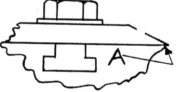

Showing how lips on cutter head should be rounded over.

ing while doing their work. They seemed to be chipping the wood instead of planing it, and this started me thinking of the way the cutting iron of a hand-smoothing plane works.

Every mechanic knows that the cover which fastens on the plane iron is rounded over at the end, and that to do proper smoothing it is necessary to have only a very small projection of the plane iron past the end of the cover. After turning these points over in my mind I stopped the planer and took the knives off. I then examined the head and found that, instead of being rounded over right to the edge, it had square edges about 1/16 of an inch wide. I filed these off to a nice round and replaced the knives, allowing them to have only a very slight projection. Since then I have been getting perfect, smooth work from my planer. When a planer is working right the shavings should be fine (unless a heavy cut is being taken) when they come from the knife. The small projection of the knife and the round edges on the planer head will make them curl. The accompanying illustration will convey my meaning quite clearly. A is where the edges were rounded off with the file.

I may say that my shavings from both sticker and planer are so fine that I have no difficulty in selling them for the same price as sawdust. For any person turning out fine work I recommend this method of adjusting the knives, but, of course, for heavy cuts on large stuff it would not work satisfactorily.

Framing a Roof From Plans and Elevations

Owners of Planing Mills Who Take Contracts for Buildings Could Use Machines for Making the Necessary Bevel Cuts

By William Graham

To frame a roof without hips or valley rafters is a very simple matter, but many good men "fall down" on hips and valleys just because they have not given a little study to the principles upon which the steel square is based. It is my intention in this article to

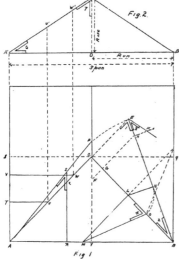

Figs. 1 and 2—Showing how bevel cuts are obtained.

endeavor to make clear these principles; and, let me say further that anyone who will give a little time to study Figs. 1 and 2 should have no difficulty in roof-framing with either hip or valley rafters. The illustrations are very simple, and the writer strongly recommends anyone wishing to become proficient in roof-framing to draw out these examples for himself. In this way one gets a thorough grasp of the thing, and it is not easily forgotten. The scale used is ¼ inch to 1 foot, the span is 16 feet, and the rise 5 feet. First draw Fig. 1, which is the plan of a hipped roof,

the outside lines representing the outside lines of wall plates, A C and B C the centre lines of plan of the two hips, and also put in the ridge line. Now draw Fig. 2, which is the front elevation of hipped end, making the rise 5 feet.

The surfaces of almost any roof are a series of triangles, and if the true shape of these triangles can be obtained then we have the bevels for all the cuts. If a hip roof is framed as shown in these illustrations—that is, with same pitch on all sides (which should always be the case if the building will permit), the plan of the hips A C and B C bisect the squares B Y C 9 and A Y C 8. This is important to remember, and will be referred to again. Now produce the line A C to E, making C E equal to the rise, 5 feet; join E B, and then E B is the length of the hip.

It will be seen here that the imaginary triangle which supports the hip has been developed, giving the length of the hip, also bevels for bottom cut and top side cut, shown at 1 and 2. Two bevels are required for the top cut of the hip rafter. To obtain bevel 3 take any point G in C B and draw F G H at right angles to C B; then from H draw H K at right angles to E B, and make H K equal to F G; join E K; then bevel 3 is for top edge of hip rafter and 6 and 7 are bevels for common rafters. To obtain the bevels for top cut of Jack rafters proceed as follows: Use B as centre and E as radius; cut the hip at P and join A P. It will be seen that here we have dropped the surface of the hipped end into the horizontal plane, thus giving us the true shape. Let R S represent a Jack rafter; then bevel 5 is for the top cut.

It will be understood that the cuts for foot and side cut at top of Jack rafters are same as for common rafters—that is, 6 and 7. If T U and V W be the plan of two Jack rafters, then A' V' and A' W' will be their true lengths respectively. It is very seldom in these days of rush that the hip rafters are, backed, or, in other words, levelled to receive the sheeting. The method of obtaining same is here given for the benefit of those who wish to go thoroughly into the drawing. Take any point L in the hip and draw L M at right angles to C B; draw L N at right angles to E B; then with L as centre and N as radius, draw the arc, cutting C B at O and join M O. Bevel 4 is for backing of hip. If this diagram were drawn to a larger scale it would be all that would be required to frame up the roof successfully.

We will now study the steel square in connection with this drawing. It will be seen what a great help a little knowledge of drawing is to the woodworker. In this case the span and rise is all that is required to commence operations. First, take the common rafter. We know that the rise is 5 feet and the span 16 feet;

therefore the run, which is half the span, is 8 feet. But before we go any further, let me say how useful it would be if the reader make a long scale rule, say, 3 feet long, with the twelfth scale marked on. He could then measure across the square much easier and with greater accuracy. Take the square and with the scale rule measure across from 5 inches on the blade to 8 inches on the tongue, using the twelfth scale. We find the length of common rafter to be 9 feet 5 inches. Put this length down on a piece of board, as Fig. 3, for reference.

To mark out the rafter, place the square on the scantling with 5 and 8 on outside edge and mark out, using the 5 side for ridge cut. Measure out the length of the rafter, 9 feet 5 inches, and place the square in same position, with 8 on the extreme length of rafter. Mark out on the 8 side of square. This gives us the foot cut Fig. 4. We will now take the hip rafter. It will be seen in Fig. 1 that C B is the diagonal of the square B Y C 9, so we must first find the length of the diagonal. Because B Y C 9 is a square we know that B Y is equal to C Y, both 8 feet. Then if we measure across the steel square from 8 to 8 we get 11 feet 3½ inches, which is the length of the diagonal. The length of the hip is the longest side of the right angled triangle B C E, and C E is 5 feet; therefore, if we measure across the steel square from 5 feet to 11 feet 3½ inches we get 12 feet 5 inches, which is the length of the hip.

Take care to put all these measurements down on the piece of board previously mentioned, as they are easily forgotten. To mark out the hip rafter take your square, place on scantling (broad side), with lengths 5 feet and 11 feet 3½ inches, and mark on 5 side for ridge cut. To mark the top side, or edge of scantling, use length of hip and diagonal—that is, 12 feet 5 inches and 11 feet 3½ inches, and mark on 12

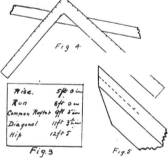

Figs. 3, 4 and 5—Getting bevel cuts on rafters with a steel square.

Rise.	5ft 0 in
Run	8ft 0 in
Common Rafter	9ft 5 in
Diagonal	11ft 3½ in
Hip	12ft 5

foot 5 inch side of square. In measuring out the length of the hip it must be thoroughly understood that in roof-framing you work from centres all the time; therefore, you must measure the length of the hip from the centre (see Fig. 5), and not from long or short point. This is the cause of so many hips being cut an inch or two long or short, as the case may be. Also, in fixing the hip rafter this centre line should strike point C in Fig. 1. Having measured the length

of the hip, use rise and diagonal for foot cut, cutting on diagonal.

For cutting Jack rafters the top side bevel and bevel at foot are the same as for common rafters. To obtain the top bevel use length of common rafter, 9 feet 5 inches, and run 8 feet, and cut on length of common rafter. There is little opportunity for a young mechanic to get practical experience in setting out hip rafters, etc., as an experienced hand is generally in charge of the work. The writer gained considerable experience and practice by making models of roofs at home, using ½ inch by 1 inch rippings for scantling, and this is strongly recommended to young mechanics.

Toys Made of Wood

The accompanying photograph gives an idea of what is being accomplished in Canada in the way of manufacturing wooden toys and novelties to replace

A group of "Made in Canada" wooden toys.

the lines that were formerly imported from Germany. A study of this collection might offer suggestions from which other articles could be evolved. There is undoubtedly a number of articles that could be designed and worked out by studying the things that children are mostly interested in.

Both girls and boys have to be provided for. The toy set of furniture shown in the photograph and the small table, doll carriage, doll bed, etc., are good examples of the class of toys that is required for girls, while the engine, building blocks, constructor toy, gun, miniature wagons, etc., are articles that boys always clamor for. Then, again, there is the quoit game (the shield-shaped board with numbers on), swinging clubs, dumb bells, etc., for the older boys and girls.

There is plenty of scope for the man with an inventive mind to work out something that could be manufactured at a profit, but we would strongly advise anyone going into the manufacture of wooden toys and novelties to study the market before going into it to any extent, as there are undoubtedly a few toys on the market for which there is very little demand. It would also be a good idea to visit the toy section at the Toronto Exhibition and examine the different exhibits. In this way an idea could be gained of what is already being made, and thus prevent overlapping.

Things that come to the man who does nothing but wait are apt to be mildewed.

Some Questions from Our Readers

Kiln for Drying British Columbia Spruce.

Namu, B.C., June 5, 1916.

Editor Canadian Woodworker:

I want to kiln-dry British Columbia spruce for boxes, and do not know whether it has ever been tried out or not. Will you kindly ask the question in the Canadian Woodworker; also how large a kiln would have to be built for a capacity of fifteen thousand boxes a day, and what is the best style of kiln for the purpose? W. S. Ball.

Answer No. 1.—We put practically all our box lumber through dry kilns. For a plant capable of producing 15,000 boxes per day I should say that the kilns required would be about three of a daily capacity of 40,000 feet each. There is no question but that a perfectly good box for any purpose can be made of kiln-dried spruce lumber.

Cameron Lumber Company,

Victoria, B.C., July 3, 1916.   J. O. Cameron.

Answer No. 2.—To dry 15,000 feet of 5/4 spruce your subscriber should figure on a five-day proposition, and for this amount of lumber per day he should have a long, progressive kiln.

It is a very easy proposition to dry spruce, but the conditions for doing it must be right. The lumber must enter into a very moist atmosphere, at a moderate temperature, and must be kept there until the board is heated through. When the moisture from the inside of the board is started out the moisture can be gradually reduced, but it must never be entirely eliminated. Spruce requires more moisture at the finishing of the drying process than any other wood.

The National Dry Kiln Company.

Indianapolis, Ind.

Manufacturers of Three-ply Alderlite.

Regina, July 19, 1916.

Editor Canadian Woodworker:

Could you kindly put us in touch with the manufacturers or distributors of "Alderlite," which we believe we have seen advertised in your columns. We would like quotations on this material.

Regina Sash and Door Company, Ltd.

Alderlite is a three-ply stock, manufactured in Russia, and previous to the war was sold by C. Dupre & Co., Toronto. Their stock has been exhausted, however, and it is impossible to secure further shipments from Russia at the present time. Would British Columbia fir not suit your requirements?—Editor.

Enquiry for Canadian Three-ply Stock.

Liverpool, July 21, 1916.

Editor Canadian Woodworker:

We have in the past imported Russian three-ply alder, prices for which at the moment are very prohibitive. We believe that Canada is now manufacturing Canadian three-ply. If you know of a responsible firm manufacturing same we should consider it a great favor if you are able to ask them to get into touch with us. There is a good demand now for three-ply stock, and after the war it will undoubtedly be greater, as, for some little time at any rate, Russia would seem to require a large proportion of her own manufacture. Trusting you will be able to put us in touch with a firm manufacturing three-ply stock and thanking you in anticipation.

The Liverpool Timber Company, Ltd.

There is some alder grown on the Pacific Coast, but not in commercial quantities. Canada was a large user of Russian three-ply Alderite before the war, but it is not obtainable now at any price. There is, however, manufactured in British Columbia a grade of three-ply Douglas fir which we understand can be obtained in sheets as large (and larger, if required) than the Russian three-ply stock and at a very moderate price. The grain of Douglas fir bears a slight resemblance to the Russian Alderlite. One difference in the product is that the plies in Russian Alderlite are secured with waterproof cement, whereas in Douglas fir they are secured with glue, but we feel sure the Canadian product would give good service if not exposed to dampness. The British Columbia Government maintain in Toronto an exhibit of British Columbia woods, and we are asking their representative to forward you a sample of the three-ply Douglas fir, and we trust that this material will, in some measure at least, meet your requirements. If at any time you should desire further information we will be glad to furnish it, if possible.—Editor.

Making Shavings Into Sawdust.

Sudbury, June 22, 1916.

Editor Canadian Woodworker:

Will you kindly inform us if there is a machine on the market that will reduce planer shavings to sawdust? We have an abundance of planer shavings, for which there is no demand, and if these shavings could be reduced to sawdust in any way we could sell the sawdust, as there is a good market for it.

Thanking you for any information you may be able to give us, we remain,

The Evans Company, Ltd.

The Williams Patent Crusher and Pulverizer Co., Old Colony Building, Chicago, manufacture a machine which they advise us will reduce shavings to the ordinary coarse grade of sawdust. They also say, however, that the horsepower consumed is high and capacity delivered small, so it would be advisable to go into the matter thoroughly before installing expensive equipment. One of our friends, who operates a woodworking factory in Toronto, recently explained to the writer how he made a slight change in the adjustment of his planer knives which resulted in the shavings being so fine that he had no difficulty in disposing of them as sawdust. A sketch showing how the adjustment was made appears elsewhere in this issue. If you try this out we would like to know the result of your experiments.—Editor.

July 22, 1916.

Editor Canadian Woodworker:

Dear Sir,—After reading your article on making pipes from maple and cherry would like to get some information in regard to making pipes. What machinery would we require, and where could the mouthpieces be bought, or could they be manufactured in the same plant? Subscriber.

The manufacture of pipes being an industry in which there is very little done in Canada, we have been unable to secure any particulars about the machinery required. It would seem to require something on the principle of a lathe, which might have to be made to order. Perhaps some of our readers could furnish us with the information desired.—Editor.

The Lumber Market

Domestic Woods

A factor that has entered into the production of lumber in Canada, and which promises to contribute something to the prices, is the shortage of labor. Sawmill owners cannot secure enough men to operate their mills and load cars, and the situation is likely to become much worse. The forestry battalions which have been recruited in Canada to cut timber in England will take such a large number of experienced busy men out of the country that it will seriously affect the cutting and getting out of logs next winter. Stocks have been sufficiently low for some time to ensure that there would be no decrease in prices, but it now looks as though they might soar if the demand from the factories continues.

Furniture factories in Canada, while not running to full capacity—on account of the shortage of labor—are nevertheless busy, and there is every prospect that they will continue to be so. Of course, if lumber prices increase very much it will mean that prices of furniture will have to increase in proportion. But manufacturers will be well advised to try and make satisfactory arrangements to keep themselves provided with sufficient lumber to prevent closing their factories down.

The shell box business is still taking care of fairly large quantities of both hard and soft woods, but lumbermen are exercising more care in selling lumber to the firms engaged in shell box making, as, judging from the prices at which some of them accept contracts, some person is going to go short when it comes to paying up. Therefore, the lumbermen are taking no chances, as they can dispose of their stocks to firms handling more satisfactory lines of business.

The building trade remains practically at a standstill. Manufacturers of mouldings have been using considerable basswood, with the result that it is now very scarce, and the price is stiff in accordance.

Prices at Toronto on car lots vary very little from last month, and are as follows:

Ash, white, firsts and seconds, 1½ and 2 inch, $60; ash, white, No. 1 common, $45; ash, brown, firsts and seconds, 4/4, $50; ash, brown, common and better. $40; ash, brown, firsts and seconds, 6/4 and 8/4, $60; birch, firsts and seconds, 4/4, $44; birch, No. 1 common and better, 4/4, $35; birch, firsts and seconds, 6/4 and 8/4, $46; basswood, firsts and seconds, 4/4, $40; basswood, No. 1 common and better, 4/4, $35; basswood, No. 1 common and better, 4/4, $35; basswood, 6/4 and 8/4, firsts and seconds, $44; basswood, 6/4 and 8/4, common and better, $38; soft elm, firsts and seconds, 4/4, $40; soft elm, No. 1 common and better, 4/4, $32; soft elm, 6/4 and 8/4, firsts and seconds, $42; rock elm, firsts and seconds, 6/4 x 8/4, $55; soft maple, firsts and seconds, 4/4, $33; soft maple, common and better, 4/4, $25; soft maple, 6/4 and 8/4, firsts and seconds, $35; hard maple, firsts and seconds, 4/4, $40; hard maple, common and better, 4/4, $33; red oak, plain, firsts and seconds, 4/4, $66; red oak, plain, No. 1 common, 4/4, $40; red oak, plain, firsts and seconds, 6/4 and 8/4, $68; white oak, plain, firsts and seconds, 4/4, $68; white oak, plain, No. 1 common, 4/4, $40; white oak, ¼ cut, firsts and seconds, 4/4, $89; white oak, ¼ cut, 5/4 and 6/4, firsts and seconds, $92; white oak, ¼ cut, No. 1 common, $57; hickory, firsts and seconds, 4/4, $70; hickory, 6/4 and 8/4, firsts and seconds, $75. White pine: 1 inch, No. 1 cuts and better, $50 to $52; 1½ inch, No. 1 cuts and better, $60 to $62; 1 x 8 box and common, $23.50 to $24.50; 1 x 10 inch box and common, $25 to $26; 1 x 12 inch box and common, $27; fir flooring, 1 x 3, edge grain, $31.50; fir flooring, 1 x 4, edge grain, $35; fir flooring, 1 x 4, flat grain; No. 1 and 2, 1 inch clear fir, rough, $44.50; No. 1 and 2, 1¼ and 1½ inch, clear fir, rough, $50.

Imported Woods

A number of United States lumber firms are advising their customers to purchase sufficient lumber now to cover their requirements for some time, as the car shortage is already beginning to make itself felt, and by September, when the railroads commence to move the crops, cars will be almost unobtainable. This appears to be good advice, and it looks to us as though Canadian manufacturers would be well advised to figure out what their lumber requirements will amount to for a few months and get in touch with the market. Furniture factories in the United States are busy. According to the American lumber journals, they are buying lumber in large volume and paying fair prices for their purchases. This has had a decided tendency to stiffen the prices of the woods in demand. Furniture manufacturers are using a great deal more red and sap gum than they have ever used before. Large quantities of gum are also being used for the manufacture of interior trim, and box factories are taking the lower grades, with the result that prices have stiffened accordingly. There has been a falling off in the demand for oak by the furniture manufacturers in the United States, but the makers of hardwood flooring are still taking it in fairly large quantities so that there is no alteration in prices. Automobile manufacturers are steady buyers of elm in 2, 2½, and 3-inch stock. Considerable hickory and ash are also being used, while low-grade cotton wood is in the same position as gum. The highest priced woods, like mahogany and walnut, are strong in price and in good demand by furniture manufacturers. Thick stock, including maple and ask, is still in demand, with prices about the same. One report from the United States is to the effect that quite a little high-grade hardwood, particularly high-grade plain and quartered oak, poplar, and gum, is being exported to England and France, to be used for war purposes.

Prices quoted on some of the best sellers at New York are as follows: Ash, firsts and seconds, 1 inch, $55 to $60; firsts and seconds, 1¼ and 1½ inch, $63 to $67; firsts and seconds, 2 inch, $65 to $70; No. 1 common, inch, $35 to $38; No. 1 common, 1¼, 1½, and 2 inch, $43 to $48; basswood, firsts and seconds, inch, $44 to $46; firsts and seconds, 1¼, 1½, and 2 inch, $45 to $48; No. 1 common, inch, $31 to $33; No. 1 common, 1¼, 1½, and 2 inch, $33 to $35; firsts and seconds, 1¼, 1½, and 2 inch, $40 to $42; No. 1 common, inch, $28; No. 1 common, 1¼, 1½, and 2 inch, $30 to $32; red birch, firsts and seconds, inch, $59; firsts and seconds, 1¼, 1½, and 2 inch, $61 to $63; No. 1 common red birch, inch, $41; No. 1 common, 1¼, 1½, and 2 inch, $43 to $48; sap birch, firsts and seconds, inch, $51; firsts and seconds, 1¼, 1½, and 2 inch, $53 to $55; No. 1 common sap birch, inch, $35; No. 1 common sap birch, 1¼, 1½, and 2 inch, $37 to $43; chestnut, firsts and seconds, inch, $48 to $52; firsts and seconds, 1¼, 1½, and 2 inch, $52 to $55; No. 1 common, inch, $36 to $38; No. 1 common, 1¼, 1½, and 2 inch, $38 to $42; red gum, firsts and seconds, inch, $40 to $44; No. 1 common, inch, $30 to $34,

Squaring and Edging on the Jointer

For factories that only have one or two jointers, and desire at times to use them for squaring or edging, the machine illustrated herewith is an especially desirable one, as the attachment can be quickly swung free from the jointer.

A jointer is a very dangerous machine, but this attachment reduces the danger to the minimum. The machine will work stock of any kind without risk to the hands of employees and at very much greater speed than an ordinary jointer. It takes the winds out of a board and planes it perfectly smooth. The stock can vary one inch in thickness without making

The attachment is driven through gears running in oil and incased in the standard of the machine, and the feed is raised and lowered by means of a hand wheel. The attachment is provided with cone pulleys, with two speeds, 25 and 40 feet per minute. The post or standard of the machine is of heavy cast iron, and is very rigid and substantial. The machine is manufactured and sold by Jackson, Cochrane & Co., Berlin, Canada.

Applying Emery Direct to Sanding Disks

While looking over a home-made sanding disk in a small shop some time ago, I found that instead of

Jointer with Attachment for Squaring and Edging Stock.

any adjustment, and for stock varying more than this the feed can be raised or lowered. It will handle stock from one inch wide to the entire width of the jointer, and it is impossible for the knives to throw the stock back. The attachment will feed fast or slow or take a light or heavy cut in hard or soft wood.

The feed is an endless chain, carrying a series of fingers, which extend the full width of the jointer. The fingers are spaced 3 inch centres, and the rows are staggered, so that in the width of the table they are only 1¼ inches apart. The fingers have coil springs, designed to exert the same pressure as a man would when feeding stock by hand. The pressure can easily be regulated by raising or lowering the attachment by means of a hand wheel. In operating, it is not necessary to exert any pressure in shoving the stock under the feed rollers, and no attention has to be paid that the stock follows closely after or in the same line as the preceding piece. It is only necessary to slide the material under the feed and the mechanical fingers do the rest.

changing the paper, as is usually done, they applied emery in powder form direct to the face of the disk.

The disk, of which a number of duplicates is kept, is made of three-ply hard maple, and so arranged that it can be taken off the shaft as easily as a rip-saw. The cement for applying the emery powder is made as follows: Use equal parts of shellac, white resin, and carbolic acid (in crystals). The shellac and resin are melted together and the carbolic acid added afterwards. The face of the disk is then covered lightly with the cement and the powdered emery spread over the surface. When the cement is dry the loose emery is brushed off, leaving an emery surface that will outlast several pieces of sandpaper, both in quantity of output and quality of sanding. When the disk becomes worn, another coat of cement and emery powder can be applied.

Cracks in the dry kiln make holes in the profit account, by an unnecessary tax on the steam plant.

THE FINISHING ROOM

Staining and Coloring Different Kinds of Woods

By James Culverwell

Reference was made in last month's issue of the Canadian Woodworker to a few ways and methods of filling hardwoods. The coarse-grained woods, as oak, ash, and chestnut, need filling with a pack filler, while others with less open grain, as maple, butternut, satinwood, hickory, bird's-eye maple, and others, need a liquid filler, while in some cases no filler is required. But in whatever form the filler takes, the work must of necessity be rubbed down before proceeding.

We will now consider the subject of staining of woods, of which a great deal might be said and, I trust, prove useful. We will first take the oak, this being perhaps the most widely known of our domestic woods. There is no wood that can be treated in so many ways as the oak. It can be oiled, wax-polished, stained in various tones, fumigated, French-polished with open or filled grain if it is desired to finish in the natural state, or filled with Russian tallow and plaster paris and polished with white shellac.

Some good stains in oil are as follows: For light oak take raw sienna, sufficient dryer to dry same hard in 12 hours; thin to proper consistency; put the stain on with a brush; let stand for a few minutes, then rub off with a waste or tow, being careful that no surplus stain is left on the wood, as it will mar its transparency and spoil the effect so much desired. This stain can be made deeper, not by applying a greater body of color, but by adding burnt umber. This being a semi-transparent color, does not make your wood look cloudy. Should you wish to get it still deeper or darker in tone use burnt umber alone. Should this prove too warm in color, as it is termed (meaning red), it can be neutralized by adding raw umber or raw umber alone; or, again, a trace of green in the umbers in some cases has a very good effect. But, whatever stain is required, endeavor to use transparent colors.

A little knowledge is required to pick the transparent colors from the opaque. Raw sienna is transparent, while ochre and chromes are opaque. Burnt umber and vandyke brown are transparent; also burnt sienna, Prussian blue, and asphaltum. Asphaltum broken down in turpentine makes a very pleasing stain. Oil added to your stains improves the work, as it insures your job being cleaner, and it is easier handled than all turpentine. It has one disadvantage, however—that of darkening the wood—but where a dark stain is used this does not matter.

Asphaltum, being of a different nature, has not the affinity with oil that the other stains have. It resembles pitch in appearance, and is insoluble in water and alcohol, but dissolves readily in turpentine or naphtha. Other methods of staining are by applying to the oak some freshly-made lime water diluted to strength required. I might say here that, when staining, there should be a few pieces of the same kind of wood handy, to try out your stains, otherwise you may easily spoil a nice piece of work by applying your stains too strong.

After trying a sample of your lime water, let same dry, then oil with linseed oil or shellac, as the case may be. This will give you a fair sample of what your piece of work will look like. A strong solution of common soda has the appearance of ageing the oak. This should be applied two or three times until the desired effect is obtained. Green walnut shell, boiled down and strained through muslin, also makes a very pleasing stain. This should be diluted with water until the proper tone is obtained, and any tone can be obtained from light brown to nearly black, according to the number of coats and strength of stain used.

It should always be remembered that two or three coats of stain give better results than one thick or strong coating, as this only mars the work, making it muddy and taking away the purity of tone it would otherwise have with two or three thin coats. The much admired bronze olive tint given to oak can be obtained by fumigating, this being the best method known to bring oak to an aged appearance. The method is simple. In the case of large pieces, as doors, paneling, or large furniture, it is advisable to build in one corner of the workshop a room large enough to take the article to be fumed, say 10 feet x 6 feet x 4 feet.

After erecting, paper up all cracks that would be likely to let out fumes; place the article in, care being taken that no glue or greasy finger marks are left on the work, as otherwise these places will remain white and will not fumigate. Care must also be taken to see that every article is free, in order that the fumes can get all around. When the stock is properly arranged, place three or four saucers about on the floor, not too closely to the work to be fumigated. Place in these saucers about three-quarters of a pint of liquid ammonia, dividing same about equal. Come out and close up the door, papering up any places where fumes are likely to escape. It is also advisable to leave a small pane of glass in one side, that one may look through and see the progress made from time to time. From ten to twelve hours would be a fair time for the quantities given. The more ammonia placed in the saucers and the longer it is confined the darker it becomes.

This becomes a permanent stain, and one can take a shaving or two off without destroying the color. When taken from the room it will require oiling, shellacing, varnishing, or waxing, as the case may be. It is then when you see its real beauty. If a spare piece is placed in the room you will have the advantage of being able to test your color without fear of doing any damage to your work; and, if not dark enough, you can place more ammonia in saucers and close it up again for a short time.

This treatment does in a few hours what nature would require years to do. The next best thing to do is to apply liquid ammonia to the article with a brush. When dry, if not deep enough, go over again, keeping the ammonia week by diluting it with water. When dry it is ready for shellacing or oiling. Another good stain is to coat the article with a solution of bichromate of potash dissolved in water. The fuming has an advantage over the liquid stains, inasmuch as it does its own work when once placed, and it does not raise the grain of the wood. It is also a pure, soft-tone of color that is otherwise unobtainable by artificial means.

The best way to avoid the raising of the grain when liquid water stains are used is to first coat the work over with hot water. While wet rub it down with sandpaper, wipe off, and apply your stain. This will obviate the trouble to a great degree. Weathered oak can be obtained in the following way: Coat the article to be stained with a solution of liquid ammonia. When dry, rub down with fine sandpaper. Always keep the way of the grain, and never paper across the grain. Clean away all dust, make a mixture of lamp black and ochre, or raw sienna with a little Japan dryer added and sufficient silica's earth to bring it to a creamy consistency. You apply this after the ammonia is dry; then let stand a few minutes, and clean off with rags, tow, or waste. Let the job stand for twelve to twenty-four hours before finishing with oil, wax, or varnish, or which ever way it is desired to finish it.

Good results are obtained by giving a thin coat of orange shellac and letting stand twelve hours; then rubbing down with very fine sandpaper; or, better still, with what is known as flour paper. Dust off and give a thin coat of rubbing varnish, let stand for twenty-four hours, rub down, dust off, and varnish with a good, reliable rubbing varnish, care being taken that a good, even coat be put on and that no runs or tears are permitted to occur. Then let the job stand two or three days, when it should be rubbed down with 0.0. ground pumice stone in oil, with a piece of felt about two inches to three inches square. This will give you that soft, greenish brown which is so pleasing to look at. There are many other stains which can be applied, but space being limited they will have to be dealt with in a later issue of the Canadian Woodworker.

The Practical Testing of Varnish

Perhaps no product which the furniture manufacturer uses causes more vexation than varnish, writes T. Alexander in "American Furniture Manufacturer." In the first place, varnish-making is not an exact science; and, in the second place, the handling of it in the finishing department is often not conducted with the highest degree of intelligence.

The varnishes in most common use are composed of resins and linseed oil, thinned with turpentine (or benzine). The larger the proportion of oil the more elastic and durable will be the varnish. The smaller the proportion of oil the harder and more brilliant it is, and the quicker to dry. Rubbing a varnished surface is done after it is quite hard, and with a pad of cloth, or, better, of felt, which is wet either with oil or water, and which has been dipped lightly in powdered pumice. The effect is to grind off the natural surface of the varnish and to remove any little prominences, and thus make it more level and uniform. It is obvious that for this purpose it is convenient to have a varnish which will dry and become very hard quickly. Rubbing varnishes contain about equal parts by weight of resin and oil, and are hard enough to rub in about two days, as a general thing. The objection to such a varnish is that it is so hard it is somewhat brittle. An interesting experiment which no doubt many finishers know is to varnish a very hard piece of board with several coats of

such varnish and, after a few weeks, bore a hole in it with a good-sized auger bit. It will splinter and crack around the hole, looking as thin glass would. And in time it will, in ordinary use, crack probably from change of temperature, and the whole surface will become cracked or crazed, and its beauty will be destroyed. To prevent this, it is customary to subsequently apply a coat of a much more elastic varnish, which has a protective effect.

Some of these finishing varnishes contain three times as much oil as resin. These, while very durable, never become hard enough to be suitable for furniture, as they are likely to be tacky (sticky) when warm or under long continued pressure. Most furniture finishers will agree that furniture may be finished with a varnish containing twice as much oil as resin, and that such a varnish will in three weeks or so get hard enough to rub.

Many finishers have complained about varnish difficulties experienced in spring and fall. The reason is that varnish must have an even temperature during the time it is hardening, and if it becomes chilled after it has been applied the drying is usually very unsatisfactory. The artificial heat of winter is more readily regulated and controlled, and the conditions are more nearly uniform then than at any other season. In the early spring and late fall artificial heat is used also, but temperature is not watched in the average furniture factory. When it becomes too hot for the comfort of the workmen, windows are opened, producing drafts, to which are attributable the difficulties experienced with varnish blooming at this time. It is a general rule that the higher the quality of the varnish the more apt it is to be sensitive to conditions of this kind.

Many furniture manufacturers make the mistake of allowing insufficient time for the drying of varnish coats. Some managers will argue that the sooner they get this stuff out of the factory and onto the dealer's floor the sooner they will get their money. It is true, as I know but too well from experience, that it is not always possible to allow the time required for drying. A dealer may say: "Unless I get this by Thursday of this week I will cancel the order." This is an old story, with which you are all familiar. In such cases use a first coater that will dry quickly, even if it is necessary to sacrifice some on the durability of the finish rather than coat on a better first coater that is not dry.

The science of varnish drying has been so thoroughly discussed in previous numbers that it is needless to again repeat what has been said. As already stated, the drying of the varnish, or rather the uncertainty of its drying in a given time at different seasons, is what worries more than anything else the manufacturer of furniture. With an up-to-date varnish drying system uncertainty is removed. It may behoove you to investigate, and see what your competitor is doing.

As a general rule, much more time is given for the varnish to dry where the work is rubbed and polished than where the work is left in the gloss. Yet the practice of recoating before the undercoats are dry is not uncommon, and should be carefully guarded against. One of the most serious results is alligator checking, which is more likely to occur where several coats of varnish are applied. As varnish dries and becomes hard it contracts slightly and the film of varnish becomes thinner.

My good friend, Dr. Clifford D. Holley, has done more laboratory experimenting in the varnish line than any man I know of. He has informed us that the time a varnish requires to dry properly is regulated according to the purpose for which the varnish is to be used. Hence the samples tested should be compared with

BUILT-UP VENEER PANELS

Ready for Immediate Shipment August 12, 1916. **Subject to Prior Sale.**

Good 2 Sides, 5 ply

39 Birch, 2s	60 x 30 x 3/8	4 Birch, 2s,	72 x 20 x 3/8
19 Pl. Oak, 2s	60 x 30 x 3/8	29 Pl. Oak, 2s,	72 x 24 x 3/8
5 Mahogany, 2s	72 x 24 x 3/8	91 Qt. Oak, 1s, Pl. Oak, 1s,	60 x 30 x 3/8
56 Qt. Oak, 1s, Birch, 1s	60 x 30 x 3/8	5 Qt. Oak, 1s, Bir. 1s. (3 ply),	60 x 24 x 1/4

Good 1 Side, 3 ply

40 Birch, 1s,	72 x 24 1/4	23 Pl. Oak, 1s,	72 x 24 x 1/4
23 Pl. Oak, 1s,	72 x 24 x 1/4	4 Qt. Oak, 1s,	72 x 24 x 1/4
104 Pl. Oak, 1s,	60 x 30 x 1/4	4 Qt. Oak, 1s,	72 x 24 x 3/16

Good 1 Side, 5 ply

46 Qt. Oak, 1s,	60 x 24 x 3/8	6 Qt. Oak, 1s,	60 x 30 x 3/8
28 Qt. Oak, 1s,	60 x 18 x 3/8	100 Qt. Oak, 1s,	60 x 30 x 3/8
8 Qt. Oak, 1s,	48 x 18 x 3/8	20 Pl. Oak, 1s	60 x 30 x 3/8
5 Birch, 1s,	72 x 24 x 3/8	4 Mahogany, 1s,	72x 24 x 3/8

Good 1 Side, 3 ply

240 Maple 48 x 48 x 3/16 195 Maple 36 x 36 x 3/16 135 Maple 42 x 42 x 3/16

Write for prices

Hay & Company, Limited, Woodstock, Ont.

"Furniture in England"

This book is an encyclopaedia of artistic suggestion—an education in itself.

Upwards of 400 exquisite half-tone and color plates on pages 14 ins. x 10 ins.

Among the subscribers to this book are Her Majesty Queen Mary, His Majesty the King of Greece, and many of the foremost manufacturers the world over.

Price $12.00

Delivered to any address in Canada

Copies may be obtained from

The Woodworker Publishing Co.
Limited
Toronto Ontario

ANGLO RUBBING and POLISHING

Works free and easy and can be rubbed in two days.

WRITE FOR SAMPLE

The
Ault & Wiborg Co. of Canada, Ltd.
Varnish Works

Montreal Toronto Winnipeg

THE GLUE BOOK

WHAT IT CONTAINS

Chapter 1—*Historical Notes*
Chapter 2—*Manufacture of Glue*
Chapter 3—*Testing and Grading*
Chapter 4—*Methods in the Glue Room*
Chapter 5—*Glue Room Equipment*
Chapter 6—*Selection of Glue*

WOODWORKER PUBLISHING CO., LIMITED
345 Adelaide St. West, Toronto

the accepted standards of those types of varnises, both on the wood test surface and on a sheet of glass, on which samples of the varnises have been placed, and then set in a dust-free but unconfined place at an angle of about 30 degrees from the perpendicular. The best results are secured by resting the glass on a couple of small hooks, which permits the varnish to drain freely. The rapidity of the drying should be noted at regular intervals. When dry, the test should be saved for further examination.

After the requisite number of coats have been applied to the test boards and the finishing coat has hardened thoroughly, a sponge made of several thicknesses of felt is thoroughly moistened and laid on the varnished surface and allowed to remain for a stated number of hours undisturbed. A higher grade varnish will either show no discoloration at all or will regain its color on drying, provided, of course, that it has been suitably applied. A varnish containing a large amount of resin will be more or less badly corroded, and will remain permanently white and discolored. With a little practice, by working with varnishes of known composition, the reader can make a pretty shrewd guess at the approximate amount of resin present by the degree of didscoloration.

Sweating of varnishes on damp, warm days is due, in many cases, to the presence of fish oils. These oils absorb moisture, which causes the varnish to become clammy and sticky to the touch, especially if the temperature is above 80 degrees Fahrenheit.

One of the most important things is that the varnish should have age, although the varnish-maker may somewhat object to this, for the reason that his money is tied up so much longer in the product. A certain large manufacturing concern is Grand Rapids has two tanks reserved for its varnish, which is made according to a formula which cost the concern, as well as the varnish-maker, quite an amount of money in the way of experimenting, etc. Plenty of time is allowed for aging. The manufacturer is glad to pay the price, and the quality of the finish justifies the wisdom of the policy.

Ragged Edges

Back of all the tragedy of Failure there is always the tragic truth of neglect and slight—edges left ragged and incomplete.

Finish up as you go.

A few years ago a young man in a Western college got restless and discouraged. He wanted to leave his course unfinished. He sought the advice of a successful man, and this was the advice: "Stick it out. Finish something. There are too many men now with ragged edges crowding the ranks." The young man finished his college course with honors. To-day he is a leader and a success.

Many a man stops work with the clock. He leaves his day's work with ragged edges. He is the man who starts his days with ragged edges, and finally rounds out an incomplete life.

There is a satisfaction and a feeling of latent strength in the breast of a man who starts a thing—and finishes it. You will find this true if you do it. The most important task is always the task at hand. Complete it. Make it stand square and clean when you leave it; Look it over. Be sure no ragged edges remain.

Make thoroughness one of your masters. Searchingly note the trifles. Get them together and know them—for out of them comes—Perfection.

Finish up as you go.—Globe-Wernicke Doings.

You Don't Have to Wait for Veneers
ORDERED HERE

Notwithstanding the scarcity of MAHOGANY Veneer we have a good stock on hand and another large consignment which will be ready for shipment in about three weeks.

We carry the largest and best stock of AMERICAN BLACK WALNUT Veneers in Canada. We have some very highly figured stump wood, good straight sliced stripe and curly figured stocks, also semi rotary and full rotary cut stock of all kinds.

QUARTERED OAK we carry in all thicknesses in both sawed and sliced. In this we have received from our mill a new cutting a few days ago, including some very choice extra wide 1/16" and 1/20" sawed Oak.

Let us show you what we have to offer in FIGURED GUM, in this we have the goods and prices.

Get our prices on WALNUT LUMBER

TORONTO VENEER COMPANY
TORONTO   93-99 Spadina Avenue   ONTARIO

Veneers AND Panels

Proper Methods of Handling Stock in the Veneer Room

By Albert Hudson

There has been a considerable change in machinery and methods in the process of producing built-up stock. This change is not only apparent in the laying of veneers, but also in the gluing up of the core stock. The extent to which veneer is being used to-day in the manufacture of high-grade furniture gives occasion for

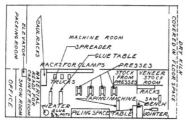

Fig. 1—Layout of veneer room.

considerable thought as to the very best conditions under which the work may be done. There are still problems to be met with in the glue room, but, given modern equipment, good stock, good mechanics, good system, a little common sense, and good judgment, the average run of veneer troubles may be overcome.

For example, what good would modern equipment be without good stock? Of what good would first-class mechanics be unless the other important factors are also of the best? Given good mechanics, generally there will be good system. Good system means increased output, and first under this heading comes the layout of the veneer room. Going through a furniture factory some time ago the writer's attention was especially drawn toward the veneer room. This room (Fig. 1) shows the layout) was located on the ground floor, close to the dry kilns. The veneers, before being used, were placed on trucks and the trucks pushed into the dry kilns.

When asked if they had any trouble with their veneered products, the reply of the management was: "No; we always dry our veneers before using. We also see that our core stock is in good condition, and give great attention to our veneer and glue department, thus ensuring good work," which is undoubtedly the proper attitude to take.

The special racks, before mentioned, are constructed as shown in Figs. 2 and 3. The veneer is piled with the grain upright, and held in place by strips running horizontal. These strips are fastened in across the frame of the rack and fixed at the ends in cut-out or

notched strips running the full length of the racks. The racks are built to fit trucks.

An important factor in the veneer room is good light. The matching of fine, fancy veneers can be done only in good light, and the man at the spreader requires good light to ensure an even spread and to detect spots if they should occur. There is a feeling in some furniture factories that veneered work can be rushed. This is a mistaken idea. Veneering is one of the big items in the manufacture of high-grade furniture, and the only way to keep the cost down is to do the work right the first time. To attempt to reduce the cost by giving little attention to the details in the laying of veneers or the preparation of the core stock, or of using inferior stock, or of rushing the work results always in failure and increases the cost in the cabinet room. This simply means moving the cost from one department to the other.

It is, unfortunately, true that the veneer room, as a department in a factory, has been found in many cases to be absorbing some of the profits made on the operations of other departments instead of paying its own way. How often one may see core stock run through the sticker and then, without having the knife marks removed, it is sent straight into the veneer room, with the idea that all the marks and defects will be hidden; or, again, the core stock is very often sent direct from the planer to the veneer room with the same idea.

In surfacing core stock it is most important that all knife marks be removed. If the core stock is allowed to go through without the knife marks being removed they will certainly show up in the finishing room, just at the time when it is impossible to remove them. Extra coats of varnish are applied and extra rubbing given to make this leading passable, but it seems to be sheer nonsense, saving 5 cents in the beginning and

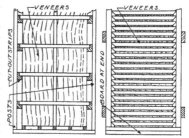

Fig. 2—Rack for drying veneers.

spending 50 cents in the finishing. All marks, whether by sticker or planer, should be removed from the core stock. It is wise to sponge all core stock that is soft to swell out any indentations that it may have received.

Of late some manufacturers have begun to tooth their core stock, and, while this takes a little longer to do, it is a wise operation. Others sand their core stock, but toothing the most important parts is the most satisfactory, and always ensures a good, flat surface. Glued up stock should never be rushed, as in gluing it up the glue swells the wood along the edge of the joint, and if not allowed time to dry out and shrink back to its normal condition before being planed or prepared it will certainly shrink below the level after preparing. After veneering, the work will show narrow depressions, extending from end to end. This should be watched very carefully, and especially when two kinds of wood are used for the core stock.

The core stock should be perfectly dry, and, after having been glued up, should be left again to become properly dried out along the joints. By observing this rule endless trouble will be avoided, as one of the most frequent causes of trouble in veneered work is badly dried core stock. Unless this is thoroughly dry trouble will follow, and you will have it in the finishing room without fail.

Passing on from the core stock to the veneers, we are confronted with the question of properly dried veneers. It is entirely wrong to think that because veneer is wrinkled it is sure to be dry. Veneer becomes wrinkled in the process of drying, fancy veneers being more liable to wrinkle than plain veneers. This is because the surface is made up of varying degrees of end wood, which dries out more rapidly than the parts that have not open pores exposed so completely to the air. When these parts begin to dry and shrink it can be plainly seen that a strain is caused on the slower drying parts, thus causing the twist. This twisting, or wrinkling, is done long before the veneer is thoroughly dry.

If veneers are badly wrinkled it would be unwise to lay them in such a condition. It would be better to moisten and place them in a redrier. If the equipment does not include a redrier, then place them between hot boards of an absorbent nature, to straighten them out. A few words might be said here about cross-banding. The question has often been asked, "Should cross-banding and face veneers be laid at the same time?" Cross-banding should be placed under pressure as soon after it touches the glue as possible, to prevent it from expanding, and this cannot be done when the two veneers are laid at the same time.

Much of the stock that is used for cross-banding contains creases where the wood broke away in the process of slicing. Now, if the cross-banding and face veneers are laid at the same time these creases will fill up with glue; then, when the stock is allowed to dry out, the glue in the creases will shrink and draw the veneer down. This is seen very often in a buffet, dresser, or dining table top. To be sure of a perfect job the cross-banding should always be laid some time before the face veneers and stripped to dry out thoroughly. This allows it to do all the shrinking possible, after which it should be sanded and then the face veneer applied.

High-grade veneers are so thin that they will not allow the amount of sanding or scraping necessary to remedy the defects of the cross-banding or core stock. There is no reason, however, why cross-banding should be laid in such a manner that the veneer is cut through on the face before a smooth surface is obtained. Overlapping of the cross-banding is a frequent source of trouble, especially where two veneers are laid at one operation. To prevent this the cross-band must be dry. Lay it first, and get it into the presses as soon as possible after it touches the glue. Veneers do not expand once pressure is applied, so that no expansion, no overlapping, and no checking of the face veneers will result if the cross-banding is applied as outlined above.

Some factories are laying four veneers at one operation. This is a very simple operation, but not always the best policy. However, in laying the four veneers at one operation everything should be at hand. Have cross-banding opened and well dried for a few days before using; see that the face veneers are dry, straight and taped at the ends. Laying four veneers at one operation certainly cuts down the cost of production, but does not always give the quality of work desired. Often the veneer man is blamed for inferior veneering when it is not his fault. The cause may be found in the rushing of the work or the preparation of the core stock, or in the veneers, or even in the glue.

While it may be a simple matter to return work to the veneer room to be reveneeres, the loss should be considered. The holding up of the different departments for these various causes is a bigger item than is usually supposed. Imagine a cabinet-maker beginning to work, say, on a batch of fifty dressers. The stock will be put into his bench-room and the different parts placed in order convenient for him to work. One of the first things he will do will be to assemble the gables for the fifty cases, but if he finds some of the gables with loose veneers, or if some require reveneering it throws him off his work, and as many cases as these gables represent cannot be assembled. Again, the reveneering of the gables will cause a delay of a week or more. In the meantime the cabinet-maker goes to work and completes the balance. The remaining parts are laid on one side, to await the return of the gables, and when these are brought into him the stock has to be handled again.

The same thing applies to all veneered goods, whether tables, tops, drawer fronts, etc. This extra work is sometimes looked upon as a matter of course, but its effect on production and profits is usually felt. Sometimes, however, these things will occur, and are unavoidable, but usually a little common sense investigation will reveal the cause. The writer has often heard the questions asked: "How thick should cross-banding be?" "What is the object of cross-banding?" I take it that cross-banding is used to add strength and durability to the veneered article. If that is so, then the amount of additional strength required will determine the thickness of the cross-banding required.

Wood is influenced by atmospheric changes, and thin wood, of course, will respond more readily than thick. However, there is this difference between them—while the thin wood will respond more readily than will the thick to the atmospheric changes, it is much more easily held in check. The thick wood, being slow to respond, moves with a force in proportion to its thickness when once it begins to move. Hence to hold it in check a force equal in strength must be opposed to it.

For example, take a panel three-eighths of an inch thick, five ply. Now, no matter what thickness of veneer or cross-banding we use, the centre, or core, will necessarily be thin, so that no great force will be required to prevent expansion, or even contraction, under atmospheric changes. Take a dresser or buffet top, say, one inch thick. A greater resisting force would be required to prevent it from yielding to the atmospheric changes. It follows, then, that in order to accomplish the purpose for which cross-banding is used a thickness must be used in proportion to the thickness of the core and the force it will have to resist.

The question often arises, "What are the best cauls for use in veneering?" There are many opinions held regarding cauls. All manner of cauls have their adherents. Personally, I believe that the wooden cauls are the best, and, if made properly, are the most satisfactory. It is quite true that more stock can be put into the press at one time with the metallic cauls. This, of course, is on account of their thinness. The writer has seen wood cauls made of three-ply stock, with cloth between the sections, and also made of three-ply rotary birch, ¼-inch core and about 3/16-inch face. These were sanded down to a smooth surface and then coated with an oil filler. Such cauls are greatly in favor in the large English factories. They are there declared to be the best and to give the most satisfaction under any system of veneering, whether hot or cold glue is used.

Wood cauls will retain heat longer than any others, and, if handled properly, will last for an indefinite period. I have seen wood cauls that have been in use for over eight years and were still in good condition. Of course, there are objections to using wood cauls—they are easily damaged; only a small amount of stock can be put into the press at one operation; they are hard to keep clean, etc. Metal cauls are favored on account of the greater facilities they offer for keeping them clean; also their cleanness, which makes it possible to get more stock into one pressing. Aluminum cauls, however, are preferable to any other kinds. Their lasting qualities are greater than any others, and, on account of them being so light it is possible to materially increase the thickness without making them any heavier than the zinc cauls. The thicker cauls hold the heat better, and this is reflected in the uniformly better setting of the glue.

Aluminum cauls do not wear thin in places, they leave the work free and clean, they are light to handle, and retain the heat well. They should, however, never be scraped, but cleaned with some soft waste. I have seen aluminum cauls that have been in use for years and were in the best of condition, while zinc cauls had been worn out. Use prepared caul wax, an application of which, before using, will ensure the cauls leaving the work free and clean.

Inferior veneer work is often caused by the cauls being too hot. They should never be so hot that they cannot be handled with the bare hands. Glue is often burned and spoiled owing to the heat of the cauls. Glue is dissolved usually at a temperature of about 150 degrees Fah., which is, however, always done under water protection. Attempt to melt glue at that temperature in the dry state and the result would be spoiled glue. Apply this to the glue that has the water forced out of it by coming into contact with hot cauls.

Common sense should be used in the heating of cauls, and whatever cauls are used they should be kept in the very best of condition to ensure good work. But for real good work use wood cauls and give them a fair trial. I

Properly laid veneers will help the output of the factory, keep down the cost of production, and will be a credit to the manufacturer in the finished product. Surely it is worth while to give a little attention to the veneer room, especially when the first cost is so trifling compared with the results.

Veneering Curved or Moulded Surfaces
By a Plane Mechanic

In addition to the ordinary troubles of laying veneers on straight surfaces there is the work of providing a properly fitting caul for the crooked job. If the core is built up or glued up of ribs and then band-sawed, excellent cauls are at once obtained. The saw-cut generally takes care of the thickness of the veneer; any difference can be adjusted, if need be.

For narrow work a lot of cheap veneers, one on top of the other to a thickness of about ¾-in., will at times be found useful as a caul, and if the first layer be a sheet of zinc it will keep the heat longer. For large work, such as panels or solid doors, where band-sawing is out of the question and the core has to be moulded or glued up of ribs and worked to shape, the cauls are made in different ways. For fairly flat work it is sufficient to take straight boards and run grooves or saw-cuts on the back deep enough and close enough together to allow of bending under pressure. A number of strong crosspieces are sawed out to correspond to the shape of both sides of the curve.

Or a number of narrow strips of exactly the same thickness are fastened to these crosspieces, close together and at right angles thereto. The sawed-out pieces should be the proper distance apart, according to the presses. Some antique furniture workers provide these narrow strips with exactly bored holes near the ends, putting a thin, flexible wire through these holes, and the cauls thus formed are laid loosely on the cleats.

For moulded rails or ogee-shaped drawer fronts and similar work to be crotch-veneered (upright), it is necessary to make well-fitting cauls. If we cannot have them molded they may be made by marking both ends and cutting them nearly to shape on jointer and rip-saw, afterwards trimming by hand. The veneers may prove to be too brittle or too thick to readily bend into the shape of the core. They must then be first given the desired shape (before gluing or laying them). Both sides of the veneer are sponged with hot water, the veneer is laid on the core, and the hot caul is clamped into place. It is understood that as many pieces of veneer are treated at the same time as are necessary to make the required width. After several hours this is repeated (without the sponging, of course), until the veneer is quite dry. Better dry it in this manner than to leave it exposed to dry air, as it may buckle up or crack. The joints are made and bound together, and binding strips are also put along the edges, as this sort of veneer (crotch) is very easily damaged.

The following precautions should be observed to insure success: The core should be fairly "dead" stuff—that is, dry material that will shrink or warp but little. It is, therefore, best glued up of narrow strips. The core should have only ordinary temperature to keep from taking up moisture from the glue, which latter ought to be, therefore, strong, as compared with glue for ordinary purposes. The glue must be allowed to thoroughly chill (that does not mean that the glue should become dry—just past the tacky state. The flat hand may be moved over it without sticking. The heat of the caul will take care of this condition of the glue, and there is little danger of overheating wooden cauls. But certainly the caul must be hot enough to make the glue flow again, which it will readily do,) before laying the veneer, so that the latter will absorb only very little moisture from the glue. The tendency of the veneer to swell is thus overcome by the heat of the caul. The first pressure should be applied quickly at the centre, working gradually towards the ends, to permit the air and superfluous glue to escape.

Last, but not least, damp weather conditions should be avoided or artificially overcome.—Veneers.

Figured American Walnut

George W. Hartzell, Piqua, Ohio, is one of the pioneer manufacturers of walnut lumber and fancy veneers in the United States. The business has been established for over forty years, and during that time many fine trees of American walnut have passed through his hands. Recently, however, he cut a tree within about forty miles of Piqua which he says is one of the most highly figured pieces of walnut that he has ever seen, and the accompanying illustration, Fig. 1

Fig. 1—View in warehouse, showing substantial crates in which veneers are shipped.

(made from a photograph), would seem to bear out the statement. The tree, which is regarded as a very rare specimen of American walnut, contained about 10,000 feet of veneer, and the figure throughout is very well represented by the illustration.

A point which is well worth taking up here is the kind of crates used by George W. Hartzell for shipping his veneers in. He believes that when a customer places an order with him for veneers he is entitled to have those veneers reach him in perfect condition

(barring accident), and for that reason he pays particular attention to his crates, using none but the very best. The photograph, Fig. 2, taken in his veneer warehouse, illustrates the point. The knowledge that they will receive their shipments of veneer in good, substantial crates no doubt bears considerable weight with the majority of veneer users when placing their orders.

Veneers

The qualities of some woods are such that the wood can be used best as a veneer. This may be due to weight, in which a soft pine core can be built up and covered with a thin layer of the heavier and more expensive wood. It may be that the wood is subject to extreme shrinkage, in which case the core readily holds it in place. It may be that the grain of the wood is best brought out by a rotary cut, taking a shaving off around the log rather than by sawing the log into flat sheets. Birch, mahogany, beech, maple, walnut, red gum, and other woods are used in this way. Other woods, notably oak, show best when cut on radial lines or "quarter-sawed." Fewer cuts can be obtained by this method, but the results usually justify the extra expense.

With some other woods it seems to make very little difference how the cut is made, but there are other qualities which influence its use.

Woodwork for Secondary Schools

The Manual Arts Press, Peoria, Ill., have recently placed on the market a book by Prof. Ira S. Griffith, bearing the above title. There are nine chapters in the book, as follows: Common Woods, Tools and Processes, Woodworking Machines, Joinery, Wood Turning, Inlaying and Wood Carving, Wood Finishing, Furniture Construction, Pattern Making. A feature of the book is the number of drawings and sketches which it contains. These go a long way in helping the student to a complete understanding of the text.

Although the book is written particularly for students in industrial schools, yet there are very few mechanics who will not find in it something that it will pay them to know. The book can be obtained from the publishers, The Manual Arts Press, for $1.75.

Fig. 2—Beautifully figured piece of American Walnut Veneer manufactured at the plant of Geo. W. Hartzell, Piqua, Ohio.

THE OHIO VENEER COMPANY

Importers and Manufacturers

Foreign and Domestic Veneers and Hardwood Lumber

We always carry a large and assorted stock of Mahogany, Circassian
Walnut, Sawed and Sliced Quartered Oak.

Send us your enquiries and orders.   We guarantee good service.

2624 to 2644 Colerain Avenue,   ⋈   ⋈   CINCINNATI, OHIO.

Indiana White Oak
Lumber and Veneers

Band sawed from the best growth of timber.
We use only the best quality of Indiana White Oak.
From four to five million feet always on hand.

Walnut Lumber

Our stock assures you good service.

Chas. H. Barnaby, Greencastle, Indiana

 Hoffman Bros. Co.

Estab. 1867, Incor. 1904

**800 West Main Street,
FORT WAYNE, INDIANA**

Manufacturers of

VENEERS and LUMBER

In the Domestic Hardwoods

ANY THICKNESS,

1/24 and 1/30 Slice Cut
(Dried flat with Smith Roller Dryer)

1/20 and thicker Sawed Veneers,
Band Sawn Lumber.

— SPECIALTY —

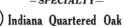 **Indiana Quartered Oak**

Veneers and Panels

PANELS

Stock Sizes for Immediate Shipment
All Woods   All Thicknesses   3 and 5 Ply
or made to your specifications

VENEERS

5 Million feet for Immediate Shipment
Any Kind of Wood   Any Thickness

J. J. NARTZIK

1966–76 Maud Ave.,   ⋈   ⋈   CHICAGO, ILL.

Veneers $_a$n^d Panels

LARGE STOCKS FOR
QUICK   SHIPMENTS

Send for Lists

Walter Clark Veneer Co.

GRAND RAPIDS, MICH.

Wood - Mosaic Co.

NEW ALBANY,   -   -   IND.

WALNUT

Lumber - Dimensions - Veneers

If you are in the market for anything in Walnut, let us figure with you. We have cut three million feet of Walnut in the last year, and have a good stock of all thicknesses dry. We can also furnish Walnut dimension—**Walnut Veneers.** We have a large stock of veneers, plain, stripe figured and stump wood.

Wood-Mosaic Co., Inc.

NEW ALBANY,   -   -   IND.

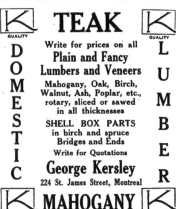

TEAK

Write for prices on all
Plain and Fancy
Lumbers and Veneers
Mahogany, Oak, Birch,
Walnut, Ash, Poplar, etc.,
rotary, sliced or sawed
in all thicknesses

SHELL BOX PARTS
in birch and spruce
Bridges and Ends
Write for Quotations

George Kersley
224 St. James Street, Montreal

MAHOGANY

DOMESTIC

LUMBER

American Walnut Veneers
Figured Wood
Sliced - Half Round - Stump Wood

5 Essential Qualifications:

1.—Unlimited supply of Walnut Logs.
2.—Logs carefully selected for figure.
3.—All processes of manufacture at our own plant under personal supervision.
4.—Dried flat in Philadelphia Textile Dryer.
5.—3,000,000 feet ready for shipment.

JOHN N. ROBERTS,   New Albany, Indiana
The Pioneer Walnut Veneer Producer of America

OAK   MAHOGANY   CIRCASSIAN   MAPLE

VENEERS

BIRCH   POPLAR   CHERRY   WALNUT   GUM

Dear Mr. Consumer:

If you're using walnut, you can't afford to take chances with anything but the best.

Properly manufactured, properly matched and properly laid, American Black Walnut is the most beautiful of woods.

But you must have the right foundation in the material.

When it comes from Penrod, it's right. Just remember that, when you need Walnut Veneers or Lumber, Plain Stock or Figured.

Penrod Walnut & Veneer Co.
"WALNUT SPECIALISTS FOR THIRTY YEARS"
KANSAS CITY, Mo.

We specialize on

Poplar

cross-banding and backing

Veneer

Highest Grade — Right Prices

The largest Poplar Veneer Mill in the world.

The
= Central Veneer Co.
Huntington, West Virginia

MAHOGANY VENEERS
Notwithstanding the scarcity, we are in a position to take care of our customers.

CIRCASSIAN
Some stocks on hand suitable for interior finish and furniture manufacturers.

FIGURED AMERICAN WALNUT
Over two million feet, nice striped and curley wood.

QUARTER SAWED WHITE AND RED OAK
1-20 to 1-8, 6 to 15-in. wide

GRAND RIM POPLAR
Cut to size. We can take care of your 1916 requirements from choice Yellow Poplar logs.

FIGURED WALNUT BUTTS
Two hundred twenty-three thousand feet newly cut Butts, some highly colored stock in this lot.

MAHOGANY LUMBER
This is something we make a specialty of, representing in Chicago and the Northwest, one of the largest importers.

HAUGHTON VENEER COMPANY
1152 W. LAKE STREET   CHICAGO, ILLINOIS

News of the Trade

The sash and door factory at Weyburn, Sask., belonging to J. Ford & Sons, was destroyed by fire last week.

The Brantford Carriage Company has purchased the stock and trade and the good-will of the Baynes Carriage Company, of Hamilton, Ont.

The stave and saw mills of the Fesserton Timber Company, Ltd., Fesserton, Ont., were totally destroyed by fire on August 8.

A large hydro-aeroplane is being built at the plant of Henry Morgan & Co., of Montreal. It is being constructed almost entirely of wood.

The Gibson Casket Company, of Mount Forest, Ont., had a quantity of lumber destroyed and their warehouse damaged by fire about two weeks ago.

The Preston Planing Mills, at Medicine Hat, Alta., are reported to be working on one of the largest orders for boxes ever placed in Western Canada.

The plant of the Quebec Shipbuilding and Repair Company. Ltd., which has been incorporated recently, with capital of $40,000, will include a woodworking shop.

Cheese factories in Western Ontario are complaining about the shortage of boxes to ship their cheese in. It is estimated that it would take 4,000 boxes to relieve the shortage at the present time.

The inability of Nasmiths Limited to secure sufficient wooden boxes to transport bread from Toronto to Borden Camp has resulted in complaints about the bread being mouldy from being transported in cardboard cartons.

The sash and door factory belonging to W. G. Scrim & Co., Vancouver, has been destroyed by fire. The loss is estimated at $9,000, of which $5,000 is stock. The stock was insured for about $3,000.

The plant of Ewing & Murphy, wood turners, at 18 Cameron Street, Toronto, was recently destroyed by fire. The machinery escaped damage, but every inch of the belting was destroyed.

The Snyder Desk and Table Company, Ltd., Waterloo, Ont., have obtained a charter. The company is capitalized at $15,000, and those interested are H. M. Snyder, A. H. Snyder, and C. H. Snyder, all of Waterloo.

The Lafleur Industrielle, Inc., Ste. Agathe des Monts, Que., advise us that their sash and door factory is operating as usual. A recent report created the impression that their factory had been destroyed by fire. This report was not correct.

The "Magog Furniture Manufacturers, Limited," are erecting a factory at Magog, P.Q. Those interested in the new company are N. C. Gendron, G. Roliand, Jr., J. D. Hamel, L. Fontaine, all of Magog, and G. H. Cotton, of Waterville, P.Q.

Mr. James Bushty and P. Villeneauve have started to manufacture fish boxes at Thessalon, Ont. They have secured contracts in advance from fishermen on the North Channel for over 10,000 boxes, which will keep them occupied for a good portion of the winter.

The "London Advertiser" states that the new planing mill being built at Harrow, Ont., for C. F. Smith, to replace the one recently destroyed by fire, will be built of cement block, and will be equipped with the latest machinery, driven by motors.

The furniture and piano trades in Ontario and Quebec are at the present time well booked up with orders, states

Mr. George Kersley, of Montreal, who specializes in veneers and hardwoods. Mr. Kersley has just returned from a visit to the principal woodworking centres of Ontario, and gives an optimistic account of the trade outlook. The general business conditions are favorable for the furniture, piano, and allied industries, and these appear likely to continue for some time.

Wilson Brothers, of Collingwood, Ont., have entered into an amalgamation with the Collingwood Hardwood Lumber Company and R. Feigehen, and will carry on business under the name of Wilson Brothers, Ltd. The business formerly conducted by Wilson Brothers, namely, the manufacture of interior finish, hardwood veneered doors to detail, stairs, and stair material and hardwood flooring in maple, beech, birch, and oak, will be continued. The hardwood flooring manufactured by Wilson Brothers is widely known for its excellent quality. As a result of the amalgamation the company will be in a position to carry on more extensively the wholesale business formerly done by their new associates. Having their own hardwood limits and sawmill, they will be in a position to quote on and supply all kinds of building material, bill stuff, and heavy hardwood lumber. Their dry kiln has a capacity of 150,000 feet, which enables them to supply kiln-dried lumber when required.

Methods of Polishing Checkers

The cheaper class of boxwood checkers are simply coated with a good quality spirit varnish, but high-grade goods are polished in the lathe. The polish that is used and the method of applying the polish differ slightly from the method that is employed in polishing flat surfaces. A bright finish on both sides and edges is only obtained after several handlings, the chief difficulty being the manipulation in the early stages, such as the provision of suitable chucks, the avoidance of the use of glass-paper, and the knack of using the polish so that it will not clog up the finer grooves. If ordinary French polish is used, it should not be applied with raw wadding. A wad made from a rubber that has been used on other work should be employed, so that there will be less loose fluff sticking on the work while the polishing is being done. The wad would not require the rag covering that is usual on flat surfaces. If a lathe is not available, very good results can be obtained by using polish for sealing up the pores of the wood and forming a smooth foundation, and then applying carefully a coating of good quality clear spirit varnish. Black goods should be stained first with French black water stain, and the polishing done with black polish. White polish made from bleached shellac, or a transparent polish, should be used in preference to polish that is made from orange or lemon shellac.

New Consignment of Mahogany Veneer

The Toronto Veneer Company, 93-99 Spadina Avenue, Toronto, have received a new consignment of mahogany veneer, totalling half a million feet, which is ready for the inspection of veneer users. The company have a large, well-lighted show room at the above address, where their stock can be seen to advantage. A cordial invitation is extended by them to all those who are interested in veneers and propose visiting the Toronto Exhibition to call and examine their stock.

It is easy enough to select a machine for one specific kind of work. The trouble arises from the fact that usually a machine is needed to do several kinds of work, and what is a help in some is a hindrance in others.

Panels and Tops

Deskbeds, Table Tops, Chiffonier and Dresser Tops
Three and Five Ply Panels

Large assortment of stock panels always on hand.
Three and five ply stock for cartridge boxes.
Your inquiry will be fully appreciated and promptly answered with full information.

WRITE US

The
Gorham Brothers Company

MT. PLEASANT   -   MICHIGAN

Long - Knight Lumber Co.

Manufacturers and
Wholesale Dealers

INDIANAPOLIS, IND., U. S. A.

HARDWOOD LUMBER

Walnut, Mahogany
and
American Walnut Veneers

SPECIAL :

5 cars 5/4 1 & 2 Pl. Red Oak

This Tests a Belt

A belt that will run continuously at a high speed, over small pulleys is truly a good belt.

The "run" illustrated is a typical instance where Amphibia Crown Planer Belting has solved a difficult belting problem.

For over 40 years we have been tanners and manufacturers of the best leather belts.

Try a sample run.

Sadler & Haworth

MONTREAL   511 William Street

TORONTO   ST. JOHN
38 Wellington St. East   149 Prince William St.

WINNIPEG   VANCOUVER
Galt Building   107-111 Water St.

"Leather like gold has no substitute."

| David Gillies | J. S. Gillies | D. A. Gillies |
| President | Vice-Pres. and Man. Dir. | Sec.-Treas. |

GILLIES BROS.
Established 1873   LIMITED

Mills and Head Office
BRAESIDE, ONT.

— Manufacturers of —

White Pine - Red Pine - Spruce

We offer Lumber for Box and Crating:

WHITE PINE MILL RUN		RED PINE LOG RUN	
4 4 x 4 10 x 6 9 0	500 M	4/4 to 8/4 x 4/5/6/7/8	
4 4 to 8/4 x 6 7 8 x		in x 10/17	1500 M
10/16	900 M	RED PINE BOX & CULLS	
WHITE PINE BOX		4/4 to 8/4 x 4 and up	
4 4 to 8 4 x 4 and up		x 6/16	1000 M
x 6/16	400 M	SPRUCE MILL RUN	
WHITE PINE MILL CULLS		2 x 4 and up x 6/22	25 M
4 4 to 8/4 x 4 and up x		SPRUCE BOX & CULLS	
4 10	200 M	4/4, 5/4, 8/4 x 4 and	
4 4 x 10 11 12 x 6/16	300 M	up x 6/16	500 M

The above stock is dry and much of it is assorted widths.

The Atlantic Lumber Co.

Manufacturers and Wholesale Dealers

American and Canadian Hardwood Lumber and Mahogany, Quartered and Plain, White and Red Oak a Specialty.

110 Manning Chambers - TORONTO
Phone Main 6386

HOME OFFICE, BOSTON MASS.

Mills: KNOXVILLE, TENN.,   WALLAND, TENN.
FRANKLIN, VA.

DARLINGS STEAM APPLIANCES

DARLING BROTHERS
LIMITED
Engineers and Manufacturers
MONTREAL, CANADA

Branches: Agents:
Toronto and Winnipeg.   Halifax, St. John, Calgary, Vancouver

Shell Box Stock

1 x 6 birch, in No. 1, 2, and 3 common.

1 x 7 and 8 birch, in No. 1, 2, and 3 common.

1 x 8¼ birch, in No. 1, 2, and 3 common

1 x 9 and wider, birch, in No. 1, 2, and 3 common.

1 x 4 and wider, spruce. 2 x 4 and wider, spruce.

For the Furniture Trade

4/4, 5/4, 6/4, 8/4, 10/4, 12/4, 16/4, birch, maple, elm, basswood and black ash.

Elm, Basswood and Black Ash

in No. 1 common and better grades—dry stock in assorted sizes.

Hart & McDonagh
Continental Life Bldg.

TORONTO   - -   ONTARIO

FOR SALE

Black Walnut Lumber

15,000	feet	4/4	No. 1 Common
20,000	"	4/4	No. 2	"
15,000	"	5/4	No. 1	"
15,000	"	5/4	No. 2	"
15,000	"	6/4	No. 2	"
40,000	"	8/4	No. 1	"
60,000	"	8/4	No. 2	"
30,000	"	8/4	No. 1	"

Will make special low prices on all of the above stock to move it quick. All Band Sawn, Equalized, and all thoroughly dry, ready for immediate shipment. Write for prices.

Geo. W. Hartzell
Piqua, Ohio

Canadian Agents Wanted

Reliable agents wanted in Canada to handle a waterproof coating for finishing concrete stucco, wood surfaces, etc., where a flat, lustreless finish is desired. Commission basis. If interested please reply to Box 20, Canadian Woodworker and Furniture Manufacturer, 345 Adelaide St. West, Toronto.　　8

Second-Hand Machines For Sale

1—one arm Cowan Sander, good as new.

1—26-in. double surface revolving bed Cowan planer.

1—6-in. four-sided Cowan moulder.

All in good working order. Operate entirely satisfactory, but replaced by different type of machines.

T. H. HANCOCK.

8　　1372 Bloor St. West, Toronto.

Machinery FOR SALE

1 Jackson-Cochrane 54-in. Resaw; 1 Cowan 42-in. Sander; 1 Jackson-Cochrane 36-in. Sander; 1 Egan 36-in. single Surfacer; 1 McGregor-Gourlay 2-drum 30-in. boss Sander; 1 Egan 30-in. 2-drum Sander; 1 C. M. C. 8-in. 4 side Sticker; 1 McGregor No. 212 30-in. Double Surfacer; 1 No. 180 Berlin 30-in. Double Surfacer; 1 Ballantyne 12-in. Sticker.

P. B. Yates Machine Co., Limited
Hamilton, Ontario

t-f

Second-Hand Woodworking Machines

Hand Saws—26-in., 36-in., 40-in.

Jointers—12-in., 16-in.

Shapers—Single and double spindle.

Re-Saw—42-in. American.

Lathes—20-in. with rests and chisels.

Surfacers—16 x 6 in. and 18 x 8 in.

Mortisers—No. 5 and No. 7 New Britain.

Rip and Cut-off Saws; Molders; Sash Relishers, etc.

Over 100 machines, New and Used, in stock.

Baird Machinery Company,
123 Water Street,

PITTSBURGH, PA.

FOR SALE

Planing Mills, Chopping Mills and Machine Shop. Bargain for someone; closing the estate of the late G. Chambers, Watford, Ont.　　8

LUMBER WANTED
For Various Shell Boxes

Spruce or other soft woods, all grades, 5/8 in. to 4 in. Hardwood 2 in.—all grades. Must be absolutely dry stock. Give width and full particulars. Prices F.O.B. Montreal. Terms cash 30 days. Address

CASTLE & SON,

200-202 Papineau Ave.,

Montreal, Que.

Toy Exhibit in London

In a recent issue of the New York Herald, the following article from their London office appeared: The idea that England was necessarily dependent on Germany for her toy supply is one of the illustrations shattered by the war. In future Hun toys will be replaced by British made toys or toys made in America or by Britain's allies in the present war. One is directly reminded of the Hun in the Toy Exhibition at the Whitehall Art Gallery only by some unsparing effigies of German notabilities and a set of ninepins surmounted by pickelhauben.

Since the war began great developments have taken place in the toymaking industry in England, and the progress made in design and craftsmanship is well illustrated in this exhibition.

Prominent among the larger toys is a set of seven models, the Gates of Old London, just as they appeared about the middle of the eighteenth century. A collection of silhouette toys, made of wood, includes a hunt in full cry. For the warlike there are knights, clad in realistic tissue armour, on mettlesome chargers of shock resisting wood.

A Tudor doll's house, the work of the Queen's Park toymakers, was bought by Queen Mary.

Some of the most attractive exhibits have been made by disabled soldiers and sailors and much is the work of children. Under the supervision they will turn out a market place with every figure keen on a bargain; or again, fired by Scott or another, a tournament in which the horses are no less desperately in earnest than the combatants.

There are, of course, many charming dolls in elegant apparel and animals innumerable. One remembers especially a stuffed elephant positively bursting with suppressed sagacity. And all are amiable save one mirthless monkey, which is harranguing a guffawing hippopotamus.

There are examples of work by men and women artists, East End working girls, cripples, cottagers, all arranged with artistic care. One is especially struck by the wit and humor, the vitality and movement of line which characterize many of the toys. In the Noah's arks, for example, cleverly chiselled animals literally prance into the ark, each with an individual and jocund gait.

PETRIE'S
MONTHLY LIST
of
NEW and USED
WOOD TOOLS
in stock for immediate delivery

Wood Planers
36" American double surfacer.
30" Whitney pattern surfacer.
26" Revolving bed double surfacers.
24" Major Harper planer and matcher.
24" Single surfacers, various makes.
20" Dundas pony planer.
18" Little Giant planer and matcher.
10" Galt Jointer.

Band Saws
42" Fay & Egan power feed rip.
38" At antic tilting frame.
30" Ideal pedestal, tilting table.

Saw Tables
Ballantine variable power feed.
Preston variable power feed.
Cowan heavy power feed.   M138.
No. 3 Crescent universal.
No. 2 Crescent combination.
Ideal variety, tilting table.
12" Defiance automatic equalizer.
MacGregor, Gourlay railway cut-off.
1½ iron frame swing.

Mortisers
No. 5 New Britain, chain.
Cowan hollow chisel.   M.190.
Galt upright, compound table.
Smart foot power.

Moulders
13" Clark-Demill four-side.
12" Cowan four side.
10" Houston four side.
0" Cowan four side.
6" Dundas sash sticker.

Sanders
24" Fay double drum.
12" C.M.C. disk and drum.
18" Crescent disk, adjustable table.

Clothespin Machinery
Humphrey No. 8 giant slab resaw.
Humphrey gang slitter.
Humphrey cylinder cutting-off machine.
Humphrey automatic lathes (6).
Humphrey double slotters (4).

Miscellaneous
Fay & Egan 12-spindle dovetailer.
Cowan dowel machine, M80.
No. 235 Wysong & Miles post boring machine.
Cowan post boring machine, M85G.
Cowan post boring machine, M23.
MacGregor Gourlay 2-spindle shaper.
Elliott single spindle shaper.
Egan sash and door tenoner.
No. 6 Lion universal woodtrimmer.
6-nail box nailer, Meyers patent.
20" American wood scraper.
16" Ideal wood lathes (3).
4-head rounding machines.
Cowan spindle carver. M03
Cowan veneer press screws.

Prices, Descriptions and Full Particulars on Request

H. W. PETRIE, LTD.
Front St. W.,　　Toronto, Ont.

USE

SIMONDS SAWS

AND

PLANER

KNIVES

We manufacture Saws of every kind and size, Circular or Band Saws, Concave, Mitre or Novelty Saws for rapid smooth cutting. Write us for our new catalogue and tell us what Saws or Knives you now require.

Simonds Canada Saw Co., Limited

Factory: St. Remi St. and Acorn Ave.,

St. John, N. B.   **Montreal**   Vancouver, B. C.

Morehead

SYSTEM

PRACTICAL POINTERS ON BOILER ROOM EFFICIENCY

IN THE NEW "Morehead" book you will find real ready-to-use information on the handling of your steam equipment which may save you thousands of dollars a year.

Besides a general discussion on the drainage of kilns and steam heated machinery generally, there are interesting

PICTURES AND DATA

which will show you how representative plants everywhere are turning their condensation into golden profit.

Write for the book today.

CANADIAN MOREHEAD MFG. CO.

WOODSTOCK   Dept. "H"   ONTARIO

GRINDING

BAND SAWS

with

26 TEETH

TO THE INCH

Your woodworking band saws can be resharpened so that all teeth come within ½ of 1/1000 of an inch alike to each other, by this

—Model A—

Automatic Band Saw Grinder

Saws 1-8 inch to 8 inches wide. Teeth up to 2 inch spacing

POINTS OF INTEREST

Bronze Bushed Grinding Wheel Spindle.
Head Suspended between Hardened Steel Pivots
No Slides to Wear. No Cams to Change.
Feed Movement obtained by a Scaled Slide
A Double Pawl, Positive Feed Movement, the quickest
and most perfect method of jointing all teeth,
can be furnished at a small extra charge.

LOWEST PRICED GRINDER MADE

Write for Details

THE WARDWELL MFG. COMPANY

Saw and Knife Sharpening Machinery

111 Hamilton Ave.   CLEVELAND, OHIO

The Glue Book

What it Contains

Chapter 1—Historical Notes

Chapter 2—Manufacture of Glue

Chapter 3—Testing and Grading

Chapter 4—Methods in the Glue Room

Chapter 5—Glue Room Equipment

Chapter 6—Selection of Glue

PRICE 50 CENTS

Woodworker Publishing Co.

Limited

345 Adelaide St. West, Toronto

Woodworkers

and

Lumbermen

The Sheldon Exhaust Fans have characteristics that adapt themselves to the service of the Planing Mill or Woodworking Plant. They are specially designed for this kind of work, having a saving in power and speed of 25% to 40%.

Write for Industrial Booklet.

SHELDONS LIMITED - Galt, Ont.

Toronto Office, 609 Kent Building

— Agents —

Messrs. Ross & Greig, 412 St. James St., Montreal, Que.
Messrs. Walker's Limited, 259-261 Stanley St., Winnipeg, Man.
Messrs. Gorman, Clancey & Grindley Ltd., Calgary and Edmonton, Alta
Messrs. Robt. Hamilton & Co., Ltd., Bk. of Ottawa Bldg. Vancouver, B.C.

50 CANADIAN WOODWORKER August, 1916

Peter Hay Machine Knives

Peter Hay knives are made of special quality steel tempered and toughened to take and hold a keen cutting edge. Hay knives are made for all purposes.
Have you our Catalogue?

Made in Canada

The Peter Hay Knife Co., Limited
GALT, ONTARIO

PRESSES

For Veneer and Veneer Drying

Made in Canada

William R. Perrin
Limited
Toronto

Flexibility

To vary the tempera-ture of a National VER-TICALLY - PIPED Dry Kiln, all
that's necessary is to turn on or turn off one or more coils. Each coil is supplied with steam *independently* of the rest—an advantage making for supreme flexibility—and found only in

Manufactured by

THE NATIONAL DRY KILN CO.
1117 E. Maryland St., Indianapolis, Ind., U.S.A.

You
CONTRACTORS
and
SMALL MILL MEN
Just the machine you need
—ANY KIND OF POWER—

A Whole Mill in Itself
A GREAT TOOL
Write for Catalogue and Prices NOW

J. L. NEILSON & CO.
Winnipeg, Man.

Acme Cold Rolled Strip Steel

STEEL BOX STRAPPING

Corrugated Joint Fasteners

SMOOTH EDGES—CONTINUOUS LENGTHS
COIL HOLDERS SUPPLIED
LACQUERED—GALVANIZED—BRIGHT
Write now for Catalog.

ACME STEEL GOODS CO.,
LIMITED
Coristine Building, MONTREAL, Quebec

Ribbon wound.—⅜" to 1½" and .013 to .035 in thickness

In bulk In Barrels and Boxes

In coils for Automatic Driving Machines.

PARALLEL SAW EDGE

DIVERGENT SAW EDGE

IT SOLVES THE LABOR PROBLEM

One boy can operate the NASH SANDER and do as much work as 5 men with sand belts. Better work too.   See the saving?

J. M. NASH, MILWAUKEE, Wis.

"OBER" signifies

DURABILITY in Construction
SIMPLICITY in Operation
EFFICIENCY in Production
of
HANDLES and other
TURNED WORK

We make LATHES for turning handles, spokes, whiffletrees, neck yokes and variety work.

SANDERS, RIPSAWS, CHUCKS, Etc.

Write for free Catalogue No. 31

THE OBER MFG. CO.
Chagrin Falls, O., U.S.A.

L OSS of power is one of the greatest drawbacks to modern manufacturing. It is constantly harassing the profits of the manufacturer. Oftentimes it is the real factor that stands between profits and losses.

Eliminate that loss of power caused by friction by installing Chapman Double Ball Bearings on your line and counter shafts.

Installation of the above bearings will reduce your power costs from 15% to 40%. The cash you paid for waste power during the past year and a half would have purchased a complete installation of Chapman Ball Bearings. Why not install them now, and cut out this unnecessary expense in the future?

No trouble to change over—our bearings suit any standard hanger

The Chapman Double Ball Bearing Co. of Canada, Limited
TORONTO   347 Sorauran Ave.   ONTARIO

The "Canadian Woodworker" Buyers' Directory

ABRASIVES
The Carborundum Co., Niagara Falls, N.Y.

BALL BEARINGS
Chapman Double Ball Bearing Co., Toronto.

BALUSTER LATHES
Garlock-Walker Machinery Co., Toronto, Ont.
Ober Mfg. Company, Chagrin Falls, Ohio.
Whitney & Son, Baxter D., Winchendon, Mass.

BAND SAW FILING MACHINERY
Canada Machinery Corporation, Galt, Ont.
Fay & Egan Co., J. A., Cincinnati, Ohio.
Garlock-Walker Machinery Co., Toronto, Ont.
Preston Woodworking Machinery Company,
Preston, Ont.
Yates Machine Co., P. B., Hamilton, Ont.

BAND SAWS
Canada Machinery Corporation, Galt, Ont.
Garlock-Walker Machinery Co., Toronto, Ont.
Jackson, Cochrane & Company, Berlin, Ont.
Petrie, H. W., Toronto.
Preston Woodworking Machinery Company,
Preston, Ont.
Yates Machine Co., P. B., Hamilton, Ont.

BAND SAW MACHINES
Canada Machinery Corporation, Galt, Ont.
Fay & Egan Co., J. A., Cincinnati, Ohio.
Garlock-Walker Machinery Co., Toronto, Ont.
Petrie, H. W., Toronto.
Preston Woodworking Machinery Company,
Preston, Ont.
Williams Machinery Co., A. R., Toronto, Ont.
Yates Machine Co., P. B., Hamilton, Ont.

BAND SAW STRETCHERS
Fay & Egan Co., J. A., Cincinnati, Ohio.
Yates Machine Co., P. B., Hamilton, Ont.

BASKET STITCHING MACHINE
Saranac Machine Co., Benton Harbor, Mich.

BENDING MACHINES
Fay & Egan Co., J. A., Cincinnati, Ohio.
Garlock-Walker Machinery Co., Toronto, Ont.
Perfection Wood Steaming Retort Company,
Parkersburg, W. Va.

BELTING
Fay & Egan Co., J. A., Cincinnati, Ohio.

BLOWERS
Sheldons, Limited, Galt, Ont.
Toronto Blower Company, Toronto, Ont.

BLOW PIPING
Sheldons, Limited, Galt, Ont.
Toronto Blower Company, Toronto, Ont.

BORING MACHINES
Canada Machinery Corporation, Galt, Ont.
Fay & Egan Co., J. A., Cincinnati, Ohio.
Garlock-Walker Machinery Co., Toronto, Ont.
Jackson, Cochrane & Company, Berlin, Ont.
Nash, J. M., Milwaukee, Wis.
Preston Woodworking Machinery Company,
Preston, Ont.
Reynolds Pattern & Machine Co., Moline, Ill.
Yates Machine Co., P. B., Hamilton, Ont.

BOX MAKERS' MACHINERY
Canada Machinery Corporation, Galt, Ont.
Fay & Egan Co., J. A., Cincinnati, Ohio.
Garlock-Walker Machinery Co., Toronto, Ont.
Jackson, Cochrane & Co., Berlin, Ont.
Neilson & Company, J. L. Winnipeg, Man.
Preston Woodworking Machinery Company,
Preston, Ont.
Reynolds Pattern & Machine Co., Moline, Ill.
Whitney & Son, Baxter D., Winchendon, Mass.
Williams Machinery Co., A. R., Toronto, Ont.
Yates Machine Co., P. B., Hamilton, Ont.

CABINET PLANERS
Canada Machinery Corporation, Galt, Ont.
Fay & Egan Co., J. A., Cincinnati, Ohio.
Garlock-Walker Machinery Co., Toronto, Ont.
Jackson, Cochrane & Co., Berlin, Ont.
Preston Woodworking Machinery Company,
Preston, Ont.
Yates Machine Co., P. B., Hamilton, Ont.

CARS (Transfer)
Sheldons, Limited, Galt, Ont.

CARBORUNDUM PAPER
Carborundum Co., Niagara Falls, N.Y.

CARVING MACHINES
Canada Machinery Corporation, Galt, Ont.
Fay & Egan Co., J. A., Cincinnati, Ohio.
Garlock-Walker Machinery Co., Toronto, Ont.
Jackson, Cochrane & Company, Berlin, Ont.

CASTERS
Foster, Merriam & Co., Meriden, Conn.

CLAMPS
Fay & Egan Co., J. A., Cincinnati, Ohio.
Garlock-Walker Machinery Co., Toronto, Ont.
Jackson, Cochrane & Company, Berlin, Ont.
Preston Woodworking Machinery Company,
Preston, Ont.
Simonds Canada Saw Company, Montreal, P.Q.

COLUMN MACHINERY
Fay & Egan Co., J. A., Cincinnati, Ohio.

CORRUGATED FASTENERS
Acme Steel Goods Company, Chicago, Ill.

CUT-OFF SAWS
Canada Machinery Corporation, Galt, Ont.
Fay & Egan Co., J. A., Cincinnati, Ohio.
Garlock-Walker Machinery Co., Toronto, Ont.
Jackson, Cochrane & Co., Berlin, Ont.
Ober Mfg. Co., Chagrin Falls, Ohio.
Preston Woodworking Machinery Company,
Preston, Ont.
Simonds Canada Saw Co., Montreal, Que.
Yates Machine Co., P. B., Hamilton, Ont.

CUTTER HEADS
Canada Machinery Corporation, Galt, Ont.
Fay & Egan Co., J. A., Cincinnati, Ohio.
Jackson, Cochrane & Company, Berlin, Ont.
Mattison Machine Works, Beloit, Wis.
Yates Machine Co., P. B., Hamilton, Ont.

DOVETAILING MACHINES
Canada Machinery Corporation, Galt, Ont.
Fay & Egan Co., J. A., Cincinnati, Ohio.
Garlock-Walker Machinery Co., Toronto, Ont.
Jackson, Cochrane & Company, Berlin, Ont.

DOWEL MACHINES
Canada Machinery Corporation, Galt, Ont.
Fay & Egan Co., J. A., Cincinnati, Ohio.
Ober Mfg. Co., Chagrin Falls, Ohio.

DRY KILNS
National Dry Kiln Co., Indianapolis, Ind.
Sheldons, Limited, Galt, Ont.

DUST COLLECTORS
Sheldons, Limited, Galt, Ont.
Toronto Blower Company, Toronto, Ont.

DUST SEPARATORS
Sheldons, Limited, Galt, Ont.
Toronto Blower Company, Toronto, Ont.

EDGERS (Single Saw)
Canada Machinery Corporation, Galt, Ont.
Fay & Egan Co., J. A., Cincinnati, Ohio.
Garlock-Walker Machinery Co., Toronto, Ont.
Simonds Canada Saw Co., Montreal, Que.

EDGERS (Gang)
Canada Machinery Corporation, Galt, Ont.
Fay & Egan Co., J. A., Cincinnati, Ohio.
Garlock-Walker Machinery Co., Toronto, Ont.
Petrie, H. W., Toronto.
Simonds Canada Saw Co., Montreal, P.Q.
Williams Machinery Co., A. R., Toronto, Ont.
Yates Machine Co., P. B., Hamilton, Ont.

END MATCHING MACHINE
Canada Machinery Corporation, Galt, Ont.
Fay & Egan Co., J. A., Cincinnati, Ohio.
Garlock-Walker Machinery Co., Toronto, Ont.
Jackson, Cochrane & Company, Berlin, Ont.
Yates Machine Co., P. B., Hamilton, Ont.

EXHAUST FANS
Garlock-Walker Machinery Co., Toronto, Ont.
Sheldons Limited, Galt, Ont.
Toronto Blower Company, Toronto, Ont.

EXHAUST SYSTEMS
Toronto Blower Company, Toronto, Ont.

FILES
Simonds Canada Saw Co., Montreal, Que.

FILING MACHINERY
Wardwell Mfg. Company, Cleveland, Ohio.

FLOORING MACHINES
Canada Machinery Corporation, Galt, Ont.
Fay & Egan Co., J. A., Cincinnati, Ohio.
Garlock-Walker Machinery Co., Toronto, Ont.
Petrie, H. W., Toronto.
Preston Woodworking Machinery Company,
Preston, Ont.
Whitney & Son, Baxter D., Winchendon, Mass.
Yates Machine Co., P. B., Hamilton, Ont.

FLUTING HEADS
Fay & Egan Co., J. A., Cincinnati, Ohio.

FLUTING AND TWIST MACHINE
Preston Woodworking Machinery Company,
Preston, Ont.

FURNITURE TRIMMINGS
Foster, Merriam & Co., Meriden, Conn.

GARNET PAPER AND CLOTH
The Carborundum Co., Niagara Falls, N.Y.

GRAINING MACHINES
Canada Machinery Corporation, Galt, Ont.
Fay & Egan Co., J. A., Cincinnati, Ohio.
Williams Machinery Co., A. R., Toronto, Ont.
Yates Machine Co., P. B., Hamilton, Ont.

GLUE
Colgnet Chemical Product Co., New York,
N.Y.
Perkins Glue Company, South Bend, Ind.

GLUE CLAMPS
Jackson, Cochrane & Company, Berlin, Ont.

GLUE HEATERS
Fay & Egan Co., J. A., Cincinnati, Ohio.
Jackson, Cochrane & Company, Berlin, Ont.

GLUE JOINTERS
Canada Machinery Corporation, Galt, Ont.
Jackson, Cochrane & Company, Berlin, Ont.
Yates Machine Co., P. B., Hamilton, Ont.

GLUE SPREADERS
Fay & Egan Co., J. A., Cincinnati, Ohio.
Jackson, Cochrane & Company, Berlin, Ont.

GLUE ROOM EQUIPMENT
Perrin & Company, W. R., Toronto, Ont.

GRINDERS (Cutter)
Fay & Egan Co., J. A., Cincinnati, Ohio.

GRINDERS, HAND AND FOOT POWER
The Carborundum Co., Niagara Falls, N.Y.

GRINDERS (Knife)
Canada Machinery Corporation, Galt, Ont.
Fay & Egan Co., J. A., Cincinnati, Ohio.
Garlock-Walker Machinery Co., Toronto, Ont.
Petrie, H. W., Toronto.
Preston Woodworking Machinery Company,
Preston, Ont.
Yates Machine Co., P. B., Hamilton, Ont.

GRINDERS (Tool)
Fay & Egan Co., J. A., Cincinnati, Ohio.
Jackson, Cochrane & Company, Berlin, Ont.

GRINDING WHEELS
The Carborundum Co., Niagara Falls, N.Y.

GROOVING HEADS
Fay & Egan Co., J. A., Cincinnati, Ohio.

GUMMERS, ETC.
Fay & Egan Co., J. A., Cincinnati, Ohio.

HAND PROTECTORS
Fay & Egan Co., J. A., Cincinnati, Ohio.

HAND SCREWS
Fay & Egan Co., J. A., Cincinnati, Ohio.

HANDLE THROATING MACHINE
Klotz Machine Company, Sandusky, Ohio.

HANDLE TRIMMING MACHINE
Klotz Machine Company, Sandusky, Ohio.

HANDLE & SPOKE MACHINERY
Fay & Egan Co., J. A., Cincinnati, Ohio.
Klotz Machine Co., Sandusky, Ohio.
Nash, J. M., Milwaukee, Wis.
Ober Mfg. Company, Chagrin Falls, Ohio.
Whitney & Son, Baxter D, Winchendon, Mass.

HARDWOOD LUMBER
American Hardwood Lumber Co., St. Louis, Mo.
Atlantic Lumber Company, Toronto.
Brown & Company, Geo. C., Memphis, Tenn.
Clark & Son, Edward, Toronto, Ont.
Kersley, Geo., Montreal, Que.
Kraetzer-Cured Lumber Co., Moorhead, Miss.
Mowbray & Robinson, Cincinnati, Ohio.
Paepcke Leicht Lumber Co., Chicago, Ill.
Penrod Walnut & Veneer Co., Kansas City, Mo.
Spencer, C. A., Montreal, Que.

HUB MACHINERY
Fay & Egan Co., J. A., Cincinnati, Ohio.

HYDRAULIC VENEER PRESSES
Perrin & Company, Wm. R., Toronto, Ont.

JOINTERS
Canada Machinery Corporation, Galt, Ont.
Fay & Egan Co., J. A., Cincinnati, Ohio.
Garlock-Walker Machinery Co., Toronto, Ont.
Jackson, Cochrane & Company, Berlin, Ont.
Knechtel Bros., Limited, Southampton, Ont.
Petrie, H. W., Toronto, Ont.
Preston Woodworking Machinery Company,
Preston, Ont.
Yates Machine Co., P. B., Hamilton, Ont.

KNIVES (Planer and others)
Canada Machinery Corporation, Galt, Ont.
Fay & Egan Co., J. A., Cincinnati, Ohio.
Simonds Canada Saw Co., Montreal, P.Q.
Yates Machine Co., P. B., Hamilton, Ont.

LATHES
Canada Machinery Corporation, Galt, Ont.
Fay & Egan Co., J. A., Cincinnati, Ohio.
Garlock-Walker Machinery Co., Toronto, Ont.
Jackson, Cochrane & Company, Berlin, Ont.
Klotz Machine Company, Sandusky, Ohio.
Ober Mfg. Company, Chagrin Falls, Ohio.
Petrie, H. W., Toronto, Ont.
Preston Woodworking Machinery Company,
Shawver Company, Springfield, O.
Whitney & Son, Baxter D., Winchendon, Mass.

LUMBER
American Hardwood Lumber Co., St. Louis, Mo.
B. C. Lumber Commissioner, Toronto.
Brown & Company, Geo. C., Memphis, Tenn.
Churchill, Milton Lumber Co., Louisville, Ky.
Clark & Son, Edward, Toronto, Ont.
Dominion Mahogany & Veneer Co., Montreal,
Que.
Gillies Bros., Braeside, Ont.
Gorham Bros. Co., Mt. Pleasant, Mich.
Hart & McDonagh, Toronto, Ont.
Hartzell, Geo. W., Piqua, Ohio.
Long-Knight Lumber Co., Indianapolis.
Paepcke Leicht Lumber Company, Chicago, Ill.

MACHINE KNIVES
Canada Machinery Corporation, Galt, Ont.
Peter Hay Knife Company, Galt, Ont.
Simonds Canada Saw Co., Montreal, P.Q.
Yates Machine Co., P. B., Hamilton, Ont.

MITRE MACHINES
Canada Machinery Corporation, Galt, Ont.
Fay & Egan Co., J. A., Cincinnati, Ohio.

MITRE SAWS
Canada Machinery Corporation, Galt, Ont.
Fay & Egan Co., J. A., Cincinnati, Ohio.
Garlock-Walker Machinery Co., Toronto, Ont.
Simonds Canada Saw Co., Montreal, P.Q.

MORTISING MACHINES
Canada Machinery Corporation, Galt, Ont.
Fay & Egan Co., J. A., Cincinnati, Ohio.
Garlock-Walker Machinery Co., Toronto, Ont.
Yates Machine Co., P. B., Hamilton, Ont.

MOULDINGS
Knechtel Bros., Limited, Southampton, Ont.

MULTIPLE BOXING MACHINES
Fay & Egan Co., J. A., Cincinnati, Ohio.
Nash, J. M., Milwaukee, Wis.
Reynolds Pattern & Machine Co., Moline, Ill.

PATTERN SHOP MACHINES
Canada Machinery Corporation, Galt, Ont.
Fay & Egan Co., J. A., Cincinnati, Ohio.
Garlock-Walker Machinery Co., Toronto, Ont.
Jackson, Cochrane & Company, Berlin, Ont.
Preston Woodworking Machinery Company,
Preston, Ont.
Whitney & Son, Baxter D., Winchendon, Mass.
Yates Machine Co., P. B., Hamilton, Ont.

PLANERS
Canada Machinery Corporation, Galt, Ont.
Fay & Egan Co., J. A., Cincinnati, Ohio.
Garlock-Walker Machinery Co., Toronto, Ont.
Jackson, Cochrane & Company, Berlin, Ont.
Petrie, H. W., Toronto, Ont.
Preston Woodworking Machinery Company,
Whitney & Son, Baxter D., Winchendon, Mass.
Williams Machinery Co., A. R., Toronto, Ont.
Yates Machine Co., P. B., Hamilton, Ont.

PLANING MILL MACHINERY
Canada Machinery Corporation, Galt, Ont.
Fay & Egan Co., J. A., Cincinnati, Ohio.
Garlock-Walker Machinery Co., Toronto, Ont.
Jackson, Cochrane & Company, Berlin, Ont.
Petrie, H. W., Toronto, Ont.
Preston Woodworking Machinery Company,
Whitney & Son, Baxter D., Winchendon, Mass.
Williams Machinery Co., A. R., Toronto, Ont.
Yates Machine Co., P. B., Hamilton, Ont.

PRESSES (Veneer)
Canada Machinery Corporation, Galt, Ont.
Jackson, Cochrane & Company, Berlin, Ont.
Perrin & Co., Wm. R., Toronto, Ont.
Preston Woodworking Machinery Company,
Preston, Ont.

PULLEYS
Fay & Egan Co., J. A., Cincinnati, Ohio.

RED GUM LUMBER
Gum Lumber Manufacturers' Association.

RESAWS
Canada Machinery Corporation, Galt, Ont.
Fay & Egan Co., J. A., Cincinnati, Ohio.
Garlock-Walker Machinery Co., Toronto, Ont.
Jackson, Cochrane & Company, Berlin, Ont.
Petrie, H. W., Toronto, Ont.
Preston Woodworking Machinery Company,
Preston, Ont.
Simonds Canada Saw Co., Montreal, P.Q.
Williams Machinery Co., A. R., Toronto, Ont.
Yates Machine Co., P. B., Hamilton, Ont.

RIM AND FELLOE MACHINERY
Fay & Egan Co., J. A., Cincinnati, Ohio.

RIP SAWING MACHINES
Canada Machinery Corporation, Galt, Ont.
Fay & Egan Co., J. A., Cincinnati, Ohio.
Garlock-Walker Machinery Co., Toronto, Ont.
Klotz Machine Company, Sandusky, Ohio.
Jackson, Cochrane & Company, Berlin, Ont.
Ober Mfg. Company, Chagrin Falls, Ohio.
Preston Woodworking Machinery Company,
Preston, Ont.
Yates Machine Co., P. B., Hamilton, Ont.

SANDERS
Canada Machinery Corporation, Galt, Ont.
Fay & Egan Co., J. A., Cincinnati, Ohio.
Garlock-Walker Machinery Co., Toronto, Ont.
Klotz Machine Company, Sandusky, Ohio.
Jackson, Cochrane & Company, Berlin, Ont.
Nash, J. M., Milwaukee, Wis.
Ober Mfg. Company, Chagrin Falls, Ohio.
Preston Woodworking Machinery Company,
Preston, Ont.
Yates Machine Co., P. B., Hamilton, Ont.

SASH, DOOR & BLIND MACHINERY
Canada Machinery Corporation, Galt, Ont.
Fay & Egan Co., J. A., Cincinnati, Ohio.
Garlock-Walker Machinery Co., Toronto, Ont.
Jackson, Cochrane & Company, Berlin, Ont.
Preston Woodworking Machinery Company,
Preston, Ont.
Yates Machine Co., P. B., Hamilton, Ont.

SAWS
Fay & Egan Co., J. A., Cincinnati, Ohio.
Radcliff Saw Mfg. Company, Toronto, Ont.
Simonds Canada Saw Co., Montreal, P.Q.

SAW GUMMERS, ALOXITE
The Carborundum Co., Niagara Falls, N.Y.

SAW SWAGES
Fay & Egan Co., J. A., Cincinnati, Ohio.
Simonds Canada Saw Co., Montreal, Que.

SAW TABLES
Canada Machinery Corporation, Galt, Ont.
Fay & Egan Co., J. A., Cincinnati, Ohio.
Garlock-Walker Machinery Co., Toronto, Ont.
Jackson, Cochrane & Company, Berlin, Ont.
Preston Woodworking Machinery Company,
Preston, Ont.
Yates Machine Co., P. B., Hamilton, Ont.

SCRAPING MACHINES
Canada Machinery Corporation, Galt, Ont.
Garlock-Walker Machinery Co., Toronto, Ont.
Whitney & Sons, Baxter D., Winchendon, Mass.

SCROLL CHUCK MACHINES
Klotz Machine Company, Sandusky, Ohio.

SCREW DRIVING MACHINERY
Canada Machinery Corporation, Galt, Ont.
Reynolds Pattern & Machine Works, Moline, Ill.

SCROLL SAWS
Canada Machinery Corporation, Galt, Ont.
Fay & Egan Co., J. A., Cincinnati, Ohio.
Garlock-Walker Machinery Co., Toronto, Ont.
Jackson, Cochrane & Company, Berlin, Ont.
Simonds Canada Saw Co., Montreal, P.Q.
Yates Machine Co., P. B., Hamilton, Ont.

SECOND-HAND MACHINERY
Petrie, H. W.
Williams Machinery Co., A. R., Toronto, Ont.

SHAPERS
Canada Machinery Corporation, Galt, Ont.
Fay & Egan Co., J. A., Cincinnati, Ohio.
Garlock-Walker Machinery Co., Toronto, Ont.
Jackson, Cochrane & Company, Berlin, Ont.
Ober Mfg. Company, Chagrin Falls, Ohio.
Petrie, H. W., Toronto, Ont.
Preston Woodworking Machinery Company,
Preston, Ont.
Simonds Canada Saw Co., Montreal, P.Q.
Whitney & Sons, Baxter D., Winchendon, Mass.
Yates Machine Co., P. B., Hamilton, Ont.

SHARPENING TOOLS
The Carborundum Co., Niagara Falls, N.Y.

SHAVING COLLECTORS
Sheldons Limited, Galt, Ont.

STEAM APPLIANCES
Darling Bros., Montreal.

SINGLE SPINDLE BOXING MACHINES
Canada Machinery Corporation, Galt, Ont.
Fay & Egan Co., J. A., Cincinnati, Ohio.
Ober Mfg. Company, Chagrin Falls, Ohio.

STAVE SAWING MACHINE
Whitney & Sons, Baxter D., Winchendon, Mass

STEAM TRAPS
Canadian Morehead Mfg. Co., Woodstock, Ont.
Standard Dry Kiln Co., Indianapolis, Ind.

STEEL BOX STRAPPING
Acme Steel Goods Co., Chicago, Ill.

STEEL SHELL BOX BANDS
Acme Steel Goods Co., Chicago, Ill.

SURFACERS
Canada Machinery Corporation, Galt, Ont.
Fay & Egan Co., J. A., Cincinnati, Ohio.
Garlock-Walker Machinery Co., Toronto, Ont.
Jackson, Cochrane & Company, Berlin, Ont.
Petrie, H. W., Toronto, Ont.
Preston Woodworking Machinery Company,
   Preston, Ont.
Whitney & Sons, Baxter D., Winchendon, Mass.
Williams Machinery Co., A. R., Toronto, Ont.
Yates Machine Co., P. B., Hamilton, Ont.

SWING SAWS
Canada Machinery Corporation, Galt, Ont.
Fay & Egan Co., J. A., Cincinnati, Ohio.
Garlock-Walker Machinery Co., Toronto, Ont.
Jackson, Cochrane & Company, Berlin, Ont.
Ober Mfg. Co., Chagrin Falls, Ohio.
Preston Woodworking Machinery Company,
   Preston, Ont.
Simonds Canada Saw Co., Montreal, P.Q.
Yates Machine Co., P. B., Hamilton, Ont.

TABLE LEG LATHES
Fay & Egan Co., J. A., Cincinnati, Ohio.
Ober Mfg. Co., Chagrin Falls, Ohio.
Whitney & Sons, Baxter D., Winchendon, Mass.
Yates Machine Co., P. B., Hamilton, Ont.

TABLE SLIDES
Walter & Co., B., Wabash, Ind.

TENONING MACHINES
Canada Machinery Corporation, Galt, Ont.
Fay & Egan Co., J. A., Cincinnati, Ohio.
Garlock-Walker Machinery Co., Toronto, Ont.
Jackson, Cochrane & Company, Berlin, Ont.
Preston Woodworking Machinery Company,
   Preston, Ont.
Yates Machine Co., P. B., Hamilton, Ont.

TRIMMERS
Canada Machinery Corporation, Galt, Ont.
Fay & Egan Co., J. A., Cincinnati, Ohio.
Garlock-Walker Machinery Co., Toronto, Ont.
Preston Woodworking Machinery Company,
   Preston, Ont.
Yates Machine Co., P. B., Hamilton, Ont.

TRUCKS
Sheldons Limited, Galt, Ont.
National Dry Kiln Co., Indianapolis, Ind.

TURNING MACHINES
Fay & Egan Co., J. A., Cincinnati, Ohio.
Garlock-Walker Machinery Co., Toronto, Ont.
Klotz Machine Company, Sandusky, Ohio.
Whitney & Sons, Baxter D., Winchendon, Mass.
Yates Machine Co., P. B., Hamilton, Ont.

UNDER-CUT SELF-FEEDING FACE
PLANER
Fay & Egan Co., J. A., Cincinnati, Ohio.
Garlock-Walker Machinery Co., Toronto, Ont.
Jackson, Cochrane & Company, Berlin, Ont.

VARNISHES
Ault & Wiborg Company, Toronto, Ont.

VENEERS
Barnaby, Chas. A., Greencastle, Ind.
Central Veneer Co., Huntington, W. Virginia.
Dominion Mahogany & Beneer Company, Mont-
   real, Que.
Walter Clark Veneer Co., Grand Rapids, Mich.
Gorham Bros. Co., Mt. Pleasant, Mich.
Hartzell, Geo. W., Piqua, Ohio.
Haughton Veneer Company, Chicago, Ill.
Hay & Company, Woodstock, Ont.
Hoffman Bros. Company, Fort Wayne, Ind.
Geo. Kersley, Montreal, Que.
Nartzik, J. J., Chicago, Ill.
Ohio Veneer Company, Cincinnati, Ohio.
Penrod Walnut & Veneer Co., Kansas City, Mo.
Roberts, John N., New Albany, Ind.
Toronto Veneer Company, Toronto, Ont.
Wood Mosaic Co., New Albany, Ind.

VENTILATING APPARATUS
Sheldons Limited, Galt, Ont.

VENEERED PANELS
Hay & Company, Woodstock, Ont.

VENEER PRESSES (Hand and Power)
Canada Machinery Corporation, Galt, Ont.
Garlock-Walker Machinery Co., Toronto, Ont.
Jackson, Cochrane & Company, Berlin, Ont.
Perrin & Company, Wm. R., Toronto, Ont.

VISES
Fay & Egan Co., J. A., Cincinnati, Ohio.
Simonds Canada Saw Company, Montreal, Que.

WAGON AND CARRIAGE MACHINERY
Canada Machinery Corporation, Galt, Ont.
Fay & Egan Co., J. A., Cincinnati, Ohio.
Ober Mfg. Company, Chagrin Falls, Ohio.
Whitney & Sons, Baxter D., Winchendon, Mass.
Yates Machine Co., P. B., Hamilton, Ont.

WOOD FINISHES
Ault & Wiborg, Toronto, Ont.

WOOD TURNING MACHINERY
Canada Machinery Corporation, Galt, Ont.
Garlock-Walker Machinery Co., Toronto, Ont.
Yates Machine Co., P. B., Hamilton, Ont.

WORK BENCHES
Fay & Egan Co., J. A., Cincinnati, Ohio.

WOODWORKING MACHINES
Canada Machinery Corporation, Galt, Ont.
Garlock-Walker Machinery Co., Toronto, Ont.
Jackson, Cochrane & Co., Berlin, Ont.
Preston Woodworking Machinery Company,
   Preston, Ont.
Reynolds Pattern & Machine Works, Moline,
   Ill.
Williams Machinery Co., A. R., Toronto, Ont.

WHEELS, Grinding, Carborundum and Aloxite
The Carborundum Co., Niagara Falls, N.Y.

THE COMPLETE LINE

ACME CASTERS are the development of long experience, they are always dependable and reliable.

We solicit your inquiries and will be pleased
to have our Canadian Representative call and
assist you in your Caster and Period Trimmings problems.

Foster, Merriam & Co.

Meriden, Conn.

Canadian Representative: F. A. Schmidt, Berlin, Ont.

"ACME"
The
RELIABLE CASTER

Making TABLE-SLIDES is a Specialty Business

For more than TWENTY-FIVE YEARS we have made TABLE SLIDES exclusively. Our Factory is equipped with Special Machinery which enables us to make SLIDES,—BETTER and CHEAPER than the furniture manufacturer.

Canadian Table makers are rapidly adopting WABASH SLIDES

Because { They ELIMINATE SLIDE TROUBLES Are CHEAPER and BETTER

BY USING

Reduced Costs

Increased Out-put

WABASH SLIDES

MADE BY

B. Walter & Company
Wabash, Ind.
The Largest EXCLUSIVE TABLE-SLIDE Manufacturers in America
ESTABLISHED-1887

Just What You Need

There's no necessity now for you to **wish** for a corrugated joint fastener machine which will rise fully to your expectations; you can get one.

The Saranac Corrugated Joint Fastener Machine

is the absolute realization of your wish. This illustration gives you an idea of what it looks like, but you can have no conception of its efficiency, economy, speed and reliability until you get it working for you.

It is substantially constructed, and the workmanship and materials are of the very best. Particularly suitable for manufacturers of sash, doors, blinds, screens, porch columns, boxes, etc., etc.

Made in various types for all conditions.

Write for further particulars and prices

Saranac Machine Co., SOLE MAKERS
Benton Harbor, Mich.

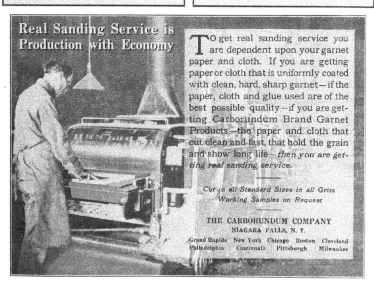

Real Sanding Service is Production with Economy

To get real sanding service you are dependent upon your garnet paper and cloth. If you are getting paper or cloth that is uniformly coated with clean, hard, sharp garnet—if the paper, cloth and glue used are of the best possible quality—if you are getting Carborundum Brand Garnet Products—the paper and cloth that cut clean and fast, that hold the grain and show long life—*then you are getting real sanding service.*

Cut in all Standard Sizes in all Grits
Working Samples on Request

THE CARBORUNDUM COMPANY
NIAGARA FALLS, N. Y.

Grand Rapids   New York   Chicago   Boston   Cleveland
Philadelphia   Cincinnati   Pittsburgh   Milwaukee

It Cuts Deep

Into the Cost of Turned Work

Who has the responsibility of supplying your turnings? Do you buy them outside and give the other fellow the profit?

Or are you dependent on the hand turner, a high-priced expert, limited in capacity, who ties up your entire production if he gets sick or leaves?

Reduce to a minimum the labor cost; keep for yourself the jobber's profits, and insure uninterrupted production.

The Mattison Automatic Shaping Lathe solves the problem. Any ordinary mechanic shapes your turnings automatically—round, square. octagon, etc.—in whatever quantities you require, and on short notice. The work is accurately formed, smooth and true to design, and at a labor cost 50 to 90 per cent. lower than is possible by hand methods.

Such a deep cut in the cost of your turned work soon pays for the lathe.

Ask to have our practical man call and go over this proposition in detail. This will not obligate you in the least.

MATTISON
AUTOMATIC
Shaping Lathe

There is only one Mattison Lathe and only Mattison builds it.

C Mattison Machine Works
283 Fifth Street, Beloit, Wisconsin

Volume Sixteen   TORONTO, SEPTEMBER, 1916   Number Nine

CANADIAN
WOODWORKER
and
Furniture Manufacturer

C.M.C.

No. 617 Variety Saw

For light, accurate work in furniture, cabinet, sash and door factories, or wherever light ripping or crosscutting or dadoing is required, the machine shown above is unequalled.

FEATURES OF CONSTRUCTION

Cast iron table equipped with removable wooden throat plate to permit use of extra thick saws or dado heads. Ripping fence of the double-faced type and can be used on either side of the saw blade. Operator can cut off stock with square, bevel or mitre ends, either to simple or compound angles, as he desires. Arbor of special high-carbon steel. **Bulletin describing this improved machine on request.**

CANADA MACHINERY CORPORATION
LIMITED
Toronto Office and Showrooms: Brock Avenue Subway.   GALT, ONTARIO.

The
Herzog Self-Feed Jointer
means

Increased Production   **Simplicity of Operation**

Small Floor Space   **Safety to Employees**

Does Four Times the Work of the Hand Jointer.

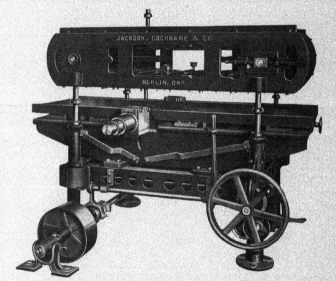

Our No. 34 Herzog Jointer, illustrated above, is one of the most efficient machines on the market to-day. It is appreciated by the manufacturer and employees alike, because, while it will produce from three to five times as much work as the hand jointer, it does not require skilled operators, but eliminates the danger so common to other makes. It can be operated by two boys. It will handle stock varying in width from 1 inch to the full width of the jointer, will feed fast or slow, takes only one-fourth the floor space of hand jointers, and requires only one-fifth of the sharpening of the knives. It is fitted with power feed raising and lowering attachment, with cylinder double belted and driven from both ends.

If interested in reducing your costs, write us.

Jackson, Cochrane & Company
KITCHENER   -   CANADA

GARLOCK-WALKER MACHINERY CO.
LIMITED
32 FRONT ST. WEST,   **TORONTO**   TELEPHONE MAIN 5346

Mr. Manufacturer:

How are you running your mouldings? Are you sticking to the old
time method of slow feed machines?

American No. 26 Fast Feed Outside Four Sided Moulder.

Here is a machine which will double your capacity—is easy of access—has all the
latest adjustments—including extra heavy spindles, large bearings, removable
heads, and carries either square or round heads.

Look it over and write us for further particulars.

Canadian Representatives

American Wood Working Machinery Co.
ROCHESTER, N.Y.

M. L. Andrews & Co.
CINCINNATI, OHIO

METAL and WOODWORKING MACHINERY of all Kinds

THE COMPLETE LINE

If your furniture is equipped with

ACME CASTERS

you are assured of satisfied customers, your most valuable asset.

Our Canadian representative is at your service and will be pleased to assist you in your CASTER and Period Trimmings problems.

Foster, Merriam & Co.

Meriden, Conn.

Canadian Representative:—F. A. Schmidt, Berlin, Ont.

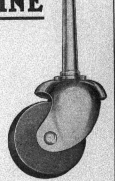

" ACME "

The
RELIABLE CASTER

Page 7 in the book on handle factory efficiency

Modern handle factories should be equipped with a machine for putting fancy knob ends on handles. A graceful, ornamental end on your handles will add much to the chances for sales. Do this work accurately and speedily on a

KLOTZ KNOBBER

It is a little machine with a mighty big purpose, and boosts the prices you can get for your handles. Honestly constructed of the same wearable material found in all of the Klotz handle factory equipment. This Knobber is builded to a "no-deficiency" standard which is protection   you and an asset to us.

Turn a Page With Us Next Issue

FOR YOU

who have a thought of manufacturing handles, or those desiring maximum production at the correct cost, get the Klotz Machine Company's idea of a perfectly modern handle factory. One with such dependable machines in it as their

BLOCKING SAW

SHAPING SAW

LATHE

THROATER

BOLTING SAW

POLISHER

DISC WHEEL OR END SHAPER

KNOBBER OR SCROLL MACHINE

All have a place in an efficient handle factory. Write

Klotz Machine Co., Sandusky, O.

Automatic Shaping Lathe
Makes Period Turnings Economically

This machine puts the cumbersome, costly, hand-turning of Period Patterns in the class of antiquated operations. Any round, oval or polygonal pattern, or any combination of these forms, can be made with little effort. Any good workman can run it. He can do twenty or thirty times as much work as a hand turner, and do it better. Cuts costs of the average pattern to less than a cent each. Its use

Enables You to Reach a Bigger Market

It will help you to so cut the cost of production that your goods can be retailed at a price within reach of all classes of buyers. In addition, the Yates Automatic Shaping Lathe is sold at a price the smallest manufacturer can afford to pay.

Send for a free sample showing what it will do and our circular "Period Patterns." You'll get some new ideas without expense or obligation to you in any way.

P.B. Yates Machine Co. Ltd.
HAMILTON, ONT. CANADA
Successors to **THE BERLIN MACHINE WORKS** — U. S. Plant: Beloit, Wis.

The "Beaver" Circular Planer Saw

The Beaver Planer Saw is made by our own special process and will cut very fast and smoothly without burning—making a clean planed or polished cut. They are already sharpened for use when sent out and are particularly adapted to toy and shell box work, where smooth clean cutting is necessary. The Beaver Planer Saw will cut quickly and is used by many of our customers on lightning cut-off or jump saw rigs with perfect satisfaction.

The Beaver Planer Saw must not be confounded with the common novelty or mitre saw sold, which burns the work and cuts very slowly.

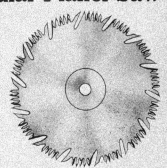

Radcliff Saw Manufacturing Co., Limited

TORONTO   550 Dundas Street   ONTARIO

Agents for L. & J. J. White Company Machine Knives

SLOW SPEED LOW POWER

BLOWER SYSTEM

The

Foster Blower

The "FOSTER" BLOWER

is specially designed for handling shavings, sawdust, excelsior, or any conceivable material. It is equipped with three bearings, shaft extending through, with bearing on inlet side. This eliminates the over-hung wheel and avoids heating, clogging and shaking. It is the only non-centre suction fan in existence. The Foster Fan is so constructed that the material entering the fan is immediately discharged without passing through or around the wheel.

Our Automatic Furnace Feeders will feed your material under boilers or into any desirable receptacle.

Consult us on your exhaust system troubles.

THE TORONTO BLOWER COMPANY

TORONTO   156 DUKE ST.   ONTARIO

Band Saw Safety and Efficiency

Read what users say about

FAY-EGAN "LIGHTNING" No. 50—36-in Band Scroll Saws

A fine, square column, thoroughly safeguarded machine like this costs a little more in the beginning, but is cheapest in the end—these and hundreds of similar letters in our files prove it.

We are getting excellent results from the No. 50 Square Column Band Saw, using it on all the varieties of work found around a railroad shop and it appears to be just what was needed. On account of the heavy construction of the machine we have very little trouble with saws breaking, as the saws are held under constant equal tension and do not buckle as in the old style "gooseneck" band saws. I also find that more work can be gotten out of the machine than from the old-style, on account of this heavier construction, as the machine can be crowded and still stand up under the strain.

BUTTE, ANACONDA & PACIFIC RY. CO.,
Feb. 23, 1915.   C. H. Spengler, M. M. Anaconda, Montana.

I do heavy work, and hardly ever break a saw; and as far as cutting ability, it is far ahead of the old gooseneck-style machine. I used a gooseneck-style before I bought the No. 50 Square Column Band Saw, and the Fay & Egan Co. No. 50 will do twice the work that the old-style gooseneck will do. With best regards, I am a friend of the No. 50 Band Saw.
H. A. SPRADLIN,
Feb. 16, 1915.   Monticello, Kentucky.

The No. 50 Square Column Band Scroll Saw which we purchased from you about two years ago has given good service, being installed in our pattern shop. We certainly prefer your machine to the "gooseneck" style which we were using before.
BAUSH & LOMB OPTICAL CO., E. Kandler,
Feb. 18, 1915.   Superintendent's Office, Rochester, N.Y.

We have had the saw in use nearly every day that our shops have been in operation during the past two years, and we do not hesitate to say that it has done at least one-third more work than the "gooseneck" style of saw, which we formerly used, could have done in the same time. We would also add that there is a considerable saving in the matter of blades.
Feb. 17, 1915.   JOHN G. ABBOTT ESTATE, Antrim, N.H.

We are using one of your No. 50 Square Column Band Scroll Saws, and consider it is the best saw we have in the factory. This saw is running fully ten hours per day and has been worked to the limit ever since it was installed, and has given us highly satisfactory results. We are using it on work as heavy as 5 in., on both oak and maple, and consider it much more satisfactory than the gooseneck-style of machine. In fact, the machine leaves nothing to be desired.
MOLINE PLOW CO., MANDT WAGON BRANCH,
Feb. 17, 1915.   Stoughton, Wisconsin.

We have three of your No. 50 Band Saws in our factory. The writer has been connected with two or three factories prior to this one and used your saws side by side with other makes, and we have always been partial to J. A. Fay & Egan Company. We keep our No. 50's running all the time. The machines have never given us any trouble. SHELL CHAIR COMPANY,
W. A. Shell, Vice-Pres. and Gen. Mgr.,
Feb. 15, 1915.   North Wilkesboro, N.C.

We have been using one of your No. 50 Square Column Band Scroll Saws about two years, and it has not given us any trouble at all toward repairs. We use this band saw for general mill work and we find it to be the best saw we ever had in our mill, both in saving blades and in sawing.
C. ARNOLD & SON,
Feb. 13, 1915.   Culman, Alabama.

Fill in and Mail Coupon Today for Free Bulletin M-4 on No. 50 Square Column Band Scroll Saw

British Columbia Douglas Fir Rotary Cut Veneer Panels

FOR
DOOR
PANELS

WALL
PANELLING

DRAWER
BOTTOMS

BUFFET,
DRESSER
and
WARDROBE
BACKS

A
BEAUTIFUL
PANEL
THAT
DOES NOT
WARP,
CHECK
OR
SPLIT

These beautiful veneer panels can be supplied in sizes and thicknesses to suit any purpose at prices that are surprisingly low

This illustration
shows a
rotary veneer
cutter
making
Douglas Fir
Veneer

Furniture
Manufacturers
can obtain
samples and
full information
on
application to

B. C. Lumber Commissioner

Excelsior Life Building, Toronto and Adelaide Streets   ·   ·   Toronto, Ontario

"Treat your machine as a living friend."

What is the problem in Canada today which is the cause of most anxiety to all manufacturers?

"THE SCARCITY OF LABOR"

It will be worse before it is better and if you are up against this proposition at all why not install one of our

No. 162 High-Speed Ball-Bearing Shapers

and do away with some of your worry?

Speed 7,000 Revolutions per minute

HERE IS WHAT WE CLAIM:—Double the amount of work can be done on this HIGH-SPEED BALL-BEARING SHAPER that can be done on an ordinary shaper, with much more ease to the operator. LET US PROVE IT TO YOU AS WE HAVE PROVED IT TO OTHERS.

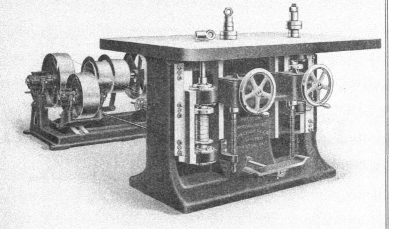

DIMENSIONS, WEIGHTS and SPEEDS

Floor Space, 120 in. x 58 in. Net Weight, 2650 lbs. Shipping Weight, 2750 lbs. Tight and Loose Pulleys on Countershaft, 10 in. x 6 in. Driving Pulleys on Countershaft, 18 in. x 5 in. Idler Pulleys, 14 in. x 5 in. R.P.M. Countershaft, 1200. Pulley on Spindle, 3½ in. x 6 in. R.P.M. Spindle, 7000. Diameter Spindle, 2 3/16 in. Spindle Centres, 28 in. Table, 58 in. x 41 in. Diameter Spindle Top, 1¼ in. Horse Power Required, 3. Front Projection, 17 in. Height, 35 in.

REPEAT ORDERS BEST PROOF OF MERIT

The Chevrolet Motor Car Company, Limited, of Oshawa, bought three of these machines November 26, 1915. On March 22, 1916, they wrote us: "The three Shapers which we purchased from you several months ago are giving every satisfaction. How soon could you supply us with two more and what allowance could you make for two old shapers?" On March 28 they ordered these two additional shapers, making five in all. They further expressed their opinion in a letter dated April 5, 1916, in which they say: "We cannot speak too highly of the design and workmanship of the machines, or the results which we are able to get from them. They are, in our opinion, one hundred per cent. perfect." Since then we have sold them still another.

We specialize in Labor Saving Machines. If you are interested in these machines, write us.

The Preston Woodworking Machinery Co. Limited, Preston, Ont.

Alphabetical List of Advertisers

Acme Steel Goods Company 50
American Hardwood Lumber Co .. 12
Atlantic Lumber Company 14
Ault & Wiborg Company 39

Baird Machinery Works 48
Barnaby, Chas. A. 43
B. C. Lumber Commissioner 8
Brown & Co., Geo. C. 12

Canada Machinery Corporation 1
Canadian Morehead Mfg. Company 51
Carborundum Company 55
Central Veneer Company 45
Chapman Double Ball Bearing Co. 54
Churchill, Milton Lumber Co. 12
Clark Veneer Co., Walter 43
Clark & Son, Edward 10
Cowan & Company 11

Darling Bros. 49
Decorators Supply Company 46
Dominion Mahogany & Veneer Co.. 16

Edwards Lumber Co., E. L. 14

Fay and Egan Company, J. A. 7
Foster, Merriam & Company 4

Garlock-Walker-Machinery Co. 3
Gorham Bros. Co. 49
Gum Lumber Manufacturers' Assn. 13

Hart & McDonagh 14
Hartzell, Geo. W. 14
Haughton Veneer Company 45
Hay & Company 39
Hay Knife Company, Peter 50
Hoffman Bros. Company 43
Huddleston-Marsh Mahogany Co. .. 13

Jackson Cochrane & Company ... 2

Kersley, Geo. 44
Klotz Manufacturing Company ... 4
Kraetzer-Cured Lumber Co. ... ..

Long-Knight Lumber Co. 13

Mattison Machine Works 56
Mowbray & Robinson 12

Nartzik, J. J. 43
Nash, J. M. 49
National Dry Kiln Company 50
Neilson & Company, J. L. 50

Ober Manufacturing Company 49
Ohio Veneer Company 43

Paepcke Leicht Lumber Co. 16
Perfection Wood Steaming Retort
  Company 51
Penrod Walnut & Veneer Company 45
Perkins Glue Company 15
Perrin, William R. 50
Petrie, H. W. 48
Preston Woodworking Mach. Co.. 9

Radcliff Saw Company 6
Reynolds Pattern & Machine Works 15
Ritchey Supply Company 47
Roberts, John N. 44

Saranac Machine Company 55
Sadler & Haworth 47
Sheldons Limited ... ...
Simonds Canada Saw Co. ... .
Smith Company, R. H. 11
Spencer, C. A. 12

Toronto Blower Company 6
Toronto Veneer Co. 40

Walter & Company, B. 55
Wardwell Mfg. Company 51
Whitney & Son, Baxter D. ... .
Wood Mosaic Company 44

Yates Machine Co., P. B. 5-48

Edward Clark & Sons
Toronto

CANADIAN HARDWOODS—winter cut **Birch, Bass, Maple**

10,000,000 ft. on hand, shipment June 1st
10,000,000 ft. sawing, shipment August 1st

ONTARIO and **QUEBEC** stocks, East or West, rail or water shipment, direct from mill to factory.

Put Up for Shell Boxes
3,000,000 ft. 4/4 Birch, No. 1 and 2 Common.
2,000,000 ft. 6/4 Birch, No. 1 and 2 Common.
1,000,000 ft. 12/4 Birch, No. 1 and 2 Common.

For Furniture Manufacturers
1,000,000 ft. each 10/4, 12/4, 16/4 Birch, 75 per cent. No. 1 and 2.

For Trim Trade
500,000 ft. 4/4 Birch, No. 1 and 2,   30 per cent. 14/16 ft., Red all in.
300,000 ft. 5/4 Birch, No. 1 and 2,   9 in. average. Red all in.
700,000 ft. 6/4 Birch, No. 1 and 2,   40 to 60 per cent. 14/16 ft. long.
500,000 ft. 8/4 Birch, No. 1 and 2,
200,000 ft. 4/4 to 8/4 Brown Ash, No. 1 and 2, 40 per cent. 14/16 ft.

We Commend to Your Notice:
Our lumber is reliable, our grades uniform, our shipments prompt, and receive personal supervision; our prices consistent. We have had long experience. We desire to serve.

R. H. Smith Co.
ST. CATHARINES, ONT.   Limited

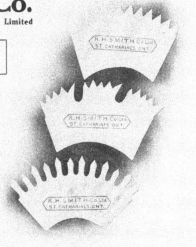

—"Arrow Head"—
Quality Saws

The Best Material is essential in
Saws and Knives.   When combined
with good workmanship, efficiency is
guaranteed. We know of no better
steel than Vanadium. If we did, we
would secure it. Our workmanship
is guaranteed. The use of a high
grade of steel has given our Mitre
Saws, Dado Heads and Woodworkers' Saws and Knives a reputation for
excellence not equalled by our competitors.

Eventually you will use
Arrow Head Saws—why
not on your next order?

Woodworking Machinery that is made for your
needs—not adapted to them

12 IN. MOULDER

Their Past and Present Records are Your Guarantee of Efficiency

Manufacturers of High Class Woodworking Machinery for Box Factories, Furniture Factories, Planing
Mills, Sash and Door Factories, etc.

We are specialists in Shell Box Equipment.   Ask the man who has installed "Cowan" Machinery.

Agents for Reynolds Screw Drivers and Morgan Nailers.

COWAN & COMPANY OF GALT, LIMITED
GALT   ONTARIO

GEO. C. BROWN & COMPANY

Manufacturers St. Francis Basin Hardwoods

Band Mills—PROCTOR, ARK. General Offices—Bank of Commerce Bldg.—MEMPHIS TENN.

STOCK FOR SALE—SEPTEMBER AND OCTOBER SHIPMENT

150 M' 4 4 No. 1 and No. 2 Com. Plain Red Oak.	200 M' 4/4 1st and 2nds Sap Gum (straight and flat).
150 M' 4 4 No. 1 and No. 2 Com. Plain White Oak.	50 M' 8/4 1st and 2nds Sap Gum (straight and flat).
70 M' 4 4 1st and 2nds. Qtd. White Oak (15 to 20 per cent. 10 in. and up).	150 M' 4/4 1st and 2nds Sel. Red Gum (straight and flat)
70 M' 4 4 No. 1 and No. 2 Com. Qtd. White Oak.	200 M' 4/4 No. 1 Common Red Gum.
20 M' 5 4 No. 1 and No. 2 Com. Qtd. Red Oak (6 in. and up).	50 M' 8/4 No. 1 Com. and Bet. Sel. Red Gum (quarter-sawed).
	60 M' 10/4 No. 1 Com. and Bet. Sel. Red Gum (quarter-sawed).

"Our Gum is all Kraetzer-Cured"—Splendid Condition.

TENNESSEE AROMATIC RED CEDAR SPECIALISTS   OUR TRAFFIC DEPARTMENT TRACES EVERY SHIPMENT

Churchill Milton Lumber Co.

Louisville, Ky.

We carry at
NEW ALBANY, IND.

Genuine Indiana Plain and Quartered Oak, Poplar

At GLENDORA, MISS.

Cypress, Gum, Cottonwood, Oak

SEND US YOUR INQUIRIES

We have dry stock on hand and can make quick shipment

Plain and Quartered Red and White Oak

and Other Hardwoods

Even Color   Soft Texture

We have **35,000,000 feet dry stock** all of our own manufacture, from our own timber grown in Eastern Kentucky

MADE (MR) RIGHT

OAK FLOORING

PROMPT SHIPMENTS

The Mowbray & Robinson Co., Inc.

GENERAL OFFICES
CINCINNATI, OHIO

Mills: Quicksand, Ky., West Irvine, Ky., Viper, Ky.

Canadian Representative:
N. H. FARNHAM

601 Elmwood Ave., BUFFALO, N. Y.

American Hardwood Lumber Co.

St. Louis, Mo.

Large stock of—
Dry Ash, Quartered Oak Plain Oak and Gum

Shipments from — NASHVILLE, Tenn.,
NEW ORLEANS, La., and BENTON, Ark.

Dry Spruce and Birch

Good Stocks, Prompt Shipments, Satisfaction

C. A. SPENCER, Limited

Wholesale Dealers in Rough and Dressed Lumber

Offices—500 McGill Building
MONTREAL   -   -   Quebec

Huddleston-Marsh Mahogany Company, Inc.

Importers and Manufacturers

MAHOGANY

Lumber and Veneer

All Grades   All Thicknesses

WE SPECIALIZE
IN
MEXICAN MAHOGANY

33 West 42nd St.   2254 Lumber St.
NEW YORK, N.Y.   CHICAGO, ILL.

—Mills and Yards—
Long Island City, New York

Long - Knight Lumber Co.

Manufacturers and
Wholesale Dealers

INDIANAPOLIS, IND., U. S. A.

HARDWOOD LUMBER

Walnut, Mahogany
and
American Walnut Veneers

SPECIAL:

5 cars 5/4 1 & 2 Pl. Red Oak

"Made in America"

RED GUM

("America's Finest Cabinet Wood")

IS THE "LIFE-SAVER"

of the furniture trade in these disturbing days. *Gum's only competitors* as a pre-eminent furniture and trim wood are the woods formerly imported but *now increasingly difficult to get.*

It is to the credit of American furniture *manufacturers, retailers* and *users* that they now *acknowledge the advantage to them* of having available (and of *relying* on) so wonderfully beautiful and dependable a furniture wood as American *Gum.*

It is to the credit of *Gum* manufacturers that they are *not* taking unfair advantage of *the BOOM IN Gum* to exact unreasonable prices.

Gum at present prices is *the one big, best buy* in the whole furniture world, for maker, seller and consumer. There's money in Gum for *You.*

A convincing fact as to RED GUM'S growing prestige is that it enjoys increasing demand among the best judges of value—artistic value, utilitarian value and economic value. (There are no other kinds.) "GUM HAS COME."

Furniture Manufacturers and Furniture Retailers who wish detailed proof as to the *salability, dependability* and *commercial desirability* of *RED GUM,* are inVited to correspond with

GUM LUMBER MANUFACTURERS ASSOCIATION
1314 Bank of Commerce Building, Memphis, Tenn.

The Atlantic Lumber Co.

Manufacturers and Wholesale Dealers

American and Canadian Hardwood Lumber and Mahogany, Quartered and Plain, White and Red Oak a Specialty.

110 Manning Chambers - **TORONTO**
Phone Main 6386

HOME OFFICE, BOSTON MASS.

Mills: KNOXVILLE, TENN., WALLAND, TENN.
FRANKLIN, VA.

Shell Box Stock

1 x 6 birch, in No. 1, 2, and 3 common.

1 x 7 and 8 birch, in No. 1, 2, and 3 common.

1 x 8¼ birch, in No. 1, 2, and 3 common

1 x 9 and wider, birch, in No. 1, 2, and 3 common.

1 x 4 and wider, spruce. 2 x 4 and wider, spruce.

For the Furniture Trade

4/4, 5/4, 6/4, 8/4, 10/4, 12/4, 16/4, birch, maple, elm, basswood and black ash.

Elm, Basswood and Black Ash

in No. 1 common and better grades—dry stock in assorted sizes.

Hart & McDonagh
Continental Life Bldg.

TORONTO - - ONTARIO

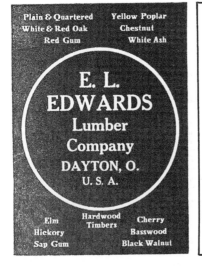

Plain & Quartered   Yellow Poplar
White & Red Oak   Chestnut
Red Gum   White Ash

E. L. EDWARDS
Lumber Company
DAYTON, O.
U. S. A.

Elm   Hardwood Timbers   Cherry
Hickory   Basswood
Sap Gum   Black Walnut

FOR SALE

Black Walnut Lumber

15,000	feet	4/4 No. 1	Common
20,000	"	4/4 No. 2	"
15,000	"	5/4 No. 1	"
15,000	"	5/4 No. 2	"
15,000	"	6/4 No. 2	"
40,000	"	8/4 No. 1	"
60,000	"	8/4 No. 2	"
30,000	"	9/4 No. 1	"

Will make special low prices on all of the above stock to move it quick. All Band Sawn, Equalized, and all thoroughly dry, ready for immediate shipment. Write for prices.

Geo. W. Hartzell
Piqua, Ohio

The Only Way to Drive Screws

There is a stage-coach method and a modern method of doing every-thing. When it comes to driving screws the only modern way is with an Automatic Screw Driving Machine, which means greater and better pro-duction and greater profit.

The Reynolds
Automatic Screw-Driving Machine

is guaranteed to produce results. It is no experiment. The No. 2 Standard Type, shown herewith, is particularly suited for placing the screws in the straps of shrapnel boxes. It may be fitted to drive, without change or ad-justment, any three consecutive sizes flat or round head, or two sizes filister head, No. 6 to 14 inclusive, wood or machine screws, or three sizes flat or round head, or two sizes filister head machine screws up to No. 20. All the above up to 1½ inches long if smaller than No. 10, or up to 1¾ inches long, No. 10 or larger. Fitted with boring attachment if desired.

Many Machines in Use in Canada.

Ask for our Catalogue.

REYNOLDS PATTERN & MACHINE CO.
101 and 103 Third Ave., Moline, Ill., U.S.A.

No. 2 Standard Type Automatic
Screw-driving Machine.

Perkins Vegetable Veneer Glue
IS A
Uniform, Guaranteed Patented Glue

¶ We have, during the past ten or more years, built up a large volume of business and maintained it, because we sell only

HIGHEST QUALITY

¶ Our list of customers proves the statement that a very large per cent. of the highest quality furniture, including office and household furniture, also doors, interior trim, etc., are built up with Perkins Glue.

¶ Be sure you have *Perkins Quality*—absolutely uniform.

¶ We will demonstrate to your satisfaction in your own factory

Ladies' Desk manufactured by The D. Hibner
Furniture Co., Limited, Kitchener, Ont., in
which Perkins Glue was used.

Originator and Patentee of Perkins Vegetable Glue.

PERKINS GLUE COMPANY, LIMITED

Address all inquiries to Sales Office.
Perkins Glue Company, SOUTH BEND, Indiana, U.S.A.

Hamilton, Ont.

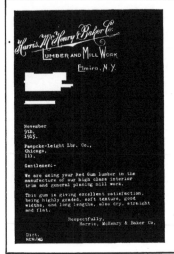

Of course it is true that

RED GUM

is America's finest cabinet wood—but
Just as a poor cook will spoil the choicest
viands while the experienced chef will turn
them into prized delicacies, so it is true that

**The inherently superior qualities
of Red Gum can be brought
out only by proper handling.**

When you buy this wood, as when you buy a new
machine, you want to feel that you have reason for
believing it will be just as represented.

We claim genuine superiority for our Gum. The
proof that you can have confidence in this claim is
shown by the letter reproduced herewith.

Your interests demand that you remember this
proof of our ability to preserve the wonderful qualities of the wood when you again want RED GUM.

Paepcke Leicht Lumber Co.

Conway Building   111 W. Washington St.
CHICAGO, ILL.

Band Mills: Helena and Blytheville Ark.; Greenville, Miss.

Buy Made in Canada Veneers

Dominion Mahogany
& Veneer Co., Limited

Head Office and Factory   Toronto Office and Warehouse
Montreal West, P.Q.   455 King St. West

Mahogany, Walnut,
and all Fancy Woods—Lumber

SAWN, SLICED and ROTARY CUT VENEERS

Mahogany, Walnut, Quartered and Plain Oak, Ash, Maple,
Birch, Elm, Poplar, Gum.

Gum and Poplar Cross Banding.   Hard Maple Pin Block Stock.
SEND US YOUR ENQUIRIES.

Canadian Woodworker

A Monthly Publication in the Interest of the Woodworking Industry.
Reaches the factories producing interior finish, doors, sash, flooring, woodenware, furniture, pianos, boxes, and general mill products.

Subscription, $1.00 a year; foreign $1.50.

Woodworker Publishing Company, Limited

Branches: Montreal, Vancouver, Chicago, New York, London, Eng.

345 Adelaide St. West, Toronto
Phone Ade. 2700

Authorized by the Postmaster General for Canada, for transmission as second class matter.
Entered as second class matter July 18th, 1914, at the Post Office at Buffalo, N.Y., under the Act of Congress of March 3, 1879.

Vol. 16	September, 1916	No. 9

Furniture Production Number

The better conditions which have characterized the furniture business of the past few months, with indications that these conditions will continue, have prompted the publishers of the "Canadian Woodworker and Furniture Manufacturer" to issue a number devoted almost exclusively to the manufacture of furniture. This number will be published in October, and will be known as a Furniture Production Number.

Recent interviews with a number of furniture dealers indicate that the demand for furniture is good. Canada as a whole has made a remarkable recovery from the European war-shock of 1914, and to-day there is an abundance of money in the country, absolute confidence prevails in all quarters, and many lines of business have returned to normal. The people are able and willing to purchase for all legitimate needs. A solid foundation has been established for future development, and a period of steady expansion would appear to be ahead of us.

Furniture manufacturers are busy, and in some factories production would be increased but for the scarcity of labor. Heavy winter sales are predicted, and the trade outlook seems to be quite bright.

Much interest has already been shown in the Furniture Production Number. The subjects which could be discussed in a number of this kind are very numerous. It will be our aim to have practical articles on a wide variety of questions pertaining to the manufacture of furniture, so that all operators of furniture factories will be sure to find much of interest and value. The invitation is here extended to our readers to contribute to this number from the results of their experiences. It is believed that such a number, devoted to prevailing styles and methods of manufacture, will be appreciated.

Trade and Commerce Convention

The Right Hon. Sir George E. Foster, Canada's Minister of Trade and Commerce, has issued a "call to action" to Canadian business men to meet in convention some time during the coming autumn to discuss ways and means of "mobilizing the business forces of Canada so as to employ our labor, increase our production, and enlarge our markets along peace lines."

The Minister has asked for the help and co-operation of boards of trade, the Canadian Manufacturers Association, bureaus of industrial and scientific research, engineering associations, labor bodies, and representative men from all the great industries, and has suggested that they give the matter as much thought and study as possible before the date of the convention, which will likely be during the latter part of October. The Minister of Trade and Commerce has also suggested that committees be selected beforehand to gather all available data, information, and ideas under the following headings: Home market possibilities and competition with imports; foreign markets and the opportunities they offer; credit information and the extension of banking facilities abroad; plans toward providing special education and training for prospective commercial representatives of Canadian firms in friendly export markets; transportation; labor; immigration and colonization; industrial equipment, plans, and processes; development of industrial research and raw materials.

After the war there will be a great struggle for the world's trade, and in view of the fact that Canada, on account of her vast natural resources is in a good position to reach out after her share of that trade, the calling of the convention as outlined above is an excellent move. Canadian manufacturers have never done a very large volume of foreign trade, and if they intend to go after it in future—as they certainly should—there is a great deal that they will require to know.

If care is used in selecting the men to act on the different committees so as to have them include men of authority, manufacturers will be able to ascertain just what they will require to know in order to make a bid for foreign trade. For instance, if they wish to do business in France they will require to know the customs of the country in regard to business.

We would like to see furniture manufacturers and the owners of woodworking factories take part in the convention, as they occupy an important position in the industrial activities of this country. They should at least be represented on the following committees: Home Market Possibilities and Competition With Imports; Foreign Markets and the Opportunities They Offer; Credit Information and the Extension of Banking Facilities Abroad; Industrial Equipment, Plants, and Processes; Development of Industrial Research and Raw Materials.

Home market possibilities and competition with imports, in particular, should receive their attention, as previous to the war there were large quantities of furniture and woodenware imported into Canada. The furniture manufacturers, as a branch of the Canadian Manufacturers Association, should be represented on several of the committees. The owners of planing mills and woodworking factories throughout the country should organize. They should then select men thoroughly conversant with woodworking to represent them at the convention. These men could find out beforehand from the Department of Trade and Commerce just what lines of manufactured woodwork were imported. They could also ascertain just what lines there is a market for in other countries. This, in turn, would enable the members of the organization to decide definitely the lines they are best able to manufacture.

Canadian Toys Outclass Those of Foreign Make

The many splendid examples of wooden toys on view in the toy section of the Government Building at the Toronto Exhibition could not fail to impress visitors with the importance of this growing industry. There was a very noticeable improvement over the toy exhibition held in the Royal Bank Building, Toronto, in April of this year. The improvement was due in part to the fact that more space was available, permitting of a much better arrangement of the exhibits and allowing a number of firms that did not exhibit at the previous exhibition to display their products. Great credit is due the Department of Trade and Commerce, who made arrangements for the necessary space and provided the carpenters to erect all the stands, and to Mr. L. G. Beebe, secretary of the Canadian Toy Association, who had the difficult and responsible task of allotting the space to the different exhibitors.

One of the outstanding features in the toy section of the exhibition was the large and varied collection of foreign-made toys and the favorable comparison made by the products of Canadian factories. In fact, one could not help thinking that the Canadian toys were superior to the foreign ones; and, judging by the interest displayed by the little folks on children's day and their evident delight in examining the many examples of wagons, sleighs, doll carriages, games, building blocks, constructor toys, juvenile furniture, rocking horses, etc., the Canadian-made toys are just as attractive to them as the foreign-made ones.

The Canadian Pacific Railway Company had a very interesting display of wooden toys and novelties in the Process Building. These had been collected from various sources and were placed on view so that Canadian manufacturers of toys and novelties could copy them or take suggestions from them. The majority of them were made in England, and they show clearly what the people of Great Britain are doing to exclude foreign-made toys.

Walnut Veneers for Furniture and Pianos

There is never very much furniture shown at the Toronto Exhibition, but what was shown this year reflected great credit on the manufacturers, it being of excellent design and workmanship. There were a great many pianos exhibited, and both in the furniture and pianos one of the predominating features was the use of fancy figured walnut veneers. There is a quiet dignity about walnut, particularly in the satin finish, that seems to "fit in" with the classic designs. It is unobtrusive, but the beauty is there for those with eyes to see it. The furniture shown was principally dining-room and bedroom suites, with the exception of davenports and couch beds. A bedroom suite in French grey enamel evoked some very favorable comments from passers-by, and some white enamel nursery furniture was the object of considerable attention. Another thing that was greatly admired was an exhibit of standard lamps—or piano lamps, as they are very often called—in wood, with silk shades. There appears to be a good sale for lamps of this kind, and some of the furniture factories are manufacturing very complete lines of them. These include floor lamps and table or reading lamps.

Western Woods at Toronto Exhibition

The manufacturers of woodwork for use in the construction of buildings, such as doors, interior trim, stair materials, newel posts, panelling, etc., no doubt found a great deal to interest them in the exhibit of the British Columbia Government. There were numerous examples of the uses to which British Columbia woods can be put. The woods shown were Douglas fir, Western soft pine, Western hemlock, Western larch, Western red. cedar, Western white pine, and Western spruce. The walls which formed part of the exhibit were divided into six sections by means of pilasters and panels of the different woods. To each section was attached doors—which could be easily swung—made from the different woods mentioned above and finished in a variety of finishes.

Besides the doors there were a number of samples of mouldings, casing, base, etc., and a large collection of fruit baskets and boxes for different kinds of fruit. There was also a rowboat, complete with oars, etc., made from Western red-cedar, which wood is said to be particularly adaptable for boat building. The exhibit was in charge of Mr. L. B. Beale, who is the British Columbia Lumber Commissioner for Eastern Canada, and who never tires of answering questions about British Columbia woods. Mr. Beale's office is in the Excelsior Life Building, Toronto.

Efficient Costing System for Planing Mills

Superintendent can tell when machines are producing full capacity—Records are valuable for estimating on jobs

In these times of keen competition and varying prices if an item is costing too much we should be able to put our fingers on the leak at once, says Mr. E. F. Duby, Superintendent Brunette Sawmill Company, Limited, New Westminster, B.C. This is only possible by a detailed LABOR COSTING SYSTEM—one that shows the cost of each operation in the process of manufacture and eliminates uncertainty as to what the cost really is.

One means of costing is by the employment of a detailed daily report of the work done, labor employed and money expended. In the planing mill business this system must be applied to each operation or department employed in the process of production. A detailed daily report system has many advantages, chief among which is the one mentioned above, viz.: That you are enabled to KNOW the cost of each operation. Another advantage of this system is that it forms a basis of comparison for different men and machines performing the same work, or a day-to-day comparison for the same men and the same machine performing the same work.

To illustrate: The writer had in the planing mill two planers of the same make and operated by the same number of men, running on 8-in. shiplap, No. 1 machine showing a meter tally of 30,000 lineal feet, while No. 2 machine showed a meter tally of 37,000 lineal feet. When this report came in the natural question was, "Why this difference?" The planer foreman was called in and asked to explain the discrepancy of 7,000 lineal feet between the two machines on identical operations. The reply is, "No. 1 machine has a defective belt, which caused several stops during the day," or "The feeder on No. 1 is a little slow and does not keep a continuous feed in the machine." What is the result? We get a new belt and eliminate the loss of time for the entire crew, or we get a new feeder, as the case may be. At any rate, the detailed report system is responsible for stopping the leak.

Again, looking at the same report we find that machine No. 1 shows waste of 15 per cent, while No. 2 shows but 8 per cent. Now, what? We again call in the foreman and ask the reason why. He replies that the trimmerman on No. 1 is reckless in his work, often cutting away two to four feet unnecessarily. We also find that the proportions in grade do not tally out as be-

Form No. 1A

BRUNETTE SAW MILL CO., LTD.

DAILY PLANER REPORT

	No.	Time	Rate	Wages
Trimmer	60	.10	.18½	1.85
Grader	72	.10	.14	1.40
Tie-up	75	.10	.13	1.30
Feeder	222	.10	.25	2.50
Total			7.05	

Meter Run	Tally	Cost per M
20,651 B.M.	17,594 B.M.	.40

Operator

Form No. 3

THE BRUNETTE SAW MILL CO., LTD.

PLANER FOREMAN'S REPORT

Date   October 7th, 1914.

No.		Meter	Tally	Waste	%	Men	Hrs.	Wages	
159	Foreman					1	10	3	46
157	Machine Man					1	10	2	75
151	Trucker					1	10	2	15
326	Clean Up					1	10	1	15
61	Rip Saw Man					1	7		95
62	Rip Saw Helper					1	7		84
	Sander								
	Sizer		43,400			1	10	3	20
	Timber Planer		52,700			6	60	12	40
1	Machine	17,130	16,200	930	5	3	30	5	10
2	Machine	18,100	17,340	760	4	3	30	6	15
3	Machine	10,130	9,620	510	5	2	16	3	20
5	Machine	35,110	23,316	1794	7	5	50	9	23
	Filer							1	12
	General Expenses							6	22
	Total		162,576					57	94

Cost per M $0.36

Signed .. .. Foreman

tween the two machines, and this knowledge (PRO-FIT) is all creditable to our detailed report, for otherwise we would have gone on in blissful ignorance, congratulating ourselves upon the fact that we are getting the most possible out of our men, machines and material.

In operations where we have but one machine on a given kind of work we watch carefully the detailed report, and if the output falls below the general average we make inquiry while the details are fresh in everyone's mind. In the event of an output above the average we make inquiry into the case, and when we find that given conditions tend toward a greater output we bend every effort toward having those conditions present at all times, at the same time trying to eliminate all unfavorable conditions.

The above methods are applied to all operations, no matter what department. Printed forms are prepared for each department, and in many departments separate forms for individual machines or operations are supplied. These forms are gotten up so as to give the greatest amount of information with the least possible extra work. For example, in the planing mill on the individual machines Form 3A is used.

This form shows the date, machine number, kind of lumber worked, grades, sizes and pieces per bundle, together with a tally of bundles in the various lengths. Note also that a meter tally is given, from which you deduct the piece tally behind the grader to get the percentage of waste.

At the bottom is shown each man employed on the operation, together with his number, corresponding to his number on the time clock and pay-roll. Also time working and rate of wages. By totalling the wages and dividing by the number of thousand feet in the tally the cost per thousand feet is gotten. This report is signed by the operator and handed to the planer foreman, and from these individual reports the planer foreman makes up his report: See Form No. 3. (This form is so drawn as to include all men connected with planing mill operations).

By the use of a system of this kind you are enabled to know when you make a price on an item whether that item will return a profit or a loss, and how much. For example, the sales manager asks what will be left at a price of $3.90 for cedar lath delivered at a 40c. point. You turn to your daily report, or still better, to your monthly report, which gives an AVERAGE COST, and find that your labor costs 42c. per thousand; to this you add cost of material, fixed and overhead expenses, together with freight, and you know to a certainty whether it is profit or loss. All guess work is eliminated.

To collect and summarize the reports and make out the general daily report in a plant employing 300 men, requires but half of the timekeeper's time; the remainder of his day being given over to his general duties. A further benefit derived from this system is that it affords a double check on the men's time, as the time clock record is checked against the time written in by the department foreman. The system is particularly advantageous in making prices on special work, as the exact LABOR COST may be readily worked out. One of the greatest benefits of the system is a tendency towards increased output. Every man's time is accounted for and his results are shown not only to his foreman, but to the superintendent and manager as well.

At the close of each month a general monthly report is made up from the general daily reports, which show the total output, time, wages and average cost of manufacture. The foremen in the various departments receive at the beginning of each month a statement of the previous month's work in their respective departments, showing the total output, average number of men working, time, total wages and average cost of manufacture, together with comments or suggestions by the superintendent.

The writer finds that the advantages of the costing system pay for the labor and trouble of carrying it on many times over, to say nothing of the satisfaction of knowing what each department is producing each day. If these systems were more generally used there would be less material sold below actual cost.

The above system is susceptible of development to any desired extent, in fact the writer has extended it to show, on the monthly report, the entire cost of production, including supplies, insurance, overhead expenses, depreciation, etc. There is no reason why the planing mill business should not have a system for determining costs and profits and losses as readily as any other manufacturing business.

With the writer it is made a part of the daily work, and is considered of just as much importance as is the output from any of the departments, for without a system of some kind we simply run along hoping things will come out right, but never knowing if they will or not.

Are Shell Box Makers Losing Money?

"Some of the makers of ammunition boxes are having a hard time trying to come out even on the prices at which they have taken the work," said a wholesale lumberman who specializes on selling lumber to box makers. "The prices at which some of them have contracted cannot possibly bring them profits, especially in view of the advances in lumber and hardware. It has happened over and over again that between the date at which they put in bids and the date at which they have taken orders lumber of the description required has gone up, while as to hardware and rope there seems to be no limit to the advances.

Box makers have placed large orders for screws in the United States rather than pay the exorbitant values asked in Canada. While the ammunition companies are being paid prices which evidently result in good profits, the Imperial Munitions Board are trying to get the uttermost cent out of the box makers, a most unfair distinction. To my personal knowledge, contracts have been offered and accepted at quotations which left a loss to the box makers. It is no doubt foolish to take work under such conditions, but a part of the responsibility must rest with the Munitions Board in offering prices which they must know are below cost.

"Then the inspection has, in some instances, been tightened up, and I believe has been made needlessly severe. But there is no redress. What between advancing costs and the keen competition, coupled with the low prices offered by the Munitions Board, the lot of many shell box makers is not a happy one. They want orders to keep their factories going, and to employ the machinery oftentimes specially purchased, but the quotations mentioned are so inadequate as to make it impossible to even break even. What are they to do?"

Adjusting the Three-Drum Sander

Making the Different Parts Work in Harmony—Hints on Putting the Paper on the Drums—Operator Should Understand Mechanism

By "W. D."

The three-drum sander should be in charge of a man that understands its mechanism. It requires accurate adjustments, and a man that does not thoroughly understand what every part of the machine is for can very soon put it in such a mess that a machinist's services will be required to give it a complete overhauling. You may say: "How is a man going to learn how to operate it if only an experienced man is allowed to touch it?" A great deal can be learned by keeping one's eyes open and watching how the experienced man goes about making the adjustments.

The writer was employed for some time in a factory where they could not get satisfactory work from their three-drum sander. Five or six different men had been given a chance to see what they could do with it. But in each case the result was the same. At last the superintendent picked out a quiet, unassuming Swede that had been engaged a few weeks previous to operate a cut-off saw and told him to see what he could do with it. That Swede was a mechanical genius. He made an excellent job of the sander and is now in charge of all the machines in the factory. The writer has acted as his assistant on a number of occasions when he was overhauling the sander, and details his method here for the benefit of readers of the Canadian Woodworker.

The first move is to closely examine the drum bearings to ascertain if there is too much play. If there is, the bearings must be tightened up, care being taken to see that they are tightened just sufficient for the shafts to turn easily. The wrench is very often applied to bearings when a good application of oil would be much better. The next part of the machine to receive attention is the bed plates. A steel straight-edge long enough to reach from end to end of the machine will be found very useful in truing these up.

If the machine has not been overhauled for a long time it may be advisable to examine the bed plates closely. If they are worn down in places they should be sent to a machine shop to be planed. This is not very often necessary, however, unless the machine is very old or has been roughly treated. Proceeding to adjust the bed plates, these will be found to be higher on the edge next the drum. This is to eliminate unnecessary friction, as otherwise the pressure rolls would have to exert too much pressure to feed the stock through the machine. The bed plates should be tested at both ends, and care taken to see that they are fastened down right.

The bottom feed rolls are the next to be lined up, and the steel straight-edge will also be found useful for them. The rolls at each end of the machine should receive attention first and then the others brought up to the straight-edge. The feed rolls should be slightly higher than the bed plates, to allow the stock to feed through the machine with a minimum amount of friction. When the feed rolls have been properly adjusted they should be firmly secured by means of the lock nuts on the adjusting screws. From the lower feed rolls we proceed to the upper feed rolls. Two short straight-edges, having both their edges parallel, should be used for adjusting the upper feed rolls. One can be used satisfactorily, but two are more convenient. If

two are used they should be set as near each end as possible, then the upper portion of the machine should be lowered until the feed rolls just touch the straight edges. The first roll on the feeding in end of the machine should be adjusted first.

Before proceeding to adjust the second roll we must remember that before reaching it the stock will pass over the first sanding drum and thus be reduced in thickness; therefore, in order to have a uniform pressure on the stock we must allow for the thickness of the cut. This is done by raising the upper portion of the machine equivalent to the thickness of the cut and adjusting the second roll to the straight edges in the same manner as the first. The third roll is adjusted in exactly the same manner as the first and second, only that we have to again raise the upper portion of the machine a distance equal to the thickness of the cut taken off by the second sanding drum. In adjusting the fourth roll it is not necessary to consider the thickness of the cut taken by the third sanding drum, as the purpose of the third drum is simply to remove the scratches made by the first two; the thickness of the cut would therefore be insignificant.

The pressure rolls come next. These can be put in parallel alignment with the other rolls by using the short straight edges before mentioned. When the bed of the machine and all the rolls are in perfect alignment it is well to give the drums some attention. These should be raised up above their normal position and the felt covering pushed back a little from the ends, so that both ends of the drum may be tried with a straight edge, to see that they are in alignment with the bed plates. This being done, the felt or canvas covering on the drum should be examined to see that there are no hard lumps or foreign matter clinging to it. A good plan is to revolve the drums slowly and brush them vigorously with a good, stiff brush.

The machine is now ready for putting on the papers, and perhaps an idea of how this was done on the machine under discussion may be of interest to readers. In the first place, the paper must be tight, and as sandpaper is not the most pleasant thing in the world to stretch tight with the bare hands, we used the following method: A piece of maple about three inches by two inches and the length of the drum was made concave on one side to fit snugly against the drum. One end of the paper was secured in the drum and the piece of maple held tightly against the paper while the drum was revolved; the other end of the paper was then secured. This resulted in the paper being good and tight.

When the papers have all been put on the machine is ready to start up. It will be wise, however, before feeding a piece of stock through, to lower the drums slightly below the bed plates. They can easily be raised to take the required cut, whereas if they were left above the bed plates it is hard to say just what thickness of a cut they would take. When a piece of stock is being fed through the machine the pressure rolls should be adjusted. These should press firmly on the stock, but they should never be so tight that they cannot be stopped from revolving by gripping them with the hand.

While the following of these instructions will not

make a man a first-class sander operator, nevertheless they will give him a good idea of the relation of one part of the machine to another, enabling him to get fairly good results when sanding solid stock. The old saving, "experience teaches," applies in this, as in all other cases, and the operator can by applying himself and observing details, soon acquire sufficient skill in the operation of the three-drum sander to be entrusted on veneered stock, which undoubtedly is the most difficult on which to secure satisfactory work.

The Change from Thick Knives to Thin Ones

Faster Feeds and Better Quality of Work Possible—Cutting Angle Varies for Different Kinds of Woods—Workmen must be Instructed
How to Handle Self-Hardening Steel

By J. Johnson

Many owners of woodworking plants are at the present time considering the advisability of changing from the old square-head thick knife planer to one having round cylinders and thin knives. The reason is that the thin knife machine has a reputation for turning out larger quantities of better planed work than the square-head machine. The man who has never had any experience with thin knife planers is naturally inclined to "nose" around other plants to find out what they are doing and whether their thin knives have been satisfactory. He will hear conflicting opinions. Some men will say: "Throw out your old machine and get a thin knife machine." Others will say: "We tried the new kind, but had to go back to the old

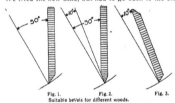

Fig. 1.   Fig. 2.   Fig. 3.
Suitable bevels for different woods.

machine, as we could not get satisfactory results." These conflicting opinions are apt to leave a man in the position where he does not know what to do for the best.

Having had considerable experience with both types of machines, I will endeavor to give some information which may assist the undecided one in making a choice. Where it is used as it should be, the thin knife planer is an unquestionable success. When the knives are properly ground and jointed it will turn out very fine work. And by properly ground knives I do not mean that it requires an expert to do the grinding. The grinding of thin knives is simple, and any workman can handle them successfully if he goes about the work in a careful manner. The machine companies that furnish the knives, or the steel companies that furnish the steel, are always willing to give complete information on how to treat their knives.

In the matter of production of a thin knife machine over a thick one it is only necessary to take a few figures, as follows: A square cutter head running at 3,000 r.p.m. with two thick knives would give 6,000 knife marks per minute. With these figures the feed would figure out at approximately 40 feet per minute. This would be a maximum speed, as the average square head planer feeds at about 30 feet per minute. If a four knife head were used the feed could be increased to about 80 feet per minute without in any way impair-

ing the quality of the finish. But with thin knives we can speed up even more than this by using six knives on the head. For thin knives I will say that it is possible to secure a finer finish than with thick knives, because instead of the chopping cut which we get with thick knives we get a scraping cut. To produce the best possible finish with a circular head and thin knives the knives should be jointed so as to form a true cutting circle.

When considering the thin knife planer as an addition to the equipment of the factory it is necessary to know what kind of woods the machine will be operating on. The reason for this is that hard and soft woods require that the knives be at different cutting angles. If the wood to be cut is all either hard wood or soft wood, then the problem becomes simple, as the cutterheads can be made so that the cutting angle will be suitable for the texture of the wood being cut. Fig. 1 shows the outline of a round cutter head, with the knife fixed at an angle of 30 degrees with the radial line (a radial line is a line drawn from the cutting edge of the knife through the centre of the cutter head). A knife set at this angle is suitable for surfacing soft woods, as basswood, cedar, pine, poplar, spruce, etc., in a dry state. Where the wood is green or in a wet state, an addition of 10 degrees to the cutting angle is advisable. If the planer is to operate on soft wood exclusively, a cutting angle of 30 degrees will give good results.

If different kinds of woods are to be planed it is advisable that the cutting angle be fixed at 30 or 35 degrees, because then for hard woods the knife can be back-bevelled to give the proper scraping cut. In Fig. 2 a knife is shown having a cutting angle of 30 degrees and back-bevelled 10 degrees. This would work satisfactorily for dressing oak, ash, elm, etc. The cutting angle would have to be more acute—in fact, it would require as much bevel as for soft wood if the wood was wet or green. Where hard woods that have been kiln dried are being operated on exclusively a cutting angle of 15 degrees, as shown in Fig. 3, should be used.

After determining the proper cutting angle for the knives the next thing to be considered is the human element. When the thin knife planer is placed in charge of a workman that has always been accustomed to handling thick knives he will require some instructions before he attempts to grind the knives. Having always been accustomed to grind his knives with a cooling flow of water, or with a vessel of water handy to dip them in from time to time to prevent the knife from burning, it is rather difficult for him to realize that the thin, self-hardening knives do not require any cooling medium other than air. Water can be used for grinding self-hardening thin knives, but when used it should be used very sparingly. In fact, you can't use too much. When grinding without water be careful to grind slow, as otherwise the knives may be heated more than is necessary. They will come back to their former state of hardness all right, but in so doing they will crack, which will mean perhaps that they will be irreparably damaged. Whether the grinding is done on the head or on an automatic knife grinder, it should be done slowly.

The knives should be jointed to a true cutting circle before they are ground. This will leave a heel or thick edge on the knife, and great care should be taken not to grind this thick edge entirely away, as otherwise there is no way of telling whether you have a true cutting circle or not. A good plan when grinding the knives on the head is to take light cuts and keep going from one knife to another, as in this way you will have a better chance to keep the cutting circle and also to keep the knives from being overheated.

It may be advisable here to say a few words about jointing the knives. Some men have the idea that the jointing apparatus was devised for the purpose of enabling machine operators to sharpen their knives without stopping the machine. If you are one of these, you are wrong. It is only for truing up the knives, and the more lightly it has to be used the better. Jointing fast or taking heavy cuts should be strictly avoided. The stone should just touch the edge of the knives and be slid back and forth slowly and evenly.

When it is desired to remove a nick from a knife, the knives should be jointed until the nick is removed and then they should be ground. In this way the true cutting circle is maintained. If proper care is given the knives on the thin knife surfacer there should be no difficulty in getting good work from it. The machine itself does not differ greatly from the square head machine. It would pay every woodworker to study up the principles of thin knives, as they have come to stay, and the thin knife machine will eventually supersede the old square head thick knife machine.

The Packing of Furniture

At a recent gathering of furniture men in the United States, Mr. R. W. Finlator, a prominent railroad man, gave some advice on the packing of furniture which may be interesting to Canadian furniture manufacturers.

"I want to say in the beginning that efforts in the past to improve the packing of property are appreciated by the carriers, and, while progress has been made, the packing has not yet reached the point where the traffic is protected as it should be.

"Little fault can be found with the packing of the finer grades; it is in the handling of the common goods where we are suffering most—rockers with sweeps attached, legs or posts of the washstands, dressers, and other goods of this character with similar protrusions in many, many cases not protected at all. You can easily imagine what happens to them. Chairs that should be crated are shipped with slight wrapping. Broken rockers are heavy contributions to our furniture damage.

"I have always felt that case goods of every description should be thoroughly wrapped, padded, and crated. The practice of shipping this class of furniture with the legs exposed is a very heavy contributor to our damage claims. Rocking chairs, in my opinion, should never be shipped with the sweeps attached; when it becomes necessary to ship in this manner they should be protected with a good, strong crate. Other chairs, except the cheaper kind, should be crated, and the cheaper grades wrapped in a manner that will protect them in transit."

Box Testing Machine

The accompanying illustration is that of a machine devised by forest service experts at the Forest Products Laboratory, Madison. Wis., for testing box materials. It consists of a hexagonal drum, with 3½-foot sides, which is lined with thin steel sheets. Pieces of scantling bolted to the bottom form what are known as "hazards." The machine is the result of experiments made to determine what is a fair test for all types of boxes. During the last year a series of box tests has been carried on in co-operation with the American Society for Testing Materials and the National Association of Box Manufacturers, with the aim to determine the strength of boxes made of various woods and different forms of construction.

In making the tests, boxes filled with cans containing water are placed in the drum, which is then rotated. For convenience in observing the results of the tests

Boxes are placed inside and the machine revolved.

the sides and ends of the box are numbered with large figures, and, in addition, other numbers are placed at specified points on each side. The "hazards" cause the boxes to be carried part way around and then dropped back to the lower level of the drum. Each fall of this sort is a pretty fair imitation of the probable treatment the box would receive in shipment. The boxes are watched carefully and notes are made of the manner in which they give way and the number of falls required to break them in pieces.

In this way, say the officials who have conducted the tests, it is possible to determine what kinds of woods are best suited for boxes. The tests showed a decided need for a standard classification of box woods, and three groups have been made, based on the data which was obtained. The Forest Products Laboratory advise us that, with the exception of a few general statements, the results of these tests have not been given out. When they are we believe they will be interesting to Canadian box makers.

Hints on the Proper Treatment of Bandsaws

Experienced Men Should be Employed—Tension of Saws Should be Uniform—Proper Alignment of Machines Necessary

By J. McDonald

The length of time that a bandsaw will last depends to a great extent on the way it is handled. It has to become adjusted to the wheels on which it is going to run before it will give its best service. A new bandsaw is somewhat similar to a new belt—it has a certain amount of stretch or elasticity, which gives it a tendency to bounce off the wheels. This extra elasticity should be removed during the first period of its use by means of the hammer and saw stretcher.

Owing to the fact that the manufacturers of saws cannot subject the saws to the same strains that they receive in the factory, it is a good policy just to run a new saw for half an hour and then take it off and go over it carefully, touching it up where necessary. Experienced sawyers know that a saw will crack under excessive speed, uneven tension, too much tension, glazing from the emery wheel, case hardening, crystallization from too severe hammering, cuts from sharp hammers, vibration, and a host of other causes. In order to have saws tensioned and fitted in the best manner and to have them operated so as to give the best results in capacity and quality of work they should be in charge of experienced men.

Inexperienced men very often ruin a number of saws before realizing that their lack of knowledge is to blame. Therefore, it pays to have experienced men, even though they should be paid higher wages. The difference in the saw bill will invariably be found to much more than offset the difference in wages. It is impossible to draft a set of rules that would be applicable in every factory, as different conditions prevail. A few of the chief causes of defective saws, however, will be enumerated here for the benefit of readers of the Canadian Woodworker:

Vibration is the first. When the speed at which the saw travels is considered and the tension in it, it can easily be understood that very little vibration will break it. To eliminate vibration it is absolutely essential that the machine be firmly secured on a good, level foundation. Assuming that this condition has been fully realized and provided for, we will deal with the machine itself. Particular attention should be given to the wheels and their shafts; also the journals and boxes. The wheels must be perfectly round and true, and the shafts must fit snug in the boxes so as to have no lost motion.

From the machine we turn to the saw itself. Uniform tension is very important. If the saw has tight and loose spots it will have a tendency to crack, due to the fact that the tensile strain is not evenly distributed throughout its length. In hammering out the "tight" and "loose" places in the saw a round-faced hammer should be used. Hammer alternately on both sides of blade and try frequently with a straight-edge. Care should be exercised at all times to keep the edges true. After the tight and loose places have been hammered out, the next thing to do is to look for and remove the twists or kinks, working from both sides of blade, and using round-faced hammer as before.

The saw should then be "opened" or tensioned until it shows the right amount of drop from the straight-edge. The most of the opening should be done in the centre of the blade, gradually decreasing to within about seven-eighths of an inch from tooth edge of saw

and about five-eighths from the back edge. The tensioning may be varied a little to suit the work in hand. Care should be taken not to get the saw too open; it should be examined from time to time with the small straight-edge. In order to have the saw travel straight on the wheels without any side motion, and to reduce the vibration of the loose side to a minimum, the tension must be perfectly uniform the entire length of the saw. The correct amount of tension varies with the amount of feed and the crown of the wheel, but there should never be so much tension that the saw will not lie flat of its own weight when placed on a table.

Some men use tension gauges, having one edge curved to the amount of tension wanted. These will be found very useful. In using a tension gauge the saw is placed on the anvil in the same position as when hammering and the tension gauge held square across the blade. If the saw has been properly tensioned the blade will conform to the curved edge of the gauge from edge to edge. If it is desired to reduce the tension, the blade should be hammered gently along the edge, care being taken not to strike nearer than about a quarter of an inch from the line of the bottom of the teeth. To increase the tension, hammer the centre of the blade and test frequently with the tension gauge.

After tensioning, the saw should be accurately fitted. The swaging and fitting of the teeth is done practically the same as in a full swaged circular saw. The swaging is side dressed or shaped to a uniform width with an under and back cut, so as to leave the extreme point of tooth a trifle the widest. The amount of swage when side dressed should not exceed No. 9 gauge in a 14-gauge saw, and in hard wood it can run with less clearance. It is always advisable to run the saw with as little swage as practicable, in order to reduce the tensile strain and also to save power. The re-sharpening of saws is something that should be taken care of, as a great many saws are ruined by running them when they are dull. No bandsaw should be run longer than two hours without being re-sharpened.

The proper amount of hook on saw teeth is governed by the stock to be re-sawn and the feed. If a saw when properly fitted and hammered runs true on the wheels out of the cut, but runs back on the wheel as soon as it enters the board, the amount of hook should be increased, as the saw should retain the same position on wheels, whether in or out of the cut. In sharpening a saw a medium sharp emery wheel should be used, and care should be taken not to crowd it on the work, or case hardening will result. Sharp gullets should be avoided, as these concentrate the bend of the saw too much at one point as it runs over the wheels. A long, round gullet should be used, having no sharp corners or abrupt angles.

The back edge of the saw should never be allowed to come in contact with the back guard wheel or any hard surface, or case hardening is sure to result. If the saw should accidentally run against the guard wheel and be case hardened, the case hardened or glazed spots should be removed at once with a soft emery wheel. It is necessary to have the tooth edge of the saw tighter than any other part, and to do this without affecting the uniform tension the saw should be rolled a little longer on the back edge. The increased length should begin at a point where the greatest tension shows, and the back edge should show about 1/32 of an inch rounding in every five feet. The upper wheel should then be tilted forward enough to make the saw have an equal pressure on the front and back edge.

A word might be said here about the guides. These should be lined with soft babbit metal or good hard end-wood, and adjusted as closely as possible to the saw without touching it. The saw must, of course, be in perfect alignment, and the guides must be watched to see that they do not throw it out of alignment. The tensile strain should be just sufficient to prevent slipping. If the operator carries out the above suggestions he should be able to keep his saws in good condition and turn out satisfactory work.

The Economy of High-speed Machines
By Frederick E. Tracey

You personally are progressive and believe in modern methods. Then why backwater so hard when it comes to putting your beliefs into practice? Isn't this what you do when the machine salesman comes around with his tale of tools of marvelous producing ability? Quite true, you say, but then you have good and sufficient reasons for not installing every giant unit that puts the "prod" in production. Let us size up the arguments both ways.

To be perfectly square and matter-of-fact I will quote the arguments used by a Grand Rapids furniture manufacturer. The machine in question was the up-to-date turning and shaping machine. It has a capacity of double that of four hand turning lathes this manufacturer used. Its work is more uniform on any given pattern than hand turning methods. Being more automatic in operation less skill is required to develop good work from it than under the old methods. This allows operators of less skill than are usually employed in hand turning to operate it, thereby saving operative expense.

These advantages were all admitted by the manufacturer. But his great objection was that the machine would stand idle most of the time because other operations in the factory could not keep up with it in capacity. Here, therefore, would be a comparatively expensive tool standing idle a greater part of the time, so that the capital invested would not be earning what it was capable of.

While this fact is distinctly true it is more than counteracted because the high-speed tool requires space, power and operator for but one machine instead of for from the two to eight units used when hand turning methods are employed. As the high-speed tool stands idle a large part of the time in small factories, its wear and tear is proportionately less. Its operators, moreover, can devote most of their time to other work.

Thus all these economies stand against the single fact that the high-speed machine may, in a sense, partially tie up a given amount of capital. The money involved amounts to about $400. If the factory is expected to earn 10 per cent. on its investment, this would amount to $40 in a year on the $400. For one working day this would be but 13c. The first hour of the operator's time spent on other work would more than pay for this total day's interest. When you add to this the saving of the entire time of the several operators needed under hand lathe methods, and the saving of floor space, power, depreciation, etc., etc., the total saving amounts to quite a few dollars a day in

favor of the high-speed tool as against a trifling maximum of 13c for interest on tied-up capital.

These considerations apply with equal force to the general run of modern tools, such as the continuous-feed glue jointer, the high-speed moulder, matcher and belt and drum sanders. This conservatism on your part is but natural and necessary—it holds things in established channels. Progress has always been fought against—but in the end it has always won. For instance, take the modern spelling. I prefer to use the old way of spelling "through" rather than "thru." The new way is shorter, better and more sensible, so that it eventually will be always spelled that way—still, the old familiar way is hard to get away from.

There you have the situation. But as factory equipment involves the loss of actual money if you cling to your old ways and preferences you can ill afford to operate any but the speediest producing units.

Useful Sanding Device
By Richard New

I recently saw a very useful sanding machine constructed, on which the head stock of a discarded turning lathe was made good use of. The firm in whose factory the machine was constructed had taken an order to turn out a number of small pieces of wood 2⅛ inches x 5¾ inches x ¼ inch thick for bellows stock for a musical instrument plant. The pieces had to be sanded on ends, edges, and sides, and, not wishing to go to the expense of purchasing a disk sander specially for this order, the foreman in charge of the shop constructed a sanding device, as shown in illustration.

Placing the head stock in a convenient place on a bench, he took three pieces of 1⅛ inch maple stock

Tail stock of turning lathe used for sanding machine.

and built up a three-ply disk 20 inches in diameter. The outer edge of disk was counter-sunk 5/16 of an inch deep to allow for a ¼-inch iron band. The iron band was fastened to the disk with four screws. When it was desired to change papers the iron band was removed, paper placed on disk, iron band replaced, and there she was ready again.

Sandpaper cloth such as is used on belt sanding machines was applied around the edge of the disk—which, it will be remembered, was about three inches thick—thus enabling two men to work at the machine at one time. Owing to the lathe cones this machine had an advantage over some machines on special work, as it was easy to change the speeds. Thus, for an outlay of less than a dollar for time and material, a device was constructed which saved the firm a pile of dollars during the time of its existence. Sometimes the most simple devices produce results far in excess of what the device itself can be valued at.

Woodworking Machines at Toronto Exhibition
Manufacturers Keep Pace With the Requirements of the Industry—Ball Bearings Wonderful Power-Savers.

When considering the wonderful strides that have been made in recent years in the manufacturing of all classes of woodwork, we must give due credit to the manufacturers of woodworking machinery. In some cases the requirements of individual factories have been outlined to them, and they have had to go to work—often with very little to work on—to devise machines to do the work. In other cases they have anticipated the wants of manufacturers, and placed machines on the market, only to find that they had to turn around and spend a great deal of time and money educating manufacturers to their advantages.

Recent observations, however, have led us to believe that manufacturers now place a great deal of confidence in machinery firms and their representatives. In the majority of cases these representatives are practical men, who, from their travels in different sections of the country, coming in touch with and helping to solve the problems of many manufacturers, possess a veritable wealth of knowledge and experience. Thus when you put your problem up to him it may not be a problem at all. It may have been solved by, or for, some other manufacturer. If the problem is new to him it may not be new to the company he represents. And so it goes. The old saying, "Profit by the experience of others," cannot be applied to better advantage than in the woodworking industry. Cultivate the acquaintance of the "machinery man" would be a good motto for the woodworker, as by keeping in touch with him and your trade paper you will be keeping abreast of the times.

The exhibit of the Canadian Machinery Corporation, Limited, of Galt, Ont., proved of great interest to visitors. Besides several standard machines well known to woodworkers, the company had some new ones. One of these was their No. 651 floor swing cut-off saw. The feature of this machine is that the saw works from below, and is operated with a foot lever, permitting the operator to have the use of both hands for manipulating the stock. The machine is equipped with a table having cross rollers to facilitate the handling of heavy material. It also has suitable stops for cutting the material to the required lengths. The other machines in the company's exhibit worthy of special mention were their No. 116 end matcher for end matching hardwood flooring; their band-saw equipped with a Newel automatic band-saw guide; a single end tenoner with anti-friction carriage and side clamp bearings; a 12 spindle dove-tailer equipped with spiral gearing, and their No. 518 chair clamp, which is made specially for clamping the side rails into the front and back posts. The No. 116 end matcher mentioned above is also equipped with spiral gearing. It is claimed for this gearing that it runs far smoother than the old cross gearing. The Newel automatic band-saw guide is a device that takes the place of the small wheel or two pieces of hardwood usually found on band-saws. The feature of this guide, as pointed out by the manufacturers, is that it does not hold the saw rigidly, but swings in such a manner that any force tending to deflect the saw from its course operates automatically to bring it back instantly to its normal path.

One of the most interesting things at the exhibition in woodworking machinery was seen at the exhibit of the Garlock-Walker Machinery Company. This was

a perfect working model of the Linderman automatic dovetail matcher. The model was placed in a position convenient for visitors to see exactly how it worked, and samples showing some of the work of which the machine is capable were on view. The Garlock-Walker Company also had in their exhibit a variety saw, manufactured by the American Woodworking Machinery Company and sold by them. The features of the machine are graduation marks on the table, which permit the operator to set the fence at the required distance from the saw without using a rule, and a graduated protractor which enables the operator to adjust either one or two fences at any desired angle. The latter feature is very useful when the stock to be cut has a different bevel on each end, as both ends can be cut at one handling. Scarcely any time is required to adjust the fences, as the graduation marks on the table are very distinct.

The exhibit of the A. R. Williams Machinery Company in the Industrial Building, featuring machines for toy-making, proved of particular interest to those already manufacturing toys and novelties and those contemplating going into the business. The fact that there was a man working on the machines actually making toys made the exhibit more real, and gave an idea of what may be accomplished in small space in the manufacture of toys, providing suitable equipment is installed. The machines exhibited were a variety saw and a band-saw, made by the Preston Woodworking Machinery Company, of Preston, Ont., for whom the A. R. Williams Company are agents. There was also a Universal woodworker and a small planer. The A. R. Williams Company also had an exhibit in their usual place in the Machinery Building, where they exhibited their Eclipse planer and several other woodworking machines, one of which was a buzz planer, equipped with Whitney self-lubricating bearings, made by the Preston Woodworking Machinery Company. The Eclipse planer is a machine that should interest the proprietors of small and medium-sized planing mills where there is not sufficient matching to warrant installing a large double planer and matcher. In using the Eclipse planer the stock is first faced or planed on one side. Two heads, which are located at the back of the machine immediately behind the last feed roll, are then adjusted to do the moulding or matching. Thus when the stock is put through the machine the top side is planed and both edges are moulded or matched. The A. R. Williams Company were demonstrating some saws made by the Radcliff Saw Company of Toronto. This company has designed a number of special saws for shell box work.

The Cowan Machinery Company, of Galt, Ont., were, as usual, among the leaders, with a display of up-to-date machines suitable for different branches of the woodworking industry. These included a power feed rip-saw, a 6-inch four-side moulder, a hollow chisel mortiser, and a chain mortiser. The Cowan Company have devoted considerable time and attention since the war started to the designing and manufacturing of special tools and equipment for shell boxes, with which they have been particularly successful.

The transmission of power by means of shafting, either from motors or engines, entails huge losses due to friction. Manufacturers who realize this and desire to save as much as possible of the power lost in this manner no doubt found much to interest them at the exhibit of the Chapman Double Ball Bearing Company. The exhibit was in charge of men fully conversant with all the details connected with the transmission of power, and many interesting examples of ball bearing equipment were on hand to demonstrate the purposes for which ball bearings can be used. To the woodworker the most important of these are line shafts, counter-shafts, and loose pulleys. To the editor the most interesting feature of the exhibit was a simple test machine, consisting of two short lengths of shafting to represent counter-shafts, arranged side by side on a table, one provided with babbited boxes and the other with ball bearings. Each shaft had a short lever arranged above it and a wire loop attached to one end, with the other end of the loop passed over a small pulley on the centre of the shaft. By exerting a pressure on the end of the lever it had the same effect on the shaft or bearings as a belt pull. If the lever on the shaft with the babbited bearings was held down with one hand the shaft could not be revolved with the free hand. By holding down the lever on the shaft with the ball bearings the shaft could be revolved easily with one finger. By this it will be seen that the amount of power required to drive a shaft equipped with babbited bearings is greatly in excess of that required to drive the same shaft equipped with ball bearings.

Ball bearings have been used in place of babbited bearings on a number of woodworking machines with a great deal of success. One of these is a shaper made by the Preston Woodworking Machinery Company. Persons familiar with the smooth working of ball bearings for shafting and loose pulleys will appreciate the quality of the work that can be turned out by a shaper equipped in the same manner.

Flooring At 300 to 400 Feet Per Minute

Knives Have to Be Ground Once a Week—Machine Requires About Two Quarts of Oil a Day.
By Charles Watson

At the present time it is common knowledge among readers of this journal that flooring can be made at 300 to 400 or more feet per minute, but I venture to say there are a good many who, like the man from Missouri, would prefer to see it done. Among other things, they would like to know how big a crew of men it takes to keep one of these machines full of lumber, how long it can run that way without shutting down or "burning up," and some facts and figures about the machine further than the off-hand statement that it has round heads and thin knives.

A short time ago I visited a planing mill where a couple of fast-feed floorers were in successful operation, turning out smooth stock at the speeds above named. Accompanied by the foreman, I looked over the details of one of the machines at lunch time, and in the afternoon watched it make flooring.

The top and bottom heads of this particular machine (there are several makes of good fast-feed floorers) are 9-in. diameter and of the round type, each head carrying eight thin-steel knives. The bottom head is just back of the top head, then come the two multiple-knife self-centering side heads, directly opposite each other. Behind these are top and bottom profile heads, each mounted in a detachable yoke. At each end of the machine are smooth feed rolls, 12 or 14-in. diameter, driven by powerful feedworks. The machine frame is massive and extra heavy. All belts are of the best leather, made endless and fitted with tighteners. At the infeeding end of the machine, and on a level with the bed, is the big 5 x 16-ft. feeding table, fitted with power-driven "spiral" rolls.

In some mills these fast-feed machines cannot be run full capacity on account of the pipes or blower being too small, but not so here. The profile heads have 6-in. blower pipes, the inside head 7-in., the outside head 8-in., the bottom head 9-in., and the top head a 10-in. pipe. The fan is a 90-in. slow-speed affair—that is, the wheel is 72-in. diameter and each of the 18 vanes 22-in. wide. It runs only 480-r.p.m., but delivers the goods.

"How fast do your cutterheads go?" I asked.

"About 3,600 or 3,700-r.p.m.," the foreman replied.

"How much lead do you give your inside guide?"

"I always give the inside guide about ⅛-in. to the foot," he said. "It is necessary on these fast-feed machines in order to keep the rough lumber up to it. That causes considerable friction, but come around here and I'll show you how we keep it cool."

He led me around to the side of the machine and showed me a ¼-in. pipe leading to and from the ends of that section of the guide extending from the front of machine to the top head. "When running air-dried stock that has grit in it we open the valve and allow a circulation of cold water through here," he explained.

"How often do you grind your knives?" I asked, "and how many times do you joint with one grinding?"

"When we are running steadily I grind about once a week and joint when necessary, sometimes four or five times on one grinding. I usually change side heads once a week, but in a pinch have run a set nine days with one grinding and two jointings, and still did good work." He agreed that the jointing must be real light and carefully done.

Continuing, he said: "After jointing the straight, thin knives two or three times, I go over them with a carborundum stone and knock off the heel. I set the top and bottom head knives with a radial knife-setting gauge, but the side heads and profile heads are set and fitted up in the grinding room on special set-up stands, and jointed on an arbor the same size and speed as the machine arbor."

"By the way," I remarked, "how much faster does your feeding table run than the machine rolls?"

"Twice as fast," he replied. "We have experimented with that and found that for such fast feeds as 300 and 400-ft. per minute the feed table should go 100 per cent. faster, to insure a steady stream of lumber going through."

Noticing a man filling up the sight-feed oil cups, I enquired of the foreman how much oil the machines took per day.

"They used to use two gallons," was the startling

reply. "but I have cut it down to something like two quarts per day. That is about the least such a machine will run with," he added.

Just then the five-minute whistle blew, and the line-shaft began to roll. Then men came in from sunning themselves and began taking their respective places, joking and laughing as they did so.

"Wait a while and watch her go," the foreman invited, when he saw me step to one side.

A truckload of 1 x 4-in. was drawn up alongside the feeding table by means of a power cable, and the machine started with a tightener. A man stood at each end of the truck, and together they began piling the stock onto the feeding table. The feed was turned on and the boards began to shoot through at 350-ft. per minute—some speed, but the finished flooring was as smooth and true as could be desired.

It certainly kept those feeders busy. They would manage to get ahead a little and have a pile on the table and the extension arms when the truck was emptied, so that one could get that truck out of the way and rustle a loaded one into place with the cable before the last board was gone. A boy at the rear table dropped the flooring onto conveyor cables as it came through. These carried it to the trimmers, the grader, the pilers, and the tyers. One man did nothing but watch the machine and occasionally test the stock with his metal flooring gauge.

"I keep a man watching the machine and stock as it goes through," the foreman explained, "because if some little thing went wrong over 1,000 feet of flooring would be spoiled in less than five minutes."

"How long does it take to change from one pattern to another?" I asked.

"About fifteen minutes," he replied, smiling. "All there is to it is to change or cut out the profile heads and possibly change the side heads. These heads are always fitted up in the grinding room, and all we do here at the machine is to take off one set and put on another. We rejoint them here, of course, after they begin to get dull."

"Well, I concluded, as I bade him good-bye and thanked him for his courtesy, "you have some machine here, but I don't want the job of feeding it."

"It's lively work, all right," he admitted.—Woodworker (Indianapolis).

Marietta Paint and Color Company Loses Its President

Hon. Charles Sumner Dana, president of the Marietta Paint and Color Company, died at his home, in Marietta, Ohio, August 1, 1916. Mr. Dana had not been well for some time, but it was only during the last few weeks of his illness that there seemed to be cause for alarm, and consequently the news of his death came as a shock to a host of friends and business associates.

To readers of this publication Mr. Dana was probably best known by his reputation as the executive head of the Marietta Paint and Color Company. He was a firm believer in the use of printer's ink, through the medium of which his company has become one of the best known in the country, especially in the furniture and piano manufacturing trades. Mr. Dana was 52 years old, and took an active part in the business of his company up till a short time before his death. In 1896 he was a member of the State Senate, but in the later years of his life his business interests left him little time for politics. He was prominent in social circles, and was a trustee and one of the leading spirits behind Marietta College, one of the oldest educational institutions in Ohio. Mr. Dana's death will not result in any radical changes of the company's policies.

Stretcher for Making Screen Doors
By N. Richard

It is not true that all unique ways of doing things are found in the large establishments. On the contrary, some of the simplest, yet most effective methods of doing things are found in the so-called one-horse plants, in which, as a rule, the majority of the equipment is of the home-made variety.

While in a small shop recently I observed them making screen doors, and noted with what ease they were able to stretch their wire cloth stiff and taut. Thinking that perhaps some reader may be called upon at some time to make screens, I will describe as nearly as I can the device they were using. The illustration shows the device in operation and the method of stretching the wire cloth. The table was built of ash 3 feet 6 inches wide and 8 feet long. The ratchet wheel and crank handle were obtained from an old chain type

Handy Machine for Stretching Wire Cloth.

water-pump, and fastened to the end of a 4-inch wooden roller. The roller was slit open in the centre to hold the end of the wire cloth. When a sufficient amount of the cloth was wound on the roll the ratchet was released and the cloth pulled back and fastened to the end of the screen frame. When this was done the ratchet was wound as tight as desired and the wire cloth secured in place all round. After being secured the cloth was cut, thus completing the job.

For quickness and taut stretching of the screen this device had any beat that I have ever seen in operation. If I were to build one myself, however, I would so arrange the roll that it could be raised or lowered as desired. This could be done by cutting a groove in legs A and B and fastening the roll in same by means of thumb-screws.

Women Box Makers in England

A review of the work now being done by women in England, which was done by men before the war, speaks as follows: "There is considerable scope for the employment of women in box and light case-making factories. In addition to the sawmill processes, women have been successfully employed on jointing, matching, tonguing and grooving, hand-holing, boring, recessing, nailing, screwing, wire stitching, corner hinging, sandpapering, buffing and finishing, printing and branding machinery. Certain kinds of ammunition box work are well within their powers, including drilling, screwing, nailing, jointing-up bottoms, and knotting and splicing handles."

Being on the job early is one good virtue, but not the only one. Being on the job persistently is one of just as much importance, and is more often overlooked.

The Stock Sawyer in the Planing Mill

Should Possess a Good Memory, and Be Able to Work from a Number of Orders at One Time Without Becoming Confused

By T. Crawford

In no part of the woodworking factory can a man with brains be placed to better advantage than at the stock sawyer's bench. There is no need to dwell on the importance of the stock sawyer. The quality of the material entering into the work and the proper utilization of the stock depends on him. A short discussion on how to secure the best results and the best order of efficiency in the stock sawyer's department may tend to bring out an exchange of ideas on the subject.

It is a case of straight figuring to determine how stock may be efficiently handled through the various machines in the mill, because it only involves certain operations on specified quantities of material. This makes it much easier to take stock of and eliminate waste energy. When it comes to the stock sawyer, it is impossible to have any definite system of motions. Also in many instances it is impossible for him to keep track of his time to record it on the time cards. If a good, reliable man is secured it pays to do away with time cards. This is because the stock can very often be cut to the best advantage by having the stock cutter working on two or three different bills of stock at the same time. To clearly understand the work of the stock sawyer it is necessary to deal with specific cases, as the work varies greatly in different factories and on different lines of work.

In this instance we will consider that the stock sawyer is employed in a jobbing planing mill. Then he will have only one single order to fill. Suppose this order is for a job of interior trim. The first item will be the window and door casings. The finished size of these will probably be 4½ inches by 77 inches for door casing and 4½ inches by 64 inches for window casing. This item presents no difficulties, as 12 foot stock will make one length of door casing and one of window casing, and 16 foot stock will make three lengths of window casing.

If the stock has been properly sorted and piled when it was unloaded at the yard, the best way is to pick out enough 6-inch stuff of the proper lengths to make the order. If the stuff varies in width, however, it should be taken to the rip-saw and ripped to a uniform width. It is always advisable to send the stock to the planer before cutting it in lengths, as it requires less labor at the planer to handle the long pieces than short ones, and they can be cut to the finished length afterwards. If the casings are a stock pattern it will be all right to have a few over, as they can be stored in a suitable place until they are required for the next order.

This part of the work is very simple, and can easily be checked up. Suppose, however, that the order calls for some other kind of wood. We will say oak, as oak is frequently included on orders received at the jobbing planing mill. The first difficulty will be in picking out the proper widths, as oak, not being purchased in as large quantities as pine, is not sorted and piled according to width and length. In the next place, oak is seldom purchased by the jobbing planing mill in clear strips and boards that can be worked up without much loss. The usual practice is to buy oak in random lengths and widths varying in grade from common to select, with possibly a few clear strips and boards for special purposes.

The order for casing might call for boards six inches wide in the rough. These may be in the lumber yard, but in most cases they will be so mixed in with other widths that it will require too much time to pick them out. It may be practical to do a certain amount of sorting on an order of this kind, but generally the best method would be to take the mixed run of lengths and widths in boards that come nearest to filling the requirements of the order.

When it comes to cutting these up, the measurements for the other items on the order of interior trim should be at hand; also any other orders on hand that call for oak. The reason for assembling all these different items is so that the stock may be cut with as little waste as possible. The same reason is responsible for doing away with time cards in the stock sawyer's department.

On the same order as the window and door casing there will be window stools, aprons, base, etc. The efficient stock sawyer will check these over and note their widths and lengths. As part of the equipment of an efficient stock-sawyer is a good memory, he will be able to memorize these widths and lengths and then just as soon as a board is placed on his swing saw table he can tell which items it will make with the least waste. If a wide board has a shake or some other defect he can tell instantly just what it can be used for. It will thus be seen that the stock-sawyer may be working on half a dozen different jobs inside of ten minutes, so that time cards are out of the question. After all, it is not records that the firm wants, it is results. Therefore, if a good reliable workman is employed it is better to let him have as free a hand as possible and not worry him with a mass of detail.

The more orders the stock-sawyer can handle at one time without becoming confused the smaller will be the waste pile. Of course, this would not apply in furniture factories, where a quantity of stock of a great many sizes is generally required for each order. There will always be a certain amount of waste that will have to be piled aside to await an order for which it can be utilized. But if the stock-sawyer cuts several orders at one time, the pile will be small in comparison with what it would be if he only cut one order at a time and laid aside all the stock which could not be used for that order.

Before leaving the subject it might be well to refer again to the cutting of oak. In cutting oak it is generally advisable to cut it to lengths at the swing saw, as if it is sent to the planer in long lengths there may be knots or other defects that will interfere when it is being cut to finished lengths. Then, again, if the sections containing knots are cut out at the swing saw it will be better for the planer, as oak knots are not good for planer knives.

To sum up, efficiency in the stock-sawyer's department depends on the man or men engaged therein settling down and studying out systematic methods of handling stock, to the end that the boards may be made to fill the orders at one handling and waste eliminated as far as possible.

Where Mahogany is Grown and How Secured*

Features in Its Production Interesting to Manufacturers—Costs More to Put Out Veneered Article than Solid

I am going to tell you a little about how mahogany is grown and then clear up some little features that puzzle some people. First of all, you know that there are three kinds of mahogany, the Island, which comes principally from Cuba. It is a small wood, very small, very crooked and very undesirable except for small work, such as chair work. Then we have the Central American mahogany and the African. Do you know that the woods that grow in the tropical zone are heavy? They sink and cannot float, and the two greatest exceptions are mahogany and cedar—the cedar that goes into cigar boxes? If it were not that they float perhaps you would not have them, because the only means of transportation are the rivers.

I would like to tell you something here that you didn't know. You think of an American tree as having a stump. Always when a tree comes down it leaves a stump, but what would you think if I told you that the mahogany tree reverses the position. Instead of having a stump at the bottom, the stump is at the top and the tree at the bottom is no larger than my thumb. The body of the tree will be three or four feet in diameter, yet no stump. As the tree grows, instead of producing a stump like an American tree, it begins to throw out fins, just narrow fins. Sometimes these fins will go out 15, 20, 25 or 30 feet, and no thicker perhaps than three to six inches, and if you stood in one of these pockets it would be like the walls of a tent.

Many people ask why they square mahogany, why not ship it round like the logs in this country? You know a mahogany tree does not grow round. It is just as likely to grow square as it is to grow round. I have seen trees where you would think it would be the thickest there would be a hole. A tree is many times as apt to be irregular in shape as it is round, and many of them are at least square on two or three sides. I remember once, a log we cut in the middle. The log when it was cut off was about 4½ inches one way and the other was about 16 or 17 inches, and the heart of that tree was within 6 inches of one edge and about 4 feet from the other edge. Now, isn't that strange? In the American National Bank you will see that very log. It is easier, therefore, to ship them square than to round them up. Besides, they have to go so far that they stack up better in the hold of the vessel than if they were round.

Then there is a little beetle. He lives to eat. This little fellow, when he starts, doesn't know when he ought to stop, so the only way to catch him is to get him when he has eaten clear through and before he starts back, and removes that piece. One year we brought some of these beach, into our plant, and we lost $30,000 worth of lumber. He wasn't content to eat the piece he was in but began on the fresh lumber.

In the tropical zone there are two seasons, the rainy and the dry; a season when practically no rain falls, then a season when all of it falls. The rainy season comes about like our June rises. When our June rise comes, the tropical June rise starts in. It lasts until about September, then eight months it is

dry. Just as soon as the rainy season is over work begins in the camps and lasts about eight months. When the dry season is over, the logs are put afloat and streams carry them out. Usually, man power is the only means of bringing the wood out. They run in teams, as many as 100 men in a team. The chief of the tribe will lead them, and they come down the corduroy roads, which have been made beforehand, so fast that somebody is killed. The wood is laid on the banks of the streams until the rainy season, when it is launched. The man who does the launching has to be exceedingly careful. If he launches his log when the water is rising, the chances are nine out of ten that he will never see it again, but if he waits until the flood tide is at its highest and then launches, the water will draw the logs to the coast. If he launces the logs as the water rises, they will go into places where they will never come out and they are lost. Many seasons the wood does not come out at all.

When the British Government (I am speaking now of Africa) took over what are now the British Colonies, they never interfered with the personal rights of the natives. The result is, the natives are their own crews, and get the wood out. Many times a white man finances them—so much so that they provision them for two years. I don't believe we get out fifty per cent. whole on the backings we give these natives. A number of them every season operate on their own account. After you have made your dicker and your trade, these men retire, and it takes from one to three days for them to divide their interests and it is an amusing sight. They will sit in a circle and all talk at once. That noise will begin early in the morning and continue until dark, and in the morning begin again. Finally it is all over; every man's interest has been talked out and the division made, and off they go. After we get the wood to the coast in this fashion, we have a time in preventing them from selling it. Where you have three or four miles of rivers and small bayous, these natives are always about, and they have to be watched.

When we get the logs we brand them on the ends and on the sides, but the natives will take out as many as they can get away with, take them up the river, clean them off as clean as when they left their mill, and then you have the privilege of purchasing them again. Once in a while the turning tide will carry a log out to sea. Whenever a log goes out to sea it will always come back. In cases where as many as 1,000 logs have gone out to sea, they have come back in a few days, and we have known where they came back into the same river where they went out. Whenever a log hits the beach, as the water comes back it will strike the log from the sea, and form a little eddy of water around one end. As it runs around it will cut under the log, then the log will lean in, and in about five days it will be buried completely out of sight. When you have to have logs, you dig under them and roll them out on the beach, so that when you are ready to load them they will be there.

We have another enemy, the teredo. You have heard about having to have copper bottoms for ships to prevent the corrosion. You know whenever salt

* Abstracted from address before Retail Furniture Dealers' Association at Indianapolis.

water mixes with fresh water it produces life. Salt water won't produce it, nor fresh water won't, but the mixing of the salt water and the fresh produces it. This life first begins about as big as a pin head. It begins to go into the log, and as he goes in he grows. If you have the eyes you can find a hole as big as the point of a cambric needle. As he goes in the hole is sealed. He eats a hole for himself and as he eats a hole he grows. One and a half inches from the outside he makes himself a hole as big as the point of a lead pencil. By the time he gets further in the hole is as big as my lead pencil. He grows faster in diameter and in length. We have taken them out of logs as long as nine or ten inches. He never bores above the water line. You would think that as soon as he makes enough holes it would make the log lighter, but he never comes out. The material is all there. You could not lay a silver-dollar on that side of the log but what it would cover one or more of these holes. They always bore straight in and below the water line.

Then you have the steamer. When the steamer comes you have to bribe the mate and captain; or if you don't they will work against you. You have to treat them generously. They won't anchor where you want them. When you remember that you are paying anywhere from $100 to $500 a day for that steamer's time, you get pretty anxious to get the logs on board. You do finally get them on board and you put about one-third of the entire cargo on deck, to bring them into New Orleans.

Have you never thought of this—why can't they imitate mahogany? After all, mahogany is merely a red wood, why is it that an imitation is not just as good when you paint it black like some types are made now? Mahogany is black in the pits, and I know of no other wood in the United States, with one exception, that is black in the pits. American walnut is black in the pits, a gum it exudes makes this black. When you take birch and try to color it mahogany, you don't have the black in the pits, and anyone who has had very much experience can tell the difference.

There is another opinion that is many times held, and that is that cheap goods are veneered. As a matter of fact, I am here to tell you that the reverse is true. It costs more to put out a good veneered article than the same article solid. It is true that veneer produces an article with a character marking so far in excess of that which you get in solid that there is no comparison. "A thing of beauty is a joy forever," and when you get a good piece of veneer you have something that you could never get in solid. It is true many cheap grades are made in veneer, but that does not prove anything.

Another thing, you know all the markings, the character in mahogany, is the stripe. How does it get that stripe? When a tree grows, like in the tropics, when the sun comes out of the equator, it will take a warp (I call it a warp, don't know how else to express it) to the right or to the left. The opposite season, when the sun comes north again, it will take that warp in the opposite direction, so if you sever a log, inside than three inches you will have about three inches where one piece of wood will be gouged out and the next piece will come out of the other side; it will represent a hollow and a ledge.

If you will look down on a mahogany table top, as you look down one side will be dark and the other side light. If you put your pencil at the light side, still holding the pencil in the same place and swing

around to the left, you will find that the place that was light at the right hand is dark. The reason is that looking in one direction you are looking into the pit end of the wood, then if you go the other way you are looking at the grain. There is some furniture put out where the veneer is almost devoid of that stripe. In order to help it, they touch it up with the brush. Now, if you want to know, and want to prove the character, you can prove it by looking at first one side and then the other side, because if you paint it with the brush it is light both ways.

Suggestions for a Drawing Table

We are indebted to Mr. C. H. Flewwelling, of St. John, N.B., for the two suggestions (illustrated herewith) on the construction of a drawing table. One of the suggestions—Fig. 1—is to have a projection all around the top, three inches wider than the table and

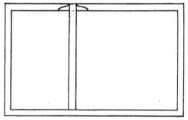

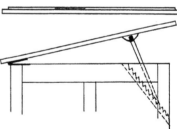

Drawing Table with Edge Grooved for T Square and having a Tilt Top.

one-half inch lower, on which to rest the T-square head. This also protects the drawing edge from knocks, thus keeping it true and square. A good method is to build a one-inch table top first and then glue half-inch clean pine boards on top.

The other suggestion—Fig. 2—is to hinge the side next the operator and put a rack on the other edge, making it easy to tilt the top and put in details on a wide drawing without stooping over.

The next issue of the "Canadian Woodworker and Furniture Manufacturer" will contain valuable information on the manufacture of furniture. Watch for it.

Original Method of Advertising a Machine

The P. B. Yates Machine Company have adopted a clever way of introducing to furniture manufacturers, their No. Seventy Automatic Turning Machine. They have issued in booklet form, a poem called "Making Good With Period Styles."

"Period styles for the masses" has been the dream of many furniture designers, but the cost of producing them has heretofore placed them out of range of those factories catering to the general public. With the advent of modern machines into the furniture industry, however, period styles have become very popular, and the general public are demanding them.

There is nothing left, then, for the manufacturer but to supply the demand. The following piece of

"I neVer dreamed these Period Styles so cussed intricate."

poetry, published in a booklet by the P. B. Yates Machine Company, besides giving one manufacturer's experience with period styles, contains some helpful hints to the man who would produce them on a commercial basis:

There was a manufacturer who had a nifty place for making Mission Furniture—he was well up in the race. He sawed it on a bandsaw, then finished it with "sand." 'Twas mostly done on machine, a little bit by hand.

He said: "These massive mission styles are used throughout the land; their value's taught in half the schools, it brings a big demand. An addition to my factory I'll build, and then I think I'll help to satisfy the call, and rear the dollars clink." His word was good, and so he did as to himself he'd said. In Mission manufacturing right soon the field he led. He sent his salesmen to the fray, with a kit of pictures plain, to show the dealers on their routes, the styles that made the gain.

They "got" the trade—those pictures did—they "sold" each dealer quick—and every single salesman soon was sending orders thick. "I'll tell you what," the old man said, as he in office sat, "It's new designs that push the sales, and you can bank on that." Then he in business went to sleep, to the tune of the dollar's clink. He'd made a "hit," what need for him to think another think?

But while he slept the public changed on furniture its poise. It didn't wake the old man up because it made no noise. He didn't need the coming change—a busy man was he, a piling up his money bags and coupon bonds in glee.

Then all at once the orders failed. The old man's eyes stuck out. He sat him down and "hollered" loud—it brought no change about. "If this keeps on," he sadly said, "my biz will go to smash. 'Twill land me in the county house to live on

beans and mash." He read, "It pays to advertise." A campaign then he run. He scattered ink throughout the land, but it didn't raise the "mon." He called his salesmen off the road and said, "What makes this slack?" "It is because," one man replied, "you've got clean off the track. Competitors are making stuff the buyers really want, and if you're going to get the biz the 'trade' you've got to haunt until you find the proper Style in Periodic Dope the Public really wants to buy.

There lies our only hope."
The old man did as he was bid. He traveled far and wide. He got the dope on Period Styles so he could well decide which one of them was suited for his factory to make and salesmen push in public mart, and healthy profits rake. "Good Lord!" said he on his return—he mopped his hairless pate—"I never dreamed these Period Styles so cussed intricate.

"I got myself all tangled up. 'Twouldn't penetrate my bean whether twisted legs meant Charles the Second or simply Jacobean. I couldn't tell the Louis tribe, at first, from old Queen Anne, or whether Hope or Hepplewhite to cane and spindles ran. But I got 'em down by labor hard to where a single glance would tell me if 'twas Empire or Italian Renaissance. I thereon held a joint debate betwixt me and myself, and chose the Bill and Mary stuff to gather in the pelf."

The factory began to hum, each lathe and back knife hand—and shaper operator, too—he worked to beat the band. But oh! the work that they turned out! Your cheeks with tears 'twould lave. 'Tis a wonder Bill and Mary didn't turn over in their grave. The turned parts looked like knob-sticks from Afric's golden strand. The finish was a calibre ne'er seen on sea or land. The workmen slept the "Monday blues," they'd stand around and swear. The office sweated drops of blood, the old man tore his hair. "I can't turn out," he wildly yelled, "these cussed Period parts, at price or grade the public wants. I see where ruin starts."

Then into all this chaos, turmoil, and language hard, a jaunty salesman briskly stepped, and handed out his card. The old man snatched it from his hand (he didn't feel his best). "Of all the things," he said, "on earth, I simply do detest, it is a drummer with the nerve to butt in on a man that's got more troubles to be solved than any mortal can." The salesman smiled. The old man's ire had missed him by a mile. He'd

"What now is going bad? I see your collar's hot." The Old Man answered, "Period Styles my Nanny's almost got."

learned the art of meeting irritation with a smile. He said, "What now is going bad? I see your collar's hot."

The old man answered, "Period Styles my Nanny's almost got. The trade demands the measly trash to furnish home and hall, and the turnings my machinery makes, they will not do at all. They're loppy, splintered, rough and bent, they wouldn't sell for cash. Next month a note of mine falls due—the business goes to smash. Wife's health is poor. She'll lose a trip she cannot do without. And John will lose his college, and Mary can't 'come out.'" The sympathetic salesman said, "Your cause is far from lost. I'll show you how to make those things at an almost-nothing cost. The P. B. Yates Machinery Plant—the firm I represent—an Automatic Turner has proceeded to invent. A wonderful machine it is that makes each Period part at rates so fast and cost so small 'twill make your eye-balls start. We call it Number 'Seventy.' I tell you what I'll do. I'll put one in and let you run the rig a day or two, and let you settle for yourself if what I say is true, about the rate and cut of cost and grade of product, too."

Disconsolate the old man said, "It won't cost anything, and maybe it will be some use and will salvation bring."

A few days later in the shop he stood a bit apart, awaiting for the "Seventy" to make her maiden start. The carriage moved in toward the knives, the chips began to fly, and one it flew and landed right in the old man's eye. He got it out—he op'ed his eye—and there beside him stood a man who held as smooth a piece as 'ere was turned in wood. Octagonal it was in shape, in William-Mary Style. Each edge was clean cut, smooth and true, yet delicate the while.

The old man scarce believed his eyes. "It made this? Tell me, when?" Before he learned, the "Seventy" had manufactured ten. "Why man alive, just look at this!" (the finish caught his eye). "No more I'll worry nights because our sanding costs are high." He kept a count throughout the day, of work that had been done, and when he cast his figures up the "Seventy" Turner won. "By Heck!" said he. "A common hand to that machine I sent, and he's making Period turnings at about one-third a cent. And every one is perfect, too, with finish smooth and neat. When salesmen show them on the road they'll bring home some bacon meat."

The weeks rolled 'round, the orders 'came—some orders—and repeat—the kind of product he turned out, one simply couldn't beat. The "note" came due—he had the cash. The boy to college went. His daughter made her social splash. Far south his wife was sent.

At last, one eve, when all was still—his men had sought their homes—he balanced up his bank account and found there many "bones." He went out in the

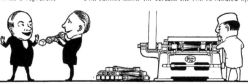

"The Old Man scarce believed his eyes. "It made this? Tell me, when?" Before he learned, the "Seventy" had manufactured ten.

factory in murky twilight dim, to where the "Seventy" Turner stood, just visible to him. And as he looked his eye grew moist—he reached his hand a bit, he softly touched the grease and chips, and thus he spoke to it: "It may seem nuts for a man like me his friend a machine to call, but—you dandy little devil! you—you saved me from the Wall."

We do not doubt that "Seventy" could fill an honored place in some part of your factory and help you win the race to gather in sufficient store of this earth's cursed gold, to mitigate the rheumatiz and chilblains when you're old, and still leave something over which your family can scrap when you are safely stowed away in Mother Nature's lap.

You cannot know for certain till you've studied up the dope we're sending out to those who must with Period Turnings cope. So quickly send us your request for "Period Patterns"—free—and details full, and photographs of little "Seventy."

You can write it up on linen bond or wrapping paper hard, or send it in by fast express or use a postal card. Use anything—just suit yourself—we do not give a d—n—just shoot 'er in—and do it now—through the mail of Uncle Sam.

The London (England) Timber News say that the United Kingdom imports yearly nearly $10,000,000 worth of manufactured woodenware and wood turnery (not joinery, doors, etc.), and of this about one-half comes from the United States and Scandinavia. Canada furnishes the bulk of the rest. The most important imports under the head of manufactured wooden articles, etc., are kitchen and house utensils, toys, dowels, skewers, chair parts, broom handles, clothes pegs, brush-backs, birch squares for spools, handles, and turned wooden boxes.

These articles, with the exception of tool and implement handles, are made almost exclusively from birch, poplar, beech, maple, basswood or spruce. The handles are made chiefly from hickory and ash. The following articles are imported in large quantities from the United States and to a small extent from Canada: Birch squares, 1¼-inch, 1½-inch, 1⅞-inch, 2-inch square, and 2½-foot, 3-foot, 3½-foot, and 4-foot long. These are used for manufacturing spools and bobbins, and for chair legs and the cabinet trade. Dowels are imported chiefly from the United States. They are all birch and maple preferably, varying from 3/16 to 1 inch in diameter, and 12 inches to 108 inches in length. The best sizes are ⅜ inch diameter, 36 inches to 48 inches long. Broom and mop handles are imported from Finland, Sweden, and the United States, the varieties being spruce, basswood, maple, or beech.

At last one eve when all was still—his men had sought their homes—he balanced up his bank account and found there many "bones."

Brazing Band Saws

By W. Martin

A great many young bandsawyers look upon brazing bandsaws as a very difficult piece of work, and even some of the older ones regard it as a very uncertain operation. These are mistaken ideas, and I will endeavor to show how easy it is to make a braze.

First cut the blade to a line squared across from the tooth edge. Be sure and square it from the front edge, as the back edge is crowned and would not do to square from. When the saw is cut, see that both ends, when laid on the leveling block, lie flat. This is essential to make a good braze, because if one corner should not fit down closely the lap could not be ground evenly all the way across, and this is necessary to make a good braze.

When the ends have been flattened down properly, joint them until they are square and show the full thickness of the blade. If a cold chisel was used to cut the saw, be sure you get all the chisel marks out, and, above all, be sure and keep the ends square. Care should be used in grinding the laps in order to avoid grinding them down to a knife edge. They should be left about 24 gauge thick on the edge, to allow for draw filing to get all the case hardening off. It is impossible to avoid case hardening to a certain extent, no matter what kind of grinding wheel is used, and the laps should always be draw filed before proceeding to make the braze.

The next thing is to make sure that both ends are alike as to thickness and width. The thickness can be calipered and the width of the laps can be made to correspond exactly by using a gauge made from a piece of band iron ¼ inch thick, ¾ inch wide, and 2 inches long. This should have a shoulder filed in it about ½ inch from the end on one of the ¾ inch sides. By placing the shoulder of the gauge against each end of the blade alternately and holding a scratch awl or some similar instrument against the end of the gauge, you make a mark exactly the same distance from the ends on both laps.

If both ends are marked in this manner they must fit when completed, providing proper care is taken to file just up to the line. The ends of the laps should always be left the thickness of about 24 gauge, as by so doing, when they are clamped together the pressure is concentrated on the part where it is needed. Another feature is that the irons do not mark or compress the blade on each side of lap, thus leaving an unsightly scar that cannot be filed out in finishing without making the blade too thin at that point and weakening the braze.

In regard to the flow, my experience has been that a good braze can be made by using good clean soft water, either clear or with a little clean borax added. The best results I ever secured, however—cleanliness and cost considered—were by using good, fresh lump borax. I had it pulverized as fine as possible, and mixed about three tablespoonfuls to half a pint of good grain alcohol. I then washed the laps off thoroughly with this mixture, using a swab made of a clean piece of cloth wrapped on a stick.

When the laps have been placed in the clamp they should be rubbed over with a clean piece of fine emery cloth before the silver solder is applied. The solder should then be washed off well with the alcohol and borax mixture, care being taken not to handle it more than is absolutely necessary to get it between the laps. This is done by prying up the top end enough to let the solder slip in easily. The pry should then be taken

out and placed under the other side, to permit the bottom iron to slip into place.

When heating brazing irons the heat should be as uniform as possible the whole length of the irons, and both irons should be at the same heat. They should be made as hot as they will stand without sparkling. When they are hot enough the scale should be scraped off. They should then be put in the clamp and the pressure applied quickly. When the bottom iron is placed the pry is removed, and care should be taken to see that the lap is centred on the bottom iron and the top one placed exactly over the bottom one, as then the pressure will bear evenly on the joint. The pressure should always be followed down, because in cooling the irons contract very considerably, and will thus release the pressure unless care is taken to keep it adjusted as tightly as possible.

Once all the red has disappeared from the irons and they have blackened, the saw can be taken out of the clamp, but it should not be disturbed any more than is necessary until it has become quite cold, when it is finished by filing and rolling. The braze should be tensioned before it receives any hammering. In filing always be careful to keep the braze as near the same thickness as the rest of the saw as possible. It is good policy to examine the tension in a new braze for a few runs, as a braze will stretch easier when it is new than when it has been running for a while.

A Complete Book for Finishers

The rapid increase of applied science in the woodworking industry has created a demand for more efficient and superior workmanship in the finishing room, so that a reliable work on modern finishing methods will no doubt be welcomed by a large number of finishers. Such a book has been written by Mr. Walter K. Schmidt, a prominent chemist and finishing expert, dealing with the whole subject of finishing from the equipment and arrangement of the finishing room up to the scientific handling of all materials. Every phase of scientific finishing room management is comprehensively discussed.

The reader is told in non-technical language how to apply the best modern methods. In one chapter there are over one hundred staining formulas. The book is 9¼ inches x 6¼ inches x 1 inch thick, handsomely bound in a black flexible cover. It is interleaved with blank pages for the entering of additional notes and formulas. The title is "Problems of the Finishing Room." The book is published by the Periodical Publishing Company, Grand Rapids, Mich., and the selling price is $5. If so desired, the book will be submitted first for inspection.

Stow Manufacturing Company's Bulletins

The Stow Manufacturing Company, of Binghamton, N.Y., have issued three bulletins—Nos. 100, 101, and 102—illustrating and describing respectively their very complete line of electric motors, electric tools, and flexible shaft. One of the features of the company's multi-speed motor is the speed control. The speed adjustments are made by means of a hand-wheel, which may be either at the top or the side of the motor. The exact speed desired can be obtained by turning the hand-wheel.

The Stow Company's line of electric tools include drills (either portable or stationary), screw drivers, hand grinders, post grinders, bench grinders, floor grinders, etc., and their line of flexible shaft is well known to the trade.

The Lumber Market

Domestic Woods

Twenty dollar hemlock before next spring is the prediction of one large wholesale lumber dealer in Toronto. In conversation with the editor recently he made the above prediction, and remarked that the proprietors of woodworking factories did not appear to realize the seriousness of the lumber situation. This was due to the fact that the majority of them had been purchasing their lumber in small quantities and had not given the subject of future supplies any serious thought. His order book showed that the most indifferent ones were those whose factories are busy. A great many of the planing mill and box men have evidently been doing some thinking, and the consequence is that they are buying ahead. Another feature is the quantities of lumber that are being purchased by large firms in the United States.

A comparison of figures showing stocks on hand a year ago and at the present time will be more convincing than words. In the matter of spruce culls, which are used in large quantities by furniture factories for crating purposes, one company had on hand at one shipping point a year ago the whole of the 1914 cut, amounting to 1,500,000 feet, together with the 1915 cut of 1,300,000 feet, being a total of 2,800,000 feet. At the present time there is on hand at that point 600,000 feet, and it is all sold, to be shipped right away. On January 1 the same company had on hand at another point 800,000 feet of the 1915 cut, of which there is only a few cars left, too green to ship.

The company only deal in hardwoods as a side line, but their figures in this connection can be taken as a pretty fair barometer of the hardwood situation. Of hardwoods they had on hand a year ago 3,000,000 feet of the 1914 cut and 2,000,000 feet of the 1915 cut. This was all shipped during the fall of 1915 and during the present summer. The 1916 cut amounted to 2,000,000, and it is all shipped that is ready to ship. At another point where the 1914, 1915, and 1916 cuts totalled 3,775,000 feet there is only 800,000 feet left, and it is going as fast as it is fit to ship.

The scarcity of hemlock is even more pronounced, as the figures will show. At one point the 1914 cut amounted to 700,000 feet, the 1915 cut 1,100,000 feet, and the 1916 cut 400,000 feet. The entire lot was sold during the fall of 1915 and spring and summer of 1916. At another point the 1915 cut of 750,000 feet was sold during the winter and spring. The 1916 cut of 400,000 feet is not ready to ship or it would have been cut into. At still another point the 1914 cut amounted to 2,000,000 feet, and the 1915 cut 7,500,000. This is all gone but 2,000,000 feet of No. 2. The 1916 cut amounted to 6,000,000 feet, the bulk of which is on hand. To further emphasize the shortage of stocks, the inventory of one of the largest hemlock manufacturers in Ontario shows only half of what it did one year ago.

Small mills that were unable to market their 1914 and 1915 cuts have sold and shipped them last spring, and at the present time a large percentage of their 1916 cut is gone. Even where lumber is available to fill orders, great difficulty is experienced in making shipments on account of the shortage of labor, and the majority of the lumber wholesalers are none too anxious to part with what stocks they have, feeling convinced that they will get much better prices by holding them.

Letters received from some of the large lumber manufacturing mills indicate that cutting in the bush will be carried on on a very small scale next winter on account of the high cost and scarcity of labor and the increased cost of all food supplies, it being estimated that it would cost from $3 to $4 per thousand more to produce lumber than it did last year. Some of the smaller mills have intimated that they do not intend to go into the bush at all.

If a large number of manufacturers were to get into the market with the intentions of stocking up it would no doubt send prices soaring, and they would not gain anything in the end, as it would simply mean that the stocks would run out that much sooner. By knowing the facts, however, they will be able to deal with the situation more intelligently.

Imported Woods

The wood-using industries of the United States are busy. The volume of lumber being consumed by the furniture, implement, and automobile industries is said to have increased considerably during the last month. This is quite in order, as the fall and winter is their busiest time. Activity in the building trades is pronounced, so that the planing mills and sash and door factories are busy, meaning, of course, that they will be consuming lumber in proportion to their activities. One of the features of the situation is the amount of hardwoods being used by the planing mills and sash and door factories.

There is a scarcity of stocks in the United States, as evidenced by the large orders that have been placed with Canadian lumber wholesalers. The scarcity is more noticeable in maple, birch, and crating lumber than in any of the other stocks, the orders placed in Canada being principally for cull spruce for crating stock and hemlock. There has been rather a heavy demand for maple and birch during recent months, which no doubt accounts for the shortage. Prices on hardwoods have not advanced materially, but they are firm. Dry stocks of poplar are rather scarce, and that wood is still in demand, at prices ranging from $60 to $63 for desirable stocks in different thicknesses.

Furniture factories are taking large quantities of oak, cherry, and gum. The present consumption of both red and sap gum is particularly noticeable. Quartered oak is still a favorite wood, despite reports to the contrary. A few of the prices on hardwoods, firsts and seconds, inch, are as follows: Basswood, $44 to $46; chestnut, $51 to $53; maple, $41 to $43; plain oak, $63 to $65; quartered oak, $89 to $91; red birch, $57 to $59; sap birch, 47 to $49; white ash, $56 to $60.

One of the American lumber journals, in speaking of the volume of orders placed recently by manufacturers, says: "For many weeks the single car order has been the rule, but reports now include orders for from one to fifteen cars, and even more."

Mahogany and walnut are being used extensively by the furniture factories, and the prices for these woods are firm. Walnut appears to be very popular at the present time for the better designs in furniture.

The one feature that may hamper Canadian manufacturers in getting shipments from the States at present is the car shortage that usually prevails at this time of the year. Taking everything into consideration, however, the situation in regard to imported woods is favorable to the manufacturers. While prices are firm, and expected to remain so, no important increase is expected.

An Unusual Boring Machine

The advent into Canada of the shell box industry has opened up new avenues of endeavor for the manufacturers of woodworking machinery, and the majority of these manufacturers were quick in grasping the many opportunities which this industry presented.

Of particular interest to the woodworking factories and to the manufacturers of shell boxes especially is

Two-spindle boring machine for shell box work.

the two spindle drop-table boring machine recently manufactured and put on the market by Cowan & Co., of Galt, Limited, Galt, Ont. The special feature of this machine is the table and gate, in which the blocks are held for the operation. Unlike other machines, this table is automatically dropped to a fixed position by means of the hand-wheel. This eliminates the necessity for removing the block to complete the boring operation. The table is again raised to its normal position by means of the foot-tread.

The spindles are two in number and of standard size. Only two operations are necessary to bore the diaphragm for the Bethlehem or fixed ammunition box.

A special feature, which is claimed by the manufacturers to be a big improvement, is the absence of all gears, the entire machine being belt-driven.

Figures Interesting to Toy Makers

Many persons do not realize the importance of the toy industry as a national asset, writes the Quebec "Telegraph." Germany and Austria have been the largest exporters of toys, with France and Japan a close second. The year before the war began Germany exported $24,600,000 worth of toys; Austria-Hungary, $1,350,000, making a total of $25,950,000, which is not equalled by the output of any of Canada's large industries listed at the last census; the clay, glass, and stone products from the Dominion reaching a value in 1910 of $25,791,000. To the United States Germany sent toys valued at over $7,736,000; to Canada about $580,000. German toys, indeed, found their way all over the world, and the industry, which is unharmed by the war, is ready to assume its place in the world of

Articles on Draughting

In the June issue of the "Canadian Woodworker" we published an article dealing with the method of teaching draughting used in technical schools, and intimated that more advanced work would be taken up in future articles if our readers were interested.

We have received a number of letters from readers in different parts of the country asking that the articles be continued. We have, therefore, asked Mr. Alexander—the writer of the article—to prepare the second one, and we have his assurance that it will be ready in time for our October issue.

It is very gratifying to know that the article was found so interesting. We are always glad to receive suggestions from our readers telling us about any special subject in connection with woodworking that they would like to see taken up.—Editor.

trade as soon as peace is declared. It is a matter of loyalty to preach the doctrine of the made-in-Canada toy, with its maple leaf stamp, as well as a matter of economical necessity.

Canada offers a good market for toys, and it has the advantage of being a market that steadily increases with the demands from a rapidly growing population. In the fiscal year 1914, which ended four months before the war began, the total value of toys imported into Canada was $1,037,000, of which $580,000 worth came from Germany and $5,600 from Austria. The United States sent toys to the value of $293,000; the United Kingdom over $91,000 worth. Imports of toys from France were valued at $33,000; those from Japan at over $26,000. For the fiscal year ending March 31, 1916, Canada imported only $624,190 of toys, of which $476,581 worth came from the United States; the United Kingdom, France, and Japan sending the bulk of the rest. In a normal year a million dollars does not pay the bill for toys demanded to satisfy Canadian children.

Unloading Mahogany Logs at New York

Many of our readers may not be acquainted with the way mahogany is secured. We are indebted to Messrs.

One of the Huddleston-Marsh Mahogany Company's boats unloading mahogany logs at port of New York.

Huddleston and Marsh Mahogany Company, New York, for the above picture, showing one of their boats unloading mahogany from Mexico at the port of New York.

Construction and Operation of a Fuming Box

By W. J. Beattie

There is no doubt as to the superiority of a fuming box, so far as results are concerned, over any method of brush fuming, or oil stain fuming. The latter never attain either the permanency or richness of effect that the former process gives to the furniture. The construction of a good fuming chamber is the first step toward a satisfactory result on fumed goods.

A room of sufficient size for most factories will be about 10 feet by 14 feet, and as high as the ceiling of the finishing room will conveniently permit; that is, not over 8 feet inside the chamber if case goods are the product, as it is not convenient to have to pile them higher. It is much better to extend the floor area if necessary.

Make a framework of 2 inch by 4 inch material, in a manner somewhat similar to the construction of a frame building. This can be roofed over with tight sheeting if you wish to utilize the space above the fuming chamber for storing purposes. The walls of the structure are covered with a good heavy duck on the outside, which is painted after completion with four coats of good heavy paint, two on each side. The doors are covered with the duck on the inside of the frames, so as not to interfere with the door fastenings, as shown.

The roof ventilator is made of galvanized iron, and must extend up through the factory roof a sufficient distance to ensure a reasonably good draft. The pipe must be 10 inches or 12 inches in diameter and fitted with a damper. The damper is faced with felt so as

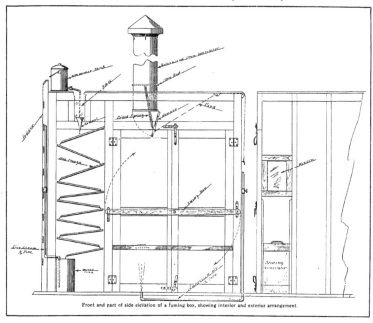

Front and part of side elevation of a fuming box, showing interior and exterior arrangement.

to close snugly against the wood facing on the lower end of the pipe, where it is held securely by the coiled spring. The coiled spring is attached to an iron rod running through the ventilator. The top end of the pipe is protected by a cap. The damper is controlled by a cord attached to it, which runs over a couple of pulleys. The end of the cord is outside the box. This damper is, of course, kept closed when the fuming operation is in progress.

The ammonia tank of one gallon capacity is placed on the roof of the chamber, preferably at the most convenient corner, and is fitted with a drip pipe, which goes through the roof. The flow is controlled by a small ammonia valve, or any iron valve will do. The liquid ammonia is run down into a funnel through which it runs into a number of iron troughs placed in zig zag fashion, and finally drips into a five gallon iron tank, in which there is a couple of gallons of water. The illustration conveys the idea in clear enough manner.

There is a pane of glass inserted so that the flow of ammonia can be seen, and any obstruction preventing its flow can be detected easily. The zig zag troughs are fastened to the inside of the room on wooden slats placed there for this purpose. A ½-inch live steam pipe is connected having its opening within two inches of the bottom of the water tank, and is controlled by a valve. The ladder, which is stationary, is used to carry up the ammonia tank and is useful also if the roof is utilized for storage purposes. The ventilator already referred to need not be in the centre as shown. If it is advisable to place it elsewhere, do so.

The doors are substantially made and covered on the inside with duck. The iron fasteners top and bottom and also the swinging wood bar across the centre, are made so that when closed they exert considerable pressure on the doors to ensure tight joints. The jamb joints and also the one where the doors meet are bevelled and covered with felt, to guard against leakage of the fumes. Part of the side elevation, front end, is shown having a window at the proper height, and also a sliding ventilator which can be raised after the top pipe damper is opened. The ventilation of the chamber after the fuming process on a batch of goods is completed, is greatly facilitated by the compressed air pipe which is shown coming through the floor just inside the doors. The air pipe has a cap on the end in which is a small hole, not over ⅛ inch in diameter. This force will raise the air out of the box in a few minutes, and it can then be opened and emptied without any inconvenience from ammonia fumes. As nearly every factory is equipped with an air compressor this feature is easily added.

The important things to remember in constructing a fuming box are, it must be snug and tight, and the only metals that will resist ammonia fumes are iron or steel. Brass will corrode in a short time; take off any brass hinges that may be on the furniture, or they will come out covered with verdigris. Case goods are piled up inside similar to what they usually are in the store room, having as little contact as possible. The framework of the structure being exposed on the inside gives plenty of opportunity for hanging supports for small parts.

The foregoing, with reference to the drawing, should be sufficiently clear to anyone contemplating the installation of a fuming box. If any other information is required it will be cheerfully furnished.

The goods to be fumed are first stained with fumed stain that has been diluted one-half by the addition of water: (Grand Rapids Wood Finishing Company's stain or Marietta Paint & Color Company's stain) this is applied freely with a brush, and as fast as the articles are stained they are piled into the fuming chamber. When the lot is completed heat the water in the tank with the live steam until it is warm enough to begin steaming, then close the doors.

The ammonia tank contains two quarts of liquid ammonia. This is plenty for a box of the size given. Open the valve so that the ammonia will drip freely. It should be all down in from fifteen to thirty minutes, and that which has not evaporated during its zig zag course will be in the water in the tank. Turn on the live steam sufficiently to cause a fairly lively commotion and raise sufficient moisture in the chamber to be noticeable on the window. But as soon as it begins to run on the glass close off the steam. The introduction of live steam irritates the ammonia (if such a term is permissible) and the fumes become very active and pungent. The humid air makes the wood easily penetrated, and the color is given a depth and permanency so desirable.

Fill the chamber so that the operation up to this point can be done before quitting time; instruct the night watchman to give another similar introduction of live steam about 12 or 1 o'clock. This helps to get all that is possible of the fuming strength of the two quarts of ammonia. At seven o'clock a.m. the pipe ventilator is opened, the compressed air valve released and in a minute or two open the sliding ventilator to allow the fresh air to rush in to replace that being driven out by the high pressure. In a very short time you can open the doors and begin emptying the box.

If the demand is sufficient, the box can be filled every morning and night, thus getting two capacities per day. Should there be at any time only a few articles to do, put in a pint or a pint and a half of the ammonia into the water tank and turn on the live steam, following the same directions as for a full box. The results will be satisfactory:

The first operation after the goods are removed from the chamber is to sand them thoroughly smooth. A pneumatic sand block is a good labor saver in this respect, if your goods present sufficient surface. After the sanding, coat with thin white shellac (be sure it is thin). When this is thoroughly dry and before sanding it, tone the sap streaks and lighter portions with spirit soluble made of nigrosine and leather brown. Have these mixtures in separate vessels so that the proper degree of shading color can be readily gotten. Use a flat camel's hair brush which can be dipped into either or both, as required. Apply the toning color in direct strokes, as the spirits soften the shellac coat and become incorporated with it. The necessity for doing this work in straight bold strokes is apparent. It is also advisable in this connection to have a wiper of cotton for the toning brush to take off the excess color, which is sure to be taken up, before applying the brush to the goods. After the toning has become thoroughly dry, sand and apply a second coat of shellac. The final sanding is followed by a coat of mineral wax, which, after it has flattened, is wiped off and fine dull gloss on a rich brown ground is the result.

If the shellac used has not been thoroughly bone dry before being cut, it retains sufficient moisture to sometimes turn the goods a light grey color in time. Hence the caution to have the shellac thin and to be sure that each coat has thoroughly evaporated its moisture before going on with the next operation.

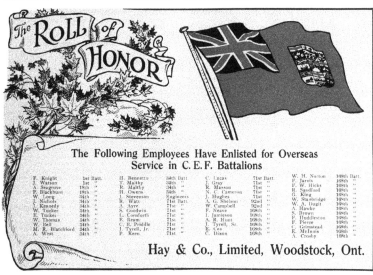

The Roll of Honor

The Following Employees Have Enlisted for Overseas Service in C. E. F. Battalions

F. Knight	1st Batt.	H. Bennetto	34th Batt.	C. Lucas	71st Batt.	W. H. Norton	168th Batt.
J. Watson	1st "	T. Maltby	34th "	J. Gray	71st "	F. Jarvis	168th "
A. Seagrove	18th "	R. Maltby	34th "	R. Masson	71st "	F. W. Hicks	168th "
P. Blackburn	18th "	H. Owens	34th "	N. C. Cameron	71st "	R. Sandford	168th "
W. Long	34th "	J. Stevenson	Engineers	J. Hughes	71st "	G. King	168th "
L. Nichols	34th "	R. Watt	71st Batt.	A. G. Shelson	92nd "	W. Stanbridge	168th "
J. Kennedy	34th "	A. Ayre	71st "	W. Campbell	92nd "	W. A. Dugit	168th "
W. Tucker	34th "	S. Goodwin	71st "	F. Neave	168th "	A. Hawke	168th "
E. Tucker	34th "	L. Cornforth	71st "	I. Jamieson	168th "	S. Brown	168th "
W. Thomas	34th "	R. Beam	71st "	A. R. Hunt	168th "	F. Huddleston	168th "
W. Bell	34th "	C. B. Priddle	71st "	J. Tyrell, Sr.	168th "	F. Pierce	168th "
M. R. Blatchford	34th "	J. Tyrell, Jr.	71st "	E. Cox	168th "	C. Grimstead	168th "
A. West	34th "	P. Keen	71st "	C. Dixon	168th "	E. McInnis	168th "
						A. Crosby	168th "

Hay & Co., Limited, Woodstock, Ont.

"Furniture in England"

This book is an encyclopaedia of artistic suggestion—an education in itself.

Upwards of 400 exquisite half-tone and color plates on pages 14 ins. x 10 ins.

Among the subscribers to this book are Her Majesty Queen Mary, His Majesty the King of Greece, and many of the foremost manufacturers the world over.

Price $12.00
Delivered to any address in Canada

Copies may be obtained from

The Woodworker Publishing Co.
Limited

Toronto　　　Ontario

The Canadian Woodworker and Furniture Manufacturer covers a specialized field. It represents maximum value to its subscribers.

Subscription price $1.00 per year.

"ENAMELS"

Our business in both White and Old Ivory Enamels and Primers has trebled during the past year

WHY?

They are the best on the market. If you are not using them, write for samples and prices.

The
Ault & Wiborg Co. of Canada, Ltd.
Varnish Works

Montreal　　　Toronto　　　Winnipeg

This danger is, of course, greater in heavier shellac, which does not dry for a considerable time.

Fumed goods can always be freshened up by an application of this wax, and it seems to me that it would be a good idea for furniture dealers to keep pint tins of it on hand, the same as they carry furniture polish, liquid veneer, etc., for varnished goods. There are floor waxes, etc., well advertised, but they are not so easily applied as the fumed mineral wax, which is very light in body and not at all sticky. One advantage of the fumed wax is that it is made of a dark brown color and so has a coloring value that a white or transparent wax does not have. The Standard Paint & Color Company, of Windsor, now make a good fuming wax which is quite equal to any imported article. Liquid ammonia is manufactured by Stuart and Foster, of Toronto.

It is quite feasible to have a dipping tank of fumed stain for small articles, such as chairs and articles that can be easily handled, though the fumes of the ammonia which enters into the composition of the stain are somewhat trying to the eyes, when disturbed in such a volume as would be contained in a dipping tank. If proper methods of ventilation are used this trouble can be greatly reduced, as it is well known that ammonia fumes are very injurious to the eyes.

The popularity of fumed finish seems to be still good for a long time, and as long as we do not perpetrate some of the atrocities that sell in the United States for fumed, we can expect it to retain its place as a desirable oak finish, particularly for library, den and dining room furniture.

When one sees a light sticker that has been running for twelve years, and is going on as smoothly as a new machine, one feels that there is one machine that, in addition to having been built for service, has been in competent hands and knows nothing of the term abuse.

Why Does Crossbanding Warp?

The following question was received from one of our readers. We would like to have some of the men who are experienced in veneering give it their attention and send us their views.—Editor.

We have experienced excessive warping of the crossbanding used in building up panels. The crossbanding is 1/28 in. poplar, and animal glue heated to 140 degrees F. is used, put on with a glue spreader. The cauls are heated to about the same temperature, and are taken directly from the heating box to the work truck. The cores used are ¾ in. to 1¼ in. thick and are chestnut. Before the pile is ready to go into the press the crossbanding on the lower layer has warped so much that the weight of the heavy cores does not hold it down, and, as a result, when the work goes into the press some of the waves lap over completely, making a lap of as much as ¾ of an inch in some cases. Of course, at this point there are three layers of crossbanding, on account of the overlapping, and this causes the face veneer to be taken off entirely when it goes through the sander.

The waves of the warping are very uniform, and naturally the crest of the wave is parallel to the grain. We cannot understand why the cores along with the cauls, which weigh about one pound to the square foot, do not hold the warping in check, unless there is some condition in the glue or applying the glue which causes it. It is glue of a good make, prepared according to instructions. Can any reader give a suggestion which might overcome this trouble, which spoils too large a percentage of the panels?   V. C.

AT YOUR SERVICE

MAHOGANY

We have a large stock of this wood on hand, in both furniture and nice mottle piano wood. Particular care has been taken in the selection of our stocks and we would appreciate an opportunity to submit samples.

AMERICAN BLACK WALNUT

We carry the largest and best stock of this wood in Veneers in Canada. We have some very highly figured stump wood, good straight sliced stripe and curly figured stocks, also semi-rotary and full rotary cut stocks of all kinds.

QUARTERED OAK

We carry this wood in all thicknesses and widths in both sawed and sliced. We have a good stock to select from.

FIGURED QUARTERED GUM

Let us show you what we have to offer in this wood. We have the goods and the prices.

We are also stocking Poplar and other rotary veneers and would appreciate your inquiries on this.

Get our prices on WALNUT LUMBER

TORONTO VENEER COMPANY

TORONTO   93-99 Spadina Avenue   ONTARIO

How to Make Perfect Flat Panels

By "Veneer Room Foreman"

In the piano factory where the writer is employed we used to have a great deal of trouble keeping our panels straight. These were sliding panels for the front of the case, and as they had to be cut in half before they were put in the case, so that one half would slide each way, it can be appreciated how straight they had to be in order to make a flush joint where they met. We were positive that by crossbanding them we would have no difficulty in keeping them straight. After events, however, proved that although they were fairly straight there was room for improvement.

We then made some experiments, and the result of these experiments was that we discovered a way of veneering our panels so as to keep them perfectly straight. Since that time—which was a few years ago—we have never had to discard a panel on account of it not having a perfectly flat surface. The results were far ahead of those obtained by the previous method. It is just possible that some reader of the Canadian Woodworker has experienced a similar difficulty, and for that reason I will tell how we veneer our panels.

In the first place, we single veneer them, but we first take great care in picking out the core stock—that is to say, we do not use a piece that has any sign of a crack or defect of any kind. This can easily be done by making a fairly large quantity of the panels and picking out the best for the ones that are required to be perfectly straight. The others can be crossbanded and used for some other purpose. We have no difficulty in this respect, because our player pianos require two or three small panels inside the case, for which we use our left-overs. After carefully choosing the core stock the next thing is to use exactly the same veneer for both sides of the panels.

The veneer on each side must exert an even pull, and this can only be accomplished by having veneer of the same texture on both sides. By exactly the same veneer, I mean that it should be cut from the same log. It does not require to be the same figure, and there is generally part of the log that is pretty near devoid of figure. After the veneer has been laid and the panels taken out of the press they are piled away to dry. Some attention should be given to the piling, as the panels must be kept perfectly flat until the glue has become thoroughly dry.

We have a kind of press for drying our panels in. It is constructed as follows: Uprights are erected from the floor to the ceiling, and about a foot from the floor horizontal pieces are nailed to the uprights. (The uprights and horizontal pieces must be at perfect right angles.) The panels are placed across the horizontal pieces, with strips between them, so that the air can get between the panels. The strips have small blocks tacked on the ends to keep them in place while the piling is being done.

When the panels are all in place, screws which are attached to the uprights on one side of the "press" are adjusted to hold the panels perfectly flat. It is not necessary to put on any great amount of pressure—just sufficient to keep the panels flat until they have been thoroughly dried out.

It must not be thought that this article is an argument against crossbanding. It is simply to point out how thin panels may be kept perfectly flat when required for a very particular job, such as sliding panels, etc.

Built-Up Veneer Panels for Walls

By J. K.

There appears to be a field for the manufacturer of built-up veneer panels that has not yet been explored. This is for wall construction in private and public buildings. There has been a tendency of late to use fibre board in large sheets in preference to lath and plaster. The sheets are about four by eight feet and one quarter inch thick. When the sheets are applied the joints are covered with mouldings to give a panelled effect. While most of this material that is used is painted or enamelled in solid colors, yet some of it is finished to represent different kinds of woods.

There is not a great deal of building going on in Canada at the present time, but there undoubtedly will be in the future; and, in view of the fact that panelled effects for the walls of parlors, living-rooms, etc., in private dwellings, as well as for the interiors of public buildings, is becoming increasingly popular, it would be wise for the makers of built-up panels to exploit their product.

One objection that might be raised to built-up panels is that moisture might enter from the wall and loosen the glue. But this objection can easily be overcome by applying asphaltum or paint to the back of the panels. This will render it positively moisture-proof, and make it impossible for dampness to have any ill-effects on the panels. People who are building good residences nowadays want the very best in regard to looks, etc., regardless of cost.

To the writer's mind, wood surpasses all other materials for the interior of a building. It is both cheering and restful. It does not require decorating like a plastered wall, and it does not show up finger marks. It is sanitary and, above all, it is durable.

There should be nothing to prevent the manufacturers of panels from figuring out and adopting certain standard sizes that could be used on any job. In different shaped rooms it might be necessary to vary the arrangement of the panels, but almost any effect could be had, even by using standard sizes, as they could be cut if necessary. One thing the panel manufacturer should do, though, is to manufacture pilasters, mouldings, cornice, and everything complete, so that he would be able to furnish all the material for the job.

It could be worked out on the same basis as the manufacture of metallic sheeting. The architect, or owner, could submit the blue prints, giving size of rooms, arrangement of doors, windows, etc., to the panel manufacturer, who would then be in a position to submit one or two sketches of different arrangements of panelling. It would also be better for him to have his own men to erect the panelling. In this way he could be reasonably sure that the work was done properly and under satisfactory conditions.

Methods of Constructing Three-Ply Panels

By J. Crow Taylor

Originally the three-ply built-up panel business was a proposition of glueing three pieces of veneer of uniform thickness together with the centre piece running crosswise of the two faces. For example, if it were a ⅜-in. panel there would be three sheets of ⅛-in. veneer, and if it were ¾-in. panel there would be three sheets of ¼-in.

With the developing of the veneer industry and the spreading out of the face veneer business there has been a gradual thinning down of face veneer, until now it ranges from 1/20 to about 1/30-in., with 1/20-in. recognized as a sort of standard for native woods and the fancy face being cut even thinner. The thinning down of face veneer has brought some questions of construction into the three-ply panel business, because manifestly there is very little structural quality in a sheet of veneer as thin as 1/20-in. To use face veneer of this kind on a three-ply panel really means now in good construction to make a five-ply job of it. This brings us to the question of three-ply construction or the question of how thin the two outer plies making the face veneer may be and still have a fair order of construction in three-ply work. To make a specific case illustrating the point under discussion, let us assume a job of three-ply ⅜-in panels. In the older days this would naturally have been made up of three sheets of ⅛-in. veneer. To-day it would more likely be made up of a sheet of ¼-in. veneer in the centre, and this faced on each side with 1/16-in.

A three-ply panel built of ¼-in. centre and 1/16-in. face veneer on each side makes a fairly good job of three-ply construction, because there is enough body and strength in the 1/16-in. face veneer to counterbalance the heavier centre body. One may even use different woods—say, gum or chestnut for the thick centre and oak for the outer face. To cut the oak into 1/16-in. for outer face is a saving of one-half in face material as compared to the old method of making a three-ply panel out of ⅛-in. stock. The trouble with this proposition, however, is that to-day the majority of oak veneer is cut as thin as 1/20-in. This would mean, in the construction of a three-ply ⅜-in. panel, that the centre piece must be even thicker when the faces have been thinned down from 1/16 to 1/20 or 1/28-in. A three-ply panel of that kind might stand; it might serve as a small door panel where it is retained on all edges, but structurally it is poor, because there is not enough body or strength in the thin outer plies to counteract any shrinking or warping tendencies of the centre body. For a panel of any great size, such as is used for the larger door panels, it would become necessary in a case of this kind to make a five-ply job, to crossband the centre or core with common veneer of whatever thickness might be desired, and then face it with the thin face veneer.

All this puts the trade to-day between two fires—that of economy in face wood on the one side and that of economy in glue and work on the other. To save the face wood by cutting it thinner and making it go farther means to use more glue on panel jobs and make five-ply instead of three-ply, and often this additional glue, with the expense involved in working it, amounts to more than the saving in face wood.

There enters, of course, for some consideration, the question of the quality of finish that it is practical to obtain from different thicknesses of face veneer. Some contend that thin face veneer makes a better job of finish, because it is not ruptured in the cutting so much as thick stock—say, stock as thick as ⅛-in. There is room for some argument on this point, and there is not much doubt but what the main influence leading to the use of thinner face veneer comes from the desire to make the face wood go farther and to save as much as possible on the expensive veneer that makes the out side face. In the larger panels it is often plainly worth while to change from three-ply to five-ply work, as this begets a better form of construction, as well as makes it practical to use the thinner face veneer.

That, however, is getting away from three-ply construction. Seemingly this question simmers down into one of whether or not we are to continue cutting faces and backs as well as fillers ⅛-in. thick to get ⅜-in. three-ply panels, or will there be a shift and more consideration given to the 1/16-in. as a face veneer, which, combined with ¼-in. centres in three-ply ⅜-in. panels, will make a very good construction and save half the quantity of fine face wood by increasing quantity of the cheaper wood used for centres.

The ideal combination, viewed from a certain angle, would be to stick to the ⅛-in., because it gives a better opportunity to use the same wood both inside and outside and get a closer utilization. For example, if one is cutting gum, the clear stock can be selected for the faces, while that showing defects which will mar it for face work, but do not impair it for centres, can be utilized in that way. This will serve to simplify the cutting of the veneer because it is all the same thickness, whereas if the faces are cut 1/16-in. and the centre is cut ¼-in. it will be necessary to sort the blocks and cut certain kinds of stock one thickness while the other kind of stock is cut another, which is not only troublesome, but a handicap to proper utilization. The sap part of a block that may have some good red underneath can be used for centres, while the red will work for faces, yet unless the centres and the faces are made the same thickness it is difficult to utilize them, because it is not practical at present to change the thickness instantly while cutting up a block. Something of the same logic enters in the consideration of oak or birch.

We must not lose sight of the fact that construction is an important item in all veneered work. In the three-ply panels it is one of the strong talking points. It gives strength and reduces the warping and shrinking tendencies as compared to solid wood; consequently we must attach a fair measure of importance to the proper construction in the making of three-ply panels. Also, to get a broad view of this subject one should study it from several angles and then seek to weigh the importance of the conclusions developed from the different angles and be governed accordingly.
—Veneers.

THE OHIO VENEER COMPANY

Importers and Manufacturers

Foreign and Domestic Veneers and Hardwood Lumber

We always carry a large and assorted stock of Mahogany, Circassian
Walnut, Sawed and Sliced Quartered Oak.

Send us your enquiries and orders. We guarantee good service.

2624 to 2644 Colerain Avenue, ‑ ‑ CINCINNATI, OHIO.

Indiana White Oak
Lumber and Veneers

Band sawed from the best growth of timber.
We use only the best quality of Indiana White Oak.
From four to five million feet always on hand.

Walnut Lumber

Our stock assures you good service.

Chas. H. Barnaby, Greencastle, Indiana

 Hoffman Bros. Co.

Reg. U.S. Pat. Off.   Estab. 1867, Incor. 1904   Reg. U.S. Pat. Off.

800 West Main Street,
FORT WAYNE, INDIANA

Manufacturers of

VENEERS and LUMBER

In the Domestic Hardwoods

ANY THICKNESS,

1/24 and 1/30 Slice Cut

(Dried flat with Smith Roller Dryer)

1/20 and thicker Sawed Veneers,
Band Sawn Lumber.

—SPECIALTY—

 Indiana Quartered Oak

Reg. U.S. Pat. Off.   Reg. U.S. Pat. Off.

Veneers and Panels

PANELS

Stock Sizes for Immediate Shipment
All Woods   All Thicknesses   3 and 5 Ply
or made to your specifications

VENEERS

5 Million feet for Immediate Shipment
Any Kind of Wood   Any Thickness

J. J. NARTZIK

1966-76 Maud Ave., ‑ ‑ CHICAGO, ILL.

Veneers and Panels

LARGE STOCKS FOR
QUICK   SHIPMENTS

Send for Lists

Walter Clark Veneer Co.

GRAND RAPIDS, MICH.

OAK   MAHOGANY   CIRCASSIAN   MAPLE

VENEERS

BIRCH   POPLAR   CHERRY   WALNUT   GUM

Wood - Mosaic Co.
NEW ALBANY,   -   -   IND.

WALNUT
Lumber ~ Dimensions ~ Veneers

If you are in the market for anything in Walnut,
let us figure with you. We have cut three million feet of Walnut in the last year, and have a
good stock of all thicknesses dry. We can als
furnish Walnut dimension—**Walnut Veneers.**
We have a large stock of veneers, plain, stripe
figured and stump wood.

Wood-Mosaic Co., Inc.
NEW ALBANY,   -   -   IND.

TEAK

DOMESTIC

Write for prices on all
**Plain and Fancy
Lumbers and Veneers**

Mahogany, Oak, Birch,
Walnut, Ash, Poplar, etc.,
rotary, sliced or sawed
in all thicknesses

SHELL BOX PARTS
in birch and spruce
Bridges and Ends

Write for Quotations

George Kersley
224 St. James Street, Montreal

LUMBER

MAHOGANY

American Walnut Veneers
Figured Wood
Sliced ~ Half Round ~ Stump Wood

5 Essential Qualifications:

1.—Unlimited supply of Walnut Logs.
2.—Logs carefully selected for figure.
3.—All processes of manufacture at our own plant
under personal supervision.
4.—Dried flat in Philadelphia Textile Dryer.
5.—3,000,000 feet ready for shipment.

JOHN N. ROBERTS,   New Albany, Indiana
The Pioneer Walnut Veneer Producer of America

OAK   MAHOGANY   CIRCASSIAN   MAPLE

VENEERS

BIRCH   POPLAR   CHERRY   WALNUT   GUM

Dear Mr. Consumer :

If you're using walnut, you can't afford to take chances with anything but the best.

Properly manufactured, properly matched and properly laid, American Black Walnut is the most beautiful of woods.

But you must have the right foundation in the material.

When it comes from Penrod, it's right. Just remember that, when you need Walnut Veneers or Lumber, Plain Stock or Figured.

Penrod Walnut & Veneer Co.

"WALNUT SPECIALISTS FOR THIRTY YEARS"

KANSAS CITY, Mo.

We specialize on

Poplar

cross-banding and backing

Veneer

Highest Grade — Right Prices

The largest Poplar Veneer Mill in the world.

The ≡ **Central Veneer Co.**

Huntington, West Virginia

We Can Fill Your Requirements in # MAHOGANY

Because we represent one of the largest importers in Chicago and the Northwest. Therefore if you need Mahogany, we are in a position to give you such service as few can render.

American Black Walnut

Particular care has been taken in the selection of our stock, which embraces some of the choicest figures we have ever seen.

Quartered White and Red Oak

Of these old favorites we carry always a large line, so varied in figure that we can suit any taste. We have many very desirable figures in flitches which will insure uniformity in quality work.

Stocks ready to ship on receipt of your order

HAUGHTON VENEER COMPANY

1152 W. LAKE STREET   **CHICAGO, ILLINOIS**

News of the Trade

The Harriston Casket Company, Harriston, Ont., recently suffered a serious loss by fire.

The death occurred recently of Mr. John Irwin, of Toronto, who for a number of years was connected with the furniture firm of Jacques & Hay.

Corporal Fred S. Leach, vice-president of the Leach Piano Company, Montreal, has been wounded in the cheek by a shrapnel bullet. He enlisted in October, 1914.

The city of Stratford will add another to its list of furniture factories. The Kindel Bed Company of Toronto have purchased a building and will move their plant from Toronto.

A new sash, door and blind factory is to be established at Fort Erie, Ont. It will be known as Hanna Bros. Planing Mills. The provisional directors are A. J. Hanna, F. Claus and A. R. Ellis.

The planing mill, etc., belonging to A. E. Pedwell, Clarksburg, Ont., was destroyed by fire, incurring a loss of over $10,000, with insurance of $4,000. A large concrete factory will be erected immediately.

Mr. George L. Orme, one of Ottawa's leading business men, died while on a visit to Vancouver, B.C. The late Mr. Orme was 74 years of age and was vice-president of the Martin-Orme Piano Company, Limited.

Mr. R. S. Gourlay, of the firm of Gourlay, Winter and Leeming, piano manufacturers, Toronto, after spending some weeks in the Western provinces, gives it as his opinion that that section of the country is going to enjoy a period of prosperity.

The Martin Aeroplanes Limited, has been incorporated with head office at Windsor, Ont., and capital stock of $100,-000, to carry on business as manufacturers of aeroplanes. The provisional directors are C. S. King, W. L. McGregor, W. Richard, O. E. Fleming and A. F. Healy, all of Windsor.

There is a possibility of a new furniture factory being established at Regina, Sask. The King Furniture Company, which has been organized to carry on a wholesale and retail business are said to be considering going into the manufacturing end. Mr. B. W. Perry is president of the concern and Mr. Robert Martin vice-president.

The Beacon Match Company, Limited, is the name of a new company that has been organized by a group of Toronto men to carry on business as manufacturers and producers of matches, etc. Those interested in the new company are R. M. Malville, F. C. Wagner, F. M. Geale, K. Campbell and James Richardson, all of Toronto.

Mr. Henry Horton Miller, ex-M.P., died at his home at Hanover, Ont. The late Mr. Miller was born at Owen Sound, Ont., on January 10th, 1861. He was a director and shareholder of the Knechtel Furniture Company, the Spiesz Furniture Company, Limited, and the Knechtel Kitchen Cabinet Company, all of Hanover.

With a view of promoting the toy-making industry in the province of Quebec, the C. P. R. arranged for a large display of toys—made in the United States—at the Quebec Exhibition. It is pointed out that there are in the province of Quebec splendid opportunities for toy-making. During the long winter months the habitant and his family are comparatively idle, and in this period they can profitably employ their time developing this new industry. Moreover, there is an abundance of suitable Canadian wood, so that the toys can be turned out at prices which will compete with imported articles.

CARVINGS FOR PERIOD FURNITURE

Composition Carvings

are now extensively used for furniture decorations.

This is a material that will interest all manufacturers of furniture.

Write for Catalog and prices.

Most durable and practical Carvings on the market.

We have thousands of stock designs in all the different styles; can also manufacture according to your special designs.

Samples sent on request.

THE DECORATORS SUPPLY CO., 2551 Archer Ave., Chicago, Ill.

Leather Belting
STANDS SUPREME

There is always one "best" material for making a thing—for belting it's leather. If you wish the best results from your machines, men and investment you will use leather belting, preferably

Amphibia Crown Planer Belting

It is made from specially tanned leather, has a minimum of stretch, and will stand the severe service of high speed work on small pulleys.

For 40 years we have made a special study of the tanning and manufacture of the best leather belts.

Ask us for prices on Amphibia Crown Planer.

| **Toronto** | **St. John, N.B.** | **MONTREAL** | **Winnipeg** | **Vancouver** |
| 38 Wellington St. E. | 149 Prince William Street | 511 William St. | Galt Building | 111 Water Street |

"TROY" GARNET PAPER

GARNET CLOTH

FINISHING PAPER

Prompt Shipments ——————— *Thoroughly Seasoned Stock*

—— RITCHEY SUPPLY COMPANY ——
126 Wellington St. W.   ·   ·   ·   TORONTO

WANTED
Woodworking Machinery

Second-hand, suitable for shell box making, all kinds, including nailers, power presses, glue spreader, hoist. Address Castle & Son, 200-202 Papineau Avenue, Montreal, P.Q.   9

Manager Open for Engagement

Man experienced in the manufacture and sale of Woodenware of all kinds, including household utensils, toys, etc., is open for position as manager or sales manager. Results guaranteed. Woodworking concerns looking for profitable lines to manufacture should communicate. Address. "Specialties," Box 36, Canadian Woodworker, Toronto, Ont.   9

FOR SALE

1 Waterous Engine, good condition, less fly-wheel; cylinder bore 12-in.; stroke 14-in.; crank shaft 5½ in.
1 Bundy Trap.
2 Detroit Traps.
2 Hot Water Receivers.
1 Baling Press.

Box No. 24, Canadian Woodworker,
9-10   Toronto, Ont.

Machinery
FOR SALE

1 Jackson-Cochrane 54-in. Resaw; 1 Cowan 42-in. Sander; 1 Jackson-Cochrane 36-in. Sander; 1 Egan 36-in. Single Surfacer; 1 McGregor-Gourlay 2-drum 30-in. Boss Sander; 1 Egan 30-in. 2-drum Sander; 1 C.M.C. 8-in. 4-side Sticker; 1 No. 108 Berlin 30-in. Double Surfacer; 1 Ballantyne 12-in. Sticker.

P. B. Yates Machine Co., Limited
Hamilton, Ontario
t-f

Second-Hand
Woodworking Machines

Hand Saws—26-in., 36-in., 40-in.
Jointers—12-in., 16-in.
Shapers—Single and double spindle.
Re-Saw—42-in., American.
Lathes—20-in. with rests and chisels.
Surfacers—16 x 6 in. and 18 x 8 in.
Mortisers—No. 5 and No. 7 New Britain.
Rip and Cut-off Saws; Molders; Sash Relishes, etc.
Over 100 machines, New and Used, in stock.

Baird Machinery Company,
123 Water Street,
PITTSBURGH, PA.

Cooperage in Canada

The National Coopers' Journal publishes a report from Mr. James Innes, of the Sutherland-Innes Company, of Chatham, Ont., which reads as follows:—

Business continues good and there is a demand for about all the stock that can be turned out at the present time with the exception of apple barrel stock, and the manufacturers have stopped producing this for the present. There is going to be a great deal more apple barrel stock used than was anticipated a month ago, but still the production will not be anything like what was expected when the trees were in bloom.

The principal trouble at the present time is labor. There is a great scarcity of men and high wages have to be paid to get even the class of men which can be obtained.

A great many of the coopers and consumers are running from hand to mouth, only getting enough stock to keep them on the ragged edge all the time. Prices are very firm, and, as the production this year will be less than it has been for years and the consumption in ordinary lines of business, outside of the barrel trade, is greater than it has been for years, we look for quite a shortage of cooperage material during the winter and spring unless the production can be increased for the balance of this season. The export demand is heavy, and, as there has been more freight offering during July and August than in the early months of the year, quite heavy shipments of both slack and tight barrel stock have been made to foreign ports during July and August.

Stability in Shaper Bearing Motion

The full significance of the perfected ball bearing as it concerns the design and work of the modern shaper is by no means generally recognized. There are two seemingly incompatible qualities which the designer of an efficient shaper wishes to combine in the high speed spindle bearings. He wants the utmost rigidity and stability, and at the same time he wants the greatest freedom of movement. And he wants these two antagonistic conditions combined in the same bearing parts without sacrificing or impairing either.

The great usefulness of the ball bearing lies in the direct application of the device to just these points. It gives support, by metal to metal contact of the most solid and positive character, of the revolving part in the stationary part; and this without interfering to any appreciable degree with freedom of rotation. This is accomplished by using, as supporting and bracing, practically perfect spheres, or balls, in practically concentric raceways. In the combination of these elements we get the realization of the paradox of a freely revolving part rigidly supported in a stationary part.

Through the medium of the rolling ball, the marvellously perfect sphere as now used, we get solid support of a rapidly revolving part without sliding contact, no movement between the parts as in a stationary bearing.

Of course, no concrete thing in this world is absolutely perfect. But the ball bearing, as some are to-day manufacturing it, is a wonderful approximation to just this thing.—Furniture Manufacturer and Artisan.

PETRIE'S
MONTHLY LIST
of
NEW and USED
WOOD TOOLS
in stock for immediate
delivery

Wood Planers
36" American double surfacer.
30" Whitney pattern surfacer.
26" Revolving bed double surfacers.
24" Major Harper planer and matcher.
24" Single surfacers, various makes.
20" Dundas pony planer.
18" Little, Giant planer and matcher.
16" Galt jointer.

Band Saws
42" Fay & Egan power feed rip.
38" Atlantic tilting frame.
30" Ideal pedestal, tilting table.

Saw Tables
Ballantine variable power feed.
Preston variable power feed.
Cowan heavy power feed. M138.
No. 3 Crescent universal.
No. 2 Crescent combination.
Ideal variety, tilting table.
12' Defiance automatic equalizer.
MacGregor, Gourlay railway cut-off.
7½' Iron frame swing.

Mortisers
No. 5 New Britain, chain.
Cowan hollow chisel. M.190.
Galt upright, compound table.
Smart foot power.

Moulders
13" Clark-Demill four-side.
12" Cowan four side.
10" Houston four side.
6" Cowan four side.
6" Dundas sash sticker.

Sanders
24" Fay double drum.
12" C.M.C. disk and drum.
18" Crescent disk, adjustable table.

Clothespin Machinery
Humphrey No. 8 giant slab resaw.
Humphrey gang slitter.
Humphrey cylinder cutting-off machine.
Humphrey automatic lathes (6).
Humphrey double slotters (4).

Miscellaneous
Fay & Egan 12-spindle dovetailer.
Cowan dowel machine, M80.
No. 285 Wysong & Miles post boring machine.
Cowan post boring machine, M85G.
Cowan post boring machine, M23.
MacGregor Gourlay 2-spindle shaper.
Elliott single spindle shaper.
No. 51 Crescent universal woodworker.
Rogers vertical resaw.
Cowan sash clamp.
Egan sash and door tenoner.
No. 6 Lion universal woodtrimmer.
6-nail box nailer, Meyers patent.
20" American wood scraper.
16" Ideal wood lathes (3)
4-head rounding machines.
Cowan spindle carver. M63.
Cowan veneer press screws.

Prices, Descriptions and Full
Particulars on Request

H. W. PETRIE, LTD.
Front St. W., Toronto, Ont.

Panels and Tops

Deskbeds, Table Tops, Chiffonier and Dresser Tops
Three and Five Ply Panels

Large assortment of stock panels always on hand.
Three and five ply stock for cartridge boxes.
Your inquiry will be fully appreciated and promptly answered with full information.

WRITE US

The

Gorham Brothers
Company

MT. PLEASANT   -   MICHIGAN

To compete successfully with other manufacturers in the production of Broom, Fork, Hoe, Rake, Mop Handles, Etc.,

YOU SHOULD USE THE

OBER NO. 10 LATHE
Simple, Rapid, Economical, Up-to-date in every way.

We also make **Sanders, Rip-Saws, Chucking Machines, Lathes** for turning irregular work, **Bolting Saws, Etc.**

Write for free Catalogue No. 31

THE OBER MFG. CO.
Chagrin Falls, O., U.S.A.

THE GLUE BOOK
WHAT IT CONTAINS

Chapter 1—*Historical Notes*
Chapter 2—*Manufacture of Glue*
Chapter 3—*Testing and Grading*
Chapter 4—*Methods in the Glue Room*
Chapter 5—*Glue Room Equipment*
Chapter 6—*Selection of Glue*

WOODWORKER PUBLISHING CO., LIMITED
345 Adelaide St. West, Toronto

LET A BOY DO IT

One boy and the NASH-SANDER will do the work of 5 men with sand belts. Don't you see the saving?

J. M. NASH, MILWAUKEE, Wis.

DARLINGS
STEAM
APPLIANCES

DARLING BROTHERS
LIMITED
Engineers and Manufacturers
MONTREAL, CANADA

Branches: Agents:
Toronto and Winnipeg   Halifax, St. John, Calgary, Vancouver

Peter Hay Machine Knives

Peter Hay knives are made of special quality steel tempered and toughened to take and hold a keen cutting edge. Hay knives are made for all purposes.

Have you our Catalogue?

Made in Canada

The Peter Hay Knife Co., Limited
GALT, ONTARIO

PRESSES

For Veneer and Veneer Drying

Made in Canada

William R. Perrin
Limited

Toronto

In a National

IN a National PROGRESSIVE DRIER, most of the pipes are confined to the discharging end of the kiln. This is but one of the many reasons why "case hardening," honeycombing and checking can be forgotten.

The National Dry Kiln Co.
1117 E. Maryland St., Indianapolis, Ind., U.S.A.

You
CONTRACTORS
and
SMALL MILL MEN
Just the machine you need
—ANY KIND OF POWER—

A Whole Mill in Itself
A GREAT TOOL
Write for Catalogue and Prices NOW

J. L. NEILSON & CO.
Winnipeg, Man.

ACME
STEEL GOODS
BOX STRAPPING
Our Specialty

Acme Galv. Twisted Wire Strap

Acme Cold Rolled Steel Hoop.

Bright—Galv.—Lacquered

WRITE FOR CATALOG

Coil of 5000-ft. Acme Twisted Wire Strap

ACME STEEL GOODS CO.—MFRS.
2840 Archer Ave., Chicago.
Coristine Bldg., Montreal, Que.

Morehead
Back to Boiler →
SYSTEM

They Saved 25% in Fuel and 50% in Repairs—

So writes Mr. William H. Turner, Secretary and Treasurer of the Easton Furniture Company, of Easton, Md., after giving the "Morehead" System a thorough trial.

Until you are draining the condensation from your kilns and returning it—every drop— to the boilers at the original temperature you are wasting heat units that cost you DOLLARS to produce.

Rejuvenate Your Steam Plant

by making the inexpensive "Morehead" System a part of it.

Save fuel by bringing the pure condensation directly back to the boilers HOT!

Save repairs by eliminating for ever the necessity for the wasteful and constantly out-of-order steam pump.

The simple, easily installed "Morehead" System will much more than pay for itself the first year of use.

The Morehead Book shows actual photographs of interesting installations. Write for your copy now.

Canadian Morehead Manufacturing Company
DEPT. H
WOODSTOCK, ONTARIO

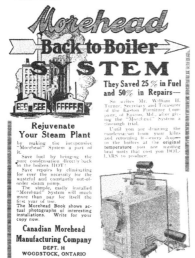

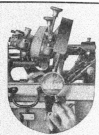

GRINDING

BAND SAWS

with

26 TEETH

TO THE INCH

Your woodworking band saws can be resharpened so that all teeth come within 1/1,000 of an inch alike to each other, by this

—Model A—
Automatic Band Saw Grinder

Saws 1-8 inch to 8 inches wide. Teeth up to 2 inch spacing

POINTS OF INTEREST

Bronze Bushed Grinding Wheel Spindle.
Head Suspended between Hardened Steel Pivots
No Slides to Wear. No Cams to Change.
Feed Movement obtained by a Scaled Slide
A Double Pawl, Positive Feed Movement, the quickest and most perfect method of jointing all teeth, can be furnished at a small extra charge.

LOWEST PRICED GRINDER MADE
Write for Details

THE WARDWELL MFG. COMPANY
Saw and Knife Sharpening Machinery
111 Hamilton Ave.   **CLEVELAND, OHIO**

Wood Steaming Retort

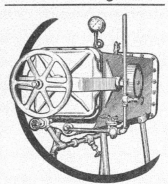

Wood Bending Manufacturers:
This is one of our

Perfection Retorts
which we guarantee will save you

50% Less Breakage

in your bending department than your present process; that your stock will dry in your forms or presses in one-third less time; that you will have no stained stock; that your stock will retain its shape much better after being bent; that it will dry in your dry-kiln in one-half less time and that your steam consumption will be reduced at least 90 per cent.

The door can be opened and closed in ten seconds, and it is steam and water tight and for this reason can be placed anywhere in your factory.

Compare this IMPROVED RETORT with your present steam boxes, then write us for our Booklet on Progressive Wood Steaming.

Made in Preston, Ontario

Perfection Wood Steaming Retort Co.
PARKERSBURG - WEST VIRGINIA

The "Canadian Woodworker" Buyers' Directory

ABRASIVES
The Carborundum Co., Niagara Falls, N.Y.

BALL BEARINGS
Chapman Double Ball Bearing Co., Toronto.

BALUSTER LATHES
Garlock-Walker Machinery Co., Toronto, Ont.
Ober Mfg. Company, Chagrin Falls, Ohio.
Whitney & Son, Baxter D., Winchendon, Mass.

BAND SAW FILING MACHINERY
Canada Machinery Corporation, Galt, Ont.
Fay & Egan Co., J. A., Cincinnati, Ohio.
Garlock-Walker Machinery Co., Toronto, Ont.
Preston Woodworking Machinery Company,
   Preston, Ont.
Yates Machine Co., P. B., Hamilton, Ont.

BAND SAWS
Canada Machinery Corporation, Galt, Ont.
Cowan & Co., Galt, Ont.
Garlock-Walker Machinery Co., Toronto, Ont.
Jackson, Cochrane & Company, Kitchener, Ont.
Petrie, H. W., Toronto.
Preston Woodworking Machinery Company,
   Preston, Ont.
Yates Machine Co., P. B., Hamilton, Ont.

BAND SAW MACHINES
Canada Machinery Corporation, Galt, Ont.
Cowan & Co., Galt, Ont.
Fay & Egan Co., J. A., Cincinnati, Ohio.
Garlock-Walker Machinery Co., Toronto, Ont.
Petrie, H. W., Toronto.
Preston Woodworking Machinery Company,
   Preston, Ont.
Williams Machinery Co., A. R., Toronto, Ont.
Yates Machine Co., P. B., Hamilton, Ont.

BAND SAW STRETCHERS
Fay & Egan Co., J. A., Cincinnati, Ohio.
Yates Machine Co., P. B., Hamilton, Ont.

BASKET STITCHING MACHINE
Saranac Machine Co., Benton Harbor, Mich.

BENDING MACHINES
Fay & Egan Co., J. A., Cincinnati, Ohio.
Garlock-Walker Machinery Co., Toronto, Ont.
Perfection Wood Steaming Retort Company,
   Parkersburg, W. Va.

BELTING
Fay & Egan Co., J. A., Cincinnati, Ohio.

BLOWERS
Sheldons, Limited, Galt, Ont.
Toronto Blower Company, Toronto, Ont.

BLOW PIPING
Sheldons, Limited, Galt, Ont.
Toronto Blower Company, Toronto, Ont.

BORING MACHINES
Canada Machinery Corporation, Galt, Ont.
Cowan & Co., Galt, Ont.
Fay & Egan Co., J. A., Cincinnati, Ohio.
Garlock-Walker Machinery Co., Toronto, Ont.
Jackson, Cochrane & Company, Kitchener, Ont.
Nash, J. M., Milwaukee, Wis.
Preston Woodworking Machinery Company,
   Preston, Ont.
Reynolds Pattern & Machine Co., Moline, Ill.
Yates Machine Co., P. B., Hamilton, Ont.

BOX MAKERS' MACHINERY
Canada Machinery Corporation, Galt, Ont.
Cowan & Co., Galt, Ont.
Fay & Egan Co., J. A., Cincinnati, Ohio.
Garlock-Walker Machinery Co., Toronto, Ont.
Jackson, Cochrane & Company, Kitchener, Ont.
Neilson & Company, J. L. Winnipeg, Man.
Preston Woodworking Machinery Company,
   Preston, Ont.
Reynolds Pattern & Machine Co., Moline, Ill.
Whitney & Son, Baxter D., Winchendon, Mass.
Williams Machinery Co., A. R., Toronto, Ont.
Yates Machine Co., P. B., Hamilton, Ont.

CABINET PLANERS
Canada Machinery Corporation, Galt, Ont.
Cowan & Co., Galt, Ont.
Fay & Egan Co., J. A., Cincinnati, Ohio.
Garlock-Walker Machinery Co., Toronto, Ont.
Jackson, Cochrane & Company, Kitchener, Ont.
Preston Woodworking Machinery Company,
   Preston, Ont.
Yates Machine Co., P. B., Hamilton, Ont.

CARS (Transfer)
Sheldons, Limited, Galt, Ont.

CARBORUNDUM PAPER
Carborundum Co., Niagara Falls, N.Y.

CARVING MACHINES
Canada Machinery Corporation, Galt, Ont.
Fay & Egan Co., J. A., Cincinnati, Ohio.
Garlock-Walker Machinery Co., Toronto, Ont.
Jackson, Cochrane & Company, Kitchener, Ont.

CASTERS
Foster, Merriam & Co., Meriden, Conn.

CLAMPS
Fay & Egan Co., J. A., Cincinnati, Ohio.
Garlock-Walker Machinery Co., Toronto, Ont.
Jackson, Cochrane & Company, Kitchener, Ont.
Preston Woodworking Machinery Company,
   Preston, Ont.
Simonds Canada Saw Company, Montreal, P.Q.

COLUMN MACHINERY
Fay & Egan Co., J. A. Cincinnati, Ohio.

CORRUGATED FASTENERS
Acme Steel Goods Company, Chicago, Ill.

CUT-OFF SAWS
Canada Machinery Corporation, Galt, Ont.
Cowan & Co., Galt, Ont.
Fay & Egan Co., J. A., Cincinnati, Ohio.
Garlock-Walker Machinery Co., Toronto, Ont.
Jackson, Cochrane & Company, Kitchener, Ont.
Ober Mfg. Co., Chagrin Falls, Ohio.
Preston Woodworking Machinery Company,
   Preston, Ont.
Simonds Canada Saw Co., Montreal, Que.
Yates Machine Co., P. B., Hamilton, Ont.

CUTTER HEADS
Canada Machinery Corporation, Galt, Ont.
Cowan & Co., Galt, Ont.
Fay & Egan Co., J. A., Cincinnati, Ohio.
Jackson, Cochrane & Company, Kitchener, Ont.
Mattison Machine Works, Beloit, Wis.
Yates Machine Co., P. B., Hamilton, Ont.

DOVETAILING MACHINES
Canada Machinery Corporation, Galt, Ont.
Cowan & Co., Galt, Ont.
Fay & Egan Co., J. A., Cincinnati, Ohio.
Garlock-Walker Machinery Co., Toronto, Ont.
Jackson, Cochrane & Company, Kitchener, Ont.

DOWEL MACHINES
Canada Machinery Corporation, Galt, Ont.
Fay & Egan Co., J. A., Cincinnati, Ohio.
Ober Mfg. Co., Chagrin Falls, Ohio.

DRY KILNS
National Dry Kiln Co., Indianapolis, Ind.
Sheldons, Limited, Galt, Ont.

DUST COLLECTORS
Sheldons, Limited, Galt, Ont.
Toronto Blower Company, Toronto, Ont.

DUST SEPARATORS
Sheldons, Limited, Galt, Ont.
Toronto Blower Company, Toronto, Ont.

EDGERS (Single Saw)
Canada Machinery Corporation, Galt, Ont.
Fay & Egan Co., J. A., Cincinnati, Ohio.
Garlock-Walker Machinery Co., Toronto, Ont.
Simonds Canada Saw Co., Montreal, Que.

EDGERS (Gang)
Canada Machinery Corporation, Galt, Ont.
Fay & Egan Co., J. A., Cincinnati, Ohio.
Garlock-Walker Machinery Co., Toronto, Ont.
Petrie, H. W., Toronto.
Simonds Canada Saw Co., Montreal, P.Q.
Williams Machinery Co., A. R., Toronto, Ont.
Yates Machine Co., P. B., Hamilton, Ont.

END MATCHING MACHINE
Canada Machinery Corporation, Galt, Ont.
Cowan & Co., Galt, Ont.
Fay & Egan Co., J. A., Cincinnati, Ohio.
Garlock-Walker Machinery Co., Toronto, Ont.
Jackson, Cochrane & Company, Kitchener, Ont.
Yates Machine Co., P. B., Hamilton, Ont.

EXHAUST FANS
Garlock-Walker Machinery Co., Toronto, Ont.
Sheldons Limited, Galt, Ont.
Toronto Blower Company, Toronto, Ont.

EXHAUST SYSTEMS
Toronto Blower Company, Toronto, Ont.

FILES
Simonds Canada Saw Co., Montreal, Que.

FILING MACHINERY
Wardwell Mfg. Company, Cleveland, Ohio.

FLOORING MACHINES
Canada Machinery Corporation, Galt, Ont.
Cowan & Co., Galt, Ont.
Fay & Egan Co., J. A., Cincinnati, Ohio.
Garlock-Walker Machinery Co., Toronto, Ont.
Petrie, H. W., Toronto.
Preston Woodworking Machinery Company,
   Preston, Ont.
Whitney & Son, Baxter D., Winchendon, Mass.
Yates Machine Co., P. B., Hamilton, Ont.

FLUTING HEADS
Fay & Egan Co., J. A., Cincinnati, Ohio.

FLUTING AND TWIST MACHINE
Preston Woodworking Machinery Company,
   Preston, Ont.

FURNITURE TRIMMINGS
Foster, Merriam & Co., Meriden, Conn.

GARNET PAPER AND CLOTH
The Carborundum Co., Niagara Falls, N.Y.
Ritchey Supply Company, Toronto, Ont.

GRAINING MACHINES
Canada Machinery Corporation, Galt, Ont.
Fay & Egan Co., J. A., Cincinnati, Ohio.
Williams Machinery Co., A. R., Toronto, Ont.
Yates Machine Co., P. B., Hamilton, Ont.

GLUE
Coignet Chemical Product Co., New York,
   N.Y.
Perkins Glue Company, South Bend, Ind.

GLUE CLAMPS
Jackson, Cochrane & Company, Kitchener, Ont.

GLUE HEATERS
Fay & Egan Co., J. A., Cincinnati, Ohio.
Jackson, Cochrane & Company, Kitchener, Ont.

GLUE JOINTERS
Canada Machinery Corporation, Galt, Ont.
Jackson, Cochrane & Company, Kitchener, Ont.
Yates Machine Co., P. B., Hamilton, Ont.

GLUE SPREADERS
Fay & Egan Co., J. A., Cincinnati, Ohio.
Jackson, Cochrane & Company, Kitchener, Ont.

GLUE ROOM EQUIPMENT
Perrin & Company, W. R., Toronto, Ont.

GRINDERS (Cutter)
Fay & Egan Co., J. A., Cincinnati, Ohio.

GRINDERS, HAND AND FOOT POWER
The Carborundum Co., Niagara Falls, N.Y.

GRINDERS (Knife)
Canada Machinery Corporation, Galt, Ont.
Fay & Egan Co., J. A., Cincinnati, Ohio.
Garlock-Walker Machinery Co., Toronto, Ont.
Petrie, H. W., Toronto.
Preston Woodworking Machinery Company,
   Preston, Ont.
Yates Machine Co., P. B., Hamilton, Ont.

GRINDERS (Tool)
Fay & Egan Co., J. A., Cincinnati, Ohio.
Jackson, Cochrane & Company, Kitchener, Ont.

GRINDING WHEELS
The Carborundum Co., Niagara Falls, N.Y.

GROOVING HEADS
Fay & Egan Co., J. A., Cincinnati, Ohio.

GUMMERS, ETC.
Fay & Egan Co., J. A., Cincinnati, Ohio.

HAND PROTECTORS
Fay & Egan Co., J. A., Cincinnati, Ohio.

HAND SCREWS
Fay & Egan Co., J. A., Cincinnati, Ohio.

HANDLE THROATING MACHINE
Klotz Machine Company, Sandusky, Ohio.

HANDLE TRIMMING MACHINE
Klotz Machine Company, Sandusky, Ohio.

HANDLE & SPOKE MACHINERY
Fay & Egan Co., J. A., Cincinnati, Ohio.
Klotz Machine Co., Sandusky, Ohio.
Nash, J. M., Milwaukee, Wis.
Ober Mfg. Company, Chagrin Falls, Ohio.
Whitney & Son, Baxter D., Winchendon, Mass.

HARDWOOD LUMBER
American Hardwood Lumber Co., St. Louis, Mo.
Atlantic Lumber Company, Toronto.
Brown & Company, Geo. C., Memphis, Tenn.
Clark & Son, Edward, Toronto, Ont.
Edwards Lumber Co., E. L., Dayton, Ohio.
Kersley, Geo., Montreal, Que.
Kraetzer-Cured Lumber Co., Moorhead, Miss.
Mowbray & Robinson, Cincinnati, Ohio.
Paepcke Leicht Lumber Co., Chicago, Ill.
Penrod Walnut & Veneer Co., Kansas City, Mo.
Spencer, C. A., Montreal, Que.

HUB MACHINERY
Fay & Egan Co., J. A., Cincinnati, Ohio.

HYDRAULIC VENEER PRESSES
Perrin & Company, Wm. R., Toronto, Ont.

JOINTERS
Canada Machinery Corporation, Galt, Ont.
Cowan & Co., Galt, Ont.
Fay & Egan Co., J. A., Cincinnati, Ohio.
Garlock-Walker Machinery Co., Toronto, Ont.
Jackson, Cochrane & Company, Kitchener, Ont.
Knechtel Bros., Limited, Southampton, Ont.
Petrie, H. W., Toronto, Ont.
Preston Woodworking Machinery Company, Preston, Ont.
Yates Machine Co., P. B., Hamilton, Ont.

KNIVES (Planer and others)
Canada Machinery Corporation, Galt, Ont.
Fay & Egan Co., J. A., Cincinnati, Ohio.
Simonds Canada Saw Co., Montreal, P.Q.
Yates Machine Co., P. B., Hamilton, Ont.

LATHES
Canada Machinery Corporation, Galt, Ont.
Cowan & Co., Galt, Ont.
Fay & Egan Co., J. A., Cincinnati, Ohio.
Garlock-Walker Machinery Co., Toronto, Ont.
Jackson, Cochrane & Company, Kitchener, Ont.
Klotz Machine Company, Sandusky, Ohio.
Ober Mfg. Company, Chagrin Falls, Ohio.
Petrie, H. W., Toronto, Ont.
Preston Woodworking Machinery Company, Preston, Ont.
Shawver Company, Springfield, O.
Whitney & Son, Baxter D., Winchendon, Mass.

LUMBER
American Hardwood Lumber Co., St. Louis, Mo.
B. C. Lumber Commissioner, Toronto.
Brown & Company, Geo. C., Memphis, Tenn.
Churchill, Milton Lumber Co., Louisville, Ky.
Clark & Son, Edward, Toronto, Ont.
Dominion Mahogany & Veneer Co., Montreal, Que.
Edwards Lumber Co., E. L., Dayton, Ohio.
Gorham Bros. Co., Mt. Pleasant, Mich.
Hart & McDonagh, Toronto, Ont.
Hartzell, Geo. W., Piqua, Ohio.
Huddleston, Marsh Mahogany Co., New York.
Long-Knight Lumber Co., Indianapolis.
Paepcke Leicht Lumber Company, Chicago, Ill.

MACHINE KNIVES
Canada Machinery Corporation, Galt, Ont.
Peter Hay Knife Company, Galt, Ont.
Simonds Canada Saw Co., Montreal, P.Q.
Yates Machine Co., P. B., Hamilton, Ont.

MITRE MACHINES
Canada Machinery Corporation, Galt, Ont.
Fay & Egan Co., J. A., Cincinnati, Ohio.

MITRE SAWS
Canada Machinery Corporation, Galt, Ont.
Fay & Egan Co., J. A., Cincinnati, Ohio.
Garlock-Walker Machinery Co., Toronto, Ont.
Simonds Canada Saw Co., Montreal, P.Q.

MULTIPLE BOXING MACHINES
Fay & Egan Co., J. A., Cincinnati, Ohio.
Nash, J. M., Milwaukee, Wis.
Reynolds Pattern & Machine Co., Moline, Ill.

PATTERN SHOP MACHINES
Canada Machinery Corporation, Galt, Ont.
Cowan & Co., Galt, Ont.
Fay & Egan Co., J. A., Cincinnati, Ohio.
Garlock-Walker Machinery Co., Toronto, Ont.
Jackson, Cochrane & Company, Kitchener, Ont.
Preston Woodworking Machinery Company, Preston, Ont.
Whitney & Son, Baxter D., Winchendon, Mass.
Yates Machine Co., P. B., Hamilton, Ont.

PLANERS
Canada Machinery Corporation, Galt, Ont.
Cowan & Co., Galt, Ont.
Fay & Egan Co., J. A., Cincinnati, Ohio.
Garlock-Walker Machinery Co., Toronto, Ont.
Jackson, Cochrane & Company, Kitchener, Ont.
Petrie, H. W., Toronto, Ont.
Preston Woodworking Machinery Company, Preston, Ont.
Whitney & Son, Baxter D., Winchendon, Mass.
Williams Machinery Co., A. R., Toronto, Ont.
Yates Machine Co., P. B., Hamilton, Ont.

PLANING MILL MACHINERY
Canada Machinery Corporation, Galt, Ont.
Cowan & Co., Galt, Ont.
Fay & Egan Co., J. A., Cincinnati, Ohio.
Garlock-Walker Machinery Co., Toronto, Ont.
Jackson, Cochrane & Company, Kitchener, Ont.
Petrie, H. W., Toronto, Ont.
Preston Woodworking Machinery Company, Preston, Ont.
Whitney & Son, Baxter D., Winchendon, Mass.
Williams Machinery Co., A. R., Toronto, Ont.
Yates Machine Co., P. B., Hamilton, Ont.

PRESSES (Veneer)
Canada Machinery Corporation, Galt, Ont.
Jackson, Cochrane & Company, Kitchener, Ont.
Perrin & Co. Wm. R., Toronto, Ont.
Preston Woodworking Machinery Company, Preston, Ont.

PULLEYS
Fay & Egan Co., J. A., Cincinnati, Ohio.

RED GUM LUMBER
Gum Lumber Manufacturers' Association.

RESAWS
Canada Machinery Corporation, Galt, Ont.
Cowan & Co., Galt, Ont.
Fay & Egan Co., J. A., Cincinnati, Ohio.
Garlock-Walker Machinery Co., Toronto, Ont.
Jackson, Cochrane & Company, Kitchener, Ont.
Petrie, H. W., Toronto, Ont.
Preston Woodworking Machinery Company, Preston, Ont.
Simonds Canada Saw Co., Montreal, P.Q.
Williams Machinery Co., A. R., Toronto, Ont.
Yates Machine Co., P. B., Hamilton, Ont.

RIM AND FELLOE MACHINERY
Fay & Egan Co., J. A., Cincinnati, Ohio.

RIP SAWING MACHINES
Canada Machinery Corporation, Galt, Ont.
Cowan & Co., Galt, Ont.
Fay & Egan Co., J. A., Cincinnati, Ohio.
Garlock-Walker Machinery Co., Toronto, Ont.
Klotz Machine Company, Sandusky, Ohio.
Jackson, Cochrane & Company, Kitchener, Ont.
Ober Mfg. Company, Chagrin Falls, Ohio.
Preston Woodworking Machinery Company, Preston, Ont.
Yates Machine Co., P. B., Hamilton, Ont.

SANDERS
Canada Machinery Corporation, Galt, Ont.
Cowan & Co., Galt, Ont.
Fay & Egan Co., J. A., Cincinnati, Ohio.
Garlock-Walker Machinery Co., Toronto, Ont.
Klotz Machine Company, Sandusky, Ohio.
Jackson, Cochrane & Company, Kitchener, Ont.
Nash, J. M., Milwaukee, Wis.
Ober Mfg. Company, Chagrin Falls, Ohio.
Preston Woodworking Machinery Company, Preston, Ont.
Yates Machine Co., P. B., Hamilton, Ont.

SASH, DOOR & BLIND MACHINERY
Canada Machinery Corporation, Galt, Ont.
Cowan & Co., Galt, Ont.
Fay & Egan Co., J. A., Cincinnati, Ohio.
Garlock-Walker Machinery Co., Toronto, Ont.
Jackson, Cochrane & Company, Kitchener, Ont.
Preston Woodworking Machinery Company, Preston, Ont.
Yates Machine Co., P. B., Hamilton, Ont.

SAWS
Fay & Egan Co., J. A., Cincinnati, Ohio.
Radcliff Saw Mfg. Company, Toronto, Ont.
Simonds Canada Saw Co., Montreal, P.Q.
Smith Co., Ltd., R. H., St. Catharines, Ont.

SAW GUMMERS, ALOXITE
The Carborundum Co., Niagara Falls, N.Y.

SAW SWAGES
Fay & Egan Co., J. A., Cincinnati, Ohio.
Simonds Canada Saw Co., Montreal, Que.

SAW TABLES
Canada Machinery Corporation, Galt, Ont.
Fay & Egan Co., J. A., Cincinnati, Ohio.
Garlock-Walker Machinery Co., Toronto, Ont.
Jackson, Cochrane & Company, Kitchener, Ont.
Preston Woodworking Machinery Company, Preston, Ont.
Yates Machine Co., P. B., Hamilton, Ont.

SCRAPING MACHINES
Canada Machinery Corporation, Galt, Ont.
Garlock-Walker Machinery Co., Toronto, Ont.
Whitney & Sons, Baxter D., Winchendon, Mass.

SCROLL CHUCK MACHINES
Klotz Machine Company, Sandusky, Ohio.

SCREW DRIVING MACHINERY
Canada Machinery Corporation, Galt, Ont.
Reynolds Pattern & Machine Works, Moline, Ill.

SCROLL SAWS
Canada Machinery Corporation, Galt, Ont.
Cowan & Co., Galt, Ont.
Fay & Egan Co., J. A., Cincinnati, Ohio.
Garlock-Walker Machinery Co., Toronto, Ont.
Jackson, Cochrane & Company, Kitchener, Ont.
Simonds Canada Saw Co., Montreal, P.Q.
Yates Machine Co., P. B., Hamilton, Ont.

SECOND-HAND MACHINERY
Cowan & Co., Galt, Ont.
Petrie, H. W.
Williams Machinery Co., A. R., Toronto, Ont.

SHAPERS
Canada Machinery Corporation, Galt, Ont.
Fay & Egan Co., J. A., Cincinnati, Ohio.
Garlock-Walker Machinery Co., Toronto, Ont.
Jackson, Cochrane & Company, Kitchener, Ont.
Ober Mfg. Company, Chagrin Falls, Ohio.
Petrie, H. W., Toronto, Ont.
Preston Woodworking Machinery Company, Preston, Ont.
Simonds Canada Saw Co., Montreal, P.Q.
Whitney & Sons, Baxter D., Winchendon, Mass.
Yates Machine Co., P. B., Hamilton, Ont.

SHARPENING TOOLS
The Carborundum Co., Niagara Falls, N.Y.

SHAVING COLLECTORS
Sheldons Limited, Galt, Ont.

STEAM APPLIANCES
Darling Bros., Montreal.

SINGLE SPINDLE BOXING MACHINES
Canada Machinery Corporation, Galt, Ont.
Fay & Egan Co., J. A., Cincinnati, Ohio.
Ober Mfg. Company, Chagrin Falls, Ohio.

STAVE SAWING MACHINE
Whitney & Sons, Baxter D., Winchendon, Mass.

STEAM TRAPS
Canadian Morehead Mfg. Co., Woodstock, Ont.
Standard Dry Kiln Co., Indianapolis, Ind.

STEEL BOX STRAPPING
Acme Steel Goods Co., Chicago, Ill.

STEEL SHELL BOX BANDS
Acme Steel Goods Co., Chicago, Ill.

SURFACERS
Canada Machinery Corporation, Galt, Ont.
Cowan & Co., Galt, Ont.
Fay & Egan Co., J. A., Cincinnati, Ohio.
Garlock-Walker Machinery Co., Toronto, Ont.
Jackson, Cochrane & Company, Kitchener, Ont.
Petrie, H. W., Toronto, Ont.
Preston Woodworking Machinery Company,
  Preston, Ont.
Whitney & Sons, Baxter D., Winchendon, Mass.
Williams Machinery Co., A. R., Toronto, Ont.
Yates Machine Co., P. B., Hamilton, Ont.

SWING SAWS
Canada Machinery Corporation, Galt, Ont.
Cowan & Co., Galt, Ont.
Fay & Egan Co., J. A., Cincinnati, Ohio.
Garlock-Walker Machinery Co., Toronto, Ont.
Jackson, Cochrane & Company, Kitchener, Ohio.
Ober Mfg. Co., Chagrin Falls, Ohio.
Preston Woodworking Machinery Company,
  Preston, Ont.
Simonds Canada Saw Co., Montreal, P.Q.
Yates Machine Co., P. B., Hamilton, Ont.

TABLE LEG LATHES
Fay & Egan Co., J. A., Cincinnati, Ohio.
Ober Mfg. Co., Chagrin Falls, Ohio.
Whitney & Sons, Baxter D., Winchendon, Mass.
Yates Machine Co., P. B., Hamilton, Ont.

TABLE SLIDES
Walter & Co., B., Wabash, Ind.

TENONING MACHINES
Canada Machinery Corporation, Galt, Ont.
Fay & Egan Co., J. A., Cincinnati, Ohio.
Garlock-Walker Machinery Co., Toronto, Ont.
Jackson, Cochrane & Company, Kitchener, Ont.
Preston Woodworking Machinery Company,
  Preston, Ont.
Yates Machine Co., P. B., Hamilton, Ont.

TRIMMERS
Canada Machinery Corporation, Galt, Ont.
Fay & Egan Co., J. A., Cincinnati, Ohio.
Garlock-Walker Machinery Co., Toronto, Ont.
Preston Woodworking Machinery Company,
  Preston, Ont.
Yates Machine Co., P. B., Hamilton, Ont.

TRUCKS
Sheldons Limited, Galt, Ont.
National Dry Kiln Co., Indianapolis, Ind.

TURNING MACHINES
Fay & Egan Co., J. A., Cincinnati, Ohio.
Garlock-Walker Machinery Co., Toronto, Ont.
Klotz Machine Company, Sandusky, Ohio.
Ober Mfg. Company, Chagrin Falls, Ohio.
Whitney & Sons, Baxter D., Winchendon, Mass.
Yates Machine Co., P. B., Hamilton, Ont.

**UNDER-CUT SELF-FEEDING FACE
  PLANER**
Fay & Egan Co., J. A., Cincinnati, Ohio.
Garlock-Walker Machinery Co., Toronto, Ont.
Jackson, Cochrane & Company, Kitchener, Ont.

VARNISHES
Ault & Wiborg Company, Toronto, Ont.

VENEERS
Barnaby, Chas. A., Greencastle, Ind.
Central Veneer Co., Huntington, W. Virginia.
Dominion Mahogany & Veneer Company, Mont-
  real, Que.
Walter Clark Veneer Co., Grand Rapids, Mich.
Gorham Bros. Co., Mt. Pleasant, Mich.
Hartzell, Geo. W., Piqua, Ohio.
Haughton Veneer Company, Chicago, Ill.
Hay & Company, Woodstock, Ont.
Hoffman Bros. Company, Fort Wayne, Ind.
Geo. Kersley, Montreal, Que.
Nartzik, J. J., Chicago, Ill.
Ohio Veneer Company, Cincinnati, Ohio.
Penrod Walnut & Veneer Co., Kansas City, Mo.
Roberts, John N., New Albany, Ind.
Toronto Veneer Company, Toronto, Ont.
Wood Mosaic Co., New Albany, Ind.

VENTILATING APPARATUS
Sheldons Limited, Galt, Ont.

VENEERED PANELS
Hay & Company, Woodstock, Ont.

VENEER PRESSES (Hand and Power)
Canada Machinery Corporation, Galt, Ont.
Garlock-Walker Machinery Co., Toronto, Ont.
Jackson, Cochrane & Company, Kitchener, Ont.
Perrin & Company, Wm. R., Toronto, Ont.

VISES
Fay & Egan Co., J. A., Cincinnati, Ohio.
Simonds Canada Saw Company, Montreal, Que.

WAGON AND CARRIAGE MACHINERY
Canada Machinery Corporation, Galt, Ont.
Fay & Egan Co., J. A., Cincinnati, Ohio.
Ober Mfg. Company, Chagrin Falls, Ohio.
Whitney & Sons, Baxter D., Winchendon, Mass.
Yates Machine Co., P. B., Hamilton, Ont.

WOOD FINISHES
Ault & Wiborg, Toronto, Ont.

WOOD TURNING MACHINERY
Canada Machinery Corporation, Galt, Ont.
Garlock-Walker Machinery Co., Toronto, Ont.
Yates Machine Co., P. B., Hamilton, Ont.

WORK BENCHES
Fay & Egan Co., J. A., Cincinnati, Ohio.

WOODWORKING MACHINES
Canada Machinery Corporation, Galt, Ont.
Cowan & Co., Galt, Ont.
Garlock-Walker Machinery Co., Toronto, Ont.
Jackson, Cochrane & Company, Kitchener, Ont.
Preston Woodworking Machinery Company,
  Preston, Ont.
Reynolds' Pattern & Machine Works, Moline,
  Ill.
Williams Machinery Co., A. R., Toronto, Ont.

WHEELS, Grinding, Carborundum and Aloxite
The Carborundum Co., Niagara Falls, N.Y.

L OSS of power is one of the greatest drawbacks to modern manufacturing. It is constantly harassing the profits of the manufacturer. Oftentimes it is the real factor that stands between profits and losses.

Eliminate that loss of power caused by friction by installing Chapman Double Ball Bearings on your line and counter shafts.

Installation of the above bearings will reduce your power costs from 15% to 40%. The cash you paid for waste power during the past year and a half would have purchased a complete installation of Chapman Ball Bearings. Why not install them now, and cut out this unnecessary expense in the future?

No trouble to change over—our bearings suit any standard hanger

The Chapman Double Ball Bearing Co. of Canada, Limited
TORONTO   347 Sorauren Ave.   ONTARIO

Making TABLE-SLIDES is a Specialty Business

For more than TWENTY-FIVE YEARS we have made TABLE SLIDES exclusively. Our Factory is equipped with Special Machinery which enables us to make SLIDES.—BETTER and CHEAPER than the furniture manufacturer.

Canadian Table makers are rapidly adopting WABASH SLIDES

Because { They ELIMINATE SLIDE TROUBLES
Are CHEAPER and BETTER

Reduced Costs
Increased Out-put
{ BY USING
WABASH SLIDES

MADE BY

B. Walter & Company
Wabash, Ind.

The Largest EXCLUSIVE TABLE-SLIDE Manufacturers in America
ESTABLISHED-1887

Just What You Need

There's no necessity now for you to wish for a corrugated joint fastener machine which will rise fully to your expectations; you can get one.

The Saranac Corrugated Joint Fastener Machine

is the absolute realization of your wish. This illustration gives you an idea of what it looks like, but you can have no conception of its efficiency, economy, speed and reliability until you get it working for you.

It is substantially constructed, and the workmanship and materials are of the very best. Particularly suitable for manufacturers of sash, doors, blinds, screens, porch columns, boxes, etc., etc.

Made in various types for all conditions.

Write for further particulars and prices

Saranac Machine Co., SOLE MAKERS
Benton Harbor, Mich.

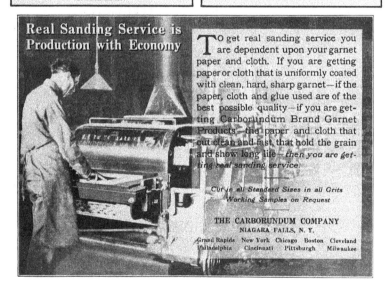

Real Sanding Service is Production with Economy

To get real sanding service you are dependent upon your garnet paper and cloth. If you are getting paper or cloth that is uniformly coated with clean, hard, sharp garnet—if the paper, cloth and glue used are of the best possible quality—if you are getting Carborundum Brand Garnet Products—the paper and cloth that cut clean and fast, that hold the grain and show long life—*then you are getting real sanding service.*

Cut in all Standard Sizes in all Grits
Working Samples on Request

THE CARBORUNDUM COMPANY
NIAGARA FALLS, N. Y.

Grand Rapids   New York   Chicago   Boston   Cleveland
Philadelphia   Cincinnati   Pittsburgh   Milwaukee

MATTISON
EDGE SANDER
With Oscillating Belt

Designed to replace
the make-shift kind

A lot of "172" Edge Sanders are being installed this year in furniture factories; and a large proportion of them are to replace out-of-date or make-shift machines. The reason is simply because the Mattison Edge Sander gives a decidedly better quality of finish, does the sanding at much larger capacity, and at less expense. Since it pays for itself in a short time through the saving of time and labor, you can well afford to scrap your old machines.

Square and Octagon Turned Work
Rapidly Sanded on This "172."

By the use of special forms, this class of work, which costs so much to finish by hand, can be sanded economically and accurately.

Machine-tool construction throughout is responsible for being able to successfully sand these pieces.

Our non-vibrating sanding forms, and the sand belt tensioning device, together with perfect mechanical construction, makes this a highly efficient Sander.

This Edge Sander will help solve
your knotty finishing problems.
Why not look into it?

C. MATTISON MACHINE WORKS
BELOIT, WISCONSIN, - 883 Fifth Street, - U.S.A.

CANADIAN
WOODWORKE
Furni ure *and* anufa urer

Furniture Production
Number

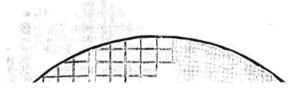

"You'll see I've amended this requisition"

"I'VE been looking into the matter of reducing our overhead charges—and I find that we can save on our power cost. For instance, I've been reading the report of a scientific test showing the enormous loss from slippage of belts over metal pulleys. It proves that the actual cost per year of running the metal pulleys, as against wood pulleys averages around $5.00 per pulley. So that I have changed your requisition from "Steel Pulleys" to read "*Dodge Wood Split Pulleys*."

"You will quite possibly find that your mechanical department has acquired a habit of specifying steel pulleys, and I would like you to see personally that *Dodge Wood Split Pulleys* have the call from now on."

We are makers of a full line of Power Transmission Machinery and Elevating and Conveying Machinery

Dodge Standard Lines

Shafting
Couplings
Collars
Iron Pulleys
Rope Wheels
Gearing
Iron Centre Wood Rim Pulleys
Wood Split Pulleys
Counter Shafts
Hangers
Pillow Blocks
Floor Stands
Mule Stands
Belt Tighteners
Friction Clutches
Rope Dressing
Sprockets
Belt Clamps
Boots
Belt Conveyors
Take Ups
Grain Shovels
Car Pullers
Trippers, etc., etc.

DODGE
WOOD SPLIT
PULLEYS

are 40% cheaper than Metal Pulleys
and give 50% greater efficiency

Sole makers in Canada of

Dodge Standard Wood Split Pulleys

With Interchangeable Bushings

DODGE MANUFACTURING CO.,
LIMITED
TORONTO

This Book mailed Free on request

A SCIENTIFIC TEST
SHOWING LOSS FROM SLIPPAGE
OF BELTS ON METAL PULLEYS

6

GARLOCK-WALKER MACHINERY CO.
LIMITED
32 FRONT ST. WEST,   TORONTO   TELEPHONE MAIN 5346

Do You Manufacture Furniture?

If so here is the very latest and most up-to-date SURFACER
for your work—the best, the cheapest in the long run.

Feeds up to 85′ per minute.

Detachable Side Clamping Cylinder Boxes.

Round Cylinder

Knife Setter and Jointer.

In Feed Rolls and Chipbreaker either solid or sectional.

Feed Rolls of large diameter—all powerfully driven.

Bed raises and lowers on long inclines.

American No. 444 Single Furniture Planer

Heavy   Strong   Durable

Does the best work at minimum expense

Let us take this up with you in detail—as well as offer you
other tools of like calibre for your work.

WRITE FOR PRICES AND DELIVERIES

Garlock-Walker Machinery Co., Limited

Canadian Representatives

American Wood Working Machinery Co.
ROCHESTER, N.Y.

M. L. Andrews & Co.
CINCINNATI, OHIO

METAL and WOODWORKING MACHINERY of all Kinds

BEAVER  BRAND

Cutting the costs of making SHELL BOXES

SPECIAL

"BEAVER"

CYLINDER

SAW

The installation of one of these saws will speed up your work so fast that you will want one for every boring machine in your plant.

The special "Beaver" Cylinder Saw is exceptionally efficient for cutting holes in shell boxes and all similar operations in the manufacture of furniture. It is made of 11 gauge steel and fitted with a heavy crucible steel spring for ejecting the cores. The shank is standard, and will fit any ordinary boring machine. It is exceptionally strong, fool proof, cuts fast and clean, and is fully guaranteed. May be had in practically any size desired. Write for prices—we can make prompt shipments.

SPECIAL

"BEAVER"

DADO

Specially designed to cut with or across the grain, leaving a smooth, clean surface. Indispensable in the manufacture of shell boxes.

This special "Beaver" Dado was designed to meet the requirements of shell box manufacturers. Its particular construction eliminates the trouble previously experienced by many manufacturers in cutting across the grain. The above illustration shows one complete order recently shipped to a shell box manufacturer. It consists of five "Beaver" Dados from 1¼ in. to 2 in. wide, also two special Planer Saws for cutting off box stock. The small groover shown in illustration is to cut slot for wire handle or for string for the cover. Get our prices on these saws—it will be worth your while.

THE RADCLIFF SAW MFG. CO., LTD., 550 Dundas Street TORONTO, ONT.

Agents for L. & I. J. White Co. Machine Knives

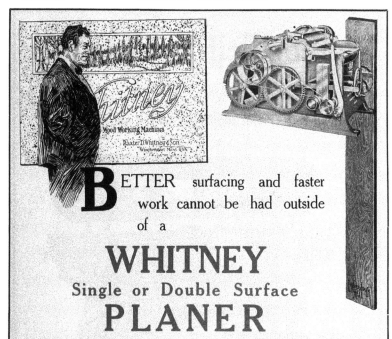

BETTER surfacing and faster work cannot be had outside of a

WHITNEY
Single or Double Surface
PLANER

The broad, clean, comprehensive statement we give you regarding Whitney Planers, when you signify your desire for facts, covers not only construction, operation and cost of maintenance, but also the performance of Whitney Planers under the varied conditions met with in planing mills, furniture and box factories.

To obtain this information places you under no obligation.

We fully believe that Whitney planers are the **best planing machines** made for rapid, smooth surfacing and lots of it, but—we assist you in comparing them with other planers so that the opinion may become mutual.

Your own opinion, formed by your investigation of facts will make you certain that buying a Whitney is your best move.

Whitney Planers are equipped either with square cylinder carrying two or four knives, or with round cylinder and four thin high speed steel knives.

BAXTER D. WHITNEY & SON,   Winchendon, Mass.

California Office:—Berkeley, California.
H. W. Petrie, Ltd., Toronto, Ont., Agents for Ontario.

Selling Representatives:—Henry Kelly & Co.
26 Pall Mall, Manchester, England.

The

Herzog Self-Feed Jointer

means

Increased Production	Simplicity of Operation
Small Floor Space	Safety to Employees

Does Four Times the Work of the Hand Jointer.

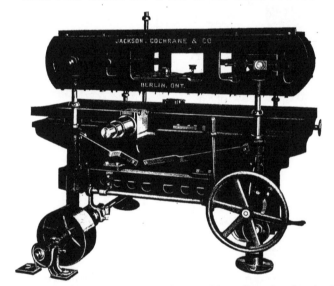

Our No. 34 Herzog Jointer, illustrated above, is one of the most efficient machines on the market to-day. It is appreciated by the manufacturer and employees alike, because, while it will produce from three to five times as much work as the hand jointer, it does not require skilled operators, but eliminates the danger so common to other makes. It can be operated by two boys. It will handle stock varying in width from 1 inch to the full width of the jointer, will feed fast or slow, takes only one-fourth the floor space of hand jointers, and requires only one-fifth of the sharpening of the knives. It is fitted with power feed raising and lowering attachment, with cylinder double belted and driven from both ends.

If interested in reducing your costs, write us.

Jackson, Cochrane & Company
KITCHENER - CANADA

Is Your Sander
Specially Adapted
To Its Work?

ENDLESS BED SANDER

MANY factories have much large surface sanding, combined with other sanding jobs of a general nature. For these conditions the proper machine is the

 ### Roll Feed Sander

It is especially adapted to sanding large quantities of wide, flat stock. The Roll Feed will handle all the stock that can be fed into it all day long. It puts on a finish ready for the stainer's brush. Changes feeds with sliding gears like an auto. Feed can be instantly reversed, preventing expensive accidents. Single handwheel adjusts drums, separately or simultaneously. Hundreds of installations all over the world.

Investigate your sanders. Make sure each is doing all that can be done on that kind of work—that it's producing the best possible finish—that it isn't eating up instead of piling up profits.

Both the Roll Feed and Endless Bed Sanders are fitted with our patented Spiral Drum, or our improved Sectional, Automatic Take-up, Straight Opening Drum. For those instances where individuals prefer the Straight Opening Drum, none gives better satisfaction than the Yates.

EACH different kind of sanding demands a special sander designed to do that work most economically. The whole Yates Sander Line is specialized—each machine perfected to do a certain kind of sanding. The

 ### Endless Bed Sander

is particularly adapted to the sanding of thin or narrow stock—no matter how long—and as short as 3 inches.

When hopper-fed the rubber discs on the travelling bed keep a machine-width stream of stock passing under the drums constantly. Its output is four to six times that of a roll feeder. Finished surfaces are left smooth as glass. Drums have a micrometer adjustment to 1.1000 of an inch. Adjustments are within easy reaching control of the operator. Feed bed has an automatic stop, both top and bottom, which prevents accident when using power hoist.

ROLL FEED SANDER

No matter what make of sander you have—or may think of having—our new Sander Book contains information you need to make a decision you won't regret. Send for it—it's free.

P. B. Yates Machine Co. Ltd.
HAMILTON, ONT. CANADA
U.S. Plant　-　-　Beloit, Wis.

No. 708 Band Resaw

A Modern Machine—
for Modern Work

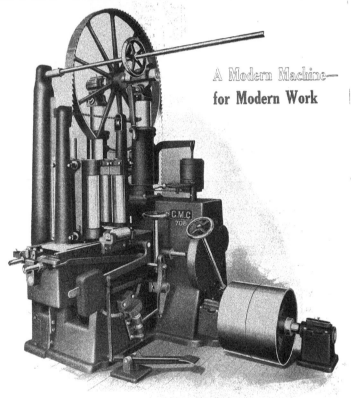

Manufactured by

CANADA MACHINERY CORPORATION
LIMITED

Toronto Office and Showrooms: Brock Avenue Subway.   **GALT, ONTARIO**

Points about the
No. 708 Band Resaw
shown on opposite page

The machine is built in two sizes, 42″ and 48″ wheels. It is designed for general work in FURNITURE FACTORIES, BOX FACTORIES, PLANING MILLS, or wherever a medium sized, high-grade Band Resaw is required. Embodies all the well-known features found in our line of Band Saws.

The 48″ machine resaws 26″ x 10″, and the 42″ machine 26″ x 8″. Frame cast in one piece and perfectly rigid.

Wheels hang between heavy vertical columns, rigidly bolted to the base. Impossible to raise the columns unevenly.

Straining device is the latest improved arrangement of our knife edge balance, which permits a high rate of speed and the use of a thin saw blade with no danger of breaking.

Six changes of speed on the 48″ machine, and five on the 42″. Feed rolls driven by spur and bevel steel gears carefully machine cut from the solid. Rolls will tilt to any angle up to 10° for making bevel siding, etc.

Rolls cannot be tilted until the feed works are drawn back by a lever convenient to the operator, thus preventing the saw from striking the rolls when tilted for bevel siding, etc.

Ask for Bulletin describing this machine

CANADA MACHINERY CORPORATION
LIMITED

GALT - ONTARIO

Toronto Office and Showrooms   -   -   Brock Avenue Subway

Woodworking Machinery that is made for your needs—not adapted to them

12 IN. MOULDER

Their Past and Present Records are Your Guarantee of Efficiency

Manufacturers of High Class Woodworking Machinery for Box Factories, Furniture Factories, Planing Mills, Sash and Door Factories, etc.

We are specialists in Shell Box Equipment.   Ask the man who has installed "Cowan" Machinery.

Agents for Reynolds Screw Drivers and Morgan Nailers.

COWAN & COMPANY OF GALT, LIMITED
GALT   ONTARIO

SLOW SPEED LOW POWER

BLOWER SYSTEM

The

Foster Blower

The "FOSTER" BLOWER

is specially designed for handling shavings, sawdust, excelsior, or any conceivable material. It is equipped with three bearings, shaft extending through, with bearing on inlet side. This eliminates the over-hung wheel and avoids heating, clogging and shaking. It is the only non-centre suction fan in existence. The Foster Fan is so constructed that the material entering the fan is immediately discharged without passing through or around the wheel.

Our Automatic Furnace Feeders will feed your material under boilers or into any desirable receptacle.

Consult us on your exhaust system troubles.

THE TORONTO BLOWER COMPANY
TORONTO   156 DUKE ST.   ONTARIO

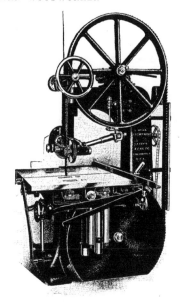

Large picture shows
No. 146 as a Band Rip
Saw. Small picture
shows how, by simply
reversing table, it be-
comes a modern Band
Resaw.

Time required to make
change 3 minutes.

Ready in a moment to either rip or resaw,
FAY-EGAN
"LIGHTNING" No. 146 is the machine you need!

If your combined ripping and resawing does not exceed 50 to 60 M lineal feet per. day on soft woods, or 20 to 25 M feet on hard woods, you do not need to put in two separate machines.

A Fay-Egan No. 146 will take care of all this work in the most economical manner.

At the very beginning you save the price of one machine.

You save the space occupied by one machine.

You realize the highest return on your investment, as the one machine is kept busy all the time.

By simply reversing the table and raising or lowering the ripping rolls, the No. 146 is changed from a rip to a resaw or from a re-saw to a rip saw—it takes but a moment to do this.

As a rip saw, it will handle material up to 24 inches wide.

As a resaw, it will cut to the center of 8 inches and up to 18 inches under the guide.

The No. 146 has all of the advantages of a separate band rip or resaw

You can find out more about this economical machine by asking for Bulletin N-3.

J. A. FAY & EGAN CO.
153-173 W. Front St.   ·   ·   ·   CINCINNATI, O.

For

Box and Shook Mfrs.

Staves and Heading
Furniture Mfrs.
Flooring Mfrs.
Shingles, Etc., Etc.

WIRES Packages any length, up to 16 ins. high and 17 ins. or more wide.

CAPACITY is 80 to 100 bundles per hour, dependent on skill of operator, the size of bundles and the number of ties. One user is tieing 1,000 box shook bundles, size 24 in. x 14 in. x 12 in., per 10 hour day, two wires per bundle.

As wire gives a stronger bundle, and ordinarily costs less than half of string cost, and as string stretches, breaks, and wastes fully 25 per cent. in using, further comment seems unnecessary.

Write for Illustrated Bulletin and Price

LOWRY'S LATEST "STANDARD"

SHOOK BUNDLER
and
BOX WIRING MACHINE

J. L. NEILSON & CO.
WINNIPEG, MAN.

FILE THIS "AD" FOR
FUTURE REFERENCE.

Our "FROST KING"

Babbitt Metal was originally manufactured more especially for large machinery makers, but it met with such universal success and gave such satisfying results that we have placed it on the market as a first-class, all-round babbitt. It is as equally efficient for heavy duty as it is for speed. It is especially adapted for planing mills, saw mills, and woodworking plants generally. Get our prices on FROST KING.

Your bearings will always be Cool and Lasting if you use HOYT BABBITT METALS

Manufactured for 40 Years by

HOYT METAL COMPANY
TORONTO - ONTARIO

New York   St. Louis   London, Eng.

Try Our New "TROJAN" Babbitt

It is especially noted for its ability to absorb and hold oil, making it unnecessary to give the bearings as much attention as is otherwise required. It is unequalled for over-head bearings in line and counter shafts and will be found an excellent babbitt for use in furniture factories, planing mills, etc., where there are numerous bearings in out of the way places. If you want a reliable, low priced babbit try our TROJAN.

Wood Steaming Retort

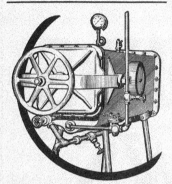

Wood Bending Manufacturers:

This is one of our

Perfection Retorts

which we guarantee will save you

50% Less Breakage

in your bending department than your present process; that your stock will dry in your forms or presses in one-third less time; that you will have no stained stock; that your stock will retain its shape much better after being bent; that it will dry in your dry-kiln in one-half less time and that your steam consumption will be reduced at least 90 per cent.

The door can be opened and closed in ten seconds, and it is steam and water tight and for this reason can be placed anywhere in your factory.

Compare this IMPROVED RETORT with your present steam boxes, then write us for our Booklet on Progressive Wood Steaming.

Made in Preston, Ontario

Perfection Wood Steaming Retort Co.

PARKERSBURG - WEST VIRGINIA

Seeing is Believing

No trouble at all to see Francis Modern Hydraulic Veneer Press Equipment in operation under conditions similar to yours. It is in use in representative panel, furniture, door, piano and other factories most everywhere. It must be seen in use to be fully appreciated.

Permanency is the Proof

Every one of the many users of Francis Outfit is as enthusiastic over it to-day as ever, because the principle is right, the construction is right.

It produces the work in the quickest, most satisfactory and most economical manner.

EASILY INSTALLED, READILY UNDERSTOOD. ANYBODY CAN OPERATE IT.

Francis Factory Address, **Rushville, Ind** **Francis**
EST. 1880.   EST. 1880.

Originators and Manufacturers of Modern Glue Room Equipment

Single Beam Retaining Clamps

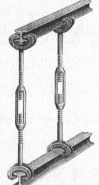

Order a Sample Lot Now

No Hydraulic Veneer Press a complete success without them, likewise advantageous for Hand Screw and other Veneer Presses.

Work done at least twice as fast.

Clamps adapt themselves to different widths of stock.

Will not "eat you up" in repairs, nor create a "scrap heap."

Made of special material throughout.

No. 1 for stock up to 18-in. wide.

No. 2 for stock up to 24-in. wide.

No. 3 for stock up to 30-in. wide.

No. 4 for stock up to 36-in. wide.

And larger sizes.

State daylight or opening of your press, and Clamp Rods will be made accordingly.

Every one of the many hundreds of users recommends these clamps.

Have You Our Catalogue?

CHAS. E. FRANCIS CO.

Francis Largest Manufacturers of Glue Room Outfits. **Rushville, Ind.** **Francis**
EST. 1880.   EST. 1880.

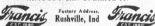

USE

SIMONDS
SAWS

AND

PLANER
KNIVES

We manufacture Saws of every kind and size, Circular or Band Saws, Concave, Mitre or Novelty Saws for rapid smooth cutting. Write us for our new catalogue and tell us what Saws or Knives you now require.

Simonds Canada Saw Co., Limited

Factory: St. Remi St. and Acorn Ave.,

St. John, N. B.   Montreal   Vancouver, B. C.

The "HARGRAVE"

Bench Clamps (any size) Cabinet Clamps, Column Clamps, Carvers' Clamps, Hand Screws

Of course you need some

J. L. NEILSON & CO.   -   -   Winnipeg, Man.

Morehead

Back to Boiler

SYSTEM

Read What The Wabash Cabinet Co. Say!

"With reference to the Morehead Trap, it is everything that Mr. McLachlan said it was. Before we made this change, our watchman was obliged to put in coal every fifteen minutes. Now he puts in a fire on an average of once every hour and WE ARE SAVING OVER 50 PER CENT. ON THE COST OF OUR FUEL."

You can do the same, The first step is to

Write for the "Back-to-Boiler" Book

In it you will find the condensation question discussed from several angles—and interesting pictures of installations under every possible condition.

The economic principle of the "Morehead" System will save you dollars. Write for the book TO-DAY.

Canadian Morehead Mfg Co.
Woodstock   Dept. "H"   Ontario

GRINDING

BAND SAWS

with

26 TEETH

TO THE INCH

Your woodworking band saws can be resharpened so that all teeth come within ½ of 1/1000 of an inch alike to each other, by this

—Model A—

Automatic Band Saw Grinder
Saws 1-8 inch to 8 inches wide. Teeth up to 2 inch spacing

POINTS OF INTEREST

Bronze Bushed Grinding Wheel Spindle.
Head Suspended between Hardened Steel Pivots
No Slides to Wear. No Cams to Change.
Feed Movement obtained by a Sealed Slide
A Double Pawl, Positive Feed Movement, the quickest and most perfect method of jointing all teeth, can be furnished at a small extra charge.

LOWEST PRICED GRINDER MADE
Write for Details

THE WARDWELL MFG. COMPANY
Saw and Knife Sharpening Machinery
111 Hamilton Ave.   CLEVELAND, OHIO

Receiving End

Discharging End

The More You Learn

about dry kilns, the more you will desire to own a NATIONAL. Its efficiency is wonderful. Its economy remarkable. Let us tell you what its users have to say about it.

Ask for our new catalog just off the press.

THE NATIONAL DRY KILN COMPANY
1117 East Maryland Street   -   -   -   INDIANAPOLIS, Indiana

DID IT EVER
OCCUR TO YOU

that possibly one of the many styles of NASH
SANDERS might be a fine thing for your work?

ASK ME ABOUT IT.

J. M. NASH, MILWAUKEE, Wis.

GRAND RAPIDS
VAPOR DRY KILN
GRAND RAPIDS
MICHIGAN

Increasing output of other kilns is part of our business.

We can arrange to utilize your present equipment and supplement with sufficient additional to bring your kilns to our basis of output.

We would suggest that you supply our engineering department with description of your kilns and lumber drying needs, so we can understandingly outline what can be done.

Your opening the matter will incur no obligation and we will give you, frankly, the results of our experience.

Woodworkers
and
Lumbermen

The Sheldon
Exhaust
Fans have
characteristics
that adapt
themselves
to the ser-
vice of the
Planing
Mill or
Wood-
working
Plant. They are specially designed for this kind of work, having a saving in power and speed of 25% to 40%.

Write for Industrial Booklet.

SHELDONS LIMITED - Galt, Ont.
Toronto Office, 609 Kent Building
— Agents —

Messrs. Ross & Greig, 412 St. James St., Montreal, Que.
Messrs. Walker's Limited, 259-261 Stanley St., Winnipeg, Man.
Messrs. Gorman, Clancey & Grindley Ltd., Calgary and Edmonton, Alta.
Messrs. Robt. Hamilton & Co., Ltd., Bk. of Ottawa Bldg, Vancouver, B.C.

If you are planning on building any clothing or gar-
ment cabinets, it will pay you to remember we are

HEADQUARTERS FOR
CLOTHING TROLLEYS

The Twentieth Century Model

In addition to the model shown above we make six other models, including the Columbia, Yukon, and Klondike; and others suitable for hotel and home use, as well as for store purposes.

We assist our customers by furnishing cabinet construction drawings free. Send for catalogue giving complete

description and prices.

HARDWARE SUPPLY CO.
Dept. D.   314 North Street, Grand Rapids, Mich.

PERFORATED VENEER TAPES

Designed for the express purpose of saving the TIME, LABOR and EXPENSE that is necessary in sanding off, or otherwise removing ordinary tapes.

"IDEAL" PERFORATED TAPES DO NOT HAVE TO BE REMOVED

Economical   *Practical*   *In General Use*

They work equally as well in 3 ply veneers as in 5 ply, and are better than open mesh cloth for taping crossbanding stock.

If YOU are not using them, send for trial coils, stating the work you desire to try them on.

IDEAL COATED PAPER COMPANY
BROOKFIELD, MASS.

NEW YORK   CINCINNATI   CHICAGO

Perkins Vegetable Glue
The Best for Built-up Stock and Veneer Laying

Perkins Vegetable Glue sticks and holds. It is unaffected by dampness, extremely dry atmosphere or severe sanding. We guarantee to effect a 20 per cent. saving in your glue room by replacing hide glue with Perkins Vegetable Glue. Along with this saving you can have a glue room comfortable as to temperature and sweet as to smell. The glue is absolutely uniform in quality. Every shipment exactly the same. Let us demonstrate in your own factory and on your own stock just what can be accomplished with Perkins Vegetable Glue.

Dinner wagon manufactured by the D. Hibner Furniture Co., Limited, Berlin, Ont., in which Perkins Glue was used.

PERKINS GLUE COMPANY, LIMITED, HAMILTON, ONT.
"Address all inquiries to Sales Office, Perkins Glue Company, South Bend, Indiana, U. S. A."

REXFORD VENEER TAPES

Combine all the qualities desired by manufacturers and users of veneers.

"Let a sample roll tell its own story"

Please state your requirements and, if possible, send sample of tape you are now using.

REXFORD GUMMED TAPE CO.

Manufacturers

Home Office, 598 Clinton St., MILWAUKEE, WIS.

St Louis, Mo. Branch, 416 Rialto Building. Olive 3529   Chicago, Ills., Branch, 20 E. Jackson Blv'd. Harrison 6107.

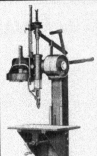

No. 2 Standard Type Automatic Screw-driving Machine.

The Only Way to Drive Screws

There is a stage-coach method and a modern method of doing everything. When it comes to driving screws the only modern way is with an Automatic Screw Driving Machine, which means greater and better production and greater profit.

The Reynolds
Automatic Screw-Driving Machine

is guaranteed to produce results. It is no experiment. The No. 2 Standard Type, shown herewith, is particularly suited for placing the screws in the straps of shrapnel boxes. It may be fitted to drive, without change or adjustment, any three consecutive sizes flat or round head, or two sizes filister head, No. 6 to 14 inclusive, wood or machine screws, or three sizes flat or round head, or two sizes filister head machine screws up to No. 20. All the above up to 1½ inches long if smaller than No. 10, or up to 1¾ inches long, No. 10 or larger. Fitted with boring attachment if desired.

Many Machines in Use in Canada.

Ask for our Catalogue.

REYNOLDS PATTERN & MACHINE CO.

101 and 103 Third Ave., Moline, Ill., U.S.A.

"Treat your machine as a living friend."

Don't Waste Time

for which you are paying big money by having a Machine in your plant which is not giving you a maximum of production every day.

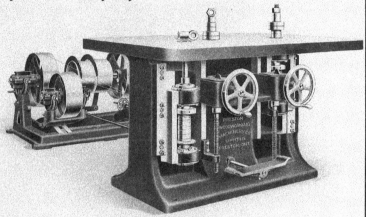

No. 162 High-Speed Ball-Bearing Shaper
SEE THIS MACHINE!

t will turn out 100% MORE WORK than the ordinary shaper which you are now using. That's our claim; twice the amount of work and much easier on the operator. Is it worth our while to cut your shaper cost in two? If so, drop us a line for more particulars.

THE CHEVROLET MOTOR CO. ARE USING 6 OF THESE MACHINES IN THEIR PLANT AT OSHAWA

DIMENSIONS, WEIGHTS and SPEEDS

Floor Space, 120 in. x 58 in. Net Weight, 2650 lbs. Shipping Weight, 2750 lbs. Tight and Loose Pulleys on Counter-shaft, 10 in. x 6 in. Driving Pulleys on Countershaft, 18 in. x 5 in. Idler Pulleys, 14 in. x 5 in. R.P.M Countershaft, 200. Pulley on Spindle, 3¼ in. x 6 in. R.P.M. Spindle, 7000. Diameter Spindle, 2 3/16 in. Spindle Centres, 28 in. Table, 8 in. x 61 in. Diameter Spindle Top, 1¼ in. Horse Power Required, 3. Front Projection, 17 in. Height, 35 in.

"The Machine you want in your plant is not the cheapest machine to buy but the machine that will save you the most money to use."

The Preston Woodworking Machinery Co. Limited, Preston, Ont.

"Treat your machine as a living friend."

Don't Waste Material

It Costs Too Much Money These Days

Our No. 130 Short Centre Power Feed Rip Saw

will reclaim short material otherwise waste.

How much lumber do you waste in a year? Lots of it could be ripped and used if you had this machine **which will take care of stock as short as 8″ singly or 6″ when ends are butted together.** It will also handle stock of any long length equally as well.

SAVE YOUR WASTE MATERIAL AND YOU WILL
SOON SAVE THE PRICE OF THE MACHINE

Send for a cut. It's well worth your consideration.

The machine is specially designed for use in Furniture Factories and will recommend itself to the user by reason of the **Great Saving in Labor** and the **Large Increase in Production** possible.

The Preston Woodworking Machinery Co., Limited, Preston, Ont.

Southern Hardwoods

Fullerton-Powell Hardwood Lumber Co.

Manufacturers and Dealers

South Bend, Indiana

Oak and Gum a Specialty

Daily Capacity 200,000 Feet

We handle Southern Hardwoods exclusively in any grade or thickness desired. Fifteen years' experience in shipping the Canadian trade successfully enables us to render the best of service.

We solicit your inquiries and orders. Remember to W-I-R-E at our expense when in a hurry.

Alphabetical List of Advertisers

Acme Steel Goods Company 82
Acme Veneer & Lumber Company 78
American Hardwood Lumber Company 97
Anderson Lumber Company, C. G. 27
Atlantic Lumber Company 22
Ault & Wiborg 100

B. C. Government 23
Baird Machinery Company 106
Barnaby, Chas. A. 75
Beach Furniture Company 106
Bennett & Witte 31
Boiler Inspection & Insurance Co. 103
Brown & Company, Geo. C. 29
Binns & Bassick 104
Burns & Knapp Lumber Company 34

Canada Glue Company 107
Canada Machinery Corporation 8-9
Canadian Morehead Mfg. Company 15
Carborundum Company 113
Central Veneer Company 79
Chapman Double Bearing Company 112
Churchill-Milton Lumber Co. 28
Clark & Son, Edward 22
Clark Veneer Company, Walter 75
Cooley & Marvin Company 102
Cowan & Company 101
Cumberland Valley Lumber Company 31

Darling Bros. 72
Decorators Supply Company 90
Dermott Land & Lumber Company 30
De Vilbiss Mfg. Company 91
Dodge Manufacturing Company 2
Dominion Mahogany & Veneer Company 36
Dulweber Company, John 84

Elgie & Jarvis Lumber Company 72
Edwards Lumber Company, E. L. 85

Farrin Lumber Company, M. B. 81
Faultless Caster Company 104
Fay & Egan Company, J. A. 11
Foster Merriam Company 100
Francis Company, Chas. E. 13
Fullerton-Powell Hardwood Lumber Co. 21

Garlock-Walker Machinery Company 3
Grand Rapids Dry Kiln Company 16
Graves, Bigwood Company 97
Ginn Lumber Manufacturers' Assn. 29
Gummed Papers, Limited 100

Hardware Supply Company 16
Harland & Son, Wm. 80
Hart & McDonagh 22
Hartzell, Geo. W. 34
Haughton Veneer Company 70
Hay & Company 72
Hay Knife Company, Peter 72
Hoffman Bros 75
Hoyt Metal Company 12
Huddleston-Marsh Mahogany Co. 28
Hutchinson Woodworker Company 111
Hyde Lumber Company 33

Ideal Coated Paper Company 17
International Time Recording Co. of Canada 95

Jackson, Cochrane & Company 6

Kentucky Lumber Company 30
Kersley, Geo. 77-83

Lamb-Fish Lumber Company 34
Long-Knight Lumber Company 32
Louisville Point Lumber Company 80

Martin-Barriss Company 90
Mattison Machine Works, C. 114
Meadows Toronto Iron Wire & Brass Works
    Company 101
Mowbray & Robinson Company 82

Nartzik, I. J. 78
Nash, J. M. 16
National Dry Kiln Company 16
National Lock Company 108
Neilson & Company, J. L. 12-15

Ohio Veneer 75
Onward Manufacturing Company 105

Paepcke-Leicht Lumber Company 36
Paasche Air Brush Company 92
Penrod, Jurden & McCowen 26
Penrod, Walnut & Veneer Co. 70
Perfection Wood Steaming Retort Co. 13
Perkins Glue Company 17
Perrin, William R. 23
Petrie, H. W. 99
Preston Woodworking Machinery Co. 19-20
Pringle Company, R. E. T. 111
Probst Lumber Company 31

Radcliffe Saw Company 4
Read Bros. 22
Rexford Gummed Tape Company 18
Reynolds Pattern & Machine Co. 18
Ritchey Supply Company 103
Roberts, John N. 77
Russe & Burgess 82

Sadler & Haworth 101
Sargeac Machine Company 113
Sheldons Limited 10
Simonds Canada Saw Company 14
Skedden Brush Company 90
Smith & Company, Alfred Z. 107
Smith Company, R. H. 28
Spencer, C. A. 97
Stevens-Hepner Company 90
Stimson, J. V. 34

Thoman-Flinn Lumber Company 22
Toronto Blower Company 10
Toronto Veneer Company 81

Waetjen & Company, Geo. L. 78
Walter & Company, B. 113
Walter & Son, J. 109
Wardwell Manufacturing Company 13
Weber-Knapp Company 103
Whitney & Son, Baxter D. 5
Wood-Mosaic Company 77

Yates Machine Company, P. B. 7-96

Edward Clark & Sons

Toronto

CANADIAN HARDWOODS—winter cut **Birch, Bass, Maple**

10,000,000 Feet, Immediate Shipment

Ontario and Quebec stocks, East or West, rail or water shipment, direct from mill to factory. Stocks in any grade or thickness, selected and graded for your requirements, selected widths, selected color.

Case Goods and Chair Manufacturers

1,000,000 ft. 4/4 Birch, mill run.
1,000,000 ft. 4/4 Birch, No. 1 common.
1,000,000 ft. 4/4 Birch, No. 2 common.
500,000 ft. each 5/4, 6/8, 8/4, No. 1 common.
500,000 ft. 4/4 Basswood.
500,000 ft. 4/4 Soft Elm.

For Trim Trade

200,000 ft. 4/4 Birch, No. 1 and 2.
300,000 ft. 5/4 Birch, No. 1 and 2.
400,000 ft. 6/4 Birch, No. 1 and 2.
300,000 ft. 4/4 to 8/4 Brown Ash.

Parlor Frame Makers

400,000 ft. 10/4 Birch, No. 1 C and B.
600,000 ft. 12/4 Birch, No. 1 C and B.
200,000 ft. 16/4 Birch, No. 1 C and B.
50,000 ft. 6/4 S. Maple, No. 1 C and B.
50,000 ft. 10/4 S. Maple, No. 1 C and B.
50,000 ft. 12/4 S. Maple, No. 1 C and B.

Shell Box Trade

300,000 ft. 1x9 and up Birch, No. 1 and No. 2C.
200,000 ft. 1x9 and up Birch, No. 2 and No. 3C.
200,000 ft. 1x6 to 8 Birch, No. 2 and No. 3C.
600,000 ft. 6/4 x 8/4 Birch, No. 2 nd No. 3C.

We Commend to Your Notice:

Our lumber is reliable, our grades uniform, our shipments prompt, and receive personal supervision; our prices consistent. We have had long experience. We desire to serve.

R. H. Smith Co.
ST. CATHARINES, ONT.　Limited

"Arrow Head"
Quality Saws

The Best Material is essential in Saws and Knives. When combined with good workmanship, efficiency is guaranteed. We know of no better steel than Vanadium. If we did, we would secure it. Our workmanship is guaranteed. The use of a high grade of steel has given our Mitre Saws, Dado Heads and Woodworkers' Saws and Knives a reputation for excellence not equalled by our competitors.

Eventually you will use Arrow Head Saws—why not on your next order?

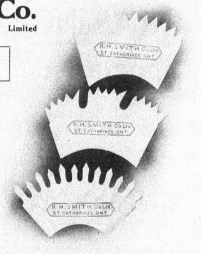

PRESSES
MADE IN CANADA
Support Home Industries by specifying
PERRIN PRESSES

Perrin Presses are built "up to a standard" and not down to a price, and even with this high standard of quality you will find them very favorable in price comparison with other Makes.

We build Hydraulic and Mechanical Presses for all kinds of mill and factory uses, in any size or type to suit your special requirements.

We can save money for you not only on your first cost, but also on your up-keep because Perrin Presses are guaranteed against defective and faulty construction.

William R. Perrin, Limited
Toronto　-　-　Ontario

CRATING
Specialists

5/8" and 1" All Grades
At Right Prices

TRY US!
It will be worth your while

Read Bros., Limited
43 Victoria Street
Toronto   -   Canada

Shell Box Stock

1 x 6 birch, in No. 1, 2, and 3 common.
1 x 7 and 8 birch, in No. 1, 2, and 3 common.
1 x 8¼ birch, in No. 1, 2, and 3 common
1 x 9 and wider, birch, in No. 1, 2, and 3 common.
1 x 4 and wider, spruce.   2 x 4 and wider, spruce.

For the Furniture Trade
4/4, 5/4, 6/4, 8/4, 10/4, 12/4, 16/4, birch, maple, elm, basswood and black ash.

Elm, Basswood and Black Ash
in No. 1 common and better grades—dry stock in assorted sizes.

Hart & McDonagh
Continental Life Bldg.
TORONTO   -   -   ONTARIO

Thoman-Flinn Lumber Co.
Write us when in the market for
WEST VIRGINIA
HARDWOODS
Such as
Chestnut
Basswood
Poplar
Maple
Beech
and soft textured — uniform colored Plain and Quartered
White and Red Oak
also
Southern Hardwoods
Red and Sap Gum—Cottonwood—Tupelo
Direct Mill Shipments—No Yard Grades
CINCINNATI   -   -   OHIO

The Atlantic Lumber Co.

Manufacturers and
Wholesale Dealers

American and Canadian Hardwood Lumber and Mahogany, Quartered and Plain, White and Red Oak a Specialty.

110 Manning Chambers
Phone Main 6386   -   TORONTO
HOME OFFICE, BOSTON MASS.
Mills: KNOXVILLE, TENN.,   WALLAND, TENN.
FRANKLIN, VA.

British Columbia Furnishes a Flag Pole for Kew Gardens, London

The Pole being towed up the Thames.

It is a far cry from a flag pole to FURNITURE, PANELS, BOXES, MOULDINGS, SASH, DOORS, and the many other products of wood working factories, but you are asked to let it reach you.

DOUGLAS FIR—which produced this 215 foot flag pole.

WESTERN SOFT PINE—which is crowding White Pine in all markets.

WESTERN LARCH—the giant of the larch family.

WESTERN RED CEDAR—which yields the famous B.C. shingles and bevel siding.

SITKA SPRUCE—which furnishes most of the Aeroplane Stock.

ENGELMANN SPRUCE—next to Sitka the largest of the Spruce genus.

WESTERN HEMLOCK—whose wood does not admit kinship to its eastern botanical cousin.

All possess qualities which make them specially suited for various kinds of manufacture.

Are you sure that some of these CANADIAN woods will not fit your requirements exactly? Space prevents a full description of all these woods, but a request for samples and information concerning their qualities addressed to the

British Columbia Lumber Commissioner

Excelsior Life Building,

will be gladly complied with.   **Toronto**

BRITISH COLUMBIA HAS A WOOD FOR EVERY USE

HARDWOOD LUMBER

Ready to Ship

We have on sticks at Brasfield, Ark., some of the finest hardwood lumber ever manufactured. Ask us for quotations on the following:

In pile at Brasfield, Ark.

Ash

67,000 ft. 4-4" Log Run.
16,000 ft. 8-4" Log Run.

Plain White Oak

37,400 ft. 4-4" 1s and 2s.
8,700 ft. 5-4" 1s and 2s.
12,000 ft. 6-4" 1s and 2s.
107,000 ft. 4-4" No. 1 Com.
9,000 ft. 5-4" No. 1 Com.
47,000 ft. 4-4" No. 2 Com.
87,000 ft. 4-4" No. 3 Com.

Plain Red Oak

39,000 ft. 4-4" 1s and 2s.
3,000 ft. 5-4" 1s and 2s.
11,000 ft. 6-4" 1s and 2s.
242,000 ft. 4-4" No. 1 Com.
64,000 ft. 6-4" No. 1 Com.
87,000 ft. 4-4" No. 2 Com.
92,000 ft. 4-4" No. 3 Com.
8,000 ft. 5-4" No. 3 Com.

Quartered White Oak

63,000 ft. 4-4" 1s and 2s.
9,000 ft. 5-4" 1s and 2s.
13,600 ft. 8-4" 1s and 2s.
134,800 ft. 4-4" No. 1 Com.
19,400 ft. 5-4" No. 1 Com.
13,700 ft. 8-4" No. 1 Com.
18,000 ft. 4-4" No. 2 Com.

Quartered Red Gum

139,700 ft. 4-4" 1s and 2s.
42,450 ft. 5-4" 1s and 2s.
20,300 ft. 6-4" 1s and 2s.
244,000 ft. 4-4" No. 1 Com.
35,000 ft. 6-4" No. 1 Com.
14,000 ft. 8-4" No. 1 Com.

In pile at Memphis, Tenn.:

Quartered Red Oak

F.A.S. No. 1 Com. No. 2 Com.
4-4" 76,000 ft. 28,400 ft.

In pile at Helena, Ark. :

Quartered White Oak

4-4" 56,000 ft. 94,000 ft.
1,800 ft.
4-4" Clear Qtd. Strips
14,800 ft.

4-4" Clear Qtd. Strips
16,200 ft.

Plain Red & White Oak

4-4" 82,000 ft. 140,000 ft.
88,000 ft.

Plain Red Gum

274,000 ft. 4-4" F.A.S.
38,700 ft. 5-4" F.A.S.
16,000 ft. 6-4" F.A.S.
13,000 ft. 3-4" No. 1 Com.
258,000 ft. 4-4" No. 1 Com.
22,000 ft. 5-4" No. 1 Com.
4,000 ft. 6-4" No. 1 Com.

Plain Sap Gum

83,000 ft. 4-4" Box Boards
13-17"
16,000 ft. 4-4" F. A. S.
6-12"
187,000 ft. 4-4" F. A. S.
13-17"
63,000 ft. 4-4" F. A. S.
18" and up.
32,000 ft. 5-4" F. A. S.
Regular.
111,000 ft. 6-4" F. A. S.
7,500 ft. 3-4" No. 1 Com.
220,000 ft. 4-4" No. 1 Com.
16,000 ft. 5-4" No. 1 Com.
96,000 ft. 6-4" No. 1 Com.
42,000 ft. 8-4" No. 1 Com.
147,000 ft. 4-4" No. 2 Com.
8,000 ft. 5-4" No. 2 Com.
3,000 ft. 6-4" No. 2 Com.
167,000 ft. 6-4" No. 2 Com.

Elm

16,000 ft. 5-4" Nos. 2 and 3 Com.
20,000 ft. 6-4" Nos. 2 and 3 Com.

Sycamore

31,000 ft. 4-4" Log Run, Plain.
22,000 ft. 4-4" Log Run, Qtd.

Ash

8-4" 48,200 ft. 34,000 ft.
10-4" 18,300 ft. 11,000 ft.
12-4" 16,400 ft. 10,900 ft.
16-4" 28,300 ft. 13,400 ft.

PENROD, JURDEN & McCOWEN

Incorporated

Memphis, Tenn.

The Most Complete

WALNUT STOCK

in America

Our stock of Walnut Lumber is comprehensive. The following lumber in stock at Logansport, Indiana:

Walnut

	Feet
4/4 F. A. S.	73,000
4/4 No. 1 Com.	117,000
4/4 No. 2 Com.	200,000
5/4 F. A. S.	39,000
5/4 No. 1 Com.	6,000
5/4 No. 2 Com.	80,000
6/4 No. 2 Com.	15,000
8/4 F. A. S.	5,000
8/4 No. 1 Com.	5,000
8/4 No 2 Com.	32,000
9/4 F. A. S.	2,000
9/4 No. 1 Com.	6,000
10/4 F. A. S.	9,000
10/4 No. 2 Com.	5,000
12/4 F. A. S.	5,000
12/4 No. 2 Com.	3,000
16/4 No. 1 Com.	2,000
5/8 No. 1 Com.	2,000

	Feet
3/4 No. 1 Com.	10,000
3/4 No. 2 Com.	8,000

We also have a nice stock of Plain and Quartered Oak, Cherry and Poplar in addition to this Walnut shown.

H. A. McCOWEN & Co., Memphis, Tenn.

Southern Rotary Veneers

of PJM Quality

Our big rotary mill at Helena, Ark., is famous for quality products. We manufacture all Southern woods, including

Oak, Gum, Ash,

Cypress, Elm, etc.

Send Us Your Inquiries, and let us explain why we are holding the trade of the most discriminating buyers in America.

PENROD, JURDEN & McCOWEN

Incorporated

Memphis, Tenn.

CANADIAN LUMBER

FOR THE

CANADIAN
FURNITURE MANUFACTURER

RIGHT PRICES	GOOD GRADES	PROMPT SHIPMENTS

Let Us Quote You on the Following Stocks:

Special

3 cars 4/4 Nos. 2 & 3 Com. Ash
4 " 4/4 No. 1 Com. & Betr. Birch
1 car 4/4 " 3 " " 9 in. & up Birch
3 cars 8/4 " 2 " " Maple

Regular Stock

Birch	-	-	-	-	-	4/4 to 12/4
Maple	-	-	-	-	-	4/4 to 12/4
Ash	-	-	-	-	-	4/4
Elm	-	-	-	-	-	4/4 to 16/4
Basswood	-	-	-	-		4/4 to 6/4
White & Red Pine	-	-				4/4 & 8/4
Jack Pine and Spruce						

The above in all grades

Manufacturers and Strictly Wholesale Dealers

C. G. ANDERSON LUMBER CO.
LIMITED

705 Excelsior Life Building

TORONTO - ONTARIO

RED GUM

WE MANUFACTURE IT—at Glendora, Miss., from Selected Virgin Forest Delta Timber. Write your requirements, as we carry stock from 1 in. to 2 ins. thick, grades First and Seconds, to Number Two Common in Plain Red, Quartered Red and Sap Gum—

SOUTHERN HARDWOODS

Also including Oak, Plain and Quartered, White and Red, Cottonwood, Ash, Elm and Cypress.

INDIANA HARDWOODS

On our yards, New Albany, Ind., from which point we will quote very attractive prices on

2 cars 1 in. 1s and 2nds Poplar.	2 cars 1 in. 1s and 2nds Plain White Oak.
2 cars 1 in. Clear Saps Poplar.	2 cars 1 in. No. 1 Common Plain White Oak.
1 car 1½ in. Select Poplar.	1 car 1½ in. 1s and 2nds Plain Red Oak.
1 car 1¼ in. No. 1 Common Poplar.	1 car 2 in. 1s and 2nds Plain Red Oak.
1 car 1¼ in. No. 2A Common Poplar.	1 car 1½ in. No. 1 Common Plain Red Oak.
	2 cars 1 in. No. 2 Common Plain Oak.

Members:

Our Shipments handled promptly.   Southern Hardwood Traffic Assn. of Memphis.
Our Gum is right.   Gum Lumber Manufacturers' Assn. of Memphis.
Our grades are right.   National Hardwood Lumber Assn. of Chicago.

CHURCHILL-MILTON LUMBER COMPANY
Sales Office: LOUISVILLE, KENTUCKY

THIS picture shows the jungle which has to be overcome in getting out Mahogany logs; also the native labor used in that business; but best of all it shows the class of Mahogany logs we have been getting out of Mexico the past year, and from which is produced our high-class lumber and veneer.

If you want the best, write us.

Huddleston-Marsh Mahogany
Company
33 West 42nd St., NEW YORK CITY

Selected Red and Sap Gum

St. Francis Basin Band Sawed Stock

KRAETZER-CURED   **STRAIGHT AND FLAT**

> "Replying to your inquiry as to what we thought of your Kraetzer-Cured Gum, we are glad to state the lumber worked better than any Gum we ever put through our factory.
>
> "We expect to handle a great deal of this treated Gum from you, and want to state that we are well pleased with your service."
>
> (Name furnished on request.)

YOU CAN OBTAIN THE SAME RESULTS

Send us a list of your requirements.   —   Delivered Prices will be cheerfully quoted.

GEO. C. BROWN & CO.

Mill—PROCTOR, Ark.   **Main Office—MEMPHIS, Tenn.**

We specialize in Tennessee Aromatic Red Cedar

RED GUM

("AMERICA'S FINEST CABINET WOOD")

Before RED GUM became recognized as a cabinet wood which manufacturers are proud to use and to tell about, and which *consumers* are proud to own, a good many high-class furniture manufacturers were obliged to call RED GUM by some sort of pretty nickname.

NOW ALL THAT IS CHANGED

The RED GUM producers have spent (and are still spending) thousands of dollars to tell the public that RED GUM is not only "America's most winsome wood," and *a thoroughly reliable wood* when properly manufactured, but also that to have a suite of "Genuine American RED GUM" in one's home is a *fact to boast of.*

As the public is being continuously educated to CALL FOR RED GUM BY NAME, isn't it good policy for all furniture manufacturers who make RED GUM furniture to call it by its RIGHT NAME and nothing else? Isn't it GOOD BUSINESS to do so? If they *don't, what will they say* when the public ASKS for "REAL RED GUM?" Are we right?

Write us for samples, particulars and general information.
Our reply will be prompt, personal and dependable.

GUM LUMBER MANUFACTURERS' ASS'N

1314 Bank of Commerce Building, Memphis, Tennessee

Dermott Land and Lumber Co.

Manufacturers of

Southern Hardwoods

80 E. Jackson Blvd.
CHICAGO   -   U.S.A.

THE DERMOTT KING
BAND SAWN SOUTHERN HARDWOODS

Quartered White Oak

1 cars 1 in. 1sts and 2nds.
5 cars 1 in. No. 1 Common.
1 car 1¼ in. 1sts and 2nds.
1 car 1¼ in. No. 1 Common.
½ car 1½ in. 1sts and 2nds.
1 car 2 in. 1sts and 2nds.
1 car 2 in. No. 1 Common.
1/3 car 2½ in. Com. and Bet.
1/3 car 3 in. Com. and Bet.

Plain Red Oak

5 cars 1 in. 1sts and 2nds.
10 cars 1 in. No. 1 Common
2 cars 1¼ in. 1sts and 2nds.
2 cars 1½ in. No. 1 Common
2 cars 2 in. 1sts and 2nds.
1 car 2 in. No. 1 Common.

Plain White Oak

1 car ⅜ in. 1sts and 2nds.
1/3 car ⅝ in. 1sts and 2nds.
1 car ¾ in. 1sts and 2nds.
10 cars 1 in. 1sts and 2nds.
10 cars 1 in. No. 1 Common.
2 cars 1¼ in. 1sts and 2nds.
4 cars 1¼ in. No. 1 Com.
2 cars 1½ in. 1sts and 2nds.
3 cars 1½ in. No. 1 Com.
2 cars 1¾ in. 1sts and 2nds.
1 car 2 in. 1sts and 2nds.
2 cars 2 in. No. 1 Common.
1 car 3 in. 1sts and 2nds.
½ car 4 in. 1sts and 2nds.

Gum

1 in. to 2 in. Plain Sawn Red.
1 in. to 2 in. Quarter Sawn Red.
1 in. to 2 in. Figured Plain and Quartered.
1 in. to 2 in. Sap.

All well cared for band sawn stock properly edged and trimmed. Nice assorted dry stock on hand at our modern band mill at Dermott, Arkansas.

KENTUCKY LUMBER COMPANY

Sales Office:
LEXINGTON, KY.

Operating Modern Band Mills at
WILLIAMSBURG, KY.
BURNSIDE, KY.

Sawing

SOFT WHITE OAK, POPLAR CHESTNUT, BASSWOOD

SULLIGENT, ALA.

Sawing

RED and SAP GUM, TUPELO WHITE and RED OAK

We cater to discriminating trade who look for Quality and Service as well as Price

An opportunity to show you is all we ask

Look Here, Mr. Consumer

WE Make our own lumber,
Make good grades,
Don't advertise much,
Have no salesmen's expense.

Why Don't You Buy from Us

and

SAVE THAT DIFFERENCE ?

The Louisville Point Lumber Company

LOUISVILLE   -   KENTUCKY

HARDWOODS

A few items we wish to move

DRY QUARTERED WHITE OAK

1 car 1 x 4" and wider 22 and 25" long clear 1 face stock.

1 car 1 x 4" and wider 31" long clear 1 face.

1 car 1 x 4" and wider 37", 49" and 55" long clear 1 face.

PLAIN OAK

1 car 2 x 2—30" long.

1 car 2 x 2—30" oak.

2 x 2—30" ash.

2 x 2—24" oak.

Sooner or later you will buy dimension. Why not get del'd prices now?

The Probst Lumber Co.
Cincinnati, Ohio

THE
M. B. FARRIN LUMBER
COMPANY
CINCINNATI   -   OHIO

Manufacturers

HARDWOODS

We have very large stocks of well assorted West Virginia Hardwoods on hand at Cincinnati and also at mills.

Plain Oak   Whitewood
Quartered Oak   Red Gum
Chestnut   Ash

KILN DRIED STOCK
A SPECIALTY

BONE DRY
HARDWOODS

Ready for Quick Shipment

SPECIALTY

PLAIN
and Well Figured

QUARTERED OAK

Even Color   Soft Texture

CUMBERLAND VALLEY
LUMBER CO.
Cincinnati   -   Ohio

BENNETT & WITTE
CINCINNATI, OHIO

We carry in stock ready for prompt shipment:

1/2", 5/8", 3/4", 4/4", 5/4", 6/4" and 8/4"

Plain Sawn Sap and Red Gum

1/2", 5/8", 3/4", 4/4", 6/4" and 8/4"

Plain Red and White Oak

especially adapted to the use of the furniture manufacturer.

All thicknesses of ASH and SOFT SOUTHERN ELM for the Automobile Body Manufacturer.

SEND US YOUR INQUIRIES

Plain and Quartered
Red and White Oak

and Other Hardwoods

Even Color   Soft Texture

We have **35,000,000 feet dry stock**
all of our own manufacture, from our own
timber grown in Eastern Kentucky

MADE (MR) RIGHT

OAK FLOORING

PROMPT SHIPMENTS

The Mowbray & Robinson Co., Inc.

GENERAL OFFICES
CINCINNATI, OHIO

Mills: Quicksand, Ky., West Irvine, Ky., Viper, Ky.

Canadian Representative:
N. H. FARNHAM
601 Elmwood Ave., BUFFALO, N. Y.

Long - Knight Lumber Co.

Manufacturers and
Wholesale Dealers
INDIANAPOLIS, IND., U. S. A.

HARDWOOD
LUMBER

Walnut, Mahogany

and

American
Walnut Veneers

SPECIAL:

5 cars 5/4 1 & 2 Pl. Red Oak

RUSSE & BURGESS
MEMPHIS, TENN.   INC.

We manufacture about 40 million feet
Hardwood Lumber
per year

Our goods are all band sawn. About 50 per cent. is
Kraetzer Cured. Let us have your inquiries for OAK,
both Quartered and Plain.

RED and WHITE ASH, also GUM in RED and SAP

HYDE LUMBER COMPANY

Sales Office—SOUTH BEND, Ind.   Band Mills—LAKE PROVIDENCE, La.

Southern Hardwoods

Dry Stock List

Plain Red Oak

	1s & 2s	No. 1 Com.	No. 2 Com.	No. 3 Com.
3/4	41,000	57,000		
4/4	67,000	133,000	104,000	54,000
5/4	29,900	16,000		

Plain White Oak

	1s & 2s	No. 1 Com.	No. 2 Com.
4/4	83,000	107,000	58,000
4/4	Sound Wormy		45,000

Quartered White Oak

Scant 4/4 No. 1 Common and Better	13,100
4/4 1sts and 2nds	46,000
4/4 No. 1 Common	186,000
5/4 1sts and 2nds	8,200
5/4 No. 1 Common	12,400
4/4 2½ to 5½ in. Strips	29,000
4/4 No. 2 Common	28,000

Quartered Red Oak

4/4 No. 1 Common and Better	26,000

Cypress

	1s & 2s	Select	No. 1 Shop	No. 1 Com.
4/4	39,000	18,400	51,000	28,000
5/4	28,000	12,000	18,000	22,000
6/4	31,000	25,000	17,000	26,000
8/4	36,000	28,000	14,000	27,000

Ash

	1s & 2s	No. 1 Com.	No. 2 Com.
4/4	46,000	72,000	24,000
5/4	17,000	35,000	11,000
6/4	42,000	29,000	14,000
8/4	38,000	36,000	23,000
10/4	28,600		
12/4	14,000		
16/4	16,600		

Cottonwood

4/4 13 to 17 in. Box Boards	33,000
4/4 9 to 12 in. Box Boards	41,000
4/4 18 in. and up Panel	28,000
4/4 13 in. and up 1sts and 2nds	29,000
4/4 No. 1, 2 and 3 Common	42,000
5/4 1sts and 2nds	22,000

Elm

4/4 No. 2 Common and Better	64,000
5/4 No. 2 Common and Better	31,000
6/4 No. 2 Common and Better	84,000
8/4 No. 2 Common and Better	67,000
10/4 No. 1 Common and Better	18,000
12/4 No. 1 Common and Better	15,000

Plain Sawn Sycamore

4/4 No. 1 Common and Better	39,000

AMERICA'S FINEST   REDGUM   CABINET WOOD

Sap Gum

	1s & 2s	No. 1 Com.	No. 2 Com.
3/4	41,000	18,000	
4/4	163,000	172,000	189,000
5/4	92,000	9,100	62,000
6/4	14,000	32,000	27,000
8/4	23,000	43,000	19,000
4/4 Box Boards, 13 to 17 in.		106,000
4/4 Box Boards, 9 to 12 in.		133,000
4/4 1s and 2s, 13 to 17 in.		102,000
4/4 1s and 2s, 18 in. and wider		37,000

Kraetzer Cured Sap Gum

Our Lake Providence, La., plant, is equipped with a Kraetzer Preparator and we solicit your inquiries for this stock.

Tupelo

4/4 1s and 2s	49,000	No. 1 Com.	62,000
4/4 No. 2 Common			23,000
4/4 13 to 17 in. Box Boards			18,000

Plain Red Gum

3/4 1s and 2s	42,000	No. 1 Com.	37,000
4/4 1s and 2s	118,000	No. 1 Com.	163,000
5/4 1s and 2s	12,000	No. 1 Com.	55,000
6/4 1s and 2s	13,000	No. 1 Com.	14,000
8/4 1s and 2s	8,000	No. 1 Com.	42,000

Quartered Red Gum

4/4 No. 1 Common and Better	43,000
5/4 No. 1 Common and Better	36,000
6/4 No. 1 Common and Better	28,000
8/4 No. 1 Common and Better	52,000

Plain Sawn—Figured Red Gum

4/4 No. 1 Common and Better	42,000

Quarter Sawn—Figured Red Gum

4/4 No. 1 Common and Better	21,000
5/4 No. 1 Common and Better	8,000
6/4 No. 1 Common and Better	9,000
8/4 No. 1 Common and Better	34,000

Write for prices and descriptive information. Our stock will please you and our prices are right.

CHARLES O. MAUS,
CANADIAN REPRESENTATIVE

J. V. Stimson & Co.

OWENSBORO, KENTUCKY

Manufacturers of

BAND SAWN
HARDWOODS

QUARTERED and PLAIN OAK
ASH
POPLAR
HICKORY
WALNUT
QUARTERED SYCAMORE

OAK and WALNUT
OUR SPECIALTY

BLACK WALNUT

Large Stock 1 to 3 inches thick ready
for immediate shipment

also

Well Assorted Stocks of

West Virginia & Kentucky

O A K

Ash, Maple, Hickory, Chestnut, Basswood and Poplar

Prices and stock list on request

Burns & Knapp
Lumber Company
CONNEAUTVILLE, PA.

FOR SALE

Black Walnut
Lumber

15,000	feet	4/4	No. 1	Common	
20,000	"	4/4	No. 2	"	
15,000	"	5/4	No. 1	"	
15,000	"	5/4	No. 2	"	
15,000	"	6/4	No. 2	"	
40,000	"	8/4	No. 1	"	
60,000	"	8/4	No. 2	"	
30,000	"	9/4	No. I	"	

Will make special low prices on all of the above
stock to move it quick. All Band Sawn, Equalized,
and all thoroughly dry, ready for immediate shipment. Write for prices.

Geo. W. Hartzell
Piqua, Ohio

We have on our Charleston
Yard a complete stock of

WHITE
and   **OAK**   PLAIN
RED   and
                        QUARTERED

RED
and   **GUM**   PLAIN
SAP   and
                        FIGURED

Can Make Immediate Shipment

Representative in Eastern Canada:
FRANK T. SULLIVAN
71 South Street
BUFFALO - New York

Lamb-Fish Lumber Co.
"The Largest Hardwood Mill in the World"
CHARLESTON, MISS.

Cable Address:
ELELCO

E. L.
EDWARDS
LUMBER
CO.

DAYTON, O.
U. S. A.

Plain and Quartered White and Red Oak
Black Walnut, Cherry, Ash
Hickory, Elm, Chestnut
Red Gum, Sap Gum
Yellow Poplar
Basswood

*Delivered Prices, any place
on the Globe reached
by rail or vessel.*

Of course it is true that

RED GUM

is America's finest cabinet wood—but
Just as a poor cook will spoil the choicest viands while the experienced chef will turn them into prized delicacies, so it is true that

The inherently superior qualities of Red Gum can be brought out only by proper handling.

When you buy this wood, as when you buy a new machine, you want to feel that you have reason for believing it will be just as represented.

We claim genuine superiority for our Gum. The proof that you can have confidence in this claim is shown by the letter reproduced herewith.

Your interests demand that you remember this proof of our ability to preserve the wonderful qualities of the wood when you again want RED GUM.

Paepcke Leicht Lumber Co.

Conway Building　　　111 W. Washington St.
CHICAGO, ILL.

Band Mills: Helena and Blytheville Ark.; Greenville, Miss.

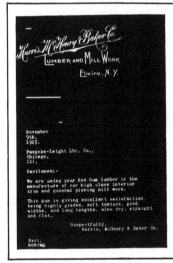

November 9th, 1915.

Paepcke-Leight Lbr. Co.,
Chicago,
Ill.

Gentlemen:-

We are using your Red Gum lumber in the manufacture of our high class interior trim and general planing mill work.

This gum is giving excellent satisfaction, being highly graded, soft texture, good widths, and long lengths, also dry, straight and flat.

Respectfully,
Harris, McHenry & Baker Co.

Dict.
REN/MG

Buy Made in Canada Veneers

Dominion Mahogany
& Veneer Co., Limited

Head Office and Factory
Montreal West, P.Q.

Toronto Office and Warehouse
455 King St. West

Mahogany, Walnut,
and all Fancy Woods—Lumber

SAWN, SLICED and ROTARY CUT VENEERS

Mahogany, Walnut, Quartered and Plain Oak, Ash, Maple, Birch, Elm, Poplar, Gum.

Gum and Poplar Cross Banding.　　　Hard Maple Pin Block Stock.

SEND US YOUR ENQUIRIES.

Canadian Woodworker

A Monthly Publication in the Interest of the Woodworking Industry.
Reaches the factories producing interior finish, doors, sash, flooring,
woodenware, furniture, pianos, boxes, and general mill products.

Subscription, $1.00 a year; foreign $1.50.

Woodworker Publishing Company, Limited

Branches: Montreal, Vancouver,
Chicago, New York, London, Eng.

345 Adelaide St. West, Toronto

Phone Ade. 2700

Authorized by the Postmaster General for Canada, for transmission as second class matter.
Entered as second class matter July 18th, 1914, at the Post Office at Buffalo, N.Y., under the Act of Congress of March 3, 1879.

Vol. 16   October, 1916   No. 10

The Outlook for the Furniture Industry

Furniture factories in Canada are busy. They have plenty of orders on hand and are operating their plants as near to full capacity as the shortage of skilled mechanics will permit. One large factory we know of has been working overtime for the greater part of the past summer. Retailers report that sales are very satisfactory. A retailer interviewed a few days ago stated that his company's sales for the months of June and July had been as good as at any time since they have been in business, and for the months of August and September their sales had been greater than during June and July. When asked as to the prospects for the future, he stated that his company were preparing to go right ahead the same as during the past few months.

These reports are distinctly encouraging. They indicate that the country is prosperous and that people as a rule are not compelled to retrench in their expenditures for furniture.

The sales are not confined to any particular grade, but are fairly evenly distributed, the medium grades, perhaps, being in greatest demand. It is not improbable that a general advance in prices will be made very shortly, as the cost of production is steadily increasing and is likely to continue to increase for some time to come.

During the fiscal year ended March 31, 1916, imports of furniture into Canada amounted to approximately $700,000. These figures are only about one-quarter of what they amounted to in previous years. The imported furniture that is most often seen is of the higher grades. Canadian manufacturers are now turning out furniture that compares favorably with any of the high class goods formerly imported, and in the near future we believe that the imports (with the exception of some special kinds) will be reduced to practically nil. Altogether the outlook is a bright one, and we anticipate that sales will be particularly heavy this fall and winter.

The Christmas trade last year was good and there is every prospect of it being much better this year. Furniture, especially living room pieces and novelty furniture, is becoming exceedingly popular for Christmas gifts. Enterprising manufacturers have seen the trend in this direction, and as a result there is a wide range of attractive pieces from which prospective purchasers can make selections. There is nothing more appropriate for Christmas gifts than furniture, and its use for that purpose should be encourged in every way.

This issue of the "Canadian Woodworker and Furniture Manufacturer" is styled a "Furniture Production Number." It is devoted almost exclusively to the manufacture of furniture. The articles cover a range of subjects of vital interest to those engaged in the industry, written by men well qualified to deal with their respective subjects. The advertising pages are equally interesting. They contain the announcements of the most progressive manufacturers and dealers, and they show where the best equipment can be obtained. Any reader who is in the market for machinery, lumber or supplies will be benefited by a study of the advertisements in this issue.

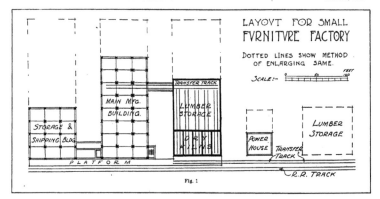

LAYOVT FOR SMALL
FVRNITVRE FACTORY

DOTTED LINES SHOW METHOD
OF ENLARGING SAME.

SCALE:-

Fig. 1

The Design and Construction of a Furniture Factory

By S. B. Lincoln*

In laying out a modern furniture factory, first consideration should be given to the relative arrangement of buildings and departments, and the relation of same to railroad tracks. Space for lumber storage should be provided at a point near the plant, but so located that it will not constitute a serious fire hazard. Lumber will be taken either from storage or directly from cars into dry kilns and these should be conveniently located for loading by either method.

It is desirable to have on hand at all times a supply of lumber which has been thoroughly dried and for this purpose it will be found convenient to have a storage building located immediately adjoining dry kilns in which the lumber may be kept on trucks previous to cutting up.

The system of tracks used in dry kilns may readily be extended through into the storage building. From the storage building, lumber, still handled in truck loads—through use of transfer cars—should be taken into the cutting department, which will in general be located on the first floor.

Beyond this point there is more or less latitude in the arrangement of machinery and processes, but in general it will be found well to have the work gradually travel toward the upper floors of the building so that painting, varnishing and finishing may be done on the top floor. From this point finished goods should be taken into separate storage and shipping building, which may be connected with main manufacturing building by a bridge. The ground floor of storage building should be used for shipping room and should have outside platform to come level with car floors.

The above outlines the general arrangement which might be followed for most plants of moderate size and a typical layout is shown by the accompanying sketch, Fig. 1. This shows ample space for the enlargement of all departments in a logical manner. In the case of a very much larger plant where manufacturing can-

*Engineer Chicago office of Lockwood, Greene & Company, Engineers for Industrial plants.

not be combined in one building some changes from this plan would have to be worked out. In such a case the main manufacturing building might be changed to consist of several buildings.

The power plant should be located in a separate building, and as wood refuse and sawdust should form a part of the fuel, a storage bin should be provided at a sufficient height so that material from bin will be fed directly into the boilers. The usual shavings collector system with collector located directly on the roof of the power house could be installed.

The number of storeys for a furniture factory would depend upon the size and nature of the product, etc., but where land is plentiful, either a three or four storey building would probably work out best for economy of construction and for manufacturing. Dry kiln and lumber storage would probably be one storey in height unless land was very scarce indeed, in which case these departments could be worked out for a two-storey building. The storage and shipping building could readily be made five or six storeys, as the case might require.

On the matter of materials of construction, the dry kilns and storage building should be built of non-combustible materials. For dry kilns, either brick or tile walls should be used, and the roof construction would best be a combination of hollow tile and concrete. This would require further insulation in the form of wood roofing over concrete, or cinder fill over concrete. The storage building should be built entirely of reinforced concrete.

For manufacturing space either concrete or heavy mill construction could be used. Some manufacturers object to concrete floors because of the necessity of changing machines and cutting holes through the floors, which does not work out so well with concrete as with plank floors. Aside from this feature, either construction may be well adapted to the work. If constructed with plank floors, beams and columns are con-

sidered better from a fire protection standpoint than
unprotected steel.

It is assumed that buildings for this purpose will
be provided with automatic sprinklers throughout and
in this case the insurance rate for concrete or for first
class mill construction is about the same. In either
case ample light and ventilation should be provided.
All stairways and toilet-room floors may be made of
concrete, as the question of cutting through floors does
not occur at these points. Stairways, elevators and
toilets are best located in fireproof towers outside of
the building proper, as in this way they do not obstruct
the main floor space.

For the space occupied by woodworking machinery
proper, and in fact for the whole manufacturing building, with the exception of toilets, the heating should
be provided by the indirect heating system. That is,
where a fan and heater located at a central point is
used to blow a supply of warm fresh air into the
buildings. This is particularly necessary for the room
in which machines are located which are connected
to shavings collector system, as this system removes
large quantities of air from the room and it is necessary
to replace this.

The storage building should be heated slightly to
prevent freezing of sprinkling pipes. Heat for dry
kilns should be provided by exhaust steam piping with
connection for live steam through reducing valve.

Elevators should be provided of ample size, and on
account of the length of some of the materials handled,
it is better to have them open at the narrow end or
ends. Elevators opening on one side only provide the
greatest safety, particularly where an operator is not
maintained constantly on the car. A size of about
8 feet by 14 feet is suggested as a good size for car
platform.

The questions of safety and welfare of employees
are receiving more and more attention in these days,
and in woodworking factories, as well as in other lines,
these should receive ample consideration when plant
is first laid out. Guards for moving belts, generous

This building has steel floor, beams and columns, with floors of plank
and maple; exterior walls of brick; windows of steel sash.

End of building shown in picture is left ready to take future extension of this building of about same length as section already constructed. This end wall will form a fire wall dividing completed building
into two sections.

Just above the sidewalk in the third bay to right of doors there may
be seen an iron box built into wall. This is provided with roller and an
iron door and is used for the purpose of unloading lumber from wagon
into first storey of building. Sidewalk is about 4 feet above building
floor at this point.

washing and toilet facilities, good light and fresh air
are all matters which should be provided for.

A small machine shop will doubtless be required
to take care of repairs, and in a small factory this can
be located in one corner of the main manufacturing
building, while for a larger plant this would work out
better in a separate building.

Provision for future expansion of plant is a matter
which should always receive consideration, and usually
this can be provided for in the original design without
much trouble.

This building is 65 feet wide with one centre row of columns spaced 20 feet apart lengthwise of building. Floor beams are of steel. Floors
of 2 x 6 in. hemlock on edge. Maple floors are laid diagonally; this providing the best trucking service.
This view shows a large amount of floor space unobstructed by columns and the excellent light obtained by large windows of steel sash.

Buying, Handling and Drying Lumber

Its Admitted Importance in the Manufacture of Furniture Entitles it to the Thoughtful Consideration of Those Responsible for the Quality of the Product

By W. J. Beattie

Mr. W. J. Beattie.

Lumber being the major portion of the raw material used in the manufacture of furniture, it deserves the greatest consideration, and plans for delivering it to the swing saws in the best possible shape are always of interest, if they have anything new to offer or suggestions which may give some reader a new thought or idea to graft on to his present methods.

As practically all lumber is bought "unsight, unseen," certain kinds and grades at varying prices, it is a case of depending largely on the honor of the lumber manufacturer to deliver the goods as per order, and to his credit, he nearly always does so.

The most fruitful source of possible tribulation is when stock is bought, not direct from the manufacturer, but from some firm who is selling the product of perhaps half a dozen mills, stock which they as a rule do not see. If the mill ships stock not up to specifications, the trouble has to be settled between the seller and producer, meanwhile the buyer may be greatly inconvenienced waiting for a settlement, as there will be a considerable proportion of the stock O.K. and which he is likely in urgent need of, to start on the way through the kiln.

As a rule the most satisfactory way is to "tie up" to a few good manufacturers of lumber, who have proved themselves able to deliver just what they promise, and if there are any favors going they go to the steady customers; this is human nature, and is not at all uncommon.

I know of one of our largest furniture manufacturers who has bought quartered oak for years, practically from the one firm. The lumber firm never gave less than the invoice showed, in fact the furniture firm's expert scaler was nearly always from two to three hundred feet ahead of the invoice, and do you know that they were always honest enough to split the difference with the lumber firm. Needless to say, the feeling of confidence established by such methods works to the mutual benefit of these concerns, and when the lumberman has any really extra choice stock the furniture firm is sure to be informed and given the first choice to avail themselves of a good investment.

Strict integrity in such matters makes business friendships that are really helpful. It is foolishness to say that friendship and business are alien to each other. There is no basis for any such contention. There is nearly always an element of uncertainty in the buying of lumber, not that the lumber will not be as represented, but as to the general figure and texture (quartered oak). It is quite within the legal limits that considerable variety will occur, long boards may be clear and sound as a bell, but hard, and showing a "stringy" figure; others will be soft and of a reddish color and have broad blotches of figure not easily matched. Others again, will be too "loudly" figured

(for present-day taste). As the grade rules do not consider figure, only to 90 per cent. of the aggregate of one face, the furniture manufacturer has to make his selection of figured stock according to his best judgment.

The conditions of soil, latitude, exposure, etc., have very much to do with the appearance and texture of the lumber. The boards off the north side of an oak tree will differ from those off the south side, and all this gives the lumber that variety of color and figure which make the manufacture of it into furniture such a fascinating occupation. If every board in a carload was almost an exact duplicate of its fellows, the handsome creations in quartered oak would lose at least fifty per cent. of their charm. It is the height of folly to expect every board in a car of quartered oak to show the same percentage of figure, either in quantity or similarity. This variation makes the working of quartered oak not merely a mechanical operation, but one requiring a degree of taste and an eye for beautiful effects.

Figure, If Too Uniform, Becomes Monotonous

Some "Philistines" think (or appear to) that the real article should look like the monotonous machine reproductions—which are duplicates of a certain part of a veneer flitch—that tire the eye on account of their uniformity, as every article of a lot is exactly similar in appearance, and as wooden looking as a thing can be made. No change, no variety, no novelty, only monotony. What is needed in quartered oak furniture is harmonious blending, or matching, of figure, so that the eye is pleased and rested, the same as looking at a well executed picture. "A thing of beauty is a joy forever," is just as true of furniture as of anything else.

These remarks so far have been confined to oak, as that material is still the premier furniture wood, and is altogether likely to remain so for some time to come. This statement is not made in any way to attempt to detract from the value or beauty of other woods, but in medium and better grade furniture oak is still the greatest quantity seller.

The buyer of the raw material has also to consider the fact of dimension cut stock, which is now being manufactured by many mills, and it will be found that in such parts as pilasters for case goods, legs for tables, drawer bottoms, case and glass backs, etc., there is quite a field for economical investigation. As the pilasters or posts for dressing cases and buffets are practically a standard height, and finish 1⅜ inches or 1¼ inches square, it is safe to assume that they are staples; a good quantity can be secured, and the wastage is practically nil.

The same is true of table legs in oak, up to 2 inches square, finished, though I have seen dimension chair legs that were 2½ inches square, finished size, that worked up perfectly. It depends on the way the stock is manufactured and dried. Every reputable manufacturer of dimension stock sells his product absolutely guaranteed.

While the lumber measure of dimension stock costs more than the same stock in planks, there is the wastage and labor to offset these items, and it will usually be found a paying proposition. The drawer bottom requirements will vary more as to lengths and widths, and glass backing still more so. Three-ply material has no rivals when it comes to these requirements, and can be bought at so much per foot face measure.

Three-ply drawer bottoms and glass backing are a good and truthful talking point in favor of your goods. They can be honestly called reinforced or laminated, and guaranteed to "stay put" and appeal to those who like good sound and dependable furniture. The rotary cut poplar makes a fair bottom for the drawer of cheaper goods, but has a "pleasant" habit of being able to shrink or swell on very slight provocation.

For glass backs rotary cut stock is a delusion and a snare, don't use it.

Paper mill board is being used quite freely for glass backing, and if of good hard calendared stock,

are as flexible as they should be; that is, that the conditions are not such that a car of lumber has to be completely cut up once it has been pushed into the factory, which is a kind of a "pot luck" style. A car of seleced stock for tops or fronts would not usually be required all at one time, and if the facilities for running the car in, cutting off the required quantity and then returning the car to the shed, are not handy, then there is no need of the greater care in separating as to width and figure before loading. If the lumber is a good grade, a good stock cutter will do as good as a man can be expected to, but he will have a lot of looking over and handling to do, which takes time, and the whole production of the factory depends more on him than any one individual in the business.

The accompanying illustration of a small factory, designed by the writer, will explain the remarks on more economical lumber handling in a fairly clear

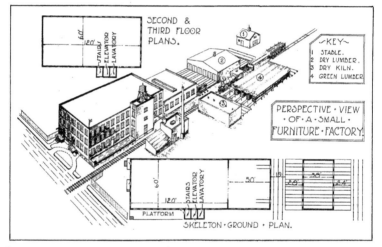

there is no objection to it, as it can be varnished and made look quite respectable, on fairly good furniture.

There are other parts of household furniture that are being made as specialties by manufacturers, such as case and table tops, drawer fronts, with flat and otherwise carved mouldings, etc., but as these are not quite in line with the present article we will get back to lumber.

No matter what kinds or grades of oak are being bought, the proper place to separate and select it into the various sizes and kinds, is while unloading it from the railroad car. There is no other time when it can be done so effectively. The requirements of the factory are known and can be provided for by the intelligent loading and labelling of the kiln cars, and thus save valuable time in the "break out" department of the factory.

The former paragraphs assume that the handling facilities for the loaded cars after they leave the kiln

manner. The lumber is separated and piled on the respective kiln cars as it is being unloaded. It can then be easily transferred into any one shed desired, to await its turn to the dry kiln.

It might be noted, in passing, that this shed, or group of sheds, are made completely of iron, having hollow iron pipes bedded into cement piers, which support skeleton iron roof frames that are covered with corrugated iron. The gutters of these roofs need not be watertight, in fact are as well left a few inches apart so that the drip falls into a cement open drain which leads to a general outlet. All the ground space under the sheds has a rough coat of cement, thus ensuring a clear and dry floor. As this air can circulate freely and the lumber is protected from the sun, an ideal out-door drying is immediately in evidence.

The rails supporting the kiln cars are carried on cement piers of the right height to meet the transfer cars at each end. Across the railway track is the storage shed for the dried lumber, having six tracks, so that it is quite convenient to get any kind of dry material quickly and easily.

The dry kiln illustrated is of the moist air, box type, though that is only suggested. The dry lumber shed is kept moderately warm during cold weather, say not less than 60 degrees. The moist air or vapor process of drying lumber has long since passed the experimental stage, and no progressive firm contemplating a change would think of installing any other.

These kilns are made in two styles, the one known as the "Progressive" and the other the "Box" type. The former would usually be chosen by the larger manufacturer, who is cutting up large quantities of several kinds of lumber of 1 inch thickness.

The "Box" kiln outfit can be easily added to as the business grows, and makes an appeal on this ground, for a factory that is making a start.

The system of drying is similar, the lumber being steamed for at least twenty-four hours, in a compartment created by dropping a curtain; the car of well saturated stock is then gradually moved along and as the air conditions surrounding the lumber are gradually growing less humid, it is in time delivered in perfect condition, straight, bright, sound, free from checks, and much easier on saws and machine knives than lumber dried in the old style hot air kiln. These kilns are not cheap affairs, but are made to last a life time. The only wood in connection with them is the doors and the roof ventilators, though the latter can be of some more durable material. There is every facility for ease of regulation, and they have reduced the drying of lumber to an almost exact science.

Advantages of Moist Air Kiln

Don't let that remark about science frighten any one, as the method is as simple as A. B. C.

The greatest advantage a moist air kiln has over the old drying methods is the fact that the sap cells of the lumber are practically freed from the substance that absorbs moisture readily and causes swelling, and also the resultant shrinkage when conditions change. The writer has furniture in his own home that has proved this contention during the past summer, when it will easily be remembered that with the great heat there was also much humidity. The furniture that was made in a factory where the lumber was dried in a moist air kiln was not much affected, while other pieces that were constructed of material dried in an ordinary hot air kiln swelled so much that it was with difficulty the drawers could be pulled out, in order to plane sufficient off the lower edges to make them run freely.

These are facts, and are not written simply to support a statement. I would not go so far as to say that lumber dried in a moist air kiln will not swell under any conditions; that would not be true; but it is a positive fact that the liability of bothersome swelling is practically eliminated, even in such weather as we have had this last summer.

To return to the kiln; reasonably well air dried 1-inch lumber is dried in from ten days to two weeks. The kiln can be regulated to take a longer time, and if the consumption of lumber is not too great for the easy capacity of the kiln, it is just as safe and requires less heat to give the lumber a longer stay in the kiln. I have seen oak that was bandsawed into chair slats dried in a vapor kiln, and there was not a check in any of them. If a kiln will do that what more can anyone ask?

The thick oak is not dried in the same kiln section

as 1-inch lumber, as it is obvious that a much longer and slower process is required to condition thick material. This applies to dimension square stock also. The methods of testing lumber for moisture, or to determine whether it is dry enough to use, will be dealt with in a future article on furniture manufacturing.

To refer again to the factory, illustration; the space between the dry lumber shed and the factory is to be roofed over, having considerable wired glass in the roof, so as not to darken the work room. This was left off in the picture to better illustrate the handling of the lumber.

The plant was designed to be run by electric power; the steam plant is for heating and dry kilns, the method of heating the factory being hot air. The cuttings are "hogged" and blown over to the fire. The plant was designed with a view to future extensions, which would make the heating apparatus central.

As other references may be made to the illustration in future articles, the above will explain apparent shortcomings in the layout.

The author of the above article, Mr. W. J. Beattie, has contributed a number of articles to the pages of the "Canadian Woodworker and Furniture Manufacturer" from time to time. Mr. Beattie has had a wide and varied experience in the furniture business. He learned his trade as a cabinet maker in Peterborough, Ont., when a great deal of the work was done by hand, and attended a drawing class in the evenings at the Mechanic's Institute. After spending several years in Peterborough, Mr. Beattie removed to Toronto, where he worked as a cabinetmaker, and incidentally learned the trade of wood carving. During the evenings and Saturday afternoons he attended the art schools of the city and Mr. Beattie now considers that this time was well spent.

After a number of years he left Toronto and sought to broaden his experience by working for some time in the factories of the United States. Returning from the States he located in Woodstock, Ontario, and accepted a position as designer and detailer for the Anderson Furniture Company of that place. When that factory passed under the control of the Canada Furniture Manufacturers Limited, Mr. Beattie was made superintendent of one of their plants in Woodstock and later in Waterloo. From Waterloo he went to Owen Sound, where he has remained as superintendent of the National Table Company. Mr. Beattie has had experience in the building and installing of machinery and in practically every department of a furniture factory. We feel sure that our readers will look forward to reading other articles from Mr. Beattie's pen.—Editor.

Inspector for Shell Boxes

The Imperial Munitions Board recently appointed Mr. C. M. Tremear to a position which required him to visit shell box manufacturers and criticize and assist them in their work. Mr. Tremear is a practical woodworker and thoroughly acquainted with everything essential in connection with such duties.

This move by the Munitions Board was an excellent one in the interests of the efficient production of shell boxes. It was appreciated by manufacturers, and we only hope that the Board will continue to place men in control of their inspection work who understand what is required, rather than permit the work to be done by men who may secure their appointment by political influence.

Making Detail Drawings and Stock Bills

The Second of a Series of Articles on This Important Subject—What the Mechanic Should Know to be Able to Interpret a Drawing

By W. Alexander

In the article "How to make and read working drawings," which was published in the June issue of the "Canadian Woodworker and Furniture Manufacturer," the writer explained the basic principles of draughting and making detail drawings. A kitchen table of simple design and construction was used as an illustration. The method of drawing plans and elevations was fully dealt with. It should therefore be safe to assume that readers have these points fixed firmly in their minds.

In this instance the object to be drawn is a small living room table. In the ordinary routine of the furniture factory the draughtsman would prepare the drawing and make out a stock bill of lumber required. After being cut roughly to size and dressed the lumber would be taken to the machines which perform the operations necessary for the fitting together of the different parts.

It is essential that the men that operate these machines and the cabinet makers that finally assemble the parts be able to interpret the drawing made by the draughtsman. They will be best able to do this if they understand how to make the drawing and take off the stock bill themselves. We will therefore go over the ground covered by the draughtsman in making the drawing.

In the first place, he would either have given to him, or decide himself, the size of the top. Then the other details would be worked out, having in mind the appearance of the completed table and economy of manufacture. There is no hard and fast rule as to which view should be drawn first. The writer generally draws the front elevation first and then the plan, but with small objects, like the one illustrated, the three different views can be worked in at the same time.

It may be noticed that there are one or two slight differences in the drawing of this table and the kitchen table shown in the former article. For instance, only half the front elevation and plan are shown. The reason for this is clear. Both ends of the table are the same and it would only be a waste of time and paper to draw the whole thing.

It might be well to point out here that draughting is a commercial proposition. The draughtsman should always bear in mind that the aim is not to draw a pretty picture, but simply to facilitate the manufacturing processes by providing something for the men to work from.

Another difference between the accompanying drawings and the one shown in the last article will be seen in the plan. The outline of the top is shown in broken lines. The really proper way would be to show the top in full lines and everything underneath it in dotted or invisible lines. The method shown here is the one generally adopted. By showing the top in broken lines it is possible to put the under framing in full lines and the case-rail, which, of course, is under the drawer, and therefore invisible, in dotted lines.

In place of the end elevation is shown a section. The only difference is that in an end elevation the end of the drawer front, case-rails, back rail, drawer

bottom and shelf and the drawer bearers would be shown in dotted lines. It will be noticed that the legs are stop chamfered. The usual way of showing this is to show a section through one of the legs, but in factory drawings it is usually just written on.

As mentioned before, the draughtsman should avoid putting any more on a drawing than is necessary. Thus the stretcher that joins the two legs at the end and supports the shelf could have been shown on the plan in dotted lines; also the shelf could have been shown on the plan in dotted lines. There was nothing to be gained by showing them, as all the measurements could be secured from the elevation and sec-

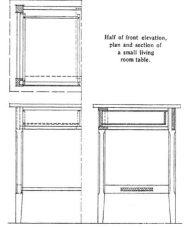

Half of front elevation, plan and section of a small living room table.

tion, so they were left out. No measurements are given on these drawings as, for the purpose of this article they are not necessary and they might have been more confusing than enlightening.

In taking off stock bills it is necessary to have a system. If the different parts are measured at random and noted on the stock bill it is difficult to check the work. The system followed by the writer is to start at the top and work down. If the same system is followed all the time it is a simple matter to find out where you left off if you happen to be interrupted in your work. In taking the length of the top, case rails and shelf from the front elevation and the length of the back rail from the plan it will, of course, be necessary to double the length shown on the drawing.

It might be mentioned that the form of construction shown is about as cheap as it is possible to use, but nevertheless, it is precisely the way the table would be constructed in the average furniture factory. The back rail, the two end rails and the bottom stretchers are dowelled into the legs. The two case rails are tenoned into the legs. Two pieces are screwed onto the end rails for the drawers to slide on. The lower case rail and the back rail are grooved on the saw to receive the centre drawer bearer. The drawers are machine dovetailed at the front and nailed at the back. The top is secured at the front by screws through the top case rail; at the ends by pieces which are screwed to the end rails and also to the top, and at the back, screw holes are bored on the angle through the back rail. The shelf is dowelled into the two end stretchers.

The Characteristics of Different Woods

Their Adaptability for Furniture Making—A Few Remarks on the Proper Handling of Gum—Use of Birch Could be Increased by Advertising its Beauty

Staff Article

There are certain staple woods which are used year after year for the manufacture of furniture. They have been tried and tested and found serviceable, and manufacturers are dubious about making experiments with any "outsiders." Then, again, there are certain woods that show up well in some particular style of design. Thus we find oak for Jacobean, mahogany for the Adam period, and so on. Walnut appears to be the favored wood for the William and Mary designs, although mahogany is also being used extensively, and even oak is finding its way into this particular style.

The suitability of mahogany for furniture making has been dealt with so often that it seems somewhat superfluous to deal with it here at any length. The varieties of mahogany that are mostly used to-day are Honduras, Mexican, Cuban, and African. Honduras is a favorite. The Cuban is used for manufacturing chairs and other articles which require small pieces of good strength. African mahogany, generally speaking, has a nice, even figure. It is rather coarse, however, and is mostly used on medium class goods. It is not very plentiful at the present time on account of the difficulty of getting ships to transport it and the high freight rates prevailing.

Mexican mahogany shows a high figure and has a lustre all its own. The consumption of Mexican mahogany has been increased since the beginning of the war, due in part to the difficulty of securing shipments of other kinds. White mahogany or primavera, as it is mostly called, is so seldom used now that it is scarcely worth considering. About the only way in which white mahogany differs from other mahogany is in the color. The texture and grain of the wood is practically the same. The features that make mahogany suitable for furniture making are its pleasing color and figure, its working qualities, and durability.

Next in importance to mahogany as a high-class furniture wood is walnut. One thing the writer has never been able to understand, although he has spent a good many years in piano and furniture factories, is why some people persist in calling this beautiful wood "black walnut." Even some mechanics that should know better continually call it by that name. This practice does walnut incalculable harm. When persons not posted on woods hear the expression it settles in their mind, and the next time they see a piece of black walnut, perhaps imitation ebony, they decide at once that it is "black walnut," and they come to the conclusion that they do not care much for it. Then the next time they are in the market for a piece of furniture and the salesman suggests walnut, they tell him they do not like walnut, and the matter drops there.

Quite a lot of walnut is very dark in color, but it most certainly is not black. If it were simply called walnut without adding the prefix "black," walnut furniture would be judged on its appearance, and thus stand a better chance of winning favor with that section of the general public who are ignorant of its beauties simply because they have not had them pointed out to them. As far as the working qualities of walnut are concerned it differs very little from mahogany. The wood is perhaps a trifle harder, but it responds readily to tools.

Oak is still used in large quantities for furniture. It is a very reliable wood, and particularly suitable for furniture that is in constant use, such as dining-room, living-room, and den furniture. The best grade of oak is quartered white oak, but of late years quartered red oak has occupied a prominent place in the manufacture of furniture. The latter is quite satisfactory when the finish is to be golden oak, but for the fumed and Jacobean finishes quartered white oak should be used if possible. For cheaper grades of furniture, however, the price of quartered white oak excludes it and quartered red oak has to be substituted. Plain sawn oak has been coming into use quite prominently of late for bedroom furniture and for certain parts of living-room chairs.

On the cheaper and medium grades of the latter the custom is to use quartered oak for the most prominent parts, such as the arms and front legs and plain oak for the other parts. For any parts, of course, which are covered by the upholstering, maple, elm, or some other common wood is used. The writer would much prefer to see a piece of furniture made entirely of plain oak in either the golden or fumed finish than a piece made in "surface oak." Oak is an open-grained wood, with a very pleasing figure. It is easily kiln dried and presents no difficulties in manufacturing. A great variety of finishes are obtainable by subjecting it to different treatments, such as staining, filling, waxing, etc.

Gum lumber has come to the front wonderfully during the past couple of years. It has several features which make it valuable for furniture making. It can be obtained in large, clear boards that cut to good advantage without waste of time on the part of the stock-cutter. Some of it is beautifully figured and makes choice furniture when finished in the natural color, or if without figure it can be stained either walnut or mahogany. The red gum tree produces both sap wood and heart wood. Commercially the term "red gum" applies to the heart wood of the red gum tree. Unselected gum or sap gum may be partially sap wood or all sap wood. Some furniture manufacturers have in the past experienced difficulty in drying gum lumber so as to prevent it from shrinking and opening up the joints after the furniture has been assembled.

The fact that gum is now being handled successfully by a great many furniture manufacturers would seem to indicate that the difficulties have been due more to unsatisfactory drying methods than anything else. It is a difficult matter to lay down an exact formula for kiln drying gum lumber, or, for that matter, lumber of any kind, for the reason that the application of the requisite heat and circulation must be properly regulated throughout the entire process or trouble is almost certain to result.

Moreover, woods of different shapes and thicknesses are very differently affected by the same treatment. The tissue composing the wood, which may vary in form and physical properties, exerts its own peculiar influence upon the behavior of the wood while drying; therefore, to kiln dry lumber of any kind properly the operation requires an experienced kilnman who knows how to reason these different questions as they arise. The following simple rules will aid in keeping the lumber in proper condition:

The lengths should be kept separate—that is, 12-foot lumber should be piled on one truck, 14-foot on another, etc. Stickers should be placed directly over each other. There should be at least seven stickers to a 12-foot length. An air space of about two inches should be left between each board. From this point on through the kiln it is difficult, as already stated, to lay down an exact formula, but an experienced kiln man can successfully handle it from this point on, and it is safe to say that about the same methods should be employed as when drying birch lumber, care always being taken not to dry the lumber too fast. Experienced men have stated that the maintenance of a high humidity in drying gum lumber is necessary for keeping it straight, and drying should start at about 90 per cent. and a temperature of 125 degrees F., and the condition when drying is completed should be a humidity of from 40 per cent. to 50 per cent. and a temperature of 175 degrees F.

After the lumber is thoroughly dry it is a good idea to take it from the trucks while it is hot and pile it in a solid pile on a firm foundation in a dry shed, and leave it for about ten days before using. Air-dried sap gum, one inch thick, should remain on sticks about six months before going into kiln. Air-dried red gum should remain on sticks about eight months before going into kiln.

Birch has been used for some years as a furniture wood, but the most of it has been finished in imitation of mahogany. During the past couple of years suites made of birch (mostly bedroom suites) have made their appearance from time to time in some of the leading Canadian furniture stores. Why they have not met with a greater measure of success is somewhat of a mystery. Birch is a splendid furniture wood, but perhaps the fact that it has masqueraded for so long under false colors has acted against it. The writer is inclined to the belief, however, that it is due more to indifference on the part of the manufacturers of birch than anything else that it has not come to the front. They have not advertised it to the public as possessing a beauty of its own which makes it particularly suitable for furniture. Birch finished to represent mahogany has not been a great success. In the first place, it is manufactured to sell at a lower price than mahogany, and for that reason no great pains are taken to give it the proper color because if the imitation were too good it might spoil the sale of higher-priced furniture. Also, the finish being inferior causes the light-colored edges to show through in a very short time.

The general public are to-day being educated in the matter of design through the "movie" films, illustrations in the advertising sections of the daily papers and popular magazines and various other sources. If the manufacturers of birch would co-operate and create a demand for birch, finished as birch, by inserting attractive advertisements in some of the journals that find their way into the homes of the people, the latter would begin to enquire for birch furniture at their dealers, and furniture manufacturers would not be slow in catering to the demand. The advertisements could be gotten up by showing illustrations of simple designs, such as modified Sheraton, etc., and pointing out the admirable qualities of the wood.

A Near Ideal Packing and Shipping Room

Staff Article

The writer was in a furniture factory some time ago which bore evidence of the fact that the man who designed it was no amateur in the furniture business. One or two features in the room where the trimming, packing, and shipping is done are worth passing on to the readers of the "Canadian Woodworker and Furniture Manufacturer." The accompanying drawing, which is a section through the lower two floors of the wing in which the packing and shipping rooms are located, together with a few remarks, will be sufficient to show how the furniture is handled once it reaches this department.

It might be well to state first, however, that the lower floor is situated below the ground level. The railway siding passes up the side of the building, and is indicated at D. The lower floor, A, is the cabinet

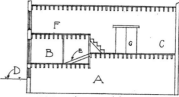

Chute from Packing Room to Shipping Room

room. The machine room is also located on the lower floor, but in the main portion of the building. The low ceilinged part of the cabinet room is used for piling drawer stock and assembled cases. The floor marked C is where the trimming and packing is done. The upholsterers and the girls who pad the chairs, etc., for the packers are located on the floor marked F. An elevated track, built on a level with the floor C, extends out into the yard, where the crating lumber is piled. The cars of lumber are run into the packing room, where a power feed ripsaw and a large swing saw, located at the end of the room near the door, convert it into the required sizes. Racks are provided, in which stocks of different sizes are kept.

When the furniture is trimmed it is crated and sent down the chute E into the shipping room B. Thus it is entirely out of the way, and the floor B, being on a level with the floor of the railway cars, it is ready for shipping just as soon as the cars are placed. There is space in the shipping room to accommodate a large stock of crated furniture, so that one shipment need not be placed in the way of another.

The chute has a partition on each side and doors at the top in the packing room, which can be shut to prevent a cold draught coming up when the shipping room doors are open for loading the furniture into the cars. Steps are provided at the side of the chute for the convenience of the shippers.

Designing Furniture to Sell at a Price

How to Obtain Best Results from Artistic Point of View—Designer Must be Conversant with Manufacturing Processes—Keeping Costs Down to a Minimum

By Walter Wrench

Walter Wrench.

All furniture is designed to a price, from the most expensive cabinet work to the cheapest grades of factory furniture, and each grade calls for careful consideration and forethought on the part of the designer, in order to obtain the best possible results and yet keep within the stipulated price at which the suite is to sell.

In commercial furniture produced in batches the designer is constantly scheming and experimenting, often in actual practice, with the sole object in view of producing good-looking furniture — attractive and styles-getting, yet economical to manufacture, both in material and labor. There are many points to watch—many labor-saving "stunts" which experience alone teaches, the mastery of which make the designer worth his salary in reduced manufacturing costs. It is a fact that one hour's thought at the drafting table will save hundreds of dollars in the machine room alone.

Nowadays furniture design seems to mean designing suites cheaper than one's rivals in all his grades, and yet designing suites which lose none of their artistic appeal by reason of this cheapness. To this end there are a number of basic principles which must be observed and the considerations of which I shall treat under separate headings.

In case goods—that is to say, dressers, chiffoniers, toilet tables, and the like, in bedroom furniture, and sideboards and china cabinets in dining-room furniture, and, in fact, in all cases where they are used—mirrors are a basic factor in design, i.e., commercial furniture design.

The Importance of Mirror Sizes.

It is upon the size and price of the mirror that all such cases are based, particularly since the enormous increase in price of mirror plate consequent on the war. The designer has received his instructions to design, let us say, a bedroom suite with a given size of mirror for his dresser. He proceeds then with the dimensions of that mirror as the base on which his design is to be built. The mirror naturally limits the size of his case to a certain extent. In order to economize on the size of his mirror and still have a well-proportioned case the designer resorts to various methods to obtain this result. Let us suppose that a 24 by 30 mirror is to be used on a medium-grade dresser. The design is to be a moderation of the Adam period and consequently the mirror frame would have to be light, certainly not more than one inch and three-eighths wide if moulded, and less if plain.

If an inch and five-eighths standard is used, and one inch allowed for the projection of the case top, a case 38 inches in length is the result. Should the desire be to have a case of more imposing length, this may be accomplished, to the improvement of the design, by the method illustrated, Nos. 1 and 2. This method not only gives the designer scope for the introduction of those little features which "make a case," but also materially add to the balance of the design.

The 24 by 30 mirror is often used the reverse way, to give height necessary to the usefulness of the dresser. When this is not done, as cited in the first case, it is necessary to give a rather high base rail to obtain the minimum height of case. The method of treating that high toilet base rail will either make or mar the design, for nothing looks worse than to see a plain,

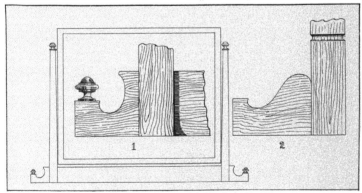

Method of lengthening a case when proportioning it from the size of the mirror.

heavy-looking base under a slender toilet-frame rail. A method of overcoming this defect is shown in No. 3.

In better grade furniture, economy can be effected in the size of the mirror in its relation to the case length by using the wall mirror method, where the mirror frame is hung without standards on the wall behind the case. This permits the case to be one-third longer than the width of the mirror. With sideboards, length of case in relation to the mirror may be regulated at will by the size of panels at either end of mirror, as shown in illustration No. 4.

The Case Itself.

The designer, of course, has the manufacturing costs in mind all the time, and when the size of the dresser or sideboard has been fixed by the dimensions of the mirror he turns to the consideration of the artistic side of the proposition. At this stage of the procedure it must be understood that the sketch design has still to be made, and it is upon this that he now works. Later, further economies must be effected in the drafting of the working designs.

The number of drawers put into a case depends upon the grade of furniture being designed. If it is a 42-inch dresser there will rarely be more than four drawers—two drawers in the first tier and two full-length drawers. Now, the designer reasons to himself if work may not be saved on the drawers without the sacrifice of any feature valuable to the selling possibilities of the case. If the two drawers at the top can be combined in one, making three full-length drawers, so much ripping, planing, and dovetailing will be saved, not to mention one drawer less to be assembled and fitted. To most people a divided single drawer will serve the purpose just as well. The money saved by this economy may be used in part to decorate the front of the top drawer, and experience has taught me that a single divided drawer, decorated by some panelled or fluted effect, is not only more economical to the manufacturer, but sells better than two plain small drawers. The decoration requires to be done intelligently, though.

Decorative Details.

The designer at work upon his sketch of a dresser, having decided the spacing of the drawers, turns his attention to the decorative details which are to add

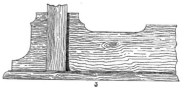

How to improve the appearance of a heavy back rail.

character to his design. Most important of all, perhaps, is the foot of the case, turned or tapered as the estimate of selling price may warrant. A few years ago designers of cheap or medium-grade furniture were practically forced to use the straight tapered foot for reasons of economy, but now that automatic turning lathes have been installed in most factories, turned legs have become almost as cheap to produce as the simple, straight-tapered leg.

If he should decide upon a turned leg, the designer will be careful to refrain from any turning which presents difficulty in sanding. When on cheap work one is limited to simple effects, for any turning containing too many members means additional cost in cleaning up. For this same reason the case-top moulding, if one is used on the cheap case, should be so designed as to eliminate, if possible, any hand-sanding. During my experience on this class of work few mouldings have been so pleasing to the eye as the simple one illustrated at No. 5. It was found that this bead—for it is no more than a bead—if carefully run past a well-balanced and sharpened shaper-cutter required little or no sanding at all. The secret of the charm of simple mouldings—and, indeed, of all mouldings—is a knowledge of the value of the shadows which the receding members will cast, and of the value of the high lights which are caught by the protruding members. The mouldings Nos. 6 and 7 well illustrate this point, and both of them have been used with satisfactory effect by the author.

It is absolutely essential to successful designing that a complete knowledge of the capabilities and fail-

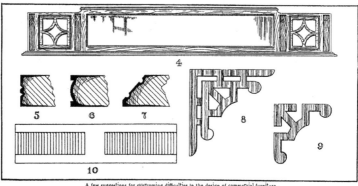

A few suggestions for overcoming difficulties in the design of commercial furniture.

ings of every machine used in the furniture factory be known to the designer. This is particularly true where one is working on competitive goods designed to a price. It has been pointed out how the decorative features of foot and case top must be modified or allowed full scope, according to the capacity of the machines which are used to obtain the desired effects. Here is an example of how this knowledge is put to account:

A dresser in the Chinese Chippendale motif is being designed, and, in accordance with the spirit of the style, a fret bracket is to be introduced at the angle formed at the junction of the base and foot. The designer then, knowing how much more expensive it is to do jigsawing than bandsawing, designs accordingly. His first idea, let us assume, is bracket No. 8, which he later modifies, as shown in No. 9, because of the cheaper cost of production of this bracket, due to the elimination of all jigsaw work. Note that while the jigsaw

Dresser designed by the Author.

work was obviated the true fret effect was obtained at the same time.

Knowledge of machines guides the designer in his choice of mirror frames and their shaping. Sometimes a mirror frame is patterned to have "round corners." The quarter circle is fixed as to its minimum radius by the designer's knowledge of the size of the smallest shaper head. Familiarity with the diameter of the sanding drums also is a deciding factor in the drafting of the sweep of curves in all bandsawed work.

Study of the capabilities of the "sticker" rewards the designer for his pains, particularly in cheaper grade work. Many pleasing effects, cheaply produced, are made possible by the use of this machine. After a few years' experience the designer of commercial furniture comes to weigh every motif of his design as to how much it will cost to produce in the machine room and

how much work a particular feature will call for in the assembling shop, even when making the first rough sketch design.

Chairs Designed to a Price.

The manufacturer of case goods who also runs chairs "en suite" takes care to standardize them as far as possible. With cheap lines the same chair has to do service with several "period" suites. This is made possible by means of slight alterations to back and legs. In the early days of my Canadian experience nothing hurt my conscience more than this same standardization of the period styles in chairs and cases, but the fact that this standardization has made possible good, cheap furniture for the working man does much to attone for what at first seemed little short of sacrilege to me.

Chairs designed to a price, then, must be simple of line and construction. The periods best adapted to cheap chairs (that is to say, as we simplify them in our modern adaptations) are, without doubt, Adam, Sheraton, and certain types of Chippendale, but emphatically not Louis XV. or XVI. The Mission style of chair is, of course, the cheapest of all to manufacture, and is excellently suited to certain purposes, as living-room, den, or club furniture.

Enrichments.

The introduction of composition and pressed fibre carvings was a boon to the designer working on furniture the price of which is fixed by competition. Many excellent carvings are now obtainable, and, used with discrimination, will add a charm to any case or chair when carefully worked in. An abuse of this type of cheap decoration is found where the "plastered-on" effect is used. It is very noticeable on some suites, and is evidently the work of inexperienced designers. Often it seems to me that a case has been taken and the carvings just piled on, regardless of balance or the rudiments of good taste. The correct way to use these fibre carvings is to design the case, as it were, around them, following out the motif of the selected carving in moulding and turning of the case so that all the various features blend in one harmonious whole. This may truly be called designing; the abuse referred to is just the "botching" of amateurs.

Where a little latitude is permissable in price it is an advantage for the designer to work out his own designs, which allows him to put his own personality into his work at a slight increase in cost. For most purposes, however, the ample range of stock designs carried by the manufacturers of this class of enrichment meet every ordinary requirement in a satisfactory manner.

The writer was interested to examine specimens of a type of composition carving which is rapidly gaining favor in Canada. This particular composition seems to be identically the same as that used by the Brothers Adam, which famous designers, by the way, were among the first to experiment with this mode of decoration. Carvings of this composition are remarkable for the sharpness of the most minute detail, and, as has been proved by the test of time in the case of those used by the Brothers Adam, improve with age.

One of the most pleasing, and at the same time inexpensive, forms of decoration, suited to drawer fronts and chair front rails and backs, is the pressed beading in panelled effect, as shown in No. 10. The pressed beading is cheap and can be obtained in any reasonable width. In the example shown, the groove was cut clean through, and the panel in the centre glued in afterwards. It was then flanked on each side by a section of beaded moulding. The effect produced is, in my estimation, better than that produced by fluting, which is sometimes used in its stead.

Economy in Curved Parts.

When curved parts, such as chair arms or pedestal bases, and mirror frame top rails are used in furniture designed to a price which allows only a small margin of profit, the designer will endeavor to plan them so that they fit the one within the other, so as to economize in lumber. While the expense thus saved in cost of material is a fraction compared to the saving effected by a motif designed to give the best effect with the minimum of machine work, yet it is attention to just such small details as this that ensure the final profit.

I have on a number of occasions been able by experimenting with my patterns to fit the top rail of the mirror frame of a shaped toilet table within the dresser top rail and the chiffonier top rail within that again, so that there was practically no waste at all. On pedestal bases where the lumber is of considerable thickness this economy is particularly worth watching.

When the sketch design has been accepted, the designer proceeds to put into practice upon his working draft all the economies which his sketch has suggested to him. Apart from the artistic side of designing furniture to a price, the method of construction is the big factor in the production of furniture of this class. As every factory has its own method of construction, differing more or less the one from the other, it would be useless for me to dilate upon my own favorite method, since every superintendent is quite sure that his particular method is the best. When all is said and done, perhaps it is, for a gang of men trained to one method would be likely to lose more in getting used to a method unknown to them than the innovation would be worth.

In closing, I can but add to my suggestions regarding the designing of furniture to a price that it is to methods of economy whereby artistic features may be added to a design without unnecessary labor expenditure, to which the aspirant to designing success should devote his attention rather than to economy in material, which is a small item by comparison. To this end, a close study of machines and methods is not only to be advised, but is absolutely essential to attain the object in view.

Making Furniture in an Up-to-date Plant

An Illustrated Article Showing Arrangement of Machinery and Describing the Methods of Converting the Raw Material into the Finished Product

Staff Article

Before proceeding to describe the operations of the George McLagan Furniture Company at Stratford, Ont., it might not be out of place to make a few remarks about Stratford as a furniture town. It is the home of a number of progressive factories representing practically every branch of the industry, and the spirit of co-operation that prevails between the different factories augurs well for the future of the city and the industry.

One of the points where this co-operation is most noticeable is in the shipping arrangements. If one factory is sending a shipment of furniture to a certain point, (mostly long distances) and there is not sufficient to fill the car, the other factories are communicated with and told how many cubic feet of space is available. In this way it is possible to make up carloads and thus exercise the greatest economy in freight charges.

The factory of the George McLagan Company is the largest in Stratford. Further than that, it is one of the largest and finest equipped factories in Canada. One wing of the building is 300 feet by 50 feet and the other 320 feet by .60 feet, all four storeyed. A study of the accompanying diagram will convey to the reader some idea of the convenient arrangement of the different parts of the factory and dry kilns, etc.

As one of the first essentials of a successful enterprise is a well and carefully designed product, we will give an outline of how this end of the business is conducted at the McLagan Company's factory. An expert designer is employed whose duty it is to work out new designs. From these the most likely ones are

Rear view showing kiln—The Geo. McLagan factory.

selected and the designer makes up a working or detail drawing. A stock bill is then made from the drawing and a single piece of furniture made and set up dry or partially finished for inspection. This is called the sample, and there is a special department in the factory that works on nothing else but samples and patterns.

If there are no changes to be made in the sample, it is finished up and stained and filled. It is then sent to the photograph gallery to be photographed, after which it is numbered and catalogued and photographs sent out to the company's travellers.

When the details of manufacture have been decided on the foreman of the sample and pattern department makes his layouts and rods, corrects the first sample stock bill and has the new ones made from it. Each department is provided with files on

which are photographs of the company's entire line. To economize on the cost of photographs the ones used in the factory are made on blue print paper, such as is used for architect's drawings. It is considerably cheaper than the usual photographic paper and as the men in the factory are not in the same position as the dealer, on whom a good impression has to be made, the blue print paper is quite satisfactory. When the stock bills are ready the jobs are billed out as required, and the lumber prepared ready to manufacture them.

From the yard the lumber is run on cars into a continuous dry kiln of 180,000 feet capacity. It enters into the spraying or steaming chamber and passes on through to the other end, when it is thoroughly dry and ready for the planer. The length of time the lumber remains in the kiln depends on what kind it is and

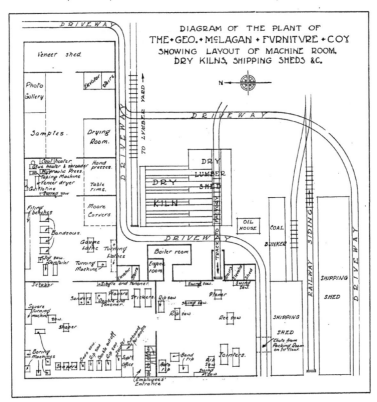

DIAGRAM OF THE PLANT OF
THE ⋆ GEO. ⋆ McLAGAN ⋆ FVRNITVRE ⋆ COY
SHOWING LAYOUT OF MACHINE ROOM,
DRY KILNS, SHIPPING SHEDS &C.

Cabinet department at the plant of the Geo. McLagan Furniture Company.

Part of the machine department in the Geo. McLagan Furniture Company's factory.

what condition it is in when it goes in. It generally requires about a week, however, or a little longer.

After being taken from the kiln the lumber is run into the factory on trucks as required, and put directly through a special planer to show the grain or placed at the several swing saws by means of overhead tracks and block and tackle pulleys.

The swing sawyers cut it roughly to the lengths marked on the stock bill, and in the case of wide stock such as tops, gables, etc., it is then ripped and matched. After being matched the stock is tongued and grooved and sent to be glued up. When the gluing has been completed and the rough glue scraped off with scrapers, the stock is passed through automatic

buzz planers which surface or true up one side ready for the planers and stickers.

The different jobs or parts of jobs are kept on separate trucks from the time they leave the swing saws until they reach the cabinet room, the stock tags which are written out for each part accompanying them.

After the different parts of the stock have received the necessary operations at the planers and stickers, their course is somewhat varied. The case tops, gables, etc., are sent to the jointers to have one edge jointed, then to the rip saws to be ripped to a width, and in the case of moulded tops, they go to the shapers to have the mould run. From there they go to the scraper and finally to the sanders. Round extension table tops are handled in pairs. They are jointed on one edge and then bandsawed, and if moulded, sent to the shapers. They are then scraped and sanded the same as the case tops.

Turnings for both the hand lathes and square turning lathes that require thickening pieces are sent to the veneer room to have these glued on. The square turnings are cleaned up in the cabinet room as much as possible by various sanders, and the rest by hand.

In the cabinet room, which is located on the second floor, the jobs are assembled, examined by the foreman and sent to the finishing room on the third floor, or if not to be finished immediately, to the stock room on the top floor.

All the goods are sent to the stock room after being finished and varnished, where they are piled away and are taken from here by the stock keeper as the orders are filled. Each order is picked out and sent to the rubbing room to be rubbed and polished, and thence to be trimmed and packed ready for shipment. The trimming and packing room is one the second floor and a chute (indicated on drawing) is provided for sending the crated or packed furniture down to the shipping sheds.

The equipment of the veneer room is very complete. The veneer is cut to the required sizes on a swing saw and "guillotine." Where it is necessary to joint narrow stock it is put through a taping machine. From there it goes to a drier and then it is ready to be put on the core stock.

The McLagan Company crossband practically all of their veneered work with 1/28 inch poplar, and lay the crossband and face veneers at the same time, the crossband being put through a glue spreader. They veneer buffet and sideboard tops 60 and 66 inches long in mahogany and walnut and have no difficulty from the crossband wrinkling. The cauls used are three-ply birch. They are made with one-quarter inch core and three-sixteenth inch outside veneers. The ebanding of the table rims is done in one corner of the veneer room as indicated on the diagram.

Mr. Geo. McLagan, the president of the company, is known and respected by furniture men all over Canada. He understands the furniture business from A to Z, and is at home in any part of the factory or the office. The active management of the business is in the hands of Mr. D. M. Wright, who has proven himself a capable executive. The manufacturing end of the business is under the supervision of Mr. F. G. Scrimgeour. Mr. Scrimgeour has had wide experience in the manufacture of furniture and has surrounded himself with an efficient working staff.

Part the Planer Plays in Reducing Costs

First Essential is a Good Machine—Stock Should be Faced Before it is Sent to the Planer—Operator Should be Man Possessing Accuracy and Speed

By S. E. S.

In taking up the above subject I am glad to inform the readers of the "Canadian Woodworker and Furniture Manufacturer" that I am a "wood butcher" myself, and have a few things to say that I have learned from experience, and which I try to practice. I do not doubt but that some readers hold opinions contrary to mine, but as I have handled men and made furniture for eighteen years my experience may be of interest to some.

Probably in no other industry will be found so many different ways of doing the same thing as in woodworking establishments. In factories turning out the same kind of goods, various methods are employed to attain the same results, the method of procedure being usually developed by the most competent person around the plant.

Now as the object in view is to reduce cost to a minimum in planer work, we must agree that there is no profit where there is not system. Many a foreman and manager—and I hope they are becoming scarce—look on a planer as a machine with knives and rollers to feed the stock through, and as long as it comes out at the back without much trouble the planer is all right. It may be dipping on the ends or the stock may be thin on one edge or thick or thin in the centre. We hear them remark "why it's all right, the sander will fix it."

Good management in a furniture factory simply means the best way of doing things. Now let us have a system to our planer work. I will take this up under three different heads, each one leading into the other, and will take up the sanding proposition in a later issue. First comes the preparation of stock for the planer; second the planer and its operator, knife grinding and setting, and operation; third, the results.

Now let me take the readers back to the breaking out department. Let us go and see this man, whoever he may be, matching up some quartered oak tops. Does he lay them on the truck haphazard? No, he will, if he has been properly trained, lay those pieces down, matching grain and color. Follow the same stock to the undercut planer, where it is to be faced before being glued and jointed. Here we see the operator again matching grain for no other reason than to have the grain all run the same way in each piece in the top.

Having the stock faced before gluing gives the jointer man a chance to do good work. If the pieces were not faced we would have tops perhaps zigzagged, hollow, or whatever way the material happened to be. The stock is, of course, worked from the face side marked by the ripper.

We will suppose that the tops are jointed for the planer. Now here is a point where I may find criticism, as I think I hear some saying that facing stock before gluing does not pay, or that one-inch lumber in its irregular thickness will not stand facing. I can only ask the reader to try it. The next thing is to examine the planer. I maintain that every modern plant should be equipped with two planers, one single cylinder for roughing and one double cylinder for use entirely as a finishing machine.

In spite of the positive claims of the various machine makers, the main difference between machines of different pattern is one of convenience rather than principle. Many of the older makes of planers were just as solid, correct in speed and accurate in the quality of the work turned out as any of the up-to-date machines. Makers have been experimenting too long with light frames in their hurry for cheapness, and are again back to the heavy frame, larger mandrels, increased friction surface in journals and pulleys. All these go to make up the vital improvements in the latest machines.

If the planer man fails to turn out satisfactory work it may be because the planer is not right, and he is being asked to do the impossible on a machine that should have been scrapped long ago. I will only deal here with a first class machine, there being several of these on the market. We all know that in a smoothing planer the greatest difficulty lies in adjusting the bottom rolls of the bed plate so as to get smooth running without getting the work so tight on the bed that it is impossible to feed it through.

This depends on the use you put your planer to. If your stock has been faced off so it will lay flat, the rollers may be worked closer to the bed. If not, you are obliged to run your bottom rollers a shade higher in order to assure a steady feed. Too many of us do not give the proper attention to the preparation of our stock and expect the planer to do everything, and if it doesn't turn out good work blame it on the machine or the operator.

Thus you will understand that if we are not careful in matching the pieces in our tops and run the grain both ways our planer, no matter how well it is running, is bound to tear a little when cutting against the grain. We will now suppose that our stock has all been carefully faced. We should then see that the cylinders are adjusted in their bearings to run free without vibration. Have all gearing and working parts carefully adjusted and oiled and then proceed to set the knives. The writer prefers to whet knives and balance them accurately each time they are ground before putting them in.

In whetting it is a good plan to turn face of knife back a little. This gives you a thicker edge and at the same time does not affect the clearance. Use a short parallel bar, say 5 or 6 inches long by 3 inches wide by 1½ inches thick, and always set your knives off the bed, moving block back and forth from end to end until both knives are touching alike. On roughing planer, set knives out ⅛ inch, as you will often have heavy cuts in roughing uneven stock, but on your cabinet surfacer set as close to cylinder as will permit your general stock to work nicely; one-sixteenth inch is plenty.

The knives being set, see that the pressure bar is the same on both ends, and, of course, use as much pressure as you can without sticking the stock. With the feeding-in rolls ⅛ inch lower than knives, and chip breaker working freely and set ⅛ inch below knives, there is no reason why we should not have a steady feed. Sometimes I have noticed the chip-breaker on some machines pinching or binding, being too thick on the heel, and instead of allowing stock to pass under freely causes incessant sticking. This is remedied by grinding off the thick part.

Many of the above suggestions only a mechanic would take a delight in figuring out. A good planer man is a mechanic having many of the symptoms of an artist. He should be a man very painstaking, yet alert and quick of motion, true in judgment, always having presence of mind, and above all things not afraid of work. In short, he should be accurate and able to turn out sufficient work to leave a reasonable profit for his employer at the close of each day. An operator possessing these two requirements is all that can be expected by any employer of labor.

Let the planer man have the first hour each day to go over his machine and touch up his knives, no matter what is waiting. Acquire the habit. Generally speaking, the planer man will perform his duties better when he sees his employer taking an interest in him and his machine by getting him good knives, good belts and providing proper working facilities, such as keeping the shop clean, well lighted and arranged, and of all things, having stock come to him in systematic rotation. We will now assume that we have a good man at our planer and a good helper, one that can detect any flaws in stock as it is being piled on his truck, have it set aside and repaired before the job goes up to the saws. The general results are these:—Clean and accurate work, stock that can be depended upon for even thickness for boring, shaping, sanding and veneering.

In closing these few remarks; does it not all depend upon the employer to supply the machine that will, when properly adjusted and taken care of, do the work? Is it not up to him if he wants profits to hire the right man and keep him?·

Wood Fibre Ornaments for Furniture

By J. Walter, of J. Walter & Sons, Manufacturers of Wood Fibre Products

Decorations for use in making furniture have been a vital point. The usual method has been hand or machine carving, which was more or less unsatisfactory from a point of cost and also design. It is very difficult to obtain a good wood carver that can turn out a good period carving at a commercial price.

Ornamentation made out of wood fibre(not pulp) is only of recent origin. The technical difficulties were many, both chemical and mechanical. The first stage in making a good ornament lies, mainly, in making a sound copper deposit over the carving by the galvanic process, which is an art in itself. The deposit is backed up in the usual way to stand very heavy pressure and wear and tear. The so-called mould ready, it is transferred to an automatic machine of special design that ejects the fibrous mass into the mould, under a pressure between six and eight tons to the square inch, according to the size of carving.

The reader might guess that there are a great many minor points which enter into the manufacture of this product which are kept shop secrets. To an onlooker it seems to be a very simple operation, but when this process was started, no mechanical or chemical data were available, and it had to be mapped out by the cut-and-dry method. Hundreds of thousands of dollars have been spent by various experimenters in making articles out of wood fibre, from a beer keg to a telephone pole, but the majority of them had to give up their wild dream when they found out what an obstreperous material wood fibre is. It does not lend itself readily for making into a plastic state successfully, as its capillary action for taking up water is so great, and consequent shrinkage in drying out after it is formed into an article makes it useless. This feature in itself has taken years of experimenting to make it commercially successful.

The wood fibre ornament as known to-day has many merits over hand carving. It can be glued and nailed and worked with any ordinary wood-working tools the same as wood. It can be used to-day or to-morrow, or in a year, and does not deteriorate with age. It is the closest resemblance to wood as to body and surface that the present day technical appliances will allow, and undoubtedly as time goes on great improvements will be made on this product as to its cost and otherwise.

Cost System for Furniture Factories

A Synopsis of the Actual Results of the Different Departments Can be Had Every Two Weeks, Enabling Manufacturers to Have Records of Output and Expenses

By Wm. Cawkell*;

The necessity of a reliable cost system is now admitted by every furniture manufacturer who has given the matter serious thought, but even many of these are too busy to go into this perhaps most important detail of their business as fully as they should. They put it off until they have more time. Other manufacturers look upon the suggestion of a system as a reflection on their knowledge of the trade, and in this they are often confirmed by their foremen.

I feel that I cannot too strongly urge upon furniture manufacturers to put in an adequate cost system based upon productive labor. The system I am going to outline, together with the check it provides will pay for itself over and over again. Remember that you don't know what the lack of a system is costing you. This is dangerous. Why not find out? A system such as described here can, after its installation, be handled by one competent girl.

First, there should be a perpetual inventory kept of lumber, much as one would keep a ledger account. The first entry is a debit of the lumber on hand. If more than one kind is used a separate account should be kept for each kind. The quantity and average cost should be stated for the first entry, and after that every purchase should be entered, giving the quantity and price per foot. All amounts used are entered as a credit, then the balance should be brought down every two weeks and the average cost price shown. Stock should be taken of lumber on hand every six months and compared with the quantity as shown by the inventory. This will prove whether the estimates of the quantities used have been substantially correct or not, as the lumber wastage varies as much as ten per cent., and again there is often a wide difference of opinion when figuring wastage.

A similar system should be followed in the case of all materials, including packing-paper and twine, and it should be remembered that these should always be figured as material cost and not as expenses.

Then in factory cost figuring, direct or productive labor should be the basis, because labor is the one thing that fluctuates the least, and this is the method that is acknowledged to be the most satisfactory.

The next thing to do is to divide the factory cost according to the several departments or stages of manufacture. In order to give a concrete example we will suppose the following divisions:

First, the Machine Department; Second, Cabinet Department; Third, Finishing Department; Fourth, Assembly Department and Packing Department; Fifth, General. Wages in the packing room should be classed as productive labor, the percentage on which will provide for the overhead expenses in that depart-

*Secretary Furniture Manufacturers' Association.

Wm. Cawkell.

ment, the same as the other departments as explained further on in the article. Some factories are not able to allocate the packing wages to any specific job, in which case the wages and overhead of this department are included in General Department.

Having decided upon the departmental division of your factory, the next is to ascertain the fixed charges, and to allocate them properly. They usually consist of: first, rent and taxes; second, insurance on property and fixtures, on stock and on machinery; third, depreciation on machinery; fourth, coal, heat and light. Some of these, such as rent, are charged according to space, others according to value. Then there are many points to be considered when these charges are allotted, which must be dealt with by the accountant putting in the system. In deciding on the amount of these fixed charges the figures for the previous years are, of course, available as a basis.

The first thing is to find the wages paid by each department for productive labor, and all other charges excepting selling expense. Then will be found what percentage on direct labor it will require to cover all the other charges. Cost of yard men and kiln expenses are often covered by adding an estimated amount to cost of lumber, but to make sure of the exact cost, I suggest that these wages be charged as non productive labor to the machine department, so that all wages in pay roll are charged either as productive or non-productive. You then get actual figures. Engineer and saw filers should be charged as non-productive labor to the machine department. Everything possible should be charged to the department which used it. The essence of this system is to departmentalize, and therefore every charge that it is possible to charge to one or more departments should be so dealt with. General expenses are provided for by a percentage on the total of all the productive labor.

Now for a continual report showing the results after every pay day. The report would read something like the example shown on the opposite page.

These reports, made out every two weeks—after each pay day—will show exactly what amount the percentage on direct labor had provided and whether the percentage allowed is sufficient to carry the various expenses. It will also keep tab on the expenses. After factory cost has been ascertained by adding together materials, direct labor, and a percentage on direct labor to cover overhead, a further percentage should be added to cover selling expense and 15 per cent. put on at this stage would about equal 10 per cent. on the selling price and that is what you have to consider. Then a further percentage will have to be added to this total to provide for profit; this percentage will vary according to what an article will carry.

Use of Time Cards

No article on cost systems would be complete that did not deal with this most important point, which no manufacturer can afford to evade, and I know of what I write. If you have not got this time card system in use, get it going, and it will open your eyes. These cards have every possible kinds of work that your men can do in your factory printed on them and each man must put a tick opposite the work he has done, mark the time taken, the number of the job, and his own number. Every workman fills in a card for every different job he works on; he may turn in

Cabinet Hardware and Upholstering Materials

There are numerous occasions in the daily routine of the average furniture factory when the foremen of the various departments present themselves at the office with the information that certain items are required in their departments.

The foreman of the trimming department may discover when the first batch of a new style dresser or buffet reaches his department that a different kind of hinge is required from those previously used. The

	Productive wages	Percentage on productive labor	Amount provided	Fixed charges	Sundry Expenses	Non-productive wages	Total	Deficit	Gain
Machine dept.									
Cabinet dept.									
Finishing dept.									
Assem. and packing dept.									
General dept.									
Totals									

REPO   No. 1   FOR TWO WEEKS   NOV. 18th, 1916

Office Mgr. Sig.

Supt. Sig.

Gen. Mgr. Sig.

Remarks

Net Gain

several cards in a day. These are assembled and checked against his registered time; then you know that this time has all been charged against some job.

After the cards are checked against the man's registered time, they are then put with the cost sheet of each job to be entered there. After you have had this system put in compare the time spent on work with the time turned in before you had this system and you will get a surprise.

When men are sometimes employed on saw sharpening or machinery repairing, a separate colored card may be used, and this card should also bear foreman's signature as a check. This time should be charged as non-productive labor to the machinery department.

It is impossible to deal with every point of accounting systems within the compass of this article, but if it has the effect of causing furniture manufacturers to install a suitable cost system based on a percentage of productive labor then my purpose in writing will have been fulfilled.

The details can usually be safely left to the cost accountant putting in the system, who will study the requirements of each particular factory, but the time cards are a very important feature and I recommend them to everyone because when they have been installed the time shown by them has been very different to the estimated time turned in before this effective and absolutely reliable method had been employed. There is no real argument against their use, and they have proved their value wherever they have been adopted.

Steamer Chairs

The Muskoka Navigation & Hotel Company, Gravenhurst, Ont., are in the market for 300 steamer chairs. Write them for particulars.

foreman upholsterer may find that he needs certain fittings which have not been used before.

The problem the office is up against is to find out just where they can purchase the desired articles. This very often necessitates a search through a pile of catalogues, and results in the loss of a great deal of time that the manager or his assistants can ill afford.

The "trade" will, therefore, no doubt appreciate an itemized list which has been prepared by Alfred Z. Smith & Company, of Winston-Salem, N.C., covering their entire line of hardware and trimmings for cases, tables, desks, kitchen safes and cabinets, also chair and upholstering supplies and finishing room and packing room materials.

The list is a convenient size, being printed on letter-size paper, and should be found very handy if placed in the desk drawer. Another feature that should appeal to the manufacturer is that Alfred Z. Smith & Company promise "prompt service at bottom prices."

Your Suggestions, Please!

Although self-satisfaction is dangerous, we assume the risk of expressing some satisfaction with this issue. We believe it will be of more than passing value to our readers.

Doubtless there are many ways in which the issue could be improved. Will each reader send us his suggestions? They will help us to make the publication more instructive, more valuable.

Read this issue thoroughly and keep it for reference as a buying guide.

Producing 100 Buffets in the Cabinet Room

Cabinet Room Should Not Start a Job Until All Parts Are Complete—Contract Work
Most Satisfactory to Manufacturer and Employee

By Albert Hudson

The cabinet room of a furniture factory is one of great importance. Reputations can be won and lost in this department and the cabinet maker should do his part faithfully and thoroughly, whether he is working by the day or by contract. All work should be completely machined before it is given out to the cabinet maker. This is an important factor, and it is the duty of the management to see that it is carried out.

Contract work in the cabinet room is the most satisfactory, both to the manufacturer and to the mechanics, and it is possible to do everything by piecework in this department. However, it is not always wise to insist that the workmen must take piecework. Some workmen would rather quit the job than work piecework. These men should be encouraged, and eventually they will begin to like the idea of piecework. I have known workmen of this class, and to-day they will not work by the day. They want piecework only.

It is not wise to give out work in the cabinet room unless everything belonging to the particular job is ready. This beginning a job before it is completely machined is one of the drawbacks to the cabinet room, as it causes delay and worry that could easily be avoided if the work was handled properly in the machine room. Skilful handling of the work in the cabinet room is also an important factor. Some factories adopt the sectional method of working in the cabinet room. Others have their men work in gangs of three or five, one of them being the leader.

For example; if a batch of 100 buffets having a low mirror back similar to that shown in the illustration, Fig. 1, is going through a cabinet room where sectional work is the practice, the different operations can be taken care of in a manner similar to that observed by the writer while visiting a furniture factory some time ago. The cases were clamped up with the case clamp and then pushed forward for the different operations in a systematic way until the buffet was turned out as the finished product as far as the cabinet room was concerned.

The layout of the cabinet room was somewhat on the principle of that shown in Fig. 2. The gables were assembled and placed in separate piles—rights and lefts. This enabled the man at the clamp to take up a pair of gables without any difficulty when assembling the cases. The case frames were cleaned up and placed in separate piles, making it possible for the man at the clamp to pick up the proper frames without groping around for them. The cases, after being assembled were pushed forward from bench to bench until they were completed. They were then taken to the finishing room.

The assembling of the drawers for these buffets was done by piecework, a certain price being paid for every hundred. All the operations, however, can be done by the piecework system, a price being paid for

Albert Hudson.

every section. It will be clearly seen that the man assembling the gables will be through and wanting a job first. It is therefore very important to have another one ready for him so that no time will be wasted.

This system can be worked out to good advantage on almost any kind of furniture. In a factory where the men work in gangs, one man being the leader, piecework is the most satisfactory, both to the manufacturer and to the men. By this method the mechanic contracts for the whole job, and is provided with helpers to assist him. He arranges the pay of his helpers, and it is up to him to get the work through satisfactorily, both to himself and to his employer. The mechanic should be able to cost the work up and lay it out in a systematic way so as to get the best results from his helpers and to pay himself, or, if working by the hour, to pay his employer.

For example, take the 100 buffets. The contract price of these are, say, $150 for the whole batch; this makes each buffet $1.50. The mechanic rates himself at 30c per hour or $3.00 per day. At that rate one buffet would have to be produced in five hours to ensure his pay. It is essential that a mechanic should study out for himself the different sections of the work in hand, and figure the time on each section. He will then be able to tell if he is working inside the price and will not have to wait until the job is finished to know how he stands.

Again, in the actual work there must be system, in order that the minimum time will be taken up. The writer would suggest that the following method be used for making the sections of the 100 buffets.

Buffet Base	Minutes
1 Assembling the gables	20
2 Cleaning up case frames	15
3 Assembling the cases	35
4 Screwing on the tops, cleaning up the case, nailing in the cupboard bottoms	45
5 Assembling the drawers and putting in the drawer bottoms	25
6 Fitting doors	30
7 Fitting drawers, nailing on the outside back, looking over the case	60
Total, buffet base	230

Buffet Back	Minutes
1 Assembling the buffet back frame	10
2 Cleaning up the square brackets and shelves on the top	25
3 Cleaning up the back frame	15
4 Screwing on the brackets and fastening on the top piece, or shelf	15
5 Looking over the whole back and screwing in place on the buffet base	5
Total, buffet and back	300

By estimating the work in sections in this manner it is always possible to know how much has been earned, even when only one section has been completed. It also enables the mechanic to know at intervals whether he is working inside the price. This is an important feature, as it is essential for him to know just how the time compares with the value of the work completed. In cases where the mechanic is provided with helpers, the system must, of course, be modified in some of its details, but it can be used, just the same.

For example: If the mechanic is provided with two helpers, it would be up to him to lay his work out to the very best advantage. This is only possible when the workman is thoroughly efficient, both in the quality of his workmanship and in his ability to lay out the work properly. If he estimated or rated himself at $3 per day and the two helpers were receiving $1.50

Fig. 1—Style of buffet to be made.

per day each, this would bring the total amount to be earned by the gang up to $6 per day.

The mechanic, having estimated how long each section should take, would be able to figure out how much work his gang would have to do each day according to the wages paid. It would then be up to him to see that $6 worth of work was done in the day. He should calculate the time of the helpers according to each section in proportion to their pay, taking his own time as a basis. The mechanic should expect the helpers to do a little better than just earn their pay, however, in order to compensate him for time lost in showing them how the work should be done, and making good their mistakes. This rule applies whether the work is being done by the day or the piece. If day

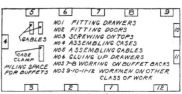

Fig. 2—Layout of cabinet room.

work, then the employer should receive the extra margin.

In Fig. 3 is shown how the benches can be arranged and the work allotted to the different workmen when making buffets.

It is wise to give the helpers similar tasks to do as far as possible. By doing so, both are in a kind of competition all the time and they will speed up themselves without feeling driven or overworked. The helpers will be through a little before the mechanic. It is then a good policy—and only fair—to give them some of the finer work to do. This is where the mechanic may lose a little time, but he need not lose by it financially, providing he handles the work systematically.

When the helpers know that there is a chance of them getting some of the fine work to do they will endeavor to get their allotted portion finished as quickly as possible. This, of course, considerably shortens the time on the job and increases the pay envelopes. The successful handling of helpers is an important feature of the work and it pays to encourage them. The costing and estimating of a job, and the systematic handling of the work so as to turn out the finished product as economically as possible can only be done by training oneself to it and by constant practice.

All kinds of furniture can be worked out on the same principle as the buffets. Cost out in sections and work accordingly.

The fire which occurred at the plant of the Hoffman Brothers Veneer Company, Fort Wayne, Ind., destroyed only one of their six veneer saws and a small quantity of veneers. The company advise us that the fire does not in any way affect the orders they have on hand or materially decrease their production. They had three new mills in course of erection before the fire and are in a position to meet the demands of the trade for veneers or domestic hardwoods, such as quartered and plain red and white oak, ash, cherry, hard maple, domestic black walnut, etc., also band sawn hardwood lumber.

Helper 2	Mechanic	Helper 1
Clean up frames.	Assemble cases.	Assemble gables.
Glue up drawers and put in bottoms at same time.	Fit drawers and doors.	Screw on tops.
Screw on top.	Nail on outside back.	Clean up buffet back and top shelf.
Clean up cases.		Fasten on shelf and fit up back.
Assemble buffet backs.		Fit drawers and back.
Fit drawers and doors.		

Fig. 3—Showing arrangement of benches.

Historic Carving Applied to Furniture

Details of Settee of Period of Louis XIV.—Decoration for Jacobean Furniture, of Rugged Type—Specimen of Grindling Gibbons' More Simple Efforts

By Lewis Clarke

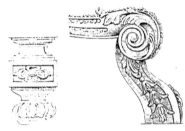

Fig. 1—Details of settee of Period of Louis XIV

To the student of furniture, be he designer or connoisseur or both, nothing is so pleasing or enlightening as the study of carving as a means of decoration, as it has developed through history. Carving is without doubt one of the earliest forms of decorative art, being employed by the Egyptians in conjunction with the painted forms of enrichment. To the savage tribes of Britain in the earliest period recorded by history, carving in its cruder form was known. The Greeks and Romans undoubtedly achieved in their carving of wood and stone an excellence and purity of form unexcelled to this day.

The master cabinet-maker, Chippendale, lays special emphasis upon the necessity of careful study of the classic Roman ornament upon which he declares the whole art of the design of furniture is based. The student of carving will readily admit the far-reaching influence which the motifs originated by the Roman and Greek sculptors and carvers have had upon the various styles of furniture, interior and exterior decorations throughout the whole of the civilized world in

with the more modern period styles, in order that this article may be of practical value to the modern manufacturer. For the benefit of those interested in design—and what manufacturer of furniture is not—the writer wishes to acknowledge his indebtedness for much of the success which has been his in the field of special order design and general commercial furniture design, to a careful study of carving. To those who wish to fit themselves for the profession of designing or to improve upon their ideas, the writer strongly advises a study of the historic examples of carving as one essential course of their education. To this end the careful sketching of authentic museum specimens is the best form of study, or failing this, a supply of books on the subject and intelligent study of them will help to give that knowledge of carving which is necessary to furniture design.

Perhaps it will be thought that too much importance is being attached to carving as it affects furniture. The writer recognizes that the all-important necessity for the success of a piece of furniture is to have it correctly proportioned, but the fact remains that most periods call for some carved decoration and this carving if badly worked in would spoil the most perfectly proportioned piece of furniture ever designed. It is a fact that many of the most highly valued pieces of antique furniture are badly proportioned, and are valued solely for the beautiful carving with which they are enriched. The manufacturer of today, having seen to it that the proportions of his case are well balanced should pay equal attention to the carved enrichments which add so much to its selling value.

It is recognized by the writer that modern factory furniture which must be sold at a price within the reach of all, must be sparingly decorated with carving, for carving is an expensive proposition. The illustrations which are given are extracted from a museum sketch book used by the author, and are of authentic carved details of three periods. These carvings will

Fig. 2— Scroll on settee of Louis XIV. period.

every country where craftsmen have wrought their creations in wood or stone. Practically every type of moulding, plain or decorated, is based upon the classic Roman orders.

After this brief reference to the classic origin of many forms of decorative carving it is as well to deal

suggest ideas for the decoration of modern reproductions of the styles they represent, if carefully considered and intelligently modified to simpler forms. In short, it is not suggested for one moment that these carvings could be used in their present form upon modern factory furniture; careful adaptation and sim-

Fig. 3
Carving by Grindling Gibbons.

plification will be necessary in order to make them practical for furniture sold at the present low prices.

The first two sketches, Fig. 1, are details of a settee of the period of special interest in view of Louis XIV., and will be of the fact that this period is rapidly finding favor in the furniture world, both in Canada and the United States. Louis XIV. is without doubt the coming period style. The drawing of the settee foot illustrates well the "bluntness" which characterizes this style. No doubt the composition c a r v i n g manufacturers could help out on the flower motif, leaving the furniture manufacturer to sink the panel into which it insets. The carved decoration of the moulding at the base could readily be simplified for shaper work.

As regards the arm, it would be essential that the outline be retained, while the carved enrichment should be modified. The scroll might well be plain; and it would look well, at that. The panelled effect beneath the scroll should be preserved, and if composition buds were used inset into the panel instead of overlapping, as in the original, practically the whole charm of the arm would be intact. It is not intended, neither is it advisable, that such a straight copy should be made. The purpose of these examples of authentic period carvings is not to encourage copying, but to provide the designer with good specimens of the various periods they represent. In designing along these lines the designer must remember, however, that to obtain the Louis XIV. spirit in his work he must observe the decorative motifs used in the period.

Louis XIV. carving is based upon the classic forms without doubt, but the interpretation is rugged, though not by any means coarse. It is a bold rendering of refined motifs—and this must be the thought uppermost in the mind of the designer when adapting or designing carving for furniture of this period. In passing it may be mentioned that the carved details illustrated were originally gilded, in common with the practice of gilding carved embellishments in vogue during the Louis XIV. period.

Fig. 2 shows a carved rail from an entirely different period. This is from a Stuart chair in the South Kensington Museum. Stuart, or as it is more commonly classified, Jacobean, is another instance of the rugged type of carved decoration as one finds it in this period. Its adaptation to modern requirements is a simple matter. Where it is desirable to use it in the form shown, all that is necessary to cheapen the cost of production is to omit the leaf motif and substitute plain scrolls. With care this carved rail could be worked up into a very attractive feature for a Stuart dining room suite. It would look well beneath a china-cabinet, sideboard or side table, where the rest of the design was blended in harmony with its simple beauty.

Fig. 3 is a direct contrast. This is a specimen of Grindling Gibbons' more simple efforts under the architect Wren. The motif is natural, no attempt being made to conventionalize the idea except at the head of the "swag." This carving lends itself readily to adaptation, and manufacturers will see in it opportunities for an attractive feature for a suite. To the writer it suggests unlimited possibilities as an enrichment and central motif for the door of a sideboard or chiffonier.

At least it will have served its purpose if it induces some enterprising manufacturer to step from the beaten track and try out a suite based on the style of Grindling Gibbons, the greatest of the historic carvers.

Composition Carvings

Furniture enriched with carvings, whether it be of classic design or otherwise will always be popular and the problem that confronts manufacturers is to satisfy the public demand and at the same time produce furniture at popular prices. The grade of furniture being manufactured, of course, governs to a certain extent the portion of the cost that can be appropriated for carvings.

On the other hand, if the cost of the carving can be kept down, more time can be spent in putting a little better finish on some other part of the furniture. Composition carvings, on account of their adaptability for furniture and their reasonable cost are rapidly supplanting wood carvings. There are two or three lines on the market. One of these, manufactured by the Decorators Supply Company, of Chicago, has been used extensively for furniture enrichment and for interior decoration in large buildings for a number of years. It is manufactured from a solution of whiting, glue and a number of other ingredients which are a trade secret. The process of making the models is inexpensive and furniture manufacturers have found it advantageous to use the product as they can carve their own original carvings—from which the models are made—to suit their individual tastes. They can thus secure an exclusive line of ornaments bearing the characteristics of their own designer.

Furniture manufacturers that do not employ a designer can submit a blue print or pencil sketch of what they require. The original carving is then made by the manufacturers at a nominal cost (no charge is made for the model). In this way the furniture manufacturer can secure artistic ornaments which he probably would not get if he were obliged to have the entire lot of goods carved by hand as he could not afford to have his carver spend proportionately as much time on all the ornaments as he would on the original carving. In other words, the furniture manufacturer or his designer can criticize the original carving and have it made almost perfect, and know that he will get the duplications exactly like the original article.

Another feature of the material is that it can be applied to curved and irregular surfaces just as satisfactorily as to flat surfaces. The ornaments are steamed slightly and glued directly to the raw wood. The steaming process is a simple matter. It is only necessary to stretch a piece of canvas across a steam box or a small galvanized pan and heat the pan sufficiently for the steam to ooze through the canvas. The canvas is then coated with thin glue and the ornaments

placed on the canvas. In a short period they will absorb sufficient moisture to make them soft, when they can be bent to almost any shape desired.

It will be seen that the composition ornaments offer advantages in the matter of applying them that cannot be had with wood carvings or others made from hard, non-pliable material. Another advantage is the matter of price. The Decorators Supply Company guarantee their product and state that the cost of any ornament averages about 10 per cent. of the cost of the same ornament carved in wood.

Large Furniture Factory in Quebec

Diagram Showing Layout of Buildings and Arrangement of Departments at Plant of
J. W. Kilgour & Bros., Ltd., Beauharnois, Que.

Staff Article

While the furniture industry in the Province of Quebec is steadily expanding, it is worthy of note that some of the oldest factories, by having continually enlarged their plants, are still to the fore in the race. The one we have in mind at present is that of J. W. Kilgour & Brothers, Ltd., at Beauharnois, which is a large plant, and probably the oldest in the province, dating back to the year 1863. At that date it consisted of a small frame structure, while to-day there stands a three-storey brick building of mill construction and having 125,000 square feet of floor space. It is exceptionally well lighted and ventilated, and is equipped throughout with an automatic sprinkler system.

There are three brick towers used for the freight elevators, one of which rises to a considerable height above the level of the roof and encloses a tank contain-

ing 25,000 gallons of water as a reserve against low pressure from the town pumping system in case of fire. There is also an auxiliary pump, capable of pumping 1,000 gallons a minute direct from the lake into the system, and always ready to start at a moment's notice.

The lumber used in the construction of the company's medium-grade dining-room, bedroom, drawing-room, and kitchen furniture is largely hardwood, and is carried in three yards—one at the railway siding, where all shipments by rail are piled; one at their wharf, to take care of that arriving by water; and a third surrounding the factory. In these yards will be found at all times between two and three million feet of lumber, air drying before it enters two spacious kilns with a capacity of about 100,000 feet. The one kiln is of the moist-air type, while the other is the

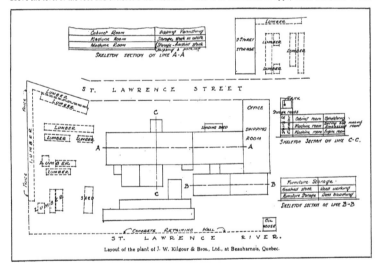

Layout of the plant of J. W. Kilgour & Bros., Ltd., at Beauharnois, Quebec.

standard dry type. After passing through the kiln the lumber is run out on transfer cars to a siding and allowed to cool for about 24 hours. After being taken into the factory it is first brought to thickness on a high speed planer, situated close to the entrance. Three swing-saws are used for cutting the stock to lengths, after which it passes on to the larger power feed band ripsaws, then to the automatic jointer, and so on till it is prepared for the cabinet makers on the third floor. The finishing and packing departments are on the same level as the cabinet room. A reference to the accompanying diagram will show the location of the different departments.

On the ground floor, and directly below the packing room is a department 140 feet by 50 feet, where all orders are assembled, after which the copy of order is handed over to the shipping clerk for attention.

The boiler room of the plant is located in a wing in the rear, and contains three boilers having a total capacity of 350 h.p. These are largely fed by cuttings and shavings, etc., the latter being conveyed from machine to fire automatically by the blower system. In close connection with this department are the machine shop and saw filing room, these being equipped as far as possible with automatic machinery.

Another very important department in connection with this plant is the rattan chair department, where all sorts of reed and fibre goods are made. The material used here is very largely imported, as there are no reed manufacturers in this country. After the reed arrives it is put through a bleaching process before working into the chairs, etc. Then there are the upholstering, spring bed, and mattress departments, any of which deserve an article by themselves.

The goods are routed through the factory on a manufacturing number, which is to be found on a ticket accompanying each batch, and from which the men register their time for the cost department (a department, by the way, which is sadly lacking in so many plants, resulting in unnecessary price-cutting and losses).

The office is in the front of the west wing, and is well fitted in dull-finished quartered oak, and does credit to the plant it represents. The whole establishment gave the representative of the "Canadian Woodworker and Furniture Manufacturer" a very pleasing impression, being situated as it is on the bank of that beautiful Lake St. Louis, with the ground rising gradually away into a small hill and commanding a view for miles along the lake.

Which are the Best—Belt or Drum Sanders?
Manufacturers of Both Put Forth Certain Claims—Work to be Sanded Should Govern the Buyer's Choice—Both Types Dealt With in an Impartial Manner
By Frederick E. Tracey

One can go into a half-dozen furniture plants and, in instances, find as many sanding methods used. Where the same types of machines are employed the operation and results vary greatly. Yet every man will swear by his way of doing things. This lack of uniformity is due in part to the conflicting claims of the tool builders. The belt sander people advertise that their machines will duplicate hand finish, and that their capacity is as great as the roll-feed or endless-bed drum sanders. The endless-bed manufacturers claim exactly the same for their tools, and so it goes.

All this sort of thing is confusing, and makes it hard for the furniture manufacturer to really gauge his sanding department efficiency unless he tries all methods. As this is impossible, I will present the unbiased opinion of one who has worked for a company which makes practically all types of sanders.

The endless-bed and the belt type of machines are capable of giving a finish that requires no hand-touching up on all furniture parts that are to be covered over and even on considerable other work. Of course, surfaces that are to be glued require the least bit of roughness, so as to afford the glue a good "foothold." However, where really fine work is necessary hand-touching up is required, in spite of claims to the contrary by the machine makers. Occasionally, though, an operator may become so expert as to eliminate the necessity of this, but I am talking about average cases.

Some manufacturers make the mistake of finishing their fine work entirely by hand. The finish is all that could be desired, but it is gained at too heavy a cost.

The big endless-bed or roll-feed machine positively has a far greater capacity than the belt-sander on all work having two opposite faces flat that can be sent through it. The smaller the pieces the greater is this ratio of capacity in favor of the endless-bed machine. This applies to pieces as small as two and three inches, which can be set in an easily-made frame and a hun-

dred or more pieces all be sent through together. The finish on the drum machine is also more uniform than that given by the belt-sander, because on the belt machine the operator varies his arm pressure, particularly towards the close of the day, when he is tired.

The big drum machines, however, require more highly skilled operators. They are very complicated tools and hard to keep everything about them going just right. Certain defects in the sanding may be caused by a dozen different things, so that the operator must have an understanding as to all this.

In one factory the operator will use paper on his drum machine that is too coarse, so that the parts sanded must be sent to the belt-sander before they are touched up by hand. In practically all cases this is unnecessary. As likely as not, in the factory across the way, the operator is using the proper combination of grit and paper to give a finish that only requires hand-finishing, with no extra machine sanding.

A great many other factors contribute to the amount of extra machine or hand-sanding necessary. The operator may have his adjustments wrong, or run the paper too long, or one and a dozen other things. But given a man who knows his machine and the drum machine is more economical usually than the belt-sander—at least, in plants that are large enough to keep the big machine busy several hours a day. Where the belt-sanders have the preference is on odd shapes—very high work, such as boxes where the parts are not sanded before being put together, etc.

Some factories object to the expense of the endless-bed machine because of the rubber pad replacements necessary. However, where this is necessary, frequently it is not the machine's fault, but is due to the fact that the operator is feeding with too heavy pressure, etc. Where the factory can afford it, both types of drum-sanders and the belt machine too should be used, as each has its advantages and economies.

Furniture Design and Prevailing Styles

Noticeable Improvement During the Last Few Years—Modern Machinery Has Played Important Part—William and Mary Period Most Popular at Present

Staff Article

Somnoe of Louis XVI. bedroom suite.

One of the most noticeable and pleasing features in connection with the furniture industry in Canada during the past few years has been the tendency on the part of manufacturers to produce furniture of good design at reasonable prices. It is now possible for the man with a moderate salary to choose furniture to his taste, and even for the workingman there is a wide range of dining room, living room, and bedroom furniture in simple yet pleasing designs, the price of which is within his means. There is, of course, and always will be, a certain amount of "gingerbread" furniture—handsawed and jigsawed and carved to death—on the market, but the quantity is fast diminishing, and there is no person better pleased about it than the manufacturers.

The designer is playing a more important part to-day in the manufacture of furniture than he has played since the days of special order and hand made furniture. No modern furniture factory of any pretentions is complete without a designer. But the man holding that position must be able to do more than draw pictures. He must possess an intimate knowledge of the manufacturing operations and be able to interpret his designs in dollars and cents. Some of the most successful designers are superintendents who handle that branch of the business as well as supervising the manufacture of the product.

One of the principal factors in placing well-designed furniture within the reach of the great middle class has been the perfecting of machines to turn out large quantities of work in a short time that formerly required tedious and endless hand labor. These are so well known to manufacturers that it would be superfluous to deal with them here. If two of them might be singled out for special mention, however, they are the improved four-sided moulders and automatic turning machines.

The designs to be seen on the floors of leading Canadian furniture stores cover a wide range. Perhaps the most popular at the time of writing is the William and Mary design. Of this style there is some splendid examples. With their excellent proportions and graceful turnings they reflect credit on their several manufacturers. Walnut appears to be the predominating wood for this style, although there are some suites in mahogany and oak. Walnut is particularly appropriate, mahogany is pleasing, but for the writer's taste oak is not suitable.

We read about Jacobean furniture having been put out of the running for the time being, but the quantity to be seen seems to indicate either that it is still popular, or that it cannot be sold.

The examples shown are mostly confined to living room and library pieces. Jacobean furniture never made a very strong "hit" with the ladies, their objection to it being that it was difficult to dust properly. That there is a certain stateliness and dignity about Jacobean when well designed and executed cannot be denied, and it will surely win its way back to

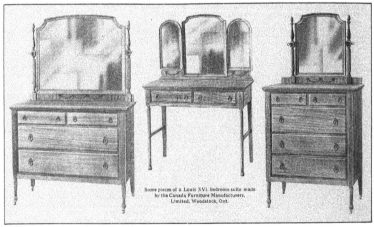

Some pieces of a Louis XVI. bedroom suite made by the Canada Furniture Manufacturers, Limited, Woodstock, Ont.

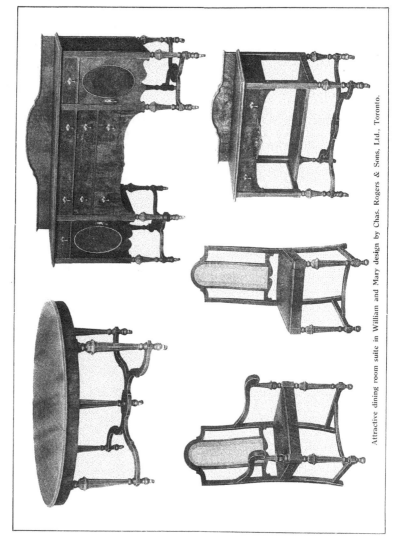

Attractive dining room suite in William and Mary design by Chas. Rogers & Sons, Ltd., Toronto.

favor. The chief factor in its downfall was undoubtedly the number of cheap and crude imitations, as some manufacturers took advantage of its popularity and adorned some of their stock designs with a couple of rope turnings and put them out as Jacobean.

Perhaps next to the William and Mary in point of popularity is the Queen Anne. There are a number of dining room and bedroom suites in this style to be seen and mahogany is the wood most in evidence. There are still a few suites of the Adam period, but this style has had quite a run and is not so much in demand at present as it has been. There will always be a certain amount of Sheraton furniture, particularly for dining room suites. Finished in mahogany, with its simple straight line designs and bands of inlay, it is a very pleasing style. There are a few suites of different styles to be seen with hand-painted designs, but this class of furniture does not appear to be as popular as it is on the other side of the border, unless in designs suitable for children's rooms.

Altogether, from the standpoint of design and quality of workmanship, Canadian furniture generally speaking is attractive and will return service equal to any that can be purchased elsewhere.

Knowing Costs in the Furniture Industry

Accountant Installing System Must be Familiar with Operations and Methods of Handling Materials—Conditions in Individual Plants Must be Considered

By M. L. Cooley*

Cost Finding is an infant industry. But a few years ago, plants which operated cost finding methods which would now be accepted as satisfactory cost systems were so rare as to be negligible in number.

This infant is blessed with a rugged constitution. Notwithstanding the blows it has received, through those who did not or would not understand, those who commenced installation but were unwilling to give it a fair trial, and those who attempted installation without being fitted for the task, it has survived, grown, and has become an absolute necessity in every line of business today.

In no business is a knowledge of the facts of cost of manufacture more necessary than in woodworking. In no other business is it more essential that the cost accountant engaged to install a cost system be familiar with the industry, with the operations and the handling of materials, parts, and finished goods. A plan which might be successful in a hardware, textile or paper business might work to great disadvantage in a woodworking plant.

The plan must fit the industry. It must not be complicated, yet experience shows that a plan so simple as to be capable of being installed by an ordinary bookkeeper is too simple to give the facts required. It demands the services of an expert, one who is not only a first-class accountant, but who can also bring to bear on the problem a knowledge of conditions in woodworking and experience in the principles of operating and managing a woodworking plant.

Further, the individual conditions existing in any one plant show wide differences from those of every other plant, and while the fundamental principles of cost finding are the same for all woodworking manufacturers, and the skeleton plan may well be uniform for all manufacturers of similar lines, the filling out of this skeleton plan, in adapting it to the needs of a particular business, takes a different aspect in every single case. Like the human race, each shows a distinct personality in its finished form and even in cases where several plants are operated under one management there will be essential differences in the application of certain parts of a skeleton uniform system.

By working on the basis of a skeleton plan whose basic principles have been found satisfactory by many woodworking manufacturers, as a uniform system, the details can be elaborated to the extent required by any manufacturer, of whatever size, yet the same basis is used in every case for arriving at costs of finished product and for setting selling prices, thereby rendering competition more intelligent.

In a recent test, a number of manufacturers in a woodworking industry were asked to give their estimate of cost for a certain article. These costs varied widely and even as much as one hundred per cent. If one manufacturer believes his cost is $25.00 and his neighbor says his is $50.00, it is manifest that somebody is going to lose when the goods are sold, and frequently both lose, one because his figures are too low, and the other because he feels he must meet that unintelligent competition.

There are three basic elements of cost of production; material, labor, and overhead expense. Each of these can and should be accurately determined and proven each month.

Material Costs

Experience shows that it is entirely practicable to use a simple plan which will give the cost of lumber delivered to the cut-off saws and determine the value of the rough parts, with the added benefit of an accurate control of investment in materials.

There are several methods of arriving at the cost of lumber delivered to the cut-off saws, such as scaling lumber as piled, determining the average measurement of courses per pile, recording the courses as taken from the piles, or even recording exact measurements; a second method is to determine the footage of material produced at the cut-off saws plus a percentage for wastage in cutting; and a third method would be to fix standards for the valuation of the various parts in rough sizes. Of these methods the first gives a closer control of the lumber inventory.

Accounts should also be kept with each other class of materials, charging each with all purchases and crediting the value of materials put into process. The following of materials from raw stores through product in process, product in the white, and finished product is simply a matter of records which quickly becomes routine work for an ordinary clerk.

The value of cost accounting as it relates to materials covers not only a control of raw materials and the investment therein, but gives comparative costs of materials used in various lots of similar parts and finished goods and leads to the reduction of waste.

Labor

Labor is divided into two classes, productive and non-productive. Direct labor is charged to the work

*Mr. Cooley is a member of the firm of Cooley & Marvin Co., Accountants and Engineers, Boston.

in process account and non-productive labor is classed as a departmental factory expense. The total direct labor charge for the month should equal the sum of the various charges to production orders, or departmental analysis.

The ascertainment of labor costs gives opportunity to compare results of both men and machines, and offers to one who understands the application of the principles of scientific management an opportunity to obtain some wonderful results in reducing costs and increasing output in conjunction with the establishment of standards, planning the work and scheduling production.

Factory Expense

This element is assembled from the general divisions of factory burden, such as design expense, superintendence and foremen, departmental non-productive labor, repairs, shipping, etc., etc. In many cost systems in other industries this account is cleared through a journal entry arising from an analysis of the production order costs of the month.

In woodworking cost finding it has been found that it is better to create a "Reserve for Factory Burden" account, to credit that account and charge work in process monthly with an amount of factory expense in proportion to the amount of direct labor incurred. The theory which we sometimes encounter, in actual practice, that the entire factory expense for the month should be apportioned to that month's production, is fundamentally wrong, as the light month's production would thus have to stand nearly as much factory burden as a month when the production is heavy, so that the costs have extreme variations and are comparatively useless as a basis for selling prices.

The account "Reserve for Factory Burden" is in effect the credit side of "Factory Expense" account. Each month charges for the general divisions of factory burden are accumulated in detail accounts covering non-productive labor, repairs, power, heat, light, and protection, insurance, etc., etc., and at the close of the year the balances in these accounts are credited and carried to the debit of "Reserve for Factory Burden." The "Reserve for Factory Burden," on the other hand, is credited each month with the amount of burden applicable to the cost of product for the month in proportion to the amount of direct labor incurred, and charged to work in process.

These charges are based on percentage rates and the accuracy of the percentage rates is determined when at the close of the year, or preferably once a month, the balance of charges applicable to factory burden is contrasted with the balance in "Reserve for Factory Burden." If the "Reserve for Factory Burden" is in excess of the charges which have been incurred for factory expense, this indicates that the percentage applied to work in process has been too high. If the amounts accumulated in the reserve for factory burden are less than the aggregate of the balance in the factory expense account it indicates that the percentage rates used in the appointment of burden have been too low, and the percentages will then be corrected for the ensuing period.

The treatment of administrative and selling expenses and interest does not occasion much trouble. So long as a manufacturer knows and proves the factory cost of material, labor and factory expense he is in a position to establish a policy, but so long as he is in the dark on these matters, working on a basis of estimates or competitors' prices only, he must expect disaster.

Monthly Balance Sheet

As a result of such a system, the figures are provided for a set of monthly statements showing the assets and liabilities of the business, the loss or gain for the month, and comparative figures of income and expenses. No one who has ever had these monthly statements would dream of going back to the old method of getting a problematical "trial balance" once a month—which means absolutely nothing to even an expert accountant unless analyzed, and then is without much value, because the amount of stock is unknown—and closing the books once a year with a physical inventory which cannot be checked. Every inventory which was ever taken twice showed variations, and it is common knowledge that a continuing book inventory as mentioned above, in connection with a sound cost system, is very much more dependable than the usual hasty and inaccurate "count of stock."

With such a cost system as previously outlined in operation, the manufacturer is in a position to be handed a set of statements each month which not only automatically relieves him of making personal researches of the books or investigation of the plant in order to obtain an estimate of current condition and trend, but supplies him with reliable information on facts regarding which he had previously no way of obtaining any data whatever.

An ideal set of monthly statements such as is in current use by certain woodworking manufacturers, provides:

A Balance Sheet showing in as much detail as may be required the exact condition of all assets and liabilities at the close of the month compared with previous months and the previous year;

A Loss and Gain Statement for the current month compared with the same month of the previous year and for the period to date compared with the same period of the previous year. This statement gives information regarding the gross sales, deductions, net sales, factory cost, gross manufacturing profit, commercial expenses, operating profit, other income, other charges, and final net profit.

Supplementary to this statement is information regarding departmental operations and percentages, such as gross profit to net sales, returns to gross sales, etc.

A statement giving an analysis of the inventory of merchandise and supplies, including raw materials, product in process, and finished product on hand.

A statement of the distribution of factory burden and a statement of various administrative and selling expenses entering into commercial expense, all of which statements are compared by month for the year.

While it is true that an ordinary commercial bookkeeping system could be made to provide certain important facts each month, such as the condition of cash, accounts and notes receivable, and accounts and notes payable, in many cases even that information is unreliable and it is a fair statement that the best work is done where a sound cost system is in operation, controlled by the accounting records, and it is only by such means that reliable information can be produced every month on matters relating to cost of sales, stock on hand, labor, material, and expense charges, and investment in plant and equipment.

Lesson in Furniture Making for Beginners

Taking up in Detail the Construction of a Sheraton Dressing Table—Accurate Drawings Assist Readers to Follow Points as They are Dealt with in The Article

By C. A. Zupann

Photograph of Sheraton dressing table.

The statement has been made that the m o d e r n workman has been turned into a machine. This may be true in the furniture factory to-day, though it seems to be an exaggeration. In the golden age of English furniture the designer and cabinet maker were quite often one and the same person. It is to be regretted that this is so seldom the case at the present time. To guard against the day when furniture workers may become machines on account of the high degree of specialization now in vogue it is essential that the workman know more about the industry than simply the part of the construction work to which he is assigned. His mind should trace the piece through every phase, even though his work be limited to only one part of the many processes involved.

Only in this way can he realize the dignity and importance of his work in the entire scheme, and only in this way can he gain the full joy from his work. A viewpoint from many angles will then be found. He will be able to appreciate the ideal of the designer at one end and the attitude of the consumer at the other, with all the intervening workers, as yard-men, machine men, cabinet makers, finishers, trimmers, packers, salesmen, and the many others who have to do with each article manufactured. Many a man bewails the trend of industry as crushing out individuality, and at the same time neglects to inform himself of the origin of the pieces upon which he is working—at times can scarcely tell the style after which they are made. When a stock number and order number are sufficient description for the workman, then intelligent enjoyment of work has surely ceased.

Among the styles being reproduced and modified to-day the Sheraton is the most simple, and the American adaptations lend themselves to easy reproduction by the amateur or the young cabinet maker. For this reason an example of this modified style is chosen as the basis of this article.

Thomas Sheraton was one of the five great Georgian designers, Chippendale, Heppelwhite, and the Adam brothers being the others. He was born in 1751 and died in 1806. The fifty-five years of his life saw a great many hardships, which, it is said, rather embittered the latter part of his life. He was primarily a designer and draftsman, usually selling his designs to other manufacturers. History, however, classifies him as being a journeyman cabinet maker as well as a designer and writer. Among other things he was a Puritan preacher.

The qualities of character shown by this last fact are undoubtedly seen in the straight and severe lines shown in most of his designs. Sheraton never reached an age of prosperity: His book of designs was purchased more by cabinet makers and producers than by the consuming public. In this respect it differed from Chippendale's book, and, owing to this fact, he cut off rather than increased the sale of his designs and furniture. As he grew older it is told by his friends that he became quite soured and bitter toward life in general, perhaps with some cause, for he died in poverty.

Sheraton's work at its best was simple, and so refined as to be called "severe," a term well given, as in some cases there was not even a moulding to give a change from the straight line. In this case simple banding or inlay was the only ornamentation. The legs of all his designs were invariably straight, and variety was obtained by having them square, tapered, turned, or fluted. His chair seats were "all over" stuffed, or caned, and the legs were seldom placed with stretchers or any form of under-framing. He employed a variety of methods of decoration, though, as said previously, these were used very sparingly. Among these methods were carving, inlaying, painting, and lacquering, and the use of veneer and wedgewood plaques. His sources of inspiration were the work of the French artisans and designers of the Louis XVI. and Empire periods and the work of Chippendale and Heppelwhite. The spade foot illustrated in the sketch shown in this article was a favorite, and was used in many of his designs.

The characteristics of the American adaptations of this style are square, tapered legs, with or without feet, inlaid bands or stripes, a predominance of straight lines with very simple curves, if any, and light construction. The chair backs of this style are rectangular in form, and quite often have a raised part on the top back rail. A lower back rail two or two and one-half inches above the seat is always employed. The pieces are usually made in mahogany, though other delicate woods may be used. A Sheraton design produced in oak would be as incongruous as Jacobean in gum or birch.

For any readers of the "Canadian Woodworker and Furniture Manufacturer" who desire to construct the dressing table shown in the accompanying photograph, which is of the modified Sheraton style, the following directions will undoubtedly be sufficient.

Get out all stock called for in the stock bill and dress and sand to the dimensions given. Saw the legs and the top rails of the toilet frames. Saw grooves

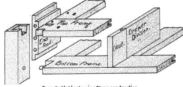

Corner of table showing frame construction.

one-fourth inch wide by one-half inch deep into the front and back rails or frames and into the cleats or drawer divisions. Saw one-fourth inch grooves one-fourth inch deep into drawer fronts and ends to receive the one-fourth inch bottom. Saw a three-eighths inch dado one-fourth inch deep into drawer ends to receive the drawer backs. Saw rabbets three-eighths by three-fourths inches into the top and bottom edges of end rail and back rail, to receive the top and bottom frames. After boring dowel holes in the stiles, rabbet the stiles and rails of toilet frames three-eighths inch by five-eighths inch. It will be impossible to saw this rabbet into the top rails of end toilets owing to the curve. In this case it must be cut out on the shaper or by one of several hand methods.

It has been taken for granted in the preceding directions that a circular saw is available. If this is not the case, however, the work described can be done quite easily with a Stanley plow and rabbet plane. Saw tenons on the ends of cross rails and end rails in frames, and also upon the ends of the drawer divisions which join the cleat. After boring the dowel holes in the toilet rails, half tenons must be sawed on the ends in order to fit them into the rabbets in the toilet stiles. Assemble and glue the five frames. Also glue the cleats on the drawer divisions.

Stock Bill—Sheraton Dressing Table.

It'm.	No.	Name.	Stock.	Long.	Wide.	Th'k.
1	1	Top rail in middle toilet	Mahogany	13	2	⅜
2	1	Bottom rail in middle toilet	Mahogany	13	1½	⅞
3	2	Stiles in middle toilet	Mahogany	23½	1¾	⅞
4	2	Top rails in end toilets	Mahogany	9	3½	⅞
5	2	Bottom rails in end toilets	Mahogany	9	1¾	⅞
6	2	Inside stiles in end toilets	Mahogany	21¼	1½	⅞
7	2	Outside stiles in end toilets	Mahogany	19½	1½	⅞
8	2	Rails under end toilets	Mahogany	11½	1½	⅞
9	1	Top	Mahogany	39¾	19¼	⅞
10	4	Legs	Mahogany	30	1⅝	1⅝
11	2	End rail	Mahogany	15	5½	⅝
12	1	Back rail	Birch	34½	5½	¾
13	2	Front rails in frames	Mahogany	36½	2½	⅝
14	2	Back rails in frames	Birch	36½	2½	⅝
15	4	End rails in frames	Birch	13¾	2½	⅝
16	4	Cross rails in frames	Birch	13¾	2½	⅝
17	2	Drawer divisions	Birch	15½	2	⅝
18	2	Cleats on drawer divisions	Mahogany	3¼	2	⅝
19	2	Drawer fronts end	Mahogany	8	3¾	⅞
20	4	Drawer ends end	Birch	13½	3¾	½
21	2	Drawer backs ends	Birch	7½	3	⅜
22	1	Drawer front	Mahogany	17½	3¾	⅞
23	2	Drawer ends	Birch	15½	3¾	½
24	1	Drawer back	Birch	16¾	3	⅜
25	2	Toilet supports	Birch	18	1½	⅞
26	2	Drawer bottoms	3 ply veneer	7½	15½	⅛
27	1	Drawer bottoms	3 ply veneer	16¾	15½	⅛
28	1	Toilet back	3 ply veneer	22	14	3/16
29	2	Toilet backs	3 ply veneer	20	10	3/16

When frames are dry, lay out the dovetail on the front rail of top frame and also the notches on the other corners of both frames to the points marked. Bandsaw these corners to the lines marked. With a chisel cut out the dovetail mortises in the tops of the front posts to receive the dovetails on top frame. Bore dowel holes in all remaining notches in the frames and corresponding holes into the posts. Also bore three dowel holes into each end of the end rails and back rails and corresponding holes into the posts. Glue up the legs and end rails in pairs. Bore holes through the frames for screws, which will hold the frames into the rabbets in the rails. Also bore holes for screws to secure the drawer divisions. The top is secured by screws passing through from the underside of the top frame. The holes for these screws must be bored at this time. Assemble the ends, back rail and frames with glue and screws. The top may be screwed in place at any time after this.

The toilet frames are held in place by a seven-eighths inch strip screwed to the back of the centre frame. The short rails under swing frames may be doweled and glued or screwed to the top. In the event of screws being used, an unglued dowel should extend

from the end of this rail into the stationary toilet stile. The end toilets are fastened with two brass butts to the stationary toilets and are kept from swinging by friction buttons in each short rail.

The simplest form of drawer construction that can be made is the plain rabbet. A slightly more complicated form or housed rabbet, which is very easily made upon the circular saw is shown in Fig. 2. Where the

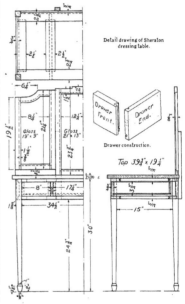

Detail drawing of Sheraton dressing table.

Drawer construction.

Top 39¾" x 19¼"

ability of the artisan permits, handmade dovetails are naturally preferable.

The piece should be thoroughly cleaned and sponged with water and sanded before starting to finish. Stain with a good mahogany stain. Fill with mahogany filler and follow with a coat of shellac and two coats of good varnish. Sand well after each coat and rub the last coat of varnish to a dead finish.

If inlay is desired on this piece, the legs may be banded one-eighth of an inch wide about one-fourth inch in from the edge of the leg. The drawers, top, and toilet frames may also be banded. It will be understood that this inlaying must take place before any of the parts are assembled.

A few years ago the application of varnish, shellac, or enamel in a furniture factory by the spray method would have been considered so expensive that it would not be thought practical. To-day every up-to-date factory has installed or has on order one or more spray outfits.

Correct Trimmings for Period Designs

Original Sketches Showing Necessity of Harmony — Styles Blend into Each Other— Manufacturers of Trimmings Have Risen to the Occasion

By James Clark

The present large and increasing demand for well-designed furniture has opened the pages of historical matter regarding the various "Period" designs, and has caused a considerable amount of study and research in order to produce styles that will be at least fairly representative of their particular time.

This has, of course, always been the case in the more expensive grades of furniture, as all the leading

Queen Anne.

manufacturers have invariably carried a limited number of designs that were quite correct and were bought by the wealthier or more discriminating class of customers, but now the demand has become quite general and in the large field of medium grade goods the call is for "Period" designs, which are being manufactured in large quantities, and it may be said, are very presentably put out.

The proper trimming for these various styles is very essential, and the manufacturers of handles, knobs, etc., have risen to the occasion and produced the goods. They have, generally, made the one mistake, of illustrating their catalogues with so-called "Period" styles and not mentioning the particular "Period" design the trimmings were for. It is quite easy for a furniture trimmer in a factory to substitute one pull for another, and so trim a Queen Anne suite with the hardware that

was intended for a William and Mary. Ever see it done? Of course, a certain number of handle and knob is specified, but these have a way of being out of stock sometimes, and the next best thing is put on. The result is a mistake, and if anyone sees it that knows about such details he or she is apt to say things very uncomplimentary about the lack of taste and knowledge of the manufacturer.

There are instances where it is very difficult to decide on the correctness of the design of the furniture mounts, as styles blend more or less into each other; that is, those styles that immediately succeed each other, and unless the trimmings carry a characteristic touch of the style required, mistakes will occur, though they are not so noticeable . If they harmonize there is no great harm done.

The developments of furniture trimming is quite interesting. The first attempts at making metal mounts were certainly crude. The ones manufactured to-day show the advance in that line, and doubtless the trimming of "Period" furniture is much better done than it was when the various styles came into being.

The writer has attempted to show the relation between the furniture and its mounts by several original sketches, using the forms of ornamentation on the various styles, in the handles, which should be a correct method, keeping in mind that the handles are metal, in most cases.

The Queen Anne buffet illustrated should look quite properly trimmed with the pulls shown, providing that the color of the trimmings harmonize with the mahogany buffet. Queen Anne furniture looks quite perfect in mahogany, and next best in walnut. This style, as the name indicates, was developed about

William and Mary buffet, showing relation of trimmings to turnings.

Queen Anne buffet with appropriate trimmings.

1702-1714, and furniture correctly made along these lines is singularly graceful and pleasing.

William and Mary

This style is having a tremendous "run" at the present time, and deservedly so. This is the style that immediately preceded the Queen Anne, and had naturally considerable influence on it, and some examples of furniture of that period show the transition quite plainly. Turned legs and under-framing came into

Adam dresser, showing how drawer pulls harmonize with general design.

vogue, and the general tendency to more strength in line is noticeable. Oak and walnut were generally used, and these are the two woods that are being drawn upon for this purpose to-day.

The trimming, in keeping, is less ornate than the Queen Anne, and the drop pull is used almost exclusively, being of a form to harmonize with the turnings. The illustration of the buffet and two drop

Jacobean.

pulls conveys the idea. Trimmings of a dead or dull brass are the best finish for this style, though on walnut bright nickel is sometimes used, but to the writer's mind it looks somewhat "tinny" and cheap.

"Adam" Style

This very popular and beautiful furniture was the work of the brothers R. and J. Adam, architects, and the period of their activities dated from 1750 to 1790 and shows the influence of Hepplewhite, particularly

Jacobean table with bobbin turnings.

with the added classical detail of similar ornamentation that had been unearthed from the ruins of the city of Pompeii.

A great deal of composition ornament was and is used on the "Adam" style, in low relief and very delicate detail, along with inlay and painted ornamentation.

The forms of urns, rosettes, floral festoons, more

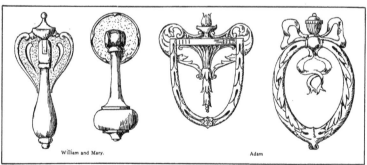

William and Mary.   Adam

Mr. J. R. Ball, Ball Furniture Co.,
Hanover, Ont.

Mr. James Baird, Baird Bros.,
Plattsville, Ont.

Mr. G. A. Gruetzner, Hespeler Furniture Co.
Hespeler, Ont.

or less conventionalized, are very prominent, and in some cases are apt to convey a funereal effect, if care is not taken. Delicate fluting is also very much in evidence, and the cross fluting on drawer fronts, figures, etc., has a pleasing effect.

The brass mounts partake of the delicacy of detail that the ornamentation possesses, and must conform to the general design. The simpler design of an "Adam" dresser and the two handles, illustrated, are in keeping with each other.

The "Adam" style is still a strong factor in the line of present sellers, and is made in walnut, mahogany and in lesser degree in oak. The latter wood must be selected for small figure, and looks well in a rich, dull fumed finish. Handles having hand painted plates can be effectively used on the better grades of "Adam" furniture. The most faithful attention to detail is necessary to the successful manufacture of this style.

The Jacobean Style.

This style has had, and is still having, a very considerable demand, and the prospects seem to be fairly bright for its continuance. It was brought into being during the English Renaissance, or revival period, which includes the styles Tudor, Elizabethan, Jacobean, William and Mary, and Queen Anne. The Jacobean division dates from 1603 to 1688, and is a development of the Elizabethan, there being no distinct line between the two styles, and they can be used together in furnishing a room quite correctly.

This style is sometimes referred to as Charles II. by some makers, but it is simply a piece of the Jacobean period, and might be called James I., Charles I., or James II. with equal right, or English Renaissance.

The Jacobean style is of good, substantial appearance, running structurally on straight lines, and more or less architectural in appearance. The trimmings partake of the same solidity of character and the drop pull is generally in evidence on modern adaptations. The twisted spiral turnings of the furniture are indicated in the designs of the pulls. Any other ornamention is in flat relief and not at all showy. The "bobbin" turned legs and stretchers are also a feature of this "period."

The illustration of the table, herewith, shows the "bobbin" turned leg, and could be also made with the spiral twist turning. The three illustrations of pulls show two designs suitable for either way.

The Jacobean style is particularly effective for libraries and the larger dining-rooms; also for rotunda furniture, where an effect of real solid worth and dignity is desired.

As in the trimming of other styles, that of the Jacobean is equally important, and the handle-makers all have fine examples of correct designs from which to choose.

Seasoning Lumber Artificially

A new process to rapidly season lumber artificially has been tested at Columbia University, New York, says the American Contractor, where, according to reports, it was demonstrated that black and tupelo gum planks, one inch thick, green from the saw, were thoroughly dried in 24 hours, the lumber continuing straight and free from cracks, the fibre of the wood at the same time becoming materially harder and stronger.

The new process, which is called a lumber vulcanizing process, it was stated, consists of drying the lumber in steel cylinders under superheated steam and in vacuum. The lumber, right from the saw, is placed on cars and the loaded cars are sealed in the cylinders, where the superheated steam is applied for a definite period. Explanation given of the working of the new method is that the steam penetrates to the heart of the green timbers, raising the sap water to boiling point and bringing it quickly to the surface of the lumber, thus eliminating the condition known to lumbermen as "case hardening."

The utilization of timbers heretofore considered practically worthless from a commercial point of view, the conversion of millions of acres of timber-bearing swamp lands of the South into sources of wealth and immense additions to the revenue-producing possibilities of the lumber and timber industry of the entire South are made possible, it is asserted, by the perfection of this new process.

Don't forget our Information Department, which is ready at all times to answer any question of interest to subscribers.

Vapor System of Drying Lumber

By E. U. Kettle, of the Grand Rapids Veneer Works

Patent office records will show to any person who takes the trouble to look them up that a great many apparatuses have been invented for drying lumber, and certain systems and processes have been evolved which, through lack of knowledge on the part of the inventors, have been found purely theoretical and have not been at all practical.

A well-known company of dry kiln engineers who control a patented process, and who have installed some very large batteries of lumber drying kilns, have records which show that in many cases where their services have been applied to the remodelling of other installations, they found apparatus had been installed with little or no consideration relative to correct conditions of heat, humidity, compression, circulation and ventilation, all of which must be duly proportioned according to certain fixed laws if good results are desired. Adaption of certain distinctive features of diverse patents are also frequently met with.

To those who desire to install lumber drying kilns, or remodel existing apparatus, I would say—Investigate; and place your contract with the concern which can produce a satisfactory record for efficiency, and do not allow the question of efficiency to be influenced by price. A cheap installation in a cheap building will assuredly prove highly expensive in the long run.

Without going into an analysis of the numerous methods of drying lumber, and passing over the several freak systems now obsolete, I may say that taking present well-known types of kilns into consideration, the vapor process has proven most efficient and economical for drying green lumber, and special dimension stock, such as gun stocks, spokes and wagon stock, shoe last blocks, squares, etc.

This process is in successful operation in large woodworking plants in Canada making sash, doors, interior trim, flooring, furniture, pianos, auto bodies, wagons, etc., all of which use lumber and dimension stock in different conditions of dryness. This is sufficient proof of its universal adaptability to any desired purpose.

The apparatus consists of a combination of horizontal radiators operated by exhaust steam, and situated below rails on which kiln cars of lumber are placed. There is a pit under these horizontal coils, which for obvious reasons is cooler than any other part of kiln when in operation. This comparatively cool pit is a vital factor of the installation; hence, the use of horizontal radiators.

The ingress and egress of air is easily controlled by special apparatus as is also the percentage of humidity and temperature, whilst a system of specially designed hoods and dampers secures the requisite amount of compression, circulation and ventilation, and prevents down drafts in ventilating stacks during variable winds. The latter is a decidedly advantageous feature, although it is often omitted by other kiln designers.

The process of drying stock by this installation is divided into three distinct periods; i.e., steaming, sweating and drying.

During the first period steam is admitted and by a unique system is circulated between the courses of lumber until the latter reaches the limit of expansion, with pores opened, moisture content equalized, and of an uniform temperature from centre to surface. This treatment liquifies the coagulated sap in the lumber, and prepares the way for removal of same by the ensuing processes of sweating and drying.

During the sweating, or preliminary evaporation period, a considerable amount of vapor exudes from the lumber, and this maintains a high percentage of humidity in the air of the kiln, keeping the lumber still expanded and enabling the interior moisture to escape to the exterior.

Then follows the third, or active evaporation period, which completes the operation, leaving the lumber set in expansion, and practically of fixed dimensions. Lumber dried in this manner is immune from the expansion and contraction due to climatic variations to which ordinary dried lumber readily responds.

It will be seen that this is a reversal of the ordinary drying process. In other words, the interior is dried first. Dried from the inside out instead of from the outside in.

The following well-known facts may be worthy of

Mr. H. A. Simpson, Manager Eastern Townships Furniture Co., Arthabaska, Que.

Mr. J. G. Hay, President and Manager North American Bent Chair Co., Owen Sound, Ont.

Mr. G. W. Gibbard, Manager Gibbard Furniture Co., Napanee, Ont.

repetition herein, inasmuch as they are the basis of the above described system:

1. The moisture content of the lumber must be equalized before drying operations commence.

2. Surface evaporation must not exceed the rate of transfusion of moisture from interior to surface.

3. The proper regulation of temperature and humidity is necessary to avoid case-hardening, honeycombing, excessive shrinkage and other defects.

4. Drying must be uniform at all points, otherwise stresses are set up in the lumber, causing defects.

5. Wood is soft and plastic whilst hot and moist and will become set in whatever shape it is in as it dries.

6. Shrinkage is greater, the higher the temperature of drying whilst still moist.

7. Brittleness is produced by overdrying.

8. The temperature and humidity within kiln are no criterion of the conditions of drying within the courses of lumber if the circulation is deficient.

9. Where stagnation of air exists, or movement of air is sluggish, the temperature will drop and humidity increases.

10. The extraction of heat from the air which comes in contact with the lumber through evaporation of moisture causes it to cool and descend. Hence, the purpose of pit unobstructed by radiators.

The Superiority of the Wooden Pulley

Exhaustive Studies by Experts Have Proved it to be More Efficient than an Iron or Steel Pulley—Other Features are its Light Weight and Moderate Price

[Editor's Note.—The most common appliance in mills, factories, and machine shops is the pulley. Wherever belt transmission is used the pulley is in service. The cost of installing and maintaining mills where considerable machinery is employed constitutes a large per cent. of operating expenses. For that reason the pulley is of direct interest to every man who uses one or more. The article which follows was prepared by the National Lumber Manufacturers' Association, and is reproduced from the Hardwood Record.]

Are you using wood pulleys or are you not? If, instead, you are using steel pulleys, are you paying some 50 per cent. more at the outset for your pulleys than if you bought wood pulleys? In this case are you getting 50 per cent. greater efficiency and durability, or are you simply favoring the steel trust with a welcome contribution, which, considering its source, it is more than glad to accept?

Wood pulleys have been on the market for over thirty years, and that is too short a time to be able to say how long they will last. Many wood pulleys built thirty years ago are still in use and as good as new. Steel pulleys have been on the market sixteen years, more than long enough to demonstrate their average but owing to the inevitable crystallization to which they are subject when in service, this period has been length of life. So much for durability and its corresponding lack in the two types of pulleys.

There is only one class of service for which a wood pulley is not suited, and that is when exposed to excessive moisture, a service in which a regular leather belt cannot be successfully used. In such service a wood pulley should not be used, nor a sheet steel pulley either. Long-continued exposure to excessive moisture will cause a wood pulley to go to pieces. Moisture penetrating the riveted joints of a steel pulley cause rust and corrosion, and in a short time the joints work loose and the pulley is gone. The pulley for wet places is the cast-iron pulley.

The following table gives a comparison in price to

Mr. J. E. Alain, Victoriaville Furniture Co., Victoriaville, Que.   Mr. J. S. Knechtel, Knechtel Furniture Co., Hanover, Ont.   Mr. D. Hibner, D. Hibner Furniture Co., Kitchener, Ont.

the user between wood and steel pulleys. Only the most popular sizes are shown, but the comparison holds equally true for all sizes:

Sizes.	Wood.	Steel.	Excess cost of steel. Per cent.
6 x 3	$ 1.16	$ 1.82	57
12 x 4	1.60	2.55	59
24 x 6	3.96	6.02	52
36 x 8	8.04	13.20	64
48 x 10	14.98	26.81	85
60 x 12	24.28	46.20	90

Iron pulleys are even more expensive. Now, if you pay that much more for iron and steel pulleys, you ought to get that much more service out of them. But do you?

Wooden Pulleys Transmit More Power.

A number of years ago Haswell, the leading American authority on mechanics, made a series of exhaustive tests of wood and iron pulleys, and the findings of these tests have never been controverted. Haswell proved that the coefficient of friction of belts running over wood pulleys was 47 and over iron pulleys 24. That is, that with the same belt tension you can transmit, even according to his conservative figures, on the average, nearly twice the power over a wood pulley that you can over an iron pulley. Over a steel pulley you cannot transmit so much even as you can over an iron pulley, for steel pulleys are never perfectly round, and their faces are covered with high and low spots, so that for both reasons the belt is prevented from coming in perfect contact with the pulley face. This saving of power transmitted, rather than waste in belt slippage, is very considerable in the course of a year.

Wooden Pulleys Weigh Less.

The following table of weights is instructive, especially in view of the claims of the steel pulley manufacturers that their pulleys are light. These figures are approximate as representing several makes of wood and steel pulleys, but the ratio of comparison is correct:

Sizes.	Wood. Pounds.	Steel. Pounds.	Excess weight of steel. per cent.
6 x 3	1¾	7	300
12 x 4	8½	19	124
24 x 6	31	41	32
36 x 8	79	121	53
48 x 10	155	245	58
60 x 12	245	384	57

What do you gain by putting this extra weight on your hangers and shafting? The answer is, nothing;

Mr. W. J. McMurtry, Gold Medal Furniture Co., Toronto, Ont.

while, in addition, you penalize yourself severely every minute that your hangers and shafting carry this weight and every working minute that your engine has to turn it.

Wooden Pulleys Run True.

Trueness of running is another advantage which wood pulleys possess over steel pulleys. The latter are forced into temporary approximate roundness by dies in the course of their manufacture, but soon spring back out of this even approximate roundness when subjected to the strains and shocks of actual service. This is why you never yet saw a steel pulley in service that did not gallop or wobble or possess a combination of both undesirable gaits. Think about this. Galloping and wobbling is destructive to belts and bad for hangers and shafting.

Wood pulleys do not crystallize, even under hard service; the hardwood from which they are made offers the maximum resistance to compression, and yet has sufficient spring to relieve itself from sudden shocks.

Wooden Pulleys Make Yearly Saving.

Last year Prof. Price of the University of Toronto conducted a series of tests for the purpose of analyzing more closely than had heretofore been done the relative efficiency of wood and iron pulleys. His report has been published, and makes interesting reading. These tests seem to have been broad and thorough enough, and made under a sufficient number of varying conditions to be really conclusive. Prof. Price considered the subject under four heads, and so arranges his results, but it is sufficient to state that he

Mr. Geo. McLagan, Geo. McLagan Furniture Co., Stratford, Ont.

found the wood pulley far more efficient than the iron pulley, and that, reduced to dollars and cents, you save on the average about $6 a year per pulley by using wood pulleys rather than iron ones. Had steel pulleys been used in this test rather than iron, there is every reason to believe that the showing would have been even more favorable to the wood pulley.

The interests of the steel trust and the steel pulley manufacturers are parallel, and they have doubtless been of great assistance to each other. The steel pulley manufacturers take advantage of the effect of the word "steel" on their hearers' mind when their salesmen are talking and on their readers' minds when they speak through their advertising. They harp on the alleged fact that this is a "steel age." Steel may be good in some places, but a pulley is not one of these places. The flexibility of steel, its tendency to distortion, and its inevitable crystallization make it utterly untrustworthy for pulleys.

Certainly lumber manufacturers ought to favor a product made from lumber if they sincerely believe in the merits of wood over substitutes. But laying aside all natural bias of this sort, by using wood pulleys you are using something that has been proven more efficient, more durable, and more economical, both in first cost and operation, than either of the substitutes on the market.

The Kew Flagstaff

When in 1913 the great flagstaff which had been erected in the Royal Botanic Gardens at Kew in the year 1861 had to be taken down the wish was expressed that another spar of the same species and from the same locality might soon be forthcoming. The original flagstaff was of Douglas fir, its height being 118 feet, and it came from British Columbia, the donor being Captain Edward Stamp. By a strange coinci-

Flag pole for Kew Gardens, England, on the towpath.

dence the Government of British Columbia had already written the authorities, offering an even larger spar for erection in some public position, not being aware that the famous flagstaff at Kew was soon to be dismantled, and when the action of the authorities became known the Government offered to replace the famous landmark. The new flagstaff, which was to be over 200 feet in length, was not difficult to find, but transportation of such a stick involved some difficulty, which was fortunately overcome.

There were but two steamers trading to the British Columbia coast which were able to handle the giant. and, although it was planned that the Royal Mail Steam Packet Company's boat, the S.S. Merionethshire should take the flagstaff in November, 1914, a washout occurred on the logging line connecting the timber limits at Gordon Pasha Lake with the salt water and shipment had to be postponed. The dimensions of the flagpole when, in August, 1915, it was finally loaded the S.S. Merionethshire were: Length, 215 feet; diame-

Flag staff furnished by British Columbia Government for Botanic Gardens at Kew, England, being towed up the Thames.

ter at butt, 33 inches; at 115 feet up, 22½ inches; and at the top, 12 inches. The steamer arrived in London in December, and before the stick reached its ultimate destination it had to be placed in the Thames, towed up-river from Limehouse to Kew, about a dozen bridges having to be negotiated, the last one of all being Kew Bridge. Arrived at Kew, the flagstaff was landed, and then began a tortuous course of over a mile across lawns and flower-beds, with clumps of trees and conservatories presenting serious obstacles. The photographs were taken by the Daily Mirror and Topical Press Agency, to whom due acknowledgement is made for the courteous permission to reproduce.

The "Onward" Sliding Furniture Shoe

An invention which made its appearance on the market a few years ago and which has met with a large sale is the sliding furniture shoe. One of the greatest advantages claimed for it is that it will not injure the floor, carpet, matting or rug, nor wrinkle the rug on a polished floor.

The illustration below is the product of the Onward Manufacturing Company, of Kitchener, Ont. This company makes a complete line of sliding shoes of different kinds and sizes suitable for different weights of furniture. One of these is made with a metal base and is adaptable for any kind of floors. For rugs, carpets and highly-finished waxed floors, however, the glass shoe is recommended. For metal beds, having either round or square posts, a shoe with a wishbone bushing is made. This bushing takes a grip of the inside of the tubing and prevents the shoe from dropping out. A more substantial shoe is made, which is suitable for pianos.

A feature claimed for the "Onward" Sliding Shoe is that a joint, on the ball and socket principle, allows it to adjust itself to uneven surfaces. The size of shoe required should be governed by the weight of the furniture—the larger the shoe the easier it slides.

Illustrated herewith are an ornamental pull and an ash tray as manufactured by the Weber-Knapp Company, of Jamestown, N.Y. This company is now manufacturing a full line of hardware for talking machine cabinets.

Faultless Caster Co., winners of gold medal at Panama-Pacific Exposition.

Casters for Furniture

The Faultless Caster Company, of Evansville, Indiana, are manufacturers of a complete line of casters, covering the requirements of the furniture industry. The important feature of their socket casters is the pivot point at the top end of the stem. The socket being rounded at the end, the stem of the caster fits snugly into it and takes the weight of the article, allowing the caster to turn easily and quickly. Another feature is the patent steel springs which prevent the casters dropping out every time the piece of furniture is lifted.

The hospital caster made by this company should be interesting to manufacturers who wish to equip furniture with noiseless casters. The wheels run on axles fitted with ball bearings in concave raceways and they have canvas-rubber composition tires. For iron and brass beds a line of casters are made which have attached sockets. These are very convenient as it is only necessary to compress the socket—which is made in the form of a spring—to allow it to enter the tube or post of the bed and the spring prevents the caster from dropping out when the bed is lifted.

The company state that their line of casters finished in oxidized finish is taking well with the furniture trade, as they are considered appropriate for period furniture. The Faultless Company have an attractive catalogue illustrating their different makes.

"Onward" sliding furniture shoe. Ash tray and ornamental drawer pull made by the Weber-Knapp Co., Jamestown, N.Y.

Save Money by Purchasing

"Made in Canada"

Veneered Panels

Maple - Elm - Basswood - Birch - Plain Oak
Quartered Oak - Mahogany - Black Walnut

Single 3, 5 or 7 Plies.

Hay & Company, Limited Woodstock Ontario

Peter Hay Machine Knives

Peter Hay knives are made of special quality steel tempered and toughened to take and hold a keen cutting edge. Hay knives are made for all purposes.

Have you our Catalogue?

Made in Canada

The Peter Hay Knife Co., Limited
GALT, ONTARIO

DARLINGS STEAM APPLIANCES

DARLING BROTHERS
LIMITED
Engineers and Manufacturers
MONTREAL, CANADA

Branches   Agents:
Toronto and Winnipeg   Halifax, St. John, Calgary, Vancouver

THE ELGIE & JARVIS LUMBER CO., LTD.

18 Toronto Street - TORONTO

White Pine, Red Pine, Spruce
for

SHELL BOXES	CRATING
500,000 ft. 1 x 4 and up white pine culls.	150,000 ft. 5/8 spruce.
300,000 ft. 1 x 4 and up M. R. white pine.	75,000 ft. 2 in. M. C. spruce.
100,000 ft. 1 x 6 M. R. spruce.	100,000 ft. 3 in. M. C. spruce.
60,000 ft. 1 x 10 M. R. spruce.	200,000 ft. 1, 2, 3 in. Wormy Pine.
300,000 ft. 3 x 6 and up spruce.	

Veneers Used for Furniture and Pianos

Staff Article

That furniture and piano factories are large users of veneers is amply shown by the trade records of our Department of Trade and Commerce. During the year ending March, 1914, when all the woodworking industries were busy, veneer to the value of $284,625 was imported into Canada. Of this $284,497 worth came from the United States. For the year ending March, 1915, the figures show a decrease. From that time on, however, there was a gradual improvement. Taking the figures by months, one gets a fair idea of how the veneer-using industries have recovered from the depression caused by the war, and the recovery would have been very much greater if it had not been for the shortage of labor, as the majority of the manufacturers have sufficient orders on hand to permit them running their factories to full capacity. The value of the veneer importations for March, 1915, amounted to $13,837, and for the corresponding month of 1916, $35,908.

The kinds of veneer used by furniture and piano manufacturers are chiefly American walnut, African and Cuban mahogany, quartered oak, sawn and sliced and poplar for crossbanding and inside veneers.

The fact that piano manufacturers are rather handicapped in the matter of design influences to a certain extent the amount of fancy veneers used. That is to say, a piano has to conform to certain definite standards which do not permit of much elaboration of the case. For instance, the top must be hinged; the top door must be removable in order that the piano may be tuned; the bottom door must be removable, or at least it is customary to make it so. Thus it will be seen that the designer of a piano case has not very much scope to vary his design. He therefore relies to a great extent on the use of fancy veneers to increase the attractiveness of the case.

For walnut pianos a great deal of butt-joint veneer is used. That is, veneer in which two short pieces are jointed together end to end, the joint being exactly in the centre. Some beautiful results can be obtained in this way. Besides the butt-joint walnut, there is a great deal of American walnut used that does not require jointing; also some Circassian walnut. Where nicely-figured American walnut can be secured of a sufficient width and length to make jointing unnecessary it is generally advisable to buy it, as jointing costs money. Circassian walnut has never been used extensively for pianos, probably because the purchaser of a piano generally wants one that will match the furniture in the living-room or parlor in which it is to be placed, and living-room or parlor furniture in Circassian walnut is seldom—if ever—seen. Then, again, Circassian walnut appears to be very scarce at the present time, and with the natural consequence that it is very expensive.

There is no particular kind or grade of mahogany used for pianos. About the only thing demanded is that it be of good figure and texture, free from wind shakes, knot holes, and other defects. In buying veneer to be cut up for piano cases a great deal of money can be saved by paying particular attention to the matter of sizes when placing orders. Veneer is purchased in logs or flitches. Some times a log contains a great many flitches of various widths and lengths. The buyer may purchase the entire contents of the log, which in some cases amounts to 40,000 feet or more, or he may only purchase one or two flitches.

If the lengths of the flitches are in multiples of five feet or thereabouts they will be found to cut to good advantage if upright pianos are the product. A certain amount of latitude can be allowed in the width, but the buyer should endeavor to secure enough veneer of sufficient width to make the gables without jointing, as if these have to be jointed it means a big bill for labor, particularly where the work is done by hand. Besides this, the gables are hardly wide enough to look well with a joint in the centre. If about one-sixth of the log or flitch measures seventeen inches wide it will take care of the gables (assuming that poplar or some similar common veneer is used for the inside). If a portion of the log or flitch measures about nineteen inches wide it will be very acceptable for the top doors.

The manufacturer, as a rule, however, does not object to jointing the top doors. For the bottom doors, which are usually about twenty-one inches wide, it is very seldom that these can be made without jointing, but when they can, it means money saved. In buying veneer for a furniture factory it is not necessary to pay so much attention to the matter of size. Furniture is manufactured in such a wide range of designs and in so many different sizes that veneer of almost any dimensions can be utilized. It is advisable, however, to have a reasonable proportion of it of a suitable size for large stock. A great deal of the dining-room and bedroom furniture being shown at the present time is veneered with figured American walnut veneer. This wood shows up to splendid advantage in the William and Mary and other period designs so popular just now.

A number of furniture manufacturers seem to prefer the African mahogany to the Cuban, saying that the African has a softer and much nicer stripe than the Cuban mahogany. African mahogany is not very plentiful just now, however, on account of the lack of ships in which to transport it from its native haunts to the veneer mills. There is now—and probably always will be—large quantities of oak veneer used in furniture factories. Some of it is sliced, but the greater proportion of it is sawn. The sawn oak undoubtedly gives the most satisfactory results.

Very little sliced oak is used in piano factories, the majority of manufacturers having raised the same objection to it. The texture of the wood is made up of hard and soft portions, and when it is being sliced the knife cuts smoothly while cutting the soft portions, but when it comes to the hard spots or flakes it does not cut so easily, thereby causing the soft portion to break away at one end of the flake. Then when the veneer is laid it is almost impossible to glue all these places down, with the result that they show up when the stock is being sanded.

Practically all of the veneered work on pianos and furniture is crossbanded. Besides making a much more substantial job, it pays to crossband. Where single veneering is the practice, checks very often develop, which cause considerable annoyance, not to mention

Unique effect obtained in matching Circassian walnut veneers at the Heintzman Piano Co.'s factory, Toronto. The veneer was purchased from the Dominion Mahogany & Veneer Co., Toronto.

the expense of repairing. Crossbanding reinforces the wood, and any minute checks which happen to be in the core stock will be held securely.

Whether the growing popularity of period designs will tend to increase the use of veneer for furniture making is pretty hard to say. One thing is almost certain, however, it will not tend to reduce it. Period designs are largely made in mahogany and walnut, and call for large quantities of these woods. There is also the Jacobean and the still popular fumed oak, for which considerable oak veneer is required. The old Colonial designs, of which there are still some examples to be seen, called for a great deal of veneering, but whether they required more than the period designs only time will tell.

There are just as many big opportunities to-day as there ever were.

Why Crossbanding Warps

Editor "Canadian Woodworker"—

In looking over the September number I notice your reader V. C. asking about his trouble with crossbanding, and as I have had some little experience with the same I have taken the liberty of sending on an account of the method we use.

Our crossbanding is thoroughly dried before use, as we put it all on top of the hot table about twelve layers thick with a heavy weight on top to keep it flat; some use the double table and have the crossbanding between the two tops. It should have at least eight hours to dry, and be taken off the hot plate just before putting through the glue spreader, but should not be put through it hotter than you can handle with comfort.

I may say that we use very few cauls, only a layer of paper on the face veneer to protect it. The stock is put into the press as quickly as possible, and very seldom are we troubled with the crossbanding overlapping.

I do not think there can be any fault with V. C.'s glue if it hardens all right otherwise.—A.B.C.

* * *

Glue Should Not be Allowed to Set

Editor "Canadian Woodworker"—

In answer to question on page 40 of the September number, I think the trouble can be overcome by having the ¾ and 1¼ inch core stock warm and not giving the crossband a chance to set in places while expanding in other places, and get it under pressure as quickly as possible.

On five-ply stock ten minutes is my limit from the time I start to glue the crossband until I start it under pressure. I get better results in handling small batches than by trying to put too many pieces in the press at a time. Crossband will wrinkle and expand until you get it pressed. Has V. C. not found that the last few glued have come out in better shape than those glued first?

I do not like to use the glue too thick. Of course, chestnut core stock will take a heavier glue than birch, but I like to have an even coat on both sides of the crossband. I re-dry all my stock, as damp crossband will cause all kinds of trouble. Does V. C. start to press his batch from centre to both ends or lock all the retainers at once?—Subscriber.

* * *

Put in Small Batches

Editor "Canadian Woodworker"—

Replying to V. C.'s question on page 40 of the September issue, "Why crossbanding warps," I would say a better name would be swelling, instead of warping, for undoubtedly that is what happens. V. C. does not state which is put through the spreader, the crossbanding or the core, but it does not make any difference in this case. It is the moisture in the glue that comes in contact with the soft porous poplar, swelling it in a very short time. If the top of the pile were left as long as the bottom, it would go the same way.

Poplar crossbanding being rotary cut, the swelling occurs more between the growths, thus causing the freshly glued piece to corrugate about every two inches. This accounts for the regularity of the waves. Remedy,—try putting in smaller batches and getting them under pressure quicker, say from 10 to 13 pieces. These should be under pressure in at least 4½ minutes. Anything over that time would be excessive,

OAK   MAHOGANY   CIRCASSIAN   MAPLE

VENEERS

BIRCH   POPLAR   CHERRY   WALNUT   GUM

THE OHIO VENEER COMPANY

Importers and Manufacturers

Foreign and Domestic Veneers and Hardwood Lumber

We always carry a large and assorted stock of Mahogany, Circassian
Walnut, Sawed and Sliced Quartered Oak.

Send us your enquiries and orders. We guarantee good service.

2624 to 2644 Colerain Avenue,   --   --   **CINCINNATI, OHIO.**

Indiana White Oak
Lumber and Veneers

Band sawed from the best growth of timber.
We use only the best quality of Indiana White Oak.
From four to five million feet always on hand.

Walnut Lumber

Our stock assures you good service.

Chas. H. Barnaby, Greencastle, Indiana

 Hoffman Bros. Co.

Estab. 1867, Incor. 1904

**800 West Main Street,
FORT WAYNE, INDIANA**

Manufacturers of

VENEERS and LUMBER

In the Domestic Hardwoods

ANY THICKNESS,

1/24 and 1/30 Slice Cut
(Dried flat with Smith Roller Dryer)

1/20 and thicker Sawed Veneers,
Band Sawn Lumber.

—SPECIALTY—

Indiana Quartered Oak

Veneers and Panels

PANELS

Stock Sizes for Immediate Shipment
All Woods   All Thicknesses   3 and 5 Ply
or made to your specifications

VENEERS

5 Million feet for Immediate Shipment
Any Kind of Wood   Any Thickness

J. J. NARTZIK

1966-76 Maud Ave.,   -   -   CHICAGO, ILL.

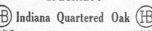

Veneers and Panels

**LARGE STOCKS FOR
QUICK SHIPMENTS**

Send for Lists

Walter Clark Veneer Co.

GRAND RAPIDS, MICH.

and the crossband would swell so as to give trouble.

In a batch of 25 pieces, make two lifts; put 12 in on bottom of press, using the other 13 pieces as packing or filling on top. After four hours, open press, remove the 13 on top, veneer and replace on top again without disturbing the bottom 12, and leave at least twelve hours. These batches should be put in conveniently in 4½ minutes each, from time of starting to glue to the time pressure is applied.

The quality of the glue has nothing to do with the swelling, it is the moisture from the warm glue.—Veneer Foreman.

* * *

Crossband Should Be Dry

Editor "Canadian Woodworker"—

In answering V. C.'s query "Why Crossbanding Warps," I have answered it only as I have read it. If I could see their veneer man at work and the veneers that are used, along with the product, I could more readily solve the problem.

It seems to me that the trouble is in the crossbanding not being dry, and as soon as it touches the hot glue it wrinkles. Then when the hot cauls are placed in position it makes matters worse. The surface begins to dry first and the moisture cannot get away. This undoubtedly causes the whole trouble.

I was foreman in a factory four years ago and we had the very same trouble. Our crossbanding overlapped and wrinkled, and some of our veneered work was just simply wrinkles and swellings which always showed through when the stock was sanded.

We experimented one way and another and as a last resort took our veneer and crossbanding into the dry kiln for drying. Our troubles then disappeared, showing plainly that the wet veneers were the cause of all the trouble. Take a thin piece of wood and wet it, place it in a warm place and it will warp, twist and swell. The same thing applies to crossband.

In the cold glue system no such trouble is experienced, but if the veneers are not dry they will peel off in a few days after leaving the presses. I have experienced all this myself, both with hot glue and cold glue.
—Foreman.

Solid or Built-up Table Tops—Which?

Leading Furniture Manufacturers Tell Experiences and Express Opinions on Interesting and Timely Subject

We publish below a number of letters received from prominent Canadian furniture manufacturers, pointing out the advantages and disadvantages of solid and built-up table tops. It is evident from the letters that both kinds have their adherents, but the concensus of opinion appears to be in favor of the built-up tops for really high-class work.

Public Demands Solid Tops.

The strongest argument that we have found in favor of the solid tops is the fact that the buying public is demanding this class of stock more and more all the time. We do not claim that a solid top is necessarily better than a veneered top. Trouble will be experienced with either if they are not properly made, and if they are properly made and handled there is no trouble with either. At the same time, a piece of furniture made of birch is just as serviceable as solid mahogany, but the people will not pay the price for it that they will for mahogany.

The fact that solid mahogany is so much preferred to veneered mahogany is a fair index of what the situation will be with regard to quartered oak as the supply is decreased. There was a time when the exponents of veneered tops claimed that solid tops could not be made so that they would not warp, joints open up, etc., but this statement is being proven wrong every day by manufacturers of high-grade furniture all over the country. For dining tables especially, we are fully convinced that the solid top is more practical than the veneered, owing to the fact that veneered stock is more readily affected by heat than solid stock.

Veneered Tops Superior to Solid

Our opinion is that a properly made built-up top is superior to a solid top, but a job of this kind adds to the expense, and is only found in a high-class article. A veneered top must have a proper core (if framed up, so much the better and more expense), properly cross-banded, and the top veneer can be as choice as required. This should give a top evenly matched and of the best quality.

A solid top is not so evenly matched as to figure, but is generally used on cheap and medium goods, owing to its cheapness in manufacture. We still consider a solid top superior to a cheaply-veneered top, as our experience has been that unless proper care is exercised in making the core and in cross-banding a veneered top is bound to give trouble. But where this care is taken the veneered top is superior to all others.

Edges Have to Be Stained.

There are a number of points in favor of both solid and built-up tops. Personally, we prefer the solid top, but in built-up tops you do not have the trouble of joints going, and you certainly secure a much better looking table by matching your veneers.

Points we have against the built-up tops are that the cores show on the edges, and as the core is always some other wood than the veneer used, the edges have to be stained. The stain in time fades. Another objection we have to the built-up top is that, even if the edges are veneered, they are liable to chip, for no other piece of furniture gets as many knocks as the edge of a table.

Time Will Tell Which Are Best.

The question of which is the better, a solid or veneered table top, has often been brought up by furniture men, and it appears to the writer that the question can only be answered by defining what kind of a top is under discussion.

If I should be asked which kind of a top would be the better for a kitchen table, I would answer, without hesitation, a solid top. For a dining table, which is subjected to frequent applications of water and other cleansing fluids, it would appear that it would be better solid, judging from a utilitarian standpoint.

A good solid top of well seasoned mahogany or walnut, jointed properly, is almost indestructible while in ordinary use. Furthermore, these woods are more cherished as they grow older. We know that the most beautiful and expensive tops are veneered, and wisely so, for how could we ever expect to develop art in the furniture world without using the beautiful inlaid patterns and fine figured veneers which have been prepared for us by skilled artisans and artists of great ability?

Our present built-up tops seem durable, but we have no laminated tops one hundred years old and older that we can use in comparison with the tops which have been left us by

Wood - Mosaic Co.

NEW ALBANY, - - IND.

WALNUT

Lumber ~ Dimensions ~ Veneers

If you are in the market for anything in Walnut, let us figure with you. We have cut three million feet of Walnut in the last year, and have a good stock of all thicknesses dry. We can also furnish Walnut dimension—**Walnut Veneers.** We have a large stock of veneers, plain, stripe figured and stump wood.

Wood-Mosaic Co., Inc.

NEW ALBANY, - - IND.

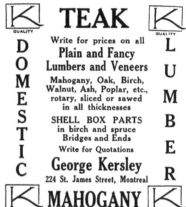

TEAK

D O M E S T I C

Write for prices on all
**Plain and Fancy
Lumbers and Veneers**
Mahogany, Oak, Birch,
Walnut, Ash, Poplar, etc.,
rotary, sliced or sawed
in all thicknesses

SHELL BOX PARTS
in birch and spruce
Bridges and Ends

Write for Quotations

George Kersley
224 St. James Street, Montreal

L U M B E R

MAHOGANY

American Walnut Veneers
Figured Wood
Sliced ~ Half Round ~ Stump Wood

5 Essential Qualifications:

1.—Unlimited supply of Walnut Logs.
2.—Logs carefully selected for figure.
3.—All processes of manufacture at our own plant under personal supervision.
4.—Dried flat in Philadelphia Textile Dryer.
5.—3,000,000 feet ready for shipment.

JOHN N. ROBERTS,   New Albany, Indiana
The Pioneer Walnut Veneer Producer of America

Mahogany and Walnut Veneer

Our recent arrival of choice logs which have been cut into veneer
now makes our stock complete in figured and plain woods, 1/28"
to ¼" in thickness, also every kind and thickness of

Domestic Veneer
and
Thin Lumber

thoroughly kiln dried. Our large and complete stock enables us
to ship promptly upon receipt of orders.

ACME VENEER & LUMBER CO.

8th, Harriet and 7th Streets   -   -   -   -   CINCINNATI, Ohio

---

200,000 Ft. ¼" Gum
12 to 36" wide, 60 to 84" long
DRAWER BOTTOMS
AND
BOX STOCK
For Quick Shipment
GEO. L. WAETJEN & CO.
MILWAUKEE   -   -   -   WISCONSIN

"Furniture in England"

This book is an encyclopaedia of artistic suggestion—
an education in itself.

Price **$12.00** Delivered to any address in Canada

Copies may be obtained from

THE WOODWORKER PUBLISHING CO., LIMITED
TORONTO   -   ONTARIO

Buyers of Veneers and Panels

will find it to their advantage to purchase from the manufacturers and dealers whose
advertisements appear in this publication. They are progressive firms—the leaders in
the business, which is a guarantee of good service and prompt attention to orders.

Give your business to the man who will spend his time and money to get in touch
with you. He deserves it—if his stock and prices are right.

Dear Mr. Consumer :

If you're using walnut, you can't afford to take chances with anything but the best.

Properly manufactured, properly matched and properly laid, American Black Walnut is the most beautiful of woods.

But you must have the right foundation in the material.

When it comes from Penrod, it's right. Just remember that, when you need Walnut Veneers or Lumber, Plain Stock or Figured.

Penrod Walnut & Veneer Co.
"WALNUT SPECIALISTS FOR THIRTY YEARS"
KANSAS CITY, Mo.

We specialize on

Poplar

cross-banding and backing

Veneer

Highest Grade — Right Prices

The largest Poplar Veneer Mill in the world.

The == **Central Veneer Co.**

Huntington, West Virginia

We Can Fill Your Requirements in **MAHOGANY**

Because we represent one of the largest importers in Chicago and the Northwest. Therefore if you need Mahogany, we are in a position to give you such service as few can render.

American Black Walnut

Particular care has been taken in the selection of our stock, which embraces some of the choicest figures we have ever seen.

Quartered White and Red Oak

Of these old favorites we carry always a large line, so varied in figure that we can suit any taste. We have many very desirable figures in flitches which will insure uniformity in quality work.

Stocks ready to ship on receipt of your order

HAUGHTON VENEER COMPANY
1152 W. LAKE STREET   CHICAGO, ILLINOIS

our forebears. It appears that time only will answer the question as to whether the present built-up tops will continue in favor.

Retail Dealer Likes Veneered Tops.

The question of the relative value of solid or veneered tops is very like the question of which is the best roadway; there are places where cobble stones give better service than asphalt and vice versa, and so with tops. A kitchen table top subject to frequent washing would be better solid, and, personally, the writer would prefer a solid dining table top to a veneered one, provided proper care was given to the selection and seasoning of the wood, etc. Should a veneered top get badly dented, it is almost impossible to restore it again, whereas the writer has planed down many an old solid top and made it as good as new.

However, in the general hurry of modern factory life and the difficulty of getting experienced labor, and with the improvements in veneering methods, the tendency is to veneer all tops. The retail dealer likes them, as generally the wood is better matched and the manufacturer finds less trouble on the whole. Veneered tops, like democracy, are still on trial, for there are thousands of solid tops still in existence that have done duty for half a century and are as good as new to-day, but it is doubtful if there will be as large a percentage of modern veneered tops in as good shape in another half century.

Built-up Tops More Desirable Than Solid.

Our experience has been that the solid top is much preferred. Built-up tops will give good satisfaction on the better grade of goods, as they receive special care and attention in use, but on the medium and cheaper grades, where they are used carelessly, by putting hot dishes on them and spilling hot liquids on them, they will not stand. If they could receive the proper attention, the built-up tops would be much preferred, as the figure in the grain is much superior to the solid wood. When properly matched the veneer makes a beautiful top, and gives an effect which it is not possible to get with the solid wood.

When library tables and parlor tables are considered the built-up tops are much superior in every way, always providing the workmanship is right. It is very particular work to make built-up stock properly, and as there has been so much of it turned out with loose methods, much of it has gone bad. In our opinion, providing the workmanship is good, built-up stock is more desirable than solid. The built-up stock costs more to make than solid, should stand more wear, and is very much more pleasing to look at.

Buyers Prefer Solid Tops.

The best argument against built-up table tops appears to be prejudice on the market against anything but a solid table top.

The built-up top, under ordinary conditions, should be as good as the solid one, but tables, as you know, owing to the use they are put to, are subjected to considerable dampness and considerable heat at times, and either of these seems to mean destruction or serious damage where veneer is used. It, therefore, seems to us as though the prejudice is well founded on the part of the consumer.

The writer has many times had to guarantee to the buyer that the table tops were all solid. Of course, a few years' time works wonders in any line, and it may be possible that built-up tops will become very popular in time to come.

Depends on Class of Job.

The use of solid or built-up table tops depends entirely upon the value and requirements of the job and the cost of material to be used. If the tables or desks for which the tops are to be used are made of quartered oak, mahogany, or some other valuable material, the job would, of course, be high grade and would warrant the use of built-up tops. In

this case the saving in expense by using veneer instead of solid material would offset the extra cost of labor in building up the tops.

But, on the other hand, if the goods are cheap or medium grade, and the material consists of some of our native woods, which are not expensive, we would hardly be warranted in going to the expense of built-up tops. The solid material, if well jointed and glued and secured to the desks or tables, will keep its place as long as the veneered top, and be less liable to damage from bruises. It would, therefore, seem to us that it is not a matter of preference, but rests entirely upon the requirements of the job, whether we use solid or built-up material.

Built-up Tops Invite Trouble.

For extension tables we prefer solid tops. Built-up ones are apt to blister when hot utensils are placed on them. Moreover, built-up tops invite at all times and under all conditions more or less trouble through veneer loosening and peeling off.

Found Built-up Tops Unsatisfactory.

Our experience with built-up extension table tops was not satisfactory. Some years ago we purchased about five dozen tops, and they gave us so much trouble that we did not try any more of them. We do not think that a built-up table top can be made to stand the different degrees of atmosphere they are subjected to. We make all our tops solid, as we have found them to give much the best satisfaction.

Manufacture Only Solid Tops.

We have only on a few occasions manufactured built-up tops and invariably have not found them as satisfactory as a solid top. At the present time in our entire line we do not list a single top other than solid, which is the best proof of our opinion on this matter.

Perforated Veneer Tapes

The Ideal Coated Paper Company, of Brookfield, Mass., have for many years specialized on veneer tapes of every description for the taping of veneers. The company state that their tapes are the only ones recommended by the builders of the Dennis taping machine for use on that machine.

With the object of reducing the cost in taping they put on the market the "Ideal" perforated tapes, designed for the express purpose of saving the time, labor, and expense entailed in the sanding off or otherwise removing ordinary tapes, as the perforated tapes can be left on the veneers, perfect joints being made with them. In three-ply veneers the perforated tape is applied on the underside of the face veneer, the perforations permitting a wood-to-wood contact directly over the joints when the veneer is glued to the centre or core stock, thus insuring a perfect joint.

For this work the large single perforation is recommended. The economy in the use of this tape can be readily appreciated. For taping crossbanding stock greater strength is required, and for this purpose the "Ideal" double perforated tape is recommended. The Ideal Coated Paper Company advise us that their tape is very much cheaper than the open mesh gummed cloth tape which was formerly used in this work, and that it is better adapted for the purpose. In fact, they say that their tapes have practically done away with the use of cloth tapes. If there are any who are not acquainted with "Ideal" perforated tapes, liberal sample coils can be obtained direct from the mills at Brookfield, Mass., or through one of the company's branch offices at New York, Cincinnati, or Chicago.

William and Mary Design Buffet
made from American Black Walnut Veneer

Cut by courtesy of
Chas. Rogers & Sons, Toronto

American Black Walnut Veneers

The recent heavy demand for walnut furniture shows no sign of letting up. We are ready to supply you in almost unlimited quantities of very highly figured stump wood, straight sliced stripe, and curly figured stock, as well as the semi and full rotary cut stock.

Mahogany Veneers

You know how hard it is to get a high-grade mahogany—or any kind of mahogany, for that matter—and, anticipating this condition some time ago, we prepared, ourselves, and have an excellent stock on hand—some of the finest figures you ever saw, too.

Quartered Oak (in all thicknesses)

We carry a large stock of this wood in all thicknesses, both in sawed and sliced veneers. It is all high-grade stock, specially selected for the furniture trade, in extra widths.

Figured Gum

This is becoming a very popular wood, and is demanding a ready sale among the trade. We can supply you with any quantity in a very large assortment of figures.

All these woods are carried in stock at our warerooms, where you are invited to call and personally make your selections.

Write for samples and prices

TORONTO VENEER COMPANY
TORONTO   93-99 Spadina Avenue   ONTARIO

Electric Glue Pots

The use of electricity has made it possible for the workman to keep his glue hot while working with it at his bench. The electric glue pot is now used extensively in furniture and piano factories. In some branches of both these industries it is necessary to have the glue pot on the bench all the time.

This is particularly true of the furniture industry,

Electric glue pot without water jacket.

where chairs are the product. In the action departments of piano factories it has been customary to have bench glue pots heated with alcohol lamps. These, besides being dangerous, in time consume considerable alcohol. The electric glue pot would therefore appear to be particularly adapted for this work. It

Electrically heated glue pot with water jacket.

would be much easier to turn the current on and off at a lamp socket than to light matches, and besides, this last is an exceptionally dangerous proceeding.

Another department in both furniture and piano

factories where the electric glue pot could be used to advantage is where the veneer is taped and matched. Every factory has not sufficient of this work to warrant installing a taping machine, and consequently, the glue has to be applied by hand. For this work it is absolutely necessary to have the glue pot on the bench. The system generally followed is to have two pots on the heater, so that they can be used alternately, and the glue kept warm. This, of course, incurs loss of time that would not be necessary if the pot could be kept on the bench.

The electric glue pot is one of the successful and important applications of electricity to the mechanical arts. It offers an efficient and practical method for the preparation of glue and is superior for some purposes

A row of benches equipped with electric glue pots.

to a pot heated by the ordinary methods—i.e., steam, coal, gas or kerosene oil—is portable and permits of extremely flexible arrangement.

There are two types of electric glue pots, viz., jacketless, having no water jacket and being heated by direct thermal contact with the walls of the glue pot, and the water jacketed type, having an interposed heating medium—water. We illustrate herewith two glue pots. Fig. 1 is the jacketless type, and is made by the Canadian General Electric Company. This company also make a water jacketed glue pot. Fig. 2 is a water jacketed pot made by the Canadian Westinghouse Company. The other illustration shows a row of benches with electric glue pots made by the Canadian General Electric Company in use.

ACME
STEEL GOODS
BOX STRAPPING
Our Specialty

Acme Galv. Twisted Wire Strap

Acme Cold Rolled Steel Hoop.

Bright—Galv.—Lacquered

WRITE FOR CATALOG

Coil of 5000-ft. Acme Twisted Wire Strap

ACME STEEL GOODS CO.—MFRS.

2840 Archer Ave., Chicago.
Coristine Bldg., Montreal, Que.

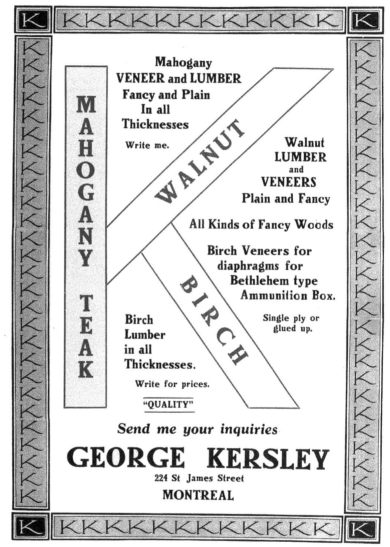

MAHOGANY TEAK

Mahogany
VENEER and LUMBER
Fancy and Plain
In all
Thicknesses

Write me.

WALNUT

Walnut
LUMBER
and
VENEERS
Plain and Fancy

All Kinds of Fancy Woods

Birch Veneers for
diaphragms for
Bethlehem type
Ammunition Box.

BIRCH

Birch
Lumber
in all
Thicknesses.

Single ply or
glued up.

Write for prices.

"QUALITY"

Send me your inquiries

GEORGE KERSLEY

224 St James Street
MONTREAL

Grasp the Opportunity

The Yard on which the lumber listed below is located is being closed out and especially attractive prices will be made on all items. *All thoroughly dry—good stock. Prompt shipment. No car shortage.*

QUARTERED WHITE OAK

4,000 ft. 3/4" (scant 4/4")
1s and 2s and No. 1 Com.
2,500 ft. 4/4" 10" and up, 1s and 2s.
10,000 ft. 4/4" 5" to 6" Clear Strips.
4,000 ft. 4/4" 2½" to 4½" Clear Strips.
40,000 ft. 4/4" 2½" to 5½" Common Strips.
25,000 ft. 4/4" No. 2 Common.
7,000 ft. 4/4" No. 3 Common.
2,000 ft. 5/4" No. 2 Common.
1,200 ft. 6/4" 2½" to 5½" Clear Strips.
10,000 ft. 6/4" No. 2 Common.
1,000 ft. 8/4" 1s and 2s.
6,500 ft. 10/4" 12/4" and 16/4" 1s and 2s and No. 1 Common.

QUARTERED RED OAK

14,000 ft. 4/4" No. 1 and No. 2 Common.
6,000 ft. 5/4" No. 1 and No. 2 Common.
7,000 ft. 6/4" No. 1 Common.
3,800 ft. 8/4" 1s and 2s.
1,500 ft. 8/4" No. 1 Common.

PLAIN OAK

7,000 ft. 3/4" (scant 4/4") 1s and 2s and No. 1 Common.
66,000 ft. 4/4" No. 1 Common.
14,000 ft. 4/4" No. 2 Common.
22,000 ft. 5/4" No. 1 Common.
33,000 ft. 5/4" No. 2 Common.
1,000 ft. 6/4" 1s and 2s.
22,000 ft. 6/4" No. 1 Common.
20,000 ft. 6/4" No. 2 Common.
32,000 ft. 8/4" 1s and 2s.
31,000 ft. 8/4" No. 1 Common.
98,000 ft. 8/4" No. 2 Common.
15,000 ft. 8/4" Sound Common.
1,000 ft. 12/4" 1s and 2s.
3,000 ft. 12/4" No. 2 Common.

POPLAR

7,200 ft. 4/4" 1s and 2s.
2,400 ft. 4/4" 1s and 2s, 11" and up Sap, no Defect.
700 ft. 4/4" 4" to 6" Clear Strips.
30,000 ft. 4/4" No. 1 Common.
5,200 ft. 4/4" No. 2 Common.
1,100 ft. 5/4" 1s and 2s, Sap no Defect.

1,100 ft. 5/4" No. 1 Common.
20,000 ft. 6/4" 1s and 2s, S.N.D. Extra Fine,
5,500 ft. 8/4" 1s and 2s, Sap no Defect.
15,000 ft. 8/4" No. 1 Common.
3,800 ft. 12/4" 1s and 2, Sap no Defect.

ASH

15,000 ft. 4/4" No. 1 Common.
12,000 ft. 4/4" No. 2 Common.
5,790 ft. 5/4" No. 2 Common.
32,000 ft. 8/4" No. 1 Common.
22,000 ft. 8/4" No. 2 Common.
2,100 ft. 10/4" No. 2 Common.
4,300 ft. 12/4" 1s and 2s.
5,000 ft. 12/4" No. 2 Common.
2,500 ft. 16/4" 12" and up 1s and 2s.
4,000 ft. 16/4" 1s and 2s.

CHERRY

800 ft. 4/4" No. 1 Common.
400 ft. 6/4" No. 1 Common.
9,700 ft. 8/4" 1s and 2s.
6,100 ft. 8/4" No. 1 and 2 Common.

CHESTNUT

25,000 ft. 4/4" 1s and 2s.
22,900 ft. 4/4" No. 1 Common.
4,100 ft. 5/4" No. 1 Common.
9,100 ft. 6/4" 1s and 2s.
9,000 ft. 6/4" No. 1 Common.

GUM

5,000 ft. 4/4" No. 1 Common, Sap.
1,200 ft. 6/4" No. 1 Common, Sap.
2,000 ft. 3/4" 1s and 2s, Red.
7,000 ft. 3/4" No. 1 Common, Red.
2,000 ft. 6/4" No. 1 Common, Red.
1,900 ft. 3/4" 1s and 2s and No. 1 Com., Qtd. Red.
5,300 ft. 6/4" 1s and 2s. Figured, Qtd. Red.

MISCELLANEOUS

3,900 ft. 8/4" 1s and 2s Basswood.
1,000 ft. 8/4" No. 3 Common Basswood.
2,000 ft. 8/4" Log Run Walnut.
500 ft. 4/4" 1s and 2s and No. 1 Com. Maple.
300 ft. 8/4" 1s and 2s and No. 1 Com. Maple.
2,100 ft. 10/4" 1s and 2s Maple.
500 ft. 6/4" 1s and 2s Red Birch.

SPECIAL—75000 ft. 8/4 in. American Black Walnut

Extra fine—band sawn and equalized, exceptionally good widths and lengths. Sawn for No. 1 Common and Better but will contain small per cent. good No 2 Common. Walnut of this quality is rare and it is bone dry.

The John Dulweber Co.
Cincinnati, Ohio

THE FINISHING ROOM

Applying Finish by Compressed Air Method

By M. A. Burt

The earliest practical method of applying wood finishes was by means of the hand brush. Then, with the demand for a cheaper and quicker process, the dipping method came into vogue. No doubt each of these methods has its peculiar advantage over the other, according to the conditions prevailing, but it is interesting to note that within the past six or eight years there has been a rapid progress made in what is called the air brush system of applying finish, and that it is fast replacing both the hand brush and the dipping methods. Under many conditions the air brush has been found more rapid and economical than either of the other two processes.

The system has been given such prominence in trade journal and direct advertising that it seems scarcely necessary to give an extensive technical description of the equipment in this article. The several manufacturers stand ready to give such information without in any way obligating the one seeking the information.

Several years ago, in a Minneapolis, Minnesota, factory, I first saw the air brush in operation and asked the foreman if he preferred that method to dipping.

"Far better than dipping or hand finishing," was the answer."   -

"Why?"

"In the first place, we are saving about 35 per cent. in labor. Then in finishing the interiors of our kitchen cabinets the operators can reach the corners and other difficult places with no trouble at all. They just stand away and 'shoot' the spray. We put on all the coats with these machines and would not think of going back to the old method."

"But how about the loss of finishing material?" I asked. "It seems to me that there is a lot of 'spray' going 'up the flue.'"

"You are right about that," he told me. "At first we used to have about 23 per cent. waste, but we have reduced this to about 16 per cent., and as the operators become familiar they are lowering that figure. Even with the greater waste, the saving in time more than makes up for that loss."

A few months later a visit was made to a case goods plant at Stevens Point, Wisconsin, and the manager was asked if he had anything of special interest. "Come to the finishing room!" he said. Here there were four booths for spraying. "I should think that the nature of your product would make the dipping process better for you," I said.

He smiled and answered, "We used to dip practically everything, but have learned better. In the dipping process, whether on panels of veneer or solid material, we always had our thin and thick places on account of the running of the finish. But our worst trouble came from the built-up material, of which we use a large quantity finished with water stains. The chief trouble with the dipping process was that frequently the stain was not thoroughly, or quickly

enough, wiped off, and the superfluous moisture was not removed before it had time to soak through the face veneer. The natural result was that the glue softened and the veneer worked loose, and open joints were one of the ills that we had too much of before we installed this spraying equipment."

He told me lots of other things, but why get all the information from one source?

Last March I visited a piano factory in Chicago, Illinois, the management of which claims the honor of building more high grade pianos than any other plant in the country. Then in June I called on one in Greater New York that makes practically the same claim. Both pianos are of national reputation, and far be it from me to take sides. The point is that the air brush equipment was in use in both plants.

I remarked in each that I thought the finish for high grade piano work required the hand brush, and in each place I was informed to the effect that such an old slow-going method was obsolete. They wanted to know why a progressive manufacturer should be annoyed with unevenness, sags, runs, streaks, fatty edges, brush marks, etc., that must eventually be rubbed down or touched up, when with a spray they could apply a full coat of material on a vertical surface and have the work uniform in thickness, thereby saving the rubbers much work.

First and last, during my visits in woodworking institutions, large and small, making interior trim, refrigerators, chairs, and every kind and quality of household furniture, and in visits that have extended from the Atlantic seaboard to the Mississippi River, I have found many installations of the air brush system, some of one make and some of another, all working successfully, and some of the information gathered is here recorded for such as may be interested.

In brushing carved or ornamental relief work the hand finisher encounters the difficulty of the low spots collecting more varnish than the high places, and the finish may be far from perfect. Using the spray on this work results in a surface that is covered uniformly and evenly, it all being accomplished with one sweep of the hand at even speed.

In order to have the finishing material of the same consistency when applied by air brush as it is when brushed, it is sometimes necessary to reduce the material slightly with turpentine or some other solvent. The matter of reducing materials is governed by conditions existing in different plants. Then, too, since the inception of the spraying system most varnish manufacturers have turned their attention to the making of varnishes particularly adapted to spraying. Generally speaking, it is safe to say that the use of a free flowing coach varnish requires no reducing; rubbing varnishes are reduced slightly, as they do not flow as readily as coach varnishes. The sprayed coat of rubbing varnish is more easily and quickly rubbed than when brushed, and the undercoat more easily sanded.

In spraying pigment primers or first coats, it is better to have the pigment content less than for brushing. Because of the greater uniformity of surface, a flat finish more closely resembles a rubbed finish than when brushed. Most finishers find it difficult to brush shellac uniformly and without laps. In the application of varnish the "air brush" shows up to great advantage. The sprayed coat of shellac is smooth and uniform.

In brushing white undercoats and enamels, several coats are necessary, while in spraying—owing to the uniformity of surface and the absence of brush marks— one or more coats may frequently be saved. Usually one coat of white primer when sprayed will cover a piece of work as well as two coats when brushed.

One factory mentioned above was successful in spraying stains and fillers. There are others, but the process is not so applicable to these materials as to others mentioned. Economy of time is greater with the shellac and primers, on work with many angles and surfaces the saving is very noticeable. As one can readily understand, varnish and enamel coats cannot be handled as rapidly as the undercoats.

Although the several types of spraying equipment offered are of fine mechanical construction they will be found fairly simple to operate, and the man of average intelligence will find a week ample time in which to become reasonably proficient.

Under normal conditions there is usually some loss of material where the air brush system is used. Users admit of a loss varying from 8 to 25 per cent. With respect to this question, however, let me present the opinion of one of the partners of a firm making white enamels and varnishes. Early in September we were discussing several questions relative to finishing room problems, and I asked what he thought of the spraying proposition.

"It is one of the most progressive things that has happened in the furniture industry for years," was the answer.

"Is it not pretty soft for the manufacturers of finishing materials?" I asked.

"You refer to the use of more material," he answered. "Yes, but you know it is not necessary to use any more material with the air brush than with the hand brush. In fact, I believe as good work could be done with less. The extra loss is caused by two things; I may say three—ignorance, maliciousness and graft. Plainly speaking, if the operator is honest and intelligent he soon learns to operate the apparatus successfully and to use a minimum amount of material. He may be spiteful at the installation of a system that reduces labor, or he may be the recipient of certain graft, more or less common, in which case he will use as much material as possible."

That last paragraph came from one of the largest manufacturers of enamels in the United States, and should give users as well as non-users of the spraying equipment something to think about. The air brush, like most things of benefit, had its difficulties in the experimental stage, to pass through, but it is now in a state in which no furniture man need fear its use in his plant, especially if the apparatus be made by a reputable concern. No good manufacturer wants to install his equipment where it will not be a benefit to the user, and the intending purchaser will be given every proof of its efficiency before it is installed.

Should the Foreman-Finisher be a Chemist?

By K. A. Skinner, in charge of Specialty Department of A. Ramsay & Son Company, Montreal

In writing this article I have tried to use phrases and terms easily understood by the foreman-finisher and those under him, down to the man who applies the stain and filler, hoping it will be of some help to the workman who is endeavoring to advance his position, i.e., one who is trying to do his work a little better— for that kind of workman of to-day is the foreman-finisher of to-morrow.

After giving the subject some thought, and judging from my past experience as foreman, I can find no reason why a foreman-finisher should be a chemist, or of what use the knowledge of chemistry would be to him in his regular work of taking the furniture from the wood department and carrying it through the different stages to completion.

The materials used in this work are, first, the stain and filler (the stain being of many different kinds— golden oak, fumed oak, early English, etc., as well as mahogany—which all contain more or less aniline dye) and then paste filler (transparent and colored), and different lines of varnish.

Let us take, for example, the varnish, which I consider the most troublesome of any one of the lines used by the foreman-finisher. And it is usually at a time when the department is full of rush orders that (to use an old expression) "the deviltries of varnish" appear. The foreman-finisher immediately seeks the cause of his trouble. He goes carefully over every detail and all the operations, and finds that each has been done according to rule and precedent. Then a complaint goes at once to the maker of the varnish, who writes that the varnish is exactly the same as the previous shipment, consequently the trouble must be in the finishing department.

Now, if our foreman-finisher has a knowledge of chemistry and a perfectly equipped laboratory and plenty of time to wait he analyzes his varnish and finds it is composed of any one of the many gums used— Congo, pontianak, manilla, or kauri—and different oils —chinawood or linseed—and thinners—turps., benzine, etc.—and after all this trouble and work he would find any varnish maker would have been pleased to tell him what the ingredients were for the asking, but he would not have told him the manipulation of the different ingredients, or the different heats that were applied, or, in fact, how to make varnish, consequently the foreman-finisher finds himself really "up against it," with no chance of getting the trouble straightened, and he then tries samples of varnishes from every maker in the country.

I would like to suggest a system which helped me a great deal. Instead of studying chemistry I had a testing room, properly equipped with everything needed to make conclusive tests. After trying out many makes of varnish I found one that seemed to answer my purpose satisfactorily. I then took a gallon sample of this varnish, sealed it up, and called it "Standard No. 1"; also my polishing varnishes—in fact, every one of the different lines used in finishing the article, and when my trouble day came along (and every foreman-finisher surely has one!) if the fault was not in the application, or that I had rushed the work, I usually had the varnish maker in a hole, for my standard was there to be compared with the last barrel which gave me so much trouble, and after a thorough investigation it was usually found that a new gum had been used, or a new thinner, or the wood-oil preparation was not just the same as last, or some one of the many different things which would cause the varnish to be a little different than the last. Insist on a well-equipped testing department. Have your tests going the year round, with records and dates, and let the varnish maker study the chemistry end of it.

Compressed Air as a Cost Reducer

The method of applying stains, fillers, shellacs, varnishes, etc., that has been handed down to us from the days of our forefathers was by means of brushes held in and manipulated by the hands of men, the quality of the work depending on the skill of the men. If one possessed a higher degree of skill than another, the result was, perhaps, superior and inferior work on different parts of the same job.

Another feature was the passing of the system of training apprentices. In recent years, in order to secure good brush hands it has been necessary to engage men and train them. This is an expensive method. In the first place, they invariably have to be paid men's wages; and during the time they are becoming proficient the work is costing the employer more than the selling price of the product will permit and net him a reasonable profit.

Increase of speed in applying finishing materials was obtained by the adoption of the dipping process, but this method proved satisfactory only on cheap grades of furniture. Even here its economy is questionable on account of the waste of materials. When the article is dipped, every portion, seen and unseen, is coated alike, and considerable varnish, or whatever material is being applied, lodges in the corners, etc.

Then the system of applying finishing materials by means of compressed air was introduced. It appears to possess all the qualities of the hand-brush method with quite a number added. The DeVilBiss Manufacturing Company, of Toledo, Ohio, have been particularly successful in their efforts to develop the compressed air system. The equipment manufactured by them is known as the Aeron system, and is in use in a number of furniture factories in Canada.

It is claimed for the Aeron system that it covers the work smoothly and uniformly without sags, fatty edges, streaks, or thin spots, whether the surface is large or small, flat or irregular. An intricately carved piece of work can be finished in approximately the same time as a flat panel of the same area and with just as much uniformity. Thin edges and corners present no difficulties. The Aeron applies stains, primers, surfacers, shellacs, varnishes, undercoats, and enamels.

There are two styles of Aerons—one with the attached fluid cup and the other without the cup, receiving the fluid from a container placed overhead. Of these two styles there are several sizes and different types, suitable for different classes of work. Then there is the air compressor and the air receiver, and also the transformer set, together with air duster for regulating the air pressure and purifying the air, and for cleaning the parts to be finished.

There is also the electric air heater attached to the Aeron at the last possible point of contact, which heats the air and keeps it heated till it reaches the work and also raises the temperature of the material. The steel fumexers, in which the aeroning is done, are made in a number of sizes, equipped with fans and exhaust pipes to take away all fumes, etc. Through the combination of the fireproof steel fumexer and the "auto-cool electric exhaust fan" all noxious, poisonous fumes and vapors arising from the work are confined to a given area and then entirely removed from the finish-

ing room. At no time does the operator come in contact with the spray or offensive, stifling, and dangerous fumes.

Some of the most important features claimed for the Aeron equipment are: The maximum volume of spray obtainable can be varied by simply turning a knurled screw or nut. The sensitiveness of the trigger control enables the operator to regulate the spray to any degree almost without effort. A slight pressure of the thumb or finger held in a natural position on the controlling lever opens the air valve and starts the spray. Releasing the pressure closes the valve. An adjustable stop limits the opening. The valve is very sensitive, and can be opened to any desired degree. The Aeron is cleaned by spraying a thinner or solvent through it after using. If the Aeron is left standing for a time, so that the material dries in the tube, a slight pressure on the cleaning needle or a movement of the control lever will remove the obstruction. At night the complete spray head of the attached cup type can be removed by turning a thumb-screw and swinging back a clamp. The spray head can then be placed

The Aeron System of finishing.

in the solvent over night. In the case of the separate container Aeron, the whole Aeron can be kept in the solvent when not in use, or if left connected with the fluid hose and container, it need not be cleaned at night.

A single motion of the clamping lever on the attached cup type Aeron releases the fluid cup, and, after refilling, it can be replaced and locked in position in an instant. The automatic locking safety block and tackle provides a means for lowering the five-gallon container, from which the separate container type of Aeron receives the material for refilling.

The DeVilBiss Company have a competent service organization which is at the disposal of users of the Aeron system. Its purpose is to co-operate with them and enable them to secure the best possible results from their installation.

How to Produce Good Enamelled Furniture

Carefully Prepared Wood the First Essential—Flat Coats Must be Followed by Gloss or Semi-gloss—Thin Coat of Zinc White Before Enamelling an Advantage

By James Culverwell*

James Culverwell.

The enamelling of woodwork requires a considerable amount of practical k n o w - ledge to get good results. There are different grades of work done, a great deal depending upon the quality and finish of the article to be enamelled; also upon the materials used, and how applied. Woodwork roughly finished cannot under any circumstances be made into a perfect job of e n a m e l l i n g without endless labor. Plane marks, scratches by sandpaper, or the raised grain of wood will show through twenty coats of paint, and more so when enamelled. It must be borne in mind that the more gloss you require the more perfect must be the foundation. Nothing is more discouraging to the finisher than to see plane marks through his work when finished.

The cheaper grades of work do not get either the proper attention of the cabinetmaker or the preparation or finish by the finisher, let it be doors, furniture or whatever the article that is to be enamelled. It is usual to shellac the knots, prime two coats (red lead should never be used as a primer when work is to be finished white) and one coat of enamel. The better grade work is first cleaned up by the cabinetmaker, care being taken to always sandpaper the way of the grain. Never allow the sandpaper to cross the grain or scratches will readily be seen to appear which will remain all through the work and give an uneven appearance. Sandpapering being finished, the work must be thoroughly dusted, care being taken to free the mouldings from dust.

Your attention is next drawn to the knots and rosinous portions, which must be carefully treated. There are many ways of treating knots. The knot must be well covered, but not thickly covered, otherwise the edge of the shellac will show through the entire job, particularly when enamelled. It will be clearly seen that two thin coats will be far better than one thick one. A good way to treat shellaced knots is to take the edges off the shellac after with the finger by spreading it completely out. It must not be thought that sandpaper will cut down the edges when dry, as this cannot be done successfully. It will therefore be seen that care in shellacing knots is an

important factor in insuring passable work. When the knots have been treated the work is primed with white lead broken up in oil, a little white dryer, and a small portion of turpentine.

This should be allowed to dry at least forty-eight hours. (Cheaper work is usually painted again after twenty-four hours). It is then lightly sandpapered with either No. 1 or No. 0 paper. When well dusted, all stopping and facing up should be done before painting again. The second coat consists of white lead, one-eighth linseed oil, seven-eighths turpentine. Let stand at least twenty-four hours; when hard it should be finely sandpapered and well dusted off. The paint must be of a different nature to the two previous coats. To use two or three coats of flatting is not right and will injure the enamel. Flat coats must be followed by gloss or semi gloss, and as oil is injurious to white, a little of the enamel can be used in the paint with advantage.

After this coat has been allowed to stand and get hard, it is again finely and carefully sandpapered and well dusted. It is then ready for the enamel, which can be applied. A thin coat of zinc white after the last coat of paint and previous to enamelling is an advantage. By the above process a fair job can be accomplished. There are many ways of making enamels, but to-day there are so many different kinds and qualities by so many different makers that it does not pay to attempt to make it.

There is yet a better grade of enamelling which is done in the following way: The woodwork must receive special attention before anything in the way of painting is done. It must be absolutely free of plane marks and scratches, or marks of any kind. As I said previously, knots can be treated in various ways. In some cases the knot is taken out. But a good and certain cure for a knot is to put on gold leaf applied as follows: place with the finger a rub of white Japan, rub it out level, then place the gold leaf on, cover the knot about one-half inch all round and let stand a few hours before painting.

(To be continued in November issue)

Air Brush Method of Finishing

The finishing of furniture is important. The average buyer invariably examines and criticizes the finish, while paying little attention to the mechanical details of construction. The reason for this is apparent. The details of construction are concealed, and about the only person who knows where to look for them is a mechanic or a person conversant with furniture manufacturing. The finish is different. It is the "face" of the piece of furniture, and, apart from the design, practically decides whether the sale will be made or not.

Then, again, in order that the reputations of the manufacturer and the dealer that handles his goods may be sustained, it is essential that the finish be durable as well as attractive. Then last, but by no means least, there is the question of manufacturing costs. The purpose of a business is essentially to make money, and the furniture factory being no exception to the

*Mr. Culverwell held the position of instructor of wood finishing in the Toronto Technical School for three years. He has had wide experience in the trade and will be pleased to answer any questions submitted by our readers on matters pertaining to wood finishing—Editor.

BY APPOINTMENT TO

HIS MAJESTY THE KING

Established 1791

Wm. Harland & Son wish to announce to the Furniture Manufacturers that in spite of the great increase in the cost of raw materials and labor, the price and quality of Harland's Varnishes are the same to-day as in July, 1914.

This has only been made possible by carefully thought-out changes in their organization, and still more, by the co-operation of the Furniture Trade itself

Send them your orders and by your cooperation Wm. Harland & Son will be able to keep the high standard of Quality which has been the Firm's watchword for One Hundred and Twenty-five years.

WM. HARLAND & SON
Varnish Makers
400 Eastern Avenue

Toronto, Ont.

Factories:—Toronto, Ont.
Buffalo, N.Y.
Merton, Eng.

Established 1895

Skedden Brush Co.

Manufacturers of

ALL GRADES OF

Brushes and Brooms

102-104 Bay St. N., Hamilton, Ont.

A Real Factory Broom

TRY IT!

Our stapled metal-cased Brooms are made specially for factory use. Have greater strength, greater sweeping surface, wear longer, and will not cut at the shoulders when sweeping about machinery.

Patented

Manufactured by

Stevens-Hepner Co., Ltd.

Port Elgin, Ont.

"ENAMELS"

Our business in both White and Old Ivory Enamels and Primers has trebled during the past year

WHY?

They are the best on the market. If you are not using them, write for samples and prices.

The
Ault & Wiborg Co. of Canada, Ltd.
Varnish Works
Montreal   Toronto   Winnipeg

general rule, it is necessary to place their product on the market having a pleasing and durable finish and yet at a cost that will enable the manufacturer to meet competition and make a profit sufficient to meet his obligations and insure the continuity of his business. The very nature of the work of applying liquids to the surface of a piece of furniture or its unassembled parts has until recently defeated all attempts to accomplish it cheaply and satisfactorily.

Furniture manufacturers will, therefore, be interested in knowing that the system of applying finishing materials by means of compressed air or the "air brush" has proved successful, and that it is now possible to secure first-class work in the finishing department at a reduction in cost that calls for the attention of all progressive manufacturers. Varnishes, paints, enamels, shellacs, stains, fillers, etc.—in fact, liquids of all kinds—can be applied with the air brush. Further, they can be applied more evenly, more economically, and more satisfactorily than by the hand-brush method. The air forces the finishing material into every crevice by the simple sweep of the hand and the pull of a trigger. This simplicity enables the operator to turn out the maximum amount of work without be-

Showing the Paasche air brush in operation.

coming fatigued, as he would in doing the work with a hand brush. Inexperienced men can be trained in a very short time to produce better work than an old-time brush hand. It is possible to keep the finishing room much cleaner, and the improved conditions are reflected in the better spirits of the workmen.

The Paasche Air Brush Company, of Chicago, have been prominent in the manufacture of finishing equipment. An illustration showing their equipment set up for use and the method of operating it is reproduced on this page. Best finishes are obtained by heating all materials as well as the air, and the Paasche Company have made provision for this contingency by providing a heat unit so constructed that any degree of heat desired may be acquired by simply manipulating a three-heat switch—low, medium, high—conveniently arranged for this purpose. As raw, damp, and cold air can frequently be traced as the primary cause of a poor finish, the advantages of this heating appliance will be appreciated.

Another feature of the equipment is the oil and water separator, which has been devised for the purpose of eliminating all moisture, dirty oil, and grease from the finish. The Paasche system of air brush finishing is acknowledged by large and progressive plants in the United States and Canada as a great time and labor-saving device, turning days of hard work into hours of greatest efficiency.

Production Days Now At Hand

Picturing installation in factory of one of the 3,000 highly satisfied Aeron System users.

The **dominant idea and main effort** in the finishing department now is **to improve** present **quality of work; to check rising** finishing **costs; to increase output,** and obtain maximum results.

As a **quick, absolute and proved solution** to any such problems that may be yours, let us suggest that you

Let the *Aeron System* Help You

Surpasses Brushing · Finishing

It makes it possible for you to accomplish these very things and more.

Actual, average performance—not "what it ought to do"—shows that the **AERON SYSTEM** of painting or varnishing with air enables 1 man to do the work of anywhere from 2 to 6, or more, men working with a brush or other less efficient methods—thereby not only checking the rise in but actually cutting finishing costs 50 per cent. and more.

—Also, it shows that the **AERON SYSTEM** cannot be equalled for applying shellacs, varnishes, lacquers, practically any materials on wood and metal products, and that it always produces the highest quality of work possibly obtainable.

—Still further, it shows that the **AERON SYSTEM** is the positive means of placing the finishing department on an efficient and economical working basis—insuring the best results at all times, and giving a speed-up to production that means the desired output, up to capacity, at any time.

We have more **FACTS** that are worthy of consideration, and a new, interesting, informative booklet, we should like to send you. SIMPLY ADDRESS—

The DeVilbiss Manufacturing Co.

1318 Dorr Street,   *Aeron System*   **Toledo, Ohio, U.S.A.**

FURNITURE
REFRIGERATOR
KITCHEN CABINET
and METAL GOODS

MANUFACTURERS

Everywhere

Who have used makes of all
Lower Priced and Less Efficient
Systems on the market prefer

These Greatest Time and Labor-Saving Devices for reason of speed, Economy, Efficiency and Quality of Results produced.

Paasche Equipments cut your finishing costs 50 to 80 per cent., trebles your output and saves hundreds of Dollars.

Find . . .

PAASCHE

Superior Finishing Equipments

Absolutely UNEQUALLED for applying

Varnishes, Paints, Enamels, Shellacs, Stains, Fillers, Bronzes, Lacquers and Liquid Materials of Every Description.

"THE PAASCHE WAY" ⬅➡ Unquestionably the Best

EASIEST TO OPERATE

Due to the Unmatchable Features :—

- Convenience of Handling Materials.
- Perfect Heating Unit.
- Efficient Water and Oil Separators
- Paasche's Ideal Pneumatic Brushes Insure Most Perfect Control and Distribution of Materials for any and all requirements.
- Superior Exhaust and Finishing Cabinet, Insuring Perfect Ventilation.
- Admitting ample light—Convenient and easy to clean—Made in all sizes for all requirements.

Advise us as to Type of your electric current, voltage and cyclage. Also state whether or no you have compressed air in your plant and we will give you complete detailed information.

Nova Scotia Steel & Coal Co., Ltd.
L. Matheson & Company, Limited
Beach Furniture Company, Limited
Fisher Motor Company
Gray Dort Motors, Limited
Geo. Gale & Sons
General Railway Signal Company
The Victoriaville Furniture Company
The Steel Company of Canada
Canadian Fairbanks-Morse Company
The Globe Wernicke Company
National Cash Register Company
Dahlstrom Metallic Door Company
and
Hundreds of equally well known American manufacturers.

The Most Progressive — Leading Plants

Throughout the Dominion of Canada of which we name a few Use

Paasche Superior Finishing Equipments Exclusively

There's a reason.   Take it up with us—
We will tell you why.   Write today.

Paasche Air Brush Co. Inc.

MANUFACTURERS

No. 88 Madison Terminal Bldg.   CHICAGO

BOILER EXPLOSION IN WOOD BENDING FACTORY, LONDON, ONT. BOILER HOUSE TOTALLY DESTROYED.
FACTORY BADLY DAMAGED

BOILER INSPECTION AND INSURANCE

PERIODICAL INSPECTION OF STEAM BOILERS BY EXPERTS

ENGINEERING ADVICE AS TO INSTALLATION AND OPERATION OF STEAM BOILERS

PLANS FOR BOILER SETTINGS, DUTCH OVENS, CHIMNEYS, ETC.

INSURANCE AGAINST ALL LOSS OR DAMAGE TO PROPERTY
AND
AGAINST LIABILITY FOR PERSONAL INJURIES RE-
SULTING FROM THE EXPLOSION OF STEAM BOILERS

ENGINE INSPECTION AND INSURANCE

INSPECTION AND INDICATION OF STEAM ENGINES BY EXPERTS

INSURANCE AGAINST LOSS OR DAMAGE TO PROPERTY CAUSED
BY BREAKDOWN OF ENGINE OR BURSTING OF FLY-WHEEL

SERVICE PROTECTION SUPPLEMENTED BY INSURANCE

THE BOILER INSPECTION & INSURANCE COMPANY
OF CANADA

CONSULTING ENGINEERS—EFFICIENCY EXPERTS
POWER PLANT SPECIALISTS
(ESTABLISHED 1875)

BRANCH OFFICE	HEAD OFFICE	BRANCH OFFICE
227 BOARD OF TRADE BLDG.	CONTINENTAL LIFE BLDG.	904 ELECTRIC RY. CHAMBERS
MONTREAL	TORONTO	WINNIPEG

News of the Trade

The McCart Novelty & Toy Company, Limited, Toronto, have been granted a charter.

The sawmill and cheese box factory of J. G. Stanley, Glencolin, Ont., suffered fire loss recently.

The Colleran Spring Bed Company, Limited, a new company with head office at Toronto, Ont., will manufacture furniture, beds and wood products.

J. C. Risteen & Company, Limited, who conduct a woodworking factory at Fredericton, N.B., were visited by fire recently. The dryhouse was entirely destroyed.

The factory of the "Magog Furniture Manufacturers, Limited," at Magog, Quebec, is well under way. The foundations are in and the superstructure is being rapidly completed.

The Imperial Rattan Company, of Stratford, Ont., have purchased the rights in Great Britain and the Colonies for the manufacture of "Old Hickory" verandah and garden furniture.

The Stratford Manufacturing Company, Limited, is stretching out after export business. Within the past month the company made a shipment of kitchen cabinets to South Africa.

A carriage factory at Sydney, N.S., owned by J. A. McCallum, was recently destroyed by fire. The damage to the building and contents amounted to $15,000 or $20,000. The stock and building were insured.

The Melotone Talking Machine Company have established a factory at 236 Fort Street, Winnipeg. They will use only high grade mahogany and oak lumber in the construction of the cabinets for their machines.

The Carola Company of Canada, Limited, has been incorporated with capital of $40,000, to manufacture talking machines. The head office of the company is to be at Toronto and the provisional directors are W. K. Cook, H. R. Peterson, F. J. Foley and D. J. Coffey.

Mr. A. E. Clark, of Edward Clark & Son, wholesale lumber dealers, Stair Building, Toronto, has been reappointed a member of the Inspection Rules Committee of the National Hardwood Lumber Association for the current year.

A by-law is to be submitted to the people of Shallow Lake, Ont., to authorize a loan of $4,000 to the Richard Ceaser Woodworking Company to establish a woodworking factory at that place. The company propose to manufacture a line of woodenware.

Mr. D. W. Karn, of the Karn & Morris Piano & Organ Company, of Woodstock, Ont., died at his home in Toronto on September 19. Mr. Karn was a well-known figure in business circles. He was treasurer of the Canadian Lilderman Company, of Woodstock.

Mr. A. P. Willis, of the Willis Piano Company, of Montreal, predicts that Canada will do considerable export business in pianos with Australia and New Zealand, and states that a large consignment of Canadian pianos has already been forwarded to those countries.

The Knechtel Furniture Company, of Hanover, Ont., have installed a machine at their Southampton, Ont., factory for applying finishing materials. It is one of the Aeron system machines, manufactured by the DeVilbiss Manufacturing Company, of Toledo, Ohio, for applying the materials by air pressure.

The Canadian Rattan Chair Company, Limited, Victoriaville, Que., have been so busy of late that they have increased

their production nearly a hundred per cent. They are working full blast every day to double their business, and with this in view have recently bought out the Victoriaville Manufacturing Company.

The Bell Furniture Company, of Southampton, Ont., have installed two Aeron system varnish sprayers with eight-foot fumexers, manufactured by the DeVilbiss Manufacturing Company, of Toledo, Ohio. The company have also installed, on the same air system, two rubbing machines, made by the Rockford Tool Company, of Rockford, Illinois.

The Kindel Bed Company, Limited, have transferred their manufacturing operations from Toronto to Stratford, Ont, where they have purchased a factory. The company advise us that they are installing a complete outfit of woodworking machinery. They also state that they expect to have one of the most up-to-date upholstering plants in Canada.

A collection of over 8,000 samples of German and Austrian manufactured products will be on view at Convocation Hall, Toronto, from October 23 to November 8. The samples have been secured by the Canadian Department of Trade and Commerce for the purpose of showing manufacturers how they can extend their home markets and capture a portion of the export trade of Germany and Austria.

The Canadian Woodenware Company, of St. Thomas, Ont., who commenced operations in July, 1915, and worked up a large business manufacturing washboards, etc., were burned out a few weeks ago. The building, which was owned by the St. Thomas Dehydration Company, was completely destroyed, with stock, machinery, etc. The Canadian Woodenware Company have acquired a planing mill, where they will carry on their business for the present. Later, when their affairs are straightened out, they may build, and will require new machinery for the entire plant.

At the office of Arthur D. Little, Limited, Montreal, the C. P. R. exhibited on October 3, 4, 5 and 6 a collection of toys, mainly made in the United States, with the object of fostering the Canadian toy industry. Taken as a whole, the toys were considerably ahead in design and finish of the German goods which were formerly so largely imported; many were of a character which will especially appeal to the children of this continent, with a notable absence of the flimsy make-up present in German toys. The designers have not merely copied the foreign articles; they have, in the main, worked along original lines. Several of the goods could not be strictly classed as toys—they were rather useful novelties, such as baskets, flower holders, coat hangers, cigarette holders, etc. One of the most attractive toys was a set of "Alice in Wonderland," 17 pieces, while some doll cots, one in birch bark, were noticeable. Practically all the goods (in which animals were conspicuous), were of wood.

Henry Morgan & Company, Limited, Montreal, are now turning out an average of 2,000 Bethlehem shell boxes per day, in addition to a large number of shooks. The box-making department has been reorganized, and practically all new machinery purchased for the manufacture of the Bethlehem box and for the shooks. The equipment includes two Canadian Lindermann automatic dovetail matchers; Morgan, Rochester, U.S.A., nailers; machinery for making three-ply, etc. A new blower system has been added, and arrangements made for the shavings not required for the boiler to be conveyed to a separate bin from which they are carted away in a motor truck. The machinery on the floor devoted to the box manufacture has been rearranged, enabling more boxes to be turned out, while the shooks are made on another floor, and conveyed direct into a motor or carts in the street below. Additional space has also been acquired for the storage of lumber, of which half a million feet can now be kept in stock.

The manager knew there was a "leak"

—somewhere

H E started in with a systematic search to locate it—he went through each department only to find that everything seemed to be running smoothly. It *worried him* as it does you. He made up his mind one day that it must be in the production end of the business. Accordingly he got after the time sheets—*and he found it.*

In all other branches of his business he had employed the most up-to-date methods and machinery, but here in this one department the old-fashioned method was still in use. He put in an up-to-date time keeping system and stopped the "leak"—*the serious leak.*

YOU can prevent any "leakage" of time with an

INTERNATIONAL DIAL TIME RECORDER

(Fully Automatic)

Each employee Records his own time and Records it correctly. No chance to make a mistake or to tamper with the time sheet. The time slips taken from the Dial Recorder at the end of the pay period gives the pay roll in complete form *and unchangeable*, a ready made up and easily filed.

There are 260 different styles and sizes of Internationals. One of them will suit your business.

Let us get together on this proposition—a card to-day will bring full particulars

The International Time Recording
Company of Canada, Limited

WINNIPEG	F. E. Mutton,	MONTREAL.
Geo. Morris,	General Manager,	W. A. Wood, Jr.,
Sales Agent,	Anderson & William Sts.,	Sales Agent,
400 Electric Railway	**TORONTO**	Cartier Bldg., McGill and
Chambers		Notre Dame Sts.

Second-Hand Machines
For Sale

We have the following to offer. Ask us about others not listed, if you are interested.

1 Egan	30-in. 2 Drum Sander.
1 McGregor Gourlay	30-in. 2 Drum Boss Sander.
1 Jackson Cochrane	36-in. 3 Drum Sander.
1 Egan	42-in. 3 Drum Sander.
1 Cowan	42-in. 3 Drum Sander.
1 Cowan	48-in. 3 Drum ,Sander.
1 C. M. C.	8-in. 4 side Sticker.
2 Ballantyne	12-in. 4 side Sticker.
1 Egan	36-in. Single Surfacer.
1 Harper	26-in. Single Surfacer.
1 Cowan No. 220	24 x 30 Single Surfacer.
1 McGregor Gourlay QV 15-in. Planer & Matcher.
1 McGregor Gourlay TV 9-in. Planer & Matcher.
1 Jackson Cochrane 54-in. Band Resaw 8-in. blade.
1 Jackson Cochrane No. 105 Rip Saw.

P. B. Yates Machine Co., Limited
Hamilton, Ontario   t-f

Second-Hand
Woodworking Machines

Band Saws—26-in., 36-in., 40-in.

Jointers—12-in., 16-in.

Shapers—Single and double spindle.

Re-Saw—42-in. American.

Lathes—20-in. with rests and chisels.

Surfacers—10 x 6 in. and 18 x 8 in.

Mortisers—No. 5 and No. 7 New Britain.

Rip and Cut-off Saws; Molders; Sash Relishers, etc.

Over 100 machines, New and Used, in stock.

Baird Machinery Company,
123 Water Street,
PITTSBURGH, PA.

Woodworking Machinery
For Sale

1—12 Spindle dovetail machine, J. & C. make.

1—16 Spindle dovetail machine, J. & C. make.

1—18 Spindle upright boring machine, Can. Machinery Corporation make.

1—Double spindle boring machine for shell boxes, J. & C. make, with countershaft.

1—Single spindle boring machine, American Machine Company make, same as Cowan M-248.

1—Dado machine with automatic feed with five heads and two saws.

1—Veneer press, 4 sections, 4 screws to each section.

1—J. & C. Single end tenoning machine.

1—J. & C. two spindle shaper with countershaft.

1—J. & C. spindle carving machine with countershaft.

1—table slide dovetail groover with countershaft.

1—rolling top saw table.

1—shell box automatic hopper feed handle cleat machine.

Beach Furniture Limited
10-12   Cornwall, Ont.

Wanted

Second hand disc and drum sander, disc 24 ins., drum 18 ins.

1 36-in. lathe for irregular work.

1 small Dowel machine.

Address: The Canadian Ski Mfg. Co.,
10   Three Rivers, Que.

FOR SALE

1 Waterous Engine, good condition, less fly-wheel; cylinder bore 12-in.; stroke 14 in.; crank shaft 5½ in.

1 Bundy Trap.

2 Detroit Traps.

2 Hot Water Receivers.

1 Baling Press.

Box No. 24, Canadian Woodworker,
9-10   Toronto, Ont.

Material for Smoking Pipes

A sale of a large quantity of mountain laurel roots from one of the United States National forests in the southern Appalachians is reported by officials in charge, who say that the roots will be used to make pipes. The mountain laurel root is similar in appearance to the French briar, which the majority of pipe smokers are said to prefer. The French briar is the root of the white heath or "bruyere." These roots are gathered in large quantities, and after being cleaned and sawed into blanks they are placed in hot water and simmered for twelve hours or more. This process gives them the rich hue for which the best pipes are noted. It is said that in 1915 the value of the blanks shipped to this country was almost $300,000, and in addition a large number of finished pipes were imported.

On account of the present scarcity and high price of French briar, a number of pipe manufacturers in this country have been on the lookout for substitutes, and the Forest Products Laboratory has conducted experiments to determine the availability of other woods. It is reported that the mountain laurel root burns out more readily than briar, but Forest Service experts are trying to find a method of hardening the wood, and have succeeded in an appreciable extent. They have also found that a number of the various kinds of chaparral which are abundant in the West give promise of yielding material which will be the equal of French briar in. every way. Other woods now widely used for pipe making are apple wood, red gum, ebony and birch, together with smaller amounts of olive wood, rosewood and osage orange.

Considerable amounts of the laurel roots are being used, and officials expect to make further sales. The lands purchased by the Government in the southern Appalachians are reported to contain unlimited quantities of laurel, which is widely known for the delicate beauty of its flowers. In places it forms extensive thickets, which are almost impenetrable. Visitors to the mountains say that in the spring these thickets, or "pink beds" as they are called by the mountaineers, are indescribably beautiful and form one of the main attractions of the region.

PETRIE'S
MONTHLY LIST
of
NEW and USED
WOOD TOOLS
in stock for immediate
delivery

Wood Planers
36" American double surfacer.
30" Whitney pattern surfacer.
26" Revolving bed double surfacers.
24" Major Harper planer and matcher.
24" Single surfacers, various makes.
20" Dundas pony planer.
18" Little Giant planer and matcher.
16" Galt jointer.

Band Saws
42" Fay & Egan power feed rip.
38" Atlantic tilting frame.
30" Ideal pedestal, tilting table.

Saw Tables
Ballantine variable power feed.
Preston variable power feed.
Cowan heavy power feed.   M188.
No. 3 Crescent universal.
No. 2 Crescent, combination.
Ideal variety, tilting table.
12" Defiance automatic, equalizer.
MacGregor Gourlay, railway cut-off.
7½" iron frame swing.

Mortisers
No. 5 New Britain, chain.
Cowan hollow chisel. M.190.
Galt upright, compound table.
Smart foot power.

Moulders
13" Clark-Demill four-side.
12" Cowan four side,
10" Houston four, side.
6" Cowan four side.
6" Dundas sash sticker.

Sanders
24" Fay double drum.
12" C.M.C. disk and drum.
18" Crescent disk, adjustable table.

Clothespin Machinery
Humphrey No. 8 giant slab resaw.
Humphrey gang slitter
Humphrey cylinder cutting-off machine.
Humphrey automatic lathes (6).
Humphrey double slotters (4).

Miscellaneous
Fay & Egan 12-spindle dovetailer.
Cowan dowel machine, M80.
No. 235 Wysong & Miles post boring machine.
Cowan post boring machine, M85G.
Cowan post boring machine, M23.
MacGregor Gourlay, 2-spindle shaper.
Elliott single spindle shaper.
No. 51 Crescent universal woodworker.
Rogers vertical resaw.
Egan sash clamp.
Egan sash and door tenoner.
No. 6 Lion universal woodtrimmer.
6-nail box nailer, Meyers patent.
20" American wood scraper.
16" Ideal wood lathes (3)
4-head rounding machines.
Cowan spindle carver. M63.
Cowan veneer press screws.

Prices, Descriptions and Full Particulars on Request

H. W. PETRIE, LTD.
Front St. W.,   Toronto, Ont.

Panels and Tops

Deskbeds, Table Tops,
Chiffonier and Dresser
Tops
Three and Five Ply Panels

Large assortment of stock panels always on hand.
Three and five ply stock for cartridge boxes.
Your inquiry will be fully appreciated and
promptly answered with full information.

WRITE US

The
Gorham Brothers
Company
MT. PLEASANT   -   MICHIGAN

Dry Spruce
and Birch

Good Stocks, Prompt Shipments, Satisfaction

C. A. SPENCER, Limited
Wholesale Dealers in Rough and Dressed Lumber
Offices—500 McGill Building
MONTREAL   -   -   Quebec

American Hardwood
Lumber Co.
St. Louis, Mo.

Large stock of—
Dry Ash, Quartered Oak
Plain Oak and Gum

Shipments from — NASHVILLE, Tenn.,
NEW ORLEANS, La., and BENTON, Ark.

- AT YOUR SERVICE -

To discriminating buyers
we offer the following
attractive list of
HARDWOODS:

1 Car 1 x 4 and up No. 1 Com. and Btr. Black Ash.
2 Cars 1 x 4 and up No. 2 Com. and Btr. Elm Ash.
4 Cars 1 x 4 and up No. 3 Com. Black Ash.
1 Car 1 x 4 and up No. 2 Com. and Btr. Beech.
5 Cars 1 x 4 and up No. 1 Com. and Btr. Birch.
2 Cars 1 x 3 and up No. 2 and 3 Com. Birch.
1 Car 5/4 in. 1st and 2nd Birch.
1 Car 3 x 3 and up Birch Hearts.
2 Cars 4 x 4 in. Birch Hearts.
2 Cars 5/4 in. No. 1 Com. and Btr. Hard Maple.
2 Cars 5/4 in. No. 1 Com. and Btr. Hard Maple.
5 Cars 1 x 4 and up No. 2 Com. and Btr. Soft Elm.
1 Car 5/4 in. 1st and 2nd Basswood.
2 Cars 5/4 in. No. 1 Com. and Btr. Basswood.
1 Car 5/4 in. No. 2 Com. and Btr. Basswood.
1 Car 5/4 in. No. 2 and 3 Com. Basswood.
1 Car 5/8 in. Log Run Basswood.

All in prime condition
and ready for prompt shipment
Your enquiries solicited
and will receive
prompt attention

Graves, Bigwood & Co.
Manufacturers
of
PINE
HEMLOCK   } LUMBER
HARDWOOD

Mills at—Byng Inlet, Ont.
and
Deer Lake, Ont.

The Lumber Market

Domestic Woods

There is no feature of importance to report in connection with the lumber situation. Prices remain practically the same as they have been for some time. The prices on hemlock are stiffening and show a tendency to carry out the prediction of the lumberman quoted in our last issue, namely, that the price would reach $20 by spring.

We understand that the Imperial Munitions Board has declined for the time being to take further deliveries on some kinds of shell boxes, but we do not expect that their action will affect the prices of lumber. Stocks are scarce and for that reason prices are not at all likely to decline. Furniture factories throughout Canada are busy and are using hardwoods in good quantities. While stocks in these lines appear to be large enough to meet the demands, yet they are fast diminishing, and prices are about as they have been, but very firm.

The most of the box factories are working as near to their capacity as the scarcity of labor will permit, and the grades used by them are higher in price than they have been for some time.

The building trade is still dead, but in some of the smaller towns the planing mills seem to get enough business to keep their wheels turning. There are always odd dwellings going up and one of the chief features about this is the increased use of hardwoods for interior trim. While it does not affect stocks to any extent, yet it is helping to keep prices firm. One thing is certain, manufacturers will not be making any mistake by stocking up their yards, as the opinion among lumbermen is that prices are bound to go up. They are, of course, in touch with the situation, and are watching with a great deal of interest to see what is going to be done in the bush this winter. With the increased cost of labor and supplies it is almost certain that the cut will not be very large, and with the decreased stocks on hand prices cannot help but advance.

Some prices quoted at Toronto on car lots are as follows: white ash, 1sts and 2nds, 1½ and 2 inch, $60; birch, 1sts and 2nds, inch, $44; basswood, 1sts and 2nds, inch, $42; soft elm, 1sts and 2nds, inch, $40; soft maple, 1sts and 2nds, inch, $33; hard maple, 1sts and 2nds, inch, $40; plain red oak and plain white oak, 1sts and 2nds, inch, $60; quartered white oak, 1sts and 2nds, inch, $92.

Imported Woods

The prices of imported woods still remain firm. Woodworking plants, automobile factories and furniture factories in the United States continue busy, and while they are not making extraordinary demands for lumber, yet they are consuming steady quantities. As long as this condition continues there will be no decrease in prices.

The car shortage which occurs nearly every year at this time is affecting shipments to some extent. This is said to be most noticeable in the Southern States. Stocks are reported to be low in some quarters. This may be true of certain grades and kinds on which there has been a run, but on the whole they appear to be fairly well balanced. The woods most in demand are all grades of gum, birch, maple, ash and walnut.

A feature of the demand for birch is its continually increasing use for interior trim, and as building activities in the United States are very pronounced, large quantities are no doubt being used for this purpose, and prices are not likely to come down. Maple is a good reliable wood and is always more or less in demand by implement factories and others requiring a wood with strength, at a moderate price.

The demand for veneers and panels is strong, and consequently prices are fairly high. Quartered oak veneer is a leader just at present, while quartered oak lumber is dropping off perceptibly. Plain oak is in pretty good demand and prices are not showing any tendency to decline; poplar and chestnut are still being used in large quantities by piano and furniture manufacturers for core stock for veneering, and prices are firm.

Ash is in about the same class as maple. It is one of the good reliable hardwoods and is in demand by the automobile manufacturers. Therefore it retains its price. Elm is used extensively in the manufacture of piano backs and for unexposed parts of furniture, and the prices quoted on it are much the same as they have been for several weeks.

The prices of red and sap gum have not shown any increase of late, but the wood being much in demand for furniture and interior trim has caused them to remain steady. Mahogany and walnut are still being used in fairly large quantities for furniture.

The prices on some of the most prominent woods are white ash, $56 to $60; quartered oak, $89 to $91; red birch, $57 to $59; chestnut, $51 to $53; plain oak, $63 to $65; sap birch, $47 to $49; maple, $41 to $43.

Sincerity in Advertising

Humor has no place in advertising.

Nor has poetry. Nor any touch of lightness.

Spending money is serious business. And most folks so regard it.

You are seeking confidence. Deserve it.

You are courting respect. Avoid frivolity.

People are not reading ads for amusement. They seek information. And they want it from a man who seems sincere.

Picture a typical customer. Consider his wants—and his ignorance—respecting what you have to sell.

Consider the importance—to him and to you—of what you ask him to do.

Write as though that man were before you.

Write as though your future depended on that sale. Your future does, when your words go to millions.

Don't pass an ad until you feel that the reader will find it resistless.

Make your case impregnable.

Make every word ring with truth.

There is nothing so winning in the world as absolute sincerity. Nothing is so abhorrent as its lack.

Don't lose sight of the fact that you must take the public as you find it—not as you think it ought to be.

CARVINGS FOR PERIOD FURNITURE

Composition Carvings

are now extensively used for furniture decorations.

This is a material that will interest all manufacturers of furniture.

Write for Catalog and prices.

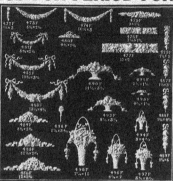

Most durable and practical Carvings on the market.

We have thousands of stock designs in all the different styles ; can also manufacture according to your special designs.

Samples sent on request.

THE DECORATORS SUPPLY CO., 2551 Archer Ave., Chicago, Ill.

Correctly Dried Hardwoods

This trade - mark stands for the best there is in hardwoods.

Martin-Barriss hardwoods are perfectly kiln-dried by the most scientific method. A period of three weeks is required, during which time the heat and humidity of the kilns are maintained at the proper degree.

We can point to many elaborate interiors where the fine woodwork and splendid carvings have been executed in correctly dried Martin-Barriss hardwoods.

We are prepared to cut into logs to obtain the particular figure desired by our customer.

Ask us for prices and particulars
of our complete Hardwood stocks.

Martin-Barriss Company
CLEVELAND, OHIO

VENEER TAPE

for all purposes

PAPER TAPE
in any width or weight.

CLOTH TAPE
Open or close mesh, gummed or ungummed.

Send for 20 Feet Free Sample of any kind you want to try out.

Buy "MADE-IN-CANADA" Goods

Gummed Papers, Limited
Brampton, Ontario

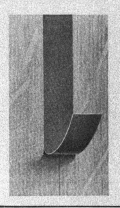

THE COMPLETE LINE

If your furniture is equipped with

ACME CASTERS

you are assured of satisfied customers, your most valuable asset.

Our Canadian representative is at your service and will be pleased to assist you in your CASTER and Period Trimmings problems.

Foster, Merriam & Co.
Meriden, Conn.

Canadian Representative:—F. A. Schmidt, Kitchener, Ont.

"ACME"
The
RELIABLE CASTER

"CR WN" Pl ner B l ing

WE CLAIM IT—WE PROVE IT

We Claim,—

That "CROWN" PLANER BELTING takes the offensive against strong competitors and holds its ground against formidable counter attacks of the critical. A belting made to stand the acid test.

That it solves the high speed heavy duty problems.

That it spells 100% possible transmission of power.

That it envolves durability against the hard test of usage.

Our belting is made from the highest grade of leather, carefully selected and tanned by ourselves—no "weak link" in "CROWN" PLANER BELTING.

Running machines with slipping belts is like walking down a moving stairway—it means a waste of time and energy. No wasted energy with "Crown" Planers—they GRIP.

Our Proof,—

To substantiate all we believe and say of "CROWN" PLANER BELTING, we are ready to place a belt on your machine for approval. We are convinced of its superiority and stand ready to convince you.

Try a sample run and let us PROVE OUR CLAIM.

"Leather like gold has no substitute."

Sadler & Haworth

Estd 1876

Tanners and Manufacturers

For 40 years Tanners and Manufacturers of the Best Leather Belts

| Toronto | St. John, N.B. | MONTREAL | Winnipeg | Vancouver |
| 38 Wellington St. E. | 149 Prince William Street | | Call Building | 1. Water Street |

Mahogany from Mexico

The picture shown here is taken from a photograph furnished to us by the Huddleston-Marsh Mahogany Company, New York City, importers and manufacturers of mahogany lumber and veneer. It seems a long way from this little stream to the Canadian consumers

Stream down which mahogany logs are floated.

of mahogany lumber, yet it is down this stream that there has floated hundreds of the mahogany logs which produced the lumber used in Canada for furniture. The stream is located in Mexico, and empties into a river, which in turn flows into the Gulf of Mexico, near Frontera, Mexico.

The climate there is such that the young man in the foreground dresses the year around about as you see him. He was invited to stand in the water in order to show its depth. The logs are brought down in the flood, when the water at this point will be at least ten feet deep.

The Huddleston-Marsh Mahogany Company are bringing up from that section about two cargoes of logs each month, and state the wood is of excellent texture, good, even color, producing lumber of good widths, as well as some beautifully figured veneer. Our readers are probably aware that Mexican mahogany is harder than African wood, and seldom has cross-breaks or shake defects.

German Woodworking Industry

Information from Berne is to the effect that Germany's woodworking industry has suffered enormously through the war. At the end of 1915, 8,300 factories and workshops had been closed, out of a total of 20,-900 existing before the war. The number of workers, too, had fallen from 260,000 to 117,000, or 55 per cent. The home trade was worse than the foreign, since with the call to arms the creation of new homes and households practically ceased.

Unlike so many other industries, the woodworking factories have found little employment in the war. At first they were engaged in making baskets for munitions, and later other employment was discovered for them. One piano factory has been turning out barbed wire, and a furniture factory has been making shells, but for the most part the wood workers have been of little use in the munitions field. Every week more factories are being shut down, and a large number are working only half-time.—Cabinet Maker (London).

COOLEY & MARVIN CO.

Room 709-712 Tremont Building.   Telephones Haymarket 3927-3928.

BOSTON - MASSACHUSETTS

Canadian Office: Toronto, Ontario, 39 Adelaide Street, East

AUDITORS   ENGINEERS
COST ACCOUNTANTS

COST ACCOUNTANTS FOR THE NATIONAL ASSOCIATION OF CHAIR MANUFACTURERS (U.S.A.)

We specialize in auditing, accounting and cost finding methods for Woodworking Manufacturers and maintain an engineering department devoted to factory organization, reduction of costs and efficiency in operation.

"TROY"
GARNET PAPER

COMBINATION and CLOTH

FINISHING
PAPER

RITCHEY SUPPLY CO.
126 Wellington St. W.
TORONTO

FURNITURE
HARDWARE

Brass Pulls in all the Period Designs

Pulls, Label Holders, Card Index Rods, etc., for Filing Cabinets

Slides for Ladies' Desks

Brass Sockets

Coat and Hat Hooks

Trimmings for Smokers' Stands

Hardware for Talking Machines

Write for catalogue and samples

Weber-Knapp Co.
Jamestown, N. Y.

Cabinet Hardware Supplies
FOR PROMPT SHIPMENT

ALFRED Z. SMITH & CO.
Winston-Salem, N. C.

CABINET   **UPHOLSTERERS'**

HARDWARE   **SUPPLIES**

Materials for Packing and Finishing Rooms

WHEN IN THE MARKET FOR SUPPLIES, CONSIDER OUR ADVANTAGE OF LOCATION

Our stock is especially complete, as follows:

CHIFFOROBE FIXTURES	TACKS—Upholstering—Packing
WIRE NAILS and BRADS	BOX STRAPS
WOOD SCREWS	BURLAP—Sewed Scrap Pieces
BRASS PLATED BUTTS	WIPING RAGS
DRAWER GLIDES	POLISHING WASTE
CASTERS	ARTIFICIAL LEATHER
BRASS LEG SOCKETS	LEATHER GIMP

METALINE and GILT UPHOLSTERY NAILS

UNIVERSAL SLIDES   SAMPLES AND PRICES

FOUR SIZES   SENT ON REQUEST

Faultless Casters

Have the bearing on a round headed pivot which allows the caster to turn easily and quickly in any direction. Only one point of contact to each set. No balls to loose or rust.

The springs in the socket fitting into the neck of the stem prevent the caster from dropping out.

These two points are worthy of YOUR consideration BEFORE you place your caster contract.

SEND FOR SAMPLES AND PRICES.

Faultless Caster Co.
Evansville, Indiana

Gold Medal, Highest Award Pan-Pacific International Exposition in 1915

Full Size C-3-5

FURNITURE TRIMMINGS

of every description, including many Period designs

New Designs in Glass Knobs

Kitchen Cabinet Trimmings
in any finish to order

CASTERS
for all kinds of furniture

Send for our samples and prices before placing orders

THE BURNS & BASSICK COMPANY
BRIDGEPORT, CONN., U. S. A.

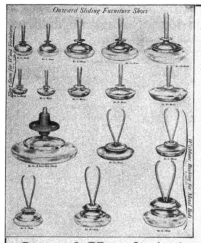

Onward Sliding Furniture Shoes

The "Onward"
Sliding Furniture Shoe

(Successor to the old fashioned wheel Caster)

The caster has had its day. Onward Sliding Furniture Shoes have come to stay. They are very much more superior in every respect. They are made in all sizes and styles, suitable for all kinds of Furniture. Made with Glass Base and Mott Metal Base. Do no damage to carpets or floors. Easily attached, no sockets required.

Special prices to manufacturers in quantities.

Free descriptive circular upon request.

MADE IN CANADA BY

Onward Manufacturing Co. - Kitchener, Ont.

To Our Trade :

We are well prepared to care for all your cabinet hardware requirements and solicit your specifications or inquiries—all of which will have careful attention.

Orders should be placed not less than 30 days in advance of your needs.

NATIONAL LOCK
COMPANY
ROCKFORD   -   -   ILLINOIS

MAKERS OF

**Cabinet Hardware   Furniture Trimmings   Locks
Butts and Hinges   Wood Screws   Nails and Brads
Hanger Bolts, etc., etc.**

The "Canadian Woodworker" Buyers' Directory

An Index to the Best Equipment and Supplies

ABRASIVES
The Carborundum Co., Niagara Falls, N.Y.

AIR BRUSH EQUIPMENT
DeVilbiss Mfg. Co., Toledo, Ohio.
Paasche Air Brush Co., Chicago, Ill.

BABBITT METAL
Hoyt Metal Co., Toronto.

BALL BEARINGS
Chapman Double Ball Bearing Co., Toronto.

BALUSTER LATHES
Garlock-Walker Machinery Co., Toronto, Ont.
Ober Mfg. Company, Chagrin Falls, Ohio.
Whitney & Son, Baxter D., Winchendon, Mass.

BAND SAW FILING MACHINERY
Canada Machinery Corporation, Galt, Ont.
Fay & Egan Co., J. A., Cincinnati, Ohio.
Garlock-Walker Machinery Co., Toronto, Ont.
Preston Woodworking Machinery Company,
Preston, Ont.
Yates Machine Co., P. B., Hamilton, Ont.

BAND SAWS
Canada Machinery Corporation, Galt, Ont.
Cowan & Co., Galt, Ont.
Garlock-Walker Machinery Co., Toronto, Ont.
Jackson, Cochrane & Company, Kitchener, Ont.
Petrie, H. W., Toronto.
Preston Woodworking Machinery Company,
Preston, Ont.
Yates Machine Co., P. B., Hamilton, Ont.

BAND SAW MACHINES
Canada Machinery Corporation, Galt, Ont.
Cowan & Co., Galt, Ont.
Fay & Egan Co., J. A., Cincinnati, Ohio.
Garlock-Walker Machinery Co., Toronto, Ont.
Petrie, H. W., Toronto.
Preston Woodworking Machinery Company,
Preston, Ont.
Williams Machinery Co., A. R., Toronto, Ont.
Yates Machine Co., P. B., Hamilton, Ont.

BAND SAW STRETCHERS
Fay & Egan Co., J. A., Cincinnati, Ohio.
Yates Machine Co., P. B., Hamilton, Ont.

BASKET STITCHING MACHINE
Saranac Machine Co., Benton Harbor, Mich.

BENDING MACHINES
Fay & Egan Co., J. A., Cincinnati, Ohio.
Garlock-Walker Machinery Co., Toronto, Ont.
Perfection Wood Steaming Retort Company,
Parkersburg, W. Va.

BELTING
Fay & Egan Co., J. A., Cincinnati, Ohio.
Sadler & Haworth, Montreal, Que.

BLOWERS
Sheldons, Limited, Galt, Ont.
Toronto Blower Company, Toronto, Ont.

BLOW PIPING
Sheldons, Limited, Galt, Ont.
Toronto Blower Company, Toronto, Ont.

BOILER INSPECTION
Boiler Inspection & Insurance Co., Toronto.

BOILER INSURANCE
Boiler Inspection & Insurance Co., Toronto.

BORING MACHINES
Canada Machinery Corporation, Galt, Ont.
Cowan & Co., Galt, Ont.
Fay & Egan Co., J. A., Cincinnati, Ohio.
Garlock-Walker Machinery Co., Toronto, Ont.
Jackson, Cochrane & Company, Kitchener, Ont.
Nash, J. M., Milwaukee, Wis.
Preston Woodworking Machinery Company,
Preston, Ont.
Reynolds Pattern & Machine Co., Moline, Ill.
Yates Machine Co., P. B., Hamilton, Ont.

BOX MAKERS' MACHINERY
Canada Machinery Corporation, Galt, Ont.
Cowan & Co., Galt, Ont.
Fay & Egan Co., J. A., Cincinnati, Ohio.
Garlock-Walker Machinery Co., Toronto, Ont.
Jackson, Cochrane & Company, Kitchener, Ont.
Neilson & Company, J. L. Winnipeg, Man.
Preston Woodworking Machinery Company,
Preston, Ont.
Reynolds Pattern & Machine Co., Moline, Ill.
Whitney & Son, Baxter D., Winchendon, Mass.
Williams Machinery Co., A. R., Toronto, Ont.
Yates Machine Co., P. B., Hamilton, Ont.

BROOMS
Stevens-Hepner Co., Port Elgin, Ont.

BRUSHES
Skedden Brush Co., Hamilton, Ont.

CABINET PLANERS
Canada Machinery Corporation, Galt, Ont.
Cowan & Co., Galt, Ont.
Garlock-Walker Machinery Co., Toronto, Ont.
Jackson, Cochrane & Company, Kitchener, Ont.
Preston Woodworking Machinery Company,
Preston, Ont.
Yates Machine Co., P. B., Hamilton, Ont.

CARS (Transfer)
Sheldons, Limited, Galt, Ont.

CARBORUNDUM PAPER
Carborundum Co., Niagara Falls, N.Y.

CARVING MACHINES
Canada Machinery Corporation, Galt, Ont.
Fay & Egan Co., J. A., Cincinnati, Ohio.
Garlock-Walker Machinery Co., Toronto, Ont.
Jackson, Cochrane & Company, Kitchener, Ont.

CASTERS
Burns & Bassick Co., Bridgeport, Conn.
Faultless Caster Co., Evansville, Ind.
Foster, Merriam & Co., Meriden, Conn.
National Lock Co., Rockford, Ill.
Onward Mfg. Co., Kitchener, Ont.
Smith & Co., Alfred Z., Winston-Salem, N.C.
Weber-Knapp Co., Jamestown, N.Y.

CLAMPS
Fay & Egan Co., J. A., Cincinnati, Ohio.
Francis Co., Chas. E., Rushville, Ind.
Garlock-Walker Machinery Co., Toronto, Ont.
Jackson, Cochrane & Company, Kitchener, Ont.
Neilson & Company, J. L., Winnipeg, Man.
Preston Woodworking Machinery Company,
Preston, Ont.
Simonds Canada Saw Company, Montreal, P.Q.

CLOTHING TROLLEYS
Hardware Supply Co., Grand Rapids, Mich.

COLUMN MACHINERY
Fay & Egan Co., J. A., Cincinnati, Ohio.

CORRUGATED FASTENERS
Acme Steel Goods Company, Chicago, Ill.

CRATING LUMBER
Read Bros., Limited, Toronto.

COST ACCOUNTANTS
Cooley & Marvin Co., Boston, Mass.

CUT-OFF SAWS
Canada Machinery Corporation, Galt, Ont.
Cowan & Co., Galt, Ont.
Fay & Egan Co., J. A., Cincinnati, Ohio.
Garlock-Walker Machinery Co., Toronto, Ont.
Jackson, Cochrane & Company, Kitchener, Ont.
Ober Mfg. Co., Chagrin Falls, Ohio.
Preston Woodworking Machinery Company,
Preston, Ont.
Simonds Canada Saw Co., Montreal, Que.
Yates Machine Co., P. B., Hamilton, Ont.

CUTTER HEADS
Canada Machinery Corporation, Galt, Ont.
Cowan & Co., Galt, Ont.
Fay & Egan Co., J. A., Cincinnati, Ohio.
Garlock-Walker Machinery Co., Toronto, Ont.
Jackson, Cochrane & Company, Kitchener, Ont.
Mattison Machine Works, Beloit, Wis.
Yates Machine Co., P. B., Hamilton, Ont.

DOVETAILING MACHINES
Canada Machinery Corporation, Galt, Ont.
Cowan & Co., Galt, Ont.
Fay & Egan Co., J. A., Cincinnati, Ohio.
Garlock-Walker Machinery Co., Toronto, Ont.
Yates Machine Co., P. B., Hamilton, Ont.

DOWEL MACHINES
Canada Machinery Corporation, Galt, Ont.
Fay & Egan Co., J. A., Cincinnati, Ohio.
Ober Mfg. Co., Chagrin Falls, Ohio.

DRY KILNS
Grand Rapids Veneer Works, Grand Rapids,
Mich.
National Dry Kiln Co., Indianapolis, Ind.
Sheldons, Limited, Galt, Ont.

DUST COLLECTORS
Sheldons, Limited, Galt, Ont.
Toronto Blower Company, Toronto, Ont.

DUST SEPARATORS
Sheldons, Limited, Galt, Ont.
Toronto Blower Company, Toronto, Ont.

EDGERS (Single Saw)
Canada Machinery Corporation, Galt, Ont.
Fay & Egan Co., J. A., Cincinnati, Ohio.
Garlock-Walker Machinery Co., Toronto, Ont.
Simonds Canada Saw Co., Montreal, Que.

EDGERS (Gang)
Canada Machinery Corporation, Galt, Ont.
Fay & Egan Co., J. A., Cincinnati, Ohio.
Petrie, H. W., Toronto.
Simonds Canada Saw Co., Montreal, P.Q.
Williams Machinery Co., A. R., Toronto, Ont.
Yates Machine Co., P. B., Hamilton, Ont.

EFFICIENCY ENGINEERS
Cooley & Marvin Co., Boston, Mass.

END MATCHING MACHINES
Canada Machinery Corporation, Galt, Ont.
Cowan & Co., Galt, Ont.
Fay & Egan Co., J. A., Cincinnati, Ohio.
Garlock-Walker Machinery Co., Toronto, Ont.
Jackson, Cochrane & Company, Kitchener, Ont.
Yates Machine Co., P. B., Hamilton, Ont.

EXHAUST FANS
Garlock-Walker Machinery Co., Toronto, Ont.
Sheldons, Limited, Galt, Ont.
Toronto Blower Company, Toronto, Ont.

EXHAUST SYSTEMS
Toronto Blower Company, Toronto, Ont.

FACTORY LOCKERS
Meadows Co., Ltd., George B., Toronto.

FACTORY TRUCKS
Francis Co., Chas. E., Rushville, Ind.

FILES
Simonds Canada Saw Co., Montreal, Que.

FITTING MACHINERY
Wardwell Mfg. Company, Cleveland, Ohio.

FLOORING MACHINERY
Canada Machinery Corporation, Galt, Ont.
Cowan & Co., Galt, Ont.
Fay & Egan Co., J. A., Cincinnati, Ohio.
Garlock-Walker Machinery Co., Toronto, Ont.
Petrie, H. W., Toronto.
Preston Woodworking Machinery Company,
Preston, Ont.
Whitney & Son, Baxter D., Winchendon, Mass.
Yates Machine Co., P. B., Hamilton, Ont.

FLUTING HEADS
Fay & Egan Co., J. A., Cincinnati, Ohio.

FLUTING AND TWIST MACHINE
Preston Woodworking Machinery Company,
Preston, Ont.

FURNITURE CARVINGS
Decorators Supply Co., Chicago, Ill.
Walter & Son, J., Kitchener, Ont.

FURNITURE SHOES
Onward Mfg. Co., Kitchener, Ont.

FURNITURE TRIMMINGS
Burns & Bassick Co., Bridgeport, Conn.
Faultless Caster Co., Evansville, Ind.
Foster, Merriam & Co., Meriden, Conn.
National Lock Co., Rockford, Ill.
Smith & Co., Alfred Z., Winston-Salem, N.C.
Weber-Knapp Co., Jamestown, N.Y.

GARNET PAPER AND CLOTH
The Carborundum Co., Niagara Falls, N.Y.
Ritchey Supply Company, Toronto, Ont.

GELATINES
Canada Glue Company, Brantford, Ont.

" Made in Canada "

JOINT and VENEER
GLUES

Also

Sounding Board and Other Gelatines

COLD WATER

VEGETABLE PASTES

Your request for samples or prices invited.

Canada Glue Co.
Limited

Head Office and Factory: BRANTFORD, ONT.

BRANCH STORES:

16 Wellington St. East, Toronto, Ont.   Head of Frontenac St., Montreal, Que.

"Canadian Woodworker" Buyers' Directory—Continued

GRAINING MACHINES
Canada Machinery Corporation, Galt, Ont.
Fay & Egan Co., J. A., Cincinnati, Ohio.
Williams Machinery Co., A. R., Toronto, Ont.
Yates Machine Co., P. B., Hamilton, Ont.

GLUE
Canada Glue Company, Brantford, Ont.
Perkins Glue Company, South Bend, Ind.

GLUE CLAMPS
Jackson, Cochrane & Company, Kitchener, Ont.
Neilson & Company, J. L., Winnipeg.

GLUE HEATERS
Fay & Egan Co., J. A., Cincinnati, Ohio.
Francis Co., Chas. E., Rushville, Ind.
Jackson, Cochrane & Company, Kitchener, Ont.

GLUE JOINTERS
Canada Machinery Corporation, Galt, Ont.
Jackson, Cochrane & Company, Kitchener, Ont.
Yates Machine Co., P. B., Hamilton, Ont.

GLUE SPREADERS
Fay & Egan Co., J. A., Cincinnati, Ohio.
Francis Co., Chas. E., Rushville, Ind.
Jackson, Cochrane & Company, Kitchener, Ont.

GLUE ROOM EQUIPMENT
Francis Co., Chas. E., Rushville, Ind.
Perrin & Company, W. R., Toronto, Ont.

GRINDERS (Cutter)
Fay & Egan Co., J. A., Cincinnati, Ohio.

GRINDERS, HAND AND FOOT POWER
The Carborundum Co., Niagara Falls, N.Y.

GRINDERS (Knife)
Canada Machinery Corporation, Galt, Ont.
Fay & Egan Co., J. A., Cincinnati, Ohio.
Garlock-Walker Machinery Co., Toronto, Ont.
Petrie, H. W., Toronto.
Preston Woodworking Machinery Company,
Preston, Ont.
Yates Machine Co., P. B., Hamilton, Ont.

GRINDERS (Tool)
Fay & Egan Co., J. A., Cincinnati, Ohio.
Jackson, Cochrane & Company, Kitchener, Ont.

GRINDING WHEELS
The Carborundum Co., Niagara Falls, N.Y.

GROOVING HEADS
Fay & Egan Co., J. A., Cincinnati, Ohio.

GUM LUMBER
Brown & Co., George C., Memphis, Tenn.
Churchill-Milton Lumber Co., Louisville, Ky.
Dulweber Co., John, Cincinnati, Ohio
Edwards Lumber Co., E. L., Dayton, Ohio.
Fullerton Powell Hardwood Lumber Co., South
Bend, Ind.
Hyde Lumber Co., South Bend, Ind.
Kentucky Lumber Co., Lexington, Ky.
Lamb-Fish Lumber Co., Charlestown, Miss.
Louisville Point Lumber Co., Louisville, Ky.
Penfod, Jurden & McCowen, Memphis, Tenn.
Russe & Burgess, Memphis, Tenn.

GUMMED PAPERS
Gummed Papers, Limited, Brampton, Ont.

GUMMERS, ETC.
Fay & Egan Co., J. A., Cincinnati, Ohio.

HAND PROTECTORS
Fay & Egan Co., J. A., Cincinnati, Ohio.

HAND SCREWS
Fay & Egan Co., J. A., Cincinnati, Ohio.

HANDLE THROATING MACHINE
Klotz Machine Company, Sandusky, Ohio.

HANDLE TRIMMING MACHINE
Klotz Machine Company, Sandusky, Ohio.

HANDLE & SPOKE MACHINERY
Fay & Egan Co., J. A., Cincinnati, Ohio.
Klotz Machine Co., Sandusky, Ohio.
Nash, J. M., Milwaukee, Wis.
Ober Mfg. Company, Chagrin Falls, Ohio.
Whitney & Son, Baxter D., Winchendon, Mass.

HARDWARE
Burns & Bassick Co., Bridgeport, Conn.
Faultless Caster Co., Evansville, Ind.
National Lock Co., Rockford, Ill.
Smith & Co., Alfred Z., Winston-Salem, N.C.
Weber-Knapp Co., Jamestown, N.Y.

HARDWOOD LUMBER
American Hardwood Lumber Co., St. Louis, Mo.
Anderson Lumber Co., C. G., Toronto, Ont.
Atlantic Lumber Company, Toronto.
Brown & Company, Geo. C., Memphis, Tenn.
Burns & Knapp Lumber Co., Conneautville, Pa.
Churchill-Milton Lumber Co., Louisville, Ky.
Clark & Son, Edward, Toronto, Ont.
Cumberland Valley Lumber Co., Cincinnati,
Ohio.
Dulweber Co., John, Cincinnati, Ohio
Dermott Land & Lumber Co., Chicago, Ill.
Edwards Lumber Co., E. L., Dayton, Ohio.
Elgie & Jarvis Lumber Co., Ltd., Toronto.
Farrin Lumber Co., M. B., Cincinnati, Ohio
Fullerton Powell Hardwood Lumber Co., South
Bend, Ind.
Graves, Bigwood Company, Toronto, Ont.
Hyde Lumber Co., South Bend, Ind.
Kersley, Geo., Montreal, Que.
Kraetzer-Cured Lumber Co., Moorhead, Miss.
Lamb-Fish Lumber Co., Charlestown, Miss.
Martin-Barriss Co., Cleveland, Ohio.
Mowbray & Robinson, Cincinnati, Ohio.
Paepcke Leicht Lumber Co., Chicago, Ill.
Penrod, Jurden & McGowen, Memphis, Tenn.
Penrod Walnut & Veneer Co., Kansas City, Mo.
Probst Lumber Co., Cincinnati, Ohio.
Read Bros., Limited, Toronto.
Spencer, C. A., Montreal, Que.
Thoman Flinn Lumber Co., Cincinnati, Ohio.

HUB MACHINERY
Fay & Egan Co., J. A., Cincinnati, Ohio.

HYDRAULIC VENEER PRESSES
Perrin & Company, Wm. R., Toronto, Ont.

JOINTERS
Canada Machinery Corporation, Galt, Ont.
Cowan & Co., Galt, Ont.
Fay & Egan Co., J. A., Cincinnati, Ohio.
Garlock-Walker Machinery Co., Toronto, Ont.
Jackson, Cochrane & Company, Kitchener, Ont.
Petrie, H. W., Toronto, Ont.
Preston Woodworking Machinery Company,
Preston, Ont.
Yates Machine Co., P. B., Hamilton, Ont.

KNIVES (Planer and others)
Canada Machinery Corporation, Galt, Ont.
Fay & Egan Co., J. A., Cincinnati, Ohio.
Simonds Canada Saw Co., Montreal, P.Q.
Yates Machine Co., P. B., Hamilton, Ont.

LATHES
Canada Machinery Corporation, Galt, Ont.
Cowan & Co., Galt, Ont.
Fay & Egan Co., J. A., Cincinnati, Ohio.
Garlock-Walker Machinery Co., Toronto, Ont.
Jackson, Cochrane & Company, Kitchener, Ont.
Ober Mfg. Company, Chagrin Falls, Ohio.
Petrie, H. W., Toronto, Ont.
Preston Woodworking Machinery Company,
Preston, Ont.
Shawver Company, Springfield, O.
Whitney & Son, Baxter D., Winchendon, Mass.

LUMBER
Acme Veneer & Lumber Co., Cincinnati, Ohio.
American Hardwood Lumber Co., St. Louis, Mo.
Anderson Lumber Co., C. G., Toronto, Ont.
Bennett & Witte, Cincinnati, Ohio.
B. C. Lumber Commissioner, Toronto.
Brown & Company, Geo. C., Memphis, Tenn.
Churchill, Milton Lumber Co., Louisville, Ky.
Clark & Son, Edward, Toronto, Ont.
Cumberland Valley Lumber Co., Cincinnati,
Que.
Dermott Land & Lumber Co., Chicago, Ill.
Dominion Mahogany & Veneer Co., Montreal.
Elgie & Jarvis Lumber Co., Ltd., Toronto.
Farrin Lumber Co., M. B., Cincinnati, Ohio.
Gorham Bros. Co., Mt. Pleasant, Mich.
Graves, Bigwood Company, Toronto, Ont.
Hart & McDonagh, Toronto, Ont.
Hartzell, Geo. W., Piqua, Ohio.
Huddleston, Marsh Mahogany Co., New York.
Hyde Lumber Co., South Bend, Ind.
Kentucky Lumber Co., Lexington, Ky.
Long-Knight Lumber Co., Indianapolis.
Louisville Point Lumber Co., Louisville, Ky.
Paepcke Leicht Lumber Company, Chicago, Ill.
Penrod, Jurden & McCowen, Memphis, Tenn.
Probst Lumber Co., Cincinnati, Ohio.

Read Bros., Limited, Toronto.
Russe & Burgess, Memphis, Tenn.
Stimson & Co., J. V., Owensboro, Ky.
Thoman Flinn Lumber Co., Cincinnati, Ohio.

HYDRAULIC PRESSES
Francis Co., Chas. E., Rushville, Ind.
Perrin & Company, W. R., Toronto.

MACHINE KNIVES
Canada Machinery Corporation, Galt, Ont.
Peter Hay Knife Company, Galt, Ont.
Simonds Canada Saw Co., Montreal, P.Q.
Yates Machine Co., P. B., Hamilton, Ont.

MITRE MACHINES
Canada Machinery Corporation, Galt, Ont.
Fay & Egan Co., J. A., Cincinnati, Ohio.

MITRE SAWS
Canada Machinery Corporation, Galt, Ont.
Fay & Egan Co., J. A., Cincinnati, Ohio.
Garlock-Walker Machinery Co., Toronto, Ont.
Simonds Canada Saw Co., Montreal, P.Q.

MOTORS
Pringle, Ltd., R. E. T., Toronto, Ont.

MULTIPLE BOXING MACHINES
Fay & Egan Co., J. A., Cincinnati, Ohio.
Nash, J. M., Milwaukee, Wis.
Reynolds Pattern & Machine Co., Moline, Ill.

PANELS
Waetjen & Co., George L., Milwaukee, Wis.

PATTERN SHOP MACHINES
Canada Machinery Corporation, Galt, Ont.
Cowan & Co., Galt, Ont.
Fay & Egan Co., J. A., Cincinnati, Ohio.
Garlock-Walker Machinery Co., Toronto, Ont.
Jackson, Cochrane & Company, Kitchener, Ont.
Preston Woodworking Machinery Company,
Preston, Ont.
Whitney & Son, Baxter D., Winchendon, Mass.
Yates Machine Co., P. B., Hamilton, Ont.

PLANERS
Canada Machinery Corporation, Galt, Ont.
Cowan & Co., Galt, Ont.
Fay & Egan Co., J. A., Cincinnati, Ohio.
Garlock-Walker Machinery Co., Toronto, Ont.
Jackson, Cochrane & Company, Kitchener, Ont.
Petrie, H. W., Toronto, Ont.
Preston Woodworking Machinery Company,
Preston, Ont.
Whitney & Son, Baxter D., Winchendon, Mass.
Williams Machinery Co., A. R., Toronto, Ont.
Yates Machine Co., P. B., Hamilton, Ont.

PLANING MILL MACHINERY
Canada Machinery Corporation, Galt, Ont.
Cowan & Co., Galt, Ont.
Fay & Egan Co., J. A., Cincinnati, Ohio.
Garlock-Walker Machinery Co., Toronto, Ont.
Jackson, Cochrane & Company, Kitchener, Ont.
Petrie, H. W., Toronto, Ont.
Preston Woodworking Machinery Company,
Preston, Ont.
Whitney & Son, Baxter D., Winchendon, Mass.
Williams Machinery Co., A. R., Toronto, Ont.
Yates Machine Co., P. B., Hamilton, Ont.

PRESSES (Veneer)
Canada Machinery Corporation, Galt, Ont.
Jackson, Cochrane & Company, Kitchener, Ont.
Perrin & Co., Wm. R., Toronto, Ont.
Preston Woodworking Machinery Company,
Preston, Ont.

PULLEYS
Fay & Egan Co., J. A., Cincinnati, Ohio.

TIME CLOCKS
International Time Recording Co., Toronto.

RED GUM LUMBER
Gum Lumber Manufacturers' Association.

RESAWS
Canada Machinery Corporation, Galt, Ont.
Cowan & Co., Galt, Ont.
Fay & Egan Co., J. A., Cincinnati, Ohio.
Garlock-Walker Machinery Co., Toronto, Ont.
Jackson, Cochrane & Company, Kitchener, Ont.
Petrie, H. W., Toronto, Ont.
Preston Woodworking Machinery Company,
Preston, Ont.
Simonds Canada Saw Co., Montreal, P.Q.
Williams Machinery Co., A. R., Toronto, Ont.
Yates Machine Co., P. B., Hamilton, Ont.

Period Carvings

They are very economical.
You can use them To-day or in Ten years from now.
No steaming required.

Have

They

Brad

Holes

?

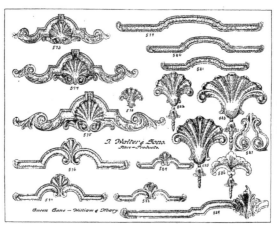

Yes

Have you a copy of our Catalogue?
If not, why not?
We have over 800 designs to choose from.
Next year's samples are ready for mailing.

We solicit your inquiries.

J. Walter & Sons

KITCHENER   -   -   **ONTARIO**

"Canadian Woodworker" Buyers' Directory—Continued

RIM AND FELLOE MACHINERY
Fay & Egan Co., J. A., Cincinnati, Ohio.

RIP SAWING MACHINES
Canada Machinery Corporation, Galt, Ont.
Cowan & Co., Galt, Ont.
Fay & Egan Co., J. A., Cincinnati, Ohio.
Garlock-Walker Machinery Co., Toronto, Ont.
Jackson, Cochrane & Company, Kitchener, Ont.
Ober Mfg. Company, Chagrin Falls, Ohio.
Preston Woodworking Machinery Company,
   Preston, Ont.
Yates Machine Co., P. B., Hamilton, Ont.

SANDERS
Canada Machinery Corporation, Galt, Ont.
Cowan & Co., Galt, Ont.
Garlock-Walker Machinery Co., Toronto, Ont.
Jackson, Cochrane & Company, Kitchener, Ont.
Nash, J. M., Milwaukee, Wis.
Ober Mfg. Company, Chagrin Falls, Ohio.
Preston Woodworking Machinery Company,
   Preston, Ont.
Yates Machine Co., P. B., Hamilton, Ont.

SASH, DOOR & BLIND MACHINERY
Canada Machinery Corporation, Galt, Ont.
Cowan & Co., Galt, Ont.
Fay & Egan Co., J. A., Cincinnati, Ohio.
Garlock-Walker Machinery Co., Toronto, Ont.
Jackson, Cochrane & Company, Kitchener, Ont.
Preston Woodworking Machinery Company,
   Preston, Ont.
Yates Machine Co., P. B., Hamilton, Ont.

SAWS
Fay & Egan Co., J. A., Cincinnati, Ohio.
Radcliff Saw Mfg. Company, Toronto, Ont.
Simonds Canada Saw Co., Montreal, P.Q.
Smith Co., Ltd., R. H., St. Catharines, Ont.

SAW GUMMERS, ALOXITE
The Carborundum Co., Niagara Falls, N.Y.

SAW SWAGES
Fay & Egan Co., J. A., Cincinnati, Ohio.
Simonds Canada Saw Co., Montreal, Que.

SAW TABLES
Canada Machinery Corporation, Galt, Ont.
Fay & Egan Co., J. A., Cincinnati, Ohio.
Garlock-Walker Machinery Co., Toronto, Ont.
Jackson, Cochrane & Company, Kitchener, Ont.
Preston Woodworking Machinery Company,
   Preston, Ont.
Yates Machine Co., P. B., Hamilton, Ont.

SCRAPING MACHINES
Canada Machinery Corporation, Galt, Ont.
Garlock-Walker Machinery Co., Toronto, Ont.
Whitney & Sons, Baxter D., Winchendon, Mass.

SCREW DRIVING MACHINERY
Canada Machinery Corporation, Galt, Ont.
Reynolds Pattern & Machine Works, Moline, Ill.

SCROLL SAWS
Canada Machinery Corporation, Galt, Ont.
Cowan & Co., Galt, Ont.
Fay & Egan Co., J. A., Cincinnati, Ohio.
Garlock-Walker Machinery Co., Toronto, Ont.
Jackson, Cochrane & Company, Kitchener, Ont.
Simonds Canada Saw Co., Montreal, P.Q.
Yates Machine Co., P. B., Hamilton, Ont.

SECOND-HAND MACHINERY
Cowan & Co., Galt, Ont.
Petrie, H. W.,
Williams Machinery Co., A. R., Toronto, Ont.

SHAPERS
Canada Machinery Corporation, Galt, Ont.
Fay & Egan Co., J. A., Cincinnati, Ohio.
Garlock-Walker Machinery Co., Toronto, Ont.
Jackson, Cochrane & Company, Kitchener, Ont.
Ober Mfg. Company, Chagrin Falls, Ohio.
Petrie, H. W., Toronto, Ont.
Preston Woodworking Machinery Company,
   Preston, Ont.
Simonds Canada Saw Company, Montreal, P.Q.
Whitney & Sons, Baxter D., Winchendon, Mass.
Yates Machine Co., P. B., Hamilton, Ont.

SHARPENING TOOLS
The Carborundum Co., Niagara Falls, N.Y.

SHAVING COLLECTORS
Sheldons Limited, Galt, Ont.

SHOOK BUNDLER
Neilson & Company, J. L., Winnipeg, Man.

STAINS
Ault & Wiborg, Toronto, Ont.
Harland & Son, Wm., Toronto.

STEAM APPLIANCES
Darling Bros., Montreal.

SINGLE SPINDLE BOXING MACHINES
Canada Machinery Corporation, Galt, Ont.
Fay & Egan Co., J. A., Cincinnati, Ohio.
Ober Mfg. Company, Chagrin Falls, Ohio.

STAVE SAWING MACHINE
Whitney & Sons, Baxter D., Winchendon, Mass.

STEAM TRAPS
Canadian Morehead Mfg. Co., Woodstock, Ont.
Standard Dry Kiln Co., Indianapolis, Ind.

STEEL BOX STRAPPING
Acme Steel Goods Co., Chicago, Ill.

STEEL SHELL BOX BANDS
Acme Steel Goods Co., Chicago, Ill.

SURFACERS
Canada Machinery Corporation, Galt, Ont.
Cowan & Co., Galt, Ont.
Fay & Egan Co., J. A., Cincinnati, Ohio.
Garlock-Walker Machinery Co., Toronto, Ont.
Jackson, Cochrane & Company, Kitchener, Ont.
Petrie, H. W., Toronto, Ont.
Preston Woodworking Machinery Company,
   Preston, Ont.
Whitney & Sons, Baxter D., Winchendon, Mass.
Williams Machinery Co., A. R., Toronto, Ont.
Yates Machine Co., P. B., Hamilton, Ont.

SWING SAWS
Canada Machinery Corporation, Galt, Ont.
Cowan & Co., Galt, Ont.
Fay & Egan Co., J. A., Cincinnati, Ohio.
Garlock-Walker Machinery Co., Toronto, Ont.
Jackson, Cochrane & Company, Kitchener, Ont.
Ober Mfg. Company, Chagrin Falls, Ohio.
Preston Woodworking Machinery Company,
   Preston, Ont.
Simonds Canada Saw Co., Montreal, P.Q.
Yates Machine Co., P. B., Hamilton, Ont.

TABLE LEG LATHES
Fay & Egan Co., J. A., Cincinnati, Ohio.
Ober Mfg. Co., Chagrin Falls, Ohio.
Whitney & Sons, Baxter D., Winchendon, Mass.
Yates Machine Co., P. B., Hamilton, Ont.

TABLE SLIDES
Walter & Co., B., Wabash, Ind.

TENONING MACHINES
Canada Machinery Corporation, Galt, Ont.
Fay & Egan Co., J. A., Cincinnati, Ohio.
Garlock-Walker Machinery Co., Toronto, Ont.
Jackson, Cochrane & Company, Kitchener, Ont.
Preston Woodworking Machinery Company,
   Preston, Ont.
Yates Machine Co., P. B., Hamilton, Ont.

RECORDERS (TIME)
International Time Recording Co., Toronto.

TRANSMISSION MACHINERY
Dodge Mfg. Co., Ltd., Toronto, Ont.

TRIMMERS
Canada Machinery Corporation, Galt, Ont.
Fay & Egan Co., J. A., Cincinnati, Ohio.
Garlock-Walker Machinery Co., Toronto, Ont.
Preston Woodworking Machinery Company,
   Preston, Ont.
Yates Machine Co., P. B., Hamilton, Ont.

TRUCKS
Sheldons Limited, Galt, Ont.
National Dry Kiln Co., Indianapolis, Ind.

TURNING MACHINES
Fay & Egan Co., J. A., Cincinnati, Ohio.
Garlock-Walker Machinery Co., Toronto, Ont.
Ober Mfg. Co., Chagrin Falls, Ohio.
Klotz Machine Company, Sandusky, Ohio.
Whitney & Sons, Baxter D., Winchendon, Mass.
Yates Machine Co., P. B., Hamilton, Ont.

**UNDER-CUT SELF-FEEDING FACE
PLANER**
Fay & Egan Co., J. A., Cincinnati, Ohio.
Garlock-Walker Machinery Co., Toronto, Ont.
Jackson, Cochrane & Company, Kitchener, Ont.

VARNISHES
Ault & Wiborg Company, Toronto, Ont.
Harland & Son, Wm., Toronto.

VARNISH SPRAYERS
DeVilbiss Mfg. Co., Toledo, Ohio.
Paasche Air Brush Co., Chicago, Ill.

VEGETABLE PASTES
Canada Glue Company, Brantford, Ont.

VENEERS
Acme Veneer & Lumber Co., Cincinnati, Ohio.
Barnaby, Chas. A., Greencastle, Ind.
Central Veneer Co., Huntington, W. Virginia.
Dominion Mahogany & Veneer Company, Mont-
   real, Que.
Walter Clark Veneer Co., Grand Rapids, Mich.
Gorham Bros. Co., Mt. Pleasant, Mich.
Hartzell, Geo. W., Piqua, Ohio.
Haughton Veneer Company, Chicago, Ill.
Hay & Company, Woodstock, Ont.
Hoffman Bros. Company, Fort Wayne, Ind.
Geo. Kersley, Montreal, Que.
Nartzik, J. J., Chicago, Ill.
Ohio Veneer Company, Cincinnati, Ohio.
Penrod Walnut & Veneer Co., Kansas City, Mo.
Roberts, John N., New Albany, Ind.
Toronto Veneer Company, Toronto, Ont.
Waetjen & Co., George L., Milwaukee, Wis.
Wood Mosaic Co., New Albany, Ind.

VENEER TAPE
Gummed Papers, Limited, Brampton, Ont.
Ideal Coated Paper Co., Brookfield, Mass.
Rexford Gummed Tape Co., Milwaukee, Wis.

VENTILATING APPARATUS
Sheldons Limited, Galt, Ont.

VENEERED PANELS
Hay & Company, Woodstock, Ont.

VENEER PRESSES (Hand and Power)
Canada Machinery Corporation, Galt, Ont.
Francis Co., Chas. E., Rushville, Ind.
Garlock-Walker Machinery Co., Toronto, Ont.
Jackson, Cochrane & Company, Kitchener, Ont.
Perrin & Company, Wm. R., Toronto, Ont.

VISES
Fay & Egan Co., J. A., Cincinnati, Ohio.
Simonds Canada Saw Company, Montreal, Que.

WAGON AND CARRIAGE MACHINERY
Canada Machinery Corporation, Galt, Ont.
Fay & Egan Co., J. A., Cincinnati, Ohio.
Ober Mfg. Company, Chagrin Falls, Ohio.
Whitney & Sons, Baxter D., Winchendon, Mass.
Yates Machine Co., P. B., Hamilton, Ont.

WIRE SCREENS
Meadows Co., Ltd., George B., Toronto.

WOOD FINISHES
Ault & Wiborg, Toronto, Ont.
Harland & Son, Wm., Toronto, Ont.

WOOD PULLEYS
Dodge Mfg. Co., Ltd., Toronto, Ont.

WOOD TURNING MACHINERY
Canada Machinery Corporation, Galt, Ont.
Garlock-Walker Machinery Co., Toronto, Ont.
Yates Machine Co., P. B., Hamilton, Ont.

WORK BENCHES
Fay & Egan Co., J. A., Cincinnati, Ohio.

WOODWORKING MACHINES
Canada Machinery Corporation, Galt, Ont.
Cowan & Co., Galt, Ont.
Garlock-Walker Machinery Co., Toronto, Ont.
Hutchinson Woodworker Co., Toronto, Ont.
Jackson, Cochrane & Company, Kitchener, Ont.
Preston Woodworking Machinery Company,
   Preston, Ont.
Reynolds Pattern & Machine Works, Moline,
   Ill.
Williams Machinery Co., A. R., Toronto, Ont.

WHEELS, Grinding, Carborundum and Aloxite
The Carborundum Co., Niagara Falls, N.Y.

A Profit Maker

This one machine will do the following work: Cutting r a f t e r s, studs, braces, Boring for dowelling, Ten-oning, Dadoing, Routing, Miter-ing, Sanding, Rip-ping, Tool-grind-ing, etc.

The HUTCHINSON
COMBINATION WOODWORKER

Not simply a saw-rig; not a big, complicated machine. Assures maximum service at minimum investment. Always ready for work; easy to set up right at the job. Embodies on one base all the equipment a contractor needs. Sold on its ability to give satis-faction. Ask for catalog and details of our free trial offer.

The Hutchinson Woodworker Co.
Limited
99 Sherbourne Street, Toronto
U.S.A.:—29 Franklin Street, Buffalo, N.Y.

—for those one-thousand-and-one odd jobs of drilling

You need one of these handy portable electric drills around your plant. You need it because it will save its cost to you in a few weeks, and then go on working for you just the same. You can take it to any part of the plant or yard and run it —any place, anywhere wherever there is an or-dinary lamp socket to attach it.

Use it on your regular drilling jobs, too; it will cut down your production costs.

Write for catalogue and prices.

The VAN DORN PORTABLE ELECTRIC DRILL
(with Universal Motor)

R. E. T. PRINGLE
LIMITED
95 King Street East
TORONTO - ONT.

809 Unity Building
MONTREAL - P. Q.

MEADOWS FACTORY EQUIPMENT

The majority of accidents may be prevented by properly guarding dangerous parts of machines.

We make guards to fit any ma-chine. Our service is at your dis-posal.

Our literature sent on request.

Up-to-date Factories have

Meadows Metal Lockers

because we specialize on factory lockers and know exactly what is required.

The GEO. B. MEADOWS Toronto Wire Iron & Brass Works Co., Limited
479 West Wellington Street, TORONTO, Canada

You Have Paid for an Installation of
Chapman Double Ball Bearings
in Your Factory, over and over again, BUT—

Have You Got Them?

Ordinary line shafting wastes from 15 per cent. to 60 per cent. of power.

Line shafting equipped with Chapman Double Ball Bearings will eliminate about 75 per cent. of the friction, thus averaging a total saving of from 15 per cent. to 30 per cent.

Over 2,000 Canadian factories are making this saving every day. They have installed Chapman Double Ball Bearings. Everyone of these factories is a working testimonial of their efficiency.

Chapman Double Ball Bearings will increase the power from your present power plant or decrease the cost of operating that same plant if extra power isn't required.

Chapman Double Ball Bearings fit any adjustable hanger and require oiling and attention only once a year. No extra equipment required to install.

Write to-day for full particulars

The Chapman Double Ball Bearing Company of Canada
Limited
Toronto   339-351 Sorauren Ave.   **Ontario**

American Branch: The Transmission Ball Bearing Co., Inc., 32 Wells St., Buffalo, N. Y.

Making TABLE-SLIDES is a Specialty Business

For more than TWENTY-FIVE YEARS we have made TABLE SLIDES exclusively. Our Factory is equipped with Special Machinery which enables us to make SLIDES,— BETTER and CHEAPER than the furniture manufacturer.

Canadian Table makers are rapidly adopting WABASH SLIDES

Because { They ELIMINATE SLIDE TROUBLES } { Are CHEAPER and BETTER }

Reduced Costs   BY USING

Increased Out-put   **WABASH SLIDES**

MADE BY

B. Walter & Company
Wabash, Ind.
The Largest EXCLUSIVE TABLE-SLIDE Manufacturers in America
ESTABLISHED-1887

Just What You Need

There's no necessity now for you to wish for a corrugated joint fastener machine which will rise fully to your expectations; you can get one.

The Saranac Corrugated Joint Fastener Machine

is the absolute realization of your wish. This illustration gives you an idea of what it looks like, but you can have no conception of its **efficiency, economy, speed** and **reliability** until you get it working for you.

It is substantially constructed, and the workmanship and materials are of the very best. Particularly suitable for manufacturers of sash, doors, blinds, screens, porch columns, boxes, etc., etc.

Made in various types for all conditions.

Write for further particulars and prices

Saranac Machine Co., SOLE MAKERS
Benton Harbor, Mich.

INCREASED production and greater economy in your sanding department can be obtained by using

CARBORUNDUM BRAND GARNET PAPER AND CLOTH

They show a saving in labor, quicker production and better results. Carborundum Brand Garnet Paper and Cloth is made from the best garnet grains. Free from all impurities. Uniformly coated upon flexible backings. The best materials united with superior methods of manufacture make Carborundum Brand Garnet Paper and Cloth the ideal abrasive for the woodworking trade.

Working samples upon request

THE CARBORUNDUM COMPANY
NIAGARA FALLS, N. Y.

New York   Chicago   Grand Rapids
    Boston   Cincinnati

urner

uc ion costs

WHY is it that more than 2,000 factories, throughout the country are using Mattison Automatic Shaping Lathes? Because they realize that turnings bought outside, or whittled out on a hand lathe, cost too much.

Do you know any reason why **you** can't make your own turnings just as cheap or cheaper than the job shop? You cannot afford to be dependent on outside sources and pay a premium for every turning you use.

And you can't make headway if you hinge your production on a hand turner. He's a high-priced expert, limited in capacity and unreliable. Factory can't run without turnings, and if he gets sick or quits, you're up against it.

Solve this problem of turning once and for all. Add a Mattison Automatic Shaping Lathe to your force. It's always on the job—efficient, dependable and minimizes your labor cost. An ordinary mechanic with this lathe automatically shapes, rounds, squares, octagons, ovals, etc., in whatever quantities you require, at a 50 to 95 per cent. reduction of hand-method costs. Pays for itself in a few months.

Don't sacrifice profits by putting off your investigation of this **standard machine turner.**

CANADIAN WOODWORKER

and Furniture Manufacturer

C. M. C.

No. 518 Chair Clamp

for pressing up the rails connecting the front posts to the back posts, and for pressing up the side rails

The Features:

Will take in any box seat or slip seat chair, either round or flat post.

Will clamp up to 19 in. x 19 in.

Pedestal is a cored casting, with flared base, and planed on top where table revolves.

The table, 24 in. x 31½ in., is made of cast iron, and is planed on top; also on bottom where it is fitted to the pedestal.

Provision is made for locking the table in position every quarter of a revolution. The operator does not require to move from his position when clamping rails and corner blocks of chairs.

The back fence or bar of clamp is adjustable, and can be swung to one side, giving the operator free access in placing chair in clamp. Both front and back bars are provided with adjustable steel holders for holding wooden blocks shaped to suit the form of chair post.

Front blocks on table are tapped in two places, so that pressure screws can be changed to accommodate different styles of chairs.

CANADA MACHINERY CORPORATION
LIMITED

Toronto Office and Showroom: Brock Avenue Subway.　　　　GALT, ONTARIO

Power!

Who Pays the Power Bill?
Who Pays for Pulleys?

YOU—the Management—pay for the power that drives your factory. And, remember—you pay for the power you *waste* as well as for the power you use. Exhaustive scientific tests prove conclusively that the actual cost per year for running *metal pulleys* as against *wood* pulleys averages *$5.00 per pulley.*

Dodge Wood Split Pulleys cost 40 per cent. less than metal pulleys and are 50 per cent. more efficient. They help you cut the cost of your power bills. Next time you require pulleys see that the "requisition" calls for *Dodge Wood Split Pulleys* and not for steel pulleys.

DODGE
WOOD SPLIT PULLEYS

DODGE MANUFACTURING CO., Limited, TORONTO

Every buyer should send for this book mailed Free on request

A SCIENTIFIC TEST

14

GARLOCK-WALKER MACHINERY CO.
LIMITED
32 FRONT ST. WEST,   TORONTO   TELEPHONE MAIN 5346

Do You Manufacture Furniture?

If so here is the very latest and most up-to-date **SURFACER** for your work—the best, the cheapest in the long run.

Feeds up to 85′ per minute.

Detachable Side Clamping Cylinder Boxes.

Round Cylinder

Knife Setter and Jointer.

In Feed Rolls and Chipbreaker either solid or sectional.

Feed Rolls of large diameter—all powerfully driven.

Bed raises and lowers on long inclines.

American No. 444 Single Furniture Planer

Heavy   Strong   Durable

Does the best work at minimum expense

Let us take this up with you in detail—as well as offer you other tools of like calibre for your work.

WRITE FOR PRICES AND DELIVERIES

Garlock-Walker Machinery Co., Limited

Canadian Representatives

American Wood Working Machinery Co.
ROCHESTER, N.Y.

M. L. Andrews & Co.
CINCINNATI, OHIO

METAL and WOODWORKING MACHINERY of all Kinds

The IMPROVED
ELLIOT WOODWORKER

The Elliot Woodworker cuts and rips over 3 inches thick, easily cross cutting plank 3 inches thick and 24 inches wide. Table tilts to any angle for ripping, bevels or compound mitres. Powerful electric motor direct connected to saw spindle. Runs from ordinary house current.

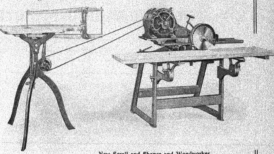

New Scroll and Shaper and Woodworker.

Illustration shows our latest machine, a combination Jig Saw and Shaper placed beside the woodworker and run direct from motor pulley. Both the jointer and rip saw can be run at the same time as the jig saw and shaper. The jig saw is made on the arm principle, without springs, and it has no dead centre, making it possible to do heavier and faster work. By loosening two screws the arms are detached and shaper head is set up in place of the saw. The shaper spindle is steel 1⅛ in. with ⅞ in. babbitted bearings and heavy enough for all classes of work.

Illustration shows the Elliot Woodworker with Jointer attached. Both jointer and rip saw can be run at the same time.

Why *you* should have one

An Elliot Woodworker is the greatest time and labor saving machine any woodworking plant can add to its present equipment. When you have ripping, cross-cutting, boring, rabbiting, tenoning, grooving, mitring, housing out of stair strings, dadoing, shaping, scroll-sawing, or tool grinding to do the combination shown above will do it all quicker and better than any other machine you have in your plant.

Both the woodworker and jig saw are portable and can be moved about at will. So compact and simple are these machines that they only take a couple of minutes to set up for any job required.

Space does not permit us to go into details here, but if you will send us your name and address we will mail you full particulars and prices. Write today.

THE ELLIOT WOODWORKER
LIMITED
Cor. College and Bathurst Streets

TORONTO   -   ONTARIO

Ready to Glue—Without Further Preparation

That's the way core stock comes from the Yates Type G-2 Edging and Jointing Saw. Smooth enough and true enough to glue at once. Saves the cost of an extra run through a jointer— money you can count right into your pocketbook. Prepares your stock for rough gluing three times as fast as it can be done on a hand jointer, and at a fraction of the cost.

The Type G-2 can be used for all kinds of edging and ripping. The saw line is kept perfectly straight, automatically permitting operation along the very edge of bad or shaky spots without danger to saw, stock, or operator. It is particularly adapted to the salvage of usable portions of low grade stock. Although before the trade but a short time the Type G-2 has dozens of well-satisfied users, some of whom have installed additional saws like the first.

We have a circular describing and illustrating this new saw, and its operation. Sent free without obligation or expense of any kind.

P.B. Yates Machine Co. Ltd.

HAMILTON, ONT. CANADA

Successors to **THE BERLIN MACHINE WORKS** -- U. S. Plant: Beloit, Wis.

The
Herzog Self-Feed Jointer

means

Increased Production   Simplicity of Operation
Small Floor Space   Safety to Employees

Does Four Times the Work of the Hand Jointer.

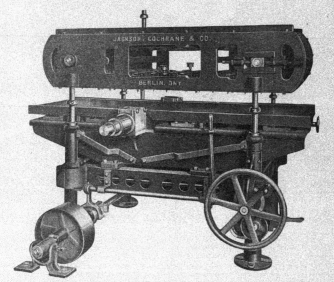

Our No. 34 Herzog Jointer, illustrated above, is one of the most efficient machines on the market to-day. It is appreciated by the manufacturer and employees alike, because, while it will produce from three to five times as much work as the hand jointer, it does not require skilled operators, but eliminates the danger so common to other makes. It can be operated by two boys. It will handle stock varying in width from 1 inch to the full width of the jointer, will feed fast or slow, takes only one-fourth the floor space of hand jointers, and requires only one-fifth of the sharpening of the knives. It is fitted with power feed raising and lowering attachment, with cylinder double belted and driven from both ends.

If interested in reducing your costs, write us.

Jackson, Cochrane & Company
KITCHENER - CANADA

S a n d e r s !

3 and 2 Drum Roll Feed	*— the Fay-Egan "Lightning" line is the most complete made.*
3 and 2 Drum Endless Bed	Put your sanding problems up to us—our line of
Belt—all kinds	sanders is complete—we can and will recommend
Single Drum	the type best suited to your individual needs.
Drum and Disc	Every Fay-Egan Sander is a unit of highest efficiency,
Double Disc	developed from a thoroughly practical design, by
Double Spindle	skilled mechanics with unlimited manufacturing facil-
Spindle and Disc	ities and highest grade materials.
Flexible Arm	Machine sanding is one of the greatest economizers
etc.	in modern woodworking—are you getting the benefit ?

*Write for Bulletins covering any
or all kinds of Sanders.*

J. A. Fay & Egan Co.

Established 1830

*World's Oldest and Largest Manufacturers of Woodworking
Machinery.*

153-173 West Front St.   -   Cincinnati, Ohio

Faultless Casters

Have the bearing on a round headed pivot which allows the caster to turn easily and quickly in any direction. Only one point of contact to each set. No balls to loose or rust.

The springs in the socket fitting into the neck of the stem prevent the caster from dropping out.

These two points are worthy of YOUR consideration BEFORE you place your caster contract.

SEND FOR SAMPLES AND PRICES.

Faultless Caster Co.
Evansville, Indiana

Gold Medal, Highest Award Pan-Pacific International Exposition in 1915

Full Size C-3-5

NEW DESIGNS in GLASS KNOBS

FURNITURE TRIMMINGS
of every description, including many new designs

Kitchen Cabinet Trimmings
in many new designs

CASTERS
for all kinds of furniture

Send for our samples and prices before placing orders

THE BURNS & BASSICK COMPANY
BRIDGEPORT, CONN., U. S. A.

"Treat your machine as a living friend."

Don't Waste Time

for which you are paying big money by having a Machine in your plant which is not giving you a maximum of production every day.

No. 162 High-Speed Ball-Bearing Shaper

SEE THIS MACHINE!

It will turn out 100% MORE WORK than the ordinary shaper which you are now using. That's our claim; twice the amount of work and much easier on the operator. Is it worth your while to cut your shaper cost in two? If so, drop us a line for more particulars.

THE CHEVROLET MOTOR CO. ARE USING 6 OF THESE MACHINES IN THEIR PLANT AT OSHAWA

DIMENSIONS, WEIGHTS and SPEEDS

Floor Space, 120 in. x 58 in. Net Weight, 2650 lbs. Shipping Weight, 2750 lbs. Tight and Loose Pulleys on Countershaft, 10 in. x 6 in. Driving Pulleys on Countershaft, 18 in. x 5 in. Idler Pulleys, 14 in. x 5 in. R.P.M. Countershaft, 1200. Pulley on Spindle, 3½ in. x 6 in. R.P.M. Spindle, 7000. Diameter Spindle, 2 3/16 in. Spindle Centres, 28 in. Table, 58 in. x 41 in. Diameter Spindle Top, 1¾ in. Horse Power Required, 3. Front Projection, 17 in. Height, 35 in.

"The Machine you want in your plant is not the cheapest machine to buy but the machine that will save you the most money to use."

The Preston Woodworking Machinery Co. Limited, Preston, Ont.

Alphabetical List of Advertisers

Acme Steel Goods Company 61
Acme Veneer & Lumber Company 46
American Hardwood Lumber Company 13
Atlantic Lumber Company 14
Ault & Wiborg 39

B. C. Government 39
Baird Machinery Company 56
Banning, Leland G. 16
Barnaby, Chas. A. 43
Beach Furniture Company 56
Brown & Company, Geo. C. 12
Burns & Bassick 57
Burns & Knapp Lumber Company 16

Canada Machinery Corporation 1
Canadian Morehead Mfg. Company 59
Carborundum Company 63
Central Veneer Company 47
Chapman Double Bearing Company 57
Churchill-Milton Lumber Co. 13
Clark & Son, Edward 10
Clark Veneer Company, Walter 43
Cooley & Marvin Company 102
Cowan & Company 11
Cumberland Valley Lumber Company 31

Darling Bros. 53
Decorators Supply Company 49
Dermott Land & Lumber Company 17
De Vilbiss Mfg. Company 40
Dodge Manufacturing Company 2
Dominion Mahogany & Veneer Company .. 18
Dulweber Company, John 84

Elliott Woodworker, Limited 4
Edwards Lumber Company, E. L. 15

Farrin Lumber Company, M. B. 13
Faultless Caster Company 8
Fay & Egan Company, J. A. 7
Foster Merriam Company ....
Francis Company, Chas. E. ....
Fullerton-Powell Hardwood Lumber Co. ...

Garlock-Walker Machinery Company 3
Gorham Brothers 53
Grand Rapids Veneer Works 59
Graves, Bigwood Company 51
Gum Lumber Manufacturers' Assn. 12

Hart & McDonagh 14
Hartzell, Geo. W. 14
Haughton Veneer Company 47
Hay & Company 30
Hay Knife Company, Peter 61
Hoffman Bros. 43
Hoyt Metal Company 57
Huddleston-Marsh Mahogany Co. 13

Jackson, Cochrane & Company 6

Kentucky Lumber Company 30
Kersley, Geo. 45

Lamb-Fish Lumber Company 15
Long-Knight Lumber Company 14

Martin-Barriss Company 10
Mattison Machine Works, C. 64
Mowbray & Robinson Company 15

Nartzik, J. J. 43
Nash, J. M. 59
National Dry Kiln Company 61
Neilson & Company, J. L. 61

Ohio Veneer 43

Paepcke Leicht Lumber Company 18
Penrod Walnut & Veneer Co. 47
Perfection Wood Steaming Retort Co. ...
Perkins Glue Company 17
Perrin, William R. 61
Petrie, H. W. 56
Preston Woodworking Machinery Co. 9
Probst Lumber Company 51

Radcliff Saw Company 55
Read Bros. 58
Rexford Gummed Tape Company 51
Reynolds Pattern & Machine Co. 62
Ritchey Supply Company 58
Roberts Veneer Company 45

Sadler & Haworth 40
Saranac Machine Company 68
Sheldons Limited ....
Simonds Canada Saw Company ....
Skedden Brush Company 39
Smith Company, R. H. 11
Spencer, C. A. 13
Stimson, J. B. 15

Toronto Veneer Company 44

Waetjen & Company, Geo. L. 46
Walter & Company, B. 63
Wardwell Manufacturing Company 50
Waymoth, A. D. 53
Whitney & Son, Baxter D. ....
Wood-Mosaic Company 45

Yates Machine Company, P. B. 5-56

Edward Clark & Sons
Toronto

CANADIAN HARDWOODS—winter cut Birch, Bass, Maple
10,000,000 Feet, Immediate Shipment

Ontario and Quebec stocks, East or West, rail or water shipment, direct from mill to factory. Stocks in any grade or thickness, selected and graded for your requirements, selected widths, selected color.

Case Goods and Chair Manufacturers

1,000,000 ft. 4/4 Birch, mill run.
1,000,000 ft. 4/4 Birch, No. 1 common.
1,000,000 ft. 4/4 Birch, No. 2 common.
500,000 ft. each 5/4, 6/8, 8/4, No. 1 common.
500,000 ft. 4/4 Basswood.
500,000 ft. 4/4 Soft Elm.

For Trim Trade

200,000 ft. 4/4 Birch, No. 1 and 2.
300,000 ft. 5/4 Birch, No. 1 and 2.
400,000 ft. 6/4 Birch, No. 1 and 2.
300,000 ft. 4/4 to 8/4 Brown Ash.

Parlor Frame Makers

400,000 ft. 10/4 Birch, No. 1 C. and B.
600,000 ft. 12/4 Birch, No. 1 C. and B.
200,000 ft. 16/4 Birch, No. 1 C. and B.
50,000 ft. 6/4 S. Maple, No. 1 C. and B.
50,000 ft. 10/4 S. Maple, No. 1 C. and B.
50,000 ft. 12/4 S. Maple, No. 1 C. and B.

Shell Box Trade

300,000 ft. 1x9 and up Birch, No. 1 and No. 2C.
200,000 ft. 1x9 and up Birch, No. 2 and No. 3C.
200,000 ft. 1x6 to 8 Birch, No. 2 and No. 3C.
600,000 ft. 6/4 x 8/4 Birch, No. 2 nd No. 3C.

We Commend to Your Notice:

Our lumber is reliable, our grades uniform, our shipments prompt, and receive personal supervision; our prices consistent. We have had long experience. We desire to serve.

R. H. Smith Co.

ST. CATHARINES, ONT.   Limited

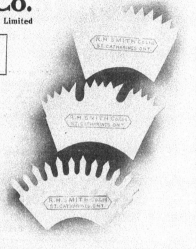

"Arrow Head"
Quality

The scarcity of labor deals directly with the manufacturer: Arrow Head Saws and Knives increase production, and save labor, made from Vanadium Steel by skilled workmen and are guaranteed to do 20% more than any other Saw. The Arrow Head Planer Tooth Saw is the smoothest and most satisfactory saw of its kind for Shell Box, Toy and other kinds both Rip and Cross Cut.

Eventually Arrow Head Saws and Knives. Why not your next order?

Shell Box Machinery

We manufacture and sell Full Equipment for manufacturing the Bethlehem Ammunition Box, including the following:—

Special Five Spindle Borers   Dado Machines

Cut Off Saws   Rip Saws   Band Resaws

Veneer Presses   Nailing Machines, Etc., Etc.

Full particulars and prices on application

COWAN & COMPANY OF GALT, LIMITED
GALT   -   ONTARIO

Selected Red and Sap Gum

St. Francis Basin Band Sawed Stock

KRAETZER-CURED   STRAIGHT AND FLAT

"The car of special width gum box boards that you shipped us that were specified 17 inches in width, we found upon their arrival that this car of gum box boards was the best car that ever came into our yards. We have found every board to have the correct width and we failed to find a warped board in the car, and the grade and measurements were perfectly satisfactory. We were highly pleased with the kind of lumber that this car contained as every board in the car will make box boards.

Thanking you very much for the kind of lumber that you shipped us in this car, we remain—"

(Name furnished on request.)

YOU CAN OBTAIN THE SAME RESULTS

Send us a list of your requirements.   —   Delivered Prices will be cheerfully quoted.

GEO. C. BROWN & CO.

Mill—PROCTOR, Ark.   Main Office—MEMPHIS, Tenn.

We specialize in Tennessee Aromatic Red Cedar

It is much to the credit of the better class of furniture manufacturers that they are now lavishing their worthiest designing on

RED GUM

("AMERICA'S FINEST CABINET WOOD")

RED GUM Furniture Makers have MADE A BIG HIT WITH THE WISER DEALERS, *just as furniture dealers have pleased their most discriminating customers* in putting on the market a wide range of beautifully designed and conscientiously manufactured pieces of RED GUM furniture. RED GUM is no longer identified only as a low price product. Its growing favor with the very best and most critical trade completely justifies the foresight of the industry. RED GUM FURNITURE *is now* THE FASHION. *Are YOU taking full advantage of this fact?*

Neither are up-to-date furniture *manufacturers* overlooking for a minute the wonderful value and applicability of SAP GUM to a hundred and one of their manufacturing requirements. When *properly* manufactured (as it is by the members of the undersigned Association) SAP GUM is *thoroughly reliable*, just as RED GUM is.

Let us send you the RED GUM literature and let us direct you to a group of buildings trimmed with RED GUM in your neighborhood.

Gum Lumber Manufacturers Association

1314 Bank of Commerce Bldg.   Memphis, Tennessee

Dry Spruce
and Birch

Good Stocks, Prompt Shipments, Satisfaction

C. A. SPENCER, Limited
Wholesale Dealers in Rough and Dressed Lumber

Offices—500 McGill Building
MONTREAL   -   -   Quebec

Churchill Milton
Lumber Co.
Louisville, Ky.

We carry at
NEW ALBANY, IND.

Genuine Indiana Plain and
Quartered Oak, Poplar

At GLENDORA, MISS.

Cypress, Gum,
Cottonwood, Oak

SEND US YOUR INQUIRIES

*We have dry stock on hand and
can make quick shipment*

American Hardwood
Lumber Co.
St. Louis, Mo.

Large stock of—

Dry Ash, Quartered Oak
Plain Oak and Gum

Shipments from — NASHVILLE, Tenn.,
NEW ORLEANS, La., and BENTON, Ark.

THE
M. B. FARRIN LUMBER
COMPANY
CINCINNATI   -   OHIO

Manufacturers

HARDWOODS

We have very large stocks of well assorted
West Virginia Hardwoods on hand
at Cincinnati and also at mills.

Plain Oak   Whitewood
Quartered Oak   Red Gum
Chestnut   Ash

KILN DRIED STOCK
A SPECIALTY

Huddleston-Marsh Mahogany
Company, Inc.

Importers and Manufacturers

MAHOGANY

Lumber and Veneer
All Grades   All Thicknesses

*WE SPECIALIZE
IN
MEXICAN MAHOGANY*

33 West 42nd St.   2254 Lumber St.
NEW YORK, N.Y.   CHICAGO, ILL.

—Mills and Yards—
Long Island City, New York

The Atlantic Lumber Co.

Manufacturers

Strictly Soft Textured

Quartered and Plain **White Oak Red Oak**

Chestnut, Poplar, Cherry

Tennessee Scented Red Cedar

High Grades—Quick Shipments

110 Manning Chambers - TORONTO
Phone Main 6386

HEAD OFFICE, BOSTON MASS.

Mills: KNOXVILLE, TENN.,   WALLAND, TENN.
FRANKLIN, ARRINGDALE and BUTTERWORTH, VA.

Shell Box Stock

1 x 6 birch, in No. 1, 2, and 3 common.
1 x 7 and 8 birch, in No. 1, 2, and 3 common.
1 x 8¼ birch, in No. 1, 2, and 3 common
1 x 9 and wider, birch, in No. 1, 2, and 3 common.
1 x 4 and wider, spruce. 2 x 4 and wider, spruce.

For the Furniture Trade

4/4, 5/4, 6/4, 8/4, 10/4, 12/4, 16/4, birch, maple, elm, basswood and black ash.

Elm, Basswood and Black Ash

in No. 1 common and better grades—dry stock in assorted sizes.

Hart & McDonagh

Continental Life Bldg.

TORONTO   -   -   ONTARIO

Long - Knight Lumber Co.

Manufacturers and
Wholesale Dealers

INDIANAPOLIS, IND., U. S. A.

HARDWOOD LUMBER

Walnut, Mahogany

and

American Walnut Veneers

SPECIAL:

5 cars 5/4 1 & 2 Pl. Red Oak

FOR SALE

Black Walnut Lumber

15,000	feet	4/4	No. 1 Common
20,000	"	4/4	No. 2 "
15,000	"	5/4	No. 1 "
15,000	"	5/4	No. 2 "
15,000	"	6/4	No. 2 "
40,000	"	8/4	No. 1 "
60,000	"	8/4	No. 2 "
30,000	"	9/4	No. 1 "

Will make special low prices on all of the above stock to move it quick. All Band Sawn, Equalized, and all thoroughly dry, ready for immediate shipment. Write for prices.

Geo. W. Hartzell

Piqua, Ohio

Plain and Quartered Red and White Oak

and Other Hardwoods

Even Color   Soft Texture

We have **35,000,000 feet dry stock**
all of our own manufacture, from our own
timber grown in Eastern Kentucky

MADE (MR) RIGHT

OAK FLOORING

PROMPT SHIPMENTS

The Mowbray & Robinson Co., Inc.

GENERAL OFFICES
CINCINNATI, OHIO

Mills: Quicksand, Ky., West Irvine, Ky., Viper, Ky.

Canadian Representative:
N. H. FARNHAM
601 Elmwood Ave., BUFFALO, N. Y.

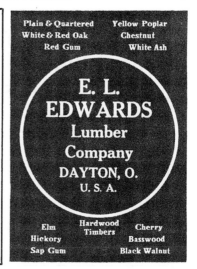

Plain & Quartered   Yellow Poplar
White & Red Oak   Chestnut
Red Gum   White Ash

E. L. EDWARDS Lumber Company
DAYTON, O.
U. S. A.

Elm   Hardwood Timbers   Cherry
Hickory   Basswood
Sap Gum   Black Walnut

J.V. Stimson & Co.
OWENSBORO, KENTUCKY

Manufacturers of

BAND SAWN HARDWOODS

QUARTERED and PLAIN OAK
ASH
POPLAR
HICKORY
WALNUT
QUARTERED SYCAMORE

OAK and WALNUT
OUR SPECIALTY

We have on our Charleston
Yard a complete stock of

WHITE and RED **OAK** PLAIN and QUARTERED

RED and SAP **GUM** PLAIN and FIGURED

Can Make Immediate Shipment

Representative in Eastern Canada:
FRANK T. SULLIVAN
71 South Street
BUFFALO - New York

Lamb-Fish Lumber Co.
"The Largest Hardwood Mill in the World"
CHARLESTON, MISS.

Handsomely

ⅯⒷ Figured

Hardwoods

Martin-Barriss hardwoods are noted for their excellent grain and handsome figuring. In order to please our particular customers we are prepared to cut into several logs in order to strike the desired figure.

The Martin-Barriss method of kiln-drying is conducted on the most scientific principles in order that perfectly dried hardwoods may be the result.

ASK US FOR PRICES

The Martin-Barriss Co.
Cleveland, Ohio

BLACK WALNUT

Large Stock 1 to 3 inches thick ready for immediate shipment

also

Well Assorted Stocks of

West Virginia & Kentucky

OAK

Ash, Maple, Hickory, Chestnut, Basswood and Poplar

Prices and stock list on request

Burns & Knapp
Lumber Company
CONNEAUTVILLE, PA.

I Manufacture
Kentucky River Quartered
and Plain White Oak

Wanted by all furniture manufacturers for its

Soft Texture and Even Color

I also manufacture

Soft Texture Plain Red Oak
Walnut and Poplar
Ash and Hickory
Gum and Chestnut
Cuban and Mexican Mahogany

Wire (at my expense) or write for prices for anything in above list

Leland G. Banning, CINCINNATI OHIO

Dermott Land and Lumber Co.

Manufacturers of

Southern Hardwoods

80 E. Jackson Blvd.

CHICAGO   -   U.S.A.

Quartered White Oak
4 cars 1 in. 1sts and 2nds.
5 cars 1 in. No. 1 Common.
1 car 1¼ in. 1sts and 2nds.
1 car 1¼ in. No. 1 Common.
½ car 1½ in. 1sts and 2nds.
1 car 2 in. 1sts and 2nds.
1 car 2 in. No. 1 Common.
1/3 car 2½ in. Com. and Bet.
1/3 car 3 in. Com. and Bet.

Plain Red Oak
5 cars 1 in. 1sts and 2nds.
10 cars 1 in. No. 1 Common.
2 cars 1¼ in. 1sts and 2nds.
2 cars 1½ in. No. 1 Common.
2 cars 2 in. 1sts and 2nds.
1 car 2 in. No. 1 Common.

Plain White Oak
1 car ¾ in. 1sts and 2nds.
1/3 car ⅝ in. 1sts and 2nds.
1 car ¾ in. 1sts and 2nds.
10 cars 1 in. 1sts and 2nds.
10 cars 1 in. No. 1 Common.
2 cars 1¼ in. 1sts and 2nds.
4 cars 1¼ in. No. 1 Com.
2 cars 1½ in. 1sts and 2nds.
3 cars 1½ in. No. 1 Com.
2 cars 1¾ in. 1sts and 2nds.
1 car 2 in. 1sts and 2nds.
2 cars 2 in. No. 1 Common.
1 car 3 in. 1sts and 2nds.
½ car 4 in. 1sts and 2nds.

Gum
1 in. to 2 in. Plain Sawn Red.
1 in. to 2 in. Quarter Sawn Red.
1 in. to 2 in. Figured Plain and Quartered.
1 in. to 2 in. Sap.

**All well cared for band sawn stock properly edged and trimmed. Nice assorted dry
stock on hand at our modern band mill at Dermott, Arkansas.**

Are Your Products
Perkins Glued?

Ladies' Desk manufactured by The Jacques Furniture Co., Kitchener, Ont., in which Perkins Glue was used.

The live dealer realizes the value of selling his customers Perkins Glued products because they never come apart. Many of Canada's largest furniture factories use Perkins Vegetable Glue. It costs 20 per cent. less than hide glue, is applied cold, antiquating the obnoxious and wasteful cooking process, is absolutely uniform in quality and will not blister in sanding.

We will be glad to demonstrate in your own factory and on your own stock what can be accomplished with Perkins Vegetable Glue.

Write us today for more detailed information.

Perkins Glue Co.

Hamilton, Ont.   Limited

Address all inquiries to Sales Office, Perkins Glue Company,
South Bend, Indiana, U. S. A.

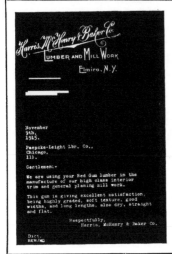

Of course it is true that

RED GUM

is America's finest cabinet wood—but .
Just as a poor cook will spoil the choicest
viands while the experienced chef will turn
them into prized delicacies, so it is true that

The inherently superior qualities
of Red Gum can be brought
out' only by proper handling.

When you buy this wood, as when you buy a new
machine, you want to feel that you have reason for
believing it will be just as represented.

We claim genuine superiority for our Gum. The
proof that you can have confidence in this claim is
shown by the letter reproduced herewith.

Your interests demand that you remember this
proof of our ability to preserve the wonderful qualities of the wood when you again want RED GUM.

Paepcke Leicht Lumber Co.

Conway Building   111 W. Washington St.
CHICAGO, ILL.

Band Mills: Helena and Blytheville Ark.; Greenville, Miss.

Buy Made in Canada Veneers

Dominion Mahogany
& Veneer Co., Limited

Head Office and Factory   Toronto Office and Warehouse
Montreal West, P.Q.   455 King St. West

Mahogany, Walnut,
and all Fancy Woods—Lumber

SAWN, SLICED and ROTARY CUT VENEERS

Mahogany, Walnut, Quartered and Plain Oak, Ash, Maple,
Birch, Elm, Poplar, Gum.

Gum and Poplar Cross Banding.   Hard Maple Pin Block Stock.

SEND US YOUR ENQUIRIES.

Canadian Woodworker

A Monthly Publication in the Interest of the Woodworking Industry.
Reaches the factories producing interior finish, doors, sash, flooring,
woodenware, furniture, pianos, boxes, and general mill products.

Subscription, $1.00 a year; foreign $1.50.

Woodworker Publishing Company, Limited

Branches: Montreal, Vancouver,
Chicago, New York, London, Eng.

345 Adelaide St. West, Toronto
Phone Ade. 2700

Authorized by the Postmaster General for Canada, for transmission as second class matter.
Entered as second class matter July 18th, 1914, at the Post Office at Buffalo, N.Y., under the Act of Congress of March 3, 1879.

| Vol. 16 | November, 1916 | No. 11 |

Export Trade for Box Makers

"For the past three years trade enquiries have been published for many firms anxious to take up the representation of Canadian box makers, and a number of trade enquiries have been published from South African merchants prepared to purchase in 100,000 lots. Business has not resulted, and it is hoped that some organization will take up the manufacturing of box shooks for export on the most modern plans, so that Canada may secure a good share of the present demand as well as that of the near future, which competent authorities say, will be in the millions of boxes for oranges only."

The above paragraph, which is somewhat in the nature of a reproach to Canadian box makers for their lack of activity in going after export business, appeared in a recent edition of the weekly bulletin issued by our Department of Trade and Commerce. Box manufacturers interviewed on the subject nearly all voiced the opinion that the lack of interest shown in export business was due to the inability of manufacturers to cope with the trade at home. As one large manufacturer put it, "our factories are operating at from one-third to one-half less than full capacity on account of the shortage of men. A large number of factories have been working on shell box contracts, and the consumers of boxes to whom the products of their factories formerly went have been forced to look to other factories for their boxes. Thus you will see that the demand being made on us at the present time is abnormal, and makes it impossible for us to consider export business."

The manufacturer quoted above, in reply to the following question, "If sufficient men were available to run all factories to full capacity would South African business look attractive " said: "In my opinion, it would be impossible for box makers in Eastern Canada to touch the business. In the first place, the boxes required will in all probability be a cheap grade, and the prices of lumber in this section of the country are too high for us to supply them." Our friend further stated that he thought if it was desired to develop

trade with England, Australia, New Zealand, or South Africa the logical ones to handle the business are the box makers of British Columbia.

We may rest assured that the box makers of the United States will put forth a supreme effort to secure a large portion of the export trade. Those manufacturers whose factories are located in the South in the districts where cheap grades of pine suitable for boxes can be secured and those who operate factories in connection with sawmills are the ones who are most favorably situated to do an export business. We believe, however, that the box makers of British Columbia are on an equal if not a better footing to secure export trade than the manufacturers of the United States. British Columbia has almost unlimited supplies of lumber, particularly suitable for box making, both as regards quality and price. She has factories that are second to none in the matter of equipment and efficient operating. In a short time she will have a fleet of merchant ships in which to transport her products to foreign markets. To us this looks an admirable combination, and we look forward to seeing the box makers of British Columbia make the most of their opportunities.

Piano Sound Boards

We wonder if it has ever occurred to any of the owners of woodworking plants in Canada that quite a nice little business might be worked up manufacturing piano sound-boards. It has been estimated by men in a position to know that there are approximately 20,000 pianos manufactured in Canada in a year, each of which requires a sound-board. These boards have never been manufactured in Canada, in spite of the fact that there is a 25 per cent. duty on the imported ones, and at the present time there is also the 7½ per cent. war tax.

As far as the making of sound-boards is concerned we believe they could be readily manufactured in any up-to-date woodworking plant. They vary in size according to the size of the piano, but, roughly speaking, they are about 4 feet 6 inches square. They are made from spruce about one-half inch thick, and it will be readily appreciated that the wood must be absolutely free from knots and other defects. The required size is obtained by jointing up narrow boards, but instead of being jointed as for ordinary work these must be jointed so that the grain in the finished board will run diagonally. This will be more readily understood by looking at the back of a piano, as the sound-board can be seen from the back.

We do not know just exactly what machinery would be required to manufacture sound-boards, but we believe that by adding a large surfacer and sander to their equipment any of the up-to-date woodworking factories would be able to manufacture them. Another method would be to surface them in two sections and finish them on the sander. The most important point to be considered, of course, would be the cost at which they could be produced.

As near as we can learn, the cost of sound-boards, including duty and war tax, vary from $2 to $3.25, according to size and quality. We believe that Canadian manufacturers could easily compete at these prices, and, as the business would probably amount to $55,000 or $60,000 a year, it should be worth going after. It should be understood by any person contemplating manufacturing them that they have only to make the board itself. The strips, or ribs as they are called, which are secured to the board and run diagonally in the opposite direction to the grain of the board, are made in the piano factory. The attaching of the ribs to the board forms part of the work of the "bellying department."

Another point which seems worth taking up is the boxes in which American manufacturers of sound-boards pack their product for shipment. These are identical in shape with the boxes used for shipping pianos. The boards are placed in the boxes in an upright position, and the shape and size of the boxes insure them being kept upright. Then, after the boxes have served their purpose for transporting the sound-boards, they are used by the piano manufacturer to ship pianos.

Demand for Ply Wood

Judging from the number of enquiries that have been received at the office of the Canadian Woodworker from manufacturers in different parts of Canada and England for three-ply stock, it looks as though there was room for some enterprising Canadian firm to start and manufacture it. There has been a tendency on the part of some men to confuse three-ply stock with built-up panels, but this is a mistake, as what is commonly known as three-ply stock or ply wood is a cheap product that has been used in large quantities for drawer bottoms, backing, and dust boards, etc., for case goods in furniture factories.

The fact that it is cheap, however, should not be taken as an indication that it is only good for cheap furniture. It is used on some of the best furniture,

and makes excellent drawer bottoms and case backs. It cannot be split; it is light and flexible, and the whole back can be put on in one piece and secured with small nails or screws, thereby saving considerable time and labor, which in turn means reduced costs.

The largest part—if not all—of the ply wood used in Canada was manufactured in Russia, and was first introduced into this country about four or five years ago. Manufacturers were rather slow to recognize its advantages at first, but once they did they purchased it in large quantities, and now that the war has stopped shipments being made from Russia, they are looking around to find out where fresh supplies can be obtained from.

The Russian product was made in two or three different thicknesses and kinds of wood, but that which found most ready sale was "alderlite," 3/16 inch thick. Alderlite is a very close-grained wood, about the color of maple, and having a texture similar to birch. We believe there is an attractive market for any person that can manufacture this material on a large scale and sell it at the right price. The whole matter would, of course, have to be gone into very thoroughly. The Russian alderlite is made with a waterproof cement, which seems to place it very highly in favor with English users. As near as we can learn, the method of procedure in manufacturing it is to cut the veneer on a rotary machine, put the plys through a machine to be coated with the cement, and then place it under pressure in huge hydraulic presses. Thus it will be seen that the veneer is not dried, but is used green.

The moisture from the veneer does not affect the waterproof cement. The outside veneers very often split, due to shrinkage, but, as the material is manufactured in sheets about 60 inches square and the splits will not occur in the same place on both sides, they can easily be avoided when the stock is being cut up for use. It will be evident that any Canadian firm starting to manufacture three-ply stock would have to go into it on much the same basis, in order to sell their product at a price that would net them a fair profit and enable them to compete with the Russian product after the war. The first thing to do would be to choose a suitable wood, and at the present moment we cannot think of a better one than birch. After deciding on the wood, it would be necessary to secure a suitable adhesive (preferably waterproof cement).

The making of a cheap, efficient waterproof cement may be a secret, but we are confident that there are men in Canada capable of finding out the secret. We believe that this line might be taken up profitably by panel factories manufacturing built-up veneer panels.

It must not be thought that the manufacturing of this material would in any way affect the built-up veneer panel business. Built-up veneer panels are a high-class product, used for exposed parts, or what may be called face work, on a wide range of cabinet work. Thus it will be seen that "ply wood" and built-up veneer panels have their distinctive fields.

Manufacture of Shell Boxes Still Active

Appointment by Munitions Board of a Practical Woodworker to Visit Factories Has Met with General Approval

Staff Article

A round of the woodworking factories in Toronto reveals the fact that there is still considerable activity in the manufacture of shell boxes. Manufacturers generally state that the prices being paid are too low, and that they are simply finishing up contracts and would not consider further orders at present prices.

Some time ago the superintendent of one factory told the editor that they would not accept further orders at any price, as they had sufficient other work to keep them busy. About a week later his firm received a large order. Some people higher up had put in a tender, and it was whispered around among the trade that he had underbid all former prices on that particular kind of box.

Undoubtedly some firms have made lots of money on shell boxes, but one cannot help thinking that the whole business has been very badly managed. In the first place, many of the drawings and specifications were prepared by men who were sadly lacking in the knowledge of how to construct a box with a view to strength and economy. They seem to have acted on the principle of "the government is paying for it; we should worry."

The writer was talking to a shell inspector a few days ago who has had experience in both the United States and Canada. He said: "You people over in this country have wasted enough money on the making of shell boxes to help up the last war loan." Continuing, he said: "You ought to see the shell boxes the Russian Government have had made for their shells. They are simply strong boxes, that can be made very cheaply, but they are perfectly satisfactory for the work for which they are intended." Referring again to the Canadian-made shell boxes, he said the samples he had seen looked as if they were to be used for jardinieres in the drawing-room.

While deductions have to be made from certain of these statements, all woodworkers will agree that there is room for a great deal of criticism in regard to the amount of unnecessary work that they were compelled to put on shell boxes, and which had to be paid for. The Imperial Munitions Board have, to a certain extent, remedied the situation by appointing a practical woodworker to visit the different factories and confer with the manufacturers in regard to any details that seem to offer room for improvement. The writer can truthfully say that he has not yet gone into a factory manufacturing shell boxes where the Munitions Board have not been highly commended for making this appointment.

The whole trouble with the Munitions Board has been that they were not acquainted with, or interested in the problems of the manufacturer. As a matter of fact, they could not interest themselves to any extent on account of looking after their own work; and, further than that, their interest would not have availed much, as they were not practical men. They appear to have realized, however, that it was necessary to appoint an intermediary who could size up the situation from the manufacturers' viewpoint as well as theirs, and, according to manufacturers, they are to be congratulated on their selection of an intermediary.

In regard to prices, undoubtedly the most satisfactory to both parties would be to establish fair prices.

based on the actual cost of production. Even here, however, there would be drawbacks, as the cost of production depends on a number of factors which cannot be controlled by the manufacturer. The principal ones are the cost of labor and cost of materials. As these changed it would be necessary to alter the established prices, and considerable difficulty would be experienced, as conditions vary in different localities.

Prices are fixed by competition in practically every industry, about the only exception being where one firm has a monopoly of the tools of production. As no firm has a monopoly on the production of shell boxes, prices will most likely continue to be fixed by competition. So that about the only thing manufacturers can hope for is that the competing will be done in an intelligent and honorable manner.

National Basket and Fruit Package Meeting

An advance announcement of the annual meeting of the National Basket and Fruit Package Manufacturers' Association stated that the meeting was to be held in New York City, Nov. 15 and 16. As this issue of the Canadian Woodworker went to press before details of the subjects discussed at the meeting were available, we publish below an outline of the proposed programme, as contained in the advance announcement:

A report from a special committee giving recommendations for standard four-quart baskets and hampers, the dimensions recommended to be as follows:

Dimensions for Standard Four-quart Basket.

5⅜ in. wide by 9¾ in. long on the bottom, 4 5/16 in. deep, to be made with a 32½ in. inside hoop.

This basket holds a full four quarts.

Dimensions for Standard Bushel Hamper.

Staves, 20 in. long.

Diameter of the top of hamper inside at the top of the inside hoop, 15 in.

Diameter inside of the hamper at the middle hoop, 12 in.

Diameter inside of the hamper at the bottom, 9 in.

Height of hamper inside, from bottom to top of inside hoop, 18⅞ in.

This hamper holds one bushel.

Dimensions for Standard One-half Bushel Hamper.

Staves, 13⅞ in. long.

Height inside, 12⅝ in.

Diameter of top inside, 12 3/16 in.

Diameter of bottom inside, 8 in.

This hamper will hold 16 quarts.

A careful analysis of a cost account system which is now being used with satisfactory results in one of the largest basket factories to be presented at the meeting, and details of this system to be supplied to those present at the annual.

Several prominent members of the basket association to give talks on prevailing conditions in the trade, and a comparison of the basket and fruit package industry with other industries to be made by one of the manufacturers of fruit packages.

Cutting Up Lumber in the Furniture Factory

Second of Series of Articles Illustrating Practical Methods by which Satisfactory and Economical Results are Obtained in the Manufacture of the Finished Product

By W. J. Beattie

Dry lumber is the first requisite of a furniture factory, also the most important. The necessity for it being dry is too obvious to need any discussion.

There are various ways of determining the condition of lumber which is about to be cut up for manufacture, the measuring method and the weighing. The latter is possibly the one most used, though either is quite reliable. The way to determine the percentage of moisture in lumber is to take a fairly wide board, cut off about two feet of it and then from the cut end take off several pieces, each piece not more than a quarter of an inch long, weigh them, and then have them so thoroughly dried that not a particle of moisture can possibly remain, weigh them again, and the percentage of moisture is then easily arrived at.

There are special scales made for such purposes, and these are so sensitive that a very small fraction of an ounce is registered. There can also be an elec-

Plate corresponding to lumber rule.

tric oven used for the purpose of extracting the moisture from the pieces of lumber, though the same work, not so well done, can be accomplished by laying the pieces on warm steam pipes, or the cylinder of the engine.

The measuring method is done by using the same kind of pieces, except that the edges of the pieces of board should be jointed before being cut across into the quarter inch lengths. The exact calipered measurement is taken before and after the "roasting" process, and the percentage of shrinkage is arrived at. Of course, it is understood that the pieces of wood submitted to this test have been out of the dry kiln at least twenty-four hours.

Another way to test without exact measurement or weighing, is to take several pieces of exactly the same width, dry one or two thoroughly, contrast the difference in width, then lay them together where they will be subject to the same air conditions as the average factory work room, and if the dried ones swell to the same width as the others it can be safely assumed that the conditions of the lumber is reasonably good, though the writer would not recommend it as being infallible.

We are now assuming that the lumber is dry and ready for the swing saw. The average quantity of lumber piled on a kiln car is about 2,000 feet and the top tiers are well above a man's head when the car is rolled into the factory. The usual method is to

climb up and throw off four or five tiers of lumber into the most convenient position, which will likely be between the car and the swing saw table, where it lies in a more or less confused manner, and is very unhandy to sort for requirements. When the first lot is cut off the man scrambles up again and dumps more off, etc., until the lumber is down low enough for him to reach without climbing.

Did you ever wonder about how many hours in, say, a year, is wasted by such useless waste of energy? The better way to overcome this is to have the car of lumber run onto a hydraulic lift that can be adjusted so that the lumber is always waist high. The difference in the ease and rapidity of cutting will repay many times the investment. Some plants plane all the lumber on one side before it goes to the swing saw. This is open to objection in many cases, as it is generally advisable to run stock through a self-feed buzz planer on the one side, then through a first planer before the matching up and jointing, and it is evident that if the lumber in the long lengths is first planed on one side there is less material left to work on.

This will apply especially to tops, drawer fronts, etc., that need to be straight and true. As there are no two factories in the country that work under exactly the same conditions, opinions will vary, and all honest ideas are worthy of consideration.

I think a fair contention is that the average factory will find, it more advantageous to cut the lumber into lengths at the swing saw, then buzz or plane it on one side before ripping and matching, for all the exterior parts of furniture. The smaller interior parts and case backing, etc., are excepted.

Percentage of Waste

The percentage of waste in cutting up lumber into furniture seems an enormous loss when you figure that it amounts to 25 to 45 per cent. Talk about conservation of resources, to estimate what that waste means each year in the furniture industry alone would be staggering. The waste, or the cause of it, starts long before the lumber has reached your swing saw. The lumber comes in lengths running from four feet to eighteen feet, and as said in a former article, the proper place to divide and select it is before it goes into the dry kiln, and thus avoid unnecessary waste in cutting up all kinds of lengths at the swing saw.

It is not good practice to try and cut out half a dozen jobs at one time, so as to utilize all kinds of different lengths as they occur. Such a method is a time waster, and adds confusion in the matter of room, and also the time-keeping on each job.

The necessity of having a real good man as stock cutter is generally admitted. A man that has a general knowledge of the needs of the business, and can see the finished article in his mind as he gets out the various requirements, besides keeping a sharp eye on the rip saws, is worth a liberal inducement.

The actual percentage of waste can only be ascertained by measuring the lumber as it is being cut at the swing saw. A quick way that involves the least loss of time is to have the end of the swing saw table converted into a lumber rule so that the board is measured before there has been anything cut from it, and the quantity put on a small tally card or bill that is in a convenient position where it takes the least time. With the job number and quantity of articles on the heading, having room for the different kinds of lumber, the amount of material used for that lot is easily arrived at. As the stock bill sizes are net, the difference is the percentage of waste, and affords a record to check from, improve upon, or contrast the difference in waste that various grades of lumber will show, which, compared with the time consumed in cutting out, will give a correct reference as to the most economical lumber to use.

The sketch of the swing saw table here shown will give the idea as to the measuring of the lumber, also the fuller sized sketch of the one-eighth inch iron plate that is let into the saw table, and which has the figures corresponding to the both sides of a lumber rule stamped into the face. The width of a swing saw table is 18 inches, and the length of this iron plate would be 18 inches. The figures representing the various board lengths are at the outside end and are always visible. It would be quite simple to have this plate wide enough so that the 6 feet and 4 feet lumber lengths could be accommodated.

The figures representing the board lengths on the back of the saw table can be painted white on a black ground, using iron plates, and turning the farther ones so that the operator can see the figures plainly. The rule at the other side of the saw is one of the gravity automatic kind that are in general use.

The saw table is constructed as much as possible of iron, and has pipe and flange supports running to the floor. The rollers are of 4-inch steam pipe filled with a wood core that carries a 1-inch or ¾-inch arbor through the centre and into the side pieces.

A swing saw table gets about the roughest possible use and cannot be built too strongly. The illustration shows one that has the maximum strength and does not look cumbersome. The two supporting pipes directly in front of the operator could be given a slant in at the bottom and would thus be clear of shoe leather.

Vapor Kilns

Twenty vapor kilns have recently been installed in one plant in Oregon, which involves, it is claimed, the largest outlay in investment of the kind on the Pacific Coast.

One unique feature of this operation, says the West Coast Lumberman, is the fact that the loads are placed in the kilns at one end only, the rear of the kilns being walled up.

This installation has proven very satisfactory according to the owners and builders, because the stock is dried to a very low moisture content with soft outer surfaces, bright in color and very light in weight. The material is manufactured into sash and door stock, as well as other items of interior trim used in the building trade.

The increasing popularity of this modern drying method is evidenced by the large number of installations that are found in all of the western states and provinces in Canada.

A very important innovation has been developed in the demonstrations of the ability of this type of kilns to dry dimension lumber by the vapor process, as this is especially important to the lumberman who desires to ship his stock during the winter months.

Drawer Bearer for Cheap Case

The construction of cheap and medium grade case goods requires a great deal of thought and study in order to produce furniture that will present a workmanlike appearance and yet not involve too much labor in the making.

For the construction of buffets and dressers having short drawers requiring a centre bearer it is nearly always advisable to use frames. For dressers having long drawers the case rails at the front can be tenoned into the front posts, and the back muntins, which, of course, have to be run longitudinally, can be tenoned into the back posts. A reference to the accompanying sketch will convey the idea.

With this form of construction it is necessary to provide drawer bearers. One or two different kinds have been used, and, while they serve the purpose, they do not show up to advantage if the party buying the furniture happens to know enough to pull the drawer right out and examine the inside of the case. A drawer-bearer that was used in a factory where the writer was employed for some time is shown at B in the sketch. It is 1¼ inches wide and the same thickness as the case rails. It is checked out to fit snugly against the front post and a trifle loose at the back one.

The man who assembles the gables puts on the drawer-bearers (being guided by the mortises in the posts), and secures them with a nail at each end. The

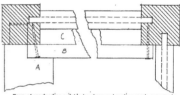

Form of construction suitable for cheap and medium grade case goods.

case rail A is then notched out as shown, with a shoulder on one side to fit against the post, and on the other to fit against the drawer bearer B. These two shoulders are cut, of course, on both ends of the case rails at one operation on the double end tenoner. The piece C is for the drawer to slide against, and can be put on with a couple of small nails at the same time as the drawer bearer. Another method would be to make the drawer bearers thicker and rebate them on the sticker.

The feature of this bearer is that it makes a neat job and gives very much the appearance of a frame without being nearly as expensive as a frame. The strips for the bearers can be run out in large quantities from cuttings. Then when a batch of cases comes through, the length of the bearer can be secured from the rod or drawing and the strips cut off and notched accordingly.

How to File a Handsaw

This is the title of an interesting booklet which we have received from the Simonds Canada Saw Company, of Montreal. and which they will send to any person upon receipt of their request. The book is intended for apprentices in the woodworking industry and for manual training students. It contains a great deal of information, clearly illustrated, on what has always been one of the most difficult things for a young mechanic to learn.

Piano Making a Highly Important Industry

Work in Average Factory Confined to Woodworking Operations— Executives Have Same Problems to Solve in Cutting Costs as Heads of Other Plants

By "Piano Maker"

In view of the fact that there are between twenty and thirty thousand pianos manufactured in Canada in a year, piano making can be regarded as one of the country's important industries.

A number of the parts used in the construction of a piano are metal. These enter the piano factory as raw material, but as they are really purchased in a finished or semi-finished state from jobbers or firms specializing in their manufacture, the piano manufacturer has very little, if anything, to do with them further than utilizing them in their proper place in the assembling of the finished product. Piano manufacturing, as carried on in the average factory, is a woodworking proposition. The superintendents and managers have the same problems to solve as those that confront the heads of other woodworking establishments.

Outsiders may imagine that because the product sells for a high price the costs of the various operations do not have to be checked as closely as in furniture factories and other woodworking plants. In this, however, they are mistaken. That bugabear of the modern manufacturing plant, "overhead expense," coupled with the high cost of labor, make it imperative for the piano manufacturer to cut down manufacturing operations to a minimum in order that the cost of the product may be kept at a point where it will be possible to sell at a profit.

Perhaps in no other branch of the woodworking industry does labor represent such a large percentage of the cost of the product as in piano manufacturing. This is partly due to the fact that the operations are so many and varied and partly because the bulk of the labor must be highly skilled.

The first step in the manufacture of a piano is to have an accurate drawing. A stock bill is then made out, showing the dimensions of each piece of lumber required. The back and the case are billed separately, as these are assembled in different departments. The manufacture of the backs calls for large quantities of ash and elm, which must be of thick stock. A reference to Figs. 1 and 2 will show how the back is constructed. Fig. 1 shows the outside and Fig. 2 the inside. It will be noticed that the illustrations are of two different kinds of backs. One is constructed with five posts and the other with six. Also, one has the handles which are provided for lifting the piano secured between the two end posts while the other has a cleat extending right across the back with hand-holes at the ends.

Apart from a few minor details, however, backs do not generally vary greatly in the method of construction. The posts are tapered and are spaced so that the iron frame or plate (shown in Fig. 3) on which the wires are strung, and which is provided with bolt holes can be bolted to them. Special bolts are made for this work. One end has a screw-thread while the other end is threaded for a nut.

The blocking which goes between the posts, top

Fig. 1—Outside view of piano back.

Fig. 5—End view of piano action.

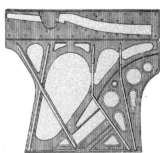

Fig. 3—Metal frame or "plate" on which wires are strung.

Fig. 4—Corner of plank into which tuning pins are driven.

Fig. 2—Inside view of piano back.

and bottom, affords an opportunity to use up a lot of odd stock, as any kind of wood can be glued up and used for this purpose. After being glued up the blocking is tapered to correspond with the taper on the posts and then cut to length.

In putting the backs together the first step is to glue size the ends of the blocks and then assemble the posts and blocks. This is done in a press having side clamps. After the glue has set, the partly assembled

Fig. 6 — Piano sound-board, showing diagonal strips or "ribs"

back is removed and levelled off with a plane. Then the two pieces (called "linings") which can be seen at top and bottom Fig. 1, are glued on, also the tapered piece (called a "wrest plank") at the top, and the notched strips at the sides and bottom Fig. 2.

Fig. 4 is a "close up" view of the corner of the wrest plank and shows the method of construction quite clearly. The wrest plank is made of hard maple, about 1¼ inches thick, with maple plywood glued on top (in this illustration nine-ply, but not always so). It will be noticed that the lower edge of the wrest plank is rebated. The rebate is flush with the strips at the sides and bottom, and on these the soundboard shown in Fig. 6 is secured. The notches in the strips are to receive the ends of the diagonal strips or "ribs" on the soundboard.

The piano action, an end view of which is shown in Fig. 5, is made principally of wood, wire, and felt, wood being most in evidence. The wood mostly used is maple. The action is partly assembled where it is manufactured, but some of the parts, such as hammers, dampers (upright piece with V-shaped piece of felt in front of hammer, which "damps" the wire to stop the echo), etc., are assembled by mechanics called "action finishers" and "action regulators." The majority of these men can do both jobs, but in large factories the work is divided. The men work piece work, and are paid so much per piano.

The construction and method of assembling the case of the piano will be dealt with in a later article.

Sizes of Fruit Boxes

Mr. A. H. Flack, the Dominion Government's chief fruit inspector for the Prairie Provinces, in discussing the sizes of fruit boxes, writes as follows:

"The apple box has frequently been discussed, and no decision has ever been reached in regard to it. We have for domestic purposes two sizes to select from, both of which are in use in British Columbia to-day. One is the box specified for export in the Inspection and Sales Act, section 325, which has inside dimensions of 10 x 11 x 20 inches, containing 2,200 cubic inches, while the other is the Oregon box, with inside dimensions of 10½ x 11½ x 18 inches, cubic contents being 2,173½ cubic inches. As far as the adaptability of

these packages for apples is concerned, I do not believe there is very much difference.

"If anything, the Oregon box has the advantage. The chief argument in favor of the Oregon box, to my idea, is that the length is identical with the pear and peach box. This is really a very important point, as it enables mills to produce cheaper by utilizing material for other packages that otherwise would be wasted, and is certainly an advantage in warehousing stock and loading mixed cars.

Peach and Apple Boxes.

"In regard to apple crates, it is quite obvious that some steps should be taken in order to have only one size in use. Undoubtedly several ideas exist as to what constitutes the best package for this purpose. In regard to the peach box we have one in use in British Columbia measuring 11 x 18 inches inside, and another measuring 11½ x 18 inches. I think it woud be advisable to agree as to which of these two packages to adopt. It will, of course, be necessary to have several different depths, and I would suggest 3½ inches, 4 inches, and 4½ inches. In regard to the width, 11½ inches would be the same as the apple and pear box suggested, with the accompanying advantage.

New Kind for Pears.

"Regarding the pear box, I would suggest the one measuring 11½ inches wide, 8½ inches deep, and 18 inches length inside. This would give us a package of the same length and width as the apple and peach box suggested. In view of the fact that a lug package is now being used for some markets, it would, I consider, be desirable to standardize one. I would favor the Pacific lug, as used by the Pacific Fruit and Produce Company. This package is used largely for cherries, and is shipped containing 2 pounds net by weight. It measures 14 inches in width, 5½ inches deep, and 16¼ inches in length, having a hand gauge on each end, with a cleat 14 x 1½ x ½ inches under the cover, which permits a free circulation of air."

Fine Furniture Sold by Auction

The accompanying illustration is taken from the "Cabinet Maker" (London, Eng.), and shows one of a set of twelve chairs of Chippendale design and a hall table that formed part of the contents of Norbury Park,

Hall table and Chippendale chair from a collection of furniture recently sold in England.

near Dorking, in Surrey, which were recently sold by auction. The chairs are in mahogany with claw and ball feet, and are upholstered in green leather. The table is also mahogany, with cabriole legs and claw feet. The top consists of a marble slab and an extra Verde marble slab 4 ft. 6 in. wide.

Standard Parts for Filing Cabinets

Installation of a White Parts Stock Room—Machine Room Manufactures for Stock Only—Orders Issued on White Parts for Cabinet Room

We reproduce herewith an article that was published in a recent issue of "Hardwood Record," under the title, "Twin-Six Efficiency vs. Filing Cabinets." The article, which contains many points of interest to woodworkers, compares the wonderful developments in the field of woodworking with the achievements of automobile manufacturers who have secured near-constant application of power by improving the one-cylinder engine until it has evolved through the two cylinder, four cylinder, six cylinder, eight cylinder, and finally developed into the twin-six cylinder engine:

Ten years ago two young men in Muskegon, Mich., wanted to get into a business for themselves. They wanted to manufacture filing cabinets. Their assets were made up, chiefly, of a PLAN, nerve, and the ability to work. So they started in a small way—they had to—putting on the market a line of office equipment similar to that of other manufacturers.

Many people thought they wouldn't last long; that their capital wasn't large enough, but they built good cabinets. their prices were right, and they worked hard—how hard they alone know. Gradually they established a reputation for quality and their business grew; still their assets remained about as before—just nerve, the ability to work, and The Plan.

In order to put the automobile within the reach of everyone. Henry Ford standardized. Not only did he confine himself to one chassis, but wherever he could make two or more parts uniform and interchangeable he did so. The result was increased output and decreased cost—efficiency.

PART No. 400..............USED IN SECTIONS

NO. PIECES	SECTION NO.	NO. PIECES	SECTION NO.	NO. PIECES	SECTION NO.	
1 Pr.	559					PART NO.
1 Pr.	253					
1 "	250					
1 "	235					
1 "	334					400
1 "	233					
1 "	232					Pilasters
1 "	231					
1 "	230					
1 "	226					
1 "	211					

Brown-Morse record card indicating number of different sections in which this one part is common. (Actual size card 4x6.)

The Browne-Morse Company, the company formed by the two young men mentioned above, standardized—that was the first part of its plan. Instead of making different sectional combinations of various sizes, as it had heretofore, and as other manufacturers were doing, it selected two heights and two depths as the standard dimensions to make its various combinations. The company build everything—with the exception of a few specials—up to and around these sizes. In either standard size end panels, tops, bases, stiles, rails were

Perpetual inventory card used by Brown-Morse Company. Three colors used—White card for oak, buff for birch, blue for mahogany. (Actual size card 5x8.)

uniform in the three woods used—birch, oak, and mahogany. No matter what the combination—letter file, document file, drawers for legal blanks, card index drawers—as many dimensions as possible were reduced to a common standard. The result was a saving in cost, an increased efficiency, and a better and more flexible file.

Filing cabinets had always been made as a whole. Orders for a certain quantity of a certain style or number would be started through the machine room. When the requisite number of panels, tops, drawer slides, pilasters, cleats, etc., were accumulated on trucks, they were taken to the assembling room, and then another style cabinet would be started through. A certain amount of congestion—which means loss of time and increased cost—on the floor and in the machine room was bound to accompany this method. Then, too, a completed cabinet occupies more than four times the space that it would if "knocked down."

Why wasn't it feasible to manufacture ahead and carry in stock at all times the various wood parts that went to make up this office equipment? Why couldn't these hundreds of parts—all cut, sanded, ready to be put together as needed—be carried in stock the same as the raw material (lumber) was carried?

Having, by this time, outgrown its first home and taken up new quarters in a large, modernly equipped factory, the Browne-Morse Company proceeded to put this other efficiency idea into effect—the second part of the plan. A white parts stock room was installed. equipped with racks and bins, which would accommodate the required number of each particular wood part. As each part carried an arbitrary number, its particular bin was correspondingly numbered; and. in addition. at the end of each bin-section an index of that section, showing the bin numbers, of what wood the part is made, and the capacity in pieces of the bin. The machine room manufactured for stock only, and orders were issued on the white parts for the cabinet room.

The successful working out of this stock room plan has been due in part to the simple, yet complete, record system used. But, as the Browne-Morse Company has been able to install card index systems for others, it was able to do the same thing for itself. The perpetual inventory, showing exactly the number and kind of the various parts carried in the stock room, is kept by the woodworking superintendent, duplicate cards of which are carried in the front office. This inventory, as can be seen from the accompanying cut, shows what the section is, its number, bin number, the number of pieces and dates when put in stock, together with the shop order numbers, disbursements, and the amount on hand.. Each section or part has its own card, different colors being used for the different woods—buff for birch, white for oak, and blue for mahogany. There is also shown on the card the "low limit"—the number of pieces below which the stock is not allowed to run. When this is reached new cutting bills are issued and the bin is then kept automatically replenished.

In addition to the inventory record is what might be called the standardization record. Each particular part is again given a card, indexed under its same arbitrary number, showing the pieces and section number to which this one part is common.

In cutting the rough lumber for stock it has been found that a saving has been effected in the raw material, the waste being less than under the old method. Edgings and trimmings, too small for the cutting bill in question, are sent back to the lumber storage, sorted, and later worked up for other items.

The proof of the pudding is in the eating. Results in cold dollars and cents are what we are all looking for these days. That the white parts room—cutting for stock—has made for efficiency is proved by the fact that the Browne-Morse Company has been able to double its output at a saving of 25 per cent. And not only that, but, operating under the old method, it would be forced to carry a stock of finished goods fully 50 per cent. greater than it does. Carrying in the white parts for every cabinet and cabinet combination enables it to be ready at any time for such rush orders as may come its way.

The twin-six motor delivers almost constant power, with a maximum of flexibility and the least tendency to stall and choke. The Browne-Morse Company has developed this same twin-six efficiency—a woodworking plant that is truly flexible, delivering quality goods in the greatest volume, at the lowest cost.

Work—Nerve—Plan—a combination that is hard to beat.

Should Manufacturers Buy Hardwoods Now?

Scarcity of Stocks on Hand and Small Number of Firms Preparing to Cut in the Bush Make the Situation Very Critical—High Prices will Undoubtedly Prevail

Staff Article

This question was put to a lumberman recently who deals in hardwoods on a large scale. His answer was, "Yes, if they can." We believe that the whole situation is summed up in those few words. Hardwoods are among the scarcest commodities on the market at the present time. The consumption during the past year and more has been greater than manufacturers generally will credit. More than that, large quantities of hardwoods have been finding their way into entirely different channels from those in which they are customarily used.

Of these undoubtedly the most important is the manufacture of shell boxes. Some manufacturers may say: "There can't be such a terrible lot of hardwood going into shell boxes." Maybe not, but one prominent lumberman informed the writer that his company had been shipping eight and ten cars a week for this purpose. If one company alone has been supplying eight and ten cars a week during the time the shell box business has been active (and the sales in all probability amounted to more than that during the earlier stages of the activity) one can form some idea of the "dent" that has been made in hardwood stocks.

To further emphasize the shortage of stocks, an interview with a lawyer who is promoting an enterprise in which considerable lumber, both hardwoods and soft woods, will be required for construction purposes might be referred to. The remark was made that he might experience some difficulty in securing lumber. He and his assistant looked at each other and smiled grimly, and his reply was somewhat as follows: "No person can tell us anything about the shortage of lumber. My assistant has covered Ontario, Quebec, and the Maritime Provinces trying to buy lumber at right prices, but there is nothing doing."

Apart from the scarcity of stocks must be considered the great scarcity of labor at loading points. As manufacturers are well aware, the stocks on hand are located at many points in different sections of the country. When orders are placed with lumber firms, points where the stocks are piled. As most of these they in turn transmit the orders to their men at the points are in outlying districts, and the rural sections of the country have, generally speaking, contributed their full quota of men to the fighting forces of the Dominion, it is quite clear that there is great difficulty in securing sufficient men to load cars. The result is that higher wages have to be offered as an inducement for men to leave the farms and other forms of employment to load lumber. And as the lumbermen have to make up the difference in wages, it is only reasonable that they should demand higher prices for their product. Manufacturers in every part of the country are feeling the labor shortage, so that it should not be difficult for them to appreciate the position of the lumbermen in this respect.

The price of any commodity in the world of business is governed by supply and demand. If the seller is anxious to sell, the buyer has the advantage, and vice versa. Interviews with leading lumbermen during the last two weeks reveal the fact that they are not "pushing" their goods. As a matter of fact, they are quite content—those who are fortunate enough to have fairly extensive stocks on hand—to let things drift, or, in other words, they are not anxious to sell.

It is gratifying to note, however, that they are not

making any attempt to hold up prices. Further, they are doing their best to furnish lumber at reasonable prices to their customers, especially those who are in the habit of giving them their business. A striking illustration of this was brought to the writer's attention the other day. A lumberman called on the manager of a factory with whom he had had a long and pleasant business association and told him he had a car of lumber that would just suit his requirements. The manager, through his farsightedness, was fairly well stocked up and wouldn't really need the lumber for some time. The price quoted, however, was such that he could not afford to miss the opportunity, so he made the purchase. Some time later the owner of the business paid a visit to the factory and the manager mentioned the extra car of lumber he had purchased. The owner, literally speaking, "went up in the air," and started to howl about the drain on the firm's bank account. The manager waited patiently until the irate gentleman had reached terra firma again, and then told him politely that the car was in the railway yard close by and that he could transfer it to another firm immediately for five dollars a thousand more than he had paid for it. The owner, who it might be stated, pays more attention to the selling end of the business than the manufacturing end, saw the point, and told him to hang on to it. This incident actually occurred, and

the three parties concerned are well-known business men.

In regard to the future supply of hardwoods, the outlook is pretty gloomy. So far very few firms have gone into the bush, and those who have are only preparing to cut on a small scale. It is now getting late in the season, and if no more go in by the end of November, hardwood prices, to quote one party, "will turn into aeroplanes." The market will simply develop into a bidding contest, and prices will be fixed according to what the most urgent buyer is willing to pay. The question that confronts manufacturers is: With the present or prospective selling prices of their product can they afford to be urgent buyers?

In any event, prices of hardwoods must advance. Even if a large number of firms go into the bush to cut, the scarcity and subsequent high price of labor and the increased cost of practically all food supplies will add considerably to the cost of their output of logs. The selling price of the lumber must cover all this additional expense.

The "Canadian Woodworker and Furniture Manufacturer" has advocated for some months past that manufacturers should stock up as much lumber as possible. The situation is becoming more acute all the time, but even now it may be possible to make provision for the inevitable coming of higher prices.

Special Machinery in the Toy Industry

Describing Two Instances Where the Inventive Genius of Manufacturers Enabled Them to Cut Costs so Low That Competition with Foreign Made Toys was Easy

By R. Newbecker

It cannot be denied by any person acquainted with the toy industry that the demand for toys constructed of wood has been steadily increasing for the past several years. The days of the trashy, cheap tin toy of foreign make are about over.

Present indications seem to point that within a short period of time millions of dollars' worth of wooden toys will be manufactured from the waste or off-fall of the woodworking plants, which formerly had apparently no other value than to be sold for firewood or burnt up under the boilers.

The reason that up to the time of the war very little, if any, attempt was made to convert the waste into toys was due principally to two causes. First, the high cost of labor as against the cheap foreign labor, and, second, the lack of native-made machinery to produce toys at a cost that would make it possible for manufacturers to compete against the foreign product. Therefore, to bring this industry to a stage of development where it will be possible to compete successfully with foreign-made toys and provide employment for native workers, special machinery combining two or three machines in one will first be necessary to bring the cost of production down to the lowest possible point.

In this regard I will cite an incident which occurred in the United States some time ago. In the manufacture of rulers and yardsticks it had been customary with the plants engaged in this line to first rip off the strips, then plane them and sand them, after which they were cut to length and stained and printed. Then along came a man who also engaged in this line, but not in the same way as the other plants were doing. He first constructed a machine about 20 feet long, in which he combined all these operations. Thus, instead

of planing, ripping, sanding, etc., he simply stuck a 6-inch board in his machine and it did the rest. The board was first ripped into pieces of the desired dimensions, while directly following the saws were the planer knives, and at the back of these were specially constructed sanding disks doing the rough sanding. Then there was another set of specially constructed emery disks doing the polishing. It will be seen that this one operation reduced the work to less than one-fourth, and it wasn't long before the manufacturer in question was able to produce rulers and yardsticks at prices that defied competition.

Another incident of a similar nature occurred about a year before the present war broke out. A certain St. Louis gentleman engaged in the manufacture of wooden toys, such as small chairs, tables, taborettes, settees, etc., the articles being carded five to a card and sold in the 5 and 10-cent stores. However, he soon realized that the foreign product was sold cheaper than what his labor would amount to, not figuring in the material, so he set to work to invent a machine with which he would be able to safely compete against the articles of foreign manufacture.

After considerable study and several reverses this genius was finally rewarded through the perfecting of a machine which, up till a short time ago, was thought impossible in the woodworking line. This consisted of a die-cutting machine capable of cutting any size or shape in thin lumber up to ⅜ inch in thickness. Thus by means of this machine one boy and a girl are able to produce more round, curved, irregular, or, in fact, any of the shaped designs so essential in toy furniture than from six to ten men would be able to produce by ordinary methods on the scroll saws. This manufacturer is not only able to compete, but is also able to underbid all foreign-made goods in his line.

The machine is built along similar lines to the ordinary stamping press, with the special patented die-cutting arrangement attached. With this machine they often cut out, in exact shape, as many as twelve different shaped pieces with one cut, all ready to be passed on and finished. The board is laid on the stamping table, then the die comes down with enormous pressure and cuts out the wood in fancy outlines, just as easily as mother used to cut out cookies by means of the ordinary tin cutter. The whole operation is completed in less time than it takes to count ten.

While the present market for wooden toys and novelties is a large one, nevertheless to be successful in this line specially constructed and planned out machinery are almost essential in order to make the line pay.

No better advice was ever given the manufacturer about to embark on toy manufacturing than the article in the August issue of the "Canadian Woodworker"—"The Practical Manufacture of Wooden Toys," by "Experience." If his advice is followed there will be little danger of going wrong when starting out to manufacture wooden toys and novelties.

Extracting Turpentine from Pine

The Chicago Post, in a recent issue, published the following interesting article on how turpentine workers in Southern forests extract "spirits" and resin from pines:

"In the Carolinas, whence most of the turpentine is derived, tracts of turpentine land are called 'orchards.' Early in the year, before the 'juice' of the pine is ready to exude, the trees are 'boxed.' The boxes are not, as the term would seem to signify, appendages attached to the outside of the tree, but are cuplike cavities cut into the trunks about a foot from the ground. They measure from 10 to 14 inches across and are 4 inches in width and 7 in depth. Each of them is presumed to hold a quart.

"Usually there are two, and sometimes even three or four, such boxes to each tree. The life of the pine, as of other trees, is drawn from the soil through the bark; hence it is necessary always to leave an uncut strip of the bark between the boxes. The number of the boxes depends, therefore, upon the size of the tree and the width of the intervening strips.

"The instrument used by the cutters is an axe made especially for the purpose, measuring three inches across the blade and nine inches in length, having a handle which is bent several degrees out of a straight line, ordinarily to the right; but if the user cuts 'left-handed' it is bent toward the left.

"To one who has never used such an axe it seems extremely awkward, and the novice who handles it is apt to come to grief. In his hands it is a kind of boomerang, and it is impossible to determine just where it will hit. An experienced cutter, however, takes his position near a pine, gives the axe a peculiar swing, and in a few moments has cut and cleaned out such a box as an inexperienced hand would hardly hope to make with an hour's labor.

"As the weather grows warmer the boxes become filled with the exudations of the tree, and the first year's product, being the most valuable, is dipped out and kept separate from that yielded by trees that have been boxed in previous seasons. After a time the running rapidly diminishes, and then another small, steel-pointed instrument—the scraper—must be brought into requisition.

"This process is repeated as often as proves necessary, and over this sacrificed surface the crude turpentine finds its way into the cup. Much of it is thickened by the heat of the sun before reaching its destination, but is scraped off, and, with the rest, subjected to the process of distillation. That in the cups, being very adhesive, is removed by means of flat scoops. The whole is placed in barrels and taken to the distillery. Here the crude product goes into the retort and comes out as 'spirits' and resin, which are converted into the numerous commercial articles with which all are familiar."

Speaker's Chair for House of Commons

Herewith we reproduce a photograph of the new chair for the Hon. A. Sevigny, Speaker of the House of Commons. It is of special design, and was entirely made by the Bromsgrove Guild (Canada), Limited, in their workshops in Montreal. The chair is of oak, upholstered in green Canadian leather, and it will be seen that the carving is very elaborate and appropriate, the national emblems and arms of Canada and the

Speaker's chair made by Bromsgrove Guild, Montreal.

national emblems of the United Kingdom being very skilfully treated in the design. Thus the central and chief subject is the arms of the Dominion, surrounded by maple leaves, while on one side are the Welsh leek and Irish shamrocks and on the other the English rose and Scotch thistle. Carrying out the idea of national emblems still further, squirrels are carved on each side of the chair, and the maple leaf is also freely introduced into the other decorations. The initial of the speaker—S— is worked out on either arm of the chair, immediately above the leather seat.

Wide Range of Upholstering Materials

Leather Still Occupies Leading Place—Tapestries Much in Evidence for Period Designs—United States Supplying Greater Part of the Demand

Staff Article

Leather still occupies the leading position as an upholstering material, although there has been a large increase in the use of tapestries and other fabrics. This has been particularly noticeable in the Jacobean and William and Mary styles. In early days, when the master craftsmen of England and France plied their trade, they used scarcely anything else but tapestries and velvets, etc., for upholstering. Some leather was used for dining room chairs, but here also tapestries and fabrics predominated. The use of tapestries for the period designs is generally regarded as an evidence of good taste.

The amount being used, however, has not in any way relieved the situation in regard to the shortage of leather. As a matter of fact the scarcity of good tapestries is just as pronounced, if not more so, than the scarcity of leather. Before the war large quantities of the former were imported from Belgium, France and England. Fairly good shipments are still being secured from England, but the manufacturers of the United States are being called upon to supply a large portion of the trade. The chief difference between the Belgian, French and English tapestries and those manufactured in the United States is in the designs and color combinations. The American manufacturers have done well when the disadvantages they have had to labor under are considered. In the first place they have been handicapped in the matter of dyes. Then again, previous to the war the custom was to look abroad for upholstering fabrics for the better grades of furniture.

This class of furniture, principally composed of reproductions and modifications of period styles, calls for designs that harmonize with the style. These are mostly dainty floral designs with carefully matched colors. Textile manufacturers abroad have been producing this class of goods for years and consequently were in a position to furnish American and Canadian furniture manufacturers with the correct thing. Textile manufacturers on this side of the Atlantic then had to confine themselves more or less to the cheaper grade goods, for which large floral designs in pronounced colors are mostly favored. Since the outbreak of war they have been called upon to manufacture the better grade goods and considering the shortage in certain materials they have achieved splendid results. While the goods being produced are mostly confined to two or three predominating shades the designs and colorings are in good taste.

The war has not by any means been the only cause of the shortage of leather. This has been brought about as the result of several economic factors which have been at work for a number of years, gradually exerting their influence until now the world's supply of hides is showing signs of becoming almost completely exhausted. The most important factor from the supply side has been the increasing proportions in which prairie lands in Western Canada and the Western States have been taken up for agricultural purposes, thereby crowding out the large cattle ranches which formerly furnished perhaps about half of the world's supply of hides.

From the demand side the principal feature has been the growth of the automobile industry. When one considers the enormous quantities of leather that have been required in this industry for upholstering purposes one can begin to realize something of the demand which has been made on the already fast diminishing supply of hides. Then again, the manufacturers of ladies' footwear have been using large quantities of leather to make the high top boots which are the prevailing style. The small amount required for this purpose may not look much, but when it runs into millions of pairs of boots it means a good many hides. The manufacture of army boots has no doubt been responsible for the consumption of hides in considerable quantities, but it can only be regarded as a contributory cause to the shortage.

Hides for the manufacture of upholstery leather, of course, have to be choice. That is to say, they must be of large size or "spready," to use a trade term. This is essential in order that the leather will cut to good advantage when the upholsterers receive it. The hides also have to be free from blemishes such as brand marks, scars, etc. After being partly tanned they are put through huge machines and split into three and four thicknesses, depending on the quality desired. If split into four pieces they are called—naming from the hair side of the hide—buffing, machine buff, main split, extra split, slab. The first cut or buffing is extremely thin and only suitable for purses, camera case coverings, etc. The machine buff which is the second cut, is the most valuable. The next two cuts, the main split and extra split are both spoken of by upholsterers as simply "splits."

The final cut or slab is of very little use unless perhaps for insoles for boots or some similar purpose. If it is desired to make one of these split pieces of better quality the extra thickness has to be allowed for and this reduces the number of splits which can be taken from the hide. After the different cuts have been subjected to certain processes of oiling and enamelling they are put through embossing machines to receive the "grain" or figure.

The quantity of artificial leather being used for upholstering is constantly growing. The question has often been debated, "which are the best, leather splits or artificial leather?" I will not attempt to answer the question at once, but will give a brief outline of the conditions to be met with in the average furniture factory.

The price at which the furniture is to sell governs to a certain extent the amount to be spent for upholstering. Practically all cheap living room and dining room chairs and a great many of the higher priced suites are upholstered in artificial leather. In the majority of the factories the upholsterers work piece work. The foreman of the department and the management as a rule fix the price.

The price being paid for some chair already in stock is usually taken as a basis, and an allowance made for any extra work that may be necessary, or, if less work is required a deduction is made. An assortment of different grades of leather and artificial leather is usually carried in stock. The grade to be used is decided on and the foreman does the cutting. The upholsterers gauge the amount of time they can afford to spend on each piece by the price (the quality of the work, of course, being subject to the foreman's inspection. Some of the cheap grades of artificial leather are expensive at any price. They stretch like a piece of elastic and it requires so much time to get the wrinkles out and make a tight job that the upholsterers demand a bigger price. On the other hand, some of the better grades are much easier to work with than "splits," and thus find favor with the workmen.

As far as the wearing qualities are concerned there is not much choice between leather splits and a good grade of artificial leather. Some of the manufacturers of leather splits claim that the dressing will wear off the artificial leather in a short time and the fabric backing begin to show through. This statement, however, is not borne out by actual facts. The final consumers, the public, are the ones that have to be satisfied, and the writer during a number of years' experience in factories manufacturing upholstered furniture has never heard any complaints from dealers about their customers being dissatisfied, and they are certainly the ones to whom the dissatisfaction would be expressed if cause existed for such.

For really high grade furniture the "Simon pure" leather is undoubtedly the best, but for medium grade furniture the writer's choice would be a good grade of artificial leather in preference to leather splits. And the public can rely on getting a serviceable grade for the simple reason that it costs the manufacturer more to bother with a poor grade than it does to buy a good grade.

The Manufacture of Office Furniture
By J. N.

I have been asked to contribute an article on the manufacture of office furniture, but as office furniture does not differ materially as far as mechanical points of construction are concerned from other furniture or woodwork, I think it would be advisable to deal with the subject from a different angle.

The manufacturers of office furniture are in a somewhat different position from the makers of household furniture. The latter have certain definite lines to work along. That is to say, requirements in the home do not change very much from year to year. Dining room suites, bedroom suites, parlor suites, library furniture and living room furniture have always been and will always be more or less the same, with the exception of certain details in the matter of design and construction which are worked out to suit the tastes and pocket books of the buying public.

The manufacturers of office furniture, on the other hand, are up against an entirely different proposition. They have to anticipate the requirements of modern business offices. This is no small problem. In the multiplicity of business enterprises that make up our commercial and industrial life there are numerous details of office routine to be provided for. In order that these may be taken care of systematically and as efficiently as possible, filing cabinets, record cabinets, card index drawers and a host of other receptacles for filing trade journals, catalogues, etc., must be designed. Office equipment cannot be designed haphazard, or at least if it is there will likely be some difficulty in convincing the business man that he needs it in his office.

There is only one logical way to design office equipment, and that is, to keep in touch with what business men are doing in different sections of the country. This can be done by reading business magazines and getting a line on how business men are thinking and how they are handling their office work. By getting their viewpoint on the filing of office records, etc., you are able to design furniture that will take care of their requirements almost before they are ready for it.

We hear a lot about metal office furniture, and while it is undoubtedly finding favor in some offices, it will never supplant wood. In fact the writer is inclined to believe that the days of metal office furniture are numbered. The chief aim in modern offices appears to be to make the surroundings as congenial for employees as possible, working on the assumption that they will do more and better work. Metal furniture certainly does not fit in with this line of reasoning. It is much more appropriate in a gaol, as it harmonizes nicely with the stone walls, cement floors and barred windows. One can scarcely imagine men accustomed to the good furniture and comfortable surroundings of their homes and clubs spending their business hours working at metal desks or sitting on metal chairs. We can therefore expect that men who equip their offices with metal furniture have been taken in by the manufacturer's or salesman's talking point that metal furniture will not burn. It certainly will not burn, but that is not to say that the contents will remain intact. To make metal furniture so that it will

Two types of filing cabinets.

protect papers and other valuables from being damaged by heat it would have to be built the same as a safe or vault. This, besides being much too cumbersome, would occupy a great deal too much space. The accompanying illustrations show two types of filing cabinets manufactured by the Office Specialty Company, of Toronto, which can be utilized for a wide range of requirements and other units can be added when necessary.

When you are dissatisfied with your job or your surroundings, it is a good idea to carefully analyze the matter, that you may understand clearly just what is wrong. It may be merely your own health or habits.

The minute a man begins to see his own shortcomings plainly, they commence to disappear.

Uniformity in Drying Lumber for Furniture

Equipment Must be Designed to Suit Individual Requirements—Air Dried Stock Deteriorates Rapidly from Surface Checking and Case-Hardening

By A. E. Krick

Drying lumber is a process which must be varied to a great extent, according to the demands of the factory in which it is to be used. Lumber dried for wagon stock requires one kind of treatment, that for automobile bodies another, and so on throughout the category of products utilizing wood as raw material.

Each branch of the woodworking industry is apt to believe that its lumber must be better dried than that of other branches. It is true that there are some peculiar requirements for each and every class of work which are not to be found in other classes, and it naturally follows that ways and means of drying must be varied to suit the different requirements.

Furniture factories are probably the ones for which the most careful and scientific work must be done. They use a great variety of stock of a great many thicknesses. The lumber must be dry, and it must be uniformly dry—that is, different sections of a board must contain exactly the same given percentage of moisture. It must be dried without checking, casehardening, or honeycombing, and there must be no warping. In order for it to work properly through the big machines now in almost universal use for building up table tops, chair bottoms, and various other wide pieces used in the manufacture of furniture, the stock must be in such condition that it will readily take and assimilate glue.

It must be as soft and mellow as possible, in order that it may be worked easily through the many knife machines to which it must go. An extensive list of requirements might be detailed to emphasize the need of exceptionally carefully dried stock for furniture factories.

Then, again, the needs of factories making furniture are not always identical. What will do for one class of furniture will not do for another. There is no panacea in the shape of a dry kiln which will cure everybody's troubles in drying regardless of his needs.

There must be careful consideration given to each manufacturer's requirements by the dry kiln designer, and his kilns must be accommodated to that manufacturer's needs. This fact is going to preclude the possibility of our ever being able to order dry kilns right off out of stock, from a shelf as it were, and without going into the matter carefully first.

As lumber becomes more valuable it becomes increasingly important to dry it properly, and to avoid having any of it to throw away. In fact, it is becoming quite well understood that air-drying is making too much cull lumber. This is particularly true of thick stock.

For a great many years we have been disposed to think that air-drying was an absolutely necessary thing. Now we know that it is not only not absolutely necessary, but that there are many things which make it advisable to do just as little air-drying as possible.

A piece of lumber is in much better condition as far as shape, strength, and general appearance is concerned right after it leaves the saw than it is after it has been air-dried for a long time. It is flat and straight, clean and bright; there are no checks, and there is no case-hardening, honeycombing, or hollow-horning.

If it can be taken while in this good condition and put into a properly designed and equipped dry kiln, the net results are much better than if it is air-dried before going to the kiln. On the other hand, if the lumber is air-dried it deteriorates rapidly from surface checking and case-hardening. The best drying apparatus in the world cannot repair the damage thus done. True, after the entire board is dry it may be that the checks will not show. The grain often closes up and conceals the checks, but the tensile strength of the piece has been shaken.

Often in working apparently good stock through the machines little pieces will break out on a crown

Furniture factory dry-kiln of modern design and construction.

cut. This is because of trouble such as that just mentioned. There are a number of large furniture manufacturers who are now equipped to dry stock in any condition in which it may be received. These people are free from the handicap of having to carry an enormous stock of varying grades and thicknesses. Likewise they have the advantage of not having to wait long for dry lumber of any kind. A few days in the kilns will suffice.

Up to date kiln manufacturers have made great strides during the last few years. Some very old and fixed ideas have been revised. Results are now being obtained which not long ago were thought impossible. Another very satisfactory thing from the manufacturer's standpoint is the fact that there are kiln designers and manufacturers now who are confident of their ability to do what they claim, so confident that they are willing to take all of the risk of making good before claiming their recompense.

A great deal might be said regarding how lumber should be treated after it leaves the kilns, but this phase of the matter is becoming well understood. It is a fact that lumber often leaves the kilns in good shape, but is afterward subjected to unfavorable atmospheric conditions which deteriorates its quality.

After leaving kilns, lumber for furniture should be kept in a dry storage room having as nearly as possible the same temperature and humidity as the shop in which it will be worked.

Improvement in Shell Box Prices

In recent contracts for shell boxes there has been, we are informed, a slight improvement in the prices paid by the Imperial Munitions Board, although the prices are still low. The general idea was that the cutting was due to many small firms, without any substantial backing, but it is now understood that the chief sinners in quoting inadequate values are large firms, mainly connected with the regular box-making industry. The Imperial Munitions Board have been blamed for letting contracts at such low figures. The official view, however, is said to be that the unremunerative prices are solely due to the manufacturers. The board merely, as a matter of business, accepted the lowest bids, and if these only represent the cost, or even a loss, this is the fault of the bidders. The board naturally try to get the work done at the cheapest rate, and have taken advantage of the mad competition to secure contracts beneficial to the Government but not to the manufacturer. On the other hand, it is freely said that this method has resulted in several firms falling down on the contracts, with the result that the boxes have had to be placed elsewhere, at enhanced values. The rises in lumber and in hardware have caused manufacturers to stiffen their quotations, besides which makers of boxes have learned by experience that former prices were not, in many instances, profitable.

Self Centering and Adjusting Turning Lathe

The self-centering and self-adjusting variety wood-turning lathe shown in the accompanying illustration is manufactured by A. D. Waymoth & Company, Newton Place, Fitchburg, Mass. This lathe is intended for turning and boring short wood turnings. It takes stock from 1 foot to 34 inches long and turns, bores and cuts one piece off the other until the stick is all turned up.

This variety turning lathe is used for such work as fuse and gain plugs, wood toys, pipe bowls and stems, knobs, drawer pulls, handles, wood faucets and numerous other turnings of a similar nature.

Some of the features claimed for the machine are that the headstocks have taper boxes with a nut on either end to take up the wear and keep them in line, is adjustable in all places where there is any wear, so that the slack can be taken up and an arrangement is provided whereby side motion occasioned by the

Handy lathe for making small turnings.

continual wear of the tail-stock and middle-piece may be taken up. The company also state that an ordinary mechanic can operate it.

The machine is equipped with a patented roughing knife holder, an adjustable front tool holder and accurately fitted tools. Parties submitting samples or drawings of the turnings they wish to make will be furnished with full information.

The company manufacture six sizes of lathes and self-lubricating countershaft is furnished with each, also certain other small parts which are necessary to complete the equipment. A special taper attachment for turning long handles can be obtained at an extra charge.

Furniture Number Appreciated

That our "Furniture Production Number" which was published last month was well received by furniture manufacturers and the men employed in furniture factories is evidenced by the numerous letters we have received complimenting us on the issue. We publish two of these below.

Exceeded Expectations

The October issue of your journal came duly to hand, and I must say that it exceeded my expectations, and it certainly does you credit.

As a Canadian, I am proud to see our trade journal becoming a worthy rival of some of the United States publications. Canadians owe it to themselves to give all the support and encouragement possible to make the "Canadian Woodworker and Furniture Manufacturer" the best trade paper on the continent.

W. J. Beattie.

Of Genuine Interest to Trade

The furniture production number for October was a very interesting one, and reflects great credit on your company, and undoubtedly is of genuine interest to the furniture trade.

Canada Furniture Manufacturers, Limited,
J. Wegenast, Purchasing-Agent,
Woodstock, Ont.

Where Will Future Mechanics Come From?

An Experienced Woodworker Makes a Plea for the Boys Employed in Modern Factories—Says They Should be Given a Chance to Learn Trade Properly

By Lewis Taylor

The need of skilled men in the woodworking industry is becoming more urgent every year, and it is about time that some means were being taken to insure a steady supply of skilled mechanics, in order that the industry may keep up to the standards of its past records. We have relied a great deal in the past on the steady flow of immigration to keep up the supply, but this is by no means a safe source to rely on. Owing to our different methods of working and different types of machinery, it is found difficult and expensive to "break them in."

By far the best method in my opinion would be the systematic training of apprentices, either by the indenture system or otherwise. The indenture system seems to be taking us a step backward to the past, but, after all, it appears to be the best solution for the problem. The trouble is that no manufacturer likes to lose valuable time instructing youths in the trade, with the chance of them jumping out at the first opportunity after becoming a little proficient, but if there was a legal tie, besides the moral one, binding them for a stated period, then it would be to the material interest of the boys to learn all they could, and the employer to teach them all he could in the shortest space of time.

In a period of four years any boy of average intelligence ought to master any trade, providing he was properly instructed by capable men, and if a boy knew that by diligence and industry he would at the expiration of that time receive journeymen's wages, it certainly would be an encouragement for him to do the best he could.

We are met everywhere by the superintendents of factories with the same complaint—that there are no capable mechanics nowadays like there were some years ago. Why is this? It is because the boys in this country do not get a chance to learn a trade. They are hired in the factories, to be sure, but they are put at sweeping or some small job, and kept there indefinitely. As a matter of fact, every factory ought to have a certain number of apprentices on the way all the time, advancing them systematically from step to step, until they become proficient mechanics. One thing is certain: there never will be any good mechanics in our factories until they are taught. Mechanical skill and knowledge comes only with experience, and by that means alone can a boy develop any mechanical skill he may possess. We speak about a certain firm having a good plant, but the firm that has not good experienced men has a poor plant indeed.

I make the plea for the better training of boys, believing that the woodworking industry in this country is only in its infancy, and with the future will come greater responsibilities in the shape of higher grade products and more of them. I believe there are no better mechanics than Canadians properly trained, and with increased training comes increased efficiency. With well-trained men greater production with the minimum of expenditure is possible, as there is no cheaper man in the factory than the well-trained one, whether he works day work or piece work.

It is an old and a very true saying that "that which we learn in our youth we never forget in our mature years." Take a boy learning his trade under the supervision of a strict mechanic, say of the old school; call him a crank if you like, but we may be sure that if there is any latent mechanical ability in that boy it will be brought out. The boy will be well ground in those little details that do not seem to amount to much by themselves, but amount to so much in the aggregate.

Many of our best workmen have to thank the fact that they were placed with competent mechanics in their early years—mechanics who were not only good workmen themselves, but good criticizers and fault-finders. It is only by finding fault with the work that the quality is improved. During my experience in factories I have known many boys who, after spending two or three years at the trade, became discontented with it owing to the poor advancement they made, not through their own fault, but through foremen who were either too careless or not energetic enough to give them a helping hand. What an asset to the individual firms those boys would have been had they received the proper training, as each factory has their own system of working, and therefore the boys would have grown up in their systems and proved valuable cogs in the industrial wheel.

Much has been heard of the specialty system of working, and, while I think it is a very efficient system of working and productive of good results, it does not produce mechanics. If we are going to work on the principle of "after us the deluge," it is perfectly satisfactory, but I do not think it is fair, either to the employer or employee, for a man to know only a small part of his trade. If he were to learn the trade right through he could then specialize on any particular branch. This is particularly true regarding the woodworking department of the piano trade, where the work is cut up into sections. If a man only knows one small branch he is helpless when thrown out of that employment and, vice versa, if he is wanted by the employer for another branch he is equally helpless.

As the case now stands regarding mechanics, everything is left to chance, and anything left to chance seldom turns out well in the long run. Sooner or later attention will have to be given to this question, for it is a certainty that without good men we cannot turn out good work. No matter how efficient our machinery may become we must always remember that a machine has no brains, and will just as soon work with poor material as not.

All must agree that it is only by thorough methods and foresight that a man or a firm succeeds in leading their fellows or competitors. It is getting things down to a scientific basis that counts, and it is the firm that is going to have the most skilled men that is going to forge ahead; for, on the same principle that "a few bad apples will spoil a barrel of them," so will a few untrained men spoil a great deal of material in any factory where they may be. As long as we neglect the boys (the future mechanics) we may expect to be up against the proposition of inferior workmanship.

Find that which you can do better than most people and it will give you a good start toward distinction.

The unbridled tongue is often a handicap to progress, as is also the habit of betraying confidences.

The Lumber Market

Domestic Woods

In our September issue we stated that a prominent lumberman had predicted $20 hemlock for next spring. Subsequent events have proved that he was right about the price, but that he was a few months out in the date. Sales have been made of hemlock during the past few weeks at as high as $22. Lumber prices are stiffening considerably on nearly every item.

This stiffening is not confined to eastern woods, either; British Columbia woods are showing the same tendency. Stocks are low and conditions all over the country are the same in regard to labor at loading points, thus making shipments difficult.

Woodworking factories are busy, especially furniture factories. A gentleman prominently connected with the furniture industry visited a few of the furniture towns in Ontario about a week ago, and on his return stated that some of the factories he had visited were turning out more furniture than they had ever done, despite the scarcity of labor. This may sound impossible, but the gentleman in question accounts for it by saying that the manufacturers are utilizing all the labor-saving equipment they can secure.

The furniture factories must therefore be using large quantities of lumber and stocks are rapidly diminishing. Despite certain statements that shell box prices are too low to warrant any firm manufacturing them, shell boxes are being turned out in fairly large quantities, and are responsible in no small feasure for the present shortage of stocks.

The situation in regard to stocks of domestic hardwoods and the prospects for future supplies is dealt with in an article published elsewhere in this issue. The majority of the users of hardwoods are alert to the situation and are snapping up any available offerings. It is difficult to quote prices as they are governed to some extent by the demand made on individual lumber firms and the stocks on hand to meet the demands. At the present time lumbermen are in the position where they do not have to do much soliciting for business and prices are firm accordingly. They are higher than they were a month ago, and likely to go higher.

Imported Woods

The brightest item on the list of imported woods at present is oak. It faded away from its usual popular position some time ago but is now one of the woods most in demand. It is an easy matter to account for its return to favor, in the furniture industry, at any rate.

A great many furniture manufacturers turn out nothing but oak furniture. If they wish to switch off oak onto mahogany or walnut, they are up against the problem of adding a veneer department to their factory. If they wish to stick to the solid furniture their choice must be oak, birch or gum. And of these, oak seems to be the most popular with the public, although gum is rapidly gaining in favor.

There is a scarcity of all grades of hardwoods in the United States, but the shortage is probably not so pronounced as in Canada. Prices are firm on all grades and are likely to continue so. Practically all

of the woodworking industries in the United States are busy, and consequently the market for hardwoods is a steady one.

Reports indicate that logging conditions in the Southern States have been fairly good, but some of the mills have had to shut down on account of the impossibility of transporting the logs.

In regard to the hardwoods which come from the northern States, it is said that the car shortage has made it exceedingly difficult to make shipments, and that the stocks are a little more favorable than they have been for some time.

These few thoughts expressed by speakers at a meeting of the Michigan Hardwood Manufacturers' Association reflects the situation in the United States:

"We want to caution our members to watch their production carefully in order that they do not overproduce on any particular size of item; a general overproduction is impossible."

"Our stocks are gradually growing smaller and we have not been getting the values we should within $1.00 to $1.50 per thousand feet."

"It has been estimated that the hardwood production for 1916 will exceed that of 1915 by at least ten per cent., and still stocks are decreased."

The following figures showing the stocks held by members of the Association speak for themselves and can be taken as a pretty fair index of the stocks throughout the country:

Stocks Compared With One Year Ago

All No. 2 common and better stocks, with the exception of ash, show a decrease as follows:

Basswood ... ..., ...	3,568 M
Beech	2,319
Birch	3,479
Rock Elm	894
Soft Elm	309
Maple	7,366
	17,835 M Decrease
Ash	31 M Increase
	17,804 M Net Decrease

No. 3 common stocks show the following increases and decreases:

Ash		257 M
Basswood		988
Rock Elm		301
Beech	2,058 M	
Birch	4,762	
Soft Elm	446	
Maple	1,604	
	8,870 M	1,546 M
	1,546 M Decrease	
	7,324 M Net Increase	

It is recognized by lumbermen in the States that the cost of production is increasing steadily and it is more than likely that they will increase their selling prices in proportion. "Hardwood Record," in speaking of the hardwood situation, says: "Sales in northern hardwoods considerably exceed production for the past year, and the general decrease in stocks was made in spite of a ten per cent. increase in production. The actual figures developing on the different hardwoods substantiate the opinions which were formerly based only on general observation, that lumber is a good bargain today and is going to be a more expensive commodity in the very near future."

Turned Wooden Articles

The list of articles of wood which may be produced on a lathe is practically without limit, and in size they range from a porch column and wagon hub down to a pill box and collar button. Among some of the articles shown in the accompanying collection are druggists' boxes, button and tassel moulds, curtain fixtures, bonnet stands, hubs and spokes for children's carriages, tent buttons, knobs, balls, lungs, tops, piano pins, dowels, organ stops, pencil cases, faucets, wheels, lather, paint, stencil, file, auger, and other handles, parlor croquet, whip sockets, wooden jewelry, checkers, pipes and pipe stems.—"Barrel and Box."

Preparing the Wood

For certain forms of finish it is not necessary to make the wood perfectly smooth, but as a general rule the better the surface is prepared the more satisfactory will the finish be, writes A. Ashmun Kelly in "Veneers." The finisher has, of course, to take the wood as it comes to him from the wood-worker, and it may or may not be as it should be. At any rate, he will see that the surface is sandpapered smooth and free from dust before he begins to apply any treatment. There are two classes of woods used in finishing, soft and hard; some are hardly to be classed in either list, being something between the two. These are birch, cherry, Circassian walnut and maple. However, for convenience they may be considered as hardwoods.

Soft woods have a close grain and require a liquid filler. Hardwoods usually have a coarse, open grain, but not always. Maple is a hardwood, though classed among the intermediate woods, and it does best with a liquid filler, shellac being used.

The very open-grain woods, like ash and chestnut, require a heavy paste filler, to fill the large openings solid, and two applications may be necessary. As some of the liquids that the paste contains will be absorbed by the wood, a coat of thin liquid is sometimes used over the first coat of paste. This satisfies the wood and prevents its taking any oil from the varnish, if that be the finish. The matter of foundation is of the very first importance in wood finishing. Even where a simple stain and a coat of wax constitutes the finish, it is necessary to have a right foundation or sur-

Collection of small wood turnings.

face, if you would get the best color effect and a uniform coloring.

Paste filler for oak is light of color, but the color may be made to agree with any kind of wood. In its usual light color it makes no noticeable change in the color of the wood, as when we fill oak with it; in this case we get what is called the "natural finish." But since most woods are not entirely uniform of color, it is usually best to add some coloring. In the case of a light wood a very little coloring suffices. In some cases oak is antiqued by means of the filler, which is made very dark, and the stain only equalizes the color or makes the surface color more uniform and pronounced. When filler is made to agree with the color finish, it should be rather darker than the stain, but as near the color of the natural wood as possible when the finish is not greatly altered from the natural.

Individuals have different methods of finishing. Some apply a coat of raw linseed oil to the wood before filling, and no doubt this is a good practice, as it satisfies the wood for liquid, as previously described.

A new dining room suite in Queen Anne design, made by the Geo. McLagan Furniture Co., Stratford, Ont.

How to Produce Good Enamelled Furniture

By James Culverwell

It might be well to say that priming of new wood-work does not mean a good heavy or thick coat, but a coat of paint the consistency of cream, well spread and rubbed into the wood. The priming is as import-ant as any of the coats, but as a rule it is treated as an unimportant process and thick and heavy paint used, trusting to the new wood to absorb it. After the priming the work should stand forty-eight hours. It will dry on the surface in less time, but that does not mean that the paint is hard through, which should be the case before further coats are applied. It is es-sential that every coat of paint have sufficient time to dry if good results are to be obtained.

Following coat after coat obtains bad and danger-ous results, causing the enamel in many instances to crack, owing to the coatings underneath being what is termed in the trade "green," meaning dry, but not hard. The priming is the only coat of paint that should have any oil added to it, apart from the oil that is in the lead when ground by the manufacturers. When the priming has stood forty-eight hours or longer, if time will permit it should be rubbed down with 0.0. sandpaper or, as it is sometimes called, "flour paper." Care should always be taken when sandpapering with fine paper that the paint does not collect on it in lumps and cause scratches that will injure the face of the work. Free all work, as ledges and crevices of dirt and keep dirt and dust from the surrounding floor and benches. Merely moving about causes the dust to rise and then pitch upon the work while wet, more or less spoiling it. Dust is one of the painter's greatest enemies and one cannot be too clean and careful in this respect when first-class work is aimed at.

Another thing that is often overlooked is care of brushes. When a job is finished for the time being, say at night, the brushes should be freed of all color by scraping them over or under a putty or pallette knife, then either placing them into turpentine, oil, or water as the case may be. Paint brushes may be carefully hung in water; varnish brushes are better hung into linseed oil; while enamel brushes are better hung in turpentine. Care of your brushes is a great asset to good work.

The water, oil or turpentine should be allowed to just come up to the stock of the brush and never above or below. Should the water or spirits evaporate, more must be added, because when the hair or any portion of it is exposed to the air it causes oxidation to set in. Thus a film forms over the hairs of the brush and finally falls on to the work, which proves most annoy-ing and unsatisfactory. Should a brush be allowed to get into that state it will be found far better and will save a lot of worry if it is laid aside for more inferior work. Better still is to wash the brushes out, but this cannot always be done when they are in use from

* This article is continued from the October issue.

day to day. A good finisher will take the greatest care of his brushes.

To body up work for a first-class job it will re-quire from six to eight or nine coats before enamell-ing. After the first two or three coats of lead, zinc white will be found better in many respects. It can be kept to a smoother surface with less labor and has another advantage of helping the work to remain white when finished. The more the finished work is exposed to the light, the whiter it remains. When covered up or kept from light, as inside of doors, it will always turn more or less yellow. In bodying up, the coats of paint must of necessity be alternate, that is, one coat in turpentine or flat, followed by a gloss or semi-gloss coat. As I remarked before, no oil must enter the paint after the priming is done, so that in the place of oil can be used either a little enamel, French oil, or any white varnish made by a reliable firm, but not a spirit varnish or a paper varnish, unless it be an oil paper varnish.

Exercise Care in Sandpapering

As the work is approaching the final coats greater care must be exercised in sandpapering. Rub the flour paper face to face before using for fear of scratch-ing, as a scratch now is fatal to a perfect finish. The first coat of enamel must be applied with as great care as the finishing one. It must be put on very evenly, not more on one place than another. The first coat should stand for two or three days at least. When it is thoroughly dry and hard and has a fine and even gloss, it is ready for the next process, that of felting down, which may be done in water or in oil. Cut a piece of felt about one-quarter or three-eighths of an inch thick and about two inches square, or larger if preferred, and use 0.0. ground pumice stone. When the whole of the work has been gone over in this way it must be carefully cleaned of all pumice dust. When it is ready for the last and final coat of enamel, which must be applied as full as possible, but not full enough to sag or tear or run at the corners of the panels, it must be constantly watched until it begins to set up.

The conditions of painting or rather preparing, painting and enamelling of furniture depend largely upon the nature of the wood. An open-grained wood requires filling first, which can be done as follows: Make a composition of bolted whiting and parchment size. The parchment size is made by simmering down in water the parchment cuttings, the quantity of water and length of time over a not too hot fire must be determined according to strength of size required. The strength should be such that when size is cold it will be about the consistency of jelly. It must not be used too strong. This preparation can be used when a job has to be prepared and painted at short notice, as several thin coats can be applied during the day.

The best results are obtained by standing the jar

containing the "compo," as it is termed, in boiling water and applying a thin coating over the furniture with a camel hair mop. Care must be exercised not to apply another coating before the previous one is thoroughly dry and hard. The whiting must be carefully sifted and just enough water put on it to soak it in. When precipitated, pour off the water and add the parchment size. This is better done the day before being used, as it will work much more smoothly. After two or three, or as many coats as are necessary to give the work a smooth solid appearance are applied, it can be carefully rubbed down.

The edges of the work, as corners of legs, arms or any edges, must be looked after, otherwise you will soon rub through, with bad results, showing a poor and unfinished line, and apart from that it rounds the edges that should be left nice and square. The job is now ready for painting. Some prefer to give it a coat of clear parchment size before painting, but care must be taken in the use of size. If too strong it will in time crack all over. It must be of the right consistency, not too strong, yet strong enough to bind and bold well and not powder off. The first coat of paint may have a little linseed oil mixed with the lead, and white Japan or patent dryers. All successive coats must be free from oil. The reason is that if you build up the ground work for enamel with soft colors (the use of oil causes the color to be too elastic, or "remain green," to use a practical term), your enamel being of a hard and brittle substance, when dry, will crack and split and apart from that it will turn the paint and enamel a bad color, especially a white.

Use Small Portion at a Time

Enamelling must be done in a warm room away from any damp or moisture. The damping of floor around the work just before enamelling is a mistake and should not be done, as dampness in the air causes enamel or varnish to "bloom." If the floor needs washing where the work stands it should be done and allowed to dry. Cold air and draughts will chill enamel and mar the effect. Should anything go wrong after applying the final coating of enamel and you wish to give it another coat, this must not be done upon a gloss surface or the result will be unsatisfactory. Under no circumstances must a gloss coat be put upon gloss; it must first be pumiced down again after being allowed to get dry and hard. To pumice with water will be found more speedy than with oil for this purpose.

When enamelling, only a small portion should be put out at one time and care taken to cork the can up at once. Then with thumb held on the cork turn the can upside down. This will exclude the air and the enamel will collect up around the cork and cause the can to be airtight, or nearly so. This prevents to a great degree the skinning of the enamel and also keeps it from getting thick or fat. Keeping the enamel free from air causes it to flow smoothly and not become ropey as it otherwise would. A real good job can be done by paying special attention to the woodwork before painting. The knots should be carefully treated with either white shellac or for preference cover them with gold leaf. Bring up the work in thin coats of lead, followed by zinc white with alternate coats of gloss or semi-gloss and flat, carefully sandpapering each coat when dry. Care should be taken in sandpapering not to rub the edges of the work. If not it will be cut off, and, to use a trade term, " it will grin," and cause the work to look far from solid.

The number of coats must be judged according to the condition as you go along, also as to the class of work you wish to turn out. Ordinary class work calls for priming, two coats of paint to follow, then one coat of enamel. A better class job requires from four to six coats, with better attention between each coat, and in its application, and so on, according to the class of work required. For the very best work, body up as stated, make a nice clean, even job of enamelling, and when hard, felt down with ground pumice stone No. 00, thoroughly clean off, then give a flowing coat of enamel and let stand for ten to twelve days. Felt down again with either putty powder or tripoli; this can be done either in oil or water.

If done in oil it is best to clean it off with a little dry flour and rub down with a piece of silk. This will clean off the work and free it from all grease. Then with a clean, damp chamois or wash leather in the left hand, damp the lower part of the palm of the right hand with the chamois, commence at the top and draw the hand briskly down the work. This will make a peculiar squeaking row; but has the advantage of bringing the work to a very high and perfect polish. This must not be repeated over the same ground too quickly. Move the hand to the right or left as the case may be. The pressure of the hand will have to be judged by practice. White enamel can, if left long enough to get perfectly hard, be French polished with white polish, but this is a thing that must not be done until it has stood a considerable time and the polish must be sparingly used or otherwise it will crack. French polishing is not necessary if the work has been carried through with the greatest care and cleanliness. A hurried up job is invariably a failure.

How to Obtain Colors

The work may be enamelled or finished in any color. The best ground colors are recommended for this purpose and can be obtained as follows:—Pearl grey, white lead 25 parts, venetian red 1 part, deep green 1 part; olive grey, white lead 40 parts, lamp black 3 parts, chrome green 1 part; opal grey, burnt sienna 1 part, Cobalt blue 2 parts, zinc white 30 parts; pine grey, white lead 40 parts, raw umber 1 part, lamp black 1 part; heliotrope, zinc white 2 parts, carmine 3 parts, ultra marine blue 4 parts; sage green, white lead 30 parts, light green 2 parts, burnt sienna 1 part; lavender, 100 parts white lead, ultramarine 3 parts, madder lake 1 part; mauve, white lead 6 parts, prussian blue 2 parts, madder lake 1 part; cream, white lead 33 parts, Italian ochre 1 part.

There are many other pleasing colors to which furniture can be painted. When you have decided in what shade you intend to finish, the last three or four coats must be of a similar color, finally staining the enamel to match the last coat. It should be a few shades lighter, however, as all paints or enamels have the habit of drying a few shades darker than when applied, the best results are obtained by painting up from the wood itself—and not using the preparation mentioned, viz., bolted whiting and parchment size, but it takes longer, owing to the time required for drying. After the work is finished and allowed to stand and harden, it can be painted with any designs, as for example, Adam design with medallions, ribbons, etc. This must be done in very thin color so as not to stand above the surface. When this is allowed to harden it can be polished with a white French polish and a pad or rubber and spirited out. This last process requires a practical French polisher.

Save Money by Purchasing

"Made in Canada"

Veneered Panels

Maple - Elm - Basswood - Birch - Plain Oak
Quartered Oak - Mahogany - Black Walnut

Single 3, 5 or 7 Plies.

Hay & Company, Limited   Woodstock Ontario

"Furniture in England"

This book is an encyclopaedia of artistic suggestion—an education in itself.

Upwards of 400 exquisite half-tone and color plates on pages 14 ins. x 10 ins.

Among the subscribers to this book are Her Majesty Queen Mary, His Majesty the King of Greece, and many of the foremost manufacturers the world over.

Price $12.00

Delivered to any address in Canada

Copies may be obtained from

The Woodworker Publishing Co.
Limited

Toronto   Ontario

Established 1895

Skedden Brush Co.

Manufacturers of
ALL GRADES OF

Brushes and Brooms

102-104 Bay St. N., Hamilton, Ont.

"ENAMELS"

Our business in both White and Old Ivory Enamels and Primers has trebled during the past year

WHY?

They are the best on the market. If you are not using them, write for samples and prices.

The Ault & Wiborg Co. of Canada, Ltd.
Varnish Works

Montreal   Toronto   Winnipeg

Should the Foreman Finisher be a Chemist?

By W. J. Chambers, Manager, Manufacturers Sales Department
Sherwin-Williams Company.

To ask the foreman finisher to be a chemist is asking too much, as it is practically a physical impossibility. Every manufacturer wants his foreman finisher to be the best in his line, and therefore, the foreman must be one who has devoted the greater part of his life to that kind of work.

A man cannot possibly devote his time to working up to be a first-class finisher and yet have sufficient time to devote to chemistry. A chemist should be a man who has had a college education, for if he is to be worth anything he must know a great deal about chemistry. This, readers will agree, requires a great deal of time for study and experimenting, which a man who is devoting his time to a trade and to become a first-class finisher cannot afford.

Besides this, knowledge of chemistry to a finisher is of very little value, as what he really requires to know is how the goods which he is working with will act when applied, and what finish he can obtain with the varnish he is using. Some of the best experts in the country place very little importance on a chemical analysis when the proper results are to be obtained. They prefer to know what the piece of goods they are handling will do and how it will act and, therefore, some of the best experts on finishing devote very little thought to chemical analysis, as in their opinion (with which I quite agree) there is nothing to be learned by analysis, excepting that in some cases it may give one thoroughly conversant with finishes some slight suggestion in regard to duplicating the goods. A general chemical analysis will not convey any idea of the essential composition of the varnish. It will show the percentage of oil to color, whether the oil is linseed, rosin or petroleum, and in a general way what the color is, but it shows practically nothing concerning the physical condition of the materials before manufacture, or how they were combined, and unless the chemist has a practical knowledge of finishing it does not show even this much.

For example:—an analysis may show the presence of linseed oil, but it will not show the consistency of the oil used, nor whether two or three different consistencies were used. Neither will it show if the oil was combined as a gum varnish in a drier, or whether it was burnt or boiled. It will not show whether the drier used was a paste drier, a Japan drier, or whether the drier was added at the time of boiling or not.

In view of these facts, the essential knowledge for a finisher is to know how to use the goods and what is to be done to obtain the best results. Chemistry, when combined with a knowledge of finishing, generally means that either one or the other of these two important subjects must suffer. A chemist must be a man who knows his work, and if he is to be such a man, must devote all his time to chemistry. A finisher cannot possibly spare the time when learning his trade, to study chemistry, and after he has learned something about it he is generally too busy to study chemistry. If a foreman finisher, therefore, is not a chemist, there is no reason whatever why he should not be the best of finishers and have the best and most thorough knowledge of his trade.

Our experts are now pointing out that there is both a distinction and difference between cost-accounting and accounting for costs, as one is purely a clerical task and the other involves analytical study.

Ever Have These TROUBLES in Your Finishing Room?

{ Quality of finish not up to standard.
Production costs creeping higher.
Delivery promises hard to meet.
Contentment of workmen on the wane.
Maintenance and operating costs much too high.

Unnecessary Troubles— each and every one of them—as several thousand progressive manufacturers in Canada and the United States gladly testify. These manufacturers experienced the same difficulties; they investigated and installed the AERON SYSTEM— NOW after long service THEY COULDN'T BE CHANGED TO ANY OTHER METHOD ······ There you have the very good reason for

The Supremacy of the Aeron System

in the industrial painting or finishing field today, and THE VERY GOOD REASON why it MERITS FIRST CONSIDERATION if you have finishing room troubles, or desire to initiate a general improvement.

In applying a coating of shellac, varnish, enamel, lacquer, bronze, japan or any liquid material —on wood or metal—the AERON SYSTEM always produces a superior uniformity and smoothness of surface . . . . The AERON SYSTEM does this at a positive saving of 50 to 90 per cent. in time and labor, which not only checks but reduces production costs, speeds-up production and makes possible quick deliveries. In this respect of completeness, the AERON SYSTEM provides for maximum exhausting results at 1-10th the usual cost, making for healthful, pleasant working conditions at almost negligible cost . . . . Practical, efficient and durable, the AERON SYSTEM calls for the minimum of up-keep expense.

Let us help you—more facts and new Booklet "J" will gladly be sent—ADDRESS—

The DeVilbiss Mfg. Co.
1318 Dorr St., Toledo, Ohio, U.S.A.

Veneers AND Panels

Costs Too Much to Joint and Tape Veneer by Hand
By a Veneer Man

The introducing of an up-to-date cost system into a factory is like turning on a searchlight. It reveals some astonishing facts. One or more of the departments in your factory, Mr. Manufacturer, may be jogging along, turning out the work in sufficient quantities to keep pace with the other departments in the factory. but if you were to have handed to you a detailed cost sheet showing in actual figures what the work in those departments is costing you, in all probability you would lose a few nights' sleep.

Have you turned the searchlight on your veneer department? Are your men still jointing and taping veneer by hand methods? The writer was recently shown through the factory of a well-known piano company who turn out on an average about twenty-five or thirty pianos a week. The factory was very well equipped and managed, with the exception of the veneer room. In this department the only thing that was modern was a hydraulic press. The jointing and taping was all done by hand, and three men and a boy were required to do it.

A few days after visiting the piano factory a visit was made to the veneer room of a large furniture factory in the same city. The contrast was almost too great for words. In the first mentioned factory as much of the veneer was cut on the bandsaw as possible and the rest by hand. To do the jointing a "shooting board" was placed on the bench, a couple of dozen pieces of veneer placed on it and held in place by a couple of bars reaching up to the ceiling, and then the edges of the veneer were planed until a straight edge fitted their accurately. After the edges were nicely jointed the veneer was tacked on boards and the joints taped and left to set. The pieces were then removed from the board, the joints rubbed down on the opposite side to the tape, and the "burr" around the tack holes hammered down. There is no doubt about a good job being obtained by this method, but it costs money.

What a difference in the other factory. The veneer was cut on a large knife machine, or "guillotine," it only being necessary for the operator to press his foot on a lever to bring the knife down and cut the veneer off like cheese. The jointing was done on a regular veneer jointing machine. This machine consists of a flat table with a jointer head and saw mandrel attached to it. and a chain feed similar to that on a continuous feed jointer.

To operate the machine the veneer is laid flat on the table and fed into the chain feed. It will be understood that this chain feed runs vertically instead of horizontally as on the continuous feed jointer. When the veneer starts through the machine, it first comes in contact with the circular saw, which cuts off any rough or ragged edges. Then about eighteen inches or two feet past the saw it comes in contact with the jointer knives, which joint the edges up nice and true. After the jointing is completed the veneer is put through a taping machine.

With the aid of these machines one man can cut, joint, and tape enough veneer in a couple of hours to keep a large staff going. The work is as well, if not better done than if it were done by hand, and there is no unsightly tack holes in the finished work. It is impossible to tell just how much your veneered work is costing you unless you have a cost system. Perhaps you are finding it difficult to meet the selling prices of your competitors. Then heed a timely warning and investigate the costs in your veneer department. You may say, "All that sounds good, but I can't afford to install three expensive machines." Can't you? Take the piano factory mentioned above as an example.

By installing cutting, jointing, and taping machines they might incur a liability of $1,200. (This figure is simply taken as an example, and may be quite a bit out.) But, to offset this, in place of the three men and a boy, as at present employed, they would only require one man, and with the firm's output of twenty-five or thirty pianos a week, in all probability he would have enough spare time to assist at the gluing machine when the presses were being put in. We will assume that the factory is working ten hours a day and that the men are being paid 30 cents an hour. This is none too high, as it requires a mechanic to joint veneer properly by hand. By eliminating two men (never mind the boy, although the writer's experience has been that, with few exceptions, boys are a nuisance in the veneer room) there would be a saving of $6 a day, or, in other words, the machines would pay for themselves in 200 days—less than a year.

Even in some factories where the costs are definitely known, veneered work is costing more than it should. When the cost and efficiency bug first made its appearance, the machine room was the department on which the "searchlight" was turned. Every operation was scrutinized to see if it could not be worked in conjunction with some other operation, thereby saving one handling of the material.

After the machine room came the cabinet room. Bench work was cut down to about one-tenth of its former volume by designing the product so that practically all of the work could be done on machines and also by insisting on accurate work by the machine operators.

Even in the finishing room the hand or brush method of applying stains, shellacs, varnishes, etc., was considered far too slow, and the speedier and cleaner method of putting them on by means of the air brush adopted. Yet in some factories, where all these other departments have been placed on an efficiency basis, the veneer is still jointed and taped by hand.

When the man at the machine becomes informed on the burden of overhead cost he has a spell of wondering why the boys in the office should have the best time and get the most money. But do they?

Glue Should be Carefully Tested

By Lewis Thompson

The old saying that "a chain is as strong as its weakest link" holds good with a piece of furniture. It will be just as strong as the glue used in its manufacture. Therefore, a few words regarding the preparation and use of glue may not be amiss. In the first place, when trying a new brand of glue care should be taken to give it a thorough test. I do not believe in soaking glue previous to boiling it, as the full strength of the glue is not conserved, whereas, in putting the raw glue on to boil steeped in water, the full strength of the glue is retained, and it can be thinned at the discretion of the user.

In testing, take a definite amount of glue and water, boil thoroughly, and, when ready, skim the fat off and divide the glue in three separate pots. Have that in each pot of a different consistency—fairly thick, medium, and thin. Having in the meantime prepared the stock to receive the glue, the next thing is to apply it and put the job under pressure, making each separate gluing according to the amount of water taken to dilute it. It is necessary to know, say after it has set for two hours, how the glue stands, how the water assimilated, and how satisfactory the results are. It seems unnecessary to say how to complete the test. Give the piece sharp strokes with a chisel fair on the joint. On breaking the joints a good idea can be had of the strength of the glue. The glue that will stand the most water and still give satisfactory results is the most profitable glue to use. If the method of testing outlined above is carefully carried out the quality of the brand which it is proposed to use can be ascertained.

When we have satisfied ourselves as to the quality of glue we are using, and have confidence in it, we should always see that we give it fair play. Boil it carefully, and do not make up too much at a time, so as to avoid using stale glue. Always have a fresh supply at hand, and, above all things, avoid diluting the glue with impure water, such as water from the glue boiler. This often contains harmful ingredients. I worked for one foreman who insisted on us putting the fresh water in the glue half an hour before we used it, explaining that owing to the density of the glue being so much greater than water, that if the glue and water were not thoroughly mixed up, the water would naturally remain on top.

I have seen men in the act of gluing, finding their glue too thick, go to the glue boiler and put a ladle or two of water into their glue, give it a hasty stir, and away went the glue on the stock. And then we find firms having trouble with their glue! I repeat, again, after the glue has been found all right in the test, it should be given fair play in every particular, from boiling to setting. In this way glue troubles will be minimized. There is one point I would like to emphasize about the process of gluing—that is, be sure that every portion of the stock to be glued, however large or small, is covered with glue. There are some men who, to save themselves the trouble of cleaning the glue off afterwards, skimp the gluing, and thus pave the way for trouble. One man whom I worked with some years ago was overburdened with this fault; he was as stingy putting on his glue as a man could very well be unless he put none on.

Of course, the inevitable happened. The stock came back just barely hanging together. The foreman, on examining the returned work, found in nearly every joint just a thin ribbon of glue, with more than half the joint absolutely bare. This bad work put the firm to great expense, besides loss of reputation, through really no fault of their own. There is no need of flooding the glue on, but care should be taken to see that every part of the joint is covered, and then if there is any "come back" through dampness etc., the glue will be under there, and the workman will be protected to that extent. Spare the glue and spoil the joint is a bad motto to follow. The question of what consistency of glue is the best to use for general purposes is often debated. I speak now of good standard glues. I prefer glue a little on the thin side. It will then work easily and perform the true function of glue—that is, to identify itself with both portions of the joint, soak in, as it were, and bind. Glue is not like mortar, but just the contrary, for we must squeeze all the surplus glue out or there will be trouble. Thus, the toothing plane is a great help to the glue, giving it a chance to get a grip on the wood fibres, and for durability a toothed joint or surface is much better than one that is not toothed.

In applying pressure to glued surfaces, care should be taken to see that the heat has properly struck the glue, so that it will run freely. This can be done by applying the pressure gradually, especially in the case of veneering. It is a bone of contention in many factories as to the length of time that should elapse before it is safe to release the stock under pressure. Of course, a great deal depends on the latent heat that remains in the stock. A press of veneered stock will hold the heat a long time, while one slender gluing is soon cooled.

Any ordinary gluing should have at least one hour before releasing the pressure. I refer, of course, to joints. It is not giving fair play to the glue to release it sooner than this. We must not do like a man whom I knew who released the pressure in twenty minutes. The joint came loose and he attempted to fix it by simply renewing the pressure. It would be hardly a safe joint to let go, I am sure. After all, glue is the furniture man's friend, and, as such, should be treated with respect, not contempt.

Will Brown Oak Become Popular?

The scarcity of mahogany, which is not merely temporary, but likely to exist for some time after the end of the war, will probably have the effect of bringing British brown oak once more to the front as a furniture wood, writes the Cabinet Maker. The demand for reproduction in mahogany drove it practically off the market for some years, notwithstanding the fact that it is in itself a very beautiful timber. We have no doubt that ere long brown oak furniture will once more make its appearance. If it does, we hope that designers will recognize the qualities of the wood and treat it accordingly. Brown oak work in the old days was usually characterized by elaborate carved and moulded detail. This, we think, was a mistake. Generally speaking, the work should be kept simple; the panels should be plain, in order that due emphasis may be given to the beautiful figure and color of the wood itself. Furniture manufacturers, in searching for timber and veneers, may also discover the merits of walnut curls. Fine effects can be obtained from these also if they are judiciously used. They could be used in "Queen Anne" style instead of mahogany, which was so popular just before the war broke out.

There are plenty of other good things to talk about besides the faults and shortcomings of others.

OAK　MAHOGANY　CIRCASSIAN　MAPLE

VENEERS

BIRCH　POPLAR　CHERRY　WALNUT　GUM

THE OHIO VENEER COMPANY

Importers and Manufacturers

Foreign and Domestic Veneers and Hardwood Lumber

We always carry a large and assorted stock of Mahogany, Circassian
Walnut, Sawed and Sliced Quartered Oak.

Send us your enquiries and orders. We guarantee good service.

2624 to 2644 Colerain Avenue,　　-　　-　　CINCINNATI, OHIO.

Indiana White Oak
Lumber and Veneers

Band sawed from the best growth of timber.
We use only the best quality of Indiana White Oak.
From four to five million feet always on hand.

Walnut Lumber

Our stock assures you good service.

Chas. H. Barnaby, Greencastle, Indiana

 Hoffman Bros. Co.

Estab. 1867, Incor. 1904

800 West Main Street,
FORT WAYNE, INDIANA

Manufacturers of

VENEERS and LUMBER

In the Domestic Hardwoods

ANY THICKNESS.

1/24 and 1/30 Slice Cut
(Dried flat with Smith Roller Dryer)

1/20 and thicker Sawed Veneers,
Band Sawn Lumber.

—SPECIALTY—

 Indiana Quartered Oak

Veneers and Panels

PANELS

Stock Sizes for Immediate Shipment
All Woods　All Thicknesses　3 and 5 Ply
or made to your specifications

VENEERS

5 Million feet for Immediate Shipment
Any Kind of Wood　　Any Thickness

J. J. NARTZIK

1966-76 Maud Ave.,　　-　　-　　CHICAGO, ILL.

Veneers and Panels

**LARGE STOCKS FOR
QUICK SHIPMENTS**

Send for Lists

Walter Clark Veneer Co.

GRAND RAPIDS, MICH.

Buying Narrow Veneer

The purchasing of narrow veneer because it can be obtained cheap is a foolish policy unless it can be used without jointing. This fact seems to have been overlooked by some buyers of veneer. In the majority of cases where narrow veneer is jointed, if the cost of jointing is added to the cost of the veneer it will be found that the job has cost more than if veneer of the required width had been bought in the first place.

There may be exceptions, as in the case of fancy burl walnut and crotch mahogany. These are very expensive woods at best, and the quantity used may not be very large, it may be advisable to buy the narrow stock. In the more common kinds, however, such as a good striped mahogany and quartered oak, it is best to buy the wide stock. The whole trouble is that the party buying the veneer has no conception of the amount of time—not counting tape and glue—that it requires to joint a pile of veneers.

There is, of course, another side to the question. Wide veneer should not be purchased when it is only narrow stock that has to be veneered. The whole thing is a question of the buyer being a practical man. If the superintendent is allowed to do the buying, or even to have a say in what shall be bought, the buying can be done on a more economical basis, because he is in touch with the requirements of the factory.

The writer was once employed in a factory where the manager was a mark for all the veneer travellers in the business that had any "junk" to sell. He held his position for a few years; and his services were apparently satisfactory, but in the long run the owners evidently woke up to the fact that they should have been making larger profits. At any rate, the manager was asked to send in his resignation. His successor was a man that knew the business from the ground up, but when his bills for lumber and veneer went to the head office (the factory was located at a different point from the owners on account of his predecessor's bills being lighter.

The new man, however, was engaged on a three years' contract, and before the three years were up the owners were perfectly satisfied. In conclusion, it might be well to mention that there is considerable of the veneer purchased by his predecessor still on the premises. A lot of it has gone to the boiler room for fuel, but as it is valued in the stock book at the purchase price it has to be got rid of slowly to prevent the owners thinking some person has eaten up $4,000 or $5,000 worth of stock.—J. S.

Strength of Glue

A series of tests to determine the strength of glue were made by Messrs. O. Linder and E. C. Frost and reported to the recent meeting of the American Society for Testing Materials. The results show a strength of from 1,100 to 1,950 pounds per square inch for a glue made of one part dry glue and three parts water, and a strength of 60 to 70 per cent. of the above figures for a 1.5 glue. Prolonged heating lowers the strength of the glue. Glue solutions heated to 150 degrees F. for 20 hours showed a loss of 30 to 45 per cent. in strength.

The average woodworker does not think of becoming a chemist, and yet those who study the subject realize chemistry is an important factor in wood.

VENEERS FROM STOCK ON SHORT NOTICE

MAHOGANY
We have a large stock of this wood on hand, in both furniture and nice mottle piano wood. Particular care has been taken in the selection of our stocks and we would appreciate an opportunity to submit samples.

AMERICAN BLACK WALNUT
We carry the largest and best stock of this wood in Veneers in Canada. We have some very highly figured stump wood, good straight sliced stripe and curly figured stocks, also semi-rotary and full rotary cut stocks of all kinds.

QUARTERED OAK
We carry this wood in all thicknesses and widths in both sawed and sliced. We have a good stock to select from.

FIGURED QUARTERED GUM
Let us show you what we have to offer in this wood. We have the goods and the prices.

ROTARY CUT VENEERS
We are now carrying in stock Poplar, Maple, Oak, Gum, and Birch.

Get our prices on WALNUT LUMBER

TORONTO VENEER COMPANY
TORONTO   93-99 Spadina Avenue   ONTARIO

Wood - Mosaic Co.

NEW ALBANY,   -   -   IND.

WALNUT

Lumber ~ Dimensions ~ Veneers

If you are in the market for anything in Walnut, let us figure with you. We have cut three million feet of Walnut in the last year, and have a good stock of all thicknesses dry. We can also furnish Walnut dimension—**Walnut Veneers.** We have a large stock of veneers, plain, stripe figured and stump wood.

Wood-Mosaic Co., Inc.

NEW ALBANY,   -   -   IND.

 TEAK

Write for prices on all
**Plain and Fancy
Lumbers and Veneers**
Mahogany, Oak, Birch,
Walnut, Ash, Poplar, etc.,
rotary, sliced or sawed
in all thicknesses

SHELL BOX PARTS
in birch and spruce
Bridges and Ends

Write for Quotations

George Kersley
224 St. James Street, Montreal

 MAHOGANY

D O M E S T I C

L U M B E R

Notice—Change in Name

Effective October 16th, 1916. The Roberts Veneer Company succeeded the Roberts and Conner Company and John N. Roberts, Veneer Manufacturers.

The Roberts and Conner Company and John N. Roberts desire to thank their many friends for their long and continued patronage and bespeak a continuance of same with the Roberts Veneer Company.

The Roberts Veneer Company will continue to produce Veneers in Walnut, Oak and Poplar, specializing on American Walnut in sliced, half-round and stump wood.

Roberts Veneer Co., New Albany, Ind.

OAK — MAHOGANY — CIRCASSIAN — MAPLE
VENEERS
BIRCH — POPLAR — CHERRY — WALNUT — GUM

| Mahogany | Figured Red Gum |
| American Walnut | Curly Birch |

Quartered White Oak
(Sawn or Slice Cut)
And all other kinds of

Domestic Veneer
AND
Thin Lumber

ACME VENEER & LUMBER CO.
8th, Harriet and 7th Streets   -   -   -   -   CINCINNATI, Ohio

---

200,000 Ft. ¼" Gum
12 to 36" wide, 60 to 84" long
DRAWER BOTTOMS
AND
BOX STOCK
For Quick Shipment
GEO. L. WAETJEN & CO.
MILWAUKEE   -   -   -   WISCONSIN

THE GLUE BOOK
WHAT IT CONTAINS
Chapter 1—*Historical Notes*
Chapter 2—*Manufacture of Glue*
Chapter 3—*Testing and Grading*
Chapter 4—*Methods in the Glue Room*
Chapter 5—*Glue Room Equipment*
Chapter 6—*Selection of Glue*
WOODWORKER PUBLISHING CO., LIMITED
345 Adelaide St. West, Toronto

Buyers of Veneers and Panels

will find it to their advantage to purchase from the manufacturers and dealers whose advertisements appear in this publication. They are progressive firms—the leaders in the business, which is a guarantee of good service and prompt attention to orders.

Give your business to the man who will spend his time and money to get in touch with you. He deserves it—if his stock and prices are right.

OAK   MAHOGANY   CIRCASSIAN   MAPLE

VENEERS

BIRCH   POPLAR   CHERRY   WALNUT   GUM

Burl Walnut Veneers

We have some exceptionally fine burl veneers in American Walnut. In figure, color and general attractiveness this stock is really remarkable, and must be seen to be appreciated. If you are interested in material which is out of the ordinary and is suitable for the highest class of cabinet work or interior trim, ask us to send samples. We know that you will be interested.

Penrod Walnut & Veneer Co.
"WALNUT SPECIALISTS FOR THIRTY YEARS"
KANSAS CITY, Mo.

We specialize on

Poplar

cross-banding and backing

Veneer

Highest Grade — Right Prices

The largest Poplar Veneer Mill in the world.

The **Central Veneer Co.**

Huntington, West Virginia

We Can Fill Your Requirements in **MAHOGANY**

Because we represent one of the largest importers in Chicago and the Northwest. Therefore if you need Mahogany, we are in a position to give you such service as few can render.

American Black Walnut

Particular care has been taken in the selection of our stock, which embraces some of the choicest figures we have ever seen.

Quartered White and Red Oak

Of these old favorites we carry always a large line, so varied in figure that we can suit any taste. We have many very desirable figures in flitches which will insure uniformity in quality work.

Stocks ready to ship on receipt of your order

HAUGHTON VENEER COMPANY

1152 W. LAKE STREET   CHICAGO, ILLINOIS

Better Methods of Edging and Joining Material

By a Member of the P. B. Yates Machine Co.'s Efficiency Staff.

All classes of successful manufacturers must be keen to sense changes in trade conditions and quick to react to them in the most advantageous manner, but there are very few classes whose continued success is so dependent on that ability as the furniture manufacturer. Changes in public taste, transportation difficulties, labor problems, even storms at his source of raw materials, react immediately on his business affairs. Right now, there are factors growing out of the war that are of foremost importance, and, of these, labor and raw materials claim attention most frequently.

The call of so many men into military service, coupled with the labor demands of increased business, has created a shortage of labor all over Canada and

A Special Edging Attachment for Sizing Long Stock Quickly and Easily.

the United States. There is no industry but feels its effects, and only a few that are as vitally affected as the furniture manufacturing business, because a vastly increasing trade has developed along with the diminishing of the labor supply.

At the same time the price of the furniture manufacturer's raw materials has considerably increased, putting up to him the immediate necessity of solving, at the same time, two problems in cost cutting, either of which is in itself cause for worry. To create more labor is impossible, as is also the reduction of prices for raw materials. It follows that the manufacturer's only recourse is to look to more efficient production for aid in his difficulty.

The word efficiency is sadly over-used these days—so much so that it attracts but little attention. The need for efficiency in practice, however, is greater than ever, and there is still much room for its application in the average furniture plant. Expressed in factory terms efficiency means the greatest possible output with a given amount of labor, and a minimum amount

of raw materials. It follows that in order to bring a working force up to its greatest efficiency, it is necessary that they be provided with the most efficient equipment with which to perform their work.

In these days, when the times demand utmost efficiency, it surprises one to find so many furniture plants using antiquated methods and machinery. Perhaps this is more noticeable during preparation of raw stock for further operations than elsewhere. Practically all plants are doing their cut-off work with up-to-date swing saws, but in the matter of edging and joining only a few are so fortunate. These processes are still carried on largely with hand feed edging saws and hand planers, with their attendant waste of time and effort.

The amount of "lost motion" through the use of old and slow edging saws is remarkable. To illustrate, the writer, not long ago, took observations of a workman, without his knowledge, at an old hand feed edger. He was edging strips for table tops 48 inches long. The average time occupied by a single cycle of operations—from taking up raw stock back to taking up raw stock again—was sixteen seconds. That sounds like pretty snappy work, but analyze the cycle, bearing in mind that the man was being paid 20 cents an hour, and that the useful part of his effort was that used in actual edging—that is, while the saw was in the cut. As the saw was in the cut seven seconds out of the sixteen it is clear that only 44 per cent. of the workman's time was actually used for production. The remaining 56 per cent. was wasted on motions necessary to produce with that type of an edging saw.

It is evident that an edging saw saving a considerable part of this lost 56 per cent. would be a vast improvement in that factory. The loss of 56 per cent. of a man's time on one operation isn't much, but examine its cumulative results. The workman in question spent about four hours a day at that operation. His daily loss of time was approximately two and a quarter hours, with a money value of 45 cents. In a year the total loss amounts to $135, a sum which is almost double the interest and depreciation charges on the most modern edging saw, which would do that same work in one-fourth the time, besides having a wider range of usefulness. The time saving capabilities alone of such a machine would make it a sensible investment, to say nothing of the increase in output and quality.

The unnecessary labor that must accompany older types of edging and joining saws is a drain on the finances of the furniture manufacturer great enough to demand careful thought. The claim is often made that better edging equipment does not increase output materially because the rate of hand feed is frequently nearly equal to automatic feed and there is nothing gained by installing such equipment. The answer is found in the innumerable instances where modern equipment has reduced the number of men

CARVINGS FOR PERIOD FURNITURE

Composition Carvings

are now extensively used for furniture decorations.

This is a material that will interest all manufacturers of furniture.

Write for Catalog and prices.

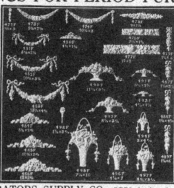

Most durable and practical Carvings on the market.

We have thousands of stock designs in all the different styles; can also manufacture according to your special designs.

Samples sent on request.

THE DECORATORS SUPPLY CO., 2551 Archer Ave., Chicago, Ill.

It Takes "CROWN PLANER" to Stand This

It takes more than ordinary leather belting to stand the service of planer work as illustrated. Observe the severe turns and double turns over the small pulley. These pulleys are carrying a heavy load and turning up at a high speed.

Crown Planer leather belting is ideal for just such runs. The choice leather, the special tannage, and the careful workmanship combine to make Crown Planer the best leather belting.

Write us to-day. Say under what conditions you want the belt to run. We will recommend a Belt for your purpose that will break all your previous records for economy in full power generated and transmitted and in low up-keep.

Sadler & Haworth

Tanners and Manufacturers

Established 1876

FOR 40 YEARS TANNERS AND MANUFACTURERS OF THE BEST LEATHER BELTS

MONTREAL, 511 William Street

| TORONTO | ST. JOHN | WINNIPEG | VANCOUVER |
| 38 Wellington Street East | 149 Prince William Street | Galt Building | 107-111 Water Street |

necessary to perform these operations. The writer knows of one where such an installation reduces the labor required by fifty per cent. and increased the output at the same time by sixty per cent. Many similar to this might be cited.

Another important fact bearing on the labor saving problem is that the most modern of the edging saws cut so smoothly that, for rough gluing, the stock needs no further preparation, thus saving the extra time and labor required to run it through the regular glue jointer. Considering that a large proportion of glued-up work at the present time is either core stock, or missible, this item of saving looms with large importance.

Much of the edging and jointing machinery in use at the present time is inadequate to efficiently meet the demands put upon it. The rate of output is below present standards and compels the employment of labor that might be better engaged on some other work which the use of machinery cannot expedite. Where old machinery is used, an excessive number of machines must be employed to meet production demands, which is at once reflected in the expense of repairs and maintenance. One can arrive at no other conclusion than that the efficiency the times so urgently demand can be injected into furniture manufacturing only by the use of superior time and labor saving machines in edging and jointing operations.

On the usual edging saw, stock must be returned

eventually to the infeed end of the machine, thus making just half the stock travel non-productive. The logical method of procedure is to keep the stock constantly moving forward until assembled. This would make the edging and sizing cuts, two distinct operations which should, logically, be performed on separate machines. The lost motion eliminated would more than offset the cost of the small added equipment necessary. The increased efficiency attained by using the saw for edging purposes only would eventually reduce the amount of labor necessary to carry on the operation.

Only a moment's reflection is necessary to note that the machines, to be better, must be radically different from the old, or no improvement would result. Such is actually the case. One of the greatest principles on which factory efficiency is based is that operative motions must be as few in number as is consistent, and through the shortest distance. Any useless motion reduces the total of the worker's energy, and by just that much reduces his producing capacity. To illustrate, the workman above mentioned walked three feet for each piece of stock put through the saw. In a year this totals a hundred and fifty miles of wasted effort. Improved machines for edging and jointing must be constructed with this principle in view.

The ultimate cost of raw materials is another factor that prompts efforts to better the edging and jointing processes. There is little prospect of an immediate re-

An Example of the Compact, Self-contained, Efficient Up-to-date Edging and Jointing Saws.

FOR SALE!

1 car 6/4 1s and 2s Dry Hickory 80 per cent. 14 ft.
1 " 6/4 No. 1 Com. and Bet. Hickory.
2 " 2 x 2—30 in. Clear Dry Oak. Sqrs.
2 " 2 x 2—19 in. Clear Dry Oak Sqrs.
1 " 2 x 2—24 and 30 in. Clear Dry Oak Squares.
1 " 1½, 2, 2½ and 3 x 3—30 in. Clear Dry Oak Squares.
1 " 2 x 2—30 in. Clear Dry Gum Sqrs.
1 " 1 x 1—5 ft. Clear Dry Oak Sqrs.
1 " 1 x 4 in. and up 22 and 25 in. Clear Dry Quartered White Oak.
1 " 1 x 4 in. and up, 31, 49 and 55 in. Clr. Dry Quartered White Oak.
1 " 3 in. No. 1 Com. and Bet. Dry Qtd. White Oak.
3 " 4/4 x 4 in. and up 4 to 8 ft. Clear Ash Shorts.
1 " 6/4 and 8/4 x 4 in. and up 4 to 8 ft. Clear Ash Shorts.
1 " 10/4 x 4 in. and up, 4 to 8 ft. Clear Ash Shorts.

Shall we send you quotations?

The Probst Lumber Co.

Cincinnati, Ohio

Rexford

Veneer Tapes

Combine all the qualities desired by manufacturers and users of veneers.

"Let a sample roll tell its own story"

Please state your requirements and, if possible, send sample of tape you are now using.

Rexford Gummed Tape Co.

Manufacturers

Home Office, 598 Clinton St., MILWAUKEE, Wis.

St. Louis, Mo. Branch, 416 Rialto Building. Olive 3529
Chicago, Ill., Branch, 20 E. Jackson Blvd. Harrison 6107

- AT YOUR SERVICE -

We submit for your interest the following choice HARDWOODS:

1 car 1 x 4 and up No. 1 Com. and Btr. Black Ash.
2 cars 1 x 4 and up No. 3 Com. and Btr. Black Ash.
4 cars 1 x 4 and up No. 3 Com. Black Ash.
2 cars 1 x 4 and up No. 3 Com. Black Ash.
1 car 5/4 No. 2 Com. and Btr. Birch.
1 car 6/4 x 10 and up No. 1 Com. and Btr. Birch.
1 car 3-inch First and Second Birch.
1 car 4 x 6 Birch Hearts.
2 cars 2-inch No. 1 Com. and Btr. Hard Maple.
1 car 6/4 No. 1 Com. and Btr. Basswood.
1 car 5/4 No. 2 and 3 Com. Basswood.
1 car 2-inch No. 1 Com. and Btr. White Oak.
6,000 ft. 4-inch First and Second Hickory.
1,500 ft. 3-inch No. 1 Com. and Btr. Hickory.
1,500 ft. 2-inch No. 1 Com. and Btr. Hickory.

All the above stock thoroughly dry and well manufactured

Write, wire or 'phone your inquiries and they will have careful attention

Graves, Bigwood & Co.

Manufacturers of

PINE
HEMLOCK | **LUMBER**
HARDWOOD |

712 Traders Bank Building
TORONTO   -   ONT.

Mills at—Byng Inlet, Ont.
and
Deer Lake, Ont.

duction of prices, therefore economy in the use of raw materials offers the most promising avenue along which to discover opportunities for the reduction of ultimate costs. More care in the cutting and edging of materials yields a slight advantage, but the method that promises most is that of purchasing a poorer grade of raw materials, and then by cutting out the unusable portions, save the sound parts for manufacture. The possibilities in this direction have long been recognized, but until recently machine developments have not permitted the economical working up of such stock.

There has been put on the market during the last few months a new machine which admirably fulfils this purpose. It is built to conform to the principles of saw edging herein explained, although when so desired it can be used for sizing as readily as any other edging saw. This saw has a travelling feed bed carrying the stock under pressure rolls, which are not in any sense feed rolls, hence never spoil the saw line when slightly worn. The stock is absolutely certain to travel through the cut in a straight line. This permits very close edging, and with it all unusable portions of stock can be eliminated, and the sound portions retained, thus saving enough of poor stock to make its use profitable.

The forethought of the designers of this saw has made it adaptable to the needs of any purchaser by providing extra wide saw tables for use where edging and sizing is to be done on the same saw, and, for exceptional conditions, a special sizing device, which, when used in conjunction with the saw, greatly expedites the work.

There can be no question that the labor situation, together with the price level of raw materials, make more necessary than ever the rigid practice of economical production. Labor and time of men and machines must be conserved to the utmost. The operations of edging and rough jointing offer the best field for reform which can best be accomplished by the use of up-to-date labor and time-saving edging and jointing saws.

Enquiries for Canadian Products

The following inquiries have been received by the Department of Trade and Commerce at Ottawa during the past couple of weeks. The names and addresses of the parties making the inquiries can be obtained from the Department by quoting the number which precedes each item. It is interesting to note the large number of inquiries that are being made and the different kinds of articles that are being asked for.

1527. **Woodenware.**—A South African firm of wholesale dealers requests catalogues and full particulars of Canadian-made woodenware of all kinds.

1528. **Handles.**—A South African firm of wholesale dealers requests samples and prices and particulars of delivery on hammer, pick, axe, tool, broom and other handles.

1465. **Brushware.**—A Durban firm asks for catalogues and price-lists of Canadian-made brushware.

1466. **Woodenware.**—A Durban firm of general merchants asks for catalogues and price-lists illustrating woodenware of all kinds, including stepladders.

1467. **Handles.**—A Durban firm of general merchants asks for samples, price-lists and full particulars from Canadian manufacturers for the regular supply of handles of all kinds.

1614. **Windsor chairs, chests of drawers, three large and two small, in knocked down form.**—A Paisley firm wishes to enter into correspondence with Canadian manufacturers of the above, with a view to future business.

1615. **Sewing machines, mangles, wringers.**—A Paisley firm wishes quotations for the above from Canadian exporters.

1616. **Wooden furniture.**—A Glasgow house with a distribution in Scotland, Northern England and Ireland wishes quotations for wooden cradles, cots and cribs; wooden furniture; wood lattice (lath thick); games and toys; woodenware. Particulars in detail will be furnished by the inquirer for manufacturers' guidance. Best references.

1617. **Furniture—Lumber.**—A South African firm of furniture manufacturers requests samples and particulars from Canadian manufacturers of 3-ply veneer and other lumbers suitable for furniture making.

1557. **Woodenware.**—A Durban firm requests catalogues and full particulars of Canadian-made woodenware of all kinds.

1558. **Handles.**—A Durban firm requests samples and prices and particulars of delivery on hammer, pick, axe. tool, broom and other handles.

1330. **Woodsplit pulleys.**—A Johannesburg firm in close connection with the mines and manufacturing interests makes request for full particulars from Canadian manufacturers of woodsplit pulleys.

1331. **Handles.**—A South African firm requests samples and prices and particulars of delivery on hammer, pick, axe, rake, tool, broom, and other handles.

1332. **Handles.**—A Johannesburg firm asks for samples and quotations on hammer, pick and tool handles.

1390. **Box shooks.**—Quotations asked for on box shooks, packed in crates, in lots of 10,000 each. Particulars of size required may be obtained from the Department of Trade and Commerce, Ottawa.

1435. **Cart and carriage wood material.**—A Bloemfontein firm of cart and carriage builders will purchase Canadian-made hubs, rims, spokes, shafts, dash-boards and lumber, such as white poplar. Full particulars with illustrations and prices requested.

1437. **Pony carts or buggies.**—A South African firm are prepared to handle for the Orange Free State a Canadian-made pony cart. Will handle at least one hundred a year of the cheaper make and forty to fifty of the better make.

1440. **Wood.**—A Bloemfontein firm requests catalogues and price-lists on doors, window frames, mantel pieces, mouldings, pine and spruce deals, and shelvings; also ash and poplar boards.

Wants Designs and Machines for Furniture Making

Victoria, B.C. Nov. 13th, 1916.

Publishers Canadian Woodworker,
    Toronto, Ont.

Gentlemen:

Have just received your October issue of the "Canadian Woodworker," with which I am well pleased. I have been thinking of making furniture in a small way, and would like to obtain a design book, if possible, on Chippendale furniture. Could you inform me as to who would be able to furnish same? Would also like to know what machine would be the best for all-round use for the making of furniture.

Thanking you in anticipation, I am,

    Yours in earnest,

        "Subscriber."

Photographic Proof of the Quality of

BEAVER BRAND BANDSAWS

The above cut shows one of our 6 in. Band saws which has been pulled off the wheels in the planing mills of Belgo Canadian Pulp Company at Shawinigan Falls, Que. The saw has been twisted into five coils less than 20 inches in diameter without a single crack—thus demonstrating the toughness of our band saw steel.

RADCLIFF SAW MFG. CO., LIMITED   550 DUNDAS ST. TORONTO, ONT.

Agents for L. & I. J. White Co., Machine Knives

The Complete Line

VICTOR CASTERS
Roller Bearing

Roll and swivel easily under great weight. Nos. 4 and 6 sizes can also be furnished with protection caps preventing the accumulation of threads around the axle when used in knitting mills, shoe factories, etc.

VICTOR CASTERS can be furnished in the following sizes :—Nos. 1, 2, 3, 4, 6, 8, 9, 10, 10½, 11, 12.

ACME CASTER—the Reliable Casters.

Our Canadian representative is at your service and will be pleased to assist you in your Hardware problems.

Victor Caster

Foster Merriam & Co.,   MERIDEN, CONN.

Canadian Representative—F. A. Schmidt, Kitchener, Ont.

News of the Trade

The United Pole Company, Amherst, N. S., have obtained a charter.

Berthiaume & Sequin, sash and door manufafcturers, Montreal, P. Q., have been registered.

It is stated that the assets of the Elmira Interior Woodwork Company, Limited, will be sold by tender.

Wm. Thomson & Sons, lumber dealers, Thurso, P. Q., recently lost their planing mill by fire. The loss was covered by insurance.

The R. Laidlaw Lumber Company's plant at Sarnia, Ont., suffered a slight loss by fire recently as the result of a fire which wiped out a neighboring plant.

E. Russell Purtle, office manager for Snyder Brothers, Waterloo, Ont., was married on October 11 at St. Mary's Church, Kitchener, to Miss Ida M. Gabel.

The box-making business formerly carried on at South Waterville, N.S., under the management of Charles H. Lloyd, will be carried on by Henry Lloyd & Sons.

The Canadian Cooperage Manufacturing Company, Limited, Smith's Falls, Ont., were burnt out recently. The loss is covered by insurance.

The Fletcher Manufacturing Company, Glencoe, Ont., who manufacture soda fountains, etc., are erecting an addition to their woodworking plant.

Fire totally destroyed the sash and door factory owned by Ed. Charbonneau at Lachute, Que. The loss amounts to about $15,000 and is partly covered by insurance.

The sash and door factory at St. Hughes, Que., owned by Adelard Paquette was recently destroyed by fire. The loss is estimated at $20,000 and the insurance is $10,000.

A new industry—the manufacture of cane furniture—has been started in Kerrobert, Sask. Work has been commenced on the line, and the promoters hope that it will develop into a business of wide scope.

At the annual meeting of the Nova Scotia Carriage Company, Limited, held in Amherst, the shareholders after hearing the report of the directors, strongly recommended that the company go into liquidation.

An industrial by-law will be submitted to the people of North Bay, Ont., on December 11 for the purpose of obtaining their verdict on the matter of providing the North Bay Toy Company with a free site and building.

Dominion Hardwoods Limited, Deseronto, Ont., will probably proceed at once to rebuild their plant which was recently destroyed by fire. The town of Deseronto has carried a by-law granting the company certain important concessions.

The death occurred recently in Toronto of Samuel May, a pioneer in the manufacture of billiard tables and accessories. Mr. May was a recognized authority on everything pertaining to billiard tables and had been in business in Toronto for over fifty years.

At a meeting of the logging, lumbering and woodworking section of the National Safety Congress in session at Detroit, October 20, Mr. F. G. Lovett, inspector for the Furniture Manufacturers' Safety Association, Toronto, was elected vice-president for the ensuing year.

Graves, Bigwood & Company, the well-known lumber merchants, have removed from No. 1026 Traders Bank Building, Toronto, to larger offices at No. 712 in the same building. The firm has always been prominently identified in the past with the pine and hemlock trade and the manufacture of box shooks, but they are now specializing in hardwood lumber of all kinds.

Fire totally destroyed the box factory of James Trevail & Son at South Middleton, Ont., recently. The amount of insurance on the building and contents was very small. The firm made a specialty of cheese boxes, and as they had a lot of orders on hand the loss will be very heavy.

The Maple Leaf Toy. Company, Limited, has been incorporated with capital of $40,000, to manufacture toys, novelties, woodenware and furniture. The company's place of business will be at Toronto, and the provisional directors are H. W. Fielden, F. J. Jackson and D. H. Morgan.

Work has started on the erection of a planing mill for Hector McLean, Orillia, Ont. The building will be one storey, 50 x 90, frame construction, brick boiler house. The owner will be in the market for engine, boiler, and planing mill equipment.

The Kindel Bed Company have commenced making shipments from the factory recently acquired by them at Stratford, Ont. It is understood, however, that the company still have some overhauling to do to the building and that they purpose installing a Grand Rapids dry kiln.

The entire plant of Haley & Son, manufacturers of doors, sash, box shooks, etc., at St. Stephen, N.B., including the office and supplies and the lumber in the yards was destroyed by fire on October 17th. The fire broke out in the glue room of the factory shortly after the whistle blew and was blazing fiercely when first discovered. The company advise us that they have already commenced to rebuild.

The Frederickhouse & Abitibi Pulp Wood Company has been organized with capital of $300,000, to exploit 180 square miles of timber limits owned by them near Cochrane, Ont. To better utilize the timber to advantage the company will erect an up-to-date woodworking plant to manufacture box shooks, cigar boxes, laths, broom and pick handles, doors and sash and wooden spools. Communications should be addressed to J. A. McAndrews, Lumsden Building, Toronto.

The Dyment carriage factory at Barrie, Ont., was the scene of a fire a few days ago which did considerable damage to the painting department and destroyed quite a lot of stock in the woodworking department that was ready to be assembled. Also a portion of the roof was destroyed. The origin of the fire is a mystery, but as it spread up the elevator shaft to the upper floor it is assumed that it started at or near the bottom of the shaft. The loss is estimated at $15,000 or $20,000, and 70 men are temporarily thrown out of employment.

The Kilgour Davenport Company of Toronto, are building an addition to their factory. The addition will be three storeys and will contain about 20,000 square feet of floor space. The company state that the great demand for their sofa beds makes it necessary to extend their operations. They are doing more than double the business they did last year and are at present three months behind with their orders. Besides davenports the company manufacture a line of den chairs, and when their new addition is completed will make a divanette folding bed of their own design.

Porcelain Tops Wanted

A Canadian manufacturer wants to get in touch with manufacturers of porcelain tops for kitchen cabinets. Information will be appreciated by the publishers of this journal.

Panels and Tops

Deskbeds, Table Tops, Chiffonier and Dresser Tops
Three and Five Ply Panels

Large assortment of stock panels always on hand.
Three and five ply stock for cartridge boxes.
Your inquiry will be fully appreciated and promptly answered with full information.

WRITE US

The

Gorham Brothers Company

MT. PLEASANT   -   MICHIGAN

"TROY"
GARNET PAPER

COMBINATION and CLOTH

FINISHING
PAPER

RITCHEY SUPPLY CO.
126 Wellington St. W.
TORONTO

Crating Lumber

⅝" and 1"   All Grades
At Right Prices

Try Us—It will be worth your while.

Read Bros., Limited

43 Victoria Street
Toronto   -:-   Canada

DARLINGS
STEAM
APPLIANCES

DARLING BROTHERS
LIMITED
Engineers and Manufacturers
MONTREAL, CANADA

Branches:
Toronto and Winnipeg   Agents:
    Halifax, St. John, Calgary, Vancouver

Many Points of Superiority
are embodied in the
WAYMOTH VARIETY
WOOD-TURNING LATHE

This machine in one operation turns, bores and cuts off wood turnings in any size up to 34 inches in length. It embodies many improvements. There is an arrangement whereby all side motion occasioned by the continual wear of the tail-stock and middle-piece is taken up; by the use of a patent rougher bed a knife is used 10 inches long instead of one about 2 inches, as formerly; a rack and gear is used instead of a cord and crank for operating the swing bed.

Our lathes are sold in Canada by
A. R. Williams Machinery Co., Toronto, Vancouver and Winnipeg.
General Supply Company of Canada, Ottawa.
Williams & Wilson, Montreal.

A. D. WAYMOTH & CO., FITCHBURG, MASS.

Second-Hand Machines
For Sale

We have the following to offer. Ask us about others not listed, if you are interested.

1 Egan 30-in. 2 Drum Sander.
1 McGregor Gourlay 30-in. 2 Drum Boss Sander.
1 Jackson Cochrane 36-in. 3 Drum Sander.
1 Egan 42-in. 3 Drum Sander.
1 Cowan 42-in. 3 Drum Sander.
1 Cowan 48-in. 3 Drum Sander.
1 C. M. C. 8-in. 4 side Sticker.
2 Ballantyne 12-in. 4 side Sticker.
1 Egan 36-in. Single Surfacer.
1 Harper 26-in. Single Surfacer.
1 Cowan No. 220 24 x 10 Single Surfacer.
1 McGregor Gourlay OV 15-in. Planer & Matcher.
1 McGregor Gourlay OV 9-in. Planer & Matcher.
1 Jackson Cochrane 54-in. Band Resaw 5-in. blade.
1 Jackson Cochrane No. 165 Rip Saw.

P. B. Yates Machine Co., Limited
Hamilton, Ontario t-f

Second-Hand
Woodworking Machines

Band Saws—26-in., 36-in., 40-in.
Jointers—12-in., 16-in.
Shapers—Single and double spindle.
Re-Saw—42-in., American.
Lathes—20-in. with rests and chisels.
Surfacers—16 x 6 in. and 18 x 8 in.
Mortisers—No. 5 and No. 7 New Britain.
Rip and Cut-off Saws; Molders; Sash Relishers, etc.
Over 100 machines, New and Used, in stock.

Baird Machinery Company,
123 Water Street,
PITTSBURGH, PA.

Woodworking Machinery For Sale

1—18 Spindle upright boring machine, Can. Machinery Corporation make.
1—Double spindle boring machine for shell boxes, J. & C. make, with countershaft.
1—Dado machine with automatic feed with five heads and two saws.
1—Veneer press, 4 sections, 4 screws to each section.
1—J. & C. Single end tenoning machine.
1—J. & C. two spindle shaper with countershaft.
1—J. & C. spindle carving machine with countershaft.
1—table slide dovetail groover with countershaft.
1—rolling top saw table.
1—shell box automatic hopper feed handle cleat machine.

Beach Furniture Limited
10-12 Cornwall, Ont.

FOR SALE

Pine, Birch, Spruce, suitable for shell box manufacturing. Apply Office Specialty Mfg. Co., Ltd., Newmarket, Ont.
11-12

FOR SALE
Continuous Feed Glue Jointer
Falls make. Practically new.

Canadian Linderman Co.
Woodstock, Ont.

Valuable Box Factory
For Sale

The undersigned will receive offers for the purchase of that valuable property known as the

Czerwinski Box Factory

situated on Logan Avenue and Tecumseh Street, Winnipeg, Canada. This property comprises what is believed to be the finest box factory in Western Canada, fully equipped with the latest and best machinery for box manufacturing and ready for immediate operation. It also includes the factory site, consisting of about 2½ acres of land close in to the business centre of the city, and possesses unexcelled trackage facilities. Street cars pass the factory door. This is a magnificent opportunity for any person wishing to engage in box manufacturing. The whole may be purchased at a very great bargain.

Further information may be obtained upon application to

H. E. Deneen,
Assignee of the Czerwinski Box Co., Limited
300 Electric Railway Chambers,
Winnipeg, Man.

A New Finish for Wood

Two of the partners of A. S. Toone and Sons, who are jacquard card manufacturers at Nottingham, England, have patented a process for finishing the varnished or painted surfaces of wood boards and veneers, cardboard, paper, and other materials, to produce a perfectly smooth and highly polished surface, claimed to be superior to that ordinarily obtained with either varnish or paint, writes "The Cabinet Maker." The surface of a piece of any of these materials is first given one or more coats of spirit varnish (resin dissolved in spirit) or paint, composed of coloring matter, mixed with vegetable oils, and as soon as the coating is dry, it is placed in contact with the polished face of a metal finishing plate; heat and pressure is applied for a short period of time, and then the material and plate are thoroughly cooled before separating them. The action of the heat and pressure causes the surface of the finishing-plate to make perfect contact, so that the polished face is transmitted to the coated face of the material, whilst the cooling process before removal causes the coating material to become thoroughly set, and renders the finish obtained permanent. The process described is most conveniently carried on in a press heated by steam, fitted for changing cold water for the steam, so that the process of cooling the press and its contents can be rapidly effected. The finishing plates are preferably made of bronze or nickel, with highly polished faces, and, if desired, they may be plated to prevent corrosion.

PETRIE'S
MONTHLY LIST
of
NEW and USED
WOOD TOOLS
in stock for immediate delivery

Wood Planers
36" American double surfacer.
30" Whitney pattern surfacer.
26" Revolving bed double surfacers.
24" Major Harper planer and matcher.
24" Single surfacers, various makes.
20" Dundas pony planer.
18" Little Giant planer and matcher.
16" Galt jointer.

Band Saws
42" Fay & Egan power feed rip.
38" Atlantic tilting frame.
30" Ideal pedestal, tilting frame.

Saw Tables
Ballantine variable power feed.
Preston variable power feed.
Cowan heavy power feed. M138.
No. 3 Crescent, universal.
No. 2 Crescent, combination.
Ideal variety, tilting table.
12" Defiance automatic, equalizer.
MacGregor Gourlay railway cut-off.
7½" Iron frame swing.

Mortisers
No. 5 New Britain, chain.
Cowan hollow chisel. M.190.
Galt upright, compound table.
Smart foot power.

Moulders
13" Clark-Demill four-side.
12" Cowan four side.
10" Houston four, side.
6" Cowan four side.
6" Dundas sash sticker.

Sanders
24" Fay double drum.
12" C.M.C. disk and drum.
18" Crescent disk, adjustable table.

Clothespin Machinery
Humphrey No. 8 giant slab resaw.
Humphrey gang slitter.
Humphrey cylinder, cutting-off machine.
Humphrey automatic lathes (6).
Humphrey double slotters (4).

Miscellaneous
Fay & Egan 12-spindle dovetailer.
Cowan dowel machine, M80.
No. 235, Wysong & Miles post boring machine.
Cowan post boring machine, M88G.
Cowan post boring machine, M23.
MacGregor Gourlay 2-spindle shaper.
Elliott single spindle shaper.
No. 51 Crescent universal woodworker.
Rogers vertical resaw.
Cowan sash clamp.
Egan sash and door tenoner.
No. 9 Lion universal woodtrimmer.
6-nail box nailer, Meyers patent.
20" American wood scraper.
16" Ideal wood lathes (3).
4-head rounding machines.
Cowan spindle carver. M63.
Cowan veneer press screws.

Prices, Descriptions and Full Particulars on Request

H. W. PETRIE, LTD.
Front St. W., Toronto, Ont.

FROST KING   *"The Standard for Forty Years"*   TROJAN

The Dependable Babbitt

Equally as good for heavy duty as speed. It is especially adapted for manufacturers of woodworking machinery, planing mills, furniture factories, box factories and woodworking plants.

Try it

Our High Grade Babbitt

Owing to its phenomenal ability to absorb oil "Trojan" babbitt will be found ideal for overhead bearings in line and counter-shafts as it is unnecessary to oil as often as with the ordinary babbitts.

Try it

HOYT METAL COMPANY   -   Toronto, Ont.

NEW YORK   ST. LOUIS   LONDON, ENG.

You Have Paid for an Installation of

Chapman
Double Ball Bearings

in Your Factory over and over again, BUT—

HAVE YOU GOT THEM ?

Ordinary line shafting wastes from 15 per cent. to 60 per cent. of power.

Line shafting equipped with Chapman Double Ball Bearings will eliminate about 75 per cent. of the friction, thus averaging a total saving of from 15 per cent. to 30 per cent.

Chapman Double Ball Bearings fit any adjustable hanger and require oiling and attention only once a year. No extra equipment required to install.

Write to-day for full particulars.

The Chapman Double Ball Bearing Co. of Canada, Limited
Toronto   339-351 Sorauren Ave.   Ontario

American Branch: The Transmission Ball Bearing Co., Inc., 32 Wells St., Buffalo, N. Y.

BOOKS FOR SALE

The following books are offered at special prices subject to previous sale:

Carpentry and Joinery, by Gilbert Townsend. Published in 1915 by the American School of Correspondence, 258 pages, illustrated. Price $1.00.

Saw Fitting Manual, a treatise on the care of saws and knives. Deals with everything in the saw and knife alphabet, from adjustments to widths. 144 pages. Price $2.00.

Common-sense Handrailing, by Fred T. Hodgson. Published by Frederick J. Drake & Company, Chicago. 114 pages, illustrated. Price 50c.

Roof Framing Made Easy, by Owen B. Maginnis. Published by The Industrial Publication Company, New York. 104 pages, illustrated. Price 50c.

Handrailing Simplified, by An Experienced Architect. Published by William T. Comstock, New York, 62 pages, illustrated. Price 50c.

How to Join Mouldings; or, The Arts of Mitering and Coping, by Owen B. Maginnis. Published by William T. Comstock, New York. 72 pages, illustrated. Price 50c.

Popular Mechanics Shop Notes. Published by Popular Mechanics, Chicago. Easy Ways to do Hard Things, etc. Years 1905, 1906. Price 40c. each.

Cabinet Making, by J. H. Rudd. Published by Grand Rapids Furniture Record Company. 210 pages, illustrated. Price $1.50.

Utilization of Wood-Waste (Second Revised Edition), by Ernst Hubbard. Published in 1915 by Scott, Greenwood & Sons. 192 pages, illustrated. Price $3.00.

Woodworker Publishing Company, Limited
345 Adelaide Street West, Toronto, Ontario

The "Canadian Woodworker" Buyers' Directory

An Index to the Best Equipment and Supplies

ABRASIVES
The Carborundum Co., Niagara Falls, N.Y.

AIR BRUSH EQUIPMENT
DeVilbiss Mfg. Co., Toledo, Ohio.

BABBITT METAL
Hoyt Metal Co., Toronto.

BALL BEARINGS
Chapman Double Ball Bearing Co., Toronto.

BALUSTER LATHES
Garlock-Walker Machinery Co., Toronto, Ont.
Whitney & Son, Baxter D., Winchendon, Mass.

BAND SAW FILING MACHINERY
Canada Machinery Corporation, Galt, Ont.
Fay & Egan Co., J. A., Cincinnati, Ohio.
Garlock-Walker Machinery Co., Toronto, Ont.
Preston Woodworking Machinery Company,
Preston, Ont.
Yates Machine Co., P. B., Hamilton, Ont.

BAND SAWS
Canada Machinery Corporation, Galt, Ont.
Cowan & Co., Galt, Ont.
Garlock-Walker Machinery Co., Toronto, Ont.
Petrie, H. W., Toronto.
Preston Woodworking Machinery Company,
Preston, Ont.
Yates Machine Co., P. B., Hamilton, Ont.

BAND SAW MACHINES
Canada Machinery Corporation, Galt, Ont.
Cowan & Co., Galt, Ont.
Fay & Egan Co., J. A., Cincinnati, Ohio.
Garlock-Walker Machinery Co., Toronto, Ont.
Petrie, H. W., Toronto.
Preston Woodworking Machinery Company,
Preston, Ont.
Williams Machinery Co., A. R., Toronto, Ont.
Yates Machine Co., P. B., Hamilton, Ont.

BAND SAW STRETCHERS
Fay & Egan Co., J. A., Cincinnati, Ohio.
Yates Machine Co., P. B., Hamilton, Ont.

BASKET STITCHING MACHINE
Saranac Machine Co., Benton Harbor, Mich.

BENDING MACHINES
Fay & Egan Co., J. A., Cincinnati, Ohio.
Garlock-Walker Machinery Co., Toronto, Ont.
Perfection Wood Steaming Retort Company,
Parkersburg, W. Va.

BELTING
Fay & Egan Co., J. A., Cincinnati, Ohio.
Sadler & Haworth, Montreal, Que.

BLOWERS
Sheldons, Limited, Galt, Ont.

BLOW PIPING
Sheldons, Limited, Galt, Ont.

BORING MACHINES
Canada Machinery Corporation, Galt, Ont.
Cowan & Co., Galt, Ont.
Fay & Egan Co., J. A., Cincinnati, Ohio.
Garlock-Walker Machinery Co., Toronto, Ont.
Jackson, Cochrane & Company, Kitchener, Ont.
Nash, J. M., Milwaukee, Wis.
Preston Woodworking Machinery Company,
Preston, Ont.
Reynolds Pattern & Machine Co., Moline, Ill.
Yates Machine Co., P. B., Hamilton, Ont.

BOX MAKERS' MACHINERY
Canada Machinery Corporation, Galt, Ont.
Cowan & Co., Galt, Ont.
Fay & Egan Co., J. A., Cincinnati, Ohio.
Garlock-Walker Machinery Co., Toronto, Ont.
Jackson, Cochrane & Company, Kitchener, Ont.
Neilson & Company, J. L., Winnipeg, Man.
Preston Woodworking Machinery Company,
Preston, Ont.
Reynolds Pattern & Machine Co., Moline, Ill.
Whitney & Son, Baxter D., Winchendon, Mass.
Williams Machinery Co., A. R., Toronto, Ont.
Yates Machine Co., P. B., Hamilton, Ont.

BRUSHES
Skedden Brush Co., Hamilton, Ont.

CABINET PLANERS
Canada Machinery Corporation, Galt, Ont.
Cowan & Co., Galt, Ont.
Fay & Egan Co., J. A., Cincinnati, Ohio.
Garlock-Walker Machinery Co., Toronto, Ont.
Jackson, Cochrane & Company, Kitchener, Ont.
Preston Woodworking Machinery Company,
Preston, Ont.
Yates Machine Co., P. B., Hamilton, Ont.

CARS (Transfer)
Sheldons, Limited, Galt, Ont.

CARBORUNDUM PAPER
Carborundum Co., Niagara Falls, N.Y.

CARVING MACHINES
Canada Machinery Corporation, Galt, Ont.
Fay & Egan Co., J. A., Cincinnati, Ohio.
Garlock-Walker Machinery Co., Toronto, Ont.
Jackson, Cochrane & Company, Kitchener, Ont.

CASTERS
Burns & Bassick Co., Bridgeport, Conn.
Faultless Caster Co., Evansville, Ind.
Foster, Merriam & Co., Meriden, Conn.
National Lock Co., Rockford, Ill.
Onward Mfg. Co., Kitchener, Ont.

CLAMPS
Fay & Egan Co., J. A., Cincinnati, Ohio.
Garlock-Walker Machinery Co., Toronto, Ont.
Jackson, Cochrane & Company, Kitchener, Ont.
Neilson & Company, J. L., Winnipeg, Man.
Preston Woodworking Machinery Company,
Preston, Ont.
Simonds Canada Saw Company, Montreal, P.Q.

COLUMN MACHINERY
Fay & Egan Co., J. A., Cincinnati, Ohio.

CORRUGATED FASTENERS
Acme Steel Goods Company, Chicago, Ill.

CRATING LUMBER
Read Bros., Limited, Toronto.

CUT-OFF SAWS
Canada Machinery Corporation, Galt, Ont.
Cowan & Co., Galt, Ont.
Fay & Egan Co., J. A., Cincinnati, Ohio.
Garlock-Walker Machinery Co., Toronto, Ont.
Jackson, Cochrane & Company, Kitchener, Ont.
Ober Mfg. Co., Chagrin Falls, Ohio.
Preston Woodworking Machinery Company,
Preston, Ont.
Simonds Canada Saw Company, Montreal, Que.
Yates Machine Co., P. B., Hamilton, Ont.

CUTTER HEADS
Canada Machinery Corporation, Galt, Ont.
Cowan & Co., Galt, Ont.
Fay & Egan Co., J. A., Cincinnati, Ohio.
Jackson, Cochrane & Company, Kitchener, Ont.
Mattison Machine Works, Beloit, Wis.
Yates Machine Co., P. B., Hamilton, Ont.

DOVETAILING MACHINES
Canada Machinery Corporation, Galt, Ont.
Cowan & Co., Galt, Ont.
Fay & Egan Co., J. A., Cincinnati, Ohio.
Garlock-Walker Machinery Co., Toronto, Ont.
Jackson, Cochrane & Company, Kitchener, Ont.

DOWEL MACHINES
Canada Machinery Corporation, Galt, Ont.
Fay & Egan Co., J. A., Cincinnati, Ohio.

DRY KILNS
Grand Rapids Veneer Works, Grand Rapids,
Mich.
National Dry Kiln Co., Indianapolis, Ind.
Sheldons, Limited, Galt, Ont.

DUST COLLECTORS
Sheldons, Limited, Galt, Ont.

DUST SEPARATORS
Sheldons, Limited, Galt, Ont.

EDGERS (Single Saw)
Canada Machinery Corporation, Galt, Ont.
Fay & Egan Co., J. A., Cincinnati, Ohio.
Garlock-Walker Machinery Co., Toronto, Ont.
Simonds Canada Saw Co., Montreal, Que.

EDGERS (Gang)
Canada Machinery Corporation, Galt, Ont.
Fay & Egan Co., J. A., Cincinnati, Ohio.
Garlock-Walker Machinery Co., Toronto, Ont.
Petrie, H. W., Toronto.
Simonds Canada Saw Co., Montreal, Que.
Williams Machinery Co., A. R., Toronto, Ont.
Yates Machine Co., P. B., Hamilton, Ont.

END MATCHING MACHINES
Canada Machinery Corporation, Galt, Ont.
Cowan & Co., Galt, Ont.
Fay & Egan Co., J. A., Cincinnati, Ohio.
Garlock-Walker Machinery Co., Toronto, Ont.
Jackson, Cochrane & Company, Kitchener, Ont.
Yates Machine Co., P. B., Hamilton, Ont.

EXHAUST FANS
Garlock-Walker Machinery Co., Toronto, Ont.
Sheldons Limited, Galt, Ont.

FILES
Simonds Canada Saw Co., Montreal, Que.

FILING MACHINERY
Wardwell Mfg. Company, Cleveland, Ohio.

FLOORING MACHINES
Canada Machinery Corporation, Galt, Ont.
Cowan & Co., Galt, Ont.
Fay & Egan Co., J. A., Cincinnati, Ohio.
Garlock-Walker Machinery Co., Toronto, Ont.
Petrie, H. W., Toronto.
Preston Woodworking Machinery Company,
Preston, Ont.
Whitney & Son, Baxter D., Winchendon, Mass.
Yates Machine Co., P. B., Hamilton, Ont.

FLUTING HEADS
Fay & Egan Co., J. A., Cincinnati, Ohio.

FLUTING AND TWIST MACHINE
Preston Woodworking Machinery Company,
Preston, Ont.

FURNITURE CARVINGS
Decorators Supply Co., Chicago, Ill.

FURNITURE TRIMMINGS
Burns & Bassick Co., Bridgeport, Conn.
Faultless Caster Co., Evansville, Ind.
Foster, Merriam & Co., Meriden, Conn.

GARNET PAPER AND CLOTH
The Carborundum Co., Niagara Falls, N.Y.
Ritchey Supply Company, Toronto, Ont.

GRAINING MACHINES
Canada Machinery Corporation, Galt, Ont.
Fay & Egan Co., J. A., Cincinnati, Ohio.
Williams Machinery Co., A. R., Toronto, Ont.
Yates Machine Co., P. B., Hamilton, Ont.

GLUE
Perkins Glue Company, South Bend, Ind.

GLUE CLAMPS
Jackson, Cochrane & Company, Kitchener, Ont.
Neilson & Company, J. L., Winnipeg.

GLUE HEATERS
Fay & Egan Co., J. A., Cincinnati, Ohio.
Jackson, Cochrane & Company, Kitchener, Ont.

GLUE JOINTERS
Canada Machinery Corporation, Galt, Ont.
Jackson, Cochrane & Company, Kitchener, Ont.
Yates Machine Co., P. B., Hamilton, Ont.

GLUE SPREADERS
Fay & Egan Co., J. A., Cincinnati, Ohio.
Jackson, Cochrane & Company, Kitchener, Ont.

GLUE ROOM EQUIPMENT
Perrin & Company, W. R., Toronto, Ont.

GRINDERS (Cutter)
Fay & Egan Co., J. A., Cincinnati, Ohio.

GRINDERS, HAND AND FOOT POWER
The Carborundum Co., Niagara Falls, N.Y.

GRINDERS (Knife)
Canada Machinery Corporation, Galt, Ont.
Fay & Egan Co., J. A., Cincinnati, Ohio.
Garlock-Walker Machinery Co., Toronto, Ont.
Petrie, H. W., Toronto.
Preston Woodworking Machinery Company,
Preston, Ont.
Yates Machine Co., P. B., Hamilton, Ont.

GRINDERS (Tool)
Fay & Egan Co., J. A., Cincinnati, Ohio.
Jackson, Cochrane & Company, Kitchener, Ont.

"We are
Saving over
40% of
our coal!"

"I need only
clean the
boiler once
in
four weeks!"

MR. O. H. HALE, of the King Furniture
Company, Warren, Ohio, is so enthusiastic about the

Morehead
⫸ Back to Boiler ⫸
SYSTEM

that he has written an account of his experiences which will profit his brother engineers
everywhere.
For the Morehead "Back-to-Boiler" Book—
with its many pages of data and interesting
illustrations—and a complete copy of Mr.
Hale's letter, just write

Canadian Morehead Mfg. Co.
WOODSTOCK Dept. "H" ONTARIO

"You never
have to think
about low
water in the
boiler!"

"I have worked
other Traps
but I never saw
one as simple as
the Morehead!"

WOULD YOU PAY $1
FOR A 15c LUNCH?

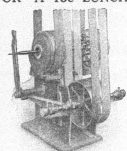

Then why pay $1 for your
round work sanding when the
Nash Sander will do it for 15c?

J. M. NASH, MILWAUKEE, Wis.

GRINDING

BAND SAWS

with

26 TEETH

TO THE INCH

Your woodworking band saws can be resharpened so that all teeth
come within ½ of 1-1000 of an inch alike to each other, by this

—Model A—
Automatic Band Saw Grinder
Saws 1-8 inch to 8 inches wide. Teeth up to 2 inch spacing
POINTS OF INTEREST
Bronze Bushed Grinding Wheel Spindle.
Head Suspended between Hardened Steel Pivots
No Slides to Wear. No Cams to Change.
Feed Movement obtained by a Scaled Slide
A Double Pawl, Positive Feed Movement, the quickest
and most perfect method of jointing all teeth,
can be furnished at a small extra charge.
LOWEST PRICED GRINDER MADE
Write for Details

THE WARDWELL MFG. COMPANY
Saw and Knife Sharpening Machinery
111 Hamilton Ave. CLEVELAND, OHIO

GRAND RAPIDS
VAPOR DRY KILN
GRAND RAPIDS
MICHIGAN

Increasing output of other kilns is part
of our business.

We can arrange to utilize your present
equipment and supplement with sufficient
additional to bring your kilns to our basis
of output.

We would suggest that you supply our
engineering department with description of
your kilns and lumber drying needs, so we
can understandingly outline what can be
done.

Your opening the matter will incur no
obligation and we will give you, frankly,
the results of our experience.

"Canadian Woodworker" Buyers' Directory—Continued

GRINDING WHEELS
The Carborundum Co., Niagara Falls, N.Y.

GROOVING HEADS
Fay & Egan Co., J. A., Cincinnati, Ohio.

GUM LUMBER
Brown & Co., George C., Memphis, Tenn.
Churchill-Milton Lumber Co., Louisville, Ky.
Edwards Lumber Co., E. L., Dayton, Ohio.
Lamb-Fish Lumber Co., Charlestown, Miss.
Peared, Jurden & McCowen, Memphis, Tenn.

GUMMERS, ETC.
Fay & Egan Co., J. A., Cincinnati, Ohio.

HAND PROTECTORS
Fay & Egan Co., J. A., Cincinnati, Ohio.

HAND SCREWS
Fay & Egan Co., J. A., Cincinnati, Ohio.

HANDLE & SPOKE MACHINERY
Fay & Egan Co., J. A., Cincinnati, Ohio.
Nash, J. M., Milwaukee, Wis.
Whitney & Son, Baxter D., Winchendon, Mass.

HARDWARE
Bunge & Bassick Co., Bridgeport, Conn.
Faultless Caster Co., Evansville, Ind.

HARDWOOD LUMBER
American Hardwood Lumber Co., St. Louis, Mo.
Anderson Lumber Co., C. G., Toronto, Ont.
Atlantic Lumber Company, Toronto.
Banning, Leland G., Cincinnati, Ohio.
Brown & Company, Geo. C., Memphis, Tenn.
Burns & Knapp Lumber Co., Connellsville, Pa.
Churchill-Milton Lumber Co., Louisville, Ky.
Clark & Son, Edward, Toronto, Ont.
Dermott Land & Lumber Co., Chicago, Ill.
Edwards Lumber Co., E. L., Dayton, Ohio.
Farrin Lumber Co., M. B., Cincinnati, Ohio.
Graves, Bigwood Company, Toronto, Ont.
Kersley, Geo. Montreal, Que.
Kraetzer-Cured Lumber Co., Moorhead, Miss.
Lamb-Fish Lumber Co., Charlestown, Miss.
Martin-Barriss Co., Cleveland, Ohio.
Mowbray & Robinson, Cincinnati, Ohio.
Paepcke Leicht Lumber Co., Chicago, Ill.
Penrod Walnut & Veneer Co., Kansas City, Mo.
Probst Lumber Co., Cincinnati, Ohio.
Read Bros., Limited, Toronto.
Spencer, C. A., Montreal, Que.
Thoman Flinn Lumber Co., Cincinnati, Ohio.

HUB MACHINERY
Fay & Egan Co., J. A., Cincinnati, Ohio.

HYDRAULIC PRESSES
Perrin & Company, W. R., Toronto.

HYDRAULIC VENEER PRESSES
Perrin & Company, Wm. R., Toronto, Ont.

JOINTERS
Canada Machinery Corporation, Galt, Ont.
Cowan & Co., Galt, Ont.
Fay & Egan Co., J. A., Cincinnati, Ohio.
Garlock-Walker Machinery Co., Toronto, Ont.
Jackson, Cochrane & Company, Kitchener, Ont.
Petrie, H. W., Toronto, Ont.
Preston Woodworking Machinery Company, Preston, Ont.
Yates Machine Co., P. B., Hamilton, Ont.

KNIVES (Planer and others)
Canada Machinery Corporation, Galt, Ont.
Fay & Egan Co., J. A., Cincinnati, Ohio.
Simonds Canada Saw Co., Montreal, P.Q.
Yates Machine Co., P. B., Hamilton, Ont.

LATHES
Canada Machinery Corporation, Galt, Ont.
Cowan & Co., Galt, Ont.
Fay & Egan Co., J. A., Cincinnati, Ohio.
Garlock-Walker Machinery Co., Toronto, Ont.
Jackson, Cochrane & Company, Kitchener, Ont.
Petrie, H. W., Toronto, Ont.
Preston Woodworking Machinery Company, Preston, Ont.
Whitney & Son, Baxter D., Winchendon, Mass.

LUMBER
Acme Veneer & Lumber Co., Cincinnati, Ohio.
American Hardwood Lumber Co., St. Louis, Mo.
Anderson Lumber Co., C. G., Toronto, Ont.
Banning, Leland G., Cincinnati, Ohio.
B. C. Lumber Commissioner, Toronto.
Brown & Company, Geo. C., Memphis, Tenn.
Churchill, Milton Lumber Co., Louisville, Ky.
Clark & Son, Edward, Toronto, Ont.
Dermott Land & Lumber Co., Chicago, Ill.
Dominion Mahogany & Veneer Co., Montreal, Que.
Edwards Lumber Co., E. L., Dayton, Ohio.
Farrin Lumber Co., M. B., Cincinnati, Ohio.
Gorham Bros. Co., Mt. Pleasant, Mich.
Graves, Bigwood Company, Toronto, Ont.
Hart & McDonagh, Toronto, Ont.

Hartzell, Geo. W., Piqua, Ohio.
Huddleston, Marsh Mahogany Co., New York.
Long-Knight Lumber Co., Indianapolis.
Probst Lumber Company, Chicago, Ill.
Probst Lumber Co., Cincinnati, Ohio.
Read Bros., Limited, Toronto.
Stimson & Co., J. V., Owensboro, Ky.
Thoman Flinn Lumber Co., Cincinnati, Ohio.

MACHINE KNIVES
Canada Machinery Corporation, Galt, Ont.
Peter Hay Knife Company, Galt, Ont.
Simonds Canada Saw Co., Montreal, P.Q.
Yates Machine Co., P. B., Hamilton, Ont.

MITRE MACHINES
Canada Machinery Corporation, Galt, Ont.
Fay & Egan Co., J. A., Cincinnati, Ohio.

MITRE SAWS
Canada Machinery Corporation, Galt, Ont.
Fay & Egan Co., J. A., Cincinnati, Ohio.
Garlock-Walker Machinery Co., Toronto, Ont.
Simonds Canada Saw Co., Montreal, P.Q.

MULTIPLE BOXING MACHINES
Fay & Egan Co., J. A., Cincinnati, Ohio.
Nash, J. M., Milwaukee, Wis.
Reynolds Pattern & Machine Co., Moline, Ill.

PANELS
Waetjen & Co., George L., Milwaukee, Wis.

PATTERN SHOP MACHINES
Canada Machinery Corporation, Galt, Ont.
Cowan & Co., Galt, Ont.
Fay & Egan Co., J. A., Cincinnati, Ohio.
Garlock-Walker Machinery Co., Toronto, Ont.
Jackson, Cochrane & Company, Kitchener, Ont.
Preston Woodworking Machinery Company, Preston, Ont.
Whitney & Son, Baxter D., Winchendon, Mass.
Yates Machine Co., P. B., Hamilton, Ont.

PLANERS
Canada Machinery Corporation, Galt, Ont.
Cowan & Co., Galt, Ont.
Fay & Egan Co., J. A., Cincinnati, Ohio.
Jackson, Cochrane & Company, Kitchener, Ont.
Petrie, H. W., Toronto, Ont.
Preston Woodworking Machinery Company, Preston, Ont.
Whitney & Son, Baxter D., Winchendon, Mass.
Yates Machine Co., P. B., Hamilton, Ont.

PLANING MILL MACHINERY
Canada Machinery Corporation, Galt, Ont.
Cowan & Co., Galt, Ont.
Fay & Egan Co., J. A., Cincinnati, Ohio.
Garlock-Walker Machinery Co., Toronto, Ont.
Jackson, Cochrane & Company, Kitchener, Ont.
Petrie, H. W., Toronto, Ont.
Preston Woodworking Machinery Company, Preston, Ont.
Whitney & Son, Baxter D., Winchendon, Mass.
Yates Machine Co., P. B., Hamilton, Ont.

PRESSES (Veneer)
Canada Machinery Corporation, Galt, Ont.
Jackson, Cochrane & Company, Kitchener, Ont.
Perrin & Co., Wm. R., Toronto, Ont.
Preston Woodworking Machinery Company, Preston, Ont.

PULLEYS
Fay & Egan Co., J. A., Cincinnati, Ohio.

TIME CLOCKS
International Time Recording Co., Toronto.

RED GUM LUMBER
Gum Lumber Manufacturers' Association.

RESAWS
Canada Machinery Corporation, Galt, Ont.
Cowan & Co., Galt, Ont.
Fay & Egan Co., J. A., Cincinnati, Ohio.
Garlock-Walker Machinery Co., Toronto, Ont.
Jackson, Cochrane & Company, Kitchener, Ont.
Petrie, H. W., Toronto, Ont.
Preston Woodworking Machinery Company, Preston, Ont.
Simonds Canada Saw Co., Montreal, P.Q.
Williams Machinery Co., A. R., Toronto, Ont.
Yates Machine Co., P. B., Hamilton, Ont.

RIM AND FELLOE MACHINERY
Fay & Egan Co., J. A., Cincinnati, Ohio.

RIP SAWING MACHINES
Canada Machinery Corporation, Galt, Ont.
Cowan & Co., Galt, Ont.
Fay & Egan Co., J. A., Cincinnati, Ohio.
Garlock-Walker Machinery Co., Toronto, Ont.
Jackson, Cochrane & Company, Kitchener, Ont.
Preston Woodworking Machinery Company, Preston, Ont.
Yates Machine Co., P. B., Hamilton, Ont.

SANDERS
Canada Machinery Corporation, Galt, Ont.
Cowan & Co., Galt, Ont.
Fay & Egan Co., J. A., Cincinnati, Ohio.
Garlock-Walker Machinery Co., Toronto, Ont.
Jackson, Cochrane & Company, Kitchener, Ont.
Nash, J. M., Milwaukee, Wis.
Preston Woodworking Machinery Company, Preston, Ont.
Yates Machine Co., P. B., Hamilton, Ont.

SASH, DOOR & BLIND MACHINERY
Canada Machinery Corporation, Galt, Ont.
Cowan & Co., Galt, Ont.
Fay & Egan Co., J. A., Cincinnati, Ohio.
Garlock-Walker Machinery Co., Toronto, Ont.
Jackson, Cochrane & Company, Kitchener, Ont.
Preston Woodworking Machinery Company, Preston, Ont.
Yates Machine Co., P. B., Hamilton, Ont.

SAWS
Fay & Egan Co., J. A., Cincinnati, Ohio.
Radcliff Saw Mfg. Company, Toronto, Ont.
Simonds Canada Saw Co., Montreal, P.Q.
Smith Co., Ltd., R. H., St. Catharines, Ont.

SAW GUMMERS, ALOXITE·
The Carborundum Co., Niagara Falls, N.Y.

SAW SWAGES
Fay & Egan Co., J. A., Cincinnati, Ohio.
Simonds Canada Saw Co., Montreal, Que.

SAW TABLES
Canada Machinery Corporation, Galt, Ont.
Fay & Egan Co., J. A., Cincinnati, Ohio.
Garlock-Walker Machinery Co., Toronto, Ont.
Jackson, Cochrane & Company, Kitchener, Ont.
Preston Woodworking Machinery Company, Preston, Ont.
Yates Machine Co., P. B., Hamilton, Ont.

SCRAPING MACHINES
Canada Machinery Corporation, Galt, Ont.
Garlock-Walker Machinery Co., Toronto, Ont.
Whitney & Sons, Baxter D., Winchendon, Mass.

SCREW DRIVING MACHINERY
Canada Machinery Corporation, Galt, Ont.
Reynolds Pattern & Machine Works, Moline, Ill.

SCROLL SAWS
Canada Machinery Corporation, Galt, Ont.
Cowan & Co., Galt, Ont.
Fay & Egan Co., J. A., Cincinnati, Ohio.
Garlock-Walker Machinery Co., Toronto, Ont.
Jackson, Cochrane & Company, Kitchener, Ont.
Simonds Canada Saw Co., Montreal, P.Q.
Yates Machine Co., P. B., Hamilton, Ont.

SECOND-HAND MACHINERY
Cowan & Co., Galt, Ont.
Petrie, H. W.
Williams Machinery Co., A. R., Toronto, Ont.

SHAPERS
Canada Machinery Corporation, Galt, Ont.
Fay & Egan Co., J. A., Cincinnati, Ohio.
Garlock-Walker Machinery Co., Toronto, Ont.
Jackson, Cochrane & Company, Kitchener, Ont.
Petrie, H. W., Toronto, Ont.
Preston Woodworking Machinery Company, Preston, Ont.
Simonds Canada Saw Co., Montreal, P.Q.
Whitney & Sons, Baxter D., Winchendon, Mass.
Yates Machine Co., P. B., Hamilton, Ont.

SHARPENING TOOLS
The Carborundum Co., Niagara Falls, N.Y.

SHAVING COLLECTORS
Sheldons Limited, Galt, Ont.

SHOOK BUNDLER
Neilson & Company, J. L. Winnipeg, Man.

STAINS
Ault & Wiborg, Toronto, Ont.

STEAM APPLIANCES
Darling Bros., Montreal.

SINGLE SPINDLE BOXING MACHINES
Canada Machinery Corporation, Galt, Ont.
Fay & Egan Co., J. A., Cincinnati, Ohio.
Ober Mfg. Company, Chagrin Falls, Ohio.

STAVE SAWING MACHINE
Whitney & Sons, Baxter D., Winchendon, Mass.

STEAM TRAPS
Canadian Morehead Mfg. Co., Woodstock, Ont.
Standard Dry Kiln Co., Indianapolis, Ind.

STEEL BOX STRAPPING
Acme Steel Goods Co., Chicago, Ill.

STEEL SHELL BOX BANDS
Acme Steel Goods Co., Chicago, Ill.

Peter Hay Machine Knives

Peter Hay knives are made of special quality steel tempered and toughened to take and hold a keen cutting edge. Hay knives are made for all purposes.

Have you our Catalogue?

Made in Canada

The Peter Hay Knife Co., Limited
GALT, ONTARIO

PRESSES

For Veneer and Veneer Drying

Made in Canada

William R. Perrin
Limited

Toronto

If You Stay in Business

you will eventually buy

National Dry Kilns

That will come in the natural progress of events. You should begin to investigate now.

We have a new catalog for you.

THE NATIONAL DRY KILN CO.
1117 East Maryland St., INDIANAPOLIS, Indiana

You

CONTRACTORS
and
SMALL MILL MEN

Just the machine you need
—ANY KIND OF POWER—

A Whole Mill in Itself
A GREAT TOOL.

Write for Catalogue and Prices NOW

J. L. NEILSON & CO.
Winnipeg, Man.

◁ ACME ▷

STEEL GOODS
BOX STRAPPING
Our Specialty

Acme Galv. Twisted Wire Strap

Acme Cold Rolled Steel Hoop.

Bright—Galv.—Lacquered

WRITE FOR CATALOG

Coil of 5000-ft. Acme Twisted Wire Strap

ACME STEEL GOODS CO.—MFRS.
2840 Archer Ave., Chicago.
Coristine Bldg., Montreal, Que.

"Canadian Woodworker" Buyers' Directory—Continued

SURFACERS
Canada Machinery Corporation, Galt, Ont.
Cowan & Co., Galt, Ont.
Fay & Egan Co., J. A., Cincinnati, Ohio.
Garlock-Walker Machinery Co., Toronto, Ont.
Jackson, Cochrane & Company, Kitchener, Ont.
Petrie, H. W., Toronto, Ont.
Preston Woodworking Machinery Company, Preston, Ont.
Whitney & Sons, Baxter D., Winchendon, Mass.
Williams Machinery Co., A. R., Toronto, Ont.
Yates Machine Co., P. B., Hamilton, Ont.

SWING SAWS
Canada Machinery Corporation, Galt, Ont.
Cowan & Co., Galt, Ont.
Fay & Egan Co., J. A., Cincinnati, Ohio.
Garlock-Walker Machinery Co., Toronto, Ont.
Jackson, Cochrane & Company, Kitchener, Ont.
Preston Woodworking Machinery Company, Preston, Ont.
Simonds Canada Saw Co., Montreal, P.Q.
Yates Machine Co., P. B., Hamilton, Ont.

TABLE LEG LATHES
Fay & Egan Co., J. A., Cincinnati, Ohio.
Whitney & Sons, Baxter D., Winchendon, Mass.
Yates Machine Co., P. B., Hamilton, Ont.

TABLE SLIDES
Walter & Co., R., Wabash, Ind.

TENONING MACHINES
Canada Machinery Corporation, Galt, Ont.
Fay & Egan Co., J. A., Cincinnati, Ohio.
Garlock-Walker Machinery Co., Toronto, Ont.
Jackson, Cochrane & Company, Kitchener, Ont.
Preston Woodworking Machinery Company, Preston, Ont.
Yates Machine Co., P. B., Hamilton, Ont.

RECORDERS (TIME)
International Time Recording Co., Toronto.

TRANSMISSION MACHINERY
Dodge Mfg. Co., Ltd., Toronto, Ont.

TRIMMERS
Canada Machinery Corporation, Galt, Ont.
Fay & Egan Co., J. A., Cincinnati, Ohio.

Garlock-Walker Machinery Co., Toronto, Ont.
Preston Woodworking Machinery Company, Preston, Ont.
Yates Machine Co., P. B., Hamilton, Ont.

TRUCKS
Sheldons Limited, Galt, Ont.
National Dry Kiln Co., Indianapolis, Ind.

TURNING MACHINES
Fay & Egan Co., J. A., Cincinnati, Ohio.
Garlock-Walker Machinery Co., Toronto, Ont.
Whitney & Sons, Baxter D., Winchendon, Mass.
Yates Machine Co., P. B., Hamilton, Ont.

UNDER-CUT SELF-FEEDING FACE PLANER
Fay & Egan Co., J. A., Cincinnati, Ohio.
Garlock-Walker Machinery Co., Toronto, Ont.
Jackson, Cochrane & Company, Kitchener, Ont.

VARNISHES
Ault & Wiborg Company, Toronto, Ont.

VARNISH SPRAYERS
DeVilbiss Mfg. Co., Toledo, Ohio.

VENEERS
Acme Veneer & Lumber Co., Cincinnati, Ohio.
Barnaby, Chas. A., Greencastle, Ind.
Central Veneer Co., Huntington, W. Virginia.
Dominion Mahogany & Veneer Company, Montreal, Que.
Walter Clark Veneer Co., Grand Rapids, Mich.
Gorham Bros. Co., Mt. Pleasant, Mich.
Hartzell, Geo. W., Piqua, Ohio.
Haughton Veneer Company, Chicago, Ill.
Hoffman Bros. Company, Fort Wayne, Ind.
Geo. Kersley, Montreal, Que.
Nartzik, J. J., Chicago, Ill.
Ohio Veneer Company, Cincinnati, Ohio.
Penrod Walnut & Veneer Co., Kansas City, Mo.
Roberts, John N., New Albany, Ind.
Toronto Veneer Company, Toronto, Ont.
Waetjen & Co., George L., Milwaukee, Wis.

VENEER TAPE
Rexford Gummed Tape Co., Milwaukee, Wis.

VENTILATING APPARATUS
Sheldons Limited, Galt, Ont.

VENEERED PANELS
Hay & Company, Woodstock, Ont.

VENEER PRESSES (Hand and Power)
Canada Machinery Corporation, Galt, Ont.
Garlock-Walker Machinery Co., Toronto, Ont.
Jackson, Cochrane & Company, Kitchener, Ont.
Perrin & Company, Wm. R., Toronto, Ont.

VISES
Fay & Egan Co., J. A., Cincinnati, Ohio.
Simonds Canada Saw Company, Montreal, Que.

WAGON AND CARRIAGE MACHINERY
Canada Machinery Corporation, Galt, Ont.
Fay & Egan Co., J. A., Cincinnati, Ohio.
Whitney & Sons, Baxter D., Winchendon, Mass.
Yates Machine Co., P. B., Hamilton, Ont.

WOOD FINISHES
Ault & Wiborg, Toronto, Ont.

WOOD PULLEYS
Dodge Mfg. Co., Ltd., Toronto, Ont.

WOOD TURNING MACHINERY
Canada Machinery Corporation, Galt, Ont.
Garlock-Walker Machinery Co., Toronto, Ont.
Waymoth & Co., A. D., Fitchburg, Mass.
Yates Machine Co., P. B., Hamilton, Ont.

WORK BENCHES
Fay & Egan Co., J. A., Cincinnati, Ohio.

WOODWORKING MACHINES
Canada Machinery Corporation, Galt, Ont.
Cowan & Co., Galt, Ont.
Elliot Woodworker Limited, Toronto, Ont.
Garlock-Walker Machinery Co., Toronto, Ont.
Jackson, Cochrane & Company, Kitchener, Ont.
Preston Woodworking Machinery Company, Preston, Ont.
Reynolds Pattern & Machine Works, Moline, Ill.
Williams Machinery Co., A. R., Toronto, Ont.

WHEELS, Grinding, Carborundum and Aloxite
The Carborundum Co., Niagara Falls, N.Y.

No. 2 Standard Type Automatic Screw-driving Machine.

The Only Way to Drive Screws

There is a stage-coach method and a modern method of doing everything. When it comes to driving screws the only modern way is with an Automatic Screw Driving Machine, which means greater and better production and greater profit.

The Reynolds
Automatic Screw-Driving Machine

is guaranteed to produce results. It is no experiment. The No. 2 Standard Type, shown herewith, is particularly suited for placing the screws in the straps of shrapnel boxes. It may be fitted to drive, without change or adjustment, any three consecutive sizes flat or round head, or two sizes filister head, No. 6 to 14 inclusive, wood or machine screws, or three sizes flat or round head, or two sizes filister head machine screws up to No. 20. All the above up to 1½ inches long if smaller than No. 10, or up to 1¾ inches long, No. 10 or larger. Fitted with boring attachment if desired.

Many Machines in Use in Canada.

Ask for our Catalogue.

REYNOLDS PATTERN & MACHINE CO.
101 and 103 Third Ave., Moline, Ill., U.S.A.

Making TABLE-SLIDES is a Specialty Business

For more than TWENTY-FIVE YEARS we have made TABLE SLIDES exclusively. Our Factory is equipped with Special Machinery which enables us to make SLIDES,— BETTER and CHEAPER than the furniture manufacturer.

Canadian Table makers are rapidly adopting WABASH SLIDES

Because { They ELIMINATE SLIDE TROUBLES / Are CHEAPER and BETTER

Reduced Costs

Increased Out-put

BY USING

WABASH SLIDES

MADE BY

B. Walter & Company
Wabash, Ind.
The Largest EXCLUSIVE TABLE-SLIDE Manufacturers in America
ESTABLISHED-1887

Just What You Need

There's no necessity now for you to wish for a corrugated joint fastener machine which will rise fully to your expectations; you can get one.

The Saranac Corrugated Joint Fastener Machine

is the absolute realization of your wish. This illustration gives you an idea of what it looks like, but you can have no conception of its efficiency, economy, speed and reliability until you get it working for you.

It is substantially constructed, and the workmanship and materials are of the very best. Particularly suitable for manufacturers of sash, doors, blinds, screens, porch columns, boxes, etc., etc.

Made in various types for all conditions.

Write for further particulars and prices

Saranac Machine Co., SOLE MAKERS
Benton Harbor, Mich.

Carborundum Brand Garnet Paper and Cloth are Uniformly Graded—Absolutely so

YOU can always depend upon Carborundum Brand Garnet products—they are uniformly coated with uniformly graded North River Garnet. There is never any coarse grain to leave heavy deep scratches on the work. The paper or cloth will not fill because it is uniformly coated with free, clean cutting grain.

Carborundum Brand Garnet Products are Uniform—Absolutely so

May we send you sample sheets for trial?

THE CARBORUNDUM COMPANY
NIAGARA FALLS, N. Y.
NEW YORK CHICAGO PHILADELPHIA BOSTON CLEVELAND CINCINNATI
PITTSBURGH GRAND RAPIDS MILWAUKEE

MATTISON
"Automatic"
The Standard Shaping Lathe
OVER 2000 IN USE

Courtesy of
Robbins Table Co.
Owosso, Mich.

This plant has been a user of the Mattison Lathe for several years, and it has been the means of saving them thousands of dollars in the manufacture of their round, square and octagonal turnings.

And Now Come the Queen Anne Styles

Hundreds of concerns have utilized the Mattison Lathe for handling their round, square and octagon Period Turnings. It is the accepted method for that line of work. Now come the Queen Anne styles to still further increase the usefulness of this versatile machine.

Perhaps you are among those who held up installing a Mattison Lathe, because you thought you would squeeze through somehow until the styles took a change, by turning your rounds in a hand lathe and possibly buying your squares and octagons outside.

But Turned Styles are still steady sellers, and you'll be making them along with the new Queen Anne designs. It's a "puttery" job to whittle out Queen Anne legs by hand methods.

Why not take full advantage of this new design by installing a Mattison Lathe, which at the same time will enable you to lower production costs on your other turned work?

Remember, the Mattison Lathe isn't a special machine for one kind of turning. It's an all 'round turning method, and equips you not only for today's but for future patterns, unthought of now.

Ask to have our practical man call and tell you how the Model "B" will solve your problems in turned work, once and for all.

There is no other Shaping Lathe like the Mattison
and it is built only by

C. Mattison Machine Works, 883 Fifth St., Beloit, Wisconsin

Volume Sixteen   TORONTO, DECEMBER, 1916   Number Twelve

CANADIAN WOODWORKER
and
Furniture Manufacturer

Watch that Output

For better surfacing and more of it buy a
Whitney Motor-Driven
Single or Double Surface Planer

"Make-it-pay" is the religion of modern business. We can tell you concisely, if you write, why a Whitney Planer is going to pay dividends in the form of easily-won profits from the day it is installed in your plant— our telling will be of double interest to you because it will be backed up with statements by leading concerns in your own line who looked for the best planing machine value and found it in the Whitney Planer.

We will describe various equipments, applicable to Whitney Planers, whether Belt Driven or Motor Driven, whether Single or Double Surfacer, that will make the machine you buy especially efficient for handling your planing—in some cases we advise square cylinders carrying two or four ordinary knives, for other work, round cylinders with four thin, high-speed steel knives, sectional rolls and steel sectional chip breakers or flexible steel chip breakers.

May we send you the details?

Baxter D. Whitney & Son, Winchendon, Mass.

California Office: Berkeley, California
H. W. Petrie, Limited, Toronto, Ont., Agents for Ontario

Selling Representatives: Henry Kelly & Co., 26 Pall Mall, Manchester, England

Power!

Who Pays the Power Bill?
Who Pays for Pulleys?

YOU—the Management—pay for the power that drives your factory. And, remember—you pay for the power you *waste* as well as for the power you use. Exhaustive scientific tests prove conclusively that the actual cost per year for running *metal pulleys* as against *wood* pulleys averages *$5.00 per pulley.*

Dodge Wood Split Pulleys cost 40 per cent. less than metal pulleys and are 50 per cent. more efficient. They help you cut the cost of your power bills. Next time you require pulleys see that the "requisition" calls for *Dodge Wood Split Pulleys* and not for steel pulleys.

DODGE

WOOD SPLIT PULLEYS

DODGE MANUFACTURING CO., Limited, TORONTO

Every buyer should send for this book mailed Free on request

A SCIENTIFIC TEST

14

GARLOCK-WALKER MACHINERY CO.
LIMITED
32 FRONT ST. WEST   **TORONTO**   **TELEPHONE MAIN 5346**

Do You Manufacture Furniture?

If so here is the very latest and most up-to-date **SURFACER** for your work—the best, the cheapest in the long run.

Feeds up to 85' per minute.

Detachable Side Clamping Cylinder Boxes.

Round Cylinder

Knife Setter and Jointer.

In Feed Rolls and Chipbreaker either solid or sectional.

Feed Rolls of large diameter—all powerfully driven.

Bed raises and lowers on long inclines.

American No. 444 Single Furniture Planer

Heavy   Strong   Durable

Does the best work at minimum expense

Let us take this up with you in detail—as well as offer you other tools of like calibre for your work.

WRITE FOR PRICES AND DELIVERIES

Garlock-Walker Machinery Co., Limited

Canadian Representatives

American Wood Working Machinery Co.
ROCHESTER, N.Y.

M. L. Andrews & Co.
CINCINNATI, OHIO

METAL and WOODWORKING MACHINERY of all Kinds

Sanders!

3 and 2 Drum Roll Feed

3 and 2 Drum Endless Bed

Belt—all kinds

Single Drum

Drum and Disc

Double Disc

Double Spindle

Spindle and Disc

Flexible Arm

etc.

—the Fay-Egan "Lightning" line is the most complete made.

Put your sanding problems up to us—our line of sanders is complete—we can and will recommend the type best suited to your individual needs.

Every Fay-Egan Sander is a unit of highest efficiency, developed from a thoroughly practical design, by skilled mechanics with unlimited manufacturing facilities and highest grade materials.

Machine sanding is one of the greatest economizers in modern woodworking— are you getting the benefit?

Write for Bulletins covering any or all kinds of Sanders.

J. A. Fay & Egan Co.

Established 1830

World's Oldest and Largest Manufacturers of Woodworking Machinery.

153-173 West Front St.   -   Cincinnati, Ohio

Ready to Glue—Without Further Preparation

That's the way core stock comes from the Yates Type G-2 Edging and Jointing Saw. Smooth enough and true enough to glue at once. Saves the cost of an extra run through a jointer—money you can count right into your pocketbook. Prepares your stock for rough gluing three times as fast as it can be done on a hand jointer, and at a fraction of the cost.

The Type G-2 can be used for all kinds of edg-ing and ripping. The saw line is kept perfectly straight, automatically permitting operation along the very edge of bad or shaky spots without danger to saw, stock, or operator. It is particularly adapted to the salvage of usable portions of low grade stock. Although before the trade but a short time the Type G-2 has dozens of well-satis-fied users, some of whom have installed additional saws like the first.

We have a circular describing and illustrating this new saw, and its operation.   Sent free without obligation or expense of any kind.

P. B. Yates Machine Co. Ltd.

HAMILTON, ONT. CANADA

Successors to THE BERLIN MACHINE WORKS -- U. S. Plant: Beloit, Wis.

New Straight Edge
Ripping and Jointing Machine

With saw cutting from below—the only mechanical way.
Double chain and idle pressure roll feed.

Examine the Special Features of this machine.

Prevents any twisting on either long or short stock.

Saw arbor easily accessible for oiling.

Double feed chain of flat milled links, each link removable and interchangeable.

Instantaneous adjustment of all rolls easily made and automatically locked at any point.

Spring counterbalance makes roll adjustment easy.

Saw blade easily accessible by removing plate.

Chain adjustable above table for rough, crooked, or finished stock.

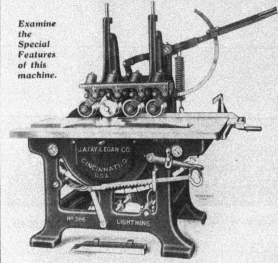

No. 386 Straight Edge Ripping and Jointing Machine.

Saves 33⅓ to 50 per cent. in Wages, 10 per cent. in lumber and also uses up your waste stock.

NOTE—**Saw cuts from below** and all saw dust and slivers are drawn down and away from the saw and do not scatter over the work and get into the operator's *eyes, ears, nose and mouth*, as in the case where saw cuts from above.

Circulars fully describing this new machine will be gladly mailed to all interested parties.

Jackson, Cochrane & Company
KITCHENER, ONTARIO

We are the Original Builders of the
Chain Feed Straight Edger and Rip Saw

of the illustrated type. Nearly one hundred of the No. 121 size have been placed throughout the United States and foreign countries.

The machine has been demonstrated to show 100% increase in production over the ordinary type of rip saws and edgers in common use, and, at the same time, do a much better grade of work. It eliminates the jointing operation entirely on core stock where gluing surfaces are required. **Can you think of a better investment to increase efficiency?** We have recently built a smaller machine of similar design, and our increased manufacturing facilities enable us to offer these machines at attractive prices. In giving you a choice of two machines at different prices we are sure to meet your requirements.

No. 121 Patent Chain Feed Straight Edger and Rip Saw.

Write for Descriptive Circular and Quotations to

The Defiance Machine Works
DEFIANCE, OHIO, U.S.A.
Everything in Woodworking Machinery — Over Five Hundred Designs.

New Five-Spindle Boring Machine

For boring the 5 holes in the diaphragms of the Bethlehem Shell Box in one operation.

Diaphragms with straight holes can be bored two in one operation.

Machine is heavily constructed, has heavy strong bearings and spindles, which are driven by spiral cut gears. It is specially designed for this work and has a very strong and steady feed. Capacity is very large and depends mostly on the operator's willingness for work.

We also make special boring cutter heads for both the straight and taper holes. They use much less power and do more and better work than the usual style of bits used.

We will mail blue prints of this machine and the heads to any interested parties.

JACKSON, COCHRANE & CO.   -   KITCHENER, ONT.

THE COMPLETE LINE

Top Bearing
PIANO CASTERS
Heavy Steel

These Casters are sturdy, reliable and superior in every way.
The Top Bearing Socket enables them to swivel easily under the heaviest pianos.

Roller Bearing
VICTOR CASTERS

The reliable
ACME CASTERS

have always been distinctive.

Our Canadian representative is at your service.

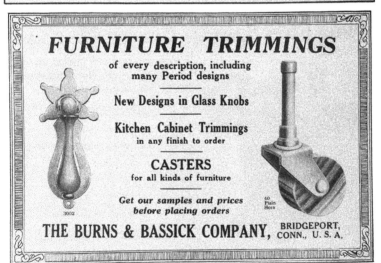

No. 161—PIANO CASTER—Top Bearing

FOSTER MERRIAM & CO.   MERIDEN, CONN.
Canadian representative—F. A. SCHMIDT, Kitchener, Ont.

FURNITURE TRIMMINGS
of every description, including many Period designs

New Designs in Glass Knobs

Kitchen Cabinet Trimmings
in any finish to order

CASTERS
for all kinds of furniture

Get our samples and prices before placing orders

3002

40 Plain Horn

THE BURNS & BASSICK COMPANY,   BRIDGEPORT, CONN., U. S. A.

"Treat your machine as a living friend."

Don't Waste Time

for which you are paying big money by having a Machine in your plant which is not giving you a maximum of production every day.

No. 162 High-Speed Ball-Bearing Shaper

SEE THIS MACHINE!

It will turn out 100% MORE WORK than the ordinary shaper which you are now using. That's our claim; twice the amount of work and much easier on the operator. Is it worth your while to cut your shaper cost in two? If so, drop us a line for more particulars.

THE CHEVROLET MOTOR CO. ARE USING 6 OF THESE MACHINES IN THEIR PLANT AT OSHAWA

DIMENSIONS, WEIGHTS and SPEEDS

Floor Space, 120 in. x 58 in. Net Weight, 2650 lbs. Shipping Weight, 2750 lbs. Tight and Loose Pulleys on Counter-shaft, 10 in. x 6 in. Driving Pulleys on Countershaft, 18 in. x 5 in. Idler Pulleys, 11 in. x 5 in. R.P.M. Countershaft, 1200. Pulley on Spindle, 3½ in. x 6 in. R.P.M. Spindle, 7000. Diameter Spindle, 2 3/16 in. Spindle Centres, 28 in. Table, 58 in. x 41 in. Diameter Spindle Top, 1¼ in. Horse Power Required, 3. Front Projection, 17 in. Height, 35 in.

"The Machine you want in your plant is not the cheapest machine to buy but the machine that will save you the most money to use."

The Preston Woodworking Machinery Co. Limited, Preston, Ont.

Alphabetical List of Advertisers

Aberdeen Lumber Company 17
Acme Steel Goods Company 61
Acme Veneer & Lumber Company 46
American Hardwood Lumber Company 13
Atlantic Lumber Company 17
Ault & Wiborg 39

Baird Machinery Company 52
Banning, Leland B. 11
Barnaby, Chas. A. 47
Beach Furniture Company 62
Brown & Company, Geo. C. 12
Burns & Bassick 8
Butns & Knapp Lumber Company 49

Canada Machinery Corporation ....
Canadian Morehead Mfg. Company 59
Carborundum Company 63
Central Veneer Company 43
Chapman Double Bearing Company 57
Churchill-Milton Lumber Co. 13
Clark & Son, Edward 10
Clark Veneer Company, Walter 46
Cowan & Company 56

Darling Bros. 55
Decorators Supply Company 51
Defiance Machine Works 7
Dermott Land & Lumber Company 14
De Vilbiss Mfg. Company 40
Dodge Manufacturing Company 2
Dominion Mahogany & Veneer Company 18

Elliott Woodworker 54
Edwards Lumber Company, E. L. 15

Farrin Lumber Company, M. B. 13
Fay & Egan Company, J. A. 4
Foster Merriam Company 8

Galbraith & Company 40
Garlock-Walker Machinery Company 3
Gorham Brothers 55
Grand Rapids Veneer Works 50
Graves, Bigwood Company 53
Gum Lumber Manufacturers' Assn. 12

Hartzell, Geo. W. 14
Haughton Veneer Company 47
Hay & Company 80
Hay Knife Company, Peter 63
Hoffman Bros. 48
Hoyt Metal Company 57
Huddleston-Marsh Mahogany Co. 13

Jackson, Cochrane & Company 6

Kersley, Geo. 46

Lamb-Fish Lumber Company 15
Long-Knight Lumber Company 14

Martin-Barriss Company 49
Mattison Machine Works, C. 64
Morden, E. P. 17
Mowbray & Robinson Company 15

Nartzik, J. J. 46
Nash, J. M. 59
National Dry Kiln Company 61
Neilson & Company, J. L. 61

Ohio Veneer 43

Paepcke Leicht Lumber Company 18
Penrod Walnut & Veneer Co. 47
Perfection Wood Steaming Retort Co. ...
Perkins Glue Company 56
Perrin, William R. 61
Petrie, H. W. 52
Pickrel Walnut Company 10
Preston Woodworking Machinery Co. 9
Probst Lumber Company 33

Radcliff Saw Company 54
Read Bros. 55
Rexford Gummed Tape Company 58
Reynolds Pattern & Machine Co. 62
Ritchey Supply Company 55
Roberts Veneer Company 45

Sadler & Haworth 51
Saranac Machine Company 64
Shafer Lumber Company, Cyrus C. 17
Sheldons, Limited 59
Signal Systems Company 52
Simonds Canada Saw Company ....
Skedden Brush Company 89
Smith Company, R. H. 11
Spencer, C. A. 13
Stimson, J. B. 15

Toronto Veneer Company 44

Waetjen & Company, Geo. L. 46
Walter & Company, B. 63
Waymoth, A. D. 55
Whitney & Son, Baxter D. 1
Wood-Mosaic Company 45

Yates Machine Company, P. B. 5

Edward Clark & Sons
Toronto

CANADIAN HARDWOODS—winter cut Birch, Basswood, Maple

Ontario and Quebec Stocks, East or West, rail or water shipment
direct from mill to factory.

Our stocks in all kinds of Hardwoods at the present moment are so thoroughly depleted that we are unable to supply a comprehensive stocklist, and for the time being we can only solicit your enquiries for what you may require in the hope that we may be able to quote you from the items we have remaining on hand.

Wishing You the Compliments of the Season

We Commend to Your Notice:

Our lumber is reliable, our grades uniform, our shipments prompt, and receive personal supervision; our prices consistent. We have had long experience. We desire to serve.

R. H. Smith Co.
ST. CATHARINES, ONT.   Limited

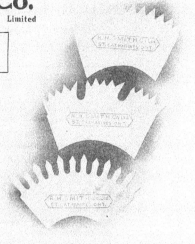

"Arrow Head" Quality

The scarcity of labor deals directly
with the manufacturer: Arrow Head
Saws and Knives increase production,
and save labor, made from Vanadium
Steel by skilled workmen and are
guaranteed to do 20% more than any
other Saw. The Arrow Head Planer
Tooth Saw is the smoothest and most
satisfactory saw of its kind for Shell
Box, Toy and other kinds both Rip
and Cross Cut.

*Eventually Arrow Head
Saws and Knives. Why
not your next order?*

I Manufacture
Kentucky River Quartered
and Plain White Oak

Wanted by all furniture manufacturers for its

Soft Texture and Even Color

I also manufacture

Soft Texture Plain Red Oak
Walnut and Poplar

*Canadian
Representative:*

Ash and Hickory

A. E. KLIPPERT

Gum and Chestnut

Cuban and Mexican Mahogany

11 Laxton Ave.,

TORONTO

*Wire (at my expense) or write for
prices for anything in above list*

Phone Parkdale 1354

Leland G. Banning, CINCINNATI OHIO

Selected Red and Sap Gum

St. Francis Basin Band Sawed Stock

KRAETZER-CURED   STRAIGHT AND FLAT

> "The car of special width gum box boards that you shipped us that were specified 17 inches in width, we found upon their arrival that this car of gum box boards was the best car that ever came into our yards. We have found every board to have the correct width and we failed to find a warped board in the car, and the grade and measurements were perfectly satisfactory. We were highly pleased with the kind of lumber that this car contained as every board in the car will make box boards.
>
> Thanking you very much for the kind of lumber that you shipped us in this car, we remain—"
>
> (Name furnished on request.)

YOU CAN OBTAIN THE SAME RESULTS

Send us a list of your requirements.   —   Delivered Prices will be cheerfully quoted.

GEO. C. BROWN & CO.

Mill—PROCTOR, Ark.   Main Office—MEMPHIS, Tenn.

We specialize in Tennessee Aromatic Red Cedar

"Made in America"

RED GUM

("America's Finest Cabinet Wood")

IS THE "LIFE-SAVER"

of the furniture trade in these disturbing days. *Gum's only competitors* as a pre-eminent furniture and trim wood are the woods formerly imported but *now increasingly difficult to get.*

It is to the credit of American furniture *manufacturers, retailers* and *users* that they now *acknowledge the advantage to them* of having available (and of *relying* on) so wonderfully beautiful and dependable a furniture wood as American *Gum.*

It is to the credit of *Gum* manufacturers that they are *not* taking unfair advantage of *the BOOM IN Gum* to exact unreasonable prices.

Gum at present prices is *the one big, best buy* in the whole furniture world, for maker, seller and consumer. There's money in Gum for *You.*

A convincing fact as to RED GUM'S growing prestige is that it enjoys increasing demand among the best judges of value—artistic value, utilitarian value and economic value. (There are no other kinds.) "GUM HAS COME."

Furniture Manufacturers and Furniture Retailers who wish detailed proof as to the *salability, dependability* and *commercial desirability* of *RED GUM,* are invited to correspond with

GUM LUMBER MANUFACTURERS ASSOCIATION

1314 Bank of Commerce Building, Memphis, Tenn.

Dry Spruce
and Birch

Good Stocks, Prompt Shipments, Satisfaction

C. A. SPENCER, Limited
Wholesale Dealers in Rough and Dressed Lumber

Offices—500 McGill Building

MONTREAL   -   -   Quebec

Churchill Milton Lumber Co.
Louisville, Ky.

We carry at
NEW ALBANY, IND.

Genuine Indiana Plain and Quartered Oak, Poplar

At GLENDORA, MISS.

Cypress, Gum, Cottonwood, Oak

SEND US YOUR INQUIRIES

We have dry stock on hand and can make quick shipment

American Hardwood Lumber Co.
St. Louis, Mo.

Large stock of—

Dry Ash, Quartered Oak Plain Oak and Gum

Shipments from — NASHVILLE, Tenn.,
NEW ORLEANS, La., and BENTON, Ark.

THE
M. B. FARRIN LUMBER
COMPANY
CINCINNATI   -   OHIO

Manufacturers

HARDWOODS

We have very large stocks of well assorted West Virginia Hardwoods on hand at Cincinnati and also at mills.

Plain Oak   Whitewood
Quartered Oak   Red Gum
Chestnut   Ash

KILN DRIED STOCK
A SPECIALTY

Huddleston-Marsh Mahogany Company, Inc.

Importers and Manufacturers

MAHOGANY
Lumber and Veneer
All Grades   All Thicknesses

*WE SPECIALIZE
IN
MEXICAN MAHOGANY*

33 West 42nd St.   2254 Lumber St.
NEW YORK, N.Y.   CHICAGO, ILL.

—Mills and Yards—
Long Island City, New York

Dermott Land and Lumber Co.

Manufacturers of
Southern Hardwoods

80 E. Jackson Blvd.
CHICAGO   -   U.S.A.

THE DERMOTT KIND
BAND SAWN SOUTHERN HARDWOODS

Quartered White Oak

4 cars 1 in. 1sts and 2nds.
5 cars 1 in. No. 1 Common.
1 car 1¼ in. 1sts and 2nds.
1 car 1¼ in. No. 1 Common.
½ car 1½ in. 1sts and 2nds.
1 car 2 in. 1sts and 2nds.
1 car 2 in. No. 1 Common.
1/3 car 2½ in. Com. and Bet.
1/3 car 3 in. Com. and Bet.

Plain Red Oak

5 cars 1 in. 1sts and 2nds.
10 cars 1 in. No. 1 Common.
2 cars 1¼ in. 1sts and 2nds.
2 cars 1½ in. No. 1 Common
2 cars 2 in. 1sts and 2nds.
1 car 2 in. No. 1 Common.

Plain White Oak

1 car ¾ in. 1sts and 2nds.
1/3 car ⅝ in. 1sts and 2nds.
1 car ¾ in. 1sts and 2nds.
10 cars 1 in. 1sts and 2nds.
10 cars 1 in. No. 1 Common.
2 cars 1¼ in. 1sts and 2nds.
4 cars 1¼ in. No. 1 Com.
2 cars 1½ in, 1sts and 2nds.
3 cars 1½ in. No. 1 Com.
2 cars 1¾ in. 1sts and 2nds.
1 car 2 in. 1sts and 2nds.
2 cars 2 in. No. 1 Common.
1 car 3 in. 1sts and 2nds.
½ car 4 in. 1sts and 2nds.

Gum

1 in. to 2 in. Plain Sawn Red.
1 in. to 2 in. Quarter Sawn Red.
1 in. to 2 in. Figured Plain and Quartered.
1 in. to 2 in. Sap.

All well cared for band sawn stock properly edged and trimmed. Nice assorted dry stock on hand at our modern band mill at Dermott, Arkansas.

Long - Knight Lumber Co.

Manufacturers and Wholesale Dealers
INDIANAPOLIS, IND., U.S.A.

HARDWOOD LUMBER

Walnut, Mahogany

and

American Walnut Veneers

SPECIAL:
5 cars 5/4 1 & 2 Pl. Red Oak

FOR SALE

Black Walnut Lumber

15,000 feet 4/4 No. 1 Common
20,000 " 4/4 No. 2 "
15,000 " 5/4 No. 1 "
15,000 " 5/4 No. 2 "
15,000 " 6/4 No. 2 "
40,000 " 8/4 No. 1 "
60,000 " 8/4 No. 2 "
30,000 " 9/4 No. 1 "

Will make special low prices on all of the above stock to move it quick. All Band Sawn, Equalized, and all thoroughly dry, ready for immediate shipment. Write for prices.

Geo. W. Hartzell
Piqua, Ohio

Plain and Quartered
Red and White Oak

and Other Hardwoods

Even Color   Soft Texture

We have **35,000,000 feet dry stock**
all of our own manufacture, from our own
timber grown in Eastern Kentucky

MADE (MR) RIGHT

OAK FLOORING

PROMPT SHIPMENTS

The Mowbray & Robinson Co., Inc.

GENERAL OFFICES
CINCINNATI, OHIO
Mills: Quicksand, Ky., West Irvine, Ky., Viper, Ky.

Canadian Representative:
N. H. FARNHAM
601 Elmwood Ave., BUFFALO, N. Y.

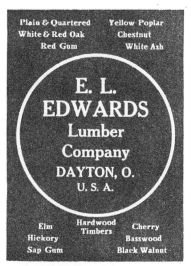

Plain & Quartered   Yellow Poplar
White & Red Oak   Chestnut
Red Gum   White Ash

E. L.
EDWARDS
Lumber
Company
DAYTON, O.
U. S. A.

Elm   Hardwood   Cherry
Hickory   Timbers   Basswood
Sap Gum   Black Walnut

J.V. Stimson & Co.

OWENSBORO, KENTUCKY

Manufacturers of

BAND SAWN
HARDWOODS

QUARTERED and PLAIN OAK
ASH
POPLAR
HICKORY
WALNUT
QUARTERED SYCAMORE

OAK and WALNUT
OUR SPECIALTY

We have on our Charleston
Yard a complete stock of

WHITE
and   **OAK**   PLAIN
RED   and
            QUARTERED

RED
and   **GUM**   PLAIN
SAP   and
            FIGURED

Can Make Immediate Shipment

Representative in Eastern Canada:
FRANK T. SULLIVAN
71 South Street
BUFFALO - New York

Lamb-Fish Lumber Co.
"The Largest Hardwood Mill in the World"
CHARLESTON, MISS.

All
American
Black
Walnut

A part of our immense stock of logs continually on hand.

We Handle Walnut
Exclusively

Let us have your inquiries on Squares, Lumber, Logs, Veneers
or any Special Stock.

Pickrel Walnut Co.
St. Louis, Mo.

CYRUS C. SHAFER LUMBER COMPANY

Box 457　　　　　　　　　　　　　　　　　　　　SOUTH BEND, INDIANA

Mills and yards — INDIANA,　TENNESSEE,　ARKANSAS,　MISSISSIPPI

WHITE ASH	Feet	PLAIN OAK	Feet
1 in. FAS and No. 1 and 2 Com.	260,666	1 in. FAS and No. 1 and 2 Com.	1,627,293
1¼ in. FAS and No. 1 and 2 Com.	98,016	1¼ in. FAS and No. 1 and 2 Com.	364,386
1½ in. FAS and No. 1 and 2 Com.	219,183	1½ in. FAS and No. 1 and 2 Com.	219,539
2 in. FAS and No. 1 and 2 Com.	746,966	2 in. FAS and No. 1 and 2 Com.	988,180
BASSWOOD		**QTR. RED OAK**	
1 in. No. 2 Common and Better	20,000	1 in. FAS and No. 1 Common	30,880
BEECH		1¼ in. FAS and No. 1 Common	8,500
1 in. No. 2 Common and Better	40,700	1½ in. FAS and No. 1 Common	22,500
ELM		2 in. FAS and No. 1 Common	3,000
1 in. FAS and No. 1 and 2 Com.	47,397	2½ to 4 in. FAS and No. 1 Common	3,500
1½ in. FAS and No. 1 Common	33,671	**QTD. WHITE OAK**	
2 in. FAS and No. 1 Common	116,782	1 in. FAS and No. 1 Common	203,528
2½ in. FAS and No. 1 Common	14,019	1¼ in. FAS and No. 1 Common	21,500
3 in. FAS and No. 1 Common	60,099	1½ in. FAS and No. 1 Common	20,500
GUM		2 in. FAS and No. 1 Common	29,500
1 in. FAS and No. 1 Common Red	26,000	**POPLAR**	
1 in. FAS and No. 1 Common Sap	169,690	⅝ in. FAS and No. 1 Common	17,200
1 in. Boxboards	38,905	1 in. FAS and No. 1 and 2 Common	739,736
HARD MAPLE		1¼ in. FAS and No. 1 and 2 Common	66,156
1 in. FAS and No. 1 Common	29,980	1½ in. FAS and No. 1 and 2 Common	238,550
SOFT MAPLE		2 in. FAS and No. 1 and 2 Common	114,874
1 in. FAS and No. 1 Common	20,700	2½ in. FAS and No. 1 Common	48,590
1½ in. FAS and No. 1 Common	9,340	3 in. FAS and No. 1 Common	58,599
2 in. FAS and No. 1 Common	12,000	4 in. FAS and No. 1 Common	12,040
		WALNUT	
		1 in. Walnut	18,750

We can surface or resaw to suit your requirements.　Our Traffic Manager traces all shipments.

E. P. MORDEN,　　　　　　　　　　　　　　PHONE NORTH 5903
CANADIAN SALES MANAGER　　　**Box 464**　　　**Toronto, Can.**

The Atlantic Lumber Co.

Manufacturers

Strictly Soft Textured

Quartered **White Oak**
and
Plain **Red Oak**

Chestnut, Poplar, Cherry

**Tennessee Scented
Red Cedar**

High Grades—Quick Shipments

110 Manning Chambers - **TORONTO**
Phone Main 6386

HEAD OFFICE, BOSTON, MASS.

Mills: KNOXVILLE, TENN.,　　WALLAND, TENN.
FRANKLIN, ARRINGDALE and BUTTERWORTH, VA.

ASH
GUM

All Grades

WIRE FOR PRICES

Quick Service Guaranteed

Seven Million Feet in our yard

Aberdeen Lumber Co.
Pittsburgh, Pa.

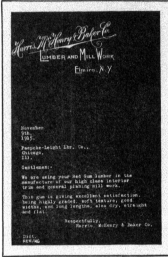

Of course it is true that

RED GUM

is America's finest cabinet wood—but
Just as a poor cook will spoil the choicest
viands while the experienced chef will turn
them into prized delicacies, so it is true that

The inherently superior qualities
of Red Gum can be brought
out only by proper handling.

When you buy this wood, as when you buy a new
machine, you want to feel that you have reason for
believing it will be just as represented.

We claim genuine superiority for our Gum. The
proof that you can have confidence in this claim is
shown by the letter reproduced herewith.

Your interests demand that you remember this
proof of our ability to preserve the wonderful qualities of the wood when you again want RED GUM.

Paepcke Leicht Lumber Co.

Conway Building   111 W. Washington St.
CHICAGO, ILL.

Band Mills: Helena and Blytheville Ark.; Greenville, Miss.

Buy Made in Canada Veneers

Dominion Mahogany
& Veneer Co., Limited

Head Office and Factory   Toronto Office and Warehouse
Montreal West, P.Q.   455 King St. West

THREE-PLY PANELS

7/8 in. Thick for Diaphragms

1/4 in., 5/16 in. and 7/24 in.

GUM and BIRCH VENEER

We Can Guarantee Prompt Shipments

Canadian Woodworker

A Monthly Publication in the Interest of the Woodworking Industry.
Reaches the factories producing interior finish, doors, sash, flooring,
woodenware, furniture, pianos, boxes, and general mill products.

Subscription, $1.00 a year; foreign $1.50.

Woodworker Publishing Company, Limited

Branches: Montreal, Vancouver,
Chicago, New York, London, Eng.

345 Adelaide St. West, Toronto

Phone Ade. 2700

Authorized by the Postmaster General for Canada, for transmission as second class matter.
Entered as second class matter July 18th, 1914, at the Post Office at Buffalo, N.Y., under the Act of Congress of March 3, 1879.

Vol. 16   December, 1916   No. 12

'Room for More Toy Manufacturers

One of the most pleasing features of Christmas
shopping this year is the "Made in Canada" signs
which decorate nearly all of the toy displays. Those
that were not made in Canada were made in the
United States and they fit in with our ideas of what
toys should be. The trashy German toys are gone—
let us hope forever. In their place we find toys that
are good to look at and durable; toys that will provide good healthy amusement and exercise for Canadian children and employment for Canadian workmen.. The assortment is large and the prices are within the reach of all.

Wooden toys and novelties occupy a leading position and manufacturers are to be congratulated on the
skill and ingenuity shown in their products. That
there is still room in this infant industry for other
manufacturers is the opinion of Mr. L. G. Beebe,
secretary of the Canadian Toy Association. Mr.
Beebe in a recent interview stated that practically all
Canadian toy manufacturers are away behind with
their orders, and that on a recent trip to New York he
found United States manufacturers in much the same
position. He also stated that there were very few
American toys coming into Canada, for the simple
reason that American manufacturers cannot produce
sufficient toys to satisfy the demand of their home
market.

When asked if the Canadian Toy Association
would hold a toy exhibition next spring similar to
the one held in the Royal Bank Building, Toronto,
in April, 1916, Mr. Beebe said he expected that they
would, but he had not had time to go into the matter very fully as yet, and that he would not be in a
position to announce the date for a few weeks at least.

Any person or firm contemplating going into the
manufacture of toys and novelties should endeavor
to have their goods on view at this exhibition. The
primary purpose of the exhibition last year was to
show toy buyers what lines were manufactured in
Canada, and to show manufacturers what lines were
made in Germany and other foreign countries, but incidentally some of the exhibitors secured some pretty
good orders.

As the majority of the manufacturers now have
their lines established, we think it would be safe to
assume that an exhibition next spring and succeeding years would be almost entirely for the purpose
of booking orders for delivery for the following Christmas. If this proved to be the case it would prove of
decided advantage to both the buyer—especially the
large buyer—and the manufacturer. The buyer of large
quantities would be sure of his deliveries and the
manufacturer would be able to distribute the work
evenly over the year, and thus be in a position to
take care of the requirements of smaller buyers who
are obliged to wait till nearer Christmas to place their
orders.

A "Safety First" Committee

The employees of a large furniture factory in Ontario recently decided that they would like to take
some steps to further the "safety first" movement in
their factory. They realized that they could do a
great deal by co-operating among themselves. They
therefore decided to hold a meeting and appoint a
safety committee to be composed of one man from
each department. No foremen or superintendents were
to be on the committee. It was to be composed entirely of working men and the idea behind the whole
movement was that each man should work as far as
possible to protect the life and limb of his fellow
workmen. When the men approached the management and told them what they had in mind they received the heartiest co-operation and were told to
hold their meeting in the company's time.

That this is a step in the right direction no person
can deny, and the men are to be congratulated on
their recognition of their duty to each other and on
their initiative in carrying the thing through. There
is also some credit due to the firm for taking the broad
viewpoint which they did. Some firms might have
been inclined to regard the movement as an interference either with their system of conducting the manu-

iacturing operations, or with the efficiency of the working force. We think, however, that there would be very little danger of any firm losing by such an arrangement. The mere fact that the men were working with the set intention of being as careful as possible would make them better workers. Also, the goodwill that a firm would receive when their employees knew that the firm had some interest in the safety of their employees would be worth something.

The human factor, or in other words, the working man, is receiving a great deal more consideration in industry today than he ever did before. For a number of years, while remarkable changes were being effected in the production of commodities by the installation of labor saving machines, manufacturers were inclined to pay too much attention to the machines and too little to the men that operated or were expected to operate them. Now that overhead costs are constantly creeping up in spite of the fact that all the latest labor saving machinery on the market has been installed, manufacturers have been looking around for some other means by which production might be increased and they have discovered that the working man is a very important factor in production.

They have therefore made observations and studies and found that he works best under certain conditions. For instance, in a bright clean well-ventilated factory with modern conveniences and where all machines and points of possible danger have been fitted with guards to insure reasonable protection for the workers, the latter will work as if work was a pleasure and thereby increase the production of the factory very considerably. Then again, the question of wages is an important one. The day of price haggling between employer and employee is fast disappearing. Remuneration of employees is being worked out on a scientific basis in numerous modern manufacturing plants. It is true that some employees think their employer should be a sort of banking institution paying them whatever money they choose to ask without expecting very much in the way of labor in return.

Also, unfortunately there are employers who are always striving to keep down the cost of their product by cutting the prices paid to their men. As a rule, however, these latter fail in their object because their good men leave them and an inferior type of workmen takes their place. This finally results in an inferior product and loss of business. The writer has often heard the statement made that "employers who attempt to reduce their costs by cutting their men's wages show their own incompetence." The reason given is "that a competent manager or employer knows that one of the first essentials of a successful enterprise is satisfied employees and that he will, therefore, endeavor to economize in some other way, preferably by speeding up production.

There is no rule in connection with the payment of wages or piece rates that would be applicable in every business. It is necessary for every firm to work out their own system. About the only thing to be borne in mind is that a broad minded policy is necessary. With up-to-date factories as mentioned above and wages fixed on a scientific basis so that capable and industrious workmen would be remunerated in accordance with their services, the working conditions would be pretty near ideal and should result in the factory producing close to one hundred per cent. efficiency.

An American Opinion

The "Furniture Manufacturer and Artisan" writes as follows: "The war is inculcating many lessons. The human race as a whole is learning that there are still many men willing to sacrifice everything to an ideal of patriotism. Medical science is face to face with new problems, in which the control of the body by something more than the mind is involved. The answer of young men to the call to arms in that great division of the British Empire adjacent to our own land presents at our very doors the two sharply contrasting phases of those who go forth rejoicing in their strength and those who return—if they come back at all—to take up occupations open to mutilated men. The Bishop of Kingston writes in the National Review of the drainage from Canada's schools and colleges. Many young men whom he questioned told him simply, 'I felt it was up to me to go.' It is the same spirit that animates our untrained volunteers who have offered themselves for service in Mexico. Into the statistics of the record it is not possible to enter. School after school sends its list of honors gained not in the quiet aula, but in the bloody arena. One theological college has sent every one of its undergraduates to the front. Canada has promised to supply 500,000 men. She has despatched 350,000. Every Canadian city is one vast house of mourning. Yet there is a pride that shines through tears and sees beyond a roadside grave somewhere in Flanders to the skies."

Get Advance Knowledge

Knowledge, without the question of a doubt, is always desirable. But what we might call advance knowledge is forty times more valuable. Nearly everyone can see beauty in an object after it is pointed out to them. That we might call knowledge. But very few people have the ability to see the beauty first—before anyone else comments on it. That is advance knowledge.

The trade magazine is constantly giving advance knowledge. It gives facts that form the foundation of future judgments—and strengthens the ability to know ahead of time.

Don't be satisfied to know things when everybody knows them. Know them a little ahead of hand. Be prepared with your knowledge.—Playthings.

Woodworking Conditions in Montreal, Que.

Staff Correspondence

"It all depends upon the war." This, in the main, is the view of Montreal firms connected with the woodworking industry, in commenting upon the outlook for 1917. The majority are just now busy—some are very busy—and the opinion generally expressed is that business will continue brisk as long as the war lasts. Take the box makers as an instance. The Imperial Munitions Board recently gave out a large number of orders which will keep the factories employed for some time, and further orders are looked for in the new year. In addition, the regular business is good, stimulated by commercial conditions, which are in a great measure affected by the higher wages paid to mechanics and laborers. The former are earning abnormal wages, consequent on munition work, while laborers' wages are nearly double those paid three years ago, and men cannot be obtained even at these figures. The ordinary box business has also been increased by the "dry" legislation in Ontario, an unlooked for result of the prohibition movement. Up to the time the "dry" legislation became operative, Ontario manufactured the boxes required for its liquor trade, but now that Montreal is the base of liquor supply for Ontario, a large and increased number of boxes is required for sending the liquor to Ontario, and these are being manufactured in Montreal.

Sash and door and interior trim manufacturers are rather quiet, but the outlook for 1917 is promising. There are signs that building will be more active next year. During the past two months the contracts awarded to Montreal contractors for work in the city and at outside points have shown a notable improvement; some of the contracts are very important, the largest being for munitions plants. Much of the work will be carried out during 1917, and will make for increased work in this branch of the woodworking industry. It is generally admitted that prices of lumber will be considerably higher, due to a scarcity arising from a short cut and from advanced operating costs. This applies to shell box lumber and that required for building purposes. Quotations have lately shown an appreciable rise, and an addition of five dollars a thousand feet next year may be looked for, if the opinions of wholesale lumbermen may be taken as a guide.

Montreal has no furniture factories in the ordinary sense of the words. A fair amount of high grade furniture is made to order; this department has been rather dull during the past year and prospects for 1917 are stated by manufacturers to be not over bright. Since the commencement of the war, business in high grade furniture has languished, while that in medium grades has lately shown a marked revival. The piano trade, too, is in excellent shape; evidently the mechanic with his larger earning power is putting a part of his money into furniture and pianos.

It may be said generally that the woodworking industry is short of suitable labor and that in certain directions the output would be larger if more men could be obtained. The prospect of securing these men is poor as long as the war lasts. Materials, too, look as if they will still further advance. Lumber, as already stated, is likely to be dearer, and hardware is pretty certain to go higher.

Heads of firms connected with the woodworking trade say that they are without definite data on which to form an opinion as to the course of business in 1917. So many prophecies have gone astray owing to unexpected happenings that it is impossible to forecast even approximately the trend of business. Many, however, believe that 1917 will witness an improvement, and those in the shell box section are hopeful of getting more benefit out of orders than they have received in 1916. Some of the contracts it is admitted have proved unprofitable, owing to the eagerness to get work, and the consequent tendering at unremunerative prices. Fortunately, conditions are now better, and there is a chance of business returning a little dividend.

Kiln-Drying Last Blocks

Among the experiments conducted by the United States Forest Products Laboratory, Madison, Wis., one has to do with the kiln-drying of maple last blocks. It takes nearly two years to properly air season maple for making shoe lasts, consequently it has been felt for some time that if methods could be developed for artificial drying that would prove commercially successful it ought to help out considerably. The experiments of the Forest Service indicate that by a comparatively slow process these last blocks can be kiln-dried—that is, they can be put in shape for use within two months—which is quite a saving in time when compared with the two years required to put them in shape by air-drying.

A Submarine for the Kiddies

Here is an illustration of a new English toy that appeared in a recent issue of "The Cabinet Maker" (London). It is a practical and substantial toy, and will undoubtedly make a strong appeal to Canadian children. It is styled a "nursery submarine," and, while it may not travel under water, it can be used either as a wagon or a rocker.

A reference to the scale in the illustration shows that the extreme length of the toy is approximately 3 feet 6 inches, and we believe that toy manufacturers wishing to try it out will have no difficulty in constructing it from the drawings. The illustration may offer a suggestion to some of our readers for a Christmas box for the kiddies or for some little one of their acquaintance.

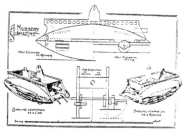

The Woodworking and Furniture Outlook for 1917

Increased Trade Anticipated throughout Canada notwithstanding Abnormal Conditions—Abundance of Money in the Country—Labor Shortage Restricting Production—Furniture Supply likely to be Unequal to the Demand—Scarcity of Canadian Hardwoods—The Cost of Materials Steadily on the Increase

The publishers of this journal asked the opinion of a number of manufacturers of general woodwork and furniture as to the outlook for 1917. Under prevailing conditions no man desires to assume the role of a prophet, but we have all given some thought to the probable trend of the particular business in which we are engaged. The letters published below may be taken as reflecting the conditions which will be likely to characterize woodworking and furniture production during 1917, subject, of course, to the influence of important war events.

* * *

The outlook seems to be bright. We feel that business will be good for us and we are making up stock so that we will be prepared for it. The unemployed are few, as may be noted by applicants for work. Wages are higher than for years past, and people seem to be purchasing freely. Everything is on the upward trend, yet the increase in our furniture business warrants us in looking forward with confidence to a busy time during 1917.

Beach Furniture, Limited,
Cornwall, Ont.

* * *

We see no reason why we should not have all the business we can take care of. The shortage of help is going to restrict the output and we think the furniture factories will be able to sell all of the goods they can make. There is also going to be trouble getting raw material. This will hold back the output some, and with ordinary sales the manufacturer will find trouble in making stock fast enough.

Eastern Townships Furniture Mfg. Co.,
Arthabaska, Que.

* * *

Regarding furniture conditions, our opinion is that trade in this particular line should be fairly good, but prices must go up in order to cover cost of production. As prices of labor and all materials going into the construction of furniture rise, the cost will rise; and in order to prevent loss to the manufacturer, prices must advance. Lumber, glass, varnishes, benzine, oil, leather, glue, nails, etc., are all advancing. Canadian lumber has been very scarce during 1916, and there are no indications of plenty during the coming year. Stocks of hardwood are very low at the present time and labor during the coming winter must inevitably be scarce and wages very high. Much of the lumber, such as elm, maple, beech and other hardwoods, has had to be brought in from the United States against a 7½ per cent. war tax. European glass has been largely stopped during the war, and American and Canadian manufacturers find it difficult to supply the demand. Gums and other commodities entering into the manufacture of varnishes are scarce and very high in price. Owing to the extensive use of gasoline and benzine, all petroleum products are likely to be high. Leather has become very high in price, owing to its extensive use during the war, also to the scarcity of hides. Glue stocks are also scarce and there may be a famine in this article. Money, of course, is plentiful, but banks and money lenders are likely to be very conservative for some time to come, especially until

peace is restored. Manufacturers of furniture are not likely to have their loans increased for some time.

We believe the market for furniture is likely to be good, but owing to cost of material and the scarcity and high price of labor, the supply may not be equal to the demand. The latter part of 1914 and the year 1915 were very hard years on manufacturers of furniture, and it will take a few years of good business to put them where they were previous to those years. For many years previous to 1916 costs of production were going up, and through the keenness in competition prices declined from year to year. During 1916 the prices became intolerably low, so that manufacturers in order to save themselves from utter ruin were obliged to put them up. The crop of 1915 put farmers, especially in the west, on their feet, and the manufacture of munitions, clothing for soldiers and other war materials has supplied Canada with large quantities of money. The purchasing power of the people, generally, has very much increased. We therefore look for 1917 to be a very fair year in the furniture business.

Ball Furniture Company,
Hanover, Ont.

* * *

Normal business for the woodworking industry in this section is not very encouraging for the year 1917. Construction work is at a standstill. Small orders are up to the standard from dealers, yet the heaviest part of our business is derived from the larger contractors and builders, and our construction department generally.

We have had to look to other sources for an output and have secured bookings for ammunition boxes, and foreign orders for the French and British governments. These are chiefly lumber requirements, and this business will diminish our yard stocks, and enable us to operate this winter on company-owned timber limits, both in hardwood and spruce. We anticipate to cut between three and four million feet.

With the present orders booked, we do not anticipate any falling off in our shop pay rolls for some months.

Rhodes, Curry Company, Limited,
Amherst, N.S.

* * *

Our plant is devoted to turning out special equipment for banks, offices, etc., in conjunction with school desks and opera chairs. In these lines the prospects for next year seem quite good, but the question of skilled labor will determine largely our output. If labor, especially skilled labor, is sufficiently available to enable us to handle to good advantage the orders we have and those in prospect, we look for a good season.

Canadian Office and School Furniture Co., Ltd.,
Preston, Ont.

* * *

It is mere conjecture to make predictions for all of 1917. This terrible war, of course, has a tremendous influence on, not only the furniture industry, but all industries, and while we know the Allies are going to win, we do not yet know what further sacrifices have to be made in the way of men and money, or how long the war will last.

With the large number of men that are still necessary

for war purposes it will mean still greater scarcity of skilled labor and a further advance in wages. In this respect it quite often happens that when there is a scarcity of men you cannot get them to work to full capacity, or get the proper production per man, therefore the cost of labor is increased proportionately.

Raw materials are continually advancing, and have already advanced very heavily in some lines. This specially applies in the case of upholstering supplies. All goods that are used for furniture coverings have increased 50 to 75 per cent. Ticking, denim and cotton filling materials 100 per cent. Everything in the way of steel and iron have fully doubled in price, and is still advancing. It means, of course, that furniture will have to be largely increased in price, unless the manufacturer is satisfied to lose money. While prices have been advancing they are still away below what they should be, as per reasons herein outlined. Lumber and veneers have advanced from 10 to 25 per cent.

The furniture industry is in a very healthy condition; there is a very heavy demand for all kinds of goods. Perhaps a portion of the demand has been due to the merchants buying in advance of their requirements, knowing that the prices are steadily advancing. The principal reason for the heavy demand is high wages and the absence of unemployment.

As there is no indication of the war being over in 1917, this condition of affairs will likely continue, as the various stores will no doubt keep heavy stocks to protect themselves, but just as soon as peace is in view the raw materials will come down in price. The dealers, knowing this, will likely reduce their stocks and there will be a lull in the furniture industry during the time of readjustment to normal conditions, which should not take long. Owing to the conditions of Europe they will still need heavy supplies from Canada, and with the renewal of immigration, Canada should continue to be very prosperous.

Gold Medal Furniture Mfg. Co.,
Toronto, Ont.

* * *

From the present outlook we believe that the supply of furniture will not equal the demand, as the manufacturers are working short-handed and with a large percentage of incompetent help. This, coupled with the shortage of raw materials, will, we think, curtail the output considerably. We believe that it will be a condition for the retailer to accept whatever the manufacturer has to offer, rather than select what he would prefer to have.

The Charles Rogers & Sons Company,
Toronto, Ont.

* * *

Relative to the outlook for the furniture industry for 1917, we scarcely feel ourselves capable of pronouncing on this subject further than to say that from our viewpoint, based on the present status of trade, and the disposition of the furniture dealers to place orders in advance for 1917 delivery, we are rather anticipating that the year of 1917 will show a marked improvement for furniture over the first six months of 1916. As most of the factories have had practically all they could handle during the last six months of the present year, we would not anticipate any improvement over this period during 1917.

The Geo. McLagan Furniture Co., Ltd.,
Stratford, Ont.

* * *

It is impossible for us to venture much of an opinion on the outlook for the woodworking industry in 1917. The building trade has been, for the past two years, in a very weak condition, although considerably helped out in recent months through the demand for additional factory accommodation. This will not continue indefinitely, and we must rely in the future more on house building for orders. It would be

foolish to predict to what extent this branch of the trade will develop, next year. Easy money conditions will exert a favorable influence, and we are hopeful that conditions will gradually improve.

Georgian Bay Shook Mills Limited,
Midland, Ont.

* * *

As yet we have not been able to form any definite opinion as to what the demand in our line will be for 1917, but do not think that it will be very active—chiefly on account of the scarcity of the supply of lumber and timber is beginning to be very apparent, and, as a consequence, the prices are advancing rapidly and further advances are in prospect. This fact, together with the labor situation, is bound to reduce the volume of building operations.

The above applies chiefly to house-building. The heavy timber department of our business has been more active this year than last, and we think the prospects for fairly active trade in this line for 1917 are good.

The Boake Manufacturing Company, Ltd.,
Toronto, Ont.

* * *

We expect the demand to be fully as great as 1916, with greater labor troubles and difficulty in getting materials. On the whole, would say the demand for furniture will exceed the supply, mainly due to labor shortage.

Knechtel Furniture Company,
Hanover, Ont.

* * *

We expect an increasing demand for furniture for 1917 at advancing prices, but as the labor conditions are very difficult, those who will buy early will be wise.

Kilgour Bros.,
Beauharnois, Que.

* * *

It is rather a hard matter to forecast the conditions of the furniture business for 1917. We believe all factories are quite busy at present, though none of them with orders very far ahead. The labor market is still very uncertain and no doubt help will be harder than ever to get. Prices, of course, are advancing, and we believe must continue to advance, so it is simply a matter of whether the public will pay the greatly increased prices that will be asked for furniture.

North American Bent Chair Co., Ltd.,
Owen Sound, Ont.

* * *

Re outlook for the furniture industry in 1917. From all indications of cost of raw materials of every description entering into our furniture line, it will be necessary to make a substantial advance in price on our product in order to make the price right, so that there is a legitimate manufacturing profit on same. The lumber, no doubt, will increase from 10 to 20 per cent., depending upon the demand. Hardware has already advanced from 10 to 25 per cent., shellac has increased about 60 per cent., stains from 35 to 50 per cent., glue from 50 to 75 per cent., and many of the articles are still on the way going up. The abnormal increase of various commodities will have a tendency to curtail the volume of the business.

The Globe Furniture Company, Ltd.,
Waterloo, Ont.

* * *

It is a little hard for us to give a correct opinion of conditions in 1917, owing to the fact that the sale of a number of our lines depends on whether we have an early or late spring, and also whether the weather is warm. Our lines are sold principally in the spring and summer, and if the weather is inclined to be cold and rainy, it means that our line does not sell as well. Therefore, a lot depends on the weather for us.

As far as building is concerned, from reports that we

have had from our Western representatives, there is practically nothing doing in this at all, therefore, as no building
is going on, no ladders are wanted. This, of course, does not
apply to the East. Taking it all in all, however, from the
way that we have received orders for 1917 shipment, we believe that business will be just as good if not better than
this year.

For future shipments Ontario is certainly keeping up to
the mark, and we believe Quebec will do the same, but we
cannot say now how the West is going to do. However,
we are making plans to sell just as much of our line for
1917 as we did for 1916, and we do not think we will be
disappointed if we can judge from the business that we
have received for the opening of spring.

Stratford Mfg. Company, Limited.
Stratford, Ont.

* * *

What is the outlook for 1917? This is a most difficult
question. Probably there is no one who can answer it. We
can only guess at what is coming in the business world. .

When the war started we adopted the slogan, "Business
as usual," in order to stimulate trade. At the present time
we need no such stimulant, as the demand for goods is
greater than the supply. Advance in price does not seem
to curtail the demand; on the contrary, it helps to increase it.

Historians tell us that great wars cause a boom, followed
by drop in values, and it is reasonable to suppose that after
this war prices will fall. The demand for luxuries will not
be so great, people will be more contented with the ordinary
things of life, and we are inclined to think that factories
will likely be occupied in making the commoner articles of
use.

Whether this war continues or whether it ceases, so
long as the Allies rule the seas, we are of the opinion that
there will be a demand for all the goods our factories can
make. Owing to the great destruction of property which
is taking place in Europe, and the large depletion of stocks
of all kinds of material in this country, we think that producers and manufacturers will have all they can do for years
to come. We do not think that local conditions will be very
different from conditions in general, but as far as we can
judge, the demand of this locality will continue good.

Labor is scarce and independent. During the winter
time it flocks to the cities. It will probably grow restless
again with the opening of spring, and will be a serious problem until the return of our soldiers from the front. '

S. Anglin & Company,
Kingston, Ont.

* * *

It is a fact that the furniture demand has materially revived during the last six months, and it is reasonably safe
to anticipate favorable conditions in our line—at least during the first six months of next year, as the surplus stock
both on the floor of the retailers and in the warehouses of
the manufacturers has been reduced to the minimum. Labor
is exceedingly scarce, so that manufacturers will be unable
to operate under normal conditions.

The situation of our materials is daily becoming more
serious, and may possibly influence the conditions considerably before we have reached the end of another year, as
costs are becoming abnormally high on some material. This
is most certain to prohibit the sale of some lines of furniture; that is, goods covered with leather and textiles.

The sum total of our idea is that conditions will be some
better in the furniture line during 1917.

The D. Hibner Furniture Company,
Kitchener, Ont.

* * *

Taking into consideration the general shortage of labor
in furniture factories of this country and the prospects that
the war will last at least another year or more, there will
be lots of means in this country for the people to continue
buying furniture of all descriptions. A continuance of the
shortage in labor will cause the production to be reduced.
The better class of furniture will always be made in Ontario
and the medium goods in the province of Quebec, as we are
closer to the raw materials and can get cheaper labor. We
are exceedingly busy at present, and are optimistic for the
year 1917, as far as the furniture manufacturers are concerned.

The Victoriaville Furniture Company,
Victoriaville, Que.

* * *

We believe the furniture industry is in a healthy condition and the outlook good for the year 1917. Canadians are
buying more in their own market, and the extra business
in our home factories which formerly went to the United
States is a big help to the industry. Shortage of labor will
no doubt curtail production, and with the increased cost of
both labor and material, furniture will have to advance in
price.

Bell Furniture Company, Limited,
Southampton, Ont.

* * *

We are of the opinion that furniture manufacturers will
get all the business that they can handle during 1917. However, owing to abnormal labor conditions, the output will
be far below factory capacity.

Baird Bros.,
Plattsville, Ont.

* * *

We submit that there never was a time in the history
of the Dominion than an opinion could be given with less
certainty of being correct. However, it appears to us that,
should the war continue, as appears likely, during the following year, the cost of labor and supplies will still further
increase; so will difficulties of transportation.

With regard to the character of the work likely to be
undertaken provided the war continues, we are of the opinion
that, notwithstanding the enormous number of factory buildings erected during the present year, a very large proportion
of the buildings erected next year will be of a similar class.
We consider it probable that high-class business blocks,
schools and residences will be erected in greater numbers
than during the present year. *

The Knight Bros. Company, Limited,
Burks Falls, Ont.

* * *

You ask me for a forecast of what conditions I think
will prevail during 1917 in our industry. It looks to me as
though our industry were in for all the trials and troubles
in the world to get out stuff. We are $30,000 or more behind on our orders today, and, with the shortage of help and
the willingness of people to buy, I see no chance of catching
up, and no hope in the immediate future of any relief. I
think you can print in your valued publication that the industries of Canada are certainly taxed to their utmost at the
present time.

F. D. Greenough,
Toronto Furniture Company, Limited,
Toronto, Ont.

A Miniature Catalogue

The Stow Manufacturing Company, of Binghamton, N.Y., are sending out to the trade a miniature of
their bulletin No. 102. This bulletin, which measured
9 x 6 inches and contained illustrations and descriptions of their very complete line of portable tools and
flexible shafting, has been photographed and reduced
to 4⅞ x 3¼ inches. The miniature is a convenient
size for quick reference, and is remarkable for the
clearness of detail of the type and illustrations.

Ontario Woodworking and Furniture Centres

Scarcity of Skilled Labor Dominating Feature—Majority of Manufacturers Behind with Orders—Building Picking Up in Smaller Towns.

A visit to several of the leading towns in Western Ontario found the woodworking and furniture factories busy. The answer invariably received from manufacturers was that they were short of skilled help and filled up with orders, but that, all things considered, they had no complaints to make. The planing mills that cater entirely to the building trade are in a little different position. Generally speaking, those that have had men leave them to enlist in the army have not found it necessary to replace them. But in one or two of the towns, particularly London and Brantford, planing mill men report that local building has been picking up considerably during the past few months.

Ham & Nott, Limited, of Brantford, are large manufacturers of refrigerators, screen doors, and window screens, and have recently added a line of kitchen cabinets. This firm are fortunate, in that they have not been affected very much by the labor shortage, despite the fact that over two hundred of their employees, or about the equivalent of the men on their payroll when the war started, have enlisted. They still have a large number of their original employees, of course, as a great many of those who have enlisted were men who were engaged from time to time after the war started. The reason why the labor shortage has not affected them to any extent is because they can use a great deal of unskilled labor, and this class of labor appears to be plentiful. The day the writer called fifteen men had applied for work. In their metal working department, where the refrigerators are lined, the winter months are the company's busiest time, as they are able to secure men who work outside in the summer and seek inside employment when the cold weather comes. They, therefore, to use their own expression, "make hay while the sun shines," and are ready with a stock to take care of spring and summer trade.

Schultz Brothers, of Brantford, are general contractors, and have a large plant, where they manufacture general planing mill products and ready-cut houses. They also make novelties, such as miniature billiard tables, crokinole boards, games, etc., and have lately added a line of poultry houses. As the company take contracts for buildings in different parts of the country, they are able, to some extent, to keep the planing mill end of their business going. The novelties and games have been much in demand for Christmas trade, and the company look forward to a fairly brisk demand for ready-cut houses for summer resorts, etc., in the spring.

John T. Grantham conducts a planing mill at Brantford. He has been doing a good business with the farmers in the surrounding territory. During the past few months local building has shown a marked improvement, owing to a scarcity of houses in the city, and he has been able to keep his staff working steadily.

Piano cases and talking machine cabinets are the products of the Brantford Piano Case Company, of Brantford. The company have been kept busy filling their orders. The manufacture of piano cases in a factory devoted exclusively to that work appears to have met with some of the prejudices that the manufacturers of built-up tops, etc., for furniture encountered during the early period of their labors. That the prejudice has been overcome to a large extent is evidenced by the fact that the Brantford Piano Case Company are making cases for a number of the leading Canadian piano manufacturers.

The Adams Wagon Company built a large addition to their plant at Brantford just previous to the beginning of the war, but as yet they have been unable to use the addition to full advantage on account of the shortage of labor. While they have had practically all the business they can handle, it has been principally for farm trade. During the last few months, however, they have noticed a decided improvement in city business.

"If you can tell us where we can get some good men we will have no complaints to make," was the answer received at the plant of the Valley City Seating Company, at Dundas, when they were asked how they found business in their line. The company make a specialty of high-grade church work and kindred lines, but since the war started they have manufactured large quantities of shell boxes. Lately, however, there has been considerable activity in their regular lines, and the labor question is their greatest trouble.

William Gerry & Sons, of London, operate a planing mill, and detail work for buildings is their chief product. Local building has picked up, but they furnish considerable material for out-of-town work. When asked if they had been affected by the shortage of labor, they said that two or three of their men had enlisted, but so far they had not found it necessary to replace them.

The Dominion Office and Store Fixture Company, of London, are going out of business, and at present are finishing up what showcase stock, etc., they have on hand. The factory which the company occupied has been sold. They may reorganize and start up again.

Store fixtures and church woodwork are made at the factory of the London Art Woodwork Company, Ltd., which is a new company, of which Mr. Keller, formerly superintendent for the Dominion Office and Store Fixture Company is part owner and manager. The company is making a good start, and has the advantage of being in the hands of a man with wide experience in the class of work being manufactured.

The principal product of the Beck Manufacturing Company, of London, is cigar boxes, and as this is a staple line, not affected by the usual conditions with which other woodworking plants have to contend, the company have been able to keep things running more or less smoothly. The majority of the staff is composed of girls and boys, so that the serious effects of the labor shortage have been escaped.

The London Box Company state that they have an order on hand for about 100,000 shell boxes, which they expect to start on about the first of January. Their plant has been so tied up with local business that they have been unable to get at them earlier. Mr. Watson, the manager, in speaking of the labor question, said that they had been able to secure enough men, but that their product is costing them more to manufacture, as they are obliged to pay about 25 per cent. more in wages and they are getting about 30 per cent. less in efficiency.

The plant of the Canada Furniture Manufacturers, Ltd., at Woodstock, has a capacity of about 400 men, but at the present time the company is unable to secure more than about half that number. They are exceptionally busy keeping up with their orders. A four-storey addition to provide room for about 100 men, which was started before the war, has been completed. It will not be equipped for the present, but when it is the machinery will be motor-driven. The company have been considering the advisability of discarding the five boilers which are at present necessary to furnish steam power and generating electric power for the entire plant.

The Crown Lumber Company, Ltd., of Woodstock, who make a specialty of bent work for wagon and carriage factories, report that they have been able to secure a steady volume of orders, and that their factory is busy, but they are in a position to handle additional business.

Passing of a Prominent Woodworker of Old Quebec

The woodworking industry of Quebec City has lost one of its most prominent figures by the death of Mr. E. T. Nesbitt, which occurred at the Montreal General Hospital on November 23. Mr. Nesbitt conducted a large contracting, planing mill and box making busi-

The late E. T. Nesbitt, who conducted a large woodworking business in Quebec.

ness. He was 62 years of age, and until very recently was able to attend to his business, although not enjoying the best of health.

Mr. E. T. Nesbitt was one of the best known constructional men in Canada—and by the same token he was one of the best liked. Perhaps his most useful work was the founding of the National Association of Builders' Exchanges, of which he was president for three years, and in which, by virtue of his excellent services, he was elected honorary president for life. Into the work of this association Mr. Nesbitt threw himself heart and soul, and both workmen and employer owe to him several much needed pieces of legislation in recent years.

Another sphere of Mr. Nesbitt's activity was the Canadian Manufacturers' Association, over whose local destinies at Quebec he presided as chairman for two years. He was a member of the council of the Quebec Board of Trade and was vice-president of a special committee appointed to enquire into the city's

system of taxation. He was a good Britisher, as evidence of which mention may be made of the fact that he was a past president of the St. George's Society of Quebec.

He had an equal command of both the English and French languages, and those who knew him can recall occasions upon which, when he was acting in a presiding capacity at some important function in the old capital, he has made an impromptu oration in the one language and followed it immediately with a perfect rendering in the other.

Strength of Lock-Cornered Boxes

The United States Forest Products Laboratory, Madison, Wisconsin, has recently completed strength tests on lock-cornered boxes. A brief report giving the results of the tests is as follows:

"Tests were made on 44 lock-cornered white pine boxes for the purpose of writing specifications for boxes of this construction. These tests indicate that lock-cornered boxes should be made of the same thickness of material as nailed boxes of the same type. It will be possible to use nails one size smaller, however, because of the greater rigidity of lock-corner construction.

"One hundred and fifty-one tests made to determine the holding power of nails in dry yellow pine showed a rapid decrease of resistance to tension as soon as the nail was overdriven; in shear rapid decrease did not occur until the nail had been overdriven one-sixteenth of an inch."

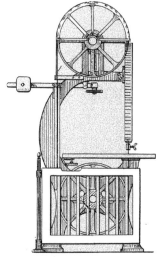

Effective home-made guards for a bandsaw. Note the arrangement on upper wheel to keep saw from pulling off; also, the piece to protect operator's face when leaning over the table. The guard for the lower wheel is attached to an upright piece of piping so that it can be swung out to remove the saw.

A Blower System for the Woodworking Plant

Discussing Its Importance as a Factor in Production—Elements to be Considered in its Installation—Cost Relatively Inexpensive

By "Superintendent"

One of the first requisites of a woodworking plant is a well-designed blower system. This point does not require emphasizing to those readers who are familiar with machine rooms where it is necessary on occasions to locate the individual machines by means of shovels. The increased cost of production where such a condition prevails may be more easily imagined than described. Machine operators or others desiring to move stock from one machine to another find on every side a pile of shavings facing them. True, in some factories laborers are engaged to remove the shavings before they accumulate to any great extent, but the expense incurred by such a method would soon pay for a good blower system. Then, again, there is the health of the employees to be considered, for it stands to reason that their health will be affected very seriously if they are obliged to work in an atmosphere thick with dust.

The designing and installing of a fan blower system should be, and usually is, placed in the hands of a reliable firm specializing on that class of work. But at the same time it is advantageous for the factory superintendent to possess some knowledge of the mechanical principles involved in the construction of a blower system. By coupling this knowledge with his knowledge of the work that the machines turn out he is often able to make suggestions about the connections for various machines. For instance, a machine that ordinarily takes light cuts may occasionally be required to take very heavy cuts, and therefore a blower pipe large enough to take away the shavings during the period of the heavy cut would be necessary. The designing of a fan blower system is based on the following principles:

The velocity with which air escapes into the atmosphere from a reservoir is dependent upon the pressure therein maintained and upon the density per unit of volume, usually designated as the "head due to the velocity." The velocity produced is that which would result if a body should fall freely through a distance equal to this head. In the case of the flow of water such a head always exists, as, for instance, when a standpipe is employed to produce the requisite pressure. Its theoretical velocity of flow from an opening at the bottom of the standpipe is determined by the formula for falling bodies, which is:

$$V = \sqrt{2gh}.$$

In which V=velocity in feet per second.

In which g=acceleration due to gravity.

In which h=head in feet, here 50 feet.

This formula, when worked out, gives 56.7 feet per second for the velocity of the flow of the standpipe mentioned above.

In the case of air, however, an actual homogeneous head never exists, but in its stead we have to deal with an ideal head, which can only be determined by dividing the pressure by the density. As the density of air

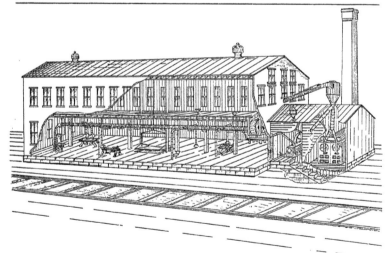

Fig. 1—Exhaust system as applied in a woodworking factory.

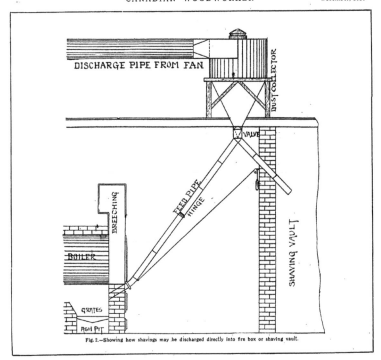

DISCHARGE PIPE FROM FAN

DUST COLLECTOR

VALVE

FEED PIPE

HINGE

BREECHING

BOILER

SHAVING VAULT

GRATES

ASH PIT

Fig. 2.—Showing how shavings may be discharged directly into fire box or shaving vault.

is so much less than that of water, it is evident that for a given pressure the head will be far greater in the case of air. But the velocity of discharge is dependent only on the distance fallen, which is represented by the head, whether real or ideal. As a consequence, air under a stated pressure escapes at vastly higher velocity than water under the same conditions.

It is evident, therefore, that the pressure created by a given fan varies as the square of its speed. That is, doubling the speed increases the pressure fourfold. The volume of air delivered is, however, practically constant per revolution, and, therefore, is directly-proportional to the speed.

It is a peculiar function of a fan blower that instead of always delivering a fixed volume of air, regardless of requirements, it automatically increases the volume as the resistances are decreased. On the other hand, if the blower be in operation with a fairly free outlet, in excess of its capacity area, and that free area be decreased, the pressure produced will immediately rise, thus tending at once to overcome the increased resistance. Therefore, if a certain maximum pressure is known to be required the fan may be so

speeded as to give this at such times as the conditions demand, while at other times, when less pressure or volume of air is required, proper manipulation of the blast gate will economize power.

The work done by a fan in moving air is represented by the distance through which the total pressure is exerted in a given time, and, as ordinarily expressed in foot pounds, is equal to the product of the velocity of the air in feet per second, the pressure in pounds per square foot, and the effective area in square feet over which the pressure is exerted. From this it is evident that the work done varies as the cube of the velocity, or as the cube of the revolutions of the fan. That is, eight times the power is required at twice the speed. The reason is evident in the fact that the pressure increases as the square of the velocity, while the velocity itself coincidentally increases; hence, the product of these two factors of the power required is indicated by the cube of the velocity.

The actual work which a fan may accomplish depends not only on its proportions, but upon the conditions of its operation and the resistances which are to be overcome. Evidently it is improper to compare fans when operating under such conditions that these resistances cannot be definitely determined, and for proper comparison of different fans, therefore, the areas through which the air is discharged should bear some constant relation to the dimensions of the wheels themselves.

It has been determined experimentally that a peripheral discharge fan, if enclosed in a case, has the ability, if driven at a certain speed, to maintain the pressure corresponding to its tip velocity over an effective area which is usually denominated the "square inches of blast." This area is the limit of its capacity to maintain the given pressure. If it be increased the pressure will be reduced, but if decreased the pressure will remain the same. As fan housings are usually constructed, this area is considerably less than that of either the regular inlet or outlet. It, therefore, becomes necessary, in comparing fans upon this basis, to provide either the inlet or outlet with a special temporary opening of the requisite area and the proper shape, and make proper correction for the contracted vien. The square inches of blast, or, as it may be termed, the capacity area of a cased fan, may be approximately expressed by the empirical formula:

$$\text{Capacity area} = \frac{D\ W}{X}$$

In which D=diameter of fan wheel, in inches.

W=width of fan wheel at circumference, in inches.

X=a constant, dependent upon the type of fan and casting.

The value of X has been determined by the leading manufacturers of blowers, but these values must be used with discretion, acquired through experience and a thorough knowledge of all the conditions liable to affect the fan in operation.

The capacity of a fan—that is, the amount of air which it will deliver—depends considerably upon the construction of the fan and upon the resistance to the flow of the air.

The following table shows a safe capacity for some leading forms of blowers and exhaust fans, operating against a pressure of 1 oz. per square inch, or $1\frac{3}{4}$ in. water, nearly:

Diameter of wheel. Feet.	Revolutions per minute.	Horsepower required.	Capacity per minute Cu. ft.
4	350	6.0	10,600
5	325	9.4	17,000
6	275	13.5	29,600
7	230	18.4	42,700
8	200	24.0	46,000
9	175	29.0	56,800
10	160	35.5	70,300

If the resistance is greater than above given, the required capacity may be obtained by increasing the number of revolutions, increasing the horsepower correspondingly. Or if the resistance is less, the required capacity may be obtained by reducing the number of revolutions, with a corresponding decrease of power. In selecting a fan it is advisable to choose a large one, so that it may run easily, quietly, and economically. A small fan, which must be run at high speed, wastes too much energy and often makes a disagreeable noise.

Fig. 1 gives a general idea of the application of a blower system in a woodworking factory. This will give the owner of the moderate sized planing mill who may be inclined to regard a blower system as something away beyond his resources an idea how simple and relatively inexpensive a proposition it really is in comparison with the present labor cost consumed in disposing of his shavings, not mentioning the improved working conditions which result when the installation is made.

Fig. 2 shows the usual method of disposing of the shavings. They can be fed direct into the firebox or into a shavings vault. This arrangement works out very satisfactorily, as when it is desired for any reason to shut off the flow of shavings into the firebox, the other pipe can be opened and the shavings permitted to flow into the shavings vault. In view of the close proximity of the vault to the firebox, it is an easy matter for the fireman to shovel the shavings into the firebox when the accumulation in the vault becomes too great.

Labor Saving Equipment

A few years ago the application of varnish, shellac, or enamel in a furniture factory by the spray method would have been considered so expensive that it would not be thought practical. To-day every up-to-date factory has installed or has on order one or more spray outfits.

Conditions as to labor have changed so that the factories were forced to adopt some other method than brushing by hand, and in most cases the spray has been surprising in results, not only as to the saving in labor, but the cost of material has not been as great as expected. The coats as applied are put on so evenly the saving in rubbing must be considered and credit given to the spray machine.

So far we have applied enamel primers, enamels, shellac, varnish stains, and wax, with good results, both as to quality of the work done and saving in both labor and material. A great deal depends on the operator, and patience should be exercised, as it is bound to cost more to break a man in than the same work could be done for by hand; but practice makes perfect, and, with the selection of a brainy operator, is sure to bring splendid results in the use of the spray.

New Five Spindle Boring Machine

Owing to the advanced price of material and labor and the price the Munition Board is paying, it is necessary to cut down the cost of production of shell boxes by installing machines that will turn out the work at the smallest cost.

The Bethlehem box requires three diaphragms with five holes in each diaphragm, and boring them at two or three operations was too slow and cost too much. Cowan & Company, of Galt, Ont., grasped the idea and have manufactured and put on the market a machine for boring the five holes at one operation and in the same time as it would take to bore one hole. The machine is self contained and driven by belt to the master spindle which drives the other spindles, with steel cut gears. The spindles are vertical, so that the diaphragms are placed on a horizontal table and are held by an eccentric clamp, convenient to the operator and raised up to the boring bits by foot lever, leaving both hands of the operator free to handle the stock. The boring bits cut out about a quarter of an inch on the circumference of the hole and a pin shoves the centre out by means of a spring in the centre of the shank of the bit. The guard protecting the cutter heads raises and lowers with the table and thus always protects the operator. It can be readily raised for changing cutter heads.

Correct Trimmings for Period Designs

A Continuation of the Article Published in the October Issue—Well Executed Drawings of Representative Pieces from Different Periods

By James Clark

Chinese Chippendale pulls.

The "Chinese Chippendale" buffet illustrated is quite bare of any rococo detail, and is on rather severe straight lines, relieved somewhat by the angle lined fret work that is planted on. Chippendale was the first great designer of the Georgian period, and so impressed his work by his masterly handling of detail, that furniture designed on the same lines has been called Chippendale ever since. Chippendale was at the height of his career about the middle of the 18th Century.

The Chinese influence in his work is the result of travel and observation. A trip to the "Flowery Kingdom" must have been quite a serious proposition at that time. While adaptations of Chinese Chippendale furniture that are just as straight in outline as the illustration are frequently trimmed with handles showing the rococo outline and detail, the result is a mismatch, and the writer suggests that handles more in keeping with the general design of the piece will look better, and submits two simple sketches along these lines.

Hepplewhite

Hepplewhite was the next most noted designer to follow Chippendale, and his designs were as plain and straight lined in general construction as possible. His style would be called chaste in general opinion, and he used ornamentation very sparingly, using inlaying and painting, which combined with the natural beauty of mahogany (the wood used). He succeeded in producing the style that bears his name, and which to this day is largely manufactured.

One of the principal features of his ornamentation is the Prince of Wales' feather, and which it must be admitted makes a pleasing crown or centre piece. Other details of the ornamentation are characterized by delicacy, and while they emphasize the severity of the outline by contrast, the combination makes the pleasing effect that satisfies the lover of good furniture.

The trimmings are in keeping, and the ornamenta-

Sheraton trimmings.

tion is in keeping with the style. Some are much plainer than any of the illustrations shown.

The dressing case shown is a modern factory adaptation of this style, and would look alright with the knob pulls or the one showing the Prince of Wales' feather.

Sheraton

The Sheraton style is the next one to consider. Thomas Sheraton was the last great furniture designer of the 18th Century, and his work, like Hepplewhite's, was of the straight lined type. He depended a great deal on inlay for the ornamental effects, and certainly produced many beautiful designs. He is ranked as the greatest designer of the 18th Century by many critics.

The plain little "everyday" buffet shown is on Sheraton lines, and of easy manufacture. The drawer

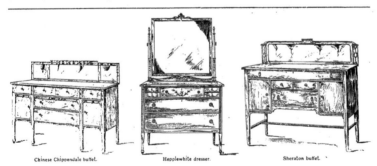

Chinese Chippendale buffet.   Hepplewhite dresser.   Sheraton buffet.

fronts and doors could have "inlaid lines" to relieve their plainness. There is not a moulding to be seen, but by careful selection of material a nice medium priced buffet can be produced.

The trimmings shown are on Sheraton lines; they of course look more ornamental in black and white

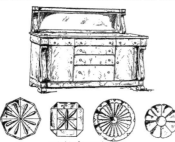

Colonial buffet and trimmings.

than in their everyday dress. They offer a fair range to cover different adaptations of the style. Sheraton and Hepplewhite are two styles that in general outline are very similar, and very often the only difference in some examples of the two styles is the Prince of Wales' plume, which was never used by Sheraton.

The Sheraton style shows to best advantage in mahogany, gum, or birch. If made in oak the material must be selected for quiet effects.

Colonial

Colonial furniture, so called, was developed from the various styles of furniture that were brought to America previous to the Revolution, and is, of course, somewhat of a misnomer, but as that is the name that has been given it, we accept it as such. The style

Hepplewhite trimmings.

of to-day known as "Colonial" is of solid and substantial construction, and of, in most cases, no ornamentation beyond the beauty of material.

In the writer's opinion, at least, the spiral post or pilaster carvings and claw feet do not look nearly so

well as the perfectly plain, band-sawn pilasters, standards and feet.

The illustration shown is of the plain and somewhat massive style. It is easy to imagine that a carved lion's paw under the same case would spoil the appearance; a ball and claw foot would not be so objectionable. All readers have, no doubt, seen just such a case as this one spoiled by putting on the former style of foot. A sideboard of this style veneered in crotch mahogany is a sight to gladden the heart of any home lover, who would have a mental picture of a dining room with a full suite installed, and a natural desire to become the posessor of such.

The trimmings for Colonial furniture are nearly always knobs; no handles that could ever be suspected of making any noise will do. On the cheaper styles plain turned wood knobs or square cones, are used; on the better examples glass knobs are greatly favored. These latter are made in a variety of shapes and designs, and are practically indestructible.

Colonial furniture is what might be known as a "roomy" style; that is, it requires a room of generous proportions to harmonize with its character, and dining room and library furniture buyers find the Colonial style a great and reliable exponent of their ideals.

While mahogany is the best medium in the material line, quartered oak nicely and tastefully selected, finished "full up" and rubbed to a dull smooth surface, is the next best and has been largely used.

Side and Swage Set for Saws

There are two alternative methods of obtaining clearance in the saw cut for the blade of a saw, viz., by bending over to right and then to left the alternate points of the teeth, known as side or spring set, or by spreading out each tooth at the point equally on each side of the saw, known as swage set, writes H. W. Durham, M.E., in the "Timber Trades Journal." In addition the clearance for the saw is sometimes obtained in circular saws used for cross-cutting by making the saw thicker at the rim than at the centre, these saws being known as "hollow ground."

For general purposes the two first-mentioned methods are used, and with both it is of the greatest importance that the teeth should be evenly and accurately set, and the extra time and trouble spent in doing this is well repaid by results, for the following reasons: First, a saw which has been properly set will cut smoother and more rapidly than one which has only been carelessly treated, or which has been set by an incompetent man.

Secondly, saws having the points of some of the teeth projecting beyond others do not cut freely and become dull quickly, because the projecting teeth do most of the work and the keen points are rapidly blunted; also the surface of the board is rougher and more uneven than if the saw had been correctly set.

Thirdly, there is always a tendency for the projecting teeth to dig sideways into the wood, causing the saw to run in the cut, with the risk also of the point becoming bent or broken.

Fourthly, if a saw has been accurately and evenly set, it will cut freely with a narrower set than is possible if the work has been badly done, thus effecting a saving of both wood and power.

Both spring and swage set have their own particular advantages and advocates, there being a great difference of opinion amongst sawyers as to whether the best results are obtained by side or by swage set.

How to Get the Accurate Cost

By F. E. Mutton, Vice-President International Time Recording Company

To obtain the actual cost of production it is the great desire of all manufacturers, for by this and this only are they able to figure out whether they are actually making a profit each month.

The cost of a manufactured article consists of three parts or items—burden expense, raw material, and time or labor. We arrive at this burden expense by adding up all of the items of fixed charges, office expense, selling expense, material consumed in manufacturing, such as oil, waste. etc., which cannot be charged direct to any specific article, non-productive labor (which should be obtained by dividing your pay roll into two parts of productive and non-productive). The total of this non-productive pay roll should go into the burden. This item of burden expense you will see amounts to a large sum, and not one dollar of it has been paid for actually producing. How are you going to get it back? You cannot add it to the cost of raw material you have purchased, sell it and get your money back. Raw material has not caused any of this expense. You cannot charge it to non-productive

labor, for that has not produced anything you can sell and get your money back. There is nothing left to charge or add this burden expense to but productive labor.

Not until you begin work on your raw material is there anything produced by or from which you can get back the money paid out for burden expense, therefore we must add the burden expense to productive labor. How can we add it and have it properly distributed? You will say perhaps a certain per cent. to the cost of material and labor. We have shown you above that raw material had nothing to do with expense, therefore you should not add a certain per cent. to the cost of that material as a burden. It was the productive labor that consumed that burden. Then you say you will add a larger per cent. to the cost of labor. But this will not do. For example, you have two men working side by side at the bench; one you pay 45 cents per hour, the other man 20 cents per hour. Each man works on his respective piece of work eighteen (18) hours. By the percentage method you charge, say, 40

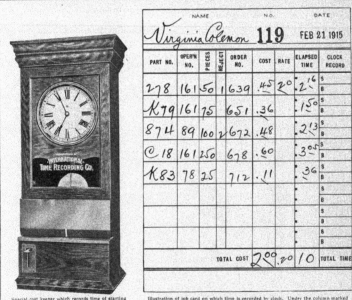

PART NO.	OPER'N NO.	PIECES	REJECT	ORDER NO.	COST	RATE	ELAPSED TIME	CLOCK RECORD	
								NAME	Virginia Colemon NO. 119 DATE FEB 21 1915
278	161	50	1	639	.45	.20	2.16		S / B
K79	161	75		651	.36		1.50		S / B
874	89	100	2	672	.48		2.13		S / B
C18	161	250		678	.60		3.05		S / B
K83	78	25		712	.11		.36		S / B
									S / B
									S / B
									S / B
			TOTAL COST	2.00	.20	10			TOTAL TIME

Special cost keeper which records time of starting and stopping jobs on the job card.

Illustration of job card on which time is recorded by clock. Under the column marked "clock record" will be seen letters S and B. B indicates when job was begun, as 7 a.m. S indicates when job was finished, as 8.16 a.m.

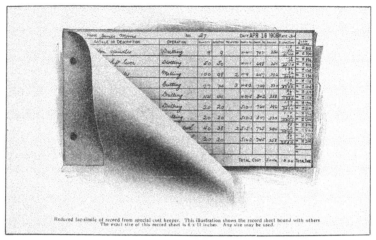

Reduced fac-simile of record from special cost keeper. This illustration shows the record sheet bound with others
The exact size of this record sheet is 8 x 11 inches. Any size may be used.

per cent. of labor cost on each man's work to cover burden expense. On this basis for the first man's share you charge $3.24 and only $1.44 for the second man, and you cannot show where the first man consumed more of the actual burden expense than the other, and yet you are charging over twice as much to the first man's labor.

To get the proper amount to add to this labor to cover this expense you should divide the total amount of the burden expense for the month by the total number of productive hours. Say your burden expense is $720 and your productive hours are 3,600. This gives you a rate of 20 cents a productive hour; therefore to the first man's wages of $8.10 you add the burden rate of 20 cents per productive hour, eighteen hours $3.60, making labor $8.10, burden cost $3.60, material $4.50, total $16.20, the absolutely accurate cost. To the second man's wages of $3.60 you add the burden hour rate of 20 cents per productive hour, eighteen hours $3.60, making labor $3.60, burden cost $3.60, material $1.50, accurate cost $8.70, and to this you add what per cent. of profit you intend to make.

By the percentage basis add 40 per cent., or $3.24, to the first man's wages of $8.10, material $4.50, making cost $15.84, a difference of 36 cents short of actual cost. To the second man's wages of $3.60 add 40 per cent., or $1.44, plus cost of material $1.50, making a total of $6.54, a difference of $2.24 short of actual cost. This is where you fall short of making the profit you had figured on. If you will give this careful thought you will see it is the only way to arrive at the accurate cost.

Take another example. Say we have three men working by the piece. The first man we pay 40 cents per 100. In six hours he turns out 475 pieces, $1.90. The second man we pay 50 cents per 1,000. In six hours he turns out 6,000 pieces, $3. The third man we pay 1 cent a piece. In six hours he turns out 225 pieces, $2.25. If you add 40 per cent. to labor cost you

will add 76 cents to the first man, $1.20 to the second man, and 90 cents to the third man, when each and every one of them took six (6) hours and consumed an equal amount of the burden expense, and should be charged an equal amount, but on a percentage basis you would charge each man a different amount for burden expense. It s necessary to keep a record of the hours that piece hands work in order to obtain the accurate burden their product should bear.

Another example: A piece worker starting to work in the morning and working only half a day, five (5) hours, earning $1. By the old method they would charge 40 per cent. to this to cover burden expense, making $1.40. This burden expense keeps right on the other half of the day, but what bears it? If she had worked the rest of the day she would have earned $2, and 40 per cent. burden expense makes $2.80. Although the burden expense went on just the same, but as she did not work and you could not, according to your methods, charge it to her labor, where does it come in? If you had obtained a productive hour rate by dividing the total burden expense by the total productive hours, the burden expense for this half day she did not work would be properly and accurately taken care of.

It is comparatively easy to obtain the cost of material that enters into your product, but the time or labor is the slippery element of cost, also the proper burden rate has been difficult to ascertain. By the use of our special cost keeper you can get the elapsed time of every operation, every job, exact labor cost to a cent on every order that goes through your works, also the productive hours, which will give you the proper burden rate and absolutely accurate distribution of your burden expense, saving you thousands of dollars a year.

If you want to keep the individual operation on one card, with this special clock you can register the elapsed time on and off as many times as you like. If when you issue an order you wish at the same time to start a job ticket, having it accompany the order through the works, accumulating the time of each operator who works on it, this special clock will record the elapsed time of each man working on the job right opposite the operation he performs, so when the order comes through it will have with it one ticket with the exact amount of elapsed time of each operator multiplied by their rate, giving you the exact and entire labor cost on this one original ticket. If your business is such that you wish a coupon tag to follow the work, each operator detaching a coupon to show what work he did, this special clock will stamp the elapsed time on his coupon, showing (if it is piecework) the comparison of piece price with an hourly rate.

If you prefer to keep a record of the various jobs a man works on in a day on one card or sheet, this clock

will give you the elapsed time of each job on that same sheet, directly opposite the order he is working on, so that the total elapsed time will foot up to the exact working hours of the day, multiplied by the rate per hour will give you the amount of wages you are paying the workman, thus proving first the productive hours and minutes, second the wages you pay, third the absolutely accurate labor cost, fourth the accurate burden hour to figure the burden expense.

With this special clock you can obtain the above results in the quickest, simplest, most accurate as well as the most inexpensive way that has ever been devised. The expense of installing this system is but a drop in the bucket as compared with what it will earn for you. You cannot afford to be without it.

We are ready to give you more information on this subject if you are interested in our product.

Practical Methods in the Furniture Factory
Third of a Series of Articles Dealing in Logical Order with the Various Manufacturing Operations—Material Tags and Factory Trucks
By W. J. Beattie

The article in the November issue of the "Canadian Woodworker" dealt principally with the installation and equipment of the swing saw. The next thing that naturally follows is by what means are we going to assist the stock cutter to get out his material intelligently. The use of billing tags for the material for a job is now fairly well used, but the readers of this magazine might be surprised if the names of some of the factories in Ontario could be given that cut up their stock without the use of any such help.

Their method is somewhat like this: A more or less correct bill of the stock is written in lead pencil on a piece of thin lumber, or paper; this is handed to the man of the swing saw and he is instructed to get out so many of the articles. This he does; they go through the rip saws, are matched up, the required

Fig. 1—Stock tag filled out as it would be written,

thickness is written on the face of one of the articles, and also what it is. Of course, the planer man often planes the instructions off, but all he has to do is to take out his rule, measure the rough size and trot over to the swing saw and find out, or try to, what it is. The stock cutter may be able to tell him, but it generally requires a return visit to settle the question, and so it goes on, more or less, until the goods are machined and ready for the cabinet room.

You can go into some machine rooms and see truck loads of stuff that it would be hard to tell what purpose or job the material is for. This is not a pipe dream though it may sound like one to many. And if

you have any doubt of it make a visit to some plants that are generally supposed to be up-to-date in all details, and there will be practical illustration of it. If you have ever heard or seen a foreman who was nearly worked or worried to death trying to keep track of the material of anywhere from 25 to 50 jobs going through a machine room, this would be quite sufficient cause for it. The unfortunate wretch is run to death doing actually nothing, and becomes a nervous wreck long before he is at the years of greatest usefulness.

The Better Way

It is so long ago that I hate to think of it when the writer first encountered the idea of tagging the material through the factory (just like so many individual parcels of freight). The stock bill having been made out from the drawing and being in accordance with the rod, the sizes being the finished dimensions in every case, and the bill for one article only, the tags are written from the stock bill. The desired quantity of each of the items, according to the number of articles to be manufactured, are written on the tags, along with the net sizes and other particulars, to guide the workmen through whose hands the material will pass.

The illustration of a tag filled out as it would be written, is here shown, so as to convey the idea clearly. See Fig. 1.

There is another form, Fig. 2, that has room for the various shop numbers of the machines. The route that the particular item of stock is to travel can be indicated when the stock tag is written. This is done in some factories that have things down "more pat" others.

The crosses show the machines or operations that these tops have to pass through, and the check marks opposite are made by the workman as he gets through with that particular part of the machine work. No. 1 is the swing saw, No. 2 the rip saw, etc., until the last operation, the belt sanders, before the bench man takes the tops into his bench room. All the various parts of the article are gotten out in the same manner.

No. 1 will be found a good serviceable form for average use, and may be preferably of a light red color with black lettering, as a red tag is more easily seen, being a decided contrast to the color of the white material. As red tags are not at present so easily procurable as a short time ago, the next best is a light yellow, having red lines and lettering which show black in the illustration, as Fig. 1.

The tags are sorted out by the stock cutter and placed in a rack, from which any tag may be removed without disturbing any other, and handed first to the rip saw. The rip saw tailer, whose duty it is to count the stock as it comes through the rip saw and pile it on the various trucks, fastens this tag by a piece of string to one of the pieces, or if the whole load is one

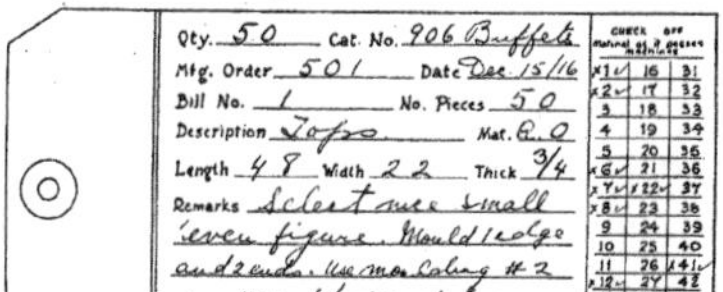

Fig. 2—Another form of tag having room for shop numbers of machines.

kind of material he may hang the tag on one of the truck stakes. The string on the tag may be dispensed with, though the tag is less likely to be lost or mislaid, and a dozen balls of twine will do a great lot of material tags.

As is readily seen, there is considerable room for written instructions on these tags, and surely it is a matter of economy for the one man who writes the tags from the bills to put on all the information, so that the men have no doubt as to what is wanted. Then they are not chasing around asking the foreman all sorts of needless questions, some of which, to answer, he would likely have to go and consult the rod or the stock bill, and besides, any conscientious workman likes to know that there is no doubt as to his being right. Therefore, the written instructions serve a twofold purpose.

Factory Trucks

Factory trucks are a very important item in the equipment of a plant, and it is a rare thing to see a factory that has too many trucks. I can at present only remember one plant that had more than they actually needed, and the reason for that was that the company that operated the plant had procured some from other plants that had, during the course of events, been closed down.

There should be sufficient room between the swing saw and the rip saws for the necessary trucks, so that the stock cutter, or his helper, can lay the different lengths of stock on their separate trucks. Also, the rule that the stock should always be on wheels until delivered to the cabinet makers is a good one, and a dozen trucks more than what you have got makes a big difference in the cost of moving stock. It lessens handling and saves time that would be spent scouting around for an empty truck.

To induce the financial man of a furniture plant to part with the necessary money to provide plenty of "material locomotion" is harder than to get him to spend ten times the amount in almost any other way. Try it! You see it is harder to convince most people that the simplest things right under their noses are the ones that are generally overlooked, and that in

the end are of very material importance. Look at the trucking facilities in the iron working plants. In many the trucks are run by electric motors, superseding the hand propelled variety. Certainly they have heavier material, but it shows progressive ideas. The costs of material handling are given as much attention as the other manufacturing expenses.

What a furniture plant needs is plenty of good serviceable four-wheel, four-staked trucks, on roller bearings, so that the truck when loaded to capacity can be easily moved.

One necessity for quantity is, that unless the material is being made up into large batches, it necessitates the piling of too many different sizes, etc., on the one truck and entails too much labor in re-arranging them as the material moves along through the machine room. Such a thing as this is sure to be overlooked by the non-practical man, who is sure to ask, "why don't you load all your trucks to capacity"? Never having wrestled with a loaded truck having twelve different items of material on it, he cannot see that any further investment in trucks is necessary.

A good factory truck costs from six to eight dollars, or at least it did so, and an investment of less than one hundred dollars means the difference between famine and plenty.

Before going on from the swing saw through the various stages of manufacture, in rotation, we will, in the next issue, take up the matter of design, rod-making, patterns, stock bills, etc., as these are really the first things to be prepared in manufacturing, though the attempt is being made to drive home the necessity of some one or two things in each number, quite independently of anything in other monthly articles.

Aniline Dyes

Two interesting facts have recently been shown in the aniline dye situation in the United States. One is, that since the war shut off supplies of coal-tar dyes, there has been a larger use made of natural dyes; and the other is the extent to which the United States has drawn upon China for synthetic indigo, which had been imported by China from Germany before the war began. Since the beginning of the war, between three and four million pounds of synthetic indigo have been re-exported from China to the United States.

In normal times the price was about 15 cents (gold) a pound; but China had to be paid so dearly to re-export these supplies that the price of indigo has ranged from 90 cents to $1.30 a pound, and for its small remaining holdings China is now asking $2 a pound. In nearly all cases weights were enormously short by reason of drying out of moisture, but the indigo paste was correspondingly concentrated above 20 per cent. So-called Chinese indigo, a natural extract containing hardly over 2 per cent. to 3 per cent. indigotine, is not considered profitable for use in the United States.

The world uses annually some 80,000,000 pounds of synthetic indigo, of which 95 per cent. was produced by Germany, which exported three-fourths of the product to the far east, no less than 50 per cent. going to China. When war broke out there were nearly 6,500,000 pounds of synthetic indigo in China, and the profits realized by the native dealers who were so fortunate as to hold large stocks have run into millions of taels.—Department of Trade and Commerce Weekly Bulletin.

The Lumber Market

Domestic Woods

Some of the leading lumber firms report that inquiries from furniture manufacturers and others have been coming in very irregularly during the last month, and that they have not experienced the steady demand that existed for some time. Practically no variation in prices has therefore been noticeable. The irregularity of the buying is accounted for by the fact that most of the manufacturers take stock at the end of the year and have postponed purchases till after that date.

The assumption is no doubt correct as present indications point to a busy period for furniture manufacturers during at least the first half of 1917, and there is no reason as yet to believe that conditions will be any different after that period. Many of the manufacturers have orders on hand and in prospect that will keep them hustling for some time.

The leading topic of discussion among lumbermen just now is to what extent logging operations are going to be carried on during the winter. In fact it is rather a difficult matter to find any members of the firms at their offices. One prominent lumberman interviewed about a week ago stated that he was leaving that night for the Ottawa Valley and would prefer to say nothing until he returned. As he did not return before this issue went to press, we were unable to ascertain what conditions prevailed in the section of the country which he visited. The majority of the lumbermen are spending a great deal of time scouring the country endeavoring to make arrangements for their future requirements.

This is good evidence that there is an unusual shortage of stocks and that higher prices will prevail. Undoubtedly the number of logs taken out this winter will be very much less than in former years. The serious shortage of labor is the principal reason, although the high cost of food supplies and other commodities have been contributing factors. Some firms have even refrained from taking out logs because they expected that the cost of manufacturing them into lumber next spring would be prohibitive.

It is estimated that the quantity of hemlock to be cut this winter will only be about 40 per cent. of that which was cut last winter, and as stocks of hemlock are already down to the last notch with lumber firms hanging on tight to anything they are fortunate enough to possess, prices will be very high. As sales of hemlock have already been made at $22 per thousand and higher, it will readily be seen that any further increase would make it prohibitive except to very urgent buyers.

Some of the important white pine manufacturers are not taking out any logs this winter, and those who are are operating on a scale very much smaller than usual.

Imported Woods

The situation at the present time is somewhat out of the ordinary and nearly every one interested in the purchase or sale of lumber is trying to solve the problem of future prices. Current prices are high; in fact about the highest they have been for years, and the indications are that they will go higher. On the other hand, to buy now and then face a lower market next spring would be a mistake. Lumbermen are nearly all of the opinion that prices will go higher and for the most part base their opinions on the facts that stocks carried over the winter will be a great deal higher than for many years, that the scarcity of labor will affect the quantity of new lumber cut and that the increased cost of logging brought about by the higher wages and higher cost of camp supplies will add considerably to the cost of the finished lumber.

The demand for southern hardwoods is said to be greater than can be taken care of with the present shortage of stocks and scarcity of cars. The big problem with United States lumber firms appears to be that of transportation. The combined efforts of the Interstate Commerce Commission, the American Railway Association, traffic departments and lumber associations have not as yet succeeded in bringing about any appreciable relief from the serious situation that has existed for some time, and lumber manufacturers are faced with the problem of trying to place their products in the hands of the consumers, who are just as anxious to secure the material which they have on order.

One of the features of the lumber situation in the United States is the amount of lumber required for the building trades. The favorable weather during the last few months, when building operations ordinarily would have ceased, has resulted in large quantities of lumber being used which would otherwise have been available for furniture manufacturers and others.

Also, furniture factories, automobile factories, and other large users of hardwoods have been unusually busy, with the consequent result that the already low stocks have been heavily taxed.

A very marked increase in the demand for oak, especially plain white oak, has been reported, and prices are accordingly high. Gum, both red and sap, is still in popular demand by furniture factories, and manufacturers of interior trim, and prices are slightly higher than they have been. Some of the southern hardwood companies are said to be executing large orders for walnut gun stocks and propeller shafts for aeroplanes, thus consuming large quantities of these woods in other than the usual channels.

In northern hardwoods, birch continues in popular demand, and in thick stocks is very hard to obtain at any price. Maple, ash and elm are in about the same position as birch. One of the leading American lumber journals makes a comparison between present prices in the Chicago market and those existing a year ago as follows: Maple clears, $49.50, as against $42.50 a year ago; No. 1, $43.50, as against $35.50; oak clears, $51, as against $48 a year ago; selects, $47.50, as against $44; beech clears, $47, as against $40 a year ago; No. 1, $40.50, as against $34 a year ago. These figures speak for themselves, and while they are quoted for a particular market, the proportions are applicable to any market.

The demand for basswood and chestnut continues, as both these woods are widely used, the former by furniture manufacturers and the latter by piano manufacturers for core stock. Some of the current quotations on leading woods at Boston are as follows: Basswood, $44 to $46; chestnut, $51 to $53; maple, $42 to $45; plain oak, $63 to $65; quartered oak, $89 to $91; red birch, $57 to $59; sap birch, $47 to $49; white ash, $56 to $60; with prices very firm.

Shading and Toning the Colors on Furniture
By C. W. Inglis

Probably there is no operation in the manufacture of furniture that gives as good returns for the time and trouble that is taken, as in the shading and toning up of the colors in the finishing room. But if you take a look around any furniture store, you will constantly be confronted with pieces of furniture that have received no attention in this line whatever.

This is not confined to the cheaper grades alone. I have frequently noticed it on some of the best grades.

I recall to mind an extension table of good quality quartered oak with a nice golden color and a beautiful polish. The top was fairly well matched with regard to grain, but each board in the top showed such a difference in shade that you could stand back and count the number of boards without any trouble whatever.

Now, this fault can very easily be overcome so that the contrast in colors will disappear altogether, or can be reduced to such an extent that they will not be remarkable.

As golden oak is still the most popular color, I will try and explain as near as I can on paper, the method of shading on this color first.

After the article has been stained and filled and has had sufficient time to dry, dissolve in a pan or plate of any description, some Vandyke brown in vinegar. Then add enough nigrosine to suit the color (as all golden oak stains are not the same it is impossible to give proportions), and apply with a pad of cotton to the light parts of the wood, previously working out the pad on the board so that the colors will be blended.

If the shading colors are used in proper proportion to match the color of the furniture it is best to use them strong. Then draw some fine sandpaper over it. In this way it blends better with the other work, and will not have the appearance of being doctored. Vandyke brown will be found to be a useful color in shading anything from a light antique to a dark walnut. In light antique or natural, raw sienna can be used with enough Vandyke to counteract a yellow tinge in the raw sienna. This will be quite effective in covering sappy or light parts in the wood.

Now that walnut is in such demand it is absolutely necessary to do a lot of shading, because the manufacturer cannot afford to discard all pieces with sappy streaks, on account of the scarcity of the lumber, and consequent high prices, and inferiority of quality.

For, as everyone knows, good walnut is not as plentiful as it was when in such demand some years ago. Walnut crystals make good shading material for walnut, and where the color required is not too red, nigrosine is added in the manner stated above.

In fumed oak, shading was never the success that it is on other finishes, simply because the colors get into the grain, which is very undesirable, and some-times produces more or less of a failure. The most effective way of dealing with sap streaks in fumed oak is to use a unifume composed of one ounce of pyrogallic acid, one ounce of tannic acid dissolved in one gallon of water. Apply this solution with a brush, and after it is dry, sandpaper with a fine paper until perfectly smooth.

The article is then ready for the fume-box, or to be stained, as the case may be, after which it will be found that sap streaks have entirely disappeared. If any are noticeable at all the colors will at least harmonize, so that they will not be objectionable.

Like everything else, shading requires practice to be perfect. In some factories men are employed who do nothing else, and they become quite expert.

It can also be overdone. But with a little care and good judgment on the part of the finisher, the result will more than compensate for the time and trouble that is spent on it.

Movable Rack for the Finishing Room

The writer had occasion a short time ago to visit the finishing department of a large furniture factory and was astonished to find that with the exception of a few racks suspended from the ceiling which were used for parts in the white, and a very inconvenient arrangement of slats up against one of the end walls, there was no place to put varnished stock, and the workmen were compelled to walk back and forth from their benches, thus wasting a great deal of time.

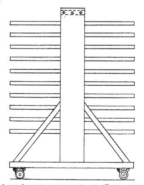

A movable rack that can be used from the filling room to the rubbing room.

Movable racks, provided they can be easily and quickly moved, are one of the handiest things that any finishing room can be equipped with. They can be wheeled up close to the worker's bench so that all he has to do is take the stock from the rack and then replace it when he has finished with it. Then when the rack full of stock has been finished it is quickly run out of the way and another one put in its place.

In finishing rooms where air brushes are used, movable racks are almost indispensable. In fact, when finishing materials are being applied to small parts it is hardly possible to operate an air brush machine to capacity without having the stock on movable racks so that it can be wheeled up to the machine and then quickly wheeled away, to be replaced by another when the parts have been coated. If this method is not adopted it requires a number of hands to carry the stuff around.

A rack built on the principal of the one shown in the accompanying illustration will be found very serviceable. The first thing to be borne in mind is that it must not be top heavy; it must be fairly well balanced, and above all, it must have good casters or wheels in order that it can be moved easily and without shaking all the stuff off. This article, of course, assumes that the factory floors are reasonably smooth.

For a factory that manufactures considerable small stuff, or for mirror frames, toilets, doors, etc., racks of the above description can be utilized, from the filling room to the last coat of varnish, and also in the rubbing room. They can be rolled to a varnish bench from the sanders, and the varnishing done with the least possible amount of time and labor. Also, if the rubbing room happens to be located on a different floor it is a very easy matter to roll the racks onto an elevator.

It is not necessary to give detailed instructions about how to make a rack of this kind, as any mechanic can make them. They should, however, be mounted on double wheel casters, having roller bearing plates, and the wheels should not be less than four inches in diameter, to insure ease of motion. Bolt the caster plates to the arms as screws cannot be depended on.

Any factory will find racks patterned from this illustration a good investment, as they are easily made and not expensive. They can be very easily moved from place to place.

Standardizing Wood Finishes
By J. B. S.

Very aggressive action has recently been taken in the United States to bring about the standardization of finishes on furniture. This subject has been discussed among manufacturers on numerous occasions, but apparently they saw too many difficulties (real or imaginary) in the way of its fulfilment. The latest action appears to have emanated from the dealers themselves, and it looks as though the manufacturers realized the necessity of carrying this matter into effect.

The woods to be standardized are American walnut, mahoganies, Jacobean oak and ivory finishes. If the results are satisfactory with these, others will be added. An authority on wood finishing has been retained to prepare sample panels of standard finishes for each wood. The veneer for the panels has been chosen by experts from the stocks of leading veneer companies. An effort was made to secure pieces of the figure and texture in common use rather than odd or fancy pieces. When the standard finishes are decided on the panels will be distributed to furniture manufacturers.

That some such step was necessary must be evident to any person that has ever been in a furniture store. Tables, chairs, and buffets are very often made in different factories, and when a customer decides on a set of chairs and then tries to find a table and buffet that matches for color and cannot find them it puts the dealer in a very disagreeable position. More than that, it reflects on the manufacturers, because it appears to be a direct challenge that they are conducting their manufacturing operations simply to make money, with no thought of the public they are serving.

In the opinion of the writer there is room for the standardization of colors in Canada. Take for instance golden oak furniture. One can go through a furniture store and see about as many different shades of golden oak as there are colors in the rainbow. Even in the product of one factory making chairs, tables and buffets, if you happen to get a set of chairs out of one batch and a buffet out of another, you will often notice a very great difference in the color. The reason is that very little attention is paid to keeping colors standard. Manufacturers fail to look at it from the standpoint of the customer or dealer. When the foreman finisher submits a sample for the inspection of the superintendent or manager, they very rarely compare it with some of their finished stock. They pass upon it, simply relying on their eye and their judgment as to whether it is a pleasing color or not.

It is not only between chairs and buffets either that this difference in color shows up. Buffet backs are usually made up in batches separate from the cases. Two, and sometimes three styles of backs are obtainable for one case. And on a number of occasions the writer has seen buffet backs shipped back from a dealer, along with a very sarcastic letter commenting on the ability of the manufacturer to match colors. That kind of thing hurts business, and it should not be a very difficult matter for manufacturers to get together and standardize colors.

There might be unwillingness on the part of some, as they no doubt feel that the color of the finishes has a great deal to do with the sale of their furniture. It is quite alright for them to have their own colors and shades in the case of individual or exclusive suites. But surely they would not lose anything by standardizing on those items which they sell to dealers by the dozen or by the hundred.

The purchasing of furniture by the dealer is conducted largely on a price basis. He may be able to buy a buffet from one manufacturer at $20 to match chairs which he buys from another manufacturer. Then again, he may be able to buy a buffet from another firm at $18 that will match the chairs equally as well. When the salesman comes around with photographs of new designs, the dealer will say, "your prices are right but what kind of finish is on your goods? Will it match my other lines?" If the salesman can come back and say "the finish is standard, therefore you can rely on our goods matching your other lines," he has some basis to work on, and the dealer knows just exactly what he is getting.

If the foreman finisher has sample boards coated with standard finishes, together with minute details of the ingredients of the finish, the manufacturer can be absolutely sure that the buffets he is shipping today are exactly the same color as those he shipped a year ago.

Save Money by Purchasing

"Made in Canada"

Veneered Panels

Maple - Elm - Basswood - Birch - Plain Oak
Quartered Oak - Mahogany - Black Walnut

Single 3, 5 or 7 Plies.

Hay & Company, Limited　Woodstock Ontario

"TROY"

GARNET PAPER

FINISHING PAPER

Extra Strong—Keen Cutting

Ritchey Supply Co.
126 Wellington St. W.
Toronto

Established 1895

Skedden Brush Co.

Manufacturers of

ALL GRADES OF

Brushes and Brooms

102-104 Bay St. N., Hamilton, Ont.

"ENAMELS"

Our business in both White and Old Ivory Enamels and Primers has trebled during the past year

WHY?

They are the best on the market. If you are not using them, write for samples and prices.

The
Ault & Wiborg Co. of Canada, Ltd.
Varnish Works

Montreal　　　Toronto　　　Winnipeg

Matching Birch with Mahogany and Walnut

To finish birch natural, if it is desired to keep it as clear and white as possible, first give it a coat of white shellac or pigment surfacer. The writer prefers the latter, as it makes a better foundation for this kind of finish than the shellac. It is better to make your own surfacer, because you can make a given quality cheaper than you can buy it, and when you make it yourself you have a better idea of what you are using. A good surfacer for a natural birch finish is made as follows: Mix together one gallon of good pale varnish, one-half gallon pale brown japan, and one pint of turpentine. Sift into this six pounds of finely ground and bolted pure silex and stir thoroughly. Let stand for twenty-four hours, then strain through book muslin, and reduce with benzine to the proper consistency.

Do not use cheap material in making this surfacer. The same grade of varnish that is used for coating on top should be used. Good varnish pays in every way. It works easier, spreads out farther, and sands better, and will hold up the top coats better than will a cheap varnish. Two coats of varnish on top of a good surfacer, such as the above, will equal three coats on top of a coat of shellac. But the surfacer should be given plenty of time to dry before sanding—at least twenty-four hours in a good drying temperature before being sanded and varnished. Natural birch finish may be rubbed dull or polished, as desired.

To finish birch used in conjunction with mahogany veneer, one should use the same stain that is used on the veneer, except that a double-strength stain is used on the birch. A good way to do is to first coat the birch with the double-strength stain, and, when dry, sand if necessary, then coat both the birch and mahogany with the regular stain. This will produce a uniform color without making the birch too dark. The birch should be filled with mahogany. To prevent the filler producing a cloudy effect, coat with a very thin shellac wash and sand lightly before filling.

To finish birch used with walnut veneer the birch should first be stained with a water stain. There are various walnut powders on the market that are soluble in water, but what are known as walnut crystals make as good a stain as any. These crystals may be obtained from any wood-finishing supply house. To dissolve these the water should be quite warm, but not hot, and should be poured on the crystals, stirring constantly until thoroughly dissolved, to prevent falling to the bottom in a gummy mass.

Unlike mahogany, the walnut veneer is not stained, but is finished almost its natural color. A little coloring matter, such as asphaltum or black japan, is frequently added to the filler, but that is all. Fill both the birch and walnut at the same time, using the same filler. When dry, shellac and body up in the usual way. Do not use a pigment surfacer on these finishes. Dull-rub or polish, as desired.—Thomas Wilde, in "Veneers."

By the Way

Usually the more a man's employer takes him into his confidence the more loyal the man is to the employer.

The man who starts in late in the morning has a hard time catching up with himself. Get started early if you would get ahead.

You should have confidence in yourself, but not too much. Put some of it into the other fellow, to keep you from getting too chesty.

Aeron System Adaptability

Whether you manufacture chairs or refrigerators, brass beds or automobiles, your finishing department has its own individual problems.

You may experience difficulty in obtaining and holding to the required quality of finish—

You may find that your production costs are climbing so high that there is little margin of profit left—

You may meet any number of finishing room perplexities in the day's work.

Your greatest need is a sure, satisfactory solution to these problems.

The practicalness and completeness of the AERON SYSTEM give it an adaptability which insures its making good under every condition—

This adaptability is proved by the superior performance and unfailing service of the AERON SYSTEM in helping 3,000 manufacturers in every line of wood and metal work to successfully solve their finishing problems.

We shall be glad to explain in detail to you how the AERON SYSTEM saves 50% and more in time and labor, produces the highest desired quality of work, and will solve your finishing room troubles—SIMPLY ADDRESS—

The DeVilbiss Manufacturing Co.
1318 Dorr Street   Toledo, Ohio, U.S.A.

For Chair Work and Narrow Surfaces
—wood or metal—Type K Aeron

cannot be equalled. It has a trigger control that regulates the spray from full capacity of nozzle down to a fine line—action is instantaneous and is accomplished by the slightest pull on the trigger. It is light and easy to handle.

Veneers AND Panels

Making Built-Up Panels in the Planing Mill

By M. A. Oliver

One place where the planing mill using veneer has a different problem from that of the panel factory, is in the matter of cores. There are various other lines of work where there are solid or heavy body cores built up from strips, instead of being built up out of veneer, but in panel factories proper built-up work in veneer means so many plies of veneer, and very seldom involves what we term a solid core made out of strips jointed together.

In mill work quite a lot of the undertakings involve the using of a body made up of lumber, or strips glued together edgewise, and again there come times when panels even for mill work are best made by built-up veneer without this lumber coring. Now and then, too, there may be a job where the planing mill man is undecided just which is best to use—to build up his work, say, with five or seven-ply veneer, or to build a centre core of strips glued together. Therefore, while it is impossible to give rules that can be followed in all cases, some effort will be made here to give suggestions that will help guide the planing mill man in deciding which to use under certain conditions.

For ordinary flat panel work, when the thickness is not great, it is better and simpler, as a general proposition, to build up the panels just as the regular panel manufacturers do it; that is, all out of veneer. This is better for making large door panels, say for single-panel doors when they are to be faced with either mahogany, oak or birch.

For the making of these panels many planing mills use a centre body or core made up of strips $\frac{1}{4}$ or $\frac{3}{8}$-in. thick, varying in width from $1\frac{1}{2}$- in. to 3-in., according to the practice in vogue in the planing mill. If these strips are used and cross-banded each side before the face veneer is put on, a job is secured against which no serious objections could be urged, except that the cost is materially increased without the construction being improved sufficiently to justify the extra expense.

A centre-piece of $\frac{1}{4}$-in. veneer will do just as well. Indeed, some of the best work that has come to notice is built up in this way, some of it even involving thicknesses of $\frac{3}{4}$-in., where the centre veneer is $\frac{1}{4}$-in. thick and the cross-banding on each side about 3/16-in., with a facing of anything from 1/16-in. to 1/28-in. This is something near the thickness required for a large single door panel, and it is much easier to put up a panel this way than to run a lot of strips through the sticker and then glue them into a thin centre sheet.

The greatest shortcoming, however, that develops when centre cores are built up of strips is the tendency to glue the face veneer right on the core without cross-banding. It has been demonstrated that it is a poor policy to take a solid core built up of strips and glue fine face veneer on it without cross-banding. When a planing mill man gets this point thoroughly

fixed in his head he will find that many times it is better to use a veneer core than a strip, for when he has to make the job three-ply anyway, he can make it all up out of common veneer with less trouble than it would take to make it out of strips.

This applies not only in the kind of panel work referred to, but in lots of other mill work. Indeed, if there is one practice the planing mill man needs to guard against it is that of making a solid body of lumber and then putting face veneer right onto it. No matter whether it is a mantel piece, door panel, wall panel, stairway panel, or what it is, if it is a job involving a thickness of something like an inch, there is a natural temptation to make it up out of $\frac{7}{8}$-in. stock and face it with veneer. This is the thing that must be guarded against. If you use stock in this way, figure on crossing it with another ply of veneer before putting on the face, and if the requirements in the way of thickness are not too much it would be better and simpler to build it up altogether of veneer.

Cores for Long Panels.

There are exceptions to all this, however. There come times when it is better, even though the thickness can be made out of veneer, to use a centre core of strips of lumber. One of these exceptional cases is when there is a panel to be made of great length; that is, when there is a panel longer or larger in any way than it is possible to get veneer in one length. Rotary-cut veneer may be had up to 8 feet; the general run of it, however, is from 6 feet down. Now, let us suppose a man is making a panel of 10 feet or more. There have been some made 16 feet, and possibly some 20 feet long without intervening supports for the centre.

In a case of this kind, if veneer is used for the centre core there would be an end joint in it that would weaken it. Say a man wants to make a panel 10 feet long and has no veneer that is over 6 feet in length. Then he should make his centre body out of strips of lumber, preferably the full length, but should it be necessary to joint a few strips here and there through the centre, the joints could be broken so that when the whole was finished it would give a great deal more strength than if it had been made of veneer that did not reach the whole length. Generally, in these large panels there is thickness enough to allow one to use strips in the centre core, say, $\frac{1}{2}$ inch thick; then to get a good stiff sheet of veneer on each side in addition to the face veneer.

In a case of this kind it is even advisable to use an exceptionally heavy centre-piece for the sake of the strength it would give, and make a variation from the ordinary practice of building up the core. Take the case of the $\frac{3}{4}$-in. core spoken of above, with the $\frac{1}{4}$-in. centre and 3/16-in. veneer on either side. If there were a large panel involving a great reach in either width or length, it would be better for the sake of strength to use more body in the centre and make the outside veneering lighter. That is, make the centre no less than ⅜-in. and put ⅛-in. cross-banding on each side, after which the face veneer could follow.

Or, if a little more thickness would be permitted, make the centre core ½ in., or even ¾ in., and use 1 16 in. for cross-banding on each side under the face veneer. This would be a matter then of studying for strength in the centre body instead of in long reaches, where it is not practical to get it in veneer. It does not add to the strength of any given section of the panel to vary it this way, but it does give you a chance to put in more strength of body running the direction the strength may be needed in this particular job.

There are a number of peculiar cases that come up in manufacturing mill work that call for the taking into consideration of various other things than the mere balancing up of the different plies. Sometimes it is the matter of big panels, like those under discussion, and again, it may be simply the facing of a plank body for a pilaster effect, or a casing around columns, or something in which it is desired to use lumber, but to get a face of some fine wood on the outside and place the back against a brick wall or framing, or something. Quite frequently it is a brick or plaster wall, and then there come other complications; the matter of swelling and shrinking caused by moisture in the wall, also the disposition of the plank to warp.

There are frequently jobs of this kind where the architect, specifying mahogany, for example, will not permit the use of solid boards; but wants the built-up body, but to safeguard this warping tendency from the moisture in the wall. It is quite a problem to build these up so that one can feel sure of the veneer holding as it should, because the moisture not only sets up strains by causing swelling of the body, but it also at times loosens the glue and causes blistering off. The usual practice is to cut grooves in the back of a solid body of this kind that goes against a wall, so as to allow the swelling to expend itself in the back. This helps some, but it is no absolute safeguard.

Another plan is to cross-band the back with veneer or with strips; another is to build such stock up out of narrow strips, and yet another one is to build up out of veneer like a veneer panel. No matter which plan is followed, one thing that the man doing the work must look after is that work of this kind is not put up against green walls.

Insist that the walls stand until not only all the mortar, but all the plaster and everything is thoroughly dried out before such finish is put up. It is better even if artificial heat can be applied, because it takes considerable time for all the moisture to evaporate out of a new wall—more time than the average builder is willing to allow. There have been many jobs of mill work practically ruined by rushing them into the building while the walls are yet green, and the safest way, if it were practical, to let the walls of a new building stand for six months before putting in the fine mill work. Such a thing is seldom heard of these days, unless in an exceptional case now and then where a man is carefully building a home and taking a lot of time about it. So the next best thing, wherever practical, is to insist upon artificial drying by the use of furnace heat, or whatever other heat is being used, or by putting stoves or something into the rooms until you are sure they are thoroughly dried out.

Another problem frequently involved in mill work is that of making curves and rounded corners without a joint or a break in the face veneer. It may happen around stairways or in a number of places in which veneering plays an important part, because nothing else can be used to make these rounded corners without a break. So, instead of avoiding these things, they should be taken up gladly as a part of the field in which veneered mill work excels. Generally, in making up the core or body for work of this kind, full-width lumber is used and kerfed for bending. Whether or not it is an improvement to use a core body of built-up strips and then kerf for bending is an open question.

It makes a better body for the sides, but it is doubtful if it will improve the corner itself. After these core bodies are carefully bent to shape, which involves exercise of judgment in kerfing, bear in mind there should be some veneering done on the back as well as on the face. This may seem a little extravagant at first, but it is really the best way to secure the job, because it insures the work holding its position and not depending on the fastening. Veneer inside and outside, just as you would a piece of straight work, but be sure you have the exact shape and make your forms very carefully before gluing up, because there is not much chance to alter the shape in general after it is once glued together.

Veneering Edges of Tops

There are many ways to reduce the cost and turn out plenty of work if we watch for ideas and put them to a test. There is one way of veneering edges on tops which will reduce the cost considerably and does not cost a great deal to install. We all know that veneering tops on the ends and edges after they have been squared up is generally a slow process, usually done with clamps. By using the following method half the time can be saved. First you need four iron rods about 5 ft. long and ¾ in. thick. Eighteen tops can be made in this form (we will call it form). Next we need two

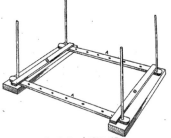

Form for Veneering Edges of Tops

pieces for each top, to be 10 in. longer than longest top made. Holes to be bored in these pieces, A, large enough to slip easily over iron rods. Allowance must be made in boring holes for two pieces, B, one at each end, 1½ in. square and long enough to cover end of top, also room enough for two wedges about 1¼ in. thick. Then we need two heavy pieces, 2½ ft. long, 1 ft. wide, 2 or 3 in. thick. Bore holes in these for iron

THE OHIO VENEER COMPANY

Importers and Manufacturers

Foreign and Domestic Veneers and Hardwood Lumber

We always carry a large and assorted stock of Mahogany, Circassian
Walnut, Sawed and Sliced Quartered Oak.

Send us your enquiries and orders.   We guarantee good service.

2624 to 2644 Colerain Avenue,   ⋈   ⋈   CINCINNATI, OHIO.

Established 1867   Incorporated 1904

QUARTERED OAK VENEERS

If you want the highest grade of quartered oak veneers, made from *northern grown* trees, you can find them at our factory. For many years we have been manufacturing specialties for manufacturers making fine furniture and musical instruments.

Reg. U.S. Pat. Office

"You Can't Mistake the Flake"

and you can't mistake the service we are prepared to render you in handling your orders. Our business has been built up on quality and service. A trial order will convince you.

Hoffman Brothers Company

800 West Main St., Fort Wayne, Ind.

WE SPECIALIZE ON—

POPLAR   CROSS-BANDING AND BACKING   VENEER

Highest Grade   —   Right Prices

The Largest Poplar Veneer Mill in the World

THE CENTRAL VENEER CO.,   HUNTINGTON, WEST VIRGINIA

rods to fit in. In making tops from 20 to 24 in. wide the holes in these heavy pieces should be 17 in. apart. These are simply to hold up rods. Place rods in position, two rods at each end, and slip over the pieces, A, bored for different lengths of tops. Then we lay two strips, 18 in. long by 1 in. wide and ¼ in. thick, across the two pieces, A, a short distance from each end of top. Strips are to allow veneer to extend beyond surface of top. Then lay top flat on these pieces, glue end of top, place veneer on and one piece, B, at each end. Slip over two more pieces, A, and wedge up top, and continue on in same way until the form is full; the pieces between the tops keep rods from spreading, therefore the last top can get just as much pressure as first one. The pieces, B, can be tapered a little on the side to be placed next to veneer. By so doing you are sure to get good pressure in centre as well as ends. In tapering, start in centre of piece and taper equally to each end.

The above described form is for ends of tops only. After ends are veneered, another form can be made to veneer the edges in the same manner. Where your tops are 30 to 50 inches long you need six rods instead of four; two for the centre. The pieces A must be bound on both ends with good, strong band iron or thin sheet steel. These forms are great time-savers where many tops are veneered on the edges. For still more efficiency in this part of the work a small motor-driven saw on an extension wire is also very necessary. Tops can be trimmed very rapidly with one of these.—Veneers.

Sometimes one spends a lot of effort humoring someone who is not worth it, but it doesn't happen often.

Three-ply Panels for Shell Boxes

The Dominion Mahogany and Veneer Company, Limited, Montreal West, P.Q., have equipped their factory for manufacturing three-ply panels for diaphragms for shell boxes. Large quantities are now being turned out, and with additional equipment now being installed, the capacity will be brought up to 20,000 per day. Birch and other Canadian hardwoods are being used, either entirely or with a hardwood core with a gum face and back, in accordance with the regulations of the Imperial Munitions Board. The company are also supplying shell box manufacturers making their own diaphragms with either gum or Canadian hardwood veneers in thicknesses of ¼ in., 9/32 in., 7/24 in., 5/16 in., or such other thicknesses as may be desired. The gum veneer used and sold by the company is supplied by the largest mills in the United States.

Oscillating Belt Sander

The C. Mattison Machine Works, of Beloit, Wisconsin, have recently sent out a circular showing a number of photographs taken in woodworking factories showing the variety of work that can be accomplished with their "172" Oscillating Belt Sander.

The company state that the "172" will sand practically any piece of straight or bent work that can be supported on the tables or held freehand against the platen, sanding rolls or special forms, and that the quality of the work is excellent because the oscillating motion does away with scratch marks.

Another feature claimed for the machine is, that it has a very much greater capacity than a small drum or spindle sander.

VENEERS FROM STOCK ON SHORT NOTICE

MAHOGANY

We have a large stock of this wood on hand, in both furniture and nice mottle piano wood. Particular care has been taken in the selection of our stocks and we would appreciate an opportunity to submit samples.

AMERICAN BLACK WALNUT

We carry the largest and best stock of this wood in Veneers in Canada. We have some very highly figured stump wood, good straight sliced stripe and curly figured stocks, also semi-rotary and full rotary cut stocks of all kinds.

QUARTERED OAK

We carry this wood in all thicknesses and widths in both sawed and sliced. We have a good stock to select from.

FIGURED QUARTERED GUM

Let us show you what we have to offer in this wood. We have the goods and the prices.

ROTARY CUT VENEERS

We are now carrying in stock Poplar, Maple, Oak, Gum, and Birch.

Get our prices on WALNUT LUMBER

TORONTO VENEER COMPANY

TORONTO   93-99 Spadina Avenue   ONTARIO

Wood - Mosaic Co.

NEW ALBANY,   - -   IND.

WALNUT

Lumber ⊷ Dimensions ⊷ Veneers

If you are in the market for anything in Walnut, let us figure with you. We have cut three million feet of Walnut in the last year, and have a good stock of all thicknesses dry. We can also furnish Walnut dimension—**Walnut Veneers.** We have a large stock of veneers, plain, stripe figured and stump wood.

Wood-Mosaic Co., Inc.

NEW ALBANY,   · - ·   IND.

D O M E S T I C

TEAK

Write for prices on all
**Plain and Fancy
Lumbers and Veneers**
Mahogany, Oak, Birch,
Walnut, Ash, Poplar, etc.,
rotary, sliced or sawed
in all thicknesses

SHELL BOX PARTS
in birch and spruce
Bridges and Ends

Write for Quotations

George Kersley
224 St. James Street, Montreal

L U M B E R

MAHOGANY

Notice—Change in Name

Effective October 16th, 1916. The Roberts Veneer Company succeeded the Roberts and Conner Company and John N. Roberts, Veneer Manufacturers.

The Roberts and Conner Company and John N. Roberts desire to thank their many friends for their long and continued patronage and bespeak a continuance of same with the Roberts Veneer Company.

The Roberts Veneer Company will continue to produce Veneers in Walnut, Oak and Poplar, specializing on American Walnut in sliced, half-round and stump wood.

Roberts Veneer Co., New Albany, Ind.

Mahogany
American Walnut
Quartered White Oak
Figured Red Gum
Curly Birch
(Sawn or Slice Cut)

And all other kinds of

Domestic Veneer
AND
Thin Lumber
ACME VENEER & LUMBER CO.
8th, Harriet and 7th Streets   -   -   -   -   CINCINNATI, Ohio

---

200,000 Ft. ¼″ Gum
12 to 36″ wide, 60 to 84″ long
DRAWER BOTTOMS
AND
BOX STOCK
For Quick Shipment
GEO. L. WAETJEN & CO.
MILWAUKEE   -   -   -   WISCONSIN

THE GLUE BOOK
WHAT IT CONTAINS
Chapter 1—*Historical Notes*
Chapter 2—*Manufacture of Glue*
Chapter 3—*Testing and Grading*
Chapter 4—*Methods in the Glue Room*
Chapter 5—*Glue Room Equipment*
Chapter 6—*Selection of Glue*
WOODWORKER PUBLISHING CO., LIMITED
345 Adelaide St. West, Toronto

Veneers and Panels
PANELS
Stock Sizes for Immediate Shipment
All Woods   All Thicknesses   3 and 5 Ply
or made to your specifications
VENEERS
5 Million feet for Immediate Shipment
Any Kind of Wood   Any Thickness
J. J. NARTZIK
1966-76 Maud Ave.,   -   -   CHICAGO, ILL.

Veneers and Panels

LARGE STOCKS FOR
QUICK   SHIPMENTS
Send for Lists

Walter Clark Veneer Co.
GRAND RAPIDS, MICH.

OAK   MAHOGANY   CIRCASSIAN   MAPLE
VENEERS
BIRCH   POPLAR   CHERRY   WALNUT   GUM

Burl Walnut Veneers

We have some exceptionally fine burl veneers in American Walnut. In figure, color and general attractiveness this stock is really remarkable, and must be seen to be appreciated. If you are interested in material which is out of the ordinary and is suitable for the highest class of cabinet work or interior trim, ask us to send samples. We know that you will be interested.

Penrod Walnut & Veneer Co.
"WALNUT SPECIALISTS FOR THIRTY YEARS"
KANSAS CITY, Mo.

Indiana White Oak Lumber and Veneers

Band sawed from the best growth of timber. We use only the best quality of Indiana White Oak. From four to five million feet always on hand.

Walnut Lumber
Our stock assures you good service.

Chas. H. Barnaby, Greencastle, Indiana

Haughton Veneer Company

Office:
122 S. Michigan Ave.

Warehouse:
1150-52 W. Lake St.

Chicago - Illinois

American Walnut

Quartered

White and Red Oak

Mahogany

LET US QUOTE YOU

Straight Edge Ripping and Jointing Machine

To combine in one operation what formerly required several is an important step towards the reduction of manufacturing costs, and nowhere in the woodworking factory does it apply with greater emphasis than in that department of the machine room where the lumber is first assembled into the required widths for the various items indicated on the stock bill, because in this department a reduction in the number of operations generally means a considerable saving in lumber.

This point can perhaps be brought out more clearly by referring to some of the methods at present in use for jointing up stock. One of these is to rip the boards to a width either on a hand feed rip saw or a power

the market a machine (illustrated on this page) for which the manufacturers, Jackson, Cochrane & Company, Kitchener, Ont.,—owners of the Canadian patent rights—claim several distinct advantages: that it will rip stock of any thickness up to 4 inches, depending on the diameter of the saw used, and leave an edge that will make a perfect joint. Also, that it will rip stock up to 18 inches wide.

The machine is of modern open type design, making all parts easily accessible. The table is 31 inches wide, 60 inches long and 33¼ inches above floor, and the table equipment consists of an adjustable ripping fence, self-locking in any position, a bronze index plate to show exact setting and an auxiliary fence to slip over the main fence for cutting narrow stock.

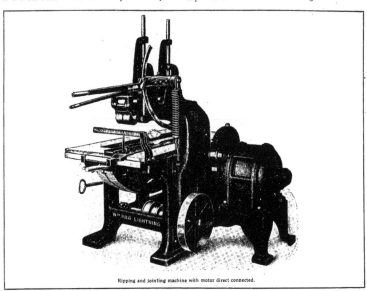

Ripping and jointing machine with motor direct connected.

feed rip and then joint the edges that are to be glued on a jointer or buzz planer. In some factories the jointing is done on a hand feed jointer where, if there is a wind in the board, it has to be run over the knives two or three times, taking off about an eighth of an inch each time. This waste, together with the excessive labor cost, makes it extremely difficult for a manufacturer using such methods to manufacture at a profit. In more modernly equipped factories the practice is to do the jointing on a self-feed jointer where the operator just feeds the stock in. This method, while it greatly reduces the amount of labor required, is still responsible for considerable waste, as an allowance in width has to be made at the rip saw for the jointing.

There has recently been designed and placed on

The saw arbor is 1 7/16 inches diameter, mounted in self-oiling bearings. The driving pulley is outside of the frame, between heavy bearings to permit the machine being driven from any direction or from individual motor. The saw is held securely by 5-inch flanges and cuts from below.

The feed works consist of two separate driven chains, composed of flat milled links which are interchangeable and removable. The chains can be adjusted to vary their projection above the table line to facilitate the feeding of dressed or rough material. Four pair of pressure rolls are provided, two at the back, and two at the front of the saw, insuring that two pair will always be acting on the stock, whether it is long or short.

Single Furniture Planer

The machine shown in the accompanying illustration is the No. 444 Single Furniture Planer made by the American Woodworking Machinery Company and sold by the Garlock-Walker Machinery Company, Limited, 32 Front Street West, Toronto. The machine is made in three widths, 24 inch, 30 inch and 36 inch, and the makers state that pieces from 4 inches long can be surfaced true. Also that stock up to 7 inches thick can be planed without difficulty.

The machine is equipped for two rates of speed, and has a patent sectional roll and chipbreaker which permits narrow strips of varying thickness to be surfaced simultaneously.

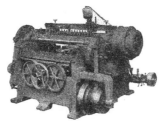

No. 444 Single Furniture Planer

The patented sectional infeed roll has eight tool steel springs 2⅜ inches long, and the construction is such that the roll can be taken apart and reassembled in a short time.

The journals are 2¼ inches diameter and the boxes are of the clamping type with automatic circulation of oil and emergency oilers. The boxes can be removed for babbiting on a jig. The chip breaker can be obtained either solid or sectional to correspond with the infeed roll.

Where motor drive is desired, a special motor is furnished with extended shaft to receive pulleys at each end to drive the cylinder. This can be located on the floor behind the planer or overhead. The company also furnish a motor which is directly attached to the cylinder, on a taper, with holding nut. The motor is bolted up to a supporting bracket attached to the machine frame.

Galbraith & Co., Ltd.
Owen Sound, Ont.
Manufacturers of
"Well-Made" Wood Specialties
Toys, Games, Turned Goods and Kitchenware

We Specialize on
CUSTOMS TURNINGS

Your order will receive prompt attention

Really Dry
Hardwoods

As evidence of the dryness of Martin-Barriss hardwoods we can refer you to many handsome interiors of elaborate detail.

Our process of drying requires three weeks during which time the heat and humidity of the kilns are kept at an efficient degree.

If our stock does not include the particular figure you want, we will gladly cut into new logs.

Have you our prices?

The Martin-Barriss Co.
Cleveland, Ohio

BLACK WALNUT

Large Stock 1 to 3 inches thick ready for immediate shipment

also

Well Assorted Stocks of

West Virginia & Kentucky

OAK

Ash, Maple, Hickory, Chestnut, Basswood and Poplar

Prices and stock list on request

Burns & Knapp
Lumber Company
CONNEAUTVILLE, PA.

News of the Trade

W. S. Anderson is erecting a spool wood mill at the Gravevard Brook, Ludlow, N.B.

The Thomas Meyer Company, of Granby, P.Q., are erecting a $5,000 addition to their box factory.

A newspaper item states that Mr. Hiram Goudey, of Port Maitland, N.S., will erect a furniture factory at that place.

Mr. Harry Brown, of the Atlantic Lumber Company, Toronto, who enlisted as a private with the 15th Battalion of the First Canadian Contingent, has secured a lieutenancy.

Mr. James Cruickshank, the founder of the Cruickshank Wagon Works, Weston, Ont., is dead. Mr. Cruickshank was actively interested in the business for many years, but lately has been living retired.

Mr. L. B. Beale, the British Columbia Lumber Commissioner for Eastern Canada, with offices in the Excelsior Life Building, Toronto, has returned to his duties after spending several months in British Columbia.

The St. John, N. B., Globe says there is a very marked scarcity of office desks in New Brunswick, and that dealers placing orders with Ontario manufacturers have to wait for upwards of four months to have them filled.

The Canadian Toy Association will likely hold another toy exhibition in Toronto early next year. Mr. L. G. Beebe, the secretary, states that the date will be announced later, when preliminary arrangements have been completed.

Mr. J. D. Langelier, of Montreal, has built a piano factory at Point aux Trembles, P.Q., at a cost of $25,000, and is also interested in the Parlor Furniture Manufacturers, Limited, which is completing a factory in the same district.

Another Canadian aeroplane industry is being launched. Canadian Aeroplanes, Limited, Toronto, has been federally incorporated with capital stock of $500,000. The company is authorized to manufacture all kinds of flying machines, etc.

The death occurred recently of Mr. T. G. Eagen, who for 46 years has been associated with Heintzman & Company, Limited, piano manufacturers, Toronto. Mr. Eagen was a stockholder in the company and for many years held the position of superintendent.

Messrs. Arnold and Whitehead, of Guelph, Ont., have made arrangements to carry on the business formerly conducted by the Dominion Casket Company, under the name, the New Dominion Casket Company. They will manufacture caskets in the factory used by the old firm.

The Beacon Match Company, a recently incorporated concern, are expected to commence operations at Bracebridge, Ont., with machinery capable of manufacturing 7,400,000 matches per day, gradually increasing the output as conditions warrant.

The Namu Box Company, Limited, has been incorporated in British Columbia with capital of $24,000, to manufacture boxes and carry on the business of sawmill proprietors, shingle manufacturers, etc. The head office of the company will be in Vancouver.

Thorold, Ont., is believed to stand a good chance of securing a new industry. It is reported that a basket factory at that place which has been vacant for some time may be purchased by a company who propose to manufacture piano sound boards and other piano parts.

Work has commenced on the rebuilding of the Canadian Cooperage Company's plant at Smith's Falls, which was destroyed by fire. The new buildings are expected to be larger than the old ones. The company are also said to be considering the erection of a heading mill.

The Goderich Organ Company, Limited, Goderich, Ont., have secured an order for 30,000 shell boxes of the "Bethlehem" type, and are proceeding to manufacture them right away. The company advises us that they are well equipped to make these boxes and will not require to install any new machines.

The ratepayers of North Bay have passed a by-law authorizing the town to grant a loan of $15,000 to L. W. Henderson, Dr. Edgar Brandon, W. G. Armstrong, H. H. Thompson, G. W. Duncan, Hon. Geo. Gordon, all of North Bay, who propose to erect a factory for manufacturing toys and woodenware in large quantities.. Work is to start on the building as soon as a site can be secured.

The Turner, Day and Woolworth Handle Company, of Louisville, Ky., advise us that the Turner, Day and Woolworth Ltd., have leased a factory at Hamilton, Ont., and will manufacture all kinds of hickory handles. Machinery is now en route to the factory. The company will only employ about 25 men for the first two or three months, but that number will be increased later.

The Toronto Veneer Company, 93 Spadina Avenue, Toronto, report that they are finding business good. They have six cars of veneer on hand and eight more on the way, and say that they are disposing of it about as fast as they get it shipped in. They find that mahogany is still the leading wood, with walnut following closely, and oak, particularly in the piano business, slackening off.

After a long illness, Mr. J. Arthur Villeneuve, president of L. Villeneuve & Company, Limited, Montreal, lumber dealers and sash and door manufacturers, died on Sunday, December 10th, aged 31 years. He was also head of the Eagle Lumber Company, Limited, Montreal, with mills and factories at Mont Laurier and St. Jerome, P. Q. Mr. Villeneuve suffered for a long period from heart disease.

A special meeting of the Kincardine Board of Trade was held on December 6th to discuss a proposal received from Mr. J. B. Watson, in which he offered to form a joint stock company with a capital of $50,000, to build a furniture factory in Kincardine. As the proposal called for several concessions from the town, the Board of Trade appointed a committee to take the matter up in detail with Mr. Watson, and report back at an early date.

The Bell Piano and Organ Company have made a request to the Guelph City Council for a fixed assessment of $10,000 a year for ten years. The company state that owing to the war and other causes they had been unable to pay the interest on the debentures, but that they are anxious to revive and extend their business. They propose to carry on the business again with a large staff of workmen as soon as arrangements can be made.

The Parlor Furniture Manufacturers, Limited, a newly incorporated company with head office at Room 35 Dandurand Building, Montreal, are erecting a factory at Pointe-Aux-Trembles, Que. The company write us that the building will be of brick construction, three storeys, with all modern improvements and conveniences, and will provide 16,500 square feet of floor space. The machinery will all be run by electric power and an electric elevator will be installed. The company own their own timber limits, which they estimate will last them for twenty-five years. They will make their own frames for medium and high grade upholstered furniture. The products of the factory will include everything from tabourets to luxurious Chesterfields. A staff of competent men will be employed, with a designer from a leading United States firm.

CARVINGS FOR PERIOD FURNITURE

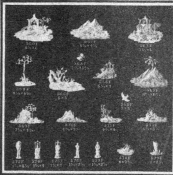

Composition Carvings

are now extensively used for furniture decorations.

This is a material that will interest all manufacturers of furniture.

Write for Catalog and prices.

Most durable and practical Carvings on the market.

We have thousands of stock designs in all the different styles; can also manufacture according to your special designs.

Samples sent on request.

THE DECORATORS SUPPLY CO., 2551 Archer Ave., **Chicago, Ill.**

CROWN PLANER

"Made Good" Here

Here is a con where real "pull" counts. Imagine the inefficiency of this excellent machine if the belting slipped on these seven pulleys! Here is a graphic argument in favor of using the very best leather belting. You get more out of your machines and men and more service per dollar invested, out of your belting by using Crown Planer Belting. On the one illustrated it is giving absolutely satisfactory service.

Send us the conditions of your most difficult run and we will send you a Crown Planer Belt that will exceed your most optimistic expectations.

Any of our offices can supply you

Made in Canada Since 1876

Sadler & Haworth

TANNERS AND MANUFACTURERS
MONTREAL, 511 William Street

TORONTO   ST. JOHN   WINNIPEG   VANCOUVER
38 Wellington Street East   149 Prince William Street   Galt Building   107-111 Water Street

For Sale

One Cowan 8-inch four-side sticker, in good condition.

Garlock-Walker Machinery Co., Ltd.,
12　　　　　32 Front Street West, Toronto.

Second-Hand Machines For Sale

We have the following to offer. Ask us about others not listed, if you are interested.

1 Egan　　　　　30-in. 2 Drum Sander.
1 McGregor Gourlay 30-in. 2 Drum Boss Sander.
1 Jackson Cochrane 38-in. 3 Drum Sander.
1 Egan　　　　　42-in. 3 Drum Sander.
1 Cowan　　　　42-in. 3 Drum Sander.
1 Cowan　　　　48-in. 3 Drum Sander.
1 C. M. C.　　　8-in. 4 side Sticker.
2 Ballantyne　　8-in. 4 side Sticker.
1 Egan　　　　　26-in. Single Surfacer.
1 Harper　　　　26-in. Single Surfacer.
1 Cowan No. 220　24 x 10 Single Surfacer.
1 McGregor Gourlay QY 16-in. Planer & Matcher.
1 McGregor Gourlay QY 9-in. Planer & Matcher.
1 Jackson Cochrane 54-in. Band Resaw 6-in. blade.
1 Jackson Cochrane No. 105 Rip Saw.

P. B. Yates Machine Co., Limited
Hamilton, Ontario　　　　　t-f

Second-Hand Woodworking Machines

Band Saws—26-in., 36-in., 40-in.

Jointers—12-in., 16-in.

Shapers—Single and double spindle.

Re-Saw—42-in. American.

Lathes—20-in. with rests and chisels.

Surfacers—16 x 6 in. and 18 x 8 in.

Mortisers—No. 5 and No. 7 New Britain.

Rip and Cut-off Saws; Molders; Sash Relishers, etc.

Over 100 machines, New and Used, in stock.

Baird Machinery Company,
123 Water Street,
PITTSBURGH, PA.

Woodworking Machinery For Sale

1—18 Spindle upright boring machine, Can. Machinery Corporation make.

1—Veneer Press, 4 sections, 4 screws to each section.

1—Jackson & Cochrane single end tenoning machine.

1—Jackson & Cochrane two spindle shaper with countershaft.

1—Jackson & Cochrane spindle carving machine with countershaft.

1—Table slide dovetail groover with countershaft.

1—Rolling Top saw table.

Beach Furniture Limited
12　　　　　　　　　Cornwall, Ont.

FOR SALE

Pine, Birch, Spruce, suitable for shell box manufacturing. Apply Office Specialty Mfg. Co., Ltd., Newmarket, Ont.
11-12

WANTED

Birch, 2 cars, dry 1 in. unselected Firsts and Seconds; red all in, 60 per cent. 14 and 16 ft.
12-1-2　　　CHARLES F. SHIELS & CO.,
Cincinnati, Ohio.

For Sale—Panel Poplar

1 Car ¾ in. 7 to 17 in. wide, mostly 12 in. and up 48,000 ft. ¾ in., 17 to 23 in. wide, 50 per cent. 16 ft. long.
32,000 ⅝ in., 24 in. and up wide, 50 per cent. 16 ft. long.
21,000 ft. ¾ in., 24 in. and up wide, 50 per cent. 16 ft. long.
12-1-2-　　　CHARLES F. SHIELS & CO.,
Cincinnati, Ohio.

Woodworking Plant
for Sale or to Rent

Is thoroughly equipped for all kinds of woodwork manufacture. Situated on three lines of railroad. List of machinery, etc., furnished to interested parties. Terms reasonable. Address Box 24, Canadian Woodworker, 345 Adelaide Street West, Toronto.　　12-1

Canadian Agents Wanted

We desire to appoint an agent in Canada to handle our enamels, etc. Will give agency for entire Dominion or for Provinces, as desired. Write in first instance to Box 25, "Canadian Woodworker and Furniture Manufacturer," Toronto.
12-1

Valuable Box Factory For Sale

The undersigned will receive offers for the purchase of that valuable property known as the

Czerwinski Box Factory

situated on Logan Avenue and Tecumseh Street, Winnipeg, Canada. This property comprises what is believed to be the finest box factory in Western Canada, fully equipped with the latest and best machinery for box manufacturing and ready for immediate operation. It also includes the factory site, consisting of about 2½ acres of land close in to the business centre of the city, and possesses unexcelled trackage facilities. Street cars pass the factory door. This is a magnificent opportunity for any person wishing to engage in box manufacturing. The whole may be purchased at a very great bargain.

Further information may be obtained upon application to

H. E. Deneen,

Assignee of the Czerwinski Box Co., Limited
300 Electric Railway Chambers,
Winnipeg, Man.

WATCHMAN'S CLOCKS
Hardinge Portable
The only unlimited System.

Patrol clock, each $39.60 complete. Stations, any number needed $4.10. F.O.B. Toronto. Write for catalogues.
Also Key Type Clocks, limited capacity.
Signal Systems Co., 110 Church St., Toronto

PETRIE'S
MONTHLY LIST
of
NEW and USED
WOOD TOOLS
in stock for immediate delivery

Wood Planers
30" Whitney pattern surfacer.
26" Revolving bed single surfacer.
26" Revolving bed double surfacers.
24" Major Harper planer and matcher.
24" Dundas surface planer.
24" Single surfacers, various makes.
22" Stationary bed surface planer.

Band Saws
42" Fay & Egan power feed rip.
38" Atlantic tilting frame.
30" Ideal pedestal, tilting table.

Saw Tables
Ballantine variable power feed.
Preston variable power feed.
Cowan heavy power feed, M188.
No. 8 Crescent universal.
No. 2 Crescent combination.
Ideal variety, tilting table.
12' Defiance automatic equalizer.
MacGregor Gourlay railway cut-off.
7½' Iron frame swing.

Mortisers
Cowan hollow chisel, M190.
Galt upright, compound table.
No. 2 Smart foot power.
No. 1 Smart foot power.

Moulders
18" Clark-Demill four-side.
12" Cowan four side.
10" Houston four side.
6" Cowan four side.
6" Dundas sash sticker.

Sanders
24" Fay double drum.
12" C.M.C. disk and drum.
18" Crescent disk, adjustable table.

Clothespin Machinery
Humphrey No. 8 giant slab resaw.
Humphrey gang slitter.
Humphrey cylinder cutting-off machine.
Humphrey automatic lathes (6).
Humphrey double slotters (4).

Miscellaneous
Fay & Egan 12 spindle dovetailer.
MacGregor Gourlay 12 spindle dovetailer.
Cowan-post boring machine, M80.
Cowan-post boring machine, M85G.
Cowan dowel machine, M80.
Cowan post boring machine, M23.
Elliott single spindle shaper.
Rogers vertical resaw.
Cowan sash clamp.
Egan sash and door tenoner.
No. 6 Lion universal woodtrimmer.
8 nail box nailer, Meyers patent.
20" American wood scraper.
16" Ideal wood lathes (3).
4-head moulding machines.
Cowan spindle carver, M63.
Cowan Veneer press screws.
Dunbar horizontal automatic shingle machine.

Prices, Descriptions and Full Particulars on Request

H. W. PETRIE, LTD.
Front St. W., Toronto, Ont.

FOR SALE!

1 car 6/4 1s and 2s Dry Hickory 80 per
    cent. 14 ft.
1 "   6/4 No. 1 Com. and Bet. Hickory.
2 "   2 x 2—30 in. Clear Dry Oak. Sqrs.
2 "   2 x 2—19 in. Clear Dry Oak Sqrs.
1 "   2 x 2—24 and 30 in. Clear Dry
    Oak Squares.
1 "   1½, 2, 2½ and 3 x 3—30 in. Clear
    Dry Oak Squares.
1 "   2 x 2—30 in. Clear Dry Gum Sqrs.
1 "   1 x 1—5 ft. Clear Dry Oak Sqrs.
1 "   1 x 4 in. and up 22 and 25 in.
    Clear Dry Quartered White Oak.
1 "   1 x 4 in. and up, 31, 49 and 55 in.
    Clr. Dry Quartered White Oak.
1 "   3 in. No. 1 Com. and Bet. Dry
    Qtd. White Oak.
3 "   4/4 x 4 in. and up 4 to 8 ft. Clear
    Ash Shorts.
1 "   6/4 and 8/4 x 4 in. and up 4 to 8
    ft. Clear Ash Shorts.
1 "   10/4 x 4 in. and up, 4 to 8 ft. Clear
    Ash Shorts.

Shall we send you quotations?

The Probst Lumber Co.

Cincinnati, Ohio

Rexford

Veneer Tapes

Combine all the qualities desired
by manufacturers and users of
veneers.

"Let a sample roll tell
its own story"

Please state your requirements
and, if possible, send sample of
tape you are now using.

Rexford Gummed Tape Co.

Manufacturers

Home Office, 598 Clinton St., MILWAUKEE, Wis.

St. Louis, Mo. Branch, 416 Rialto Building. Olive 3529
Chicago, Ill., Branch, 20 E. Jackson Blvd. Harrison 6107

- AT YOUR SERVICE -

*We submit for your interest
the following choice*
HARDWOODS:

2 cars 4/4 No. 2 Com. and Bet. Soft Elm.
2 cars 6/4 No. 1 Com. and Bet. Soft Elm.
3 cars 8/4 No. 1 Com. and Bet. Soft Elm.
2 cars 16/4 No. 1 Com. and Bet. Soft Elm.
1 car   4/4 No. 1 Com. and Bet. Black Ash.
2 cars 4/4 No. 3 Com. Black Ash.
15,000 ft. 6/4 No. 1 Com. and Bet. Birch.
1 car 6/4 x 10 and up No. 1 Com. and Bet. Birch.
33,000 ft. 10/4 No. 1 Com. and Bet. Birch.
1 car 16/4 x 6 Sound Birch Hearts.
5 cars 4/4 No. 3 Com. Beech and Maple.
7,000 ft. 16/4 First and Second Hickory.
  200 ft. 12/4 No. 1 Com. and Bet. Hickory.
2,500, ft. 8/4 No. 1 Com. and Bet. Hickory.
50,000 ft. 4/4 x 3 Spruce Crating.
50,000 ft. 4/4 x 4 and up Spruce Crating.

*All the above stock thoroughly
dry and well manufactured*

*Write, wire or 'phone your
inquiries and they will have
careful attention*

Graves, Bigwood & Co.

Manufacturers
of

PINE
HEMLOCK } **LUMBER**
HARDWOOD

712 Traders Bank Building
TORONTO - ONT.

Mills at—Byng Inlet, Ont.
and
Deer Lake, Ont.

Put this machine on *your* pay roll

There are hundreds of places in your plant where you can use this machine in cutting down cost of production. Its many uses are unlimited and for convenience it is unexcelled.

The

Elliot Woodworker

It cross cuts, rips, miters, dadoes, grooves, bores, grinds, etc. A 16 foot stair string can be housed out in 30 minutes. No plant is too small or too large to find constant use for one or more machines.

Let us send you illustrated circular and prices, and suggest how it can be used in YOUR plant.

THE

ELLIOT WOODWORKER, Ltd.

Cor. College and Bathurst Sts.   TORONTO, ONT.   Patented 1912—14—15

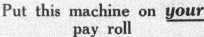

"BEAVER" Cylinder Saws for "Bethlehem" Shell Box Diaphragms

These Saws will cut straight holes in veneer diaphragms rapidly and without tearing the different plys asunder. They can be set and filed as any ordinary saw, thus holding their clearance and size. Our latest type saw is fitted with new style core ejector. This kicker bears on exact centre of wood eliminating all friction.

We also supply special cutter heads for taper nose blocks in 4.5″ and 18 P. R. Shell Boxes.

RADCLIFF SAW MFG. CO., LIMITED

550 DUNDAS ST.
TORONTO, ONT.

Agents for L. & I. J. White Co., Machine Knives

Panels and Tops

**Deskbeds, Table Tops,
Chiffonier and Dresser
Tops
Three and Five Ply Panels**

Large assortment of stock panels always on hand.
Three and five ply stock for cartridge boxes.
Your inquiry will be fully appreciated and
promptly answered with full information.

WRITE US

The

Gorham Brothers
Company

MT. PLEASANT - MICHIGAN

"TROY"
GARNET PAPER

COMBINATION and CLOTH

FINISHING
PAPER

RITCHEY SUPPLY CO.
126 Wellington St. W.
TORONTO

Crating Lumber

⅝" and 1" All Grades
At Right Prices

Try Us—It will be worth your while.

Read Bros., Limited
43 Victoria Street
Toronto   -:-   Canada

**Fuse Plugs, Gain Plugs, and
Wooden Toys**

can be produced most perfectly, economically, and conveniently by
the use of the

NEW WAYMOTH
WOOD-TURNING LATHE

This machine is most suitable for the production of a wide
variety of turnings. In one operation it turns, bores, and cuts off
wood turnings of any size up to 24 inches in length.

Let us tell you more about it.

Our lathes are sold in Canada by
A. R. Williams Machinery Co., Toronto, Vancouver and Winnipeg.
General Supply Company of Canada, Ottawa.
Williams & Wilson, Montreal.

A. D. WAYMOTH & CO., FITCHBURG, MASS.

DARLINGS
STEAM
APPLIANCES

**DARLING BROTHERS
LIMITED**
Engineers and Manufacturers
MONTREAL, CANADA

Branches:
Toronto and Winnipeg   Agents:
Halifax, St. John, Calgary, Vancouver

COWAN'S
Woodworking Machinery is built for Labor-Saving, High Efficiency and Long Service

Tenoner

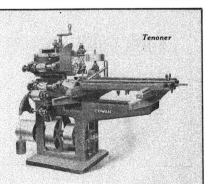

We can supply you with any kind of a woodworking machine — One of our regulars or a special machine to suit your requirements.

COWAN & COMPANY OF GALT
LIMITED
GALT - ONTARIO

Ask the Man Who is Using "Cowan" Equipment

Perkins Vegetable Veneer Glue
IS A
Uniform, Guaranteed Patented Glue

¶ We have, during the past ten or more years, built up a large volume of business and maintained it, because we sell only

HIGHEST QUALITY

¶ Our list of customers proves the statement that a very large per cent. of the highest quality furniture, including office and household furniture, also doors, interior trim, etc., are built up with Perkins Glue.

¶ Be sure you have *Perkins Quality*—absolutely uniform.

¶ We will demonstrate to your satisfaction in your own factory

Ladies' Desk manufactured by The D. Hibner Furniture Co., Limited, Kitchener, Ont., in which Perkins Glue was d.

Originator and Patentee of Perkins Vegetable Glue.

PERKINS GLUE COMPANY, LIMITED
Address all inquiries to Sales Office, Perkins Glue Company, SOUTH BEND, Indiana, U.S.A.

Hamilton, Ont.

FROST KING   *"The Standard for Forty Years"*   TROJAN

The Dependable Babbitt

Equally as good for heavy duty as speed. It is especially adapted for manufacturers of woodworking machinery, planing mills, furniture factories, box factories and woodworking plants.

Try it

Our High Grade Babbitt

Owing to its phenomenal ability to absorb oil "Trojan" babbitt will be found ideal for overhead bearings in line and counter-shafts as it is unnecessary to oil as often as with the ordinary babbitts.

Try it

HOYT METAL COMPANY   -   Toronto, Ont.

NEW YORK   ST. LOUIS   LONDON, ENG.

You Have Paid for an Installation of

Chapman
Double Ball Bearings

in Your Factory over and over again, BUT—

HAVE YOU GOT THEM ?

Ordinary line shafting wastes from 15 per cent. to 60 per cent. of power.

Line shafting equipped with Chapman Double Ball Bearings will eliminate about 75 per cent. of the friction, thus averaging a total saving of from 15 per cent. to 30 per cent.

Chapman Double Ball Bearings fit any adjustable hanger and require oiling and attention only once a year. No extra equipment required to install.

Write to-day for full particulars.

The Chapman Double Ball Bearing Co. of Canada, Limited
Toronto   339-351 Sorauren Ave.   Ontario

American Branch: The Transmission Ball Bearing Co., Inc., 32 Wells St., Buffalo, N. Y.

BOOKS FOR SALE

The following books are offered at special prices subject to previous sale:

Carpentry and Joinery, by Gilbert Townsend. Published in 1913 by the American School of Correspondence. 258 pages, illustrated. Price $1.00.

Saw Fitting Manual, a treatise on the care of saws and knives. Deals with everything in the saw and knife alphabet, from adjustments to widths. 144 pages. Price $2.00.

Common-sense Handrailing, by Fred T. Hodgson. Published by Frederick J. Drake & Company, Chicago. 114 pages, illustrated. Price 50c.

Roof Framing Made Easy, by Owen B. Maginnis. Published by The Industrial Publication Company, New York. 104 pages, illustrated. Price 50c.

Handrailing Simplified, by An Experienced Architect. Published by William T. Comstock, New York. 52 pages, illustrated. Price 50c.

How to Join Mouldings; or, The Arts of Mitering and Coping, by Owen B. Maginnis. Published by William T. Comstock, New York. 72 pages, illustrated. Price 50c.

Cabinet Making, by J. H. Rudd. Published by Grand Rapids Furniture Record Company. 210 pages, illustrated. Price $1.50.

Utilization of Wood-Waste (Second Revised Edition), by Ernst Hubbard. Published in 1915 by Scott, Greenwood & Sons. 192 pages, illustrated. Price $1.50.

Woodworker Publishing Company, Limited
345 Adelaide Street West, Toronto, Ontario

The "Canadian Woodworker" Buyers' Directory
An Index to the Best Equipment and Supplies

ABRASIVES
The Carborundum Co., Niagara Falls, N.Y.

AIR BRUSH EQUIPMENT
DeVilbiss Mfg. Co., Toledo, Ohio.

ANNUNCIATORS
Signal Systems Company, Toronto, Ont.

BABBITT METAL
Hoyt Metal Co., Toronto.

BALL BEARINGS
Chapman Double Ball Bearing Co., Toronto.

BALUSTER LATHES
Garlock-Walker Machinery Co., Toronto, Ont.
Whitney & Son, Baxter D., Winchendon, Mass.

BAND SAW FILING MACHINERY
Canada Machinery Corporation, Galt, Ont.
Fay & Egan Co., J. A., Cincinnati, Ohio.
Garlock-Walker Machinery Co., Toronto, Ont.
Preston Woodworking Machinery Company,
Preston, Ont.
Yates Machine Co., P. B., Hamilton, Ont.

BAND SAWS
Canada Machinery Corporation, Galt, Ont.
Cowan & Co., Galt, Ont.
Garlock-Walker Machinery Co., Toronto, Ont.
Jackson, Cochrane & Company, Kitchener, Ont.
Petrie, H. W., Toronto.
Preston Woodworking Machinery Company,
Preston, Ont.
Yates Machine Co., P. B., Hamilton, Ont.

BAND SAW MACHINES
Canada Machinery Corporation, Galt, Ont.
Cowan & Co., Galt, Ont.
Fay & Egan Co., J. A., Cincinnati, Ohio.
Garlock-Walker Machinery Co., Toronto, Ont.
Petrie, H. W., Toronto.
Preston Woodworking Machinery Company,
Preston, Ont.
Williams Machinery Co., A. R., Toronto, Ont.
Yates Machine Co., P. B., Hamilton, Ont.

BAND SAW STRETCHERS
Fay & Egan Co., J. A., Cincinnati, Ohio.
Yates Machine Co., P. B., Hamilton, Ont.

BASKET STITCHING MACHINE
Saranac Machine Co., Benton Harbor, Mich.

BENDING MACHINES
Defiance Machine Works, Defiance, Ohio.
Fay & Egan Co., J. A., Cincinnati, Ohio.
Garlock-Walker Machinery Co., Toronto, Ont.
Perfection Wood Steaming Retort Company,
Parkersburg, W. Va.

BELLS, ELECTRIC
Signal Systems Corporation, Toronto, Ont.

BELTING
Fay & Egan Co., J. A., Cincinnati, Ohio.
Sadler & Haworth, Montreal, Que.

BLOWERS
Sheldons, Limited, Galt, Ont.

BLOW PIPING
Sheldons, Limited, Galt, Ont.

BORING MACHINES
Canada Machinery Corporation, Galt, Ont.
Cowan & Co., Galt, Ont.
Fay & Egan Co., J. A., Cincinnati, Ohio.
Garlock-Walker Machinery Co., Toronto, Ont.
Jackson, Cochrane & Company, Kitchener, Ont.
Nash, J. M., Milwaukee, Wis.
Preston Woodworking Machinery Company,
Preston, Ont.
Reynolds Pattern & Machine Co. Moline, Ill.
Yates Machine Co., P. B., Hamilton, Ont.

BOX MAKERS' MACHINERY
Canada Machinery Corporation, Galt, Ont.
Cowan & Co., Galt, Ont.
Fay & Egan Co., J. A., Cincinnati, Ohio.
Garlock-Walker Machinery Co., Toronto, Ont.
Jackson, Cochrane & Company, Kitchener, Ont.
Neilson & Company, J. L., Winnipeg, Man.
Preston Woodworking Machinery Company,
Preston, Ont.
Reynolds Pattern & Machine Co., Moline, Ill.
Whitney & Son, Baxter D., Winchendon, Mass.
Williams Machinery Co., A. R., Toronto, Ont.
Yates Machine Co., P. B., Hamilton, Ont.

BRUSHES
Skedden Brush Co., Hamilton, Ont.

CABINET PLANERS
Canada Machinery Corporation, Galt, Ont.
Cowan & Co., Galt, Ont.
Fay & Egan Co., J. A., Cincinnati, Ohio.
Garlock-Walker Machinery Co., Toronto, Ont.
Jackson, Cochrane & Company, Kitchener, Ont.
Preston Woodworking Machinery Company,
Preston, Ont.
Yates Machine Co., P. B., Hamilton, Ont.

CARS (Transfer)
Sheldons, Limited, Galt, Ont.

CARBORUNDUM PAPER
Carborundum Co., Niagara Falls, N.Y.

CARVING MACHINES
Canada Machinery Corporation, Galt, Ont.
Fay & Egan Co., J. A., Cincinnati, Ohio.
Garlock-Walker Machinery Co., Toronto, Ont.
Jackson, Cochrane & Company, Kitchener, Ont.

CASTERS
Burns & Bassick Co., Bridgeport, Conn.
Faultness Caster Co., Evansville, Ind.
Foster, Merriam & Co., Meriden, Conn.
National Lock Co., Rockford, Ill.
Onward Mfg. Co., Kitchener, Ont.

CLAMPS
Fay & Egan Co., J. A., Cincinnati, Ohio.
Garlock-Walker Machinery Co., Toronto, Ont.
Jackson, Cochrane & Company, Kitchener, Ont.
Neilson & Company, J. L., Winnipeg, Man.
Preston Woodworking Machinery Company,
Preston, Ont.
Simonds Canada Saw Company, Montreal, P.Q.

COLUMN MACHINERY
Fay & Egan Co., J. A., Cincinnati, Ohio.

CORRUGATED FASTENERS
Acme Steel Goods Company, Chicago, Ill.

CRATING LUMBER
Read Bros., Limited, Toronto.

CUSTOMS TURNINGS
Galbraith & Co., Ltd., Owen Sound, Ont.

CUT-OFF SAWS
Canada Machinery Corporation, Galt, Ont.
Cowan & Co., Galt, Ont.
Fay & Egan Co., J. A., Cincinnati, Ohio.
Garlock-Walker Machinery Co., Toronto, Ont.
Jackson, Cochrane & Company, Kitchener, Ont.
Ober Mfg. Co., Chagrin Falls, Ohio.
Preston Woodworking Machinery Company,
Preston, Ont.
Simonds Canada Saw Co., Montreal, Que.
Yates Machine Co., P. B., Hamilton, Ont.

CUTTER HEADS
Canada Machinery Corporation, Galt, Ont.
Cowan & Co., Galt, Ont.
Fay & Egan Co., J. A., Cincinnati, Ohio.
Jackson, Cochrane & Company, Kitchener, Ont.
Mattison Machine Works, Beloit, Wis.
Yates Machine Co., P. B., Hamilton, Ont.

DOVETAILING MACHINES
Canada Machinery Corporation, Galt, Ont.
Cowan & Co., Galt, Ont.
Fay & Egan Co., J. A., Cincinnati, Ohio.
Garlock-Walker Machinery Co., Toronto, Ont.
Jackson, Cochrane & Company, Kitchener, Ont.

DOWEL MACHINES
Canada Machinery Corporation, Galt, Ont.
Fay & Egan Co., J. A., Cincinnati, Ohio.

DRY KILNS
Grand Rapids Veneer Works, Grand Rapids,
Mich.
National Dry Kiln Co., Indianapolis, Ind.
Sheldons, Limited, Galt, Ont.

DUST COLLECTORS
Sheldons, Limited, Galt, Ont.

DUST SEPARATORS
Sheldons, Limited, Galt, Ont.

EDGERS (Single Saw)
Canada Machinery Corporation, Galt, Ont.
Fay & Egan Co., J. A., Cincinnati, Ohio.
Garlock-Walker Machinery Co., Toronto, Ont.
Simonds Canada Saw Co., Montreal, Que.

EDGERS (Gang)
Canada Machinery Corporation, Galt, Ont.
Fay & Egan Co., J. A., Cincinnati, Ohio.
Garlock-Walker Machinery Co., Toronto, Ont.
Petrie, H. W., Toronto.
Simonds Canada Saw Co., Montreal, P.Q.
Williams Machinery Co., A. R., Toronto, Ont.
Yates Machine Co., P. B., Hamilton, Ont.

END MATCHING MACHINES
Canada Machinery Corporation, Galt, Ont.
Cowan & Co., Galt, Ont.
Fay & Egan Co., J. A., Cincinnati, Ohio.
Garlock-Walker Machinery Co., Toronto, Ont.
Jackson, Cochrane & Company, Kitchener, Ont.
Yates Machine Co., P. B., Hamilton, Ont.

EXHAUST FANS
Garlock-Walker Machinery Co., Toronto, Ont.
Sheldons Limited, Galt, Ont.

FILES
Simonds Canada Saw Co., Montreal, Que.

FILING MACHINERY
Wardwell Mfg. Company, Cleveland, Ohio.

FLOORING MACHINES
Canada Machinery Corporation, Galt, Ont.
Cowan & Co., Galt, Ont.
Fay & Egan Co., J. A., Cincinnati, Ohio.
Garlock-Walker Machinery Co., Toronto, Ont.
Petrie, H. W., Toronto.
Preston Woodworking Machinery Company,
Preston, Ont.
Whitney & Son, Baxter D., Winchendon, Mass.
Yates Machine Co., P. B., Hamilton, Ont.

FLUTING HEADS
Fay & Egan Co., J. A., Cincinnati, Ohio.

FLUTING AND TWIST MACHINE
Preston Woodworking Machinery Company,
Preston, Ont.

FURNITURE CARVINGS
Decorators Supply Co., Chicago, Ill.

FURNITURE TRIMMINGS
Burns & Bassick Co., Bridgeport, Conn.
Faultless Caster Co., Evansville, Ind.
Foster, Merriam & Co., Meriden, Conn.

GARNET PAPER AND CLOTH
The Carborundum Co., Niagara Falls, N.Y.
Ritchey Supply Company, Toronto, Ont.

GRAINING MACHINES
Canada Machinery Corporation, Galt, Ont.
Fay & Egan Co., J. A., Cincinnati, Ohio.
Williams Machinery Co., A. R., Toronto, Ont.
Yates Machine Co., P. B., Hamilton, Ont.

GLUE
Perkins Glue Company, South Bend, Ind.

GLUE CLAMPS
Jackson, Cochrane & Company, Kitchener, Ont.
Neilson & Company, J. L., Winnipeg.

GLUE HEATERS
Fay & Egan Co., J. A., Cincinnati, Ohio.
Jackson, Cochrane & Company, Kitchener, Ont.

GLUE JOINTERS
Canada Machinery Corporation, Galt, Ont.
Jackson, Cochrane & Company, Kitchener, Ont.
Yates Machine Co., P. B., Hamilton, Ont.

GLUE SPREADERS
Fay & Egan Co., J. A., Cincinnati, Ohio.
Jackson, Cochrane & Company, Kitchener, Ont.

GLUE ROOM EQUIPMENT
Perrin & Company, W. R., Toronto, Ont.

GRINDERS (Cutter)
Fay & Egan Co., J. A., Cincinnati, Ohio.

GRINDERS, HAND AND FOOT POWER
The Carborundum Co., Niagara Falls, N.Y.

GRINDERS (Knife)
Canada Machinery Corporation, Galt, Ont.
Fay & Egan Co., J. A., Cincinnati, Ohio.
Garlock-Walker Machinery Co., Toronto, Ont.
Petrie, H. W., Toronto.
Preston Woodworking Machinery Company,
Preston, Ont.
Yates Machine Co., P. B., Hamilton, Ont.

GRINDERS (Tool)
Fay & Egan Co., J. A., Cincinnati, Ohio.
Jackson, Cochrane & Company, Kitchener, Ont.

Morehead
Back to Boiler
SYSTEM
PRACTICAL POINTERS ON BOILER ROOM EFFICIENCY

IN THE NEW "Morehead" book you will find real ready-to-use information on the handling of your steam equipment which may save you thousands of dollars a year.

Besides a general discussion on the drainage of kilns and steam heated machinery generally, there are interesting

PICTURES AND DATA

which will show you how representative plants everywhere are turning their condensation into golden profit.

Write for the book today.

CANADIAN MOREHEAD MFG. CO.
WOODSTOCK   Dept. "H"   ONTARIO

ARE YOU INTERESTED

in a Saving of $3.00 to $4.00 a day ?

If you are, ask me about the No. 5 Sander for plain, swelled or taper round work.

J. M. NASH, MILWAUKEE, Wis.

Woodworkers
and
Lumbermen

The Sheldon **Exhaust Fans** have characteristics that adapt themselves to the service of the Planing Mill or Woodworking Plant. They are specially designed for this kind of work, having a saving in power and speed of 25% to 40%.

Write for Industrial Booklet.

SHELDONS LIMITED - Galt, Ont.
Toronto Office, 609 Kent Building
— Agents —
Messrs. Ross & Greig, 112 St. James St., Montreal, Que.
Messrs. Walker's Limited, 200-261 Stanley St., Winnipeg, Man.
Messrs. Gorman, Clancey & Grindley Ltd., Calgary and Edmonton, Alta.
Messrs. Robt. Hamilton & Co., Ltd., Bk. of Ottawa Bldg, Vancouver, B.C.

GRAND RAPIDS
VAPOR DRY KILN
GRAND RAPIDS
MICHIGAN

Increasing output of other kilns is part of our business.

We can arrange to utilize your present equipment and supplement with sufficient additional to bring your kilns to our basis of output.

We would suggest that you supply our engineering department with description of your kilns and lumber drying needs, so we can understandingly outline what can be done.

Your opening the matter will incur no obligation and we will give you, frankly, the results of our experience.

"Canadian Woodworker" Buyers' Directory—Continued

GRINDING WHEELS
The Carborundum Co., Niagara Falls, N.Y.

GROOVING HEADS
Fay & Egan Co., J. A., Cincinnati, Ohio.

GUM LUMBER
Brown & Co., George C., Memphis, Tenn.
Churchill-Milton Lumber Co., Louisville, Ky.
Edwards Lumber Co., E. L., Dayton, Ohio.
Lamb-Fish Lumber Co., Charlestown, Miss.
Penrod, Jurden & McCowen, Memphis, Tenn.

GUMMERS, ETC.
Fay & Egan Co., J. A., Cincinnati, Ohio.

HAND PROTECTORS
Fay & Egan Co., J. A., Cincinnati, Ohio.

HAND SCREWS
Fay & Egan Co., J. A., Cincinnati, Ohio.

HANDLE & SPOKE MACHINERY
Defiance Machine Works, Defiance, Ohio.
Fay & Egan Co., J. A., Cincinnati, Ohio.
Nash, J. M., Milwaukee, Wis.
Whitney & Son, Baxter D., Winchendon, Mass.

HARDWARE
Russ & Bassick Co., Bridgeport, Conn.
Faultless Caster Co., Evansville, Ind.
Foster, Merriam & Co., Meriden, Conn.

HARDWOOD LUMBER
Aberdeen Lumber Co., Pittsburg, Pa.
American Hardwood Lumber Co., St. Louis, Mo.
Atlantic Lumber Company, Toronto.
Banning, Leland G., Cincinnati, Ohio.
Brown & Company, Geo. C., Memphis, Tenn.
Burns & Knapp Lumber Co., Conneautville, Pa.
Churchill-Milton Lumber Co., Louisville, Ky.
Clark & Son, Edward, Toronto, Ont.
Dermott Land & Lumber Co., Chicago, Ill.
Edwards Lumber Co., E. L., Dayton, Ohio.
Farrin Lumber Co., M. B., Cincinnati, Ohio.
Graves, Bigwood Company, Toronto, Ont.
Kersley, Geo., Montreal, Que.
Lamb-Fish Lumber Co., Charlestown, Miss.
Martin-Barriss Co., Cleveland, Ohio.
Mowbray & Robinson, Cincinnati, Ohio.
Paepcke Leicht Lumber Co., Chicago, Ill.
Penrod Walnut & Veneer Co., Kansas City, Mo.
Pickrel Walnut Co., St. Louis, Mo.
Probst Lumber Co., Cincinnati, Ohio.
Read Bros., Limited, Toronto.
Shafer Lumber Co., Cyrus C., South Bend, Ind.
Spencer, C. A., Montreal, Que.
Thoman Flinn Lumber Co., Cincinnati, Ohio.

HUB MACHINERY
Fay & Egan Co., J. A., Cincinnati, Ohio.

HYDRAULIC VENEER PRESSES
Perrin & Company, Wm. R., Toronto, Ont.

JOINTERS
Canada Machinery Corporation, Galt, Ont.
Cowan & Co., Galt, Ont.
Defiance Machine Works, Defiance, Ohio.
Fay & Egan Co., J. A., Cincinnati, Ohio.
Garlock-Walker Machinery Co., Toronto, Ont.
Jackson, Cochrane & Company, Kitchener, Ont.
Petrie, H. W., Toronto, Ont.
Preston Woodworking Machinery Company, Preston, Ont.
Yates Machine Co., P. B., Hamilton, Ont.

KNIVES (Planer and others)
Canada Machinery Corporation, Galt, Ont.
Fay & Egan Co., J. A., Cincinnati, Ohio.
Simonds Canada Saw Co., Montreal, P.Q.
Yates Machine Co., P. B., Hamilton, Ont.

LATHES
Canada Machinery Corporation, Galt, Ont.
Cowan & Co., Galt, Ont.
Defiance Machine Works.
Fay & Egan Co., J. A., Cincinnati, Ohio.
Garlock-Walker Machinery Co., Toronto, Ont.
Jackson, Cochrane & Company, Kitchener, Ont.
Petrie, H. W., Toronto, Ont.
Preston Woodworking Machinery Company, Preston, Ont.
Whitney & Son, Baxter D., Winchendon, Mass.

LUMBER
Aberdeen Lumber Co., Pittsburg, Pa.
Acme Veneer & Lumber Co., Cincinnati, Ohio.
American Hardwood Lumber Co., St. Louis, Mo.
Anderson Lumber Co., C. G., Toronto, Ont.
Banning, Leland G., Cincinnati, Ohio.
B. C. Lumber Commissioner, Toronto.
Brown & Company, Geo. C., Memphis, Tenn.
Churchill, Milton Lumber Co., Louisville, Ky.
Clark & Son, Edward, Toronto, Ont.
Dermott Land & Lumber Co., Chicago, Ill.
Dominion Mahogany & Veneer Co., Montreal.
Edwards Lumber Co., E. L., Dayton, Ohio.
Farrin Lumber Co., M. B., Cincinnati, Ohio.
Gorham Bros. Co., Mt. Pleasant, Mich.
Graves, Bigwood Company, Toronto, Ont.

Hart & McDonagh, Toronto, Ont.
Hartzell, Geo. W., Piqua, Ohio.
Huddleston, Marsh Mahogany Co., New York.
Long-Knight Lumber Co., Indianapolis.
Paepcke Leicht Lumber Company, Chicago, Ill.
Pickrel Walnut Co., St. Louis, Mo.
Probst Lumber Co., Cincinnati, Ohio.
Read Bros., Limited, Toronto.
Shafer Lumber Co., Cyrus C., South Bend, Ind.
Stimson & Co., J. V., Owensboro, Ky.
Thoman Flinn Lumber Co., Cincinnati, Ohio.

MACHINE KNIVES
Canada Machinery Corporation, Galt, Ont.
Peter Hay Knife Company, Galt, Ont.
Simonds Canada Saw Co., Montreal, P.Q.
Yates Machine Co., P. B., Hamilton, Ont.

MITRE MACHINES
Canada Machinery Corporation, Galt, Ont.
Fay & Egan Co., J. A., Cincinnati, Ohio.

MITRE SAWS
Canada Machinery Corporation, Galt, Ont.
Fay & Egan Co., J. A., Cincinnati, Ohio.
Garlock-Walker Machinery Co., Toronto, Ont.
Simonds Canada Saw Co., Montreal, P.Q.

MULTIPLE BOXING MACHINES
Fay & Egan Co., J. A., Cincinnati, Ohio.
Reynolds Pattern & Machine Co., Moline, Ill.

PANELS
Waetjen & Co., George L., Milwaukee, Wis.

PATTERN SHOP MACHINES
Canada Machinery Corporation, Galt, Ont.
Cowan & Co., Galt, Ont.
Fay & Egan Co., J. A., Cincinnati, Ohio.
Garlock-Walker Machinery Co., Toronto, Ont.
Jackson, Cochrane & Company, Kitchener, Ont.
Preston Woodworking Machinery Company, Preston, Ont.
Whitney & Son, Baxter D., Winchendon, Mass.
Yates Machine Co., P. B., Hamilton, Ont.

PLANERS
Canada Machinery Corporation, Galt, Ont.
Cowan & Co., Galt, Ont.
Fay & Egan Co., J. A., Cincinnati, Ohio.
Garlock-Walker Machinery Co., Toronto, Ont.
Jackson, Cochrane & Company, Kitchener, Ont.
Petrie, H. W., Toronto, Ont.
Preston Woodworking Machinery Company, Preston, Ont.
Whitney & Son, Baxter D., Winchendon, Mass.
Yates Machine Co., P. B., Hamilton, Ont.

PLANING MILL MACHINERY
Canada Machinery Corporation, Galt, Ont.
Cowan & Co., Galt, Ont.
Fay & Egan Co., J. A., Cincinnati, Ohio.
Garlock-Walker Machinery Co., Toronto, Ont.
Jackson, Cochrane & Company, Kitchener, Ont.
Petrie, H. W., Toronto, Ont.
Preston Woodworking Machinery Company, Preston, Ont.
Whitney & Son, Baxter D., Winchendon, Mass.
Yates Machine Co., P. B., Hamilton, Ont.

PRESSES (Veneer)
Canada Machinery Corporation, Galt, Ont.
Jackson, Cochrane & Company, Kitchener, Ont.
Perrin & Co., Wm. R., Toronto, Ont.
Preston Woodworking Machinery Company, Preston, Ont.

PULLEYS
Fay & Egan Co., J. A., Cincinnati, Ohio.

RED GUM LUMBER
Gum Lumber Manufacturers' Association.

RESAWS
Canada Machinery Corporation, Galt, Ont.
Cowan & Co., Galt, Ont.
Fay & Egan Co., J. A., Cincinnati, Ohio.
Jackson, Cochrane & Company, Kitchener, Ont.
Petrie, H. W., Toronto, Ont.
Preston Woodworking Machinery Company, Preston, Ont.
Simonds Canada Saw Co., Montreal, P.Q.
Williams Machine Co., A. R., Toronto, Ont.
Yates Machine Co., P. B., Hamilton, Ont.

RIM AND FELLOE MACHINERY
Defiance Machine Works, Defiance, Ohio.
Fay & Egan Co., J. A., Cincinnati, Ohio.

RIP SAWING MACHINES
Canada Machinery Corporation, Galt, Ont.
Cowan & Co., Galt, Ont.
Fay & Egan Co., J. A., Cincinnati, Ohio.
Garlock-Walker Machinery Co., Toronto, Ont.
Jackson, Cochrane & Company, Kitchener, Ont.
Preston Woodworking Machinery Company, Preston, Ont.
Yates Machine Co., P. B., Hamilton, Ont.

SANDERS
Canada Machinery Corporation, Galt, Ont.
Cowan & Co., Galt, Ont.
Fay & Egan Co., J. A., Cincinnati, Ohio.
Garlock-Walker Machinery Co., Toronto, Ont.
Jackson, Cochrane & Company, Kitchener, Ont.
Nash, J. M., Milwaukee, Wis.
Preston Woodworking Machinery Company, Preston, Ont.
Yates Machine Co., P. B., Hamilton, Ont.

SASH, DOOR & BLIND MACHINERY
Canada Machinery Corporation, Galt, Ont.
Cowan & Co., Galt, Ont.
Fay & Egan Co., J. A., Cincinnati, Ohio.
Garlock-Walker Machinery Co., Toronto, Ont.
Jackson, Cochrane & Company, Kitchener, Ont.
Preston Woodworking Machinery Company, Preston, Ont.
Yates Machine Co., P. B., Hamilton, Ont.

SAWS
Fay & Egan Co., J. A., Cincinnati, Ohio.
Radcliff Saw Mfg. Company, Toronto, Ont.
Simonds Canada Saw Co., Montreal, P.Q.
Smith Co., Ltd., R. H., St. Catharines, Ont.

SAW GUMMERS, ALOXITE
The Carborundum Co., Niagara Falls, N.Y.

SAW SWAGES
Fay & Egan Co., J. A., Cincinnati, Ohio.
Simonds Canada Saw Co., Montreal, Que.

SAW TABLES
Canada Machinery Corporation, Galt, Ont.
Fay & Egan Co., J. A., Cincinnati, Ohio.
Garlock-Walker Machinery Co., Toronto, Ont.
Jackson, Cochrane & Company, Kitchener, Ont.
Preston Woodworking Machinery Company, Preston, Ont.
Yates Machine Co., P. B., Hamilton, Ont.

SCRAPING MACHINES
Canada Machinery Corporation, Galt, Ont.
Garlock-Walker Machinery Co., Toronto, Ont.
Whitney & Sons, Baxter D., Winchendon, Mass.

SCREW DRIVING MACHINERY
Canada Machinery Corporation, Galt, Ont.
Reynolds Pattern & Machine Works, Moline, Ill.

SCROLL SAWS
Canada Machinery Corporation, Galt, Ont.
Cowan & Co., Galt, Ont.
Fay & Egan Co., J. A., Cincinnati, Ohio.
Garlock-Walker Machinery Co., Toronto, Ont.
Simonds Canada Saw Co., Montreal, P.Q.
Yates Machine Co., P. B., Hamilton, Ont.

SECOND-HAND MACHINERY
Cowan & Co., Galt, Ont.
Petrie, H. W.
Williams Machinery Co., A. R., Toronto, Ont.

SHAPERS
Canada Machinery Corporation, Galt, Ont.
Fay & Egan Co., J. A., Cincinnati, Ohio.
Garlock-Walker Machinery Co., Toronto, Ont.
Jackson, Cochrane & Company, Kitchener, Ont.
Petrie, H. W., Toronto, Ont.
Preston Woodworking Machinery Company, Preston, Ont.
Simonds Canada Saw Co., Montreal, P.Q.
Whitney & Sons, Baxter D., Winchendon, Mass.
Yates Machine Co., P. B., Hamilton, Ont.

SHARPENING TOOLS
The Carborundum Co., Niagara Falls, N.Y.

SHAVING COLLECTORS
Sheldons Limited, Galt, Ont.

SHOOK BUNDLER
Neilson & Company, J. L. Winnipeg, Man.

STAINS
Ault & Wiborg, Toronto, Ont.

STEAM APPLIANCES
Darling Bros., Montreal.

SINGLE SPINDLE BOXING MACHINES
Canada Machinery Corporation, Galt, Ont.
Fay & Egan Co., J. A., Cincinnati, Ohio.
Ober Mfg. Company, Chagrin Falls, Ohio.

STAVE SAWING MACHINE
Whitney & Sons, Baxter D., Winchendon, Mass.

STEAM TRAPS
Canadian Morehead Mfg. Co., Woodstock, Ont.
Standard Dry Kiln Co., Indianapolis, Ind.

STEEL BOX STRAPPING
Acme Steel Goods Co., Chicago, Ill.

STEEL SHELL BOX BANDS
Acme Steel Goods Co., Chicago, Ill.

Peter Hay Machine Knives

Peter Hay knives are made of special quality steel tempered and toughened to take and hold a keen cutting edge. Hay knives are made for all purposes.

Have you our Catalogue?

Made in Canada

The Peter Hay Knife Co., Limited
GALT, ONTARIO

PRESSES

For Veneer and Veneer Drying

Made in Canada

William R. Perrin
Limited
Toronto

NATIONAL EFFICIENCY

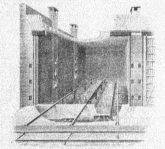

stands for *real merit*.

"Efficiency" is a comparative word and often loosely used. The efficiency of

NATIONAL DRY KILNS

is way above the ordinary. They will dry more lumber—do it better—and with less steam than any others.

Write us for catalog

THE NATIONAL DRY KILN CO.
1117 East Maryland St., INDIANAPOLIS, Indiana

"GOODYEAR"
LOAD BINDER

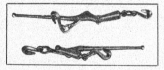

For
LUMBERMEN, LOGGERS, SASH, DOOR and BOX MILLS
CONTRACTORS, DRAYMEN
For
ANYBODY who hauls loads of material requiring binding.

Order NOW

J. L. NEILSON & CO.
WINNIPEG, MAN.

STEEL GOODS
BOX STRAPPING
Our Specialty

Acme Galv. Twisted Wire Strap

Acme Cold Rolled Steel Hoop.

Bright Galv.—Lacquered

WRITE FOR CATALOG

Coil of 5000-ft. Acme Twisted Wire Strap

ACME STEEL GOODS CO. OF CANADA, LTD.—Mfrs.
20 St. Nicholas St., Montreal, Que.

"Canadian Woodworker" Buyers' Directory—Continued

SURFACERS
Canada Machinery Corporation, Galt, Ont.
Cowan & Co., Galt, Ont.
Fay & Egan Co., J. A., Cincinnati, Ohio.
Garlock-Walker Machinery Co., Toronto, Ont.
Jackson, Cochrane & Company, Kitchener, Ont.
Petrie, H. W., Toronto, Ont.
Preston Woodworking Machinery Company,
Preston, Ont.
Whitney & Sons, Baxter D., Winchendon, Mass.
Williams Machinery Co., A. R., Toronto, Ont.
Yates Machine Co., P. B., Hamilton, Ont.

SWING SAWS
Canada Machinery Corporation, Galt, Ont.
Cowan & Co., Galt, Ont.
Fay & Egan Co., J. A., Cincinnati, Ohio.
Garlock-Walker Machinery Co., Toronto, Ont.
Jackson, Cochrane & Company, Kitchener, Ont.
Preston Woodworking Machinery Company.
Preston, Ont.
Simonds Canada Saw Co., Montreal, P.Q.
Yates Machine Co., P. B., Hamilton, Ont.

TABLE LEG LATHES
Fay & Egan Co., J. A., Cincinnati, Ohio.
Whitney & Sons, Baxter D., Winchendon, Mass.
Yates Machine Co., P. B., Hamilton, Ont.

TABLE SLIDES
Walter & Co., B., Wabash, Ind.

TENONING MACHINES
Canada Machinery Corporation, Galt, Ont.
Fay & Egan Co., J. A., Cincinnati, Ohio.
Garlock-Walker Machinery Co., Toronto, Ont.
Jackson, Cochrane & Company, Kitchener, Ont.
Preston Woodworking Machinery Company,
Preston, Ont.
Yates Machine Co., P. B., Hamilton, Ont.

RECORDERS (TIME)
International Time Recording Co., Toronto.
Signal Systems Company, Toronto, Ont.

TRANSMISSION MACHINERY
Dodge Mfg. Co., Ltd., Toronto, Ont.

TELEPHONES—PRIVATE SYSTEMS
Signal Systems Company, Toronto, Ont.

TRIMMERS
Canada Machinery Corporation, Galt, Ont.
Fay & Egan Co., J. A., Cincinnati, Ohio.

Garlock-Walker Machinery Co., Toronto, Ont.
Preston Woodworking Machinery Company,
Preston, Ont.
Yates Machine Co., P. B., Hamilton, Ont.

TRUCKS
Sheldons Limited, Galt, Ont.
National Dry Kiln Co., Indianapolis, Ind.

TURNING MACHINES
Defiance Machine Works, Defiance, Ohio.
Fay & Egan Co., J. A., Cincinnati, Ohio.
Garlock-Walker Machinery Co., Toronto, Ont.
Whitney & Sons, Baxter D., Winchendon, Mass.
Yates Machine Co., P. B., Hamilton, Ont.

**UNDER-CUT SELF-FEEDING FACE
PLANER**
Fay & Egan Co., J. A., Cincinnati, Ohio.
Garlock-Walker Machinery Co., Toronto, Ont.
Jackson, Cochrane & Company, Kitchener, Ont.

VARNISHES
Ault & Wiborg Company, Toronto, Ont.

VARNISH SPRAYERS
DeVilbiss Mfg. Co., Toledo, Ohio.

VENEERS
Acme Veneer & Lumber Co., Cincinnati, Ohio.
Barnaby, Chas. A., Greencastle, Ind.
Central Veneer Co., Huntington, W. Virginia.
Dominion Mahogany & Veneer Company, Mont-
real, Que.
Walter Clark Veneer Co., Grand Rapids, Mich.
Gorham Bros. Co., Mt. Pleasant, Mich.
Hartzell, Geo. W., Piqua, Ohio.
Haughton Veneer Company, Chicago, Ill.
Hay & Company, Woodstock, Ont.
Hoffman Bros. Company, Fort Wayne, Ind.
Geo. Kersley, Montreal, Que.
Nartzik, J. J., Chicago, Ill.
Ohio Veneer Company, Cincinnati, Ohio.
Penrod Walnut & Veneer Co., Kansas City, Mo.
Roberts, John N., New Albany, Ind.
Toronto Veneer Company, Toronto, Ont.
Waetjen & Co., George L., Milwaukee, Wis.

VENEER TAPE
Rexford Gummed Tape Co., Milwaukee, Wis.

VENTILATING APPARATUS
Sheldons Limited, Galt, Ont.

VENEERED PANELS
Hay & Company, Woodstock, Ont.

VENEER PRESSES (Hand and Power)
Canada Machinery Corporation, Galt, Ont.
Garlock-Walker Machinery Co., Toronto, Ont.
Jackson, Cochrane & Company, Kitchener, Ont.
Perrin & Company, Wm. R., Toronto, Ont.

VISES
Fay & Egan Co., J. A., Cincinnati, Ohio.
Simonds Canada Saw Company, Montreal, Que.

WAGON AND CARRIAGE MACHINERY
Canada Machinery Corporation, Galt, Ont.
Fay & Egan Co., J. A., Cincinnati, Ohio.
Whitney & Sons, Baxter D., Winchendon, Mass.
Yates Machine Co., P. B., Hamilton, Ont.

WATCHMAN'S CLOCKS
Signal Systems Company, Toronto, Ont.

WOOD FINISHES
Ault & Wiborg, Toronto, Ont.

WOOD PULLEYS
Dodge Mfg. Co., Ltd., Toronto, Ont.

WOOD TURNING MACHINERY
Canada Machinery Corporation, Galt, Ont.
Defiance Machine Works, Defiance, Ohio.
Garlock-Walker Machinery Co., Toronto, Ont.
Waymoth & Co., A. D., Fitchburg, Mass.
Yates Machine Co., P. B., Hamilton, Ont.

WORK BENCHES
Fay & Egan Co., J. A., Cincinnati, Ohio.

WOODWORKING MACHINES
Canada Machinery Corporation, Galt, Ont.
Cowan & Co., Galt, Ont.
Elliot Woodworker Limited, Toronto, Ont.
Garlock-Walker Machinery Co., Toronto, Ont.
Jackson, Cochrane & Company, Kitchener, Ont.
Preston Woodworking Machinery Company,
Preston, Ont.
Reynolds Pattern & Machine Works, Moline,
Ill.
Williams Machinery Co., A. R., Toronto, Ont.

WHEELS, Grinding, Carborundum and Aloxite
The Carborundum Co., Niagara Falls, N.Y.

No. 2 Standard Type Automatic Screw-driving Machine.

The Only Way to Drive Screws

There is a stage-coach method and a modern method of doing everything. When it comes to driving screws the only modern way is with an
Automatic Screw Driving Machine, which means greater and better production and greater profit.

The Reynolds
Automatic Screw-Driving Machine

is guaranteed to produce results. It is no experiment. The No. 2 Standard
Type, shown herewith, is particularly suited for placing the screws in the
straps of shrapnel boxes. It may be fitted to drive, without change or adjustment, any three consecutive sizes flat or round head, or two sizes filister
head, No. 6 to 14 inclusive, wood or machine screws, or three sizes flat or
round head, or two sizes filister head machine screws up to No. 20. All the
above up to 1½ inches long if smaller than No. 10, or up to 1¾ inches long,
No. 10 or larger. Fitted with boring attachment if desired.

Many Machines in Use in Canada.

Ask for our Catalogue.

REYNOLDS PATTERN & MACHINE CO.
101 and 103 Third Ave., Moline, Ill., U.S.A.

Making TABLE-SLIDES is a Specialty Business

For more than TWENTY-FIVE YEARS we have made TABLE SLIDES exclusively. Our Factory is equipped with Special Machinery which enables us to make SLIDES,—BETTER and CHEAPER than the furniture manufacturer.

Canadian Table makers are rapidly adopting WABASH SLIDES

Because { They ELIMINATE SLIDE TROUBLES Are CHEAPER and BETTER

BY USING

Reduced Costs

Increased Out-put

WABASH SLIDES

MADE BY

B. Walter & Company
Wabash, Ind.

The Largest EXCLUSIVE TABLE-SLIDE Manufacturers in America
ESTABLISHED-1887

Just What You Need

There's no necessity now for you to wish for a corrugated joint fastener machine which will rise fully to your expectations; you can get one.

The Saranac Corrugated Joint Fastener Machine

is the absolute realization of your wish. This illustration gives you an idea of what it looks like, but you can have no conception of its **efficiency, economy, speed** and **reliability** until you get it working for you.

It is substantially constructed, and the workmanship and materials are of the very best. Particularly suitable for manufacturers of sash, doors, blinds, screens, porch columns, boxes, etc., etc.

Made in various types for all conditions.

Write for further particulars and prices

Saranac Machine Co., SOLE MAKERS
Benton Harbor, Mich.

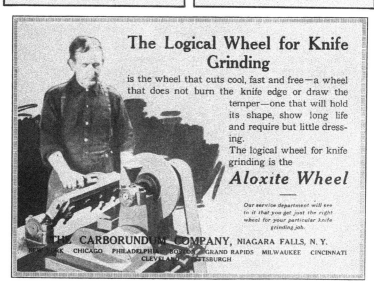

The Logical Wheel for Knife Grinding

is the wheel that cuts cool, fast and free—a wheel that does not burn the knife edge or draw the temper—one that will hold its shape, show long life and require but little dressing.

The logical wheel for knife grinding is the

Aloxite Wheel

Our service department will see to it that you get just the right wheel for your particular knife grinding job.

THE CARBORUNDUM COMPANY, NIAGARA FALLS, N. Y.

NEW YORK   CHICAGO   PHILADELPHIA   BOSTON   GRAND RAPIDS   MILWAUKEE   CINCINNATI
CLEVELAND   PITTSBURGH

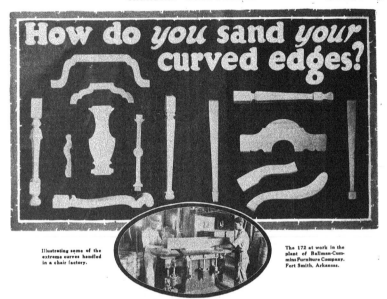

How do *you* sand *your* curved edges?

Illustrating some of the
extreme curves handled
in a chair factory.

The 172 at work in the
plant of Ballman-Cum-
mins Furniture Company,
Fort Smith, Arkansas.

Many furniture factories are replacing the clumsy handmade spindles, edge sanders of rickety, "sewing-machine" construction and slow, tedious hand-methods with the Mattison Oscillating Belt Sander, because they realize that for accurate and rapid sanding of curved edges, it is the only practical method on the market.

MATTISON
EDGE SANDER
With Oscillating Belt

Employs
Stationary Forms

Employs
Stationary Forms

Because of machine tool design, rigid and accurate construction, the Mattison Edge Sander is in a class by itself.

It handles an enormous variety of work. The picture shows some of the work which is met in the chair factory. But any cabinet shop having edge sanding to do, can use one or more of these machines with profit because it successfully employs stationary forms.

If any vibration is imparted to the sandbelt as it passes over the narrow form, it is bound to mark the edge being sanded. No homemade or rickety sewing-machine affair can be used with any degree of success for form work.

A construction which is rigid and heavy, combined with non-vibrating sanding forms and a perfect belt-tensioning device, make the Mattison Edge Sander the only practical mechanical method for every class of edge-work.

Can you afford to handicap production by using makeshift edge-sanding equipment?   Write today for our trial proposition.

C. Mattison Machine Works, 883 Fifth St., Beloit, Wis, U.S.A.

Lightning Source UK Ltd.
Milton Keynes UK
UKHW012008070219

336896UK00012B/927/P